P9-CDX-071

DICTIONARY OF SCIENTIFIC BIOGRAPHY

PUBLISHED UNDER THE AUSPICES OF
THE AMERICAN COUNCIL OF LEARNED SOCIETIES

The American Council of Learned Societies, organized in 1919 for the purpose of advancing the study of the humanities and of the humanistic aspects of the social sciences, is a nonprofit federation comprising forty-six national scholarly groups. The Council represents the humanities in the United States in the International Union of Academies, provides fellowships and grants-in-aid, supports research-and-planning conferences and symposia, and sponsors special projects and scholarly publications.

MEMBER ORGANIZATIONS

AMERICAN PHILOSOPHICAL SOCIETY, 1743
AMERICAN ACADEMY OF ARTS AND SCIENCES, 1780
AMERICAN ANTIQUARIAN SOCIETY, 1812
AMERICAN ORIENTAL SOCIETY, 1842
AMERICAN NUMISMATIC SOCIETY, 1858
AMERICAN PHILOLOGICAL ASSOCIATION, 1869
ARCHAEOLOGICAL INSTITUTE OF AMERICA, 1879
SOCIETY OF BIBLICAL LITERATURE, 1880
MODERN LANGUAGE ASSOCIATION OF AMERICA, 1883
AMERICAN HISTORICAL ASSOCIATION, 1884
AMERICAN ECONOMIC ASSOCIATION, 1885
AMERICAN FOLKLORE SOCIETY, 1888
AMERICAN DIALECT SOCIETY, 1889
AMERICAN PSYCHOLOGICAL ASSOCIATION, 1892
ASSOCIATION OF AMERICAN LAW SCHOOLS, 1900
AMERICAN PHILOSOPHICAL ASSOCIATION, 1901
AMERICAN ANTHROPOLOGICAL ASSOCIATION, 1902
AMERICAN POLITICAL SCIENCE ASSOCIATION, 1903
BIBLIOGRAPHICAL SOCIETY OF AMERICA, 1904
ASSOCIATION OF AMERICAN GEOGRAPHERS, 1904
HISPANIC SOCIETY OF AMERICA, 1904
AMERICAN SOCIOLOGICAL ASSOCIATION, 1905
AMERICAN SOCIETY OF INTERNATIONAL LAW, 1906
ORGANIZATION OF AMERICAN HISTORIANS, 1907
AMERICAN ACADEMY OF RELIGION, 1909
COLLEGE ART ASSOCIATION OF AMERICA, 1912
HISTORY OF SCIENCE SOCIETY, 1924
LINGUISTIC SOCIETY OF AMERICA, 1924
MEDIAEVAL ACADEMY OF AMERICA, 1925
AMERICAN MUSICOLOGICAL SOCIETY, 1934
SOCIETY OF ARCHITECTURAL HISTORIANS, 1940
ECONOMIC HISTORY ASSOCIATION, 1940
ASSOCIATION FOR ASIAN STUDIES, 1941
AMERICAN SOCIETY FOR AESTHETICS, 1942
AMERICAN ASSOCIATION FOR THE ADVANCEMENT OF SLAVIC STUDIES, 1948
METAPHYSICAL SOCIETY OF AMERICA, 1950
AMERICAN STUDIES ASSOCIATION, 1950
RENAISSANCE SOCIETY OF AMERICA, 1954
SOCIETY FOR ETHNOMUSICOLOGY, 1955
AMERICAN SOCIETY FOR LEGAL HISTORY, 1956
AMERICAN SOCIETY FOR THEATRE RESEARCH, 1956
SOCIETY FOR THE HISTORY OF TECHNOLOGY, 1958
AMERICAN COMPARATIVE LITERATURE ASSOCIATION, 1960
MIDDLE EAST STUDIES ASSOCIATION OF NORTH AMERICA, 1966
AMERICAN SOCIETY FOR EIGHTEENTH-CENTURY STUDIES, 1969
ASSOCIATION FOR JEWISH STUDIES, 1969

DICTIONARY
OF
SCIENTIFIC BIOGRAPHY

FREDERIC L. HOLMES
Yale University

EDITOR IN CHIEF

Volume 17

Supplement II

LEASON HEBERLING ADAMS – FRITZ H. LAVES

CHARLES SCRIBNER'S SONS · NEW YORK

First publication in an eight-volume edition 1981.

Library of Congress Cataloging in Publication Data

Main entry under title:

Dictionary of scientific biography.

"Published under the auspices of the American Council of Learned Societies."
Includes bibliographies and index.
1. Scientists—Biography. I. Gillispie, Charles Coulston.
II. American Council of Learned Societies
Devoted to Humanistic Studies.
Q141.D5 1981 509'.2'2 [B] 80-27830
ISBN 0-684-

ISBN 0-684-16963-0 Vols. 1 & 2
ISBN 0-684-16964-9 Vols. 3 & 4
ISBN 0-684-16965-7 Vols. 5 & 6
ISBN 0-684-16966-5 Vols. 7 & 8
ISBN 0-684-16967-3 Vols. 9 & 10
ISBN 0-684-16968-1 Vols. 11 & 12
ISBN 0-684-16969-X Vols. 13 & 14
ISBN 0-684-16970-3 Vols. 15 & 16
ISBN 0-684-19177-6 Vol. 17
ISBN 0-684-19178-4 Vol. 18

Published simultaneously in Canada
by Collier Macmillan Canada, Inc.

1 3 5 7 9 11 13 15 17 19 20 18 16 14 12 10 8 6 4 2

Printed in the United States of America.

Editorial Staff

PREFACE

The sixteen volumes of *Dictionary of Scientific Biography* published between 1970 and 1980 included articles on scientists representing "all periods of science from classical antiquity to modern times." For reasons summarized in the preface to volume I, subjects for the twentieth century were limited to "relatively important figures" and excluded anyone still living when the original lists were established. The principal objective of Supplement II has been to enrich the coverage of scientists who have been active during the twentieth century. Some were subjects overlooked in the original volumes, but the majority are scientists whose lives ended between 1970 and 1981, the time at which planning for the present volume began.

An early decision made by the editorial board was to restrict coverage to the same areas of science—mathematics, astronomy, physics, chemistry, biology, and the earth sciences—that were covered in the main volumes. This decision was based on the practical limitation that the editors were then empowered to plan for only a single supplementary work. Aware of the proliferation of specialty areas not only within the natural sciences, but in the social sciences and medicine, the editors concluded that it would be impractical at this point to expand the original conception of the *DSB* to include them. The board expresses its hope that the auspicious success of the existing volumes will encourage similar ventures in the areas of science that have not been incorporated into it.

Its focus on twentieth-century science confronted the board in acute form with a limitation already noted in the Preface to Volume I. Historians of science have yet to subject twentieth-century science to scholarship as intense as that which they have devoted to earlier eras. For many outstanding recent scientists, there exist only obituaries. For some of them, historians are currently engaged in scholarship the results of which are represented in articles appearing within this volume. In some cases historians have taken on subjects outside their previous scholarly experience, and we are appreciative of their venturesome spirit. A larger proportion of the articles in this volume than in previous volumes necessarily relies upon the willingness of colleagues in the fields of the subjects to write on their lives and work. What is sometimes lost in the detached perspective that a historian can bring is often gained in the intimate glimpses of the personality and working style of a scientist that can come only from someone who knew him or her well.

Knowledgeable readers will notice that there are major figures eligible by the above criteria for inclusion in *Supplement II* who do not appear. In some cases it was not possible to identify an author qualified and willing to undertake the task. In other cases it was not possible for an author who had undertaken the task to complete it before the publication deadline. It is envisioned that future supplementary volumes will remedy these omissions.

Readers will also notice that there are relatively few women represented among the subjects of *Supplement II*. The board has received urgent requests to ensure that a large proportion of women be included. Sympathetic though we have been to the viewpoint represented in this advice, we have concluded that we should not engage in retrospective affirmative action. The small representation of women in *Supplement II* is a realistic reflection of the very real barriers that hindered entrance of women into science during the periods in which the careers of the individuals included in this volume began. We look forward to future volumes to reflect the dramatic improvements in the career prospects for women in science that have taken place so recently that most of those who have benefited from it are still active scientists.

Frederic L. Holmes
Yale University

Contributors to Supplement II

ANDRÁS ADÁM
Hungarian Academy of Sciences, Budapest
KALMAR

MARK B. ADAMS
University of Pennsylvania
AGOL; ASTAUROV; CHETVERIKOV; FILIPCHENKO; KARPECHENKO; LEVIT; LEVITSKII; LYSENKO; OPARIN; SEREBROVSKII

SERGEI ADIAN
Soviet Academy of Sciences, Moscow
NOVIKOV

KATHLEEN AHONEN
Finland
HEISKANEN

RICHARD C. ANDERSON
Augustana College
LEVERETT

RICHARD L. ANDERSON
Lexington, Kentucky
COCHRAN

LUIS CARLOS ARBOLEDA
Universidad del Valle, Colombia
KURATOWSKI

MICHAEL F. ATIYAH
Oxford University
HODGE

FRANCISCO AYALA
University of California at Irvine
DOBZHANSKY

P. C. BAAYEN
Stichting Mathematisch Centrum
DE GROOT

WILLIAM BACK
United States Geological Survey
ATWOOD

LAWRENCE BADASH
University of California at Santa Barbara
FAJANS

CARL JAY BAJEMA
Grand Valley State College
BUMPUS

GIOVANNI BATTIMELLI
Università di Roma
TAYLOR

WILLIAM BECHTEL
Georgia State University
DAKIN; FILDES; NEUBAUER

JAMES D. BERGER
The University of British Columbia
SONNEBORN

PIERRE DE BÉTHUNE
Université de Paris-Sud
FOURMARIER

J. MICHAEL BLISS
University of Toronto
BEST; MINKOWSKI

MARY BOGIN
Cornell University
KING, H. D.; NEWMAN

PHILIP K. BONDY
Yale University School of Medicine
LONG

WILLIAM M. BOOTHBY
Washington University
WANG

MERRILEY BORELL
University of California at San Francisco
RUŽIČKA; STEENBOCK

KENNARD B. BORK
Dennison University
MATHER

WALLACE A. BOTHNER
University of New Hampshire
KNIGHT

J. BOUCKAERT
Belgische Geologische Dienst
VAN LECKWIJCK

JOANNE BOURGEOIS
University of Washington
KUENEN

A. BRACHNER
Deutsches Museum, Munich
GAEDE

DAVID BRANAGAN
University of Sydney
EDWARDS

WILLIAM R. BRICE
University of Pittsburgh
CAMPBELL

GUNNAR BROBERG
Uppsala University
DAHLBERG

WILLIAM H. BROCK
University of Leicester
RIDEAL

C. KEITH BROOKS
University of Papua
WAGER

AART BROUWER
Rijksmuseum van geologie en mineralogie, Leiden
VAN DER VLERK

LAURIE M. BROWN
Northwestern University
SAKATA; TOMONAGA; YUKAWA

RICHARD W. BURKHARDT, JR.
University of Illinois
CUNNINGHAM; HEINROTH; KOHLER; SELOUS; VON FRISCH

JOHANNES BÜTTNER
Medizinische Hochschule Hannover
WEBER

PAUL L. BUTZER
Technische Hochschule Aachen
HELLY

SIMONNE CAILLÈRE
Université de Paris-Sud
ORCEL

DOUGLAS E. CAMERON
University of Akron
ALEKSANDROFF

GEOFFREY CANTOR
University of Leeds
STONER

ELOF AXEL CARLSON
State University of New York, Stony Brook
PAYNE

ERNST CASPARI
University of Rochester
BELAR; HADORN; STERN

J. W. S. CASSELS
University of Cambridge
MORDELL; HEILBRONN

DAVID C. CASSIDY
State University of New York, Stony Brook
HEISENBERG; PLACZEK; SAUNDERS

LUIGI CERRUTI
Universita di Torino
NATTA

A. B. CHAPMAN
University of Wisconsin
COLE

GUSTAVE CHOQUET
Université de Paris
DENJOY

FREDERICK B. CHURCHILL
Indiana University
EIMER; MELDOLA

MARGARET J. CLARKE
College of St. Scholastica
WILHELM

SEYMOUR S. COHEN
State University of New York, Stony Brook
BAWDEN; KUNITZ; MIRSKY; STANLEY

EDWIN H. COLBERT
Museum of Northern Arizona
GREGORY

MICHEL COLCHEN
Université de Paris-Sud
PRUVOST

DAVID L. COLTON
University of Delaware
ERDÉLYI

KENNETH W. COOPER
University of California at Riverside
SCHRADER

CONTRIBUTORS TO SUPPLEMENT II

THOMAS D. CORNELL
Rochester Institute of Technology
TUVE

ALBERT B. COSTA
Duquesne University
CONANT; FIESER; WINSTEIN; WOLFROM

T. G. COWLING
University of Leeds
CHAPMAN

OLIVIER DARRIGOL
Université de Paris
BRILLOUIN; DIRAC

JOSEPH W. DAUBEN
Graduate Center of the City University of New York
ROBINSON, A.

MANSEL DAVIES
University of Wales at Aberystwyth
BURY

JACQUES DEBELMAS
Université de Grenoble
GIGNOUX; MORET

CLAUDE DEBRU
Centre National de la Recherche Scientifique
ARTHUS; HENRI; TREFOUEL

JEAN DIEUDONNÉ
Université de Paris
DELSARTE; EHRESMAN; MONTEL; SIEGEL

D. L. DINELEY
University of Bristol
WHITTARD

J. L. DOOB
University of Illinois at Urbana-Champaign
FELLER

SIGALIA DOSTROVSKY
Antioch College
HUNT, F. V.

ROBERT DOTT, JR.
University of Wisconsin
KAY

J. J. DOZY
Netherlands
ESCHER

ELLEN T. DRAKE
Oregon State University
BANDY; CHANEY; KING, W. B. R.

CLAUDE DUPUIS
Muséum National d'Histoire Naturelle
HENNIG

JOHN T. EDSALL
Harvard University
LINDERSTRØM-LANG; RITTHAUSEN; ROUGHTON

HAROLD M. EDWARDS
New York University
HASSE; TAKAGI

CHURCHILL B. EISENHART
The National Institute of Standards and Technology
SHEWHART

STEVEN ELLIS
NASA: Ames Research Center
BÉKÉSY

WOLF VON ENGELHARDT
Universität Tübingen
CORRENS

PAUL ERDÖS
Hungarian Academy of Sciences, Budapest
TURÀN

C. W. FRANCIS EVERITT
Stanford University
SCHIFF

LINDA N. EZELL
Smithsonian Institution
DRYDEN

BERNADINO FANTINI
Università di Roma
MONOD

FREDERICK FELLOWS
Norwood, New York
VAN VLECK

MARTIN E. FINK
ALBRIGHT

RACHEL FINK
Mount Holyoke College
HYMAN

JEAN-LOUIS FISCHER
Université de Paris
ROSTAND

CHARLES FLEMING
Wellington, New Zealand
MARSDEN

ERIK FLÜGEL
Universität Erlangen-Nürnberg
KUHN

W. FORSTER
The University of Southampton
SCHAUDER

HENRY FRANKEL
University of Missouri
BULLEN; EWING; HESS

V. J. FRENKEL
Soviet Academy of Sciences, Leningrad
ALIKANOV; ANDREEV; FOK; FRUMKIN; KHOKHLOV; KRAVETZ; KRUTKOV; LEIPUNSKII; LIFSHITS; LUKIRSKII; PAPALESKI; POMERANCHUK; STEPANOV; TIMOSHENKO; UMOV; VERNOV

HANS FREUDENTHAL
Rijksuniversiteit Utrecht
KOEBE; VAN DANTZIG

GERALD M. FRIEDMAN
Rensselaer Center of Applied Geology
ILLING

ROBERT MARC FRIEDMAN
University of California at San Diego
BJERKNIS; SIEGBAHN; VEGARD

JOSEPH S. FRUTON
Yale University
HOFMEISTER; SCHOENHEIMER; MICHAELIS, L.; ZERVAS

W. BRUCE FYE
Marshfield Clinic, Wisconsin
MARTIN

JEAN GAUDANT
Université de Paris-Sud
ARAMBOURG

C. STEWART GILLMOR
Wesleyan University
BERKNER

BENTLEY C. GLASS
American Philosophical Society
DEMEREC; DUNN; TIMOFÉEF-RESSOVSKY

STANLEY GOLDBERG
Smithsonian Institution
GOUDSMIT

JUDITH GOODSTEIN
California Institute of Technology
DUMOND; EPSTEIN

STEPHEN JAY GOULD
Harvard University
CRAMPTON

JAMES A. GREEN
University of Warwick
BRAUER

BRYAN GREGOR
Wright State University
NIEUWENKAMP

JOSEPH T. GREGORY
University of California at Berkeley
JEPSEN

MIRKO D. GRMEK
École Pratique des Hautes Études
GLEY

ROBERT C. GUNNING
Princeton University
BOCHNER

MARTIN GUNTAU
Wilhelm-Pieck-Universität Rostock
VON BUBNOFF

E. TEN HAAF
Rijksuniversiteit Utrecht
RUTTEN

WILLIAM J. HAGAN, JR.
Saint Anselm College
VINOGRAD

HEINI HALBERSTAM
University of Illinois at Urbana-Champaign
DAVENPORT

WALTER K. HAYMAN
University of York
NEVANLINNA

JOHN HENDRY
London Business School
CHADWICK; FRISCH; THOMSON

BRUCE HEVLY
Stanford University
WEBSTER

EDWIN HEWITT
University of Washington
NAIMARK

CONTRIBUTORS TO SUPPLEMENT II

BRUCE M. HILL
University of Michigan
SAVAGE

MORRIS HIRSCH
University of California at Berkeley
BOWEN

LILLIAN HODDESON
University of Illinois at Urbana-Champaign
SLATER

HELMUTH HÖLDER
Universität Münster
BENTZ; SCHÄFER

FREDERIC L. HOLMES
Yale University
KREBS

GERALD HOLTON
Harvard University
EHRENHAFT

RODERICK HOME
University of Melbourne
MARTYN

JOHN HORVATH
University of Maryland
RIESZ

KARL HUFBAUER
University of California at Irvine
ZWICKY

I. IBRAGIMOV
Mathematics Institute, Leningrad
LINNIK

AARON J. IHDE
University of Wisconsin
DANIELS; HEILBRON; MAGNUS-LEVY; SCHMIEDEBERG; THOMAS; VICKERY

NATASHA X. JACOBS
Indiana University
BAUR

HAROLD L. JAMES
BUDDINGTON

KAREN E. JOHNSON
Saint Lawrence University
MAYER

DANIEL P. JONES
National Endowment for the Humanities
JULIAN

JEAN-PIERRE KAHANE
Université de Paris-Sud
SALEM

WALTER KAISER
Johannes-Gutenberg-Universität Mainz
ECKART; EWALD

GEORGE B. KAUFFMAN
California State University
COSSA; MARVEL; SCHWARZENBACH

JAMES KEESLING
University of Florida
BORSUK

S. T. KEITH
Haringey College
GABOR

RALPH H. KELLOGG
University of California at San Francisco
BOHR

DAVID G. KENDALL
University of Cambridge
RENYI

WILLIAM C. KIMLER
North Carolina State University
ALLEE; BATES; CARSON; McATEE; POULTON; SHELFORD

SHARON E. KINGSLAND
Johns Hopkins University
LACK

ROBERT E. KOHLER, JR.
University of Pennsylvania
STEPHENSON

SALLY GREGORY KOHLSTEDT
Syracuse University
GOLDRING

KENKICHIRO KOIZUMI
Toyo University
TANAKADATE

FRITZ KRAFFT
Johannes-Gutenberg-Universität Mainz
STRASSMANN

HELGE KRAGH
Niels Bohr Institut
DORGELO; LEMAÎTRE; SCHERRER

EMIL KUHN-SCHNYDER
Universität Zürich
DIETRICH; EDINGER; HUENE; PFANNENSTIEL

KEITH J. LAIDLER
University of Ottawa
EYRING

ERICH L. LEHMANN
University of California at Berkeley
NEYMAN; SCHEFFE

ULRICH LEHMANN
Universität Hamburg
HÄNTZSCHEL

DERRICK H. LEHMER
University of California at Berkeley
VANDIVER

TIMOTHY LENOIR
Stanford University
HÖBER

JOHN E. LESCH
University of California at Berkeley
GULICK; KING, H.; THIERFELDER; VERNEY

STUART W. LESLIE
Johns Hopkins University
BRIGGS

ALBERT C. LEWIS
McMaster University
MOORE, R. L.

R. BRUCE LINDSAY
Brown University
KNUDSEN

MOSHE L. LIVŠIC
Ben Gurion University of the Negev
AKHIESER

JOEL LLOYD
GILLULY

STEFAN MACHLUP
Case Western Reserve University
ONSAGER

KARL MAGDEFRAU
Ludwig-Maximilians-Universität München
ZIMMERMANN

JANE MAIENSCHEIN
Arizona State University
RIDDLE; STEVENS, N.

MARJORIE MALLEY
Bartlesville, Oklahoma
AYRTON

URSULA B. MARVIN
Center for Astrophysics
LANE; RAMSDELL

BRIAN MASON
Smithsonian Institution
FOSHAG; SCHAIRER

JOHN GWYN MAY
University of California at Berkeley
SNOW

ERNST MAYR
Harvard University
CHAPMAN; DAVIS; JORDAN, K.; NOBLE; ROMER; SCHMIDT; STRESEMANN

PAULINE MAZUMDAR
University of Toronto
MARRACK

MACLYN McCARTY
Rockefeller University
LANCEFIELD

D. P. McKENZIE
University of Cambridge
BULLARD

DUNCAN McKIE
University of Cambridge
TILLEY

HERBERT MEHRTENS
Technische Universität Berlin
MOUFANG

CHRISTOPH MEINEL
Universitat Hamburg
FREUDENBERG; ZIEGLER

KARL VON MEŸENN
Universitad Autónoma de Barcelona
JORDAN, P.; KLEIN

ALEXIS MOISEYEV
France
BARRABÉ

GREGORY N. MOORE
McMaster University
BERNAYS; GÖDEL; MOSTOWSKI; TARSKI

PAUL B. MOORE
University of Chicago
LAVES; ZACHARIASEN

CONTRIBUTORS TO SUPPLEMENT II

NEIL D. MORGAN
University of Bristol
CHICK; DUBOIS; HALLIBURTON; HARRINGTON; KLENK; KNORR; PURDIE

ALBERT MOYER
Virginia Polytechnic Institute and State University
WEBSTER

FREDERICK NEBEKER
American Philosophical Society
LEFSCHETZ

ALLAN A. NEEDELL
Smithsonian Institution
A. C. OFFORD

KARL-HEINZ NITSCH
Universität Göttingen
WINKLER

MARY JO NYE
University of Oklahoma
BECQUEREL

A. C. OFFORD
Oxford University
LITTLEWOOD

ROBERT C. OLBY
University of Leeds
DARLINGTON; JONES, W. J.; MACLEOD

JANE M. OPPENHEIMER
Bryn Mawr College
HERBST

DONALD E. OSTERBROCK
University of California at Santa Cruz
BOWEN

LARRY OWENS
University of Massachusetts at Amherst
BUSH

JOHN PARASCANDOLA
National Library of Medicine
HUNT, R.

KAREN V. H. PARSHALL
University of Virginia
ALBERT

PHILIP J. PAULY
Rutgers University
OSTERHOUT

CLAUDINE PETIT
Université de Paris
TEISSIER

GIULIANO PICCOLI
Università di Padova
BIANCHI; DAL PIAZ; GORTANI

LEWIS PYENSON
Université de Montréal
BARNES; LAUB; MCLENNAN

BARBARA REEVES
Ohio State University
CORBINO; GARBASSO

NATHAN REINGOLD
Smithsonian Institution
BRONK

ROBIN E. RIDER
University of California at Berkeley
EVANS

MARK RIDLEY
Animal Behavior Research Group, Oxford
DEBEER; GARSTANG;

JOHN S. RIGDEN
American Institute of Physics
DENNISON

ENDERS A. ROBINSON
University of Tulsa
LEVINSON

W. P. DE ROEVER
Amstelveen, the Netherlands
BROUWER

PIERRE ROUTHIER
Paris, France
BARRABÉ

JAMES SANGSTER
Montreal, Canada
BARNES

MENACHEM M. SCHIFFER
Stanford University
BERGMANN

RUDOLF SCHLATTER
Museum zu Allerheiligen, Switzerland
PEYER

ERHARD SCHOLZ
Gesamthochschule Wuppertal
TEICHMÜLLER

H. SCHÖNEBORN
Technische Hochschule Aachen
KRULL

WILFRIED SCHRÖDER
Bremen, West Germany
ERTEL

HELMUT SCHUBERT
Patentstelle für die Deutsche Forschung
MEISNER

HANS-PETER SCHULTZE
University of Kansas
GROSS

MAX SCHWAB
Martin-Luther-Universität Halle-Wittenberg
WALTHER

SILVAN S. SCHWEBER
Brandeis University
WENTZEL

KARL SEEBACH
Ludwig-Maximilians-Universität München
TIETZE

ROBERT W. SEIDEL
The Laser History Project
BIRGE; BRODE

JOHN W. SERVOS
Amherst College
BADGER; SLICHTER

I. I. SHAFRANOVSKII
Plekhanov Mining Institute, Leningrad
BELOV; SHUBNIKOV

ALEXEI N. SHAMIN
Institute of the History of Science and Technology, Moscow
BELOZERSKII; KARGIN; LEBEDEV; PALLADIN; SHEMIAKIN

ELIZABETH NOBLE SHOR
University of California at San Diego
HELLAND-HANSEN; HUBBS; RITTER

D. SIGOGNEAU-RUSSELL
Muséum National d'Histoire Naturelle
LEHMAN, J. P.

BRIAN SKINNER
Yale University
BATEMAN; KNOPF

CYRIL STANLEY SMITH
Cambridge, Massachusetts
MEHL

EMIL L. SMITH
University of California at Los Angeles
STEIN

H. A. M. SNELDERS
Rijksuniversiteit Utrecht
DAM

GEORGE D. SNELL
Jackson Laboratory, Maine
LITTLE

MICHAEL STANLEY
England
TAYLOR, H. S.

G. LEDYARD STEBBINS, JR.
University of California at Davis
ANDERSON; BABCOCK; CLAUSEN; GOODSPEED; RENNER; TURESSON

ERICH E. STEINER
University of Michigan
CLELAND

ANTHONY N. STRANGES
Texas A & M University
GIAUQUE; SCATCHARD

WILFRIED STUBBE
Universität Dusseldorf
MICHAELIS, P.

ROGER H. STUEWER
University of Minnesota
ALLISON

FRANK SULLOWAY
Harvard University
ALLEN, J. A.

CHARLES SUSSKIND
University of California at Berkeley
WATSON-WATT

JOHN C. SWANN
Smithsonian Institution
CHAIN; RICHARDS

CARL P. SWANSON
University of Massachusetts at Amherst
SAX

CONTRIBUTORS TO SUPPLEMENT II

FERENC SZABADVARY
National Museum for Science and Technology, Budapest
POLÁNYI

KRYSZTOF SZYMBORSKI
Skidmore College
LARK-HOROWITZ; WILSON, H. A.

JOSEPH N. TATAREWICZ
Smithsonian Institution
UREY

LEONID TATARINOV
Soviet Academy of Sciences, Moscow
ORLOV

ANGUS E. TAYLOR
University of California, Berkeley
University of California, Los Angeles
FRÉCHET

S. JAMES TAYLOR
University of Virginia
BESICOVITCH

JÜRGEN TEICHMANN
Deutsches Museum, München
POHL

MARIE THARP
(Retired) Lamont Doherty Geological Observatory
HEEZEN

CHRISTIAN THIEL
Universität Erlangen-Nürnberg
LÖWENHEIM

V. V. TIKHOMIROV
Soviet Academy of Sciences, Moscow
BOGDANOV; FEDOROV; NIKOLAEV; STRAKHOV

KEITH J. TINKLER
Brock University
RUSSELL

A. S. TROELSTRA
Universiteit Amsterdam
HEYTING

RUDOLPH TRÜMPY
Eidgenössische Technische Hochschule, Zurich
STAUB

ALBERT W. TUCKER
Princeton University
LEFSCHETZ

JOHN R. G. TURNER
University of Leeds
KETTLEWELL; SHEPPARD

MELVYN C. USSELMAN
University of Western Ontario
ROSSITER

THOMAS G. VALLANCE
University of Sydney
BROWNE; STILLWELL

V. S. VARADARAJAN
University of California at Los Angeles
HARISH-CHANDRA

F. J. VINE
University of East Anglia
HILL, M. N.

A. P. A. VINK
University of Amsterdam
EDELMAN

ROBERT P. WAGNER
University of Texas at Austin
PATTERSON

RAYMOND L. WALKER
Oak Ridge National Laboratory
CAMERON

BYRON H. WAKSMAN
National Multiple Sclerosis Society
WAKSMAN

JOHN H. WARNER
Yale University
FENN; HARVEY

SPENCER R. WEART
American Institute of Physics
COTTON; CREW; KOWARSKI; PEGRAM

LIONEL WEISS
Cornell University
WOLFOWITZ

TREVOR WILLIAMS
Oxford University
BATES

EDWARD O. WILSON
Harvard University
MACARTHUR

LEONARD G. WILSON
University of Minnesota
GARROD

GEORGE WISE
General Electric Corporation
COOLIDGE

WILLIAM R. WOODWARD
University of New Hampshire
STEVENS, S. S.

XIA XIANGRONG
Habei Geological Survey
LI

ERI YAGI
Toyo University
NISHINA

KENZO YAGI
Tokyo University
KUNO

ELLIS L. YOCHELSON
Smithsonian Institution
DUNBAR; MOORE, R. C.

HATTON S. YODER, JR.
Carnegie Institution
ALLEN, E. T.; ADAMS

JOSEF ZEMANN
Universität Wien
MACHATSCHKI

DICTIONARY OF SCIENTIFIC BIOGRAPHY SUPPLEMENT II

ADAMS–ZWICKY

ADAMS, LEASON HEBERLING (*b.* Cherryvale, Kansas, 16 January 1887; *d.* Bethesda, Maryland, 20 August 1969), *geophysics.*

Adams was the son of William Barton Adams and Katherine Heberling Adams. Educated in a one-room schoolhouse in the farm belt of central Illinois, Adams entered the University of Illinois at the age of fifteen and graduated in 1906 with a B.Sc. degree in chemical engineering.

After two years as a chemist with Morris and Company of Chicago and the Missouri Pacific Railroad in St. Louis, Adams became a member of the Technologic Branch of the U.S. Geological Survey in Pittsburgh. In 1910 he began a career at the Geophysical Laboratory of the Carnegie Institution of Washington, becoming its acting director (1936–1937) and director (1938). After retirement in 1952, Adams continued research as a consultant to the director of the National Bureau of Standards, then accepted an invitation from the University of California at Los Angeles to become visiting professor (later professor in residence) of geophysics at the Institute of Geophysics and Planetary Physics. He returned to Washington in 1965, when failing eyesight made further experimental work difficult.

Adams married Jeanette Maude Blaisdell of St. Louis in 1908; they had four children: Leason Blaisdell, William Muirhead, Madeline Jeanette, and Ralston Heberling. After forty-six years of "harmonious partnership," Jeanette Adams died. Adams married Freda R. Ostraw in July 1956; they lived in Pacific Palisades, California, until 1965.

Adams was active in formal organizations such as the Philosophical Society of Washington, the Chemical Society of Washington, the Geological Society of Washington, the Washington Academy of Sciences, and the American Geophysical Union, serving as president of each, as well as more socially oriented scientific clubs, such as the Metachemical Club, the Pick and Hammer Club, and the Cosmos Club. He was an early radio enthusiast, building receivers as the technology advanced. He also designed and supervised the building of two homes, managed a sizable formal flower garden, played golf regularly, built his own sailboat, and contributed to the blazing of a portion of the Appalachian Trail with his colleague J. Frank Schairer.

From 1910 to 1916 Adams was the research associate of John Johnston, who was studying the effects of pressure and temperature on physicochemical systems of importance to geology. He helped design and construct an apparatus for making measurements at pressures up to 2 kilobars and 400°C, the highest conditions achieved simultaneously at that time. They measured the effect of pressure on the melting points of metals, on density, and on the physicochemical behavior of simple solids. In the course of these experiments Adams and Johnston produced new calibration curves for the copper–constantan and the platinum–platinum–rhodium thermocouples that were in "remarkable concordance" with the resistance thermometer measurements of the National Bureau of Standards and the nitrogen-gas thermometer at the Geophysical Laboratory.

With the onset of World War I, the sources of optical glass in Germany were cut off; and the Geophysical Laboratory, because of the expertise of its staff in phase equilibria studies on glass-forming silicate systems, accepted the task of developing optical glasses. One of the principal problems to be solved was the annealing of the formed glasses. Adams and Erskine D. Williamson invented an annealing method based on heat transfer theory that was particularly useful for large blocks. The procedure was successfully applied to the fabrication of the two-hundred-inch mirror for the Mount Palomar telescope. His interest in cooling problems was eventually applied to the earth itself, and in 1924 he published "Temperatures at Moderate Depths Within the Earth." Adams also laid the foundation for the important conclusion that the earth, while solidifying as a magma (silicate liquid), crystallized from the bottom upward, thereby furnishing a reasonable explanation for the segregation of the elements abundant in the crust.

After World War I, Adams turned his attention to the measurement of the elastic constants of rocks at 25°C under high (2–12 kilobars) pressures. He solved the porosity problem in natural rocks by sealing the specimens in a thin metal jacket and immersing the jacketed specimen in an organic fluid. The volume change, measured by piston displacement, was used to calculate the compressibility. More important, Adams, again with Williamson, related the compressibility and density of the rock to the seismic velocities of the primary and secondary waves. By comparing the velocities deduced in the laboratory with those measured in the earth, Adams was able to identify the kinds of rocks with the appropriate elastic properties (for instance, peridotite and eclogite) that might occur deep in the crust of the earth.

A self-taught student of thermodynamics, Adams was persuaded that the validity of those relationships could be tested only in the laboratory. In a series of papers with Ralph E. Gibson he accurately measured the thermodynamic properties of systems such as $NaCl–H_2O$, $K_2SO_4–H_2O$, and $NH_4NO_3–H_2O$. In addition, as a result of his investigations on equilibria of aqueous solutions under pressure with various solid phases, in 1936 he provided the first complete definition of the activity function proposed by Gilbert N. Lewis.

With his acceptance of the duties of the director of the Geophysical Laboratory of the Carnegie Institution of Washington in 1936, Adams gave full attention to administration. During World War II he was appointed chairman of Division I (ballistics) of the Office of Scientific Research and Development, directed by Vannevar Bush. The task was to investigate the erosion of gun barrels resulting from the action of hot propellant gases under high pressure. The team he assembled produced significant results that led to improvements in ordnance. Some of the metals developed were later used in pressure vessels by staff members of the Geophysical Laboratory to investigate hydrothermal systems important to petrology. At the close of World War II he instituted a comprehensive review of the Geophysical Laboratory's programs and greatly strengthened those involving controlled experimental studies.

After retiring from the Geophysical Laboratory, Adams resumed experimental studies at the University of California at Los Angeles, where his principal interests were the kinetics of transitions and polymorphic changes under pressure. He never lost his interest in thermodynamics; his last paper was titled "Enthalpy Changes as Determined from Fusion Curves in Binary Systems."

Recognition of Adams' scientific abilities include the Edward Longstreth Medal of the Franklin Institute in 1924, for his research on the annealing of optical glass; an honorary Sc.D. from Tufts University in 1941; the Presidential Medal for Merit in 1948, for his contributions to the war effort; the Bowie Medal, the highest award of the American Geophysical Union, in 1950; and election to the National Academy of Sciences in 1954. In addition, he held honorary memberships in the Royal Astronomical Society (London), Sigma Xi, and Phi Lambda Upsilon.

BIBLIOGRAPHY

I. ORIGINAL WORKS. Adams' writings include "The Influence of Pressure on the Melting Points of Certain Metals," in *American Journal of Science*, 4th ser., **31** (whole no. 181) (1911), 501–517, written with John Johnston; "On the Effect of High Pressures on the Physical and Chemical Behavior of Solids," *ibid.*, **35** (whole no. 185) (1913), 205–253, written with John Johnston; "The Annealing of Glass," in *Journal of the Franklin Institute*, **190** (1920), 835–870, written with Erskine D. Williamson; "On the Compressibility of Minerals and Rocks at High Pressures," *ibid.*, **195** (1923), 475–529, written with Erskine D. Williamson; "Temperatures at Moderate Depths Within the Earth," in *Journal of the Washington Academy of Sciences*, **14** (1924), 459–472; "The Compressibilities of Dunite and of Basalt Glass and Their Bearing on the Composition of the Earth," in *Proceedings of the National Academy of Sciences*, **12** (1926), 275–283, written with Ralph E. Gibson; "The Melting Curve of Sodium Chloride Dihydrate. An Experimental Study of an Incongruent Melting at Pressures up to Twelve Thousand Atmospheres," in *Journal of the American Chemical Society*, **52** (1930), 4252–4264, written with Ralph E. Gibson; "Equilibrium in Binary Systems Under Pressure. I. An Experimental and Thermodynamic Investigation of the System $NaCl–H_2O$ at 25°," *ibid.*, **53** (1931), 3769–3813; "A Method for the Precise Measurement of Optical Path-Difference, Especially in Stressed Glass," in *Journal of the Franklin Institute*, **216** (1933), 475–504, written with Roy W. Goranson; "A Note on the Stability of Jadeite," in *American Journal of Science*, **251** (1953), 299–308; and "Enthalpy Changes as Determined from Fusion Curves in Binary Systems," *ibid.*, **264** (1966), 543–561, written with Lewis H. Cohen.

An autobiographical article is "Adams, Leason Heberling," in *Modern Men of Science*, **I** (New York, 1966), 1–2.

II. SECONDARY LITERATURE. A complete bibliography of Adams' works is in Ralph E. Gibson, "Leason Heberling

Adams," in *Biographical Memoirs. National Academy of Sciences*, **52** (1980), 3–33. See also Merle A. Tuve, "Twelfth Award of the William Bowie Medal Citation," in *Transactions of the American Geophysical Union*, **31**, no. 3 (1950), 341–343.

H. S. YODER, JR.

AGOL, IZRAIL' IOSIFOVICH (*b.* Bobruisk, Russia, 20 November 1891; *d.* probably in Lubianka prison, Moscow, U.S.S.R., 10 March 1937), *genetics, philosophy of biology*.

The son of a carpenter, Agol was born into a large Jewish family. His sister became involved in revolutionary activities during her teenage years, and he soon followed her example. After his graduation from a Vilnius gymnasium about 1909, he became active in radical politics. He made his living tutoring, composed poetry, and wanted to become a writer. In 1915 he joined the Russian Social Democratic Party. Shortly thereafter, he was drafted into the Russian army and fought in World War I.

At the time of the October 1917 revolution, Agol was in Belorussia, where he immediately got involved in party activities. He first worked as secretary of the local Bolshevik Party committee in Bobruisk, then in a party regional committee in Minsk, and after the formation of the Belorussian Soviet Republic (1919) he became a member of its Central Executive Committee. In 1919 in Vilnius he served as secretary of finance in the government of the short-lived Lithuanian-Belorussian Soviet Republic. During the civil war (1919–1921) he fought on the side of the Reds and served as assistant manager in the supply section of the 16th Army, stationed on the western front.

In 1921 Agol moved to Moscow, where he studied medicine at Moscow University and wrote articles for *Pravda* and later *Trud*. Upon graduating from the university in 1923, he worked as a psychiatrist, and at the end of 1924 he entered the Institute of Red Professors, where he first studied in the department of philosophy and then transferred into the department of natural science, graduating in 1927. From 1925 to 1928, Agol worked in the laboratory of B. M. Zavadovskii at the Ia. M. Sverdlov Communist University. There, focusing on morphogenesis and the possible inheritance of acquired characteristics, he learned surgical techniques for removing the thyroid glands of hens and attempted to repeat the experiments of Guyer and Smith on the inheritance of immune blood characteristics. In 1926 he wrote sympathetic obituaries of Paul Kammerer, characterizing him as a victim of the machinations of Western capital but expressing reservations about his views on the inheritance of acquired characteristics.

Agol's fascination with Marxist philosophy permeated his organizational, scholarly, and scientific work. In the mid 1920's Soviet dialectical materialists agreed in their condemnation of "idealist" positions, but were beginning to split into two camps over whether the emphasis should be on materialism (the mechanists) or dialectics (dialecticians, or Deborinites—followers of A. M. Deborin). Agol sided with the Deborinites and published many articles, books, and pamphlets criticizing vitalistic and mechanistic approaches to embryology, heredity, and evolution.

Since 1926 Serebrovskii had been advocating Morganist genetics as the only theory of inheritance fully consonant with dialectical materialism. Initially Agol did not entirely support Serebrovskii's position, but he became a wholehearted convert after H. J. Muller's demonstration of X-ray mutagenesis in 1927. Agol attached himself to Serebrovskii's genetics laboratory at the Moscow Zootechnical Institute and participated in his replication of Muller's results, together with N. P. Dubinin, V. N. Slepkov, and V. E. Al'tshuler.

In late 1928 Agol was appointed to head the Timiriazev Biological Institute, a center of Marxist Lamarckian research under Glavnauka, the Central Administration of Scientific Research Institutes of the Commissariat of Education. In 1929 he was elected to the Communist Academy and, at its meeting of 8–13 April 1929, he argued that genetics was the Deborinite approach to inheritance, criticized Lamarckism as "mechanistic," and called for its official condemnation and its expulsion from the academy. Although this position was not supported by A. M. Deborin and O. Iu. Schmidt, the organizers of the meeting, its final resolution instructed Agol to subordinate the work of the Timiriazev Institute to the Communist Academy, to discharge certain Lamarckians, and to organize within the Timiriazev Institute a genetics laboratory, to be headed by A. S. Serebrovskii.

Agol became an important member of the staff of the laboratory, where Serebrovskii gathered a group of young biologists that included N. I. Shapiro. Together with N. P. Dubinin, B. N. Sidorov, and others from Serebrovskii's laboratory at the Moscow Zootechnical Institute, the group developed (1929–1931) a theory of gene structure known as step-allelism, based on complementation maps of the various alleles of the *scute* region, located at the tip of the X chromosome in *Drosophila melanogaster*, which influences the fly's head and thoracic bristles. Although many of their interpretations proved unsound, theirs was one of the earliest attempts to analyze gene structure through complementation mapping. The theory is more fully discussed by Carlson (1966).

Agol gained an international reputation for his work on *scute*[4], which overlapped other alleles and disproved the hypothesis that the various "step-alleles" were subgenes that could be linearly arranged. This work (and his party membership) led Glavnauka to nominate him for a Rockefeller Foundation International Fellowship to study in the United States. Together with Solomon G. Levit, Agol spent 1931 working with Muller at the Zoology Department of the University of Texas in Austin. Agol's descriptions of communism and the Soviet Union may have influenced Muller's politics; in 1933 he moved to the Soviet Union and worked there until 1937.

While Agol was in Texas, ideological shifts were taking place in his homeland. At the time of the first Five-Year Plan (1929–1932), the Communist Party began to exert direct control over all aspects of Soviet life, and a new and more rigid party line was beginning to be imposed. In particular, the views of Deborin, which had apparently triumphed in 1929, were now considered an unacceptable ideological deviation. As a militant Deborinite, Agol came under attack as a "Menshevizing idealist." No information is available about his activities in the months following his return to the Soviet Union in early 1932. By the fall, however, the ideological storm had abated. Agol was posted to Kiev and worked in the All-Ukrainian Association of Marxist-Leninist Scientific Research Institutes, serving as its vice-president (1933–1934). In 1934 he was elected to the Ukrainian Academy of Sciences in genetics. Within that academy's Institute of Zoology and Biology he organized a genetics division, which he directed (1934–1936).

In 1936 Agol returned to Moscow to accept a high administrative post in Glavnauka. Later that year, however, the purges began in earnest, and party members who stood accused of earlier ideological deviations and had worked abroad were especially at risk. On 19 December 1936 *Pravda* announced that Agol had been arrested as a "Trotskyite bandit," a "Menshevizing idealist," and an "enemy of the people." He was shot on 10 March 1937—the day after his mentor, Muller, left Russia. Thereafter Agol officially became an "unperson" in the Soviet Union, and censors deleted his name even from lists of scientific references. Although he was rehabilitated in the 1950's, he is rarely mentioned in Soviet sources.

BIBLIOGRAPHY

I. Original Works. Agol's publications on the philosophy of biology include *F. Engel's* (Moscow, 1920); "Dialektika i metafizika v biologii" (Dialectics and metaphysics in biology), in *Pod znamenem marksizma,* 1926, no. 3, 118–150; "Pamiati professora P. Kammerera" (In memory of Professor P. Kammerer), in *Antireligioznik,* 1926, no. 11, 29–30; "Predislovie" (Preface), in Paul Kammerer, *Zagadka nasledstvennosti* (Moscow and Leningrad, 1927), iii–viii (translation of *Das Rätsel der Vererbung*); *Dialekticheskii metod i evoliutsionnaia teoriia* (Dialectical method and evolutionary theory; Moscow and Leningrad, 1927); "Problema organicheskoi tselesoobraznosti" (The problem of organic purposiveness), in *Estestvoznanie i marksizm,* 1930, no. 1(5), 3–20; "Neovitalizm" (Neovitalism), in *Bol'shaia sovetskaia entsiklopedia,* XI (Moscow, 1930), 255–288; "Darvin i darvinizm" (Darwin and Darwinism), *ibid.*, XX (Moscow, 1930), 442–470; and *Vitalizm i marksizm* (Vitalism and Marxism), 3rd ed. (Moscow, 1932).

For examples of his publications on genetics, see "Poluchenie mutatsii rentgenovskimi luchami u Drosophila melanogaster" (Mutations induced in *Drosophila melanogaster* by X rays), in *Zhurnal eksperimental'noi biologii* ser. A, **4,** no. 3–4 (1928), 161–180, written with A. S. Serebrovskii, N. P. Dubinin, V. N. Slepkov, and V. E. Al'tshuler; "Stupenchatyi allelomorfizm u *Drosophila melanogaster*" (Step-allelomorphism in *Drosophila melanogaster*), *ibid.*, **5,** no. 2 (1929), 86–101; and "K voprosu o zarodyshevom puti u Drosophila melanogaster" (Concerning the embryonic development of *Drosophila melanogaster*), *ibid.*, **6,** no. 4 (1930), 369–372.

For Agol's account of his childhood, see *Khochu zhit': Provest'* (I want to live: A narrative; Moscow, 1936).

II. Secondary Literature. There are no secondary works on Agol, but there are occasional references to him in Elof Axel Carlson, *The Gene* (Philadelphia, 1966) and *Genes, Radiation, and Society* (Ithaca, N.Y., 1981); A. E. Gaissinovitch, *Zarozhdenie i razvitie genetiki* (The birth and development of genetics; Moscow, 1988) and "The Origins of Soviet Genetics and the Struggle with Lamarckism, 1922–1929," Mark B. Adams, trans., in *Journal of the History of Biology,* **13,** no. 1 (1980), 1–51; and David Joravsky, *Soviet Marxism and Natural Science 1917–1932* (New York, 1961).

Mark B. Adams

AKHIEZER, NAUM IL'ICH (*b.* Cherikov, Belorussia, 6 March 1901; *d.* Kharkov, U.S.S.R., 3 June 1980), *mathematics.*

Akhiezer graduated from the Kiev Institute of People's Education in 1923, then continued his studies for the candidate degree at the University of Kiev under D. A. Grave, who was well known in the field of algebra. In his candidate's thesis, on his aerodynamic research, he applied methods of complex analysis to aerodynamic problems. He was the first to obtain a formula for the conformal mapping of a double-connected polygonal domain onto a ring.

Akhiezer was very much involved in problems of approximation theory. In 1928 he solved a difficult problem: Among all the polynomials of degree n with three fixed senior coefficients, find the least deviation from the zero polynomial in a given interval of the real axis. Akhiezer showed that the solution of this problem can be obtained with the help of Schottky's functions. This result gave impetus to further development of the classical theory of the least deviation from the zero polynomials of P. L. Chebyshev, N. E. Zolotarev, and V. A. Markov. Akhiezer later showed that the problem of the least deviation from the zero polynomial in the case where k senior coefficients are fixed can be reduced to the problem of finding some domain that is the complex plane with k segmental sections along the real axis, and to the construction of the Green function of this domain.

In 1933 Akhiezer moved to Kharkov, where he was head of the complex analysis department and president of the Kharkov Mathematical Society for many years. In 1934 he was elected corresponding member of the Ukrainian Academy of Sciences. At that time Akhiezer and M. G. Krein began to investigate the L-moments problem: Find a density $\rho(t)$ $(-\infty \leq a \leq t \leq b \leq \infty)$ of a mass distribution with given moments that satisfies an additional condition $0 \leq \rho(t) \leq L$. Their researches in this field were summarized in *Some Questions in the Theory of Moments* (1933). They also found the precise value of a constant in the theorem of Dunham Jackson concerning the approximation of a periodic function by trigonometric polynomials (this value was obtained independently by Jean Favard).

During World War II, Akhiezer was at the Alma-Ata Mining Institute (1941–1943) and the Moscow Power Engineering Institute (1943–1947). He returned to Kharkov in 1947, the year in which his book on approximation theory was published. In it he combined new ideas of functional analysis with classical methods and presented his own results. In 1949 Akhiezer was rewarded by the Chebyshev Prize for this book.

Further investigations were inspired by S. N. Bernstein, one of the founders of approximation theory, who in 1924 had formulated the following problem: Let $\phi(x)(-\infty\ x < \infty)$ be a function such that $\inf_{-\infty<x<\infty}\phi(x) > 0$ and $\lim_{|x|\to\infty} x^{2n}/\phi(x) = 0$ $(n = 0,1,2,\cdots)$. Let us introduce the space C_ϕ of continuous functions $f(t)$ such that $\lim_{|x|\to\infty} f(x)/\phi(x) = 0$ and let $\| f \| = \sup_{-\infty<x<\infty} \frac{|f(x)|}{\phi(x)}$ be the norm in C_ϕ. Find the necessary and sufficient conditions on ϕ such that polynomials will form a dense set of C_ϕ. In their joint work on this problem, Akhiezer and K. I. Babenko studied an important class M_ϕ of polynomials $P(x)$ satisfying the inequality $|P(x)| \leq \phi(x)$ and introduced the expression

$$J_\phi = \sup_{Q\in M_\phi} \int_{-\infty}^{+\infty} \frac{\ln|P(x)|}{1+x^2}dx.$$

In 1953 Akhiezer and Bernstein found that the condition $J_\Psi = \infty$, where $\Psi = (1 + x^2)^{1/2}\phi \times$, is a sufficient and necessary condition for the completeness of the set of polynomials in C_ϕ. This result established the complete solution of Bernstein's problem. In another group of works Akhiezer studied the following problem: Among all the entire functions of a finite degree with given values or derivatives at a finite set of given points in the complex plane, find the least deviation from the zero entire function. Akhiezer also obtained a generalization of Bernstein's inequality for the derivative of an entire function of a finite degree. Later he and B. Ia. Levin extended these results to important classes of many-valued functions. In 1961 another important book, *The Classical Moment Problem*, was published. At the same time Akhiezer worked on problems connected with "continual analogues" of the classical problem of moments. He also further developed a work by Mark Kac on the Fredholm determinants of the Wiener-Hopf equation with a hermitian kernel.

Akhiezer investigated the orthogonal polynomials with respect to a weight function on a set of arcs of a circle or on a set of intervals of the real axis; he then applied these methods to inverse problems in spectral analysis. Let

$$L = -\frac{d^2}{dx^2} + q(x)(x\geq 0, y(0) + hy'(0) = 0)$$

be a Sturm-Liouville operator, where $q(x)$ is a real continuous function and h is a real constant. Let us assume that the spectrum of L has g gaps. Akhiezer constructed and studied a hyperelliptical Riemannian surface of genus g, associated with the operator L; the corresponding Bloch function $E(\lambda;x)$ on this surface is single-valued with respect to λ. The set $\{P_1(x),\cdots,P_g(x)\}$ of its zeros plays an important role; B. A. Dubrovin later showed that the potential $q(x)$ can be expressed explicitly with the help of the functions $\{P_1(x),\cdots,P_g(x)\}$. These results are presented in the appendix to the third edition of *Theory of Operators in Hilbert Space*. Using the above results, S. P. Novikov, B. A. Dubrovin, and more recently V. A. Marchenko obtained solutions of remarkable classes of nonlinear partial differential equations of the Korteweg–de Vries type in an explicit and effective form.

BIBLIOGRAPHY

I. ORIGINAL WORKS. Akhiezer wrote 130 papers and 8 books. The latter include *Theory of Approximation*, Charles J. Hyman, trans. (New York, 1956); *Some Questions in the Theory of Moments*, W. Fleming and D. Prill, trans. (Providence, R.I., 1962), written with M. G. Krein; *The Classical Moment Problem*, N. Kemmer, trans. (New York, 1965); and *Theory of Operators in Hilbert Space*, 2 vols., E. R. Dawson, trans. and W. N. Everitt, ed. (Boston, 1981; 3rd ed., 1987), written with I. M. Glassman.

II. SECONDARY LITERATURE. M. Berezanskii, A. N. Kolmogorov, M. G. Krein, B. Ia. Levin, B. M. Levitan, and V. A. Marchenko, "Naum Il'ich Akhiezer (k semidesiatiletiiu so dnia rozhdeniia" (. . . to the seventieth birthday), in *Uspekhi matematicheskikh nauk*, **26**, no. 6 (1971), 257–261; M. G. Krein and B. Ia. Levin, "Naum Il'ich Akhiezer (k shestidesiatiletiiu so dnia ruzhdeniia)" ([. . . to the sixtieth birthday]), *ibid.*, **16**, no. 4 (1961), 223–232; and, V. A. Marchenko, *Nonlinear Equations and Operator Algebra*, V. I. Rublinetskii, trans. (Dordrecht and Boston, 1988).

MOSHE LIVŠIČ

ALBERT, ABRAHAM ADRIAN (*b.* Chicago, Illinois, 9 November 1905; *d.* Chicago, 6 June 1972), *mathematics*.

Adrian Albert was the second of three children of Elias and Fannie Fradkin Albert. Albert's parents, originally from Russia, both immigrated to the United States, but his father's route to America had begun long before he met and married Fannie Fradkin, who was twenty years his junior. At the age of fourteen Elias ran away from his home in Vilnius for a new life in England. On his arrival he abandoned his Russian name (which remains unknown) and assumed the English surname Albert in honor of the prince consort. Although Elias taught school in England, on coming to the United States he worked as a retail merchant. This allowed him to provide a reasonably comfortable life for the family he insisted on raising in a formally orthodox Jewish, if not deeply religious, atmosphere.

Except for the years 1914 to 1916, when his family lived in Iron Mountain, Michigan, Adrian received all of his elementary, secondary, and university education in Chicago. In 1922 he entered the University of Chicago, obtaining his B.S. degree in 1926, his M.S. in 1927, and his Ph.D., under Leonard E. Dickson, in 1928. On 18 December 1927 Albert married Frieda Davis, and they took his National Research Council fellowship to Princeton for the academic year 1928–1929. Following a two-year instructorship at Columbia University, the Alberts returned to the University of Chicago permanently in 1931. Beginning as an assistant professor of mathematics, Albert moved steadily through the academic ranks, rising to the rank of full professor in 1941. In 1943 he was elected to the National Academy of Sciences, and in 1960 his university honored him with the E. H. Moore Distinguished Service Professorship. At Chicago, Albert served as the chairman of the mathematics department from 1958 to 1962 and as dean of the Division of Physical Sciences from 1962 to 1971. Albert was recognized for his leadership and service to mathematics in 1965 when he was elected president of the American Mathematical Society (AMS), a position he held through 1966. After he stepped down from his deanship in 1971, at the mandatory retirement age of sixty-five, he swiftly succumbed to the diabetes that had plagued him for many years.

Albert's contributions to mathematics ranged over three related algebraic areas: associative and nonassociative algebras and Riemann matrices. In the late 1920's Dickson's presence at the University of Chicago made it a world center for the study of algebras. In his 1923 book *Algebras and Their Arithmetics*, and in its more influential German translation of 1927, Dickson extended the theory of algebras that Joseph H. M. Wedderburn had so elegantly set up in his paper "On Hypercomplex Numbers." Wedderburn showed that the classification of finite-dimensional associative algebras over a field essentially reduced to a classification of the division algebras. From 1928 to 1932 researchers in this area pushed toward the classification of the finite-dimensional division algebras over the field of rational numbers $\boldsymbol{Q}$. Albert was edged out in the race for this result in 1932 by the German team of Richard Brauer, Helmut Hasse, and Emmy Noether after independently hitting upon many of their ideas, most notably that of the Brauer group. At Hasse's urging, he and Albert coauthored a paper for the *Transactions of the American Mathematical Society* later in 1932 detailing Albert's contributions to this result.

Putting this setback behind him, Albert continued to work actively on associative algebras and focused most notably on determining whether all finite-dimensional central division algebras are crossed products, a question finally settled by Shimshon Amitsur in 1972. Albert's book, *Structure of Algebras*, published as a colloquium volume by the AMS in 1939, remains a definitive work on the theory of algebras and testifies to his achievements in this field.

In spite of his interest in associative algebras, however, Albert profited perhaps more from his contact with the geometer Solomon Lefschetz than with the increasingly reclusive Wedderburn during his year at Princeton. Lefschetz introduced Albert to multiplication algebras of Riemann matrices, constructs from algebraic geometry that Hermann Weyl made formally algebraic in his 1934 paper "On Generalized Riemann Matrices," published in the *Annals of Mathematics*. In a series of articles that also appeared in the *Annals* in 1934 and 1935, Albert gave necessary and sufficient conditions for a division algebra over Q to be the multiplication algebra of a Riemann matrix, the main research problem in the area. For this work he was awarded the Cole Prize in algebra by the AMS in 1939.

During the war years Albert continued his pure research efforts, concentrating on nonassociative algebras. In 1932 the physicist Pascual Jordan had defined the so-called Jordan algebra over a field for use in quantum mechanics. The algebra J over a field F of characteristic unequal to two is a Jordan algebra provided that for a, b in J, $ab = ba$ and $(a^2b)a = a^2(ba)$. Thus Jordan algebras are commutative but nonassociative. In the spirit of Wedderburn before him, Albert developed the basic structure theory of these algebras and published his main results in 1947. In addition Albert contributed to the war effort through his participation in the Applied Mathematics Group at Northwestern University, serving as its associate director in 1944 and 1945. He was also interested in the interrelations between pure mathematics and cryptography and lectured on this subject at the AMS regional meeting in 1941 at Manhattan, Kansas.

Throughout the 1950's and 1960's Albert continued his research on nonassociative algebras and frequently returned to associative questions. It was also during these decades that he contributed significantly to mathematics at the political level. He was instrumental in the late 1940's in securing government research grants for mathematics commensurate with those awarded in the other sciences. In 1950 he served on the committee to draft the mathematics budget for the newly formed National Science Foundation, and from January 1955 to June 1957 he chaired the "Albert Committee," which evaluated training and research potential in the mathematical sciences in the United States. He acted as consultant to the Rand Corporation and to the National Security Agency, and as trustee to both the Institute for Advanced Study and the Institute for Defense Analysis, after having directed the latter's Communications Research Division at Princeton from 1961 to 1962. At the international level Albert was elected vice president of the International Mathematical Union in 1970 to serve a four-year term beginning on 1 January 1971.

Although Albert's research on Riemann matrices and especially on nonassociative algebras was important, he made his primary contribution to mathematics in the field of associative algebras. His work in that area completed a major chapter in the history of algebra that had begun in 1907 with the foundational results of Joseph H. M. Wedderburn.

BIBLIOGRAPHY

I. Original Works. Albert wrote or edited eight books and 141 papers during his career. Many American graduate students in mathematics received their first introduction to advanced algebraic concepts from Albert's classic text, *Modern Higher Algebra* (Chicago, 1937), while specialists consulted his treatise *Structure of Algebras* (Providence, R.I., 1939), a work that remains one of the standard sources on the theory of algebras. Indicative of his deep interest in teaching, five of his other books were textbooks on algebra and geometry aimed at college students of various levels of sophistication.

Albert's original research focused primarily on three algebraic topics: associative and nonassociative algebras and Riemann matrices. Among his most noteworthy contributions to these areas are "A Determination of All Normal Division Algebras in Sixteen Units," in *Transactions of the American Mathematical Society*, **31** (1929), 253–260; "On Direct Products," *ibid.*, **33** (1931), 690–711; "Normal Division Algebras of Degree Four over an Algebraic Field," *ibid.*, **34** (1932), 363–372; "A Determination of All Normal Division Algebras over an Algebraic Number Field," *ibid.*, 722–726, written with Helmut Hasse; "On the Construction of Riemann Matrices: I," in *Annals of Mathematics*, 2nd ser., **35** (1934), 1–28; "Normal Division Algebras of Degree 4 over F of Characteristic 2," in *American Journal of Mathematics*, **56** (1934), 75–86; "A Solution of the Principal Problem in the Theory of Riemann Matrices," in *Annals of Mathematics*, 2nd ser., **35** (1934), 500–515; "On the Construction of Riemann Matrices: II," *ibid.*, **36** (1935), 376–394; "Normal Division Algebras of Degree p^e over F of Characteristic p," in *Transactions of the American Mathematical Society*, **39** (1936), 183–188; "Simple Algebras of Degree p^e over a Centrum of Characteristic p," *ibid.*,

40 (1936), 112–126; "Non-associative Algebras: I, Fundamental Concepts and Isotopy," in *Annals of Mathematics*, 2nd ser., **43** (1942), 685–707; "A Structure Theory for Jordan Algebras," *ibid.*, 2nd ser., **48** (1947), 546–567; "A Theory of Power-Associative Commutative Algebras," in *Transactions of the American Mathematical Society*, **69** (1950), 503–527; "A Construction of Exceptional Jordan Division Algebras," in *Annals of Mathematics*, 2nd ser., **67** (1958), 1–28; and "On Exceptional Jordan Division Algebras," in *Pacific Journal of Mathematics*, **15** (1965), 377–404.

II. SECONDARY LITERATURE. Four obituary notices appeared in American journals. Nathan Jacobson's article in the *Bulletin of the American Mathematical Society*, n.s. **80** (1974), 1075–1100, systematically details Albert's most significant research and concludes with a complete bibliography of Albert's works. Irving Kaplansky contributed a less mathematical and more biographical sketch to *Biographical Memoirs. National Academy of Sciences*, **51** (1980), 3–22, which included a list of Albert's twenty-nine doctoral students. One of these students, Daniel Zelinsky, wrote a more personal memoir for the *American Mathematical Monthly*, **80** (1973), 661–665. Finally, the issue of *Scripta Mathematica*, **29** (1973), which was to have been dedicated to Albert on his sixty-fifth birthday, served instead as a memorial volume and was prefaced by a short tribute to his life and work by his longtime friend and colleague I. N. Herstein.

See also Shimshon Amitsur, "On Central Division Algebras," in *Israel Journal of Mathematics*, **12** (1972), 408–420; David Kahn, *The Codebreakers: The Story of Secret Writing* (London, 1967); and Joseph H. M. Wedderburn, "On Hypercomplex Numbers," in *Proceedings of the London Mathematical Society*, 2nd ser., **6** (1907), 77–118.

KAREN HUNGER PARSHALL

ALBRIGHT, FULLER (*b.* Buffalo, New York, 12 January 1900; *d.* Boston, Massachusetts, 8 December 1969), *medicine, endocrinology.*

Albright was among a select group (including Herbert M. Evans, Edward C. Kendall, Robert Loeb, John P. Peters, and Lawson Wilkins) who forged the approach to clinical research in medicine and endocrinology during the middle third of the twentieth century. For more than twenty years he dominated the field of parathyroid physiology, until his career was terminated by illness in 1956. He held the rank of associate professor of medicine at Harvard Medical School and was physician in medicine at Massachusetts General Hospital at the time of his retirement. Albright was president of the American Society for Clinical Investigation (1943–1944) and the Association for the Study of Internal Secretions (1946–1947). He was elected to the National Academy of Sciences in 1955.

In 1897 Albright's father, John Joseph Albright, an affluent widower with three children, married Susan Fuller, a Smith College graduate. Fuller Albright, the third of their five children, was reared in a happy, close-knit family amid the advantages of wealth. He attended the Nichols School in Buffalo (founded by his father) and entered Harvard at age sixteen, graduating cum laude three years later. He enrolled at Harvard Medical School in 1920 and was an "exceptionally good" student. In his final year he gave a paper to the Boylston Society titled "The Physiology and Physiological Pathology of Calcium," with Joseph C. Aub as his sponsor. Aub's patronage was fortunate, for aside from interests in lead poisoning and calcium metabolism, he had trained in the Cornell division's metabolism unit at Bellevue, a prototype for the future Ward 4 research unit at Harvard. Albright graduated in 1924 and interned at Massachusetts General Hospital in 1925, staying another year as Aub's research fellow to work on calcium metabolism. Also in 1925, Ward 4 research opened and James B. Collip isolated parathormone, the principal ligand of calcium regulation.

Following a year as assistant resident at Johns Hopkins (1927), his interest in parathyroid function led Albright to join Jacob Erdheim in Vienna (1928–1929). Erdheim, a prolific investigator and brilliant bone pathologist, had noted tetany in parathyroidectomized animals as early as 1906. Albright frequently commented in later years that Erdheim "knew more about disease processes than any other living man." Erdheim's influence undoubtedly routed any thoughts of private practice, for in 1929 Albright joined the Harvard medical faculty and the staff of Massachuchusetts General Hospital, where he remained until his retirement.

Albright's marriage in 1933, to Claire Birge of New York, was of significance to his professional life. In 1935 he developed signs of Parkinson's disease, and it is doubtful that he could have maintained the strenuous pace of his later activities without her extraordinary courage, as well as the loyal support and physical care given by their sons, Birge and Reed. Though he was hampered by illness, his output was prolific in spite of the relatively short span of years in which he had to work.

In the late 1920's endocrinology was still a descriptive area of medicine—something of a freak show. Albright's work brought biochemical techniques and a rigorous scientific discipline into the field, compelling others to restructure their qualitative method of thought.

Albright began calcium balance studies with Walter Bauer and Aub in the new research unit (later renamed the Mallinckrodt Ward). Although the Cornell group originated the modern metabolic unit, Albright perfected it through his exploitation of the routine. The accuracy of his studies remained unchallenged until the cumbersome technique of that day yielded to newer methodology (isotopic dilution techniques). These 1929 balance studies with Aub, along with the first case report of idiopathic hypoparathyroidism, by Albright and his close friend Read Ellsworth, were the earliest analyses using controlled calcium diets. Clarifying parathormone's action on calcium and phosphorus excretion (1930), Albright highlighted the role of bone as an available reserve supply of calcium and underscored total protein as a cardinal measurement in the assessment of serum calcium. In 1931 Albright, with J. R. Cockrill and Ellsworth, showed that the inverse relationship of serum calcium and phosphorus was coupled to parathormone activity, the former varying directly with parathormone levels and the latter inversely with increased renal excretion.

In 1934, on the basis of a large series of patients with primary hyperparathyroidism, Albright gave a detailed description of the complicating bony lesions (osteitis fibrosa generalisata) and nephrocalcinosis (Albright's expression), further differentiating between "diffuse hyperplasia of all parathyroid glands rather than a solitary adenoma of one." He organized the Stone Clinic—later dubbed "the quarry"—that uncovered eighty-nine cases of hyperparathyroidism by 1947.

Albright in 1937 reported a nonheritable condition with seemingly unrelated findings: areas of cystic bone changes (without metabolic derangement), distinctive skin pigmentations, and the precocious onset of puberty in girls—now cataloged as "Albright's syndrome." His comment—"If pathological manifestations which at first seem to be totally disconnected are found to occur together in a sufficient series of patients, some relation between them is apparent"—forecast an eponymic kismet for Albright's co-workers: Albright coauthored at least eight descriptions of new syndromes that were named for himself and various colleagues.

Between 1930 and 1945 Albright evolved a classification of metabolic bone disease (Albright's coinage) that charted the differences between osteoporosis, osteomalacia, and osteitis fibrosa, and emphasized that osteoporosis was not a calcium derangement but rather a defect in protein metabolism resulting in decreased formation of bone matrix. Osteomalacia was characterized as a failure of mineralization, and osteitis fibrosa as bone destruction. Albright's name became inseparable from the condition known as postmenopausal osteoporosis (his label) because he correctly related the osteopenia to estrogen withdrawal. Before 1940 neither physiological menopause nor artificial withdrawal of estrogen (oophorectomy) had been a recognized cause of decreased bone mass.

In 1942, with C. H. Burnett, P. H. Smith, and W. Parson, Albright reported a group of patients with short stature, hypocalcemia, ectopic calcifications, mental retardation, short metacarpal and metatarsal bones, and insensitivity to normal parathormone—a constellation Albright designated in his milestone paper "Pseudohypoparathyroidism [PHP]—An Example of Seabright-Bantam Syndrome" [sic]. In it he erected a theoretical scaffold based on four observations: (1) the Sebright bantam rooster has female feathering with normal androgen production; (2) the American male Indian is beardless with normal testosterone; (3) there existed "patients with low basal metabolic rates without other evidence of hypothyroidism"; and (4) patients with PHP were unresponsive to normal parathormone. Albright concluded that all four phenomena were examples of end organ failure and inferred that renal cell receptors in these patients were unresponsive to parathormone.

This was the first instance in which blunting of a normal response (hormonal resistance) had been contemplated. An accepted mechanism today, at the time this was an inspired orchestration of ideas. Albright assembled a new series of patients in 1952 (variations on a theme of PHP) with the same phenotypic findings but lacking calcium and phosphorus changes. He tagged the disorder "pseudopseudohypoparathyroidism," theorizing that both conditions were expressions of different mutations of the same gene. Questions persist about these reciprocally related genetic conditions, and the specific gene awaits identification. Present data concerning end organ resistance in patients with PHP show a reduction of regulatory protein in the receptor cell membranes coupled to the adenylate cyclase system of the kidney and other tissues. Albright's analogy of the Sebright bantam reconciles with present data only insofar as it is an example of hormonal resistance—the defect in these cockerels is related to the conversion of androgen to estrogen (testosterone to estradiol) by their skin cells rather than an end organ defect, that is, the defect is proximal to the end organ.

In 1941 Albright hypothesized that a metastatic kidney tumor (nonendocrine) was capable of secreting a parathormone-like substance. More than an improvisational abstraction, this theory was monumentally prescient: at least a dozen or more polypeptide messengers have been involved with ectopic hormonal secretion. In 1948 Albright, with Edward C. Reifenstein, published the monograph *Parathyroid Glands and Metabolic Bone Disease*, culminating more than twenty years of research on calcium metabolism. Encyclopedic in scope, it detailed many syndromes first described by Albright, including vitamin D–resistant rickets, milk-alkali syndrome, and osteomalacia secondary to steatorrhea. Its text still provides a footing for most of what is accepted about parathyroid physiology.

With the increasing availability of biochemical information in the 1930's, Albright initiated studies on the pituitary-gonad and pituitary-adrenal axes. Pathophysiological information on menstrual dysfunction was sparse; however, by 1936 he had partially resolved the nature of amenorrhea in women of childbearing age, discriminating between primary ovarian failure and pituitary hypofunction. His biological assay methods (crude by present standards) for measuring pituitary gonadotropins and, later, follicle-stimulating hormone (FSH) in the urine (1943), led to further study of menopause and of testicular dysfunctions. Using these estimations as "hormonal measuring sticks," he defined menopause as a "primary condition in the ovary, and not secondary to hypofunction of the anterior pituitary," with the demonstration of an increase in FSH after menopause, that is, "menopause is a physiological ovarian amenorrhea."

Albright's convincing evidence that menorrhagia was related to hypoprogesteronism ended an era of inappropriate hysterectomy for functional uterine bleeding. In 1938 he introduced the phrase "medical curettage" (suggested by J. S. L. Browne), advocating progesterone administration for five to six days every six weeks, thus restoring a normal menstrual cycle.

Albright's classic description, written with his research fellow, Henry E. Klinefelter, Jr., and with E. C. Reifenstein (1942), of males with gynecomastia, aspermatogenesis, and increased secretion of FSH (Klinefelter's syndrome) signaled a new generation of research (chromosomal mapping) directed to the study of genetic disorders. In 1956 the discovery of the presence of superfluous X-chromatin material in these patients (47XXY karyotype) clearly established a genetic transmission of the syndrome.

In 1943 Albright delivered his landmark Harvey Lecture "Cushing's Syndrome: Its Pathological Physiology, Its Relationship to the Adrenogenital Syndrome, and Its Connection with the Problem of the Reaction of the Body to Injurious Agents (Alarm Reaction of Selye)." The conceptual fusion of ideas and their source, relating the rationale for appropriate corticosteroid therapy and its pitfalls, have been forgotten; however, the alpha lessons can be abstracted from this disquisition. Using elegant carbohydrate balance studies, Albright elaborated upon some of the fundamental distinctions between the "S" (sugar) hormone and the "N" (nitrogen-sparing, testosterone-like) hormone of the adrenal cortices. His theory, while oversimplified, is fundamentally correct: excess production of the former causes Cushing's syndrome, and excess secretion of the latter causes adrenogenital syndrome. Overabundant "S" hormone (glucocorticoid intoxication) was properly viewed by Albright as an antianabolic syndrome that was adrenal in origin, whether secondary to anterior pituitary stimulation (basophilic adenoma) or primary in the adrenals (cortical hyperplasia or adenoma).

As Albright anticipated, other ectopic hormonal syndromes have come to light. Adrenocorticotropic substances from both endocrine and nonendocrine tumors may produce a syndrome of hypercortisolism. The subtle paradigms relating female secondary sexual characteristics to adrenal origin, and the view that the counterpart (adrenogenital syndrome) to Cushing's syndrome is overvirilization by the adrenal cortices, are Albrightian in origin. The endocrine rhetoric of hyper-this or hypo-that, the concatenation of circular diagrams bristling with modifying arrows—all shorthand expressions of hormonal circuitry—emanate from Albright's chorographic style.

Convinced that he was a burden to those who loved him, Albright reached an agonizing decision in 1956—to undergo brain surgery (chemothalamectomy). The results were catastrophic. He hemorrhaged into his brain on the third postoperative day and spent the last thirteen years of his life an akinetic mute, bound to a wheelchair at Massachusetts General Hospital.

Acknowledged at the time of his death as the doyen of parathyroid physiology, Albright bequeathed a rich legacy. He left behind a dynamic model of bone resorption and remodeling that paired with a working conceptualization of the pituitary-ovarian and pituitary-adrenal axes, provided a rational approach to corticosteroid therapy and a solid foundation for clinical research in endocrinology. Of deep concern to Albright was the peril of segregating clinical practice from clinical investigation. He repeatedly stressed that the care of the sick was the primary objective of the clinical investigator, and "only secondarily, if at all, the study of laboratory animals." Moreover, he was acutely aware that scientific theory ought not to be written in concrete, often making the point that each of his hypotheses was "subject to change tomorrow."

BIBLIOGRAPHY

I. ORIGINAL WORKS. All of Albright's research papers and articles, including material relevant to the preceding text, as well as his appointments, honors, and awards, are listed in A. Leaf, "Fuller Albright," in *Biographical Memoirs. National Academy of Sciences*, **48** (1976), 3–22. Not included is the monograph written with Edward C. Reifenstein, *The Parathyroid Glands and Metabolic Bone Disease* (Baltimore, 1948). Albright's remarks on ectopic hormonal (parathyroid hormone) production from a nonendocrine tumor are in "Case Records of the Massachusetts General Hospital (Case 27461)," in *New England Journal of Medicine*, **225** (1941), 789–791.

Albright's presidential address to the American Society for Clinical Investigation (May 8, 1944), "Some of the 'Do's' and 'Do-Not's' in Clinical Investigation," was printed in *Journal of Clinical Investigation*, **23** (1944), 921–926. His presidential address to the Association for the Study of Internal Secretions (June 6, 1947), "A Page out of the History of Hyperparathyroidism," is printed in *Journal of Clinical Endocrinology*, **8** (1948), 637–657. Both contain reproductions of his illustrations.

Albright's extensive slide collection dealing with endocrine topics and metabolic studies has been restored and cataloged by his longtime associate Anne P. Forbes; copies are available by arrangement with the Countway Library of the Department of Medicine, Harvard Medical School.

II. SECONDARY LITERATURE. No full-length biography of Albright has been published. For brief biographical sketches and panegyric materials, see Lloyd Axelrod, "Bones, Stones and Hormones: The Contributions of Fuller Albright," in *New England Journal of Medicine*, **283** (1970), 964–970; C. Frederic Bartter, "Fuller Albright," in *Birth Defects, Original Article Series*, **7**, no. 6 (1971), 3–4; S. Gilbert Gordan, "Fuller Albright and Postmenopausal Osteoporosis: A Personal Appreciation," in *Perspectives in Biology and Medicine*, **24** (1981), 547–560; P. H. Henneman, "Fuller Albright (1900–1969)," in *New England Journal of Medicine*, **282** (1970), 280–281; and John E. Howard, "Fuller Albright: The Endocrinologists' Clinical Endocrinologist," in *Perspectives in Biology and Medicine*, **24** (1981), 374–381.

Photos and illustrations are in E. B. Pyle, "Fuller Albright's Inimitable Style," in *Harvard Medical Alumni Bulletin*, **56** (1982), 46–51. An account of the circumstances surrounding Albright's brain surgery is in Irving S. Cooper, *The Vital Probe* (New York, 1981), 161–170. Appraisals of Albright's influence on twentieth-century clinical investigation and endocrinology are in James Bordley III and Abner M. Harvey, *Two Centuries of American Medicine* (Philadelphia, 1976), 223, 553–555; Abner M. Harvey, *Science at the Bedside* (Baltimore, 1981), 255–257; and in V. C. Medvei, *A History of Endocrinology* (Lancaster, England, 1982), 398–399, 486–489, 524–530, 700–701, 730, with page facsimiles of Albright's introduction to the chapter "Diseases of the Ductless Glands," taken from Russell L. Cecil and Robert F. Loeb, eds., *A Textbook of Medicine*, 9th ed. (Philadelphia, 1955), and his address to the Society for Clinical Investigation cited above. A detailed history of the Mallinckrodt Ward, with commentaries on Albright's work, is in James H. Means, *Ward 4* (Cambridge, Mass., 1958).

MARTIN E. FINK

ALEKSANDROV (OR ALEXANDROFF), PAVEL SERGEEVICH (*b.* Bogorodsk [formerly Noginsk], Russia, 7 May 1896; *d.* Moscow, U.S.S.R., 16 November 1982), *mathematics*.

Pavel Sergeevich Aleksandrov was the youngest of six children (four sons, two daughters) of Sergei Aleksandrovich Aleksandrov, a rural government doctor, and of Tsezariia Akimovna Aleksandrova (née Zdanovskaia), whose main concern was the education of her children. Both parents instilled in him an intense interest in science and music. His mother taught him German, in which he was as proficient as in his native Russian, and French. Aleksandrov's early education was in the public schools of Smolensk, where his family moved in 1897 when his father became senior doctor in the Smolensk state hospital. The development of Aleksandrov's mathematical abilities and interest in the fundamental problems of mathematics were encouraged by his grammar school mathematics teacher, Aleksandr Romanovich Eiges. (In 1921 Aleksandrov was married for a brief time to Ekaterina Romanovna Eiges, sister of his teacher.)

Aleksandrov matriculated in the mathematics department of Moscow University in September 1913, intending to become a teacher. In the fall of 1914 he attended a lecture given by the brilliant young mathematician Nikolai Nikolaevich Luzin and became his first student. In 1915 Aleksandrov obtained his first mathematical result on the structure of Borel sets. When it was initially explained to him, Luzin doubted that the method Aleksandrov employed would work and suggested that another approach be taken. Aleksandrov persisted, however. The result he obtained may be stated as follows: every nondenumerable Borel set contains a perfect subset.

Enjoying the success of this first project, Aleksandrov energetically embarked on his second project: the continuum hypothesis. It is now known that Cantor's famous hypothesis cannot be proved or disproved within the framework of the theory of sets, so that Aleksandrov's efforts to obtain a definitive result were doomed to failure. His lack of complete success led him to conclude that his mathematical career was ended, and he left the university to go to Novgorod-Severskii, where he worked as a producer in the local theater, and then to Chernigov, where he helped to establish the Chernigov Soviet Dramatic Theater in the spring of 1919.

Life during the years 1918 to 1920 was filled with turmoil following the October Revolution of 1917. Aleksandrov was arrested and jailed for a brief time in 1919 by a group opposing the new Soviet government, but he was released when the Soviet army reoccupied Chernigov. In addition to his work in the theater, he embarked upon a series of public lectures on literature and mathematics. In December 1919, following a six-week illness, he decided to go to Moscow University and resume his study of mathematics.

After returning to Moscow in September 1920, Aleksandrov prepared for his master's examinations by studying with Pavel Samuilovich Uryson. Their association blossomed into a deep friendship. During the summer of 1922 the two P.S.'s (as they were referred to by their fellow students) and several friends rented a dacha on the banks of the Klyaz'ma. The two young mathematicians embarked on the study of topology, a recently formed field of mathematics. Their work of that summer and fall was guided by a mere handful of articles, among them the pioneering work of Maurice Fréchet (1906) and Felix Hausdorff's monumental *Grundzüge der Mengenlehre* (1914). Their primary concern was to obtain necessary and sufficient conditions for a topological space to be metrizable. The outcome of their research was the lengthy and authoritative paper "Mémoire sur les espaces topologiques compacts," which, because of various problems, was not published until 1929.

Their search for a metrization theorem was successful, but their formulation made its application difficult. The search for a workable result continued until one of Aleksandrov's students, Yuri M. Smirnov, as well as J. Nagata and R. H. Bing, independently achieved a workable formulation (1951–1952). Using modern terminology, the condition that Aleksandrov and Uryson derived may be stated as follows: A topological space is metrizable if and only if it is paracompact and has a countable refining system of open coverings.

Encouraged by their work, the two young men visited Göttingen, the intellectual hotbed of German mathematics. (To finance their journey, they gave a series of lectures in and around Moscow on the theory of relativity.) During the summer of 1923, the two presented their results, which were enthusiastically received by such mathematicians as Emmy Noether, Richard Courant, and David Hilbert. This summer not only marked the first time since the revolution that Soviet mathematicians had traveled outside their country, but it also set the stage for mathematical exchanges between Moscow and Göttingen. In fact, Aleksandrov returned to Göttingen every summer until 1932, when such exchanges became impossible because of the restrictiveness of regulations imposed by the German government.

Aleksandrov and Uryson returned to Göttingen in the summer of 1924; they also visited with Felix Hausdorff in Bonn and with L. E. J. Brouwer in Holland. After their time with Brouwer, the two young Soviets went to Paris and then to the Atlantic coast of France for a period of work and relaxation. In Batz, France, their tour ended tragically on 17 August 1924 when Uryson drowned.

The death of his friend seemed to intensify Aleksandrov's interest in topology and in the seminar which the two had begun organizing in the spring of 1924. One of the first students in this seminar was the first of Aleksandrov's students to make substantial contributions to mathematics in general and to topology in particular. Andrei Nikolaevich Tikhonov developed the concept of the product of an infinite number of topological spaces (at least for infinitely many copies of the closed unit interval $[0,1]$). He developed the concept to solve a problem posed by Aleksandrov: Is every normal space embeddable as a subspace of a compact Hausdorff space?

After the death of Uryson, Aleksandrov returned to Moscow and made plans to spend the academic year 1925–1926 in Holland with Brouwer. One of the reasons was that during their visit with Brouwer in 1924, Aleksandrov and Uryson had been persuaded by him to have their topological work published in *Verhandelingen der Koninklijke akademie van wetenschappen*. Due to a series of delays, this monumental work was not published until 1929. The original version was in French; since then it has appeared in Russian three times (1950, 1951, 1971), each time with footnotes by Aleksandrov updating contributions by mathematicians answering questions posed by the work.

Aleksandrov formed lifelong friendships during his summers in Göttingen, the most important of which, with Heinz Hopf, began in 1926. Their friendship grew during the academic year 1927–

1928, which they spent at Princeton University. When they returned to conduct a topological seminar at Göttingen during the summer of 1928, they were asked by Richard Courant to write a topology book as part of his Yellow Collection for Springer. This request resulted in a seven-year collaboration that culminated in 1935 with the publication of *Topologie*, a landmark textbook on topology. Two additional volumes had originally been planned, but the war prevented completion of the project.

Aleksandrov loved to swim and to take long walks with his students while discussing mathematical problems. His athletic inclinations were restricted by his eyesight, which had been poor from his youth. (He was totally blind during the last three years of his life.)

In 1935 Aleksandrov and his close friend Andrei Nikolaevich Kolmogorov acquired a century-old dacha in the village of Komarovka, outside Moscow. They shared the house and its surrounding garden until death separated them. It not only became a convenient and popular place for Aleksandrov and his students to gather but also sheltered many renowned mathematicians who came to meet and work with Aleksandrov or Kolmogorov.

Aleksandrov's achievements were not limited to pure mathematics. From 1958 to 1962 he was vice president of the International Congress of Mathematicians. He held the chair of higher geometry and topology at Moscow State University, was head of the mathematics section of the university, and served as head of the general topology section of the Steklov Institute of Mathematics of the Soviet Academy of Sciences. For thirty-three years Aleksandrov was president of the Moscow Mathematical Society; he was elected honorary president in 1964. In addition to serving as editor of several mathematical journals, he was editor in chief of *Uspekhi matematicheskikh nauk* (*Russian Mathematical Surveys*). In 1929 he was elected a corresponding member of the Soviet Academy of Sciences, and a full member in 1953. He was a member of the Göttingen Academy of Sciences, the Austrian Academy of Sciences, the Leopoldina Academy in Halle, the Polish Academy of Sciences, the Academy of Sciences of the German Democratic Republic, the National Academy of Sciences (United States), and the American Philosophical Society, and an honorary member of the London Mathematical Society. He was awarded honorary doctorates by the Dutch Mathematical Society and Humboldt University in Berlin.

In his last years Aleksandrov supervised the editing of a three-volume collection of what he considered his most important works: *Teoriia funktsii deistvitel'nogo peremennogo i teoriia topologicheskikh prostranstv* ("The Theory of Functions of Real Variables and Theory of Topological Spaces"; 1978), *Teoriia razmernosti i smezhnye voprosy: Stat'i obshchego kharaktera* ("Dimension Theory and Related Questions: Articles of a General Nature"; 1978), and *Obshchaia teoriia gomologii* ("The General Theory of Homology"; 1979).

In addition Aleksandrov wrote his autobiography, portions of which appeared in two parts under the title "Stranitsii avtobiografii" ("Pages from an Autobiography") in *Russian Mathematical Surveys* along with papers presented at an international conference on topology in Moscow (June 1979) of which he was the prime organizer.

Aleksandrov's mathematical results were substantial and diverse. Through his joint work with Uryson he is credited with the definition of compact spaces and locally compact spaces (originally described as bicompact spaces). In 1925 he first formulated the modern definition of the concept of topological space. The concept of compact space undoubtedly led to the definition of a locally finite covering of a space that he used to prove that every open cover of a separable metric space has a locally finite open cover, or, in modern terminology, that every separable metric space is paracompact. This idea appeared later in the result of A. H. Stone that every metric space is paracompact, a fact used by Smirnov, Nagata, and Bing in their metrization theorem.

In the period 1925 to 1929 Aleksandrov is credited with laying the foundations of the homology theory of general topological spaces. This branch of topology is a blend of topology and algebra, his study of which was inspired by Emmy Noether during his Göttingen summers and during a visit she made to Brouwer (while Aleksandrov was working with him) in the winter of 1925 to 1926. It was during this visit that Aleksandrov became interested in the concept of a Betti group, which he was to use in his work. (He coined the term "kernel of a homomorphism," which appeared in print for the first time in an algebraic supplement to *Topologie*.)

Aleksandrov's arguments used the concept of the nerve of a cover that he had introduced in 1925. The nerve of a cover ω of a topological space X is a simplicial complex N_ω whose vertexes are in a one-to-one correspondence with the elements of ω, and any vertexes $e_1, \ldots, e_k$ of N_ω form a simplex in N_ω if and only if the elements of ω corresponding to these vertexes have a nonempty intersection.

Based upon this, one may define a simplicial transform $\Pi_{\omega}^{\omega'}$ (ω' being a cover contained in or succeeding ω), which is called the "projection" of the nerve N_{ω}' into N_{ω}.

For a compact space X, the collection of all such projections formed by letting ω range over the directed family of all the finite open covers of X is the projective spectrum S of X. This projective spectrum is the directed family of complexes N_{ω}, which are linked by the projections $\Pi_{\omega}^{\omega'}$. The limit space of the projective spectrum is homeomorphic to X, which implies that the topological properties of the space X may be reduced to properties of the complexes and their simplicial mappings. Among other results, this work led to Aleksandrov's theorem that any compact set of a given dimension lying in a Hilbert space can, for any $\epsilon > 0$, be transformed into a polyhedron of equal dimension by means of an ϵ-deformation, that is, a continuous deformation in which each point is displaced by at most ϵ.

These concepts led to the creation of the homological theory of dimension in 1928 to 1930. Aleksandrov's works frequently seemed to be springboards for other mathematicians, including many of his own students: A. N. Tikhonov, L. S. Pontriagin, Y. M. Smirnov, K. A. Sitnikov, A. V. Arkhangel'skii, V. I. Ponomarev, V. I. Zaitsev, and E. V. Shchepin, to name only a few.

In his autobiography Aleksandrov broke down his mathematical life and the associated papers into six periods:

1. The summer of 1915—the structure of Borel sets and the A-operation
2. May 1922–August 1924—basic papers on general topology
3. August 1925–spring 1928—the definition of the nerve of a family of sets and the establishment of the means of the foundations of homology theory of general topological spaces by a method that permitted him to apply the methods of combinatorial topology to point-set topology
4. The first half of 1930—the development of homological dimension theory, which built upon his spectral theory
5. January–May 1942—Because of World War II, Aleksandrov, Kolmogorov, and other scientists were sent to Kazan in July 1941. Although Aleksandrov returned to Moscow for the start of the fall session at the university, he was told to go back to Kazan. There, during the winter of 1941, he wrote a work devoted to the study of the form and disposition of a closed set (or complex) in an enveloping closed set (or complex) by homological means. One important by-product of this paper was the concept of an exact sequence, an important algebraic tool used in many branches of mathematics
6. The winter of 1946–1947—duality theorems for nonclosed sets; he regarded the paper containing these results as his last important work

In the late 1940's and early 1950's, Aleksandrov and his pupils built upon this last work with the construction of homology theory for nonclosed sets in euclidean spaces. At all times his works seemed motivated and guided by geometric ideas undoubtedly stemming from his youthful fascination with the subject.

One of the pervasive elements in Aleksandrov's works is the theory of continuous mappings of topological spaces, beginning with his theory of the continuous decompositions of compacta, which led to the theory of perfect mappings of arbitrary, completely regular spaces. Included in this development is Aleksandrov's theorem on the representation of each compactum as a continuous image of a perfect Cantor set. This result gave rise to the theorem that every compactum is a continuous image of a zero-dimensional compactum of the same weight, and is part of the foundation of the theory of dyadic compacta.

Most of Aleksandrov's life was spent in university teaching and research, and in many ways was structured around education and his students. It included visits to Kamarovka, musical evenings at the university, public talks, and private concerts. The last twenty-five years of his life seem to have been devoted to his students and education in general, as indicated by the survey articles he wrote during this period. He seemed to radiate the same kind of magnetism and contagious fervor for mathematics and life that first drew Aleksandrov, Uryson, and other young students to cluster around Luzin in their student years.

BIBLIOGRAPHY

I. ORIGINAL WORKS. "Some Results in the Theory of Topological Spaces, Obtained Within the Last Twenty-five Years," in *Russian Mathematical Surveys*, **15**, no. 2 (1960), 23–84; *Teoriia funktsii deistvitel'nogo peremennogo i teoriia topologicheskikh prostranstv* ("The Theory of Functions of Real Variables and Theory of Topological Spaces"; Moscow, 1978); *Teoriia razmernosti i smezhnye voprosy: Stat'i obshchego kharaktera* ("Dimension Theory and Related Questions: Articles of a General Nature"; Moscow, 1978); "The Main Aspects in the Development of Set-Theoretical Topology," in *Russian Mathematical Surveys*, **33**, no. 3 (1978), 1–53, with V. V. Fedorchuk; *Obshchaia teoriia gomologii* ("The General Theory of Homology"; Moscow, 1979); "Pages from an Autobiography," in *Russian Mathematical Surveys*, **34**, no. 6 (1979), 267–302, and **35**, no. 3 (1980), 315–358.

II. SECONDARY LITERATURE. A. V. Arkhangelskii *et al.*, "Pavel Sergeevich Aleksandrov (On His 80th Birthday)," in *Russian Mathematical Surveys*, **31**, no. 5 (1976), 1–13; and *Russian Mathematical Surveys*, **21**, no. 4 (1966), an issue dedicated to Aleksandrov, with a biographical introduction by A. N. Kolmogorov *et al.*

DOUGLAS EWAN CAMERON

ALIKHANOV, ABRAM ISAAKOVICH (*b.* Elisavetpol [now Kirovobad, Azerbaijan S.S.R.], Russia, 4 March 1904; *d.* Moscow, U.S.S.R., 8 December 1970), *physics*.

Alikhanov's father, Isaak Abramovich Alikhanian, was a railway engineer; his mother, Iulia Artemevna, was a housewife. There were four children: two sons (the younger, Artem Alikhanian, became a well-known physicist) and two daughters. In 1912 the family moved to Aleksandropol (now Leninakan, Armenian S.S.R.), where Alikhanov began his studies at a commercial school. He graduated from another commercial school, in Tiflis (now Tbilisi, Georgian S.S.R.), in 1921 and the same year enrolled in the Georgian Polytechnical Institute. After a few months, however, he moved to Petrograd (now Leningrad), where he continued his studies at the Faculty of Physics and Mechanics of the Petrograd Polytechnical Institute, from which he graduated in 1930. It was at that time that he russified his name from Alikhanian to Alikhanov.

In 1927 Alikhanov started to work at the Physical-Technical Institute. Combining studies at the Polytechnical Institute with scientific research at the Physical-Technical Institute was common practice among the best students of the faculty.

Alikhanov's first research work at the Physical-Technical Institute, devoted to X-ray physics, was carried out in Petr Ivanovich Lukirskii's laboratory under Lukirskii's supervision. Alikhanov x-rayed aluminum in the temperature range of 550–600°C and proved that in this range aluminum retains its structure and there is no allotropic transformation. This result contradicted the data obtained by others. Subsequently, X rays became not a tool but the subject of Alikhanov's research. The most important of these investigations was a study of the total internal reflection of X rays from thin layers and the estimation of the depth of their penetration into the medium. Alikhanov also proved that the laws of classical optics can be applied to the reflection of hard X rays. This research formed the basis for the monograph published in 1933.

Research on nuclear physics began at the Physical-Technical Institute in late 1932. In 1934 Alikhanov was appointed head of the laboratory for positron physics. The work of his group was devoted to the investigation of pair production and of the resultant positron spectrum, especially the latter's dependence on the γ radiation released and on the atomic number of the irradiated material. For observation of positrons, Alikhanov, his brother, and his student M. S. Kozodaev used an original combination of a magnetic spectrometer and two contiguous Geiger-Müller counters making coincidence counts. This work became a starting point for the application of radio engineering to experimental nuclear physics in the Soviet Union. Another series of investigations by Alikhanov and collaborators was concerned with the β-decay of artificially radioactive nucleides. This work used not a Wilson cloud chamber but the Alikhanov-Kozodaev spectrometer, which enabled the authors to obtain more precise data concerning the form and bounds of the β-spectrum and its dependence on the atomic number of the source of radiation. The scattering and slowing of relativistic electrons (those with velocities comparable with the velocities of photons, keeping in mind use of the relativistic corrections) in matter were investigated as well.

Two more of Alikhanov's prewar investigations must be mentioned. A method of determining the rest mass of the neutrino, using decay of the nuclei of Be^7, was suggested in his laboratory in 1938. A secondary nucleus of Li^7 gains momentum from the neutrino created as a result of capture of an orbital electron by the nucleus of Be^7. The discrete energy spectrum of the neutrino and, as a consequence, of the nucleus facilitates the experimental solution of the difficult problem of determining the rest mass of the neutrino. The information about the experiments planned at Alikhanov's laboratory under his leadership appeared in the Soviet physical literature in 1940 but the work was not carried out because of the war. In 1942 the experiment was carried out by James Allen.

Another aspect of Alikhanov's work (which he began in the late 1930's) was associated with cosmic rays. By the summer of 1941 an expedition to the Pamir Mountains under his leadership had been prepared, but the project was not realized because of the war. In 1942, however, as well as later, observations of cosmic rays were carried out during an expedition to the Armenian Mountains. The experiments demonstrated the presence of a stream of fast protons in cosmic radiation, but also included a serious mistake. The conclusion that there exist in cosmic radiation particles (called by Alikhanov and his group "varitrons") possessing a broad spectrum of masses was erroneous. Alikhanov was sharply criticized because the possibility of the existence of particles with a broad mass spectrum was perceived as nonrealistic at that time. Its reality was determined later, through a quite different type of experiment.

At the end of 1942 Alikhanov joined the research on uranium. He became the head of investigations, conducted in a special laboratory, of a heavy-water technique in reactor engineering, and in 1945 he established a research center for these investigations, the Heat Engineering Laboratory (now the Institute of Theoretical and Experimental Physics) at Moscow. He served as its director almost until the end of his life. In 1949 the first heavy-water reactor went into operation at the institute. Before the war Alikhanov (with Igor Vasilevich Kurchatov) had participated in the design and construction of the first cyclotrons in the Soviet Union and Europe (at Kurchatov's laboratory in 1944). In 1952 he returned to this work, and in 1961 a proton accelerator with strong focusing and an energy of 7 GeV was completed at the institute. Alikhanov and his collaborators also participated in the design of a 70 GeV accelerator at Serpukhov. During the last years of his life he worked in high-energy physics and the physics of mesons, having been stimulated by the discovery of nonconservation of parity in weak interactions.

Alikhanov was married twice. He and his first wife, A. Prokofieva, had a son, Ruben, who became a physicist. They were divorced about 1942. Alikhanov's second wife, Slava Roshal', was a violinist. Their son, Tigran, and their daughter, Evgenia, both became musicians.

Painting also played a great role in Alikhanov's life. His friend the Armenian artist Martiros Sergeevich Sarian once painted a portrait of him.

Alikhanov's work was highly recognized: in 1939 he was elected corresponding member, and in 1943 full member, of the Soviet Academy of Sciences. He was three times awarded the State Prize, and in 1954 he became a Hero of Socialist Labor.

During the last two years of his life, Alikhanov suffered from the effects of a cerebral hemorrhage. Nevertheless, he did not retire. In 1968, however, he resigned his directorship, and from then on, he limited his work to scientific research in his laboratory at the institute.

BIBLIOGRAPHY

I. ORIGINAL WORKS. "Polnoe vnutrennee otrazhenie rentgenovykh luchei ot tonkikh sloev" (The total internal reflection of X rays from thin layers), in *Zhurnal eksperimentalnoi i teoreticheskoi fiziki*, **3**, no. 2 (1933), 115, written with Lev. A. Artsimovich; *Optika rentgenovskikh luchei* (X-ray optics; Leningrad and Moscow, 1933); "A New Type of Artificial β-Radioactivity," in *Nature*, **133** (1934), 871–872, written with Artem I. Alikhanian and B. S. Dzhelepov; "Emission of Positrons from a Thorium-Active Deposit," *ibid.*, **136** (1935), 475–476, written with Artem I. Alikhanian and M. S. Kozodaev; "Zakon sokhraneniia impulsa pri annigiliatsii positronov" (The conservation of momentum law in positron annihilation), in *Doklady Akademii nauk SSSR*, **1**, no. 7 (1936), 275, written with Artem I. Alikhanian and Lev A. Artsimovich; "Obrazovanie par pod deistviem γ-luchei" (Pair creation under γ-ray influence), in *Izvestiia Akademii nauk SSSR, seriia fizicheskaia*, (1938), no. 1/2, 33; "Forma β-spektra RaE vblizi verkhnei granitsy i massa neitrino" (The spectrum mode of RaE near the upper bound and the mass of the neutrino), in *Doklady Akademii nauk SSSR*, **19**, no. 5 (1938), 375; "O poteriakh energii bystrymi elektronami pri prokhozhdenii cherez veshchestvo" (On energy losses of fast electrons during passage through matter), *ibid.*, **25**, no. 3 (1939), 193, written with Artem I. Alikhanian; "Novii dannii o prirode kosmicheskikh luchei" (New data on the nature of cosmic rays), in *Uspekhi fizicheskikh nauk*, **27**, no. 1 (1945), 22, written with Artem I. Alikhanian; "Tiazhelovodnyi energeticheskii reaktor s gazovym okhlazhdeniem" (The heavy-water energy reactor with gas cooling), in *Atomnaia energiia*, **1** (1956), 5, written with V. V. Vladimirskii, P. A. Petrov, and P. I. Khristenko; *Slabye vzaimodeistviia: Noveishie issledovaniia β-raspada* (Moscow, 1960), trans. by William E. Jones as *Recent Research on Beta-Disintegration* (New York, 1963); "Dalneishie poiski ($\mu \rightarrow e + \gamma$)-raspada" (Further searches of ($\mu \rightarrow e + \gamma$)-decay), in *Zhurnal experimentalnoi i teoreticheskoi fiziki*, **42** (1962), 630–631, written with coauthors; *Izbrannye trudy* (Selected works), S. I. Nikitin *et al.*, eds. (Moscow, 1975).

II. SECONDARY LITERATURE. A. P. Aleksandrov, B. V. Dzelepov, S. I. Nikitin, and I. B. Khariton, "Pamiati Abrama Isaakovicha Alikhanova" (In memory of A. I. Alikhanov), in *Uspekhi fizicheskikh nauk*, **112**, no. 3 (1974), 725–727; B. G. Gasparin. A. P. Grinberg, and V. J. Frenkel, *A. I. Alikhanov v Fiziko-tekhnicheskom institute* (A. I. Alikhanov in the Physical-Technical Institute; Leningrad, 1986); A. P. Grinberg, "Gipoteza neitrino i novie podtverzhdaiushchii ee dannie" (Neutrino hypothesis and the new data that confirm it), in *Uspekhi fizicheskikh nauk*, **26**, no. 2 (1944), 189; and "Positive Electrons from Lead Ejected by γ-Rays," in *Nature*, **133** (1934), 581.

V. J. FRENKEL

ALLEE, WARDER CLYDE (*b.* near Bloomingdale, Indiana, 5 June 1885; *d.* Gainesville, Florida, 18 March 1955), *ecology*.

Clyde Allee, professor of zoology at the University

of Chicago (1921–1950), was a leader of American ecology during the years in which it was effectively established as a subdiscipline of biology. His large number of students and his part in producing the textbook *Principles of Animal Ecology* (1949) were his most lasting influences on the field, although he was a pioneer of ecological studies of social behavior in animals.

Born on a farm to Mary Newlin and John Wesley Allee, Clyde grew up in the strong Quaker (Society of Friends) community of southern Indiana. He attended the Friends Academy in Bloomingdale and Earlham College in Richmond. Upon graduation in 1908, Allee studied with Victor Shelford at the University of Chicago, where he received the Ph.D. in 1912. That same year he married Marjorie Hill, also a member of the Society of Friends; they had three children. Profoundly religious (in an ethical, not mystical, manner), Allee became intellectually devoted to the theoretical problems of social behavior, cooperation in nature, and the evolution of human ethics. Science held a significant role in his philosophy, and he felt science and religion could mutually profit from an interchange of values. Also reflecting his values was the time he gave to scientific societies, especially the Ecological Society of America and the American Society of Zoologists, editorship of *Physiological Zoology* (1928–1954), collaboration with students and colleagues, and teaching. He was also a trustee of Earlham College and the Woods Hole Marine Biological Laboratory. In 1938, following a series of operations on a spinal tumor, he was paralyzed from the waist down and confined to a wheelchair for the rest of his life. The death in 1945 of Marjorie Allee, a successful author of children's books and helpful collaborator on Allee's books, also deeply affected him emotionally, yet he continued an extremely active academic life. He retired from Chicago in 1950, but took the chairmanship of the biology department at the University of Florida. He married Ann Silver in 1953. Allee died in 1955 shortly after entering the hospital with a kidney infection.

Allee came under two strong influences in his doctoral education, and the manner in which he combined them remained the mark of his career and contribution to ecology. As Shelford's student he was in the circle of biologists making Chicago a center of research on ecological succession; their descriptive studies were tied in part to Shelford's attempt to create a physiology of environmental factors. Another influence was F. R. Lillie, departmental chairman at the University of Chicago and also director of the laboratory at Woods Hole, Massachusetts, where Allee spent his summers. There he learned the techniques of the experimental zoologists such as Lillie, C. O. Whitman, and J. Loeb. The experimentalists' dependence on tropisms as the new paradigm for animal behavior was attractive because it was a program for laboratory progress, and Allee began a long-running series of experimental investigations into what physical factors affect the movements and behavior of a small aquatic crustacean. He taught at a succession of universities for the next decade, always returning to Woods Hole for summer research and teaching. The synthesis he produced was to use the lab-derived results about physical factors to understand the animal's behavior and place in nature. He also argued that tropisms were insufficient to account for adaptive peculiarities in nature, thus moving away from strictly physiological and behaviorist explanations. Because he designed experiments in the laboratory to shed light on the larger ecological setting, he became an influential figure in a young science.

By 1921 Allee was back at Chicago and deeply involved in showing how distribution of animals is a response to physical factors. His work was distinguished within ecology by his emphasis on the behavior that produces the distribution. Thus he began his most famous work, on animal aggregation, by opening a specifically ecological research program, described in *Animal Aggregations: A Study in General Sociology* (1931), with the search for the general principles that lead to aggregation and, ultimately, the structure of biotic communities. He saw all animals existing in a social network, reproducing and interacting ecologically, manifesting a level of sociality more loosely organized than the highly specialized insect societies that dominated earlier theories. Extending the classifications of societies by A. V. Espinas, P. Deegener, and especially W. M. Wheeler, Allee organized the subject around the type or degree of integration of the social group, and considered a far greater range of aggregations to have social significance.

In his career he directed the research of about forty graduate students, who themselves took important posts in American ecology. Together, their physiological experiments demonstrated survival and growth benefits for members of groups, and through the 1930's he expanded into social effects on learning. Following the impetus of T. Schjelderup-Ebbe's far-reaching theory of "peck-order," Allee's group by the 1940's centered more on the conditions and ecology of social hierarchies and less on the simpler physiological questions.

Allee himself became more and more concerned

with principles and the practice of ecology, producing several well-known books. *The Social Life of Animals* (1938) presented his theory that cooperation, by which he meant automatic mutual interdependence, is a fundamental trait of life, a principle as important as struggle for existence. Claiming that cooperation is found throughout the biological world, he extrapolated the development of innate tendencies to human altruism in an attempt to use the biology of sociality to improve human society. He denied the supremacy of competition as the law of nature. Within behavioral ecology, however, his grand principle of universal tendencies did not lead to a productive research program. Instead, after 1950, Niko Tinbergen became the major figure in the field with his emphasis on the individual advantage found within social behaviors, including cooperation and apparent altruism.

Allee's great institutional contribution was leading the group that became an identifiable Chicago school of ecology. Together with A. E. Emerson, O. Park, T. Park, and K. P. Schmidt, he produced a standard textbook, *Principles of Animal Ecology*, that presented the heart of their view that the ecological community is a metaphorical organism, with homeostasis and an evolutionary history. The distinctive Allee stamp was the weight given physical factors, organized around the idea that shared principles structure all communities. It was an intentional extension of the earlier descriptive approaches and the classifying of types and biogeographical regions; and it was distinct from the Eltonian school's concentration on food webs, competition, and predation.

BIBLIOGRAPHY

A detailed biography appears in Karl P. Schmidt, "Warder Clyde Allee, 1885–1955," in *Biographical Memoirs. National Academy of Sciences*, **30** (1957), 2–40, which includes a complete bibliography of Allee's works. The Joseph Regenstein Library, University of Chicago, has a small collection of Allee's papers, including research notes, notebooks, correspondence of the Committee on the Ecology of Animal Populations of the National Research Council, and manuscripts. A larger collection of his correspondence is in the Lilly Library, Earlham College, Richmond, Indiana. The Northwestern University Archives, Evanston, Illinois, holds correspondence concerning his lectures for the Norman Wait Harris Lecture Series, which became *The Social Life of Animals*.

WILLIAM C. KIMLER

ALLEN, EUGENE THOMAS (*b.* Athol, Massachusetts, 2 April 1864; *d.* Arlington, Massachusetts, 17 July 1964), *geochemistry*.

The son of Frederick Allen, a merchant and manufacturer, and of Harriet Augusta Thomas Allen, Eugene attended Amherst College, from which he graduated in 1887 with the B.A. degree. He then entered Johns Hopkins for graduate study in chemistry, receiving the Ph.D. in 1892. During his graduate years he served as associate in chemistry at Women's College of Baltimore (now Goucher College). Following a year as acting professor of chemistry at the University of Colorado, he studied at Harvard for two years (1893–1895) and then was professor of chemistry at the Missouri School of Mines (now University of Missouri, Rolla) from 1895 to 1901. At the end of his first year in Missouri he married Harriet Doughty of Arlington, Massachusetts, on 26 August 1896.

In 1901 Allen joined the U.S. Geological Survey in Washington, D.C. On 1 January 1907 he transferred to the Geophysical Laboratory of the Carnegie Institution of Washington when that department was formed from the geophysical laboratory of the U.S. Geological Survey, whose appropriations had been curtailed. There he served primarily as chemist until his retirement on 1 May 1932. On several occasions, during absences of the director, Arthur L. Day, Allen was officially designated acting director of the Geophysical Laboratory. He continued as research associate (1932–1933) following his retirement.

Although Allen had already contributed to geochemistry through a careful analytical study of a sedimentary rock containing native iron in the coal measures of Missouri, it was the influence of William F. Hillebrand while he was at the U.S. Geological Survey that set him on the road to becoming a superior analytical geochemist. The devising of a "clear and logical course to follow" in making a complete analysis, double precipitation as a routine procedure, and the continual search for refinement of technique were lessons he applied well. In turn Allen inspired others to seek the highest precision in their analytical work—at least one assistant, Emanuel G. Zies, became an important geochemist in his own right. And a colleague at the U.S. Geological Survey, Arthur L. Day, became his mentor as director of the Geophysical Laboratory and a lifelong collaborator in research, as well as a close friend.

The experimental study of the plagioclase feldspars, the most abundant of the rock-forming minerals, by Day, Allen, and Joseph P. Iddings set the format for the principal line of geochemical research at the Geophysical Laboratory. The work was done at the U.S. Geological Survey, funded for the most part by the Carnegie Institution of Washington through special grants, and published in 1905. In recognition of the role of the Carnegie Institution of Washington in supporting the work, the director of the U.S. Geological Survey authorized its distribution as Carnegie Institution of Washington publication no. 31, which later was designated *Publication of the Geophysical Laboratory* no. 1.

After the development of equipment for maintaining and measuring relatively constant high temperatures, scientists were able to determine the melting points of the compositional range of plagioclase feldspars with considerable accuracy. (The nitrogen-gas thermometer developed then remains the fundamental thermodynamic scale for temperature measurement and calibration.) Allen was also responsible for the analyses of high-purity metals used as secondary standards for melting-point determinations. The analyses showed that the plagioclases consisted of a continuous series of solid solutions that melted in a simple way, the beginning-of-melting curve and the all-liquid curve forming a simple loop (Roozeboom's type I). The methods he helped develop for measuring the physicochemical behavior of minerals made from exceptionally pure chemicals have served as the foundation for all subsequent phase equilibria studies.

Other works, now recognized as classics, were written in collaboration with John K. Clement, James L. Crenshaw, Clarence N. Fenner, John Johnston, Esper S. Larsen, Jr., Herbert E. Merwin, Robert B. Sosman, Walter P. White, Frederic E. Wright, and Emanuel G. Zies, all leaders in their respective fields. They include studies on mineral polymorphism, the processes of ore mineral deposition, the chemical analysis of glass, and the composition of volcanic gases and hot springs.

Allen, White, and Wright placed an upper limit on the stability of wollastonite ($CaSiO_3$), an important metamorphic mineral, by showing that it inverted to another form at 1190°C. Allen established with Wright and Clement that $MgSiO_3$, a common mineral in igneous rocks and meteorites, also exhibited polymorphism, in which the transitions produced measurable heat effects. The synthetic compound diopside, a principal end-member of the pyroxenes, which in terms of composition lies between wollastonite and enstatite, was shown by Allen, White, Wright, and Larsen to melt congruently—that is, to a liquid of its own composition (though it was demonstrated by Kushiro and Schairer in 1963 to melt incongruently—to a diopside solid solution and liquid), to exhibit different extents of solid solution toward $MgSiO_3$ and $CaSiO_3$, and to have optical and physical properties close to those of natural diopside. In the course of this investigation, a high-temperature heating stage for the petrographic microscope was constructed and another polymorph of $MgSiO_3$ (now called protoenstatite) was discovered. Allen and Clement were the first to show that water is an essential, structurally bound, constituent of the amphibole tremolite.

During the period from 1910 to 1917, Allen was mainly concerned with the determination of sulfur in metallic sulfides and sulfates. He viewed the genesis of ores as primarily a chemical problem and believed the conditions of formation could be obtained by forming the ore minerals in the laboratory, thereby determining the range of conditions of stability in a systematic fashion. Showing great insight into the pitfalls of mineral synthesis, he stated that it is necessary for a chemist working in this field to keep in constant touch with geologists. His sulfide research culminated in the classic study with Zies and Merwin on secondary copper sulfide enrichment (1916). By quantitative study of the reactions of natural sulfides with copper sulfate solutions, they showed that cuprous sulfate reacts more readily than cupric sulfate in replacing the original sulfide. The order of stability of the sulfide enrichment products was determined to be chalcopyrite, covellite, chalcocite, and eventually metallic copper and sulfuric acid. The project was carried out in cooperation with Louis C. Graton and his colleagues at Harvard, and with many copper companies, as part of the Secondary Enrichment Program.

The entry of the United States into World War I resulted in the loss of access to the Jena optical glasses. The Geophysical Laboratory did the basic research for the production of high-quality optical glasses in the United States. Determining the composition of the Jena glasses fell to Allen and Zies. The glasses contained combinations of elements that were more difficult to separate and determine accurately than those found in rocks. The two scientists concentrated on the problem of determining boron and arsenic accurately, and in the course of their work devised a series of rapid analytical methods. Another classic study resulted (1918).

For the remainder of his life's work, Allen took as his text a statement from Thomas C. Chamberlin and Rollin D. Salisbury's *Geology* (I, 1904): "It is one of the outstanding problems in geology to determine the origin of the volcanic gases." Inspiration no doubt was also gained from his colleagues Day and Ernest S. Shepherd, as well as from the enthusiasm of Thomas A. Jaggar, but his experiences in the field with volcanism on the Katmai expedition of 1919 probably determined his interest. Four studies now recognized as classics resulted: (1) on the fumaroles of the Katmai region, Alaska, in the Valley of Ten Thousand Smokes, written with Zies (1923); (2) on the hot springs of Lassen National Park, California, written with Day (1924); (3) on the steam wells at "The Geysers," California, written with Day (1927); and, perhaps the most famous, (4) on the hot springs of Yellowstone National Park, Wyoming, written with the help of Day (1934). These pioneering studies made available a vast amount of field data and chemical analyses, collected over sufficient time periods to serve as a guide not only to the understanding of geysers, hot springs, and fumaroles, but also to the production of geothermal energy.

Allen was a kindly, modest man always helpful to his younger colleagues. He was a superb conversationalist and an informative letter writer. His holidays, usually spent in New England, involved hiking and botanical tours of the countryside.

A member of the American Chemical Society, Allen served in 1904 as president of its Washington, D.C., section, the Chemical Society of Washington. In addition, he was a fellow of the Geological Society of America and the American Geophysical Union. The National Academy of Sciences elected him a member in 1930.

BIBLIOGRAPHY

I. Original Works. Allen's writings include "Native Iron in the Coal Measures of Missouri," in *American Journal of Science*, **4** (1897), 99–104; "Some Reactions Involved in Secondary Copper Sulfide Enrichment," in *Economic Geology*, **11** (1916), 407–503, written with Emanuel Zies and Herbert Merwin; "A Contribution to the Methods of Glass Analysis . . . ," in *Journal of the American Ceramic Society*, **1** (1918), 739–786; "The Condition of Arsenic in Glass and Its Role in Glassmaking," *ibid.*, 787–790; "A Chemical Study of the Fumaroles of the Katmai Region," in *National Geographic Society Contributed Technical Papers*, **1** (1923), 75–155, written with Emanuel Zies; "The Source of the Heat and the Source of the Water in the Hot Springs of the Lassen National Park," in *Journal of Geology*, **32** (1924), 178–190, written with Arthur L. Day; "Steam Wells and Other Thermal Activity at 'The Geysers,' California," *Carnegie Institution of Washington Publication* 378 (1927), written with Arthur L. Day; and "Hot Springs of the Yellowstone National Park," in *Proceedings of the Fifth Pacific Science Congress*, III (Victoria and Vancouver, B.C., Canada, 1934), 2275–2283, written with Arthur L. Day.

II. Secondary Literature. Charles A. Anderson, "Eugene Thomas Allen," in *Biographical Memoirs. National Academy of Sciences*, **40** (1969), 1–17, with a bibliography of Allen's publications; and Margaret D. Foster, "CSW Past-President Celebrates 100th Birthday," in *The Capital Chemist*, **14** (1964), 95. There is information on Allen in the personnel files of the Geophysical Laboratory at the Carnegie Institution of Washington, covering the period 1907–1964.

H. S. Yoder, Jr.

ALLEN, JOEL ASAPH (*b*. Springfield, Massachusetts, 19 July 1838; *d*. Cornwall-on-Hudson, New York, 29 August 1921), *zoology*.

Allen's ancestors, who were of English descent, settled in New England in the 1630's and 1640's, and became farmers. His father, Joel Allen, had learned the carpenter's trade but later turned to farming. His mother, Harriet Trumbull Allen, taught school before their marriage. Joel, the first of the couple's four sons (one of whom died in infancy) and one daughter, grew up on the family farm in Springfield. His early training was very puritanical; and his parents, who were Congregationalists, were rigid in their religious practices.

Allen showed an early interest in the wildlife around the family farm. At the age of thirteen he began collecting birds, which he measured, weighed, described, and sketched. A new world was opened up to him when he discovered that people actually wrote books on birds and that names for them existed in English and Latin. He avidly read ornithological works by John James Audubon, Alexander Wilson, Thomas Nuttall, and others, and aspired to write a book on the birds of New England. During winters he attended a district school, where he was far ahead of the other pupils and was largely self-taught. Summers were spent on the farm helping his father, who did not share his son's interest in natural history. When not doing his chores, Allen spent every spare moment preparing specimens and making notes on his natural history observations. His constant disappearances naturally annoyed his father, who could not appreciate his absorption in such unpractical affairs. Allen's mother was more supportive and often intervened in his favor. As a result he was eventually allowed one day a week to pursue his zoological collecting and researches.

At the age of twenty Allen enrolled at the Wilbraham Academy in Springfield, which he attended for three winter terms. In 1861 he reluctantly sold his extensive zoological collection, which included hundreds of mounted specimens, to the academy. The sale gave him enough money to enroll in the Lawrence Scientific School at Harvard, where he studied under Louis Agassiz, the "revered teacher" to whom Allen later dedicated his autobiography. Agassiz's first goal as a teacher was to train his pupils to observe. Allen was provided with hand lenses and several genera of corals, and told to find out their methods of growth and development. "After a few hours of application we were asked 'Well, what have you seen?' and the same query was daily repeated. We reported what we thought we had discovered and if we had seen aright we were encouraged with a few words of approval; if we were mistaken the reply was 'You are wrong; you must look again; you must learn to see'" (*Autobiographical Notes*, 8–9). Agassiz's lectures, which were well attended by notable Bostonians, supplemented these observational exercises. Allen also took courses from Jeffries Wyman and Asa Gray; his fellow students included Alpheus Hyatt, William James, Edward Sylvester Morse, Alpheus Spring Packard, Jr., Nathaniel Southgate Shaler, and Addison Emery Verrill.

In 1865 Allen was invited to accompany Agassiz as one of fifteen assistants during his expedition to Brazil. Allen was detailed to a smaller party that was to make collections in the provinces north of Rio de Janeiro. He remained in Brazil for eight months, doing extensive inland collecting as far north as Bahia. When he returned to Boston, a chronic intestinal ailment acquired in Brazil forced his retirement to the family farm, where he remained for a year. He recovered and was able to spend several months collecting in Indiana, Illinois, and Iowa before returning to Harvard as curator of mammals and birds at the Museum of Comparative Zoology in October 1867.

Allen spent the next eighteen years at the museum. During this period he made collecting expeditions to eastern Florida (1868–1869), the Great Plains and Rocky Mountains (1871–1872), Yellowstone, Montana (1873), and Colorado (1882). Some measure of his collecting zeal may be gained from the record of specimens taken over a nine-month period in the Great Plains and Rocky Mountains: 200 animal skins, 60 skeletons, 240 skulls, 1,500 bird skins, 100 birds in alcohol, many nests and eggs, and numerous fish, mollusks, insects, and crustaceans. While in the Yellowstone area in 1873 he was frequently in danger from hostile Sioux Indians. His military escort was Custer's Seventh Cavalry, which was to perish three years later at the Little Big Horn, sixty miles south of where Allen had been collecting.

From 1874 to 1882 Allen concentrated largely on research and publication. His monograph *The American Bisons* (1876) was followed by one on North American pinnipeds—walruses, sea lions, and seals (1880). By 1882 Allen had so overworked himself that he was incapacitated by a nervous breakdown; in later years his health remained delicate.

In 1885 Allen entered the second phase of his career. Financial restrictions at the Museum of Comparative Zoology compelled him to accept the post of curator of birds and mammals at the American Museum of Natural History in New York, where he remained until his death. There he presided over a twentyfold increase in the museum's specimens of birds and mammals, which required his constant efforts in cataloging, labeling, and tending the collections. Only in 1888 was he assisted by Frank M. Chapman, who was to become a distinguished ornithologist in his own right.[1] The following year Allen became editor in chief of all the museum's zoological publications, of which nearly sixty volumes appeared under his supervision.

Allen was a cofounder (with Elliott Coues and William Brewster) of the American Ornithologists' Union in 1883, and he was elected its first president (1883–1890). He played an active role in drafting this organization's Code of Nomenclature (1886), which later became the basis for the International Code of Zoological Nomenclature and has had a major influence in standardizing nomenclatural rules throughout the world. In 1910 Allen was elected a member of the International Commission on Zoological Nomenclature. For twenty-eight years he edited the A.O.U.'s journal, *The Auk* (1884–1912), as he had the eight volumes of its predecessor, *Bulletin of the Nuttall Ornithological Club* (1876–1883). His publications include 1,470 titles, mainly dealing with birds and mammals but also treating nomenclature, biogeography, and evolution. Among his honors were an honorary Ph.D. from Indiana University (1886), the Walker Grand Prize from the Boston Society of Natural History (1903), and the Linnaean Society (London) medal (1916). He was also a member of the National Academy of Sciences (1876) and president of the Linnaean Society of New York (1890–1897).

Allen's personality was marked by a tireless zeal for work, a scrupulous and sometimes pedantic attention to detail, and a sincere consideration for others. He was critical, however, of careless work or of conclusions based on insufficient evidence. His delicate health was probably linked to his extreme shyness, which prevented him from lecturing publicly and made even the delivery of scientific papers distasteful.

Allen's first wife, Mary Manning Cleveland, whom he had married on 6 October 1874, died in April 1879; they had a son, Cleveland Allen. On 27 April

1886 Allen married Susan Augusta Taft, who survived him. They had no children.

Besides his prodigious labors in descriptive natural history and taxonomy, Allen's most notable scientific contributions were in the domains of biogeography and nomenclature. His first major publication, which appeared in 1871, dealt with the mammals and winter birds of eastern Florida. In this article Allen set forth a number of general views on the nature of climatic variation among species and the relationship of climate to the primary biogeographic zones of the world. He held that temperature, analyzed according to isothermal lines, and humidity were the two main determinants of geographic range among species. He also affirmed the generalization that increased humidity and rainfall are associated with darker coloration in most species (Gloger's rule).[2] Another of his biogeographical generalizations—that the extremities become shorter in colder climates—has become known as Allen's rule (pp. 230–233).[3] (Allen's rule is actually a special case of Bergmann's rule, which states that in colder climates, races of species are larger in body size—and hence proportionally smaller in surface area—thus reducing heat loss.[4])

Allen's 1871 publication also proposed a revised terminology and conceptualization for major faunal areas of the globe. Objecting to the use of such zoologically inappropriate terms as "Old World" and "New World," Allen proposed eight major biogeographic zones: (1) an Arctic Realm; (2) a North Temperate Realm; (3) an American Tropical Realm; (4) an Indo-African Tropical Realm; (5) a South American Temperate Realm; (6) an African Temperate Realm; (7) an Antarctic Realm; and (8) an Australian Realm. Within the American portion of the North Temperate Realm, he recognized eastern and western provinces. Allen's 1871 paper received the Humboldt Scholarship of the Lawrence Scientific School and brought him recognition as a major American naturalist.

Allen's biogeographic views inspired subsequent conflict with Alfred Russel Wallace, who endorsed Philip Lutley Sclater's (1858) somewhat differing faunal boundaries. In his revised biogeographic subdivisions Allen (1878) emphasized the importance of geographic isolation in addition to climatic influences; and, following Wallace and Sclater, he allotted Madagascar and the Mascarene Islands to a separate "Lemurian Realm." He also stressed (1876) that the most aberrant or specialized forms of zoological groups are consistently found at the periphery of the range, a view that was important for later advocates of geographic isolation as a crucial mechanism in speciation.

Although Allen believed that Darwin had overemphasized the importance of natural selection, he specifically rejected mutationism as a mechanism of evolution (1877). Like Darwin in his later years, he thought the environment played an important role in inducing adaptive changes that are then inherited. In this respect he was basically a neo-Lamarckian, as exemplified by his theory of bird migration. On this topic he argued that "what was at first a forced migration [owing to climatic changes] would become habitual, and through the heredity of habit give rise to that wonderful faculty we term the instinct of migration" ("Origin of the Instinct of Migration in Birds," 153). Similarly, he understood what has become known as Allen's rule in neo-Lamarckian terms (as the result of decreased circulation in the extremities in colder climates, presumably impeding growth) rather than in terms of natural selection favoring heat conservation (1872, 215).

In the area of biological nomenclature Allen was a conservative thinker who nevertheless adapted to change. He at first resisted the notion of using trinomials to distinguish geographic races; but eventually, under the weight of ever growing museum collections and greater evidence for meaningful geographic variants, he accepted this revision of the binomial system. He was a master at coping with the tangle of problems created by zoological synonyms, and he exerted a major influence in establishing rules to decide upon proper nomenclature. Not a brilliant thinker or a bold generalizer, Allen nevertheless possessed a patient love of detail that underlay his prolific achievements as a descriptive naturalist.

NOTES

1. The American Museum of Natural History, which had been founded in 1869 by A. S. Brickmore, one of Agassiz's students, grew rapidly in Allen's lifetime. In 1908 the department of birds and mammals was separated, with Chapman taking over the department of birds and Allen continuing as head of the department of mammals. By 1921, the year of Allen's death, the combined staff of these two departments had grown to seventeen (Chapman, "Biographical Memoir," 5).
2. Constantin Lambert Gloger, *Das Abändern der Vögel durch Einfluss des Klima's* (Breslau, 1833).
3. See also Allen's "Geographical Variation in North American Birds," 213–215; and "The Influence of Physical Conditions in the Genesis of Species," 116–117.
4. Carl Georg Bergmann, "Über die Verhältnisse der Wärmeökonomie der Thiere zu ihrer Grösse," in *Göttinger Studien*, pt. 1 (1847), 595–708. See also Ernst Mayr, *Animal Species and Evolution* (Cambridge, Mass., 1963), 319–323.

BIBLIOGRAPHY

I. ORIGINAL WORKS. A complete bibliography of Allen's publications up to 1916 is in his *Autobiographical Notes and a Bibliography of the Scientific Publications of Joel Asaph Allen* (New York, 1916). Frank M. Chapman's biographical memoir of Allen (see below) completes Allen's bibliography. The most important of Allen's publications are "On the Mammals and Winter Birds of East Florida, with an Examination of Certain Assumed Specific Characters in Birds, and a Sketch of the Bird Faunae of Eastern North America," in *Bulletin of the Museum of Comparative Zoology*, **2** (1871), 161–450; "Geographical Variation in North American Birds," in *Proceedings of the Boston Society of Natural History*, **15** (1872), 212–219; *The American Bisons, Living and Extinct, Memoirs of the Geological Survey of Kentucky*, **1**, pt. 2 (1876) and *Memoirs of the Museum of Comparative Zoology*, **4**, no. 10 (1876); "Geographic Variation Among North American Mammals, Especially in Respect to Size," in *Bulletin of the United States Geological and Geographical Survey of the Territories*, **2** (1876), 309–344; "The Influence of Physical Conditions in the Genesis of Species," in *Radical Review*, **1** (1877), 108–140; "The Geographical Distribution of the Mammalia, Considered in Relation to the Principal Ontological Regions of the Earth, and the Laws That Govern the Distribution of Animal Life," in *Bulletin of the United States Geological and Geographical Survey of the Territories*, **4** (1878), 313–377; *History of North American Pinnipeds: A Monograph of the Walruses, Sealions, Sea-bears and Seals of North America* (Washington, D.C., 1880); "Origin of the Instinct of Migration in Birds," in *Bulletin of the Nuttall Ornithological Club*, **5** (1880), 151–154; and *The Code of Nomenclature and Check-List of North American Birds Adopted by the American Ornithologists' Union* . . . (New York, 1886; rev. ed., 1908; 3rd ed., 1910). Fifteen of Allen's publications were reprinted in *Selected Works of Joel Asaph Allen* (New York, 1974).

II. SECONDARY LITERATURE. Other than Allen's *Autobiographical Notes*, the most complete treatment of his life and work is Frank M. Chapman, "Biographical Memoir. Joel Asaph Allen 1838–1921," in *Biographical Memoirs. National Academy of Sciences*, **21** (1927), 1–20; a shorter version was published as "In Memoriam: Joel Asaph Allen," in *The Auk*, **39** (1922), 1–14. See also Henry Fairfield Osborn, "Joel Asaph Allen," in *Dictionary of American Biography*, I, 197–198; Keir B. Sterling, "Introduction," in *Selected Works of Joel Asaph Allen* (New York, 1974); and Witmer Stone, "Dr. J. A. Allen," in *The Auk*, **38** (1921), 490–492.

FRANK J. SULLOWAY

ALLISON, SAMUEL KING (*b.* Chicago, Illinois, 13 November 1900; *d.* Oxford, England, 15 September 1965), *physics*.

Samuel K. Allison, the son of Samuel Buell Allison, an elementary school principal, and Caroline King Allison, attended elementary school and Hyde Park High School in Chicago. In 1917 he entered the University of Chicago. Doing honors work in chemistry and mathematics and studying physics under R. A. Millikan, he received his B.S. degree in 1921. Two years later he earned his Ph.D. degree in chemistry (with a thesis on a problem in experimental physics) under William D. Harkins. He married Helen Catherine Campbell on 28 May 1928, and they had two children, Samuel and Catherine.

Allison then received a National Research Council Fellowship to pursue postdoctoral research in William Duane's laboratory at Harvard University. He arrived there in the fall of 1923, just as G. L. Clark, another National Research Council Fellow, and Duane challenged the validity of the Compton effect, discovered by Arthur H. Compton at Washington University in St. Louis in late 1922. Duane and Compton (who transferred to the University of Chicago in mid 1923) debated the issue vigorously at a meeting of the American Physical Society in December 1923 and during an exchange of visits to each others' laboratories in early 1924. Allison was drawn into the dispute, and his initial experiments seemed to support Clark and Duane's position. Only after Compton and Duane engaged in a second debate at a meeting of the British Association for the Advancement of Science in Toronto in the summer of 1924 did Allison find persuasive evidence for the validity of the Compton effect, which compelled Duane to withdraw his objections at the December 1924 meeting of the American Physical Society.

Allison left Harvard in mid 1925 for a third year of postdoctoral research at the Carnegie Institution in Washington, D.C., after which Leonard B. Loeb was instrumental in bringing him to the University of California in Berkeley. He remained in the Berkeley physics department four years, until mid 1930, rising from the rank of instructor to associate professor. Continuing his X-ray researches, soon with the help of his graduate student, John H. Williams, Allison designed and constructed a new high-resolution double-crystal X-ray spectrometer, used it to measure precisely the intensities and widths of various X-ray lines, and in this way confirmed the dynamical theory of X-ray diffraction that had been developed independently by C. G. Darwin in 1914 and by P. P. Ewald in 1917. This collaboration cemented a lifelong friendship between Allison and Williams. When Arthur Compton persuaded Allison to return to Chicago in 1930, Williams followed as a National Research Council Fellow from 1931 to 1933 and then joined the faculty of the University of Minnesota. In subsequent years, the two spent several weeks together each summer on fishing and canoe trips in Wisconsin and Canada.

Allison remained on the faculty of the University of Chicago for the rest of his life. Until 1935 his research continued in the field of X rays; it culminated in the publication of his and Compton's authoritative treatise, *X-Rays in Theory and Experiment* (1935), a work which, as Compton acknowledged in the preface, Allison was primarily responsible for writing. After its completion, Allison redirected his research into the growing field of experimental nuclear physics. He spent six months in Rutherford's Cavendish Laboratory in Cambridge, England, learning techniques and working with Cockcroft and Walton's accelerator. On his return, he built a machine of this type, thereby introducing experimental nuclear research into the University of Chicago's physics department. He and his students used it to make precision measurements of the masses of various light isotopes by means of proton-induced reactions.

Allison's experience in experimental nuclear research placed him in a position to play a key role in the Manhattan Project. In January 1941 he obtained a contract from the National Defense Research Committee to study the feasibility of using beryllium as a neutron reflector and as a moderator in a pile. Thus, when Compton organized the Metallurgical Laboratory a year later in February 1942, bringing Enrico Fermi's Columbia team and Eugene Wigner's Princeton team to Chicago, Allison already had an active and knowledgeable group working there. He was promoted to full professor, appointed head of the chemistry division, and joined Fermi and others in the construction of the first pile under the West Stands of Stagg Field. When it achieved criticality on 2 December 1942, Allison was in charge of the "suicide squad" that was to dump a huge jug of cadmium solution over the pile to quench the chain reaction in the remote possibility of an uncontrolled runaway.

The principal mission of the Metallurgical Project, with plants at Chicago, Oak Ridge, Tennessee, and Hanford, Washington, was to develop methods for the large-scale production of plutonium. As Compton's responsibilities as overall director grew, he decided to appoint Allison, in 1943, director of the "Met Lab" in Chicago. Allison's deep scientific and technical knowledge, his singleness of purpose and exceptional administrative abilities, and his personal patience, sound common sense, and dry sense of humor made him a natural and logical choice for this position. As director he had to solve a host of problems of all kinds, and he had to work fruitfully with people of many backgrounds, including engineers and executives from the Du Pont Company, whom he assisted in the design and construction of the reactors at Hanford.

In November 1944 Allison transferred to Los Alamos, where he served as Robert Oppenheimer's associate director and as chairman of the Technical and Scheduling Committee, which in early 1945 became the nerve center for coordinating experiments, facilities, and materials for the construction of the atomic bomb. He also became a key member of the "Cowpuncher Committee" charged to "ride herd" on the development of the implosion method of detonation. It was Allison who called the countdown of the Trinity test of the plutonium bomb in the early morning hours of Monday, 16 July 1945.

Three weeks after the war ended, on 1 September 1945, Allison sounded another call. Some sixteen scientists, including Fermi and Harold C. Urey, joined Allison at a news conference in the Shoreland Hotel on the South Side of Chicago to announce their affiliation with the newly established Institute for Nuclear Studies of the University of Chicago. Allison, as its director-designate, used this occasion to make an eloquent plea for the return of free and unhampered research. He remarked, bluntly, that if secrecy was imposed, all first-rate scientists would work on subjects as innocuous as the colors of butterflies' wings. "Sam's butterfly speech" became a clarion cry for civilian control of nuclear research after the war. Allison became a founding sponsor of the *Bulletin of the Atomic Scientists*, and he continued to courageously argue the case against secrecy in government-supported research even after the shocking arrest of Alan Nunn May in early 1946 and of Klaus Fuchs in early 1950 as Soviet spies.

In 1946 Allison received the Medal of Merit for his wartime contributions, with a special citation from President Truman, and he was also elected to the National Academy of Sciences. As director of Chicago's Institute for Nuclear Studies (later renamed after Enrico Fermi), he assembled a distinguished group of physicists, chemists, and astrophysicists, and he fostered a spirit of scientific collaboration among them. He also personally carried out and supervised a good deal of research in experimental low-energy nuclear physics using a new 400-keV Cockcroft-Walton accelerator that he constructed in 1949. A number of his students from these years, as from earlier and later ones, became distinguished scientists in their own right.

In 1958 Allison resigned as director of the Enrico Fermi Institute to be able to devote more time to his research. The following year he was named Frank P. Hixon Distinguished Service Professor. In 1960 he traveled to Egypt, and in 1961 to Argentina, to serve as an adviser to those scientifically developing countries. In 1963 he completed the construction of a new 4-MeV Van de Graaff accelerator and again agreed to become director of the Enrico Fermi Institute. Two years later, on 6 September 1965, while serving as the United States delegate to the Plasma Physics and Controlled Nuclear Fusion Research Conference in Culham, England, he suffered an aortic aneurism and died following surgery at the Radcliffe Infirmary in nearby Oxford.

BIBLIOGRAPHY

I. ORIGINAL WORKS. Allison's papers are preserved in the University of Chicago Library. His doctoral thesis was published as "The Absence of Helium from the Gases Left after the Passage of Electrical Discharges: I, Between Fine Wires in a Vacuum; II, Through Hydrogen; and III, Through Mercury Vapor," in *Journal of the American Chemical Society*, **46**, no. 4 (1924), 814–824, written with W. D. Harkins. Some representative later publications are: "Experiments on the Wave-lengths of Scattered X-Rays," in *Physical Review*, 2nd ser., **26**, no. 3 (1925), 300–309, written with W. Duane; "The Resolving Power of Calcite for X-Rays and the Natural Widths of the Molybdenum $K\alpha$ Doublet," *ibid.*, 2nd ser., **35**, no. 12 (1930), 1476–1490, written with J. H. Williams; "The Masses of Li^6, Li^7, Be^8, Be^9, and B^{10} and B_{11}," *ibid.*, 2nd ser., **55**, no. 7 (1939), 624–627; "Passage of Heavy Particles Through Matter," in *Reviews of Modern Physics*, **25**, no. 4 (1953), 779–817, written with S. D. Warshaw; and "Experiments on Charge-Changing Collisions of Lithium Ionic and Atomic Beams," in *Physical Review*, 2nd ser., **120**, no. 4 (1960), 1266–1278, written with J. Cuevas and M. Garcia-Munoz.

II. SECONDARY LITERATURE. Insightful obituary notices of Allison were written by Alvin Weinberg in the *Bulletin of the Atomic Scientists*, **22**, no. 1 (1966), 2, and by H. L. Anderson in *Nature*, **209** (19 February 1966), 758–759. Later biographical accounts were prepared by the University of Chicago (undated pamphlet on the Frank P. Hixon Distinguished Service Professorship, 32 pp.), and by R. S. Shankland, *Dictionary of American Biography, Supplement Seven 1961–1965* (New York, 1981), 10–11.

For Allison's role in the Compton-Duane controversy, see Roger H. Stuewer, *The Compton Effect: Turning Point in Physics* (New York, 1975), 249–273. For information on Allison's role in the Manhattan Project and during the postwar period, see Richard G. Hewlett and Oscar E. Anderson, Jr., *A History of the United States Atomic Energy Commission*, I, *The New World, 1939–1946* (University Park, Penn., 1962); Daniel J. Kevles, *The Physicists: The History of a Scientific Community in Modern America* (New York, 1978); Alice Kimball Smith, *A Peril and a Hope: The Scientists' Movement in America: 1945–1947* (Chicago, 1965); Henry De Wolf Smyth, *Atomic Energy for Military Purposes: The Official Report on the Development of the Atomic Bomb Under the Auspices of the United States Government, 1940–1945* (Princeton, 1945).

ROGER H. STUEWER

ANDERSON, EDGAR (*b.* Forestville, New York, 9 November 1897; *d.* St. Louis, Missouri, 18 June 1969), *plant genetics*.

The son of A. Crosby Anderson, a private school administrator who later became a professor of dairy husbandry at Michigan Agricultural College (now Michigan State University), and Inez Evora Shannon Anderson, Edgar was from the age of three brought up in a college environment. Throughout his childhood he was surrounded by cultivated houseplants, which his mother carefully tended, in which he took great interest, and on which he experimented while a child. At Michigan Agricultural College, which he entered at the age of sixteen, almost his only extracurricular activity was the Horticultural Society. After graduation he served briefly during World War I in the Naval Reserve as Gunner's Mate Second Class. In 1919 he became a graduate student at Bussey Institution of Harvard University.

In graduate school his interests and character were molded by life among a small group of brilliant, carefully selected graduate students who had almost continuous association with a few of the most outstanding life scientists of their day. His capacity for making precise, detailed observations of plants was developed by association with his major professor, Edward M. East. The deeply religious attitude of friendliness and humanity that he acquired from his parents was broadened when he left the Methodist Church and adopted the Quaker faith, which he maintained with lifelong devotion. For two years he courted Dorothy Moore, a laboratory assistant who had graduated in botany at Wellesley College, and he married her in 1923, the year after he received his Ph.D. During the remaining forty-six years of his life she gave him constant encouragement in his work. They had one daughter, Phoebe. During his graduate years frequent walks in the countryside made him aware of native plants in addition to those in gardens.

In 1923 Anderson was appointed geneticist at the Missouri Botanical Garden and assistant professor of botany at neighboring Washington University. Except for a return for four years to Harvard and the Bussey (1931–1935), his entire career was spent in these two St. Louis institutions.

His association with Jesse Greenman, a leading taxonomist, made him aware of the complexity of variation in plant genera and species. He decided to investigate plants not by the relatively casual, subjective methods that were then in vogue, but by precise measurements of living individuals in populations. Studying the native species of blue flag *Iris*, he quickly found that variation within each of two species, *I. versicolor* and *I. virginica*, is of a different nature from that which makes up the differences between the species. He finally concluded that the origin of *I. versicolor* could be explained only by assuming that it is a stabilized hybrid or allopolyploid containing the entire set of chromosomes belonging to *I. virginica* plus another set that belongs to an Alaskan species, *I. setosa* var. interior. He thus became a leading investigator of hybridization as a source of variation within species.

During his return to Harvard and the Bussey his association with another botanist from Missouri, Robert Woodson, strengthened his belief in the importance of hybridization. It was fully confirmed by his collaboration with Karl Sax on the genus *Tradescantia*. Sax, formerly a fellow graduate student who returned to the Bussey in 1929, was a leading authority on plant chromosomes and did much to make Anderson aware of their importance for investigating hybridization. The research on *Tradescantia* was continued after his return to St. Louis in 1935 and led to his concept of introgression, a term that he gave to a succession of three processes: interspecific hybridization, backcrossing to one of the parental species, and natural selection of highly adaptive gene combinations generated by the increased added variation derived from the nonrecurrent species.

The years 1935–1954 were Anderson's most productive period. His position at the Missouri Botanical Garden stimulated his desire to produce new and improved cultivated plants for midwestern gardens. A trip to the Balkans in 1934 convinced him that, because of its similarity in climate to the American Midwest, this region was the most promising source of new varieties. Several that he produced, particularly in ivy and boxwood, are now widely grown. Throughout thirty-four years as the garden's geneticist, including three years as its director, he produced a flood of short, popular articles on gardening.

Throughout conversations and collaboration with botanists of cultivated plants, particularly Thomas Whitaker, his former Bussey roommate Paul Mangelsdorf, and R. G. Reeves of Texas, Anderson came to realize that the precise methods for recording variation that he had developed would be particularly valuable for investigating crop plants such as maize. The series of papers published on this crop between 1940 and 1954 were original and stimulating. Two of his graduate students, Hugh Cutler and William L. Brown, became leading botanists devoted to maize.

In his two books, *Introgressive Hybridization* (1949) and *Plants, Man, and Life* (1952), Anderson distilled his research accomplishments and philosophy of life. The latter book, reprinted in 1967 by the University of California Press, has gained renewed popularity.

He was named director of the Missouri Botanical Garden in 1954, but resigned in 1957. He was not fitted to be an administrator either by ability or temperament. Anderson was a member of the National Academy of Sciences and the American Academy of Arts and Sciences. He served as president of the Botanical Society of America, the Herb Society of America, and the Society for the Study of Evolution.

During the last twelve years of his life he suffered recurrent illness and made no more contributions to science. Nevertheless, to the end he remained devoted to the Botanical Garden, and he continued to publish popular articles on horticulture. On 18 June 1969 he died of a heart attack while writing one of these papers.

BIBLIOGRAPHY

I. Original Works. A complete bibliography, prepared by Erna R. Eisendrath, is in *Annals of the Missouri Botanical Garden*, **59** (1972), 346–351. Some of Anderson's most significant publications include "The Problem of Species in the Northern Blue Flags, *Iris versicolor L.* and *Iris virginica L.*," in *Annals of the Missouri Botanical Garden*, **15** (1928), 241–332; "The Species of *Tradescantia* Indigenous to the United States," in *Contributions from the Arnold Arboretum, Harvard University*, **9** (1935), 1–132, written with Robert E. Woodson; "A Cytological Monograph of the American Species of *Tradescantia*," in *Botanical Gazette*, **97** (1936), 433–476, written with Karl Sax; "An Experimental Study of Hybridization in the Genus *Apocynum*," in *Annals of the Missouri Botanical Garden*, **23** (1936), 159–168; "The Species Problem in *Iris*," *ibid.*, 457–509; "Hybridization in *Tradescantia*. III. The Evidence for Introgressive Hybridization," in *American Journal of Botany*, **25** (1938), 396–402, written with Leslie Hubricht; and "Recombination in Species Crosses," in *Genetics*, **24** (1939), 668–698.

"Races of *Zea mays*. I. Their Recognition and Classification," in *Annals of the Missouri Botanical Garden*, **29** (1942), 69–88, written with Hugh C. Cutle ; "Maize in Mexico—A Preliminary Survey," *ibid.*, **33** (1946), 147–247; "The Northern Flint Corns," *ibid.*, **34** (1947), 1–28, written with William L. Brown; "The Southern Dent Corns," *ibid.*, **35** (1948), 225–268, written with William L. Brown; *Introgressive Hybridization* (New York, 1949); "Origin of Corn Belt Maize and Its Genetic Significance," in John W. Gowen, ed., *Heterosis* (Ames, Iowa, 1952), 124–148; *Plants, Man, and Life* (New York, 1952; repr. Berkeley, 1967); "Introgressive Hybridization," in *Biological Reviews, Cambridge Philosophical Society*, **28** (1953), 280–307; "The Role of Hybridization in Evolution," in Willis H. Johnson and William C. Steere, eds., *This Is Life: Essays in Modern Biology* (New York, 1962), 287–314.

II. SECONDARY LITERATURE. *Annals of the Missouri Botanical Garden*, **59** (1972), a volume dedicated to Anderson, contains a biographical sketch, recollections and reminiscences by friends and students, and a complete bibliography. See also the article by G. Ledyard Stebbins in *Biographical Memoirs. National Academy of Sciences*, **49** (1978), 3–23.

G. LEDYARD STEBBINS

ANDREEV, NIKOLAI NIKOLAEVICH (*b.* Kurmani, Russia, 28 July 1880; *d.* Moscow, U.S.S.R., 31 December 1970), *physics.*

Andreev's father, Nikolai Fedorovich Andreev, was a minor official; his mother, Alexandra Nikitichna Konvisarova, was a housewife. When Andreev was five, his parents died in a fire at the family home, and the boy was then brought up by relatives. A special role in his upbringing was played by his paternal aunt, an actress at the Malyi Theater in Moscow. From 1890 to 1892 Andreev studied at the classical gymnasium in Moscow, and then enrolled in a military school. There he received a strong physical-mathematical education, studied European languages, and developed his musical talent (he played piano, flute, and oboe). In 1898 he was admitted to the Higher Technical Institute in Moscow, but the next year was expelled for his participation in the student movement. Then he studied at Moscow University as an extern. His mathematical abilities attracted the attention of Nikolai Vasilievich Bugaev. In 1901 he married Nadezhda Nikolaevna Nikolskaia, the mother of his best friend. They had one daughter. In 1903 Andreev moved to Göttingen to enter the university there. Three years later he transferred to Basel, Switzerland, where he worked under the guidance of August Hagenbach. By the time Andreev defended his doctoral dissertation in 1908, he was an original scholar. His interests focused on optics, both experimental and theoretical. He worked out the method for determining the number of optically active components in solutions of substances of high molecular weight by their rotation of the plane of polarization, and he developed Paul Drude's theory of the dispersion of light. The latter work was the basis of his dissertation, for which he was awarded the doctorate cum laude.

After defending his dissertation, Andreev returned to Moscow. The same year he was elected a member of the Russian Physical-Chemical Society. In Moscow he soon came in contact with the best physical school of prerevolutionary Russia, that of Petr Nikolaevich Lebedev, and worked in Lebedev's laboratory from 1910 to 1912. His interests at that time included X-ray physics, acoustics, and radio engineering. He taught at a number of institutions of higher learning, among them Moscow University (1912–1917). There in 1913 he defended his M.Sc. thesis, which opened the way to a professorship. The following year Andreev married Maria Gerasimovna Larionova; they had a son and three daughters. During World War I, Andreev was involved in defense work, developing an acoustical method for locating artillery batteries. The same problem was investigated by Max Born and others for the Germans.

From 1918 to 1920 Andreev was professor of physics at the Agricultural Institute in Omsk; and in 1920 he returned to Moscow. By that time his interests were concentrated on acoustics and the theory of vibrations. Andreev was a founder of nonlinear acoustics and of the acoustics of musical instruments, which he continued to develop after moving to Leningrad in the mid 1920's, at the invitation of Abram Fedorovich Ioffe. There, at the Physical-Technical Institute, Andreev established an acoustical laboratory and in 1932 he founded the world's first institute of musical acoustics. He also set up departments of acoustics at a number of institutions of higher education. His investigations on piezoelectricity stimulated works by Igor Vasilievich Kurchatov and Pavel Pavlovich Kobeko on ferroelectricity.

In 1940, at the invitation of Sergei Ivanovich Vavilov, Andreev moved to Moscow, where he began to work at the P. N. Lebedev Physics Institute. There he founded the acoustical laboratory that in 1953 became the Acoustical Institute of the Soviet Academy of Sciences. Andreev worked at this institute for the rest of his life; and in 1978 the institute was named for him.

In acoustics Andreev developed the method of

calibrating the amplitudes of oscillating membranes and the design of telephones and microphones, as well as the theory of oscillations of piezocrystals (which had applications in radio engineering). He investigated architectural acoustics, hydroacoustics, and the theory of propagation of sound waves, and was much interested in problems of biological acoustics (human voices, noises of animals or insects).

Andreev also considered philosophical problems: when he was still quite young, he translated and published at his own expense *Science and Hypothesis* by Henrí Poincaré. After the revolution he published popular books on Einstein's theory of relativity and on quantum mechanics, as well as a number of textbooks on mechanics, acoustics, and the theory of vibrations. He was a founder (in 1923) and the first editor of the popular scientific magazine *Iskra* ("Spark"), the editor in chief of *Zhurnal eksperimentalnoi i teoreticheskoi fiziki* (1950–1955), and of *Akusticheskii zhurnal* (1954–1970). But above all, Andreev was the founder and recognized leader of the Soviet acoustical school, which has included academicians L. M. Brekhovskikh, B. P. Konstantinov, and A. A. Kharkevich, and professors A. I. Belov, G. A. Ostroumov, and A. V. Rimskii-Korsakov.

Andreev's scientific and organizational accomplishments were highly appreciated by the scientific community and the Soviet government. In 1933 he was elected corresponding member, and in 1953 full member, of the Soviet Academy of Sciences. In 1945 and 1953, he received the Lenin Prize, and in 1970 he was awarded the title Hero of Socialist Labor.

Throughout his life, music was important to Andreev. He was an excellent pianist.

BIBLIOGRAPHY

I. Original Works. Andreev's works are listed in *Uspekhi fizicheskikh nauk*, **44**, no. 3 (1951), 472, **71**, no. 3 (1960), 525, and **101**, no. 4 (1970), 773; A bibliography is also in Glekin (see below). *Theoretische und experimentelle Untersuchungen über den Einfluss der Temperatur auf die Dispersion des Lichtes* (Ph.D. diss., Basel, 1909); "K dispersii zatukhaiushchikh voln" (On the dispersion of attenuated waves), in *Zhurnal Russkogo fiziko-khimicheskogo obshchestva*, **41**, no. 1 (1910), 46–56; *Elektricheskie kolebaniia i ikh spektri* (Electrical oscillations and their spectra; M.S. thesis, Moscow, 1917); "Ostrota slukha" (Sharpness of hearing), in *Zhurnal prikladnoi fiziki*, **1**, no. 1–2 (1924), 252–263; "Tekhnicheskii amplitudometr" (Technical amplitude meter), *ibid.*, **2**, no. 3–4 (1925), 205–212); "O privedennom uravnenii struny" (On the reduced string equation), *ibid.*, **4**, no. 1 (1927), 21–26; "O kolebaniiakh kvartservoi plastinki po tolshchine" (On the distribution of a quartz plate's vibrations by its thickness), in *Zhurnal tekhnicheskoi fiziki*, **2**, no. 1 (1932), 119–124; *Akustika dvizhushcheisia sredy* (Moving-medium acoustics; Moscow and Leningrad, 1934), written with I. G. Rusakov; "O dereve dlia muzykalnykh instrumentov" (On the wood for musical instruments), in *Trudy NII muzykalnoi promyshlennosti*, I (Moscow and Leningrad, 1938), 13–28; "O golose moria" (On the voice of the sea), in *Doklady Akademii nauk SSSR*, **23**, no. 7 (1939), 625–628; "Piezoelektricheskie kristally i ikh primenenie" (Piezoelectric crystals and their use), in *Elektrichestvo* (1947), no. 2, 5–13; "Ob organakh slukha u nasekomykh" (On the hearing organs of insects), in *Problemy fiziologicheskoi akustiki*, III (Moscow and Leningrad, 1955), 89–94; "Einige Fragen der nichtlinearen Akustik," in *Proceedings of the Third International Congress on Acoustics, Stuttgart*, I (Amsterdam and London, 1961), 304–306.

II. Secondary Literature. G. V. Glekin, *Nikolai Nikolaevich Andreev* (Moscow, 1980), and *Materialy k biobibliografii sovetskikh uchennykh* (Materials for the biobibliographies of Soviet scientists; Moscow, 1963).

V. J. Frenkel

ARAMBOURG, CAMILLE LOUIS JOSEPH (*b.* Paris, France, 3 February 1885; *d.* Paris, 19 November 1969), *paleoanthropology, vertebrate paleontology*.

Arambourg was the son of Victor Arambourg and Clarisse Pfeifer Arambourg. His parents, who belonged to the upper middle class, had lived in Lyons for many years, but moved to Paris before his birth. His father, who was passionately interested in photography, was a pioneer in that art and a friend of the manufacturers and inventors Louis and Auguste Lumière, and of the photographer Nadar (Félix Tournachon). Arambourg's mother was an accomplished musician.

He was educated at a private school, Sainte-Croix de Neuilly, where he prepared his baccalaureate degree in 1903. Later he studied at the Institut National Agronomique, where he took his degree in agricultural engineering in 1908.

Since his father had settled as a farmer near Oran, Algeria, Arambourg joined him there and helped him to improve the water supply of the vineyards. During this work, deep plowings unearthed well-preserved fossil fishes from the Upper Miocene (the "Sahelian" of Augûste Pomel). Arambourg married Julie Marie Froget at Oran on 16 June 1910. In 1912 he undertook excavations near Oran to collect more material. In order to identify and further study these fishes, he frequently visited the geological laboratory of the Algiers Faculty of Sciences.

When World War I broke out, Arambourg had to leave Algeria to serve in the French army. He fought in the Dardanelles campaign and in Serbia in 1915, and in Macedonia from 1916 to 1918. L'armée d'Orient had established its positions near Salonika, along the east bank of the Vardar River, in an area where Upper Miocene ("Pontian") lacustrine deposits, rich in vertebrate remains, outcrop widely. While the soldiers dug trenches, Arambourg collected numerous bone fragments. As a result, the lieutenant colonel of his regiment allowed him to undertake paleontological excavations and to gather an important collection of mammalian remains belonging to species already known from the deposits at Pikermi in Greece, remains that had been described fifty years earlier by Albert Gaudry. Moreover, the army staff transported his collection to Salonika. When the war was over, these fossils were shipped to Algiers and, later, to the Museum of Natural History in Paris.

Since he had received geological training as part of his studies at the Institut National Agronomique, Arambourg used some of the time spent in Macedonia surveying for a geological map (1:50,000) of the vicinity of Guevgueli, in the Vardar Valley. This map was printed in 1919 by the Service Géographique de l'Armée d'Orient.

After the war Arambourg returned to Algeria. In 1920, when his father decided to sell the farm, Arambourg applied for a professorship of geology at the Institut Agricole d'Alger. During his ten years in Algiers, he continued collecting Upper Miocene fossil fishes and began investigations of the Pleistocene vertebrate fauna. He also paid many visits to Paris to study in the Museum of Natural History, where Marcellin Boule was in charge of the paleontological collections. Shortly after publishing his important monograph "Les poissons fossiles d'Oran" (1927), Arambourg became a skillful specialist in fossil mammals.

Although he settled in Paris in 1930, having been appointed professor of geology at the Institut National Agronomique (he succeeded Lucien Cayeux, under whom he had studied), Arambourg continued his paleontological interest in Africa, undertaking a number of scientific expeditions. He organized an expedition to the Omo Valley in Ethiopia (October 1932–May 1933), where he collected Pleistocene vertebrates in the fossiliferous localities discovered thirty years earlier by Émile Brumpt north of Lake Rudolf. During the same expedition, excavations were made in the Miocene of Losodok (southwest of Turkana Lake, Kenya). Four and a half tons of rocks and fossils were collected and shipped to the Museum of Natural History in Paris.

At this time Arambourg studied the geology of the northern and western borders of Lake Rudolf. He also published a short paper on the volcanic formations of Turkana in 1934 and prepared a geological map (1:500,000) of this area in 1943. Three volumes of scientific reports on geology, paleontology, and anthropology were published under his direction by the Museum of Natural History (1935–1948). The second of these contains his detailed geological description of Turkana and of the lower valley of the Omo River (1943). Four color plates of landscapes and of human types encountered during the expedition were printed in this volume. These plates exhibit a series of fine autochromes made by Arambourg himself.

In November 1936 Arambourg succeeded Marcellin Boule as professor of paleontology at the Museum of Natural History. He soon organized a scientific expedition, lasting from 1938 to 1939, to collect fossil fishes in the Cretaceous localities of Lebanon and in the Oligocene deposits of Iran. In 1946 he returned to Lebanon. The Maghreb increasingly fascinated him, however. He undertook excavations in the Villafranchian of northern Africa, especially at Aïn Hanech, near Sétif, in northern Algeria, where he discovered spheroid artifacts (1947–1948), and at Ternifine, near Mascara, Algeria, where three mandibles and a parietal of pithecanthropus were exhumed (1954–1955).

Although he had to retire from his professorship at the Museum of Natural History in the fall of 1955, Arambourg retained his passionate interest in paleontology, as demonstrated by his third visit to the Cretaceous fish deposits of Lebanon in 1961 and by his desire to return to the Omo Valley of Ethiopia. He was head of the French team of the tripartite International Omo Research Expedition and participated in its first three field seasons (1967–1969), during which numerous australopithecine remains (including several mandibles, one maxillary, and one parietal) were collected. He had begun organizing the fourth field season when he died suddenly.

Arambourg was deeply convinced that paleontologists had much to learn from explorations in the field; he collected a great quantity of vertebrate fossils during his life and generously offered this collection to the Museum of Natural History in Paris. Among the fossils were more than 1,500 specimens of Upper Miocene fish from Oran and its vicinity. Moreover, for his monograph on the fossil

ARAMBOURG

vertebrates from the phosphate deposits of northern Africa, he gathered about 100,000 items with the assistance of field geologists interested in his study.

Arambourg's paleontological work covers mainly three specialties: paleoichthyology, paleomammalogy, and paleoanthropology. Occasionally he also developed original views on the evolution of living forms.

Paleoichthyology. In paleoichthyology Arambourg was interested primarily in the study of Cretaceous and Cenozoic fish, since he could, according to the actualistic principle, use the distribution of living fishes as a guide for the interpretation of the bathymetric and paleoclimatic significance of the fossil localities under study. He was able to establish that the fossil remains of Upper Miocene fish fauna from Licata, Sicily, were indicative of bathypelagic conditions (1925). Arambourg was also the first to recognize the occurrence of scales modified into light organs in fossil myctophids (1920). Moreover, he established that the Upper Miocene fish faunas from Licata and Oran had lived in subtropical conditions. Finally, he considered that the composition of the paleomediterranean fish fauna (*paleomediterranean* was coined by him to qualify that which lived in the Mediterranean during the Upper Miocene) is reminiscent of an older one that had a Tethysian extension. This interpretation was confirmed when Arambourg studied the Oligocene fish fauna of Iran (1939, 1967) and identified among them several species belonging to genera whose recent distributions are divided between the Atlantic and the Indo-Pacific oceans.

In his thorough study of the fossil vertebrates (mainly sharks) from the phosphate mines of Morocco, Arambourg was able to demonstrate that these economically important marine sediments had been deposited during three successive stratigraphic stages ranging from Upper Cretaceous to Middle Eocene. He also established that all were deposited in shallow waters under tropical-to-subtropical conditions.

Arambourg was especially interested in Cretaceous fish faunas and, more particularly, in those from the Lebanese deposits. Unfortunately, he never found time to describe this exceptionally rich and significant material. As a result, except for the Maastrichtian fauna from the phosphate deposits of northern Africa, his main studies on Cretaceous fossil fishes are a short memoir on the Lower Cretaceous of Gabon (1936) and a monograph on the Cenomanian of Jebel Tselfat, Morocco (1954), in which he relied on the composition of this fish fauna to establish its stratigraphic age.

Arambourg generally paid little attention to fossil reptiles, although he described some of their remains in his study of the vertebrates from the phosphate mines of northern Africa. Nevertheless, he identified a giant pterosaur, *Titanopteryx philadelphiae*, in the Maastrichtian phosphate deposits of Roseifa, Jordan (1954, 1959). Although the anatomical identification of the largest and most significant bone fragment remained somewhat problematical, he was able to ascertain, through methods commonly used in histology, that it had come from one of the largest pterosaurs.

Paleomammalogy. As a paleomammalogist Arambourg was mostly interested in the period from the Miocene to the Pleistocene. His excavations in the "Pontian" of the Vardar Valley, near Salonika, resulted in "Les vertébrés du Pontien de Salonique" (1929), written with Jean Piveteau. His interest in Miocene mammals is further exemplified in his "Mammifères miocènes du Turkana (Afrique-Orientale)" (1934) and "Vertébrés continentaux du Miocène Supérieur de l'Afrique du nord" (1959), in which he described mainly the material he had collected at Oued el Hammam, near Mascara. He also established that the first appearance of the Hipparion fauna clearly antedated the end of the Miocene and that the term "Pontian" is no longer useful as a stratigraphic stage defined by its mammalian remains. He emphasized the endemic character of the Upper Miocene mammals of northern Africa and considered that the life forms of Oued el Hammam included mainly genera whose recent species live under tropical conditions.

The study of the Plio-Pleistocene vertebrates of northern and eastern Africa was Arambourg's most constant preoccupation for forty years. In 1929 he published "Les mammifères quaternaires de l'Algérie" and in 1932 "Révision des ours fossiles de l'Afrique du nord." In 1938 he published his study "Mammifères fossiles du Maroc," in which he relied on the composition of the mammalian fauna from the Paleolithic of Morocco to interpret the climatic changes in northern Africa during Pleistocene times. In his masterly "Contribution à l'étude géologique et paléontologique du bassin du Lac Rodolphe et de la basse vallée de l'Omo" (1948), Arambourg described the vertebrates collected fifteen years earlier and determined that the Omo formation belongs to the Kagerian stage (Lower Pleistocene). In addition, he believed that Africa had been a center of mammalian evolution and dispersal.

Finally, after having collected for thirty years (1931–1961), Arambourg undertook a wide synthesis of the Villafranchian vertebrates of northern Africa.

Just before his death he completed a comprehensive monograph that was published posthumously in two parts with slightly different titles: "Les vertébrés pléistocènes de l'Afrique du nord" (1969–1970) and *Vertébrés villafranchiens d'Afrique du nord* (1979).

Paleoanthropology. Arambourg's major achievement was unquestionably his masterly contribution to the improvement of paleoanthropological knowledge. When he received the Gaudry Prize from the Société Géologique de France, he said that his interest in paleoanthropology began in secondary school as a result of the controversies generated by the discovery of the first remains of *Pithecanthropus erectus*—he was given the nickname "Fossil Man" by his schoolmates.

Arambourg's first discovery in the field of paleoanthropology was an Upper Paleolithic ossuary near Bejaïa, in northeastern Algeria, from which he exhumed seven skeletons and many skulls of Cro-Magnons in 1934. In 1949 he noted the occurrence of calcareous spheroid artifacts in the Lower Villafranchian of Aïn Hanech, near Sétif, Algeria. Although he was initially somewhat uncertain about their human origin, he compared them with the Kafuen industry (pebble culture) of Uganda, and also with the Soan of northwest India and the Patjitanian of Java. A still more important discovery was pithecanthropic remains (*Atlanthropus mauritanicus*) associated with primitive Acheulean bifaces and large Clactonian flakes in the Early Middle Pleistocene (Kamasian) of Ternifine. Arambourg stressed in his published studies the importance of this discovery, which enabled him to demonstrate that the Acheulean industry had been created by the pithecanthropi.

This interpretation was rapidly confirmed in 1955, when a pithecanthropic mandibular fragment of Rissian age was found near Casablanca in association with an evolved Acheulean industry. Shortly afterward Arambourg was concerned with the discovery of a Neanderthaloid skull associated with a Mousterian industry at Jebel Irhoud, Morocco. Except for the australopithecines who probably fashioned the spheroid artifacts of Aïn Hanech, the Maghreb had finally produced a complete human paleontological series associated with lithic industries arranged in a regular stratigraphic sequence.

Arambourg was more than eighty years old when he first discovered australopithecine remains, during the first field season of the International Omo Research Expedition. His death deprived him of the full benefit of the foresight that had led him to return to Ethiopia, where, he was convinced, he would uncover the remains of modern humanity's remote ancestors.

Evolutionary Theories. Arambourg's theories on the evolution of living forms are scattered throughout many publications. In 1935 he emphasized, in his "Contribution à l'étude des poissons du Lias supérieur," that the evolutionary process proceeded regularly—albeit in discrete or discontinuous manifestations—in a definite and clearly evident direction. He emphasized that fossil faunas remained almost constant during relatively long periods, while striking renewals might suddenly occur between two successive stratigraphic stages. In this respect the tempo of evolution appears to have been essentially variable, as shown by the fact that even in the more classical series of orthogenetic specializations, the transitional forms that are postulated to have connected the different stages through almost imperceptible gradations are generally lacking (*Notice . . . Arambourg*, 1936). Moreover, in his comments in "L'évolution des vertébrés" (1937), Arambourg demonstrated that the evolutionary discontinuities can be correlated with the major events of geological history and suggested that the structure and the morphology of living forms resulted from a balance between their "inner medium" and the physicochemical factors of their external environment. In this respect he believed himself to be a Lamarckian, although this interpretation is neo-Lamarckian.

Nevertheless, Arambourg distinguished two types of evolutionary processes: a general one that is tied directly to the increasing entropy of the universe and determines the global orthogenesis of the living world, and peculiar processes induced by the major geological events that have periodically modified the earth's surface. These peculiar processes would be responsible for the rapid emergence of new organic types. In this way he explained the extreme scarcity of transitional forms between the major groups.

In the first edition of *La genèse de l'humanité* (1943), Arambourg still emphasized the importance of the successive discontinuities that suddenly produced evolutionary novelties and gave rise to temporarily stable types. Later, in *Le gisement de Ternifine* (1963), he objected that the macro- and megaevolutionary discontinuities evidenced during the development of living forms seriously contradicted the Darwinian and neo-Darwinian conceptions of evolution. He argued that organic evolution proceeds from continual interactions between living beings and their physical environment, a view very similar to the cytochemical neo-Lamarckism of Paul Wintrebert.

Then, taking into account the evolutionary history

of humanity, Arambourg considered that it had proceeded by successive and progressive steps, strictly distributed through time, each corresponding to a temporarily stable human type characterized by its lithic industry. He also noted that no gradual transition can be recognized between the human fossil types or between the successive lithic industries. He had been even clearer in his "Considérations sur l'état actuel du problème des origines de l'homme" (1956), in which he asserted: "Australopithecines, Pithecanthropines, Neanderthalians, and Homo sapiens are much more than morphological stages, and . . . they really correspond to the successive steps of the series that ends with Homo sapiens" (p. 146).

Even though this point of view rapidly became obsolete, Arambourg's interpretation of the evolutionary process seems to have been somewhat prophetic. In fact, he believed that "macroevolution truly constitutes the elementary mechanism of transformism," so that in every systematic category, evolution has proceeded by successive steps, or quanta. In this respect his view of evolution, seen as "a continuous series of discontinuities" ("L'évolution transformiste des hominiens," 1965), anticipates the punctualist conception of life history, made up of geologically instantaneous events of morphological transformations.

His exceptionally thorough research brought Arambourg many honors. He was president of the Société Géologique de France (1950), the Société Préhistorique Française (1956), and the fourth Pan African Congress on Prehistory, held in Leopoldville, Belgian Congo (now Kinshasa, Zaïre), in 1959. The same year, he received the Gaudry Prize from the Société Géologique de France. In 1961 he was elected to the Académie des Sciences.

BIBLIOGRAPHY

I. Original Works. Among more than 230 publications, the most significant works of Arambourg are "Traces d'organes lumineux observés chez quelques Scopélides fossiles," in *Comptes rendus sommaires des séances de la Société Géologique de France* (1920), 167–168; "Révision des poissons fossiles de Licata," in *Annales de Paléontologie*, **14** (1925), 39–132; "Les poissons fossiles d'Oran," in *Matériaux pour la carte géologique de l'Algérie*, 1st ser., Paléontologie, no. 6 (1927); "Les mammifères quaternaires de l'Algérie," in *Bulletin de la Société d'Histoire Naturelle d'Afrique du Nord*, **20** (1929), 63–84; "Les vertébrés du Pontien de Salonique," in *Annales de Paléontologie*, **18** (1929), 59–138, written with Jean Piveteau; "Révision des ours fossiles de l'Afrique du Nord," in *Annales du Musée d'Histoire Naturelle de Marseille*, **25** (1932–1933), 247–301; "Mammifères miocènes du Turkana (Afrique-Orientale)," in *Annales de Paléontologie*, **22** (1934), 123–146; "Les grottes paléolithiques des Beni-Segoual (Algérie)," Archives de l'Institut de Paléontologie Humaine, memoir no. 13 (Paris, 1934), written with Marcellin Boule, Henri Vallois, and René Verneau; "Contribution à l'étude des poissons du lias supérieur," in *Annales de Paléontologie*, **24** (1935), 1–32; "Historique et itinéraire de la mission," in *Mission scientifique de l'Omo*, I, fasc. 1 (Paris, 1935), 1–8, written with P. A. Chappuis and R. Jeannel; "Esquisse géologique de la bordure occidentale du Lac Rodolphe," *ibid.*, 9–16; "Les poissons fossiles du bassin sédimentaire du Gabon," in *Annales de Paléontologie*, **24** (1936), 137–159, written with D. Schneegans.

Other works are *Notice sur les travaux scientifiques de M. Camille Arambourg* (Paris, 1936); "L'évolution des vertébrés," in *L'Encyclopédie Française*, V (Paris, 1937), chap. 2; "Mammifères fossiles du Maroc," in *Mémoires de la Société des Sciences Naturelles du Maroc*, **46** (1938), 1–72; "Sur des poissons fossiles de Perse," in *Comptes rendus hebdomadaires des séances de l'Académie des Sciences*, **209** (1939), 898–899; *La genèse de l'humanité* (Paris, 1943; 8th ed., 1969); "Contribution à l'étude géologique et paléontologique du bassin du Lac Rodolphe et de la basse vallée de l'Omo. Première partie: Géologie," in *Mission scientifique de l'Omo*, I, fasc. 2 (Paris, 1943), 157–230, and I, fasc. 3 (1948), 231–562; "Sur la présence, dans le Villafranchien d'Algérie, de vestiges éventuels d'industrie humaine," in *Comptes rendus hebdomadaires des séances de l'Académie des Sciences*, **229** (1949), 66–67; "Les vertébrés fossiles des gisements de phosphates (Maroc-Algérie-Tunisie)," in *Notes et mémoires du Service géologique du Maroc*, no. 92 (1952), written with J. Signeux; "Les poissons crétacés du Jebel Tselfat," *ibid.*, no. 118 (1954); "Sur la présence d'un Ptérosaurien gigantesque dans les phosphates de Jordanie," in *Comptes rendus hebdomadaires des séances de l'Académie des sciences*, **238** (1954), 133–134.

Later works include "Le gisement de Ternifine et l'*Atlanthropus*," in *Bulletin de la Société Préhistorique Française*, **52** (1955), 94–95; "Découverte de vestiges humains acheuléens dans la carrière de Sidi-Abd-er-Rahmann," in *Comptes rendus hebdomadaires des séances de l'Académie des Sciences*, **240** (1955), 1661–1663, written with Pierre Biberson; "Considérations sur l'état actuel du problème des origines de l'homme," in *Colloques internationaux du Centre National de la Recherche Scientifique*, **60** (1956), 135–147; "Vertébrés continentaux du Miocène Supérieur de l'Afrique du nord," in *Publications du Service de la Carte Géologique de l'Algérie*, n.s., Paléontologie, **4** (1959); "Gisements de phosphates maëstrichtiens de Roseifa (Jordanie). *Titanopteryx philadelphiae nov. gen., nov. sp.*, ptérosaurien géant," in *Notes et mémoires sur le Moyen-Orient*, **7** (1959), 229–234; *Le gisement de Ternifine*. I, Archives de l'Institut de Paléontologie Humaine, memoir no. 32 (Paris, 1963), written with R. Hoffstetter; "L'Évolution transformiste des hominiens," in *Revista da Faculdade de letras, Universidade de Lisboa*, 3rd ser., no. 9 (1965), 3–15; "Les poissons oligocènes de l'Iran," in *Notes et mémoires sur le Moyen-Orient*, **8** (1967), 1–247; "Sur la découverte, dans le Pléistocène Inférieur de la vallée de l'Omo (Éthiopie), d'une mandibule d'Australopithécien," in *Comptes rendus hebdomadaires des séances de l'Académie des Sciences*, **265D** (1967), 589–590, written with Yves Coppens; "Les vertébrés du Pléistocène de l'Afrique du Nord (première partie: proboscidiens et périssodactyles)," in *Archives du Muséum National d'Histoire Naturelle*, **7**, 7th ser., no. 10 (1969/1970), 1–126; and *Vertébrés villafranchiens d'Afrique du Nord (artiodactyles, carnivores, primates, reptiles, oiseaux)* (Paris, 1979).

II. SECONDARY LITERATURE. The most complete biography of Arambourg is Robert Courrier, *Notice sur la vie et les travaux de Camille Arambourg (1885–1969), membre de la section de minéralogie et géologie: Ses recherches sur la genèse de l'humanité* (Paris, 1974), with a bibliography and portrait. See also Yves Coppens, "Camille Arambourg and Louis Leakey ou 1/2 siècle de paléontologie africaine," in *Bulletin de la Société Préhistorique Française*, **76** (1979), 291–314, with a complete bibliography and portrait. Further information on Arambourg's life and work is in E. Ennouchi, "Camille Arambourg (1885–1969)," in *Bulletin de la Société des Sciences Naturelles et Physiques du Maroc*, **50** (1970), 1–7; Jean Gaudant, "Camille Arambourg (1885–1969), précurseur du ponctualisme," in *Revue d'histoire des sciences*, **39** (1986), 31–34; B. Gèze, "Camille Arambourg (1885–1969)," in *Comptes rendus des séances de l'Académie d'Agriculture de France*, **56** (1970), 101–104; and J.-P. Lehman, B. Gèze, L. Balout, and R. Heim, *Hommage à Camille Arambourg à l'occasion de son 80ᵉ anniversaire* (Paris, 1965).

JEAN GAUDANT

ARTHUS, NICOLAS MAURICE (*b.* Angers, France, 9 January 1862; *d.* Fribourg, Switzerland, 24 February 1945), *physiology*.

The son of Nicolas Arthus, a leather merchant, and of Marie Adélaïde Manuelle, Arthus prepared for the entrance examination for the École Polytechnique, then studied physiology, physics, and medicine at the Sorbonne. He became a pupil of the physiologist Albert Dastre and his technical assistant from 1887 to 1895. His first research papers (1889) concerned glycogenesis and were written with Dastre. In November 1890, he presented a doctoral dissertation in natural science on blood clotting, *Recherches sur la coagulation du sang*. Drawing a close analogy between blood and milk, Arthus discovered that calcium was a necessary component of milk clotting, a fact that had recently been shown for blood clotting. Both blood and milk clotting also require a ferment. Arthus worked on the chemistry of milk and blood proteins, and on the physiology of milk digestion and of milk and blood clotting. In the summer of 1892, he visited the physiological chemistry laboratory of Wilhelm Friedrich Kühne in Heidelberg.

In 1893 Arthus presented a doctoral dissertation in physics, *Recherches sur quelques substances albuminoïdes. La classe des caséines, la famille des fibrines*. He studied the comparative chemistry of casein and fibrin and showed that they had different coagulation properties. In 1894 he published a book on these topics, *Coagulation des liquides organiques*. He continued to work on milk clotting and discovered in 1903 that the secretion of the milk-clotting ferment is triggered by the presence of milk in the stomach. But he concentrated mainly on blood coagulation, a topic to which he made several valuable contributions. In 1901 and 1902 he studied the coagulation-preventing properties of sodium citrate, a subject that acquired much therapeutic importance. He also studied the time variations of coagulation speed and fibrin-ferment secretion. His blood-clotting studies, covering the period from 1890 to 1908, are of equal interest for physiology and for surgery.

Arthus also studied the activity of such proteolytic enzymes as trypsin. In 1896 he presented a medical dissertation, *Nature des enzymes*. Pointing out that the analytical chemistry of proteins was in a state of utmost confusion, Arthus advocated the view, already formulated by L. de Jager, that enzymes were not chemical compounds but "properties" or agents resembling physical processes (such as light or magnetism) that trigger chemical effects on their substrates—a concept that turned out to be misleading.

After serving as a lecturer in physiology at the Sorbonne (1890–1895), Arthus was appointed professor of physiology, physiological chemistry, and general microbiology at the University of Fribourg in 1895. In 1900 he was appointed laboratory director at the Pasteur Institute in Lille, which was headed by Albert Calmette. At that time he did mainly studies on blood and on toxic proteins. In 1903 he was given the position of lecturer at the medical school of the University of Marseilles and in 1907 was appointed professor of physiology and director of the Physiological Institute at the University of Lausanne, a position he held until his retirement in 1932.

In 1902 Charles Richet and Paul Portier reported to the Société de Biologie their discovery of anaphylaxis, the increased sensitivity of an organism to several contacts with poisonous substances, which were given in low, nontoxic doses. They coined the word *anaphylaxie*, meaning that the organism's defense was lowered in this state. The new phenomenon aroused much interest and some controversy. Independently, Arthus was working on an organism's reaction to proteins of other animal species. In order to recognize these heterologous proteins, he needed a precipitating serum, which he prepared by using horse serum repeatedly injected into rabbits, thus producing the precipitation reaction in the rabbit's serum. In spite of rigorous asepsis, he observed local edema and necrosis. In order to avoid these problems, he used intravenous injections in the same rabbits, which died after a few minutes. Intravenous injections of horse serum into unsensitized rabbits never created such problems. Arthus then compared his fortuitous discovery with Richet and Portier's recent report of anaphylaxis. The phenomenon of local anaphylaxis, or Arthus' phenomenon, was reported to the Société de Biologie in 1903. Arthus' use of Richet's terminology for a slightly different phenomenon led him to the concept of serum anaphylaxis.

In Sir Henry Dale's view, Arthus' studies, together with other studies on the guinea pig, contributed to a better interpretation of anaphylaxis, which does not consist of a decrease in the organism's defense but of an increased sensitivity. Arthus tried to answer the question of whether the tissue lesions resulted from a precipitate within the organism. In 1909 he described the anaphylactic properties of blood serum proteins and of other proteins, such as ovalbumin. He debated with Richet about the specificity of the anaphylactic reaction and noted that the local reaction is not specific in the rabbit, while in species that do not present any local reaction, such as the dog, the anaphylactic syndrome is specific. From the study of a rather peculiar case, the rabbit's anaphylaxis, Arthus came to the conclusion that anaphylaxis and immunity are different phenomena. In contrast, in 1910 Pierre Nolf maintained that anaphylaxis and immunity were expressions of the same state of the organism. In Arthus' view, the rabbit's local reaction did not result from a blood precipitate. In 1910 Arthus made use of more specific tools, the snake venoms, in order to better analyze the relationships among anaphylaxis, protein poisoning, and immunity. In his view, anaphylaxis was a phenomenon of poisoning by heterologous proteins rather than a sign of immunity.

Snake venoms, Arthus thought, were able to produce both anaphylactic and immune reactions. He tested all sorts of venoms, did thorough physiopathological studies, and quantitatively studied the neutralization of venoms by serums. He distinguished between several properties of the venoms and demonstrated that cobra venom acts like curare. He showed that other venoms have coagulating properties, or depressing properties on cardiac rhythm or blood pressure. But since Arthus was primarily interested in the distinction that he thought had to be established between anaphylaxis and immunity, he concentrated on the anaphylactic syndrome, or protein poisoning, and on immunity, two phenomena that may be produced by several injections of small doses of venoms. In the rabbit Arthus observed the increased sensitivity to protein poisoning that is characteristic of anaphylaxis and the absence of more specific symptoms like curare paralysis, which is a sign of immunity. Discussing Nolf's experiments, he reviewed the body of evidence in his 1921 book *De l'anaphylaxie à l'immunité*. Arthus, who strongly distrusted theories, was perhaps the victim of the peculiarities of the phenomenon he had discovered in the rabbit. Later, Arthus continued to work on venoms. He showed that attenuated venoms, which are no more toxic, can still produce anaphylactic reactions.

Along with his basic research in experimental physiology, Arthus was deeply committed to teaching, which in his view included the writing of textbooks. A tireless experimentalist and a philosopher of the experimental method, he was also a tireless writer. His textbook *Éléments de chimie physiologique* (subsequently *Précis de chimie physiologique*) had eleven editions between 1895 and 1932. It was translated into German and Spanish. His textbook *Éléments de physiologie* (later *Précis de physiologie*) had seven editions between 1902 and 1927. He also published *La physiologie* in 1920 and *Précis de physiologie microbienne* in 1921.

Arthus retired from his chair in Lausanne in 1932. Since working facilities were no longer available to him, he returned to Fribourg, where he was given the directorship of the Institute of Bacteriology and Hygiene, a position he held until 1942. He died in his eighty-fourth year.

BIBLIOGRAPHY

I. ORIGINAL WORKS. *Recherches sur la coagulation du sang* (Paris, 1890); *Recherches sur quelques substances albuminoïdes. La classe des caséines, la famille des fibrines* (Paris, 1893); *Coagulation des liquides organiques. Sang, lymphe, transsudats, lait* (Paris, 1894); *Éléments de chimie physiologique* (Paris, 1895), continued as *Précis de chimie physiologique* (11th ed., Paris, 1932); *Nature des enzymes* (Paris, 1896); *La coagulation du sang* (Paris, 1899); *Élé-*

ments de physiologie (Paris, 1902), continued as *Précis de physiologie* (7th ed., 1927); "Injections répétées de sérum de cheval chez le lapin," in *Comptes-rendus de la Société de biologie* **55** (1903), 817–820; "Lésions cutanées produites par les injections de sérum de cheval chez le lapin anaphylactisé par et pour ce sérum, en collaboration avec Maurice Breton," *ibid.*, **55** (1903), 1478–1480.

"Sur la séroanaphylaxie du lapin," *ibid.*, **60** (1906), 1143–1145. "La séroanaphylaxie du lapin," in *Archives internationales de physiologie*, **7** (1909), 471–526; "Sur la séro-anaphylaxie," in *Presse médicale*, **5** (1909), 305–306; "Le venin de cobra est un curare," in *Archives internationales de physiologie*, **10** (1910), 161–191; "De la spécificité des sérums antivenimeux," *ibid.*, **11** (1912), 265–338; "Études sur les venins de serpents," *ibid.*, **12** (1912), 162–177, 271–288, 369–394; "Anaphylaxie-immunité," in *Comptes-rendus de la Société de biologie*, **82** (1919), 1200–1202; *La physiologie* (Paris, 1920); *De l'anaphylaxie à l'immunité* (Paris, 1921); *Précis de physiologie microbienne* (Paris, 1921); and "Pour mieux connaître les anatoxines," in *Bruxelles médical* (1930), 1–13.

II. SECONDARY LITERATURE. Obituary by Léon Binet, in *Bulletin de l'Académie de médecine*, **129** (1945), 374–376, and by Henri Roger, in *Presse médicale*, **53** (1945), 261–262. See also Daniel Bovet, *Une chimie qui guérit. Histoire de la découverte des sulfamides* (Paris, 1988); Henry Dale, "Mécanisme de l'anaphylaxie," in *Presse médicale*, **60** (1952), 680–682; and Guy Saudan, "La physiologie à la haute école de Lausanne: Le premier demi-siècle (1881–1932)," in *Gesnerus*, **45** (1988), 263–270.

CLAUDE DEBRU

ASTAUROV, BORIS L'VOVICH (*b.* Kazan, Russia, 27 October 1904; *d.* Moscow, U.S.S.R., 21 June 1974), *zoology, cytogenetics.*

Astaurov was a leading Russian cytogeneticist. In the 1920's, as a member of Sergei Chetverikov's group at Kol'tsov's Institute of Experimental Biology in Moscow, he helped to originate *Drosophila* population genetics. In 1930 he switched to silkworm research, for which he won his candidate (1936) and doctoral (1939) degrees, and subsequently developed techniques for artificial parthenogenesis, intraspecific and interspecific androgenesis, and the production of the first artificially produced polyploid animal species. As a corresponding (1958) and later full (1966) member of the U.S.S.R. Academy of Sciences, he played a central role in reestablishing genetics as a legitimate science in the Soviet Union. Except for the period 1930–1935, Astaurov spent his entire career in Moscow at the Kol'tsov institute.

Astaurov was born into a prerevolutionary Russian medical family. His mother, née Ol'ga Andreevna Tikhenko, earned a doctorate in medicine from the University of Paris (Sorbonne); his father, Lev Mikhailovich Astaurov, studied medicine at the universities of Moscow and Kazan. Astaurov was born in Kazan (where his parents were working during a cholera epidemic) but was almost immediately taken to Moscow, where he lived for most of his life. Upon completing his secondary education in 1921, he entered Moscow University and studied in the biological division of the physicomathematical faculty, earning his diploma in zoology in 1927 for a study of the hereditary variation of balancers in *Drosophila melanogaster*.

Drosophila Work. At the university, Astaurov took courses with M. A. Menzbir and A. N. Severtsov, but his primary interest was experimental biology as taught by N. K. Kol'tsov and his colleagues M. M. Zavadovskii (dynamics of development), A. S. Serebrovskii (genetics), S. S. Chetverikov (genetics, entomology), and S. L. Frolova (cytology). Astaurov was an undergraduate at Moscow University (1921–1927), a graduate student at its Zoological Institute (1927–1930), and a staff member of the Moscow branch of the Genetics Division of KEPS (Commission on the Study of Natural Productive Forces of the U.S.S.R. Academy of Sciences; 1926–1930), participating in summer expeditions to Kazakhstan (1928) and Turkmenistan (1929). However, most of his research from 1924 through 1929 was conducted at Kol'tsov's Institute of Experimental Biology (IEB), under the auspices of the People's Commissariat of Public Health (Narkomzdrav), where he worked first as a laboratory technician and then as a scientific associate in its genetics section, headed by S. S. Chetverikov.

In 1925, together with N. K. Beliaev, E. I. Balkashina, and S. M. Gershenzon, Astaurov undertook the first genetic study of natural populations of various species of *Drosophila*. The results formed the basis of Chetverikov's famous papers of 1926 and 1927, and, together with the study conducted in Berlin by two other members of Chetverikov's laboratory, N. W. and E. A. Timofeeff-Ressovsky, initiated the long series of genetic investigations of natural *Drosophila* populations that formed the core of population genetics over the next four decades. His *Drosophila* work involved Astaurov in three problem areas which were to help shape his research career.

In the mid 1920's the Chetverikov group discovered that *Drosophila obscura* Fall. from the environs of Moscow was reproductively isolated from the flies brought to Russia from T. H. Morgan's laboratory which had been named *Drosophila obscura* Fall. by Sturtevant. Working with S. L. Frolova on an investigation of the two forms, which were morphologically almost indistinguishable, Astaurov established that they were distinct species and coined the name *pseudoobscura* for the American form, which was to become a mainstay of Dobzhansky's later investigations. This work led Astaurov to contemplate the cytogentic basis of speciation, a theme he developed in his later work.

Astaurov was the first to identify the *tetraptera* mutation in *Drosophila melanogaster*, which develops the fly's rudimentary halteres (balancers) into a second pair of wings. He established that the penetrance and expressivity of the mutation showed great variation depending on its genotypic milieu. However, even in inbred strains with a common genotypic milieu raised in a constant environment, the trait remained highly inconstant. In particular, there was no correlation between the trait's appearance on the left and on the right sides of flies. Astaurov concluded that some mutations have high degrees of developmental instability, but that they can be found only in artificial laboratory conditions and almost never in nature, since natural selection works toward a definite and precise norm of reaction and a narrow range of spontaneous variability.

Finally, in the 1925 survey of natural populations, Astaurov noticed several female *Drosophila phalerata* which produced only females in subsequent generations. These puzzling strains were given to A. E. Gaisinovich for analysis, and he published a cytogenetic explanation with which Astaurov was not wholly satisfied. The genetic, cytological, developmental, and environmental factors which govern sex determination would subsequently become a central motif of Astaurov's researches.

The period of the first Five-Year Plan involved ideological denunciations, arrests, institutional disruptions, and a new emphasis on the immediate utility of research. In 1929 Chetverikov was arrested and exiled, and his research group was disbanded and dispersed. As a result, in 1930 Astaurov abruptly dropped his work on *Drosophila* and began his lifelong study of the silkworm.

Silkworm Studies. In the late nineteenth and early twentieth centuries, Russian investigators had carried out important work on the biology of silkworms at the Caucasus Station (later the Tbilisi Institute) of Sericulture, the Central Sericulture Station in Moscow, and the Moscow Sericulture Society, headed by the renowned breeder A. A. Tikhomirov.

By the late 1920's work on silkworm breeding had intensified for both practical and theoretical reasons. Practically, the research base had been expanded by the creation of several new institutions, notably SANIISH (the Central Asian Scientific Research Institute of Sericulture) in Tashkent and ZAKNIISH (the Caucasian Scientific Research Institute for Sericulture and Silk Production) in Tbilisi, and in 1929 the Moscow section of KEPS—headed by Kol'tsov—was given responsibility for coordinating silkworm research. In addition to the obvious practical importance of the silkworm, it was also of theoretical interest because its eggs and larvae were especially useful in studying developmental mechanics, parthenogenesis, and sex determination. Since males produced considerably more silk than females, such work also promised to have important consequences for production. Beginning in 1928, the silkworm became an important object of research for several important Soviet biologists who had worked with *Drosophila*, notably Kol'tsov, Frolova, Efroimson, Beliaev, and (after 1937) Chetverikov.

In 1928 N. K. Beliaev became a consultant of SANIISH and worked there full-time beginning in 1929, transferring to ZAKNIISH in 1932 (where he worked until his arrest and execution in 1937). In 1930, on Kol'tsov's advice and Beliaev's urging, Astaurov moved to Tashkent to join SANIISH, and in 1932 began to study artificial parthenogenesis in the domesticated silkworm, *Bombyx mori* L. His initial studies drew upon the report of H. Sato (1931), who had managed to produce female adult parthenogenetic moths by treating unfertilized eggs of *Bombyx mori* with hydrochloric acid. Astaurov's analysis demonstrated that it was not the acid, but rather the high temperature used in the treatment, which produced parthenogenesis. He found that by soaking unfertilized eggs in water at 46°C. for eighteen minutes, up to 82 percent of the eggs would produce parthenogenetic adult moths. Astaurov conducted general studies on the effects of temperature and radiation on egg development.

Kol'tsov managed to have Astaurov reappointed to the IEB in late 1935 and he returned to Moscow, where he was awarded the degree of Candidate of Biological Sciences, without a dissertation, for his work. He did not rejoin the genetics section, however, which was then headed by N. P. Dubinin, but rather the institute's laboratory of developmental mechanics, directed by D. P. Filatov. Although the other investigators in the lab were studying development in amphibians and reptiles, Filatov warmly supported Astaurov's work on the development of silkworms. In this context, drawing on the work by Karpechenko on plant polyploidy, Astaurov developed techniques for producing triploid and tetraploid silkworms by mixing x-irradiation, temperature shock, and hybridization (1936–1955). In 1940 Astaurov suggested that animal speciation through polyploidy might occasionally occur in nature indirectly by parthenogenesis and hybridization; this view, which he developed in subsequent publications, remains controversial.

In 1937 Astaurov developed a technique of artificial androgenesis. He found that high temperature applied to newly fertilized eggs disturbs the female nuclear apparatus and makes it possible to produce parthenogenetic male moths in which the cytoplasm comes entirely from the female, whereas the nucleus results from the fusion of two haploid sperm pronuclei. In 1938 he defended his dissertation (published as a book in 1940) on artificial parthenogenesis in the silkworm, for which he was awarded the degree of Doctor of Biological Sciences in 1939. With V. P. Ostriakova-Varshaver, he later developed a technique (1956) of interspecific androgenesis by fertilizing irradiated eggs of *Bombyx mori* (L.) with the sperm of nonirradiated wild silkworms, *Bombyx mandarina* (Moore) and then applying heat shock (40°C., 120–135 minutes). Eventually he was able to produce a fertile tetraploid bisexual species, which he called *Bombyx allotetraploidus*, by parthenogenesis, selection, and hybridization.

Lysenkoism. Astaurov's research career coincided with the rise of T. D. Lysenko, the struggle between his so-called Michurinist biology and Soviet genetics, and the reestablishment of genetics as a legitimate scientific field after 1964. Astaurov played an important and unique role in these events.

In January 1939 Astaurov's teacher and patron, Kol'tsov, was vilified in the press by Lysenko's supporters, and a special meeting at the institute was held in which Dubinin and others condemned Kol'tsov for his earlier support of eugenics. In April, Kol'tsov was removed as director of the IEB, which was transferred into the U.S.S.R. Academy of Sciences and renamed the Institute of Cytology, Histology, and Embryology. Kol'tsov died of a heart attack in December 1940. Although Astaurov refused to denounce his teacher and wrote a laudatory obituary, he remained at his post in Filatov's laboratory, and several years after the latter's death in 1943 was appointed head of the laboratory of developmental mechanics (1947).

Following the August 1948 session of the Lenin All-Union Academy of Agricultural Sciences at which Lysenko's biology was endorsed by the Communist Party, geneticists were fired from the Kol'tsov institute and from the Severtsov Institute of Evolutionary Morphology, and the two institutes were united into the new Severtsov Institute of Animal Morphology. In late 1948 Astaurov was replaced as laboratory director and briefly gave up his work on the silkworm, but remained a senior member of its staff and was one of the few Soviet geneticists to maintain his scientific research post—this despite the fact that, unlike a number of his colleagues, he never supported Michurinist biology or renounced genetics.

Astaurov's staying power was probably related to several factors. He was apolitical and had maintained a low profile in the controversies over genetics. Officially he worked in the institute's embryology laboratory rather than in its genetics division. Perhaps most important, his work involved the application of heat and soaking at early developmental stages in an agriculturally important organism, in order to deliberately alter its biological and hereditary character in desirable and potentially useful ways. This approach fit the contemporary ideological emphasis on the "transformation of nature" and the control of living processes to improve agricultural production. In 1951 Astaurov was made a member of the scientific council of his institute. In 1953 he was awarded the Order of the Red Banner of Labor for his researches on the silkworm. In 1953–1954 he attended the Moscow Party Committee's night school in Marxism-Leninism and earned a degree, which may have conferred a certain legitimacy.

During the period 1950–1974 Astaurov devoted considerable effort to combating Lysenkoism and reestablishing Soviet genetics. With the advent of de-Stalinization in 1955, he resumed his former post as head of the Filatov laboratory of developmental mechanics. In 1956 he was elected a member of the Moscow Society of Naturalists, subsequently becoming the head of its genetics section (1960–1966) and a member of its editorial board (1961–1969). In 1958 Astaurov was awarded a Soviet patent for his techniques of controlling the sex of silkworms. That same year he was elected corresponding member of the U.S.S.R. Academy of Sciences in cytology, and was subsequently appointed to various editorial boards, national scientific commissions, and the Soviet certification board for scientific degrees, positions where he worked on behalf of genetics.

In 1958 Astaurov was scheduled to be the only member of the Soviet delegation to the Tenth International Genetics Congress in Montreal who was not a Lysenko supporter. In a letter to the Communist Party Central Committee, Astaurov resigned from the delegation. After citing health and family reasons, he added that it would hurt his international scientific reputation to be in such company, and stated his belief that "the views of all the other members of the delegation verge on absurdity" (Berg, 1979). At a time when international travel was a coveted privilege, Astaurov's resignation contributed to his moral authority.

Astaurov's most telling arguments against Lysenko during the period came from his own silkworm research. Although Anglo-American genetics had long held that heredity is governed by the cell nucleus, in France and elsewhere on the Continent, and among Lysenkoists, a strong belief in cytoplasmic inheritance persisted. Astaurov's technique of artificial intraspecific and interspecific androgenesis provided a way of demonstrating the differential effects of nucleus and cytoplasm on development. Using his techniques, he was able to generate hybrids of *Bombyx mori* and *Bombyx mandarina* in which the cytoplasm came entirely from one type and the nucleus entirely from the other. In each case, all the hybrid's characteristics, and those of its subsequent parthenogenetic generations, were identical to those of the type providing the nucleus. In another experiment, he demonstrated that an unirradiated enucleated cell dies if combined with a lightly irradiated nucleus; but even a heavily irradiated enucleated cell was viable when combined with an unirradiated nucleus. Astaurov has indicated that although he realized such experiments were of little interest in the West, in "vast regions of the world" where the Lamarckian doctrine "strongly prevailed," "such experiments retained their value and played rather a great role in the ultimate triumph of contemporary evolutionary genetics" (letter to Ernst Mayr, 4 May 1974).

Rebirth of Soviet Genetics. Following Khrushchev's ouster in October 1964, Astaurov played an active role in the vindication of Soviet genetics, publishing articles popularizing its achievements and its history. In 1966 he was elected full member of the U.S.S.R. Academy of Sciences and president of a newly created professional organization, the N. I. Vavilov All-Union Society of Geneticists and Selectionists (1966–1974). On 15 June 1967 the former Kol'tsov and Severtsov institutes (amalgamated since 1948) again separated, and Astaurov became director of the newly established Institute of Developmental Biology, which largely through his efforts was named after Kol'tsov in 1976. Also in 1967 Astaurov was awarded his second Order of the Red Banner of Labor for introducing scientific discoveries in agriculture.

Astaurov played a key role in creating a new historiography for Soviet genetics, editing volumes of early classics, some never before published, and writing biographies of his teachers Chetverikov, Filatov, and especially Kol'tsov. When the biochemist and gerontologist Zhores A. Medvedev was confined to a mental hospital on 29 May 1970 for his writings critical of Lysenko and the Soviet government, Astaurov played a central role in orchestrating the successful efforts by leading members of the U.S.S.R. Academy of Sciences to obtain his release. Astaurov also became a strong advocate of the development of research in human genetics, a subject which had been suppressed since 1937. In various popular and philosophical articles, he raised sociobiological issues and explored the hereditary component of human moral and altruistic behavior.

Astaurov's defense of Soviet geneticists and the Kol'tsov legacy involved him in a controversy that dominated the final years of his career. Astaurov and Dubinin had parted company in the late 1930's over the latter's condemnation of Kol'tsov's views on human heredity. Dubinin, who was also elected a full member of the U.S.S.R. Academy of Sciences in 1966, assumed control of the Institute of Genetics from Lysenko in 1965, and it was transformed into the Institute of General Genetics; but Dubinin insisted on maintaining the Lysenkoists at the institute, and established an ideological and administrative tone some of his fellow geneticists found unpleasant. As a result, in 1967 a number of outstanding geneticists of the older generation who had recently joined the institute's staff—notably V. V. Sakharov, B. N. Sidorov, and N. N. Sokolov—transferred their laboratories to Astaurov's institute.

Subsequently Dubinin joined the party, attacked some trends in human genetics as racist, and published an autobiography which criticized Kol'tsov for being a counterrevolutionary and a eugenicist. Astaurov defended human genetics work, much of which was being pursued by former students of Kol'tsov, and worked to establish the legacy of his teachers. Dubinin helped launch an ideological attack on Astaurov's institute when one of its members failed to return home from an international congress in Italy. Astaurov left the hospital to defend his institute, but died almost immediately thereafter of a heart attack on 21 June 1974. He was buried in the most prestigious cemetery in the Soviet Union, located on the grounds of the former Novodevichii monastery in Moscow.

For his work on experimental genetics and developmental biology, Astaurov won two Orders of the Red Banner of Labor (1953, 1967), a Soviet patent (1958), the Lomonosov Prize (1968), the Mechnikov Gold Medal (1970), and the Mendel Memorial Medal of the Czech Academy of Sciences

(1965). He was also elected to various foreign and international bodies, including the International Institute of Embryology (Utrecht, 1956), the International Society of Cell Biology (Liège, 1957), the American Zoological Society (1960), the International Entomological Society (1966), and the Finnish Academy of Sciences (1974).

BIBLIOGRAPHY

I. ORIGINAL WORKS. Astaurov is the author of more than 200 works, including 5 books and approximately 130 scientific articles (20 in English), 50 popular articles and interviews, 20 book reviews, and 20 historical works. His first publication was "Issledovanie nasledstvennogo izmeneniia galterov u *Drosophila melanogaster* Schin." (A study of the hereditary variation of halteres in *Drosophila melanogaster* Schin.), in *Zhurnal eksperimental'noi biologii*, ser. A, **3**, no. 1–2 (1927), 1–61.

For his work on silkworms, see "Artificial Mutations in the Silkworm (*Bombyx mori* L.)," in *Genetica*, **17**, no. 5–6 (1935), 409–460; "Iskusstvennyi partenogenez i androgenez u shelkovichnogo chervia" (Artificial parthenogenesis and androgenesis in the silkworm), in *Biulleten' VASKhNIL*, 1936, no. 12, 47–52; *Iskusstvennyi partenogenez u tutovogo shelkopriada: Eksperimental'noe issledovanie* (Artificial parthenogenesis in the silkworm: An experimental study; Moscow and Leningrad, 1940); *Tsitogenetika razvitiia tutovogo shelkopriada i ee eksperimental'nyi kontrol'* (Cytogenetics of the development of the silkworm and its experimental control; Moscow, 1968); *Nasledstvennost' i razvitie: Izbrannye trudy* (Heredity and development: Selected works; Moscow, 1974); and *Partenogenez, androgenez, poliploidiia* (Parthenogenesis, androgenesis, and polyploidy; Moscow, 1977).

Astaurov also wrote important works in the history of Soviet biology, notably "Dve vekhi v razvitii geneticheskikh predstavlenii" (Two landmarks in the development of genetic concepts), in *Biulleten' Moskovskogo obshchestva ispytatelei prirody*, biol. sec., **70**, no. 4 (1965), 25–32; *Nikolai Konstantinovich Kol'tsov* (Moscow, 1975), written with P. F. Rokitskii; and "Nauchnaia deiatel'nost' N. K. Beliaeva: K istorii sovetskikh geneticheskikh issledovanii na shelkovichnom cherve" (The scientific activity of N. K. Beliaev: On the history of Soviet genetic studies of the silkworm), in *Iz istorii biologii* (On the history of biology), V (Moscow, 1975), 103–136, written with Z. S. Nikoro, V. A. Strunnikov, and V. P. Efroimson. Astaurov's incomplete autobiographical sketch was published posthumously as "K itogam moei nauchnoi deiatel'nosti v oblasti genetiki" (Summing up my scientific activity in the field of genetics), in *Istoriko-biologicheskie issledovaniia* (Historical biological studies), VI (Moscow, 1978), 116–160.

II. SECONDARY LITERATURE. Bibliographies of Astaurov's publications, together with biographical articles by P. F. Rokitskii, are available in *Boris L'vovich Astaurov*, Materialy k biobibliografii uchenykh SSSR, ser. biologicheskikh nauk, Genetika, no. 2 (Moscow, 1972), and *Nasledstvennost' i razvitie* (Moscow, 1974). See also D. K. Beliaev's introduction to the posthumous Astaurov festschrift, *Problemy eksperimental'noi biologii* (Problems of experimental biology; Moscow, 1977). Rokitskii published a brief biography in *Vydaiushchiesia sovetskie genetiki* (Leading Soviet geneticists; Moscow, 1980), 77–87. In English, see the obituary by Gaisinovitch in *Folia Mendeliana*, **10** (1975), 247–252; and the informative survey by Raissa L. Berg, "The Life and Research of Boris L. Astaurov," in *Quarterly Review of Biology*, **54**, (1979), 397–416.

MARK B. ADAMS

ATWOOD, WALLACE WALTER (*b*. Chicago, Illinois, 1 October 1872; *d*. Annisquam, Massachusetts, 24 July 1949), *geography, geomorphology, geology*.

Atwood was the son of Adelaide Adelia Richards Atwood and Thomas Green Atwood, the owner of a planing mill and a descendant of an old Massachusetts family. He entered the University of Chicago in December 1892 and received the bachelor's degree in 1897. At Chicago he came under the strong influence of Rollin D. Salisbury and T. C. Chamberlin. His academic work under these scholars led to his distinguished and varied career as a Rocky Mountain geologist for the U.S. Geological Survey, an exceptionally talented teacher, president of Clark University in Worcester, Massachusetts, and founder and director of that institution's Graduate School of Geography. Atwood also was author or coauthor of numerous geography textbooks for students ranging in age from grammar school through college. On 22 September 1900 he married Harriet Towle Bradley; they had two daughters and two sons. The latter, Rollin and Wallace, Jr., became a glaciologist and a geologist, respectively.

A member of numerous professional societies, Atwood was a fellow of the Geological Society of America, the American Academy of Arts and Sciences, and the American Antiquarian Society. He also was president of the National Parks Association, the National Council of Geography Teachers, and the Pan-American Institute of Geography and History.

Atwood was a true conservationist and environmentalist. His consistent philosophy regarding the significance of geography is clearly articulated in his presidential address to the Association of American Geographers in 1935. He felt that the task of the geographer was to sensitize schoolchildren, teachers, college and university students, government officials, and businessmen to the importance of geographical conditions, and to explain the economic and social setting of human activity. He stated, "Nature has determined through the variety in soils, in landscapes, in climate, and in peoples, the interdependence of one part of the earth upon another and of one people upon the activities of another." He stressed the importance of geographical understanding for "the establishment of goodwill [*sic*] among the peoples of the earth" and thus as part of a peaceful solution to world problems.

Atwood also was a strong advocate of the protection and wise use of natural resources, and he developed a deep commitment to the National Park System from its earliest days. Throughout his career he demonstrated the vital role that science must have in establishing public policy. He taught that man does not conquer nature, but "He may discover the laws of nature and accomplish better and better adjustments to the natural conditions of this earth." This theme was lost sight of for nearly three decades, until the strong environmental movement caused a reawakening in the middle 1960's.

Atwood's strong belief in the need for field observation is clearly evident in his classic geologic study with Kirtley F. Mather on the San Juan Mountains (1932). In this work he describes his own extensive fieldwork and explains the mountains and scenery in terms of the sequence of events in the physiographic history of the region that produced the particular environment. Indeed, he had the opportunity to do extensive fieldwork throughout his life. Even before he received his doctorate from the University of Chicago in 1903, Atwood had worked with the New Jersey Geological Survey, the Wisconsin State Geological Survey, and the Illinois State Geological Survey, and had begun his long association with the U.S. Geological Survey (1901–1946). Atwood began his academic career in 1903 as instructor in physiography and general geology. He had risen to the rank of associate professor by 1913, when he succeeded William Morris Davis as professor of physiography at Harvard. While at Harvard, Atwood continued training students in physiography and began the preparation of his geography textbooks. He remained at Harvard until he was appointed president of Clark University in 1920, with the primary responsibility of developing the Graduate School of Geography and becoming its director. Under his directorship, that school became a world center of excellence at which many leaders in the field obtained their graduate training. In 1946, after being named president emeritus of Clark University, he continued his interest in world travel and geologic studies of the Rocky Mountains. He died of cancer at his summer home and is buried at Cambridge, Massachusetts.

BIBLIOGRAPHY

I. ORIGINAL WORKS. A bibliography of Atwood's works, prepared by George B. Cressey, is in *Annals of the Association of American Geographers*, **39**, no. 4 (1949), 296–306. Some of his principal works are "Physical Geography of the Evanston-Waukegan Region," in *Illinois State Geological Survey Bulletin*, **7** (1908), written with James W. Goldthwait; "The Interpretation of Topographic Maps," *U.S. Geological Survey Professional Paper* no. 60 (1908), written with Rollin D. Salisbury; "Glaciation of the Uinta and Wasatch Mountains," *ibid.*, no. 61 (1909); "Mineral Resources of Southwestern Alaska," in *U.S. Geological Survey Bulletin* no. 379 (1909), 108–152; "Geology and Mineral Resources of Parts of the Alaska Peninsula," *U.S. Geological Survey Bulletin* no. 467 (1911); *New Geography: Book II* (Boston, 1920); *Home Life in Far-away Lands* (Boston, 1928), written with Helen G. Thomas, who was coauthor of a number of other textbooks.

Other works include "Physiography and Quaternary Geology of the San Juan Mountains, Colorado," *U.S. Geological Survey Professional Paper* no. 166 (1932), written with Kirtley F. Mather; *The Physiographic Provinces of North America* (Boston, 1940); *The Rocky Mountains* (New York, 1945); and *Our Economic World* (Boston, 1948), written with Ruth E. Pitt.

II. SECONDARY LITERATURE. There is no critical biographical study of Atwood, but additional information about his life is in Kirtley F. Mather, "Memorial to Wallace Walter Atwood," in *Proceedings of the Geological Society of America* (1949), 106–112. A later, somewhat anecdotal biography is William A. Koelsch, "Atwood, Wallace Walter," in *Dictionary of American Biography*, Supp. 4 (1974), 31–33. Another biography by Koelsch, with a good bibliography and an outline of Atwood's career, is "Wallace Walter Atwood, 1872–1949," in *Geographers Biobibliographical Studies*, **3** (1979), 13–18.

WILLIAM BACK

AYRTON, HERTHA (*b.* Portsea, England, 28 April 1854; *d.* New Cottage, North Lancing, Sussex, England, 26 August 1923), *engineering*.

Ayrton, born Phoebe Sarah Marks, was the third of six children of Levi and Alice Theresa (Moss) Marks. Her father, a Polish-born clockmaker and jeweler, died in 1861, leaving the family in debt. Sarah received her early education through the beneficence of an aunt. While she was in her teens she adopted the name Hertha, after the Teutonic goddess eulogized by Swinburne in a popular poem. From the age of sixteen she considered herself an agnostic, but she always remained proud of her Jewish heritage.

Hertha Marks first supported herself by tutoring and embroidery work, sending much of her earnings to her impoverished family. Her dream of a university education was made financially possible largely through the efforts of Barbara Leigh-Smith Bodichon, one of the founders of Girton College, Cambridge. Hertha entered Girton in 1876 after having passed the Cambridge University Examination for Women in 1874 with honors in English and mathematics. She was coached by Richard T. Glazebrook and completed the Cambridge Tripos in 1881. At that time women could not receive a Cambridge degree.

Marks's invention and patenting of a line divider in 1884 led her to consider a scientific career. She began studies that year at Finsbury Technical College, London, under the professor of physics and noted electrical engineer William Edward Ayrton. They were married on 6 May 1885.

After the birth of their daughter Barbara, Ayrton began experiments with the electric arc, at first assisting her husband but soon taking over the researche completely.

In 1893 the direct-current arc was widely used for lighting, and was therefore of significant commercial and industrial interest. Arc lamps were plagued with problems. They hissed, sputtered, hummed, and rotated, producing unsteady illumination in a changing array of colors. Their heat melted most materials, a challenge for those who wished to devise suitable insulators. Since the arc electrodes were consumed during operation, the arc length continually changed, which required adjustments to be made in the circuit. It could be difficult to maintain the arc light, particularly with long arcs or low currents. The functional dependence of potential on time and position within the arc and the way arc resistance varied were all unknown. Scientists had debated the merits of various types of carbon electrodes but had reached no consensus to guide the manufacturers.

Ayrton investigated the relations in the direct-current arc between power supplied, potential across the arc, current, and arc length. Her results, which she showed agreed with data of other observers, were that (1) power used in the arc is a linear function of the current when arc length is held constant; (2) power is a linear function of arc length when current is held constant; (3) potential and current are inversely proportional when arc length is held constant. These results (for solid carbons and silent arcs) can be expressed by the equation $V = a + bl + (c + dl)/A$, where V is the potential difference between the electrodes, l is the arc length, A is the current, and a, b, c, and d are constants that depend on the electrodes.

Ayrton next showed that the potential required to send a given current through a fixed arc length depends principally upon the nature of the surface of the depression (crater) that forms on the tip of the positive carbon (or is performed during manufacture). By casting the arc image onto a screen, she was able to describe and explain both the arc's appearance and the changes that occurred in the carbons during operation.

Ayrton's tour de force was her analysis of the hissing arc, the instability of which presented a baffling engineering problem. She found that this undesirable condition resulted from oxidation of the positive carbon. (In the stable arc only vaporization of the carbon took place.) The proximate cause of hissing was the positive crater's spread from the tip of the carbon to its sides. The resulting fissure allowed air to rush into the arc, causing the light to rotate and producing a noise similar to that created by wind blowing through a door frame. Obviously the then-common practice of manufacturing carbons with grooves along their sides led to the very condition the engineer wished to avoid. Moreover, electrodes with flat ends would be able to withstand higher currents than the usual tapered carbons before they developed grooves.

This paper established Ayrton's reputation, and in 1899 the Institution of Electrical Engineers permitted her to read the paper herself (which was unusual for a woman) and awarded her £10 for it. In May 1899 she was elected the first woman member.

In 1902 Ayrton published *The Electric Arc*, a comprehensive study based mainly on her published papers and including a useful historical survey. In the book she also showed that the common assumption by engineers of a large "back E.M.F." or a negative resistance in the arc was not necessary, and that short arcs were more efficient than long arcs.

Ayrton patented a number of improvements in searchlight carbons that she developed for the British Admiralty. She also designed improved cinema projectors.

During 1901, while she cared for her ailing husband at the shore, Ayrton analyzed sand ripple patterns formed on the beach by the sea. She showed that a succession of water vortices originated from each ripple in turn, thus creating the patterns. These studies found a practical application in the Ayrton fan, a hand-operated device she designed during

World War I to clear poisonous gases from the trenches by means of air vortices. A variety of hindrances prevented the fan from being widely used, a failure that pained her deeply.

In 1906 the Royal Society awarded Ayrton the Hughes Medal for her experimental investigations of the electric arc, and also for her work on sand ripples. In spite of this and other honors (she apparently was the first woman to read her own paper to the Royal Society, in 1904), the society declined to elect her a fellow, deciding that, as a married woman, she was not qualified for election.

Ayrton and Marie Curie met in 1903 and were friends until Ayrton's death. During 1912 Ayrton provided a refuge for Curie and her daughters, enabling the famous physicist to recuperate anonymously from stress and illness.

For much of her life Ayrton was plagued by ill health. She nevertheless was active in charitable causes and especially in the suffrage movement. She left the considerable sum of £8,160 to the Institution of Electrical Engineers (*The Electrician*, **91** [1923], 469), the organization that had welcomed her without prejudice and helped launch her career.

BIBLIOGRAPHY

I. Original Works. Ayrton's most important work is *The Electric Arc* (London, 1902), which contains references to most of her earlier papers on the arc. Other significant papers are "The Uses of a Line-Divider," in *Philosophical Magazine*, **19** (1885), 280–285 (by Sarah Marks); "The Mechanism of the Electric Arc," in *Philosophical Transactions of the Royal Society of London*, **A 199** (1902), 299–336; "On the Non-periodic or Residual Motion of Water Moving in Stationary Waves," in *Proceedings of the Royal Society of London*, **A80** (1908), 252–260; "The Origin and Growth of Ripple-Marks," *ibid.*, **84** (1910), 285–310, (read 1904); "Local Differences of Pressure near an Obstacle in Oscillating Water," *ibid.*, **91** (1915), 405–510; and "On a New Method of Driving off Poisonous Gases," *ibid.*, **96** (1919–1920), 249–256.

II. Secondary Literature. A biography is Evelyn Sharp, *Hertha Ayrton, 1854–1923* (London, 1926). See also the unsigned "Mrs. Ayrton," in *The Electrician*, **91** (1923), 211, 227; Henry Armstrong, obituary in *Nature*, **112** (1923), 800–801, with a response by T. Mather, *ibid.*, 939; the article on Ayrton in Marilyn Bailey Ogilvie, *Women in Science* (Cambridge, Mass., 1986); and A. P. Trotter, "Mrs. Ayrton's work on the Electric Arc," in *Nature*, **113** (1924), 48–49.

Marjorie Malley

BABCOCK, ERNEST BROWN (*b.* Edgerton, Wisconsin, 10 July 1877; *d.* Berkeley, California, 8 December 1954), *botany, genetics*.

The son of Emilius Welcome Babcock and Mary Eliza Brown Babcock, Ernest spent his childhood and early youth in Wisconsin; after a year (1895–1896) at Lawrence College in Appleton, Wisconsin, he moved to southern California. There he attended the State Normal School, Los Angeles (the parent institution of the University of California, Los Angeles), from which he graduated in 1898. After three years Babcock entered the University of California, Berkeley, from which he received a B.S. degree in 1906 and an M.S. in 1911. He accepted an appointment to the Berkeley faculty in 1907 and remained there until his retirement in 1947. Babcock married Georgia Bowen, a childhood friend, on 24 June 1908; they had no children.

As a young man Babcock was greatly impressed with the contributions that the exact study of inheritance, following Mendelian principles, could make both to the more efficient improvement of cultivated plants and to a deeper understanding of the processes of organismic evolution.

In 1913 Thomas F. Hunt, dean of the College of Agriculture, founded Berkeley's department of genetics and appointed the thirty-six-year-old Babcock as its first chairman, with the rank of full professor. In 1916 Roy E. Clausen was appointed assistant professor. He and Babcock began the collaboration that led to the publication in 1918 of the textbook *Genetics in Relation to Agriculture*. The first work to unite these two fields, *Genetics* became widely popular throughout the United States and appeared in a second edition, revised and enlarged, in 1927. Along with Babcock's twenty-five years as teacher of the introductory course on principles of plant and animal breeding at Berkeley, it firmly established his preeminence as a teacher in his chosen field.

Babcock's research proved to be the first successful attempt to unite the disciplines of conventional systematics, plant geography, chromosome cytology, and the genetics of interspecific hybrids into an integrated analysis of species relationships and the origin of species. He selected the large genus *Crepis*, related to the dandelion in the aster family, because of the low chromosome number of one of its species, *C. capillaris*, on which Russian investigators, particularly Michael S. Navashin, had already conducted cytogenetic investigations. During the early stages of the project, Navashin spent almost two years at Berkeley to collaborate with Babcock.

Although Babcock originally conceived of cytogenetic research on *C. capillaris* that would be a plant counterpart of the outstanding investigations of Thomas Hunt Morgan and his associates on *Drosophila*, he quickly ran into technical difficulties that revealed the impractical nature of this approach. At this point his close friend and scientific collaborator, Harvey M. Hall, honorary curator of the herbarium and a leading member of Berkeley's botany department, suggested that a project designed to analyze cytogenetically the relationships and evolution of the numerous species of *Crepis* would prove to be exceptionally valuable. Babcock accepted Hall's advice and revised his plans. With the aid of collaborators and students, he achieved notable success. The joint review of cytogenetics and evolution in *Crepis*, published in 1930 by Babcock and Navashin, was the first of its kind.

Babcock realized the importance of delimiting the genus *Crepis* by investigating related genera. To develop this part of the project, he added J. A. Jenkins and G. Ledyard Stebbins, Jr., to the group supported by his research grant. Their collaboration was close for six years. It resulted in two monographs, one on the Asiatic genus *Youngia* and the other on the endemic American species of *Crepis*. The latter set forth the concept of the agamic (asexual) complex, which placed in a rational evolution-based system the relationships between a group of distinct sexual diploid species and the vigorous hybrids formed between them, hybrids that would have been sterile were it not for the fact that they had acquired the capacity to produce seeds by asexual means, parthenogenesis or apomixis.

During the early 1940's Babcock gathered together all of the knowledge acquired during the previous twenty-two years of investigations. In 1947, his seventieth year, he published his two-volume monograph *The Genus* Crepis. This monograph still stands as a pioneer analysis. In it a number of varied and specific topics, such as conventional morphological differences, floral anatomy, and chromosomal differences, as well as geographical distribution and its relation to geological changes and plant migrations in general, are all brought to bear on two problems: the phylogenetic relationships of species and species groups, and the cytogenetic processes that underlie the origin of species.

Most of the general principles that emerged from Babcock's research had already been recognized. Two of them, however, were new: (1) the concept of the agamic complex, already mentioned, and (2) the concept of phylogenetic reduction in chromosome number and size, worked out in collaboration with Navashin. The latter concept has gained renewed interest as DNA has become recognized as the major component of chromosomes. When noncoding or "spacer" DNA was identified as constituting the major portion of this substance in the nuclei of many plant species, an analysis of *Crepis* species with these facts in mind showed that differences between them involved not active genes but noncoding DNA. In appendix 1 of part 1 of his monograph he set forth his ideas about future research on the genus.

Babcock is remembered for his pioneering achievements in teaching genetics in relation to agriculture and in the integration of facts from numerous disciplines into a concerted effort to reveal trends and processes of plant evolution. He belonged to the Genetics Society of America, the American Genetic Association, and the Botanical Society of America. He was elected to the National Academy of Sciences (1946), became vice president of the American Society of Naturalists in 1934, president of the California Botanical Society in 1940, and president of the Society for the Study of Evolution in 1952. The University of California awarded him its highest honor, faculty research lecturer, in 1944.

During his final years (1953–1954) Babcock suffered from cancer; he died of it in December 1954.

BIBLIOGRAPHY

I. ORIGINAL WORKS. Published works by Babcock include *Genetics in Relation to Agriculture* (New York, 1918; 2nd ed., rev. and enl., 1927), with Roy E. Clausen; "The Genus *Crepis*," in *Bibliographia genetica*, **6** (1930), 1–90, with Michael S. Navashin; *The Genus* Youngia, Carnegie Institution of Washington Publication no. 484 (1937), with G. Ledyard Stebbins, Jr.; *The American Species of* Crepis*: Their Interrelationships and Distribution as Affected by Polyploidy and Apomixis*, Carnegie Institution of Washington Publication no. 504 (1938), with G. Ledyard Stebbins, Jr.; *The Genus* Crepis, 2 vols. (Los Angeles, 1947).

II. SECONDARY LITERATURE. G. Ledyard Stebbins, Jr., "Ernest Brown Babcock," in *Biographical Memoirs. National Academy of Sciences*, **32** (1958), 50–66.

G. LEDYARD STEBBINS, JR.

BADGER, RICHARD MCLEAN (*b*. Elgin, Illinois, 4 May 1896; *d*. Pasadena, California, 26 November 1974), *physical chemistry, spectroscopy*.

The son of Joseph Stillman Badger and Carrie Mabel Hewitt, Badger spent part of his youth in Brisbane, Australia, returning to Elgin to complete his secondary education. He attended Northwestern University (1916–1917), but his studies were interrupted by World War I, during which he served in France in the 311th Field Signal Battalion. Demobilized in 1919, Badger resumed his college career at the California Institute of Technology. In 1921 he was awarded a bachelor's degree in chemistry. His undergraduate thesis, "The Effect of Surface Conditions on the Intensity of X-ray Reflections from Crystal Planes," was among the first studies completed at Caltech in what soon became the burgeoning field of X-ray crystallography.

After graduating, Badger remained at Caltech, working closely with Richard Chace Tolman and Arthur Amos Noyes. In 1924 he completed a dissertation on the free energy of formation of hydrogen cyanide and received his Ph.D. Although his dissertation reflected a new interest in chemical thermodynamics, Badger soon returned to the study of molecular structure, becoming an authority on the spectroscopic study of molecules in the infrared, visible, and ultraviolet regions. Aside from a year (1928–1929) as a National Research Council fellow at Göttingen and at Bonn, Badger spent his entire career at Caltech, where he was successively research fellow (1924–1928), assistant professor (1929–1938), associate professor (1938–1945), professor (1945–1966), and professor emeritus. He married Virginia Alice Sherman on 8 July 1933; they had two children, Anthony Sherman and Jennifer Hewitt.

Badger is best known as the author of "Badger's rule," an equation that relates the force and internuclear distance in a diatomic molecule. Formulated in 1933 on the basis of Badger's empirical study of relevant data on diatomic molecules, the equation has the form

$$r_l = (C_{ij}/k_l)^{1/3} + d_{ij}$$

where r_l is the internuclear distance, k_l is the force constant of the bond, and C_{ij} and d_{ij} are constants with values determined by the nature of the bonded atoms. Subsequent research by Badger and others showed that the equation is applicable to polyatomic as well as diatomic molecules. With the help of Badger's rule, it became possible to calculate approximate values of interatomic distances within molecules from spectroscopic data alone. This was of considerable value in cases where interatomic distances could not be determined by electron diffraction experiments or other direct means.

Although Badger was alert to theoretical issues, he was primarily an experimentalist with a passion for spectroscopy. During the 1930's and 1940's, he and his students derived from their spectroscopic studies a wealth of information on the structure of such simple molecules as ethylene, ammonia, hydrogen cyanide, and ozone. Badger was among the first to use spectral data to determine values for internuclear distances, moments of inertia, and bond angles. Especially notable were his contributions to the study of the hydrogen bond. Here he used spectroscopic techniques to explore the conditions under which hydrogen bonds form and to study the contributions such bonds make to the stability of molecular structures. Badger's experimental results constituted a valuable resource for his co-worker at Caltech, Linus Pauling.

After World War II, Badger extended his investigations to proteins and other large molecules. Although spectroscopic methods were inadequate to specify their structures, Badger found that spectral data were of some use in eliminating otherwise plausible configurations. During the postwar years Badger also made a number of improvements in the design of infrared spectrometers.

In 1952 Badger was elected to the National Academy of Sciences, and in 1961 he was awarded the Manufacturing Chemists' Association Medal for excellence as a college chemistry teacher.

BIBLIOGRAPHY

I. Original Works. Badger published nearly one hundred scientific papers, many of which are listed in Poggendorff, *Biographisch-literarisches Handwörterbuch*, VIIb, 182–183. His most important writings are "A Relation Between Internuclear Distances and Bond Force Constants," in *Journal of Chemical Physics*, **2** (1934), 128–131; "The Relation Between the Internuclear Distances and Force Constants of Molecules and Its Application to Polyatomic Molecules," *ibid.*, **3** (1935), 710–714; and "The Spectrum Characteristic of Hydrogen Bonds," *ibid.*, **5** (1937), 369–370, written with Simon H. Bauer.

II. Secondary Literature. There is a brief obituary in *Engineering and Science*, **38** (December 1974–January 1975), 24. A biographical memoir, with selected bibliography, is Oliver L. Wulf, "Richard McLean Badger," in *Biographical Memoirs. National Academy of Sciences*, **56** (1987), 3–20.

John W. Servos

BANDY, ORVILLE LEE (*b.* Linden, Iowa, 31 March 1917; *d.* Inglewood, California, 2 August 1973), *geology, paleontology, micropaleontology, stratigraphy.*

Bandy was the son of Alfred Lee and Blanche Meacham Bandy. When he was four years old, his family moved to Corvallis, Oregon, where Bandy attended local schools and Oregon State University (OSU). There he came under the influence of Ira Allison and Earl Packard; the former taught him structural geology, sedimentology, and geomorphology; the latter, stratigraphy and paleontology. He received his B.S. in geology in 1940 and his M.S. in 1941, both from OSU. Cenozoic stratigraphy and paleontology became his main interest in his later professional life; he then focused on micropaleontology, specifically on the study of foraminifera.

From 1942 to 1946, Bandy was a communications officer in the U.S. Air Force. On 10 June 1943 he married Alda Ann Umbras. They had two children, Janet Lee and Donald Craig. After the war he was employed by the Humble Oil and Refining Company in Houston to work on the stratigraphy of the Gulf Coast. He soon realized the need for further education, especially in micropaleontology, and therefore left Houston to attend Indiana University, where he studied under the distinguished miropaleontologist J. J. Galloway. Bandy received his Ph.D. in 1948, the year he also held the Shell Oil fellowship. His dissertation, on the Eocene and Oligocene foraminifera from Little Stave Creek, Clarke County, Alabama, was published in 1949.

After receiving his Ph.D., Bandy joined the faculty of the University of Southern California (USC), where he had a distinguished career, rising through the ranks to become chairman of the department of geological sciences in 1967. He was a member of the American Association for the Advancement of Science, the American Society of Limnology and Oceanography, and the Pacific Branch of the Paleontological Society (vice president, 1954; president, 1955). He was also a member of the board of directors of the Cushman Foundation for Foraminiferal Research, serving as president from 1966 to 1967, and a fellow of the Geological Society of America. He belonged to the Swiss Geologische Gesellschaft, the subcommission on the Plio-Pleistocene Boundary of the International Geological Union, the Asociación Mexicana de Geólogos Petroleros, and Revista Española de Micropaleontología.

In spite of Bandy's involvement in many organizations, in administration, and in teaching, he published profusely throughout his professional life: 133 papers and several more in press when he died. Some of these papers are considered classics by micropaleontologists. Bandy was among the first of the geologists trained in the era before plate tectonics to recognize the vast changes in the science of geology and to adapt to these changes rather than resist them. As chairman of the USC department of geological sciences, he added programs in geochemistry, geophysics, and chemical oceanography, realizing the importance of these disciplines to the new geology.

His association with Kenneth O. Emery at USC and the latter's work charting the physical properties of the waters off southern California provided Bandy with a basis for his landmark paper on the ecology and paleoecology of some California foraminifera (1953). His frequency distribution diagrams showing the distribution of foraminifera with respect to such environmental parameters as depth, temperature, oxygen content, and salinity were soon dubbed "Bandygrams" by his students and colleagues. In his work on the San Joaquin Valley, he demonstrated (with Robert E. Arnal) how fossil foraminifera could be used to interpret the paleoenvironmental history of continental margins (1969). He used the concept of convergent evolution and the similarity of morphology as a guide to interpret environmental preference of extinct taxa. Some of his paleoecological principles and methods of correlating foraminifera structure with environment are summarized in a report of the Twenty-first International Geological Congress (1960).

After his initial concentration on benthic foraminifera, Bandy shifted his interest to planktonic forms and their zonation, using them to interpret biostratigraphy, especially the Neogene, and paleoclimatic and paleo-oceanographic conditions. Long before such methods became generally adopted in micropaleontology, he made quantitative analyses of living and fossil fauna, using physical, geochemical, mineralogic, and isotopic ratios, as well as radiometric data, to make his correlations. He contributed to the taxonomy and coiling ratios of foraminifera and to studies in problems of pollution and in interpretation of the history of sedimentary basins, by studying changes in their populations. He also used foraminifera to define the boundaries of the Pleistocene (1967). Toward the end of his career, in keeping with his awareness of the emergence of a new era in geology, he studied the relationship of paleomagnetism and magnetic reversals to foraminifera zonation (1972).

Bandy's contributions to the theory and methods of micropaleontology were closely related to the economic benefits of recognizing the types of environments where oil and gas could accumulate. While his career was mostly an academic one, he also had a strong interest in the oil and gas industry that was enhanced through his association with the American Association of Petroleum Geologists. His approach to reconstructing environmental models and his results—for example, his work on the San Joaquin Basin—are still widely used by petroleum geologists.

Bandy was the type of scientist about whom legends are told. For example, when he first joined the USC faculty, the laboratory space assigned to him was formerly a chemistry stockroom; it became known as "Bandy's broom closet." In it he produced work on the morphology, classification, and distribution of foraminifera resulting in his first dozen publications. Later, with more stature and authority as chief scientist on an oceanographic cruise, he relieved the captain of the ship of his command and ordered him set ashore because he could not abide the disruption to science resulting from the constant conflict between the captain and the scientists on board.

In the science of micropaleontology Bandy was highly original and innovative; he built the foundation for the study of paleoecology and biostratigraphy. His confidence in his quantitative methods was such that he was able to apply his profound knowledge of the tiny foraminifera to solving problems in geology and oceanography and, furthermore, to extrapolate on a grand scale. At the same time, he was also meticulous and careful in all his work. According to Professor James C. Ingle, Jr., "Bandy relished demonstrating the power of careful micropaleontologic analysis to address large geologic and oceanographic questions. He was, in fact, doing paleooceanographic research some ten years before the advent of the Deep Sea Drilling Project and the blooming of this discipline in the 1970's."

BIBLIOGRAPHY

I. ORIGINAL WORKS. Bandy's works are listed in W. H. Easton, "Memorial to Orville L. Bandy, 1917–1973," in *Geological Society of America. Memorials*, **5** (1977); in Edith Vincent's memorial to Bandy in W. V. Sliter, ed., *Studies in Marine Micropaleontology and Paleoecology: A Memorial to Orville L. Bandy*, Cushman Foundation for Foraminiferal Research, Special Publication no. 19 (1980), 7–13.

Bandy's library and samples are in the Micropaleontology Laboratory at the University of Southern California.

II. SECONDARY LITERATURE. Besides the memorials mentioned above, see a second memorial by Easton in *Journal of Paleontology*, **48** (1974), 422–424.

Personal communications and/or interviews with the following individuals were of great value in assessing Bandy's contributions: Ira Allison, Art Boucot, John V. Byrne, Robert G. Douglas, W. H. Easton, D. S. Gorsline, James C. Ingle, Jr., and James P. Kennett.

ELLEN T. DRAKE

BARNES, HOWARD TURNER (*b*. Woburn, Massachusetts, 21 July 1873; *d*. Burlington, Vermont, 4 October 1950), *physics*.

One of three children born to William Sullivan Barnes and Mary Alice Turner, Howard Barnes moved to Montreal in 1879 when his father became minister of the Unitarian church there. He received private elementary schooling and also attended the High School of Montreal. Barnes entered McGill University in 1889, when the English-language institution projected a high profile in the natural sciences. He remained affiliated with McGill for the rest of his career. Barnes married Ann Kershaw Cunliffe in 1901. She died in 1912, leaving him with four small children. Their two sons, William Howard Barnes (1903–1980) and Thomas Cunliffe Barnes (*b*. 1904), enjoyed notable careers in science. Wilfred Molson Barnes, Howard's brother, was a distinguished artist.

Barnes ascended the academic ladder at a time of great expansion in physics. He received a Bachelor of Applied Sciences degree in 1893, became demonstrator in chemistry in 1894, picked up a Master of Applied Sciences degree in 1896, and then advanced to a demonstratorship in physics. He fell into the tradition of making precision electrical and physical measurements, which was then represented by one of the two Macdonald professors of physics, Cambridge-trained Hugh L. Callendar. In 1898 Ernest Rutherford arrived from Cambridge to succeed Callendar. The following year Barnes was rewarded for his diligence in following Callendar's lessons by winning his way to Britain as a Joule Scholar of the Royal Society of London. He returned to become lecturer in physics in 1900. McGill awarded him a D.Sc. in the same year—an honorific title that paved the way for his being named assistant professor in 1901 and associate professor in 1906. In 1907 Barnes succeeded to Ernest Rutherford's Macdonald professorship, and in 1909 he took over responsibility for the Macdonald Physics Building when the director, John Cox, the second Macdonald Professor of Physics, retired to England. At a time when subatomic physics and quantum theory were making great advances, Barnes persuaded his colleagues of the significance of traditional calorimetry. Once he carried Macdonald's name after his own, the expected designations fell his way: fellow of the Royal Society of Canada in 1908 and fellow of the Royal Society of London in 1911. In 1912 he reached the pinnacle of his career as Tyndall Lecturer at the Royal Institution in London.

All Barnes's published research may be traced to his initial work under Callendar on constant-flow precision calorimetry. Although in principle easy to grasp, the technique requires much experimental ingenuity. In essence, a known amount of electrical energy is added to a mass of flowing liquid, whose specific heat is to be determined by measuring its rise in temperature. Attaining a precision of one part in 10^5 required accurate knowledge of potentials of standard electrochemical cells and resistances, and, as well, use of extremely sensitive platinum resistance-thermometers. Barnes perfected the technique and measured the specific heat of water from the supercooled state to 100° C with unprecedented precision. The constant-flow calorimeter that he pioneered is today a compact and self-contained instrument routinely used by chemists. Barnes branched out from calorimetry to consider problems in physical chemistry, such as turbulence and the nature of electrolytes, and he collaborated with Rutherford on measuring the heat effects of radium in equilibrium with its radioactive decay products.

Early in his career Barnes measured the temperature of the nearby St. Lawrence River and Lachine Rapids under winter conditions. He found that very small ice crystals are formed in cold weather, especially in rapids or swiftly moving water under cloudy skies. This form of natural ice, called frazil, can attach itself to existing ice and rapidly precipitate ice jams. Barnes spent much of his later career studying frazil and other forms of ice. On the basis of temperature and depth profiles of the St. Lawrence River between the Great Lakes and Quebec City, he proposed to channel river water, relatively warm from the lakes, through the shallower parts of the system to retard freezing (a procedure that was not adopted in construction of the St. Lawrence Seaway). His observations on thermal gradients in ice masses led to a simple and successful technique for breaking up ice jams by setting off small, intense chemical explosions.

Funding for physics at McGill evaporated following the decision of the discipline's principal Maecenas, Sir William Macdonald, to turn his philanthropic interests toward funding McGill's agricultural college, which bears his name. In 1914 Barnes resolved to accept a chair of physics at the new University of British Columbia, a satellite of McGill then about to declare its institutional independence, but shortly thereafter he changed his mind. From 1915 to 1917 he was under medical care for a nervous breakdown, although his health remained good enough during part of the war years for him to engage in some military research. He resigned his Macdonald chair in 1919, receiving an annual pension of $1,700; the Macdonald Physics Building purchased his library and apparatus. Between 1923 and 1926 he received an annual honorarium of $500 for his ice research, and from around 1924 he once more held the title of professor. The research arrangement ended in 1932, and the following year he became an emeritus professor.

Barnes's life illustrates the scientific calling in Canada during the first part of the twentieth century. In his research he contributed to a number of areas of physics, although his discoveries cannot qualify as epoch making. In administration he demonstrated no extraordinary cunning. His honors were garnered within a network of "old boys." Although he did turn his skills toward investigating a problem of great practical concern in a cold climate—ice formation and how to deal with it—no school followed in his wake. His career began and ended in a British imperial setting, but he came into and left the world in the United States.

BIBLIOGRAPHY

I. ORIGINAL WORKS. Barnes's publications are listed in standard bibliographies, such as Poggendorff. Several boxes of administrative correspondence are located in the McGill University Archives; Record Group 2, carton 63, folders 1107 and 1109 have provided biographical information for the present entry. The private archives of the Barnes family are held by Thomas W. Barnes of Beaconsfield, Quebec. The Unitarian Church of Montreal holds material concerning the Reverend William Sullivan Barnes.

II. SECONDARY LITERATURE. Obituaries include J. S. Foster in *Obituary Notices of Fellows of the Royal Society of London*, **8** (1952–1953), 25–35; and A. Norman Shaw in *Transactions of the Royal Society of Canada*, ser. 3, **45** (1951), 77–81. See also John L. Heilbron, "Physics at McGill in Rutherford's Time," in Mario Bunge and William R. Shea, eds., *Rutherford and Physics at the Turn of the Century* (New York, 1979), 42–73; and Lewis Pyenson, "The Incomplete Transmission of a European Image: Physics at Greater Buenos Aires and Montreal, 1890–1920," in *Proceedings of the American Philosophical Society*, **122**, no. 2 (1978), 92–114.

LEWIS PYENSON
JAMES SANGSTER

BARRABÉ, LOUIS (*b.* Bonneval, Eure-et-Loir, France, 16 March 1895; *d.* Paris, France, 13 February 1961), *economic geology, geology of continents.*

Barrabé was the son of a distinguished amateur naturalist who was a member of the Société Linnéenne de Normandie. His studies were interrupted by World War I, in which he fought, receiving the Croix de Guerre. From 1919 to 1921 he attended the École Normale Supérieure, and in 1921 he passed a national competitive examination to receive the degree Agrégation des sciences naturelles. During the 1920's he was active as a field geologist, even though his legs were paralyzed following an unexplained poisoning in Martinique in 1927. From 1931 to 1961, Barrabé taught at the Faculty of Sciences of the University of Paris.

Barrabé made a profound and lasting impact in his field. No narrow specialist, he was one of the last of a breed, a geologist in the broadest sense of the word, able to carry out the most detailed analyses as well as the most far-reaching syntheses. He was also the main founder of the teaching of applied geology in the French universities.

Barrabé's scientific career began with the structural deciphering of the Corbières region in southern France, located between the southern drop of the Montagne Noire and the eastern tip of the Pyrenees, where he began work in the early 1920's. By drawing remarkably accurate contacts on a poor topographic background, he was able to show the presence of a stack of nappes separated by slices of Triassic rocks. He published his findings in 1922 and 1923. Thirty years later this reconstruction was confirmed by oil drillings.

In Barrabé's comprehensive investigation of northwest Madagascar—which he expounded upon in his doctoral dissertation at the University of Paris (1929)—and in his geologic synthesis of the French Caribbean islands of Martinique and Guadeloupe, he showed himself to be a complete geologist, equally versed in petrology and stratigraphy.

Barrabé's scientific contributions in the fields of applied geology are equally impressive. With his teacher Léon Bertrand and his colleagues Pierre Viennot and Daniel Schneegans, Barrabé was one of the discoverers of hydrocarbons in France, first the small oil deposit at Gabian in Languedoc (1924) and more important, in the summer of 1939, the large natural gas deposit at St. Marcet in the Mesozoic of the North Pyrenean trough. The first drill hole—made after long resistance from adversaries who had estimated that the French subsurface was devoid of oil—produced a significant strike on the eve of World War II. Later hydrocarbon discoveries and petroleum geology in France have their roots in this discovery.

The first discoveries of uranium ore deposits in France were made by geologist-prospectors trained by Barrabé and his associate Jean Orcell. On the basis of their predictions, the first deposits were found in the Massif Central, an area that later became a major source of uranium, particularly the pitchblende vein of La Crouzille, discovered in 1948.

Barrabé's teaching emphasized three guiding principles:

1. Integrate all regional studies within a geologic synthesis. In this spirit he offered courses on the geology of the continents and the French possessions overseas. These courses had a profound influence on his students and his followers, especially on the work of François Ellenberger. Unfortunately, the course material was never published.

2. Consider mineral deposits as normal products of the geologic evolution that must be integrated into the local geologic history so that more can be discovered. This view has left a lasting imprint on French metallogeny through the works of geologist-prospectors and the books of Pierre Routhier. Of Barrabé's teaching there remains but one publication, his work on the geology of coal and coal basins, collected and completed by Robert Feys in 1965.

3. Never hold to theories, and be alert to anything new. He heeded his own advice and regarded Alfred L. Wegener's continental drift theory with respect at a time when it was being decried by many leading scientists. He was also the spiritual father of the first group of French academic geochemists.

Barrabé was self-denying, completely oblivious to questionable compromises, mindful of others' arguments, most solicitous about his students and colleagues, and a pioneer in academic unionization. During the German occupation, he was a member of the French resistance in the University of Paris. Despite his contributions, however, Barrabé never received the honors he deserved. For example, he was not elected to the Academy of Sciences, which at that time paid little attention to the applied sciences.

BIBLIOGRAPHY

I. Original Works. From 1922 to his death, Barrabé published more than one hundred scientific papers. He also wrote twelve unpublished reports and prepared several geological maps. Among these numerous works are "Sur la présence de nappes de charriage dans les Corbières orientales," in *Comptes rendus hebdomadaires des séances de l'Académie des sciences*, **175** (1922), 1081–1083; "Tectonique des Corbières orientales," in *Bulletin des Services de la carte géologique de la France*, **27**, no. 151 (1923), 21–31; "Sur la découverte d'un gisement pétrolifère à Gabian (Hérault)," in *Comptes rendus hebdomadaires des séances de l'Académie des sciences*, **179** (1924), 1179–1181, written with Pierre Viennot; "Rapport sur les résultats d'une mission effectuée en 1927 dans le sud et l'est de la Martinique," in *Annales de l'Office national des combustibles liquides*, **3** (1928), 7–42; "Contribution à l'étude stratigraphique et pétrographique de la région médiane du pays sakalave (Madagascar)" in *Mémoires de la Société géologique de France*, n.s. 5, no. 12 (1929), 1–270, his dissertation; "Sur la récente découverte d'un important gisement d'hydrocarbures dans les Petites Pyrénées au nord de Saint-Gaudens," in *Comptes rendus hebdomadaires des séances de l'Académie des sciences*, **209** (1939), 399–401, written with Léon Bertrand; and *Géologie du charbon et des bassins houillers*, collected and completed by Robert Feys (Paris, 1965).

II. SECONDARY LITERATURE. Pierre Bellair, "Louis Barrabé (1895–1961)," in *Bulletin de la Société géologique de France*, 7th ser., **4** (1962), 227–235, with bibliography and portrait.

ALEXIS MOISEYEV
PIERRE ROUTHIER

BATEMAN, ALAN MARA (*b.* Kingston, Ontario, Canada, 6 January 1889; *d.* New Haven, Connecticut, 11 May 1971), *geology*.

Bateman was one of three sons and a daughter born to Elizabeth Janet Mara and George Arthur Bateman. When he entered Queen's University, Kingston, in 1906, he turned to geology, in which he excelled, graduating in 1910. Summers found him mapping the Canadian Shield under the direction of Alfred E. Barlow; the complex geology he encountered fired his imagination and led him, in 1910, to enter Yale, from which he graduated with a Ph.D. in geology in 1913. With the exception of a two-year period (1913–1915), he was a resident of New Haven for the remainder of his life. He became a U.S. citizen in 1915 and married Grace Hotchkiss Street on 3 June 1916.

Among professional honors, Bateman was a fellow of the Geological Society of America and a member of the Society of Economic Geologists (president in 1940); the American Association of Petroleum Geologists; the Mining and Metallurgical Society of America (president in 1956); the American Institute of Mining, Metallurgical, and Petroleum Engineers; the American Geophysical Union; the Washington Academy of Sciences; and the American Academy of Arts and Sciences. He received the Penrose Medal from the Society of Economic Geologists in 1962, and in 1970 Queen's University awarded him an honorary doctorate.

Although Yale started awarding Ph.D. degrees in 1861, forty years passed before a formal curriculum of instruction for the degree in geology was begun with the appointment of John Duer Irving as professor of economic geology in 1907 and Joseph Barrell as professor of structural geology in 1908. Irving brought with him to Yale the editorial office of the recently founded journal *Economic Geology*. It was Irving's reputation that attracted Bateman to Yale for graduate studies, and it was *Economic Geology* that came to play a major role in his life. Bateman's field studies for the Geological Survey of Canada in the summers of 1911 and 1912 led to his dissertation "Geology and Ore Deposits of the Bridge River District, British Columbia," for which he received his Ph.D. in 1913. Bateman was then offered a post as a member of the Secondary Enrichment Investigation, an extraordinary honor for one who had just received the doctorate, and one that can be seen in retrospect as having also played a major role in his professional life.

Prospecting successes in the western United States had uncovered many ore deposits in which a rich, but shallow, blanket of copper sulfide minerals overlay a larger, but much less rich, volume of iron sulfide and copper-iron sulfide minerals. The rich cappings were apparently related to the modern topography and thus postdated the formation of primary ore deposits. But how the cappings formed and what controls were exerted by topography and climate were not understood. The enriched cappings were nevertheless the key to successful development of many of the large, western copper deposits, and it was essential that their origin be understood.

Directed by Louis Caryl Graton of Harvard and funded by a group of mining companies, the Secondary Enrichment Investigation combined theoretical and experimental studies at the Geophysical Laboratory in Washington, D.C., with field studies under the direction of Augustus Locke, Edgard H. Perry, and, from 1913 to 1915, Alan Bateman. The work brought Bateman into close contact with the leading economic geologists of the day and allowed him to study many of the classic ore deposits of North America. Through much of his professional life he worked on problems first encountered during these early years. The deposit that most influenced Bateman was that at Kennecott, Alaska. Although aspects of its geology remain enigmatic to the present day, the Kennecott ore led Bateman to become skilled in the study of reflected light microscopy and, in later years, to long-term consulting arrangements with the Utah Copper Company when it absorbed the Kennecott Company.

Bateman returned to Yale as an instructor in economic geology in the fall of 1915. In 1916, Irving took leave to enter officer's training, leaving Bateman to carry both teaching duties and editorial responsibility for *Economic Geology*. Irving left for overseas duty in 1917 and died in France in 1918. Bateman was promoted to assistant professor in 1916, associate professor in 1922, and professor in 1925 (Silliman professor in 1941). More important, he was formally appointed editor of *Economic Geology* in March 1919. His stewardship of the journal was his most important contribution. As the journal grew in stature and influence, so did Bateman; indeed, it is hard to tell whether the man molded the journal or the journal molded the man. By the time he stepped down as editor in July 1969, Bateman had selected and published many of the papers that led studies of mineral deposits into the mainstream of modern scientific geology, and in the process he played a major role in the science itself.

Many of Bateman's papers appeared in *Economic Geology*; two deserve comment. In 1930, Bateman reported on work performed at the behest of the Rhodesian Selection Trust on the recently discovered ores of the Zambian Copperbelt (then the Rhodesian Copperbelt). The deposits are enclosed in sedimentary rocks and contain the same minerals that Bateman had encountered during the Secondary Enrichment Investigation. Bateman's studies led him to conclude that secondary enrichment had not significantly modified the Zambian ores and that the observed mineral assemblages are primary.

The second paper was published in 1942. Bateman presented argument and evidence in favor of a magma having the approximate composition of ilmenite ($FeTiO_3$). He reached this conclusion from studies of layered iron-titanium oxide deposits. The idea of an oxide magma was not widely accepted at that time, but subsequent studies have demonstrated that such magmas may indeed play an important role in certain igneous processes.

Bateman's career spanned the years during which geology passed from a largely outdoor, observational activity to a theoretical, experimental, and predictive science. His main contributions lay in the role he played in effecting that transition.

BIBLIOGRAPHY

I. ORIGINAL WORKS. Bateman's published writings include "Geologic Features of Tin Deposits," in *Economic Geology,* **7** (1912), 209–262, with Henry G. Ferguson; "Magmatic Ore Deposits, Sudbury, Ont.," *ibid.,* **12** (1917), 391–426; "Geology of the Ore Deposits of Kennecott, Alaska," *ibid.,* **15** (1920), 1–80, with Donald H. McLaughlin; "Primary Chalcocite, Bristol Copper Mine, Connecticut," *ibid.,* **18** (1923), 122–166; "Geology of the Beatson Copper Mine, Alaska," *ibid.,* **19** (1924), 338–368; "Some Covellite-Chalcocite Relationships," *ibid.,* **24** (1929), 424–439; "The Rhodesian Copper Deposits," in *Bulletin of the Canadian Institute of Mining and Metallurgy,* no. 216 (1930), 477–513; "The Ores of the Northern Rhodesia Copper Belt," in *Economic Geology,* **25** (1930), 365–418; "The Rhodesian Copper Deposits," in *Transactions of the Canadian Institute of Mining and Metallurgy,* **33** (1931), 173–213; "Notes on a Kennecott Type of Copper Deposit, Glacier Creek, Alaska," in *Economic Geology,* **27** (1932), 297–306; *Economic Mineral Deposits* (New York, 1942; 2nd ed., 1950; repr. 1958, 1979); "Magmas and Ores," in *Economic Geology,* **37** (1942), 1–15; "The Formation of Late Magmatic Oxide Ores," *ibid.,* **46** (1951), 404–426; *The Formation of Mineral Deposits* (New York, 1951); and as editor, *Economic Geology, Fiftieth Anniversary Volume, 1905–1955,* 2 vols. (Urbana, Ill., 1955).

II. SECONDARY LITERATURE. Walter Stanley White, "Memorial to Alan Mara Bateman, 1889–1971," in *Geological Society of America, Memorials,* **3** (1974), 15–23.

BRIAN J. SKINNER

BATES, LESLIE FLEETWOOD (*b.* Bristol, England, 7 March 1897; *d.* Nottingham, England, 20 January 1978), *physics.*

Bates was the eldest of six children of William Fleetwood Bates and Henrietta Anne Pearce. He was brought up in the Kingswood district of Bristol, where his father, an ardent pacifist, was a clerk in a boot factory. Although his parents were Church of England, at the age of seven Bates attached himself to the Moravian church. Later, in India, he joined the Scottish Presbyterian church (probably for lack of a Moravian mission there). But upon returning to England, Bates suddenly eschewed organized religion, though he continued to quote the Bible freely. A colleague described him as a "secular Christian."

Bates was first educated at a small local council school, from which in 1909 he won a scholarship to the Merchant Venturers Secondary School—a good though not distinguished institution. In 1913 Bates won a scholarship to Bristol University, which he entered with the intention of gaining an honors degree in physics. The outbreak of war frustrated this goal, so he took a pass degree (B.Sc.) in physics and mathematics in 1916 and then qualified as a radiographer in the Royal Army Medical Corps, in which he served as a lieutenant in India until 1920.

On returning home he sought a grant to enable him to qualify in medicine. When this was refused, he returned to the physics department at Bristol as a demonstrator. There A. P. Chattock and W. Sucksmith were engaged in fruitful research on ferromagnetism, and Bates collaborated with them on measurements of the Richardson gyromagnetic ratio. Thus began his lifelong fascination with this field of physics.

Briefly, however, Bates's attention turned to the investigation of long-range α particles. In 1922 he was awarded a state grant to work for a Ph.D. in the Cavendish Laboratory at Cambridge under Ernest Rutherford.

In 1924 Bates returned to research on ferromagnetism following his appointment as assistant lecturer in physics at University College, London; by 1930 he had been promoted, rapidly, to the level of reader. Encouraged by the enthusiastic but sometimes irascible experimentalist Edward N. da Costa Andrade, who was appointed professor in 1928, Bates began a long investigation of the ferromagnetic properties of manganese compounds, especially those with phosphorus and arsenic, that have Curie points around room temperature. He measured a range of physical properties, including susceptibility, electrical resistivity, and specific heats. To make the pure manganese he required, he developed a technique of preparing the amalgam by electrolysis. This led him to investigate the properties of amalgams of ferromagnetic metals.

Apart from launching Bates on his research career, the London appointment gave him financial security. In 1925 he married Winifred Frances Furze Ridler, a graduate in botany of Bristol University whom he had met there after returning from India; they had a son and a daughter. They had a common interest in science—Bates had read quite deeply in biology. The marriage was long and happy, lasting more than forty years.

In 1936 Bates transferred his research activities to Nottingham, where he was appointed Lancashire-Spencer professor of physics in its University College, then affiliated with London University (it was granted full university status in 1948). There he continued his work on amalgams and, from about 1940, extended it into the field of thermomagnetic measurements to determine temperature changes during the hysteresis cycle. During the war Bates's deep knowledge of magnetic phenomena was utilized by the Inter-Services Research Bureau, especially in the degaussing of ships to protect them from magnetic mines. After the war (1946–1956) he was consultant to the Admiralty Compass Laboratory. From 1957 to 1966 he was secretary of the Magnetism Commission of the International Union of Pure and Applied Physics.

About 1950 Bates turned his attention to the powder pattern technique developed by Francis Bitter at M.I.T. This was a valuable method of delineating magnetic domain structure, and over the period 1950 to 1965 he published nearly fifty papers on this subject alone. From about 1954 he investigated, in association with the Atomic Energy Research Establishment at Harwell, the magnetic and other physical properties of a number of metals of interest to the British atomic energy program, notably uranium and thorium. In the postwar years Bates remained a keen and active research worker, but he was increasingly involved in administrative work in his department and the university, of which he was deputy vice-chancellor from 1953 to 1956.

Bates's comprehensive knowledge of ferromagnetic phenomena was distilled in 1939 into his *Modern Magnetism*, of which four editions were printed. His standing in this field was recognized by the award of the Holweck Prize and Medal (jointly by the Physical Society of London and the Société Française de Physique) in 1949. He was elected fellow of the Royal Society in 1950.

Bates's scientific merit has been variously assessed by his contemporaries: probably a fair consensus is that he fell short of brilliance but within his chosen field achieved, by patience and dedication, a wealth of results that have stood the test of time. Rutherford described him as somebody "who would probably not fly over hedges but who would nevertheless get there in the end." However, it must be said that this opinion was given early in his career and on the strength of his α particle research, which Bates did not find very congenial.

BIBLIOGRAPHY

A complete bibliography is in N. Kurti's memoir on Bates in *Biographical Memoirs of Fellows of the Royal Society*, **29** (1983); it lists 160 publications from the period 1920 to 1977. His *Modern Magnetism* was first published in 1939 and went through four editions (last printing 1963). After his death Bates's voluminous papers were cataloged by the Contemporary Scientific Archives Centre at Oxford; they are now at Nottingham University Library.

TREVOR I. WILLIAMS

BATES, MARSTON (*b.* Grand Rapids, Michigan, 23 July 1906; *d.* Ann Arbor, Michigan, 3 April 1974), *ecology, epidemiology*.

Marston Bates's career as an ecologist of diverse experience fell into two periods. From 1928 to 1948 his entomological and ecological researches were novel in bringing ecological field and laboratory methods into the study of epidemic diseases, and he made fundamental discoveries about the ecology of the mosquito vectors of malaria and yellow fever. In the second period, 1948–1970, he turned to analysis of broad environmental and social problems, with the library as his research site, and wrote about human ecology, as well as natural history. He was an author of talent who applied ecological principles in iconoclastic analyses of modern problems and attitudes. Bates's writing and his effective and popular teaching took precedence over field research in this second period. Despite the success of his books, some ecologists dismissed his later work as superficial or uncritical. Because of his literary popularity, however, his was an effective public voice for the application of ecology to environmental problems and the management of resources.

The only child of Glenn F. and Amy Mabel (Button) Bates, he grew up in Florida, where his father was a horticulturist of exotic plants. Attracted to natural history from adolescence, Bates studied biology at the University of Florida (B.S., 1927) and immediately took a job that would take him to tropical America. As an entomologist from 1928 to 1931 for the Servicio Técnico de Cooperación Agrícola, part of the United Fruit Company, he performed and directed ecological researches on insects affecting banana and coffee plantations in Honduras and Guatemala. In 1931 he entered Harvard University. Field research on Caribbean islands yielded a number of publications on the distribution, ecology, and taxonomy of insects, and he took the A.M. (1933) and Ph.D. (1934). His thesis on the butterflies of Cuba analyzed their taxonomy, life histories, and biogeographical relations. A few months after becoming an assistant in the Museum of Comparative Zoology at Harvard, Bates joined the Rockefeller Foundation's malaria investigations in Albania.

With a laboratory at Tirana, the foundation's international health division had begun to study the biology of the mosquito vectors of endemic malaria. Bates formally joined the staff in 1937. His mosquito work in Albania was characteristic of his research strategy, combining various approaches to produce ecological observations on distribution and seasonality, behavioral experiments on mating and egg laying, genetics of the various strains, techniques for rearing mosquitoes, and determinations of taxonomic distinctions. Out of this novel mixture he was able to produce what he called the natural history of disease, by which he meant a coherent picture of the relations among mosquito vector, environmental conditions, and incidence of the human disease. This was essential for a strategy of control for malaria.

Bates finished this project in 1939 and married Nancy Bell Fairchild while on home leave. She was the daughter of the tropical botanist and explorer David Fairchild, a family friend in Florida with connections to prominent families in the Midwest. The couple went to Egypt to establish a malaria research station, which was abandoned when World War II started; they settled instead in Villavicencio, Colombia, where Bates directed the recently established Rockefeller Foundation yellow-fever research station. Nancy Bell Bates, herself a naturalist and often research assistant, bore them three daughters and a son.

Although yellow-fever vaccine had been developed and mosquito control attempted since the early 1900's, epidemics broke out in South America in the 1930's. The Villavicencio laboratory was part of the Rockefeller Foundation's broad-based efforts to eliminate the disease. Previous control strategies centered on the cities, whose dense populations appeared to be a necessary factor in epidemics. But new outbreaks were in rural areas, and the Villavicencio station was set strategically on the edge of the forest and the sparsely peopled llanos of eastern Colombia. Bates again attacked all aspects of mosquito biology and uncovered the essentials of the transmission cycle of the virus. He found new vectors, *Haemagogus* mosquitoes, that carried yellow fever among a diverse pool of forest mammal hosts, including humans. The usual cause of outbreaks was the felling of forests, as the reservoir for virus was monkeys and mosquitoes in the upper tree canopy.

Using their ecological and epidemiological findings, Bates's team set up a diagnosis and treatment program with local doctors and a mosquito control program, and they worked on vaccine. Their efforts were dramatically successful even if Bates did conclude that elimination of the disease would be impossible. He returned to the United States in 1948 and wrote *The Natural History of Mosquitoes* (1949). Summing up his broad knowledge of mosquito biology and applying it to the ecology of epidemic diseases, the book became a classic text for epidemiology.

It also started his prolific writing career. Assigned by the Rockefeller Foundation to study human demography, Bates served as special assistant to the head of the foundation, then moved in 1952 to a professorship in the University of Michigan zoology department, continuing his study of human ecology. Meanwhile he had published *The Nature of Natural History* (1950), about scientific methods and their usefulness, and *Where Winter Never Comes* (1952), an analysis of human ecology in the tropics. The results of his analysis of human population dynamics appeared as *The Prevalence of People* (1955), an early warning about the ecological consequences of unrestrained population growth. His other most significant books were two analyses of human ecology, *The Forest and the Sea* (1960) and the textbook *Man in Nature* (1961).

In all of these books his distinctively broad application of standard ecological principles to human social and environmental problems was presented with a disarming wit, clarity, and apparent nontechnicality, which masked the sometimes sharp divergence from previous scholars' conclusions. From the complex web of biological interrelations into which humans, as organisms, necessarily had to fit in the world, he argued for the value of diversity in itself and as a key factor in preserving the balance of nature. He also argued for a more rational fitting of cultural practices to environmental conditions. His goal was moral or social wisdom based on a standard of objectivity and scientific knowledge.

BIBLIOGRAPHY

I. Original Works. Bates summarized the major findings of his more than 100 entomological and epidemiological papers in his thesis, "The Butterflies of Cuba," in *Bulletin of the Museum of Comparative Zoology at Harvard College*, **78**, no. 2 (1935), 63–258, and in *The Natural History of Mosquitoes* (New York, 1949). His other books were *The Nature of Natural History* (New York, 1950), *Where Winter Never Comes* (New York, 1952), *The Prevalence of People* (New York, 1955), *The Darwin Reader*, edited with P. S. Humphrey (London, 1957), *Coral Island*, with D. P. Abbott (New York, 1958), *The Forest and the Sea* (New York, 1960), *Man in Nature* (Englewood Cliffs, N.J., 1961), *Animal Worlds* (New York, 1963), *The Land and Wildlife of South America* (New York, 1964), *Gluttons and Libertines* (New York, 1968), and *A Jungle in the House* (New York, 1970).

His papers, including journals, correspondence, manuscripts, and research materials, are in the Bentley Historical Library (Michigan Historical Collections), University of Michigan, Ann Arbor.

II. Secondary Literature. Nancy Bell Bates described life and research at the Villavicencio laboratory in *East of the Andes and West of Nowhere* (New York, 1947).

William C. Kimler

BAUR, ERWIN (*b.* Ichenheim, Germany, 16 April 1875; *d.* Berlin, Germany, 2 December 1933), *botany, genetics, applied genetics, population genetics.*

Life. Born into a family of distinguished amateur botanists, Baur was elder of the two sons of Wilhelm Baur, a pharmacist, and Anna Siefert Baur, an innkeeper's daughter. The father was a founding member of the Baden Botanical Society (1881) and an authority on the mosses of the Black Forest region. He played a significant role in shaping Baur's early interest in botany by including him on numerous walking tours and field studies. This youthful interest was reinforced and cultivated to a passion by an uncle, Ludwig Leiner of Constance, with whom Baur lived during the first three years of gymnasium study (1885–1888). In 1888 Baur's family moved from the small rural town of Ichenheim to Karlsruhe, where both sons attended the gymnasium. Although not an outstanding student, Baur thrived in the company of amateur and professional botanists he met through his father. By the age of seventeen he was an accomplished amateur botanist.

When Baur completed the *Abitur* in 1894, he was spiritually and intellectually prepared to study botany at the university. But at his father's insistence he enrolled as a premedical student at the University in Heidelberg in the winter semester of 1894. After a few semesters he moved on to study at Freiburg im Breisgau, then Strassburg, settling finally in 1897 at the University of Kiel. Throughout his student years Baur faithfully attended lectures in botany and biology, coming under the influence of Friedrich Oltmanns, August Weismann, and Otto Reinke. But his awakened interest in theories of descent, inheritance, and development did not lessen the obligation to his father. At Kiel, Baur completed his medical studies, passed the medical *Staatsexamen*, and received the M.D. degree in March 1900. After a brief interlude as a ship's doctor and as an assistant physician at psychiatric clinics in Kiel and Emmendingen, he abandoned medicine and psychiatry for doctoral work in botany at Freiburg. Under the supervision of Oltmanns, he completed the doctorate in 1903 with a study of the developmental aspects of fructification in lichens.

In 1903 Baur became first assistant to Simon Schwendener at the University Botanical Institute in Berlin. He qualified as *Privatdozent* in 1904 with a work on the fungal bacteria, an area to which he never returned, having discovered by then the snapdragon *Antirrhinum majus*, his door to experimental genetics. The poor facilities and working conditions at Schwendener's institute made a lasting impression on Baur and became a source of anxiety as he turned steadily toward the study of inheritance and variation. His position at the institute, moreover, was undermined by Schwendener's retirement in 1910. With little prospect of obtaining a university chair, in 1911 Baur accepted a professorship in botany at the Landwirtschaftliche Hochschule at

Berlin. In 1914 he opened its Institute for Genetic Research at Berlin-Friedrichshagen; this modest facility served as Germany's first center for experimental and applied genetics. It also marked the beginning of an institutional campaign for Baur, whose massive and costly breeding studies precluded a return to the troubled bosom of the German universities.

World War I and its aftermath postponed Baur's plan for a Kaiser Wilhelm Institute for Genetic Research distinct from the Kaiser Wilhelm Institute for Biology (headed by Carl Correns since its opening on 17 April 1915). The Kaiser-Wilhelm-Gesellschaft did, however, provide Baur with some assistance. As a result, in 1923, he opened a new Institute for Genetic Research at Berlin-Dahlem. Intended to narrow the existing gap between genetic research and its application to agriculture, the facility represented a major institutional breakthrough for Baur. At the same time, it enabled him to fill another important deficiency: With Paula Hertwig and Hans Nachtsheim, he galvanized at Dahlem a German research nucleus that utilized the principles and methods of classical genetics outlined by the Morgan school at Columbia. He remained at Dahlem until 1928, when, in response to his demand for a large-scale program in applied genetics, he became director of the new Kaiser Wilhelm Institute for Plant Breeding and Genetic Research at Müncheberg. He headed the Müncheberg institute until his death.

Baur preferred rugged outdoor activity and rustic environs to urban life. After his marriage in 1905 to Elizabeth Venedey, they and their son and daughter lived on a small farm on the outskirts of Berlin. The family later settled on a larger farm at Müncheberg. Baur's zeal for fieldwork took him to three continents and led to his mastery of a half-dozen languages. Although self-contained and somewhat of a loner, he bridged the distance between himself and others with his optimism, equity, and complete lack of vanity.

Professional Career. A relentless advocate of the intellectual and practical value of genetics, Baur was the first to offer, at a German university, lectures on Mendelian inheritance in conjunction with a practicum in experimental breeding. The novel program he developed while a *Privatdozent* at Berlin led in 1911 to the publication of an immensely influential textbook, *Einführung in die experimentelle Vererbungslehre*. An innovative and much beloved teacher, he was also the principal architect and editor of *Zeitschrift für induktive Abstammungs- und Vererbungslehre*, the first scientific journal devoted to genetics, from its beginning in 1908. In 1911 Baur represented Germany at the Fourth International Congress of Genetics in Paris. That same year the newly established Kaiser-Wilhelm-Gesellschaft zur Förderung der Wissenschaften (Berlin) decreed the construction of the first of its scientific research institutes. Within a decade this scientific funding agency became the principal patron of new specialties and fields, including genetics. Accordingly, geneticists such as Baur, Carl Correns, and Richard Goldschmidt found a permanent home outside the traditional German university system.

World War I not only radically affected Baur's scientific and institutional activities, it also disrupted his plan for a Berlin meeting of the Fifth International Congress of Genetics in 1916. He also was prevented from serving as exchange professor at the University of Wisconsin at Madison for the academic year 1914–1915. But although he suffered profound difficulties and disappointments, Baur was not a spiritual casualty of the war. Optimistic rather than disillusioned, he expanded both his scientific program and his professional activities, most notably in applied genetics. During the Weimar period he trained as many as thirty students a year, published a major textbook in applied genetics, *Die wissenschaftlichen Grundlagen der Pflanzenzüchtung* (1921), and established the first German journal for theoretical and applied genetics, *Der Züchter* (1929). He also attended, and frequently headed, the meetings of the three major German agricultural and breeding societies, and he served on numerous national and international agricultural planning committees.

Baur was honored in Germany as the founder of applied genetics, but he performed yet another important service for his science. On behalf of the Deutsche Gesellschaft für Vererbungswissenschaft (1921), a society he founded in concert with Correns and Goldschmidt, Baur resumed responsibility for planning the Fifth International Congress of Genetics. For six years he acted as goodwill ambassador of the scientifically isolated postwar German genetics community. When the congress met at Berlin in 1927, Baur served as its president. Two years later he attended the All-Russian Congress of Genetics and Breeding, where he met many of the Russian population geneticists. In 1931 he delivered lectures in London, Sweden, and South America, speaking on evolution, applied genetics, and eugenics. That same year he organized and presided at the Berlin meeting of the Congress of the International Plant Breeders Association. Baur remained professionally active until his death from heart failure.

Scientific Work. Baur's memorable caveat to students and co-workers was "More experimentation, less speculation!" No crude empiricist, he saw in Mendelian principles the surest bridge between heredity and evolution, but for Baur surety entailed rigorous experimentation. Unlike William Bateson, who shared his view, Baur was a supremely pragmatic experimentalist. Gifted with an uncanny ability to recognize theoretical problems amenable to experimental testing, he exploited any discovery, idea, or method likely to yield concrete results; yet he was laboriously thorough and cautious of overinterpreting evidence. In this sense he strongly resembled Thomas H. Morgan. But whereas Morgan concentrated on the mechanisms of Mendelian inheritance, Baur focused increasingly on the genetic mechanisms of speciation.

Because of his methodological flexibility and his interest in the species question, Baur worked with equal ease and success in every major problem area of classical genetics. His diagnostic ability and investigative skill were evident at the outset of his career, soon after his arrival at Schwendener's institute in 1903. While engaged in a study of the viral and bacterial causes of leaf mottling, he discovered a type of leaf variegation that seemed to be inherited rather than microbially induced. In 1905 he began to study inherited variegation in the *aurea* variety of *Antirrhinum*. Unaware of the irony of following rather too closely in the footsteps of Hugo de Vries (*Oenothera*) and Carl Correns (*Mirabilis jalapa*), Baur likewise fixed on an organism whose abnormal genetic mechanisms played havoc with Mendel's laws. Over the next five years his work on *Antirrhinum* closely paralleled Correns' study of non-Mendelian inheritance in *Mirabilis*.

But Baur's investigation of variegation was far more exhaustive than Correns', encompassing plant chimeras and mosaic oddities in general. Surpassing the earlier work of the botanists Hans Winkler and Hermann Vöchting, Baur distinguished clearly among the four principal categories and causes of mosaic variegation: diseased plants, somatic mutation, graft hybrids, and hybrid seedlings. Only after extensive breeding experiments with hybrid seedlings and comparative cytological studies did Baur explain inherited variegation in terms of his theory of extranuclear plastid inheritance. Fully elaborated in his *Untersuchungen über die Vererbung von Chromatophorenmerkmalen bei Melandrium, Antirrhinum und Aquilegia* (1910), the "Plastom theory," as it was called, argued for the presence of individual hereditary particles, the plastids, in the cytoplasm of germ cells. These were of two types: normal color-bearing and mutant color-deficient. Because of their random location in the cytoplasm, plastids could not pair (fertilization) or segregate (cell division) in the manner of nuclear chromosomes, but were passed on individually to the cytoplasm of daughter cells in a random, non-Mendelian fashion.

The clarification of plastid inheritance was subsequently regarded by many botanists as Baur's most significant contribution to genetics. In contrast with Correns' interpretation of the same phenomenon, Baur did not attribute a crucial role in the hereditary process to the cytoplasm as a whole. Moreover, he suggested that mutant plastids resulted from a nuclear, chromosomal mutation, chromosomes being the true arbiters of inheritance.

The relationship among heredity, variation, and speciation had preoccupied Baur's thoughts since 1908, the year he established contact with William Bateson, Wilhelm Johannsen, and Herman Nilsson-Ehle. To a large extent Baur modeled his hybridization experiments on those of these three earliest and most influential acquaintances. He had long since established that with the exception of leaf coloration, *Antirrhinum*'s traits Mendelized. In 1908 he provided the first clear demonstration of a lethal gene, and in the following years he corroborated the phenomena of linkage, multiple alleles, and reversion. With the move to the Landwirtschaftliche Hochschule in 1911, Baur turned permanently to the problem of speciation. In particular he was on the alert for any evidence suggesting the inheritance of acquired traits.

Baur's prewar hybridization experiments led him to conclude that chromosomal mutations alone provided the material for evolution through natural selection. Accordingly, he directed his inquiries to the nature and transmission process of mutations. But it was not until the opening of the Institute for Genetics at Dahlem in 1923 that he had both the facilities and the theoretical framework needed to attack the mutation problem at the chromosomal level. Whereas Correns and Goldschmidt openly challenged the monopoly of nuclear chromosomes in inheritance, Baur not only adopted Morgan's gene theory of inheritance and the experimental methodology of the *Drosophila* school but also engaged in much the same work. Employing *Antirrhinum majus*, as the botanical counterpart of *Drosophila*, he and his co-workers mapped chromosomes, studied the rearrangement of the chromosomal material, and identified the loci of genic mutations. Baur's group also pioneered research on X-ray–induced mutations; success came some years later, largely through the efforts of his students Emmy Stein and Hans Stubbe.

While he was at Dahlem, Baur's findings on mutations compared in quality and significance with those of his American contemporary and admirer, Hermann J. Muller. Baur was one of the first geneticists to determine the frequency of small point-mutations and to note the occurrence of the same mutation in individuals of the same race. After extensive study of both lethal and nonlethal mutations, he concluded that most dominant mutations were adverse but infrequent, while the recessive small point-mutations were of neutral survival value but relatively frequent. He also investigated polyploidy and other nonlethal mutations affecting the reproductive capacity of closely related individuals.

Baur concentrated on mutation research and genetic analysis of inheritance in *Antirrhinum* until his move to Müncheberg in 1928. The vastly improved facilities at the new institute allowed him to take the final step in an experimental study of evolution begun twenty years earlier. It was a step that brought him full circle, back to the naturalist tradition and training of his youth. In 1928 and 1929 he studied and collected specimens of geographic populations of snapdragon in Spain, Portugal, and France. His collection included thousands of domestic and wild varieties, and with these he began large-scale breeding of interspecific populations of the genus *Antirrhinum*. The work of the Müncheberg period represents one of the earliest major genetic studies of geographic populations known.

After successful breeding of fertile species crosses and detection of multiple alleles responsible for small specific differences in *Antirrhinum*, Baur synthesized the whole of his genetic and mutation research into a theory of progressive evolution. In the articles "Evolution" (1931) and "Artumgrenzung und Artbildung in der Gattung *Antirrhinum*" (1932), he argued that small point-mutations are the material for evolution. Of negligible selective value on their own, these relatively frequent mutations accumulate and combine through interbreeding to yield and enhance several new characters in individuals of a certain population. Selective pressure, which acts on all members of the population, may or may not favor those individuals carrying combinations of multiple alleles controlling new characters. The stability of environmental conditions, he added, is a critical factor in the evolutionary process.

Baur died before completing a massive breeding experiment involving more than 50,000 second- and third-generation crosses of *Antirrhinum*; his records, too, remained incomplete and fragmented. It was known, however, that he was attempting to effect speciation through increased isolation of successive groups of related individuals. Whether Baur began to think in "populationist" terms has not been determined, but his work on small mutations and his interpretation of their significance for evolution did not escape the notice of evolutionary biologists working in the 1930's and 1940's.

Applied Genetics. As a close friend of the agricultural geneticist Herman Nilsson-Ehle, Baur was well aware of the discoveries and advances yielded by applied research. It was this awareness, as much as his social conscience, that enabled him to synthesize genetics and practical life into a new field. During the war years, when he first engaged in applied research, Baur mastered breeding methods that served him for the rest of his life. These combined the massive cultivation of crops, methodical measurement of the botanical characters of population and individual, establishment of pure lines (constant pedigrees) through self-fertilization, and cross-breeding of pure lines, with the application of Mendelian principles, pure-line theory, and evolution theory to practical and experimental breeding problems. Like Mendelian principles, however, these methods were largely ignored or unknown in the conservative German agricultural community. Confronting the need for education and reform in this quarter, Baur provided the rational basis of practical breeding in *Die wissenschaftlichen Grundlagen der Pflanzenzüchtung* (1921). He then proceeded to demonstrate the utility of the principles and methods outlined in his book through his work at Dahlem and Müncheberg.

Baur dealt with virtually every type of economically important plant. Among the most vital and threatened crop plants he worked on were grapes, which were struck in 1925 by an epidemic of yeast infection. At Dahlem he began hybridization experiments that by the time of his death yielded more than seven million disease-resistant varieties of grape. These were subsequently cultivated in the Rhine and Moselle regions. Baur's success with grapes guaranteed the survival of German viticulture, but no less vital to German agricultural productivity was his work on grains, potatoes, and lupines. Here his approach truly complemented and reflected his experimental and theoretical interests.

As early as 1913 Baur had proposed the study and utilization of wild types of grains, potatoes, and other food crops. After 1918 he amassed large collections of specimens from as far afield as Asia Minor (1926) and the Andes of Peru and Chile (1931). As in the case of *Antirrhinum*, he studied regional populations, appraised environmental conditions, and collected samples for genetic analysis and hybridization experiments. His methods of testing the viability of new varieties in large and small populations, by allowing environmental conditions to weed out weak members of successive generations, were likewise employed with *Antirrhinum*. After comparative analysis of genotypic composition and experiments with interspecific crosses, Baur successfully bred a wheat-rye hybrid suited to cultivation in light, sandy soils and a wild-domestic cross of potato that was both disease- and cold-resistant. Toward the end of his career he also began to study the effects of X-ray–induced mutations in wheat, hoping to produce new, improved strains of this staple.

Baur's most noteworthy agricultural triumph, and one that attracted a great deal of scientific attention, was his discovery in 1930 of a sweet, nonalkaline variety of lupine, a wild member of the bean family. Some years earlier, when he first considered lupines as a possible source of fodder, he had predicted the appearance and immediate consumption by wild animals of a sweet, mutant variety of this otherwise very bitter plant. Certain of his instincts, he bred more than two million wild lupines and found the anticipated mutants; their descendants were cultivated thereafter as fodder crops. While not a test of progressive evolution, the lupine experiment provides an apt illustration of Baur's readiness and ability to put theory into practice.

Eugenics. In 1907, while still a relatively new convert to the study of inheritance, Baur joined the Gesellschaft für Rassenhygiene, a eugenic society founded in 1905 by the physician Alfred Ploetz. For the remainder of his life Baur was an active member of this society and a regular contributor to its journal, *Archiv für Rassen- und Gesellschaftsbiologie*. Like most eugenicists of the period, Baur was not indifferent to the issue of hereditary racial differences. On the contrary, his interest in the evolution and genetic composition of the races of man led him to publish jointly with Eugen Fisher and Fritz Lenz the extremely popular but controversial *Grundlagen der menschlichen Erblichkeitslehre und Rassenhygiene* (1921).

From the beginning of his involvement with the German eugenics movement, however, Baur regarded the application of genetics to human heredity largely from the standpoint of medical science. Formerly a physician, he saw in eugenics the means for controlling and eradicating human hereditary defects and diseases. Optimism rather than morbid anxiety lay at the heart of his commitment to rational human breeding. Responding to the cultural pessimism of Oswald Spengler's *Decline of the West* (1918), Baur argued that disease was the primary cause of the degeneration of human society. Through a process he called "reversed selection," human culture interferes with the weeding-out function of natural selection to promote the spread of inherited debility. He maintained that genetics, medicine, and science in general could unite to overcome the degenerative effects of reversed selection.

Given the relatively new and unexplored status of human genetics, Baur cautioned against immediate widespread implementation of eugenic programs. He repeatedly emphasized the need for more research, especially of the type conducted by Agnes Blühm and others at the Kaiser Wilhelm Institute for Biology at Dahlem. Throughout the Weimar era Blühm investigated the teratogenic and mutagenic effects of alcohol and other chemicals on living organisms. Toward the end of his life Baur conducted similar investigations, convinced of their medical value—others, however, were skeptical. In 1931 Hermann J. Muller severely criticized Baur's position concerning the possible mutagenic effects of chemical agents. But in this very novel area of research, Baur's instincts proved once again to have been correct. They failed him, however, in one important respect. Although science continued to serve the practical needs of society, it did not guide society in the pursuit of rational humanitarian policies. In his last years at Müncheberg, the institute experienced serious financial difficulties. These were overcome soon after his death, but not without a change in the character and function of both his institute and other Kaiser Wilhelm Institutes when they came under the jurisdiction of the National Socialist government.

BIBLIOGRAPHY

I. Original Works. Baur produced a total of 109 scientific publications in his lifetime. These covered problems ranging from bacterial infections in plants to the emergence of new species. The most important of his publications are "Untersuchungen über die Vererbung von chromatophoren Merkmalen bei *Melandrium, Antirrhinum* und *Aquilegia*," in *Zeitschrift für induktive Abstammungs- und Vererbungslehre*, **4** (1910), 81–102; *Einführung in die experimentelle Vererbungslehre* (Berlin, 1911); *Grundlagen der menschlichen Erblichkeitslehre und Rassenhygiene* (Munich, 1921), with Eugen Fischer and Fritz Lenz; *Die wissenschaftlichen Grundlagen der Pflanzenzüchtung; ein Lehrbuch für Landwirte, Gärtner und Forstleute* (Berlin, 1921); "Die Bedeutung der Mutation für das Evolutionsproblem," in *Zeitschrift für induktive Abstammungs- und Vererbungslehre*, **37** (1925), 107–115; "Untersuchungen über Faktormutationen. I. *Antirrhinum majus* mut. *phantastica*," *ibid.*, **41** (1926), 47–53; "Untersuchungen über Faktormutationen. II. Die Häufigkeit von Faktormutationen in verschiedenen Sippen von *Antirrhinum majus*. III. Über das gehäufte Vorkommen einer Faktormutation in einer bestimmten Sippe von *A. majus*," *ibid.*, 251–258; *Menschliche Erblichkeitslehre und Rassenhygiene* (Munich, 1928), with Eugen Fischer and Fritz Lenz; *Handbuch der Vererbungswissenschaft* (Berlin, 1928), with Max Hartmann; "I. Evolution, II. Scope and Methods of Plant Breeding Work in Müncheberg," in *Journal of the Royal Horticultural Society*, **56** (1931), 176–190; "Artumgrenzung und Artbildung in der Gattung *Antirrhinum*, Sektion *Antirrhinastrum*," in *Zeitschrift für induktive Abstammungs- und Verer-bungslehre*, **63** (1932), 256–302.

A complete chronological bibliography of Baur's works, including articles written for newspapers and a list of the scientific journals that he edited, is available in Elisabeth Schiemann's "Nachrufe Erwin Baur" (see below).

The Erwin Baur papers are located at the Bibliothek und Archiv zur Geschichte der Max-Planck-Gesellschaft in West Berlin. A description of the contents of the collection is provided in Jonathan Harwood, "The History of Genetics in Germany," in *Mendel Newsletter*, **24** (October 1984), 1–3. The collection consists primarily of documents and papers pertaining to the establishment of Baur's Kaiser Wilhelm institutes at Dahlem and Müncheberg; it does, however, include the manuscript of Karl von Rauch's 300-page unpublished biography of Baur as well as biographical material from other sources. Little of Baur's correspondence with other botanists and geneticists survives, but a number of very important letters are available in two archival collections, Nachlass Carl Steinbrink and Sammlung Ludwig Darmstaedter, at the Staatsbibliothek Preussischer Kulturbesitz in West Berlin.

II. SECONDARY SOURCES. The most comprehensive treatment of Baur's life and career is Elisabeth Schiemann, "Nachrufe Erwin Baur," in *Berichte der Deutschen botanischen Gesellshaft*, **52** (1934), 4–114. Additional perspective on institutional developments that had an impact on Baur's career can be gained from Lothar Burchardt, *Wissenschaftspolitik im Wilhelminischen Deutschland* (Göttingen, 1975); and Günther Wendel, *Die Kaiser-Wilhelm-Gesellschaft 1911–1914* (Berlin, 1975). Three works reviewing the development of genetics in Germany during Baur's lifetime are H. Friedrich-Freska, "Genetik und biochemische Genetik in den Instituten der Kaiser-Wilhelm-Gesellschaft und der Max-Planck-Gesellschaft," in *Naturwissenschaften*, **48** (1961), 10–22; Jonathan Harwood, "The Reception of Morgan's Chromosome Theory in Germany: Inter-war Debate over Cytoplasmic Inheritance," in *Medizin historisches Journal*, **19** (1984), 3–31; and Hans Nachtsheim, "Die Entwicklung der Genetik in Deutschland von der Jahrhundertwende bis zum Atomzeitalter," in Hans Leussink, ed., *Studium Berolinense* (Berlin, 1960), 858–867.

NATASHA X. JACOBS

BAWDEN, FREDERICK CHARLES (*b.* North Tawton, Devon, England, 18 August 1908; *d.* Rothamsted, England, 8 February 1972), *plant pathology, biochemical virology*.

Bawden was the son of George Bawden and Ellen Balment, both of whom supervised a Poor Law institution (a workhouse) in Okehampton during much of his youth. It has been suggested that his early experiences helped to develop a continuing concern for the human condition. The large garden of the institution is presumed to have evoked an interest in agriculture in general, and in the growth of potatoes in particular. Bawden studied botany at the Okehampton Grammar School and some chemistry at the Crediton Grammar School, after which he entered Cambridge as a scholarship student. He has been described as sports loving, cheerful, and active in the social life of his college. As a senior scholar at Emmanuel College, he worked with Frederick T. Brooks on cereal rusts, the subject of his M.A. thesis (1930) for the diploma in agricultural science. Fungus infections were the subject of many of his later writings.

In 1927, Redcliffe N. Salaman established the Potato Virus Research Station at Cambridge with Kenneth M. Smith as senior research assistant. In the summer of 1930, Bawden was appointed an assistant in the poorly equipped station and began work on the potato and its viruses. In this period he was concerned with the detection of virus infection and developed a long-term interest in serology.

The year 1934 was important for Bawden. On 6 September he married Marjorie Elizabeth Cudmore, who had been a fellow student at both Okehampton and Cambridge; they had two sons. He also began a long collaboration with the biochemist Norman W. Pirie, who had been working in the department of pathology, where Bawden was learning serological technique with E. T. C. Spooner. Pirie and Bawden attempted to isolate the potato virus—a virus difficult to purify—and in 1936 reported the association of its infectivity with material sensitive to various proteinases. Their enthusiasm was heightened by the 1935 report of Wendell M. Stanley at the Rockefeller Institute, who had described the isolation and crystallization of the virus of the tobacco mosaic disease.

In 1936 Bawden moved to the Rothamsted Experimental Station in Hertfordshire as virus physiologist in the department of plant pathology, a position that permitted him to work with viruses other than those of potatoes. He continued his collaboration with Pirie at Cambridge. By the end of 1936, they were able to confirm, correct, and extend Stanley's report. In the next five years, the two men made many exciting discoveries of the properties of numerous plant viruses, and Pirie moved to Rothamsted in 1940. Bawden became head of his department in 1940 and director of the station from 1958 until his death.

The station, which celebrated its centenary in 1943, was for its first fifty years a site of closely monitored field experiments on the effects of fertilizers on plant growth. Supported from 1911 by a governmental development fund for the rehabilitation of British farming, the station thenceforth became a leading agency for agricultural research in England and for the development of world agriculture. This role was bolstered by the work of Bawden and his associates in plant virology and pathology. The growth of the station under his directorship was facilitated by his visible and lasting concern for the improvement of British agriculture and food production during World War II and afterward. He wrote numerous articles on problems of agriculture, attempting to communicate with the general public. His major scientific and administrative contributions established his national and international prestige. He was elected to the Royal Society in 1949 and was given many awards and responsibilities.

The distinction between nonfilterable bacteria and the filterable tobacco mosaic virus had been established by Dmitrii I. Ivanovskii and Martinus W. Beijerinck in 1892 and 1898, respectively, and these results pointed to the possible existence of very small noncellular pathogens, a conclusion affirmed by the development of more sophisticated filtration techniques. By the mid 1930's an extensive body of work had appeared on the properties of tobacco mosaic virus and on efforts to purify it from the sap of infected plants. The startling work of Stanley, based on the advances of the previous decade in protein isolation, required rapid confirmation. This was soon provided by Bawden and Pirie.

Their first paper on this subject, published in 1936 with the X-ray crystallographers John D. Bernal and Isidor Fankuchen, also described the isolation of crystallizable infectious nucleoproteins from tobacco and tomato plants infected with several distinctive strains of tobacco mosaic virus. Each virus preparation reproduced the characteristic disease in suitable plants. The authors emphasized that the small needles obtained should be described as "liquid crystalline," possessing only two-dimensional regularity. Bernal and Fankuchen later described these needles as "tactoids." When they were sufficiently purified, concentrated preparations spontaneously separated into a liquid crystalline birefringent lower layer and an upper layer showing anisotropy of flow as a result of the alignment of soluble viral particles. The soluble rods were readily sedimented, and the dried pellets, crystals, or oriented liquid layers were easily examined by X-ray crystallography. The preparations were shown to contain rods of constant diameter aligned in close-packed hexagonal array. The rods were comprised of small, identical protein subunits along their length, but it was not clear that the rods were of identical length. The crystallographic data were similar to those Ralph W. G. Wyckoff and Robert B. Corey obtained earlier on Stanley's material, establishing the virtual identity of the virus preparations obtained by the two laboratories.

The English preparations possessed nitrogen contents more characteristic of proteins than did the nitrogen content reported initially by Stanley; but more important, for the first time the virus was found to contain phosphorus and carbohydrate within a nucleate of the ribose type (RNA), isolable after disruption of the virus. Bawden and Pirie soon demonstrated the presence of RNA in numerous other plant viruses. At about this time Max Schlesinger, working in London, found that preparations of bacteriophages contained a nucleate of the deoxyribose type (DNA). In the next decades many other viruses infecting plants, animals, and microbes were also found to carry a nucleate of either RNA or DNA, which was proved by the early and mid 1950's to determine the genetic continuity and capacity for duplicability of the particular virus. In the 1970's, some infectious diseases were found to be caused by a "viroid," a small, stable infectious nucleate of the ribose type, devoid of protective protein. In the mid 1980's it was reported that the etiologic agents of a few difficult-to-study animal and human diseases are free of nucleic acid and consist largely of protein. The unequivocal proof of the existence of such agents, termed "prions," has been difficult to establish.

In their early papers and for decades thereafter, Bawden and Pirie were concerned with the rigor of the evidence that the infectious tobacco mosaic virus could be described as a nucleoprotein rod of a single length, and that the particles isolated were present as such in infected sap. They had shown that the conditions of purification facilitated aggregation. In subsequent years it was shown, mainly in Stanley's laboratory, that the infectivity and distinctive monomeric rods possessed essentially identical sedimentation constants in an analytical ultracentrifuge, that older infected plants contained an increased percentage of end-to-end dimers, and that the preparation of viruses for visualization in the electron microscope led to breakage of the rods. It is now generally believed, consistent with the known internal position of the single genetic element (RNA, lengthwise within the viral rod), that the tobacco mosaic virus must be of a distinctive length, within very narrow limits.

In 1938 the tomato bushy stunt virus was found by Bawden and Pirie to give rise to true three-dimensional isotropic crystals comprised of essentially spherical nucleoprotein particles. These properties and the sedimentation pattern of the virus very early suggested an identity of the infectious virus and the crystallizable spherical particle. Indeed, in 1942 Max Lauffer determined the size distribution function of a virus preparation by studying the diffusion of sedimenting particles and demonstrated that the diameters of the particles could not deviate from the mean by more than 1 percent.

The enlarging study of viruses isolated from various diseased plants revealed many new biological phenomena. Some of these new phenomena were facilitated by Bawden's skills in serology. For example, although the tobacco mosaic virus and cucumber viruses 3 and 4 do not have any common hosts, the virus preparations were similarly liquid crystalline. Antisera prepared against each gave significant serological cross sections that were evidence of the previously unsuspected relatedness of the proteins of the respective infectious entities.

In another group of investigations, various tobacco necrosis viruses were proved to be serologically unrelated. One isolate, derived from a single necrotic lesion and massive infection of leaves purified by Pirie and his collaborators, was found to consist of two fractions of noninterconvertible viruses of different sizes. Bawden showed these to be serologically unrelated. In 1960 one of these complex viral systems was shown, by Basil Kassanis and H. L. Nixon at Rothamsted, to lead to the production of two nucleoproteins of different sizes. These consisted of an infectious viral nucleoprotein of fairly large size and a smaller noninfectious nucleoprotein; the latter multiplied only in the presence of the former. These observations led to the recognition of dependent satellite viruses. The discovery of this phenomenon facilitated the dissection of numerous multicomponent virus systems in which virus particles containing different RNA components were unable to multiply unless all of the RNA components were present.

Although Bawden and Pirie had discovered RNA in the plant viruses by 1936, they did not undertake characterization of its physical properties or testing the biological activity of this material before 1956, focusing instead on the possibly crucial roles of the viral proteins. After the initial reports of the infectivity of the RNA of tobacco mosaic virus in 1956, they showed that their early methods of disrupting the virus permitted the isolation of infectious material. Despite the roles of ribonucleases in destroying the infectivity, the very low efficiency of infection by viral RNA prevented Bawden and Pirie's unequivocal acceptance of the RNA alone as the agent of infectious heredity. Bawden's last experimental papers, written with Pirie in 1972, were on the problems of isolating infectious viral nucleic acid directly from infected leaves.

Bawden wrote numerous reviews on virus disease as well as the text *Plant Viruses and Virus Diseases* (1939). He participated in many symposia, including some on virus multiplication, a subject that made important advances in the postwar period. His papers and discussions demonstrate a clear understanding of the centrality of the infected cell and the crucial use of such stripped-down biological materials in the study of the multiplication of bacterial and animal viruses. Nevertheless, work on similar systems of infected plant cells did not develop at Rothamsted during his tenure. His former collaborator Kassanis began such work there in 1972.

Despite Bawden's interest in the development of academic virology and his continuing research activities, his major concerns and achievements after World War II were related to improvements in agricultural science and technique. He supported measures and studies for the control of infectious disease in England. Bawden was very pleased with the results achieved at Rothamsted in the late 1950's and 1960's in the development of the virus-free, high-yielding potato stocks now used widely in that country.

Bawden received the Research Medal of the Royal Agricultural Society in 1955 and was president of the Society for General Microbiology in the year 1959–1960. He visited many developing countries to advise on agricultural problems and served as chairman of the Agricultural Research Council of Central Africa from 1964 to 1967. He was knighted in 1967 and in the following year served as president of the First International Congress of Plant Pathology.

Bawden believed it is not possible to distinguish sharply between basic and applied research in agriculture. In a 1972 article in *Nature*, writing as director of the station and as senior member of the Agricultural Research Council, he objected bitterly to proposals that long-term agricultural research be supported by contractual arrangements with a government ministry. Unfortunately, he died of heart disease before he could bring his ideas and personality to bear decisively on this important problem of scientific policy.

BIBLIOGRAPHY

I. Original Works. All of Bawden's publications are listed chronologically in an article by N. W. Pirie in *Biographical Memoirs of Fellows of the Royal Society*, **19** (1973). Key publications include "Experiments on the Chemical Behaviour of Potato Virus 'X,'" in *British Journal of Experimental Pathology*, **17** (1936), 19–63, with N. W. Pirie; "Liquid Crystalline Substances from Virus-Infected Plants," in *Nature*, **138** (1936), 1051–1052, with N. W. Pirie, J. D. Bernal, and I. Fankuchen; "The Isolation and Some Properties of Liquid Crystalline Substances from Solanaceous Plants Infected with Three Strains of Tobacco Mosaic Virus," in *Proceedings of the Royal Society of London*, **B 123** (1937), 274–320, with N. W. Pirie; "The Relationships Between Liquid Crystalline Preparations of Cucumber Viruses 3 and 4 and Strains of Tobacco Mosaic Virus," in *British Journal of Experimental Pathology*, **18** (1937), 275–291, with N. W. Pirie; "Crystalline Preparations of Tomato Bushy Stunt Virus," *ibid.*, **19** (1938), 251–263, with N. W. Pirie; "The Serological Reactions of Viruses Causing Tobacco Necrosis," *ibid.*, **22** (1941), 59–70; *Plant Viruses and Virus Diseases* (Leiden, 1939; 4th ed., New York, 1964); "The Separation and Properties of Tobacco Mosaic Virus in Different States of Aggregation," in *British Journal of Experimental Pathology*, **26** (1945), 294–312, with N. W. Pirie; "Studies on the Importance and Control of Potato Virus X," in *Annals of Applied Biology*, **35** (1948), 250–265, with B. Kassanis and F. M. Roberts; "Virus Multiplication Considered as a Form of Protein Synthesis," in *The Nature of Virus Multiplication: Symposium of the Society for General Microbiology*, II (Cambridge, 1953), 21–45, with N. W. Pirie; "Physiology of Virus Diseases," in *Annual Review of Plant Physiology*, **10** (1959), 239–256; "The Infectivity and Inactivation of Nucleic Acid Preparations from Tobacco Mosaic Virus," in *Journal of General Microbiology*, **21** (1959), 438–456, with N. W. Pirie; "Bringing Rothschild Down to Earth," in *Nature*, **235** (1972), 7; and "The Inhibition, Inactivation and Precipitation of Tobacco Mosaic Virus Nucleic Acid by Components of Leaf Extracts," in *Proceedings of the Royal Society of London*, **B 182** (1972), 319–329, with N. W. Pirie.

II. Secondary Literature. Geoffrey C. Ainsworth, *Introduction to the History of Plant Pathology* (Cambridge, 1981); Theodore O. Diener, *Viroids and Viroid Diseases* (New York, 1979); Joseph S. Fruton, *Molecules and Life* (New York, 1972); Max A. Lauffer, "Contributions of Early Research on Tobacco Mosaic Virus," in *Trends in Biochemical Science*, 9 (1984), 369–371; and A. P. Waterson and Lise Wilkinson, *An Introduction to the History of Virology* (New York and Cambridge, 1978).

Seymour S. Cohen

BECQUEREL, JEAN ANTOINE EDMOND MARIE (*b*. Paris, France, 5 February 1878; *d*. Sainte-Marguerite, near Pornichet, Brittany, France, 4 July 1953), *physics*.

Jean was the fourth Becquerel to hold the chair of physics at the Museum of Natural History in Paris. He was raised in the house at the Jardin des Plantes that had been the home, in turn, of his great-grandfather, Antoine-César Becquerel (1788–1878); his grandfather, Edmond (1820–1891); and his father, Henri (1852–1908). His mother, Lucie Jamin, was the daughter of J. C. Jamin, professor of physics at the Sorbonne. The distinguished line of physicists ended with Jean Becquerel, who married in 1921 but had no children.

It is hardly surprising that Becquerel's first toys included magnets and electroscopes or that he entered the École Polytechnique (1897) after graduating from the Lycée Louis-le-Grand. In 1903, the year that Henri Becquerel shared the Nobel Prize for physics with Marie and Pierre Curie, Jean Becquerel completed his education at the École des Ponts et Chaussées and became an assistant in physics at the Museum of Natural History, where he succeeded his father in 1909. During World War I, Becquerel directed construction of fortifications and trenches and carried out research on optical signals at sea. After the war he commuted to Paris from Fontainebleau, continuing to work in the physics laboratory established in 1938 in the house where Georges Cuvier had lived.

In addition to his research in physics, Becquerel became known for his lectures not only at the Museum of Natural History but also at the École Polytechnique, where he was tutor, examiner, and, from 1919 to 1922, temporary professor. With Paul Langevin, Becquerel was one of the few French physicists teaching quantum theory and Einstein's relativity theory in the 1920's. In addition, Becquerel was innovative in his teaching of thermodynamics by employing the new physics of atomism and kinetic theory. Had he lectured regularly at the École Polytechnique, the Sorbonne, or the Collège de France rather than at the Museum of Natural History, his teaching might have exerted greater influence on the new physics in France.

Becquerel's principal researches were in spectroscopy and magneto-optics, areas of traditional strength in French science. For these researches he received the Hughes Prize (1913) and the La Caze Prize (1936), and he was elected to the Academy of Sciences (1946). He continued his father's 1888 doctoral researches on the variations of the absorption spectra in crystals and on magnetic rotatory polarization, as well as his family's long-standing interest in phosphorescence. In 1908 he coauthored a paper with his father on the photoluminescence of uranyl salts at low temperatures.

Becquerel's fundamental line of research began in 1906 with an attempt to study the effect of a magnetic field on the optical properties of crystals. Building upon knowledge of the Zeeman effect for gases and vapors, Becquerel found a superficially similar effect in rare-earth crystals having a very fine spectrum of absorption bands. Analyzing doublets produced in a magnetic field, Becquerel argued for the existence of both positive and negative electrons. Although his interpretation proved erroneous, he later cited the discovery of the positron as a partial vindication of his earlier views.

Influenced by Pierre Curie's experimental work and Paul Langevin's classical electron theory relating magnetic susceptibility to temperature, Becquerel reasoned that the purely magnetic properties of rare-earth crystals could be more easily identified at very low temperatures, when intramolecular thermal motions decline. Placing crystalline films or solutions inside a vacuum tube containing liquid air, Becquerel found that some absorption bands weakened or disappeared, while others became better defined and more intense than at ordinary temperatures. His work caught the attention of Heike Kamerlingh Onnes, whose laboratory at Leiden was assuming leadership in low-temperature physics. Collaboration between Becquerel and Kamerlingh Onnes began in early 1908 and continued with other Dutch physicists; Becquerel moved his spectrograph to Leiden and spent one or two months there each year. He thus became a regular contributor to the most influential and largely English-language journal of low-temperature physics, *Communications from the Physical Laboratory at the University of Leiden*.

At Leiden, Becquerel studied spectra at the temperatures of liquid hydrogen and liquid helium, describing what he called the "fundamental spectrum" of rare-earth crystals, rubies, emeralds, and other substances in the range of 4.2°–1.3°K. In the period 1925–1930 he studied paramagnetic rotatory polarization, explaining these effects by the hypothesis of an internal electric field and interesting himself in a quantum interpretation. Becquerel's collaboration with Wander J. de Haas and Hendrik A. Kramers in the late 1920's and early 1930's was especially useful for determining properties associated with electron spin. Investigations of iron-containing crystals that exhibit magnetic rotation at very low temperatures in a magnetic field revealed properties of hysteresis that Becquerel christened "metamagnetism."

Becquerel's earliest published researches occasioned some notoriety because of their support for the existence and properties of "N rays," a spurious discovery made by René Blondlot in 1903. Becquerel and André Broca, professor of medicine at the University of Paris, investigated the effects of anesthetics in suppressing the supposed emission of N rays from living bodies. He also studied the deviability of N rays in a magnetic field, looking for the kinds of effects associated with the uranium rays discovered by his father. Becquerel later attributed his mistakes in these investigations to inexperience and to self-delusion resulting from the influence of older physicists who were family friends.

BIBLIOGRAPHY

I. ORIGINAL WORKS. Among Becquerel's writings are "Absorption de la lumière et phénomènes magnéto-optiques dans les composés de terres rares aux très basses températures," in *Het natuurkundig laboratorium der Rijksuniversiteit te Leiden in de jaren 1904–1922* (Leiden, 1922), 228–361; *Le principe de relativité et la théorie de la gravitation* (Paris, 1922); *Cours de physique à l'usage des élèves de l'enseignement supérieur et des ingénieurs*, 2 vols. (Paris, 1924–1926); and *Notice sur les travaux scientifiques de Jean Becquerel*, 2 vols. (Paris, 1934).

II. SECONDARY LITERATURE. Louis de Broglie, "Notice sur la vie et l'oeuvre de Jean Becquerel," in *Institut de France. Académie des Sciences. Notices et discours*, **5** (1963–1972), 1–20, has a portrait; Yves Le Grand, "Jean Becquerel (1878–1953)," in *Archives du Muséum national d'histoire naturelle*, 7th ser., **3** (1954–1955), v–xviii, includes a portrait and a bibliography.

MARY JO NYE

BÉKÉSY, GEORG VON (*b.* Budapest, Hungary, 3 June 1899; *d.* Honolulu, Hawaii, 13 June 1972), *biophysics, psychology*.

Georg von Békésy was born to Alexander and Paula Mazaly von Békésy. From an early age he had an avid interest in science and the fine arts. Since his father was for a time a Hungarian diplomat, the family moved frequently and Békésy received his early education in Budapest, Munich, Constantinople, and Zurich. He studied chemistry at the University of Bern from 1916 to 1920 and received his doctorate in 1923 from the University of Budapest for the development of a fast method of determining molecular weight through the diffusion coefficients of fluids. Afterward he worked primarily for the

Hungarian Post, Telephone, and Telegraph Laboratory (1923–1946), in the meantime becoming a *Privatdozent* at the University of Budapest in 1932 and professor of experimental physics in 1940. Békésy chose to work for the telephone company because it had the best-equipped laboratory in the area. His research interests shifted from physics to biophysics after he was asked to design a better telephone earphone. In 1946 he went to Sweden, where he spent a year at the Karolinska Institute (Stockholm) working with Yngve Zotterman. He did not return to Hungary, because he felt political conditions would make it difficult for him to continue his research.

Mainly due to the efforts of Stanley Smith Stevens, Békésy went to the United States to work at Harvard University. In 1949 he accepted an appointment in the psychology department as senior research fellow in psychophysics. During his nineteen years at Harvard, Békésy received many honors and awards from a wide variety of governments, universities, and professional societies, including the 1961 Nobel Prize in physiology or medicine. In 1967, having reached Harvards' mandatory retirement age, he went to the University of Hawaii as professor of sensory sciences, to head a laboratory built specifically for him with the assistance of the Hawaiian Telephone Company.

During his final years in Hawaii, as during his earlier career, Békésy was a solitary researcher totally engrossed in his scientific work and in his main avocation, art. He never married. In an era of collaborative group research, he rarely worked with colleagues on experimental projects. Indeed, his personal isolation may have contributed to the originality of his major achievements.

Békésy's most important scientific contributions, for which he received the Nobel Prize, were his observations clarifying "the physical mechanisms of stimulation in the cochlea of the inner ear," published in 1928. These observations arose from an interest of the Hungarian telephone system in the design of a better earphone. Békésy sought to determine the mechanical impedance of the ear so that an earphone could be correctly matched to it.

From his observations of the cochlea, Békésy realized that the human capacity for sharp pitch discrimination might not be adequately reflected in the vibration patterns he observed. His search for a neural explanation led to an interest in sensory inhibition and the neural sharpening processes attributed to inhibition by Ernst Mach in 1866. Mach's research and discoveries concerning the physiological effect of spatially distributed light stimuli on visual perception—an effect without physical basis—had a great influence on Békésy in his work with sensory inhibition. Though research into the transduction properties of the inner ear has shown that sharpening due to inhibitory processes may not be as important as Békésy thought, his work on inhibition was especially important. It was Békésy who first suggested to Floyd Ratliff that the inhibitory interaction discovered by Haldan Keffer Hartline (Nobel Prize recipient in physiology or medicine, 1967) in the eye of the horseshoe crab, *Limulus*, was similar to the phenomena earlier described by Mach.

Initially Békésy studied sensory inhibition with auditory, visual, and tactile stimuli. But he soon realized that similar inhibitory phenomena were present in virtually all sensory modalities. He then broadened his research interests to study other similarities among other senses, especially taste, thermal sensation, temporal discrimination, and color vision.

An appreciation of Békésy's fundamental discovery is provided by a brief review of the study of the vibration patterns on the basilar membrane in the inner ear. In the 1920's, when his attention was directed toward a study of the mechanical properties of the ear, it was generally accepted that the cochlea was the organ where sound vibrations were transduced into nerve impulses. Because the cochlea is completely encased in temporal bone, dissection of a normally hydrated, intact specimen had not been possible. Accordingly, there was controversy concerning the nature of the transduction. Hermann von Helmholtz, for example, had suggested in 1863 that structures in the cochlea were arranged like the tuned wires of a harp and that each element resonated at different frequencies, analyzing a tone into its harmonic components. Albert von Kolliker and Karl E. Hasse later correctly directed attention to the basilar membrane, a structure suited for this analysis because it varies in width along the length of the cochlea.

Békésy was probably the first to realize that the modes of vibrations ascribed to the basilar membrane are all members of a family of vibrations that can be produced by continuously varying the thickness and lateral stress on the membrane. Indeed, simply poking the membrane to observe the shape of the resulting depression was sufficient to determine the mode of vibration.

Using superb and original microdissection techniques, Békésy was able to grind away the bone encasing the cochlea and make this key observation. He overcame the problem of dehydration by conducting the operation under a stream of water, and overcame the problem of the membrane's transparency by sprinkling fine silver crystals on it. The depression he saw corresponded to that for traveling waves, a mode of vibration different from Helmholtz's original suggestion. He was able to verify their presence by observing different sections of the vibrating membrane under stroboscopic light. The peak of the envelope of these vibrations was seen to move from the apex toward the base of the cochlea as the frequency of the driving tone increased, thus providing one physical basis for the transduction of pitch, especially for tones above 200 herz.

At Harvard, Békésy was able to build a variety of large mechanical models illustrating this pattern of vibration. The most famous of these is a slotted brass tube filled with water and covered with a membrane varying in thickness. When the water in the tube is vibrated at different frequencies, the movement of the point of maximum vibration can readily be felt along the slot. This ear model exemplifies Békésy's basic experimental approach of direct observation of phenomena made visible by intricate laboratory techniques or calibrated mechanical models.

Though the mechanical nature of his major scientific contribution and his training in experimental physics would seem to qualify him as a biophysicist, Békésy is probably more accurately described as a psychologist. In some respects this description may seem inappropriate. He taught no formal courses in the field. He had no graduate students. And his training in physical science led him to disdain the field's rhetorical aspects.

Yet in more important respects he was very much a psychologist. Most of his work was motivated by psychological problems, such as the best design for a telephone earphone or an audiometer; the basic similarities among all the senses; and the hearing of complex sounds. Underlying all these matters was the general psychophysical question with roots in psychology going back to Gustav T. Fechner and in physical science to Isaac Newton: How do our subjective sensory experiences of touch, sound, pain, beauty, and so on arise from the interaction of physical objects and their associated energy patterns with the sensitive receptors of our body?

Békésy sought the answer to this question in art as well as in science; for him the two were never far apart. It was the beauty of the inner ear seen through a dissection microscope that inspired him to pursue an understanding of its transduction mechanism, just as it was the beauty of a work of art that persuaded him to add it to his collection.

BIBLIOGRAPHY

I. Original Works. Békésy's most important observations on the vibration patterns of the basilar membrane were first reported in "Zur Theorie des Hörens, die Schwingungsform der Basilarmembran," in *Physikalische Zeitschrift*, **29** (1928), 793–810. Summaries of his work and collections of his most important papers are in *Experiments in Hearing*, E. G. Wever, ed. and trans. (New York, 1960); and *Sensory Inhibition* (Princeton, 1967). An autobiographical work is "Some Biophysical Experiments from Fifty Years Ago," in *Annual Review of Physiology*, **36** (1974), 1–16. See also *Nobel Lectures, Physiology or Medicine 1942–1962* (Amsterdam, 1964), 719–748.

Békésy's experimental apparatus is in a small museum at the Békésy Laboratory of Neurobiology, University of Hawaii. Most of his papers and other materials are at the Library of Congress. His art collection is at the Nobel Foundation in Stockholm.

II. Secondary Literature. C. G. Bernhard, "Georg von Békésy and the Karolinska Institute," in *Hearing Research*, **22** (1986), 13–17; Floyd Ratliff, "Georg von Békésy," in *Biographical Memoirs. National Academy of Sciences*, **48** (1976), 25–49, with complete list of Békésy's writings; Jürgen Torndorf, "Georg von Békésy and His Work," in *Hearing Research*, **22** (1986), 3–10; and Jan Wirgin, eds., *The Georg von Békésy Collection* (Malmö, Sweden, 1974), a catalog of his art collection that includes a biographical sketch.

Stephen R. Ellis

BĚLAŘ, KARL (*b*. Vienna, Austria, 14 October 1895; *d*. Pasadena, California, 24 May 1931), *cytology, protozoology*.

Bělař was the son of a jurist. While still a student he developed a strong interest in biology and carried out observations in his brother-in-law's private laboratory. After graduation from gymnasium he studied zoology at the University of Vienna, specializing in protozoology. His studies were interrupted by service in the Austrian army during World War I. Bělař resumed his studies at Vienna and obtained a Ph.D. in zoology in 1919. He then moved to Berlin, where he became an assistant in Max Hartmann's department of protozoology at the Kaiser-Wilhelm-Institute (now the Max-Planck-Institute) at Berlin-Dahlem. Subsequently he was promoted to scientific member of the Institute. He also taught a course in evolution at the University of Berlin, where he became professor in 1930. In 1929 he was invited by Thomas H. Morgan to serve as a visiting professor in the newly founded department at the California Institute of Technology. He was killed in an automobile accident.

A year after graduation from gymnasium, Bělař published his first scientific paper, the description of a new species of protozoan that he called, in honor of his teacher, *Prowazekia josephi*, and its nuclear division. For many years after, his work was concerned with Protozoa, the central aspect being the nature of mitosis in that subkingdom. At the time he started his work, most zoologists believed that they could distinguish two types of cell division: "direct" or amitotic, in which the nucleus simply divided in two, and "indirect," the mitotic division in which the chromatin appears organized into chromosomes that divide individually and are distributed equally to the daughter cells. It was assumed that direct division was more primitive and that most divisions in Protozoa, Algae, and Fungi were direct, while the more efficient mitotic mode evolved in higher animals and plants.

Through a careful study of cell division in living and stained preparations of several unicellular organisms, Bělař demonstrated that all the divisions are actually mitotic. In this work he defined mitosis as involving the appearance of chromosomes and of an achromatic apparatus that may consist of spindle fibers, a centrosome, asters, and centrioles. He showed that most of the nuclear divisions previously described as amitotic are actually mitotic but are hard to recognize as such because the chromosomes are small, numerous, and frequently hard to identify; the achromatic apparatus is highly variable; and the nuclear membrane does not break down. Bělař's work on mitosis in unicellular organisms was summarized in his classical monograph *Der Formwechsel der Protistenkerne* (1926). In this work he not only reviews his own work but also great amounts of published material, reinterpreting many aspects and testing them on his own preparations.

As a result Bělař makes it convincingly clear that the nuclear divisions of almost all Protozoans are mitoses. Only the division of the macronucleus of ciliates is admitted as being a direct division, and Bělař hints that he did not regard the ciliates as closely related to the other Protozoans. His book is not simply a compilation of published material; instead, the material is used to build up a consistent picture of the changes the nucleus undergoes during the cell cycle. His view that, in spite of many variations between different species and orders, the fundamental process is always the same is put forth with great ingenuity and critical understanding, and is completely convincing. The book has stood the test of time, and Bělař's conclusions are still generally accepted.

His work on protozoan mitosis includes some important studies on the physiology of asexual reproduction and sexual processes in the heliozoan *Actinophrys sol*. These studies were suggested by Hartmann, the director of the laboratory at the Kaiser-Wilhelm-Institute, whose work centered on the physiology of sexual processes. In *Actinophrys*, Bělař studied both the cell division and the sexual process, in this case a type of autogamy. An encysted cell divides and the two daughter cells undergo a typical meiosis. The meiosis is identical with that of an animal ovum, resulting in a gamete and polar bodies. The chromosomes, in this case, are very distinct though large in number. The two daughter cells (gametes) then fuse, reconstituting the diploid cell. In further experimentation Bělař discovered that the sexual process can be induced at any time by environmental conditions, such as starvation. Asexual reproduction by cell division, on the other hand, can go on indefinitely—in Bělař's case for almost three years (at the rate of one to three divisions per day). Bělař pointed out that the problem of aging and death does not apply to *Actinophrys* because its life cycle is completely determined by environmental variables. Contrary to previous assumptions, based primarily on investigation of ciliates, continued asexual reproduction does not lead to aging.

Bělař next studied the mechanism of mitosis in the cells of higher plants and animals, particularly the formation and activity of the achromatic apparatus and its function in the separation of the chromosomes. His work was based on live observation of dividing cells, primarily spermatocytes of a grasshopper *(Chorthippus)* and mitosis in the hair cells of the anthers of the spiderwort, *Tradescantia*. In the spindle he distinguishes chromosomal fibers that are attached at one end to the centromere of the chromosomes and at the other end to the spindle poles. These fibers shorten considerably during anaphase and thus pull the chromosomes toward the spindle poles. The other component is the middle region of the spindle, which connects the two poles and in anaphase lengthens and pushes the spindle poles further apart, thus aiding the further separation of the chromosomes attached to the poles. The action of the middle region was suggested by experiments in which dividing cells were put into hypertonic solutions. In this treatment the cytoplasm shrinks due to loss of water, finally forming a thin layer around the spindle. The spindle, on the other hand, lengthens considerably, particularly in anaphase cells, and frequently gets bent on itself because it meets the resistance of the reduced cytoplasm.

Bělař's findings and conclusions formed the basis of later discussions of the nature of mitosis. The reality of the spindle fibers and the longitudinal structure of the central spindle was in doubt for twenty years because, though quite conspicuous in fixed preparations, they cannot be seen in living cells. Bělař supported their existence by indirect observations and conclusions. Only in 1953 was Inoue able to demonstrate their existence in living cells by means of polarization microscopy and to show the fibrous nature of the middle region. Since that time the picture has become more concrete by the employment of the electron microscope, which has shown both chromosomal fibers and central spindle fibers, and by biochemical studies of the behavior of tubulins, the main protein component of the spindle. The fibers of the middle region, which extend from one pole to the other, lengthen during anaphase, as Bělař envisioned; but it is now generally agreed that sliding of fibers against each other is probably the source of spindle elongation.

Bělař worked very hard, efficiently, and fast, and thus was able to accomplish much in his short life. He did not have graduate students or collaborators; all his publications bear only his name. He was very concerned with techniques and was admired for his ability to culture Protozoans. Bělař was an excellent microscopist, using both stained and live preparations, and a superior draftsman; he did not only the drawings for his own papers but many drawings in the publications of his friends and colleagues as well. He was a pioneer in the use of photomicrography, particularly of living cells. He postulated that ideally every cell discussed in a paper should be shown both as a photomicrograph and as a drawing, since the photomicrograph is more objective while the drawing includes interpretation by the author.

Bělař had a very sharp and critical mind, and was famous and feared for his severe criticisms. He had little patience with inexact technique and sloppy thinking, and had the courage to attack even the most influential biologists of his time—for instance, Rudolf Flick, the head of the Anatomy Department in Berlin; the protozoologist Franz Doflein; the zoologist Alfred Kühn; and even his supervisor Hartmann, who complains about Bělař's sharply critical remarks in the foreword he wrote for Bělař's monograph on protozoan nuclei. But Bělař used the same ruthless criticism against his own work and his earlier publications.

Besides his immersion in his work, Bělař shared aesthetic interests with in wife, Gertrud Bengelsdorff. He had great knowledge of the arts, and drew, sketched, and painted with watercolors. He was much attracted in his last two years by the charm and grandeur of the American landscape, particularly the desert and the mountains, which he spent much time sketching and painting.

Bělař's significance in the history of cytology consists in the fact that he mercilessly destroyed many assumptions that were based on poor observations and wrong interpretations, and replaced them with carefully considered and solidly based ideas. Most of these are still valid, though superior methods have made them more concrete.

BIBLIOGRAPHY

Bělař's writings include "Untersuchungen an *Actinophrys sol Ehrenberg* I," in *Archiv für Protistenkunde*, **46** (1923), 1–96; "Untersuchungen an *Actinophrys sol Ehrenberg* II," *ibid.*, **48** (1924), 371–434; *Der Formwechsel der Protistenkerne*, Ergebnisse und Fortschritte der Zoologie, no. 6 (Jena, 1926); "Die Technik der Zytologie," in Tibor Péterfi, ed., *Methodik der wissenschaftliche Biologie*, II (Berlin, 1928); *"Untersuchungen der Protozoen," ibid.*, I (Berlin, 1928); "Die zytologischen Grundlagen der Vererbung," in E. Baur and M. Hartmann, eds., *Handbuch der Vererbungswissenschaften* (Berlin, 1928); "Beiträge zur Kausalanalyse der Mitose II. Untersuchungen an den Spermatocyten von *Chorthippus (Stenobothrus) lineatus Pans*," in *Wilhelm Roux's Archiv für Entwicklungsmechanik der Organismen*, **118** (1929), 359–484; "Mitotic Spindle Analyzed by Dr. Bělař in Lecture," in *The Collecting Net*, **4**, no. 8 (1929), 8; and "Zur Teilungsautonomie der Chromosomen," W. Huth, ed., in *Zeitschrift für Zellforschung und Mikroskopische Anatomie*, **17** (1933), 51–66, a posthumous summary of Bělař's work at Pasadena on the eggs of the marine worm *Urechis caupo*. An obituary is Curt Stern, "Karl Bělař zum Gedächtnis," in *Naturwissenschaften*, **19** (13 November 1931), 921–923.

ERNST CASPARI

BELOV, NIKOLAI VASIL'EVICH (*b.* Janów, Poland, 14 December 1891; *d.* Moscow, U.S.S.R., 6 March 1982), *crystallography, crystal chemistry, geochemistry*.

Belov was the son of Vasili Vasil'evich Belov, a district physician, and of Olga Andreevna Belova. After graduating from the Warsaw gymnasium with a gold medal in 1910, he entered the metallurgical department of the Petrograd Polytechnical Institute, from which he received a degree in electrochemistry in 1921. While a student he married Aleksandra Grigorievna; they had two daughters.

After graduation Belov went to work in the central chemical laboratory of the Leningrad Tanning Trust. While there, he published short notes on the latest scientific achievements in *Priroda*, a popular magazine. The mineralogist and geochemist A. E. Fersman took notice of Belov and encouraged him to do research on minerals from Khibiny, nepheline and apatite. Belov subsequently produced works on the use of nepheline in the tanning, textile, papermaking, and woodworking industries.

In 1933 Belov was named a senior scientific worker in the Lomonosov Institute of Geochemistry, Mineralogy, and Crystallography of the Academy of Sciences of the U.S.S.R. He translated Odd Hassel's *Kristallchemie* into Russian (1936); his commentary doubled the length of the book, and the number of illustrations was increased from six to sixty. For several years this book served as the main manual for Soviet crystallographers and mineralogists. He subsequently translated the fundamental works on the structure of silicates by F. K. L. Machatschki, W. L. Bragg, Ernst Schiebold, and W. H. Taylor.

In 1936 Belov moved from Leningrad to Moscow in connection with the transfer of the Academy of Sciences to the capital of the Soviet Union. Here he joined the crystallography department founded by A. V. Shubnikov. He became absorbed in the problems of crystal chemistry and structural crystallography. During the following years Belov created the generalized crystal chemical picture of the structure of inorganic compounds on the basis of the theory of the closest packing of spheres. Earlier, Linus Pauling had shown that besides the two classical closest spherical packings—cubic and hexagonal—there is an infinite multitude of such packings. Belov demonstrated in a simple and clear way that this multitude obeys only eight laws of symmetry: eight of the space groups of E. S. Fedorov and Arthur Schoenflies (the total number of such groups is 230). The theory of close packings classified the symmetry laws and showed how to use them to interpret crystal structures.

Belov presented his findings in his monograph *Struktura ionnykh kristallov i metallicheskikh faz* (The structure of ionic crystals and metallic phases, 1947). This study, widely known as the "blue book" (for the color of its cover) became a sort of bible for research workers in structural crystallography and crystal chemistry.

During World War II, Belov remained in Moscow although the Institute of Crystallography had been evacuated to the Urals. During this time he completed his dissertation and received his doctorate in 1943. In 1946 Belov was elected a corresponding member of the Academy of Sciences of the U.S.S.R. and became a professor in the department of crystallography of the Physical-Mathematical Faculty at the University of Gorky. Although he headed this department as a commuter from Moscow, he was able to train a great number of specialists in crystallography. In *Strukturnaia kristallografiia* (1951) he presented an extremely simple form of the derivation of the 230 space groups. Belov encouraged his pupils to determine the structures of numerous minerals. In 1952 he was awarded the State Prize, First Class, in physics for his works on atomic structures of minerals.

In 1953 Belov became an academician in the department of geological and geographical sciences. At that time he was mainly interested in silicates. By then W. L. Bragg and his school had created what seemed to be an exhaustive theory of silicate structure. Its basic unit was a silicon-oxygen tetrahedron, SiO_4, consisting of a silicon atom surrounded by four oxygen atoms. Joined by common apices, such tetrahedrons form chains, ribbons, sheets, and three-dimensional frameworks. Comparatively small cations lodge between them.

From 1953 on, Belov and his pupils determined the structure of a number of silicates with large cations, such as calcium and sodium. Their basic building unit proved to be a pair of silicon-oxygen tetrahedrons with one common apex, Si_2O_7. The main role in the structures of corresponding silicates is played by combinations of large cations to which inert silicon-oxygen radicals are attached. In the Soviet Union two chapters stand out in the crystal chemistry of silicates. The first was written by W. L. Bragg. The second was created by Belov. Proceeding from the principles he had formulated, Belov and his pupils determined more than 200 different structures of minerals.

Remarkable results he had obtained in the crystal chemistry of silicates were awarded the Lenin Prize (1974) and were summarized in Belov's fundamental monograph *Ocherki po strukturnoi mineralogii* (Essays on structural mineralogy, 1976). Belov did not, however, confine himself to structural mineralogy. His works devoted to geochemical processes of mineral formation, the separation of minerals from magma, and problems of isomorphism are widely known. His structural conclusions were widely employed in industry. Belov also obtained a number of fundamental results in the development of the modern theory of symmetry (the derivation of 1,651 space groups of black-white symmetry [antisymmetry], the development of color symmetry).

In 1953 Belov became a professor, in 1961 head of the department of crystallography and crystal chemistry of the Geological Faculty at Moscow University. For many years he was a member of the International Union of Crystallographers (vice president 1957–1963, president 1966–1969). The mineralogical societies of the United States, Great Britain, and France, and the Geological Society of the German Democratic Republic, elected Belov honorary member. He was a foreign member of the Polish Academy of Sciences and held an honorary doctorate from Wroclaw University. The Soviet Union gave him the title Hero of Socialist Labor (1969), and the Academy of Sciences of the U.S.S.R. presented him with its highest award, the Lomonosov Gold Medal (1966).

Among his colleagues and friends Belov was noted for his exceptional kindness, sympathy, and unassuming manner. He possessed great erudition in history, art, and the humanities as well as in exact sciences. A phenomenal memory allowed Belov to learn foreign languages with ease. Humor and wit enlivened his scientific reports and public addresses.

BIBLIOGRAPHY

I. Original Works. Belov published more than 500 works. Among the most important are *Struktura ionnykh kristallov i metallicheskikh faz* (The structure of ionic crystals and metallic phases; Moscow, 1947); *Strukturnaia kristallografiia* (Structural crystallography; Moscow, 1951); and *Ocherki po strukturnoi mineralogii* (Essays on structural mineralogy; Moscow, 1976). His personal reminiscences are in Peter P. Ewald, ed., *Fifty Years of X-ray Diffraction* (Utrecht, 1962), 520–521.

II. Secondary Literature. "Nikolai Vasil'evich Belov (1891–1982)," in *Bibliografiia uchenykh SSSR* (Bibliographical index of the U.S.S.R.; Moscow, 1987), 1–10; "Nikolai Vasil'evich Belov (on His Seventieth Birthday)," in Belov's *Crystal Chemistry of Large-Cation Silicates* (New York, 1963); and B. K. Vainshtein and V. I. Simonov, "Nikolay Vasilyevich Belov, 14 December 1891–6 March 1982," in *Acta crystallographica*, **A38** (1982), 561–562.

I. I. Shafranovsky

BELOZERSKII, ANDREI NIKOLAEVICH (*b.* Tashkent, Russia, 29 August 1905; *d.* Moscow, U.S.S.R., 31 December 1972), *biochemistry*.

Belozerskii was born into a family of lawyers. His parents died when he was young, and he was raised first by relatives and then in an orphanage at Gatchina, where he finished high school in 1921. From 1922 to 1927 Belozerskii studied at the Faculty of Physics and Mathematics of Central Asian University (now Tashkent University). While there he became interested in the physiology and biochemistry of plants, and after graduation he undertook advanced studies in biochemistry.

During his graduate studies Belozerskii traveled to Moscow for practical training. There he met A. R. Kiesel, who invited him to work at the biochemical laboratory of the Polytechnical Museum in Moscow upon completing his graduate work. Belozerskii went to work in the laboratory in 1930, and by the end of the year he and Kiesel had organized the department of plant biochemistry at Moscow State University. He became an assistant there, and two years later he was named an associate professor. In 1943 he defended his dissertation, "Nukleoproteiny inukleinovye kisloty rastenii" (Nucleoproteins and polynucleic acids of plants). In the same year Belozerskii was appointed professor at the Faculty of Biology of Moscow University. Between 1946 and 1960, in addition to his work at the department of plant biochemistry, he headed the laboratory (which he had organized) of microorganic biochemistry at the A. N. Bach Institute of Biochemistry of the Academy of Sciences of the U.S.S.R. In 1960 he became chairman of the department of plant biochemistry, replacing Aleksandr I. Oparin.

In 1962 Belozerskii was elected a full member of the Academy of Sciences. The following year he organized and headed the department of virology at Moscow University, and in 1965 he established at the university the Interdepartmental Laboratory of Bioorganic Chemistry and Molecular Biology (later named for him), which, along with the M. M. Shemyakin Institute of Bioorganic Chemistry and the Institute of Molecular Biology of the Academy of Sciences, forms the major center for the development of physicochemical biology in the Soviet Union. In 1971 Belozerskii was elected vice-president of the Academy of Sciences of the U.S.S.R. and led its research work in chemistry and biology.

Belozerskii's major research interests focused on the chemistry and biology of proteins, primarily of nucleic acids. It was through the study of these substances that Belozerskii hoped to decipher the chemical structure of the genetic mechanism. Therefore, his first works were aimed at obtaining the chemical characteristics of plant proteins of definite types and patterns. By the 1930's he had predicted that, in addition to proteins, nucleic acids could be substances that possess both chemical and biological characteristics simultaneously.

The study of plant nucleic acids led Belozerskii to one of his most important discoveries: in 1936 he uncovered DNA in plants through separating thymine from the nucleic acids of the horse chestnut. Until this time DNA was considered to be a specific form of animal nucleic acid and RNA of plant nucleic acid. Continuing this research, by the end of the 1930's Belozerskii had shown the universal occurrence of DNA and RNA in higher and lower plants and in microorganisms. He traced the regularities in the changes of DNA and RNA during the ontogenesis of plants and proved the species specificity of DNA in microorganisms. This important research formed the basis for the principles of the systematics of organisms.

After Erwin Chargaff developed more sophisticated methods that permitted a precise analysis of DNA and RNA in the cell (early 1950's), the systematic study of the nucleotide composition of DNA and RNA in various groups of microorganisms began. The nucleotide composition of bacteria was determined to be directly connected to their evolutionary systematics. It was also shown that despite enormous variations in the composition of DNA, the nucleotide composition of RNA in bacteria changed fairly little from type to type. In 1957 Belozerskii and A. S. Spirin discovered the fraction of RNA that corresponded to the composition of cellular DNA. They proposed that it was this fraction that carried information from DNA to the proteins in the process of protein biosynthesis. This was viewed as the beginning of the complete scheme of the biosynthesis of proteins and the clarification of the "central dogma" of molecular biology. Consequently the existence of messenger RNA (m-RNA) was proved experimentally. (It was isolated in 1961.)

The research on the nucleotide composition of various species of organisms led Belozerskii to posit the possibility of using data on the chemical composition of the genome to establish a natural classification of living organisms. He and his students developed a new branch of biology, called genosystematics or molecular phenogenetics.

Belozerskii was elected a member of the Leopoldina Academy (German Democratic Republic). In 1948 he was awarded the Lomonosov Prize and in 1969, the title Hero of Socialist Labor, the highest honor of the Soviet Union.

BIBLIOGRAPHY

I. Original Works. Belozerskii's writings include "On the Nucleoproteins and Polynucleotides of a Certain Bacteria," in *Nucleic Acids and Nucleoproteins*, vol. 12 of Cold Spring Harbor Symposia on Quantitative Biology (Cold Spring Harbor, N.Y., 1947), 1–6; *Nukleoproteidy i nukleinovye kisloty rastenii i ikh biologicheskoe znachenie* (Nucleoproteins and nucleic acids of plants and their biological significance; Moscow, 1959); Instituts Solvay, Institut International de Chimie, *Nucleoproteins* (New York, 1960); *Prakticheskoe rukovodstvo po biokhimii rastenii* (Practical handbook on the biochemistry of plants; Moscow, 1961), written with N. I. Proskuriakov; and *Biokhimiia nukleinovykh kislot i nukleoproteidov* (Biochemistry of nucleic acids and nucleoproteins; Moscow, 1976), which consists of selected works.

II. Secondary Literature. Short articles are in *Modern Men of Science*, I; and *Modern Scientists and Engineers*, I (1980), 73–75. See also *Andrei Nikolaevich Belozersky*, Materialy k biobibliografii uchenykh SSSR (Materials toward the biobibliography of scientists of the U.S.S.R.), ser. biokhimii, no. 7 (Moscow, 1968)

Aleksei Nikolaevich Shamin

BENTZ, ALFRED THEODOR (*b.* Heidenheim, Germany, 26 July 1897; *d.* Hanover, Germany, 11 June 1964), *petroleum geology, paleontology*.

Bentz, son of Karl Alfons Bentz, a confectioner, and Pauline Adelheid Keller Bentz, was for decades the most important German petroleum geologist; he greatly influenced the exploitation of oil and natural gas in Germany and in other countries. When he was young, the great variety of fossils of the Swabian Juras aroused his interest in geology and paleontology, subjects that he studied, after his military service in World War I, at the universities of Tübingen and Munich between 1918 and 1922. Among his teachers were Edwin Hennig at Tübingen and Ferdinand Broili at Munich. His doctoral dissertation dealt with the stratigraphy of the Middle Jurassic and the tectonics at the western rim of the Nördlingen Ries at a time when its origin through the impact of a meteor was not yet known. (Bentz always remained skeptical of the impact theory.)

After serving as an assistant to Hennig at Tübingen (1922–1923) and then joining the Prussian State Geological Institute at Berlin in 1923, Bentz continued his biostratigraphic studies of Bajocian ammonites, dealing intensively with paleontological taxonomy and nomenclature.

In 1925 Bentz began the geological mapping of northwestern Germany and, at the same time, became familiar with petroleum geology. This was a twofold task that was to prove decisive for his further work. His careful study of the stratigraphy and tectonics of the German-Dutch border region—where, between 1904 and 1907, there had been unsuccessful attempts to drill for oil—contributed much to the later exploitation of hydrocarbons.

Bentz's geological and paleontological experience provided him with the thorough knowledge that enabled him to accept a position in the Section of Petroleum Exploration at the Prussian State Geological Institute in 1929. Numerous drillings during the 1920's, as well as the first productive oil fields near Hanover, offered Bentz the opportunity to investigate the complex problems connected with the origin and the deposits of petroleum. During these years the rotary boring method, developed in the United States, and applied seismics—for which Ludger Mintrop, the German originator, had found an interest only in the United States—became known in Germany.

In 1932 Bentz undertook the first comprehensive survey of the geological conditions leading to the formation of German petroleum deposits and areas that seemed likely to yield petroleum. A 1934 visit to the oil fields of the United States strengthened his decision to begin a systematic search for petroleum in Germany. His work in 1934 led to the establishment of the Institute for Petroleum Geology, of which he was a director. In 1940 the German government incorporated this institute—as a section for oil exploration—into the State Geological Institute. Owing to his organizational skills, Bentz achieved a nonpolitical collaboration between private oil firms and the state institute. Even during World War II he was successful in giving research priority over exploitation.

During the 1930's it was found that the most favorable preconditions for the formation, as well as discovery and exploitation, of petroleum deposits existed in the northwestern German basin (56,000 square kilometers, or 21,616 square miles) with its 8,000 meters (about 26,000 feet) of marine sediments composed of clay and sand from the Rhaetic to the Lower Cretaceous. Here, sandstone proved to be potentially oil bearing. Oil was found especially in the flanks and the apex of numerous salt domes, the rise of which, beginning in the Jurassic, intensified during the neo-Cimmerian orogenesis. With increased drilling, oil was also found in anticlines that had formed in a purely tectonic way, and in other tectonic structures, such as petroleum traps at faults and edges of sedimentary troughs.

Oil deposits were assumed to have originated at the place of their occurrence. In 1913 Johannes Stoller supposed that in the neighborhood of the salt domes, water of high salinity had caused masses of plankton to die out, and from those plankton oil had formed. The tectonic conditions and the preferential binding of oil to sandstone, however, testified to the fact that oil had moved secondarily to the present deposits. Bentz insisted that one had to differentiate between primary (mother) rocks and secondary (storage) rocks. An oil deposit discovered in the Permian salt of Thuringia (1930) and gas deposits found in 1934 in the Upper Permian of the Ems region pointed to the Upper Permian as the most important mother rocks. This, however, conflicted with the fact that no oil had been found in the upper Bunter strata, through which the oil rising from the Upper Permian would have had to pass.

For a few years after World War II German petroleum research was directed by Bentz—now head of the State Geological Institute of Lower Saxony. He began a reciprocal relationship with geologists outside Germany. He wrote numerous reports, publishing in German and foreign periodicals and in collected works. In 1953 Bentz was the first to draw attention to a potential deposit of hydrocarbons in the subsoil of the North Sea off Germany. In 1960 he announced at the World Power Conference in Madrid that the German output met 30 percent of Germany's domestic requirements.

After the war, German petroleum research results were increased under the influence of Bentz in five ways: (1) by more detailed information about the stratigraphy and structure of the northwest German basin with its submarine troughs and sills; (2) by the discovery that the origin of petroleum is not confined to the Upper Permian, but can also be in numerous horizons of the Jurassic and the Lower Cretaceous sediments; (3) by the recognition that petroleum occurs, owing to its buoyancy, mainly in secondary deposits (edges of troughs, salt domes, anticlines); (4) by the awareness that apart from tectonic and salt-tectonic oil traps, there are other traps conditioned stratigraphically and by the horizontally changing facies (stratigraphic and depositional oil traps, requiring a paleontologically exact stratigraphy); and (5) by the knowledge that the oil deposits of the Upper Rhine rift valley and of the molasse of the Alpine rim exploited after 1945 originated in early Tertiary rocks.

Successful petroleum exploration depends on the collaboration of the natural sciences, technology, the initiative of the entrepreneur, and the international exchange of methods, techniques, and expertise. Bentz was skillful in harmonizing and coordinating these factors.

During the last years of his life, Bentz was among the experts concerned with future energy supplies. Indeed, he was most optimistic about the availability of petroleum and natural gas to future generations.

Numerous institutions in many countries sought his advice about prospecting for hydrocarbons, advice that he gave on the spot. Bentz was the editor of *Lehrbuch der angewandten Geologie* (1961–1969) and *Beiträge zur regionalen Geologie der Erde* (published since 1961). In *Geologisches Jahrbuch* and in the accompanying supplements published by his institute, he attached great value to a well-balanced representation of all disciplines, including paleontology. He also devoted numerous talks and papers to the dissemination of geological knowledge.

BIBLIOGRAPHY

I. ORIGINAL WORKS. Bentz's writings include "Über Dogger und Tektonik der Bopfinger Gegend," in *Jahresbericht und Mitteilungen des oberrheinischen geologischen Vereins*, n.s. **13** (1924), 1–45, his dissertation; "Die Garantianschichten von Norddeutschland mit besonderer Berücksichtigung des Brauneisenoolith-Horizontes von Harzburg," in *Jahrbuch der preussischen geologischen Landesanstalt*, **45** (1924), 119–193; "Die Entstehung der 'Bunten Breccie,' das Zentralproblem im Nördlinger Ries und Steinheimer Becken," in *Zentralblatt für Mineralogie, Geologie, und Paläontologie*, sec. B, *Geologie und Paläontologie* (1925), 97–104, 141–145; "Über das Mesozoikum und den Gebirgsbau im preussisch-holländischen Grenzgebiet," in *Zeitschrift der deutschen geologischen Gesellschaft*, **78** (1926), 381–500; "Zur Entstehung des Erdöls in Nordwestdeutschland," in *Zeitschrift des Internationalen Bohrtechniker-Verbands*, **16** (1927), 307–308; "Tektonische Untersuchungen in hannöverschen Erdölgebieten," in *Zeitschrift der Deutschen geologischen Gesellschaft*, **79** (1927), 241–254; "Salzstöcke und Erdöllagerstätten," in *Petroleum*, **24** (1928), 1157–1164; "Über Strenoceraten und Garantianen, insbesondere aus dem Mittleren Dogger von Bielefeld," in *Jahrbuch der preussischen geologischen Landesanstalt*, **49** (1928), 138–206; "Geologische Voraussetzungen für das Auftreten von Erdöllagerstätten in Deutschland," in *Zeitschrift der Deutschen geologischen Gesellschaft*, **84** (1932), 369–389.

"Die verschiedenen Erdölhorizonte Norddeutschlands, deren primäre oder sekundäre Entstehung," in *Jahrbuch des deutschen nationalen Komittees fur die internationalen Bohrkongresse* (Berlin, 1932), 21–88; "Geologische Studienreise in nordamerikanischen Erdölfeldern," in *Petroleum*, **30** (1934), 1–43; "The History of the German Geological Survey," in *Geological Magazine*, **84** (1947), 169–177; "Die Entwicklung der Erdölgeologie," in *Zeitschrift der deutschen geologischen Gesellschaft*, **100** (1950), 188–197; "Die Entwicklung der deutschen Erdölproduktion—Rückblick und Ausblick," in *Erdöl und Kohle*, **6** (1953), 823–827; "Über die Herkunft des Erdöls in Deutschland," in *Roemeriana*, **1** (1954), 361–384; "Geophysik und Erdölerschliessung in Deutschland," in *Erdöl und Kohle*, **9** (1956), 278–280; "Relations Between Oil Fields and Sedimentary Troughs in Northwest German Basin," in Lewis G. Weeks, ed., *Habitat of Oil: A Symposium* (New York, 1958), 1054–1066; "Results and Prospects of Oil and Natural-Gas Research in Western Germany," in *World Power Conference, Sectional Meeting, Madrid* (Madrid, 1960), 1–21, written with H. Boigk; "Bildung und Erschliessung der Energiequellen der Welt aus geologischer Sicht," in *Brennstoff-Wärme-Kraft: Zeitschrift für Energietechnik und Energiewirtschaft*, **14** (1962), 569–579; and, as editor, *Lehrbuch der angewandten Geologie*, 2 vols. in 3 (Stuttgart, 1961–1969), the two parts of vol. 2 completed by H. J. Martini; and *Beiträge zur regionalen Geologie der Erde*, vols. 1–4.

II. SECONDARY LITERATURE. On Bentz's life and work, see Wolfgang Schott, "Alfred Bentz als Wissenschaftler," in *Geologisches Jahrbuch*, **83** (1965), xxix–xlviii, with an extensive bibliography.

HELMUT HÖLDER

BERGMAN, STEFAN (*b.* Czestochowa, Poland, 5 May 1895; *d.* Palo Alto, California, 6 June 1977), *mathematics*.

Stefan Bergman was the son of the Jewish merchant Bronislaw Bergman and his wife, Tekla. He graduated from the local gymnasium in 1913 and studied in the schools of engineering in Breslau and Vienna, receiving a degree as *Diplomingenieur* in 1920.

In 1921 Bergman entered the Institute for Applied Mathematics at the University of Berlin. Richard von Mises, the founder and director of the institute, was a leading theoretician in fluid dynamics and probability, and influenced Bergman during his whole career. Bergman worked on various problems of potential theory as applied to electrical engineering, elasticity, and fluid flow. To obtain a large number of harmonic functions in space, he applied and generalized the Whittaker method to create such functions by means of integrals over analytic functions. Using algebraic-logarithmic analytic functions as generators in the integral, he created harmonic functions that are multivalued in space and have closed branch lines. This led him further to a general theory of integral operators that map arbitrary analytic functions into solutions of various partial differential equations. He devoted many years of work to this topic, producing a monograph in 1969.

The decisive influence on Bergman's scientific development came from Erhard Schmidt, who, with David Hilbert, had developed an elegant and seminal approach to the theory of integral equations with symmetric kernel. The eigenfunctions $\phi_\nu(x)$ of such equations over an interval $\langle a, b\rangle$ form an orthonormal system, that is, one has

$$\int_a^b \phi_\nu(x)\phi_\mu(x)dx = \delta_{\nu\mu} = \begin{cases} 0 \text{ if } \nu \neq \mu \\ 1 \text{ if } \nu = \mu \end{cases}.$$

Bergman generalized this concept in a very original manner. Let D be a domain in the complex z-plane ($z = x + iy$) and consider analytic functions $f(z)$ in D such that

$$\iint_D |f(z)|^2 dxdy < \infty.$$

Consider a system $\{\phi_\nu(z)\}$ of the type such that

$$\iint_D \phi_\nu(z)\overline{\phi_\mu(z)}\,dxdy = \delta_{\nu\mu}$$

and such that every $f(z)$ can be written as

$$f(z) = \sum_{\nu=1}^{\infty} a_\nu \phi_\nu(z)$$

where the series converges uniformly in each closed subdomain of D. Such a system is called complete and orthonormal.

To a given domain D there exists an infinity of possible systems of this kind, but Bergman made the surprising discovery that the combination

$$K(z, \overline{\zeta}) = \sum_{\nu=1}^{\infty} \phi_\nu(z)\overline{\phi_\nu(\zeta)}$$

converges uniformly in each closed subdomain of D and is independent of the particular system used in its construction. For each $f(z)$ we have the identity

$$f(z) = \iint_D K(z, \overline{\zeta}) f(\zeta) d\xi d\eta.$$

Therefore, Bergman called $K(z, \overline{\zeta})$ the reproducing kernel of the domain; it is now called the Bergman kernel. It has a simple covariance behavior under conformal mapping and is a very useful tool in the theory of analytic functions. Bergman's thesis in 1922 summarized his researches and led to his doctor's degree.

He soon realized that his method worked equally well when applied to analytic functions $f(z_1, z_2, \cdots, z_n)$ of n-complex variables. In the early 1920's the theory of such functions was in its initial stages, and Bergman was forced to do much pioneering work in this field. He may be considered as one of the founders of this theory, which is today an important field of research. The kernel function plays a very useful role in the theory of "pseudo-conformal" mapping that carries a domain D in the space of $z_1, \cdots, z_n$ into a domain Δ in the space $w_1, \cdots, w_n$ by the relation $w_i = f_i(z_1, \cdots, z_n)$, $i = 1, \cdots, n$. Again the kernel function has the same covariance behavior, and one can construct invariants from it and introduce a metric in D that is invariant under pseudo-conformal mapping. It is a special case of an important class, called "Kähler metrics," which was much later defined to deal with Riemannian manifolds.

Several important concepts in the theory of analytic functions of n-complex variables are due to Bergman. He discovered, for example, that for a large class of domains, an analytic function in it is completely determined by its values on a relatively small part of its boundary. He called it the "distinguished boundary" of D, and it now goes by the name "Bergman-Shilov boundary." Bergman used this concept to give a generalization of the Cauchy integral formula and to represent the value of a function at an interior point of D in terms of its values on the distinguished boundary.

In 1930 Bergman became a *Privatdozent* at the University of Berlin with a habilitation thesis on the behavior of the kernel function on the boundary of its domain. In 1933 the Nazi seizure of power forced him out of his position and out of Germany. From 1934 to 1937 Bergman taught in Russia (Tomsk, 1934–1936; Tbilisi, 1936–1937). From 1937 to 1939 he worked at the Institut Henri Poincaré in Paris, where he wrote a two-volume monograph on the kernel function and its applications in complex analysis.

Just before the outbreak of World War II, Bergman moved to the United States. He taught at MIT, Yeshiva College, and Brown University. In 1945 he joined his old teacher and friend von Mises at Harvard Graduate School of Engineering. He worked on various problems of fluid dynamics, using his methods on orthonormal developments, on integral operators, and on functions of several complex variables.

He settled into a more leisurely life. In 1950 he married Adele Adlersberg. He found time to summarize his results on the kernel function in a monograph. He started a collaboration with M. Schiffer, who has shown the close connection of the kernel function of a plane domain with its harmonic Green's function. They extended the kernel function concept to the case of elliptic partial differential equations. One orthonormalizes solutions of such equations in an appropriate metric, forms the kernel function in an analogous way, and constructs from it the fundamental solutions of the equation for the given domain. This was summarized in *Kernel Functions and Elliptic Differential Equations in Mathematical Physics* (1953). In 1952 Bergman accepted a position as professor at Stanford University, where he taught and did active research until his death.

BIBLIOGRAPHY

I. ORIGINAL WORKS. Bergman's writings are given in *Poggendorff*. Important books are *Sur les fonctions orthogonales de plusieurs variables complexes avec les applications à la théorie des fonctions analytiques* (Paris, 1947); *Sur la fonction-noyau d'un domaine et ses applications dans la théorie des transformations pseudo-conformes* (Paris, 1948); *Kernel Functions and Elliptic Differential Equations in Mathematical Physics* (New York, 1953), with M. Schiffer; *Integral Operators in the Theory of Linear Partial Differential Equations*, 2nd rev. ed. (New York, 1969); and *The Kernel Function and Conformal Mapping*, 2nd rev. ed. (Providence, R.I., 1970).

II. SECONDARY LITERATURE. Obituaries are in *Applicable Analysis*, **8**, no. 3 (1979), 195–199 (by Menahem Schiffer and Hans Samelson); and *Annales polonici mathematici*, **39** (1981), 5–9 (by M. M. Schiffer).

M. M. SCHIFFER

BERKNER, LLOYD VIEL (*b*. Milwaukee, Wisconsin, 1 February 1905; *d*. Washington, D.C., 4 June 1967, *science administration, geophysics, radio engineering*.

Berkner was the son of Henry Frank Berkner and Alma Julia Viel Berkner. He and his two brothers grew up in the rural towns of Perth, North Dakota, and Sleepy Eye, Minnesota. Like many boys of his time, he was fascinated by radio, aviation, and polar exploration. He achieved distinction in all three activities, although he is remembered most for his administration of scientific and technical activities.

After graduating from high school, Berkner enrolled in a radio operators' school and then spent a year as a shipboard radio operator. He enrolled in the University of Minnesota in 1923. While there he enlisted in the U.S. Naval Reserve and was commissioned as an ensign aviator in 1927, the same year he received a B.S. in electrical engineering, his only earned degree. He married Lillian Frances Fulks in 1928; they had two daughters. His health was good until his later years, when a heart condition caused his early retirement and, subsequently, his death.

Berkner worked briefly for the U.S. Bureau of Lighthouses installing radio navigation equipment, and then joined the U.S. National Bureau of Standards in Washington, D.C., in 1928. Soon, however, he became one of the radio operators on Richard E. Byrd's first antarctic expedition (1928–1930). A severe cutback in 1933 caused the Radio Section of the Bureau of Standards to release many of its employees, including Berkner. During his five years with the bureau, Berkner published several papers. The best known are a collaborative note published in 1933 and a paper published in 1934 that both first revealed the existence of the F1 region of ionization in the ionosphere. These works soon led to his involvement in a priority dispute with Edward V. Appleton, whose own work on radio signals and the upper atmosphere led to a Nobel Prize in physics in 1947. E. O. Hulburt publicly argued for the Bureau of Standards team.

From the bureau, Berkner moved to the Department of Terrestrial Magnetism of the Carnegie Institution of Washington. There he modified and extended the work of T. R. Gilliland, who had built the first automatic multifrequency ionospheric sounder, basically a vertically directed radar that measured electron density as a function of height. Berkner constructed an improved model operating over a frequency range from about 0.5 to 15 megahertz. This type of sounder remained the major ionospheric research instrument for more than thirty years. During the next several years Berkner installed these sounders in Washington, D.C.; at Huancayo, Peru (near the geomagnetic equator); and at Watheroo, in the state of Western Australia. He did important observational work at all three sites.

In 1938 Berkner was coauthor of two papers arguing for the inclusion of the Lorentz polarization term in the magneto-ionic theory applied to the ionosphere. Through this work Berkner entered—on the losing side—a debate that had begun about 1929 and continued until 1943. In 1940 the military asked the Carnegie Institution to establish a full ionospheric and geophysical station at the Agricultural College near Fairbanks, Alaska; Berkner set up the station. Thus, with the entrance of the United States into World War II, most of the Allies' radio-propagation data were coming from stations established by Berkner and the Carnegie. Just before the war Berkner assisted Merle A. Tuve in early work on the radar proximity fuse.

Berkner's organizational ability was revealed during World War II in his work in naval aviation electronics as head of the Radar Section (1941–1943) and as director of electronic material for the Navy Bureau of Aeronautics (1943–1945). This entailed the supervision of the introduction of several important and sometimes revolutionary techniques and equipment into U.S. Navy aviation and fleet operations: very-high-frequency radio, airborne radar, aircraft electronic identification, and electronic navigation and bombing. During more than forty years in the U.S. Naval Reserve, Berkner served nearly twelve years on active duty and rose from second-class seaman to rear admiral and senior officer. In 1946 Vannevar Bush appointed Berkner first executive secretary of the Joint Research and Development Board (now the Office of Research and Engineering) of the Defense Department. He subsequently was involved in the formation of the North Atlantic Treaty Organization (NATO) as a State Department consultant and chaired the committee that established the posts of science attachés at U.S. missions abroad. In 1951 Berkner became president of Associated Universities; his duties included management of the Brookhaven National Laboratory, of which he remained director until 1960. In 1954 he began organizing the National Radio Astronomical Observatory of Green Bank, West Virginia, and in 1959 he chose Otto Struve as director of the observatory.

Berkner served on the President's Science Advisory Committee from 1956 to 1959. During this time he organized and led the Panel on Seismic Improvement, which studied the problem of detecting underground nuclear tests and led to the organization of Project Vela: the monitoring of nuclear tests to determine their effects on land, air, and space. He assisted in drafting the 1959 Antarctic Treaty, which has been widely studied and is often cited as a precedent in international environment and space issues. It can be said that Berkner was among the most powerful technical advisers during the Eisenhower years. At the end of the Eisenhower administration, he moved to Dallas, Texas, to become first president of the Graduate Research Center of the Southwest, which later was incorporated into the University of Texas system. He served from 1960 until a heart attack forced his retirement in 1965.

His experience in Antarctica and his lifelong concern with ionospheric and upper atmospheric physics led Berkner to propose a third International Polar Year for 1957–1958. It was modeled on the first (1882–1883) and second (1932–1933) International Polar Years, but it greatly expanded upon them. He proposed the idea in early 1950 at the home of James Van Allen, in the presence of Van Allen, Sydney Chapman, and other geophysicists. It was endorsed by the International Union of Radio Science (URSI) and by the International Council of Scientific Unions (ICSU), which established the Special Committee for the International Geophysical Year (IGY), with Chapman as president and Berkner as vice president. The idea for a third International Polar Year broadened into a world program for the International Geophysical Year of eighteen months, to occur during the period of maximum sunspot activity, from 1957 to 1958. (Later, IGY was extended to the end of 1959.) This gigantic geoscientific program with over thirty-thousand participants was the most extensive in history and involved Berkner in national and international planning meetings from 1950 through 1958, during almost the entire period when he directed Associated Universities. The largest single element of the U.S. IGY program, in terms of cost, was the rocket and satellite portion. (Indeed, the United States entered the space age through the IGY.) The next largest program was the ionospheric portion. Significant programs also were established in meteorology, geomagnetism, oceanography, solar physics, seismology, gravitation and geodesy, cosmic rays, aurora, and airglow. Berkner directly supervised the rocket and satellite program. He organized the Space Science Board of the National Academy of Sciences, was its chairman from 1958 to 1962, and worked in the planning of the National Aeronautics and Space Administration (NASA).

Berkner received many honorary degrees and awards, and was a member of several scientific organizations. The honorary degrees followed his work in the IGY and his presidency of the American Geophysical Union, the Institute of Radio Engineers, URSI, and ICSU. He was also a member and treasurer of the National Academy of Sciences.

BIBLIOGRAPHY

I. Original Works. The Berkner papers are deposited in the manuscript collection of the Library of Congress in Washington, D.C. Berkner was coeditor of two books: *Manual on Rockets and Satellites, Annals of IGY*, VI (London, New York, and Oxford, 1958), with Gilman Reid, John Hanessian, Jr., and Leonard Cormier; and *Science in Space* (New York, 1961), with Hugh Odishaw. He also wrote *The Scientific Age* (New Haven, 1964).

His most important scientific and engineering papers were published between 1928 and 1941. They concern the ionosphere, radio-wave propagation, and the construction and operation of ionospheric equipment. Some of the most significant of them are "Radio Observations of the Bureau of Standards During the Solar Eclipse of August 31, 1932," in *Journal of Research, National Bureau of Standards*, **11** (1929), 829–845, written with S. S. Kirby, T. R. Gilliland, and K. A. Norton; "Studies of the Ionosphere and Their Application to Radio Transmission," in *Proceedings, Institute of Radio Engineers*, **22** (1934), 481–521, written with S. S. Kirby and D. M. Stuart; "Constitution of the Ionosphere and the Lorentz Polarization Correction," in *Nature*, **141** (1938), 562–563, written with H. G. Booker; and "A Fundamental Problem Concerning the Lorentz Correction to the Theory of Refraction," in *Science*, **87** (1938), 257–258, written with H. G. Booker. Some of Berkner's papers are abstracted in Laurence A. Manning, *Bibliography of the Ionosphere: An Annotated Survey Through 1960* (Stanford, Calif., 1962). Most of Berkner's papers written after World War II concern broader questions of science planning and organization and involve the editing of collective efforts.

II. SECONDARY LITERATURE. An obituary appeared in the *New York Times* (5 June 1967), 43. For additional biographical information on Berkner, see *Current Biography 1949* (New York), 41–43; and Walter Sullivan, "Profile of Lloyd V. Berkner," in *ICSU Review*, **3** (1961), 208–211.

C. STEWART GILLMOR

BERNAYS, PAUL ISAAC (*b*. London, England, 17 October 1888; *d*. Zurich, Switzerland, 18 September 1977), *mathematical logic, set theory*.

Bernays came from a distinguished German-Jewish family of scholars and businessmen. His great-grandfather, Isaac ben Jacob Bernays, chief rabbi of Hamburg, was known for both strict Orthodox views and modern educational ideas. His grandfather, Louis Bernays, a merchant, traveled widely before helping to found the Jewish community in Zurich, while his great-uncle, Jacob Bernays, was a *Privatdozent* at the University of Bonn. In 1887 his father, Julius Bernays, a businessman, married Sara Brecher, who had likewise descended from Isaac Bernays. Their first child, Paul, born in London as a Swiss citizen and a *Bürger* of Zurich, was followed by a brother and three sisters.

After living in Paris for a time, the family settled in Berlin, where Paul Bernays had what he later described as a happy childhood. From 1895 to 1907 he attended the Köllnisches Gymnasium in Berlin. At an early age his talent as a pianist attracted attention, and he began to try his hand at composing. While at school, he added to his musical interests a growing attraction to ancient languages and mathematics. At eighteen, after hesitating between a career in music and one in mathematics, he opted for engineering (which his parents regarded as an eminently practical way to use his mathematical talent) and spent the summer semester of 1907 studying that subject at the Technische Hochschule in Charlottenburg. This experience made it clear to him that his future lay in pure mathematics rather than in its applications. Consequently, in the winter he transferred to the University of Berlin. For the next two years he studied there the three subjects of central interest to him: mathematics, mainly under Issai Schur, Edmund Landau (including lectures on set theory), Leo Frobenius, and Friedrich H. Schottky; philosophy under Alois Riehl, Carl Stumpf, and Ernst Cassirer; and physics under Max K. E. L. Planck. Then, from 1910 to 1912, he attended lectures at Göttingen, chiefly by David Hilbert, Landau, Hermann Weyl, and Felix Klein in mathematics; by Woldemar Voigt and Max Born in physics; and by Leonard Nelson in philosophy. Göttingen would always remain his spiritual home.

In 1912 he received his doctorate at Göttingen with a thesis, supervised by Landau, on the analytic number theory of binary quadratic forms. After his *Habilitationsschrift* on modular elliptic functions was accepted at the University of Zurich the following year, he served as assistant to Ernst F. F. Zermelo (who was then professor there) and as a *Privatdozent* until the spring of 1919. When Zermelo left the University of Zurich in 1916 (both for reasons of health and because of disagreements with the university), Bernays took over his courses. Although Bernays lectured primarily on topics in analysis, during his last year in Zurich (probably influenced by Hilbert), he gave courses on the foundations of geometry and on set theory. While at Zurich, he became both a friend and a colleague of Pólya, had conversations with Einstein, and was received socially at the home of Hermann Weyl.

When Hilbert came to Zurich to lecture on "Axiomatisches Denken" in the fall of 1917, he invited Bernays to come to Göttingen as his assistant and to help him to resume investigations of the foundations of arithmetic. During 1918 Bernays quickly wrote for Göttingen a second *Habilitationsschrift*, establishing in it the completeness of propositional logic. But he continued to teach at Zurich during both the summer semester and the winter semester of 1918. When he moved to Göttingen in 1919, Bernays received the *venia legendi*, which permitted him to lecture. He served as a *Privatdozent* until he was made untenured extraordinary professor in March 1922. As at Zurich, he lectured chiefly on topics in analysis, but, beginning in 1922, he also gave courses on the foundations of geometry and (jointly with Hilbert) on the foundations of arithmetic. During the winter semester of the year 1929–1930 he first lectured on mathematical logic, giving at the end a version of his axiomatization for set theory. Believing that he learned better orally than through reading, he attended lectures by Emmy Noether, van der Waerden, and Herglotz. During the academic holidays he traveled regularly to Berlin to visit his family. His father died in 1916.

On 28 April 1933 the dean at Göttingen ordered Bernays, as a "non-Aryan," to stop teaching pending a final decision on his official status by the minister of education. In August he was relieved of his position as an assistant at the Mathematical Institute, and a month later his right to teach was officially withdrawn by the minister. For six months during this period Hilbert employed Bernays privately as his assistant. Finally Bernays and his family, having remained Swiss citizens, returned to Zurich.

From time to time, beginning with the summer semester of 1934, Bernays held a temporary teaching position at the Eidgenössische Technische Hochschule (E.T.H.) in Zurich. Meanwhile, he visited the Institute for Advanced Study at Princeton during the academic year 1935–1936 (as he was to do again during the academic year 1959–1960). In October 1939 the E.T.H. granted him the *venia legendi* for four years, renewing it in 1943. At last, in October 1945, he received a half-time appointment as extraordinary professor at the E.T.H., a position that he continued to hold until 1959, when he became professor emeritus. During the spring semester of 1956, and again during 1961 and 1965, he was visiting professor at the University of Pennsylvania and gave lectures at several American universities. Although some mathematicians thought that the E.T.H. had not granted Bernays a position commensurate with his abilities, he himself always remained grateful to the E.T.H. for its support under difficult circumstances.

Bernays, who remained mathematically active until the end of his life, held a variety of positions: corresponding member in the Academy of Sciences of Brussels and in that of Norway, president of the International Academy of the Philosophy of Science, and honorary chairman of the German Society for Mathematical Logic and Foundational Research in the Exact Sciences. In addition he served on the editorial board of *Dialectica* and was a coeditor for both the *Journal of Symbolic Logic* and *Archiv für mathematische Logik und Grundlagenforschung*. In 1976 he received an honorary doctorate from the University of Munich for his work in proof theory and set theory. After a brief illness he died of a heart condition at the age of eighty-eight.

Bernays' earliest publications, beginning in 1910, were devoted to philosophy. They showed the direct influence of his teacher Leonard Nelson, head of the neo-Friesian school, which had revived the philosophy of Jacob Fries and extended it to ethics. Except for his doctoral thesis and his Zurich *Habilitationsschrift*, Bernays published no mathematical articles until 1918. His attempt during that period to extend the special theory of relativity was preempted when Einstein introduced general relativity.

Then, answering Hilbert's call to Göttingen to collaborate on foundational (especially proof-theoretic) questions, Bernays began the work in mathematical logic that molded his career. The first product was his 1918 *Habilitationsschrift*, "Beiträge zur axiomatischen Behandlung des Logik-Kalküls," which was devoted to the metamathematics of the propositional calculus (the heir of George Boole's logic) and was a contribution to Hilbert's program. In contrast to most earlier logicians, Bernays had a firm grasp of the difference between syntax and semantics, and he distinguished carefully between provable formulas and valid formulas. By establishing the completeness theorem for propositional logic, he showed these two notions to be equivalent in that context. Further, he gave a partial solution to Hilbert's decision problem (*Entscheidungsproblem*) by stating a decision procedure for validity in this part of logic. He also demonstrated that one of the axioms for propositional logic in *Principia Mathematics* was redundant while the four remaining axioms were independent. In these independence proofs, many-valued logic was utilized for the first time. Although in 1926 his *Habilitationsschrift* of 1918 was published in part, Post had independently published the same completeness result in 1921—with the effect that, outside the Hilbert school, Post was usually given credit for it.

The most enduring achievement of the collaboration between Hilbert and Bernays was their *Grundlagen der Mathematik*, published in two volumes in 1934 and 1939, which for decades remained the standard work on proof theory. It appears that, while the overall approach was due to Hilbert, the specific contents and the actual writing came from Bernays. There Bernays developed the ε-calculus and the ε-theorems for eliminating quantifiers so as to give a decision procedure for various theories. Moreover, he provided the first detailed proof of Gödel's second incompleteness theorem and supplied the first correct proof for Herbrand's theorem (Herbrand's proof was faulty). Finally, he demonstrated a proof-theoretic version of Gödel's completeness theorem for first-order logic.

In 1937, when Bernays published the first of his seven-part article on his axiomatization of set theory, he pointed out that his aim was to modify von Neumann's axiom system (based on function and argument rather than on set and membership) so as to make it resemble more closely Zermelo's original system, and thereby to use some of the set-theoretic concepts of Schröder and Whitehead-Russell, while expressing the theory in first-order logic. Bernays chose his groups of axioms in order to render them analogous, whenever possible, to Hilbert's (1899) groups of axioms for geometry. As Hilbert had done, Bernays explored (in 1941 and 1942) the consequences of various axioms within a group, and showed which of his axioms were needed to develop number theory on the one hand and analysis (up to Lebesgue measure) on the other. While doing so, he formulated the principle of dependent choices (the form of the axiom of choice sufficient for analysis), later independently rediscovered by Tarski. In the last two sections (published in 1948 and 1954) he investigated the independence of his axioms, mainly by using number-theoretic models in the spirit of Ackermann. Bernays' axiom system, slightly modified by Gödel, is now generally known as Bernays-Gödel set theory.

"In philosophy," Bernays wrote in his autobiography about his return to Zurich in 1934, "I came into closer contact with Ferdinand Gonseth. . . . Because of my interior dialogues on the philosophy of Kant, Fries, and Nelson, I had come very close to Gonseth's views, and so I joined his school of philosophy." Bernays was sympathetic, in particular, to Gonseth's "open philosophy," with its emphasis on dialogue between opposing viewpoints, and encouraged tolerance between diverse foundational positions (such as Platonism and intuitionism). In 1946 Bernays, Gonseth, and Karl Popper founded the Internationale Gesellschaft zur Pflege der Logik und Philosophie der Wissenschaft, which started the journal *Dialectica* the following year. Bernays' philosophical writings, so refreshingly undogmatic, are models of clarity.

Bernays directed a number of doctoral theses. At Göttingen he was involved with Haskell Curry's and Gerhard Gentzen's, and even supervised Saunders Mac Lane's (1934), though Weyl conducted Mac Lane's oral examination since the Nazis had already dismissed Bernays. At the E.T.H. in Zurich he directed theses by Martin Altwegg (1948), Hugh Ribeiro (1949), J. Richard Büchi (1950), Walter Strickler (1955), Erwin Engeler (1958), and Hersz Wermus (1961), as well as serving as *Korreferent* for six other dissertations.

Bernays' precise contributions are often difficult to ascertain because of his preference for collaboration (especially with Hilbert) and his modesty. Engeler, his former student, described him as a "great scholar and kind man." Yet Hilbert's 1922 letter recommending Bernays for an extraordinary professorship at Göttingen remains the most fitting tribute to his life's work:

> Bernays' publications extend over the most diverse fields of mathematics . . . [and] are all marked by thoroughness and reliability. . . . He is distinguished by a deep-seated love for science as well as a trustworthy character and nobility of thought, and is highly valued by everyone. In all matters concerning foundational questions in mathematics, he is the most knowledgeable expert and, especially for me, the most valuable and productive colleague.

BIBLIOGRAPHY

I. ORIGINAL WORKS. A bibliography of Bernays' publications up to 1976 can be found in Gert H. Müller, ed., *Sets and Classes: On the Work by Paul Bernays* (New York, 1976) and is supplemented in Müller's 1981 article mentioned below. The same book, which contains a photo of Bernays, reprints his seven-part article giving his axiom system for set theory and includes an English translation of his 1961 article on strong axioms of infinity. Fourteen of his philosophical essays (eight of them on logic or foundational questions) are reprinted in his *Abhandlungen zur Philosophie der Mathematik* (Darmstadt, 1976). He also edited the later editions of David Hilbert's *Grundlagen der Geometrie* (Stuttgart, 1977) and was an editor of Leonard Nelson's *Gesammelte Schriften*.

The *Wissenschaftshistorische Sammlungen* at the Eidgenössische Technische Hochschule in Zurich has a rich collection of material on Bernays, including many of his unpublished lectures and a voluminous correspondence, as well as his lecture notes for courses (generally in Gabelsberger shorthand). Some of Bernays' letters and manuscripts are located at the *Niedersächsische Staats- und Universitätsbibliotek* in Göttingen.

II. SECONDARY LITERATURE. Bernays wrote a brief autobiography for the Müller volume as well as the *Lebenslauf* at the end of his 1912 dissertation. Obituaries are by Ernst Specker and Erwin Engeler, in *Neue Zürcher Zeitung* (26 September 1977); and Gert H. Müller, in the *Mathematical Intelligencer*, **1** (1978), 27–28. More extensive discussions of his life and work can be found in Erwin Engeler, "Zum logischen Werk von Paul Bernays," in *Dialectica*, **32** (1978), 191–200; Abraham Fraenkel, "Paul Bernays and die Begründung der Mengenlehre," in *Dialectica*, **12** (1958), 274–279; Henri Lauener, "Wissenschaftstheorie in der Schweiz," and "Paul Bernays (1888–1977)," in *Zeitschrift für Allgemeine Wissenschaftstheorie*, **2**, no. 2 (1971), 294–299, and **9**, no. 1 (1978), 13–20 Gert H. Müller, "Framingham Mathematics," in *Epistemologia*, **4** (1981), 253–285; Andrés R. Raggio, "Die Rolle der Analogie in Bernays' Philosophie der

Mathematik," in *Dialectica*, **32** (1978), 201–207; Ernst Specker, "Paul Bernays," in Maurice Boffa *et al.*, eds., *Logic Colloquium '78* (1979), 381–389; Gaisi Takeuti, "Work of Paul Bernays and Kurt Gödel," in L. Jonathan Cohen *et al.*, eds., *Logic, Methodology, and Philosophy of Science*, VI (1982), 77–85.

GREGORY H. MOORE

BESICOVITCH, ABRAM SAMOILOVITCH (*b.* Berdyansk, Russia, 24 January 1891; *d.* Cambridge, England, 2 November 1970), *mathematics*.

Besicovitch was the fourth child in the family of four sons and two daughters of Samuel and Eva Besicovitch. The family had to live frugally. All the children were talented and studied at the University of St. Petersburg, the older ones in turn earning money to help support the younger. From an early age Besicovitch showed a remarkable aptitude for solving mathematical problems.

Besicovitch graduated in 1912 from the University of St. Petersburg, where one of his teachers was Andrei A. Markov. When, in 1917, he became professor in the School of Mathematics at the newly established University of Perm, his intention was to work in mathematical logic. He abandoned this idea because the library was inadequate, and as a result continued to work on fundamental problems in analysis. In order to obtain a counterexample to a plausible conjecture about repeated Riemann integrals in $\mathbb{R}^2$, he was led to construct a compact plane set F of zero Lebesque measure but containing unit line segments in every direction (1919–1920). Using a suggestion of J. Pál, Besicovitch later (1928) used F to show that zero is the lower bound of the area of plane sets in which a unit segment can be continuously turned through two right angles. The solution of this "Kakeya problem" was the subject of a lecture filmed by the Mathematical Association in 1958. The set F has turned out to be useful in many contexts. For example, R. O. Davies applied the Besicovitch method to a construction of Otton Nikodym to produce a plane set of full measure with continuum many lines of accessibility through each point. Charles Fefferman used F to obtain a negative solution to the multiplier problem for a ball, and C. R. Putnam obtained a characterization of the spectra of hyponormal operators using Davies' work.

In 1916 Besicovitch married Valentina Vietalievna, a mathematician older than himself. Three years later, during the civil war, the University of Perm was destroyed and partially reestablished at Tomsk. Besicovitch locked books in cellars and preserved much of the valuable property of the Faculty of Mathematics, then worked with A. A. Fridman to reestablish the university after the liberation of Perm. In 1920 he returned to Leningrad as professor in the Pedagogical Institute and lecturer in the university. The political powers forced him to teach classes of workers who lacked the background to understand the lectures, so there was little time for research.

Besicovitch was offered a Rockefeller fellowship to work abroad, but repeated requests for permission to leave the country were refused. In 1924 he escaped and made his way to Copenhagen, where the Rockefeller fellowship enabled him to work for a year with Harald Bohr, who was then developing the theory of almost periodic functions. This contact resulted in several papers, in the most important of which ("On Generalized Almost Periodic Functions") he showed that the analogue of the Riesz-Fischer theorem is false for Bohr's almost periodic functions and developed a new definition of "almost periodic" for which Riesz-Fischer is valid. His *Almost Periodic Functions* became the standard work on the classical theory of that subject.

In 1925 Besicovitch visited Oxford. Godfrey H. Hardy quickly recognized his analytical abilities, securing for him a position as lecturer at the University of Liverpool in 1926–1927. He moved to Cambridge in 1927 as a university lecturer and was elected a fellow of Trinity College in 1930. Besicovitch's wife remained in Russia, and the marriage was dissolved in 1926. While in Perm he had befriended a woman named Maria Denisova and her children. He brought them to England and in 1928 married the elder daughter, Valentina Alexandrovna, then aged sixteen.

At about the time of his arrival in England, Besicovitch started his deep analysis of plane sets of dimension 1 that has become the foundation of modern geometric measure theory as developed by Herbert Federer and his school. He obtained the fundamental structure theorems for linearly measurable plane sets—the regular sets are a subset of a countable union of rectifiable arcs and have a tangent almost everywhere ("On the Fundamental Geometrical Properties of Linearly Measurable Plane Sets of Points"), while the irregular sets intersect no rectifiable arc in a set of positive length, have a tangent almost nowhere, and project in almost all directions onto a set of zero linear Lebesque measure. Besicovitch also considered sets with noninteger dimension ("On Linear Sets of Points of Fractional Dimension," "On Lipschitz Numbers") and showed that these could not have nice geometric properties of regularity. These sets occur naturally in many physical situations where there is an element of randomness (for an extensive recent account see Mandelbrot, who calls such sets "fractals"). The basic Besicovitch contribution to geometric measure theory is carefully developed by Federer, and a simpler version is given by Falconer.

Besicovitch had shown that the study of local density was fundamental to an understanding of the geometric properties of small sets. John Manstrand, 1954, showed that a strict density cannot exist for any set of non-integer dimension. Claude Tricot developed a new packing measure and showed by considering density that only for surfacelike sets with integer dimension can Hausdorff and packing measures be equal and finite positive.

In 1950, on his fifty-ninth birthday, Besicovitch was elected to the House Ball chair of mathematics at Cambridge. Although he retired in 1958, he remained active in teaching and research and spent eight successive years as visiting professor at various universities in the United States. Besicovitch developed a new interest in the definition of area for a parametric surface in $\mathbb{R}^3$ starting about 1942. He produced a beautiful example of a topological disk with arbitrarily small Lebesque-Fréchet area (defined by approximating polyhedral) but arbitrarily large (three-dimensional) Lebesque measure. He concluded that the only satisfactory concept of area is a two-dimensional Hausdorff measure and undertook a program of solving anew many of the classical problems of surface area.

Besicovitch exhibited an open mind in all of his work. When solving a problem, most mathematicians make a commitment to the nature of the solution before that solution has been found, and this commitment interposes a barrier to the consideration of other possibilities. Besicovitch seems never to have been troubled in this way. He therefore obtained results that astounded his contemporaries and remain surprising today.

In recognition of his outstanding talent, Besicovitch received the Adams Prize of the University of Cambridge in 1930, was elected a fellow of the Royal Society in 1934, was awarded the Morgan Medal by the London Mathematics Society in 1950, and received the Sylvester Medal of the Royal Society in 1952. His many mathematical contributions remain a stimulus to research activity.

In the decade 1980–1989 there was an explosion of interest in fractals as a tool for modeling phenomena from a wide variety of different contexts. Dynamical systems involving the interation of a transformation produce fractals for a critical set—see the book by Heinz-Otto Peitgen and P. H. Richter *The Beauty of Fractals*. In physics there are many critical phenomena where fractals provide a helpful insight, and each year has seen international symposia focusing on this area. It is fair to say that the geometrical insights of Besicovitch's work laid the foundation for this development.

[The material in this biography is condensed from an obituary notice published in the Bulletin of the London Mathematical Society, with some comments about recent developments.]

BIBLIOGRAPHY

I. Original Works. A complete list of Besicovitch's works is in J. C. Burkill, "Abram Semoilovich Besicovitch," in *Biographical Memoirs of Fellows of the Royal Society*, **17** (1971), 1–16. Papers by Besicovitch cited in the text are "Nouvelle forme des conditions d'intégrabilité des fonctions," in *Journal de la Société de physique et de mathématique* (Perm), **1** (1918–1919), 140–145; "On Generalized Almost Periodic Functions," in *Proceedings of the London Mathematical Society*, 2nd ser., **25** (1926), 495–512; "Fundamental Geometric Properties of Linearly Measurable Plane Sets of Points," in *Bulletin of the American Mathematical Society*, **33** (1927), 652; "On Kakeya's Problem and a Similar One," in *Mathematische Zeitschrift*, **27** (1928), 312–320; "On the Fundamental Geometrical Properties of Linearly Measurable Plane Sets of Points," in *Mathematische Annalen*, **98** (1928), 422–464; "On Linear Sets of Points of Fractional Dimension," *ibid.*, **101** (1929), 161–193; "On Lipschitz Numbers," in *Mathematische Zeitschrift*, **30** (1929), 514–519; *Almost Periodic Functions* (Cambridge, 1932; repr. New York, 1955); "On the Fundamental Geometrical Properties of Linearly Measurable Plane Sets of Points (III)," in *Mathematische Annalen*, **116** (1939), 349–357; "On the Definition and Value of The Area of a Surface," in *Quarterly Journal of Mathematics* (Oxford), 1st ser., **16** (1945), 88–102; "Parametric Surfaces, III. On Surfaces of Minimum Area," in *Journal of the London Mathematical Society*, **23** (1948), 241–246; "Parametric Surfaces, I. Compactness," in *Proceedings of the Cambridge Philosophical Society*, **45** (1949), 5–13; Parametric Surfaces, II. Lower Semicontinuity of the Area," *ibid.*, 14–23; "Parametric Surfaces, IV. The Integral Formula for the Area," in *Quarterly Journal of Mathematics* (Oxford) 1st ser., **20** (1949), 1–7; and "The Kakeya Problem," in *American Mathematical Monthly*, **70** (1963), 697–706.

II. Secondary Literature. R. O. Davies, "On Accessibility of Plane Sets and Differentiation of Functions of Two Real Variables," in *Proceedings of the Cambridge Philosophical Society*, **48** (1952), 215–232; K. J. Falconer, *The Geometry of Fractal Sets* (Cambridge, 1985); Herbert Federer, *Geometric Measure Theory* (New York, 1969); Charles Fefferman, "The Multiplier Problem for the Ball," in *Annals of Mathematics*, 2nd ser., **94** (1971), 330–336;

Benoit B. Mandelbrot, *The Fractal Geometry of Nature* (San Francisco, 1982); J. M. Marstrand, "Some fundamental geometrical properties of plane sets of fractional dimensions," Proceedings of the London Mathematical Society, **15** (1954), 257–302; Otton Nikodym, "Sur la mésure des ensembles plans dont tous les points sont rectilinéairement accessibles," in *Fundamenta mathematicae*, **10** (1927), 116–168; H.-O. Peitgen and P. H. Richter, *The beauty of fractals* (Berlin, 1986); Pietronera and Tosatti, eds., *Proceedings of Sixth International Symposium on Fractals in Physics* (North Holland, Amsterdam, 1986); C. R. Putnam, "The Role of Zero Sets in the Spectra of Hyponormal Operators," in *Proceedings of the American Mathematical Society*, **43** (1974), 137–140; S. J. Taylor and C. Tricot, Packing measure and its evaluation for a Brown path, Trans. Amer. Math Soc **288** (1985) 679–699.

S. James Taylor

BEST, CHARLES HERBERT (*b.* West Pembroke, Maine, 27 February 1899; *d.* Toronto, Canada, 31 March 1978), *physiology, endocrinology.*

Best was the son of Herbert Huestis Best, a Canadian-born general practitioner, and of Luella May Fisher Best. The highlight of Best's medical research was his involvement in the discovery of insulin in the years 1921 and 1922. What began as a summer research job for a bright student turned into one of the most exciting and controversial medical adventures of modern times. After the discovery, Best finished his education and embarked on a long, productive career as a physiologist at the University of Toronto. He received much acclaim and many honors for the insulin work and considered himself one of the codiscoverers of insulin. On 3 September 1924 he married Margaret Hooper Mahon; they had two sons.

Best completed his bachelor's program in physiology and biochemistry at the University of Toronto in May 1921 and was hired for the summer as a research assistant to John J. R. Macleod, a professor of physiology. In May, Macleod asked Best and the other student assistant, Edward Clark Noble, to help test Frederick G. Banting's hypothesis that ligation of the pancreatic ducts in living animals (dogs) would cause selective degeneration of pancreatic cells and permit the isolation of the much-sought antidiabetic internal secretion. Best won a coin toss to see who would work first with Banting; Noble later decided it would be convenient for Best to finish the assignment.

Best did all the chemical tests for Banting, measuring blood, urinary sugars, and urinary nitrogen in the series of experiments on depancreatized, duct-ligated, and normal dogs. Late in July, Banting and Best began injecting extracts of degenerated pancreas into depancreatized diabetic dogs and recorded frequent declines in blood sugar after the injections. These experiments were repeated and extended at intervals through October and November and were reported in Banting and Best's first paper, "The Internal Secretion of the Pancreas," presented on 30 December 1921 at the American Physiological Society meeting at Yale and published in the *Journal of Laboratory and Clinical Medicine* in February 1922.

Historical re-creation of Banting and Best's experiments, based on their complete notebooks, indicates that their achievement was similar to that of other researchers, including E. L. Scott, Israel S. Kleiner, and Nicolas C. Paulesco, all of whom had been able to produce pancreatic extracts that often reduced hyperglycemia or glycosuria. Lacking both training and expertise, though possessing boundless enthusiasm, Banting and Best understood neither the advances made by co-workers nor the limitations of their own papers. Their papers were marred by factual errors and misinterpretations, a serious misrepresentation of Paulesco's work (based on Best's translation error), and failure to assess Banting's original, erroneous assumption that the external and internal secretions are antagonistic within the pancreas.

Macleod, whose instructions for preparing saline extract of chilled pancreas were vital to Banting and Best's early successes, was impressed by their pattern of favorable results and by Banting's stubborn optimism. New research was begun, under Macleod's direction, involving Banting, Best (now a master's student), and an experienced biochemist, James B. Collip. In December 1921 the team made rapid advances in extractive technique—utilizing fresh pancreas in acid alcohol—and in the exploration of the extracts' antidiabetic properties.

Banting's temperament clashed with Macleod's in virtually every way. He began to feel that he and Best were being shunted aside in the work and insisted that they should be responsible for preparing the first extract to be used in clinical tests. Macleod consented. Assisted by Banting, Best made extract tested on a patient, fourteen-year-old Leonard Thompson, in Toronto General Hospital on 11 January 1922. The extract was not effective enough to justify further administration. Tests on Thompson resumed on 23 January, using purified extract made by Collip. The spectacular antidiabetic effects of Collip's extract demonstrated to the Toronto group that they had made a very big discovery. In May

1922 they announced their discovery of insulin and its therapeutic benefits for diabetics, in "Effect Produced on Diabetes by Extracts of Pancreas," in *Transactions of the Association of American Physicians*.

Best studied the extract's effect on the respiratory quotient of human diabetics for his master's thesis. As part of the collaborating group, he was a joint author of all their major papers, his name following Banting's in the agreed-upon alphabetical order. In the spring of 1922, he returned to the extraction problem because technical difficulties had caused Collip to lose the ability to make insulin. Best was instrumental in the group's rediscovery of an extractive technique, and when Collip left Toronto in June 1922, to return to the University of Alberta, Best replaced him as director of insulin production at Toronto University's fledgling Connaught Laboratories. From this time, however, production advances were made principally by Eli Lilly and Company, with whom the Canadians had decided to collaborate.

When Banting learned that he and Macleod had been awarded the 1923 Nobel Prize in physiology or medicine for the discovery of insulin, he immediately announced that he was sharing his half of the prize money equally with Best. His generous allocation of credit to Best stemmed largely from the moral support his assistant had given him at several points during both the research and the vicious quarreling with Macleod and Collip. Best was disappointed not to have received more formal recognition for his insulin work.

Best completed his medical training in Toronto in 1925 and did postgraduate research under Henry H. Dale at the National Institute for Medical Research and the University of London, receiving the D.Sc. in 1928. He was given an appointment in the School of Hygiene at the University of Toronto, and in 1929 he succeeded Macleod as professor of physiology. In 1941 he succeeded Banting as head of the Banting and Best Department of Medical Research. Best's later research centered on explorations of the lipotropic effects of choline and on various insulin-related problems. In the late 1930's Best's laboratory pioneered in the isolation and production of heparin, which soon found an important clinical application as an anticoagulant in vascular surgery. Other research ranged from early studies in exercise physiology (using 1928 Olympic marathon runners) to wartime work on night vision and seasickness. With Norman B. Taylor, Best coauthored a widely used textbook, *The Physiological Basis of Medical Practice* (Baltimore, 1st ed., 1937, 10th ed., 1979). He wrote and reminisced often about his role in the discovery of insulin, but his memory was too selective to make his accounts entirely reliable.

BIBLIOGRAPHY

I. Original Works. The Fisher Rare Books Library at the University of Toronto and the university archives contain several major collections of documents relating to the discovery of insulin. The Banting papers include the Banting and Best notebooks. The Macleod papers and the records of the university's Insulin Committee are also important. Most of the documents in the extensive Best collection are from a later period, but many important autobiographical transcripts are contained in the papers of Best's chosen biographer, W. R. Feasby. The Best-Dale correspondence in the H. H. Dale papers at the Royal Society of London is also illuminating. Documents relating to the discovery of insulin that Dale and Best deposited in the Wellcome Institute are duplicated at Toronto.

Best's most important account of his role in the discovery, written in September 1922, was published as part of "Banting's, Best's, and Collip's Accounts of the Discovery of Insulin," in *Bulletin of the History of Medicine*, **56** (1982), 554–568. All of his important scientific papers, plus many reminiscences and fragments of autobiography, are in *Selected Papers of Charles H. Best* (Toronto, 1963). For the most complete bibliography of his publications and a list of his honors, see his obituary, by Sir Frank Young and C. N. Hales, in *Biographical Memoirs of Fellows of the Royal Society*, **28** (1982), 1–25.

II. Secondary Literature. The standard account of the discovery of insulin, the first to utilize the documents cited above, is Michael Bliss, *The Discovery of Insulin* (Chicago, 1982). Further detail is in Bliss, *Banting: A Biography* (Toronto, 1984). Earlier critical appraisals of Banting and Best's researches are Ffrangcon Roberts, "Insulin," in *British Medical Journal*, no. 4833 (16 December 1922), 1193–1194, and Joseph H. Pratt, "A Reappraisal of Researches Leading to the Discovery of Insulin," in *Journal of the History of Medicine*, **9** (1954), 281–289. The Royal Society obituary is somewhat incomplete and inaccurate, relying too heavily on Best's later memories. Several versions of W. R. Feasby's unpublished biography of Best are in the Feasby papers at the University of Toronto.

Michael Bliss

BIANCHI, ANGELO (*b*. Casalpusterlengo, Italy, 20 December 1892; *d*. Padua, Italy, 24 September 1970), *mineralogy, petrography*.

Bianchi was the youngest of the nine children of Giacomo Bianchi, the district doctor of Casalpusterlengo, and of Maria Platner. He studied at the classical lyceum in Lodi, and in 1910 he entered the University of Pavia to take a degree in natural sciences. He studied mineralogy under his uncle Luigi Brugnatelli and became his assistant while still a student. He graduated in natural sciences in 1915.

During World War I, Bianchi, an artillery officer in the Italian army, was wounded in battle on 3 September 1917; he was awarded the Silver Medal and the War Cross. Also in 1917 his father and his brother Camillo, an engineer, died in the influenza epidemic.

In the year 1922–1923 Bianchi taught mineralogy at the University of Sassari, and at the University of Padua from 1923; in 1926 he succeeded Ruggero Panebianco as professor of mineralogy at Padua and director of the Mineralogical Institute, which he reorganized. Also in 1926 he married Camilla Gallo, from Pavia; they had four children.

Bianchi's first scientific interest was mineralogy. Through extensive collecting in the field and a rigorous analytic method, he studied the cleft minerals from Val Devero (Ossola, Piedmont) and from Miage (Mont Blanc), publishing classical descriptions beginning in 1914. His last contribution on the subject was published posthumously.

The collaboration, started in Padua, with Giorgio Dal Piaz, professor of geology, shifted Bianchi's scientific interest to petrography. The collaboration continued from 1928 with Giambattista Dal Piaz, Giorgio's son, who succeeded his father at Padua in 1942. The two scientists became brothers-in-law when Giambattista married Graziella Gallo, a sister of Bianchi's wife.

In 1934 Bianchi and Dal Piaz published a long monograph on the crystalline formations of the eastern Alto Adige and nearby areas. The petrographic analysis by Bianchi and the geologic interpretation by Dal Piaz are closely connected, a method of research they continued in later studies on Adamello and in the Western Alps. In their monograph they established the pre-Alpine age of the bodies of granite metamorphosed into orthoschists, dating confirmed by radiometric methods, and they interpreted the crystalloblastic texture and the petrographic facies in the framework of the tectonics of the Alps.

Bianchi took part in two scientific missions in East Africa (then ruled by Italy) headed by Michele Gortani (1936, 1938). He studied the granitic and migmatitic crystalline basement, and the volcanics associated with the rift system, in a region not scientifically explored until then, and the results were published between 1937 and 1974. He had previously studied the volcanics of the Dodecanese (1929–1930).

Other major research in collaboration with Giambattista Dal Piaz, on the Tertiary Adamello pluton, began in 1937. It involved several collaborators and was not completed when Bianchi died. A main contribution by him on the Adamello petrographic types appeared in 1970. Bianchi also investigated problems of ores and quarried stones.

Italian geological science is much indebted to Bianchi for his organizational activity. He headed the geologic section of the Italian National Research Council (CNR) from 1956 to 1968, founded and directed the CNR National Center for the Geologic and Petrographic Study of the Alps (1962–1968), and was president of the commission for the completion of the geologic map of Italy at the scale 1:100,000 (1960–1965), which supervised the surveying and mapping of 140 sheets (over 250 for the whole nation) and completed its work in 1970, in time for the centenary of the Geological Survey of Italy. Bianchi was president of the Italian Geological Society (1937) and of the Italian Mineralogical Society (1949–1951). He was twice dean of the Faculty of Sciences of Padua University (1941–1943, 1943, 1949–1952), was prorector (1949–1958), and was a member of the Administrative Council for many years.

Bianchi ceased teaching in 1963 and became emeritus in 1969. He combined a great ability for scientific synthesis and excellent teaching skills. His textbook of mineralogy for university students is still widely used. He continued to work until the day before his sudden death. Bianchi did much to aid needy students: Italian refugees from Istria, Fiume, and Italian Dalmatia, which were sent to Yugoslavia after World War II; and Hungarian refugees in 1956.

Bianchi was an honorary member of the Swiss Society for Natural Sciences; a national member of the Accademia Nazionale dei Lincei and of the Accademia Nazionale dei XL; and of several regional cultural institutions. He won the Royal Prize for Mineralogy of the Accademia dei Lincei in 1932.

Bianchi left his rich collection of minerals to the Institute of Mineralogy and Petrography at Padua. The sulfate bianchite $(Zn,Fe)SO_4 \cdot 6H_2O$, was named for him.

BIBLIOGRAPHY

I. ORIGINAL WORKS. Bianchi's more than 100 publications include the following: "Ilmenite di Val Devero (Ossola)," in *Rendiconti dell'Accademia nazionale dei Lincei*, 5th ser., **23** (1914), 722–727; "I minerali del Miage

(Monte Bianco)," in *Atti della Società italiana di scienze naturali*, **64** (1925), 132–174, with A. Cavinato; "La provincia petrografica effusiva del Dodecaneso (Mar Egeo). Riepilogo e conclusioni," in *Memorie dell'Istituto geologico dell'Università di Padova*, **8** (1930), 122–22; "La Val Devero e i suoi minerali. Atlante 1932," *ibid.*, alleg. **10** (1932), 1–15; "Monografia sull'Alto Adige orientale e regioni limitrofe. Studi petrografici," *ibid.*, **10** (1934), 1–243; "Atlante geologico-petrographico dell'Adamello meridionale, regione fra lo Stabio e il Caffaro," *ibid.*, **12** (1937), 1–16, with Giambattista Dal Paz; "Osservazioni geologiche e petrografiche sulla regione di Harar," in *Bollettino della Società geologica italiana*, **56** (1937), 499–516, with M. Gortani; *Corso di mineralogia* (Padua, 1968); "I tipi petrografici fondamentali del plutone dell'Adamello. Tonaliti, quarzodioriti, granodioriti e loro varietà leucocrate," in *Memorie degli Istituti di geologia e mineralogia dell'Università di Padova*, **27** (1970), 1–148, with E. Callegari and P. Jobstraibizer; "Itinerari geologici nella Dancalia meridionale e sugli altipiani Hararini, Etiopia, 1937–1941," in Accademia Nazionale dei Lincei, *Missione geologica dell'Azienda generale italiana petroli (A.G.I.P.) nella Dancalia meridionale e sugli altipiani Hararini (1936–1938)*, I (Rome, 1973), 1–237, with M. Gortani; and "Sulla presenza di sölvsbergiti in Africa orientale," *ibid.*, II (Rome, 1974), 201–231.

II. SECONDARY LITERATURE. The studies on eastern Alto Adige were reviewed by F. Karl and G. Morteani in *Tschermaks mineralogische und petrographische Mitteilungen*, **7** (1960), 290–316; the scientific results on Adamello pluton, by C. Exner in *Mitteilungen der Geologischen Gesellschaft in Wien*, **54** (1961), 261–265.

Obituaries include Conrad Burri, in *Verhandlungen der Schweizerischen naturforschenden Gesellschaft* (1974), 233–236; Ezio Callegari, in *Bollettino dell'Associazione mineraria subalpina*, **7** (1970), 13–31; Giambattista Dal Piaz, in *Celebrazione Lincee*, **50** (1971), 131,–31, and in *Bollettino della Società geologica italiana*, **92** (1973), 3–24; Roberto Malaroda, in *Atti dell'Accademia della scienze di Torino*, **106** (1972), 661–674; Giuseppe Schiavinato, in *Rendiconti dell'Istituto lombardo, Accademia di scienze e lettere*, **105** (1971), 3–10; and Bruno Zanettin, in *Atii dell'Istituto veneto dell'Accademia di scienze, lettere ed arti*, **129** (1971), 3–13, and in *Atti e memorie dell'Accademia patavina di scienze, lettere ed arti*, **84** (1972), 39–65.

GIULIANO PICCOLI

BIRGE, RAYMOND THAYER (*b.* Brooklyn, New York, 13 March 1887; *d.* Berkeley, California, 22 March 1980), *physics*.

Birge was a pivotal figure in introducing modern quantum physics in the United States and an architect of one of the most prestigious departments of physics. He was the son of John Thaddeus Birge, who was in the water transport and laundry machine businesses, and of Carolyn Raymond Birge. He married Irene Adelaide Walsh on 12 August 1913; they had two children, Carolyn Elizabeth and Robert Walsh. His son also became a physicist.

Because of the father's business reverses in 1905, the family moved to his hometown of Troy, New York, where Birge entered business school. An engineer uncle, Charles Raymond, rescued him the following year by financing his education at the University of Wisconsin, where another uncle, the distinguished limnologist Edward Asahel Birge, helped guide his professional training. His teachers included Charles E. Mendenhall, Leonard R. Ingersoll, and Benjamin W. Snow. Birge received the A.B. in 1909, the M.A. in 1910, and the Ph.D. in 1914. Most of his professional career was spent at the University of California, Berkeley, where he chaired the physics department (1933–1953). Birge was a member of American Philosophical Society, the National Academy of Sciences (to which he was elected in 1933), the American Physical Society (of which he was president in 1955), the Optical Society of America, the American Association for the Advancement of Science, and the Astronomical Society of the Pacific. He was named faculty research lecturer at the University of California in 1946, and the physics laboratory constructed there in 1963 was named for him.

Influenced by some early experiments of Robert W. Wood, which he followed up in his dissertation work under Mendenhall, Birge early became adept in the techniques of molecular spectroscopy and a "missionary for the Bohr atom," as he styled himself. After serving as instructor (1913–1915) and assistant professor (1915–1918) at Syracuse University, he joined the faculty of the University of California at Berkeley, where Gilbert Newton Lewis, the dean of the College of Chemistry, had developed the chemist's alternative to the Bohr atom.

Birge's advocacy of the Bohr theory and his work in molecular spectroscopy attracted the attention of the National Research Council, which nominated him to the Committee on Radiation in Gases. This committee prepared the report "Molecular Spectra in Gases" between 1922 and 1926, a crucial time in the evolution of the quantum theory. Birge's contribution to the report, on electronic bands, was the most extensive chapter. Synthesizing data from his dissertation research, from astrophysical studies at the Mount Wilson Observatory, and from his own research group at the University of California, Birge reinterpreted them in light of the Bohr theory. Appearing in 1926, just when quantum mechanics was being formulated, the report had less impact, perhaps, than John H. van Vleck's *Quantum Principles and Line Spectra* (1926) but played a similar role in educating the American community of physicists in quantum theory.

As a result of this work, Birge and his research group acquired a worldwide reputation. International research fellows like Gerhardt Dieke and Hertha Sponer (later Hertha Franck) came to his laboratory, then the largest academic physics laboratory in the country, to work with him. One of his own graduate students, Edward U. Condon, was led to the Franck-Condon principle through his association with Sponer in this laboratory. Sponer and Birge extended the technique developed by Sponer to measure heats of dissociation in iodine, ordinarily masked by excited states, to other nonpolar molecules and developed the Birge-Sponer method for measuring these quantities. With John J. Hopfield, Birge discovered the Birge-Hopfield systems of bands. Birge was not, however, capable of reinterpreting this work in the light of quantum mechanics, which he found uncongenial to his imagination: "I still believe in pictures, for the sake of the experiments, whether they are true or not," he wrote to Edwin C. Kemble. "I think an approximately true picture may be of more use to the advancement of science than a more accurately true mathematical structure that is too abstruse for many investigators to profit by."

A by-product of the molecular spectra work was the discovery of several important isotopic species, including two new heavy isotopes of oxygen, O^{18} and O^{17}, by W. F. Giauque, using data and analysis supplied by Birge, who also calculated the relative masses of the new isotopes. Their discovery had radical consequences for chemistry, for atomic weight determinations had been based upon the atomic weight of oxygen. Recalculating the values of atomic weights showed a discrepancy with the atomic weight of hydrogen, which Birge and Donald Menzel, a Lick Observatory astronomer, suggested might be due to a heavy hydrogen isotope composing about 1 part in 4,500 in hydrogen gas of mass 1. This suggestion contributed to the discovery of deuterium by Harold Urey in 1931. Birge also discovered carbon 13 through an analysis of a faint band in vacuum electric furnace spectra from Mount Wilson that he confirmed by using his own carbon monoxide absorption spectra.

Birge built not only a research group but also a department. Under department chairmen E. Percival Lewis and Elmer E. Hall, Birge and Leonard Loeb were given extraordinary latitude in building up departmental research through recruiting, special funds from the university's Board of Research, and active patronage from G. N. Lewis' College of Chemistry and University of California President W. W. Campbell. Together they brought the department such luminaries as Ernest Lawrence, Robert Brode, and Harvey White. Succeeding Hall as chairman in 1933, Birge sustained this tradition throughout the Great Depression. Edwin M. McMillan and Luis Alvarez, both of whom were later Nobel Prize winners, were added to the department in these years. The University of California Radiation Laboratory was an offshoot of the department.

Although many department chairmen sacrifice research to administration, Birge found a new field in which he distinguished himself: the establishment, by the latest experimental, calculational, and statistical techniques, of the values of the general physical constants. This work began with his studies of the most probable value of the Planck constant and became a focus of his activity when he compiled a table of constants for diatomic molecules for the *International Critical Tables* in 1928. In order to synthesize the reported experimental observations of these values, Birge developed the Birge-Bond diagram to plot them and introduced the method of least squares into error analysis. He was led from this to a study of the fundamentals of statistical error analysis in collaboration with W. Edwards Deming of the Fixed Nitrogen Laboratory. His interest in statistics extended to his administrative relations with the Berkeley mathematics department, where he was instrumental in securing statistician Jerzy Neyman in 1938.

Birge was an important institution builder in an age when the institutions of science were growing as never before. The patronage he gave to Ernest Lawrence and others who built the first of the modern academic accelerator laboratories, now the Lawrence Berkeley Laboratory, was crucial to the emergence of modern "big science."

BIBLIOGRAPHY

I. ORIGINAL WORKS. Birge's writings include "The Most Probable Value of the Planck Constant h," in *Physical Review*, 2nd ser., **14** (1919), 361–368; "Electronic Bands," in Edwin C. Kemble *et al.*, "Molecular Spectra in Gases; Report of the Committee on Radiation in Gases," in *Bulletin of the National Research Council*, **11** (December 1926), 69–259; "The Heat of Dissociation of Non-Polar Molecules, as Determined from Band Spectra, and from Other Sources," in *Physical Review*, 2nd ser., **27** (1926), 640–641, written with Hertha Sponer; "The Heat of Dissociation of Nonpolar Molecules," *ibid.*, **28** (1926), 259–283, written with Hertha Sponer; "Further Evidence of the Carbon Isotope, Mass 13," in *Nature*, **124** (1929), 182, and in *Physical Review*, 2nd ser., **34** (1929), 376; "An Isotope of Carbon, Mass 13," in *Nature*, **124** (1929),

127, and in *Physical Review*, 2nd ser., **34** (1929), 376; "Molecular Constants Derived from Band Spectra of Diatomic Molecules," in *International Critical Tables*, V (1929), 409–418; "Probable Values of the General Physical Constants," in *Physical Review Supplement*, **1** (July 1929), 1–73; "Recent Work on Isotopes in Band Spectra," in *Transactions of the Faraday Society*, **25** (1929), 718–725; "Evidence from Band Spectra of the Existence of a Carbon Isotope of Mass 13," in *Astrophysical Journal*, **72** (1930), 19–40, written with Arthur S. King; "Precision Determination of the Mass Ratio of Oxygen 18 and 16," in *Physical Review*, 2nd ser., **37** (1931), 233, written with H. D. Babcock; "The Relative Abundance of the Oxygen Isotopes, and the Basis of the Atomic Weight System," *ibid.*, 1669–1671, written with Donald Menzel; "On the Statistical Theory of Errors," in *Reviews of Modern Physics*, **6**, no. 3 (1934), 119–161, written with W. Edwards Deming; "Physics and Physicists of the Past Fifty Years," in *Physics Today*, **9**, no. 5 (1956), 20–28, which describes his educational experiences; "History of the Physics Department, University of California, Berkeley," 5 vols. (Berkeley, 1966), typescript; and "Physics," in Verne A. Stadtman, ed., *The Centennial Record of the University of California* (Berkeley, 1967), 97.

Birge's papers are collection 73/79C, Bancroft Library Manuscripts Division, University of California at Berkeley. Some of his correspondence is in Sources for History of Quantum Physics, Berkeley.

II. SECONDARY LITERATURE. On Birge see Edna Tartual Daniel, "Interview with Raymond Thayer Birge, Physicist" (Berkeley, 1960), in the Regional Oral History Office, Bancroft Library; and Robert W. Seidel, "The Origins of Physics Research in California," in *Journal of College Science Teaching*, **6**, no. 1 (1976), 10–23, and "Physics Research in California: The Rise of a Leading Sector in American Physics" (Ph.D. diss., University of California, 1978).

Writings related to Birge's work include Harold D. Babcock, "Some New Features of the Atmospheric Absorption Bands, and the Relative Abundance of Isotopes O^{16}, O^{18}," in *Proceedings of the National Academy of Sciences*, **15** (1929), 471–477, and "Relative Abundance of the Isotopes of Oxygen," in *Physical Review*, 2nd ser., **34** (1929), 540–541; Harold D. Babcock and G. H. Dicke, "The Structure of the Atmospheric Absorption Bands of Oxygen," in *Proceedings of the National Academy of Sciences*, **13** (1927), 670–678; W. F. Giauque, "An Isotope of Mass 17 in the Earth's Atmosphere," in *Nature*, **123** (1929), 831; W. F. Giauque and H. L. Johnston, "An Isotope of Oxygen, Mass 18," in *Nature*, **123** (1929), 318; and Harold C. Urey, F. G. Brickwedde, and G. M. Murphy, "A Hydrogen Isotope of Mass 2," in *Physical Review*, 2nd ser., **39** (1932), 164–165.

ROBERT W. SEIDEL

BJERKNES, JACOB AALL BONNEVIE (*b.* Stockholm, Sweden, 2 November 1897; *d.* Los Angeles, California, 7 July 1975), *meteorology, oceanography*.

Bjerknes's father, Vilhelm, has often been called the father of modern meteorology. This title is doubly appropriate, for not only did Vilhelm's vision and research programs lead to establishing new directions, methods, and the conceptual foundation for the science, but Vilhelm's son also played a fundamental role in creating this new era in atmospheric science. Science was very much part of the world into which the younger Bjerknes was born. His father was professor of mechanics and mathematical physics at the Stockholm Högskola (later University of Stockholm); his father's father, Carl Anton, was professor of pure mathematics at the Royal Frederick University (later University of Christiana [Oslo]); his mother, Honoria Bonnevie, studied natural science; and an aunt, Kristine Bonnevie, zoologist and embryologist, became Norway's first woman professor.

Bjerknes was a reserved child. At home he had to compete for his parents' attention with a sickly older brother and twin younger brothers; at school he was harassed by Swedish classmates, especially when Norway broke away from the union with Sweden in 1905. The family moved home to Norway in 1907 after Vilhelm received a personal professorship in mechanics and mathematical physics at the University of Christiania. His research, which eventually proved crucial for Jacob, entailed trying to establish a physics of atmospheric change. When Vilhelm moved to the University of Leipzig in 1913 to head its new Geophysical Institute, Jacob first continued his schooling in Norway, but in 1916 enrolled at Leipzig. Here he began research as one of his father's two assistants, funded by the Carnegie Institution. In 1917 the family moved to Bergen, where as part of a plan to expand the Bergen Museum into a university, Vilhelm became professor at a new geophysical institute. Although a university was not established until 1948, Vilhelm Bjerknes established a weather forecasting service in Bergen and Jacob became its head.

As chief forecaster at Bergen (1918–1931), Jacob, along with other young Norwegian and Swedish scientists there, effected a major transformation in meteorological practice and theory. In 1931 he assumed the professorship of meteorology at the Bergen Museum. Now able to devote considerably more time to research and also to rely on increasingly sophisticated technologies for obtaining observations in the upper troposphere, he began path-breaking studies relating lower atmospheric weather phenomena to wave patterns in air currents higher up. During the time spent at Bergen, on 11 July 1928,

Bjerknes married Hedvig Borthen. They had a son and a daughter.

When the Germans invaded Norway in April 1940, Bjerknes was in the United States, where he was helping to introduce the Norwegian forecasting methods and research programs. Not able to return home, he accepted a professorship of meteorology at the University of California at Los Angeles, where Jørgen Holmboe, a former assistant to his father, was already working. The two developed a major academic department of meteorology and during the war trained a large number of military meteorologists. Bjerknes initiated several pioneering research programs related to upper atmospheric patterns, atmospheric general circulation, and interactions between the atmosphere and the oceans. Although he remained in Los Angeles until his death, he kept close contact with Norway and Norwegian meteorology, both by visiting there and by inviting Norwegians to come as guest researchers to the new "Bergen" on the Pacific coast.

Jacob Bjerknes is said to have discovered in 1918 that cyclones are composed of weather fronts, that is, three-dimensional surfaces of discontinuity separating air masses of different origin and physical characteristics. Although use of the term "discovered" enabled the Bergen scientists to legitimize their findings by imputing that they were in nature, waiting to be seen, the actual process of constituting new concepts and models was more complex and reveals considerable ingenuity and insight. At Leipzig, Jacob had inherited the research of Herbert Petzold, the Geophysical Institute's first doctoral student. Before being called into the army, Petzold was given the problem of studying the kinematics and dynamics of line squalls, which are long, narrow bands of thunderstorms and showers accompanied by strong winds.

These puzzling storms had become increasingly significant because of the danger they posed to aeronautical operations. Vilhelm Bjerknes understood that aviation held the greatest potential for meteorology's growth and renewal as a discipline; the Leipzig institute's origin and mission was bound to German aeronautical development. His research program at the time entailed finding graphic methods for directly applying known hydrodynamic and thermodynamic principles to the atmosphere and oceans. Jacob began studying line squalls by relating them to the kinematics of wind flow in proximity to so-called lines of convergence, in which, on a horizontal plane, wind converges from two directions.

In Bergen, Vilhelm Bjerknes sought to establish resources and markets for his meteorology. Taking a cue from the massive buildup of meteorological services among the warring nations, he first tried to interest the military in creating a field weather service for Norway's growing air force. When a threat of famine confronted Norway in 1917 and 1918, government intervention in all aspects of agricultural production prompted Bjerknes to propose implementing, during the summer of 1918, an experimental weather service for western Norway and an expansion of the existing Norwegian Meteorological Institute in Christiania.

Without any weather observations from Britain and Iceland during the war, forecasting for western Norway seemed hopeless if one used traditional methods based on delineating patterns of atmospheric pressure. In order to overcome the lack of observations and to experiment with what he called rational precalculations of atmospheric changes, Vilhelm adapted the structure of the German military field weather service: an extremely tight network of observation stations using telephone and wireless telegraphy for rapid exchanges of detailed weather data. Jacob's study of lines of convergence in the wind field offered a possible cognitive foundation for forecasting, in that these lines were associated with rain patterns. It was hoped that a dense network of observation stations on the many skerries and islands off the coast might allow the early detection of lines of convergence, which in turn might permit calculation of their motion for the next few hours according to a formula developed by Jacob while he was at Leipzig.

Although Jacob found that the data did not permit calculations, during the course of the forecasting he concluded that moving cyclones (midlatitude, low-pressure systems) are composed of two such lines of convergence that separate a tongue of warm air from surrounding colder air. The cyclone's rain pattern and other physical characteristics seemed to be linked to these lines in the horizontal wind field. During the next several months Jacob and Vilhelm Bjerknes, along with Halvor Solberg, modified the model of 1918 several times. They attained a satisfactory version when they abandoned the original kinematic analysis of wind flow and replaced it with a more physical approach stressing the actual weather. Consequently, the geometric two-dimensional lines of convergence were replaced in their thinking and model by three-dimensional surfaces of discontinuity separating air masses of differing physical properties.

Although rapid transitions or discontinuities in temperature, pressure, and wind velocity had been noted in Jacob's first model—and even by nineteenth-century meteorologists—and although such discontinuities had been theoretically postulated, the 1919 Bergen cyclone model represented the introduction of discontinuities into meteorology as the major focus for practice and, eventually, theory. These "fronts," as they later were termed, were endowed with scientific reality through the introduction of major changes in forecasting practice. In Norway, and especially abroad, the advent of commercial aviation prompted revolutionary changes in the types of weather data and the frequency of observation; when analyzed according to the methods devised in Bergen, the data allowed the regular reproduction of fronts in weather forecasting practice.

The cyclone model based on fronts entailed the ability to specify in time and three-dimensional space important weather phenomena associated with cyclones to a much greater degree than earlier models could achieve. During the next few years the Bergen group of Vilhelm and Jacob Bjerknes, Ernst Calwagen, Halvor Solberg, and Tor Bergeron, among others, expanded this preliminary cyclone model and accompanying forecasting methods to a comprehensive system for explaining midlatitude weather changes. The polar front, the occlusion process, an evolving cyclone model, and air mass analysis were the conceptual foundation for a new era in meteorology.

Although Jacob's cyclone model was three-dimensional, the vertical structure of the fronts was inferred indirectly from cloud observations and from Max Margules' earlier theoretical model of the character of boundary surfaces between air masses. Observations from the upper air were still too sporadic and sparse to reveal the exact nature of the upper portions of fronts. From 1922 to 1923 Jacob spent a year in Zurich, where, in addition to trying to introduce the Bergen methods into the Swiss weather service, he studied weather data from Alpine observatories (some up to 4,000 meters) to confirm the sloped character of the frontal surfaces. This work resulted in his doctoral dissertation at the University of Christiania (1924).

In the late 1920's and 1930's improvements in aerological instrumentation permitted better upper-air observations. Using the 1928 observations obtained with balloons and recording instruments by the Belgian P. Jeaumotte, Jacob proposed in 1932 a number of extraordinarily insightful ideas on the relation between upper-air currents and cyclonic and anticyclonic (high-pressure systems) activity in the lower atmosphere. He analyzed the flow of air over and under the sloping frontal surfaces in great detail, showing how the surface of discontinuity of a cold front eventually extends over and behind the cold air mass, where it becomes a warm front. By pointing out the wavelike character of the westerly air current in the upper atmosphere above the frontal surfaces, he had the beginnings of a theory associating the upper- and lower-level phenomena that also accounted for the wavelike pattern of the upper wind current.

After publishing in the 1930's a series of articles with the Finnish meteorologist Erik Palmén on the three-dimensional structure of cyclones, Jacob returned to the problem of the theory of waves in the upper westerly wind current, now including as a dynamic factor the so-called beta-effect, the Coriolis parameter's gradient. Using minimal mathematics and theoretical apparatus, but with extraordinary physical insight and ability to simplify in order to clarify, Jacob laid down the foundational ideas upon which the next decade's theoreticians, such as Carl-Gustaf Rossby, would develop a new perspective on atmospheric motions. In 1944 he and Holmboe presented a theory linking cyclonic development to pressure and vorticity changes arising from horizontal divergence in a baroclinic air current. When in 1947 Jule Charney, who had been awarded the first UCLA Ph.D. in meteorology in 1946, added boundary conditions to this model, a mature theory of cyclogenesis arising from baroclinic instability was completed.

In the 1950's Jacob Bjerknes continued research on upper-level air currents, including the newly discovered jet stream. Toward the end of the decade he began investigating a subject that held his attention for the rest of his life: interactions between the ocean and the atmosphere, and their impact on weather, climate, and ocean currents. His work included an explanation of the "Niño" effect in the Pacific Ocean. More important, his efforts provided a stimulus to studying the two geophysical fluids—atmosphere and ocean—as a single system, a project that has become increasingly significant for science and society.

BIBLIOGRAPHY

I. Original Works. Bjerknes's publications are listed in Poggendorff, VIIb, 404–405. His most important articles are in *Selected Papers of Jacob Aall Bonnevie Bjerknes*, M. G. Wurtele, ed. (North Hollywood, Calif., 1975). Unpublished papers and correspondence have been deposited in the University of Oslow Library and the District State Archive (Statsarkiv) in Bergen.

II. Secondary Literature. An obituary by Arnt Eliassen is "Jacob Bjerknes og hans livsverk. Minneforelesning holdt ved Universitet i Oslo, 25 September 1975," in *Det Norske videnskaps-akademi, Arsbok 1976*, 142–156. Bjerknes's Leipzig and early Bergen work is analyzed in Robert Marc Friedman, *Appropriating the Weather: Vilhelm Bjerknes and the Construction of a Modern Meteorology* (Ithaca, N.Y., 1989). Discussions by meteorologists on various aspects of Bjerknes's work are in his *Selected Papers*.

Robert Marc Friedman

BOCHNER, SALOMON (*b.* Cracow, Austria-Hungary [now Poland]), 20 August 1899; *d.* Houston, Texas, 2 May 1982), *mathematics*.

Bochner, a mathematician noted for the breadth and originality of his work in analysis, was born in a highly orthodox Jewish community. His father, Joseph, a small businessman, and his mother, Rude Haber, were self-educated beyond grade school but were both assiduous readers—she especially of Shakespeare and Ibsen, and he especially of works of Hebrew scholarship. Bochner had a younger sister, Fanny.

When a Russian invasion threatened in 1914, the family fled to Berlin. Although they were poorer after this move, young Salomon had much greater educational opportunities in the more liberal atmosphere of Berlin. He attended an outstanding gymnasium, brilliantly passing his entrance examinations only a few months after arriving in the city. He was strongly drawn to history and the humanities, but his mathematical talents had long been evident and the limited financial resources of his family impelled him to pursue a surer career in mathematics. He earned a Ph.D. from the University of Berlin in 1921, with a dissertation on orthogonal systems of complex analytic functions.

Inflation was then devastating Germany, so to help his family Bochner abandoned mathematics for a few years and went into the import-export business, in which he was successful but quite unhappy. His real interest was still mathematics; and at the urging and with the full support of his family, he returned to that field.

Bochner's dissertation topic had been treated independently by S. Bergman, so Bochner moved on to other topics. At that time Harald Bohr was developing the theory of almost periodic functions, which he described in short notes published in 1923. Bochner read these and was inspired to work out much of the theory on his own, in the process developing a highly original method of summation quite different from, and in some ways better than, that used by Bohr. This so impressed Bohr that he promptly invited the younger mathematician to visit Copenhagen. Bochner also discovered an alternative characterization of almost periodic functions in terms of the compactness of the sets of translates of the functions; this was the basis for future generalizations of the notion of almost periodicity by Bochner and others. From 1924 to 1926 Bochner was a fellow of the International Education Board, and in addition to Copenhagen he visited Oxford to work with G. H. Hardy and Cambridge to work with J. E. Littlewood. That led to an interest in yet another area, the theory of zeta functions and their functional equations, to which he would return later.

Bochner spent the years from 1926 to 1933 as a lecturer at the University of Munich. This was a very productive period, during which he began the research on Fourier analysis that was perhaps his greatest achievement. The culmination of this work at Munich was the publication in 1932 of *Vorlesungen über Fouriersche Integrale*, an influential book that has become a mathematical classic. Among much else the book contains Bochner's most famous theorem, characterizing the Fourier-Stieltjes transforms of positive measures as positive-definite functions; this result was the cornerstone of the subsequent development of abstract harmonic analysis. The book contained the seeds of what developed in other hands into the theory of distributions, another major mathematical tool.

In 1937 Bochner published his generalization of the Lebesgue integral to functions with values in an infinite-dimensional normed linear space, the Bochner integral. In a quite different area, during his earlier years at Munich he studied the continuation of Riemann surfaces; his paper on that subject (1928) contained, incidentally, a result about mathematical logic that was discovered independently some seven years later by Max Zorn and is usually called Zorn's lemma. Bochner was an influential figure in Munich, not just among the mathematicians but also among the physicists. He published papers on X-ray crystallography with H. Seyfarth.

The rise of Nazism in Germany impelled Bochner to move in 1933, when he accepted a position at Princeton University. He became a naturalized citizen of the United States in 1938. Except for a visiting professorship at Harvard in the spring of 1947 and one at the University of California at Berkeley in the spring of 1953, he was away from Princeton only for brief trips and summers during the thirty-five years before his retirement. Until World War II broke out, he returned to Europe every summer to visit his family, to whom he was devoted. He urged them to move to England; and when they did so, he helped to arrange for his sister Fanny's children to attend good schools there. During one of these trips he met Naomi Weinberg, whose father was a New York real estate entrepreneur and founder (and for fifty years publisher) of the Jewish newspaper *The Day*. He and Naomi were married on 1 November 1938, with John von Neumann as best man. They had a daughter, Deborah.

Bochner was a major figure at Princeton and influential in the world of mathematics. During the 1930's a number of National Research Council fellows came to Princeton to work with him: Ralph Boas, R. H. Cameron, Norman Levinson, William T. Martin, and Angus Taylor, among others; K. Chandrasekharan and Kentaro Yano followed later. He had thirty-five doctoral students in a wide variety of fields, almost a quarter of the Ph.D.s in mathematics produced during his years on the Princeton faculty. When he was awarded the Leroy P. Steele Prize of the American Mathematical Society in 1979, he was cited for the cumulative impact of his mathematical work and for his influence on mathematics through his students. He was elected to the National Academy of Sciences in 1950 and served as vice-president of the American Mathematical Society from 1957 to 1958.

Bochner's research continued unabated at Princeton, in part as an extension of the work on almost periodic functions and Fourier analysis he had begun in Germany. He undertook the first profound investigation of the summation of Fourier series in several variables, discovering that Riemann's localization theorem, a basic result for Fourier series in one variable, fails in the case of several variables. John von Neumann had extended the theory of almost periodic functions to functions on arbitrary groups, and he and Bochner jointly extended the theory still further to functions with values in complete normed linear spaces. Bochner also introduced various generalized notions of almost periodicity.

With characteristic originality and facility Bochner extended his research in a number of new directions. In the late 1930's he began investigations in the theory of functions of several complex variables by determining the envelopes of holomorphy of tube domains. He continued in the 1940's by developing extensions of the Cauchy integral formula, including the Bochner-Martinelli formula that has been basic in the subject, and by using these formulas to characterize the boundary values of complex analytic functions. In this work, in the guise of conglomerate functions and their saltuses, there appeared the germ of the theory of cohomology groups with coefficients in holomorphic vector bundles, a theory that was extensively developed by others over the next quarter of a century as a basic tool in complex analysis. Bochner's outlook in this field was summarized in *Several Complex Variables* (1948), written with W. T. Martin. Also in the late 1930's Bochner began to work in differential geometry, showing that Riemann surfaces admit real-analytic embeddings into Euclidean spaces.

This research continued in the 1940's in joint work with Deane Montgomery, in which they proved that the group of holomorphic automorphisms of a compact complex manifold is a complex Lie group. At the Princeton Bicentennial Conference in 1946, Bochner presented a proof showing that the additive Cousin problem always has a solution on a compact Kähler manifold with positive Ricci curvature, thus inaugurating the theory of curvature and Betti numbers that he developed further in the 1940's and 1950's. This theory has been of fundamental importance, with a wide variety of applications—for instance, in the magisterial work of Kunihiko Kodaira characterizing algebraic manifolds as those complex manifolds that admit a Hodge metric. During the 1940's and 1950's Bochner also embarked on investigations in mathematical probability, summarized in his 1955 book *Harmonic Analysis and the Theory of Probability*. In addition he wrote a number of papers on partial differential equations, on zeta functions, and on gamma factors, among other topics.

In the mid 1960's Bochner turned in an altogether new direction, devoting himself wholeheartedly to the history and philosophy of science, particularly of mathematics. His books *The Role of Mathematics in the Rise of Science* (1966) and *Eclosion and Synthesis* (1969), as well as papers of that period, contain fascinating, idiosyncratic, and often provocative observations. They fuse his lifelong interest in history, literature, and philosophy, along with his undoubted gift for languages, with an understanding of mathematical creativity. He participated in the seminars of the history of science program at Princeton during this period, and was the only scientist member of the editorial board that supervised the publication of the five-volume *Dictionary of the History of Ideas*.

In 1968 Bochner retired from the Henry Burchard Fine professorship of mathematics at Princeton, which he had held since 1959, and moved to Rice University as the first Edgar Odell Lovett professor of mathematics. He was as influential at Rice as he had been at Princeton, if not more so, for he also served as chairman of the mathematics department there from 1969 to 1976 and took a great interest in developing the department. In addition to teaching mathematics he participated in the history of science program, was responsible for the foundation of an interdisciplinary institute for the history of ideas, and gave university-wide public lectures on the history of science. He died after a brief illness.

BIBLIOGRAPHY

I. ORIGINAL WORKS. Many of Bochner's writings were collected in *Selected Papers of Salomon Bochner* (New York, 1969). His books include *Several Complex Variables* (Princeton, 1948), written with William T. Martin; *Fourier Transforms* (Princeton, 1949), written with K. Chandrasekharan; *Curvature and Betti Numbers* (Princeton, 1953), written with Kentaro Yano; *Harmonic Analysis and the Theory of Probability* (Berkeley, 1955); *Lectures on Fourier Integrals*, Morris Tenenbaum and Harry Pollard, trans. (Princeton, 1959); *The Role of Mathematics in the Rise of Science* (Princeton, 1966); and *Eclosion and Synthesis* (New York, 1969).

II. SECONDARY LITERATURE. On Bochner's role in the "prehistory" of the theory of distributions, see J. Lutzen, *The Prehistory of the Theory of Distributions* (New York, 1982).

R. C. GUNNING

BOGDANOV, ALEKSEI ALEKSEEVICH (*b.* Sukhumi, Russia, 25 February 1907; *d.* Moscow, U.S.S.R., 18 September 1971), *geology, tectonics.*

Soon after Aleksei's birth the family moved to Geneva, where they lived until 1916. His father, Aleksei Alekseevich, studied and worked at the University of Geneva. After returning to Russia in early 1916, he was an engineer-chemist in the oil industry. Bogdanov's mother, Tatiiana Gennadievna Kartsova, studied and worked at the University of Geneva and at the Pasteur Institute in Paris. A physician bacteriologist, at the beginning of World War I she went to the Serbian front and from there was transferred to the Caucasus in 1916. She died of typhus in 1919. In 1929 Bogdanov married Irina Vladimirovna Gudkevich; they had two sons, Nikola and Aleksei.

Bogdanov was first educated at home, then at a secondary school in Yaroslavl', from which he graduated in 1925. He next studied at the geological prospecting faculty of the Moscow Academy of Mines until 1931. While a student, Bogdanov spent the summers on field expeditions. In 1941 he received his candidate degree, and in 1945 his doctorate.

In 1935 Bogdanov started teaching at the Moscow Geological Prospecting Institute: first as an assistant, then assistant professor and professor (1946). He taught general geology, geotectonics, geological mapping, and structural geology. In 1951 he transferred to Moscow State University, where he lectured on the geology of the U.S.S.R. Under his supervision a great number of young geologists received their doctorates.

Bogdanov's early geological investigations were done under the guidance of such prominent Soviet geologists as A. D. Arkhangel'skii and N. S. Shatskii. His contact with them helped to form his interest in sedimentary rocks, their stratigraphy, and especially tectonic structures.

During the 1930's and in the early 1940's Bogdanov's investigations dealt with the geological structure of regions that appeared to contain oil and gas deposits. These works contributed to the knowledge of tectonics of the regions studied, particularly the salt domes of the northern Caspian area.

In the first half of the 1940's Bogdanov headed a large expedition working in the southern Urals that found several gas- and oil-bearing geological structures. Later, for some years he supervised geological investigations in the Carpathians. During his later life Bogdanov and a large group of geologists studied the geology of central Kazakhstan. All of this work contributed much to knowledge of the stratigraphy, magmatism, and tectonics of the Soviet Union.

Though almost all of Bogdanov's regional geological investigations were applied in character, they also served as the basis for a number of important theoretical conclusions. Bogdanov revealed the peculiarities of the tectonic structure of the southeastern part of the Russian platform (the East European platform) and elucidated the history of formation of oil-bearing structures developed there. Being profoundly interested in the boundaries of the East European platform, and having analyzed the latest data, Bogdanov concluded that it stretches much further west than had been assumed and extends at an acute angle up to the Gulf of Bristol in England.

Bogdanov noted the expediency of distinguishing in the history of each platform the stage of its cratonization, that is, the beginning of the platform proper. Pursuing this idea, he suggested two stages in a platform's development. The first, the avlakogene, is characterized by the existence of vast salients of the basement (shields) separated by deep, grabenlike basins (avlakogenes or aulacogenes). The second is syneclises or plates during formation of which the avlakogenes are involved in general subsidence; this results in the formation of syneclises and anteclises, structures peculiar to all platforms. Bogdanov's concept of the relationship of the platform to its fold framework aroused appreciable scientific interest. He also showed that marginal depressions are tectonic elements arising in boundary areas between platforms and between other depressions attributed to geosynclinal structures.

During his trips abroad to participate in international geological congresses, and to keep abreast of the compilation of maps in many countries, Bogdanov always went on excursions to geologically interesting regions. He visited most of Europe, southern Asia, North Africa, and the Americas. The geological excursions in these areas provided material that enabled him to compile regional tectonic essays in which he amended the previous, or suggested quite new, treatment of the tectonic structures of these territories.

Bogdanov's interest in tectonics led him to take an active part in Shatskii's compilation of the tectonic map of the Soviet Union. It was published, on a scale of 1:4,000,000, in 1953 and was twice republished (1956, 1961) with supplementary data.

At the Twentieth International Geological Congress, held in Mexico City in 1956, Bogdanov presented the tectonic map of the U.S.S.R. It was unanimously approved, and the congress passed a resolution organizing the Subcommission on the Tectonic Map of the World. Bogdanov was elected its secretary general and enlisted the cooperation of a great number of geologists from around the world.

In the process of preparing the tectonic maps it became necessary to unify plotted images. Since not all geologists drew and understood particular structures in the same way, Bogdanov had to prepare the uniform legend that was unanimously approved and subsequently used in compilation of all international tectonic maps.

Through Bogdanov's efforts the tectonic map of Europe was published in fifteen sheets (1964). He then began to work on a tectonic map of Earth. He died before completing it.

While studying Earth's crust structure, especially in relation to the compilation of tectonic maps, Bogdanov suggested the principles of distinguishing the major structural elements, their zonation, and the periodization of Earth's tectonic history. His interpretation of peculiarities of the Paleozoic folding in the U.S.S.R. is a striking example of such an analysis.

Bogdanov was a corresponding member of the German Academy of Sciences in Berlin (1962); a foreign member of the geological societies of Belgium, France, Czechoslovakia, and Sweden; an honorary member of the Geological-Mineralogical Society of the Czechoslovakian Academy of Sciences; and an honorary doctor of the University of Paris (Sorbonne). He was also awarded a gold medal by Charles University (Prague) and the Leopold von Buch gold medal by Göttingen University.

For many years Bogdanov suffered from heart disease but, being a man of striking energy, surprising capacity for work, and exceptional cheerfulness, he paid no attention to the state of his health. After conducting field investigations under the climatically severe conditions of semidesert Kazakhstan, he died in Moscow of a ruptured aorta.

BIBLIOGRAPHY

I. Original Works. *Rukovodstvo k prakticheskim zaniatiiam po kursu obshchei geologii* (Instructions for practical studies on general geology), 2nd ed., M. M. Zhukov, ed. (Moscow, 1946), written with M. M. Zhukov, E. V. Milanovskii, and V. N. Pavlinov; "Paleozoiskie tektonicheskie struktury iuzhnoi chasti Karagandinskoi oblasti i Chu-Balkhashskogo vodorazdela" (Paleozoic tectonic structures of the southern part of Karaganda Province and Chu-Balkhash water parting), in *Tektonika SSSR*, I (Moscow and Leningrad, 1948), 79–144; "La carte tectonique de l'URSS," in *Relaciones entre la tectónica y la sedimentación*, I (Mexico City, 1957), 267–287; *Tektonicheskaia karta SSSR i sopredel'nykh stran v masshtabe 1:5000000* (Tectonic map of the U.S.S.R. and contiguous countries, scale 1:5,000,000), compiled by Bogdanov and N. S. Shatskii, explanatory note edited by Shatskii (Moscow, 1957), also translated into Chinese, edited by Ezhan Wen-yu (Peking, 1958); "Traits fondamentaux de la tectonique de l'URSS," in *Revue de géographie physique et géologie dynamique*, 2nd ser., **1**, fasc. 3 (1957), 134–165; *Problem vztahu Kaledonské a variské tektogeneze v centrálním Kazachstánu* (Prague, 1961); "Über einige allgemeine Fragen der Tektonik alter Tafeln am Beispiel der östeuropäischen Tafel," in *Geologie*, **1**, no. 9 (1965), 1017–1038; *Die tektonische und territoriale Gliederung der Palaozoiden Zentralkazachstans und des Ti-an-Schan anlässlich der 200. Jahr-Feier der Bergakademie in Freiberg, I* (Freiberg, 1966), 33–83; "Époques tectoniques et chronologiques de la subdivision et périodes de l'histoire tectonique de la craute terrestre," in *Bulletin de la Société géologique de France*, **7** (1969), 717–728; and *Tektonika platform i skladchatykh oblastei* (Tectonics of platforms and folded areas; Moscow, 1976).

II. Secondary Literature. "A. A. Bogdanov (k 60 letiiu so dnia rozhdeniia" (To the sixtieth birthday of A. A. Bogdanov), in *Vestnik Moskovskogo gosudarstvennogo universiteta*, ser. 4, *Geologiia*, 1967, no. 1; V. E. Khain, "A. A. Bogdanov i problemi tektoniki Evropi"

(A. A. Bogdanov and the problems of European tectonics), in *Biulletin Moskovskoga obshchestva ispitatelei prirodi*, **77**, otd. geol., **47**, no. 5 (1972), 17–24; E. E. Milanovskii, "Rol Alekseia Alekseevcha Bogdanova v razvitii geologicheskogo faculteta Moskovskogo universiteta" (A. A. Bogdanov's role in the development of the Geological Faculty of Moscow State University), in *Vestnik Moskovskoga universiteta*, ser. geol., no. 2, (1977), 3–8; V. A. Varsanofiev, G. S. Vartanian, and E. A. Kuznetsov, "Pamiati Alekseia Alekseevicha Bogdanova. (Geolog. 1907–1971)" (To the memory of Aleksei Alekseevich Bogdanov. [Geologist. 1907–1971]), in *Biulletin Moskovskoga obshchestva ispitatelei prirodi*, **77**, otd. geol., **47**, no. 5 (1972), 5–16; and Ia. A. Zaitsev, "Idei A. A. Bogdanova v izuchenii geologii Kazakhstana" (The ideas of A. A. Bogdanov in the resourcing of the geology of Kazakhstan), *ibid.*, 25–29.

V. V. TIKHOMIROV

BOHR, CHRISTIAN HARALD LAURITZ PETER EMIL (*b*. Copenhagen, Denmark, 14 February 1855; *d*. Copenhagen, 3 February 1911), *respiratory physiology*.

Bohr was the son of Henrik G. C. Bohr, the headmaster of a school, and of Augusta L. C. Rimestad. He entered Copenhagen University in 1872, became assistant to Peter L. Panum in 1878, and received the M.D. degree in 1880. He married Ellen Adler, daughter of a prominent Jewish financier, in 1881 in a civil ceremony. Their sons were Niels and Harald. After postdoctoral work with Carl Ludwig at Leipzig, Bohr settled in Copenhagen in 1883 and succeeded Panum as professor of physiology in 1886.

Bohr's first publication (1876) concerned the effect of salicylic acid on digestion of meat in dogs, his M.D. dissertation (1880) dealt with fat globules in milk, and his first publications from Ludwig's laboratory (1881, 1882) related the strength of muscle contraction to the frequency and strength of stimulation. His subsequent research stemmed from Ludwig's interest in respiration. Under Ludwig, Bohr devised an absorptiometer and began a lifelong study of the absorption and dissociation of respiratory gases in various media, including hemoglobin solutions, whole blood, and intact animals. His 1885 paper on deviations of oxygen from the Boyle-Mariotte law indicates his meticulous physical approach. Characteristically he strove for accurate quantitation under physiological conditions. He trusted his observations rather than generally accepted theory. This policy led him to some erroneous conclusions when methods were not as accurate as he thought, but even his mistaken ideas sometimes led to contributions of lasting importance.

To measure the pressures of oxygen and carbon dioxide in arterial blood in vivo, Bohr devised a tonometer that formed an arteriovenous shunt in experimental animals (1891). To compare these pressures with those in alveolar gas, he crystallized previously hazy notions of respiratory dead space into an equation for calculating alveolar gas composition from the compositions of inspired and mixed expired gas, the total volume of the breath, and the anatomical dead space volume, which he measured by filling the tracheobronchial tubes with water at autopsy. This "Bohr equation" remains fundamental to gas-exchange theory and is still used (in reverse) to calculate the dead space volume of patients from their alveolar gas composition, which can now be estimated independently. The data Bohr obtained in these experiments, however, indicated that arterial blood sometimes had an oxygen pressure higher (and carbon dioxide pressure lower) than simultaneous alveolar gas (1891). This led him to accept the theory that the lungs could act like a gland to secrete gases against a pressure difference when necessary. John Scott Haldane of Oxford provided strong support, but many others disagreed. Throughout his life Bohr and his students sought better evidence.

Their study of blood reached its peak when Bohr and two students, Karl A. Hasselbalch and August Krogh, using a precise microtonometer that Krogh devised, first measured in vitro the entire equilibrium curve of oxygen and whole blood (1904). They found that the curve was sigmoid, not hyperbolic (as the generally accepted theory required). Bohr continued to suppose that hemoglobin existed in multiple forms with different affinities for oxygen; in retrospect the finding opened the door to the modern concept of four interacting binding sites within a single molecule.

The same experiments also revealed that the affinity of hemoglobin for oxygen was inversely affected by carbon dioxide pressure. Their 1904 paper pointed out that this "Bohr effect" greatly facilitates oxygen uptake in pulmonary capillaries, where carbon dioxide is lost, and release of oxygen in peripheral capillaries, where carbon dioxide is gained. Bohr was unable to demonstrate the converse effect of oxygenation on carbon dioxide binding; that effect was established by an Oxford team that included a former student, Johanne Christiansen (1914).

Bohr's last important contributions to gas transport resulted from an attempt to distinguish the rate of supposed oxygen secretion from the simultaneous physical diffusion of oxygen down its pressure gradient (1909). Although the oxygen pressure in alveolar gas could be calculated by the Bohr equation, the oxygen pressure on the other side of the alveolar membrane could not be measured directly because it changed continuously and alinearly along the capillaries as oxygen entered the moving blood and reacted with hemoglobin. From his empirical oxygen-hemoglobin equilibrium curve of 1904, Bohr devised a graphical integration to calculate the mean oxygen pressure in the pulmonary capillaries. "Bohr integration" is still used to estimate the change in blood oxygen concentration along the pulmonary capillaries.

In this 1909 paper Bohr also described a method to determine the capacity of lungs in vivo for physical diffusion of oxygen despite the supposed presence of simultaneous oxygen secretion. Carbon monoxide is not secreted, and its diffusing capacity can be measured directly because that gas binds so tightly to hemoglobin that its pressure in pulmonary capillary blood remains negligible. The diffusing capacity for oxygen can then be calculated from the solubilities and molecular weights of the two gases. Bohr concluded that the diffusing capacity, which he calculated from measurements made at rest, was inadequate to account for oxygen uptake in heavy exercise: secretion seemed necessary.

August and Marie Krogh, while also looking for better evidence for the secretion theory, had already found in 1906 that the diffusing capacity for carbon monoxide increased in exercise enough to make oxygen secretion unnecessary, but they kept this observation secret. They hesitated to contradict their teacher without more evidence. They quietly continued looking until 1910, when their accumulated evidence against secretion became too convincing to be kept concealed. The carbon monoxide principle outlasted the secretion theory and is now used routinely to evaluate diffusion in pulmonary disease.

Initiating a new line of research, Bohr published in 1906 his first report on total lung volumes measured by hydrogen dilution and, in 1910, a significant study of pathological lung enlargement in pulmonary emphysema. The latter was his last research publication before his sudden death.

Despite his errors Bohr was the major contributor of his day to present understanding of respiratory gas transport. He made another lasting contribution as a teacher. Like Ludwig, he worked closely with his students in the laboratory, inspiring them to pursue lofty goals. He founded the Danish school of respiratory physiologists, which has continued to flourish.

BIBLIOGRAPHY

I. Original Works. Obituaries that list and summarize Bohr's publications include Carl J. Salomonsen, *Festkrift udgivet af Københavns Universitet i anledning af universitetets årsfest Nov. 1911* (Festschrift published by Copenhagen University for the university's anniversary . . .; Copenhagen, 1911), 53–58, which cites original Danish publications and earlier obituaries; and Robert Tigerstedt, "Christian Bohr: Ein Nachruf," in *Skandinavisches Archiv für Physiologie*, **25** (1911), ix–xviii, which cites the German versions. For English translations of his papers on Bohr's equations, Bohr's effect, and Bohr's integration, see John B. West, ed., *Translations in Respiratory Physiology*, I. S. Levij, trans. (Stroudsburg, Penn., 1975). Archival material preserved in the Institute of Medical Physiology of the University of Copenhagen includes his extensive correspondence with Ludwig.

II. Secondary Literature. Genealogies and biographies of Bohr and many of his relatives and associates are included in S. C. Bech, ed., *Dansk biografisk leksikon* (Danish biographical dictionary), 3rd ed. (Copenhagen, 1979–1984). N. Zuntz expressed the view of a scientific opponent in *Medizinische Klinik* (Berlin), **7** (1911), 434. L. S. Fridericia published a tribute in V. Meisen, ed., *Prominent Danish Scientists* (Copenhagen and Oxford, 1932). Recent evaluations of his work in historical perspective include John T. Edsall, "Blood and Hemoglobin . . .," in *Journal of the History of Biology*, **5** (1972), 205–257; and P. Astrup and J. W. Severinghaus, *The History of Blood Gases, Acids and Bases* (Copenhagen, 1986). Bodil Schmidt-Nielsen summarized his influence on Krogh in "August and Marie Krogh and Respiratory Physiology," in *Journal of Applied Physiology*, **57** (1984), 293–303. His personal life and character are recalled in Stefan Rozental, ed., *Niels Bohr* (New York, 1967).

RALPH KELLOGG

BORSUK, KAROL (*b.* Warsaw, Poland, 8 May 1905; *d.* Warsaw, Poland, 24 January 1982), *mathematics*.

Borsuk was the son of Marian Borsuk and Zofia Maciejewska. His father was a well-known surgeon in Warsaw. After receiving a master's degree in 1927 and a doctorate in 1930, both from the University of Warsaw, he became *Privatdozent* there in 1934. Borsuk married Zofia Paczkowska on 26 April 1936; they had two daughters.

After joining the faculty at Warsaw in 1929, Borsuk advanced to professor of mathematics in 1946 and director of the Mathematical Institute from 1952 to 1964. He was at the Institute for Advanced Study

BORSUK

at Princeton from 1946 to 1947, and later visiting professor at the University of California at Berkeley (1959–1960) and at the University of Wisconsin at Madison (1963–1964).

Borsuk was vice director of the Institute of Mathematics of the Polish Academy of Sciences in 1956. He was corresponding member of the Polish and Bulgarian academies of sciences.

Borsuk worked primarily in the area of geometric topology. Although he is known for widespread contributions in topology, a particularly important discovery was the distillation of a central topological feature of polyhedra and its generalization to a larger class of spaces. This concept, that of absolute neighborhood retract, was introduced in Borsuk's doctoral dissertation at the University of Warsaw under S. Mazurkiewicz, "Sur les rétractes" (published in 1931), and permeated a great deal of his work. It has greatly influenced the direction of research in topology throughout the world.

It will be helpful to define an absolute neighborhood retract. Let X be a metric space and A a subset of X. A continuous function r from X to A is said to be a retraction provided it has the property that $r(x) = x$ for all x in A. If X has the property that whenever X is embedded as a closed subset of a metric space Y, there is a retraction of Y onto X, then X is said to be an absolute retract (AR). The unit interval and the real line are examples of AR's. If whenever X is embedded in Y as a closed subset, there is a retraction of a neighborhood of X in Y onto X, then X is said to be an absolute neighborhood retract (ANR). Prime examples of ANR's are metric polyhedra and manifolds. If the polyhedron or manifold is contractible, then it is in fact an AR. The abstracting of this property of polyhedra is one of the most remarkable accomplishments of geometric topology. Although the concept was due to Borsuk, the entire community of topologists contributed to its full development.

John H. C. Whitehead showed in 1950 that an ANR is homotopy equivalent to a polyhedron. Thus this property virtually characterizes spaces that are homotopy equivalent to polyhedra. Borsuk asked in 1954 whether a compact ANR is homotopy equivalent to a compact polyhedron. This query was answered in the affirmative by J. E. West. In 1974 Robert Edwards was able to bring together the results of many researchers in topology to show that the product of a compact ANR with the Hilbert cube is a Hilbert cube manifold and the product of a compact AR with the Hilbert cube is the Hilbert cube. Complete proofs of these results, together with references, are given in Thomas Chapman's *Lectures on Hilbert Cube Manifolds* (1976). Borsuk had been intrigued by the Hilbert cube at an early stage in his career. In *The Scottish Book* in 1938 he posed the following questions: Is it true that the product of a triod with the Hilbert cube is homeomorphic to the Hilbert cube? Is it also true that the infinite product of triods is homeomorphic to the Hilbert cube? An affirmative answer to these questions was given by Richard D. Anderson in 1964. A published proof of this result, together with the fact that the product of any compact polyhedron with the Hilbert cube is a Hilbert cube manifold, was given by Anderson's student J. E. West. The remarkable theorem by Edwards would not have been possible without a thorough investigation into infinite dimensional topology, of which the Anderson-West result was the preliminary essential step. Borsuk showed exceptional insight in conceiving of this conjecture so early in his career.

For most of his career Borsuk was connected with the University of Warsaw. During the Nazi occupation of Poland, he labored at keeping intellectual life in Poland alive through the "underground university." This and other "illegal" activities led to his imprisonment, escape, and hiding until the end of the war.

When Poland began to rebuild, Borsuk and Kazimierz Kuratowski began the work of restoring mathematics in Warsaw, and Borsuk's disrupted career came back into focus. He continued his studies in topology. In the mid 1960's he came across another fundamental concept in topology, the notion of shape. An ANR is a very nice space with many convenient local properties. Unfortunately, there are many mathematical spaces that do not have nice local properties. It was Borsuk's idea that such spaces could be "smoothed" by embedding them in AR's, in particular by embedding them in the Hilbert cube. One could then study the original space by studying the system of neighborhoods of the space in the Hilbert cube. Since these neighborhoods are ANR's, one is thus studying an arbitrary compact metric space by approximating it by a system of ANR's. Borsuk's first publication in shape theory was in 1968 in *Fundamenta mathematicae*. At about the same time he gave several talks in Europe and the United States disseminating his ideas.

Shape theory has had tremendous influence in topology. There are, however, complications in giving credit to Borsuk for its discovery because it has been shown that it is equivalent to several earlier theories and constructions. In particular, etale homotopy theory and the Kan and Čech extension of the homotopy functor on the category of polyhedra are equivalent to Borsuk's theory of shape. Several mathematicians discovered these equivalences independently at about the same time. Although these constructions are in a technical sense the same, it can certainly be said that Borsuk was the first to use these ideas as he did. His motivation was always to understand the geometry of separable metric spaces, and he showed how shape theory could be an effective tool for this purpose.

In recent times many different areas of topology have begun to merge. The theory of manifolds, the theory of CW-complexes and polyhedra, the theory of ANR's, combinatorial topology, homotopy theory, algebraic topology, shape theory, infinite-dimensional topology, and geometric topology have had considerable interaction. Borsuk played a significant role in developing several of these areas and in making them fit coherently into the whole.

Borsuk was much honored by Poland, receiving many decorations to honor his contributions to mathematics, education, and political life. He was also widely honored in the mathematical community. He participated in some twenty conferences and delivered major addresses at many of them. He gave numerous talks on his work at centers of learning. In 1978 Borsuk organized the International Conference on Geometric Topology, held in Warsaw. This conference demonstrated his widespread and profound influence on topology and the high regard in which he was held.

BIBLIOGRAPHY

I. ORIGINAL WORKS. Many of Borsuk's writings are brought together in *Karol Borsuk: Collected Papers*, 2 vols. (Warsaw, 1983). Among his books are *Foundations of Geometry* (Amsterdam and New York, 1960), written with Wanda Szmielew, rev. trans. by Erwin Marquit; *Theory of Retracts* (Warsaw, 1967); *Multidimensional Analytic Geometry*, trans. Halina Spalinska (Warsaw, 1969); and *Theory of Shape* (Warsaw, 1975). With A. Krikor, Borsuk edited *Proceedings of the International Conference on Geometric Topology* (Warsaw, 1980).

II. SECONDARY LITERATURE. Writings related to Borsuk's work are Thomas A. Chapman, *Lecture Notes on Hilbert Cube Manifolds*, Conference Board of the Mathematical Society, Regional Conference Series in Mathematics, no. 28 (Providence, R.I., 1976); Jerzy Dydak and Jack Segal, *Shape Theory: An Introduction*, Lecture Notes in Mathematics no. 688 (Berlin and New York, 1978); Sibe Mardešić and Jack Segal, eds., *Shape Theory and Geometric Topology*, Lecture Notes in Mathematics no. 870 (Amsterdam and New York, 1982); D. Mauldin, ed., *The Scottish Book* (Stuttgart and Boston, 1981). (*The Scottish Book* was originally a notebook of mathematical problems started by a group of Polish mathematicians in Lwow in 1935. This edition is an English translation, together with a commentary on the problems.) See also J. E. West, "Infinite Products Which Are Hilbert Cubes," in *Transactions of the American Mathematical Society*, **150** (1970), 1–25, and "Mapping Hilbert Cube Manifolds ANR's: A Solution of a Conjecture of Borsuk," in *Annals of Mathematics*, **106** (1977), 1–8.

JAMES KEESLING

BOWEN, IRA SPRAGUE (*b*. Seneca Falls, New York, 21 December 1898; *d*. Hollywood, California, 6 February 1973), *astronomy, physics*.

Bowen was the son of James H. Bowen, Wesleyan Methodist minister, who could trace his ancestry to Welsh settlers who had emigrated to Massachusetts in 1643, and of Philinda Sprague Bowen. Both his parents were graduates of Geneseo State Normal School. James Bowen died in 1908; and Ike, as he was called by his friends, was brought up by his widowed mother, who became a teacher at the Wesleyan Methodist Seminary in Houghton, New York.

Bowen, who was interested in science from boyhood, graduated as valedictorian of his high school class in 1915. His first three years of college were at Houghton Seminary's junior college, where all the courses in mathematics, physics, and astronomy were taught by the president, J. S. Luckey. For his senior year Bowen transferred to Oberlin College, where he received the A.B. in 1919.

He then entered the University of Chicago as a graduate student in physics, studying for two years under A. A. Michelson and Robert A. Millikan. Bowen was Millikan's assistant in his spectroscopy laboratory. When Millikan moved to the California Institute of Technology as chairman of its executive council (1921), Bowen went with him as a junior faculty member. He did the laboratory work and analysis for their series of papers on the ultraviolet spectra and energy levels of highly ionized atoms. Bowen also taught physics to many of the top undergraduates at Caltech and received his Ph.D. in 1926. In 1929 he married Mary Jane Howard, a child psychologist. They had no children.

His great discovery was the identification of the "nebulium" lines in the spectra of gaseous nebulae. These two emission lines in the green spectral region, the strongest emitted in these nebulae, were of completely unknown origin. By 1926 physicists understood the periodic table and knew there were no unknown elements such as the hypothetical "nebulium." Henry Norris Russell speculated that these lines might be emitted only in gases of very low density if, for example, it took a long time for atoms to make the relevant transitions. In 1927 Bowen, who had been thinking about this problem for years, suddenly realized that the nebulium lines might actually be emitted as "forbidden" transitions in common ions, such as O^{++}, that occur too infrequently to be observed in laboratory sources. His accurate data on the energy levels allowed him to check this hypothesis at once, and he found it to be correct. Rapidly following up his discovery, he identified many of the remaining lines in the spectra of nebulae with similar forbidden transitions in O^{+}, N^{+}, and other common ions. Bowen quickly developed the essentials of the physical picture of gaseous nebulae as we understand them today.

In 1934 W. H. Wright, at Lick Observatory, obtained the first good ultraviolet spectra of planetary nebulae and noticed that the permitted O^{++} emission lines in this region had quite abnormal relative intensities. He sent his data to Bowen, who was able to explain them immediately as resulting from fluorescence. He had noticed years before that an accidental coincidence between the strongest resonance line of He^{+} and an O^{++} line from the ground term would lead to selective excitation of a single energy level of the latter ion. This process would lead through a long chain of radiative decays to the emission of just those lines that Wright had observed to be strong in the planetary nebulae, thus observationally confirming Bowen's physical idea.

Bowen spent the summer of 1938 at Lick Observatory and observed the spectra of planetary nebulae in collaboration with Arthur B. Wyse. They identified many more forbidden lines and from their quantitative intensity measurements were able to show that the abundances of the elements in the nebulae are approximately the same as in the sun and stars.

As a physicist deeply interested in astronomy, Bowen was a member of the Advisory Committee to the Observatory Council that guided the construction of the Palomar 200-inch telescope. In 1946 he was appointed director of Mt. Wilson Observatory. He supervised the completion of the 200-inch, then became director of Palomar Observatory in 1948. Bowen's superb knowledge of optics was essential in bringing the 200-inch Hale telescope and the coudé spectrograph he designed for it into successful operation. He directed the research staff of Mt. Wilson and Palomar observatories during some of the most productive years of American observational astronomy. With the coudé spectrograph and the 200-inch, Bowen measured with high accuracy the wavelengths of many nebular lines and, inverting his procedure of thirty years before, used them to determine more precisely the energy levels of the ions that emit them.

Bowen applied his knowledge and experience of optics to optimizing the design of astronomical spectrographs. His papers on this subject, on the 200-inch telescope, on Schmidt cameras, and on telescopes of the future were all highly influential in affecting the course of astronomy.

After he retired as director in 1964, Bowen devoted himself to telescope design. His masterpiece was the 100-inch Du Pont telescope at Las Campanas Observatory, Chile, a wide-field, very-high-definition instrument. He thus made important contributions to astronomy as a research worker, as an instruments maker, and as a scientific leader.

BIBLIOGRAPHY

I. Original Works. Horace W. Babcock, "Ira Sprague Bowen," in *Biographical Memoirs. National Academy of Sciences*, **53** (1982), 83–119, contains a full bibliography of Bowen's published papers and review articles. Some of his papers are in the California Institute of Technology Archives; the remainder are in the director's files of Mt. Wilson and Las Campanas observatories, still closed but no doubt ultimately to be deposited in the observatories' archives in Pasadena.

Donald E. Osterbrock

BOWEN, RUFUS (ROBERT) (*b.* Vallejo, California, 23 February 1947; *d.* Santa Rosa, California, 30 July 1978), *mathematics*.

The son of Marie De Winter Bowen, a schoolteacher, and Emery Bowen, a budget officer at Travis Air Base in California, Robert Bowen attended public schools in the small city of Fairfield. He published his first mathematical paper at the age of seventeen, and four more before he was twenty-one. The University of California at Berkeley awarded him a bachelor's degree, with prizes for scholarship, in 1967, and the doctorate in 1970. He was appointed assistant professor of mathematics at Berkeley in 1970 and was promoted to professor in 1977. On 6 March 1968 Bowen married Carol Twito of Hayward, California; they had no children. Deeply concerned with social problems, as were many of his generation, he was active in organizations devoted to preventing nuclear war. Bowen belonged to Phi Beta Kappa and the American Mathematical Society. In 1970 he changed his first name to Rufus. He died of a cerebral aneurysm.

The subject of Bowen's doctoral dissertation, and of all his later work, was dynamical systems theory. Originated by Henri Poincaré in the 1880's, this field was developed intensively in the 1960's, largely under the leadership of Bowen's dissertation supervisor, Stephen Smale. The latter's seminal paper, "Differentiable Dynamical Systems" (*Bulletin of the American Mathematical Society*, **73** [1967], 747–817), was much admired by Bowen and strongly influenced the direction of his work.

As studied by Smale, a dynamical system comprises a manifold M and a smooth mapping $f: M \to M$ (usually a diffeomorphism); the goal is to describe the limiting behavior of the trajectories $f^n(x)$ of points $x \in M$ as n goes to infinity. As Poincaré emphasized, there is no general procedure for this, and therefore one must resort to describing average, typical, or most probable behavior. Bowen's work is an important part of the program of expressing these vague ideas in mathematically precise and useful ways.

Smale singled out what he called "axiom A systems" as being simple enough to study and complex enough to include many interesting examples. His "spectral decomposition theorem" states that there is a finite number of indecomposable subsystems, called "basic sets," to which all trajectories tend, and that the dynamics of the basic sets is not too wild. Most of Bowen's work is concerned with the dynamics in a basic set of an axiom A system.

Bowen's early papers give useful estimates for the topological entropy $h(f)$ of a dynamical system. This topological invariant, which had recently been discovered by others, measures the complexity of the system. For axiom A systems Bowen proved that

$$h(f) = \limsup_{n\to\infty} n^{-1} \log N_n(f),$$

where $N_n(f)$ is the number of fixed points of f^n. Other results give lower bounds for $h(f)$ in terms of the automorphisms induced by f on the fundamental and homology groups of M.

Most of Bowen's subsequent work concerns invariant measures associated to dynamical systems. First used by Poincaré and J. Willard Gibbs, these measures are intimately related to statistical mechanics and other branches of mathematical physics. A major achievement was Bowen's construction of an invariant, ergodic probability measure μ_X for any basic set X, for which periodic points are uniformly distributed.

In this work Bowen developed new methods in the symbolic dynamics pioneered by Jacques Hadamard and Marston Morse. He constructed a certain covering of X by closed sets $R_1, \cdots, R_n$, called a Markov partition. To any x in M one associates a doubly infinite sequence $y = (y_i)$ such that $y_i \in \{1, \cdots, n\}$, and $f^i x \in R_j$ if $y_i = j$. The Markov partition has the property that each x corresponds to at most n^2 sequences, and most x to exactly one sequence. The space Σ of all such sequences is readily described; it carries the "shift" homeomorphism $Ty = z$ defined by $z_i = y_{i+1}$. There is a continuous map P from Σ onto X such that $P(Ty) = f(Py)$. This means that T and f have practically the same dynamics. Bowen was able to infer much about f from the easily analyzed dynamics of T. In particular he showed that (f, X), considered as an automorphism of the measure space (X, μ_X), is a Markov chain. This means that most points of X have trajectories that are randomly distributed over X.

Bowen then turned to analogous but much more subtle problems for continuous-time systems, called flows: here one has a one-parameter group of maps indexed by the real numbers. An important technical achievement was the description of suitable analogues, for flows, of symbolic dynamics. An important theorem (proved with David Ruelle) states that every attracting basic set Z of a twice-differentiable, axiom A flow on M carries an invariant probability measure μ_0 with the following property: For any continuous function g on M and almost every point x in a neighborhood of Z (in the sense of Lebesgue measure), the time average of g over the forward orbit of x equals the μ_0 average of g over Z. This measure is now called the Bowen-Ruelle measure. Bowen applied his results to the geodesic flow on the unit tangent bundle of a compact manifold of constant negative curvature, showing that periodic geodesics are equidistributed in the Riemannian measure as the periods tend to infinity.

In his last, posthumous paper, Bowen applied his measure-theoretic methods in a novel way to classical problems in the geometry of Fuchsian groups. Besides its considerable intrinsic interest, this paper demonstrates that Bowen's methods have application far beyond the axiom A systems for which they were invented. The Bowen-Ruelle measure, for example, has already proved significant in other dynamical settings. His papers, models of clarity, simplicity, and originality, have a permanent importance in dynamical systems theory.

BIBLIOGRAPHY

The best sources for an overview of much of Bowen's work are the following surveys written by him: "Symbolic Dynamics for Hyperbolic Flows," in *Proceedings of the*

International Congress of Mathematicians (Vancouver, 1974), 299–302; *Equilibrium States and the Ergodic Theory of Anosov Diffeomorphisms*, Lecture Notes in Mathematics no. 470 (New York, 1975); and *On Axiom A Diffeomorphisms* (Providence, R.I., 1978). His work with David Ruelle is "The Ergodic Theory of Axiom A Flows," in *Inventiones Mathematicae*, **29** (1975), 181–202. His last paper is "Hausdorff Dimension of Quasi-circles," in *Publications mathématiques de l'Institut des hautes études scientifiques*, no. 50 (1978), 259–273. Papers covering other aspects of his work are "Entropy for Maps of the Interval," in *Topology*, **16** (1977), 465–467; and "A Model for Couette Flow-data," in P. Bernard and T. Ratin, eds., *Turbulence Seminar, Proceedings 1976/77* (New York, 1977), 117–133.

MORRIS HIRSCH

BRAUER, RICHARD DAGOBERT (*b*. Berlin, Germany, 10 February 1901; *d*. Boston, Massachusetts, 17 April 1977), *mathematics*.

Brauer was the youngest of three children of Max Brauer, an influential and wealthy businessman in the wholesale leather trade, and his wife, Lilly Caroline Jacob. He attended the Kaiser-Friedrich-Schule in Berlin-Charlottenburg and had an interest in science and mathematics as a young boy, an interest that owed much to the influence of his gifted brother Alfred, who was seven years older.

In February 1919 Brauer enrolled in the Technische Hochschule in Berlin but soon realized that, in his own words, his interests were "more theoretical than practical." He transferred to the University of Berlin after one term. He took his Ph.D. there in 1925, under the guidance of the algebraist Issai Schur.

On 17 September 1925 Brauer married Ilse Karger, a fellow student and the daughter of a Berlin physician. They had two sons, George Ulrich and Fred Günther, both of whom became research mathematicians.

Brauer's first academic post was at the University of Königsberg (now Kaliningrad), where he remained until dismissed by Hitler's 1933 decree banning Jews from university teaching. He spent 1933 and 1934 at the University of Kentucky at Lexington, and 1934 and 1935 at the Institute for Advanced Study at Princeton, where he was assistant to Hermann Weyl. After this Brauer held professorships at the University of Toronto (1935–1948), the University of Michigan at Ann Arbor (1948–1952), and Harvard (1952–1971). He lived near Boston for the rest of his life. Weakened by aplastic anemia, he died of a generalized infection.

Brauer was an elected member of the Royal Society of Canada, the American Academy of Arts and Sciences, the National Academy of Sciences, the London Mathematical Society, the Akademie der Wissenschaften (Göttingen), and the American Philosophical Society. He was president of the Canadian Mathematical Congress (1957–1958) and of the American Mathematical Society (1959–1960). He received the Guggenheim Memorial Fellowship (1941–1942), the Cole Prize of the American Mathematical Society (1949), and the National Medal for Scientific Merit (1971).

Brauer was one of the most influential algebraists of the twentieth century. He built on the foundations of the representation theory of groups that were laid by Georg Frobenius, William Burnside, and Issai Schur in the years 1895–1910; and over his long career he brought the representation theory of finite groups, in particular, to a remarkable depth and sophistication. His first important research, however, was concerned with the representations of a continuous (topological) group.

By a representation of a given linear group Γ is meant a homomorphism $H:\Gamma \rightarrow GL(N,\mathbf{C})$, whereby each element s of Γ is represented by a nonsingular complex matrix or linear transformation $H(s)$ of some finite degree N. If Γ is a topological group, it is assumed that H is continuous, that is, that each matrix coefficient of $H(s)$ is a continuous function of s. Among topological groups the most important are the classical linear groups, such as the group $O(n)$ of all real orthogonal transformations of n variables, or its subgroup $SO(n)$ (often called the rotation group) consisting of the orthogonal transformations of determinant 1.

In 1897 Adolf Hurwitz introduced a new and fundamental idea into the study of such groups, that of an invariant integral. He defined such an integral for $SO(n)$ and used it to calculate polynomial invariants for this group. Schur realized that his own treatment of Frobenius' character theory of a finite group Γ could be extended to a continuous linear group Γ on which an invariant integral could be defined. In a series of papers published in 1924, he used Hurwitz's integral to find the irreducible characters of $O(n)$. Brauer was attending Schur's seminar at this time, and Schur suggested to him that it might be possible to find a purely algebraic treatment of this work, that is, one that did not rely on the analytic notion of an integral. Brauer found such a treatment, and with it calculated the irreducible characters of the groups $O(n)$ and $SO(n)$; this became his dissertation, for which he was awarded the Ph.D. summa cum laude in 1926.

While Brauer was writing his dissertation, Hermann Weyl was working on his papers on the representations of semisimple Lie groups (these include the classical linear groups). This work of Weyl's has claim to be the finest single mathematical achievement of the twentieth century. It is based on Schur's methods with the invariant integral and Élie Cartan's construction of representations of a semisimple Lie group Γ by means of representations of its Lie algebra $\mathfrak{g}$; Schur's and Brauer's results on $O(n)$ and $SO(n)$ come out as special cases. Weyl's results and methods have been the starting point of a huge amount of research in pure mathematics and in quantum physics. By contrast Brauer's algebraic treatment for the orthogonal groups is little known—it uses difficult and (now) unfashionable techniques from the theories of determinants and invariants, and was published only as his Ph.D. dissertation.

Brauer greatly admired Weyl, and during his year as Weyl's assistant at the Institute for Advanced Study, he briefly returned to the classical linear groups. From this period date a joint paper with Weyl on spinors and a paper in which Brauer calculates, by purely algebraic means, the Poincaré polynomials of the classical groups (unitary, symplectic, and orthogonal). Brauer's last paper on continuous groups, published in 1937, hints at a general representation theory for continuous groups, strictly algebraic in nature and based on invariant theory. A promised sequel never appeared.

Much of Brauer's work while he was in Königsberg (1925–1933) was concerned with simple algebras and rooted in Schur's theory of splitting fields. Suppose k is a given field, K an algebraically closed field that contains k, and that $H:\Gamma \rightarrow GL(f,K)$ is an irreducible representation of some group Γ. Then each element s of Γ is represented by a nonsingular $f \times f$ matrix $H(s)$ whose coefficients lie in K; the condition that H be irreducible means that the set of all K-linear combinations of the $H(s)$, $s \in \Gamma$, is the full matrix algebra K_f of *all* $f \times f$ matrices over K. We shall assume that the trace $X(s)$ of each matrix $H(s)$ lies in the ground field k. Then a field L ($k \subseteq L \subseteq K$) is a splitting field for H (or for its character X) if L is a finite extension of k and there exists a matrix $P \in GL(f,K)$ such that all the coefficients of all the matrices $P^{-1}H(s)P$, $s \in \Gamma$, lie in L. The least degree $m_k(H) = m_k(X)$ over k, among all such splitting fields L, is called the Schur index of H or X over k; Schur had initiated the study of splitting fields (in the case where k is an algebraic number field and $k = \mathbb{C}$) in the early 1900's, and had proved a number of facts about the Schur index.

Brauer and Emmy Noether (who was then at Göttingen) showed, in a paper published in 1927, how the splitting fields of H are determined by the algebra A of all k-linear combinations of the matrices $H(s)$, $s \in \Gamma$. A is a finite dimensional simple algebra over k that is central; its center consists only of the scalar multiples of the identity. A given field L of finite degree $(L:k)$ is a splitting field for H if and only if $L \otimes_k A$ is isomorphic to the full matrix algebra L_f over L. This last condition depends only on the algebra A, and a field L that satisfies it is called a splitting field for A. Brauer and Noether's main result (proved under certain restrictions on the ground field k, which were later shown by Noether to be unnecessary) was that the splitting fields of a given central, simple k-algebra A are (up to isomorphism) the same as the maximal subfields of the algebras B in the same algebra class as A. The algebra class $[A]$ of A is defined as follows. By Joseph Wedderburn's structure theorem (1907), A is isomorphic to the algebra D_t of all $t \times t$ matrices over a certain central division algebra D and for a certain positive integer t; the class $[A]$ then consists of all central simple algebras over k that are isomorphic to D_s, for any positive integer s. The set of all such algebra classes, with given ground field k, forms a group $B(k)$, now known as the Brauer group of k; the product of classes $[A]$, $[B]$ is defined to be the class $[A \otimes_k B]$, and the identity element of $B(k)$ is the class $[k]$. It has turned out that $B(k)$ is a fundamentally important invariant of the field k. Brauer studied it with the help of a technique of factor sets, and from this beginning has grown the theory of Galois cohomology and its many uses in number theory. In another direction, M. Auslander and O. Goldman showed in 1960 how to define $B(R)$ for an arbitrary commutative ring R, thereby beginning a new chapter in commutative algebra.

Brauer's work with Emmy Noether brought him into contact not only with this influential algebraist and her school in Göttingen but also with the famous conjecture of L. E. Dickson that every central simple algebra A over an algebraic number field k contains a maximal subfield L that is a Galois extension of k with cyclic Galois group. This conjecture was proved in 1931 in a joint paper by Brauer, Noether, and H. Hasse that is the culmination of a long development in the theory of algebras. Brauer's association with this work secured his reputation as one of the best mathematicians of the rising generation in Germany.

Brauer was abruptly dismissed in 1933, along with all other Jewish university teachers in Germany. The disadvantages and disruptions of a forced emigration at the age of thirty-two were offset to some

BRAUER

extent by the new contacts Brauer made, not only with American mathematicians but also with other German scientists who found refuge in America in the 1930's. The year 1934–1935, when Brauer was Weyl's assistant at the Institute for Advanced Study, saw an extraordinary gathering of mathematicians and physicists of the first rank: J. W. Alexander, Albert Einstein, John von Neumann, Oswald Veblen, and Weyl were permanent professors at the Institute; and the mathematics faculty at Princeton University included Salomon Bochner, S. Lefschetz, and Joseph H. M. Wedderburn. Among the visiting members of the Institute that year were, besides Brauer, W. Magnus, C. L. Siegel, and Oscar Zariski.

In 1935 Brauer took up a post as assistant professor at the University of Toronto, where he remained until 1948, becoming in due course associate and then full professor. Here he developed his modular representation theory of finite groups, which will probably continue to be regarded as his most original and characteristic contribution to mathematics.

Representation theory began with Frobenius' paper "Über Gruppencharaktere," published in 1896. In Frobenius' theorem the irreducible characters of a group were defined in terms of what can now be described as the representations of the center of the group algebra $\mathbb{C}G$ ($\mathbb{C}G$ is the linear algebra over the complex field $\mathbb{C}$ having the elements of G as basis). Schur (who had been Frobenius' pupil) reformulated Frobenius' character theory in a beautiful paper published in 1905 that became the basis of all subsequent expositions of the subject. Schur defined the character X_H of an arbitrary representation $H:G \rightarrow GL(n,\mathbb{C})$ of G to be the complex-valued function $X_H:G \rightarrow \mathbb{C}$ given by $X_H(s)$, = trace $H(s)$, $s \in G$. If H is an irreducible representation, then X_H is said to be a simple, or irreducible, character. Schur rederived Frobenius' orthogonality relations for the irreducible characters, from which follows the fundamental theorem: Two representations H, H' of the group G are equivalent if, and only if, their characters are equal, $X_H = X_{H'}$.

It was recognized very soon that Frobenius' arguments do not hold when $\mathbb{C}$ is replaced by an algebraically closed field k of finite characteristic p, although L. E. Dickson proved in 1902 that the orthogonality relations for the characters are still true, provided p does not divide the order $|G|$ of G. In two later papers Dickson considered the case where p divides $|G|$. In this case the group algebra kG is not semisimple. A representation $F:G \rightarrow GL(n,k)$ is no longer determined up to equivalence by the natural character X_F = trace F. After Dickson's papers were published in 1907, little more was done on representations over fields of finite characteristic until the mid 1930's, when Brauer laid the foundations of his modular representation theory in three fundamental papers, the last two of which were written with C. Nesbitt, who took his Ph.D. at Toronto under Brauer's supervision.

Let G_0 denote the set of all p'-elements (or p-regular elements) of G; these are the elements whose order is prime to p. Let $|G| = p^a m$, where $a \geq 0$ and p does not divide m. Then each element $\mathfrak{g}$ of G_0 satisfies the equation $\mathfrak{g}^m = 1$, so if $F:G \rightarrow GL(n,k)$ is a representation, the eigenvalues $\alpha_1, \ldots, \alpha_n$ of $F(\mathfrak{g})$ are mth roots of unity in the field k. The set $U_m(k)$ of all mth roots of unity in k forms a cyclic group of order m, the group operation being multiplication in k. But the multiplicative group $U_m(\mathbb{C})$ of all complex mth roots of unity is also cyclic of order m. Therefore one can find a multiplicative isomorphism $\mathbf{c}:U_m(k) \rightarrow U_m(\mathbb{C})$; in general this can be done in many ways. Choose any such isomorphism $\mathbf{c}$. Brauer defines the modular character (later known as the Brauer character) of the representation F to be the function $\beta_F:G_0 \rightarrow \mathbb{C}$ given by

$$\beta_F(\mathfrak{g}) = \mathbf{c}(\alpha_1) + \cdots + \mathbf{c}(\alpha_n), \mathfrak{g} \in G_0. \quad (1)$$

This gives a complex valued function in place of the k-valued natural character X_F, for which

$$X_F(\mathfrak{g}) = \operatorname{trace} F(\mathfrak{g}) = \alpha_1 + \cdots + \alpha_n, \quad (2)$$

by the daring—almost impudent—device of complexifying the eigenvalues $\alpha_1, \ldots, \alpha_n$. That definition (1), and not the natural definition (2), is the correct basis for a modular character theory soon becomes clear. If $F_1, \cdots, F_l$ is a full set of irreducible representations of G over k, then their Brauer characters $\beta_1, \cdots, \beta_l$ are linearly independent functions on G_0. For any representation $F:G \rightarrow GL(n,k)$ one has $\beta_F = \Sigma n_i(F)\beta_i$, where $n_i(F)$ is the multiplicity with which F_i appears as a composition factor in F. Brauer characters (like natural characters) are class functions, that is, $\beta_F(\mathfrak{g}) = \beta_F(\mathfrak{g}')$ whenever $\mathfrak{g},\mathfrak{g}'$ belong to the same conjugacy class of G. Frobenius had shown that the number of irreducible ordinary characters of G is equal to the number of conjugacy classes of G; Brauer showed that the number l of irreducible modular characters is equal to the number of p-regular classes, that is, to the number of conjugacy classes lying in G_0.

From the beginning, Brauer saw the modular character theory—which is based on representations of G over a field k of finite characteristic p—as a source of information on the ordinary characters—which is based on representations on $\mathbb{C}$ or some other field of characteristic zero. If $X_1, \cdots, X_s$ are the irreducible ordinary characters of G, and $\beta_1, \cdots, \beta_l$ are the irreducible modular characters, there exist nonnegative integers $d_{\sigma i}$ such that the equations

$$X_\sigma(\mathfrak{g}) = \sum_{i=1}^{l} d_{\sigma i}\beta_i(\mathfrak{g}) \qquad (3)$$

hold for all elements $\mathfrak{g}$ in G_0. To explain these equations, we need some technical preliminaries. Let L be a field of characteristic zero that is a splitting field for G; this means that for each $\sigma = 1, \cdots, s$, the character X_σ can be obtained from a matrix representation X_σ, such that all the coefficients of the matrices $X_\sigma(\mathfrak{g})$ lie in L. This is to say that L is a splitting field for all the X_σ, in the sense we used earlier. We assume that L has a subring R that is a principal ideal domain, such that L is the field of quotients of R; moreover, R should have a prime ideal $\mathfrak{p}$ containing the integer p. We identify $\overline{R} = R/\mathfrak{p}$ (which is a field of characteristic p) with a subfield of the field k. The matrix representations X_σ can be chosen so that the matrix coefficients of $X_\sigma(\mathfrak{g})$ all lie in R.

Taking these mod $\mathfrak{p}$, we get a modular representation $\overline{X}_\sigma:\mathfrak{g} \rightarrow \overline{X_\sigma(\mathfrak{g})}$ of G, that is, a representation over the field k. The Brauer character of $\overline{X}_\sigma$ is (suitably identifying a part of L with a part of $\mathbb{C}$, so that the character values $X_\sigma[\mathfrak{g}]$ can be regarded as elements of $\mathbb{C}$) identical with the restriction of X_σ to G_0. Equations (3) therefore say that $d_{\sigma i}$ is the multiplicity of F_i as composition factor of $\overline{X}_\sigma$: Brauer called the $d_{\sigma i}$ decomposition numbers; they record the decomposition that the ordinary irreducible representations X_σ of G undergo when they are reduced mod $\mathfrak{p}$. But these numbers have another, quite separate interpretation. Brauer found that the modular irreducibles $F_1, \cdots, F_l$ are in natural correspondence with the indecomposable direct summands $U_1, \cdots, U_l$ of the regular representation of the group algebra $A = kG$. (He showed, in fact, in joint work with Nesbitt, that this holds for any finite-dimensional k-algebra A. U_i is what is now called the projective cover of F_i. Brauer's ideas from this period—the late 1930's—pervade much modern research on algebras and their representations.) If splitting field L is taken to be complete with respect to a suitable discrete valuation, with R as the ring of valuation integers, then each U_i can be lifted to a representation $\tilde{U}_i$ of G over R, that is, to a representation in characteristic zero, which therefore has an ordinary character η_i, say.

There hold then the remarkable equations

$$\eta_i = \sum_{\sigma=1}^{s} d_{\sigma i}X_\sigma, \quad i = 1, \cdots, l \qquad (4)$$

that are in some sense dual to (3). From (3) and (4) one deduces equations

$$c_{ij} = \sum_{\sigma=1}^{s} d_{\sigma i}d_{\sigma j}, \quad i, j = 1, \cdots, l, \qquad (5)$$

where the c_{ij} are the Cartan invariants of the algebra $A = kG$; c_{ij} may be defined as the multiplicity (as composition factor) of F_j in U_i. Cartan invariants exist for any algebra A, but equations (5) show that for a group algebra $A = kG$, the $l \times l$ matrix $C = (c_{ij})$ has special properties: it is symmetric and positive definite. Brauer also discovered the deep theorem that det C is a power of the characteristic p of k. This was published in the first of a remarkable series of papers appearing in 1941 and 1942. In these Brauer introduced new and sophisticated methods for the study of group characters, and began to give applications of his theory to the structure theory of finite groups.

Fundamental to this work was the idea of a block. Blocks are most easily defined by taking a decomposition $1 = e_1 + \cdots + e_t$ of 1 as sum of orthogonal primitive idempotents of e_τ of the center $Z(kG)$ of kG. This can be lifted to a corresponding decomposition $1 = \hat{e}_1 + \cdots + \hat{e}_t$ in $Z(RG)$. An ordinary (or modular) irreducible character ψ of G is said to belong to the block B_τ of G if $\hat{e}_\tau$ (or e_τ) is represented by the identity matrix in a representation corresponding to ψ. In this way both sets $\{X_1, \cdots, X_s\}$ and $\{\beta_1, \cdots, \beta_l\}$ are partitioned among the t blocks $B_1, \cdots, B_t$ of G. Block theory aims to give information about the ordinary characters of a given block B_τ of G, in terms of information available for a block b in some p-local subgroup H of G (H is usually the normalizer or centralizer in G of some p-subgroup of G).

In the most favorable cases, Brauer's methods show that a part of the character table for H is almost identical with a part of the character table for G; this is now much used in the computation of character tables for known finite groups. Brauer saw in his theory a potential tool for studying finite simple groups. (We shall return to this later.)

The main facts about blocks of a group G and their relation to blocks of subgroups H of G, as well as the important refinement to equations (3) involving coefficients called the generalized decomposition numbers, had been published (sometimes without proofs, which appeared years later) by 1947. In that year Brauer also published a solution to Emil Artin's conjecture that the sums of the Artin L-series are entire functions. Brauer had held a Guggenheim fellowship in 1941 and 1942 and spent a part of this time visiting Artin in Bloomington,

Indiana. They both knew that the solution of Artin's conjecture rested on the validity of a certain statement about group characters; and at some point in the next few years Brauer realized that this statement could be proved, using methods he had developed for his modular character theory.

By 1953 he embodied the essential idea from these methods in a theorem—perhaps his most widely known—that if Θ is a complex valued class function on a finite group G, then Θ is a generalized character of G, if its restriction $\Theta|_E$ is a generalized character of E, for every elementary subgroup E of G. (If $X_1, \cdots, X_s$ are the irreducible ordinary characters of a group G, then functions of form $z_1X_1 + \cdots + z_sX_s$ with the z_i integers—not necessarily positive—are called generalized characters of G. A group E is elementary if, for some prime p, E is the direct product of a p-group and a cyclic group.) From this theorem Brauer also obtained a new proof of a much older conjecture (which he had first solved a few years earlier): that if ε is a primitive $\mathfrak{g}$th root of unity, where $\mathfrak{g}$ is the order of G, then $\mathbf{Q}(\varepsilon)$ is a splitting field for G (that is, for all the irreducible characters of G).

In 1948 Brauer moved from Toronto to the University of Michigan at Ann Arbor, and four years later to Harvard. From this period can be dated the first systematic attack on the problem of describing (or classifying) all finite simple groups. In a paper written with his pupil K. A. Fowler, Brauer proved by very elementary means a striking fact: Let G be a simple group of even order, and let x be an involution in G (that is, an element satisfying the conditions $x^2 = 1$, $x \neq 1$). Let $H = C_G(x)$ be the centralizer of x and let n be the order of H. Then G has a proper subgroup of index less than $\frac{1}{2}n(n + 1)$. This gives hope to a general program announced by Brauer in 1954: Given an abstract group H with an involution x in its center, to find all simple groups G containing H as a subgroup in such a way that $H = C_G(x)$. For the theorem above shows that (with H and x given) the number of isomorphism types of such groups G is finite—although it does not provide practical means of constructing them. It is natural to take for H the centralizer of an involution in some known simple group G_0. It may happen that G_0 is—up to isomorphism—the only simple group having an involution x such that $C_G(x) \cong H$; we have then a characterization of G_0 by an internal property.

By 1954 Brauer (and independently M. Suzuki and G. E. Wall) had found characterizations of this kind for groups of type $PSL(2,q)$. The method in such problems is to build up knowledge of the ordinary character table of G from what is given about H. This is exactly the kind of work for which Brauer's modular methods were designed, and he used these methods successfully on many simple group characterizations during the next twenty-five years, constantly refining and developing his modular theory. But now new actors began to appear in the drama of the simple groups. Suzuki and Wall used only ordinary character theory in their work on $PSL(2,q)$, and their methods—still following Brauer's program—led eventually to the discovery of new simple groups. Around 1960 J. G. Thompson began a powerful attack on the internal structure of simple groups, using new group-theoretical methods. In 1963 he and W. Feit verified the long-standing conjecture that every simple (noncyclic) group G has even order—which showed that G must possess involutions, so that Brauer's general program will always apply. The Feit-Thompson paper, remarkable for its length and difficulty, again used only ordinary character theory, based on an old theorem of Frobenius.

In the period 1960–1980 dozens of people joined the common effort to find all finite simple groups. Many new simple groups were found, some by application of Brauer's program and others from quite different sources. Brauer was deeply involved in this effort right up to the end of his life but did not see its final success, of which he must be counted one of the chief architects.

BIBLIOGRAPHY

I. Original Works. Facsimile reprints of most of Brauer's mathematical publications are in *Richard Brauer: Collected Papers*, Paul Fong and Warren J. Wong, eds., 3 vols. (Cambridge, Mass., and London, 1980), which includes an autobiographical preface by Brauer, a complete bibliography, and a reprint of Green's biographical article (see below).

II. Secondary Literature. Articles on Brauer are W. Feit, "Richard D. Brauer," in *Bulletin of the American Mathematical Society*, n.s. **1** (1979), 1–20; and J. A. Green, "Richard Dagobert Brauer," in *Bulletin of the London Mathematical Society*, **10** (1978), 317–342.

J. A. Green

BRIGGS, LYMAN JAMES (*b.* Assyria, Michigan, 7 May 1874; *d.* Washington, D.C., 25 March 1963), *physics, metrology*.

Briggs's forty-nine years of public service (thirty-eight with the National Bureau of Standards, including twelve as director) measured, both literally and symbolically, the rise of federal science in the United States. The son of Chauncey L. and Isabella (McKelvey) Briggs, he grew up on a midwestern family farm like so many other government scientists of his generation. Intellectually precocious, he completed high school by examination and entered Michigan State College at the age of fifteen. Although he complied with his parents' wishes and majored in agricultural science, earning his bachelor's degree in 1893, it was physics that intrigued and excited him. "From the moment I saw the great glass cases in the physical laboratory filled with marvelous apparatus," he later recalled, "I knew I wanted to be a physicist." In 1895, after finishing a master's degree in physics at the University of Michigan, he went to Johns Hopkins to study for his doctorate with Henry Rowland. Briggs for a time investigated the recently discovered X-ray phenomena but wrote his dissertation on soil physics, earning his Ph.D. in 1901. In the meantime, in 1896 he had taken a job as a physicist with the Department of Agriculture in Washington, D.C. On 23 December 1896 he married Katherine E. Cook, the daughter of one of his college professors; they had a son and a daughter.

Over the next twenty-one years Briggs established himself as a leading agricultural scientist and virtually founded the modern science of soil physics. Among other accomplishments, he developed the standard soil classification known as moisture equivalent (a measure of how much water a soil sample can hold against a force a thousand times greater than gravity) and wilting coefficient (a measure of how much water in a soil sample is available for plant growth). He published dozens of technical papers, including farsighted ecological studies of water in the West, and organized and headed the Department of Agriculture's first biophysical research laboratory (1906).

Briggs's career took an abrupt turn in 1917 when, as part of the national mobilization, he was transferred from the Department of Agriculture to the Bureau of Standards, where he was put to work designing a wind tunnel. Although he knew almost nothing about aeronautics, Briggs enjoyed the project and stayed on with the bureau after the war as chief of the mechanics and sound division (1920). He surrounded himself with an outstanding staff, including Hugh L. Dryden (a teenage prodigy who joined the bureau in 1919 and in 1920 earned a Johns Hopkins Ph.D.; he later became deputy administrator of NASA), and throughout the 1920's made significant contributions to the field. With funds transferred from the National Advisory Committee for Aeronautics, Briggs and Dryden undertook pioneering studies of airfoil behavior at and above the sound barrier. Briggs also invented (with Paul R. Heyl) an earth induction compass for aircraft (1922). In recognition of his achievements, Briggs was named the bureau's assistant director for research and testing in 1927. He was appointed acting director following the death of George Kimball Burgess in 1932, and President Roosevelt appointed him director the following year.

Just after the Senate confirmed Briggs's appointment, however, the Bureau of the Budget sliced his operating funds from a 1931 peak of more than $4 million to less than $2 million. "It was a bitter experience for us," Briggs recalled. "More than a third of our staff was dropped on a month's notice." Trying to keep the bureau moving forward during the Depression, Briggs emphasized programs with direct economic relevance, including research in building materials and low-cost housing. He even tried selling the administration on economic recovery through research. "Is there not an *essential* place in Government for basic research in physics and chemistry in order to provide the foundations for new industries?" he asked in a speech entitled "The Place of Government in Research" (1938). He testified on behalf of several bills giving the bureau broad responsibilities for "the further development of industry and commerce . . . through business research," but none of them made it out of Congress. The bureau limped through the Depression, managing somehow to fulfill its essential mission and even occasionally to produce significant research, including the first preparation and separation of heavy water.

It took another war to redirect Briggs's career. In the fall of 1939, following some alarming revelations from Albert Einstein and Leo Szilard about the military implications of nuclear fission, President Roosevelt appointed Briggs chairman of the supersecret Advisory Committee on Uranium, which included Szilard, Edward Teller, Eugene Wigner, and representatives from the armed forces and the White House. Characteristically, Briggs proceeded deliberately and frugally, doling out small grants for Enrico Fermi's studies of neutron absorption in graphite and for related investigations of uranium isotope separation. The fact that Briggs seemed far more intrigued by the long-range possibilities of nuclear power than by the short-term potential of nuclear bombs was upsetting to other committee members, who were more swept along by the urgency of the bomb issue. Briggs's conservative approach, which reflected the limitations of his legal mandate as well as his general reluctance to rush ahead on a long-shot project, exasperated aggressive young physicists like Ernest O. Lawrence, who urged National Defense Research Committee (NDRC) member James B. Conant to "light a fire under the Briggs committee. What if German scientists succeed in making a nuclear bomb before we even investigate possibilities?" As outside pressures mounted, Briggs did speed up certain studies, but apparently not fast enough; his committee was absorbed by the NDRC in June 1940. Briggs was replaced by Conant as chairman in June 1942, and shortly thereafter the committee's responsibilities were turned over to the U.S. Army's Manhattan Project under General Leslie R. Groves.

Even without the atomic bomb project, there were plenty of other important wartime projects for Briggs and the National Bureau of Standards. Besides contributing analytical research for the Manhattan Project, the bureau developed proximity fuses for the U.S. Army (as did Johns Hopkins University's Applied Physics Laboratory for the U.S. Navy) and guided missiles for the U.S. Navy and undertook radio propagation studies for the joint services. There were, in addition, many smaller projects in critical materials, optics, and other areas. By 1944 the bureau's budget reached a record $13.5 million (two-thirds of that consisted of funds transferred from the military), and staff rose to an all-time high of two thousand. Beyond the bureau's direct contributions to the war effort, Briggs's mobilization set the stage, scientifically and financially, for its rapid postwar expansion into electronics, computers, nuclear physics, and materials science.

Already past normal retirement age by the end of the war, Briggs stepped down in November 1945, and was succeeded as bureau director by physicist Edward U. Condon. Briggs kept an office and a laboratory at the bureau, however, and continued to do research virtually until his death. Returning to some previous interests, he conducted significant studies on the negative pressure of water in plants and attracted national attention with a scientific investigation of the curveball (he had played outfield in college). He was active in the National Geographic Society, even leading its solar eclipse expedition to Brazil in 1947, and wrote popular science articles for its magazine. He received a number of professional awards and honorary degrees, and in 1967 Michigan State named its liberal arts college for him.

BIBLIOGRAPHY

I. ORIGINAL WORKS. Briggs's complete bibliography includes more than one hundred articles ranging across soil physics, aerodynamics, metrology, and popular science writing. Wallace R. Brode, "Lyman J. Briggs: Recognition of His Eightieth Birthday," in *Scientific Monthly*, **78** (May 1954), 269–274, lists about half of them; many of the rest were published as technical bulletins for the U.S. Department of Agriculture, the National Advisory Committee for Aeronautics, or the National Bureau of Standards. Briggs also published *NBS War Research: The National Bureau of Standards in World War II* (Washington, D.C., 1949). Briggs's papers from his years with the bureau are held by the National Archives in Washington, D.C.

II. SECONDARY LITERATURE. In addition to Brode's article (see above), there are brief portraits and appreciations of Briggs by his two successors at the bureau: Edward U. Condon, "Lyman James Briggs, 1874–1963," in *American Philosophical Society Year Book* (1963), 117–121; and Allen V. Astin, "Lyman James Briggs, 1874–1963," in *The Cosmos Club Bulletin*, March 1974, 2–6. Rexmond C. Cochrane, *Measures for Progress: A History of the National Bureau of Standards* (Washington, D.C., 1966; repr. New York, 1976), gives a detailed account of Briggs's long career with the bureau. Carroll W. Pursell, Jr., "A Preface to Government Support of Research and Development: Research Legislation and the National Bureau of Standards, 1935–41," in *Technology and Culture*, **9** (1968), 145–164, skillfully places the bureau and its programs into the context of New Deal politics. Richard Rhodes, *The Making of the Atomic Bomb* (New York, 1986), offers a valuable, though somewhat unflattering, portrait of Briggs's part in the atomic bomb project.

STUART W. LESLIE

BRILLOUIN, LÉON NICOLAS (*b.* Sèvres, France, 7 August 1889; *d.* New York City, 4 October 1969), *quantum theory, solid-state physics, electronics, information theory*.

Brillouin was the son of Charlotte Mascart and Marcel Brillouin. His father held the chair of general physics and mathematics at the Collège de France; his maternal grandfather, Éleuthère Mascart, had held the chair of experimental physics in the same institution; and his great-grandfather, Charles Briot, had been a professor of mechanics at the Sorbonne. Nevertheless, the physics taught at the lycée failed to interest him until, following his father's suggestion, he read some of Pascal's letters on atmospheric pressure.

From 1908 to 1912, Brillouin joined the elite of French students at the École Normale Supérieure, where he attended the best physics and mathematics lectures available in France at that time. The most influential of his teachers were Paul Langevin, Henri Poincaré, and Jean Perrin. Unlike most of their French colleagues, they were exploring the newest and still controversial fields of physics, such as quantum theory and relativity, and maintained strong intellectual ties with the leading foreign physicists.

Brillouin's early research was inspired by some of Einstein's successes in molecular and quantum physics. In 1911, under Perrin's supervision, he extracted a new value of Avogadro's number from his measurements of the blue light of the sky. He did not publish his results (they depended too much on the state of the weather), but in his subsequent work he returned to the basic process involved, the scattering of light in a thermally fluctuating medium.

In 1912 Brillouin began an academic year in Munich under Arnold Sommerfeld, to observe the spectacular progress made there in X-ray studies and, above all, to learn more about the methods of theoretical physics. One of his first publications (1914) resulted from a fruitful application of a mathematical tool taught by Sommerfeld, the method of steepest descents, to the propagation of light in dispersive media. This study generalized Rayleigh's results on group and phase velocities, and explained such subtle physical effects as the fast and weak undulations preceding the "front" of a light signal (which travels at group velocity).

Back in Paris, Brillouin started his dissertation work on the quantum theory of solids but was soon halted by World War I. Like Maurice de Broglie, he joined the military research on radio transmission in the laboratory of General Ferrié. His main contributions, a new type of amplifier (the "resistance amplifier") and a remote-control system for ships and planes, won him the Legion of Honor. This work was not a mere interlude; it determined his lifelong interest in radio engineering problems.

In the two years following the war, Brillouin completed his dissertation work, mainly built on the idea that Einstein's theory of the quantum solid (or the improved versions by Peter Debye and by Max Born and Theodor von Kármán) should be to real solids what the theory of the ideal gas is to real gases. As the counterpart of gas pressure, he defined a pressure (in tensor form) of the solid vibration (also in analogy with the pressure of electromagnetic waves). The main interest of this work was a derivation of several important characteristics of the perfect solid independent of the quantum hypothesis (which restricts the energy of a monochromatic component of the vibrations to an integral multiple of its frequency times h). For instance, the form $F(\nu/T)$ of the thermal average of the energy per vibration of frequency ν at temperature T results from a purely classical theorem on abiabatic transformations by Ludwig Boltzmann and Paul Ehrenfest. The quantum hypothesis, Brillouin emphasized, remained compatible with many classical arguments; it just completed them.

In harmony with this judgment, the main new effect that Brillouin predicted, the "Brillouin doublet," was essentially classical: the frequency of the light scattered in a given direction by a vibrating transparent medium is shifted negatively or positively, according to the modulation by the reflecting elastic plane wave. Although the experimental confirmation of this effect had to wait until 1932, it is now commonly observed in laser experiments for both supersonic and thermal vibrations of the medium. In the first case the absolute value of the frequency shift equals the supersonic frequency; in the second case it is close to the Debye cutoff frequency.

In 1922 Brillouin published *La théorie des quanta et l'atome de Bohr*, the first textbook covering all quantum topics to be written by a French scholar. It shows some of his strongest qualities: an exceptional clarity of exposition, an analysis in depth of the physical foundation of reasoning, and a thorough consideration of the mathematical methods involved. The study of Einstein's solid had inspired Brillouin to consolidate the foundation of statistical mechanics; the calculation of the propagation and diffraction of light had caused him to develop the method of steepest descent; and the calculation of the pressure of solid vibrations brought him to extend the applications of tensor algebra to physics. Langevin recognized his student's accomplishments, and from 1923 to 1928 kept him as assistant director of his laboratory at the Collège de France.

In 1925 Brillouin was the only French theoretician to react competently to Werner Heisenberg's new matrix mechanics. In two papers published in 1926, he contributed to the exploration of the mathematical content of Heisenberg's theory. As an expert in tensor algebra, he was not intimidated by matrices. In his book on quantum theory, he had speculated, as had Born, on a new discontinuous mechanics of quantum jumps. Nevertheless, as soon as he knew of Erwin Schrödinger's theory and its formal equivalence to matrix mechanics, he favored the wave point of view, which he could easily connect to his earlier studies on wave propagation.

Brillouin's first contribution in this field was important. Through a new method of semiclassical approximation, he discovered the relation between Schrödinger's mechanics and the quantum theory of Niels Bohr and Sommerfeld. Presumably inspired by de Broglie's early analogies between mechanics and optics, he found this approximation as the quantum mechanical counterpart of the approximation of geometrical optics (in respect to the more exact wave theory). In this procedure, stationary solutions of the Schrödinger equation are sought in the form $e^{iS/\hbar}$ (as in the eikonal approximation of optics). In the first approximation ($\hbar$ small), S must be a solution of the Hamilton-Jacobi equation of classical mechanics, and the Bohr-Sommerfeld conditions ($S = 2\pi n\hbar$ on a closed trajectory) must be satisfied for the ψ function to be defined and single-valued in all space. Subsequent corrections are proportional to successive powers of $\hbar$. They intermix the various Bohr trajectories, thereby reintroducing the complex interplay of quantum states found in matrix mechanics. This method, published by Brillouin in July 1926—anticipated by Harold Jeffreys in 1923 in a purely mathematical context, reinvented by Gregor Wentzel in September 1926, and perfected by Hendrik Kramers in November 1926—is now called the (J)BWK method and is widely used in many quantum mechanical problems.

In spite of a sympathy for de Broglie's pilot waves, Brillouin did not dwell much on the interpretation of quantum mechanics. Instead, he labored to apply it to concrete situations, especially to the many-particle statistical problems with which he was most familiar. For instance, in 1927 he derived (as Wolfgang Pauli did independently) the quantum mechanical expression for paramagnetic susceptibility. In this expression appear the "Brillouin functions," generalizing the function earlier introduced by Langevin for the same problem.

In statistical matters the most novel result brought by quantum theory was the new method of treating identical particles, the Bose-Einstein and the Fermi-Dirac statistics. In 1930 Brillouin published *Les statistiques quantiques et leurs applications*, soon translated into several languages. He first demystified the new statistics, arguing that, from the elementary point of view of the distribution of identical particles over quantum cells, they were no less natural than Boltzmann's statistics. Then he fitted them into a common formal framework, starting from the expression $1\text{-}pb$ for the probability of a particle to go into a cell already occupied p times; $b = -1$ gives the Bose statistics, and $b = +1$ (never more than one particle in one cell) gives the Fermi-Dirac statistics.

More profoundly innovative were applications to the electron theory of metals. A brief survey of the work of other pioneers in this field will help in the understanding of Brillouin's contribution. In 1927, through a clever application of the Fermi-Dirac distribution, Sommerfeld managed to remove the main paradoxes in the old model of a free electron gas (for instance, the fact that only a small proportion of the "free" electrons contribute to the thermal and electrical conductivity). But his expressions for the thermal and electric conductivity still contained the mean free path, which required further quantum mechanical calculations of the electron-lattice interaction. In 1928, as a first step toward these calculations, Felix Bloch proved a very important theorem, according to which any Schrödinger eigenfunction for an electron in a periodic potential has the form $e^{i\mathbf{k}\cdot\mathbf{r}}u_{\mathbf{k}}(\mathbf{r})$, where $u_{\mathbf{k}}$ has the periodicity of the potential, and he described the first band of the energy spectrum in the strong-binding approximation (wherein the interaction with the lattice is regarded as a small perturbation to atomic bound states). In the same year Yakov Frenkel and Hans Bethe discovered the energy gaps of total reflection of electron waves by a periodic potential (in analogy with the diffraction of X rays by crystals). In 1929 Rudolf Peierls analyzed the properties of the energy bands in the approximation of weakly bound electrons, and M. J. O. Strutt discussed the exact solution of the one-dimensional problem (through the Mathieu-Hill equation). In early 1930 Philip M. Morse explained the relation between the gaps of total reflection and the band spectrum.

In the first edition of his book on quantum statistics, Brillouin corrected several mistakes in calculation and in physical interpretation made by his predecessors. Then, in the course of the preparation of the German edition (published in 1931), he greatly advanced the discussion of the band structure. Since, according to Morse, the bands are complementary to the gaps of total reflection, the energy of one-electron states as a function of Bloch's $\mathbf{k}$ will experience discontinuities whenever $\mathbf{k}$ satisfies the Bragg condition for total reflection, $\mathbf{k} = |\mathbf{k} - \mathbf{g}|$, where $\mathbf{g}$ belongs to the reciprocal lattice. Whereas his predecessors had been satisfied with one-dimensional simplifications, Brillouin took pleasure in describing the set of mutually intersecting planes in $\mathbf{k}$-space defined by this condition (see Figure 1). These planes (the bisecting planes of the vectors of the reciprocal lattice) delimit a sequence of interlocked zones, there being an energy discontinuity at the border between two consecutive zones. Originally Brillouin imagined that these nicely shaped zones would play a role in the calculation of conductivities, for Bragg's reflections had to interfere with the free propagation of electrons (the correspondence between vanishing conductivity and complete band filling was not known before A. H. Wilson's work of 1931). This intuition was confirmed when the zones were used to determine the position

BRILLOUIN

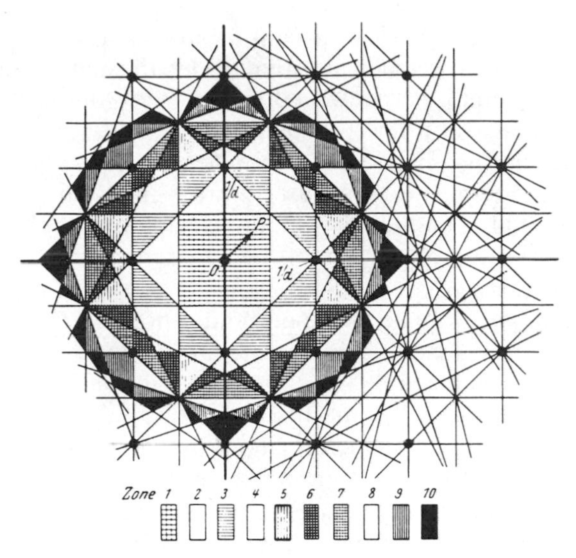

FIGURE 1. The first drawing of Brillouin's zones (from "Les électrons libres dans les métaux et le rôle des réflexions de Bragg," p. 384).

and shape of the Fermi surface (delimiting the occupied states in **k**-space), on which most relevant electronic properties of a crystal depend. The "Brillouin zones" now play a central part in any account of the electronic properties of solids.

In 1932 the Collège de France certified Brillouin's excellence by giving him the chair of theoretical physics. During the four preceding years he had taught in the same field at the Institut Henri Poincaré (as a Sorbonne professor). With the benefit of this experience, his lectures at the Collège de France focused on ongoing developments with a great mathematical clarity. His main originality was perhaps an insistence on mathematical methods common to quantum and classical physics. As he had proved in his own research, the transposition of mathematical methods from one subject to the other could help organize and develop both fields. For example, in his hands matrices were helpful in the theory of radiation pressure, in quantum mechanics, and in studies of wave propagation and electric filters (around 1936).

Meanwhile, Brillouin continued research on his favorite topics—conduction theory and wave propagation—with a growing interest in technical applications, for instance, to the telephone. In pure theoretical research some of his best (but little-known) work consisted in comparative studies of the various approximations used in the quantum theory of metals. In 1933 Brillouin introduced a new perturbation method well suited to the calculation of the corrections to the Hartree-Fock approximation (in the latter approximation the electron-electron interaction is replaced with the interaction of an electron with an average "self-consistent" field created by the other electrons). The resulting "Brillouin-Wigner" formula (which is obtained by replacing the unperturbed energy in the energy denominators of ordinary perturbative expansions with the exact energy) later proved useful in other many-particle problems, mainly in quantum chemistry and nuclear theory. Applying this method to his problem of electrons in a lattice, Brillouin could demonstrate the superiority of the Dirac-Fock approximation, a generalization of the Hartree-Fock approximation giving the best possible antisymmetrized product of individual wave functions for the global wave function. He also proved the "Brillouin theorem": the energy variation in the Dirac-Fock approximation when increasing the electron number from n to $n + 1$ is given as the $(n + 1)$th energy level of the one-electron problem in the self-consistent field. This theorem is important, for it explains the success of the most widely used approximation, that of weakly bound electrons.

In July 1939 Brillouin accepted the directorship of the French national radio broadcasting system. In the spring of 1940, Germany invaded northern France. In absence of specific military directions, Brillouin ordered the destruction of the broadcasting stations that would fall into enemy hands. He also had to organize the move of his employees to the unoccupied zone. Under the Vichy government he could not immediately resign from his position, and he had the unpleasant experience of being present at anti-Semitic and collaborationist propaganda events, incidents that he could not control. From August 1940 on, he was trying to find a position at an American university and to obtain a border pass; he did not leave France until March 1941, after his life had been threatened by the accusation in French newspapers that he had destroyed public property without orders.

Brillouin had lectured in several American universities in the 1920's and 1930's. In the academic year 1941–1942, he taught at the University of Wisconsin, and the next academic year at Brown. In 1943 he joined the military radar project at Columbia University. His research there, with the exploitation of the "Brillouin flow" of electrons in a magnetron (that is, a global rotation, as if it were a fluid mass, of the space charge in the magnetic field) won him a prize of the National Electronic Conference (1957) and later proved to be relevant to plasma theory. He also found time to promote Franco-American exchanges, as vice president of the American Institute of France, and professor and vice president of the newly created École Libre des Hautes Études de New York.

At the end of the war, uninformed about Brillouin's patriotic and scientific activities, the French Commission d'Épuration published a decree forbidding him access to any profession related to radio broadcasting. He and his friends protested vigorously and successfully: not only had he not collaborated in any political actions of the Vichy government, but he had clandestinely—and competently—sabotaged German radio-jamming devices directed against London.

In 1945 and 1946 Brillouin resumed his teaching at the Collège de France, but in 1947 he decided to emigrate to the United States. There he had a brilliant career, first as a Harvard professor (1947–1949), then as a research director at IBM (1949–1954, partly to organize a new teaching program in electronics), and as a professor at Columbia University from 1954 until his death. He was elected to the National Academy of Sciences in 1953. He had become a United States citizen in 1949.

Brillouin's postwar activities extended from consulting in radio engineering to theoretical criticism, for instance, regarding the structure of classical dynamical systems. His most important work dealt with information theory and its application to various subjects, including thermodynamics.

This last preoccupation originated in Brillouin's broader interest in such philosophical questions as the epistemological status of physical theories and the relation between physics and life. A joint meeting of physicists and philosophers that he had organized in 1938 at the Collège de France gave a first stimulus. In 1946, at Harvard, he had the opportunity for two other excursions to the frontiers of physics: he marveled at the new mathematical machines built there, and he participated in discussions on thermodynamics and life. Percy W. Bridgman impressed him with the operationalist argument forbidding the use of the concept of entropy in analyzing living organisms. On the one hand, Brillouin later commented, the usual definition of entropy through reversible transformations could not be applied in this case. On the other hand, the energy transformations produced by living organisms did not violate the second principle of thermodynamics (once it was properly applied to the organism plus its surroundings); they just suggested an intimate relation between life and the ability to exploit the energy from metastable reservoirs in complex catalytic processes. Like Schrödinger in *What Is Life?* (1945), Brillouin believed that a proper understanding of life would require a new principle, different from the known principles of physics but still a physical principle. Unable to say much positive about the nature of this principle, he contented himself with the hasty elimination of a few possibilities: It could not be a molecular principle, access to molecular constitution being complementary (in Bohr's sense) to life, nor could it be an algorithmic principle, mathematical machines (computers) being irremediably unintelligent.

Brillouin was better when he did not venture so far from mathematical physics, for instance, in his widely acclaimed *Science and Information Theory* (1956). His deepest insights concerned the relation between entropy and information. In two remarkable works of that time, Norbert Wiener's *Cybernetics* (1948) and Claude E. Shannon's *Mathematical Communication Theory* (1949), this relation had already appeared in contexts of particular interest to Brillouin: control mechanisms in the behavior of men and machines (Wiener), and electronic transmission of information (Shannon). Brillouin made the relation quantitative and general by the following argument.

Consider a physical system in a given macrostate corresponding to P equiprobable microstates (complexions). Entropy is defined as $S = k\ln P$, while the information that would be needed to completely specify the microstate is, according to Shannon's general definitions, $I = C\ln P$ (the choice of the constant C being conventional). Therefore, entropy is identical to the lack of information about the physical system under consideration, if only information is measured in proper units ($C = k$). In other words, negative entropy is identical to $-I$, the information "bound" to a physical system; this is Brillouin's "negentropy principle of information."

As a corollary, a simple way to diminish the entropy of a system is to increase information about it through more detailed measurement of its configuration. This procedure cannot be used to build cyclic machines violating the second principle of thermodynamics (as anticipated by Leo Szilard in 1929). Brillouin gave many instructive illustrations of this impossibility, the most famous of which concerned Maxwell's demon. By hypothesis, the demon is able to get information on the speeds of gas molecules, and to use this information to operate a trapdoor to separate fast molecules from slow ones. In this process the total entropy of the gas is indeed decreased. But the necessary information—here is the essential point—cannot be obtained by the demon without perturbing the gas in a way that increases the total entropy (of the system gas + demon + observation device) more than the information about the state of the gas. In general, Brillouin proved, the minimal amount of entropy increase required by an observation is $k\ln 2$, and the entropy increase during any observation is always higher than the amount of information obtained (expressed in entropy units). From this fundamental law Brillouin also derived new limits for measurements of lengths and times beyond Heisenberg's uncertainty relations.

Altogether Brillouin published some two hundred papers and two dozen books of high quality. Except for his tendency to work alone, his way of doing physics was strikingly modern. More than most of his French colleagues, he was open to new ideas and theories, and did not hesitate to cross disciplinary borders, addressing mathematical, engineering, biophysical, even philosophical problems. He constantly fought national isolationism. Against some deleterious trends of French theoretical physics before World War II, he stood out as an inspiring example.

BIBLIOGRAPHY

I. ORIGINAL WORKS. An extensive bibliography is in the Brillouin archive and in the biography by Hilleth Thomas (see below). See also Brillouin's *Titres et travaux scientifiques de Léon Brillouin* (Niort, 1931). Important books are *La théorie des quanta et l'atome de Bohr* (Paris, 1922); *Les statistiques quantiques et leurs applications aux électrons* (Paris, 1930); *Les tenseurs en mécanique et en élasticité* (Paris, 1937; 2nd ed., 1949), trans. by R. Brennan as *Tensors in Mechanics and Elasticity* (New York, 1963); *Wave Propagation in Periodic Structures* (New York, 1946; 2nd ed., 1953); *Science and Information Theory* (New York, 1956; 2nd ed., 1961); *Vie, matière et observation* (Paris, 1959).

Papers of special importance are "Über die Fortpflanzung des Lichtes in dispergierenden Medien," in *Annalen der Physik*, 4th ser., **44** (1914), 203–240; "La théorie des solides et les quanta," in *Annales scientifiques*, **37** (1920), 357–459, his doctoral dissertation; "Thermodynamique et probabilité: Révision des hypothèses fondamentales," in *Journal de physique et le Radium*, 6th ser., **2** (1921), 65–84; "Les amplificateurs à résistance," in *L'onde électrique*, **1** (1922), 7–17, 101–123; "Diffusion de la lumière et des rayons X par un corps transparent homogène. Influence de l'agitation thermique," in *Annales de physique*, **17** (1922), 88–122; "Les lois de l'élasticité sous forme tensorielle valable pour des coordonnées quelconques," *ibid.*, **3** (1925), 251–298; "Sur les tensions de radiation," *ibid.*, **4** (1925), 528–586; "La nouvelle mécanique atomique," in *Journal de physique et le Radium*, 6th ser., **7** (1926), 135–160; "La mécanique ondulatoire de Schrödinger: Une méthode générale de résolution par approximations successives," in *Comptes rendus de l'Académie des sciences*, **183** (1926), 24–26; "Les moments de rotation et le magnétisme dans la mécanique ondulatoire," in *Journal de physique et le Radium*, 6th ser., **8** (1927), 74–84; "Comparaison des différentes statistiques quantique appliquées au problèmes des quanta," in *Annales de physique*, **7** (1927), 315–331.

"Les électrons dans les métaux et le rôle des conditions de réflexion selective de Bragg," in *Comptes rendus de l'Académie des sciences*, **191** (1930), 198–200; "Les électrons libres dans les métaux et le rôle des réflexions de Bragg," in *Journal de physique et le Radium*, 7th ser., **1** (1930), 377–400; "Champs self-consistents et électrons métalliques III," *ibid.*, 7th ser., **4** (1933), 1–9; "Les bases de la théorie électronique des métaux et la méthode du champ self-consistent," in *Helvetica physica acta*, **7** supp. 2 (1934), 33–46; "La théorie des matrices et la propagation des ondes," in *Journal de physique et le Radium*, 7th ser., **7** (1936), 401–410; "La théorie du magnétron," *ibid.*, 8th ser., **1** (1940), 233–241; "Influence of Space Charge on the Bunching of Electron Beams," in *Physical Review*, **70** (1946), 187–196; "Les grandes machines mathématiques américaines," in *Annales de télécommunications*, **2** (1947), 329; "Life, Thermodynamics and Cybernetics," in *American Scientist*, **37** (1949), 554–568; "Thermodynamics and Information Theory," *ibid.*, **38** (1950), 594; "Maxwell's Demon Cannot Operate: Information and Entropy. I," in *Journal of Applied Physics*, **22** (1951), 334–337; and "Poincaré and the Shortcomings of the Hamilton-Jacobi Method for Classical and Quantized Mechanics," in *Archive for Rational Mechanics and Analysis*, **5** (1960), 76–94.

For the American Institute of Physics, Brillouin wrote extensive "Notes sur une carrière scientifique," included in the Brillouin papers deposited at the Center for History of Physics in New York City. Most of Brillouin's prewar papers have disappeared, except for the documents preserved at the Collège de France.

II. SECONDARY LITERATURE. The most detailed published biography is L. Hilleth Thomas, "Léon Nicolas Brillouin," in *Biographical Memoirs. National Academy of Sciences*, **55** (1985), 69–92. See also the introduction by A. Georges in Brillouin's *Vie, matiére et observation*; D. W. Harding, "Names in Physics: Brillouin," in *Physics Education*, **4** (1969), 46–48; and A. Kastler, "La vie et l'oeuvre de Léon Brillouin," in *L'onde électrique*, **50** (1970) 269–280. Information on Brillouin's contributions to solid-state physics are in L. Hoddeson, G. Baym, and M. Eckert, "The Development of the Quantum Mechanical Electron Theory of Metals, 1928–1933," in *Review of Modern Physics*, **59** (1987), 287–327.

OLIVIER DARRIGOL

BRODE, WALLACE REED (*b.* Walla Walla, Washington, 12 June 1900; *d.* Washington, D.C., 10 August 1974), *organic chemistry*.

Brode came of age in science at a time when the secrets of its synthetic chemical industry had been wrested from Germany through war and the bases of the quantum behavior of light were being divulged by physicists to chemists. He matured as a scientific spokesman at a time when scientific policy was at the forefront of national concern. He made important contributions in all these areas.

Brode was a son of Howard S. Brode, professor of zoology at Whitman College in Walla Walla, and of Catherine Bigham. He and his brothers (he was a triplet) all became professional scientists. Brode was trained in the sciences at Whitman College (B.S., 1921) and then at the University of Illinois, where he received the Ph.D. (1925). His work was done in Roger Adams' department, then distinguished for its work in organic chemistry. He married Ione Sundstrom on 19 March 1941.

Brode was swept up in the program to investigate synthetic dyes that Adams had originated, and took advantage of personal connections in the National Bureau of Standards to prosecute his dissertation work there, using spectrophotometry apparatus not available at the University of Illinois. He spent the three years after earning the B.S. (1921–1924) as a chemist at the Bureau of Standards.

The receipt of a Guggenheim Fellowship (1926–1928) permitted Brode to study at Leipzig with Arthur Hantzsch, a pioneer in the study of geometric isomerism and of the structure of nitrogenous organic compounds; at Zurich with Victor Henri, a pioneer in photochemistry who had discovered predissociation as a result of the absorption of light quanta; and at Liverpool with Edward C. C. Baly, an early student of the chemistry of photosynthesis. Brode's career as a government scientist was interrupted not only by his fellowship but also by an academic career spanning some twenty years (1928–1948) at Ohio State University, where he carried out a program in spectroscopy of organic compounds and was professor from 1939.

Among other things, Brode and his students clarified the influence of substituents on the absorption spectra of azo dyes, phototropic effects in the conversion of isomers of azo dyes to isomers of indigo dyes, absorption spectra of cobaltous and ferric compounds, and the relations between the optical rotation or absorption spectra and the chemical constitution of dyes. He built automatic recording spectrophotometers and spectropolarimeters needed in this work, offered a course in chemical spectroscopy, and wrote a textbook, *Chemical Spectroscopy* (1939).

Brode made broader contributions to pedagogy by devising the molecular models that came to be a mainstay of organic chemistry instruction. The laboratory manual that accompanied these models went through several editions.

During World War II, Brode worked for the Office of Scientific Research and Development. He set up a control laboratory for the Defense Plant Corporation, and served as a project leader for the War Metallurgy Committee and as a consultant on infrared filters in Division 16 of the National Defense Research Committee. In 1944 he was posted to London to serve as head of the OSRD liaison office, and served in the same capacity in Paris the following year.

After the war Brode became chief scientist at the Naval Ordnance Test Station (now the Naval-Weapons Center) at Inyokern, California, where he planned the Michelson Laboratory and recruited its civilian staff. Although he returned briefly thereafter to Ohio State, he was soon remobilized by Vannevar Bush to head the scientific branch of the Central Intelligence Agency (CIA). Because of the secret nature of the undertaking, Brode solicited the "cover" title of associate director of the National Bureau of Standards from its director, Edward Condon. When he sought to return to Ohio State after a year spent building up the scientific branch at the CIA, Condon insisted he actually assume the position of associate director.

At the National Bureau of Standards, Brode resumed studies of the photochemistry of organic compounds and made comparative studies of commercial spectrophotometers that were replacing the home-built ones he and others had long used. His experience overseas and with the CIA made him a prime candidate for the post of science adviser to the Department of State, which was created in the wake of Sputnik. Appointed by Secretary of State John Foster Dulles in 1958, he served for two years, building up a staff of science attachés for American embassies.

Brode's contributions to science and policy were recognized by his election to the presidency of the American Association for the Advancement of Science in 1958, of the Optical Society of America in 1961, and of the American Chemical Society in 1969. He became a member of the National Academy of Sciences in 1954. In his role as spokesman for science, Brode questioned the uneven support for science characteristic of the pluralistic postwar scientific establishment, and argued for the creation of a department of science to coordinate government research. He was the editor of the *Journal of the Optical Society of America* from 1950 to 1960 and of the Science in Progress series of Sigma Xi lectures from 1962 to 1966.

BIBLIOGRAPHY

I. Original Works. Among Brode's more than 200 scientific articles are the following summary reviews of his work: "Chemical Spectroscopy," in *Frontiers in Chemistry*, **4** (1945), 69–95; "Spectrophotometry and Colorimetry," in W. G. Berle, ed., *Physical Methods in Chemical Analysis* (New York, 1950), 194–253; "Color and Chemical Constitution," in George A. Baitsell, ed., *Science in Progress*, 9th ser. (New Haven, 1955), 293–323; and "Steric Effects in Dyes," in *The Roger Adams Symposium . . . University of Illinois September 3 and 4, 1954* (New York, 1955), 8–59. His textbook, *Chemical Spectroscopy* (New York, 1939; 2nd ed., 1943, 1949), and laboratory manual, *Laboratory Outlines and Notebook for Organic Chemistry* (1949), also reflect this work.

II. Secondary Literature. The most informative articles on Brode are in *McGraw-Hill Modern Men of Science*, II (New York, 1968), 53–54, and *McGraw-Hill Modern Scientists and Engineers*, II (New York, 1980), 141–143; and "Spectroscopist of the Month: Dr. Wallace R. Brode," in *Arcs and Sparks*, **9** (March 1963), 17–19. The research program at Illinois is discussed in Robert Kohler, "Adams, Roger," in *Dictionary of Scientific Biography*, XV (New York, 1980), 1–3. Brode discussed his work with defense agencies in A. B. Christman, "Interview: Dr. Wallace Brode, May 1969 re Early NOTS," NWC/75201-S-66.

Robert Seidel

BRONK, DETLEV WULF (*b.* New York City, 13 August 1887; *d.* New York City, 17 November 1975), *biophysics*.

The son of Mitchell Bronk, a Baptist minister, and Cynthia Brewster Bronk, Detlev Wulf Bronk was a pioneer biophysicist in the United States and, after World War II, an influential leader of and spokesman for the national scientific establishment. He was foreign secretary of the National Academy of Sciences (1945–1950) and simultaneously chairman of the National Research Council (1946–1950). From 1950 to 1962 Bronk was president of the Academy; president of the American Association for the Advancement of Science in 1950; chairman of the National Science Board of the National Science Foundation from 1956 to 1964; and a member of the President's Science Advisory Committee from 1956 to 1964. Bronk was president of the Johns Hopkins University from 1949 to 1953; in the latter year he became president of the Rockefeller Institute for Medical Science, which he converted to a graduate university, Rockefeller University. He retired in 1968. Bronk's pre-World War II experiences in research and administration provided the opportunities for his later ascent, as well as much of the style and viewpoints evident in his various administrative roles.

Bronk grew up in New Jersey and New York and remained a Baptist throughout his life. After enrolling in Swarthmore College, he served in the U.S. Navy's air arm during World War I, acquiring an interest, manifested during World War II, in scientific mobilization. In 1920 he graduated from Swarthmore with a B.S. in electrical engineering. In September 1921 Bronk married Helen Alexander Ramsey, who had been a fellow student at Swarthmore. Turning to physics, he received an M.S. in 1922 from the University of Michigan. By 1924 he was intent on applying physics and mathematics to physiology, receiving a Ph.D. in 1926 from the University of Michigan. Bronk's career until 1949 was linked to the University of Pennsylvania and, to a lesser extent, Swarthmore, with the exception of the academic year 1940–1941, when he joined the faculty of the Cornell Medical School in New York City. From 1939 to 1951 Bronk was editor of *Journal of Cellular and Comparative Physiology* (now *Journal of Cellular Physiology*).

In 1927 Bronk defined his goal thus: "I do not wish to become a physicist working in physiology but rather a well-rounded physiologist with a physical and mathematical background."[1] To attain this goal, he obtained a National Research Council fellowship and studied at Cambridge with Edgar D. Adrian and at London with Archibald V. Hill in 1927–1928. He collaborated with Adrian on the successful isolation of an individual motor nerve, devising an electrode to transmit the signals to the amplifier. Bronk later described his research as encompassing the "properties and activity of the single neuron, explained in terms of the laws of physics and chemistry" and how the nerve cells are integrated into a whole.[2] Bronk's most notable contributions were in the study of the regulation of the cardiovascular system. He was especially concerned with the development of electrical and optical instrumentation for physiological research.

After his return from England, Bronk organized the Eldridge Reeves Johnson Foundation for Medical Physics at the University of Pennsylvania, well endowed by the standards of that era. Here he did experimental work until 1938 or 1939, when administration claimed his energies. Under Bronk's leadership the Johnson Foundation had a significant role in expanding the field. Two future Nobelists, Ragnar A. Granit and H. Keffer Hartline, worked there. Fiscal stringencies at Pennsylvania caused a one-year move to Cornell by Bronk and his coworkers in neurophysiology. Similarly, Bronk took his closest Johnson Foundation colleagues to the Johns Hopkins University and, subsequently, to the Rockefeller Institute. In 1938 Bronk had written that progress in neurophysiology resulted from the work of a "small group of friends" at the Johnson Foundation and elsewhere.[3] The concept of the essentiality of small elite groups of interacting researchers would greatly influence his science policy views.

In the decade before World War II, Bronk became increasingly visible in the scientific community as investigator, administrator, and spokesman on the conduct and purpose of research. These and later oral and written pronouncements usually displayed the degree of his reading, a shallow but wide-ranging erudition. Given his research, Bronk defined life as adaptation, or "a modification of cellular structure by the environment," which occurred periodically or at least not continuously.[4] From the experience of his field, a merger of the physical and the biological, Bronk decried overspecialization, calling for a correlation—better still, "a synthesis of the sciences" particularly to include broader cultural values. The last required the presence of the humanities. In these writings Bronk was curiously old-fashioned, justifying the study of the single nerve cell as evading the mind-matter dichotomy and justifying the search for the "essential unity of science [as] . . . the final objective of every natural philosopher."[5]

Since man was part of the system of science and its data, Bronk saw biophysics as important for understanding research itself as well as the devices increasingly ubiquitous in a modern society. Biophysics also helped adaptation to changing environments. Repeatedly he decried the image of man conquering nature, favoring an adaptation to what research disclosed. This he identified with the coming biotechnic civilization of the writings of Patrick Geddes and Lewis Mumford. He declared that the goal of a humanitarian society was to "satisfy the biological requirements of the individual," a rather paternalistic, conservative conclusion.[6] His later career, in an environment of "big science" and large-scale military support for research, largely differed in substance from the language of his early policy pronouncements and did not quite match later statements.

When World War II broke out, Bronk moved to the national stage. From 1942 to 1946, he was coordinator of research for the air surgeon general, serving in 1945 with Theodor von Karman's group planning the U.S. Air Force's peacetime research posture. In 1945–1946 he held executive positions of both the National Academy of Sciences and the National Research Council. Bronk's elevation to presidency of the Academy in 1950, in the only contested election in that organization's history, enabled him to sit at the center of national science policy for a dozen years afterward.

In 1947 Academicians at Indiana University, dissatisfied with the conservative policy of President Frank B. Jewett and fearing a continuation in the management of the body, polled the Academy's members on who should succeed Jewett. They and others resented aspects of the wartime leadership and wanted a greater voice for members outside the traditional leading institutions. They believed the times called for activist leadership. However, the leader in the poll, Linus Pauling, withdrew his name after nomination from the floor, clearing the way for the election of Alfred N. Richards.

In 1950 the members of the Section of Chemistry of the Academy decided to oppose the official nomination of James B. Conant to succeed Richards by supporting Bronk. Contemporary sources indicate they were incensed with Conant's "authoritarian" behavior during the war. Refusing to honor Bronk's withdrawal from the floor and joined by similarly motivated Academicians, particularly among the mathematicians and geneticists, the insurgent chemists, led by Harold C. Urey, Wendell Latimer, Victor La Mer, and Joel Hildebrand, gained seventy-seven votes for Bronk to seventy-one for Conant on the ballot. Reached by phone, Conant withdrew and afterward blamed Bronk for the "treachery." The evidence known so far can support Bronk's denial of complicity. Bronk's successes as foreign secretary and as chairman of the National Research Council were in his favor, as was his straddling of the physical-biological science divide.[7]

In his multiple roles within the national science establishment, Bronk had only modest success in deflecting policies to give greater support to nonmilitary research, particularly in biological fields. His was a policy of reacting against threats to the autonomy and integrity of the research community while taking care that basic research obtained a slice of whatever programs current exigencies brought into being. Bronk also tried to expand the support of research to new and rising institutions while eschewing geographical distribution of funds.

Bronk's belief in supporting the elect few had greater success in his role as educator. At Johns Hopkins he tried unsuccessfully to create an environment where individuals could advance from high school to real research at their own pace without regard to formalities of courses and credits. In transforming the Rockefeller Institute for Medical

Science into Rockefeller University, Bronk was re-creating the ideal of the prewar Johnson Foundation on a much expanded fiscal scale and with a deliberate attempt to go far beyond the traditional biomedical fields. In the absence of undergraduates, he partially succeeded. Fiscal stringencies, however, forced his successor, Frederick Seitz, to narrow the scope of the Rockefeller University, although not back to its pre-Bronk limits.

NOTES

1. Frank Brink, Detlev Wulf Bronk," p. 16. From original in box 82, of Bronk Papers, 303-U in Rockefeller Archive Center, North Tarrytown, N.Y.
2. D. W. Bronk, "The Cellular Organization of Nervous Function," in *Transactions and Studies of the College of Physicians of Philadelphia*, 4th ser., **6** (1938–1939), 103–147, esp. 103 f.
3. *Ibid.*, p. 103.
4. *Ibid.*, p. 106; Bronk, "The Physical and Chemical Basis of Nerve Action," 1941 Priestley Lecture at Pennsylvania State College. Unpublished, edited from stenographic record.
5. Bronk, "Relation of Physics to the Biological Sciences," in *Journal of Applied Physics*, **9** (March 1938), 139–142.
6. Bronk, "The Nervous Regulation of Visceral Processes." In, Maurice B. Visscher, ed., *Chemistry and Medicine* (Minneapolis, 1940; repr. Freeport, N.Y., 1967), 261–275; Bronk, "Relation of Physics . . . ."
7. MacInness diary, April 27, 1950, in D. A. MacInness Papers, Rockefeller Archive Center; Edwin Bidwell Wilson to Bronk, May 1, 1950, Bronk Papers, box 84 of 303-U; and Rexmond C. Cochrane, *The National Academy of Sciences: The First Hundred Years* (Washington, D.C., 1978, 515–516. Speculation that Conant's defeat came about because of the H-bomb controversy seems most unlikely in view of the evidence of long-standing dissent within the Academy arising from other causes.

BIBLIOGRAPHY

The best single introduction to Bronk's life and works is Frank Brink, Jr., "Detlev Wulf Bronk," *Biographical Memoirs. National Academy of Sciences*, **50** (1979), 3–87. Brink was one of the neurophysiologists from the Johnson Foundation who accompanied Bronk to the Cornell Medical School, to Johns Hopkins, and then to the Rockefeller Institute. He played an active role in the latter's transformation into a university. (See his "Detlev Bronk and the Development of the Graduate Education Program," in *Institute to University, a Seventy-fifth Anniversary Colloquium, June 8, 1976* [New York, 1977], 69–78. The biographical memoir has an excellent bibliography of Bronk's writing that lacks the Priestley lectures (see note 4). There are a sprinkling of minor errors. More consequential is the uncritical, hagiographic tone intruding from time to time. Britton Chance's necrology in *Year Book of the American Philosophical Society* (1978), 54–66, is excellent for conveying the personal characteristics of the man. It has the unintended ironic virtue of preceding the necrology for James B. Conant.

The best sources for Bronk's career are the various collections in the Rockefeller Archive Center, North Tarrytown, N.Y. Besides the official files of his service as president of Rockefeller University, they range backward in time to his ancestors in the past century and up to his last days. One collection (303.1) is an extensive correspondence file stretching from before World War II to the 1970's. The largest and richest is 303-U, which includes extensive files of his days at the Johnson Foundation, records of his World War II activities, and personal and subject files covering much of his career. Bronk also saved drafts and notes for papers and talks, and a set of laboratory records. There are runs of correspondence with Adrian and Hill, as well as other worthies. Although an effort was made after his death to return NAS and NRC files, broadly defined, to the Archives of the National Academy, much survives in 303-U on his actions on the national level from 1945 to 1962. The various series of Central Policy files of NAS–NRC from 1946 to 1962, in the Archives of the National Academy of Sciences, are replete with evidences of Bronk's activities. No separate Bronk archive record exists in the Johns Hopkins Archives. Records of his tenure are dispersed through the single block of records of the Office of the President covering the presidencies of Remsen through Bronk.

NATHAN REINGOLD

BROUWER, HENDRIK ALBERTUS (*b.* Medemblik, Netherlands, 20 September 1886; *d.* Bloemendaal, Netherlands, 18 September 1973), *geology*.

Brouwer was the third son of Egbertus Luitzen Brouwer and Hendrika Poutsma; his eldest brother was Luitzen Egbertus Jan Brouwer, a mathematician of world note. Brouwer's father, a schoolteacher, ended his career as head of a secondary school at Haarlem. In this city, where Brouwer received his secondary education, the sight of the mineral collection at Teyler's Foundation led him to abandon his plans to become a naval officer. He studied mining engineering at the Technical University at Delft under Gustaaf A. F. Molengraaff (1903–1908). For his doctoral dissertation Brouwer investigated Molengraaff's collection of South African nepheline syenites, in part guided by Alfred Lacroix of Paris. He obtained his doctorate at Delft in 1910, his dissertation supervised by Molengraaff.

Brouwer was not active in any church but had a profound interest in mystical affairs. For instance, he firmly believed in astrology, and checked the starting date of his 1937 expedition to the Netherlands East Indies. On 2 June 1909 he married Louise Betsy van der Spil; they had three children. His elder son, Luitzen Egbertus Jan, had an education similar to his father's and became an executive in the Royal Dutch-Shell Group. Brouwer's first marriage ended in divorce (1930), as did his second marriage, to Olga Marianne Labouchere (1941–1958).

Late in 1910 Brouwer went to the Netherlands East Indies as a government official in the Mining Department. Here his main activities concerned the exploration and detailed description of the geology and petrography of the eastern part of the archipelago. The results of this exploration brought him fame among geologists the world over. One of the points Brouwer stressed was that, in spite of distinct indications of vertical movements, the island arcs mainly moved horizontally. In 1911, he was a member of Molengraaff's expedition to the island of Timor.

Brouwer returned to the Netherlands in 1916 and was appointed professor of historical geology and paleontology at Delft Technical University in 1918. He started fieldwork with his students in southern Spain and, being strongly influenced by Émile Argand's ideas, proposed a nappe structure for the Betic Cordilleras in 1924.

In 1922 Brouwer was an exchange professor at the University of Michigan at Ann Arbor. While there he delivered the lecture series "The Geology of the Netherlands East Indies" (published as one of the University of Michigan Studies in 1925). Also in 1922 he was appointed member of the Royal Netherlands Academy of Sciences and Letters, an honor he valued highly.

In 1925 Brouwer accepted a subsidiary position teaching tectonics at the State University at Utrecht. He retained this function after leaving Delft in 1928, when he became professor of general and practical geology and petrology at the University of Amsterdam, where the authorities had decided to build a new geological institute. He retired in 1957. Brouwer acquired beautiful and instructive specimens from all over the world for his institute. He started with only a few geology students but their number rose to more than a hundred. Many of his students became university professors.

In 1936 the civil war in Spain made fieldwork there impossible, and Brouwer shifted his European fieldwork from Italy and Spain to Lapland and Corsica. With a number of advanced students, he made an expedition to the Lesser Sunda Islands of the Netherlands East Indies in 1937, visiting his beloved island of Timor again. During World War II he and some of his staff members described the collections made during an expedition led by him on the island of Sulawesi (Celebes) in 1929.

In Brouwer's life almost everything was subordinate to his professional work. He was an inspiring teacher, unmatched in the communication of his enthusiasm for geology. In tackling geological problems he often turned to his well-developed intuition. His strong sense of scientific responsibility made him averse to speculation and theories.

Brouwer had great personal charm but was a loner, difficult and exacting of himself and of others. He could be militant and aggressive and was involved in many conflicts; yet he was extremely vulnerable to critical remarks.

For many years Brouwer was vice president of the Geologische Vereinigung at Bonn, which awarded him its Gustav Steinmann Medal in 1956. In 1938 he was vice president of the Société Géologique de France, and in 1954 he became an honorary member of the Société Géologique de Belgique. He was also a foreign fellow of the Geological Society of London. In the Netherlands he held many posts, mainly in the Geological and Mining Society, which awarded him its Van Waterschoot van der Gracht Medal of Honor in 1957. From 1923 to 1948 he was associate editor of the *Journal of Geology* (Chicago).

BIBLIOGRAPHY

I. Original Works. Among Brouwer's works are "On the Geology of the Alkali Rocks in the Transvaal," in *Journal of Geology*, **25** (1917), 741–778; "Geologische onderzoekingen in den oostelijken Oost-Indischen Archipel, I–V," in *Jaarb. Mijnw. Ned. Oost-Indië, Verh.* (1920–1926); *Geology of the Netherlands East Indies* (Ann Arbor, 1925); *Geological Expedition . . . to the Lesser Sunda Islands . . . Under Leadership of H. A. Brouwer*, 4 vols. (Amsterdam, 1940–1942); and *Geological Explorations in the Island of Celebes . . . Under the Leadership of H. A. Brouwer* (Amsterdam, 1947). With Karl Andrée and Walter H. Bucher he edited a new series *Regionale Geologie der Erde* (Leipzig, 1938–1941).

II. Secondary Literature. A biography and bibliography are in a memorial volume issued on the occasion of Brouwer's seventieth birthday, *Verhandelingen Koninkl. Nederl. Geol. Mijnbouwk. Genootschap*, **16** (1956). See also *Geol. & Mijnbouw.*, **52** (1973), 253–255.

W. P. de Roever

BROWNE, WILLIAM ROWAN (*b.* Lislea, County Derry, Ireland, 11 December 1884; *d.* Sydney, Australia, 1 September 1975), *geology*.

Browne was the sixth of eight children of National School teachers James Browne and Henrietta Rowan Browne, on both sides descended from loyalist, Anglican families. His paternal grandfather, presumed to have been a farmer, fought as a volunteer against the uprising in 1798; his mother's father was an architect and contractor. In October 1903 Browne entered Trinity College, Dublin, having taken first place at matriculation for all Ireland. He began a classical arts course but soon had to withdraw, suffering from tuberculosis.

Advised to seek a congenial climate, he set out in February 1904 for Sydney, Australia, where by late 1906 he was able to resume studies. But his plan to go on with classics changed after he heard about a remarkable professor of geology in Sydney, T. W. Edgeworth David. In 1910 Browne graduated B.Sc. from the University of Sydney with first-class honors in both geology and mathematics. After a year as assistant at the Adelaide Observatory, he gained a junior teaching post in geology at the University of Sydney, which thereafter became his main base. He retired as university reader in geology in 1949. In 1915 Browne married Olga M. Pauss; they had two daughters. She died in 1948, and in 1950 Browne married Ida A. Brown, a paleontologist, who died in 1976.

As Edgeworth David brought Browne to geology, so he remained Browne's inspiration. Loyalty to David shaped his career, a major part of it being devoted to realizing the synthesis of Australian geology that David had hoped to achieve. But Browne was no tame follower. He earned his place as David's true successor, as the leading Australian geologist of his day.

Browne shared David's broad geological interests, handling with equal command problems of petrology, stratigraphy, tectonics, and geomorphology; but at first he had a particular rapport with petrology. His study of carboniferous volcanism in the Hunter River region of New South Wales, begun in 1911, led to detailed and, for Australia, innovative research on secondary alteration phenomena and to concern with late Paleozoic environments and stratigraphy of the region. By 1914, at Cooma, New South Wales, he had pioneered study of regional metamorphism in that state. Later, during university vacations from 1919 to 1921, Browne joined a group led by E. C. Andrews of the Geological Survey of New South Wales that was examining the Broken Hill area. Browne's regional reconnaissance established the main petrological style of the metamorphosed Willyama Complex there. His report appended to Andrews' memoir of 1922 was also a dissertation for which he gained the D.Sc. from the University of Sydney in the same year.

During the 1920's Browne served on various research committees of the Australasian (now Australian and New Zealand) Association for the Advancement of Science set up to coordinate geological information. The experience helped focus his attention on issues beyond the local and observational—for instance, on relations between tectonic events and igneous action and metamorphism. In a presidential address of 1929 he reviewed these themes for New South Wales to the end of Paleozoic time. Another such address in 1933 took igneous action through the Mesozoic and Tertiary and for the first time showed evidence from Australia of contrasted basalt magma types. Meanwhile, in 1931 Browne had published a work, still cited, on batholiths and their time relations with tectonism. Plutonic rocks also had been his concern since working at Encounter Bay, South Australia, in 1912.

It was as if, unaware, he had been serving an apprenticeship. Without warning, in March 1934 David asked Browne to finish the book that for so long had been his last ambition. After David's death that year, the supposed manuscript was bought by the government of New South Wales, which formally repeated the author's request to Browne. He accepted but found, to his dismay, that David had left little more than a mass of jottings and some reviews written years before. He had no choice but to start afresh. Given leave by the university, he worked at the task; but at the outbreak of war in 1939, the job far from done, he returned to academic duty, with work on the book relegated to his spare time. He had already retired when *The Geology of the Commonwealth of Australia* appeared in 1950. Though credited to his mentor, it was Browne's masterly synthesis, now a classic in the literature of geology.

On the many topics he embraced while preparing the book, none became more absorbing to Browne in his active retirement than Quaternary history, in particular the consequences of Pleistocene glacial action on mainland Australia. He wrote extensively on the subject, which—as he saw it, and to his regret—was being left by geologists to geographers with inadequate geological training. He became also an ardent and effective advocate for conservation of rare natural environments. Yet he did not oppose developmental works, if sensitively handled; indeed, he served as geological consultant to various major engineering projects in New South Wales with notable success.

Disciplined, erudite, and articulate, Browne was a scholar among Australian geologists. With most he communicated by publishing his work locally, a course that matched his staunch support of Australian scientific societies. If as a teacher Browne lacked Edgeworth David's gift for fine rhetoric, his skill in orderly presentation and informed argument made an effective substitute. Of the many students set by him on distinguished careers, the list begins with C. E. Tilley, who remained his lifelong friend.

The many awards and honors, mostly Australian, received by Browne are on record in the printed memorials. Only a few are noted here. He became a fellow of the Australian Academy of Science in 1954, at the inaugural election of that body. In 1966 the Royal Society of New South Wales issued volume 99 of its *Journal and Proceedings*, consisting of papers offered in his honor, as the W. R. Browne volume. The senior award of the Geological Society of Australia, the Browne Medal, was instituted as a memorial. He died of a heart attack.

BIBLIOGRAPHY

I. ORIGINAL WORKS. Among Browne's writings are "The Geology of the Cooma District, Part I," in *Journal and Proceedings of the Royal Society of New South Wales*, **48** (1914), 172–222; "The Igneous Rocks of Encounter Bay, South Australia," in *Transactions of the Royal Society of South Australia*, **44** (1920), 1–57; "Report on the Petrology of the Broken Hill Region . . .," in *Memoirs of the Geological Survey of New South Wales*, **8** (1922), appendix, 295–353; "Presidential Address. An Outline of the History of Igneous Action in New South Wales till the Close of the Palaeozoic Era," in *Proceedings of the Linnean Society of New South Wales*, **54** (1929), ix–xxxix; "Notes on Bathyliths and Some of Their Implications," in *Journal and Proceedings of the Royal Society of New South Wales*, **65** (1931), 112–144; "Presidential Address. An Account of Post-Palaeozoic Igneous Activity in New South Wales," *ibid.*, **67** (1933), 9–95; and *The Geology of the Commonwealth of Australia*, by T. W. Edgeworth David, edited and supplemented by Browne, 3 vols. (London, 1950).

II. SECONDARY LITERATURE. A biography is T. G. Vallance, "William Rowan Browne, 1884–1975," in *Proceedings of the Linnean Society of New South Wales*, **102,** pt. 2 (1977), 76–84. A bibliography is appended to T. G. Vallance and E. S. Hills, "William Rowan Browne," in *Records of the Australian Academy of Science*, **4**, no. 1 (1978), 65–81.

T. G. VALLANCE

BUBNOFF, SERGE NIKOLAEVICH VON (*b*. St. Petersburg, Russia, 15 July 1888; *d*. Berlin, German Democratic Republic, 16 November 1957), *geology*.

Bubnoff was the son of Nikolai von Bubnov, a Russian physician, and of his German wife, Maria Türstig von Bubnoff. After their father's death in 1889, Serge and his two brothers were brought up by their mother. Despite a hearing defect that he did not overcome until later in life, with the use of a hearing aid, Bubnoff graduated from the First St. Petersburg Gymnasium with a gold medal in 1906. He had a perfect command of both German and Russian. When the family moved to Heidelberg, he enrolled at the University of Freiburg im Breisgau, from which he graduated in 1910. Bubnoff was then an assistant at the Geological-Paleontological Institute at the University of Freiburg (1910–1911) and worked for the Baden Geological Survey (1911–1913). In 1912 he obtained the doctorate with a dissertation titled "Die Tektonik der Dinkelbirge bei Basel." From 1913 to 1914 he went on a study tour to Italy and Russia. Bubnoff studied petrography at the St. Petersburg School of Mines with the crystallographer E. S. Federov and became acquainted with prominent Russian geologists. From 1914 to 1920 he was Wilhelm Salomon-Calvi's assistant at Heidelberg. In 1921 he married Eleonor Schmitt; they had two daughters.

In 1921 Bubnoff went to Breslau, following his college friend Hans Cloos. There he qualified as *Privatdozent* with a *Habilitationsschrift* on Hercynian breaks in the Black Forest ("Die hercynischen Brüche im Schwarzwald," 1922). He became a lecturer in geology and paleontology, and in 1925 he was appointed assistant professor at Breslau. In May 1929 he accepted a post at Greifswald University, where he was professor and head of the Institute of Geology and Paleontology until 1950. Bubnoff was elected corresponding member of the Prussian Academy of Sciences (Berlin) in June 1941, and of the Geological Society of America in August 1948. In 1950 he went to Berlin, where he became head of the Institute (and Museum) of Geology and Paleontology of Humboldt University. He also succeeded Hans Stille as professor of geology and head of the Institute of Geotectonics at the German Academy of Sciences at Berlin.

Bubnoff devoted his life to the geological sciences, and was especially influenced by the works of Russian geologists. His work was honored by academic societies as well as by the government of the German Democratic Republic (he received the National Prize, first class, in 1953). In 1958 the GDR Geological Society endowed the Serge von Bubnoff Medal, which is awarded to scholars for outstanding scientific achievements in the investigation of the earth.

Bubnoff made fundamental contributions to the expansion of the regional geological knowledge of Germany and of Europe. His first studies of the tectonics of the Jura plateau in southern Germany and of the granites in the Black Forest and in Thuringia were followed by works on the geology and

BUBNOFF

tectonics of the coal-mining areas in Lower Silesia. His book on the theory of nappes of the Alps was published in 1921. During his years at Greifswald, he took a particular interest in the geology of Bornholm and southern Sweden. Bubnoff's method was characterized by constant endeavor to discover greater spatial and temporal connections on the basis of concrete and detailed geological facts. It was his aim to elucidate the history of an eruptive body with respect to its substance and structure by combining geological, petrographic, and structural analyses. His great capacity for synthetic geological thought becomes evident in his *Geologie von Europa* (1926–1936), published in four volumes. His *Fennosarmatia* (1952) represents a summary of several works on the regional geological structure of central Europe written in the first half of the twentieth century.

In his book *Grundprobleme der Geologie* (1931) Bubnoff made important theoretical contributions to the investigation of the earth. Working from epistemological ideas, he came out against ill-balanced interpretations of geological documents, discussed the concept of time in geology, and systematized the geotectonic components. He developed his own tectonic conception ("Die Gliederung der Erdrinde," 1923) by subdividing the earth's crust into (1) permanent continental blocks, (2) permanent oceans, (3) weakly mobile plates, and (4) strongly mobile geosynclines. Important criteria for his characterization of the components of the earth's crust were mobility of the structural zones, oscillating amplitudes of their surfaces, tendencies of their vertical movements, the character of their tectonic deformation, and their thickness. Using these criteria, Bubnoff discussed the mechanisms of orogenesis and examined the different aspects of both the fixistic and mobilistic theories. To a certain extent he supported the ideas of mobilism, although he believed that the hypothesis that America drifted from the Old World could hardly be maintained in its original version. His explanation of orogenesis was based on theories of magmatic currents that go back to ideas developed by Otto Ampferer in his "Über das Bewegungsbild von Faltengebirgen" (1906). According to Bubnoff, the diverse manifestation of tectonic movements

> is hardly compatible with the theory of contraction, because this theory claims a simultaneity of movements all over the earth, which actually leaves no room for explaining phase differences and phase divergences, that is, different times of motion in different parts of the crust. On the other hand, phase differences quite agree with the theory of currents [*Grundprobleme der Geologie* (Haller, 1949), 220].

Bubnoff also tried to systematize general processes concerning the exogenous dynamics of the earth. From the Cambrian to the present he distinguished six major cycles, each characterized by specific facies and spatial features and concluded by an orogenetic phase. For each cycle he specified six facies phases according to the depth of the sea and the distance from the shore: first transgression (rough to medium clastic), second transgression (fine clastic), inundation, differentiation, regression, and emersion. The spatial specification, which was above all due to a particular orientation of the direction of transgression, was deduced from the geological conditions in western Europe. According to Bubnoff's calculations there was an acceleration of the sequence of cycles in the course of the earth's history. The successive cycles with their six phases become shorter and shorter.

Bubnoff explained the history of the earth as a natural process of evolution. In 1948 he drew the following conclusion from his analysis of endogenous and exogenous processes:

> In this way inorganic evolution in the history of the earth is characterized by "vergence," that is, a linear-oriented component that results in differentiation and complication of the structure of the crust, and this makes the history of the earth a historical process that cannot be repeated. The development of the earth can be described neither as a circle nor as a line but rather as a combination of both, namely, a spiral ["Der Rhythmus der Erde," in *Universitas* **3** (1948), 966].

On the basis of detailed observation of nature, Bubnoff expounded his ideas about fundamental questions of geotectonics, paleogeography, and the methodology of geological sciences in 181 publications.

BIBLIOGRAPHY

I. Original Works. A full bibliography of Bubnoff's works is A. Illner, "Wissenschaftliche Arbeiten von Serge von Bubnoff," in *Geologie*, **7** (1958), 251–256. A work in English is *Fundamentals of Geology*, W. T. Harry, trans. and ed. (Edinburgh, 1963).

II. Secondary Literature. Writings about Bubnoff include *Gedenkschrift Serge von Bubnoff zu seinem 70. Geburtstag*, a special issue of *Geologie*, **7** (1958), 237–860. See also F. Deubel, "Nachruf für Serge von Bubnoff," in *Geologie*, **7** (1958), 100–101; Yevgenii Yevgenevich Malinovskii, "Znamenityi nemetskii geolog S. N. Bubnoff v Moskovskom Universitete" (Outstanding German geologist S. N. Bubnoff at Moscow University), in *Vestnik Moskovskogo universiteta*, **1** (1957), 236–258; and Yevgenii Yevgenevich Malinovskii and Günter Möbus, "Serge N. von Bubnoff und seine Bedeutung für die Entwicklung der deutsch-sowjetischen Beziehungen auf dem Gebiet der geologischen Wissenschaften," in *Zeitschrift für geologische Wissenschaften*, **4** (1976), 457–467.

MARTIN GUNTAU

BUDDINGTON, ARTHUR FRANCIS (*b*. Wilmington, Delaware, 29 November 1890; *d*. Cohasset, Massachusetts, 25 December 1980), *geology*.

Buddington was the second child of Osmer Gilbert and Mary Salina Wheeler Buddington. When Arthur was fourteen, the family returned to the parents' native Connecticut, where the elder Buddington combined serving as pastor to a country Baptist church with operation of a small produce and poultry farm. Arthur attended public schools in Wilmington and in Connecticut and nearby Rhode Island, graduating from Westerly (Rhode Island) High School in 1908. He then entered Brown University, majoring first in chemistry and botany, then in geology. After graduating second in his class in 1912, he went on to receive the M.S. degree in 1913. His graduate work was completed at Princeton University with the granting of the Ph.D. degree in 1916.

Buddington's course was erratic for the next several years: after briefly considering a career in the petroleum industry, he began fieldwork in the Adirondacks, then returned to Brown as an instructor in 1917, only to leave within the year to teach military aerial observation at Princeton. In 1918 he enlisted in the army as a private and was assigned (because of his technical background) to the Chemical Warfare Service. Mustered out at war's end with the rank of sergeant, he returned briefly to Brown, then in 1919 accepted appointment to the Geophysical Laboratory of the Carnegie Institution. In 1920 his career finally stabilized with his appointment as assistant professor at Princeton, where he was to remain until (and after) formal retirement in 1959. He was married in 1924 to Jene Elizabeth Muntz; they had one child, Elizabeth Jene.

Aside from Princeton, Buddington was closely associated throughout his career with two scientific organizations: the New York State Museum, which sponsored his early fieldwork in the Adirondacks, and the United States Geological Survey, for which he carried out extensive field studies in Alaska and in Oregon, and—particularly during World War II—on iron deposits of the Adirondacks and adjacent regions. A member of many scientific societies, Buddington served as president of the Mineralogical Society of America in 1942; president of the Volcanology Section of the American Geophysical Union from 1941 to 1944; vice president of the Geological Society of America in 1943 and 1947; and chairman of the Geology Section of the National Academy of Sciences from 1954 to 1957. He was elected to the National Academy of Sciences in 1943 and to the American Academy of Arts and Sciences in 1947. He received the Penrose Medal of the Geological Society of America in 1954, the Roebling Medal of the Mineralogical Society of America in 1956, the André H. Dumont Medal of the Geological Society of Belgium in 1960, and the Distinguished Service Award of the U.S. Department of the Interior in 1963.

Within the broader realm of geology, Buddington was a petrologist, with a prime aim of interpreting the mineral assemblages of rocks and ores in terms of chemical principles and theory. More specifically, he considered himself a field petrologist, dedicated to the mapping and study of rocks in actual outcrop; he once estimated that he had examined and mapped 50,000 outcrops over a 44-year period, involving about 35,000 miles of travel on foot and 5,500 miles in small boat. Much of this great effort is recorded in geologic maps and documentary-type reports, most of which were published by the U.S. Geological Survey and the New York State Museum. These fact-laden reports, recognized as contributions of long-term value to the national geologic data base, also served as bases for Buddington's better-known topical papers, in which he advanced concepts of more general import, all of which stem directly from or are rooted in these field studies. Several examples follow:

Buddington's widely cited 1959 paper delineating and explaining differences in the nature of igneous intrusives in terms of depth of emplacement is based on his perceptive observations made in the course of mapping such bodies in the greatly different geologic environments of Newfoundland, the Alaska Coast Ranges, the Oregon Cascades, and the Adirondack Mountains of New York.

Definition of a "xenothermal" (shallow-depth, high-temperature) class of ore deposits (1935)—one of the first formal breaks with the then-accepted classification in which temperature of formation and depth of emplacement were assumed to vary sympathetically—clearly stemmed from his field study of shallow intrusives of the Oregon Cascades and their related ores.

Buddington's classification of anorthosites (plagioclase-rich igneous rocks) into a Grenville type characterized by massif habit and a crystal-settled type occurring within layered gabbroic complexes (1960), which brought out the profound differences in origin and significance of these mineralogically similar rocks, is based on his field studies in the Grenville terrane of the Adirondacks, coupled with observations on the Stillwater Complex of Montana made during his supervision of several Ph.D. studies.

Recognition that the mineralogic and compositional variations in the Fe-Ti-O system could serve as a measure of initial temperature and oxygen fugacity (Buddington and Lindsley, 1964)—a contribution that has led to development of increasingly sophisticated geologic thermometers and oxygen barometers—was a culmination of a long series of studies on magnetite-hematite-ilmenite ore deposits of the Adirondack region that began in the field and continued in the laboratory.

Buddington was a key member of the geology department at Princeton for nearly forty years, fourteen of which were spent as chairman (1936–1950). In his teaching, as in his research, he ranged widely over the geologic spectrum; two of his students—Harry H. Hess and J. Tuzo Wilson—were to play key roles in the "plate tectonics" revolution of geologic science in the 1960's. In his petrology courses, particularly at the graduate level, he stressed the application of theoretical and experimental chemistry to natural systems. Because of his intimate knowledge of rocks as they actually occur, however, he repeatedly noted the limitations of present knowledge of chemical theory and principles; he taught, therefore, not a developed set of organized conclusions but a method of approach designed to outlast the concepts of any given date.

BIBLIOGRAPHY

I. ORIGINAL WORKS. A small selection of Buddington's extensive writings is "The Binary System Akermanite-gehlenite," in *American Journal of Science*, **199** (1920), 131–140, written with J. B. Ferguson; *Geology of the Lake Bonaparte Quadrangle*, New York State Museum Bulletin no. 269 (1926), written with C. H. Smyth, Jr.; "Coast Range Intrusives of Southeastern Alaska," in *Journal of Geology*, **35** (1927), 224–246; *Geology and Mineral Deposits of Southeastern Alaska*, U.S. Geological Survey Bulletin no. 800 (1929), written with Theodore Chapin; "The Adirondack Magmatic Stem," in *Journal of Geology*, **39** (1931), 240–263; "Correlation of Kinds of Igneous Rocks with Kinds of Mineralization," *Ore Deposits of the Western States* (New York, 1933), 350–385; "High-Temperature Mineral Associations at Shallow to Moderate Depths," in *Economic Geology*, **30** (1935), 205–222; *Metalliferous Mineral Deposits of the Cascade Range in Oregon*, U.S. Geological Survey Bulletin no. 893 (1938), written with Eugene Callaghan; *Adirondack Igneous Rocks and Their Metamorphism*, Geological Society of America Memoir no. 7 (1939); "Some Petrological Concepts and the Interior of the Earth," in *American Mineralogist*, **28** (1943), 119–140.

Later writings include "Correlation of Reverse Remanent Magnetism and Negative Anomalies with Certain Minerals," in *Journal of Geomagnetism and Geoelectricity*, **6** (1954), 176–181; written with J. R. Balsley; "Thermometric and Petrogenetic Significance of Titaniferous Magnetite," in *American Journal of Science*, **253** (1955), 497–532, written with Joseph Fahey and Angelina Vlisidis; "Discussion," in *American Journal of Science*, **254** (1956), 511–515; "Granite Emplacement with Special Reference to North America," in *Geological Society of America Bulletin*, **70** (1959), 671–747; "The Origin of Anorthosite Re-evaluated," in *Records of the Geological Survey of India*, **86** (1960), 421–432; *Microintergrowths and Fabrics of Iron-titanium Oxide Minerals in Some Adirondack Rocks* (Hyderabad, India, 1961), 1–16, written with J. R. Balsley; *Regional Geology of the St. Lawrence County Magnetite District*, U.S. Geological Survey Professional Paper no. 376 (New York, 1962), written with B. F. Leonard; *Ore Deposits of the St. Lawrence County Magnetite District, Northwest Adirondacks*, U.S. Geological Survey Professional Paper no. 377 (New York, 1964), written with B. F. Leonard; "Iron-titanium Oxide Minerals and Synthetic Equivalents," in *Journal of Petrology*, **5** (1964), 310–357, written with D. H. Lindsley; "Sulfur Isotopes and Origin of Northwest Adirondack Sulfide Deposits," in *Geological Society of America Memoir*, 115 (1969), 423–451, written with M. L. Jensen and R. C. Mauger; *Geology of the Franklin and Part of the Hamburg Quadrangles N.J.*, U.S. Geological Survey Professional Paper no. 638 (1970), written with D. R. Baker; "Anorthosite Bearing Complexes: Classification and Parental Magmas," in C. Naganna, ed., *Studies in Precambrians* (Bangalore, India, 1975), 115–141.

II. SECONDARY LITERATURE. R. B. Hargraves, "Memorial to Arthur Francis Buddington," in *Geological Society of America. Memorials*, **14** (1984); and Harold L. James, "Arthur Francis Buddington," in *Biographical Memoirs. National Academy of Sciences*, **57** (1987), 3–24. Each has a bibliography.

HAROLD L. JAMES

BULLARD, EDWARD CRISP (*b.* Norwich, England, 21 September 1907; *d.* La Jolla, California, 3 April 1980), *geophysics*.

Sir Edward Bullard, known as Teddy, was the most distinguished and best-known British geophysicist of his generation; his experimental and theoretical work contributed to every aspect of the subject. He was one of the major figures in the development of the earth sciences during the twentieth century, both for his own contributions and for his influence on his colleagues and students. Bullard was the eldest of four children of Edward John Bullard, whose prosperous family produced Bullard's Ales, and of Eleanor Howes Crisp. After an unhappy childhood he went to Cambridge in 1926 to read natural sciences. Though he obtained a first, he found the lectures in physics very disappointing. There was a lack of generality, and all but the simplest calculus was avoided.

In the summer of 1929 Bullard became a research student at the Cavendish. His supervisor was Patrick Blackett, who suggested that he might follow up the work of Carl Ramsauer, who had shown that the total scattering cross section for electrons scattered in gases decreased as the energy was reduced below about 3 eV. This was quite inexplicable by classical theories of scattering, and Blackett suggested that Bullard should study the corresponding change in angular distribution, which could be expected to be more informative than the variation in total cross section. Bullard spent the long vacation in the "attic" of the Cavendish, where beginning graduate students worked for a month or so learning experimental techniques. He worked on the vapor pressure of tap grease, and then started to build the apparatus for electron scattering, which was a bird's nest of glass tubing for handling and circulating gases. Before he had got very far, Harrie Massey, who had started research at the same time, asked if he could join him. Bullard was delighted; the job would be much easier with two people.

After a few months Bullard and Massey had learned the tricks of vacuum electronics, and how to get reliable results. They found that there was a peak in the scattered current at an angle of about 90°. This behavior was quite inexplicable by the classical theory of collisions, but was obviously analogous to the diffraction rings around a street lamp in a fog. The explanation of Ramsauer's findings was also apparent: they resulted from diffraction around a spherical atom. The results agreed excellently with wave mechanics, and Bullard and Massey quickly wrote them up for publication (1930, 1931). This work gave Bullard a great feeling of confidence: he had carried out some elegant experiments that agreed with the quantum mechanical calculations. The room in which this work was done was close to Rutherford's, and Bullard saw more of Rutherford than he did of his supervisor, Blackett, who was on a long visit to Germany. The experience of working at the Cavendish during its most successful period made a deep impression on Bullard, and his descriptions of Rutherford's Cavendish strongly resemble the department of geodesy and geophysics at Cambridge when Bullard was its head during the 1960's.

Though the electron scattering work was very successful, the end of Bullard's studentship was approaching in 1931, and there were no jobs for physicists. That year he married Margaret Ellen Thomas; during their marriage they had four daughters. They were divorced in 1974.

Academic Career. In 1931 Bullard became a demonstrator in the department of geodesy and geophysics at Cambridge, which at the time of its formation in 1921 consisted of only one person, Sir Gerald Lenox-Conyngham. By himself Lenox-Conyngham was unable to do much. By 1931 he had persuaded the university that he needed help, and had been given funds for a junior post. On the advice of Rutherford he appointed Bullard to this position. At the same time Harold Jeffreys was appointed to a readership in geophysics. In the next eight years this small group of people had a quite remarkable impact on geophysics.

When Bullard became a demonstrator in the department, the only scientific instrument it possessed was a pendulum apparatus, and he at once set to work to improve its performance. This development is described in his Ph.D. dissertation, written in 1932. He then wished to use this technology to address a major geophysical problem. The East African Rift is one of the largest geological structures on the continents, and Bullard mounted an expedition to measure the gravity field in its neighborhood throughout central Africa. Such an ambitious approach was possible only because of the improvements he had made to the pendulum. The account of the results (1936) shows the care and thoroughness with which every part of the experiment was planned. The principal conclusion of this work was that the rift was formed by compression, an idea that Bullard later recognized as completely incorrect. Nonetheless, his gravity work on the East African Rift was of great and lasting importance. It showed how simple geophysical measurements could be used to investigate the origin of major geological structures. To do so, Bullard had to overcome the difficulties of making accurate measurements under primitive conditions in the field.

The work on gravity established Bullard's reputation, and in 1936 he was awarded the Smithson fellowship of the Royal Society. With the support of Lenox-Conyngham, he then became interested in other types of geophysical measurement of relevance to geological problems. He designed a short-period seismometer with which he and others carried out a survey of the depth to the basement beneath southeast England, using seismic refraction (1940). He also started to work on marine geophysics.

The other major project that Bullard started at this time was the measurement of heat flow from the earth's interior. Though this problem had been of great interest fifty years earlier, and had led to the famous controversy between Lord Kelvin and the geologists, it had since then largely been neglected, though a large number of temperature measurements from boreholes had been accumulated. But the conductivity of the rocks through which the holes passed was not known. Bullard and A. E. Benfield adapted an existing technique to measure the conductivity. In the winter of 1938–1939 Bullard visited South Africa, where detailed temperature measurements had been made by L. J. Krige in several deep boreholes through hard rocks of uniform lithology. He showed that the difference of a factor of about two between the geothermal gradients in South Africa and Britain was caused by a corresponding difference in the thermal conductivity (1939).

Measurement of the heat flow on land was Bullard's last major project before the outbreak of the war. In the eight years in which he had been working in geophysics, he had made important contributions to three of the four branches of the subject—gravity, seismology, and heat flow—principally from an experimental point of view. Bullard's theoretical investigations during this period were experimentally motivated. As an experimentalist he was outstanding, and his work showed the strong influence of his training in the Cavendish. The international attention his experiments generated led to his election to the Royal Society in 1941.

In November 1939 Bullard became an experimental officer attached to H.M.S. *Vernon*, which was a laboratory of the Admiralty concerned with mine warfare. The Germans had developed a new and very effective magnetic mine that aircraft could lay in shallow water. These mines were used from the start of the war in September 1939, and by the end of December had sunk sixty ships. Bullard saw what needed to be done, and after a sharp struggle with the naval scientific establishment, was put in charge of the development of methods of protecting ships from magnetic mines. Soon afterward he became head of the group concerned with sweeping all kinds of mines. He quickly developed methods of dealing with magnetic mines, and soon had time to think about other mechanisms the Germans might develop for triggering mines, such as acoustical and pressure sensors, and to develop methods of sweeping them before they came into use. In this Bullard and his group were very effective. When the Germans deployed a new type of mine to protect the beaches during the Normandy landings, they were being swept within twelve hours of the first casualty (Hepworth, p. 72). After eighteen months the losses of ships from mines had been reduced to only 10 percent of those from submarines. Bullard then decided he should move to London to join Blackett, as assistant director of naval operational research. Here he worked on a variety of problems concerned with British mines and submarine attacks on German and Italian shipping. Like most of the scientists in their twenties and thirties who worked on wartime problems, Bullard was profoundly affected by this experience.

When Bullard returned to Cambridge after the war, he found the place in a sorry state. He began by scrubbing the floor and energetically collecting surplus equipment to provide the basic needs of the department. But both he and his wife became restless, and in 1947 he agreed to become head of the physics department in Toronto. Bullard moved to Toronto in the spring of 1948 and was a very successful head of department, starting many new projects. He encouraged the development of geochemistry, especially radiometric dating and heat production, and for this purpose installed a mass spectrometer. He also continued his work on heat flow, and became interested in computers, whose development he had followed from their beginnings at Bletchley during the war. The university had recently installed FERUT, constructed by the Ferranti corporation, which Bullard used for his calculations. He worked on the generation of the earth's magnetic field and organized a visit to Scripps Institute of Oceanography to build an instrument to measure the heat flux through the seafloor.

In the spring of 1949 Bullard was offered the directorship of the (British) National Physical Laboratory on the understanding that he would stay for a decent period, which was informally agreed to be five years. In June he accepted, and resigned his professorship from the end of the year. He then left Toronto for Scripps, where he spent the summer building a heat flow probe and analyzing the evolution of the earth's magnetic field to determine the rate of westward drift. He had a very successful summer, so successful that President Sproul of the University of California offered him the directorship at Scripps, which he declined.

At the end of 1949 Bullard returned to Britain to be the director of the National Physical Laboratory (NPL), which had been founded in 1900 by the

Royal Society and the British government. Its first director, Sir Richard Glazebrook, had seen the importance of providing a measurement service for British industry, which was still a major part of its work when Bullard became director. The laboratory was large, employing more than 1,000 people by 1955. Bullard was a well-liked and effective director, and was knighted for his services. He particularly encouraged Louis Essen's work on atomic frequency standards, and the group working on electronic computations on the ACE computer, which he used extensively for his dynamo calculations (1954). Bullard remained involved in military problems and served as the joint chairman of the Anglo-American Ballistic Missile Commission. The most surprising feature of this period of Bullard's life was its scientific productivity; he wrote several of his most important and influential papers ("The Flow of Heat Through the Floor of the Atlantic Ocean," "Homogeneous Dynamos and Terrestrial Magnetism," "Heat Flow Through the Deep Sea Floor") at the NPL. Indeed, it was probably the most effective period of his whole career. Bullard made use of the workshops to construct his marine heat-flow equipment. He also used the laboratory to determine the thermal conductivity of rocks. He later remarked that it was less time-consuming to administer the NPL than the department at Cambridge, because the NPL provided effective assistance.

When he was appointed, Bullard had originally agreed to stay at the NPL for five years; in fact he stayed for six. In 1954 he began to sound out his friends at Cambridge about the possibility of his returning. James Chadwick, now master of Caius College, arranged for him to be appointed to a Bye fellowship at the college in the summer of 1955. This appointment was for three years and was not a university post. On the strength of this agreement, Bullard resigned as director of NPL. But before he could take up the Caius post, Keith Runcorn accepted the professorship of physics at Newcastle. Runcorn's departure left a vacancy as an assistant director of research, and Ben Browne, who was then reader and head of the department of geodesy and geophysics, immediately appointed Bullard to this post.

Bullard made the department into one of the best places to study geophysics in the world. He attracted excellent students and made sure they had the facilities they needed. He set them off in directions where their abilities would be best displayed. He was generous to colleagues who would otherwise have been overshadowed by his great distinction and fame. He used his contacts in the universities, industry, and government in the United Kingdom and the United States to make sure that people did excellent work. Most of the successes of the department were a direct consequence of his insight and encouragement.

The largest group that remained at the department was led by Maurice Hill. It was concerned with marine geophysics and continued the work Bullard had started before the war. Hill had developed seismic refraction at sea in order to understand the structure of the ocean basins. Bullard broadened this effort and encouraged students to build new oceangoing instruments. They constructed a proton precession magnetometer that was towed behind the ship, and the heat flow through the seafloor was measured with Bullard's heat-flow instrument. Bullard also encouraged the development of land seismology, particularly because of his interest in the detection of underground nuclear explosions. Several of his students worked on the dynamo problem, using the excellent computer facilities available in the mathematics laboratory. He himself was one of the few people in the university who was allowed to use the machine at night, when the operators were not present; he used it to develop methods of automatic data collection and reduction. Bullard also took an interest in geochemical instrumentation and in the design of sensitive mass spectrometers that Jack Miller was developing.

During this period the problem of the evolution of the ocean basins slowly came to dominate work in the department. The observations of Runcorn and his co-workers, especially that of Edward Irving on the magnetization of Australian rocks, could be understood only if large relative movements between continents occurred. But none of the marine geophysicists, including Hill, could understand how such displacements were taken up by the structures in the ocean basins. The key suggestion was made by Harry Hess of Princeton (1962), who proposed that new seafloor was created only on ridge axes. In making this suggestion Hess was strongly influenced by the discovery of heat flow anomalies associated with ridges, which had been made by Bullard, A. E. Maxwell, and R. R. D. Revelle (1956). Bullard invited Hess to talk about his ideas at a conference in Cambridge in January 1962, and many people who later worked on this problem were in the audience. When Fred Vine became a research student at the department, and found a reversely magnetized seamount in the Indian Ocean, Bullard was well aware of the reality of reversals because of his interest in the dynamo problem. He encouraged Vine and his supervisor, Drummond Matthews, to publish their ideas on how seafloor spreading and reversals could together account for the oceanic magnetic lineations. In 1965 Hess and J. Tuzo Wilson of Toronto spent their sabbaticals at the department. During this period Wilson wrote a number of papers that proposed many of the concepts of what is now called plate tectonics. Hess persuaded Vine to go to Princeton, where he used his ideas to interpret magnetic profiles from all parts of the world. Under Bullard the department at Cambridge played a major role in the construction of the new theory. His personal contribution was the use of Euler's theorem to describe the motions of continents on the earth's surface, but his influence on others involved in the work was very extensive.

Bullard's period as head of department was his most successful as a scientific administrator. He remained at Cambridge for eighteen years, during fourteen of which he was head of department. This was a much longer period than he spent at Toronto or the NPL. Because the organization was small, with a rapid turnover of students, Bullard influenced a large number of people who now occupy senior positions in the United Kingdom and North America. He also was head at a particularly important period in the development of the earth sciences. He gave considerable thought to what would happen when he retired, and was involved in discussions whose aim was the formation of a single department of earth sciences. Though Bullard strongly supported this idea, his buccaneering style frightened the other departments, and amalgamation became possible only in 1980, after he had retired. The department of geodesy and geophysics was then renamed the Bullard Laboratories in his honor.

Though most of Bullard's time was spent on the affairs of the department, he remained involved in government matters, though to a lesser degree than when he was at NPL. He attended the Geneva discussions on nuclear disarmament in 1958 as an adviser to the British government. He also became chairman of the committee in charge of British space research. Bullard enjoyed his connections with industry. He took an active interest in the family brewery, especially after his uncle died, until it was taken over by Watneys. He became a director of IBM U.K., and helped to persuade the university to buy its first IBM computer. He made a large collection of scientific books, and was especially interested in Halley, whose ability to analyze data of variable reliability he found particularly impressive.

When Bullard retired from Cambridge, his health was failing, but he was determined to make the most of his remaining time and started a new life with great bravery. In 1974 he married Ursula Cooke Curnow and moved to Scripps. He also became a geophysical consultant to the University of Alaska. Only months before he died, a group of his former students and colleagues organized a meeting in his honor at Scripps that he bravely attended. Though he was physically frail and in great pain, his comments and questions to the speakers showed his usual grasp of the essentials. Bullard died in his sleep the night after finishing his last paper.

Besides being a fellow of the Royal Society, Bullard was foreign honorary member of the American Academy of Arts and Sciences (1953), foreign associate of the U.S. National Academy of Sciences (1959), Bakerian Lecturer of the Royal Society (1967), and foreign member of the American Philosophical Society (1969). Among his awards were the Hughes Medal of the Royal Society (1953), the Chree Medal of the Physics Society (1956), the Day Medal of the Geological Society of America (1959), the gold medal of the Royal Astronomical Society (1965), the Agassiz Medal of the U.S. National Academy of Sciences (1965), the Wollaston Medal of the Geological Society of London (1967), the Vetlesen Prize of Columbia University (1968), the Bowie Medal of the American Geophysical Union (1975), the Royal Medal of the Royal Society (1975), and the Ewing Medal of the American Geophysical Union (1978).

Major Scientific Work. *Heat Flow on Land and at Sea.* The methods now used to measure the heat flow through the earth's surface both on land and at sea were devised by Bullard before and shortly after the war. On land a number of accurate measurements of temperatures in boreholes were available before he started work, but no systematic measurements of the thermal conductivity of rocks had been carried out. Furthermore, the heat flow was obtained by determining the temperature gradient from temperature differences, which Bullard (1939) showed was not the most accurate method of estimating the heat flux. But the principal error was caused by the uncertainty in the thermal conductivity of the rocks through which the boreholes passed. Bullard systematically measured its variation with rock type, and showed that the low temperature gradient in the South African gold mines compared with British coal mines resulted from the high thermal conductivity of the quartzite compared with shales. His work in this area established standards and techniques that have remained largely unchanged. His borehole observations established a reliable average value for the heat flux through the continents, but he did not discover any of the major regional variations in heat flux that are now known.

Bullard thought about the problems of measuring heat flow at sea when he first became interested in marine geology before the war. He thought the best way to measure it was to drive a spike containing the thermometers into the sediment on the seafloor and measure the temperature gradient. He built such an instrument at Scripps in 1949, using thermojunctions to measure the temperature, with a galvanometer and a camera as the recording system. The thermal conductivity of the sediments had to be measured later in the laboratory, using samples obtained from a core in the vicinity. The interior of the instrument contained air at atmospheric pressure, and the case was sealed with O-rings. Bullard knew about this method of sealing from its use in airplane hydraulic systems during the war, but this was the first time O-rings had been used in marine geophysics. They are now universally used in all deep-sea instrumentation. Bullard wrote an extensive account of the instrument in 1954, and in 1956 reviewed the results from both the Atlantic and the Pacific with Revelle and Maxwell. The observations in the Pacific showed the band of elevated heat flow that is now known to be a universal feature of spreading ridges. This paper contains a discussion of mantle rheology and of how the heat flow anomaly could be maintained by mantle convection.

As so often has happened in marine geophysics, most of the major results were obtained with the first instrument. The only major discovery that has since been made is the importance of heat transfer by movement of seawater into and out of the ocean crust. Water at temperatures as great as 350° C emerges through vents in the seafloor and transports heat by advection. Bullard's work established that the conductive heat transport through oceans and continents was the same. When the advected heat is included, the heat loss through the ocean floor considerably exceeds that through the continents.

The Earth's Magnetic Field. Bullard first became interested in the main magnetic field of the earth during the early part of the war. In connection with his work on magnetic mines he read Sydney Chapman and Julius Bartels' book on geomagnetism (1940). After the war he started a general investigation of how motions in the core might maintain the magnetic field. His first concern was with the energetics of the core motions ("The Magnetic Field Within the Earth") and the observed secular changes of the magnetic field ("Electromagnetic Induction in a Rotating Sphere," "The Westward Drift of the Earth's Magnetic Field"). He showed that the various astronomical effects, such as tidal retardation and changes in the length of day, could not generate enough energy to maintain the field against ohmic dissipation, and that the most likely energy source was the release of gravitational potential energy through some form of convection in the core. This view is now widely accepted, though little has been added to Bullard's order-of-magnitude arguments. Meanwhile, Walter Elsasser had investigated the kinematic problem of how a uniform conducting sphere could act as a self-exciting dynamo. His discussion was concerned with the general problem of how a complicated velocity field interacted with both poloidal and toroidal fields to produce a dynamo. Starting at Toronto, Bullard attempted to construct a model for a real homogeneous dynamo using Elsasser's approach. To do so he employed the digital computers at Toronto and later at NPL. This work is one of the first attempts to obtain numerical solutions to fluid mechanical problems that are analytically intractable. Such numerical experiments are now a powerful and widely used method of investigating complicated nonlinear problems. Bullard's work on the dynamo problem was one of the first uses of this approach for nonmilitary purposes. However, the model he proposed is now known not to maintain a dynamo. Nonetheless, the approach he pioneered has been widely adopted, and self-exciting kinematic dynamos similar to the one he proposed with H. Gellman have since been discovered.

Continental Fits. In the late 1950's Bullard became convinced that large horizontal displacements of parts of the earth's surface occurred. The two observations (1964) that particularly affected his views were the demonstration of offsets in the magnetic anomaly pattern of more than 1,000 kilometers off the west coast of North America and the paleomagnetic observations from Australia, which were not compatible with polar wandering and required relative motion between Australia and Europe. Bullard decided to fit the continents round the Atlantic together geometrically. He wished to demonstrate how excellent the fits were, and to separate this question from the problem of the mechanism that had caused the displacements. The confusion between these two separate problems had bedeviled the hypothesis of continental drift since A. L. Wegener's influential work. Bullard decided to produce the fits by minimizing the misfit between the continental margins rather than that of the coastlines. Both Wegener and A. L. Du Toit had argued that this was the correct procedure. S. W. Carey had previously attempted to make geometrically correct reconstructions in the same way, but he had used them to support his physically implausible idea of an expanding Earth. Bullard and his colleagues were principally interested in producing convincing fits, and the maps they generated (1965) have been widely reproduced in modern textbooks. Bullard needed a convenient description of the movement of a rigid continent on the surface of the earth. He was the first to use Euler's theorem for the purpose, and obtained the pole and rotation angle required by minimizing the misfit. Though he regarded this description simply as a convenient way of describing the motions, this theorem later became the cornerstone of plate tectonics, and was not widely known to geophysicists before his work.

BIBLIOGRAPHY

I. ORIGINAL WORKS. "Remarks on the Scattering of Electrons by Atomic Fields," in *Proceedings of the Cambridge Philosophical Society*, **26** (1930), 556–563, with H. S. W. Massey; "The Elastic Scattering of Slow Electrons in Argon," in *Proceedings of the Royal Society*, **A130** (1931), 579–590, with H. S. W. Massey; "The Elastic Scattering of Slow Electrons in Gases, II," *ibid.*, **A133** (1931), 637–651, with H. S. W. Massey; "Gravity Measurements in East Africa," in *Philosophical Transactions of the Royal Society*, **A235** (1936), 445–531; "Heat Flow in South Africa," in *Proceedings of the Royal Society*, **A173** (1939), 474–502; "Seismic Investigations on the Paleozoic Floor of East England," in *Philosophical Transactions of the Royal Society*, **A239** (1940), 29–94, with T. F. Gaskell, W. B. Harland, and C. Kerr-Grant; "Submarine Seismic Investigations," in *Proceedings of the Royal Society*, **A177** (1941), 476–499, with T. F. Gaskell; "The Protection of Ships from Magnetic Mines," in *Proceedings of the Royal Institution*, **33** (1946), 554–566; "The Time Necessary for a Bore-hole to Attain Temperature Equilibrium," in *Monthly Notices of the Royal Astronomical Society*, geophysical supp., **5** (1947), 127–130; "Electromagnetic Induction in a Rotating Sphere," in *Proceedings of the Royal Society*, **A199** (1949), 413–433; "The Magnetic Field Within the Earth," *ibid.*, **A197** (1949), 433–453.

"The Westward Drift of the Earth's Magnetic Field," in *Philosophical Transactions of the Royal Society*, **A243** (1950), 67–92, with C. Freedman, H. Gellman, and J. Nixon; "The Flow of Heat Through the Floor of the Atlantic Ocean," in *Proceedings of the Royal Society*, **A222** (1954), 408–429; "Homogeneous Dynamos and Terrestrial Magnetism," in *Philosophical Transactions of the Royal Society*, **A247** (1954), 213–278, with H. Gellman; "The Stability of a Homopolar Dynamo," in *Proceedings of the Cambridge Philosophical Society*, **51** (1955), 744–760; "Heat Flow Through the Deep Sea Floor," in *Advances in Geophysics*, **3** (1956), 153–181, with A. E. Maxwell and R. Revelle; "Continental Drift," in *Quarterly Journal of the Geological Society of London*, **120** (1964), 1–34; "The Fit of the Continents Around the Atlantic," in *Philosophical Transactions of the Royal Society*, **A258** (1965), 41–51, with J. E. Everett and A. C. Smith; "Reversals of the Earth's Magnetic Field," *ibid.*, **A263** (1968), 481–524; "The Origin of the Oceans," in *Scientific American*, **221** (September 1969), 66–75; and "Electromagnetic Induction in the Oceans," in A. Maxwell, ed., *The Sea*, IV, pt. 1 (New York, 1970), 695–730, with R. L. Parker.

II. SECONDARY LITERATURE. T. Hepworth, "On the Atlantic Shelf: A Trawler's Cruise with a Purpose," 2 pts., in *Yachting Monthly*, **81** (1946), 70–75 and 148–151; and D. P. McKenzie, "Edward Crisp Bullard," in *Biographical Memoirs of Fellows of the Royal Society*, **33** (1987), 67–97.

D. P. MCKENZIE

BULLEN, KEITH EDWARD (*b.* Auckland, New Zealand, 29 June 1906; *d.* Auckland, 23 September 1976), *seismology, applied mathematics, geophysics.*

Bullen's parents were George Sherrar Bullen, a journalist, and Maud Hannah Burfoot. Both his parents were born in 1875 and lived to be ninety-nine. His father was born in New Zealand and his mother came from England. Bullen's sister, Jean Maud Bullen, twelve years younger, became a member of the research staff of the Ionosphere Section of the Department of Scientific Research, Christchurch, New Zealand. In 1935 Bullen married Florence Mary Pressley; they had a son and a daughter.

Bullen gained an international reputation as a pioneer in modern seismology. He was a superb applied mathematician, particularly adept at constructing mathematical models of the earth's interior that enabled him to draw inferences about the numerical values for its various properties, especially density. Bullen's reputation was secured with the publication in 1940 of the Jeffreys-Bullen Seismological Tables. His work on these tables initiated his lifelong concern with density determinations of various layers of the earth's interior, which, in turn, led to his construction of Earth Model A (1940–1942) and Model B (1950), his compressibility-pressure hypothesis (1946), his hypothesis of a solid inner core (1946), and his various models for the interior of the moon and of other planets.

Bullen became a fellow of the Royal Society in 1949, a foreign associate of the United States National Academy of Sciences in 1961, and a fellow of the Pontifical Academy of Science in 1968. He was also a fellow of the Australian Academy of Science, and an (honorary) fellow of the Royal Society of New Zealand and the Royal Society of New South Wales. He received numerous scientific prizes, including the William Bowie Medal of the American Geophysical Union (1961), the Day Medal of the Geological Society of America (1963), the Research Medal of the Royal Society of Victoria (1965), and the gold medal of the Royal Astronomical Society (1974).

Bullen devoted considerable time to scientific organizations. With the advent of the International Geophysical Year he became, in 1955, chairman of its Australian National Committee; was later appointed convener of the Australian National Committee for Antarctic Research (1958–1962); and, in 1959, became vice president of the International Special Committee for Antarctic Research. From 1955 to 1957 he was president of the International Association of Seismology and Physics of the Earth's Interior and was vice president of the International Union of Geodesy and Geophysics from 1963 to 1967. He also served on the governing council of the *International Seismological Summary* during the 1960's.

Bullen's other interests included numismatics, cricket, and tennis. He was wont to visit coin shops when traveling abroad and served as president of the Auckland University Cricket Club during the 1939–1940 academic year. History of science was also one of his interests, and he contributed to the *Dictionary of Scientific Biography*. Among his articles are those on such early workers in seismology as Cargill Knott, Horace Lamb, Augustus Love, John Milne, Andrija Mohorovičić, Richard Oldham, Herbert Turner, and Emil Wiechert. He also wrote the entry on Alfred Wegener.

Bullen's early schooling was typical for a New Zealander. His primary education was at Bayfield School, Herne Bay, Auckland (1912–1918), and his secondary education was at Auckland Grammar School (1919–1922). His brilliance in mathematics expressed itself at an early age; he used to work out his mathematics homework while cycling five miles home. According to his sister, he was an extremely active boy, participating in tennis, swimming, surfing, rock climbing, and hiking with survey parties. She attributed his later increasing deafness, which he began to experience in his late twenties, to his diving. Bullen also took boat trips to White Island, the site of an active volcano; and he later suspected that these trips, along with the famous Hawke's Bay earthquake of 2 February 1931, kindled his interest in seismology and geophysics.

In 1922 Bullen earned an entrance scholarship to Auckland University College. In 1925 he became senior university scholar in pure and applied mathematics at the University of New Zealand, and he obtained a B.A. degree in 1926. He also took courses in Latin, French, philosophy, and education. He received an M.A. degree from the University of New Zealand in 1928 with a first class in mathematics, and took a B.Sc. in physics in 1930 with chemistry as a minor. During much of this time he worked at various teaching posts and was employed as a part-time lecturer at Auckland University College. In 1928 he became a full-time senior lecturer and, except for the period 1931 to late 1934, spent in England and continental Europe, he remained at this post until 1940, when he became senior lecturer in mathematics at the University of Melbourne. He stayed at Melbourne for five years, then left to become professor of mathematics at the University of Sydney. In 1946 he was awarded the Sc.D. from Cambridge.

Bullen entered St. John's College, Cambridge, in 1931 as an advanced student, intending to take the mathematical tripos in two years. He decided, however, to become a research student and "had the outstanding good fortune" to be taken in hand by Sir Harold Jeffreys, who, in Bullen's words, "literally brought me down to earth and rescued me from a pure mathematical fate." Indeed, Bullen's career as a research scientist began on 12 January 1932, during his first meeting with Jeffreys, who asked him to collaborate with him on the enormous task of constructing earthquake travel-time tables of greater accuracy than those then available. Bullen, who saw all the previous issues of the *International Seismological Summary*—the quarterly journal of the International Seismological Association, containing masses of seismological data collected throughout the world—spread upon the floor before him, accepted the offer, even though he later confessed that "on that first day, my impression of the *I.S.S.* was of a very dull-looking collection of numerical entries and strange place-names that conveyed to me little more than would, say, a collection of horse-racing booklets or the financial columns of a foreign newspaper."

Revision of the travel-time tables was not completed until 1940. Bullen and Jeffreys worked together at Cambridge until near the end of 1934. Around this time they decided it would be best for Bullen to return to Auckland to work alone since, as far as Jeffreys knew, there was no precedent for the whole of the work of a Ph.D. candidate being done jointly with a supervisor. After several months visiting a number of countries in continental Europe, Bullen returned home and began mailing his own contributions to the tables.

Earthquake travel-time tables give the length of time, T, it takes bodily seismic waves, P (longitudinal) and S (transverse) waves and various multiphased P and S waves, to travel from the focus of an earthquake through the earth's interior to a point on the surface in terms of Δ, the angular distance subtended at the center of the earth by the arc from the earthquake's epicenter to the designated point on the surface. The first travel-time tables were constructed by Karl Zöppritz, Jr., of Göttingen in 1907. They were amended and extrapolated to the antipodes by Herbert H. Turner of Oxford. The resulting Zöppritz-Turner Tables were adopted by the International Seismological Association in 1918 and used in preparing the *International Seismological Summary*.

Beginning around 1922, a number of seismologists, including Turner, Perry Byerly, and James B. Macelwane, came to the realization that the Zöppritz-Turner Tables were in need of revision. Although these researchers made considerable improvements, their work was not consistent; and about 1930 Jeffreys decided that a total revision of the tables was required. Jeffreys and Bullen published their preliminary set of tables in 1935. These tables were used by the International Seismological Association from December 1936 until near the end of 1939, when it began using a more refined set of Jeffreys and Bullen tables that were published in 1940. In 1947 these became the official tables of the International Seismological Association. Beno Gutenberg and Charles F. Richter constructed their own set of tables while Jeffreys and Bullen were busy with theirs. Although both teams used a tremendous amount of data, Jeffreys and Bullen used more. The two sets of tables compared quite well with each other: the greatest differences were less than one-tenth of the greatest correction that had to be made to the Zöppritz-Turner tables.

The Jeffreys-Bullen travel-time tables treat Earth as spherically symmetrical. Consequently the calculated travel time, T, for any given phase of a P or S wave is the same for any pair of points on the surface of the earth with the same Δ, regardless of the longitude and latitude of the epicenter and seismological recording station. In 1933 Leslie J. Comrie, a New Zealander, and Gutenberg and Richter independently suggested that further improvements in travel times would be possible if the effect of the earth's ellipticity upon earthquake travel times were taken into consideration. They pointed out that the use of geocentric instead of geographic latitudes, which had been used previously, would further reduce such errors. (A geocentric latitude of a point on the earth's surface is the angle between the radius vector from the earth's center to the point of the surface and the plane of the equator; therefore its use in seismological tables eliminates the need for making any corrections in travel time due to the ellipticity of the earth's outer surface.)

Jeffreys showed, however, that although this eliminated errors in the travel time of earthquakes that were due to the ellipticity of the earth's surface, it failed to eliminate all the effects of the earth's ellipticity, since it did not take into account errors due to the ellipticity of different layers. The travel times of seismic rays are affected not only by the surface ellipticity bulges at the end of rays but also by the ellipticity of each internal surface of constant seismic velocity encountered by the ray. Bullen took on the task of making these internal ellipticity corrections. He soon realized that in order to solve the problem, he needed to determine the density distribution at the different layers transversed by seismic rays.

At the time Bullen began working on the density problem, earth scientists generally agreed there were two major discontinuities in the earth's interior: one corresponding to the crust-mantle boundary, the other to the mantle-core boundary. In 1909 Andrija Mohorovičić of Zagreb, Yugoslavia, argued in favor of a seismological discontinuity corresponding to the crust-mantle boundary. Three years earlier Richard D. Oldham suggested a mantle-core boundary on the basis of seismological evidence. He overestimated its depth, however, placing it at 3,800 kilometers. In 1912 Gutenberg, again on seismological grounds, placed the boundary at 2,900 kilometers, extremely close to the current estimate.

There also was general agreement as to the fluidity of the earth's core. In 1926 Jeffreys was able to show, contrary to what most had believed, that the average rigidity of the earth deduced from tidal motion and Chandler wobble could be made consistent with the seismic velocities of P and S waves in the mantle only by assuming the existence of a core of very low rigidity. (The discovery of the fluidity of the earth's core, contrary to popular accounts, was not based simply upon the "fact" that seismologists had not observed the propagation of S waves through the core. In general, negative existential claims are problematical, and in this case, at least one seismologist, J. B. Macelwane, thought that he had detected the transmission of an S wave through the core. Rather, the fluidity of the earth's core depended upon Jeffreys' demonstration that an acceptable average rigidity of the earth required a liquid core.)

Despite the discovery of these two basic discontinuities, Bullen did not have too much to draw upon when he turned to the density problem, and he quickly realized that he would have to determine the answer himself. The most recent work (1923) had been done by two American earth scientists, E. D. Williamson and Leason H. Adams. Bullen began by applying their method, which contains equation (1) as its key formula, where r is the distance of a point from the earth's center, p is the density at r, γ is the universal gravitational constant, m is the mass of the matter within the sphere of radius r, and a and B are the velocities of P and S waves at level r:

$$\frac{dp}{dr} = \frac{-\gamma m p}{r^2\left(a^2 - \frac{4}{3}B^2\right)}. \qquad (1)$$

Equation (1) is applicable only to regions where constitution is essentially uniform, provided the deviation from adiabatic conditions is minimal.

Beginning with a density of 3.32 at a depth of 35 kilometers for the density and depth of the outside of the mantle, and taking the velocities of P and S waves from Gutenberg and Richter (none of this being very controversial), Bullen used equation (1) to calculate the density of the mantle at 100-kilometer intervals. This also gave him the mass of the earth's mantle, which allowed him to determine the mass of the core. At this point Bullen decided to check his results by combining them with the known moment of inertia of the core. The check proved to be effective, but not in a manner that gave him cause for immediate happiness, for it showed that his results implied the implausible hypothesis that the core was much denser near the outside than at its center. Bullen escaped the impasse by making a major discovery: that the density of the mantle does not vary continuously and, therefore, that the Williamson-Adams formula cannot be applied to the mantle as a whole.

Bullen first published his results in 1936, the same year that Inge Lehmann, a Danish seismologist, published her paper on the existence of an inner core. In his paper Bullen tentatively placed the jump in density at a depth between 300 and 400 kilometers, since he and Jeffreys, and Lehmann, independently had reported a jump in the rate increase of the velocity of P and S waves that they took to indicate a first-order discontinuity. This discontinuity came to be called the 20° discontinuity, since it shows up as a rapid change in the gradient of the P and S travel-time curves around epicentral distances of 20°. Bullen also pointed out that his calculation of the density of the earth's core, which was lower than previous estimates, was consistent with the idea of a core composed entirely of molten iron. This was important because previous estimates required the supposition that the core contained a proportion of heavier elements.

Bullen continued to work on his "density-jump" hypothesis during the next two years. In 1937 he placed the discontinuity at 481 kilometers ± 21 kilometers in light of more refined seismological estimates of the velocity of S and P waves, the boundary of the 20° discontinuity, and the depth of the mantle-core boundary. In 1939 he argued that his hypothesis gained support from findings in terrestrial magnetism and petrology. The suggestion had been made by Sydney Chapman and Albert T. Price that certain aspects of the secular variation of the geomagnetic field suggested a significant increase in the electrical conductivity of the mantle beginning at around 150 kilometers and becoming more pronounced at about 700 kilometers. Bullen noticed a report by Price and B. N. Lahiri that offered some confirmation of the idea. He proposed that if the 20° discontinuity is gradual, it might extend to a depth of 700 kilometers. This, he added, coincided with the depth of the deepest earthquakes. He then referred to Jeffreys' recent suggestion, expanded by John D. Bernal, that his "density discontinuity" might correspond to a change in the crystal structure of olivine due to tremendous pressure that could also account for increasing electrical conductivity.

While Bullen was developing his analysis of Earth's density distribution, he found time to construct earthquake time-travel tables for Earth's ellipticity, the reason he had begun working on the density problem; published a number of articles analyzing New Zealand earthquakes; and offered an estimate of twenty-six kilometers for the thickness of the upper layer of the oceanic crust in the Pacific through analysis of the Rayleigh waves from an earthquake in the Bering Sea. Moreover, he pursued the work on planetary density problems begun in 1937. In 1940 Bullen published a lengthy article on density distribution in which he summarized his previous work and introduced part of his nomenclature for layers of the earth's interior: A for the crustal layer; B, C, and D for the mantle, with C being the inhomogeneous layer containing the jump in density; E for the outer core; G for the inner core; and F for a transitional layer between E and G. He did not, however, use his nomenclature for the core because he thought it premature to offer density estimates for the inner and outer portions of the core.

In 1942 Bullen published an expanded version of his 1940 treatment of density to the layers E, F, and G. In essence this was his first presentation of

BULLEN

a complete version of his Earth Model A. He didn't refer to it by such a designation until 1950, when he needed to contrast such models with B-type models.

Although Francis Birch, a geophysicist from Harvard, was the first to suggest (1940) a solid or "frozen" inner core, Bullen was the first to offer extensive arguments for its solidity and to propose a seismological test for its existence. He first advanced the hypothesis in 1946. The idea grew out of his work on the density problem and his construction of Earth Model A. He realized that the results of Model A suggested (1) that even though the density ρ and rigidity μ greatly change at the mantle-core boundary, there was only a 5 percent change in the incompressibility k; and (2) that when values of k were plotted against values of pressure p, there was no significant change in dk/dp on either side of the mantle-core boundary.

In 1950 Bullen began referring to (2) as the compressibility-pressure hypothesis. It is tantamount to the claim that the compressibility of a substance, at least at high pressure and temperature, is largely independent of chemical composition. With (2) Bullen was able to infer the existence of a solid inner core. Beginning with the observed jump in the velocity of P waves, vp, from the outer to the inner core, and the formula for vp,

$$vp = \sqrt{[k + (4/3)\mu]/\rho},$$

he reasoned that the jump in velocity could be explained by a drop in density, a jump in incompressibility, or a jump in rigidity across the outer-inner core boundary. A sharp increase in k was ruled out by his compressibility-pressure hypothesis, while a decrease in ρ approached the impossible. Therefore he concluded that the jump in the velocity of P waves is the result of a tremendous jump in μ from an outer core of little rigidity (a fluid outer core) to an inner core of sizable rigidity (a solid inner core). He then proceeded to relate his hypothesis to Birch's solution to the origin of Earth's magnetic field and suggested that one way to test the idea was to find a PKJKP seismic wave. This is a wave that passes through the mantle and outer core as a P type, is transformed into an S type in the inner core, and then back into a P type through the outer core and mantle.

Although Bullen continued working on a number of other problems—for example, the problem of planetary densities—much of his later research was devoted to confirmation of a solid inner core. In 1961 he told a reporter from the *New York Times* that "if I can see part of the Earth proved solid before I die, I'll die happy." In 1949, starting with the compressibility-pressure hypothesis and knowing Jeffreys' values for the velocities of P waves in the inner core, he estimated that the velocity of S waves through a solid inner core should fall somewhere between 4.9 and 6.0 kilometers/second. The following year he began presenting his Earth Model B, which, unlike Model A, directly incorporates into its formalization (1), (2), and the hypothesis of a solid inner core. He also derived a set of theoretical travel-time tables for PKJKP waves, hoping that seismologists would put the tables to use and thereby confirm his hypothesis of a solid inner core. He suggested that PKJKP waves would most likely be detected over an epicentral distance Δ ranging from 130° to 155°.

In 1951 Bullen pointed out that it would be extremely difficult to observe PKJKP waves, "that they would be on the border of observability" because their amplitude would be extremely low—0.04 to 0.2 of the amplitude of their companion PKIKP waves, in which I represents the passage of a P wave through the inner core. He also hinted that matters might be even worse because the amplitude might be even less than the above estimates if the transition layer, whose character was unknown, absorbed some of the energy used in the conversion of a P wave into an S wave at the outer-inner core boundary.

In 1952 a colleague of Bullen's, T. N. Burke-Gaffney of Riverview College Observatory in Sydney, examined the observatory's seismological records back to 1909 in an attempt to identify PKJKP waves. He and Bullen concluded that although a few of the recorded impulses were found to agree with the predicted travel times of PKJKP waves, the readings were too ambiguous to warrant positive identification, and that a prerequisite for identification of PKJKP waves was that the accompanying PKIKP wave have an amplitude of at least 20μ. They also further restricted the epicentral range for observing PKJKP waves to $130° \leqslant \Delta \leqslant 142°$ and $145° \leqslant \Delta \leqslant 155°$ because it was difficult to estimate the amplitude of PKIKP waves from 142° to 145°.

Bullen suggested in 1953 that perhaps he had overestimated the rigidity of the inner core—not that it was not solid but that its rigidity might not be as great as he had previously supposed. He had used Jeffreys' estimate of the increase in the velocity gradient of P waves passing into the inner core in his former estimate of its rigidity. Gutenberg's later and somewhat lower estimate of the jump in the velocity of P waves implied a less rigid inner core. In addition, if Gutenberg's estimate were correct, it was quite likely that PKJKP waves would have insufficient amplitude for detection, since his estimate implied the use of even more energy in the P-S wave conversion at the inner core boundary.

Observation of PKJKP waves continued to be unsuccessful, and in 1955 R. O. Hutchinson argued that because he failed to observe them on two "ideal" occasions, the likelihood of their existence was quite remote. Bullen promptly replied that the "ideal" cases were not ideal. Fairly strong seismological confirmation of the existence of a solid inner core came in 1960, albeit in an unsuspected form: free oscillations of the earth brought about by a major earthquake in Chile on 22 May. In 1962 Bullen eagerly reported an analysis of the free oscillations put forth by Chaim L. Pekeris, an Israeli mathematician, showing that, among the available models, only Bullen's Earth Model B, distinguished from the other models by its postulations of a solid inner core, could explain some of the periods of the free oscillations. Additional evidence was provided by an analysis of the March 1964 earthquake in Alaska, and by the 1970's the existence of a solid inner core was well established.

To this day observation of PKJKP waves remains a matter of dispute. In 1972 Bruce Julian, David Davies, and Robert Sheppard claimed that they had identified such a wave, but it had a velocity of only about half the value predicted by Bullen. Since its velocity, even by today's standards, is too low, others have suggested that these researchers probably had observed a SKJKP wave, whose predicted velocity is much closer to what was observed. However, because both interpretations contain a J phase, either of them provides additional evidence for the solidity of the inner core, but one of less rigidity than originally predicted by Bullen. To the end of his life he believed that additional confirmation of the solidity of the inner core was needed.

Bullen was involved in an extrascientific controversy of some interest. In 1954 he was able to determine the precise time when the United States exploded a number of hydrogen bombs, and he quickly realized the scientific value that could be gained by planned nuclear explosions: that they could be used by seismologists to learn more about Earth's interior. This led him to propose their detonation for scientific purposes during his presidential address at the 1955 annual meeting of the International Association of Seismology and Physics of the Earth's Interior. Needless to say, his suggestion failed to gain universal applause.

Bullen died in 1976 of a heart attack.

BIBLIOGRAPHY

I. Original Works. Bullen published five books and more than 280 articles and papers. The most complete listing of his publications is in Sir Harold Jeffreys, *Biographical Memoirs of Fellows of the Royal Society*, **23** (1977), 19–39. Among his books are *An Introduction to the Theory of Seismology* (Cambridge, 1947; 2nd ed., Cambridge, 1953; 3rd ed., Cambridge, 1963); *An Introduction to the Theory of Dynamics* (Sydney, 1948); *An Introduction to the Theory of Mechanics* (Sydney, 1949; 8th ed., Cambridge, 1971); and *The Earth's Density* (London, 1975).

Some of the key articles, particularly relevant to his work on the density distribution problem and the solidity of the earth's inner core, are "The Variation of Density and the Ellipticities of Strata of Equal Density Within the Earth," in *Monthly Notices of the Royal Astronomical Society, Geophysical Supplement*, **3** (1936), 395–401; "Note on the Density and Pressure Inside the Earth," in *Transactions and Proceedings of the Royal Society of New Zealand*, **67** (1937), 122–124; "Composition of the Earth at a Depth of 500–700 km.," in *Nature*, **142** (1938), 671–672; "The Problem of the Earth's Density Variation," in *Bulletin of the Seismological Society of America*, **30** (1940), 235–250; "Density Variation of the Earth's Central Core," *ibid.*, **32** (1942), 19–29; "A Hypothesis on Compressibility at Pressures of the Order of a Million Atmospheres," in *Nature*, **157** (1946), 405; "Compressibility-Pressure Hypothesis and the Earth's Interior," in *Monthly Notices of the Royal Astronomical Society, Geophysical Supplement*, **5** (1949), 355–368; "An Earth Model Based on a Compressibility-Pressure Hypothesis," *ibid.*, **6** (1950), 50–59; "Theoretical Travel-Times of *S* Waves in the Earth's Inner Core," *ibid.*, 112–118; "Note on the Phase PKJKP," in *Bulletin of the Seismological Society of America*, **46** (1956), 333–334; "Oscillations of the Earth and the Earth's Deep Internal Structure," in *Australian Journal of Science*, **24** (1962), 303–307; and "Free Earth Oscillations and the Internal Structure of the Earth," in *New Zealand Mathematical Chronicle*, **5** (1976), 17–45.

Several of the best notes and articles by Bullen on his own work include his 1969 Matthew Flinders lecture, "Researches on the Internal Structure of the Earth," in *Records of the Australian Academy of Sciences*, **1** (1969), 39–58; and "Some *International Seismological Summary* Reminiscences," in *Geophysical Journal of the Royal Astronomical Society*, **20** (1970), 359–365.

II. Secondary Literature. Besides that by Jeffreys, a commemorative article is Bruce A. Bolt, *Bulletin of the Seismological Society of America*, **67** (1977), 553–557. The best historical treatment of the discovery of Earth's core that includes a section on Bullen's hypothesis of a solid inner core and its eventual confirmation is Stephen G. Brush, "Discovery of the Earth's Core," in *American Journal of Physics*, **48** (1980), 705–724.

Henry Frankel

BUMPUS, HERMON CAREY (*b.* Buckfield, Maine, 5 May 1862; *d.* Pasadena, California, 21 June 1943), *zoology, evolution, biometry, natural history education.*

One of the pioneers in the use of statistics to measure ongoing evolution, Bumpus was the second of three sons born to Laurin Bumpus and the eldest child born to Laurin's second wife, Abbie Eaton, a former schoolteacher. His father was a cabinet-maker who moved to Boston and became a missionary and social worker when Hermon was six years old.

Although he was a frail youth, Bumpus developed a strong love for nature and became an avid collector of animals. He entered Brown University in 1879 and was known as the student "who had shot, skinned, stuffed, and eaten every living animal." He studied natural history under John Jenks, a defender of the religious doctrine of special creation, and Alpheus Packard, an advocate of Lamarckian evolution. Bumpus not only collected and prepared specimens but also drew illustrations of animals, some of which were published. After graduating from Brown University in 1884, he remained there as assistant in zoology until 1886. He taught zoology at Olivet College from 1886 to 1889, during which time he married Lucy Ella Nightingdale (28 December 1886). They had two sons. Bumpus entered Clark University in 1889, wrote his Ph.D. thesis on the embryology of the American lobster, and received his Ph.D. in 1891. He then returned to Brown, where he taught zoology for the next ten years.

Bumpus established a summer school for biology students at the one-year-old Marine Biological Laboratory at Woods Hole in 1889. He served as assistant director of the Marine Biological Laboratory (1893–1895) and on its board of trustees (1897–1942). Bumpus was scientific director of the laboratory of the U.S. Fish Commission at Woods Hole from 1895 to 1900. In addition to stimulating a revival of scientific research into practical problems of the fishing industry, he published papers on the fate of flatfish fry liberated along the coast, the peregrinations of lobsters, and the rediscovery of great numbers of tilefish at the edge of the continental shelf, where they had not been reported for twenty years. His successful efforts in reviving scientific research in fisheries led to his election as president of the fourth International Fisheries Congress in 1908.

Bumpus pioneered the use of biometry to analyze biological data in America. He stimulated numerous investigators at Woods Hole to use biometrical methods in their biological studies. Bumpus published four papers summarizing his studies of variation. In one of them he described variation in the number of vertebrae and the position of the pelvic girdle relative to the vertebral column in the mud puppy *Necturus.*

Bumpus also compared samples of eggs of the English sparrow from the United States and from England. He found that the eggs from the populations of sparrows that were descendants of birds introduced into the United States were more variable than the eggs from England. Bumpus concluded that the greater variability recurred because more variant individuals had a greater chance to survive in the new American environment, where the forces of natural selection no longer operated as stringently as they did in England. In a similar study he found that populations of the small tidal-zone snail *Litorina littorea* that had been introduced into the United States were more variable than those from Europe.

On 1 February 1898 a severe winter storm with snow, sleet, and rain in Providence, Rhode Island, caused many English sparrows wintering in the vines of the old athenaeum at Brown University to be blown down. Bumpus noticed the sparrows lying exhausted or dead on the ground and quickly perceived that this could be "a possible instance of the operation of natural selection, through the process of the elimination of the unfit." He brought 136 of these sparrows to his anatomical laboratory, made quantitative measurements of nine morphological traits of each downed sparrow, and compared the measurements of the seventy-two birds that survived with those of the sixty-four birds that died. Bumpus published the results of his statistical analysis of the natural selection generated by the storm as "The Elimination of the Unfit as Illustrated by the Introduced Sparrow, *Passer domesticus*" (1899), in which he concluded, "The process of selective elimination is most severe with extremely variable individuals, no matter in what direction the variations may occur."

Bumpus' study of the selective mortality in English sparrows generated by a winter storm is considered to be a classic study of "natural selection in action" and has been summarized in many biology texts. It also has been reprinted several times. By publishing the nine morphological measurements for each of the 136 sparrows and arranging them by survivors, nonsurvivors, sex, and age of males, Bumpus enabled later investigators to reanalyze the storm-induced selective mortality in *Passer domesticus.*

Bumpus became assistant to the president and curator of invertebrate zoology at the American Museum of Natural History in 1900. He was promoted to director in 1902. In his museum activities, as in his previous university work, Bumpus contended that both teaching and research were necessary obligations of the institution. He viewed museums not merely as storehouses for natural history specimens to be used in research but also as educational institutions with roles to perform that are different from those of schools or colleges. He encouraged the replacement of "dreary series of stuffed animals" with "realistic groups in an outdoors atmosphere" and established a department of education at the museum. Bumpus actively promoted natural history in numerous official capacities, including president of the American Morphological Society (1902), the American Society of Zoologists (1903), and the American Association of Museums (1906, 1924).

Bumpus left the American Museum of Natural History in 1911 to become the business manager at the University of Wisconsin, a position he held until accepting an offer to become president of Tufts University in 1914, remaining there until he retired in 1919. Bumpus became a leader in outdoor education, as environment education was then called, during the 1920's, serving as chairman of the Special Advisory Board of the National Park Service from its inception in 1924 to 1931. The Advisory Board of the National Park Service was created in 1931, and Bumpus continued as chairman until 1940, when he resigned because of poor health. He pioneered the development of "trailside museums," contending that "the real museum is outside the walls of the building and the purpose of the museum work is to render the out-of-doors intelligible." Bumpus' trailside museums contained exhibits that explained the surrounding environment. For his creation and popularization of trailside museums in national parks, Bumpus received the Cornelius Amory Pugsley Gold Medal of the Scenic and Historic Preservation Society for 1940. The American Association of Museums awarded the Kent Diploma to him in 1941, "in recognition of distinguished service rendered to the cause of museum education," an honor that led him to write in the margin of the diploma, "More appreciated than any other testimonial."

Bumpus experienced a severe heart attack in early June 1943 and died on the twenty-first of that month.

BIBLIOGRAPHY

I. ORIGINAL WORKS. Bumpus' classic paper "The Elimination of the Unfit as Illustrated by the Introduced Sparrow, *Passer domesticus*" was published in *Biological Lectures from the Marine Biological Laboratory, Wood's Holl, Mass., 1898*, VI (Boston, 1899), 209–226. Additional papers by Bumpus include "The Embryology of the American Lobster," in *Journal of Morphology*, **5** (1891); "The Importance of Extended Scientific Investigation," in *Bulletin of the U.S. Fish Commission* (1897); "On the Reappearance of the Tile-fish, *Lopholatilus chamaeleonticeps*," in *Science*, n.s. **8**, no. 200 (1898); "Work at the Biological Laboratory of the U.S. Fish Commission at Woods Holl," *ibid.*, no. 186 (1899); "The Results Attending the Experiments in Lobster Culture Made by the U.S. Commission of Fish and Fisheries," *ibid.*, n.s. **14**, no. 365 (1901); "The Museum as a Factor in Education," in *Independent*, **61** (2 August 1906), 269–272; and "Objectives of Museum Work in National and State Parks," in *Museum News*, **15**, no. 4 (1937). A facsimile reprint of this paper, along with bibliographical references to six subsequent statistical reanalyses of Bumpus' data, is in C. J. Bajema, ed., *Natural Selection Theory: From the Speculations of the Greeks to the Quantitative Measurements of the Biometricians* (New York, 1983), 300–301, 348–365.

II. SECONDARY LITERATURE. A bibliography of Bumpus' papers and addresses is in Hermon Carey Bumpus, Jr., *Hermon Carey Bumpus: Yankee Naturalist* (Minneapolis, 1947). A. O. Mead wrote an obituary in *Science*, **99** (14 January 1944), 28–30.

CARL JAY BAJEMA

BURY, CHARLES RUGELEY (*b.* Henley-on-Thames, England, 29 July 1890; *d.* Chichester, England, 30 December 1968), *physical chemistry*.

Bury was the eldest of the five children of a solicitor. Although the family lived in Gloucestershire, he spent much time with a grandmother at Leamington. After graduating from Malvern College in 1908, he won a scholarship to Trinity College, Oxford, where his tutor was D. H. Nagel and his research supervisor in electrochemistry was Harold Hartley. In 1912 Bury graduated from Oxford with first-class honors. He then spent six months at Göttingen in the period 1912–1913.

His appointment as assistant lecturer in chemistry at University College of Wales, Aberystwyth (1913), was interrupted when Bury volunteered in August 1914 and served on the western front and in Mesopotamia; later, as a captain, he led troops through Iran to the Caspian Sea. In 1919 he returned to Aberystwyth, where he remained until 1943. At I.C.I., Billingham, Bury led a phase rule group from 1943 until 1952. In 1922 he had married Margaret Adams, an agricultural botanist. They had a son and a daughter.

Bury's first paper (1921) is an exceptional achievement. Its contents provide the basis of what

BURY

is taught to college chemistry students as the interpretation of the periodic table in terms of the electronic structure of the atoms. This lucid statement was for many years overlooked because of a common assumption that Niels Bohr was responsible for it. In the same year Bohr published a communication in which he stated: "The application of the correspondence principle . . . suggests that, after the first two electrons are bound in one-quantum orbits, the next eight electrons will be bound in two-quanta orbits, the next eighteen in three-quanta orbits and the next thirty-two in four-quanta orbits. . . ." For the atoms of the inert gases he proposed the following constitutions: helium (2_1), neon ($2_1 8_2$), argon ($2_1 8_2 8_2$), krypton ($2_1 8_2 18_3 8_2$), xenon ($2_1 8_2 18_3 18_3 8_2$), [niton] ($2_1 8_2 18_3 32_4 18_3 8_2$). He went on to say that in the rare earths we may assume the successive formation of an inner group of thirty-two electrons and, similarly, that we may suppose that the appearances of the iron, palladium, and platinum families are witnessing stages in the formation of groups of eighteen electrons. Bohr's letter leads one to conclude that he arrived at the inert gas structures largely by intuitive insight.

Bury submitted his paper some three weeks after Bohr's letter appeared. He wrote, "During the course of preparation of this paper, structures similar to those suggested by the author for the inert gases have been proposed by Bohr." He expressed his own views thus:

> Successive layers can contain 2, 8, 18, and 32 electrons. Groups of 8 and 18 electrons in a layer are stable, even when that layer can contain a larger number of electrons. The maximum number of electrons in the outer layer of an atom is 8: more than 8 electrons can exist in a shell only when there is an accumulation of electrons in an outer layer. During the change of an inner layer . . . there occurs a transition series of elements which can have more than one structure.

On these bases Bury summarizes in seven pages the valence properties of all elements up to uranium, accounting for the atomic numbers of transition elements and rare earth elements. The latter he correctly starts with cerium and finishes with lutecium. Apart from essential features reproduced in introductory accounts of the chemical elements, there are two special details.

First, Bury states: "Between lutecium and tantalum an element of atomic number 72 is to be expected. This would have the structure (2,8,18,32,8,4) *and would resemble zirconium*" (emphasis added). In January 1923 Dirk Coster and Gyorgy von Hevesy reported their identification of hafnium (atomic number 72) in zirconium minerals. Hevesy, Mary E. Weeks, and others who recount the history of the discovery of hafnium do not mention Bury but give credit to Bohr, whose first mention of element 72 only follows after Bury. Later, Bohr acknowledged Bury's priority.

Second, under the heading "The Last Period," Bury writes:

> In this period a second 18–32 transition series may be expected. . . . Little resemblance [between the actinides and lanthanides] . . . is to be expected. Possibly an element, not yet discovered, of atomic number 94 . . . is the first of a series of 7 transition elements . . . something like the ruthenium group but more electropositive.

These comments are perhaps the first reasoned predictions on transuranic chemistry.

Bury had mathematical inclinations and a very clear grasp of thermodynamics. He used partial molar values to study the state of solute molecules. In the early 1920's McBain postulated multimolecular aggregates in soap solutions where Bury anticipated complications from hydrolysis. Bury established and first explained the critical concentration for micelle formation in aqueous solutions of butyric acid:

$$\begin{aligned} nA_1 \rightleftharpoons A_n : K_{eq} &= [1/K^n] \\ &= [A_n]/[A_1]^n : \text{and}\, [A_n] \\ &= \text{micelle concentration} = [A_1/K]^n, \end{aligned}$$

where K is the Haber form of the constant and n is the number of molecules in a micelle. As $[A_1]$ = monomer concentration exceeds K, a very rapid increase in $[A_n]$ occurs. Bury followed such molecular aggregations by freezing point, specific heat, density, conductivity, and viscosity measurements.

In 1935 Bury wrote another seminal paper, in which he related the appearance of color in nearly all the well-known families of organic dyestuffs to the presence of resonance (as electronic delocalization was described) between two often strictly equivalent molecular electronic structures, for instance, in Döbner's violet:

$H_2\overset{+}{N}{=}C_6H_4{=}C(C_6H_5){-}C_6H_4{-}NH_2 \leftrightarrow H_2N{-}C_6H_4{-}C(C_6H_5){=}C_6H_4{=}\overset{+}{N}H_2$

In three pages he offers equivalent formulations for fifteen dyestuff families. Surprisingly, he instances indigo as a noncomplier (Kuhn had anticipated him in this instance). Bury did not know of Kuhn's paper.

G. N. Lewis and Melvin Calvin conducted extended studies in this area. Others who immediately used Bury's indications on dyestuff structures included Schwarzenbach, Hamer and Mills, and L. G. S. Brooker.

BIBLIOGRAPHY

I. Original Works. Bury's writings include "Langmuir's Theory of the Arrangement of Electrons in Molecules and Atoms," in *Journal of the American Chemical Society*, **43** (1921), 1602–1609; "The Densities of Butyric Acid-Water Mixtures," in *Journal of the Chemical Society* (London) (1929), pt. 1, 679–684, written with John Grindley; "The Electrical Conductivity of Butyric Acid-Water Mixtures," *ibid.* (1930), pt. 2, 1665–1668, written with John Grindley; "The Partial Specific Volume of Postassium n-Octate in Aqueous Solution," *ibid.*, 2263–2267, written with D. Gwynne Davies; "Auxochromes and Resonance," in *Journal of the American Chemical Society*, **57** (1935), 2115–2117; and "The Duhem-Margules Equation and Raoult's Law," in *Transactions of the Faraday Society*, **36** (1940), 795–797.

II. Secondary Literature. Biographical notices are Mansel Davies, "C. R. Bury: His Contributions to Physical Chemistry," in *Journal of Chemical Education*, **63** (1986), 741–743, and "Charles Rugeley Bury and His Contributions to Physical Chemistry," in *Archive for History of Exact Sciences*, **36** (1986), 75–90.

Mansel Davies

BUSH, VANNEVAR (*b.* Everett, Massachusetts, 11 March 1890; *d.* Belmont, Massachusetts 28 June 1974), *science and government, electrical engineering.*

Vannevar Bush was a statesman of twentieth-century American science. The son of Richard Perry Bush, a Universalist minister, and of Emma Linwood Paine Bush, he rose from modest beginnings in the working-class suburbs of Boston to become a noted engineer and architect of scientific institutions in the years after World War II. Between the world wars, as a professor of electrical engineering at the Massachusetts Institute of Technology, he promoted the causes of engineering education, graduate research, and applied mathematics, and developed important computational machinery. In 1932 he became MIT's first vice president and dean of engineering.

Bush moved to Washington in 1939 to assume the presidency of the Carnegie Institution; within a year the war emergency turned his attention from basic science to national preparedness. Capitalizing on friendships formed over the years, and strategically situated in the nation's capital, Bush took the initiative in mobilizing the nation's powerful communities of science and engineering for war. As chairman of the Office of Scientific Research and Development, he presided over a far-flung network of wartime laboratories from which emerged radar, the proximity fuse, amphibious landing craft, penicillin, and the atomic bomb—accomplishments that catalyzed a new public appreciation for science in the nation's service and led to the establishment of the Atomic Energy Commission and the National Science Foundation. After the war Bush continued to advise the government in matters of science and defense, enthusiastically resumed his work at the Carnegie Institution, joined the boards of directors of Merck and AT&T, and wrote several works dealing with science and the lessons of war.

Bush's roots reached deeply into the soil of New England. The descendant of a long line of Yankee sea captains, he retained something of the salty independence of the sea. His father, Perry Bush, had left the ancestral home in Provincetown to escape sectarian controversy, moving to the suburbs of Boston, where he abandoned his traditional Methodism and studied for the Universalist ministry at Tufts College. In Everett and later for many years in Chelsea, Perry Bush was a widely known and much-loved pastor with a penchant for turning mundane occasions to religious ends. A liberal Protestant, he was on the periphery of orthodox society, excluded, as were Jews and Catholics, from local institutions like the YMCA. Marginal status provoked his sympathy for the underprivileged, however, and he proved an inspired mixer. In these ways especially, Perry Bush deeply influenced his son, who throughout his life maintained a fondness for putting the ordinary to extraordinary use, was proud of his ability to move between circles powerful and ordinary, and, profoundly affected by his father's commitment to his pastoral profession, was wont to describe the vocation of engineering as itself a ministry.

In 1909 Bush enrolled at Tufts College. There he acquired a firm foundation in the art and science of engineering and entered into the vigorous extracurricular life of a turn-of-the-century college. He joined the engineering fraternity, trekked to the theater in Boston, served as the president of his junior class, and delivered speeches at banquets. He turned out for track, ran the middle distances, and managed the football team. Like other Tufts undergraduates, he attended daily chapel—often attentively, no doubt, and at other times scribbling mathematical diagrams on the back of his seating ticket.

Bush studied with a vengeance, compensating for illness that lost him a year of high school and part of his sophomore year in college. In four years Bush learned his way around the generators and currents that characterized early electrical engineering and began a lifelong romance with invention that along the way generated an abiding interest in the workings of the United States patent system. As a senior he received his first patent for the profile tracer, a surveying instrument that automatically recorded profiles of elevation. The tracer contained in embryonic form the disk integrators that were to become an important part of his later analog computers. In 1913, when he received both the B.S. and the M.S., he addressed his graduating class on "the poetry of mathematics."

The professional programs at Tufts were sheltered within an undogmatic Universalism that stressed a common humanism and a commitment to public service. Gardner Anthony, the dean of engineering and a noted educator, promoted the graphic elements in engineering, defending mechanical drawing as a universal language that distinguished the competent man from the incompetent. William Ransom, a mathematician, stressed practical problem solving while encouraging more speculative and playful topics that introduced his students to creative mathematical thinking. The influence of Anthony and Ransom, alloyed with the professional ethos that pervaded the curriculum at Tufts, left an indelible mark on Bush's career in engineering.

After receiving his master's degree in 1913, Bush was hired as a "test man" by General Electric at Schenectady, New York. Fired after a year, he talked himself into a job at Tufts teaching mathematics to the women students of Jackson College; he also taught physics to premed students. A year as an instructor whetted his appetite for further study. That summer he worked in the New York Navy Yard as an electrical inspector before enrolling, in the fall of 1915, in the graduate program at Clark University to study mathematical physics with Arthur Gordon Webster. He quickly abandoned Clark and a $1,500 scholarship, and found himself back in Chelsea, headed toward MIT and a career in engineering.

Bush's passage through the graduate program at MIT was meteoric. With the help of Oliver Heaviside's operational calculus, he attacked as a dissertation problem the oscillatory behavior of electrical power lines. In one hectic year that tested the mathematical resources of the department as well as his health—curing him, somewhat surprisingly, of the illnesses that had dogged him from childhood—Bush completed his graduate work and received one of the few doctorates jointly awarded by MIT and Harvard. His doctorate, the fifth granted by MIT, symbolized the growing importance of engineering research in the decades after 1900.

In the fall of 1916 Bush returned to Tufts as an assistant professor of electrical engineering; married Phoebe Davis, the daughter of a Chelsea merchant, on 5 September (they had two sons); and settled into a dual role as teacher and consultant for the newly established American Research and Development Corporation. As the head of AMRAD's small research laboratory, Bush discovered the world of the small, innovative, science-based firm. After the war he and Charles G. Smith developed an early rectifying tube, thereby displacing one of the cumbersome batteries that were a component of early radios and enabling the radio to use household current. AMRAD sparked Bush's entrepreneurial ambitions and involved him in the establishment of a number of small firms, notably the Raytheon Company, which became one of the nation's largest electronics firms and defense contractors.

In 1917, when the United States was drawn into World War I, Bush went to work on the problem of the German submarine, inventing a sub detector that worked but was never used. Rebuffed by Robert Millikan and the National Research Council, and dismayed by the organizational chaos that surrounded Edison's Naval Consulting Board, Bush concluded that inadequate access to the centers of power and dispersal of authority had crippled the efforts of scientists to contribute to the war. He would remember the lesson in a future and greater war.

In the fall of 1919, when the academic market for engineering turned bullish, Bush moved from Tufts to MIT as associate professor of electrical power transmission; he became professor in 1923. In his early years there he assumed direction of the electrical engineering department's research and graduate programs and helped Dugald Jackson, the department's chairman, modernize the curriculum. In 1932 Bush became MIT's first vice president and dean of engineering, serving as Karl Compton's strong right arm. With growing influence Bush spoke out on such issues as the nature of engineering education, the social responsibilities of the engineering profession, and the relationships among engineering, industry, and government.

Bush's major technical work was inspired by the need of American scientists and engineers for more adequate tools of applied mathematics. Over two decades he worked to establish a strong program in mechanical analysis at MIT. With the aim of exploring machine methods for the solution of the difficult mathematical problems confronting engineers, the program took shape in the 1920's in attempts to deal with the behavior of long-distance transmission lines. In 1925 Bush encouraged his student Herbert Stewart to mechanize the integration of the Carson equation. The Stewart device was successful but limited, and by 1931 Bush and his students had developed the differential analyzer. Comprising several disk integrators interconnected by a variety of gearings and set in motion by motor-driven shafts, the analyzer kinetically reproduced the changing terms of differential equations, tracing the desired solutions onto drawing boards. The 1931 analyzer enjoyed tremendous success, first in electrical engineering and then, in short order, in geophysics, cosmic-ray studies, and quantum mechanics, fields where progress was slowed by difficult mathematical calculations. Within a few years machines modeled upon the analyzer had multiplied within the United States and abroad.

The differential analyzer was only one of a battery of machines, including the network analyzer and the optical integraph, that were invented by Bush and his colleagues. The success of the analyzer encouraged Bush to expand his plans to make MIT an international center for machine analysis. After 1935, with the help of the Carnegie Institution of Washington and especially the Rockefeller Foundation, he built the larger, faster, more flexible machine known as the Rockefeller differential analyzer and established MIT's Center of Analysis. Put to work during the war calculating ballistics tables and the curvature of radar antennae, the Rockefeller analyzer was the most important calculator of its time. Although quickly superseded after the war by a new generation of electronic digital computers, the analyzer clearly revealed the possibilities of machine computation. Moreover, in its decisively rational, instrumental approach to problem solving it symbolized early-twentieth-century engineering.

Bush's horizons widened in the 1930's. He was elected to the National Academy of Sciences in 1934, worked with Compton's Science Advisory Board, and testified before Congress on the patent system in the monopoly hearings of 1939. He became acquainted with men who would become major actors in the events of World War II, among them Warren Weaver, an applied mathematician and officer of the Rockefeller Foundation; Frank Jewett, the head of Bell Laboratories; Conway Coe, the commissioner of patents; James Conant, the president of Harvard; and, of course, Karl Compton. In 1939 Bush left MIT to become president of the Carnegie Institution of Washington and, shortly thereafter, chairman of the National Advisory Committee for Aeronautics. Strategically positioned in the public and private spheres, Bush hoped to continue the peaceful, cooperative efforts in science and engineering that had characterized the 1920's.

As it happened, however, war, and not peace, provided the future focus of Bush's life. For months he had worried about conditions in Europe, and when Germany invaded France in May 1940, he went to work. With Compton, Conant, Richard Tolman (a Caltech physicist), Frank Jewett (who had become the president of the National Academy of Sciences), and two Roosevelt advisers, Oscar Cox and Harry Hopkins, Bush devised the National Defense Research Committee to facilitate scientific assistance to the government in the development of new weapons. The NDRC was established by Roosevelt on 27 June 1940, with Bush as chairman. His responsibilities were expanded two years later with the creation of the Office of Scientific Research and Development, which included NDRC, the Committee on Medical Research, the Office of Field Service, the Scientific Personnel Office, and a liaison office to centralize scientific exchange among the Allies' governments. With OSRD, Bush took the leading role in wartime science. Such an action was justified, Bush felt, by the times and by the failures of a previous war.

During the next five years, OSRD worked wonders. From contracting laboratories came a host of electronic devices, including microwave radars and the proximity fuse, that changed the nature of warfare. Antisubmarine warfare was improved; rockets, guided missiles, explosives, and fire control benefited from OSRD involvement, as did amphibious warfare with the development of the Dukw and the Weasel. From pharmaceutical laboratories came improved antimalarial drugs, blood substitutes, and the large-scale production of penicillin. Not least was the involvement of mathematicians in logistics in the new field of operations research. The atomic bomb was the most notorious achievement, and thwarted Bush and Conant's early hope that such a weapon would prove impossible or impractical.

As important as the weapons themselves were innovations in strategies and structures. Bush's greatest contribution to the war was, in fact, OSRD itself, an institutional device that successfully coordinated a vast network of university, industrial, foundation, and government laboratories employing tens of thousands and spending, by war's end, over $500 billion. The Radiation Lab at MIT, the Johns Hopkins' Applied Physics Laboratory, the Los Alamos Laboratory, the Clinton Works at Oak Ridge,

Tennessee, and the Hanford Works on the Columbia River at Richland, Washington, were all built and operated for OSRD projects.

OSRD's success was a consequence of several factors. First, whenever possible, Bush utilized existing private facilities rather than special-purpose government laboratories in the fashion of the earlier war. To accomplish this, OSRD contracted with universities and companies for research performed on an "actual cost" basis. Second, Bush delegated technical decisions to the scientists and engineers of OSRD's divisional committees while retaining overall responsibility for policy and liaison in his central office. With OSRD thus divided into line and staff functions, and freed from routine technical decisions, Bush was able to capitalize on administrative talents honed over the years.

With OSRD war forced a marriage of science and government unprecedented in American history. Enormously successful, it was also deeply disturbing, for it challenged traditional divisions of labor between the federal and private sectors. The irony of Bush's success is that an achievement that drew upon national strengths developed during more conservative decades should have provided the patterns for a new age in which science would never again be left largely to the devices of private enterprise. Much of Bush's postwar career was spent accommodating new demands and traditional values.

The direst change in this new age was forged by the atomic bomb. Even before Trinity, the first test of the bomb, Bush and Conant sought to raise the issue of postwar consequences. Bush was deeply afraid that a shortsighted attempt to monopolize atomic knowledge would only fuel an arms race, create a dangerous world, and undermine the political values central to American democracy. Obliged to work in secrecy, through confidential memorandums and in the interim committee created to provide advice on atomic policy to President Truman, Bush argued strongly that a stable world required the sharing of atomic knowledge under the auspices of an international organization with access to national laboratories. He shaped the Truman-Attlee-King declaration of November 1945 and influenced the Baruch Plan presented to the United Nations in June 1946. The ultimate collapse of the Baruch Plan and the emergence of the cold war confirmed many of Bush's apprehensions.

Bush's domestic efforts were more successful, and they found fruition in the Atomic Energy Commission and the National Science Foundation. Given the urgent need to control atomic energy, the AEC was enacted in July 1946, but only after bitter struggles within the scientific and political communities that revolved around the degree of control over atomic research and the role of the military. Bush had called for the National Research Foundation in his 1945 report for Roosevelt, *Science: The Endless Frontier*. Yet legislation establishing the NSF encountered even more difficulties, despite universal agreement by war's end that a national foundation had become essential to national security. Only after five difficult years, complicated by worries over political manipulation and the degree of presidential control, did the NSF become a reality in 1950.

In both developments Bush had been in at the beginning, frequently in conflict with presidential advisers seeking to protect executive authority, on the one hand, and caught between a scientific community divided over the proper relationship of science and government, on the other. Accused by critics of elitism (often justly) and of Machiavellian ambition (unjustly), Bush was forced in the heat of battle to compromise on points of debate. But in his general beliefs, he remained firm. War had made irrevocable the linking of science, prosperity, and national security. Given that fact, massive federal involvement with science was unavoidable, and it should be provided in a fashion that protected science from political interference while respecting presidential and congressional prerogatives. Bush was optimistic that both the NSF and the AEC would serve the purpose.

After the war Bush continued to serve the government in various capacities. Between 1947 and 1949 he chaired the Research and Development Board established to coordinate R&D in the newly reorganized National Military Establishment. He consulted with the Patent Office on its efforts to mechanize the searching of patent literature and testified before Congress on the reform of the patent system. Yet after Roosevelt's death, Bush found the corridors of power increasingly closed. He attempted and failed to delay the H-bomb test of November 1952, believing that it seriously compromised the prospect of international control. His frustrations were aggravated by the Oppenheimer security hearings in 1954. In a short-tempered appearance before the AEC's Personnel Security Board, he lashed out at board members, accusing them of un-American behavior in appearing to condemn the wartime director of Los Alamos for his 1949 opinion—that the H-bomb was inadvisable.

But if Bush's advisory efforts in government were increasingly frustrated, he found satisfaction in renewed involvements in the private sector. With his unparalleled knowledge of the political economy of postwar science, he proved a valuable addition as a director of AT&T and of Merck, at Merck pursuing interests in medical research cultivated during the war. He also became a successful author, writing books dealing with science, national policy, and the lessons of war that included the important *Modern Arms and Free Men* (1949). Happiest of all was his renewed involvement with the Carnegie Institution. This prestigious private foundation had been profoundly affected by the war. Given its location, its accomplished scientists, and its programs in relevant disciplines, and with its president taking the leading part in the mobilization of science, the Carnegie Institution had been transformed in short order into a virtual government laboratory for the conduct of war-related applied research, its central offices playing a dual role as OSRD headquarters. The transition from pure science to defense-related R&D was abrupt and disconcerting, and captured on a smaller stage changes remaking American science in the large. Between 1945 and Bush's retirement in 1955, the Carnegie Institution provided him a privileged location from which to reflect on these changes.

In the decade following the war, the Carnegie Institution enthusiastically resumed projects long held in check. Under Bush's leadership it continued its studies of solar storms, the earth's magnetic field, and residual magnetism in rocks; the photochemistry of chlorophyll and the formation of hybrid grasses; and bacterial resistance to antibiotics. The 200-inch Hale telescope was inaugurated at Palomar and began to provide new insights into the size and age of the universe; Barbara McClintock studied the unstable genes of maize; and Alfred Hershey revealed the secrets of the reproduction of the T-bacteriophages.

The Carnegie Institution could not, however, simply return to its prewar agenda. The new linkage between science and war assured that it had a significance greater than the sum of its programs. Several factors, Bush felt, made this so. The first was a consequence of the new relationship between the federal government and the scientific community. If national security mandated increased federal support for science, it also threatened to impose an onerous bureaucratization that would stifle innovation, militarize research, exaggerate applications, and subject the direction of science to the inexpert and the politically minded. While Bush had a deep respect for the instincts of the American people, he also feared that "creeping socialism" and the uncontrolled growth of science fueled by federal money would undermine the traditional role of the private sector. In this situation the Carnegie Institution served as both refuge and symbol—a refuge for the untrammeled pursuit of fundamental science and a symbol of the vitality of private enterprise in the search for knowledge. Moreover, during a decade when the successful development of the hydrogen bomb, increasing Soviet hostility, and the Korean War made the future appear grim, the Carnegie Institution came to represent for Bush the idealism and faith of free science.

In 1955, when Bush retired from the Carnegie Institution, he cut back on public involvements and went home to New England. There he busied himself once again with the affairs of MIT, serving as chairman of its governing corporation from 1957 to 1959 and honorary chairman from 1959 to 1971. He remained a director of AT&T until 1962 and served as chairman of Merck's board of directors from 1957 until 1962. The most durable of Bush's occupations, however, was invention. In 1912 he applied for his first patent; in 1974, the year of his death, he was granted his last—all in all, forty-nine over sixty-two years. In many ways Bush was most at home in the inventor's workshop, and it was there that his character ws clearly revealed. In all his varied roles, he was above all inventive, a rational instrument of engineering, bringing order out of chaos and solutions to problems.

Bush may have been the outstanding example of the expert whose role at the hub of an increasingly complex society captured the American imagination in the first half of the twentieth century. These were years when the figure of the engineer became not only a necessary fact of life but also a value-laden symbol that presaged the contributions of science and technology to social progress. If the consequences of this turning to science, especially in the light of its linkage with national security after the war, seemed to him ambiguous blessings, Bush never lost his optimism. The inventive spirit had helped Americans conquer difficulties in the past, he wrote in *Pieces of the Action* (1970). Given a chance, it would do so again.

BIBLIOGRAPHY

I. ORIGINAL WORKS. Among the most important of Bush's writings are *Science, the Endless Frontier: A Report to the President* (Washington, D.C., 1945); *Modern Arms and Free Men: A Discussion of the Role of Science in Preserving Democracy* (New York, 1949); and his autobiographical *Pieces of the Action* (New York, 1970). A full list of Bush's writings can be found in the short biography by Jerome Wiesner in the *Biographical Memoirs of the National Academy of Sciences*, **50** (1979), 89–117.

There is abundant material on Bush in several archival collections, notably the Vannevar Bush Papers at the Library of Congress, which emphasize the years from 1938 on; the Bush Papers, the Compton-Killian Papers, and the Dugald Jackson Papers at MIT; and the deposits in the National Archives of the various federal agencies with which he was involved, especially those of the OSRD.

II. SECONDARY LITERATURE. There is as yet no book-length treatment of Bush, although the outlines of his life and career are well treated in Wiesner's National Academy of Science memoir cited above. There is, however, substantial information about Bush in a number of other sources. For his general importance to the institutionalization of American science, refer to Daniel Kevles, *The Physicists: The History of a Scientific Community in Modern America* (New York, 1978). His contributions to electrical engineering are detailed in Karl Wildes and Nilo Lindgren, *A Century of Electrical Engineering and Computer Science at MIT, 1882–1982* (Cambridge, Mass., 1985); his work on the differential analyzer is dealt with in Larry Owens, "Vannevar Bush and the Differential Analyzer: The Text and Context of an Early Computer," in *Technology and Culture*, **27** (1986), 63–95. Bush's political and administrative activities during his Washington years receive considerable attention in numerous books dealing with the period: among them J. Merton England, *A Patron for Pure Science: The National Science Foundation's Formative Years, 1945–57* (Washington, D.C., 1983); and Alice Kimball Smith, *A Peril and a Hope: The Scientists' Movement in America, 1945–47* (Chicago, 1965). James Conant's autobiography, *My Several Lives: Memoirs of a Social Inventor* (New York, 1970), tells much about his teamwork with Bush in the NDRC and OSRD. See also Daniel Kevles, "The National Science Foundation and the Debate over Postwar Research Policy, 1942–1945: A Political Interpretation of Science—The Endless Frontier," in *Isis*, **68** (1977), 5–26, and "Scientists, the Military, and the Control of Postwar Defense Research: The Case of the Research Board for National Security, 1944–46," in *Technology and Culture*, **16** (1975), 20–47; also Nathan Reingold, "Vannevar Bush's New Deal for Research: Or the Triumph of the Old Order," in *Historical Studies in the Physical and Biological Sciences*, **17** (1987), 299–344. James Killian's *The Education of a College President: A Memoir* (Cambridge, Mass., 1985) contains several passages dealing with Bush.

LARRY OWENS

CAMERON, ANGUS EWAN (*b*. Sylvania, Pennsylvania, 14 October 1906; *d*. Oak Ridge, Tennessee, 28 September 1981), *physical chemistry, mass spectroscopy*.

Cameron was known and highly respected by his colleagues as a physical chemist and mass spectroscopist. His ideas and perceptions, which he did not hold back, strongly influenced associates in the research laboratory and in scientific meetings. He possessed a critical judgment that few ever attain. His specialty and expertise focused on the measurement of isotopic abundance of the elements. He was particularly interested in the determination of lithium and its isotopic variability in the earth's crust, as well as in the age and origin of the universe. He was a member of Alpha Chi Sigma, Sigma Xi, the American Chemical Society, and the American Society for Mass Spectrometry.

The son of Alexander George Cameron and Jennie Hoover, Cameron received his undergraduate education at Oberlin College, graduating magna cum laude in 1928. He received his doctorate in physical chemistry from the University of Minnesota in 1932. In 1933 he was awarded a one-year National Research Council postdoctoral fellowship to the School of Optics at the University of Rochester (New York). In the same year he married Jane Williams Gray. They had three sons: Allan, Douglas, and Alexander.

Cameron's first position (1934) was with the Eastman Kodak Company in Rochester, New York. Except for brief employment with other laboratories, he spent the best part of ten years with Eastman Kodak, doing physical chemical research concerned mainly with photographic processing. Cameron transferred to the Manhattan Project at Oak Ridge, Tennessee, in 1943. It was there that the fissionable uranium isotope (U^{235}) was first separated from normal uranium by an electromagnetic process. The first atomic bombs were made from the product of these separations. Cameron, with Alfred Nier, Roger Hibbs, William Harman, and others, set up the first mass spectroscopy laboratory to measure separated isotopes of uranium. His knowledge of optics and his talents in mass spectroscopy were vital to the success of the electromagnetic separation process and, ultimately, to the whole Manhattan Project.

In 1947 Cameron returned to the Eastman Kodak laboratory, where he did research on high-vacuum distillation techniques. A year later he went back to the Oak Ridge area and settled down to research and administration at the Oak Ridge Gaseous Diffusion Plant (K-25) for the Carbide and Carbon Chemicals Company, the operating contractor. At this site large process buildings were being built to house the new gaseous-diffusion separation units. The new method of separation, using UF_6, was much cheaper and easier to scale up than the electromagnetic separation method it was replacing. Cameron's leadership and knowledge furthered vacuum distillation, mass spectroscopy, and isotope separation technology for six years. With D. F. Eggers, Jr., he made an important contribution to mass spectrometry in 1948 when they developed what Cameron called an "ion velocitron," a form of the present-day time-of-flight instrument, which displayed the entire mass spectrum at one time.

In 1954 Cameron took a year's leave of absence to accept a Fulbright fellowship to the Max Planck Institute of Chemistry at Mainz, Germany. There he met and collaborated with such research leaders in mass spectroscopy as Josef Mattauch and Heinrich Hintenberger. During this association he became further engrossed in the measurement of atomic weights of the elements and the age of the universe.

His zeal for the measurement of atomic weights led to Cameron's participation in and involvement with the Commission on Atomic Weights of the International Union of Pure and Applied Chemistry (IUPAC), on which he served for more than twenty years. He continually prodded his colleagues to report their results when they made isotopic composition measurements on elements that are variable in nature or have large uncertainties, because they could be quite valuable to the scientific community and to the IUPAC as it sought to determine best values.

Cameron returned to Oak Ridge (K-25) in 1955 and a short time later transferred to the Oak Ridge National Laboratory (ORNL), where he served as an assistant director of the Analytical Chemistry Division and headed the mass spectrometry department from 1957 until his retirement in 1971.

After his retirement from ORNL, Cameron did consulting work for ORNL, Oak Ridge Associated Universities, and Litton Systems. While with the latter he was involved with the design and construction of the analytical instrument package for the Mars Viking lander that identified and measured the composition of the atmosphere of Mars in 1975.

Cameron received the Distinguished Alumni Award from the University of Minnesota in 1955, and a certificate of appreciation from the U.S. Department of Energy for his many contributions to the classified work done at the Oak Ridge plants.

BIBLIOGRAPHY

I. Original Works. Cameron wrote more than one hundred technical papers and reports and was awarded three patents for improved components for mass spectrometers. The following articles reflect his research and interests in measurements of atomic weights and isotopic abundances: "An Ion Velocitron," in *Review of Scientific Instruments*, **19** (1948), 605–607, written with D. F. Eggers, Jr.; "Variation in the Natural Abundance of the Lithium Isotopes," in *Journal of the American Chemical Society*, **77** (1955), 2731–2733; "Isotopic Composition of Bromine in Nature," in *Science*, **121** (1955), 136–137, written with E. L. Lippert, Jr.; "Report of the International Commission on Atomic Weights (1961)," in *Journal of the American Chemical Society*, **84** (1962), 4175–4197, written with Edward Wichers; and "Mass Spectrometry of Nanogram-Size Samples of Lead," in *Analytical Chemistry*, **41** (1969), 425–426, written with D. H. Smith and R. L. Walker.

II. Secondary Literature. See Richard Smyser, "Angus Ewan Cameron," in *The Oak Ridger*, 30 September 1981; and Raymond L. Walker, J. A. Carter, and W. H. Christie, "In Memoriam of A. E. Cameron," in *International Journal of Mass Spectroscopy and Ion Physics*, **42** (1982), 1–2. See also *Determination of the Isotopic Composition of Uranium*, U.S. Atomic Energy Report, TID-5213 (1950); and Clement J. Rodden, ed., *Analysis of Essential Nuclear Reactor Materials* (Washington, D.C., 1964), ch. 13.

Raymond L. Walker

CAMPBELL, IAN (*b.* Bismarck, North Dakota, 17 October 1899; *d.* San Francisco, California, 11 February 1978), *geology*.

Campbell, of Scottish ancestry, was the son of Dugald and Agnes Gilkison Campbell. He grew up first on a sheep ranch in North Dakota and then among cherry orchards in Eugene, Oregon. After attending local schools, he matriculated at the university of Oregon. His education was interrupted when he volunteered for military service in World War I as a wagoneer (361st Ambulance Company, 91st Division) from 1917 to 1919. He received the A.B. degree in 1922 and the A.M. degree in 1924. Campbell went to Northwestern University as a university fellow in 1923 and continued his graduate work in 1924 at Harvard, where he was a teaching fellow (1924–1925). He was an assistant professor of geology at Louisiana State University (1925–1928) before returning to Harvard as instructor in mineralogy and petrology (1928–1931) and receiving the Ph.D. in economic geology in 1931. On 16 September 1930 Campbell married Catherine (Kitty) Robbins Chase shortly after she received her M.A. degree in geology from Radcliffe. She received her Ph.D. from Radcliffe in micropaleontology in 1932. They had one son, Dugald. Campbell's lifelong interest in and love for nature led him to lead an active life despite two bouts with cancer, the last of which was fatal.

Campbell was a member and/or fellow of almost every major geologic and academic organization in the United States and served as an officer in many of them. A partial list includes Geological Society of America (president, 1968), Mineralogical Society of America (president, 1962), Association of American State Geologists (president, 1965–1966), American Geological Institute (president, 1961), Branner Geologic Society (president, 1938), Le Conte Geologic Society (president, 1963–1964), American Association of Petroleum Geologists (Public Service Award, 1973), American Institute of Professional Geologists (Ben H. Parker Memorial Medal, 1970), American Association for the Advancement of Science (Pacific Division president, 1957–1958), American Institute of Mining, Metallurgical and Petroleum Engineers (Hardinge Award, 1962).

If one word can describe a person's life, then for Ian Campbell the word must be "active." From childhood he was always involved with many different, frequently concurrent, projects. Not long after he was discharged from military service, he embarked on a solo motorcycle journey from Portland, Maine, to Portland, Oregon, which, given the state of the nation's highways at that time, was no mean accomplishment. In recognition of his efforts and the publicity he brought the sport of motorcycling, he received a medal from the Harley-Davidson Company. But he devoted most of his energy to geology and excelled in the roles of teacher and scholar, state geologist, and professional geologist.

Campbell's academic career began when he accepted a position as assistant professor of geology at the California Institute of Technology in 1931. In 1934 he became an associate professor and a research associate of the Carnegie Institution of Washington, D.C.; three years later he led the Carnegie-Caltech geological expedition on a pioneering trip through the Grand Canyon. While associated with Caltech, Campbell was known as a very competent teacher who demanded and received excellence from his students, requiring them to think beyond simple facts and consider their meaning. He insisted that his students master the fundamentals and techniques of geology, especially when it came to the petrographic microscope and crystallography. At Caltech he is said to have given a young Ph.D. student a sample for identification that turned out to be a kidney stone from his Norwegian elkhound.

Campbell's administrative career began at Caltech, where he served as associate chairman, had a brief tenure as acting chairman, and in 1952 became the executive officer of the Division of Geological and Planetary Sciences. He was promoted to professor of petrology in 1946, but he left the academic world and Pasadena in 1959 to become state mineralogist and chief of the State Division of Mines of California. Two years later, after some political persuasion on his part, the title became state geologist and chief of the Division of Mines and Geology. At his departure from Caltech, Campbell retained the courtesy title of research associate, and he was granted the rank of professor emeritus in 1970 after his formal retirement in 1969.

His state position demanded all of Campbell's administrative and diplomatic skills; one of his greatest achievements was the compromise he was able to negotiate on the legislation that provides for registration of geologists in California. After the Board for Registration of Geologists was organized, Campbell served as a member (1969–1978) and as its president (1972–1974). While he was state geologist, his professionalism and stature as a geologist enhanced the reputation of the Division of Mines and Geology within the state and nationally. He started the state program in "geological hazards" shortly after his arrival, beginning with a small effort in mapping geologic hazards in the Los Angeles urban area and then moving to other regions of the state. Campbell saw the need to expand the division's efforts beyond the traditional areas of geology, and he encouraged the early programs in geophysics and geochemistry. One result of this is the state Bouguer anomaly map on a 1:250,000 scale geologic map base.

Campbell left the division, from September 1966 to February 1967, to serve as director of the State Department of Conservation, then returned to complete his career in state government with the Division of Mines and Geology. During the later years with the division he turned some of his energies to the California Academy of Science and even moved his office there after his retirement in 1969. He was a member of its board of trustees from 1960 to 1976, secretary from 1970 to 1971, and president from 1971 to 1976.

Campbell's influence ranged far and wide. As a teacher he helped train and nurture a generation of Caltech geologists, at the Mines Division he helped to bring the division to the forefront among state geological departments, as a professional geologist he was a leading authority on nonmetallic mineral deposits (especially magnesium sources), and he acted as a consultant to numerous mining, oil, and utility companies.

To honor his memory, Caltech established a graduate fellowship in petrology, and the American Geological Institute set up the Ian Campbell Scholarship Fund and the Ian Campbell Medal, to be awarded "in recognition of singular performance in and contribution to the profession of geology."

BIBLIOGRAPHY

I. ORIGINAL WORKS. Among Campbell's numerous papers on the Archean of the Grand Canyon area is "Types of Pegmatites in the Archean at Grand Canyon, Arizona," in *American Mineralogist*, **22** (1937), 436–445. On economic mineral deposits, see "Magnesium Metasomatism in Dolomite from Lucerne Valley, California," in International Geologic Congress, 18th, Great Britain, *Report*, pt. 3 (London, 1950), 118–124. His reports while state geologist include "Preparedness for Disaster—The Geologist's Role," in *California Division of Mines and Geology Mineral Information Service*, **18** (1965), 51–53. A more philosophical writing after retirement is "Search for the Philosopher's Stone," in *The Professional Geologist*, **9**, supp. (December 1972), 101–104.

II. SECONDARY LITERATURE. Richard H. Jahns, "Memorial to Ian Campbell, 1899–1978," in *Geological Society of America. Memorials*, **12** (1982), with selected bibliography; and Gordon B. Oakeshott, "Ian Campbell (1899–1978)," in *Bulletin of the American Association of Petroleum Geologists*, **63** (1979), 1977–1979.

WILLIAM R. BRICE

CARSON, RACHEL LOUISE (*b*. Springdale, Pennsylvania, 27 May 1907; *d*. Silver Spring, Maryland, 14 April 1964), *ecology, natural history, marine biology*.

Carson, the third and youngest child of Maria Frazier McLean Carson and Robert Warden Carson, grew up in semirural Pennsylvania. Her father was a moderately successful insurance agent who augmented his income by the piecemeal selling of the family farm. Despite financial difficulties she was able to attend Pennsylvania College for Women (now Chatham College) in Pittsburgh, graduating magna cum laude in English and biology in 1929. Zoology professor Mary Scott Skinker encouraged Carson to face the daunting prospects for a woman in professional zoology, and she entered graduate studies at the Johns Hopkins University after summer research at the Woods Hole Marine Biological Laboratory in Massachusetts. Her thesis research, under R. P. Cowles, was detailed microscopy of the embryological development of the kidney system in a catfish species. Morphology was standard training in 1930 even for the field of fisheries biology, which she wished to enter.

Financially constrained, Carson worked for a year as laboratory assistant to the geneticist Raymond Pearl, took a position in 1931 as assistant in zoology at the University of Maryland, and continued there after taking an M.A. in marine zoology in 1932 at Hopkins. With employment opportunities restricted and the need to care for her widowed mother, she began part-time writing for the U.S. Bureau of Fisheries public information projects in 1935. Thus, out of necessity Carson began her career as a science writer. When a civil service opening arose in 1936, she entered the bureau as a junior aquatic biologist under Elmer Higgins of the Division of Scientific Inquiry, who had requested her appointment to his office.

Carson wrote informative radio broadcasts, pamphlets, and reports on fisheries research and oceanography. To supplement her income she produced articles on the sea and marine ecology, including "Undersea" in *Atlantic Monthly* (1937), which led the popular science writer Hendrik Willem van Loon and Quincy Howe, editor in chief at Simon & Schuster, to encourage her to write *Under the Sea-Wind*, a natural history of sea life with a sharp ecological focus but also a touch of drama. It appeared to good reviews in 1941, but sales were low.

In 1940 Fisheries merged with the Bureau of Biological Survey to create the Fish and Wildlife Service; Carson moved into its information section. Respected for her analytical abilities and her literary skills, she advanced to editor in chief by 1949. Significant projects that she oversaw were the promotion of applied oceanography, conservation efforts, and the new wildlife refuge network. Although she sailed on research vessels, Carson's scientific work for the Fish and Wildlife Service was as a synthesizer of information, not a field research zoologist.

She pursued her writing career with the help of the Eugene F. Saxton Memorial Fellowship, which allowed her to take a leave of absence and write *The Sea Around Us*. Using recent research, she traced the origins and history of the oceans and explained the ecological relations of the physical environment to marine and human life. Chapters were serialized in the *New Yorker* and other magazines to much acclaim, and when the book appeared in 1951, it was an immediate best-seller. It was on the *New York Times* best-seller list for eighty-six weeks, joined by a reissue of *Sea-Wind*, and Carson became a literary celebrity. Among the awards, medals, and honorary doctorates following this book was a Guggenheim fellowship for another book, which allowed her to take a leave of absence in 1951 and finally resign a year later. Her subsequent financial success gave her the opportunity to travel and to do research on her next book, a guide and natural history of seashore life in the intertidal zone. *The Edge of the Sea* appeared in 1955 and also became a best-seller.

Shy and reserved, Carson did not relish the attention bestowed on her as a popular author, although she answered the many letters she received. Rumored to be a recluse, she lived a quiet life of scholarship with her mother and her grandnephew Roger Christie, whom she had adopted when he was orphaned in 1957. She stayed in touch with a close circle of friends, was active in the Audubon Society, and corresponded with a number of professional scientists and writers.

In January 1958 one of these correspondents, Olga Owens Huckins, brought to Carson's attention environmental problems caused by indiscriminate pesticide use. Plagued by poor health, and slowed by the death of her mother in 1958 and the removal of a cancerous tumor in 1960, Carson nonetheless uncovered in four years of research a remarkable amount of evidence, never before brought together, that profligate spraying had led to unexpected damage throughout the natural environment. Applying the same principles that had informed her previous work, she explained in *Silent Spring* how the ecologically interconnected "web of life" was affected by chemical disruption, including groundwater contamination, the concentration of residues in animals high in the food chain, and the death of birds. Pest control problems had increased even through the loss of natural enemies and the evolution of resistance.

The public controversy started in 1962, even before publication of the best-selling *Silent Spring* that same year, when the *New Yorker* serialized chapters. Not opposed to all pesticide use, Carson objected to the application of a technology without sensitivity to the natural balance maintained by the complex interactions of organism and environment, or to the loss of wild habitat and species in the drive to eradicate pests. Nonetheless, she was attacked strongly by spokespersons for the agricultural chemical industry and portrayed as a fanatic opposed to all pest control. In the unfavorable reviews, innuendo and claims of error outweighed actual refutations of her carefully documented argument. Her popularity and the lyrical quality of her writing were used to cast her as a nonprofessional, sentimental nature lover, a crank in a balance-of-nature cult. The debate that raged around her, as favorable reviews pointed out, was fundamentally between ecological theorists and economic entomologists more tied to simplistic agricultural practice than to the science of complex environmental relationships. Carson's ecological theory was not radical. She tied Charles Elton's standard views on food chains and interconnectivity to the holistic views on energy and materials flow of such ecologists as Aldo Leopold. If she stressed the harmony of nature, she also had a Darwinian view of the ceaseless struggle forcing adaptation both to the environment and among organisms. She was, more pointedly, successful in transmitting the paradigmatic ecological theory of the 1950's to a wide audience.

The debate reached national proportions, and Carson became a figure of headlines, editorials, and popular culture. In 1963 the President's Science Advisory Committee report on pesticides vindicated her argument, and federal environmental protection laws were one eventual outcome. In all her books, but especially in *Silent Spring*, Carson brought the relatively new science of ecology to the public's attention, and helped change the relations of applied science to government through public environmental policy. Carson died of cancer two years after the publication of *Silent Spring*.

BIBLIOGRAPHY

I. ORIGINAL WORKS. Carson's major works are *Under the Sea-Wind* (New York, 1941); *The Sea Around Us* (New York, 1951); *The Edge of the Sea* (Boston, 1955); and *Silent Spring* (Boston, 1962).

Her library and some of her correspondence are held at the Rachel Carson Council, Chevy Chase, Maryland. Correspondence and literary papers are in the Beinecke Rare Book and Manuscripts Library, Yale University.

II. SECONDARY LITERATURE. Numerous obituaries and biographical sketches have been published, including several biographies. The most definitive of these is Paul Brooks, *The House of Life: Rachel Carson at Work* (Boston, 1972). The pesticide controversy and some of its political repercussions are discussed in Frank Graham, Jr., *Since Silent Spring* (Boston, 1970).

WILLIAM C. KIMLER

CHADWICK, JAMES (*b*. Bollington, near Macclesfield, England, 20 October 1891; *d*. Cambridge, England, 24 July 1974), *physics*.

Chadwick was the son of J. J. Chadwick, who had a laundry business in Manchester, and of Ann Mary Knowles. After attending the Manchester Municipal Secondary School he won a scholarship to Manchester University, where he studied physics under Ernest O. Rutherford. He was awarded a first-class degree in 1911, then was accepted by Rutherford as a research student for the M.Sc. At this time the department of physics at Manchester was at its height, for besides Rutherford its staff included Hans Geiger, Ernest Marsden, Charles Galton Darwin, György Hevesy, and Henry G. J. Moseley, as well as, for a while, Niels Bohr. The

Rutherford and Bohr atoms both date from this period. In 1913 Chadwick went to work with Geiger in Berlin and was still there when war broke out the following year. He was interned until the end of the war in 1918.

Internment did not prevent Chadwick from pursuing scientific interests—he was even allowed to visit German scientific colleagues—but the materials available were basic and the literature nonexistent: the science was more of an aid to survival than anything else. In 1918 he returned to Manchester and a job with Rutherford, moving to Cambridge with him when he was appointed Cavendish Professor in 1919. In 1921 Chadwick was elected to a research fellowship at Gonville and Caius College, and the following year he was appointed assistant director of research under Rutherford at the Cavendish Laboratory, a post funded by the department of scientific and industrial research to take some of the load off Rutherford. For the next thirteen years Chadwick took day-to-day charge of all the research at what was then the leading laboratory in experimental atomic and nuclear physics. He also contributed significantly to this research, often in collaboration with others. Because of his administrative duties he had no teaching load. In 1925 he married Eileen Stewart-Brown; they had twin daughters.

Chadwick's relationship with Rutherford seems to have been generally very good, but in the early 1930's the development of nuclear physics brought with it the prospect of a quarrel. Chadwick believed that the cyclotron particle accelerator invented by Ernest Lawrence would rapidly become an essential tool for nuclear physics research, and he wanted one at Cambridge. Rutherford refused to have one. In 1935, deciding it was time to move, Chadwick accepted the Lyon Jones chair of physics at Liverpool University. Over the next few years he built up the physics department, which had virtually ceased to exist as a research center, with a cyclotron as its centerpiece.

When World War II broke out in 1939, Chadwick again found himself in Europe, but this time he was able to return to England. For the next four years he divided his attention between the university and government service, with the latter increasingly predominant. Late in 1943 he moved to the United States to take charge of the British part of the atomic bomb project. Chadwick returned to Liverpool in 1946 and resumed the work of building up the physics department. In 1948 he was offered the mastership of Gonville and Caius College, Cambridge, which he decided to accept. He seems to have felt that his debt to the college, which had been very kind to him when he first arrived at Cambridge, outweighed his preference for remaining active in physics. The decision may not have been a wise one, however, for college politics led to his resignation in 1958. He retired to North Wales but returned to Cambridge in 1969 to be near his daughters.

Chadwick's early research, assigned to him by Rutherford, was concerned with gamma-ray absorption: first with its use as a precision test of radium standards and then with applications of the method devised for standardization. He investigated the excitation of gamma rays by beta rays (electrons) and then by alpha rays (helium nuclei), the latter in collaboration with the radiochemist A. S. Russell. In both cases the excitation was confirmed. In Berlin with Geiger, Chadwick set out to determine by direct observation, using a primitive Geiger point counter, the relative intensities of the discrete lines observed by Rutherford and Robinson in radioactive beta-ray spectra. Although he was able to identify a few of the most intense of the observed lines, he also found a continuous spectrum alongside the discrete one. He tried changing the detection apparatus, but this merely confirmed the conclusion. The result came as a complete surprise and could not readily be explained theoretically, but it was a clear indication of Chadwick's experimental skill. Both spectra, and the relation between them, became an important problem in atomic and nuclear physics.

After moving with Rutherford to Cambridge, Chadwick resumed research begun before the war. He worked, as before, under Rutherford's direction, effectively supplying his own solutions to the master's problems. One of his first assignments was to use the determination of alpha-ray scattering probabilities to confirm van den Broek's hypothesis that the nuclear charge of an atom on the Rutherford-Bohr model was the same as the chemical atomic number. Employing an axially symmetric scattering arrangement and a much improved optical arrangement for the counting of deflected alpha-particle scintillations, Chadwick confirmed the hypothesis for platinum with an accuracy within 1 percent and for silver and copper with slightly less accuracy. In 1921, working with E. S. Bieler, he applied the same experimental arrangement to the study of the scattering of alpha particles by hydrogen in sheets of paraffin wax. Using hydrogen gas, Rutherford had already noted discrepancies between theory and experiment; the more sophisticated analysis of Chadwick and Bieler confirmed this, leading them to propose an asymmetric model of the alpha particle. The same experimental setup was used by Chadwick and P. H. Mercier for an analysis of beta-ray scattering.

Chadwick also collaborated during this period with C. D. Ellis, whom he had met in the internment camp in Germany, on a continuation of the analysis of radioactive beta spectra and with K. G. Emeléus on the cloud chamber analysis of alpha-particle collisions. His main research throughout the 1920's, however, was in direct collaboration with Rutherford. Following up Rutherford's discovery of the artificial transmutation of nuclei under alpha-ray bombardment (they called it artificial disintegration, thinking wrongly that the alpha particles were not absorbed), they demonstrated transmutations in a range of elements besides the nitrogen of the original experiment. Chadwick and Ellis investigated the properties of the disintegration particles, confirming that they were protons. After demonstrating the existence of disintegration particles moving in different directions, they used this to eliminate the effects of hydrogen contamination (which gave spurious protons) and thereby demonstrated transmutations in still more elements. When workers in Vienna claimed to have found transmutations of elements for which Rutherford and Chadwick had found no effect, including carbon and oxygen, Chadwick's experimental skill was called on, quite successfully, to uphold the Cambridge view. Other work with Rutherford in this period, on radioactively emitted alpha particles of unusually long range, was also done with a view to the Vienna group, who had reported that the particles didn't exist.

In the second half of the 1920's Rutherford and Chadwick turned to the problem of nuclear structure raised by the earlier experiments on alpha-particle scattering by hydrogen. In 1925 they first looked at scattering by a range of other elements: magnesium, aluminum, gold, and uranium. They then turned to helium scattering, in which scattered and scattering particles were identical (alpha particles are helium nuclei), so that there was only one nuclear structure to contend with. Once again they concluded that some asymmetry in the structure would be required. They were assuming, however, that the scattering predicted by quantum mechanics in this case was the same as that predicted by classical mechanics. In 1928 Nevill Mott showed that this was not true for identical particles, and in 1930 Chadwick showed that the results of helium scattering could in fact be interpreted by quantum mechanics without any need for asymmetries.

Apart from his work on beta spectra using Geiger point counters and his one foray into cloud chamber techniques, all of Chadwick's published research in the 1920's was based on scintillation counting (the optical observation of the scintillations produced when a proton or alpha particle hits a screen of zinc sulfide). This technique had its limitations, however, and by the end of the decade electrical techniques able to surpass it were coming into use. In 1928 Geiger and Walther Müller improved Geiger's earlier point counter to make what is generally known as a Geiger counter, a very sensitive detector of beta and gamma rays. The counter was rather unreliable, in that it was subject to spurious counts in the wake of genuine ones; but it could be used reliably for coincidence counting. Chadwick responded to the new invention by quickly building some for use in the Cavendish Laboratory. Meanwhile, H. Greinacher in Bern had succeeded in detecting individual alpha particles and protons by linearly amplifying the ionization currents produced by the particles in a small ionization chamber.

By 1928 Walther Bothe and Johannes Fränz in Berlin had applied the new technique to the study of the transmutation of boron, using a polonium source of alpha rays in place of the traditional radium active deposit. Rutherford's experience had always been that polonium alpha rays did not produce transmutations; but since the new counting technique was sensitive to background gamma radiation, the use of radium active deposits, with very high gamma-ray outputs, was ruled out. Bothe and Fränz's work showed that polonium alpha rays did indeed produce transmutations despite their very low energies, a phenomenon that was soon explained by the new quantum mechanics.

Under Chadwick's direction the new counting technique was quickly taken up and developed by C. E. Wynn-Williams and others at the Cavendish. In 1930 Chadwick, J. E. R. Constable, and E. C. Pollard used the electrical linear amplification of ionization currents and, for the first time, a polonium source to study the relationship between the energies of the incident alpha rays and emitted protons in nuclear transmutations. A year later Chadwick and Constable, with an improved polonium source and an improved ionization chamber, were able to give a detailed quantitative analysis of atomic transmutations.

Meanwhile, interest had been mounting in the production of gamma radiation under alpha-particle bombardment. Gamma rays were known to be emitted along with the radioactive alpha rays. Looking at the energy spectra of the alpha rays, George Gamow suggested in 1930 that when an alpha particle was emitted from a radioactive source with less than the maximum possible energy, a gamma-ray

CHADWICK

quantum would subsequently be emitted to restore the energy balance. It had become increasingly evident in the Cavendish experiments and elsewhere that the protons emitted from nuclear transmutations were not all of the same energy, and it was therefore natural to look for gamma rays in that context as well.

In 1930 Bothe and H. Becker detected penetrating radiation, assumed to be gamma rays, emitted when light elements were bombarded with polonium alpha rays. They also noted a surprising effect for beryllium: the intensity of the penetrating radiation from this element was nearly ten times that for any other element, and the radiation was exceptionally penetrating. Soon after, H. C. Webster, working under Chadwick's direction on the same subject, observed a similar phenomenon. In June 1931 Chadwick and Webster considered the possibility that the extreme penetrating radiation from beryllium might be not gamma rays, as was generally assumed, but neutrons.

The possible existence of a neutron, envisaged as a bound state of proton and electron, had been suggested by Rutherford in 1920, and over the intervening years there had been a number of attempts made in the Cavendish to detect such particles. Chadwick himself had looked for evidence of neutrons in hydrogen in 1923 and again, with the new Geiger counters, in 1928, and throughout all the work on nuclear transmutations the possibility of neutron emissions had been kept in mind. Beryllium in particular was seen as a promising source of neutrons, because it did not emit protons under alpha-ray bombardment and, through a false argument, because naturally occurring beryl was known to contain a great deal of helium: this suggested that under cosmic radiation the beryllium nucleus might split into two helium nuclei and one neutron. Chadwick had looked for neutrons from beryllium on and off for a number of years, and his interpretation of Webster's observation was a natural one. The energy of the extremely penetrating particles was related to their direction in a way that suggested they might be material particles rather than gamma rays, and their penetrating power suggested that if this was the case, they must be uncharged. Attempts to observe their passage through an ionization chamber failed, however, and the problem was put aside.

Early in 1932 Irène Joliot-Curie and Frédéric Joliot in Paris reported that the radiation from beryllium was even more penetrating than had been thought. They still assumed it to be gamma radiation; but when Chadwick read the report, he saw, as did Rutherford, that the energy arithmetic of the collisions producing it did not add up. By now Chadwick was convinced that the radiation must be something new and might well be neutrons. Using the ionization chamber and linear amplifier of his recent investigations, together with a new and improved polonium source, he investigated the effects of collisions between the penetrating rays and a range of various substances, measuring the energies of the recoil atoms in each case. He quickly showed that the results accorded completely with the theory that the penetrating radiation was composed of neutral particles of roughly the mass of the proton, and required implausible assumptions if they were supposed to be gamma rays. A short paper announcing the discovery of the neutron was submitted in February 1932. Detailed papers by Chadwick, by Norman Feather, and by Philip Dee, who used cloud chamber techniques to further analyze the neutron's properties, followed in May.

In 1933 Chadwick did some work with Patrick Blackett and Giuseppe Occhialini, who had just demonstrated the existence of the positron. The idea was that positrons might be produced in neutron interactions, but it transpired that the observed effects were in this case due to gamma rays. The team then concentrated on the quantitative analysis of the gamma-ray production of positrons. With D. Lea, Chadwick also conducted a search of the neutrino postulated by Wolfgang Pauli to account for the continuous spectra of beta rays first demonstrated by Chadwick. Unable to detect any particles, they showed, using a very-high-pressure ionization chamber, that if the neutrino did exist, it could not produce more than one ionization in 150 kilometers of air at normal pressure.

Chadwick's last major work before leaving Cambridge for Liverpool was with Maurice Goldhaber, who joined him as a personal assistant in 1934. Following up a suggestion of Goldhaber's, they demonstrated the nuclear photoelectric effect in the form of the disintegration of deuterium under gamma-ray illumination. This work also led to the first accurate figure of the mass of the neutron, and to speculation as to the significance of slow neutrons. It was not published, however, and a few months later Enrico Fermi observed and realized the significance of the same phenomenon. Following Fermi's work, Chadwick and Goldhaber investigated slow-neutron-induced transmutations of lithium, boron, and nitrogen. After moving to Liverpool in 1935, Chadwick did some further work on the photodisintegration of deuterium with N. Feather and E. Bretscher, although he concentrated his attention

CHADWICK

on the construction of a cyclotron and the building up of the physics department there. So far as scientific publications were concerned, his career was effectively over. He still had one major contribution to make as a scientist, however, and that was to the wartime atomic energy program.

Chadwick's first response to the discovery of fission was to reproach himself for not having done it himself earlier; he had studied uranium under slow neutron bombardment with Goldhaber, but in filtering out alpha-particle releases they had also filtered out any fission products that might have been present. Chadwick did not at first respond to fission with any experimental work of his own; but once G. P. Thomson, who did respond in this way, had alerted the authorities to the possibilities of a fission bomb, Chadwick was consulted. Like Thomson, he at first saw no real prospect of a bomb—the critical mass would be enormous, and the reaction would be too slow to go far before the uranium expanded to stop it. His having read Bohr and J. A. Wheeler's analysis, in which fission was attributed to the relatively rare isotope uranium 235, led him to decide, at the end of 1939, that the possibilities could not be completely dismissed and that more information was needed. Using the Liverpool cyclotron, he set out to obtain this information.

Following the memorandum of Otto Frisch and Ronald Peierls (April 1940), in which it was estimated that a bomb might be made with just a few pounds of pure uranium 235, Chadwick was made a member of the M.A.U.D. Committee on military use of uranium and took on the coordination of relevant scientific work at British universities. By the end of the year he was thoroughly involved in this work and convinced that the development of a bomb was inevitable. As work continued through the early years of the war, Chadwick played an increasingly major role in discussions. When the British finally decided to abandon efforts at a bomb project of their own and transfer their scientists to the American project, Chadwick was appointed technical adviser to the British representatives on the Combined Policy Committee—the only scientist in the British group to have full access to all project information. The British had wanted Wallace Akers, who had been in charge of their project, to hold this post, but the Americans were suspicious of his commercial connections (he was seconded from Imperial Chemical Industries). Chadwick commanded the highest respect as a scientist and, a naturally discreet man, was completely trusted. He also had exceptional diplomatic skills.

Chadwick's abilities as a scientist and diplomat ensured that the Anglo-American collaboration proceeded well. And although he could not always stop British politicians and civil servants from upsetting the Americans, his sober advice prevailed sufficiently for the latter not to give up the joint effort. Even after the war in Europe ended, Chadwick insisted that the British put all their effort into the American project. Although the British came out of the war with less information than they would have liked, what they had, they owed substantially to Chadwick.

At the end of the war an exhausted Chadwick let it be known that he was not interested in the post of director of the planned Atomic Energy Research Establishment at Harwell, preferring to return to university life. He continued, however, to play a major consultative role in the British atomic energy program. Following up an earlier concern with medical physics, he was instrumental in the establishment of the Radiochemical Centre at Amersham for the production of radioisotopes.

Chadwick won the Nobel Prize for physics in 1935. He was knighted in 1945 and made a companion of honor in 1970. Chadwick was elected a fellow of the Royal Society in 1927; he was awarded its Hughes (1932) and Copley (1950) medals, and served as a vice president in the year 1948–1949. He also received a wide range of other scientific honors and awards.

BIBLIOGRAPHY

I. Original Works. A bibliography of Chadwick's writings is in the obituary by Massey and Feather (see below). His discovery of the neutron was reported in "Possible Existence of a Neutron," in *Nature*, **129** (1932), 312, and "The Existence of a Neutron," in *Proceedings of the Royal Society*, **A136** (1932), 692–708.

A substantial collection of Chadwick's papers and correspondence is in the archives of Churchill College, Cambridge, which also has a transcript of an interview with Chadwick conducted by C. Weiner in 1969.

II. Secondary Literature. The principal published source of information on Chadwick's life is Sir Harrie Massey and Norman Feather, "James Chadwick," in *Biographical Memoirs of Fellows of the Royal Society*, **22** (1976), 11–70. Articles dealing with the discovery of the neutron and other aspects of Chadwick's work are collected in John Hendry, ed., *Cambridge Physics in the Thirties* (Bristol, 1984), which also contains an extensive bibliography of related secondary literature; see also Norman Feather, "Chadwick's Neutron," in *Contemporary Physics*, **15** (1974), 565–572. Chadwick's wartime career is documented in Margaret M. Gowing, *Britain* *and Atomic Energy, 1939–1945* (London and New York, 1964), and *Independence and Deterrence* (London and New York, 1974).

JOHN HENDRY

CHAIN, ERNST BORIS (*b.* Berlin, Germany, 19 June 1906; *d.* Mulranny, Ireland, 12 August 1979), *biochemistry.*

Chain was the son of Russian-born Michael Chain, an industrialist, and Margarete Eisner, both of Jewish descent. The Chains also had a daughter, Hedwig. Michael Chain studied chemistry in Berlin, where he later established a firm that produced sulfates. Ernst Chain developed an early interest in chemistry, most likely because of this close industrial connection. Chain also had a strong interest in a career in music; he was an accomplished pianist and music critic for a Berlin paper.

Chain graduated in chemistry and physiology from the Friedrich-Wilhelm University in 1930 and received a D.Phil. degree from the chemical department of the Institute of Pathology at the Charité Hospital for his work on enzymes. Three months after Hitler assumed the chancellorship of Germany in January 1933, Chain immigrated to England, alone, and became a naturalized citizen six years later. His mother and sister remained in Germany, presumably for financial reasons, since the family had fallen into debt after Michael Chain's death in 1919. Chain tried to assist them, although his own financial circumstances in England were difficult. Margarete and Hedwig eventually died in the Nazi holocaust. Chain began work under Frederick Gowland Hopkins in the department of biochemistry at Cambridge in October 1933 and received a Ph.D. for research on phospholipids. In 1935 Howard W. Florey, professor of pathology at Oxford, invited Chain (on Hopkins' recommendation) to organize a biochemical section in the Sir William Dunn School of Pathology.

Chain pursued a wide range of research topics during his career, from general enzymology during his early days in Berlin and England, to phytotoxins, carbohydrate metabolism, fermentation production of ergot alkaloids, and other subjects later in life. His most significant work—the research for which he received a Nobel Prize in medicine with Florey and bacteriologist Alexander Fleming in 1945—concerned penicillin. In the late 1930's, Florey and Chain began their investigation of this mold broth filtrate from *Penicillium notatum* as part of their general interest in systemic antibacterial agents. Fleming had isolated penicillin in 1928. He found it was a promising external antiseptic against some pathogenic bacteria, but its chemical instability (among other reasons) soon led him to abandon penicillin as a therapeutic agent.

By May 1940 Chain had produced enough penicillin for Florey to conduct tests on mice infected with deadly hemolytic streptococci. These early tests yielded very promising results. Chain, in collaboration with Edward Abraham, spent the rest of the year improving the extraction and purification of penicillin for clinical tests, which began early in 1941.

Serious problems with the fermentation production of penicillin led the British and American governments, by early 1944, to coordinate hundreds of scientists for the development of a commercially feasible synthesis of penicillin. Chain and his collaborators at Oxford, however, had shed much light on the chemistry of the drug by this time. They had determined the structures and syntheses of the two compounds produced by the acid hydrolysis of the penicillin molecule. Also, they had discovered the structure of the intermediate compound between these two degradation products and penicillin. By October 1943, they proposed what turned out to be the correct structure of penicillin. That the massive penicillin project failed to produce a practical synthesis of the drug was academic, because workers succeeded in producing penicillin by fermentation, on a mass scale. Yet this development did not diminish the significance of the contributions of Chain and his colleagues. They established a fundamental understanding of the chemistry of penicillin. This aided fermentation work on the drug in the 1940's and served as the foundation for the development of semisynthetic penicillins from the mid 1950's on.

In 1948 Chain left Oxford to organize the International Centre for Chemical Microbiology at the Istituto Superiore di Sanità in Rome. Before moving, he married biochemist Anne Beloff, with whom he collaborated on many projects during the remainder of his career. (They had two sons and a daughter.) The center developed into a training ground for researchers interested in production of antibiotics. Thus, a microbiologist and a biochemist from the British pharmaceutical firm Beecham spent a year at the center (1956–1957) to learn how to produce penicillin. Chain, who was a consultant to Beecham, had the visiting scientists acquaint themselves with the production of a penicillin that he believed would lend itself to chemical modification, so that new penicillins could be tailored to meet specific therapeutic needs. Chain's later involvement with this work is not clear, but by 1957 Beecham scientists in England identified the nucleus of the penicillin molecule in their fermentation broth. Beecham soon learned how to produce this compound in large quantities. The ability to produce the penicillin nucleus on a mass scale, coupled with the earlier discovery, by chemist John C. Sheehan at MIT, of a method to add chemical side chains to this nucleus, created the framework for a new era in the history of penicillin—the era of semisynthetic penicillins.

Chain contributed to a number of fields in addition to penicillin research. Soon after arriving in Rome, he and his colleagues at the center began a project on carbohydrate metabolism. This research illuminated the role of insulin in some metabolic pathways in various tissues. Chain also advanced fermentation methods to produce therapeutically useful alkaloids from the ergot fungus in high yield.

Chain continued this work and initiated some new projects after he moved back to England in 1964 to become chairman of the department of biochemistry at Imperial College of Science and Technology in London. He and associates in the biochemistry and chemistry departments investigated the action of fusicoccins, plant toxins that cause wilting, and they established the chemical structures of several of these toxins. Chain also helped clarify the biochemistry of the myocardium, particularly in cases where the heart was deprived of oxygen, as in cardiac arrest. He and his colleagues in the biochemistry department found that unless the arrested heart was perfused with a glucose solution, other cellular energy reserves, such as adenosine triphosphate and creatine phosphate, would be depleted to levels that would hamper the heart's recovery. In the 1970's Chain and a group of chemists and biochemists investigated a new antibiotic from the soil microorganism *Pseudomonas fluorescens,* which they called pseudomonic acid. The antibiotic was active against several pathogens, but it was too toxic for clinical application.

Chain was active in many fields, but his greatest contribution to science and medicine was the introduction and development of penicillin. For this work he received many honorary degrees, prestigious scientific prizes, and a knighthood.

BIBLIOGRAPHY

I. Original Works. On the contributions of Chain and his Oxford colleagues to the introduction of penicillin, see "Penicillin as a Chemotherapeutic Agent," in *Lancet* (1940), **2**, 226–228, with H. W. Florey *et al.*; "Further Observations on Penicillin," *ibid.* (1941), **2**, 177–189, with E. P. Abraham *et al.*; "Purification and Some Physical and Chemical Properties of Penicillin," in *British Journal of Experimental Pathology*, **23** (1942), 103–115, with E. P. Abraham; and "The Earlier Investigations Relating to 2-Pentenylpenicillin," in Hans T. Clarke, John R. Johnson, and Robert Robinson, eds., *The Chemistry of Penicillin* (Princeton, 1949), 10–37, with E. P. Abraham *et al.* Chain coauthored several papers with scientists from Beecham Laboratories on the isolation of the nucleus of penicillin and its conversion to new penicillins, including "Penicillin Derivatives of *p*-Aminobenzylpenicillin," in *Nature*, **183** (1959), 180–181, with A. Ballio *et al.*; and a series of six articles in *Proceedings of the Royal Society of London*, **B154** (1961), 478–531.

Among Chain's principal publications on carbohydrate metabolism and the action of insulin are two series of papers that he published with his associates at the Istituto Superiore di Sanità in *Selected Scientific Papers from the Istituto Superiore di Sanità*, **1** (1956), 293–535, and **2** (1959), 109–149; and "Recent Studies on Carbohydrate Metabolism," in *British Medical Journal* (1959), **2**, 709–719. He reported his work on the production of ergot alkaloids in "Production of a New Lysergic Acid Derivative in Submerged Culture by a Strain of *Claviceps paspali* Stevens & Hall," in *Proceedings of the Royal Society of London*, **B155** (1962), 26–54, with F. Arcamone *et al.*; and "Biosynthesis of Ergotamine by *Claviceps purpurea* (Fr.) Tul.," in *Biochemical Journal*, **134** (1973), 1–10, with R. A. Bassett and K. Corbett.

Chain's research on the production and chemical resolution of phytotoxins from *Fusicoccum* appeared in "Fusicoccin: A New Wilting Toxin Produced by *Fusicoccum amygdali* Del.," in *Nature*, **203** (1964), 297, with A. Ballio *et al.*; and a series of four articles he published with his colleagues in chemistry and biochemistry at Imperial College in *Journal of the Chemical Society: Section C* (1971), 1259–1274; and *Journal of the Chemical Society: Perkin Transactions I* (1973), 1590–1599, and (1975), 877–883.

Signal publications by Chain on cardiac metabolism include "Effects of Insulin on the Pattern of Glucose Metabolism in the Perfused Working and Langendorff Heart of Normal and Insulin-Deficient Rats," in *Biochemical Journal*, **115** (1969), 537–546, with K. R. L. Mansford and L. H. Opie; "The Role of Glucose in the Survival and 'Recovery' of the Anoxic Isolated Perfused Rat Heart," *ibid.*, **128** (1972), 1125–1133, with D. J. Hearse; and "Recovery from Cardiac Bypass and Elective Cardiac Arrest: The Metabolic Consequences of Various Cardioplegic Procedures in the Isolated Rat Heart," in *Circulation Research*, **35** (1974), 448–457, with David J. Hearse and David A. Stewart.

On Chain's pseudomonic acid research, see "Pseudomonic Acid: An Antibiotic Produced by *Pseudomonas fluorescens*," in *Nature*, **234** (1971), 416–417, with A. T. Fuller *et al.*; and a series of three papers that he published with co-workers from Imperial College in *Journal of the Chemical Society: Perkin Transactions I* (1977), 294–322.

Also useful are "Penicillinase-Resistant Penicillins and the Problem of the Penicillin-Resistant Staphylococci," in Anthony V. S. de Rueck and Margaret Cameron, eds.,

Resistance of Bacteria to the Penicillins (Boston, 1962), 3–24; and "A Short History of the Penicillin Discovery from Fleming's Early Observations in 1929 to the Present Time," in John Parascandola, ed., *The History of Antibiotics: A Symposium* (Madison, Wis., 1980), 15–29.

II. SECONDARY LITERATURE. The most comprehensive biography of Chain is Ronald W. Clark, *The Life of Ernst Chain: Penicillin and Beyond* (New York, 1985). More useful on Chain's scientific work is Edward Abraham, "Ernst Boris Chain," in *Biographical Memoirs of Fellows of the Royal Society*, **29** (1983), 43–91. Several sources discuss Chain's work on penicillin, the most detailed of which is Howard W. Florey *et al.*, *Antibiotics*, 2 vols. (London, 1949). Also useful is John Patrick Swann, "The Search for Synthetic Penicillin During World War II," *British Journal for the History of Science* **16** (1983), 154–190. John C. Sheehan, *The Enchanted Ring: The Untold Story of Penicillin* (Cambridge, Mass., 1982), discusses the emergence of the semisynthetic penicillins.

JOHN PATRICK SWANN

CHANEY, RALPH WORKS (*b.* Brainerd, Illinois, 24 August 1890; *d.* Berkeley, California, 3 March 1971), *paleobotany, paleoecology, conservation.*

Chaney was the son of Fred A. Chaney and Laura J. Works Chaney. He received the B.S. in geology in 1912 and the Ph.D. in geology in 1919, both from the University of Chicago. On 1 June 1917 he married Marguerite Seeley; they had two sons and a daughter. Chaney was president of the Paleontological Society of America in 1939; he also was a fellow of the Geological Society of America and served as its vice president in 1940. He received an honorary D.Sc. from the University of Oregon in 1944. In 1947 Chaney was elected to the National Academy of Sciences and to the American Philosophical Society. He was a member of the Paleontological Society of Japan and the Botanical Society of Japan. In 1969 he was an honorary vice president of the Eleventh International Botanical Congress, which awarded him a Congress Medal. In 1970 he received the fifth Paleontological Society Medal from the Paleontological Society of America.

Partly to finance his field trips while he was a graduate student (1914–1917), Chaney taught science courses at the Francis W. Parker School in Chicago. From 1917 to 1922 he taught geology at the University of Iowa. In 1918 Chaney met John C. Merriam, to whom he had sent a draft of his dissertation on the Miocene Eagle Creek flora of the Columbia River gorge. Merriam was impressed by the paleoecological extensions Chaney made in his study, and in 1920 invited him to join his team at the University of California, Berkeley, to investigate the rich fossil-bearing Tertiary John Day beds of Oregon.

In 1922, through the influence of Merriam, who was then president of the Carnegie Institution of Washington, Chaney was appointed research associate; he maintained his affiliation with that institution until he retired in 1957. In 1931 Chaney became professor of paleobotany and chairman of the paleontology department at the University of California at Berkeley. He was also curator of paleobotany at the Museum of Paleontology. From 1943 to 1945 Chaney served as assistant director of the Lawrence Radiation Laboratory.

The eminent Johns Hopkins paleobotanist Edward W. Berry, whom Chaney had met in 1916, wrote in the 1918 Annual Report of the Smithsonian Institution that paleobotany was not merely the history of "the endless succession of plants which have inhabited the earth since life first came into existence, but it aims to understand and interpret these in terms of the evolution of individuals and of floras, their interactions with an ever-changing environment, and the transmutation of these facts into terms of ancient geography, topography, rainfall, temperature, and distribution" (p. 289–290). In the same philosophical vein, Chaney's approach to the study of fossil plants was a holistic one; he was not concerned merely with taxonomy and the ages of fossils but with whole assemblages in relation to their environment. The strength of his contributions, therefore, lay in paleoecology rather than in systematics. In following this course of research, Chaney was influenced by the ecologist Frederic E. Clements.

In 1917 Merriam, with Henry Fairfield Osborn and Madison Grant, founded the Save-the-Redwoods League. Chaney became associated with the League in the late 1920's, and from 1961 to 1971 he served as its president. Besides his conservation efforts for the redwoods—the League preserved tens of thousands of acres as parks—Chaney was involved with the National Park Service, serving on its Advisory Board from 1943 to 1954.

Chaney published more than 150 papers, writing well into his seventies. He has been lauded by his students and colleagues as a wise and understanding teacher with a sense of humor. In the gardens of his home in the Berkeley Hills he planted, by epochs, assemblages of many of the Tertiary floras that had survived only in Asia, thus providing his students with live representations of Tertiary vegetation. In his research Chaney emphasized the recognition of similarities rather than differences between fossil plant species and between fossil plants and living species. His comparison (1925) of the Tertiary Bridge

CHANEY

Creek flora of the John Day basin with the modern redwood forest of California enabled him to interpret living conditions of the Bridge Creek in terms of the modern forest. He also showed (1925) that the Mascall assemblage so closely resembled the modern oak-madrona forest of the west that he felt the rainfall of the Mascall epoch must have been essentially the same, thirty inches annually.

Chaney traveled extensively in search of similar assemblages, both fossil and living. He was a member of the Roy Chapman Andrews dinosaur-collecting expedition to Mongolia in 1925. In 1937 Chaney worked with H. H. Hu on the Shangwang flora of Shantung Province in northeastern China. In 1948 he visited western Hupeh Province to see a living stand of *Metasequoia*, a species of deciduous conifer that had been believed extinct. The discovery of the living *Mestasequoia* led him to undertake a revision of several members of the Taxo-diaceae. In the 1960's, under a project funded by the National Science Foundation, Chaney coordinated paleobotanical studies of the Tertiary floras of Japan.

His environmental approach to the study of Tertiary flora, involving the search for similarities, required the collection of vast quantities of specimens so that statistical studies could be made; the studies yielded such information as species dominance. Chaney also hypothesized on the transport and deposition of fossil plants by examining the transport of plant parts in modern streams of California. Harvard botanists Irving W. Bailey and Edmund W. Sinnott introduced (1915) the idea of studying fossil floras on the basis of the morphological character of leaves; this offered a basis for gauging past environments independent of phylogenetic considerations. Chaney extended and applied the method to the study of the early Tertiary Goshen flora of west-central Oregon. His 1933 monograph, written with Ethel I. Sanborn, showed that the Goshen flora indicated an environment typical of low-latitude subtropical forests such as those in Mexico and Central America today.

In 1936 Chaney introduced the idea that plant distribution can be used as a guide to age determination, since changes in it are caused by changes in climatic and other environmental conditions through time. Like most geologists working in North America in the 1930's and 1940's, Chaney denied the possibility of continental drift, although he established a very close tie between the modern vegetation of eastern Asia and the fossil vegetation of modern North America.

Chaney coined the term "geoflora" for groups of land plants in mass migration. His Arcto-Tertiary geoflora, for example, shifted its distribution from Alaska southward. The Neotropical-Tertiary geoflora then moved from Washington and Oregon into Mexico and Central America. His concepts of geofloras, plant migrations, and paleoclimates were original and highly attractive because they were presented in global terms, but they have been criticized as simplistic.

Later developments in angiosperm paleobotany have shown that mosaic evolution takes place, that various organs of a species of plant do not evolve together at the same rate, so that a fossil leaf may be similar to a modern species but other organs of the plant may be quite different, making some of Chaney's assignment of systematic affinities erroneous and thereby invalidating some of his conclusions.

Chaney's ecological approach influenced a school of thought, and he was widely recognized as the most distinguished Tertiary paleobotanist. The focus of paleobotany today, with the aid of new techniques not available to Chaney, centers more on the details of plant anatomy than on the broad concepts of Chaney. Equipped with more complete knowledge of individual plants, however, paleobotanists could well return to Chaney's imaginative approach in forming paleoecological syntheses.

BIBLIOGRAPHY

I. ORIGINAL WORKS. Chaney's professional correspondence and manuscripts (1917–1966) are in the Special Collections of the University of Oregon Library, Eugene. His papers are listed in the bibliographies of Jane Gray, "Ralph Works Chaney," in *Biographical Memoirs. National Academy of Sciences*, **55** (1985), 135–161; and Jane Gray and Daniel I. Axelrod, "Memorial to Ralph Works Chaney 1890–1971," in *Memorials of the Geological Society of America*, **3** (1974), 60–68. Writings not listed in these articles are *Redwoods of the Past* (Berkeley, Calif., 2nd ed., 1937) and "Paleobotany," in *McGraw-Hill Encyclopedia of Science and Technology*, IX (New York, 1982), 752–760.

II. SECONDARY LITERATURE. Besides the memorial articles mentioned above, see Daniel I. Axelrod, "Ralph Works Chaney (1890–1971)," in *Year Book of the American Philosophical Society* for 1971 (1972), 115–120; and "Ralph Works Chaney," in *McGraw-Hill Modern Scientists and Engineers*, I (New York, 1980), 194–195. Personal communications from the following individuals were of great value in assessing Chaney's contributions: Harlan P. Banks, Theodore Delevoryas, David L. Dilcher, and Jack A. Wolfe.

ELLEN T. DRAKE

CHAPMAN, FRANK MICHLER (*b*. West Englewood, New Jersey, 12 June 1864; *d*. New York, New York, 15 November 1945), *ornithology, conservation*.

Chapman was born into a gifted, well-to-do family. His father, Lebbens Chapman, was senior member of a New York law firm, and his mother, Mary Augusta Parkhurst, came from a family of physicians; both were of English descent. The family home was in the midst of a large and prosperous farm, in a region at that time abounding in woods, ponds, orchards, and wide fields. Endowed with his mother's musical talent, Chapman was perhaps inevitably to become an ardent bird lover in these favorable surroundings. After graduating at age sixteen from the Englewood Academy, he chose not to go to college but to work as a bank clerk (his father having died four years earlier) and to pursue his nature studies as an amateur. Early in 1884 he took part as a volunteer in a survey of the spring bird migrations organized by the American Ornithologists' Union and the United States Biological Survey. Despite the demands of his job, Chapman spent an average of two and a half hours in the field on each of sixty-nine days of a seventy-five-day period (getting up at daybreak). A. K. Fisher, who was in charge of the project, declared Chapman's report to be the best for eastern North America.

Two years later Chapman resigned from the bank to devote himself entirely to bird study. He assembled a fine collection of Florida birds, having learned in the meantime how to make bird skins, and started to work at the American Museum of Natural History as a volunteer. He impressed the curator of mammals and birds at the museum, J. A. Allen, so favorably that Allen offered him (1 March 1888) an assistantship at the princely salary of $50 a month. "[I had only] a beginner's knowledge of local birds," said Chapman later, "and had everything to learn concerning the more technical side of ornithology. There was no one under whom I could have worked and studied more profitably [than Dr. Allen]." By 1908 Chapman became curator, and when the department was divided in 1920, he became chairman of the bird department, where he remained until his retirement on 30 June 1942.

Chapman married Fannie Bates Embury on 24 February 1898. A widow with four children, she often assisted her husband on his expeditions. Together they had one son.

We have an excellent sketch of Chapman's personality from R. C. Murphy (1950), but his character can also be sensed by reading between the lines in his autobiography. Versatility, enthusiasm, warmth, and intelligence were among his strengths. A weakness for people with money and titles, occasional aberrations of judgment, and an extraordinarily one-sided anglophilia were among his foibles.

During the time he chaired the department, Chapman made it the best bird department in the world. Each of the curators he appointed became the leading world authority in his specialty: R. C. Murphy (sea birds), James P. Chapin (birds of Africa), John T. Zimmer (birds of South America), and Ernst Mayr (birds of Australia and the South Seas). Simultaneously Chapman built up the collection to the second largest in the world. His greatest acquisition was the Rothschild collection in 1932, with 280,000 specimens.

Chapman entered enthusiastically into all aspects of a museum career. He completely revitalized the exhibits and was one of the originators of habitat groups. A great friend of bird artists such as Louis Agassiz Fuertes and Francis Lee Jaques, he did much for bird art. He, more than any of his contemporaries, was responsible for the spread of bird study in the United States by his numerous popular books and by founding in 1899 the journal *Bird-Lore* (now *Audubon Magazine*), of which he was editor until 1934. He was one of America's first bird photographers, and the choicest products of his harvest of bird portraits were published in two books in which he recounted his adventures as a photographer in many lands.

As a scientific ornithologist Chapman had a much broader vision than some of his contemporaries at other museums, who seemed to be interested mainly in the "making" of new genera and subspecies. He was always vitally interested in the living bird, its habits and distribution. His essays on the biogeography and ecology of the birds of Colombia (1917), Ecuador (1926), and the Venezuelan highlands (1929–1931) were pioneering contributions in which he not only successfully classified South American birds into ecological associations but also advanced bold hypotheses on the history of South American birds. Chapman concentrated on the vertical zonation of these faunas and asked how the biota of the higher altitudes had evolved, what connections had formerly existed between the faunas of the temperate southern parts of the continent and those of the temperate zones at the higher altitudes (paramo) of the Andes, and, finally, how the faunas of the isolated mountains had originated and evolved. (By contrast, the faunas of the subtropical and temperate humid zones seemed to have originated from the humid tropical zone.) By his innovative researches Chapman had a great impact on the history of South American biogeography. To obtain the material for these zoogeographic researches he undertook a series of expeditions to South America, beginning in 1911 and continuing over many years. In addition he sent a number of collectors to Peru, Ecuador, and Colombia to enrich the material on which he based his conclusions.

When he was well into his sixties, Chapman astounded everyone by starting on an entirely new line of research, the detailed life history of tropical bird species. He spent many winters at Barro Colorado Island in Gatun Lake, Panama, which resulted in two volumes of natural history essays (1929, 1938) and several technical monographs on the habits of certain tropical bird species, such as Wagler's oropendola (1928) and Gould's manakin (1935).

Chapman was deeply interested in geographic variation and speciation. In his study of polymorphism in birds he was, like Stresemann, influenced by Hugo de Vries, and he considered such "mutations" as something very different from gradual variation, which he ascribed in the 1920's to "observable environmental factors (chiefly climatic) on the species." Like most other naturalists, he retained a belief in an inheritance of acquired characters.

Chapman's contributions to biogeography, particularly to vertical zonation, were pioneering. In the museum he championed more than any contemporary the bird as a living creature with its own behavior and environment. He played an important role in the popularization of bird study, became one of the leaders of the conservation movement in America, and, finally, built a bird department with a balanced collection and an outstanding staff. He thus made a major contribution toward raising American ornithology to its position of leadership.

BIBLIOGRAPHY

I. Original Works. Chapman's fourteen books include *Handbook of Birds of Eastern North America* (New York, 1895); *Camps and Cruises of an Ornithologist* (New York, 1908); *The Distribution of Bird-life in Colombia* (New York, 1917); *The Distribution of Bird-life in Ecuador* (New York, 1926); *Autobiography of a Bird-lover* (New York, 1933); and *Life in an Air Castle* (New York, 1938). He published about two hundred articles, many in popular journals, but most devoted to the description of new species and subspecies of birds and to reports on collections made on expeditions.

II. Secondary Literature. Obituaries were published by William K. Gregory in *Biographical Memoirs. National Academy of Sciences*, **25** (1949), 111–145, with bibliography; and by Robert Cushman Murphy in *Auk*, **67** (1950), 307–315, with bibliography listing eleven other biographies and obituaries.

Ernst Mayr

CHAPMAN, SYDNEY (*b*. Eccles, near Manchester, England, 29 January 1888; *d*. Boulder, Colorado, 16 June 1970), *geophysics, natural philosophy*.

Chapman was the second son of Joseph Chapman, chief cashier of a textiles firm, and of his wife, Sarah Gray. Their family was a nonconformist one of strict principles, but Sydney's attitude to religion grew steadily more relaxed with time. In 1922 he married Katharine Nora Steinthal; they had three sons and one daughter. He enjoyed good health, partly because of daily exercise (cycling, swimming, walking). He was elected fellow of the Royal Society of London in 1919.

Initially Chapman's father did not envisage his son as a candidate for higher education. However, aided by the advice of friends who recognized his ability, and by scholarships, Sydney's horizons expanded until he attained an M.Sc. in mathematics at Manchester University (1908) and a first class in the mathematical tripos at Cambridge University (1911). Even before he had completed the residence requirements for the tripos, he began research; his first paper was in pure mathematics, although Joseph Larmor suggested as a suitable subject one of the unsolved problems of gas theory.

What proved decisive for Chapman's future research, however, was his acceptance of an offer by Frank Dyson, the astronomer royal, of a post as senior assistant at the Greenwich Observatory. There Dyson set him to supervise the installation of new instruments in the magnetic observatory. Chapman noted that "magneticians," though assiduous in collecting data about variations in the geomagnetic field, spent little time seeking their interpretation. Encouraged by Dyson and Arthur Schuster, he set to work to rectify this. Thus began his lifelong study of the influences of the sun and the moon on terrestrial phenomena.

Chapman enjoyed meeting the distinguished people who came to visit Greenwich but found the observatory routine restrictive. Thus in 1914 he left Greenwich and returned to Cambridge as a college lecturer. During World War I his religious principles made him a pacifist. He was given exemption from military service but was asked to help the depleted Greenwich staff (1916–1918).

Despite his moves, the years 1912 to 1919 were a period of intense scientific activity for Chapman.

In 1917 he, and independently David Enskog in Sweden, gave a solution to the hitherto unsolved problems of the viscosity, thermal conductivity, and diffusion of gases. They identified the hitherto overlooked phenomenon of thermal diffusion, about which Chapman always showed a paternal pride.

In four massive papers (1913–1919) Chapman (following Schuster) examined the regular variations in the geomagnetic field arising from tidal flows in the ionosphere. He gave an improved estimate of the total conductivity of the ionosphere that depended on properties of the ionosphere not yet accessible to observation. He also showed that the lunar geomagnetic tide depends on the variable ionizing effect of solar radiation. In 1918 Chapman isolated a lunar atmospheric tide at Greenwich from sixty years of barometric records, a tide that George Airy had been unable to identify because of its smallness. He followed this up during the next thirty years with determinations of lunar tides at numerous stations scattered over the surface of the globe.

Also in 1918 Chapman made a first attempt at a theory of magnetic storms, which are sudden, irregular changes in the geomagnetic field followed by a slower recovery pattern. The theory was soon abandoned as unsound, but the preliminary analysis of storm morphology was to prove invaluable.

In 1919 Chapman was appointed to succeed a former teacher, Horace Lamb, as professor of mathematics at Manchester. He was there until 1924, when he accepted an invitation to become chief professor of mathematics at Imperial College, London. He was succeeded at Manchester by E. Arthur Milne, who had worked with him on upper-atmosphere problems at Cambridge.

After 1920 Chapman's next period of intense research activity was in the years 1928 to 1932. In 1922 and 1928 he extended his gas-theory methods to estimate the electrical conductivity of plasmas. In 1930 he joined with a junior colleague, Albert T. Price, to study what could be inferred about the earth's interior from surface magnetic variations. With a research student, V. C. A. Ferraro, he produced the first satisfactory theory of the initial phase of magnetic storms (1931)—that they were due to compression of the geomagnetic field by plasma streams emanating from the sun. The Chapman-Ferraro theory explained only one feature of magnetic storms; it treated neither the later phases nor the auroras that regularly accompany the storms. For this reason it was strongly criticized in 1939 and 1940 by Hannes Alfvén, who produced a rival theory that was especially concerned with auroras. The most significant feature of the Chapman-Ferraro theory was its recognition, twenty years before the postwar explosion of interest in plasmas, that the behavior of plasmas is essentially different from that of single charged particles.

In the 1931 Bakerian Lecture to the Royal Society, Chapman gave a trailbreaking discussion of the effects of solar ultraviolet radiation on the earth's upper atmosphere. He provided a standard theory of ionized layer formation in the lower ionosphere with which later observations could be compared and gave an enlightening account of photochemical reactions in the upper atmosphere. At that time experimental data were at best only partially available; his aim was to point the way to the observers.

In the years 1933 to 1940 Chapman, always an internationalist, made no secret of his distaste for Hitler and did all he could to get refugee scientists settled in suitable posts. He maintained friendly relations with German scientists, however, and with one of them, Julius Bartels, he worked sporadically, beginning in 1929, on a treatise called *Geomagnetism* that was published in 1940. A second book, on which Chapman had begun work much earlier, *The Mathematical Theory of Non-uniform Gases*, had been published in 1939, with T. G. Cowling as coauthor. This book expounded the 1917 gas theory of Chapman and Enskog along with later developments, especially some for which David Burnett was responsible. Both books became standard texts.

In 1933 Chapman and Ferraro completed the presentation of their theory, and in 1937 Chapman drew his work on atmospheric tides to an end. In 1938, with E. Harry Vestine, he calculated the current system of geomagnetic disturbance, on the (inexact) assumption that the currents flow only in the ionosphere. This, and work on the books, left him time only for minor pieces of work before the war. During the war, no longer a pacifist, he undertook civilian war work, ultimately (1943–1945) working for the Army Council on problems of military operational research.

In 1946 Chapman became Sedleian professor of natural philosophy at Oxford. He did not appreciate the relatively secondary status he found allocated to science there and sought to improve that status by giving general science lectures to nonscientists. He also supervised research students who were later to make their mark, including Franz D. Kahn and K. C. Westfold. His status, however, was becoming that of elder statesman and counselor rather than that of producer of exciting new ideas.

Determined not to let himself be retired when he reached the official age, in 1953 Chapman resigned his Oxford chair and took up a research post in

Alaska. In 1955 he added a similar post at the High Altitude Observatory in Boulder, Colorado, sharing his time between these two and institutes throughout the world. He also made short visits to secure the cooperation of groups in many countries in the work of the International Geophysical Year (IGY) of 1957 and 1958, of whose organizing committee he was president. His own especial contribution to the work of the IGY was as reporter of the results of auroras, in which he had now become deeply interested.

Soon after the IGY, Chapman was joined by Syun-Ichi Akasofu, a young geophysicist who had already done good work in his native Japan. Their cooperation over the decade 1960 to 1970 was remarkably fruitful, Chapman's great experience being admirably complemented by Akasofu's freshness of approach. They worked together on geomagnetic storms and auroras, drawing on recent discoveries about the van Allen belts, the magnetosphere, and the solar wind. Other collaborators during those years were Lawrence H. Aller, Peter C. Kendall, Joseph C. Cain, Marahira Sugiura, J. C. Gupta, and S. R. C. Malin.

In his last years Chapman tried, through books and review papers, to make generally available the wide knowledge he had accumulated during his career. Two of his longer papers were republished as books: *Solar Plasma, Geomagnetism and Aurora* (1963) and (written with Richard S. L. Lindzen) *Atmospheric Tides* (1970). At his death Chapman had largely completed his share of a new and comprehensive book (written with Akasofu), *Solar-Terrestrial Relations*, which was published in 1972. He died on 16 June 1970, after a few days' illness.

Chapman's most distinctive personal characteristics were kindliness, persistence, integrity, and simplicity. Those who penetrated his surface reserve found him always ready to help, and his many collaborators were made to feel increased in stature by working with him. He often returned to particular topics time and again to extend a partial solution. Chapman had a strong sense of duty and encouraged high standards in others. If convinced that a course of action was right, he would take it, even if it meant defying convention. When convinced he had made an error, he was always ready to acknowledge it. A simple directness pervaded both his writings and his way of life.

BIBLIOGRAPHY

I. ORIGINAL WORKS. Complete bibliographies of Chapman's work are in the articles by T. G. Cowling and V. C. A. Ferraro (see below). Among his writings are *The Mathematical Theory of Non-uniform Gases* (Cambridge, 1939; 2nd ed., 1952), written with T. G. Cowling; *Geomagnetism*, 3 vols. (Oxford, 1940–1967), vols. I and II written with J. Bartels, vol. III written with Marahira Sugiura; *Solar Plasma, Geomagnetism and Aurora* (London, 1963), also in *Geophysics—The Earth's Environment* (London, 1963), 373–502; *Atmospheric Tides* (1970), written with Richard S. Lindzen; and *Solar-Terrestrial Physics* (1972), written with Syun-Ichi Akasofu.

II. SECONDARY LITERATURE. Memoirs of Chapman include T. G. Cowling, "Sydney Chapman," in *Biographical Memoirs of Fellows of the Royal Society*, **17** (1971), 53–89; and V. C. A. Ferraro, "Sydney Chapman," in *Bulletin of the London Mathematical Society*, **3** (1971), 221–250. A volume that includes many interesting anecdotes in Syun-Ichi Akasofu, Benson Fogle, and Bernhard Haurwitz, eds., *Sydney Chapman, Eighty: From His Friends* (Boulder, Colo., 1968).

T. G. COWLING

CHETVERIKOV, SERGEI SERGEEVICH (*b.* Moscow, Russia, 6 May 1880; *d.* Gorki, U.S.S.R., 2 July 1959), *population genetics, evolutionary theory, entomology.*

A specialist in butterfly systematics, Chetverikov took up biometrics and genetics after the 1917 revolution. His work at the Institute of Experimental Biology in Moscow during the 1920's led to one of the earliest statements of the evolutionary synthesis (1926) and to the first systematic studies of the genetics of wild populations of *Drosophila* (1925–1926). Together with R. A. Fisher, J. B. S. Haldane, and Sewall Wright, Chetverikov is now regarded as one of the founders of population genetics and modern evolutionary theory.

Chetverikov was a hybrid of two important Moscow merchant families. His father was Sergei Ivanovich Chetverikov (1850–1929), a second-generation entrepreneur who manufactured fine cloth and was active in the Progressive Party. Chetverikov's mother, Maria Aleksandrovna, came from the Alekseev family, owners of a large firm renowned for its high quality gold braid and fine gold-laced fabrics. Her brother was Nikolai Alekseev (1852–1893), the mayor of Moscow from 1885 until his death. Her cousin was the famous actor and director Konstantin Alekseev (Stanislavsky), founder of the Moscow Art Theater. Chetverikov's scientific patron, Nikolai Konstantinovich Kol'tsov, was his distant cousin on his mother's side.

Chetverikov had three siblings. His older brother Ivan helped his father manage the factory. His younger brother Nikolai (1889–1973) studied mathematics with A. A. Chuprov and became prominent in Soviet mathematical statistics. His younger sister, Maria, left Russia after the revolution with his parents.

Butterflies. After receiving his elementary education at home, Chetverikov attended the Voskresenskii *Realschule* in Moscow to prepare for a career in engineering. From his geography teacher Vladimir Pavlovich Zykov, a docent at Moscow University, Chetverikov learned about Darwinism and natural history. Around 1895 he began to collect butterflies and decided to become a zoologist. His father opposed this decision, and after his graduation in 1897 sent him to a *Technikum* in Mittweid, a small town west of Dresden. After less than a year, however, his father relented. In order to qualify for university admission, Chetverikov spent a year and a half in Kiev preparing for the eighteen examinations required to obtain the certificate from the Fifth Kiev Gymnasium.

In September 1900 Chetverikov entered the natural sciences division of the physicomathematical faculty of Moscow University. At the beginning of his second semester he became involved in demonstrations supporting student protests in Kiev and was briefly under arrest in early 1901. In the fall he decided to repeat his first-year studies, and soon became part of a natural history group headed by the zoologist and anthropologist N. I. Zograf (1851–1922). Over the next four years he received excellent zoological training, specializing in invertebrates and especially Lepidoptera. In 1905, as a result of his participation in student disturbances at Moscow University, he was again arrested and held in the Butyrka prison for two months. He used the time to study for his examinations and passed them in the spring of 1906, earning the degree of candidate of natural sciences (equivalent to a B.S. degree).

Chetverikov was accepted as a graduate student in the department of comparative anatomy of Moscow University, headed by M. A. Menzbir (1855–1935), and spent the period 1907–1909 working on a graduate degree. For his dissertation, Chetverikov undertook a painstaking study of the anatomy of the carapace of *Asellus aquaticus* L., a widely distributed isopod crustacean. Arguing that the external skeleton of the crustaceans provides some of the most interesting cases of external anatomy in the animal kingdom, Chetverikov described the various parts of the exoskeleton and analyzed their adaptive functional significance in the struggle for existence. Published in German in 1910, the work demonstrated both a commitment to Darwinism and a mastery of the literature, problems, and techniques of invertebrate morphology.

During the period 1900–1920 Chetverikov emerged as one of Russia's leading butterfly systematists. As an undergraduate, he did work on Lepidoptera for the Imperial Society of Natural History, Anthropology, and Ethnography and its commission on the fauna of Moscow Province, headed by G. A. Kozhevnikov (1866–1933), a zoologist active in conservation. This work led to his first publication, an article on how to collect butterflies (1902), and to two notes adding to the fauna of Moscow Province (1902, 1905). Using the society's collection of butterflies and other materials, he established that what had previously been regarded as three species were in fact one (1903).

Chetverikov's third-year essay described and discussed seasonal and irregular radical fluctuations in the appearance of various Lepidoptera species, based on observations near his family's dacha in the summer of 1903. It was published in 1905 as "Volny zhizni" (Waves of life), a title drawn from William Henry Hudson's *The Naturalist in La Plata* (1892). This essay was one of the first works to draw attention to the implications of fluctuating population size in nature.

In his student years Chetverikov became involved in the collecting work of the ornithologist and biogeographer Petr P. Sushkin (1868–1928), then a privatdocent at the university. On the basis of material from Sushkin's 1902 expedition to Minusinsk and the western Sayan mountains, Chetverikov published two notes in German, one naming a new species (1903) and the other establishing four new species and two subspecies (1904). In 1904 he participated in Sushkin's expedition to the Tarbagatai mountain range. Subsequently, with Sushkin and Sushkin's young wife Anna Ivanovna, he participated in expeditions to the Minusinsk region, the western Sayans, and Lake Zaisan. As a graduate student, he classified the butterfly material collected on Leo S. Berg's expedition to the Aral Sea in 1900–1902 (1907), and on Sushkin's various expeditions. In 1910, when Sushkin left Moscow to become a professor at Kharkov University, Anna Ivanovna divorced him to marry Chetverikov.

Political disputes at Moscow University beginning in 1909 caused Kol'tsov to move his teaching to the Shaniavskii Municipal University and to the zoology department of the physicomathematical faculty of the Beztuzhev Advanced Courses for Women. On 12 November 1909 Chetverikov joined him as his laboratory assistant in the zoology department of the Beztuzhev courses, organizing an insect room and opening his own specialized course in entomology. In 1911 he published an account of the Lepidoptera collected on B. M. Zhitkov's 1908 expedition to the Iamal Peninsula. In 1913 he traveled to the Crimea and subsequently reported on its butterflies (1915). In 1916 he published a long article on butterflies for an anthology, edited by Kol'tsov, of observations of seasonal variation in the flora and fauna of the Moscow area.

Chetverikov's professional expertise led to scientific recognition. In 1913 he was elected to the Moscow Society of Naturalists. The next year he founded the Moscow Entomological Society and became editor of its journal. At its opening meeting on 1 March 1914, Chetverikov presented his first major theoretical paper, "Osnovnoi faktor evoliutsii nasekomykh" (The fundamental factor of insect evolution, 1915; American translation, 1920). Noting that the early fossil insects are among the largest known, he asked why insects have evolved toward smaller forms, in contrast to vertebrates, which have evolved toward larger ones. On the basis of calculations of the relative strengths of internal and external skeletons of different sizes, he argued that the "fundamental cause" of the evolution of insects was their outer chitinous skeleton,

> . . . thanks to which they were in a position, by continuously diminishing the size of their body, to conquer for themselves an entirely independent place among the other terrestrial animals, and not only to conquer it, but to proliferate in an endless variety of forms and thereby acquire a tremendous importance in the general economy of nature. *Thus, their smallness became their strength.* (1915, p. 24)

On 11 October 1916 Chetverikov was elected to the Russian Entomological Society of St. Peterburg.

During the civil war (1918–1922), the exigencies of the time led most Russian scientists to take on multiple jobs. Chetverikov managed to become preparator and temporary director of the zoological section of the State Polytechnical Museum in Moscow, where he worked part-time from 1920 to 1924. He published two popular manuals on how to collect butterflies (1919, 1925) and a study of the Lepidoptera from the Minusinsk region. The Zoological Museum of the Academy of Sciences declared his collection of butterflies to be one of the best in the country and urged that special measures be taken to ensure its preservation. He was invited to write the article on butterflies for the *Great Soviet Encyclopedia* (1926); it was his last publication on butterflies for thirty years.

The Kol'tsov Institute. Chetverikov would later characterize the decade 1919–1929 as the most productive and interesting period of his creative life. During these years, in connection with new teaching and research responsibilities, Chetverikov moved from butterfly systematics into biometry and genetics, developing a highly original view of the evolutionary process that led to pathbreaking studies of the genetics of natural populations.

After 1917, when the Beztuzhev Advanced Courses for Women became the Second Moscow University, Chetverikov was promoted from laboratory assistant to a lecturer in Kol'tsov's department. On 1 November 1919 the two Moscow universities were united and Chetverikov became a lecturer there, teaching a course on general entomology. In addition, he opened a new course, "Introduction to Theoretical Systematics," which he taught from 1919 to 1929. Initially the course treated the nature of species, interspecific and intraspecific variation, and quantitative methods. Through his brother Nikolai, Chetverikov had become interested in biometrics, and the course applied statistical and biometric techniques to the analysis of variation in species and populations. Then in 1921, Kol'tsov invited Chetverikov to join the staff of his Institute of Experimental Biology.

Since 1900 Kol'tsov had planned a research institute of laboratory biology. In 1916 he was finally able to obtain sufficient philanthropic funding, and he set up the Institute of Experimental Biology in a house in the Moscow merchant quarter. After the revolution, the loss of the institute's endowment forced a change in its research program: no longer able to afford expensive imported laboratory equipment and forced to cultivate new patrons, Kol'tsov obtained support from the Commissariat of Agriculture (for animal breeding studies) and the Commissariat of Public Health (for work on blood chemistry and eugenics). As a butterfly taxonomist without experience in agricultural or experimental research, Chetverikov did not figure in these plans.

In 1919, when word reached Russia of the remarkable developments in genetics by William Bateson (on poultry), W. E. Castle (on guinea pigs), and T. H. Morgan and his school (on *Drosophila*), Kol'tsov undertook studies, with V. N. Lebedev, on the inheritance of coat color in guinea pigs, and set his student Aleksandr Serebrovskii to work on the genetics of poultry. The same year, under Kol'tsov's direction, Dmitrii Romashov attempted to produce mutations in *Drosophila* through the use of X rays, but the experiments were unsuccessful, owing largely to the institute staff's inexperience with insects in general and with *Drosophila* in particular. It seems likely that Chetverikov was invited to join the institute in 1921 because he was the one figure in Kol'tsov's network who knew a great deal about insects.

Upon his appointment to the Institute of Experimental Biology (IEB), Chetverikov began to work with a number of Kol'tsov's graduate and undergraduate students. The original group (1921–1922) consisted of his wife, Anna Ivanovna; Nikolai Vladimirovich Timoféeff-Ressovsky (1900–1981); Elena Aleksandrovna Fiedler (1898–1973), who subsequently married Timoféeff-Ressovsky and adopted his name; Dmitrii Dmitrievich Romashov (1899–1963); Sergei Romanovich Tsarapkin (1892–1960); and Aleksandr Nikolaevich Promptov (1898–1948), an ornithologist. In the next two years, five bright undergraduates from Moscow University joined the group: Boris L'vovich Astaurov (1904–1974); Elizaveta Ivanovna Balkashina (1899–1981), who subsequently married Romashov; Nikolai Konstantinovich Beliaev (1899–1937); Sergei Mikhailovich Gershenzon (1906–); and Petr Fomich Rokitskii (1903–1977).

Originally the group focused on general zoological questions of systematics, evolution, biometrics, and genetics. In August 1922 Hermann J. Muller visited Kol'tsov's institute and its Anikovo genetics station, headed by Serebrovskii, and brought with him *Drosophila* cultures containing the famous mutants that had been the basis for the researches of the Morgan school. After consultations it was decided that the flies would go to Chetverikov's group, and that, in addition to pursuing their own research, all members would work together with the *Drosophila*.

From 1922 to 1924 most of the group's work was directed at mastering the theory and practice of Morganist genetics. The summers were spent at the IEB's hydrobiological station at Zvenigorod, headed by the hydrobiologist S. N. Skadovskii, which had been commissioned by Glavryba (the central administration of fisheries) to inventory the fauna of lakes in the region. During the winters the group worked with *Drosophila* in the laboratory at the IEB and held informal seminars at the apartments of group members. The discussion format was set by Chetverikov, who perused newly arrived issues of Western scientific journals for articles on genetics and assigned them to seminar members to report on. In this way, Chetverikov and his students learned genetics.

Around 1925 Chetverikov dropped his entomology course at Moscow University and opened its first genetics course. At about the same time, he organized his seminar on a more formal basis. It became known as the *So-or*, an acronym for *sovmestnoe oranie* (literally, "concerted cacophony" or, more loosely, "screeching society"); it is sometimes referred to as the "genetic *so-or*" or the *Droz-So-or* (for Drosophilists). In order to keep the group small and its discussions focused, its membership was restricted, although the institute's senior staff attended when they could. Candidates for membership were admitted only by a unanimous vote of the group.

The 1926 Paper. In the fall of 1925, Chetverikov completed a paper that has come to be regarded as one of the founding documents of the evolutionary synthesis. It was published the next year in *Zhurnal eksperimental'noi biologii*, edited by Kol'tsov, under the title "O nekotorykh momentakh evoliutsionnogo protsessa s tochki zreniia sovremennoi genetiki" (On certain aspects of the evolutionary process from the viewpoint of modern genetics). Chetverikov defined the paper's goal as "clarifying certain questions on evolution in connection with our current genetic concepts" (1961, p. 169).

The first section of the paper, dealing with mutations in nature, seeks to establish the relevance of genetics to evolutionary theory. At the time, such relevance was far from obvious: many genetic variations appear in the laboratory setting and under domestication, but natural populations appear remarkably uniform phenotypically. Lamarckians argued that such variations were probably produced by changed conditions of existence and an inheritance of acquired characteristics. Orthogeneticists who believed that there were no "random" variations regarded their appearance as part of an internally unfolding process of species development.

Chetverikov pointed out that, contrary to what would be expected if Lamarckians were right, at the present time "*we are not only completely unable to produce desirable mutations artificially, but are even unable to influence the frequency of their occurrence*" (1961, p. 170). Furthermore, contrary to what might be expected if the orthogeneticists were right, most mutations observed in the laboratory are harmful. This fact, he wrote, explains why we do not observe such variations in the wild, for "in the severe struggle for survival, which reigns in nature, the majority of these less viable mutations, originating among normal individuals, must perish very quickly, usually not leaving any descendants" (1961, p. 171).

Chetverikov also argued against the widely held view advocated by Leningrad's leading geneticist, Iurii A. Filipchenko, that microevolution and macroevolution are qualitatively different. Observing that "along with 'harmful' freakish mutations, there is also a series of 'neutral' ones, not having any biological significances and therefore not subject to selection" he emphasized that "some of these 'biologically neutral' mutations occurring randomly in the normal populations of some species or other sometimes correspond to the 'normal' features of neighboring species *or even genera and families*" (1961, p. 172). Drawing on his expertise in insect taxonomy, he gave a string of examples from the systematics of Diptera and Lepidoptera, emphasizing that "alongside the least salient traits, such as the color of the body, such important characters of *Drosophila* are changed as venation, wing structure, etc., which are fundamental in the modern systematics of insects for distinguishing the higher systematic categories" (1961, p. 173).

The second section of the paper relates biometrics to the problem of the species. Chetverikov's species concept is that of a *Paarungsgemeinschaft* (freely crossing community). Drawing on G. H. Hardy's law of equilibrium under free crossing (1908) and Karl Pearson's law of stabilizing crossing (1904), Chetverikov pointed out that new variations would not be "swamped" through crossing, as Fleeming Jenkin had argued. To the contrary, in the absence of selection, mutation, or migration, new variations would be maintained in a sufficiently large population at a constant frequency by the process of free crossing.

This fact led Chetverikov to two important deductions. The first was that *"a species, like a sponge, soaks up heterozygous* [recessive] *mutations, while remaining from first to last externally (phenotypically) homogeneous"* (1961, p. 178). The second, harkening back to Chetverikov's 1905 essay on "waves of life," concerned the effect of population size on the phenotypic expression of mutations. Chetverikov calculated that "on the one hand, the *more numerous* the population, the greater are the chances for *origin* of new mutations in it . . . the *less numerous* . . . the greater is the probability of *manifestation* in it in homozygous form of mutations absorbed by it earlier" (1961, p. 178), because the probability of homozygous expression through inbreeding is greatly increased. This led him to conclude that

> isolation entirely automatically leads to a *differentiation* within a species, to the fact that the colonies of one species, isolated from each other, begin, with time, to manifest differences in individual characters, which may be detected either by direct morphological study, or by biometric evaluation of their means and variabilities. *And so, isolation, under the conditions of a process of continuous accumulation of mutations becomes, by itself, a cause of intraspecific (and consequently, eventually, also of interspecific) differentiation.* (1961, p. 179)

Chetverikov argued that this "genetic" effect of isolation helped to explain the geographical races and insular faunas described by A. P. Semenov-Tian-Shanskii, J. T. Gulick, A. Garrett, A. G. Mayer, and H. E. Crampton.

The third section of Chetverikov's paper, "Natural Selection," analyzes the role of selection under conditions of free crossing with genotypical variability. He based his analysis on a table prepared by H. T. J. Norton in R. C. Punnett's *Mimicry in Butterflies* (1915), a book Chetverikov knew because of his work on Lepidoptera systematics. The table gives numbers indicating the relationship between different selection intensities, the frequencies of dominant and recessive mutations, and their rate of change in generations.

From the table Chetverikov drew three important conclusions. First, any advantageous mutation entering such a population will not be swamped, but has a definite chance of spreading throughout the population: hence, "Darwinism, in so far as natural selection and the struggle for existence are its characteristic features, received a completely unexpected and powerful ally in Mendelism" (1961, p. 183). Second, all other things being equal, the advantageous mutant will eventually be present in all the members of the species—hence, in the absence of isolation, never "does the species give rise to a new species, never will there be a subdivision of the species into two, never will speciation occur" (1961, p. 184). Chetverikov documented actual cases of this replacement, drawing principally on his own knowledge of Lepidoptera systematics. If isolation occurs or selection ceases, he noted, the species will remain polymorphic. Third, because Norton's table showed that dominant mutations are eliminated or incorporated much more quickly than recessives, Chetverikov concluded that natural selection promotes the accumulation of recessive genes in the population.

Chetverikov used these conclusions to address two central evolutionary problems: adaptation and speciation. Nonadaptive evolutionary processes, he argued, are possible in nature: "Systematics knows thousands of examples where the species are distinguished not by adaptive but rather by neutral (in the biological sense) characters, and to try to ascribe adaptive significance to all of them is work which is as little productive as it is unrewarding." The actual process of speciation, in Chetverikov's view, may be such a process, since *"Not selection, but isolation is the actual source, the real cause of the origin of species*" (1961, p. 188). For Chetverikov, evolution consists of two processes: "the process of differentiation, of splitting-up, leading in the end to *speciation*—isolation is its basis; the other leads to *adaptation*, to the progressive evolution of organic life, and its cause lies in the struggle for existence and the resulting *natural selection*" (1961, p. 189).

The fourth section of the paper, "Genotypic Milieu," emphasizes the creative role of natural selection in the evolutionary process. On the basis of data on gene interaction, expression, and pleiotropy developed by the Morgan school and also during the period 1922–1925 by N. V. Timoféeff-Ressovsky, a member of the *So-or*, Chetverikov argued that all traits are "a complex result of the manifold interaction of all the genes comprising the genotype of the organism" (1961, p. 189), and that "*The very same gene will manifest itself differently, depending on the complex of the other genes in which it finds itself*" (1961, p. 190). From this Chetverikov concluded that there is no difference in principle between qualitative and quantitative variation; that observed "correlative variability" is a result of gene interaction; and that "In selecting one trait, one gene, selection indirectly also selects a definite genotypic milieu, a genotype, most favorable for the manifestation of the given character . . . . *Selection results in the enhancement of the trait, and in this sense it actively participates in the evolutionary process*" (1961, p. 191).

Chetverikov's paper closes with an agenda for the future:

> . . . it is still too early to speak of a synthetic formulation of the evolutionary process. Only after we have disentangled the basic principles and regularities underlying the evolution of organisms in the widest sense of the word, as well as the phenomena of speciation, only then will we finally be able to attempt a reconstruction of the definitive structure of evolution and a consideration of its separate parts and finer details. (1961, p. 193)

This view of a coming evolutionary synthesis proved prophetic, and Chetverikov's paper helped to bring it about. By establishing the relevance of genetics to the problem of the origin and evolution of species in nature, he helped to transform population genetics into evolutionary genetics.

Genetics of Natural Populations. The logic of Chetverikov's 1926 paper led to the prediction that, despite their apparent phenotypic uniformity, natural populations should contain large numbers of hidden recessive mutants that can be revealed by genetic analysis. In the summer of 1925, members of Chetverikov's group conducted the first studies of natural populations of *Drosophila*. In exchange for his help in the posthumous study of Lenin's brain, the director of the Kaiser Wilhelm Institute for Brain Research, Oscar Vogt, had requested that a young Soviet geneticist be sent to help establish a genetics laboratory at his institute. Kol'tsov recommended Nikolai Timoféeff-Ressovsky and in 1925 he and his wife, Helena, were posted to Buch, near Berlin. Late that summer they collected seventy-eight pregnant female *Drosophila melanogaster* on the grounds of the Buch research station and mated the F_1s brother x sister. Such inbreeding revealed in the F_2s the presence of a number of mutations that had been hidden in heterozygous condition.

In Moscow, parallel studies using the same technique began in the summer of 1925 on populations of four *Drosophila* species found in the environs of the Zvenigorod station. Beliaev studied twenty-two inbred lines of *Drosophila phalerata* Meig.; Astaurov studied twenty-seven lines of *Drosophila phalerata* Meig. and twenty-two lines of *Drosophila transversa* Fall.; Balkashina studied the nine lines of *Drosophila vibrissina* Duda; Gershenzon studied nineteen lines of *Drosophila obscura* Fall. However, these four species proved difficult to cultivate in the laboratory, so in 1926 the work at Zvenigorod continued on *Drosophila melanogaster*.

The use of this species involved advantages and drawbacks. It was easy to cultivate and study in the laboratory, and its genes had been thoroughly mapped and studied by the Morgan school. However, the appearance of a recognizable mutant in wild populations might be interpreted as the result of their contamination by laboratory stocks. For this reason, in 1926 Gershenzon and Rokitskii captured over 300 fertilized females in Gelendzhik, a remote mountainous region in the Caucasus. From them 239 inbred lines were established and studied, and 32 clearly identifiable mutations were discovered in their progeny. Surprisingly, the *polychaeta* recessive appeared in approximately 50 percent of the lines. Chetverikov reported preliminary results from the Berlin and the Gelendzhik populations at the Fifth International Congress of Genetics (Berlin, September 1927), and his German summary was published in 1928.

By 1929 Chetverikov was preparing to publish his results more widely. In 1928 the journal *Nauchnoe slovo* announced the forthcoming publication of an article by Chetverikov titled "The Role of Mutations in the Evolution of Organisms." Most significantly, in the Chetverikov archives are two handwritten versions of an English text of his 1926 paper, probably dating from early 1929, with additional sections discussing the discovery of the mutagenic effect of X rays (1927) and summarizing his group's studies of natural populations.

This English version was never sent abroad, however, and some of the investigations of the Chetverikov group were not published until the mid 1930's (Gershenson, 1934; Balkashina and Romashov, 1935). According to Astaurov, "By far the most abundant material was collected on *Drosophila melanogaster* in 1926, and to this very day it has not been published in any complete form" (1974, p. 64). The principal reason was Chetverikov's arrest.

Arrest and Exile. In the spring of 1929 Chetverikov was detained by agents of the secret police (OGPU) and spent several months in its Butyrka prison in Moscow, after which he was sent into exile for six years by administrative order. No formal charges were ever brought against him and there was no trial. Chetverikov never learned the exact reason for his arrest.

Two related scenarios have been suggested. The first concerns the "Kammerer affair." In the mid 1920's there was strong support for the inheritance of acquired characteristics in the Communist Academy, which invited Paul Kammerer to Moscow to head a laboratory. A famous Viennese biologist and an outspoken socialist, Kammerer claimed to have demonstrated Lamarckian inheritance in salamanders. After articles by G. K. Noble and Hans Przibram in the 7 August 1926 issue of *Nature* suggested that his results were fraudulent, Kammerer killed himself. His obituary in *Izvestiia* (7 October 1926) mentioned a postcard, apparently sent by a Professor Chetverikov, "congratulating the Academy on Kammerer's suicide" and castigated Chetverikov as "one of the reactionary obscurantists left behind in the U.S.S.R." The next day *Izvestiia* published a letter from Chetverikov stating that he had sent no such postcard and that it was a malicious forgery. Chetverikov later conjectured that the same person who had forged his signature on the postcard in 1926 might have denounced him in 1929.

Second, Chetverikov's arrest may have had something to do with the *Droz-So-or*. In a letter dated 28 July 1929 to Maxim Gorky seeking aid in obtaining Chetverikov's release, Kol'tsov explained:

> Early this year at the time of my stay in Paris, a stupid incident occurred among the graduate students at my Institute. . . . Tempers flared, and some of the students took it all out on S. S. Chetverikov. At the University they drew up an indictment of him consisting of ten charges. . . . The charges were published, and several days later Chetverikov was arrested. . . . (Shvarts, 1975, p. 256)

The nature of the "stupid incident" remains unclear. Chetverikov may have been denounced by a student who had been refused admission to the *Droz-So-or;* or by a member of the group whose wife had been blackballed; or by a colleague or rival who was envious of his popularity and influence. Until his secret police dossier is examined, we cannot know.

Whatever the cause of Chetverikov's arrest, its consequences are clear. Within a year his entire group had been dispersed. Beliaev left in 1929 to work at the Central Asian Silkworm Breeding Institute in Tashkent; Astaurov joined him in 1930; Gershenzon accepted posts at the Zoological Institute of Moscow University and in Serebrovskii's laboratory at the Timiriazev Institute in 1930; Promptov joined the staff on the Second Moscow Medical Institute; Rokitskii joined the Institute of Animal Breeding of the newly created Lenin All-Union Academy of Agricultural Sciences. Chetverikov's wife stayed in Moscow but ceased to work at the IEB. Only Balkashina and Romashov remained. In the words of Astaurov (1974), "by 1930 the laboratory had effectively ceased to exist."

By administrative order of the OGPU, Chetverikov was exiled to Sverdlovsk in the Urals for three years (1929–1932). During the first year and a half he was unemployed. In 1931 he became a scientific consultant to Gorkomkhoz on the planning and organization of a zoo at Sverdlovsk's Park of Culture and Recreation. On 12 April 1932 his period of internal exile was extended for three years. In 1932 he moved to Vladimir, where he worked from 1 October as a teacher of entomology and biometrics at the Ministry of Agriculture's *Uchkombov* (the scientific collective for the fight against agricultural and forest pests). After it closed in the fall of 1934, he taught mathematics at the ministry's agricultural technikum in Vladimir.

Gorki and Silkworms. Following the expiration of his term of exile in 1935, Chetverikov reestablished a biological career in the city of Gorki. In 1933 I. I. Puzanov, who had known Chetverikov at Moscow University, accepted the chair in vertebrate zoology at Gorki University and became dean of its biological faculty; his assistant was Zoia Safronovna Nikoro, a docent in genetics. In 1935, on the initiative of Nikoro, Puzanov invited Chetverikov to head the genetics department of Gorki University beginning that fall. He accepted, and lived in a dormitory in Gorki until 1936, when he obtained an apartment.

Thereupon his wife joined him, bringing his enormous butterfly collection.

As head of the genetics department, Chetverikov gave a course in entomology and a course in genetics (1935–1948). He also gave popular lectures; one of them, on developments in cytogenetics during the previous decade, was issued as a pamphlet (1936). In 1936 Puzanov was replaced as dean by A. D. Nekrasov (1874–1960) who, like Puzanov, was a friend of Chetverikov's from Moscow University. The three Muscovites formed a Gorki branch of the Moscow Society of Naturalists. In the late 1930's, despite Chetverikov's open opposition to the views of Lysenko, he was strongly supported at the university by Puzanov, Nekrasov, Nikoro, the botanist S. S. Stankov, and others.

In 1938 Chetverikov was awarded the degree of Candidate of Biological Sciences without a dissertation and was made a full professor. In April 1940 he took over from Nekrasov as dean of the biological faculty, serving in that post through 1947. He was nominated for a doctorate of biological sciences and was awarded the degree in June 1945.

While in Gorki, Chetverikov devoted himself to research on silkworms. In the winter of 1937 he was approached by the scientific secretary of the Ministry of Agriculture with the proposal that he engage in breeding the silkworm for Russian conditions: silk was needed for parachutes, and Japan was no longer regarded as a reliable supplier. A small experimental base for this work was built outside of Gorki. The saturnid silkworm (*Antherea pernyi* Guér.-Mén.) is bivoltine, with two generations per year, the second hatching in late autumn. In its native habitat, October is mild; in Russia, the silkworms emerging in the fall freeze before they can spin cocoons. Chetverikov set about trying to develop a monovoltine strain.

In early 1940 Chetverikov reported on his progress at a conference on sericulture and published a preliminary account in its proceedings (1941). By 1944, at a scientific conference at Gorki University and at the seventh plenary meeting of the sericulture commission of the Lenin All-Union Academy of Agricultural Sciences, he could proudly report that his strain "Gorki Monovoltine No. 1" was under trials and produced silk of superior quality. In 1944 a monograph summarizing his six years of research was sent to the agriculture academy for publication. On 4 November of that year, the presidium of the Supreme Soviet awarded him a medal of excellence for his work.

Final Years. In the late 1940's Chetverikov's life became difficult for both personal and political reasons. On 2 July 1947 his wife died. Soon thereafter Chetverikov developed serious heart problems, leading him to resign as dean. The triumph of Lysenkoism in August 1948 led to Chetverikov's separation from the university and the loss of his pension: on 13 August 1948 the Minister of Higher Education issued an order demanding "the dismissal of incorrigible Mendelist-Morganists" that mentioned Chetverikov by name. Although it was in press, Chetverikov's silkworm monograph was never published and the manuscript may no longer exist. His "Gorki Monovoltine No. 1" was mixed with normal stock and the breed was lost.

After suffering a heart attack in September 1948, Chetverikov spent four months bedridden. He became partially blind, and by 1950 could not distinguish yellow from brown or gray from green. This made work with butterflies very difficult, so he gave his collection of more than 300,000 specimens to the Zoological Museum of the Zoological Institute of the U.S.S.R. Academy of Sciences in Leningrad. In 1956 Chetverikov published his last scientific article, a description of a new species of *Cucullia*. That year he was visited by two former students whom he had not seen in thirty years, Helena and Nikolai Timoféeff-Ressovsky, the latter just emerging from the Gulag after a decade of imprisonment and exile.

An invalid during his final years, Chetverikov was looked after by his brother Nikolai, friends, and students. With their help he managed an active correspondence. Around 1958 he dictated his memoirs, dealing mainly with his student days, to Valery N. Soyfer. In 1959 he received word that he was to be a recipient of the Darwin Medal of the (East) German Academy of Sciences (Leopoldina), one of twenty-eight awarded to those who had made major contributions to the development of Darwinism and evolutionary theory. Because of ill health, however, he was unable to attend the award ceremonies.

On 13 June 1959 Chetverikov suffered a stroke and lapsed into a coma. He died during the night of 2 July, on the twelfth anniversary of his wife's death.

Historical Significance. Immediately after Chetverikov's death, his historical significance began to be recognized. In 1959 Theodosius Dobzhansky published an abbreviated English translation of Chetverikov's classic paper; a full translation was published in 1961 by I. Michael Lerner, a Berkeley geneticist of Russian extraction. Thanks to these and other works, Chetverikov has come to be regarded, together with Haldane, Fisher, and Wright, as a founder of population genetics and the evolutionary synthesis.

Chetverikov's evolutionary approach stands apart from that of his three cofounders. Unlike them, he had not been involved in agricultural breeding or eugenics; he did not begin with genetics and biometrics and deductively extrapolate their implications for evolution, given certain unrealistic, but mathematically necessary, assumptions. Rather, he began with a systematist's concrete knowledge of natural populations and species, and sought to determine how recent findings in genetics and biometrics could illuminate their evolution in nature. Chetverikov used far less sophisticated mathematics than did Fisher, Haldane, and Wright, of course, but he spoke more directly to the concerns of practicing zoologists and naturalists.

In this sense, Chetverikov's 1926 paper set the agenda for the second phase of the evolutionary synthesis (1935–1950). His treatment of the genetic dimension of interspecific variation, speciation, the role of isolation, and the implications of population size anticipated later works by Timoféeff-Ressovsky, Dobzhansky, Julian Huxley, and Ernst Mayr. His discussion of the genetic implications of small colonies and the evolutionary significance of the genotypic milieu foreshadowed the "founder principle" and the concept of the "good mixer" developed by Mayr and others in the 1940's and 1950's. His views on the selective neutrality of many genes and his critique of adaptationism anticipated the "neutral gene hypothesis" and the "punctuated equilibrium" theories of the 1970's.

Although Chetverikov never resumed research in population genetics after his arrest in 1930, his work in the 1920's exerted a powerful formative influence over its subsequent development. Romashov and Dubinin established a new group in population genetics at the Kol'tsov institute in 1932, and its internationally renowned research constituted a conscious extension of Chetverikov's program. In Europe, population genetics and the evolutionary synthesis developed around Chetverikov's students H. A. and N. V. Timoféeff-Ressovsky, who worked in Germany from 1925 to 1945. Dobzhansky had obtained *Drosophila* stocks from Chetverikov before he left Russia for the United States in 1927; a decade later, he drew upon Chetverikov's perspective and reasoning in *Genetics and the Origin of Species* (1937), from which the evolutionary synthesis in the West is often dated. The same year, following Chetverikov's lead, Dobzhansky began his series of forty-six articles on the genetics of natural populations that would form the core of American population genetics.

BIBLIOGRAPHY

I. Original Works. Chetverikov's surviving papers (including lecture outlines, notebooks, typescripts, and manuscripts) are collected in fund 1650 of the Moscow branch of the Archive of the U.S.S.R. Academy of Sciences. In German his name is usually spelled Tschetwerikoff; in French, Tshetverikov, Tshetverikoff, or Tchetverikoff. During his lifetime he published more than thirty works. Ten have been republished in his *Problemy obshchei biologii i genetiki (vospominaniia, stat'i, lektsii)* (Problems of general biology and genetics [memoirs, articles, lectures]), Z. S. Nikoro, ed. (Novosibirsk, 1983). Nine others are in his *Fauna i biologiia cheshuekrylykh* (Fauna and biology of butterflies and moths), G. S. Zolotarenko and Z. S. Nikoro, eds. (Novosibirsk, 1984).

Chetverikov's earliest publication, and two others, are handbooks on how to collect insects: "Rukovodstvo k kollektsionirovaniiu cheshueskrylykh (Lepidoptera)" (How to collect butterflies and moths [Lepidoptera]), in G. A. Kozhevnikov, ed., *Rukovodstvo k zoologicheskim ekskursiiam i sobiraniiu zoologicheskikh kollektsii* (Handbook for zoological expeditions and making zoological collections; Moscow, 1902); *Kratkoe nastavlenie k sobiraniiu nasekomykh* (Manual on collecting insects; Moscow, 1919); and *Sbor i prigotovlenie zoologicheskikh kollektsii* (Collecting and preparing zoological collections; Moscow and Leningrad, 1925), written with N. A. Bobrinskii. In addition, he published some twenty notes on butterfly systematics.

His three most important prerevolutionary works were "Volny zhizni (Iz lepidopterologicheskikh nabliudenii za leto 1903 goda)" (Waves of life [Observations of Lepidoptera in the summer of 1903]), in *Izvestiia Imperatorskogo obshchestva liubitelei estestvoznaniia, antropologii i etnografii*, **98**, *Dnevnik zoologicheskogo otdeleniia*, **3**, no. 6 (1905), 1–5 (106–110), republished in Chetverikov, 1983, 76–83; his dissertation, "Beiträge zur Anatomie der Wasserassel (*Asellus aquaticus* L.)," in *Biulleten' Moskovskogo obshchestva ispytatelei prirody*, 1910, no. 4, 377–509; and "Osnovnoi faktor evoliutsii nasekomykh," in *Izvestiia Moskovskogo entomologicheskogo obshchestva*, **1** (1915), 14–24, published in English as "The Fundamental Factor of Insect Evolution," in *Smithsonian Report for 1918*, no. 2566 (Washington, 1920), 441–449. After the revolution he published two articles for *Bol'shaia sovetskaia entsiklopediia* (Great Soviet encyclopedia): "Babochki" (Butterflies), in IV (Moscow, 1926), 264–275, and "Biometriia" (Biometry), in VI (Moscow, 1927), 338–344. See also his letter to the editor of the newspaper *Izvestiia*, 8 October 1926.

His classic theoretical paper was "O nekotorykh momentakh evoliutsionnogo protsessa s tochki zreniia sovremennoi genetiki," in *Zhurnal eksperimental'noi biologii*, ser. A, **2**, no. 1 (1926), 3–54; summary in German, *ibid.*,

CHETVERIKOV

no. 4 (1926), 237–240; excerpts translated by Theodosius Dobzhansky in *Cold Spring Harbor Symposia in Quantitative Biology*, **24** (1959), 27–30; complete English trans. by Malina Barker, ed. by I. Michael Lerner, "On Certain Aspects of the Evolutionary Process from the Standpoint of Modern Genetics," in *Proceedings of the American Philosophical Society*, **105,** no. 2 (April 1961), 167–195. The Russian original was republished, with notes dictated by Chetverikov to Valery Soyfer, in the mid 1950's, in *Biulleten' Moskovskogo obshchestva ispytatelei prirody, Otdel biologi*, **70,** no. 4 (1965), 33–74, and in *Klassiki sovetskoi genetiki 1920–1940* (Classics of Soviet genetics 1920–1940; Leningrad, 1968), 133–170. A French translation was published as *Les Lois de l'Hérédité* (Mont-Pelerin, Switzerland, 1970).

The work of Chetverikov and his group on the genetics of natural populations was published only in abbreviated form. Before his arrest Chetverikov published only résumés of three conference papers: "Teoreticheskaia predposylka geneticheskogo analiza vidov roda *Drosophila*" (The theoretical premise of genetic analysis of *Drosophila* species), in *Trudy Vtorogo S"ezda Zoologov, Anatomov i Gistologov SSSR v Moskve 4–10 maia 1925 g.* (Proceedings of the Second Congress of Zoologists, Anatomists, and Histologists of the U.S.S.R., at Moscow, 4–10 May 1925; Moscow, 1927), 163–164; "Eksperimental'noe reshenie odnoi evoliutsionnoi problemy" (Experimental solution of one evolutionary problem), in *Trudy Tret'ego Vserossiiskogo S"ezda Zoologov, Anatomov i Gistologov v Leningrade 14–20 dekabria 1927 g.* (Proceedings of the Third All-Russian Congress of Zoologists, Anatomists, and Histologists at Leningrad, 14–20 December 1927; Leningrad, 1928), 52–54; and "Über die genetische Beschaffenheit wilder Populationen," in *Verhandlungen des Fünften internationalen Kongresses für Vererbungswissenschaft, Berlin, 1927*, II (Leipzig, 1928), 1499–1500, repr. in *Zeitschrift für induktive Abstammungs- und Vererbungslehre*, **46** (1928), 38–39.

Due to Chetverikov's arrest in 1929, all the longer treatments, though still partial, were by his students: H. A. Timoféeff-Ressovsky and N. W. Timoféeff-Ressovsky, "Genetische Analyse einer freilebenden *Drosophila melanogaster* population," in *Wilhelm Roux' Archiv für Entwicklungsmechanik der Organismen*, **109** (1927), 70–109; S. M. Gershenson, "Mutant Genes in a Wild Population of *Drosophila obscura* Fall.," in *American Naturalist*, **68,** no. 719 (1934), 569–571; and E. I. Balkashina and D. D. Romashov, "Geneticheskoe stroenie populiatsii Drosophila. I. Geneticheskii analiz zvenigorodskikh (Moskovskoi oblasti) populiatsii *Drosophila phalerata* Meig., *transversa* Fall. i *vibrissina* Duda" (The genetic structure of *Drosophila* populations. I. Genetical analysis of populations of *Drosophila phalerata* Meig., *transversa* Fall. and *vibrissina* Duda of Zvenigorod, Moscow Province), in *Biologicheskii zhurnal*, **4,** no. 1 (1935), 81–106.

After 1930 Chetverikov published only three works: *Tsitologiia nasledstvennosti za poslednie desiat' let* (The cytology of inheritance during the last decade; Gorki, 1936); "Selektsiia na monovol'tinnost' u Kitaiskogo dubovogo shelkopriada *Antherea pernyi* Guér.-Mén." (Selection for monovoltinism in the Chinese saturnid silkworm *Antherea pernyi* Guér.-Mén.), in *Selektsiia i akklimatizatsiia dubovykh shelkopriadov* (Selection and acclimatization of the silkworm; Moscow, 1941), 16–22; and "Novyi vid roda *Cucullia* Schrk. (Lepidoptera, Noctuidae) iz Iuzhnogo Priural'ia" (A new species of the genus *Cucullia* Schrk. [Lepidoptera, Noctuidae] from the foothills of the southern Urals), in *Russkoe entomologicheskoe obozrenie*, **35,** no. 4 (1956), 927–928.

A brief autobiography by Chetverikov, in German, appears in *Nova acta Leopoldina*, n.s. **21,** no. 143 (1959), 308–310. His memoirs, dictated to Valery Soyfer in 1958, have been published as "Iz vospominanii" (From the memoirs), in *Priroda*, 1974, no. 2, 68–69; "Pervyi god v Moskovskom universitete" (First year at Moscow University), *ibid.*, 1980, no. 5, 50–55; and "Vospominaniia" (Memoirs), *ibid.*, 1980, no. 11, 88–94, and no. 12, 76–85, and reprinted in Chetverikov (1983), 41–75.

II. Secondary Literature. For discussions in English specifically devoted to Chetverikov, see the articles by Mark B. Adams, "The Founding of Population Genetics: Contributions of the Chetverikov School, 1924–1934," in *Journal of the History of Biology*, **1,** no. 1 (1968), 23–39, "Towards a Synthesis: Population Concepts in Russian Evolutionary Thought, 1925–1935," *ibid.*, **3,** no. 1 (1970), 107–129, and "Sergei Chetverikov, the Kol'tsov Institute, and the Evolutionary Synthesis," in Ernst Mayr and William B. Provine, eds., *The Evolutionary Synthesis* (Cambridge, Mass., 1980), 242–278; by Theodosius Dobzhansky, "Evolution of Genes and Genes in Evolution," in *Cold Spring Harbor Symposia on Quantitative Biology*, **24** (1959), 15–30, and "Sergei Sergeevich Tshetverikov 1880–1959," in *Genetics*, **55,** no. 1 (January 1967), 1–3; by Hisao Kaneko and Kuniyoshi Ohta, "The Case of a Biologist: The Life of S. S. Chetverikov," in *Journal of the Humanities and Natural Sciences (Tokyo Keizai University)*, no. 60 (1982), 1–50, and "Supplement," *ibid.*, no. 62 (1982), 209–229 (in Japanese, summaries in English); and by I. Michael Lerner, "Introductory Note," in *Proceedings of the American Philosophical Society*, **105,** no. 2 (April 1961), 167–169.

For sources in Russian, see the articles by B. L. Astaurov, "Dve vekhi v razvitii geneticheskikh predstavlenii" (Two landmarks in the development of genetic concepts), in *Biulleten' Moskovskogo obshchestva ispytatelei prirody, Otdel biologi*, 1965, no. 4, 25–32, and "Zhizn' S. S. Chetverikova" (The life of S. S. Chetverikov), in *Priroda*, 1974, no. 2, 57–67, and its revised version, "Sergei Sergeevich Chetverikov (6 maia 1880–2 iiulia 1959): Zhizn' i tvorchestvo" (. . . Life and work), in Chetverikov (1984), 78–98; by V. V. Babkov, "Tsentral'naia problema genetiki populiatsii" (The central problem of population genetics), in Chetverikov (1983), 6–40, and *Moskovskaia shkola evoliutsionnoi genetiki* (The Moscow school of evolutionary genetics; Moscow, 1985); by P. F. Rokitskii, "S. S. Chetverikov i evoliutsionnaia genetika" (S. S.

Chetverikov and evolutionary genetics), in *Priroda*, 1974, no. 2, 70–74, and "S. S. Chetverikov i razvitie evoliutsionnoi genetiki" (S. S. Chetverikov and the development of evolutionary genetics, in *Iz istorii biologii* (From the history of biology), no. 5 (Moscow, 1975), 63–75; by V. N. Soyfer, "Neizvestnaia rabota S. S. Chetverikova po populiatsionnoi genetike" (An unknown work of S. S. Chetverikov on population genetics), in *Iz istorii biologii* (From the history of biology), no. 3 (Moscow, 1971), 177–192; by Anatolii Shvarts, "Dve sud'by" (Two fates), in *Novyi zhurnal* (New York), no. 121 (December 1975), 248–269; by N. V. Timoféeff-Ressovsky and N. V. Glotov, "Sergei Sergeevich Chetverikov, 1880–1959," in *Vydaiushchiesia sovetskie genetiki* (Leading Soviet geneticists; Moscow, 1980) 69–76; and by G. S. Zolotarenko, "Posleslovo" (Afterword), in Chetverikov (1984), 71–77.

MARK B. ADAMS

CHICK, HARRIETTE (*b*. London, England, 6 January 1875; *d*. Cambridge, England, 9 July 1977), *biochemistry*.

Harriette Chick was the fifth of the seven daughters and four sons of Samuel Chick, a lace merchant, and of Emma Hooley. All of them were brought up as strict Methodists. Harriette was educated at Notting Hill High School. Five of the Chick girls were university graduates, four of them in science or medicine. In 1896 Harriette graduated B.Sc. from University College, London, where she studied botany, obtaining the D.Sc. in 1904. After further work in Germany and England, she was appointed to the staff at the Lister Institute, where she worked for many years, first on the mechanism of disinfection and later, following the arrival of Casimir Funk in 1911, in the field of nutrition. She was made a fellow of University College, London, in 1918.

Also in 1918 the Medical Research Council and the Lister Institute appointed the Accessory Food Factors Committee to investigate the role of vitamins in metabolism. Chick was made secretary (and served on the committee until 1945), and Frederick Gowland Hopkins was chairman. On its behalf Chick made classic studies of childhood rickets in postwar Vienna, in collaboration with the medical staff of the University Children's Clinic. In 1933 she was awarded an honorary D.Sc. by the University of Manchester. From 1934 to 1937, Chick was secretary of the League of Nations health section committee on the physiological basis of nutrition. She was made C.B.E. in 1932, and D.B.E. in 1949 for her studies on nutrition. She was a founder member of the Nutrition Society (1941) and was president from 1956 to 1959. She had become an honorary member in 1949. Chick remained active as a member of the governing body of the Lister Institute and in the Nutrition Society after her retirement in 1945. She was in good health until her sudden death in 1977, at age 102. She never married.

After receiving the D.Sc. at London, partially for research on green algae in polluted water, Chick was awarded an 1851 Exhibition to work at the hygiene institutes at Vienna and at Munich under Max von Gruber. Upon returning to England she worked under the direction of Rubert Boyce at Liverpool University, an expanding center for pathology and bacteriology in England in the first decade of the century. Charles Sherrington, at that time professor of physiology at Liverpool, steered her toward the next stage in her career by suggesting that she apply to the Lister Institute of Preventive Medicine for the Jenner Memorial Research Studentship, which she was awarded in 1905. Her career there was a long and productive one. At the Lister she collaborated with the director, Charles Martin, in a pioneering study of the mechanism of disinfection.

Chick showed that disinfection is a chemical process and obeys the law of mass action. The Chick-Martin test replaced the one then in general use to measure the strength of disinfectants, which had been devised by Samuel Rideal and J. T. Ainslie Walker. These disinfection studies led them to examine the heat coagulation of proteins, the physical chemistry of protein separation by salting out, and the role of protein-water interaction in maintaining protein solubility at different pH values.

During World War I, Chick took over the preparation at the Lister Institute of agglutination serum for diagnosing typhoid and other diseases. After the war ended, she returned to the study of "vitamins," a term coined at the Lister by Casimir Funk in 1912. As a member of the Accessory Food Factors Committee, Chick contributed to a series of books on vitamins. *The Present State of Knowledge Concerning Accessory Food Factors (Vitamins)* appeared in a series of monographs from 1919 to 1932. In Vienna, after the war, Chick investigated opposing theories of the cause of rickets. About 1918 Edward Mellanby and others had advocated a vitamin deficiency, while others, notably in Glasgow, supported a "domestic" theory that linked the bony deformities to confinement, lack of exercise, and poor hygiene. Chick was able partly to reconcile these views by showing that treatment with either cod liver oil or sunshine can cure rickets. Although the issue was not entirely resolved until the later discovery of ergosterol and its conversion by sunlight to vitamin D, Chick's studies had a significant impact on further research and probably were the most lasting of her contributions to science.

BIBLIOGRAPHY

I. Original Works. "The Principles Involved in the Standardisation of Disinfectants and the Influence of Organic Matter upon Germicidal Value," in *Journal of Hygiene*, **8** (1908), 654–697, with C. J. Martin; "On the 'Heat Coagulation' of Proteins. Part I," in *Journal of Physiology*, **40** (1910), 404–430, ". . . Part II," *ibid.*, **43** (1911), 1–27, ". . . Part III," *ibid.*, **45** (1912), 61–69, and ". . . Part IV," *ibid.*, 261–295, all with C. J. Martin; "The Density and Solution Volume of Some Proteins," in *Biochemical Journal*, **7** (1913), 92–96, with C. J. Martin; "The Precipitation of Egg Albumin by Ammonium Sulphate: A Contribution to the Theory of the 'Salting-out' of Proteins," *ibid.*, 380–398, with C. J. Martin; *Studies of Rickets in Vienna, 1919–1922* (London, 1923); article in *Report on the Present State of Knowledge of Accessory Food Factors (Vitamins)*, 2nd ed., rev. and enl. (London, 1924); *Reports on Biological Standards* (London, 1932); *Diet and Climate* (London, 1935); and *War on Disease: A History of the Lister Institute* (London, 1971), with Margaret Hume and Marjorie Macfarlane.

II. Secondary Literature. Lesley Hall and Neil Morgan, "Illustrations from the Wellcome Institute Library. The Archive of the Lister Institute of Preventive Medicine," in *Medical History*, **30** (1986), 212–215.

NEIL MORGAN

CHRISTOFILOS, NICHOLAS C. (*b.* Boston, Massachusetts, 16 December 1916; *d.* Livermore, California, 24 September 1972), *engineering, particle accelerator design, magnetic fusion, nuclear physics.*

The only child of Greek nationals, Constantine and Eleni Christofilos, Nicholas Constantine Christofilos was born in Boston, where his father operated a small cafe. From the age of seven he was raised in Athens, his father having returned to his native country to open and operate a coffeehouse. His penchant for argument and his later willingness to devote his considerable creative talent and energy to American national security programs were forged in the increasingly tumultuous atmosphere of the Greek capital. In Nicholas' youth Greece experienced the establishment of a dictatorship in 1936, occupation by German troops during World War II, and an increasingly bitter civil war in which first the British and then the Americans intervened to prevent the triumph of leftist forces against a government that looked to the West for political and economic guidance and support.

As a child in Athens, in addition to argument and politics, Christofilos was exposed to music, astronomy, electrical devices, amateur radio, and science. He felt a strong attraction to science, although as a practical matter he selected engineering as a career. After earning an advanced degree in electrical and mechanical engineering from the National Technical University of Athens in 1938, Christofilos accepted a position with a company that installed elevators in office and apartment buildings. When the German occupation forces redirected the company's efforts toward the repair of military vehicles, Christofilos devoted much of the abundant spare time afforded by his supervisory position to the study of nuclear physics, especially readily available works in German.

After the war Christofilos established his own elevator installation business. He also continued his scientific interests and began designing atom smashers for nucler research. Christofilos brought an extraordinary inventiveness and entrepreneurial spirit to this work. In 1946, and with more detail in 1947, he applied for Greek and American patents for a particle accelerator of his own design, one he later learned incorporated many of the features of the synchrotron, which had been developed by Edward McMillan in the United States and Vladimir I. Veksler in the Soviet Union.

Pursuing the possibility that his accelerator work might lead to a career in one of the newly created U.S. Atomic Energy Commission (AEC) laboratories, in 1948 Christofilos wrote to Ernest O. Lawrence at Berkeley and enclosed his latest patent application. Laboratory scientists replied that the focusing scheme he described would not work. Taking the criticism to heart, Christofilos refined his invention. He realized that the magnets used in existing particle accelerators were limited by the perceived need to focus the particle beam simultaneously in the two directions perpendicular to the direction of the beam's path. Christofilos saw that simultaneous focusing was not required and proposed a magnet design that would produce alternating regions in which the gradient of the guiding field would strongly focus and then defocus the beam in a given direction (focusing in one direction while defocusing in the other). The net effect, he claimed, would be far more powerful than what was possible with any arrangement that attempted to focus the beam in both directions simultaneously. In a patent application dated 10 March 1950 Christofilos proposed magnetic fields whose gradients varied sinusoidally along the direction of the beam. Devices that use this scheme, or a similar scheme invented independently in 1952 in the United States at Brookhaven

National Laboratory, are called "strong" focusing machines. The discovery made feasible a new generation of particle accelerators with energies an order of magnitude higher than was practical with much larger and more expensive "weak" focusing magnets.

Again Christofilos sent his scheme with a letter to Lawrence's laboratory in Berkeley. In the laboratory's response it was suggested that Christofilos avail himself of a specific mathematical text so that his ideas could be expressed in a formalism familiar to American scientists. In 1952 Christofilos finally obtained the recommended text; upon mastering it he realized that his focusing scheme could indeed be demonstrated with full mathematical rigor. Rather than rely on correspondence, early in 1953 he booked passage to America to press his discovery.

One of the first stops Christofilos made upon returning to the land of his birth was the New York Public Library, where he scanned the latest issue (volume 88) of *Physical Review*. In it were articles announcing the focusing scheme that had been invented at Brookhaven National Laboratory. Christofilos immediately recognized the Brookhaven invention as essentially the same as his own and arranged to travel to the Long Island laboratory to discuss his priority. Brookhaven scientists were surprised by his revelations and the dated patent application he carried. After consulting with Berkeley colleagues, the Brookhaven scientists set up a series of meetings with AEC patent officers. The result was that, in return for a $10,000 payment, a license and agreement were granted for the use by the United States government and its contractors of the "strong focusing" principle. Christofilos also accepted a position at Brookhaven National Laboratory. He was awarded U.S. patent 2,736,799 on 28 February 1956.

Though Christofilos maintained an interest in the problems of accelerator design, his work on the interactions of magnetic fields and moving ions had already triggered an interest in using magnetic fields to contain high-energy plasmas. The goal was to produce a controlled thermonuclear reaction. Christofilos had already filed for a separate patent on a device in which a high-energy beam of electrons would orbit within a closed cylindrical container, forming a current layer that would in turn produce closed magnetic field lines. The magnetic field theoretically could contain a plasma, which would be heated to ignition temperatures by interactions with the circulating electrons.

In 1953 the subject of controlled fusion was highly classified and of great interest to the AEC. As soon as he started work at Brookhaven, the process of investigating Christofilos and obtaining the necessary security clearances for him began. Meanwhile, at Brookhaven he continued to develop his ideas. He chose the name Astron for his proposed thermonuclear reactor.

In 1954 Christofilos married the daughter of a Greek physician. Elly and Nicholas Christofilos had one son and were divorced in 1960. That same year Christofilos married Joan Jaffray; they had one son.

In 1956 the AEC established a research program based on the Astron concept at the Lawrence Radiation Laboratory in Livermore, California. Christofilos moved to California to direct the program. One of a number of competing ideas, the Astron program required the design and building of a new, high-energy, high-intensity electron accelerator as a source for the relativistic electrons that would make up the so-called E-layer. With funding from both the AEC and the Department of Defense, Christofilos set out to design and build the accelerator as well as the Astron device. Defense Department interest in the project was no doubt related to its long-term concern with the practicality of using intense particle beams for military purposes. In fact, the electron accelerator designed by Christofilos has played a major role in the free-electron laser program at the Lawrence Livermore Laboratory, an important component of the Reagan administration's Strategic Defense Initiative.

As the Astron project ran into several technical and scheduling problems, Christofilos fought extraordinarily hard to maintain support. Frequently reviewed and criticized, the project was always given high marks for its ingenuity and potential. However, already in severe trouble, it was formally canceled in December 1972, following Christofilos' sudden and unexpected death from a heart attack.

The Department of Defense took up several other ideas generated by Christofilos. Most remarkable was a suggestion he made in October 1957 (just after the Soviet Union's successful launch of Sputnik) that the earth's magnetic field could serve to trap ionized particles injected at the correct altitudes, for example, by the detonation of nuclear devices. As a potential hazard to missiles and satellites and to manned space flight, as a possible means of disrupting essential military communications, and as a way of understanding the magnetic and radiation environments of near-earth space, Christofilos' idea interested the Department of Defense's Advanced Projects Research Agency (created in the aftermath of the Sputnik launch), which decided to pursue it.

Soon the concept was independently supported by James Van Allen's discovery of natural belts of radiation trapped by the earth's magnetic field. Van Allen had made his discovery using instruments on the Explorer 1 and Explorer 3 satellites, so in April 1958, when it was decided to proceed with a test of Christofilos' idea, Van Allen was brought in to instrument a special satellite to monitor the expected artificial radiation belts. Code-named Project Argus, the effort produced trapped electrons as a result of three separate high-altitude detonations in August 1958. The Argus tests were revealed by the *New York Times* in March 1959. A great deal of attention was given to Christofilos, who was portrayed as the lone-wolf genius behind the idea.

The Astron project and related particle-beam studies occupied most of Christofilos' efforts until his death. Nevertheless, he found time to propose and take part in several other defense-related projects. One of these ideas of special importance was a proposal for a practical means of global radio communications using extremely-low-frequency (ELF) radio waves. A more complete account of this and other important aspects of Christofilos' career will have to wait until relevant records are declassified.

Among Christofilos' honors was the Elliot Cresson Medal of the Franklin Institute (1963).

BIBLIOGRAPHY

I. ORIGINAL WORKS. Christofilos published infrequently in the open literature. His strong-focusing patent (2,736,799) is available through the U.S. Patent Office, as are his numerous other patents and applications. Unclassified technical reports written after 1965 can be obtained through the U.S. National Technical Information Service (NTIS). Publications in the open literature include "Astron Thermonuclear Reactor," in *Progress in Nuclear Energy, 11th ser., I, Plasma Physics and Thermonuclear Research* (New York, 1959), 576–603; "The Argus Experiment," in *Journal of Geophysical Research*, **64** (1959), 869–875; and "High Current Linear Induction Accelerator for Electrons," in *Review of Scientific Instruments*, **35** (1964), 886–890, with R. E. Hester, W. A. S. Lamb, D. D. Reagan, W. A. Sherwood, and R. C. Wright.

II. SECONDARY LITERATURE. For biographical information see *Current Biography 1964* (1965), 82–84; John S. Foster, T. Kenneth Fowler, and Frederick E. Mills, "Nicholas C. Christofilos," in *Physics Today*, **26**, no. 1 (1973), 109–115; William Trombley, "Triumph in Space for a 'Crazy Greek,' " in *Life*, 30 March 1959, 31–34; "Up from the Elevator," in *Time*, 30 March 1959, 70–71; "Volatile Scientist Nicholas Constantine Christofilos," in *New York Times*, 19 March 1959, 16; and Burton H. Wolfe, "That Crazy Greek," in *Coronet*, **46** (December 1958), 177–180.

Accounts of Christofilos' scientific work are scarce. On the invention of strong focusing, see E. D. Courant, M. S. Livingston, H. S. Snyder, and J. P. Blewett, "Origin of the 'Strong-Focusing' Principle," in *Physical Review*, 2nd ser., **91** (1953), 202–203. On the Astron project see Joan Lisa Bromberg, *Fusion: Science, Politics, and the Invention of a New Energy Source* (Cambridge, Mass., 1982), 118–123, 201–204. On Project Argus see Walter Sullivan, *Assault on the Unknown: The International Geophysical Year* (New York, 1961), 108–163.

ALLAN A. NEEDELL

CLAUSEN, JENS CHRISTEN (*b.* Eskilstrup, Denmark, 11 March 1891; *d.* Palo Alto, California, 22 November 1969), *plant genetics*.

The son of Christen Augustinus and Christine (Christensen) Clausen, Jens was brought up on a farm. From the ages of fourteen to twenty-two, he read extensively in genetics and plant breeding, and learned the principles of physics and chemistry through self-organized experiments. A schoolmaster named Thorgille interpreted Mendelism and Darwin's *Origin of Species* for him. While a student at the University of Copenhagen (1913–1920), Clausen supported himself by teaching part-time. After receiving the master of science degree in 1920 and being appointed assistant professor in 1921, he began research on ecological genetics under Wilhelm Johannsen, a leading Danish geneticist. He investigated two species of wild pansy (*Viola arvensis* and *Viola tricolor*) found in northern Europe. His monograph (1926) was the first publication to analyze simultaneously, for any group of plant species, variation within populations, ecological adaptations, chromosome numbers, and the behavior of interspecific hybrids and their progeny.

Clausen was invited to spend the period from 1927 to 1928 in California as a Rockefeller fellow, to collaborate with Ernest B. Babcock on investigations of chromosome behavior in hybrids of the genus *Crepis*. Harvey M. Hall, director of a research program organized by the Carnegie Institution of Washington that was designed to explore the ecological genetics of plants native to California, had become much interested in Clausen's work and, after their meeting in 1927, in Clausen himself. In 1928 Clausen returned to Copenhagen, where Hall visited him and saw his research firsthand. Realizing that cytological data such as Clausen had obtained were essential to his program, Hall added him to his team, which included the taxonomist David Keck and the physiologist William Hiesey.

In the autumn of 1931, Clausen had hardly arrived at Stanford, where the team was established in a recently constructed laboratory, when Hall was taken ill (he died in March 1932). Clausen became the senior member (in age) of the team and its leader. They continued the technique of cloning plants belonging to several species in distantly related genera and growing clonal divisions in the three gardens that Hall had already established: at Stanford, at Mather, and at Timberline (the latter two in the Sierra Nevada, at 4,600 and 10,000 feet, respectively). The results of this research, published in 1940, showed clearly that adaptation to different altitudes and climates had been accompanied in each species by profound changes that established a series of genetically different races or ecotypes.

Continuing this line of research, Clausen hybridized timberline with lowland races of *Potentilla glandulosa* and analyzed progeny of first and second generations by the same clonal method. The team showed (1) that each character difference between ecotypes is governed by several different gene pairs; (2) that partial correlations between many characters exist in the F_2 generation; and (3) that genetic recombination confers upon some of the F_2 segregates adaptive properties not found in either parent. The latter results were reported in 1958.

Clausen, Keck, and Hiesey simultaneously continued a second line of research, also begun by Hall, on the nature of plant species. They analyzed progeny from about three hundred hybridizations between the approximately eighty-five annual species and named varieties of California tarweeds (Compositae subtribe Madiinae). Although their results were equal in importance to those obtained by the transplant method in *Potentilla glandulosa* and other perennial species, they are less known. This is because Clausen was too much of a perfectionist. He did not publish short papers that reported results as they were obtained, for fear that they might later need to be qualified, but almost exclusively long monographs in which a whole series of experiments were presented and analyzed together.

Since he never felt that the data on Madiinae were complete enough to warrant this kind of publication, much of the information was published only in summary fashion, or not at all. Its scope and the major conclusions drawn from it are in a slim volume, *Stages in the Evolution of Plant Species* (1951), presented at Cornell University as his Messenger Lectures. He showed (1) that a few characteristics used by taxonomists to differentiate species are inherited in simple Mendelian fashion and vary within populations, so that they are not diagnostic of biological species; (2) that most of the species are separated from each other by hybrid sterility based upon gene-controlled barriers of reproductive isolation, but that similar barriers may exist between two populations that are genetically connected via other populations with which they are interfertile; (3) that strong genetic barriers sometimes exist between populations that are virtually indistinguishable morphologically or ecologically; (4) that evolution is often reticulate below the level of the genus, but branching or furcate above this level; and (5) that some (but not all) reticulation is due to hybridization accompanied by chromosome doubling or polyploidy.

The final line of research by Clausen and his team was in response to World War II. He tried to apply his methods to native grasses in order to produce strains that could be used to revegetate overgrazed rangelands and thus increase the supply of native feed available for livestock. This goal was not achieved. Nevertheless, the investigations by Clausen and Hiesey of asexual (apomictic) seed production in bluegrasses of the genus *Poa* were of considerable scientific value.

Clausen's life was devoted to research. He never taught but did direct the research of graduate students, the most notable of whom was Robert K. Vickery. Deeply religious, he was active in the First Baptist Church of Palo Alto and a trustee of the Berkeley Baptist Divinity School. On 21 October 1921 he married Anna Hansen; they had no children. The Clausens came to the United States in 1931 and were naturalized in 1943. Clausen was named professor of biology at Stanford in 1951 and retired five years later.

Chief among Clausen's honors were a certificate of merit from the Botanical Society of America (1956); an honorary degree of Doctor of Agronomy from the Royal Agricultural College of Sweden; honorary membership in the Botanical Society of Edinburgh; election to the U.S. National Academy of Sciences, the Royal Swedish Academy of Sciences, and the Royal Danish Academy of Sciences and Letters; and knighthood in the Order of Dannebrog, conferred on him in 1961 by King Frederick IX of Denmark.

BIBLIOGRAPHY

I. ORIGINAL WORKS. Clausen's writings include "Genetical and Cytological Investigations on *Viola tricolor* L. and *V. arvensis* Murr.," in *Hereditas*, **8** (1926), 1–156; *Experimental Studies on the Nature of Species*, I, *Effect of Varied Environments on Western North American Plants*, Carnegie Institution of Washington Publication no. 520 (Washington, D.C., 1940), written with David D.

Keck and William M. Hiesey; *Experimental Studies on the Nature of Species*, II, *Plant Evolution Through Amphiploidy and Autoploidy, with Examples from the Madiinae*, Carnegie Institution of Washington Publication no. 564 (Washington, D.C., 1945), written with David D. Keck and William M. Hiesey; *Experimental Studies on the Nature of Species*, III, *Environmental Responses of Climatic Races of Achillea*, Carnegie Institution of Washington Publication no. 581 (Washington, D.C., 1948), written with David D. Keck and William M. Hiesey; *Stages in the Evolution of Plant Species* (Ithaca, N.Y., 1951); and *Experimental Studies on the Nature of Species*, IV, *Genetic Structure of Ecological Races*, Carnegie Institution of Washington Publication no. 615 (Washington, D.C., 1958), written with William M. Hiesey.

II. SECONDARY LITERATURE. For an obituary, see *Carnegie Institution of Washington Year Book*, **69** (1969–1970).

G. LEDYARD STEBBINS, JR.

CLELAND, RALPH ERSKINE (*b*. LeClaire, Iowa, 20 October 1892; *d*. Bloomington, Indiana, 11 June 1971), *botany, genetics*.

The concept of mutation as the basis of evolutionary change had its origin in Hugo de Vries's studies of *Oenothera*, the evening primrose. The validity of the mutation theory was challenged, however, when it was realized that the breeding behavior of *Oenothera* is atypical. Although *Oenothera* was much studied by the early geneticists, its genetic nature remained largely a puzzle until Ralph Cleland elucidated the unique cytogenetic mechanism that lay at its basis.

Cleland was the son of Edith Collins Cleland and Charles Samuel Cleland. When he was eighteen months old, his father, a United Presbyterian minister, accepted a call from a church in Philadelphia. Cleland grew up in a low-income urban neighborhood but spent the summers at a cottage in rural Pennsylvania, where he developed his interest in botany and natural history.

After being educated in the Philadelphia schools, Cleland received a four-year scholarship to the University of Pennsylvania, where he majored in the classics and took several courses in botany. Upon receiving the A.B. in 1915, he accepted the offer of a graduate assistantship in the department of botany, where he pursued the Ph.D. under the direction of B. M. Davis. His dissertation, completed in 1918, was a cytological study of the life history of the red alga *Nemalion multifidum*. Immediately after submitting his dissertation for publication, Cleland was inducted into the army and sent to France with a field artillery unit. In a matter of weeks the armistice was signed, and he returned to the United States. He was discharged from the army in April 1919. He married Elizabeth P. Shoyer on 11 June 1927; they had three sons.

After obtaining an instructorship at Goucher College that began in the fall of 1919, Cleland was at loose ends for the summer and offered to assist Davis with his *Oenothera* work. He chose *Oenothera franciscana*, a strain that happened to be at hand, to determine the best methods of fixation and staining for cytological study. His cytological preparations turned out to be of greater interest than a mere test of technique, revealing that four of the fourteen chromosomes regularly formed a closed circle at meiosis. This observation launched a series of studies that were of major importance in solving the genetic enigma presented by *Oenothera*.

The meticulous genetic analysis of *Oenothera* published by Otto Renner in 1917 first demonstrated that many species of *Oenothera* are permanent heterozygotes, so maintained by balanced lethal factors. Cleland's initial study of *Oenothera franciscana* was the first step toward explaining the physical basis of the mechanism that Renner's analysis had revealed. Cleland's subsequent studies during the 1920's showed that chromosome circles were widespread in the various strains of *Oenothera* and that the atypical breeding behavior of *Oenothera* could be understood in terms of its unique chromosomal mechanism.

In the academic year 1927–1928, a Guggenheim fellowship enabled Cleland to spend a year in collaborative efforts with the German *Oenothera* workers Friederich Oehlkers and Otto Renner. This resulted in convincing evidence that the number of linkage groups in various races of *Oenothera* was precisely correlated with the number of pairs and/or circles of chromosomes at meiosis. Here was rigorous proof of the correlation between gene and chromosome behavior.

When John Belling explained circle formation in *Datura stramonium* as the result of exchanges of segments between nonhomologous chromosomes, Cleland recognized that this concept of segmental interchange could be applied to *Oenothera*. He developed a scheme in which the chromosome complement of each *Oenothera* strain could be identified by the arrangement of its fourteen chromosome ends. He showed that if the arrangement of two genomes was known, when they were combined in a hybrid, the chromosome configuration at meiosis of the hybrid could be predicted.

Further, Cleland recognized that similarity in the segmental arrangement of the genomes in different strains of *Oenothera* could serve as an index of phylogenetic relationship. Thus, during the remainder of his research career he carried out a cytogenetic analysis of more than three hundred collections from natural populations of *Oenothera*. This led to the characterization of the different groups of North American *Oenothera* species and their evolutionary relationships. These studies, extending over nearly thirty years, are summarized in *Oenothera: Cytogenetics and Evolution* (1972), which Cleland completed just before his death in 1971.

In 1938 Cleland left Goucher College to assume the chairmanship of the department of botany at Indiana University, where he remained for the rest of his life. From 1950 to 1958 he served as dean of the Graduate School. Cleland's research career of some fifty years was characterized by consistently significant publications. His distinction in research brought him national recognition and election to leadership positions in professional scientific societies. He was a member of the National Academy of Sciences, the American Philosophical Society, and the American Academy of Arts and Sciences. He served as president of the Genetics Society of America, the Botanical Society of America, the American Society of Naturalists, and the Indiana Academy of Science. Cleland held honorary degrees from Hanover College, the University of Pennsylvania, and Indiana University. Other honors included the John F. Lewis Award of the American Philosophical Society, the Golden Jubilee Merit Citation of the Botanical Society of America, and honorary membership in the Genetics Society of Japan and the Botanical Society of Korea.

Essentially a modest man, Cleland had a calm, unruffled approach to problems that inspired confidence in his judgment. He was a conscientious teacher who set high standards for his students, and willingly taught at the introductory as well as at the graduate level. In spite of a busy schedule he was never impatient or short of time when students consulted with him. Retirement brought little change in Cleland's activities. He remained active in research and continued to participate in academic and scientific affairs until his death.

BIBLIOGRAPHY

I. Original Works. Cleland's scientific writings include "The Reduction Division in the Pollen Mother Cells of *Oenothera franciscana*," in *American Journal of Botany*, **9** (1922), 391–413; "Chromosome Behavior During Meiosis in the Pollen Mother Cells of Certain Oenotheras," in *American Naturalist*, **59** (1925), 475–479; "The Genetics of *Oenothera* in Relation to Chromosome Behavior, with Special Reference to Certain Hybrids," in *Zeitschrift für induktiv Abstammungs- und Vererbungslehre*, supp. vol. **1** (1928), 554–567; "Erblichkeit und Zytologie verschiedener Oenotheren und ihrer Kreuzungen," in *Jahrbuch der wissenschaftliche Botanik*, **73** (1930), 1–124, written with Friedrich Oehlkers; "Interaction Between Complexes as Evidence for Segmental Interchange in *Oenothera*," in *Proceedings of the National Academy of Sciences*, **16** (1930), 183–189, written with Albert F. Blakeslee; "Analysis of Wild American Races of *Oenothera (Onagra)*," in *Genetics*, **25** (1940), 636–644; "Phylogenetic Relationships in *Oenothera*," in *Proceedings of the Eighth International Congress of Genetics*, *Hereditas* (1949), SV 173–188; and *Oenothera: Cytogenetics and Evolution* (New York, 1972). A personal memoir is *Recollections: The Early Years* (Bloomington, Indiana, 1974).

II. Secondary Literature. The most complete account of Cleland's life and work is Erich Steiner, "Ralph Erskine Cleland," in *Biographical Memoirs. National Academy of Sciences*, **53** (1982), 120–139, with a complete bibliography.

Erich Steiner

COCHRAN, WILLIAM GEMMELL (*b.* Ruthglen, Scotland, 15 July 1909; *d.* Orleans, Massachusetts, 29 March 1980), *statistics*.

Cochran was the son of Thomas and Jeannie Cochran. Thomas Cochran, the eldest of seven children, at age thirteen had to take a job with a railroad company. The Cochrans moved several times, finally settling in Glasgow, where in 1927 William was first in the Glasgow University Bursary Competition; this award enabled him to finance his studies at the university, from which he received an M.A. with first-class honors in mathematics and physics in 1931. He shared the Logan Medal for being the most distinguished graduate of the Arts Faculty. As a result he secured a scholarship for graduate work in mathematics at Cambridge.

John Wishart had transferred from the Rothamsted Experimental Station to Cambridge in 1931; fortunately for statistics, Cochran elected to take Wishart's course in mathematical statistics, followed by his practical statistics course in the School of Agriculture. Cochran wrote his important paper presenting "Cochran's theorem" (1934) under Wishart. In the same year he was offered a position at Rothamsted that had become available when R. A. Fisher left to accept the Galton chair in eugenics at University College, London, and Frank Yates had moved up to become head of the statistics department. Cochran had to decide whether he would complete his doctorate at Cambridge or accept the

Rothamsted position. He later confided that it was not a difficult decision, because Great Britain (like the rest of the world) was in the throes of the Great Depression and few positions of this caliber were open. He did receive an M.A. from Cambridge in 1938. In his biographical sketch on Cochran, G. S. Watson states that Yates remarked, ". . . it was a measure of his good sense that he [Cochran] accepted my argument that a Ph.D., even from Cambridge, was little evidence of research ability, and that Cambridge had at that time little to teach him in statistics that could not be much better learnt from practical work in a research institute."

Cochran stayed at Rothamsted for five years. During that time he worked closely with Yates on experimental designs and sample survey techniques and had many opportunities to discuss problems with Fisher, who continued to spend much time at Rothamsted. By the time he left, Cochran had published twenty-three papers and had become a well-known statistician. One of his most exhaustive projects was a review of the long-term series of field experiments at the Woburn Experimental Station. Cochran and Yates collaborated on research on the analysis of long-term experiments and groups of experiments; here Cochran initiated his illustrious research on the chi-squared distribution and the analysis of count data. On 17 July 1937 Cochran married Betty I. M. Mitchell, who had a Ph.D. in entomology. The Cochrans were a popular couple, participating in many social activities. They had two daughters and a son. Cochran visited the Iowa State Statistical Laboratory in 1938 and accepted a position there in 1939 to develop a graduate program in statistics (it was part of the mathematics department until 1947). There he and Gertrude Cox initiated their collaboration that culminated in their famous book *Experimental Designs* (1950).

Late in 1943 Cochran took leave from Iowa State to join S. S. Wilks's Statistical Research Group at Princeton University as a research mathematician working on army-navy research problems for the Office of Scientific Research and Development. Much of his work there was devoted to an analysis of hit probabilities in naval combat that utilized little of his statistical background. In 1945 he was asked to serve on a select team of statisticians to evaluate the efficacy of the World War II bombing raids.

In 1946 Cochran joined the newly created North Carolina Institute of Statistics (directed by Gertrude Cox) to develop a graduate program in experimental statistics at North Carolina State College (now University); Harold Hotelling was to develop a graduate program in mathematical statistics at the University of North Carolina at Chapel Hill. Cochran was a member of the organizing committee for the International Biometric Society, which was founded in 1947 at Woods Hole, Massachusetts. His major contribution at North Carolina State was setting a firm foundation for a graduate program balanced in theory and practice and well coordinated with the more theoretical program at Chapel Hill.

In January 1949 the Cochrans moved to Baltimore, where he chaired the biostatistics department in the School of Hygiene and Public Health at the Johns Hopkins University. Since he was faced with medical rather than agricultural problems there, he had to develop procedures to obtain reliable information from observations rather than from experimental data, an area that became his dominant interest for the rest of his life. In 1963 he published *Sampling Techniques*.

Cochran remained at Johns Hopkins until 1957, when he joined the faculty at Harvard University to help Fred Mosteller and others develop the department of statistics. He continued to work closely with research workers at the Medical School and School of Public Health but also did his own research on a variety of topics. In 1967 Cochran was coauthor with G. W. Snedecor of the sixth edition of the latter's *Statistical Methods*. He retired from Harvard in 1976. Despite a dozen years of serious health problems, Cochran continued a wide range of professional activities and was working on the seventh edition of *Statistical Methods* and a book on observational studies until shortly before his death.

Cochran was president of the Institute of Mathematical Statistics (IMS) in 1946 and the American Statistical Association (ASA) in 1953; he served as editor of the *Journal of the ASA* (1945–1950). He was president of the Biometric Society (1954–1955) and of the International Statistical Institute (1967–1971); he was vice president of the American Association for the Advancement of Science (AAAS) in 1966. Cochran was elected to the National Academy of Sciences in 1974. He was a fellow of the ASA, the IMS, the AAAS, and the Royal Statistical Society, and was a Guggenheim fellow (1964–1965). He served on a number of scientific investigatory panels, including those concerned with the Kinsey Report, the efficacy of the Salk polio vaccine, the effects of radiation at Hiroshima, and the surgeon general's report on smoking. He wrote more than one hundred scientific articles (which are classified in Anderson).

Probably Cochran's greatest contributions to the scientific community were his guidance of students (he directed more than forty Ph.D. dissertations) and his textbooks. He had the ability to present complicated material in a format that could be understood by anyone who had an interest in collecting and analyzing data. He could explain where the usual assumptions might fail and take steps to ameliorate the effects of these failures. Cochran was quite willing to modify his statistical techniques when faced with such contingencies and often advocated approximate procedures, even though they might violate some of the accepted norms. Although he realized that there are many imperfections in the collection of data, the fact that an ad hoc statistical technique might modify (in an unknown way) the accepted probability levels did not deter him from using it. This pragmatic approach to the collection and analysis of data is held by many to be his most important contribution.

Cochran was that rarity, a man with both a keen mind and the desire to use it for the benefit of mankind. His office was always open to the struggling student, nonplussed scientist, or inquiring citizen.

BIBLIOGRAPHY

I. ORIGINAL WORKS. A bibliography of Cochran's works is part of Morris Hansen and Frederick Mosteller, "William Gemmell Cochran," in *Biographical Memoirs. National Academy of Sciences*, **56** (1987) 61–89. His books are *Fifty Years of Field Experiments at the Woburn Experimental Station* (London, 1936), written with E. J. Russell and J. A. Voelcker; *Experimental Designs* (New York, 1950; 2nd ed., 1957), written with Gertrude Cox; *Sampling Techniques* (New York, 1963; 3rd ed., 1977); *Statistical Problems of the Kinsey Report* (Washington, D.C., 1954), written with Frederick Mosteller and John W. Tukey; *Statistical Methods* (6th ed., Ames, Iowa, 1967; 7th ed., 1980), written with G. W. Snedecor; *William G. Cochran: Contributions to Statistics*, Betty I. M. Cochran, comp. (New York, 1982); and *Planning and Analysis of Observational Studies*, Lincoln E. Moses and Frederick Mosteller, eds. (New York, 1983).

Cochran's papers include "The Distribution of Quadratic Forms in a Normal System," in *Proceedings of the Cambridge Philosophical Society*, **30** (1934), 178–191; "Long-Term Agricultural Experiments," in *Journal of the Royal Statistical Society*, supp. **6** (1938), 104–148; "The Use of the Analysis of Variance in Enumeration by Sampling," in *Journal of the American Statistical Association*, **34** (1939), 492–510; "Some Consequences When the Assumptions for the Analysis of Variance Are Not Satisfied," in *Biometrical Bulletin*, **3** (1947), 22–38; "The X^2 Test of Goodness of Fit," in *Annals of Mathematical Statistics*, **23** (1952), 315–345; "The Combination of Estimates from Different Experiments," in *Biometrics*, **10** (1954), 101–129; "Some Methods for Strengthening the Common X^2 Tests," *ibid.*, 417–451; "The Planning of Observational Studies of Human Populations," in *Journal of the Royal Statistical Society*, **128** (1965), 234–265; and "Errors of Measurement in Statistics," in *Technometrics*, **10** (1968), 637–666.

II. SECONDARY LITERATURE. Works on Cochran and/or his contributions are Richard L. Anderson, "William Gemmell Cochran, 1909–1980. A Personal Tribute," in *Biometrics*, **36** (1980), 574–578; Theodore Colton, "Bill Cochran: His Contributions to Medicine and Public Health and Some Personal Recollections," in *American Statistician*, **35** (1981), 167–170; Arthur P. Dempster, "Reflections on W. G. Cochran, 1909–1980," in *International Statistical Review*, **51** (1983), 321–322; Arthur P. Dempster and Frederick Mosteller, "In Memoriam William Gemmell Cochran (1909–1980)," in *American Statistician*, **35** (1981), 38; H. O. Hartley, "In Memory of William G. Cochran," in D. Krewski, R. Platek, and J. N. K. Rao, eds., *Current Topics in Survey Sampling* (New York, 1981); Poduri S. R. S. Rao, "Professor William Gemmell Cochran: Pioneer in Statistics, Outstanding Scientist and a Noble Human Being," *ibid.*; Poduri S. R. S. Rao and J. Sedransk, eds., *W. G. Cochran's Impact on Statistics* (New York, 1984); and G. S. Watson, "William Gemmell Cochran 1909–1980," in *Annals of Statistics*, **10** (1982), 1–10.

RICHARD L. ANDERSON

COLE, LEON JACOB (*b.* Allegany, New York, 1 June 1877; *d.* Madison, Wisconsin, 17 February 1948), *genetics*.

Leon Jacob Cole, the son of Elisha Kelly Cole and Helen Marion Newton, was a pioneer in the development and promotion of theoretical and applied genetics and a prominent leader in biology.

Cole was raised in a city but his love for animals and plants drew him to the country, where he worked on farms in the summers during his elementary and high school years. He spent the years 1894–1895 and 1897–1898 at Michigan Agricultural College, then took his A.B. at the University of Michigan in 1901 and the Ph.D. at Harvard in 1906. He married Margaret Belcher Goodenow on 28 August 1906; they had a daughter and a son.

Cole's abilities were recognized early in his academic career. He was appointed assistant in zoology at Michigan (1898–1902) and Austin teaching fellow at Harvard (1903–1905) and selected as a member of the Harriman expedition to Alaska (1899), on which occasion he associated with eminent ornithologists, mammalogists, and botanists. During summers he was an investigator at the Bermuda Biological Station (1903); and in other summers between 1901 and 1906, and in 1909, he worked at the Woods Hole Marine Laboratories for the U.S. Bureau of Fisheries. He also was a member of the

Yucatan Zoological Expedition (1904), on which occasion he assisted in identifying 128 species of birds.

In 1901 Cole wrote a paper titled "Suggestions for a Method of Studying the Migration of Birds," in which he proposed the use of leg bands in studying migration and other behavior of birds. This appears to be the first time that this procedure had been recommended for such studies.

Cole became chief of the Division of Animal Breeding and Pathology at the Rhode Island Experiment Station in 1906 and started his lifelong research on pigeons. He also published a monograph for the U.S. Fish Commission, "The German Carp in the United States" (1905). From 1907 to 1910 he was instructor in zoology at the Sheffield Scientific School at Yale, where he continued his research on pigeons.

Cole joined the faculty of the University of Wisconsin in 1910 and remained there until his death, except during 1923–1924, when he was chief of the Division of Animal Husbandry, Bureau of Animal Industry, U.S. Department of Agriculture. He came to the university to establish and chair a department of experimental breeding (in 1918 renamed department of genetics), the first of its kind in the United States. He continued as chairman until 1939, by which time sixty-two Ph.D. and sixty-one M.S. degrees had been conferred on students majoring in genetics. By the late 1930's, most of the leading American, and many of the foreign, workers in this field had been his students or had been trained by his students.

Cole believed in the desirability of maintaining programs in fundamental and applied research within the department. In the early years he supervised students in both plant and animal genetics, and gave graduate students wide latitude of choice in their research problems. He was inherently a naturalist, but at the same time was an experimentalist with broad interests. His early love for birds continued throughout his life, and his research on pigeons and doves and their crosses was the basis for about half of his scientific papers.

The early research of Cole and his students was responsible for establishing the genetic background of many of the color phases in pigeons, including an early example of sex-linked inheritance. He produced a new variety of domesticated dove by an interspecific transfer of a gene for intensity of feather pigmentation from a cross of a wild species of dove with the ring dove, followed by backcrossing to the ring dove. The basis for morphological differentiation of New World from Old World pigeons was furnished by a study of an extensive collection of material at the Field Museum of Natural History in Chicago.

Cole's collection of pedigreed pigeons and doves and their crosses supplied research animals for the extensive immunogenetic studies of M. R. Irwin and his students. In 1912 Cole and his colleagues in the poultry husbandry department started what was probably the first systematic inbreeding experiment with poultry. Other genetic studies on domestic fowl and ducks were undertaken by his students.

Cole was very much aware of the importance of dealing with problems facing the livestock industry. In 1912 he began an experiment involving crosses of Angus and Jersey, and later Angus and Holstein, cattle. The primary purpose was to study the genetic basis for milk and beef production. Many of the caveats that are now taken into account in designing experiments dealing with quantitative traits had not been established by 1912. The large variability within the groups and the small numbers of individuals in the experiment allowed very few unequivocal conclusions to be drawn, but the experiment did lay the groundwork for future research.

Cole's genetic analysis of red calves in "black" breeds, of "seedy cut" in bacon, of epithelial defects in cattle, and of defects in hair and teeth in cattle provided explanations for conditions of practical importance.

Cole was largely responsible for developing the eugenic point of view in the group promoting the birth control movement led by Margaret Sanger.

During the latter part of his life, Cole's research with R. M. Shackelford supplied mink and fox breeders with interpretations of the genetic bases for color phases and many other important traits of these species.

Cole's contributions to science and to society were recognized by the American Society of Animal Production as their Guest of Honor in 1939, at which time his portrait was hung in the Saddle and Sirloin Club, Chicago. In 1945 he was awarded an honorary Sc.D. by Michigan State College. His breadth of knowledge, research contributions, administrative accomplishments, and willingness to cooperate brought requests for numerous talks and articles on broadly gauged topics as well as appointment and election to high positions in many organizations.

The following quotation from one of his most distinguished former graduate students, Ivar Johansson, undoubtedly reflects the opinion of Cole held by all of his discerning students: "Dr. Cole's graduate students all carry with them the memory of a scientist and teacher with a clear and penetrating mind, always seeking the truth honestly and straightforwardly without preconceptions and without any poses of personal authority, always exceedingly generous and willing to give the help and advice that was needed."

The faculty counterpart of this appraisal is found in the following statement by his longtime departmental colleague, R. A. Brink: "He was recognized by [the staff] as able, stimulating, actively unselfish and ready to renounce personal ambition for the general good. [He] had a most charitable view of human limitations, not excluding his own, and so, with his other qualities, enjoyed the friendship and confidence of many kinds of people. Thus, he was enabled to build up relationships with others in the college on a complementary basis whereby the possibilities which genetics held for agriculture could be realized in practice."

BIBLIOGRAPHY

I. ORIGINAL WORKS. Cole's papers after 1910 (except "The Early History of Bird Banding") are available in bound reprint files in the department of genetics, University of Wisconsin, Madison.

Selected papers are "Suggestions for a Method of Studying the Migrations of Birds," in *Michigan Academy of Science*, **3** (1902), 67–70; "The German Carp in the United States," in *Report of the U.S. Bureau of Fisheries* for 1904 (1905), 523–641; "A Case of Sex-Linked Inheritance in the Domestic Pigeon," in *Science*, n.s. **36** (1912), 190–192; "Studies on Inheritance in Pigeons. I. Hereditary Relations of the Principal Colors," *Rhode Island Experiment Station Bulletin* no. 158 (1914); "A Defect of Hair and Teeth in Cattle—Probably Hereditary," in *Journal of Heredity*, **10** (1919), 303–306; "The Occurrence of Red Calves in Black Breeds of Cattle," *Wisconsin Agricultural Experiment Station Bulletin* no. 313 (1920), written with S. V. H. Jones; "The Early History of Bird Banding in America," in *The Wilson Bulletin*, **34** (1922), 108–115; "The Inbreeding Problem in the Light of Recent Experimentation," in *Proceedings of the American Society of Animal Production* for 1921 (1922), 30–32; "The Wisconsin Experiment in Crossbreeding of Cattle," in *Proceedings of the World's Dairy Congress*, II (1924), 1383–1388; "Inherited Epithelial Defects in Cattle," *Wisconsin Agricultural Experiment Station Research Bulletin* no. 86 (1928), written with F. Hadley; "A Triple Allelomorph in Doves and Its Interspecific Transfer," in *Anatomical Record*, **47** (1930), 389 (abstract).

See also " 'Seedy Cut' as Affecting Bacon Production," *Wisconsin Agricultural Experiment Station Research Bulletin* no. 118 (1933), written with J. S. Park and Alan Deakin; "Immunogenetic Studies of Species and Species Hybrids from the Cross of *Columba livia* and *Streptopelia risoria*," in *Journal of Experimental Zoology*, **73** (1936), 309–318, written with M. R. Irwin; "The Origin of the Domestic Pigeon," in *Proceedings of the 7th World's Poultry Congress and Exposition* (1939), 462–466; "A Test of Sex Control by Modification of the Acid-Alkaline Balance," in *Journal of Heredity*, **31** (1940), 501–502, written with Emanuel Waletzky and R. M. Shackelford; "Differentiation of Old and New World Species of the Genus *Columba*," in *American Naturalist*, **76** (1942), 570–581, written with Russell W. Cumley; "Genic Control of Species-Specific Antigens of Serum," in *Journal of Immunology*, **47** (1943), 35–51, written with Russell W. Cumley and M. R. Irwin; "The Genetic Sex of Pigeon-Ring Dove Hybrids as Determined by Their Sex Chromosomes," in *Journal of Morphology*, **72** (1943), 411–439, written with T. S. Painter; "Immunogenetic Studies of Cellular Antigens: Individual Differences Between Species Hybrids," in *Genetics*, **30** (1945), 439–447, written with M. R. Irwin; "Inheritance in Crosses of Jersey and Holstein-Friesian with Aberdeen-Angus Cattle," in *American Naturalist*, **82** (1948), 145–170, 202–233, 265–280, written with Ivar Johansson; and "Hybrids of Pigeon by Ring Dove," *ibid.*, **84** (1950), 275–308, written with W. F. Hollander.

II. SECONDARY LITERATURE. Biographies of Cole are R. A. Brink, "Early History of Genetics at the University of Wisconsin-Madison" (Madison, Wis., 1974), mimeographed; W. C. Coffey, "Dr. Cole and Animal Breeding Science," in *Proceedings of the 32nd Meeting of the American Society of Animal Production* (1939), 454–463, with a response from Cole, 463–466; Ivar Johansson, "Leon Jacob Cole 1877–1948," in *Genetics*, **46** (1961), 1–4; E. W. Lindstrom, "The Influence of Wisconsin's Genetics Department," in *Proceedings of the 32nd Meeting of the American Society of Animal Production* (1939), 451–454; R. A. McCabe, "Wisconsin's Forgotten Ornithologist: Leon J. Cole," in *Passenger Pigeon*, **41** (1979), 129–131; and H. L. Russell, "Establishing a Department of Genetics," in *Proceedings of the 32nd Meeting of the American Society of Animal Production* (1939), 448–450.

A. B. CHAPMAN

CONANT, JAMES BRYANT (*b.* Dorchester, Massachusetts, 26 March 1893; *d.* Hanover, New Hampshire, 11 February 1978), *organic chemistry, science policy*.

Conant was the third child and only son of James Scott Conant and Jennett Orr Bryant. His mother was deeply involved in transcendental religious movements. His father was a photoengraver who had a shop laboratory that he used to master the chemical procedures in this new field. In 1903 James's interest in chemistry caused his parents to enroll him in Roxbury Latin School, which had a strong program in physics and chemistry. Its gifted science teacher, Newton Henry Black, encouraged Conant to do college-level work, enabling him to enter Harvard in 1910 with advanced standing (B.A., 1913; Ph.D., 1916). During World War I, Conant was a major in the Chemical Warfare Service. From 1919 to 1933 he was a member of the Harvard chemistry department. On 17 April 1921 he married Grace Thayer Richards, the daughter of Theodore William Richards, Harvard's leading chemist and Nobel laureate; they had two sons. His research career ended in 1933 with his unexpected selection as president of Harvard.

An innovative leader, Conant created new types of scholarships and professorships, and new procedures for appointments, promotions, and tenure to ensure faculty excellence. He fostered coeducation by eliminating separate classes for Radcliffe and Harvard students and by admitting women to the law and medical schools. Following World War II he ordered the cessation of all classified research as inimical to the mission of a university. The 1945 Harvard report *General Education in a Free Society* addressed itself to the need to maintain a general liberal education at the college level and proposed a required common core of courses in the humanities, social sciences, and natural sciences that stressed general relationships and values. Conant himself taught in the natural science part of the program for three years and developed the case history method of teaching science. He was the general editor of the *Harvard Case Histories in Experimental Science* (1948; reissued 1957). In 1950 he was a principal advocate before Congress for federal support of science and the formation of the National Science Foundation.

In 1953 Conant resigned the Harvard presidency to become U.S. high commissioner to West Germany and, when the Federal Republic attained full sovereignty, U.S. ambassador until 1957. His last major activity was as reformer of the American secondary schools. His studies, financed by the Carnegie Corporation, yielded ten widely read books (1959–1967) on public education and served as a catalyst for extensive reforms in schools. He was a member of the National Academy of Sciences and the Royal Society of London.

Conant intended to follow Theodore W. Richards into physical chemistry. In 1912, however, Elmer P. Kohler joined the Harvard faculty. Through Kohler, Conant became enchanted with organic chemistry and did a double doctoral dissertation under these men. His major interest thereafter was organic chemistry from the viewpoint of physical chemistry. His two major influences were physical chemists: the Dane Johannes N. Brønsted and the American Gilbert N. Lewis. Conant transferred their conceptions in thermodynamics, acids and bases, the transition state, and the shared electron-pair bond to organic chemistry. During his fifteen-year research career (1919–1934) he was a major contributor to the flourishing of physical organic chemistry in the United States; his many gifted students became leaders in several areas of the field.

Conant's interest in reaction mechanisms was evident from his first studies and in the 1920's grew into a series of important studies on the relative reactivities of quinones and organic halides. By 1924 he had established quantitative relationships between the thermodynamic and kinetic factors in these reactions and had shown how changes in structure affected the reaction rate. His work with organic halides also refuted the theory of alternating polarity along the carbon chain. Before Lewis' idea of the shared electron-pair bond (1916), dualistic theories of valence prevailed. The alternating polarity theory claimed that the carbon atoms in a saturated chain were alternately more and less positive. Therefore, the reactivity of alkyl halides should rise and fall with the lengthening of the carbon chain. Conant measured the reaction rates of organic halides and found no support for the theory. Instead, Lewis' theory, with its possibility of partial charges due to the degree of sharing in the electron-pair bond, served as a means to predict the relation between structures and reaction rates in the organic halide series.

Conant's studies on carbonyl addition reactions were important as a means to learn about reaction mechanisms. The Lewis electron-pair sharing theory of valency suggested that the carbonyl group had a polar character:

$$\mathrm{R_2C{=}O} \leftrightarrow \mathrm{R_2C^{+}{-}O^{-}}$$

Conant saw that this dipolar character could explain various carbonyl addition reactions, and he successfully explored the idea in several such reactions. His 1932 paper on semicarbazone formation from carbonyl compounds, written with his student Paul D. Bartlett, became a classic example of acid-base catalysis and of the way in which the distinctive thermodynamic and kinetic factors control the intermediate reactions and the products.

Conant also applied Lewis' ideas to the triphenylmethyl free radicals discovered by Moses Gomberg in 1900. He made reactivity studies of them and introduced a new method of preparing them by reduction of the corresponding carbonium ion by vanadous chloride (1923). Conant's papers on free-radical chemistry showed that bulky aliphatic groups in themselves cannot make a free radical stable, but they can enhance the stability of an already stabilized radical.

Among Conant's most important studies were those using the ideas of Brønsted on acids and bases. In 1927 he demonstrated the existence of "superacids," revealing that acids in a nonaqueous solvent may have an acidity enormously higher than in similar concentrations in water. This discovery provided a useful means to titrate bases that could not otherwise be titrated, and led to a deeper understanding of acids and bases in nonaqueous solvents. Conant utilized this method in his last researches on chlorophyll (1929–1934). He revealed that chlorophyll is a dihydroporphyrin, found the site where phytol is attached to the carbon skeleton of the molecule, and determined the relative basicities of the four pyrrole-nitrogen atoms, using potentiometric titration with the superacid perchloric acid in glacial acetic acid.

Although his career as a chemist was over by 1934, Conant's influence on American science continued, especially during World War II, when he was a major organizer of scientists. He was a member of the National Defense Research Committee, a science policy group formed in 1940 to mobilize scientists for the war effort. In 1941 he went to Great Britain to establish the exchange of scientific information with the British; there he learned about radar and established a London NDRC office, and the trickle of information soon became a flood.

From 1941 to 1945 Conant was chairman of NDRC, responsible for thousands of projects, including explosives, smoke generators, microwave radar, electronic countermeasures, and uranium fission. He also served as adviser to the successful synthetic rubber program. As the uranium fission project increased in size and variety, he devoted more and more time to it, overseeing its transformation from an energy development program into a weapons program. As member of the cabinet-level policy group supervising the project, he favored the use of the atomic bomb.

Conant was an architect of postwar atomic energy policy. In 1945 he went on a mission to Moscow, where he advocated international control of atomic weapons, with free access to all information and the right to inspect; he served as adviser to the American delegation at the United Nations that presented the plan for international controls vetoed by the Soviets at the end of 1946. President Truman offered Conant the chairmanship of the Atomic Energy Commission, but he declined because of the need to concentrate on his duties at Harvard as well as his awareness of the opposition of some scientists because of his approval of the bombing of Japan and his support of the May-Johnson Bill (to create the Atomic Energy Commission) in Congress. He was a member of the General Advisory Committee of the AEC (1947–1952). As chemist, educational reformer, head of a major university, and advocate of federal support for science, Conant played a major role in the rise of American science to its postwar position of prestige and high accomplishment.

BIBLIOGRAPHY

I. ORIGINAL WORKS. Important papers by Conant are "Addition Reactions of the Carbonyl Group Involving the Increase in Valence of a Single Atom," in *Journal of the American Chemical Society*, **43** (1921), 1705–1714; "The Formation of Free Radicals by Reduction with Vanadous Chloride. Preliminary Paper," *ibid.*, **45** (1923), 2466–2472, written with A. W. Sloan; "Reduction Potentials of Quinones. II. The Potentials of Certain Derivatives of Benzoquinone, Naphthoquinone, and Anthraquinone," *ibid.*, **46** (1924), 1858–1881, written with Louis Fieser; "The Relation Between the Structure of Organic Halides and the Speed of Their Reaction with Inorganic Iodides. I. The Problem of Alternating Polarity in Chain Compounds," *ibid.*, 232–252, written with W. R. Kirner; "A Study of Superacid Solutions. I. The Use of the Chloranil Electrode in Glacial Acetic Acid and the Strength of Certain Weak Bases" and "II. A Chemical Investigation of the Hydrogen-Ion Activity of Acetic Acid Solutions," *ibid.*, **49** (1927), 3047–3061 and 3062–3070, written with Norris F. Hall; "Studies in the Chlorophyll Series. I. The Thermal Decomposition of the Magnesium-free Compounds," *ibid.*, **51** (1929), 3668–3674, written with J. F. Hyde; "A Quantitative Study of Semicarbazone Formation," *ibid.*, **54** (1932), 2881–2899, written with Paul D. Bartlett; "The Study of Extremely Weak Acids," *ibid.*, 1212–1221, written with G. W. Wheland; and "Studies in the Chlorophyll Series. XIV. Potentiometric Titration in Acetic Acid Solution of the Basic Groups in Chlorophyll Derivatives," *ibid.*, **56** (1934), 2185–2189, written with B. F. Chow and E. M. Dietz.

The Harvard University Archives have a collection of Conant's correspondence. Papers relating to his World War II and postwar activities are in the National Archives.

II. SECONDARY LITERATURE. The most essential reference is Conant's autobiography, *My Several Lives: Memoirs of a Social Inventor* (New York, 1970). Fuller accounts of his work in chemistry are in Paul D. Bartlett, "James Bryant Conant,"in *Biographical Memoirs. National Academy of Sciences*, **54** (1983), 91–124; and George B. Kistiakowsky and Frank H. Westheimer, "James Bryant Conant," in *Biographical Memoirs of Fellows of the Royal Society*, **25** (1979), 209–232. Both articles contain a bibliography of his works. Brief but solid interpretations of his chemical career are Martin Saltzman, "James Bryant Conant and the Development of Physical Organic Chemistry," in *Journal of Chemical Education*, **49** (1972), 411–412; and Frank H. Westheimer, "James Bryant Conant," in *Organic Syntheses*, **58** (1978), vii–xi.

ALBERT B. COSTA

COOLIDGE, WILLIAM DAVID (*b*. Hudson, Massachusetts, 23 October 1873; *d*. Schenectady, New York, 3 February 1975), *physics*.

Coolidge spent most of his career doing and managing research in industry, making inventions of major commercial importance in the fields of lighting and X rays, and later directing the General Electric Research Laboratory. He was the only child of Albert Edward Coolidge, a farmer and shoe factory worker, and Martha Alice Shattuck Coolidge, a dressmaker. On 30 December 1908 he married Ethel Woodward; they had two children, Elizabeth and Lawrence. His wife died in 1915, and on 29 February 1916 he married Dorothy Elizabeth Machaffie.

Coolidge attended the public schools of Hudson and earned a state scholarship to the Massachusetts Institute of Technology, graduating in 1896 with a degree in electrical engineering. Summer work in a factory inclined him against an industrial career, and he chose the position of assistant in physics at M.I.T. He won a fellowship to study physics at the University of Leipzig in 1897 and 1898, under Gustav Wiedemann and Paul Drude, and earned his doctorate in 1898 with a dissertation on the determination of the dielectric constant of liquids.

Returning to M.I.T. in 1899, Coolidge taught physics for a semester before accepting a position as assistant to Professor Arthur A. Noyes, a physical chemist. When Noyes established the Research Laboratory of Physical Chemistry at M.I.T. in 1902, Coolidge joined him, focusing on the ionic theory of solutions. Coolidge's contributions were primarily experimental; for example, he designed a pressure vessel capable of sustaining high temperatures and pressures that Noyes and Coolidge used to study the properties of solutions at high temperatures.

A chemist and M.I.T. colleague, Willis R. Whitney, also served as director of the Research Laboratory of the General Electric Company, founded at Schenectady, New York, in 1900. After unsuccessfully trying to recruit Coolidge in 1902, he succeeded in 1905. Lures included a doubling of Coolidge's M.I.T. salary (he was still in debt from his years of graduate study) and a promise that he could bring his pressure vessel to Schenectady and spend one third of his time there doing pure research of his own choosing.

Coolidge brought the vessel, but immersed himself full-time in the most pressing problem facing the GE laboratory, the development of higher efficiency incandescent lamps. One route was by raising the melting point of the lamp's filament (hotter wires emit more visible light than cooler ones). Researchers in Europe had developed processes for mixing the refractory but brittle metal tungsten with a binder to make a ductile mixture, and forcing it through a diamond die to form a filament. The resulting filaments were usable but brittle after heat treatment had driven the binder off.

In 1906 Coolidge invented an improved version of this process, employing metallic rather than organic binders. Then, in an arduous empirical effort over the years 1907 to 1910, he developed a new continuous process for making tungsten wire. Blocks of hot sintered tungsten passed through a series of swaging, rolling, and drawing steps at gradually reduced temperatures. The tungsten grains gradually deformed from cubes to extended fibers, which yielded a wire that was ductile at room temperature. The great majority of all the incandescent lamps made in the world today are made by this "Coolidge process," which was one of the first inventions made by a scientist in a U.S. industrial laboratory to achieve large commercial success.

In 1911 Coolidge used tungsten as a heat-resistant target for bombardment by a high-voltage discharge to produce X rays. In 1913 he combined this with discoveries by GE colleague Irving Langmuir in electron physics to invent an X-ray tube based on a tungsten target bombarded in high vacuum with a discharge consisting overwhelmingly of electrons, rather than the previous mixture of electrons and ions. This made possible much more precise control over the frequency of X rays produced than in the previous tubes and also facilitated development of higher-voltage tubes. The improvements were so pronounced that this "Coolidge tube" became and remains the main type used in medical diagnostics.

During World War I, Coolidge helped develop a portable X-ray unit for field use and invented a shipboard acoustic device for locating submarines (the "C tube") that saw combat use. Over the next twenty years, he made many improvements in the technology of generating X rays and electron beams at high voltage, carrying those technologies up to millions of volts. Out of this work came the majority of his total of eighty-three patents, including techniques widely used in early experimental nuclear physics.

In 1932 Coolidge succeeded Whitney as director of the GE Research Laboratory. His reticence, modesty, formality, and quiet authority contrasted with his predecessor's outgoing enthusiasm. He continued Whitney's policy of putting most of the laboratory's effort into work of immediate commercial importance while maintaining a few small-scale efforts in purely scientific research of possible long-range industrial impact, such as electron and high-energy physics, and polymer and surface chemistry. The polymer work, in particular, paid off with later business successes in engineering plastics and silicones. Coolidge also served on several government advisory bodies, including the federal government's 1940–1941 Advisory Committee on Uranium, which concluded (though with less urgency than the recommendations of subsequent committees, which led to the Manhattan Project) that research on the possibility of an atomic bomb be pursued. He postponed retirement to direct the GE Research Laboratory throughout World War II, during which it was devoted to war-related efforts ranging from electronic devices for radar countermeasures to silicone rubber gaskets for battleship searchlights. After retiring from GE in 1945, he devoted the remainder of a long and active life to such hobbies as travel and photography.

BIBLIOGRAPHY

I. ORIGINAL WORKS. Coolidge published more than fifty papers in scientific journals, of which the most important describe his two major inventions: "Ductile Tungsten," in *Transactions of the American Institute of Electrical Engineers*, **29**, pt. 2 (1910), 961–965; and "A Powerful Röntgen Ray Tube with a Pure Electron Discharge," in *Physical Review*, 2nd ser., **2** (1913), 409–430. Other significant scientific papers include "Electrical Conductivity of Aqueous Solutions at High Temperatures," in *Proceedings of the American Academy of Arts and Sciences*, **39** (1903), 163–219, with Arthur A. Noyes; and "High Voltage Cathode Ray and X-Ray Tubes and Their Operation," in *Physics*, **1** (1931), 230–244, with L. E. Dempster and H. E. Tanis, Jr.

His papers are in the possession of his daughter, Elizabeth Coolidge Smith, in Portland, Oregon.

II. SECONDARY LITERATURE. Biographies of Coolidge include Herman A. Liebhafsky, *William David Coolidge: A Centenarian and His Work* (New York, 1974), which focuses on the invention of ductile tungsten; John Anderson Miller, *Yankee Scientist: William David Coolidge* (Schenectady, N.Y., 1963), the most complete; and C. Guy Suits, "William David Coolidge," in *Biographical Memoirs. National Academy of Sciences*, **53** (1982), 141–157. Coolidge's work is put into the context of the historical development of industrial research at GE and in the United States by Kendall A. Birr, *Pioneering in Industrial Research: A History of the GE Research Laboratory* (Washington, D.C., 1957); Leonard A. Reich, *The Making of American Industrial Research: Science and Business at GE and Bell, 1876–1926* (New York, 1985); and George Wise, *Willis R. Whitney, General Electric, and the Origins of U.S. Industrial Research* (New York, 1985).

GEORGE WISE

CORBINO, ORSO MARIO (*b*. Augusta, Sicily, 30 April 1876; *d*. Rome, Italy, 23 January 1937), *experimental physics*.

Corbino was the second son of seven children of Vincenzo Corbino, owner of a small pasta-making business, and Rosaria Imprescia. His father had been taught by the Franciscans, receiving an unusually good classical humanistic education; his mother, though unschooled, had a lively native intelligence. Their youngest son, Epicarmo, did not take a university degree yet became a university professor of economics and held cabinet posts in three governments after the fall of Mussolini in 1943. In 1901 Mario (as he was known) married Francesca Camilleri; they had a daughter and a son. Among Corbino's many honors were the Royal Prize for Physics for 1912 of the Accademia dei Lincei, the Mussolini Prize of the Accademia d'Italia (1933), and appointment as senator of the kingdom (1920).

Corbino's intellectual abilities were recognized early. While still in secondary school (*liceo*), he began to assist Adolfo Bartoli, professor of physics at the University of Catania. Corbino enrolled there in 1892 but transferred to Palermo when Bartoli accepted a post at Pavia in 1893. With a *laurea* in physics (1896), he began to teach in a *liceo* at Palermo while an assistant at the university. In 1902 Corbino was declared qualified to hold university professorships in both experimental physics and electrical engineering. In January 1906 he was appointed, after winning a national competition, to teach experimental physics at the University of Messina. In 1907 he won the competition for full professor. In 1909 he was moved to Rome by the minister of public instruction to teach a special physics course that the professor of experimental physics, Pietro Blaserna, had fostered to complement the laboratory work of advanced students. Corbino remained in that post at Rome until 1918, when, following the death of Blaserna, he was named professor of experimental physics and director of the physics institute, positions he held until his death.

Corbino's research may be divided into three major categories: magneto-optics; variable currents in inductive circuits, sound recording, and its radio transmission; and the electron theory of metals and the Corbino effect. Other research included the properties of nitroglycerine; the specific heats of metals at very high temperatures, which indicated large deviations from the Dulong-Petit law of constant atomic heats; and elastic deformations in solids by means of the accidental double refraction of transparent models under specific stresses, verifying the theory of Vito Volterra. Similar photoelastic studies have become widespread in engineering through the work of Ernest G. Coker and L. N. G. Filon.

The researches in magneto-optics were begun at Palermo, where, with Blaserna's early student, Professor Damiano Macaluso, Corbino published studies of what came to be called the Macaluso-Corbino effect. Related to the Faraday effect and the inverse Zeeman effect, it involves the anomalous rotatory power of sodium vapor in the vicinity of its absorption lines when placed in a magnetic field. A report presented to the Paris Academy of Sciences excited discussion abroad, and further research by Corbino yielded results in conflict with Woldemar Voigt's theory. In 1902 Pieter Zeeman undertook experiments that yielded Corbino's results under conditions not foreseen by Voigt. Corbino thus won an international reputation while in his twenties.

From 1899 on, Corbino undertook extensive research on oscillatory and otherwise variable currents in inductive circuits, Duddell's musical arc, the Wehnelt electrolytic interrupter, the Rühmkorff coil, magnetic hysteresis, and generators and motors, making early use of the Braun tube as an oscilloscope. His electrical investigations eventually encompassed high-voltage power supplies for X-ray tubes, triodes, and high-fidelity sound recording and its radio transmission. At his initiative the Istituto Nazionale di Elettroacustica was established in 1936; it was named for him after his death.

In 1911 Corbino began to publish on magnetoresistance in metals, especially bismuth, and revealed a phenomenon related to the Hall effect. The Corbino effect occurs when a circular metal disk with inner and outer concentric ring contacts is placed in a magnetic field along its axis; the current through the disk no longer moves radially but spirals around it, changing the resistance in the circuit. The 1972 proposal of G. P. Carver to use the Corbino disk arrangement in direct measurement of Hall mobilities in amorphous semiconductors has been broadly applied.

Of all Corbino's research, that close to electrical engineering paved the way for his public career. In 1916 he was an organizer of the national scientific-technical committee designed to promote links between the sciences and their related industries. Beginning in 1911, Corbino became a board member of electric utility companies, and afterward also of banks and telephone companies; eventually he became chairman of several boards and moved easily in the highest Italian industrial and banking circles. In 1921 and 1922 he served for seven months as minister of public instruction, the first experimental scientist to do so since Carlo Matteucci in 1862. In 1923 he became minister of the economy in a coalition cabinet led by Mussolini, although he was never a Fascist party member and had voted against Mussolini's first government in 1922. Corbino strongly favored the use of foreign technology and capital in Italian energy development, but an agreement he worked out with the Sinclair Oil Company aroused such vigorous opposition in both industrial and government circles, and was so tainted with rumors of collusion between Fascist bigwigs and Sinclair, that he resigned in June 1924.

Corbino's ambitious vision for the revival of Italian physics was unique among his contemporaries. From 1909 on, he propagandized for a new research tradition based on a close collaboration between theorists and experimenters, encouraged ties between science and national technical and economic planning in order to concentrate intellectual and financial resources, and worked to overcome endemic provincialism and backwardness. His great success was Enrico Fermi's group, whose quantum mechanical "experiments in theoretical physics" and prowess in nuclear physics were the local manifestation of what was intended to be a program for a national renaissance. The postwar international successes of Italian nuclear and particle physics are the legacy of both Fermi's achievements and Corbino's vision.

BIBLIOGRAPHY

I. ORIGINAL WORKS. Some of Corbino's public lectures and Senate speeches were collected in *Conferenze e discorsi di O. M. Corbino* (Rome, [1938]), which also contains a nearly complete list of his scientific and popular scientific writings. A few additional items, especially political and economic speeches, may be gleaned from the *Catalogo generale della libreria italiana*, compiled by Attilio Pagliaini and later by Arrigo Plinio Pagliaini; from *Catalogo cumulativo 1886–1957 del Bollettino delle pubblicazioni italiane*; and from the *Dizionario biografico degli italiani* article noted below. From 1908 to 1910 he published a physics text for secondary schools that went through eight editions by 1924. His 1929 speech asserting the exhaustion of traditional researches except for applications, claiming that no new forces or significant new phenomena would be found outside the atomic nucleus, and announcing the move of the Rome group into nuclear physics, has been translated by Fausta Segrè as "The New Goals of Experimental Physics," in *Minerva*, **9** (1971), 528–537, with an introductory note by Emilio Segrè.

Letters written by Corbino are in the Vito Volterra Papers and the Tullio Levi-Civita Papers, both at the Accademia Nazionale dei Lincei, Rome, and the Augusto Righi Papers, at the Accademia Nazionale dei Quaranta, Rome.

II. SECONDARY LITERATURE. The most recent comprehensive treatment of Corbino's life and multiple careers is Edoardo Amaldi and Luciano Segreto, "Corbino, Orso Mario," in *Dizionario biografico degli italiani*, XXVIII, 760–766; it contains extensive references not available elsewhere for particular aspects of his political career. An elaborate treatment of his program for the revival of Italian physics is given by Carlo Tarsitani, "La fisica italiana tra vecchio e nuovo: Orso Mario Corbino e la nascita del gruppo Fermi," in Giovanni Battimelli, Michelangelo De Maria, and Arcangelo Rossi, eds., *La ristrutturazione delle scienze tra le due guerre mondiali*, I (Rome, 1984), 323–346.

Indispensable for his childhood and family life is the autobiography of Epicarmo Corbino, *Racconto di una vita* (Naples, 1972). Obituaries may be traced through extracts reprinted in *Conferenze e discorsi* and the list in Poggendorff, VIIb. Other useful sources include Gerald Holton, "Striking Gold in Science: Fermi's Group and the Recapture of Italy's Place in Physics," in *Minerva*, **12** (1974), 159–198, reprinted as "Fermi's Group and the Recapture of Italy's Place in Physics," in Gerald Holton, *The Scientific Imagination: Case Studies* (Cambridge, 1978), 155–198; and Emilio Segrè, *Enrico Fermi, Physicist* (Chicago, 1970).

See also G. P. Carver, "A Corbino Disk Apparatus to Measure Hall Mobilities in Amorphous Semiconductors," in *Review of Scientific Instruments*, **43** (1972), 1257–1263; and D. A. Kleinman and A. L. Schawlow, "Corbino Disk," in *Journal of Applied Physics*, **31** (1960), 2176–2187.

BARBARA J. REEVES

CORRENS, CARL WILHELM (*b*. Tübingen, Germany, 19 May 1893; *d*. Göttingen, Germany, 29 August 1980), *sedimentary petrology, mineralogy, geochemistry*.

Correns was the son of Carl Erich Correns and Elisabeth Widmer. His father, a botanist, was one of the founders of modern genetics. After serving as assistant professor at Leipzig and as professor at Münster, he was the first director of the Kaiser Wilhelm Institute of Biology at Berlin-Dahlem. His mother, a native of Switzerland, was also a botanist.

Correns began his studies in natural sciences at the universities of Tübingen and Münster in 1912. After the war he went to Berlin, where he received a Ph.D. in 1920. His dissertation, supervised by J. F. Pompeckji, on limestone lenses in a Middle Devonian slate, marked the beginning of his lifelong preoccupation with the petrogenesis of sedimentary rocks.

On 30 December 1921 Correns married Agnes Ballowitz, daughter of Emil Ballowitz, professor of anatomy at the University of Münster. They had a son and a daughter. After an assistantship at the Geological Institute of the University at Munich under Erich Kaiser, Correns in 1921 entered the Prussian Geological Survey. Besides geological mapping of a Paleozoic area in Hesse, he refreshed his knowledge of chemistry by working in the laboratory of Herbert Freundlich at the Kaiser Wilhelm Institute for Physical Chemistry in Berlin-Dahlem. As a result of his *Habilitationsschrift* on the formation of sedimentary ore deposits through adsorption of metal ions by clays, Correns was appointed *Privatdozent* in mineralogy and petrography at the University of Berlin.

In 1926 and 1927 Correns took part in the German Atlantic Expedition on the ship *Meteor*. He was charged with the investigation of deep-sea sediments collected at several profiles across the South Atlantic. On this occasion Correns studied diamond occurrences in the state of Minas Gerais, Brazil. Contacts with Brazilian colleagues led to another voyage to Brazil in 1930 and a guest professorship at São Paulo in 1958.

Correns was appointed associate professor of mineralogy and geology at the University of Rostock in 1927 and full professor in 1930. In 1938 he accepted the posts of professor of mineralogy and director of the new Institute of Sedimentary Petrography at the University of Göttingen. From 1941 until his retirement in 1961, Correns was also director of the Institute of Mineralogy and Petrography at Göttingen.

Correns was one of the founders of modern sedimentary petrology, which deals with the identification of sedimentary minerals and the explanation of the phenomena of weathering, sedimentation, and diagenesis by physicochemical processes. While in Rostock, he did pioneering work in applying X-ray diffraction to the investigation of fine-grained sediments. It was found that clays are not amorphous substances, as formerly believed, but mixtures of well-defined crystalline minerals such as kaolinite, halloysite, montmorillonite, and mica minerals. Granulometric methods were developed for dividing sediments into grain size classes that could be investigated by optical microscopy, X-ray diffraction, and chemical analysis. The *Meteor* samples from the bottom of the South Atlantic were mineralogically and chemically investigated by these methods. The main types of deep-sea sediments and their areal distribution were established. For the first time the mineralogical composition of several clays and soils was determined. Observations on the decomposition of primary minerals in soil profiles gave rise to laboratory investigations on the reactions of minerals such as mica and feldspar with aqueous solutions under weathering conditions. It was found that silicates dissolve into ionic solutions at velocities depending on temperature and acidity.

In acknowledgment of Correns' achievements, the Institute of Sedimentary Petrography was established for him at Göttingen in 1938. There he continued investigations on the mineralogical composition of sedimentary rocks and conducted further experiments on weathering of silicates. Investigations on the influence of pressure on crystal growth and dissolution were important for the explanation of the diagenetic consolidation of buried sediments. Continuing Victor M. Goldschmidt's geochemical work at Göttingen, Correns and his students studied the geochemistry of halogens, chromium, carbon, sulfur, zinc, nitrogen, and titanium, with special attention to sedimentary cycles. In the Laboratory for Stable Isotopes, installed at Göttingen under his leadership in 1959, Correns initiated investigations on isotope geochemistry.

Correns was an inspiring and popular teacher who by his lectures and his readiness to communicate inspired many geology and mineralogy students. At Rostock and Göttingen he supervised sixty-one doctoral dissertations.

Correns was a member of the Deutsche Akademie der Naturforscher Leopoldina (Halle), the Akademie der Wissenschaften in Göttingen, the Romanian Academy of Sciences, the Austrian Academy of Sciences, the Accademia delle Scienze of Bologna, the Swedish Academy of Sciences, and the Serbian Academy of Sciences. He received honorary doctorates from the universities of Tübingen and Clausthal, and was elected to honorary memberships of geological and mineralogical societies in Germany, the United States, Spain, and Switzerland. Several societies awarded Correns their medals: the Göteborgs, Kungliga Vetenskaps- och Vitterhets Samhälle, the (silver) Albatross Medal (1955); the Geologische Vereinigung, the Steinmann Medal (1957); the Deutsche Geologische Gesellschaft, the Stille Medal (1963); and the Mineralogical Society of America, the Roebling Medal (1963).

BIBLIOGRAPHY

I. Original Works. Correns' writings include "Über die Bestandteile der Tone," in *Zeitschrift der Deutschen Geologischen Gesellschaft*, **85** (1933), 706–712; *Die Sedimente des äquatorialen Atlantischen Ozeans*, vol. III of *Wissenschaftliche Ergebnisse der Deutschen Atlantischen Expedition* (Berlin and Leipzig, 1935), 1–42 and (1937), 135–298, trans. by George M. Griffin as *The Sediments of the Equatorial Atlantic Ocean* (Houston, Tex., 1958); "Die Tone," in *Geologische Rundschau*, **29** (1938), 201–208; "Die Sedimentgesteine," in Thomas W. Barth, C. W. Correns, and Pentti Eskola, *Die Entstehung der Gesteine* (Berlin, 1939; repr. 1970), one of the first textbooks on sedimentary petrology; "Die chemische Verwitterung der Silikate," in *Naturwissenschaften*, **28** (1940), 369–375; "Die Stoffwanderung in der Erdrinde," *ibid.*, **31** (1943), 35–42; "Crystal Growth: Growth and Dissolution of Crystals Under Linear Pressure," in *Discussions of the Faraday Society*, **5** (1949), 267–271; *Einführung in die Mineralogie, Kristallographie und Petrologie* (Berlin, 1949; 2nd, rev. ed., with Josef Zeman and Sigmund Koritnig, 1968), 2nd ed. trans. by William D. Johns as *Introduction to Mineralogy, Crystallography and Petrography* (London and New York, 1969); "Zur Geochemie der Diagenese I. Das Verhalten von $CaCO_3$ und SiO_2," in *Geochimica and Cosmochimia Acta*, **1** (1950), 49–54; "The Geochemistry of the Halogens," in L. H. Ahrens, Kalervo Rankama, and S. K. Runcorn, eds., *Physics and Chemistry of the Earth*, I (New York, 1956), 181–233; "The Experimental Weathering of Silicates," in *Clay Minerals Bulletin*, **4** (1961), 249–265; and "The Discovery of the Chemical Elements. The History of Geochemistry," in K. H. Wedepohl, ed., *Handbook of Geochemistry*, I (Berlin and New York, 1969), 1–11.

II. Secondary Literature. A complete bibliography of Correns' publications is part of Wolf von Engelhardt, "Carl Wilhelm Correns 1893–1980," in *Fortschritte der Mineralogie*, **59** (1981), 1–12. See also Wolf von Engelhardt, "Carl Wilhelm Correns," in *Geologische Rundschau*, **70**, no. 3 (1981); Joachim Hoefs, "Memorial of Carl Wilhelm Correns," in *American Mineralogist*, **67** (1982), 399–400; P. Schneiderhön, "In Memoriam Carl W. Correns," in *Der Aufschluss* (1981); W. Schott, "Carl Wilhelm Correns," in *Nachrichten der Deutsche Geologische Gesellschaft*, **24** (1980), 4; and K. H. Wedepohl, "Carl Wilhelm Correns," in *Almanach der Österreichischen Akademie der Wissenschaften*, **130** (1980).

Wolf von Engelhardt

COSSA, ALFONSO (*b.* Milan, Italy, 3 November 1833; *d.* Turin, Italy, 23 October 1902), *chemistry, mineralogy.*

Alfonso Cossa was the son of Giuseppe Cossa, librarian of Milan's famous Biblioteca di Brera and an authority on paleography and diplomacy, and his wife, Maria Bagnacavallo. After completing his classical studies in Milan, in 1852 Alfonso was sent to Pavia, where he studied at the Collegio Borromeo. In November 1857 he received the M.D. degree from the University of Pavia with a dissertation on the history of electrochemistry. Before receiving his degree he had translated into Italian two books by Justus Liebig: *Die Grundsätze der Agricultur-Chemie mit Rücksicht auf die in England angestellten Untersuchungen* (1855) and *Zur Theorie und Praxis in der Landwirtschaft* (1856). He married Giovanna Panizza; they had two daughters, Maria and Eugenia.

Cossa was one of those rare scholars who enter fields for which they have no formal training. After his education as a physician, he successively became a botanist, an agricultural and plant chemist, a mineralogical chemist, and an inorganic and coordination chemist. When mineralogical chemistry, his favorite branch of chemistry, underwent a number of changes as a result of new physicochemical theories, Cossa, although a mature scholar with a substantial reputation, did not hesitate to return to school, along with his students, to learn from his colleagues and friends the mathematical ideas that are essential to modern chemistry. This continual intellectual development was characteristic of Cossa throughout his life.

From an early age Cossa demonstrated a striking predilection for chemistry, especially applied chemistry. He possessed unusual scientific and didactic talents, but at the beginning of his career his interests did not exhibit a precise trend, and he was forced to learn much on his own. In the mid-nineteenth century most university positions in Italy were occupied by men whose teaching was purely theoretical, with no experimental content or practical application, and who failed to consider the great progress in chemistry that was being made in other countries. In short, the teaching of chemistry in Italy was comparable with that in other countries sixty years earlier.

At the time Cossa received his medical degree, there was a great shortage of teachers in Italy. He therefore remained at the University of Pavia, where, despite his lack of specialized training, he was *assistente di medicina legale e polizia medica* (1857–1860). In May 1860 he was appointed *assistente stabile di chimica generale*, and later that year he was named *farmacista*. He remained at the university as professor of chemistry and director of the Technical Institute (1861–1866), where his outstanding abilities as a researcher and teacher were recognized by Quintino Sella, the statesman and crystallographer who helped place the new national government on a firm footing after the unification of Italy. In 1866, when Venice was united to the newly formed kingdom of Italy, Sella commissioned Cossa to organize and found a technical institute at Udine, where he remained until 1871 as professor of chemistry and director.

After a short stay (1871–1873) at the Royal High School of Agriculture at Portici, in 1873 Cossa was appointed director of the agricultural station and teacher of general and mining chemistry at the Royal Industrial Museum, both in Turin. In 1882 he succeeded Ascanio Sobrero, the discoverer of nitroglycerin, as professor of assaying and mining chemistry at Turin's Royal School of Applied Engineering, of which he was director from 1887 to 1902.

Cossa was a member of the Turin Academy of Sciences (elected 1871, president 1901–1902), the Accademia dei Lincei, the Berlin Academy of Sciences, the Mineralogical Society of St. Petersburg, and other Italian and foreign societies.

The subject of Cossa's early works was largely agricultural and plant chemistry. He worked on absorption by roots, the chemistry of plant seeds, sugarcane, plant respiration, soils, water supplies, manures, sugar-beet roots, grape must, ash of lemon leaves and fruit, vetches, the action of sulfur on calcium carbonate, and the decomposition of chlorophyll by the light of burning magnesium. Like Pasteur, Cossa was interested in wine making, and he proposed several methods for destroying the plant lice (*Phylloxera*) that attack the roots and leaves of grapevines. From 1872 to 1882 he was editor of the journal *Le stazioni sperimentali agrarie italiane*, which he founded and published in Turin. He also carried out a number of studies in medical-legal chemistry, especially on alkaloids such as veratrine. He suggested the use of dialysis for the detection of poisonous alkaloids.

Among Cossa's researches in inorganic chemistry, carried out between 1867 and 1892, were work on magnesium, ozonometry, and sulfur; one of the first observations of the behavior of aluminum in contact with metallic solutions, which helped to establish the position of aluminum in the electrochemical series; studies of aluminum amalgam; the first synthesis of hydrogen sulfide; work on boron; work on vanadium; and studies of the action of nitric acid on zinc.

Cossa's friendship with Quintino Sella and Bartolomeo Gastaldi led him to devote himself to mineralogical chemistry and petrography. Throughout his career the former was Cossa's primary field of interest, with almost half of his works being devoted to this subject. Although petrography was already flourishing in Germany, largely as a result of the efforts of Ferdinand Zirkel and Harry Rosenbusch, Cossa apparently was the first person in Italy to pursue such studies. On Cossa's urging, in 1889 the Italian government established in Rome, as part of the Royal Geological Office, a petrographic laboratory directed by Ettore Mattirolo, an engineer who was a former student and collaborator of Cossa's.

Cossa classified, characterized the properties, determined the compositions, and proposed mineralogical formulas for a variety of minerals and rocks, not only from Italy but also from many other sources. One of the minerals that he studied, a variety of paragonite—a sodium mica of composition $H_2NaAl_3(SiO_4)_3$—was named cossaite in his honor. Cossa went beyond microscopic investigations and analyses of minerals; he also prepared artificial minerals.

Closely related to his mineralogical studies were Cossa's investigations of lanthanum, cerium, and didymium. He found these elements to be widely distributed in nature, not only in many rocks but also in bones; ashes of rice, tobacco, beech trees, and other plants; and human urine. Because the knowledge of the chemistry and nature of the rare earths was obscure at the time, some of Cossa's interpretations of his data, particularly with respect to oxidation states, were incorrect; but his observations and data are still valid today.

Long interested in biography and the history of chemistry, Cossa wrote articles on the lives and work of a number of chemists. Of these the most important was his study of Angelo Sala (1576–1637), who was one of the first chemists to describe fermentation; in his emphasis on experimentation, Sala was a true precursor of Joseph Black, Joseph Priestley, and Carl Wilhelm Scheele. Cossa's study of his friend Quintino Sella led him to a new field of research. In 1885 the Accademia dei Lincei asked Cossa to commemorate the life and work of Sella, who had died the year before. In the course of his bibliographical research, Cossa found Sella's *Sulle forme di alcuni sali di platino a base di platinodiamina* (1856–1857) and decided to carry out experimental studies on platinum-ammine compounds. Beginning in 1885 and continuing until 1897, he made a long series of investigations—his most important chemical contributions. Although Cossa interpreted his data in terms of the now obsolete Blomstrand-Jørgensen chain theory and named his compounds according to Per Cleve's obsolete nomenclature system, these practices in no way invalidate his experimental results.

In attempting to reconcile some discrepancies between the work of Jules Reiset and Cleve, Cossa discovered a fifth isomer of Magnus's green salt, $[Pt(NH_3)_4][PtCl_4]$, which he formulated as $Pt(NH_3)_4Cl_2, PtCl_2$. He formulated the new yellow isomer $[Pt(NH_3)_4][PtCl_3(NH_3)]_2$ as $2(PtCl \cdot NH_3Cl)$, $Pt(NH_3)_4Cl_2$, that is, he regarded it as a compound of one molecule of platosodiammine chloride, $(Pt(NH_3)_4Cl_2$—modern $[Pt(NH_3)_4]Cl_2$—and two molecules of the chloride of a new base, which he called platosemiammine $(PtCl \cdot NH_3Cl)$. The orange-red "potassiochloride" of this new base is now known as Cossa's first salt, $K[PtCl_3(NH_3)] \cdot H_2O$.

By oxidizing $K[PtCl_3(NH_3)] \cdot H_2O$, Cossa also obtained the corresponding yellow platinum (IV) salt, $K[PtCl_5(NH_3)] \cdot H_2O$, now known as Cossa's second salt, which he called platinosemiammine potassium chloride and formulated as $Pt(NH_3)Cl_4,KCl$. Cossa also confirmed his view that one atom of platinum (II) could combine with one molecule of a base to form compounds containing pyridine (py; C_5H_5N) and ethylamine analogous to his monoammine compounds.

The Scottish chemist Thomas Anderson found that when solutions of $(pyH)_2[PtCl_6]$ and the corresponding compounds of pyridine derivatives are boiled, hydrogen chloride is released with formation of *cis*-$[PtCl_4py_2]$ (Anderson's platinic compound) or the corresponding compounds of pyridine derivatives, a reaction known as Anderson's reaction.

BIBLIOGRAPHY

I. Original Works. Cossa wrote several books and about 130 articles, many of which were published separately as well as in Italian journals such as *Gazzetta chimica italiana*, *Atti della Reale Accademia delle scienze, Torino*, *Stazioni sperimentali agrarie italiane*, *Atti della Reale accademia dei Lincei*, and *Ricerche chimiche*. An almost complete bibliography of Cossa's works is found in *In Memoria di Alfonso Cossa nel primo anniversario della sua morte* (Turin, 1903), 91–108. His most important article on platinum ammines, describing Cossa's first and second salts, is "Sopra un nuovo isomero del sale verde del Magnus," in *Gazzetta chimica italiana*, **20** (1890), 725–753, translated in *Berichte der Deutschen chemischen Gesellschaft*, **23** (1890), 2503–2509.

A number of Cossa's petrographic and mineralogical works on Italian rocks and minerals were published in *Ricerche chimiche e microscopiche su roccie e minerali d'Italia* (Turin, 1881). The collection of thousands of these specimens, ordered and classified by Cossa, is now housed in the School of Applied Engineering at Turin.

II. SECONDARY LITERATURE. *In Memoria di Alfonso Cossa nel primo anniversario della sua morte*, cited above, reprints several earlier obituaries (pages in parentheses denote the pages in this book): Luigi Gabba, in *Annali della Società chimica di Milano*, **8** (1902), 184ff. (37–43); Icilio Guareschi, in *Memorie della Reale Accademia delle scienze, Torino*, 2nd ser., **53** (1903), 79–92 (58–79); Augusto Piccini, in *Rendiconti della Reale Accademia dei Lincei, Classe di scienze fisiche, matematiche, e naturali*, **11**, no. 2 (1902), 235ff. (33–37); and M. Zecchini, in *La chimica industriale*, **4**, no. 21 (1902), 321–322 (44–52). The only article in English is George B. Kauffman and Ester Molayem, "Alfonso Cossa (1833–1902), a Self-Taught Italian Chemist," in *Ambix*, submitted for publication.

GEORGE B. KAUFFMAN

COTTON, AIMÉ-AUGUSTE (*b.* Bourg-en-Bresse, France, 9 October 1869; *d.* Sèvres, France, 16 April 1951), *physics*.

Cotton's mother died when he was two years old, and he was raised by his father, Eugène, a secondary school mathematics teacher, in comfortable rural surroundings. Following his father's wishes, he prepared for a teaching career and in 1890 entered the École Normale Supérieure in Paris. The École Normale was then the chief cradle of French scientists, and Cotton made lifelong friendships with many future leaders, for example, Élie Cartan. In 1893 he passed the *agrégation* in physics and in 1895 became a teacher at a lycée in Toulouse. He received favorable reports from his superiors—"intelligent and hard-working," "upright and friendly," "assiduously frequents the laboratory"—and in 1900 was recalled to the École Normale to serve as *suppléant* to Jules Violle.

Cotton quickly became a member of the circle of the École Normale and other scientists that centered on Pierre Curie. Among them was Eugénie Feytis, a teacher and physicist, whom he married in 1913. Cotton was promoted in 1910 to adjoint professor and in 1920 moved to the Sorbonne as professor of theoretical physics and celestial physics; in 1922 he took the chair of general physics. Elected to the Academy of Sciences in 1923, he was its president in 1938. Usually modest and reserved, Cotton revealed a bold inner spirit on alpinist and hunting excursions, and became almost loquacious at congresses devoted to Esperanto, which he saw as an important means toward international understanding.

In his research Cotton, inspired particularly by Louis Pasteur and Curie, carried forward the French tradition of working toward the essential nature of matter through simple experiments illuminated by the plainest logic. This program used symmetries, notably those found in optical investigation, as the key to discovering the most characteristic properties of materials. Although it was a different approach, through spectroscopy, that would lead in the 1920's to quantum mechanics, the classical path that Cotton followed eventually became no less important for the rise of condensed-matter physics.

In his doctoral dissertation, undertaken at the École Normale and defended in 1896, Cotton began with the familiar facts of birefringence and dichroism—the properties of certain crystals to refract and absorb light differently in different planes of polarization—and discovered a similar dichroism, in colored liquids, for opposite circular polarizations. He also discovered anomalous rotatory dispersion: the rotation of the plane of polarization showed a predictable variation with frequency in the neighborhood of a color absorption band.

In 1903 Cotton joined his École Normale classmate Henri Mouton in improving the new ultramicroscope, which could reveal submicroscopic objects as points of diffracted light. They used it for a wide variety of studies, taking a particular interest in colloids. Imposing a magnetic field on certain colloids, they found not only birefringence (previously seen by Ettore Majorana) but also circular dichroism; moreover, by freezing the tiny colloidal particles in gelatin, the pair proved that such optical effects were connected with the tendency of the anisotropic particles to line up with the magnetic field. The next step was to invisible molecules. Jean Perrin, another member of the Curie circle, was just then working up definitive proofs of the atomic theory of matter, which was still under a cloud in France; to connect optical properties with the orientation of molecules, seen as real physical entities, was less simple than it would later appear. In 1907 Cotton and Mouton closed the question by demonstrating magnetic birefringence in pure liquids such as nitrobenzene.

Meanwhile the discovery of the Zeeman effect in 1896 offered a hope that by altering spectral lines with magnetic fields, such effects as polarization could lead a researcher into the heart of atoms. Cotton had worked on the subject from the beginning; his most useful product was an instrument, later widely used, that could accurately measure magnetic fields by their influence on a current element in one arm of a balance. Further results came when he visited Zurich to collaborate with Pierre Weiss, a close friend and fellow École Normale graduate, who had built an electromagnet larger than any in France. Their 1907 measurement of the electron's charge-to-mass ratio (e/m), using the Zeeman effect, was the best then available. Cotton began to campaign vigorously for a large magnetism laboratory for France, and in 1912 the University of Paris gave fifty thousand francs to begin work.

World War I interrupted the project. Cotton and Weiss collaborated to develop a system of locating enemy cannon by listening for the blasts with widely spaced microphones, and with the help of other physicists they installed teams at the front to guide counter-battery fire. This sound-ranging work was one of the chief contributions physics made to the military effort.

After the war fund-raising for the magnetism laboratory had to begin anew. With no little effort Cotton got support from the university and the government, from funds the Academy of Sciences raised through public subscription and bequests, and directly from industrialists. In 1928 he completed a giant electromagnet at Bellevue, outside Paris. Weighing more than one hundred tons, it generated seventy thousand gauss and required water cooling to handle its nearly one hundred kilowatts of power.

In its day the magnet was among the largest experimental devices in any country. It proved its worth in Solomon Rosenblum's study of alpha rays, which revealed the fine structure of their differing velocities; in Louis LePrince-Ringuet's studies of the "penetrating component" of cosmic rays, later known as mu-mesons; in Pierre Jacquinot's Zeeman studies confirming and extending the nonlinear effects that Peter Kapitza had found in England; in the pioneering work of Franz (Francis) Simon and his collaborators, who obtained low temperatures by adiabatic demagnetization; and in many studies of the magnetic properties of condensed matter. As director of the Bellevue laboratory Cotton assigned research time at the magnet to physicists from numerous nations. He also built up auxiliary facilities such as a major spectroscope.

A firm believer in the French republic and an outspoken antifascist, Cotton was arrested along with several other professors in October 1941 and held for a month, probably to forestall any attempt at organizing an Armistice Day demonstration against the German occupation. He was arrested again for three days in April 1942 but otherwise worked unmolested at Bellevue through the war. He continued to be active into his late seventies, when failing health removed him from the laboratory life he loved.

BIBLIOGRAPHY

I. ORIGINAL WORKS. Cotton's writings include *Le phénomène de Zeeman* (Paris, 1899); *Les ultramicroscopes et les objets ultramicroscopiques* (Paris, 1906), written with Henri Mouton; *La théorie de Ritz du phénomène de Zeeman* (Paris, 1911), an extract from *Le radium, journal de physique; La polarimétrie en lumière ultraviolette* (Paris, 1934); and *Quelques instruments nouveaux de la section de physique du Palais de la découverte* (Paris, 1943), a pamphlet. Many of Cotton's writings are collected in *Oeuvres scientifiques d'Aimé Cotton* (Paris, 1956), which includes a bibliography. Some correspondence is held by the physics laboratory of the École Normale Supérieure, Paris, and bureaucratic particulars of Cotton's career are in the Archives Nationales, Paris, F^{17}24,865.

II. SECONDARY LITERATURE. Biographical works are Eugénie Cotton, *Aimé Cotton* (Paris, 1967); and Jean Rosmorduc, "Aimé Cotton: Le savant, l'homme, le citoyen" (thèse de 3me cycle, University of Western Brittany, 1971).

SPENCER R. WEART

CRAMPTON, HENRY EDWARD (*b*. New York, New York, 5 January 1875; *d*. New York, New York, 26 February 1956), *evolutionary biology*.

Crampton was the son of Henry Edward Crampton, Sr., a physician, and Dorcas Matilda Crampton (née Miller). He attended the College of the City of New York, switching to Columbia University, where he received his A.B. in 1893 and his Ph.D. in zoology in 1899. While pursuing graduate studies, he served as an assistant in biology at Columbia (1893–1895) and as instructor at the Massachusetts Institute of Technology (1895–1896). He married Marion M. Tully on 27 October 1896; they had one daughter and one son.

Crampton resumed teaching at Columbia in 1897 and remained there for the rest of his career, primarily instructing women at Barnard College. He achieved the rank of professor in 1904 and retired in 1943. Crampton taught the embryology course at the Marine Biological Laboratory in Woods Hole (1895–1902) and was in charge of embryology at the Cold Spring Harbor Laboratories (1903–1906). He also served as curator for invertebrate zoology at the American Museum of Natural History (1909–1920) and was associated with the Carnegie Institution of Washington and the Bishop Museum (Honolulu). Crampton was a fellow of the New York Academy of Sciences and served as its president in 1926. He also served as vice president of the American Society of Naturalists (1921), vice president of the American Society of Zoologists (1911), and secretary-treasurer of the Eugenics Society of the United States (1922–1925).

Crampton's career and its development serves as a fine epitome of general trends within evolutionary biology during his lifetime, from the strong interest in embryology that pervaded late-nineteenth-century work to the rediscovery of Mendelism and the non-adaptationist consensus (for differences among geographic variants and closely related species) that prevailed into the 1940's, before the Modern Synthesis reasserted the shaping power of selection at these scales. But Crampton was no passive reflector of these trends; his three massive, meticulously detailed, and beautifully illustrated monographs on the land snail genus *Partula* from Pacific Islands (1917, 1925, 1932) were primary documents in the development of this consensus.

Crampton studied at Columbia when its biology department, led by H. F. Osborn, E. B. Wilson, B. Dean, and A. Willey was a center for embryological and evolutionary studies, the two foci of Crampton's career. Crampton's early research was predominantly embryological; he also taught embryology at Woods Hole and Cold Spring Harbor. He performed important experiments on cleavage in isolated blastomeres of gastropods and on cross-species fusion in insect larvae and pupae. C. B. Davenport, in nominating Crampton (unsuccessfully) for the National Academy of Sciences, wrote that "nobody has been quite so successful in grafting insect larvae as [Crampton]."

In 1899 Crampton began a series of selection experiments on pupae of cynthia moths (published in *Biometrika* in 1904 and in the *Journal of Experimental Zoology* in 1905). He followed the work of Bumpus and Weldon in focusing upon selective elimination and differences in character means between pupae that died and those that survived to produce an imago. He was puzzled that he could demonstrate selection, but on characters that could not have influenced mortality directly (for example, on antennae not used by the pupae). He decided that selection must work for a harmonious pattern of correlation among characters, not usually on individual traits themselves, and that internal factors of development are as important to success as the match of traits to environmental dictates. He wrote in 1905: "Selection is not regarded as in any way *originative* but only as *judicial*, so to speak. As the members of any species present themselves at the bar, 'selection' decides the question of survival or destruction on the basis of the condition of correlation that is exhibited." This balance of emphasis on internal and external forces was characteristic of late-nineteenth- and early-twentieth-century evolutionary thought.

Crampton became dissatisfied with a purely laboratory approach to evolution and decided to study a case of diversification and speciation in nature. Beginning in 1906 and extending through the mid-1930's, Crampton undertook numerous expeditions to islands of the Pacific, where he studied the protean and prolific land snail *Partula* (primarily on Tahiti and Moorea). *Partula* presented many advantages for such work: it exhibits remarkable variation in color and coiling, often with distinct local races in each valley; and parents brood offspring, which can be dissected out, facilitating the understanding of inheritance in shell traits.

Crampton's three monographs on *Partula*, published by the Carnegie Institution, are among the greatest works of twentieth-century American natural history. Following themes established by J. T. Gulick in his 1905 monograph (also published by the Carnegie Institution) on Hawaiian land snails of the genus *Achatinella*, Crampton explored the roles of environment, isolation, mutation, and selection in the origin of species and geographic races. In his first monograph on *Partula* in Tahiti (published in 1917, though the title page reads 1916), Crampton studied and presented statistical summaries (all done by hand or with the simplest calculating machines) for data on 80,000 adult shells from more than 200 valleys.

Against Lamarckism (then current among many naturalists), Crampton decided that environment plays no originative role in the evolution of new variants. But since Crampton could find no correlations between shell variation and environment, he also concluded that natural selection is not an originative force, but can only act to eliminate inadaptive features. Adaptation, which Crampton regarded as universal (see his popular books of 1911 and 1931), is not built gradually by natural selection at these small scales, but arises fortuitously and "congenitally" (Crampton's favorite term) and is then preserved by selection acting in an eliminative role. Crampton therefore emphasized the primary roles of isolation, as a precondition for divergence (geography produces nothing directly but establishes a situation in which new features can spread), and mutation, as a source of fortuitous variation that can then become established in isolated populations.

The resulting pattern of differences among valleys is largely nonadaptive. Every local race must avoid elimination by natural selection (and is adapted in this sense), but its particular features represent but one among hundreds of workable possibilities, and the initiation of one or another among the numerous possible solutions is set by isolation and congenital variation (mutation), not by natural selection. This nonadaptive interpretation of small-scale variation represented a consensus among field taxonomists before the 1940's, and Crampton's monographs provided powerful support. "The role of the environment," Crampton wrote, "is to set the limits to the habitable areas or to bring about the elimination of individuals whose qualities are otherwise determined, that is, by congenital factors" (1917, p. 299).

Crampton regarded his views as within the Darwinian orbit. He wrote two popular books extolling the Darwinian worldview (1911 and 1931). Still, his insistence that natural selection can be only an eliminative force, and that everything about the origin of entire new traits (not only of small-scale favorable mutants) must be ascribed to "congenital factors," placed him outside the stricter Darwinism of the Modern Synthesis as it developed during the 1940's and 1950's. In his last monograph Crampton summarized his notion of natural selection: "'Congenital variation must not be eliminative' is the correct statement of the categorical imperative of organic differentiation, and I do not regard it as contradictory to the contentions of Darwin in any essential respect" (1932, p. 188). Nonetheless, Crampton's vision of natural history, with its large role for nonadaptive differences among local races and species, lost its popularity late in his lifetime, and Crampton himself became a forgotten figure. This eclipse is most unfortunate because Crampton's *Partula* monographs were very influential and represent an important era in evolutionary thought. Moreover, many of his ideas are getting a rehearing during renewed modern debates about evolutionary mechanisms. W. Provine, for example, demonstrates how important Crampton's views were to Sewall Wright as he developed his shifting balance theory of evolution (*Sewall Wright and Evolutionary Biology*, 1986). Students of land snails (the author of this article is one) continue to hold Crampton's work in highest regard—and to use his numerical data, surely a sign of ultimate respect for a scientist.

BIBLIOGRAPHY

The crux of Crampton's career and the source of his importance lie in his three great monographs on *Partula*, published by the Carnegie Institution of Washington, *Studies on the Variation, Distribution, and Evolution of the Genus* Partula, I, *The Species Inhabiting Tahiti* (1917), II, *The Species of the Mariana Islands, Guam, and Saipan* (1925), and III, *The Species Inhabiting Moorea* (1932). Crampton set out his views on evolution in two books intended for general audiences: *The Doctrine of Evolution: Its Basis and Scope* (New York, 1911) and *The Coming and Evolution of Life* (New York, 1931). See also "Experimental and Statistical Studies upon Lepidoptera," in *Biometrika*, **3** (1904), 113–130; and "On a General Theory of Adaptation and Selection," in *Journal of Experimental Zoology*, **2** (1905), 425–430.

STEPHEN JAY GOULD

CREW, HENRY (*b*. Richmond, Ohio, 4 June 1859; *d*. Evanston, Illinois, 17 February 1953), *physics, astrophysics, education.*

Crew was descended from devout Quaker farmers and merchants in Jefferson County, Ohio. His father, William Henry Crew, who ran a successful general store, died when Henry was eleven, providing in his will for his son and two daughters to receive a "classical education." Henry's mother, Deborah Ann Hargrave Crew, devoted herself to the children's upbringing; and it was more from her than from the local school that he learned to read and write. His first physics lessons came from activities like harnessing horses, winnowing wheat, preserving fruit, and visiting the gunsmith's shop or the sawmill. Crew's heritage gave him good health and a vigorous interest in the world around him along with intelligence and mechanical ingenuity. A career in research, however, also needed opportunities, which were in short supply at the time. In his old age Crew noted that "I have made no contribution to knowledge which is of either first-rate or second-rate importance." Nevertheless, among other honors he became president of the American Physical Society in 1909–1910, of the American Association of University Professors in 1929, and of the History of Science Society in 1930.

His education was so backward that Crew was two years older than most of his classmates when he entered high school in Wilmington, Ohio, and when he went to Princeton in 1878. Following his father's wishes, he assiduously studied Latin, Greek, and mathematics, although as a boy he had already found them dull and irksome. What did excite him were concrete subjects such as geology and botany; he enjoyed camping and fishing, collected thousands of fossils during the summers, and analyzed flowers and vegetables in the garden he kept.

In his junior year at Princeton, Crew took a physics course using the then-standard textbook of Ganot, translated from the French. The next year came laboratory work and, particularly inspiring, a course in general astronomy under Charles A. Young. On graduation Crew won a fellowship in physics and stayed on, but, as he later wrote, "In the absence of any regular graduate course in physics, I browsed in the library, played in the laboratory, and deteriorated intellectually." In the fall of 1883 he therefore went to study in Europe.

At Hermann von Helmholtz's laboratory in Berlin, Crew met other American students, who told him about the graduate school that Henry Rowland had created at the Johns Hopkins University in Baltimore. Crew went there in the spring of 1884 and became an assistant to Rowland, helping with measurements of the solar spectrum and certain physical constants. In 1887 he received his Ph.D. for a dissertation that used Doppler spectroscopy to determine the rotation period of the sun at various heliocentric latitudes; it was a hard task with the limited optical equipment at hand, and Crew later learned that his measurements were without value.

From 1888 to 1891 Crew was instructor in physics and then department head at Haverford College. Along with his teaching duties he developed a clever way to make a constant-temperature bath. This was not a route that could lead far, and in 1891—shortly after he married Helen C. Coale of Baltimore—he accepted an invitation to go to Lick Observatory.

At Lick, Crew found a tiny, isolated group of astronomers irreconcilably split by a struggle between two factions. He took the first chance to leave that appeared, becoming professor of physics at Northwestern University in the fall of 1892. At Lick he had measured some stellar spectra and had devised an ingenious way around a problem of laboratory spectroscopy: when an arc was struck to study metals, the tips tended to oxidize or weld together. Crew solved the problem by replacing one of the tips with a spinning disk.

Such a combination of telescope and laboratory studies was characteristic of a new field, astrophysics, promoted especially by George Ellery Hale. Crew served through 1941 as an associate editor of Hale's *Astrophysical Journal*. But Evanston was not one of the few places with enough people and instruments to form an astrophysical research community. As the years passed, Crew attended less and less to spectroscopy and more to pedagogy.

A friendly and conscientious teacher, Crew prepared lectures and demonstrations that could be exciting without sacrifice of precision. He was especially concerned to show the connectedness of physics, the way its branches dovetailed. In 1899 he published a textbook, *The Elements of Physics*, which went through a number of printings and revisions and was joined in 1908 by the more advanced *General Physics*, which also found wide use. These texts were more approachable than Ganot's, skipping many details of instrumentation and algebraic manipulation while retaining the key ideas. Like all textbook authors of the time, Crew hesitated to introduce new theory, although in the 1927 revision of *General Physics* (the fourth and last) he added a simple introduction to the Bohr atom and touched on relativity.

Crew taught physics not as an abstraction but as a concrete human achievement, and here he found history useful. In 1914 he and Alfonso de Salvio published a translation of Galileo's *Dialogues Concerning Two New Sciences* that was frequently reprinted. In 1919 Crew began to give lectures on the history of science; in 1928 he published *The Rise of Modern Physics*, covering much the same ground as his textbooks but in a historical mode. As a historian he called himself an amateur, for here too his main interest was pedagogical.

As a teacher Crew reached his summit when he took a leave of absence from 1930 to 1933 to become chief of the Division of Basic Sciences at the Century of Progress International Exposition in Chicago. He was in charge of the Hall of Science, a central attraction of the fair, and also organized symposia, popular books, and so forth. Many of the exhibits he inspired later found a more permanent home in the Museum of Science and Industry in Chicago.

After his retirement from Northwestern in 1933, Crew remained vigorous for many years. Besides his two daughters he had a son, William Henry Crew, who also was a physicist, as was a grandson, Henry Crew III.

BIBLIOGRAPHY

I. ORIGINAL WORKS. In 1935 Crew deposited a ten-page personal history in the Archives of the National Academy of Sciences. His papers are in the Archives of Northwestern University, Evanston, Illinois, with a microfilm copy at the Niels Bohr Library of the American Institute of Physics in New York City. Among his books are *The Elements of Physics* (New York, 1899); *General Physics* (New York, 1908); *The Principles of Mechanics* (New York, 1908); a translation, with Alfonso de Salvio, of Galileo's *Dialogues Concerning Two New Sciences* (New York, 1914; repr. 1933, 1939, 1946, 1952); *The Rise of Modern Physics* (Baltimore, 1928; 2nd ed., 1935); and his translation of Maurolycus's *Photismi de lumine* (New York, 1940).

II. SECONDARY LITERATURE. Obituaries include A. A. Knowlton, "Henry Crew," in *Isis*, **45** (1954), 169–174; and William F. Meggers, "Henry Crew," in *Biographical Memoirs. National Academy of Sciences*, **37** (1964), 33–54, with bibliography.

SPENCER R. WEART

CUNNINGHAM, JOSEPH THOMAS (*b.* London, England, 4 April 1859; *d.* London, 5 June 1935), *zoology, marine biology.*

Cunningham was an embryologist and marine biologist who from 1890 to the early 1930's was one of the leading English exponents of neo-Lamarckian evolutionary theory.

The son of William Henry Cunningham, a solicitor, Cunningham attended St. Olave's Grammar School in Southwark. He then went to Balliol College, Oxford, as Brackenbury science scholar (1878–1881), obtaining first-class honors in mathematical moderations and natural science. In 1882 he was elected to a fellowship at University College, Oxford.

Cunningham's career began at a time when British biologists were seeking to establish a marine biological laboratory comparable with Anton Dohrn's celebrated zoological station at Naples. The British perceived that a marine station of their own would be the ideal place for studying (1) the embryological development of marine organisms and (2) the food, habits, and life histories of British food fishes and mollusks. Cunningham studied at the Naples zoological station in the winter of 1882–1883. In 1883 he became assistant to the Regius professor of natural history at Edinburgh, James Cossar Ewart, who was interested in promoting marine biology.

When in 1884 a small-scale floating marine laboratory was set up by John Murray at Granton, near Edinburgh, under the auspices of the Scottish Meteorological Society, Cunningham was made its director. The Marine Biology Association of the United Kingdom was founded in the same year, and three years later, when it created a position for a naturalist at the marine laboratory being established at Plymouth, Cunningham was selected for the job. He served as naturalist to the Marine Biology Association for ten years (eight years at Plymouth and two at Grimsby), then took a position with the Cornwall Technical Instruction Committee, lecturing to fishermen about scientific means of improving their livelihood.

Cunningham moved to London in 1902 and taught zoology at Chelsea Polytechnic. In 1909 he conducted a study of the fisheries of St. Helena. He was appointed lecturer in zoology at East London College (Queen Mary College), University of London, in 1917, a position he held until his retirement in 1926. In the course of his career he was elected a fellow of the Royal Society of Edinburgh and of the Zoological Society and the Linnean Society of London. Cunningham married Sophie Crossfield in 1886; they had one son, who was killed in World War I.

Cunningham's early papers consisted primarily of studies of (1) the embryology of teleostean fishes, a subject he pursued in hopes of shedding light on the primitive ancestry of the vertebrates, and (2) the occurrence and reproduction of British food fishes. His first book, *A Treatise on the Common Sole* (Solea vulgaris), appeared in 1890.

Early in his career Cunningham concluded that the primary explanation of evolution was to be found in the origin of individual variations rather than in their selection. His study of flatfishes led him to the Lamarckian view that adaptive variations arose as the result of stimulations or irritations to the body brought about by habits or external conditions. He first made his position explicit in the preface to his translation of Gustav H. T. Eimer's *Organic Evolution as the Result of the Inheritance of Acquired Characters According to the Laws of Organic Growth* (1890). Cunningham believed that such phenomena as the distortion of the skull and the asymmetry of the eyes in flatfish and the lack of pigment on the fishes' lower sides could be readily explained as the result of the inherited effects of habits maintained over many generations.

In letters to *Nature* in the 1890's, Cunningham attacked the views of August Weismann and the English neo-Darwinists, and declared himself a disciple of Herbert Spencer. Though he believed there was ample inductive evidence to support neo-Lamarckism, he undertook experimental studies of pigment change in flatfishes to test whether the whiteness of the fishes' lower sides was due to selection or to their lack of exposure to light. The results he obtained tended, in his view, to support the latter hypothesis.

In his book *Sexual Dimorphism in the Animal Kingdom* (1900), Cunningham maintained that secondary sexual characters could best be explained as the result of habits and the inheritance of acquired characters. In an article in Roux's *Archiv* (1908) he suggested that the role provisionally ascribed by Darwin to "gemmules" in his "provisional hypothesis of pangenesis" was actually performed by hormones. Hormones were the substances through which somatic modifications affected the gametes. The findings of Thomas Hunt Morgan and other geneticists in the next two decades—and his own Mendelian experiments on fowls in the 1910's—did not change Cunningham's mind about neo-Lamarckism. He believed that the mutations Morgan and others were finding in *Drosophila* and other organisms had nothing to do with adaptation. In his book *Hormones and Heredity* (1921), he maintained that there were two kinds of variation in evolution, one somatogenic, due to the influence of external stimuli on body cells, and the other gametogenic, due to changes in chromosomes. Only the somatogenic variations were adaptive.

Opposing the notion that mutation and selection explained the evolution of all characters, and also opposing the notion that all the characters that distinguish species from one another are adaptive, Cunningham stressed the importance of the effects of external stimuli upon the development of living tissue, regarding this to be an essentially Lamarckian position. At the same time, he was skeptical of some of the claims offered on behalf of Lamarckism, including some of those set forth by the Austrian biologist Paul Kammerer. Cunningham's final book, *Modern Biology* (1928), showed him to be a thoughtful critic of contemporary biological concepts and theories.

His years in retirement were not restricted to theoretical efforts, however. In 1930 Cunningham embarked on an expedition to the island of Marajó, at the mouth of the Amazon, to study the function of the external vascular filaments on the pelvic limbs of male lungfish (*Lepidosiren*) that develop during the breeding season. True to his Lamarckian position, he interpreted these filaments and their function as an adaptation that was explained better by the direct influence of environmental conditions than by mutation and selection.

BIBLIOGRAPHY

I. Original Works. No complete bibliography of Cunningham's writings has been collected. Most of his papers published up to 1900, however, are listed in the Royal Society's *Catalogue of Scientific Papers*, **IX** (1891), 618; **XIV** (1915), 429–431.

His books are *A Treatise on the Common Sole* (Solea vulgaris), *Considered Both as an Organism and as a Commodity* (Plymouth, 1890); *The Natural History of the Marketable Marine Fishes of the British Islands* (London, 1896); *Sexual Dimorphism in the Animal Kingdom. A Theory of the Evolution of Secondary Sexual Characters* (London, 1900); *The Preservation of Fishing Nets* (London, 1902); *Hormones and Heredity: A Discussion of Adaptations and the Evolution of Species* (London, 1921); and *Modern Biology. A Review of the Principal Phenomena of Animal Life in Relation to Modern Concepts and Theories* (London, 1928).

Cunningham's articles through 1900 include "Review of Recent Researches on Karyokinesis and Cell Division," in *Quarterly Journal of Microscopical Science*, **22** (1882), 35–49; "Note on the Structure and Relations of the Kidney in *Aplysia*," in *Mittheilungen aus der zoologischen Station zu Neapel*, **4** (1883), 420–428; "The Zoological Station in Naples," in *Nature*, **27** (1883), 453–455; "Critical Note on the Latest Theory in Vertebrate Morphology," in *Proceedings of the Royal Society of Edinburgh*, **12** (1884), 759–765; "The Significance of Kupffer's Vesicle, with Remarks on Other Questions of Vertebrate Morphology," in *Quarterly Journal of Microscopical Science*, **25** (1885), 1–14; "On the Relations of the Yolk to the Gastrula in Teleosteans, and in Other Vertebrate Types," *ibid.*, **26** (1886), 1–38; "Dr. Dohrn's Inquiries into the Evolution of Organs in the Chordata," *ibid.*, **27** (1887), 265–284; "Studies of the Reproduction and Development of Teleostean Fishes Occurring in the Neighbourhood of Plymouth," in *Journal of the Marine Biological Association of the United Kingdom*, **1** (1889–1890), 10–54; and "An Experiment Concerning the Absence of Colour from the Lower Sides of Flat-fishes," in *Zoologische Anzeiger*, **14** (1891), 27–32.

Cunningham's articles after 1900 include "The Heredity of Secondary Sexual Characters in Relation to Hormones, a Theory of the Heredity of Somatogenic Characters," in *Archiv für Entwicklungsmechanik der Organismen*, **26** (1908), 372–428; "On the Marine Fishes and Invertebrates of St. Helena," in *Proceedings of the Zoological Society of London* (1910), 86–131; "Mendelian Experiments on Fowls," *ibid.* (1912), 241–259; "Results of a Mendelian Experiment on Fowls, Including the Production of a Pile Breed," *ibid.* (1919), 173–202; "On the Nuptial Callosities of Frogs and Toads from the Lamarckian Point of View," in *Journal of the Linnean Society of London, Zoology*, **36** (1927), 431–437; "The Vascular Filaments on the Pelvic Limbs of *Lepidosiren*, Their Function and Evolutionary Significance," in *Proceedings of the Royal Society of London*, **B105** (1929), 484–493; "Adaptive Evolution, with Special Reference to Metamorphosis and Sex-Limited Characters," in *Proceedings of the Linnean Society of London* (1929–1930), 165–186; "Experimental Researches on the Emission of Oxygen by the Pelvic Filaments of the Male *Lepidosiren*, with Some Experiments on *Symbranchus marmoratus*," in *Proceedings of the Royal Society of London*, **B110** (1932), 234–248, written with D. M. Reid; "Experiments on the Interchange of Oxygen and Carbon Dioxide Between the Skin of *Lepidosiren* and the Surrounding Water, and the Probable Emission of Oxygen by the Male *Symbranchus*," in *Proceedings of the Zoological Society of London* (1932), 875–887.

II. Secondary Literature. On Cunningham's life and career, see E. J. A., "Mr. J. T. Cunningham," in *Nature*, **136** (1935), 13; and D. M. Reid, "Joseph Thomas Cunningham (1859–1935)," in *Proceedings of the Linnean Society of London* (1935–1936), 205–207.

Richard W. Burkhardt, Jr.

DAHLBERG, GUNNAR (*b.* Lofta, Småland, Sweden, 22 August 1893; *d.* Uppsala, Sweden, 25 July 1956), *genetics*.

Dahlberg was the son of Henning Dahlberg, a vicar, and of Gertrud Jaede Dahlberg. He graduated from secondary school at Västervik in 1911 and studied medicine at Uppsala University under the anatomist J. V. Hultcrantz. He received his M.D. in 1926, and until 1936 he was assistant professor of medical genetics and medical statistics at Uppsala. From 1936 until his death he was director of the Swedish State Institute for Race Biology. In 1919 he married Stina Westberg, a physical therapist who during the 1920's was employed at the institute; they had one daughter.

Dahlberg's importance is threefold: he was a creative scientist, an influential teacher, and an outspoken critic of German *Rassenhygiene*. His doctoral dissertation dealt with a classic subject in the nature-nurture debate: *Twin Births and Twins from a Hereditary Point of View* (1926). It discusses the frequency of twin births in different nations and among different classes, and describes in terms used by physical anthropologists differences in size, hair texture, and so on. Then and later the amount of his laboratory work was slight, and statistical methods were used to an unusual extent. After his dissertation and several shorter papers, Dahlberg edited Herman Lundborg's comprehensive description of the Swedish Lapp population (1941), purging it, however, of ethnographical detail. *Mathematische Erblichkeitsanalyse von Populationen* (1943) is the prime example of Dahlberg's ability to criticize the optimism of *Rassenhygiene* through statistical analysis. For example, he believed there is no point in sterilization for eugenic reasons, since hereditary illnesses often make the carrier incapable of reproduction from an early age.

Shortly after graduation, Dahlberg started an informal "doctoral school" in which medical undergraduates were trained mainly in the statistical method and the fundamentals of genetics. Even before he had become director of the Swedish State Institute for Race Biology, he had been the adviser for some twenty doctoral dissertations. By the time of his death the number had quadrupled. Dahlberg's main contribution was expert information on medical statistics, about which he published two introductory books (translated into English as *Statistical Methods for Medical and Biological Students* [1940] and *Race, Reason, and Rubbish* [1942]). He profoundly influenced Swedish medical education by establishing the necessity of statistical training.

Dahlberg joined the Swedish State Institute when it opened in 1922. However, he soon came to disagree on most questions with its director, Herman Lundborg. When Lundborg was due to retire, he did his best to stop Dahlberg from succeeding him. The selection committee seemed to favor Dahlberg's rival, but after two years of strife Dahlberg was appointed to the chair. He fought his way, at times perhaps using somewhat dubious means, but there were fundamental human principles involved. Lundborg was sympathetic to the German *Rassenhygiene*, whereas Dahlberg had written against it. It is no wonder, then, that Dahlberg wanted to see Nazi supporters disqualified as judges in this affair.

This competition turned Swedish race biology and genetics away from their German leanings. Dahlberg continued to criticize the unscientific character of *Rassenhygiene*, and during World War II he constantly challenged its Swedish supporters. He had good relations with British scientists, among them the medical statistician Lancelot Hogben, who shared his socialist views and also translated his works. When Herman Muller left the Soviet Union in 1937, Dahlberg offered him a position at the Institute of Race Biology. Dahlberg's active opposition to *Rassenhygiene* during the war and his genetic expertise made him a natural choice for the UNESCO committees of 1949 and 1951 that reformulated the race concept to stress the unity of mankind. To Dahlberg "race" had no real biological meaning but merely obscured constantly changing statistical entities; hence he viewed "race biology" as a medical rather than an anthropological science.

Dahlberg had obvious problems in completing extensive investigations, preferring to toy with ideas. His format was the essay. Dahlberg's survey of the state of genetics, *Arv och ras* (*Race, Reason and Rubbish*), was a brilliant synthesis, while others of his books were more casual. Something of his attitude toward himself and toward science is revealed in the title of one of his volumes, *Dit min tanke nått* ("As Far As My Thought Has Reached," 1943). He won a relatively wide audience for modern genetics through his published works but never involved himself in legislative committees or party politics. His ideas, however, had great influence on his friend Gunnar Myrdal, who, with his wife, Alva, transformed the race question into a concern of the modern welfare state (*Kris i befolkningsfrågan*, 1934).

Dahlberg was very much a man of thought rather than of action. Despite his socialist convictions, he was an aristocrat at heart, and despite his training as a geneticist, he was no specialist but a generalist. A time-consuming aspect of his wide interests was his lexicographic work. With Herbert Tingsten, the leading liberal intellectual of his day, he edited and wrote *Svensk politisk uppslagsbok* (Swedish Political Encyclopedia, 1937), and alone he compiled *Tidens lexikon* (1950), a work of three thousand pages.

BIBLIOGRAPHY

I. ORIGINAL WORKS. Dahlberg's writings include *Twin Births and Twins from a Hereditary Point of View*, Vilgot Hammarling, trans. (Stockholm, 1926); as ed., Herman Lundborg, *The Race Biology of the Swedish Lapps*, 2 vols. (Uppsala, 1932–1941); as ed., *Svensk politisk uppslagbok* (Stockholm, 1937), edited with Herbert Tingsten; *Statistical Methods for Medical and Biological Students* (London, 1940); *Race, Reason and Rubbish*, Lancelot Hogben, trans. (London, 1942); *Mathematische Erblichkeitsanalyse von Populationen* (Uppsala, 1943); *Dit min tanke nått* (Stockholm, 1943); *Diabetes Mellitus in Sweden* (Lund, 1947); and *Tidens lexikon* (Stockholm, 1950).

II. SECONDARY LITERATURE. Dahlberg's scientific work is cataloged in L. C. Dunn, Lancelot Hogben, T. Kemp, and J. A. Böök, eds., "In Honour of Gunnar Dahlberg," *Acta Genetica*, **4** (1953); and in *Genus*, **12** (1956), with an obituary by C. Gini. A full bibliography is in *Uppsala universitets matrikel 1937–1950* (Uppsala, 1953), supplemented in *Uppsala universitets matrikel 1951–1960* (Uppsala, 1975).

GUNNAR BROBERG

DAKIN, HENRY DRYSDALE (*b*. London, England, 12 March 1880; *d*. Scarborough-on-Hudson, New York, 10 February 1952), *biochemistry*.

The youngest of eight children of Thomas Burns Dakin and Sophia Stevens, Dakin grew up in Leeds, where his father owned an iron and steel business. After finishing public school, he apprenticed with the Leeds city analyst, for whom he tested sulfur content at gasworks. Dakin subsequently obtained a B.Sc. (1901) from Yorkshire College, Leeds (then part of Victoria University, Manchester, now the University of Leeds). Dakin also received the D.Sc. from Leeds in 1907. At Leeds, Dakin worked extensively with Julius B. Cohen, an organic chemist who was investigating the optical activity of biological compounds, a topic that figured in Dakin's later work. He remained as an unofficial demonstrator in Cohen's laboratory for a year after graduation and in 1902 was awarded an 1851 Exhibition Prize, which enabled him to do research at the Lister Institute in London and to spend time at Karl Kossel's laboratory in Heidelberg.

In 1905 Dakin accepted an invitation to join the private laboratory of Christian A. Herter in New York City, of which he became director upon Herter's death in 1910. In July 1916 he married Herter's widow, Susan Dows Herter, and two years later they moved to Scarborough-on-Hudson, where he reestablished the laboratory. Dakin never held an academic appointment but was in regular contact with the research community through frequent trips to England and through his service, from 1910 to 1930, as one of four editors of the *Journal of Biological Chemistry*. (The journal had been founded by Herter.) Later in life Dakin was science adviser to the Merck Institute of Therapeutic Research and a director of Merck and Company.

Dakin received numerous honors, including honorary degrees from Yale University (1918), the University of Leeds (1936), and the University of Heidelberg (1938); the French Legion of Honor; and the Conné Medal of the New York Chemists Club. He was elected a fellow of the Royal Society in 1917. He suffered from extreme shyness and thus did not attend public meetings or conferences; he is not known ever to have given a lecture. Dakin therefore declined an invitation to give the Croonian Lecture; and when he was awarded the Davy Medal in 1941, it was presented privately by Henry Dale in Dakin's library.

Dakin's early work focused on enzymes, beginning with a study, done while still a student of Cohen's, on the mode of action of lipase. While at Heidelberg he and Kossel discovered arginase, and at the Lister Institute he examined the action of esterases on the esters of mandelic acid. His discovery of the stereochemical specificity of these enzymes led him to propose that the first step in hydrolysis consists in the ester's combination with the enzyme.

When he went to Herter's laboratory, Dakin focused on the intermediary (β) metabolism of amino acids and fatty acids. As an oxidizing agent he employed a solution of hydrogen peroxide with a trace of ferrous sulfate. With this oxidation technique Dakin provided confirming evidence for Knoop's demonstration of β oxidation of fatty acids when he showed that those fatty acids with an odd number of carbon atoms produced phenylacetic acid, while those with an even number yielded benzoic acid. Dakin continued this research in an attempt to identify the various intermediaries of fatty acid oxidation and proposed an account in which the fatty acids are oxidized at the β carbon, generating ketone bodies as intermediates. Dakin hypothesized that successive β oxidations continued until one molecule of acetoacetic acid remained. This account of the intermediate steps in fatty acid oxidation was widely accepted until the 1940's.

After taking over Herter's laboratory, Dakin frequently worked alone. A significant exception was his collaboration with Harold W. Dudley. One of their major investigations resulted in the discovery and characterization of glyoxalase, an enzyme that converts glyoxals into corresponding acids. Another joint endeavor concerned the effect of alkali on the properties of proteins. Dakin's interest in this stemmed from his observation, made while working with Kossel, of the effect of alkalies on the optically active hydantoins. He ascribed this change in optical activity to tautomeric change and hypothesized that similar effects should be observed in the polypeptide chains of proteins.

With Dudley, Dakin undertook a study of the amino acids produced from casein treated with alkali that demonstrated that the interior amino acids, but not the terminal ones, were racemized. Thus the position in the protein affected whether the amino acid was racemized. In a later extension of this work, Dakin and Henry Dale showed that crystalline albumins from hen and duck eggs, although very similar in their quantitative composition, behave as distinct antigens in anaphylactic reactions and, upon alkali treatment, yield amino acids with different optical activity (as a result of different sequencing of the amino acids).

During the early part of World War I, Dakin worked in a hospital at Compiègne, France, and there developed a buffered hypochlorite solution for treatment of infected wounds that was named for him. Later in the war he made several trips to the Lister Institute, where the research departments of the Medical Research Committee were housed, and, under its auspices, issued numerous publications (including, with Edward K. Dunham, *A Handbook on Antiseptics*) on disinfecting and sterilizing wounds.

After the war Dakin worked on developing techniques for isolating amino acids from proteins and studied the oxidative catabolism of fatty acids, unsaturated acids, hydroxyacids, and amino acids by animal tissues and yeast. The last major project he undertook was an investigation begun in the 1930's of the hematopoietic factor in liver. Although he failed to isolate the factor, he and Randolph West developed a method for producing the factor in a purer state than was previously available. This made possible the commercial production of the hematopoietic factor that could be used intravenously with safety in the treatment of pernicious anemia.

In addition to Dakin's original contributions to biochemical research, he wrote two highly influential reviews. *Oxidations and Reductions in the Animal Body* surveyed the already vast literature on biological oxidations up to 1912, and "Physiological Oxidations" provided an update to 1920.

BIBLIOGRAPHY

I. ORIGINAL WORKS. Dakin's major writings include "The Oxidation of Amino Acids with the Production of Substances of Biological Importance," in *Journal of Biological Chemistry*, **1** (1906), 171–176; *Oxidations and Reductions in the Animal Body* (London and New York, 1912; 2nd ed., 1922); "Studies on the Intermediary Metabolism of Amino Acids," in *Journal of Biological Chemistry*, **14** (1913), 321–333; "On Glyoxalase," *ibid.*, 423–431, written with Harold W. Dudley; "The Racemization of Proteins and Their Derivatives Resulting from Tautomeric Change. Part II. The Racemization of Casein," *ibid.*, **15** (1913), 263–276, written with Dudley; "On the Use of Certain Antiseptic Substances on the Treatment of Infected Wounds," in *British Medical Journal* (1915), **2**, 318; *A Handbook on Antiseptics* (New York, 1917), written with Edward K. Dunham; "Chemical Structure and Antigenic Specificity. A Comparison of the Crystalline Egg-Albumins of the Hen and the Duck," in *Biochemical Journal*, **13** (1919), 248–257, written with Henry H. Dale; "Physiological Oxidations," in *Physiological Reviews*, **1** (1921), 394–419; and "Observations on the Chemical Nature of a Haematopoietic Substance Occurring in Liver," in *Journal of Biological Chemistry*, **109** (1935), 489–522, written with Randolph West.

II. SECONDARY LITERATURE. Obituaries of Dakin are H. T. Clarke, "Henry Drysdale Dakin: 1880–1952," in *Journal of Biological Chemistry*, **198** (1952), 491–494; and Percival Hartley, "Henry Drysdale Dakin," in *Obituaries of the Royal Society*, **8** (1952), 129–148.

WILLIAM BECHTEL

DAL PIAZ, GIORGIO (*b.* Feltre, Italy, 29 March 1872; *d.* Padua, Italy, 20 April 1962), *geology, paleontology*.

Dal Piaz was the son of Basilio Dal Piaz and Corona d'Alberto. His father's brother Clemente, recognizing his nephew's intelligence, helped to finance his secondary education. Dal Piaz took his degree in pharmacy at the University of Padua in 1896.

An interest in geology, awakened by his love for the mountains surrounding his native town, led Dal Piaz to frequent the laboratory of Giovanni Omboni, professor of geology at Padua University. He graduated in natural sciences at Padua in 1898 and soon became Omboni's assistant. He married Francesca (Fanny) Pontil; they had two children.

Dal Piaz's first scientific contributions concerned glacial deposits in the province of Belluno, of which Feltre is part (1895), and the geology and petrography of the Euganean Hills, a group of magmatic origin near Padua (1896, 1897).

The deep interest of Dal Piaz in paleontology and stratigraphy, accompanied by his detailed investigations in the field, soon made him a leader in Jurassic stratigraphy of Venetia and Trentino-Alto Adige. In 1907 he published "Le Alpi feltrine," a fundamental stratigraphic work based on a rich fossil fauna of ammonites; in later publications he clarified many regional stratigraphic problems, among them heteropic facies.

Dal Piaz patiently recovered fossil bones by the hundreds, mostly of cetaceans, from the hard sandstone near Belluno, which was quarried for grinding wheels and hones; his careful study of them was a masterly contribution to odontocete (dolphin) paleontology. His last work on the subject was published posthumously. In 1908 Dal Piaz succeeded Omboni as professor of geology at Padua and as director of the Geological Institute. He greatly improved the library and the paleontological museum, which moved to a new headquarters in 1932.

Dal Piaz made a major contribution to the structural study of the Alps in 1912 by extending the new tectonic theories to the southern Alps. This work interpreted the Alpine chain as a huge structure of superimposed folds and overthrusts, and demonstrated that the faults are minor elements in a sequence of folds, arranged parallel to the chain and leaning toward the Venetian plain and the Adriatic basin. It was published in the first volume of the *Memorie dell'Istituto geologico della R. università di Padova* (now *Memorie di scienze geologiche*), founded by Dal Piaz with a large personal financial investment and directed by him for many years.

Other main contributions by Dal Piaz are the proof of the existence of Pliocene marine deposits in Venetia (1912); of Paleolithic man (1922); of Tertiary tonalites along the Alpine-Dinaric tectonic boundary, as it was then called (1926); and of pre-Würmian glacial moraines in the same area (1946, 1949). In 1922, for the seventh centenary of the University of Padua, Dal Piaz wrote a masterpiece on the development of geologic ideas in Italy since the Middle Ages.

The main scientific enterprise of Dal Piaz was the organization of the geologic mapping of northeastern Italy (the present political subdivisions Trentino-Alto Adige, Venetia, and Friuli-Venezia Giulia) on the scale 1:100,000. Started in 1923, it was completed in 1962. The last (forty-second) sheet was published just after Dal Piaz died, but he saw the printer's proofs. In this work he collaborated with many scientists; for instance, on the crystalline formations he had the cooperation of Angelo Bianchi, professor of mineralogy at Padua University. The new survey spurred much scientific research and increased knowledge of regional and applied geology (especially hydrogeology).

Dal Piaz dealt with problems of engineering geology in dam construction and ship canal excavation, of petroleum geology (Albania, Italy), of water supply (Istria, Venetia), and of thermal springs, particularly around the Euganean Hills. In the latter he demonstrated the subvolcanic, laccolithic nature of trachytic bodies (1935–1953).

Dal Piaz taught until 1942, and in 1943 he became professor emeritus. Many scientists were his pupils and followed his prejudice-free method of research. He was succeeded in the chair of geology and the direction of the Geological Institute at Padua by his son Giambattista, formerly professor of geology at Turin (who was in turn succeeded in Padua by his son Giorgio).

Dal Piaz was a member of the Accademia Nazionale dei Lincei, the Pontifical Academy of Sciences, the Accademia Nazionale dei XL, and various regional institutions. He was awarded the Royal Prize for Geology of the Accademia dei Lincei in 1916, was twice president of the Italian Geological Society, and represented Italy on the International Commission for a Stratigraphic Lexicon.

After retiring, Dal Piaz went daily to the Geological Institute at Padua. A few days after his ninetieth birthday, he died suddenly. Some of his unpublished manuscripts were printed by his followers after his death. The lyceum of Feltre is named for Dal Piaz, as are a shelter in the mountains he studied and a prize awarded by the Italian Geological Society, funding for which was raised through a subscription by his former students and admirers.

BIBLIOGRAPHY

I. Original Works. Dal Piaz's works number more than one hundred. Among the more important are "Note sull'epoca glaciale nel Bellunese," in *Atti della Società veneto-trentina di scienze naturali*, 2nd ser., **2** (1895), 336–347; "Studi geologico-petrografici intorno ai Colli Euganei. I e II," in *Rivista di mineralogia e cristallografia italiana*, **16** (1896), 49–69, and **17** (1897), 74–80; "Le Alpi feltrine," in *Memorie del R. istituto veneto di scienze, lettere ed arti*, **27** (1907), 1–176; "Studi geotettonici sulle Alpi orientali . . . ," in *Memorie dell'Istituto geologico della R. università di Padova*, **1** (1912), 1–195; "Sull'esistenza del Pliocene marino nel Veneto," in *Atti dell'Accademia scientifica veneto-trentino-istriana*, **5** (1912), 212–215; "L'università di Padova e la scuola veneta nello sviluppo e nel progresso delle scienze geologiche," in *Memorie dell'Istituto geologico della R. università di Padova*, **6** (1922) 1–41; "Il confine alpino-dinarico dall'Adamello al massiccio di Monte Croce nell'Alto Adige," in *Atti dell'Accademia scientifica veneto-trentino-istriana*, 3rd series, **17** (1926), 3–7; "Avanzi morenici prewürmiani allo sbocco della valle del Brenta . . . ," in *Acta della Pontificia accademia delle scienze*, **10**, no. 5 (1946), 155–166; and "Sui depositi morenici prewürmiani dei Colli Berici . . . ," *ibid.*, no. 29 (1949), 339–354.

II. SECONDARY LITERATURE. Biographical articles on Dal Piaz include Angelo Bianchi, "Commemorazione del socio Giorgio Dal Piaz," in *Rendiconti dell'Accademia nazionale dei Lincei*, 8th ser., **42** (1967), 112–123; Giuseppe Biasuz, "Ricordo del prof. Giorgio Dal Piaz," in *El Campanon*, **1** (1972), 1–4; Piero Leonardi, "Commemorazione del membra effettivo prof. Giorgio Dal Piaz," in *Atti dell'Istituto veneto di scienze, lettere ed arti*, **121** (1963), 1–15; and Roberto Malaroda, "Giorgio Dal Piaz paleontologo (1872–1962," in *Bollettino della Società paleontologica italiana*, **4** (1965), 3–8.

GIULIANO PICCOLI

DAM, [CARL PETER] HENRIK (*b.* Copenhagen, Denmark, 21 February 1895; *d.* Copenhagen, 17 April 1976), *biochemistry.*

Henrik Dam, son of Emil Dam, a pharmaceutical chemist, and Emilie Peterson, a teacher, was educated at the Copenhagen Polytechnic Institute. After graduating with the M.Sc. in chemistry (1920), he became an instructor in chemistry at the Royal School of Agriculture and Veterinary Medicine in Copenhagen (1920) and an instructor in biochemistry under Valdemar Henriques at the physiological laboratory of the University of Copenhagen (1923). On 15 July 1924 Dam married Inger Marie Martha Sophie Olsen. He studied microchemistry with Fritz Pregl in Graz, Austria (1925). In 1928 he was appointed assistant professor, and the following year promoted to associate professor, at the Institute of Biochemistry at the University of Copenhagen. Here he made his first observations that led by stages to the discovery of the blood-coagulation factor, vitamin K, and other bioquinones.

During the years 1932 and 1933 Dam worked at the laboratory of Rudolf Schoenheimer in Freiburg, Germany, thanks to a Rockefeller fellowship he obtained to further his studies of the metabolism of sterols. In 1933 he received the D.Sc. in biochemistry at the University of Copenhagen with a thesis entitled *Nogle undersøgelser over sterinernes biologiske betydning* (Some Investigations on the Biological Significance of the Sterols). The next year Dam worked with Paul Karrer in Zurich. He continued to hold his post as associate professor until his appointment as professor of biochemistry at the Polytechnic Institute in Copenhagen. This appointment, however, was in absentia, because since 1940 Dam was on a lecture tour to the United States and Canada under the auspices of the American-Scandinavian Foundation. The tour was planned before the occupation of Denmark by German troops in April 1940.

Dam spent the war years in the United States and returned to Denmark in June 1946. During this period he did research work at the Woods Hole Marine Biological Laboratories during the summer and autumn of 1941 and was a senior research associate at the Strong Memorial Hospital of the University of Rochester (1942–1945) and an associate member of the Rockefeller Institute for Medical Research in New York City (1945–1946). In the spring of 1949 Dam made a three-month lecture tour of the United States and Canada that was arranged by the American-Scandinavian Foundation. His topic was the role of peroxidation in vitamin E deficiency.

Until his retirement in 1965 Dam was professor at the Polytechnic Institute in Copenhagen. In 1950 his chair was changed to that of biochemistry and nutrition. From 1956 to 1963 he also was head of the biochemical division of the Danish Fat Research Institute.

During the years 1928 to 1930 Dam studied the cholesterol metabolism of chicks at the Biochemical Institute of the University of Copenhagen. It was already known that rats, mice, and dogs can synthesize cholesterol, but some experiments seemed to show that chicks could not thrive on a sterol-free diet. Dam repeated experiments on cholesterol content of chicks that had been reported by John Addyman Gardner at the St. George's Hospital Medical School at London in 1914. Dam used artificial, practically sterol-free diets consisting of casein, starch, marmite, salts, and paper (to supply indigestible matter) to which vitamins A and D were added. He found that chicks are able to synthesize cholesterol and to break it down. During his researches Dam observed that some of the chicks unexpectedly exhibited a marked bleeding tendency resembling that of scurvy. They displayed hemorrhages under the skin, in muscle, or in other tissues, and the blood coagulated more slowly than normally. Dam reported his findings in two articles in the German periodical *Biochemische Zeitschrift* (1929, 1930).

The observation of the hemorrhagic disease in chicks was also reported by William Douglas McFarlane and his associates William Richard Graham, Jr., and Frederik Richardson at the Ontario Agricultural College at Guelph, Canada (1931). During experiments to determine the need of chicks for fat-soluble vitamins, they related blood-clotting time to an ether-soluble substance in fish meal. The disease was also observed by Walther F. Holst and Everett R. Halbrook at the University of California at Berkeley (1933). They found that it could be prevented by fresh cabbage and drew the wrong conclusion that the protective factor in the cabbage was vitamin C.

After Dam returned from Freiburg to Copenhagen in the fall of 1933, he took up the research again. In June 1934 he published the detailed "Haemorrhages in Chicks Reared on Artificial Diets: A New Deficiency Disease." Dam included gizzard erosions as a characteristic of the disease, but later it was found that the cause of this was unrelated to the blood-clotting factor. In his experiments Dam showed that parenteral injections of ascorbic acid (vitamin C) failed to prevent the disease, but that an addition of hempseed to the food prevented the bleeding. He came to the conclusion that the hemorrhagic disease in chicks was owing to the lack of a hitherto unrecognized deficiency factor in the diet occurring in seeds and cereals.

With his associate Fritz Schønheyder, Dam examined a number of animal organs and plant materials for their ability to protect against hemorrhagic disease. They found that the cause of hemorrhages could not be the lack or low content of cholesterol in the diet, nor the low amount of fat, nor a lack of vitamin C, but a dietary lack of antihemorrhagic vitamin. It appeared that green leaves and hog liver were among the most potent sources of this factor (1934). It was also found that the factor was fat soluble. In 1935 they characterized the antihemorrhagic factor as a new fat-soluble vitamin, which was not identifiable with vitamins A, D, or E, that had been found to be essential in trace quantities in the diet. Dam and Schønheyder designated it as vitamin K because K was the first letter of the Scandinavian and German word *Koagulation* (coagulation). The existence of this vitamin K was promptly confirmed by Herman James Almquist and Evan Ludvig Stokstad at the University of California (1935).

Dam and his associates next carried out studies in order to discover how the new vitamin acted on blood clotting, how it could be concentrated, and whether it played any part in other animals and in humans. They found the new nutrient to be present in large amounts in hog liver fat and also in a number of cereals, seeds, vegetables, fruits, animal organs, and fats.

In 1935 Schønheyder showed that blood plasma from normal chicks accelerated the clotting of the plasma from K-avitaminous chicks. It appeared that calcium, fibrinogen, and thromboplastin were not reduced in quantity in the K-avitaminous chicks. Dam, Schønheyder, and Erik Tage-Hansen (1936) were also able to explain the mechanism by which vitamin K produces its results. They did experiments that greatly extended the knowledge of the nature of hemorrhagic disease. The amount of prothrombin, a substance formed in the liver, was below normal in the blood of birds fed diets deficient of vitamin K. The blood of chicks became depleted of prothrombin in about three weeks on vitamin K–deficient diets. Three days after feeding with vitamin K–rich preparations, the chicks' prothrombin returned to normal, as did the clotting time of the blood. Deficiency of vitamin K leads to a lack of prothrombin and thus to a lack of thrombin, which forms fibrinogen in the blood. In consequence, fibrinogen cannot be formed into the fibrin necessary for the coagulation of the blood.

Immediately after the discovery of vitamin K, its chemical properties were studied mainly by Dam and his associates and by Almquist and his group. With Schønheyder, Johannes Glavind, Liese Lewis, and Tage-Hansen, Dam developed a method for the preparation of very strong concentrates from alfalfa (1938). Having improved the method further, Dam worked on isolating and characterizing vitamin K. This investigation, carried out in collaboration with Karrer, resulted in a joint publication from the Chemical Institute of the University of Zurich and the Biochemical Institute of the University of Copenhagen, "Isolierung des Vitamin K in hochgereinigter Form" (1939). Dam, Karrer, and their associates prepared pure vitamin K from dried plant materials, in particular from dried alfalfa, by means of fat solvents. It was concentrated by the removal of chlorophyll through selective adsorption, the removal of crystalline nonactive substances by chilling of an acetone solution of the remaining material, by fractional distillation at a pressure of 10^{-3} millimeters mercury and a temperature of 115–140°C. Through this process the vitamin moved up the condenser and was finally purified by repeated chromatographic adsorption. The vitamin was obtained in the form of a light yellow oil. The elementary composition was found to be 82.2 percent carbon, 10.7 percent hydrogen, and 7.1 percent oxygen. The substance gave a characteristic color reaction with sodium ethylate and showed an adsorption spectrum in the ultraviolet region that pointed to a connection with 1,4-quinone.

The pure vitamin K was isolated nearly simultaneously by Edward Adelbert Doisy's group in the laboratory of biological chemistry at St. Louis University School of Medicine, by Louis Frederick Fieser's group at the Conserve Memorial Laboratory at Harvard University, and by Almquist (1939). They succeeded in the elucidation of its constituents. It was established that the structure of vitamin K from green plants (vitamin K_1) was phylloquinone or 2-methyl-3-phytyl-1,4-naphthoquinone. (Phytyl is the radical of the twenty-carbon alcohol phytol with four isoprene units.)

Doisy's group also isolated a vitamin K by the putrefaction of ether-extracted fish meal, which chemically differs slightly from vitamin K as reported by Dam, Karrer, and associates. Its structure appears similar to vitamin K, but it has a side chain of seven isoprene units (vitamin K_2, menaquinone, or 2-methyl-3-polyprenyl-1,4-naphthoquinone). The interest in the chemistry of vitamin K followed from the fact that the July 1939 issue of the *Journal of the American Chemical Society* contained six preliminary communications on vitamin K compounds, and the September 1939 issue, seven.

Dam made no investigations on the structure of vitamin K. He was mainly interested in experiments with vitamin K and vitamin K–active compounds using rats and rabbits as experimental animals. With Glavind (1938) he found that vitamins K are synthesized by plants and by certain microorganisms. The synthesis appears to be greatly influenced by sunlight. Peas grown in the dark contain only small amounts, while control plants raised in light contain considerably more. *Coli* bacteria synthesize vitamin K on an artificial medium that has only glucose, citrate, and asparagin as organic constituents (1941). Later Dam found vitamin K in the green alga *Chorella vulgaris* and in a lesser amount in some photosynthesizing bacteria (1944).

In studies on other vitamin K compounds, Dam, Glavind, and Karrer (1940) found that if vitamins K_1 and K_2 are converted into the diacetates of the corresponding hydroquinones, the activity is somewhat reduced. Hydrogenation of the double bond in the phytyl group of vitamin K_1 results in a lowering of the activity and replacement of the methyl group. Vitamin K activity was also found in a number of simpler but related substances, such as 2-methyl-1,4-naphthoquinone of menadione, and numerous esters of derived quinols. In 1949 Dam found that 5-methyl-4,7-thionaptenequinone, an isostere of menadione, has about 3 percent of the vitamin K activity of menadione.

Dam further studied vitamin K not only with respect to its occurrence and biological function in animals and plants, but also with respect to its application in human medicine. In 1938 he and Glavind clinically demonstrated the usefulness of vitamin K for hemorrhagic diseases in man. The first hemorrhagic condition in man that was recognized as owing to lack of vitamin K was the cholemic bleeding tendency that poses a great danger in surgery on patients with obstructive jaundice. Suitable administration of vitamin K eliminates the risk of fatal bleeding, which was shown to be owing to insufficient absorption of vitamin K from the intestine in the absence of bile.

The most frequent form of the K-avitaminosis in man is that of the infant during the first few days of life. Dam, Tage-Hansen, and P. Plum studied neonatal bleeding disorders (1939). They found that this bleeding tendency can be prevented by the administration of vitamin K to the infant immediately after birth or by administration of an excess of vitamin K to the mother a suitable time before delivery. This type of bleeding tendency results from limitation of the passage of vitamin K from mother to fetus.

The mechanism of the coagulation of the blood was envisaged as a complex but ordered succession of processes. The coagulation of the blood is owing to the transformation of the protein fibrinogen into a close network of fibrin in whose meshes the blood corpuscles are caught. This transformation is caused by the action of the enzyme thrombin, which is derived from prothrombin with the aid of thromboplastin, a substance liberated from the injured cells. The lack of vitamin K was found to cause a drop in the prothrombin content, while a rapid increase occurred after the vitamin K had been added. Dam and Glavind (1940) showed that the effect of a given dose of vitamin K on the prothrombin level was quantitatively the same whether the disease was produced by a vitamin K–free diet or by choledochus ligature. Obviously vitamin K exerts a direct influence on the formation of the prothrombin, the precursor of the enzyme that is required for the coagulation. The relation of vitamin K to the two components of prothrombin was studied by Dam in 1948.

In 1939 Dam was a recipient of the Christian Bohr Award in physiology for his discovery of vitamin K. In 1943 he was awarded the Nobel Prize in physiology or medicine for his discovery of vitamin K. The prize was shared with Doisy for his discovery of the chemical nature of vitamin K. On 27 October 1944, while war was still raging in Europe, the Caroline Institute in Stockholm made its decision about the Nobel Prizes in physiology or medicine for 1943 and 1944. Owing to the war conditions, it would have been impossible for the winners to come to Sweden. The American-Scandinavian Foundation was, therefore, asked to hold an awards ceremony at New York City in December 1944, at which the Swedish ambassador presented the prizes. On 12 December 1946, at Stockholm, Dam delivered his Nobel Prize Lecture, "The Discovery of Vitamin K, Its Biological Functions and Therapeutical Applications."

From Dam's studies in vitamin K arose the observation of some new symptoms in experimental animals, such as increased capillary permeability and coloration of adipose tissue, which turned out to be results of the ingestion of certain fats in the absence of vitamin E. From 1937 Dam worked on vitamin E both in Copenhagen and in the United States. With Glavind he found that lack of vitamin E in chicks leads to a vascular disease causing alimentary exudative diathesis followed by increased capillary permeability and muscular dystrophy (1939). Lack of vitamin E will also cause encephalomalacia. In 1931 A. M. Pappenheimer and M. Goetsch had found that a high-fat, E-deficient diet caused the development in chicks of nutritional encephalomalacia, but in ducklings on the same dietary regimen muscular dystrophy resulted. Dam and Glavind (1939) found in chicks the development of exudative diathesis, characterized by the accumulation of plasma under the skin. The alimentary encephalomalacia, which was thought to be caused by lack of vitamin E, was shown in 1958 by Dam and his associates to be specifically dependent upon dietary fatty acids of the linoleic acid series concomitantly with the absence of vitamin E.

Other manifestations of vitamin E deficiency in chicks and rats were investigated with respect to their interrelationship to dietary fat and to the role of vitamin E as an antioxidant. In 1950 Dam, with H. Granados and Erik Aaes-Jørgensen, reported that addition of sulfaguanidin to a diet with high cod-liver oil protected rats against peroxidation and yellow-brown coloration of the adipose tissue, but not against depigmentation of the incisors. They interpreted the influence on fat peroxidation as an in vivo autoxidant effect of sulfaguanidin and not as the result of the antibacterial action.

In the 1950's Dam studied gallstones in his laboratory at Copenhagen. In hamsters he found that rearing on certain artificial diets induces formation of gallstones. When the diet is deficient in polyunsaturated fatty acids and carbohydrate is furnished as an easily absorbable sugar, cholesterol gallstones are formed abundantly. Other dietary combinations lead to the formation of amorphous pigmented gallstones.

After his retirement the main subjects treated by Dam were nutritional studies in relation to gallstone formation. During his scientific career Dam and his research school produced more than three hundred publications concerning vitamin K, vitamin E, and cholesterol. Dam wrote a number of review articles that are models of fairness and acumen as well as of deep scholarship. In collaboration with Aaes-Jørgensen, Dam wrote two textbooks: *Biokemi* (1950) and *Grundrids af biokemi og ernaering* (1955). His accomplishments were honored with membership in the Danish Academy of Technical Sciences (1947) and the Royal Danish Academy of Sciences and Letters (1948), and election as foreign correspondent of the Académie Royale de Médecine de Belgique (1951) and as honorary fellow of the Royal Society of Edinburgh (1953). He was a fellow of the American Institute of Nutrition. In 1965 he received an honorary degree in sciences from St. Louis University.

BIBLIOGRAPHY

I. ORIGINAL WORKS. Bibliographies of Dam's writings are in S. Veibel, *Kemien i Danmark*, II, *Dansk kemisk bibliografi 1800–1935* (Copenhagen, 1943), 119–122; and in *Aarsberetninger for den Polyteknisk Laeranstalt 1945–1972* (Copenhagen).

His early writings include "Cholesterinstoffwechsel in Hühnereiern und Hünchen," in *Biochemische Zeitschrift*, **215** (1929), 475–492; "Über die Cholesterinsynthese im Tierkörpern," *ibid.*, **220** (1930), 158–163; "A Deficiency Disease in Chicks Resembling Scurvy," in *Biochemical Journal*, **28** (1934), 1355–1359, with Fritz Schønheyder; "Haemorrhages in Chicks Reared on Artificial Diets: A New Deficiency Disease," in *Nature*, **133** (1934), 909–910; "The Antihaemorrhagic Vitamin of the Chick," in *Biochemical Journal*, **29** (1935), 1273–1285; "The Occurrence and Chemical Nature of Vitamin K," *ibid.*, **30** (1936), 897–901, with Fritz Schønheyder; and "Studies on the Mode of Action of Vitamin K," *ibid.*, 1075–1079, with Fritz Schønheyder and Erik Tage-Hansen.

Later writings include "Anti Encephalomalica Activity of de-α-Tocopherol," in *Nature*, **142** (1938), 1157–1158, with Johannes Glavind, Ole Bernth, and Erik Hagens; "Vitamin K in Plant," in *Biochemical Journal*, **32** (1938), 485–487, with J. Glavind and I. Svendson; "Clotting Power of Human and Mammalian Blood in Relation to Vitamin K," in *Acta medica scandinavica*, **96** (1938), 108–128, with J. Glavind; "Vitamin K Lack in Normal and Sick Infants," in *Lancet* (1939), 2, 1157–1161, with E. Tage-

Hansen and P. Plum; "Alimentary Exudative Diathesis, a Consequence of E-Avitaminosis," in *Nature*, **143** (1939), 810–811, with Johannes Glavind; "Isolierung des Vitamin K in hochgereinigter Form," in *Helvetica chimica acta*, **22** (1939), 310–313, with P. Karrer, A. Geiger, J. Glavind, W. Karrer, E. Rothschild, and H. Salomon; "Die biologische Aktivität der natürlichen K-Vitamine und einiger verwandter Verbindungen," in *Helvetica chimica acta*, **23** (1940), 224–233, with Johannes Glavind and Paul Karrer; "Bildung von Vitamin K in Colibakterien auf synthetischem Substrat," in *Die Naturwissenschaften*, **29** (1941), 287–288, with J. Glavind, S. Orla-Jensen, and A. D. Orla-Jensen; "Vitamin K in Unicellular Photosynthesizing Organisms," in *American Journal of Botany*, **31** (1944), 492–493; "Dicumarol Poisoning and Vitamin K Deficiency in Relation to Quick's Concept of the Composition of Prothrombin," in *Nature*, **161** (1948), 1010–1011; "Observations on Experimental Dental Caries. Effect of Certain Quinones with, and without Vitamin K Activity," in *Acta pathologica et microbiologica scandinavica*, **26** (1949), 597–602, with H. Granados and J. Glavind; and "Influence of Sulphaguanidin on Certain Symptoms of Vitamin E Deficiency in Rats," in *Experientia*, **6** (1950), 150–152, with H. Granados and E. Aaes-Jørgensen.

Dam wrote a number of review articles, including "Vitamin K, Its Chemistry and Physiology," in *Advances in Enzymology and Related Subjects*, **2** (1942), 285–324; "Vitamin K," in *Vitamins and Hormones. Advances in Research and Application*, VI (New York, 1948), 27–53; "Fat-soluble Vitamins," in *Annual Review of Biochemistry*, **20** (1951), 265–304; "The Biochemistry of Fat-soluble Vitamins," in R. T. Holman, W. O. Lundberg, and T. Malkin, eds., *Progress in the Chemistry of Fats and Other Lipids*, III (London and New York, 1955), 153–212; and "The Determination of Vitamin K," in Paul Gÿorgy and W. N. Pearson, eds., *The Vitamins. Chemistry, Physiology, Pathology, Methods*, 2nd ed., VI (New York, 1967), 245–260.

Dam's Nobel Lecture is in *Nobel Lectures: Physiology or Medicine, 1942–1962* (Amsterdam, London, and New York, 1964), 8–24.

II. SECONDARY LITERATURE. *Current Biography 1949* (New York, 1950), 135–136; "Henrik Dam," in *McGraw-Hill Modern Men of Science* (New York, 1966), 125–126; R. A. Morton, "Obituary," in *Nature*, **261** (1976), 621; E. Rancke-Madsen, "Carl Peter Henrik Dam," in *Dansk biografisk leksikon*, III (Gyldendal, 1979), 556; and H. H. Ussing, "Henrik Dam. 21. februar 1895–17. april 1977," in *Oversigt over det K. danske videnskabernes selskabs forhandlinger* (1976–1977), 98–103.

H. A. M. SNELDERS

DANIELS, FARRINGTON (*b.* Minneapolis, Minnesota, 8 March 1889; *d.* Madison, Wisconsin, 23 June 1972), *physical chemistry.*

Daniels' career in physical chemistry was highly visible during the middle of the twentieth century and encompassed major contributions to teaching and research, including work on electrochemistry, chemical kinetics, nitrogen fixation, nuclear energy, thermoluminescence, solar energy, and the social impact of science. His active career was closely associated with the chemistry department at the University of Wisconsin in Madison.

Daniels was the son of Franc Burchard Daniels, superintendent of an American Express office, and Florence Louise Farrington Daniels, both eighth-generation descendants of English settlers who had mostly been successful in the world of business. He studied chemistry at the University of Minnesota, where he earned the B.S. in 1910 and the M.S. a year later. His Ph.D. was completed in 1914 at Harvard, where he studied the electrochemistry of thallium amalgams in the laboratory of Theodore William Richards. Since the outbreak of World War I prevented him from accepting a postdoctoral fellowship in Fritz Haber's laboratory in Berlin, Daniels took an instructorship at Worcester Polytechnic Institute (1914–1917) and was later assistant professor (1917–1918). During American participation in the war he spent a year with the Chemical Warfare Service, seeking a way to prevent fogging of the lenses in gas masks while they were being worn. On 15 September 1917, Daniels married Olive Miriam Bell; they had two sons and two daughters.

When the war ended, Daniels spent a year (1919–1920) at the Fixed Nitrogen Research Laboratory in Washington. In 1920 he accepted an assistant professorship at the University of Wisconsin, where he rapidly gained recognition as a leading physical chemist, becoming full professor in 1928. He pursued the fixation of atmospheric nitrogen for more than two decades, at first in connection with the properties and reactions of oxides of nitrogen, particularly the kinetics of such reactions. By 1940 Daniels was deeply involved in developing a heat-exchange process for preparing nitric oxide from air as an alternative to the uneconomic electric-arc process for preparing nitric acid and nitrates. However, the Wisconsin process failed to become competitive with the well-established Haber process and was ultimately abandoned after some pilot plant trials.

Before 1944, Daniels became involved in wartime research involving the kinetics of rocket propulsion, and on the detection of oil films left on water surfaces by submarines, using luminescence caused by exposure to ultraviolet light. These studies were abandoned after he was called to the Metallurgical Laboratory at the University of Chicago to direct the research being pursued toward understanding the chemical isotopes related to the development of the first atomic bombs (1945–1946). He remained as director of the laboratory after the war ended, when the activity was moved outside Chicago to become the Argonne National Laboratory of the new Atomic Energy Commission.

Daniels now sought to direct the laboratory's research toward planning and design of a power-producing nuclear reactor. The AEC, however, soon decided to slow down development of a power reactor and concentrate more intensively on development of nuclear energy for military purposes. At this point Daniels resigned his position and resumed his work at Wisconsin.

Daniels' renewed research program now focused on a survey of resources of fissionable minerals by use of thermoluminescence as an analytical tool. After these studies revealed that resources of uranium ores were clearly limited, he lost faith in placing major emphasis on development of power-producing reactors. He now became an enthusiast for developing techniques and programs for utilization of solar energy, particularly in underdeveloped nations. The last two decades of his life were devoted to research on and promotion of solar energy as the principal energy source of the future. Daniels traveled extensively around the world, particularly after retirement from Wisconsin, to develop interest in solar energy and to encourage research and development in India, Mexico, and elsewhere. He also published extensively on the subject.

Daniels was more than a dedicated research scientist, however, and contributed greatly to teaching and administration. Besides his management duties at the Metallurgical Laboratory, he served for seven years (1952–1959) as chairman of the Wisconsin chemistry department, during a period of rapid increase in the number of students and faculty, when there was a need for planning expansion of the physical plant.

In spite of many diversions Daniels remained a dedicated teacher who was constantly striving to strengthen scientific instruction, not only with regard to his own classroom performance but also with respect to content and methodology in the profession. He was author of a laboratory manual and a leading textbook of physical chemistry. Because of dissatisfaction with his own limitations in advanced mathematics, he introduced a course in mathematical preparation for physical chemistry and developed a textbook for the course. He later made the course obsolete by persuading his colleagues to incorporate a calculus requirement into the chemistry major. Daniels was also influential in adding new faculty members with a strong mathematical background.

While he was chairman of his college's Curriculum Committee in 1939, Daniels' leadership resulted in three innovations that greatly improved the breadth of curricular offerings in the university: (1) the Freshman Forum, which introduced students to various fields of learning and gave them a greater appreciation of the nature of college education as well as of various ideas awaiting exploration; (2) a history of science department to deal with the growth of scientific concepts and their impact; and (3) a course for seniors in contemporary trends, dealing with the impact of science and technology on world problems. All were implemented, the third only after war's end, when Daniels returned to the campus and took an active part in the course.

Daniels was elected to the National Academy of Sciences in 1947. The American Chemical Society honored him with three of its most prestigious awards—the Willard Gibbs Medal (1955), the Priestley Medal (1957), and the James Flack Norris Award (1957). The University of Wisconsin awarded him an honorary Sc.D. in 1966. He also received honorary Sc.D. degrees from the Universities of Rhode Island (1956), Minnesota (1959), Dakar (1960), and Louisville (1964). He was president of the American Chemical Society in 1953 and also served as president of the Geochemical Society in 1958, the Solar Energy Research Society in 1964 and 1965, and Sigma Xi in 1965, and as vice president of the National Academy of Sciences from 1959 to 1961.

Daniels was outwardly humble but possessed an inner confidence that caused him constantly to undertake difficult tasks. He was deeply religious and had a passion for social improvement. Daniels possessed a compulsive drive coupled with tireless energy that enabled him to pursue a multitude of goals that frequently had social as well as scientific objectives. An optimist by nature, he was rarely deterred by failure but quickly turned to alternative hypotheses and goals. He remained active until a few months before his death, which was caused by an inoperable cancer.

BIBLIOGRAPHY

I. Original Works. Daniels was the author of approximately 300 publications, including seven books, two of them—*Experimental Physical Chemistry* and a physical chemistry textbook—updated through seven editions with the aid of faculty colleagues. The seventh edition of the textbook was titled *Outlines of Physical Chemistry*, by F. H. Getman and Farrington Daniels (New York, 1943).

The eighth edition was published by Daniels alone under the same title. In 1955 Daniels and Robert A. Alberty published the first edition of *Physical Chemistry*. Daniels' name continued to be associated with the book through the fifth edition in 1979.

The seventh edition of *Experimental Physical Chemistry* was published in 1970 under the authorship of F. Daniels, R. A. Alberty, J. W. Williams, Paul Bender, C. D. Cornwell, and John E. Harriman.

Daniels' research papers were published primarily in the principal American Chemical Society periodicals. Papers on more general subjects were scattered widely. For a full list of his titles, see Olive Bell Daniels (below). His first book was *Mathematical Preparation for Physical Chemistry* (New York, 1928). *Chemical Kinetics* (Ithaca, N.Y., 1938) developed out of his semester at Cornell as George Fisher Baker nonresident lecturer in 1935. *Challenge of Our Times*, edited with Thomas M. Smith (Minneapolis, 1953), is a collection of lectures given by selected faculty in the early years of the Contemporary Trends course. His last book was *Direct Use of the Sun's Energy* (New Haven, 1964). Daniels' collected papers are in the University of Wisconsin Archives. Six bound volumes of collected reprints are in the archives and in the chemistry department library.

II. SECONDARY LITERATURE. There are short biographies of Daniels in *Current Biography* (1965), 106–108; *McGraw-Hill Modern Men of Science*, I (New York, 1980), 126–127; W. Miles, ed., *American Chemists and Chemical Engineers* (Washington, D.C., 1976); and James C. Spaulding, "Our Sunlit Statesman: Prof. Farrington Daniels," in *Saturday Review*, 4 April 1959, 66–67. See also Olive Bell Daniels, *Farrington Daniels: Chemist and Prophet of the Solar Age* (Madison, Wis., 1978).

AARON J. IHDE

DANTZIG, DAVID VAN (*b.* Rotterdam, Netherlands, 23 September 1900; *d.* Amsterdam, Netherlands, 22 July 1959), *mathematics, statistics, logic, philosophy, history of science.*

At the age of thirteen, Van Dantzig wrote his first mathematical paper, which was published in a Dutch mathematical periodical. After high school he studied chemistry, which he did not like; he soon stopped, owing to family circumstances that obliged him to take various odd jobs to support himself. At night he prepared for a sequence of state examinations in mathematics, which he passed in 1921, 1922, and 1923. After a short time at Amsterdam University he passed the *doctoraalexamen* (roughly equivalent to a master's degree). In 1927 he became an assistant to Jan A. Schouten at Delft Technical University. After a brief stay at a teacher training institution, Van Dantzig returned to Delft, where he became a lecturer in 1932, an extraordinary professor (roughly equivalent to associate professor) in 1938, and an ordinary professor (roughly equivalent to full professor) in 1940. He had received a Ph.D. degree at Groningen in 1932.

During the German occupation he was dismissed and obliged to move with his family from the Hague to Amsterdam. In 1946 he was appointed a professor at Amsterdam University and was a cofounder of the Mathematisch Centrum, a research and service institution. Until his death he played a leading role at this institution while retaining his chair at the university.

During his short period of study at Amsterdam University, Van Dantzig was strongly influenced by one of his mathematics professors, Gerrit Mannoury (1867–1956), whose personality had a great impact on many people (including L. E. J. Brouwer). With Mannoury, Van Dantzig shared the disbelief in mathematical certainty—intuitionist or formalist—and even more than Mannoury he stressed the social responsibility of the mathematician as a teacher and a researcher, which he expressed in a number of publications.

As an assistant to and collaborator of Schouten, Van Dantzig took up Schouten-style differential geometry and its applications, in particular projective and conformal differential geometry, and electromagnetism and thermodynamics, independent of Riemannian geometry. Unfortunately, he never elaborated on his idea of a statistical explanation of Riemannian metrics.

After the war Van Dantzig turned to probability and statistics, mainly by stimulating research in this field. After the flood of 1953 he took a part in the research preparing the now completed "Delta Works."

The most important part of Van Dantzig's work lies in topological algebra, a term coined by him. Although published in the 1930's, it was probably conceived in the late 1920's. His Groningen Ph.D. dissertation, "Studiën over topologische Algebra" (1931), is a fine example of mathematical style: it consists of a concise string of definitions and theorems organized in such a way that in this context each theorem is obvious and none needs a proof. He elaborated on this theme in a series of papers titled "Zur topologischen Algebra" that dealt with questions of metrization and completion of groups, rings, and fields, and eventually classified the fields with a nontrivial locally compact topology. In the course of these studies Van Dantzig discovered the solenoids as completions of the additive group of real numbers. These strange homogeneous spaces led to a problem on connected metric homogeneous spaces in general, solved by Van Dantzig and B. L. van der Waerden, showing that conjugacy classes of the fundamental group of such spaces must be finite.

BIBLIOGRAPHY

I. Original Works. Van Dantzig's writings include "Studiën over topologische Algebra," (Ph.D. diss., Groningen, 1931), "Zur topologischen Algebra. I: Komplettierungstheorie," in *Mathematische Annalen*, **107** (1932), 587–626, " . . . II: Abstrakte b_v-adische Ringe," in *Compositio mathematica*, **2** (1935), 201–223, and " . . . III: Brouwersche und Cantorsche Gruppen," *ibid.*, **3** (1936), 408–426; and "Über metrisch homogene Räume," in *Abhandlungen der mathematische der Seminar Hamburgischen Universität*, **6** (1928), 367–376, written with B. L. van der Waerden. A complete bibliography is in *Statistica neerlandica*, **13** (1959), 422–432.

II. Secondary Literature. On Van Dantzig's work see Hans Freudenthal, "L'algèbre topologique, en particulier les groupes topologiques et de Lie," in *Revue de synthèse*, 3rd ser., **89** (1968), 223–243. Memorial articles are Hans Freudenthal, "Levensbericht van David van Dantzig," in *Jaarboek Kon. Ned. Akademie van wetenschappen* (1959–1960), 295–299, and "In memoriam David van Dantzig," in *Nieuw archief voor wiskunde*, 3rd ser., **8** (1960), 57–73 (in Dutch); and J. Hemelrijk, "In memoriam Prof. Dr. D. van Dantzig," in *Statistica neerlandica*, **13** (1959), 416–421 (in Dutch).

Hans Freudenthal

DARLINGTON, CYRIL DEAN (*b*. Chorley, England, 19 December 1903; *d*. Oxford, England, 26 March 1981), *cytogenetics, evolution*.

Life. Darlington spent the first eight years of his life in Lancashire, where his father, William H. R. Darlington, was a schoolmaster. His mother was Ellen Frankland Darlington. On his father's appointment as secretary to the chief chemist of Crossfields Soap Ltd., the family moved to Ealing in west London. Cyril attended Mercer's School until the award of a Foundation Scholarship in 1917 led to his enrollment in one of London's famous public schools, St. Paul's. Darlington was no academic prodigy, and he hated sports. At home there was little time for fun and relaxation under the powerful influence of his puritanical father.

In 1920, intending to become a farmer in Australia, Darlington entered the South Eastern Agricultural College at Wye in Kent, where he took courses in chemistry, botany, geology, and zoology, as well as in such practical subjects as "seed testing, horse-doctoring, the analysis of milk and soil, the construction of farm implements, the study of rents and wages and agricultural law."[1] Although he appreciated the applied subjects, the introduction to the sciences attracted him to research, and in 1923, the year of his graduation, he applied (unsuccessfully) for an Empire Cotton Corporation scholarship.

In 1922, while still at Wye, Darlington read the celebrated textbook of Thomas H. Morgan, Alfred H. Sturtevant, Hermann J. Muller, and Calvin B. Bridges, *The Mechanism of Mendelian Heredity* (1915). Later he recalled how struck he had been "by the contrast between Morgan's methods and arguments and those I heard discussed in connection with the breeding of crops and livestock."[2] At the suggestion of E. S. Salmon, he applied in 1923 to work under the geneticist William Bateson at the John Innes Horticultural Institution in Merton (referred to throughout this article as the J.I.). Despite a cool reception from the great man, Darlington persuaded Bateson to accept him as an unpaid voluntary worker for three months, extended to six months. In 1924 Bateson hired him. Darlington spent the next thirty years of his life at J.I. After five years on a meager studentship, Darlington was appointed cytologist, succeeding W. C. F. (Frank) Newton (*d*. 1927), who had been his mentor in cytology. In 1937 the J.I. was reorganized on departmental lines, and Darlington became head of the cytology department. Two years later he was appointed director, thus succeeding to the post first held by Bateson.

Darlington was married three times. His first marriage was short-lived and never referred to. His second wife was Margaret Blanche Upcott, a volunteer worker in his laboratory, whom he married in 1939; they had two sons and three daughters. In June 1950 he married Gwendolen Ashead Harvey.

In 1953 Darlington accepted the Sherardian chair of botany at Oxford. By this time the most creative phase of his theoretical development was over. His reputation had been securely established with the reception of the epochal *Recent Advances in Cytology*, published in 1932. Consequently, he was able to build up a research tradition in cytogenetics at the school of botany, where physiological, taxonomic, and ecological botany had hitherto been the chief interests of the staff. In this period he wrote many articles and book reviews on aspects of human genetics, and many historical and biographical studies that followed his text on the history of genetics.[3] Despite a serious heart attack in 1964, Darlington's energy and enthusiasm continued unabated. He initiated the first Oxford Chromosome Conference in 1964[4] and completed his magnum opus, *The Evolution of Man and Society*, published in 1969.

Darlington was one of the leading biologists of his time, but honors were not heaped upon him, for he was always a controversial figure—a corrosive skeptic whose attitude was antiestablishment and antiauthoritarian throughout his career. Nevertheless, Darlington's early achievements were recognized by his election as fellow of the Royal Society in 1941, at the age of thirty-seven; five years later he received the Royal Society's Royal Medal. He was president of the Genetical Society of Great Britain from 1943 to 1946, and in 1950 became a corresponding member of the Accademia dei Lincei and of the Danish Royal Academy of Sciences. On his retirement from Oxford in 1971, Magdalen College elected him to an honorary fellowship.

Scientific Method. Darlington was a theoretician with a deep interest in the fundamental questions of evolution, but at the same time he had a great love of garden plants and an absorbing curiosity concerning their origins. The modern chromosome theory, of which he was perhaps the chief architect, was important for him in both these contexts, and he was able to promote both fundamental and applied research in cytogenetics during his thirty years at the J.I. As a theoretician, however, he scorned pure empiricism. Unaware of the writings of his contemporary Karl R. Popper, Darlington stressed his view that hypotheses should be put forward in order to attempt to refute them.[5] Long before the work of Thomas S. Kuhn, he argued for the theory-laden nature of facts in science. The distinction between fact and hypothesis, he explained, "becomes too naive when applied to the data of a new study. Every 'fact' implies a hypothesis."[6] Quoting from Byron's Don Juan—"Nothing more true than not to trust your senses and yet what are your other evidences?"—Darlington went on to attack the simplistic view that what we observe under the microscope can be characterized as either an artifact or not. "No appearance of treated material is definitely free from artifact, nor is any appearance 'pure' artifact," he declared.[7] From this it followed that one could not impute "a morphological value to almost anything seen under the microscope." Instead, the data of observation had to be compared against other such data and shown to be in harmony with the findings of genetics, so that "if we accept the chromosome theory, genetics and cytology have become one system." Provocatively he declared:

> . . . I attach greater importance to an inference from the comparison of a number of observations than to the direct evidence of a single observation. I prefer the hypothesis to the "fact". An observation of fixed material I regard, not as an inescapable fact of life, but as an experiment with life from which one may draw various conclusions depending on its agreement with other similar experiments.[8]

The reason for Darlington's concern with methodology is not difficult to determine. Cytology was riven with controversy; interpretations of the same material differed, but the ensuing conflict did not depend, he explained, "merely on details of observations or individual prejudices," but on different habits of thought. The "older habit" that he castigated was taken from histology and placed the same confidence in the interpretation of microscopic structures as it placed in macroscopic structures: "Facts are facts and have to be accepted, however improbable." Its unity rested on "a learned terminology, such as 'matrix,' 'chromonema' . . . . Behind this verbal screen a body of esoteric literature has developed, so well protected from understanding that the most critical observers on the outside have often been fain to dismiss the whole study as an imaginative imposture." But this was no longer necessary. A new system of chromosome study could now be built if we discard the old myth terminology and apply working hypotheses "founded on data derived from every available method of study . . . ."[9]

Meiosis. Although the gross features of the process by which a nucleus divides into two daughter nuclei (mitosis) had been established in the nineteenth century, controversy continued over the "individuality of the chromosomes." Theodor Boveri championed the persistence of the chromosomes as distinct structures throughout that mysterious "resting stage" between cell divisions when the nucleus takes on the appearance of a tangled skein of threads. John Bretland Farmer, the author of the term "meiosis," and many others held that the chromosomes lost their individuality in the resting stage as a result of joining up end to end to form the "continuous spireme." At the onset of nuclear division this thread was segmented to yield the chromosomes anew. It was agreed, however, that the doubling of the number of chromosomes to provide an equal set for each daughter cell was achieved by the longitudinal splitting of the chromosomes or of the spireme from which they were subsequently formed.

No such consensus existed over the procedure by which the chromosome number in the nuclei of the germ cells was halved. This special form of cell division (meiosis) had been called a "reduction division" by August Weismann in 1887 because, he

 DARLINGTON

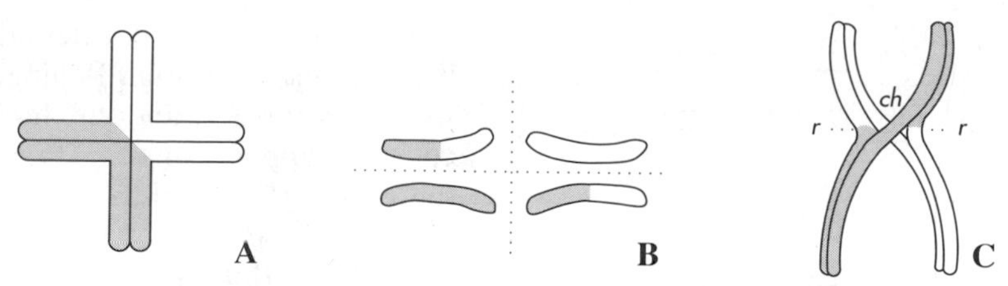

FIGURE 1. Janssens' interpretation of the cross figures. *A*. Schematized cross figure. *B*. The constituent chromatids showing, two having exchanged segments. *C*. Paired chromosomes showing points of breakage and recombination (*r*). The "chiasma," or point of crossing over, is marked *ch*. (Edmund B. Wilson, "Chiasmatype and Crossing over," in *American Naturalist,* **54** [1920], Fig. 6, p. 207)

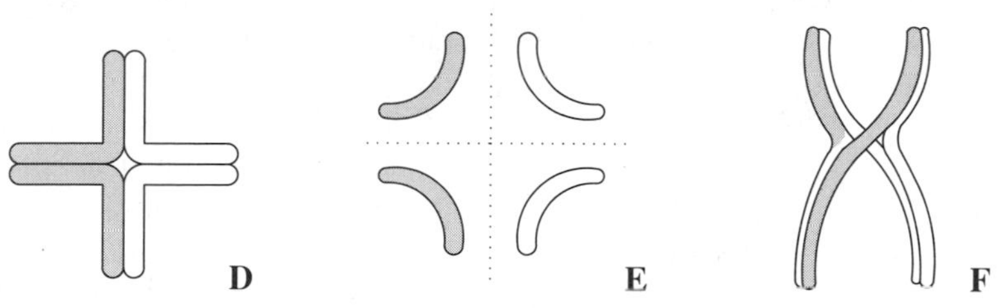

FIGURE 2. Interpretation according to the "classical theory." The chromatids in *D* and *E* show no exchange of segments. There is no breakage and reunion at the chiasma in *F*. (Edmund B. Wilson, "Chiasmatype and Crossing over," in *American Naturalist,* **54** [1920], Fig. 6, p. 207)

claimed, its function was to reduce the amount of hereditary material in each germ cell both qualitatively and quantitatively. Just how the chromosome number was halved, whether a qualitative reduction did occur, and by what means were questions that long continued in dispute. Added to these was the discovery from genetics that hereditary characters belonging to one linkage group are capable of being transferred to another. Assuming that each linkage group represents the hereditary units attached to a single chromosome, a cytological mechanism was needed whereby such block transfers could occur.

In 1909 Frans Alfons Janssens suggested that the crosslike figures of meiotic chromosomes, which he called chiasmata, resulted from the exchange of portions of neighboring threads. Like chromosomes associated in pairs, one maternal with one paternal, each chromosome split longitudinally into two chromatids. Under the strain of chromosome movements one chromatid from each chromosome broke and exchanged segments (Figure 1). Thomas Hunt Morgan saw that Janssens' "chiasmatype" theory offered the required mechanism for block transfer of hereditary factors. Unfortunately the cytological evidence was unclear, and as late as 1925 Edmund B. Wilson considered it inadequate. Opposed to the chiasmatype theory was what Darlington dubbed the "classical theory," championed by Clarence E. McClung, according to which a chiasma resulted from the exchange of pairing chromatids but without breakage and reunion, and therefore without exchange of segments (Figure 2). Consequently it did not account for recombination between genes belonging to different linkage groups, but cytologists were content that it preserved the individuality of the chromosomes and made the minimum number of assumptions about cytological processes, for neither the act of crossing-over nor the breakage of chromatids could be observed.

These rival groups were agreed on one point: the meiotic chromosomes normally paired side by side; they were parasynaptic. Yet other cytologists claimed that meiotic chromosomes paired end to end; they were telosynaptic. It was alleged that they were formed by the segmentation of the spireme, but neighboring segments remained attached end to end. The classic case was in the ring chromosomes of the evening primrose (*Oenothera*). Such a form of meiosis did not account for the genetic data and entailed the belief either that meiosis was not a uniform process throughout the animal and plant kingdoms or that all meiosis was really telosynaptic.

Although Bateson had reluctantly accepted the chromosome theory of heredity in 1922, he continued to seek evidence that would undermine the unique association of Mendelian segregation with meiosis. Darlington recalled receiving "encouragement, abundant meat and drink, but no positive direction" from Bateson, "for he had at that time lost his nerve and seemed to know that he was off the main track of enquiry"[10] (a judgment that a study of Bateson's later papers supports).

DARLINGTON

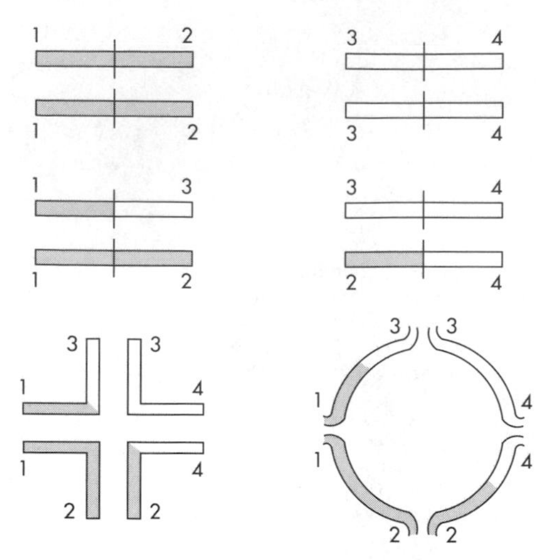

FIGURE 3. Diagram showing Belling's hypothesis of segmental interchange. Nonhomologous chromosomes 1·2 and 3·4 exchange segments to produce 1·3 and 2·4. In an individual who receives 1·2 and 3·4 from one parent and 1·3 and 2·4 from the other, pairing of corresponding segments in synapsis will produce a circle of four chromosomes. [Ralph E. Cleland, Oenothera, *Cytogenetics and Evolution* (London and New York, 1972), Fig. 5.1, p. 58. Reproduced by permission of Academic Press, Inc.]

Fortunately, in 1922 Bateson had appointed to the J.I. a cytologist, Frank Newton, who became Darlington's mentor. Newton was convinced that chromosomes that appeared to be paired end to end had begun their paired state parasynaptically. It was only the subsequent process of separation that gave the semblance of telosynapsis. All forms of meiosis were therefore fundamentally the same. His study of triploid tulips offered a crucial test of parasynapsis.

In 1926 Newton fell ill, and in 1927 he died. Darlington took over the preparation of his work for publication. The results confirmed Newton's claims for parasynapsis. Yet Newton had remained skeptical about Janssens' identification of genetic crossing-over with chiasma formation. Here Darlington went beyond his mentor; he was cautiously optimistic.

If crossing-over was to be explained in terms of chromosome behavior, then maternal and paternal chromosomes had to pair in a parasynaptic manner. Parasynapsis was established by Newton's study of triploid tulips and Darlington's study of triploid hyacinths and diploid *Tradescantia,*[11] where it was clear that the ring formation of chromosomes resulted from parasynapsis. Applying his *Tradescantia* results to *Oenothera,* Darlington concluded that the chromosome rings in this genus were composed of chromosomes that had exchanged segments with nonhomologous chromosomes. Pairing was therefore by segments rather than by whole chromosomes (Figure 3).[12]

The hypothesis of segmental interchange had first been advanced by John Belling, working with Jimson weed (*Datura*),[13] and he had conjectured that it could be applied to *Oenothera* as well. Darlington followed this up enthusiastically. He argued that as a result of segmental interchange, any one chromosome in the ring was homologous at one end with one chromosome and at the other end with another chromosome. Pairing extended only a short distance back from the ends. Proximal portions did not synapse.

This hypothesis was promptly tested experimentally by others. The result was that *Oenothera* was brought into line with the parasynaptic interpretation of meiosis in other genera. Ralph E. Cleland later remarked upon the drastic change in the picture that followed. Previously *Oenothera* was thought to offer "the most outstanding example of telosynapsis. . . ." Cleland quoted Edmund B. Wilson's judgment of 1925: "In the case of *Oenothera* . . . a 'telosynaptic' association of the chromosomes in early diakinesis seems indubitable."[14] Wilson cited the many supporters of telosynapsis in plants, but he felt they had "lost ground in recent years."[15] During the 1930's the claims for telosynapsis became increasingly implausible. Parasynapsis emerged as the rule in the meiosis of plants and animals, thus removing one of the barriers to a cytological theory of crossing-over.

The Evolution of Genetic Systems. Darlington's study of nuclear division led him to the conclusion that the most important process involved was the pairing of homologous segments of either the split halves of chromosomes—chromatids—or of whole chromosomes. This conclusion was central to his theory of meiosis, which can be summarized as follows. The pairing process satisfies the attractive forces that operate during nuclear division. In mitosis, nuclear division begins after the chromosomes have split into chromatids. Pairing is therefore between chromatids. In meiosis, however, the division process is "precocious"; the chromosomes have not yet split when nuclear division begins. Pairing is therefore between homologous chromosomes instead of between chromatids.

When chromatids are formed, however, the paired chromosomes are no longer held together by attractive forces, which are now operating between sister chromatids. The chromosome pairs would fall apart were it not for the chiasmata that have formed between them. The chiasmata maintain the unity of the quadruple structure of the four chromatids into which the two homologous chromosomes have divided and ensure regular segregation of the four to the tetrad of cells produced at the end of meiosis. The result is equational and reductional—equational because each member of the tetrad receives the same number of chromosomes, reductional because homologous chromosomes (or homologous segments) are distributed to different members of the tetrad. Because the chromosomes have split into chromatids only once, yet four cells have been formed, meiosis has effected a reduction of the chromosome complement, so that the germ cells contain half the number of chromosomes found in all other cells of the organism.

Darlington was not slow to explore the biological implications of this conception of meiosis. In the summer of 1930, encouraged by J. B. S. Haldane, he sent to *Biological Reviews* a systematic treatment of the subject the aim of which was to establish that a uniformity of principle underlay the external diversity of meiosis. He distinguished three theories of meiosis: one morphological, one cytological, and one genetic. His precocity theory was a cytological theory. He suggested that meiosis arose as a self-perpetuating aberration of mitosis, and as the only such aberration it must be the oldest observable mechanism after mitosis. It has "made possible the recurrence of fertilisation. It has inaugurated sexual reproduction as a self-repeating process or habit."[16] Janssens' theory of breakage and exchange of chromatid segments was the only genetic theory of meiosis, since it alone accounted for the genetic data from diploids and polyploids.

Darlington's suggestion about the evolution of meiosis from mitosis was one of his many speculations concerning the evolution of genetic systems. This subject formed the last chapter of his *Recent Advances in Cytology* [afterward cited as *Recent Advances*], and subsequently of a whole book, *The Evolution of Genetic Systems* (1939). His aim was "to consider the organic world from a new point of view, the point of view not of the organism itself but of its hereditary materials and mechanism." These two points of view differed profoundly, both at the level of analysis and with respect to their mutual relations. At the level of the genetic system there existed a remarkable uniformity in the constitution and organization of the genes, which had led to "parallelism in groups as widely separated as insects and flowering plants." Genotypic systematics therefore transcends phenotypic classifications. Changes believed "to have occurred at remote, chiefly pre-Cambrian periods," he claimed, can still be inferred "from current studies of the behaviour of the genetic material in hybridization." The resulting inferences were often "less speculative than those drawn from the comparative morphology of the phenotype." As for the claims of mysterious internal tendencies for different forms to evolve in similar ways—Gustav Eimer's orthogenesis and Lev Berg's nomogenesis—an adequate explanation was provided by the nature and "balance" of the genes and the role of the environment in eliminating less adaptive gene combinations.[17]

The last chapter of *Recent Advances* presents a vista of evolution beginning with the "naked gene"—some primitive bacteria may indeed be "single-gene organisms."[18] Mutation, aggregation, and physical association have led to the formation of chromosomes; mitosis, to their ordered replication; and meiosis, to sexual reproduction and sexual differentiation. The genetically determined suppression of meiosis has led to polyploidy, a trick by which hybrids can escape sterility. Other causes of the failure of fertilization could be overcome by apospory, apogamy, and parthenogenesis. Darlington described mutations and structural and numerical changes in the chromosomes as primary and secondary sources of variation, respectively. He judged numerical changes (polyploidy) to be more important as immediate agents of variation because they involved tested materials (that is, genotypes of proven merit), whereas mutations changed the genes, and such changes were usually disadvantageous. Numerical changes, however, depended upon the frequency of hybridization.[19]

Further, Darlington argued that heredity could follow two tracks. On one track the genes were separated to different chromosomes and recombined on the same chromosome by the formation of chiasmata. On the other track, genes once brought together on the same chromosome were not separated again because chiasma formation was genetically suppressed, as in the Y chromosome of *Drosophila*. The frequency of separation of autosomal genes was also reduced, he suggested, by suppression of chiasma formation in all chromosomes of one of the sexes—the male *Drosophila*, for instance. (This suggestion may have played a part in the development of the concept of a "supergene," in which a cluster of genes was preserved indefinitely.)

The Impact of *Recent Advances in Cytology*. Darlington's conclusions in *Recent Advances in Cytology* came under vigorous attack at the Sixth International Congress of Genetics (1932) and at the Sixth International Botanical Congress (1935), when, according to Darlington, C. L. Huskins "spoke publicly against me in all meetings and privately offered to come across the Atlantic and punch my head if ever I wrote any more footnotes about him."[20] Reviewing

Recent Advances for *Nature*, Sturtevant wrote admiringly of the "logical and detailed marshalling of an almost overwhelming body of evidence. . . ." Of Darlington's precocity theory Sturtevant confessed, "[It] has an attractive logical simplicity that makes it difficult to think of the phenomena in any other terms." However, he dissented from the theory that considered the segmental exchange of chromatids as the cause of chiasma formation. How was it that male fruit flies showed no chiasmata, yet the chromosomes remained in pairs after division into chromatids? Sturtevant also sounded a note of displeasure at the speculative character of much of the text. "There are many," he concluded, "who will hope that we are not to see a revival of the fashion of constructing elaborate hypothetical histories that must for ever remain hypothetical."[21] At the University of Pennsylvania, Hampton L. Carson recalled, the older cytologists

> . . . received the Darlington book [*Recent Advances*] with stiff attitudes of outrage, anger, and ridicule. The book was considered to be dangerous, in fact poisonous, for the minds of graduate students. It was made clear to us that only after we had become seasoned veterans could we hope to succeed in separating the good (if there was any) from the bad in Darlington. Those of us who had copies kept them in a drawer rather than on the tops of our desks.[22]

Darlington wrote for the expert rather than for the university student. His style in books like *Recent Advances* is precise but dense. Few concessions are made to the uninitiated, with the result that students were directed to Theodosius Dobzhansky, Michael J. D. White, and, more recently, H. L. K. Whitehouse in order to master cytogenetics without tears.[23] The clearest account of the complex cytogenetics of the genus *Oenothera* was supplied by Cleland.[24] Despite his renown, Darlington did not have as great an impact on evolutionary thinking as Dobzhansky, Ernst Mayr, or Ronald A. Fisher. Musing on this lack of impact, he identified two features of his conception of evolution that he thought responsible. First, "like some concepts in physics it reversed the common sense priorities of the observer," for the organism had become the vehicle for propagating the chromosomes, and the "genetic system had become more important than the body of the individual." Second, the genetic system "was obscure and for the mathematician it was unmanageable." In contrast with

> . . . the gene pool or the genetic code which vary in two or three dimensions and can be treated deductively, the genetic system operates in many dimensions. It integrates several interacting incommensurables. It has no single dynamic focus. It depends on a complex of individuals, populations and generalisations which all interact at the same time as they are being subjected to mutation, selection and recombination.[25]

The tidy models of the population geneticists were too simplistic for Darlington. Once or twice he tried to suggest to Fisher "that the principles of selection were not in fact going to operate with the absolute rigour he expected, but I could never get him to discuss it."[26] When Richard Dawkins' *The Selfish Gene* (Oxford, 1976) appeared, Darlington condemned it as a fantasy; he scorned Dawkins' predilection for models predicting animal behavior based on entirely selfish genes, a concept that Darlington saw as diametrically opposed to his own conception of interaction and feedback at all levels and between all levels of the genetic hierarchy.[27]

Pioneer and Controversialist. Darlington's description of his scientific biography was always amusing, often dramatic, sometimes melodramatic, and occasionally almost paranoid. As a theoretician he was bound to encounter opposition from empirically preoccupied scientists distrustful of speculation. As a hereditarian his views on the implications of genetics for man and society did not strike a sympathetic chord in a political climate that was becoming increasingly egalitarian and antieugenic. The continuity he saw between the evolution of genetic systems and the evolution of man and society yielded a historical unity in which the hereditary material was the prime determinant of events. A gene might succeed or fail in a given environment, but it might also change that environment by action of the organism upon it. Such extreme material determinism was, and has remained, anathema to many.

Darlington's debut in 1953 as an Oxford professor was stormy, but he succeeded in dragging the Oxford botany curriculum into the twentieth century. He established what is probably the first genetic garden in the world. Through his initiative Oxford University bought the magnificent arboretum at Nuneham Courteney. However, it was John W. S. Pringle, professor of zoology at Oxford, who was chiefly responsible for the unification of core teaching in botany and zoology, for the establishment of the chair of genetics, and for the introduction of the human sciences degree scheme.

NOTES

1. C. D. Darlington, "Biographical Notes, Articles, and Cvs Compiled by Darlington," in *Catalogue of Papers*, A4.

2. *Ibid.*
3. C. D. Darlington, *The Facts of Life* (London, 1953); two valuable texts on cytogenetics are his *Chromosome Atlas of Flowering Plants* (London, 1956), written with A. P. Wylie; and *Chromosome Botany* (London, 1956).
4. *Chromosomes Today*, **1** (1966)—the entire issue is the proceedings of the conference.
5. See, for example, his "Meiosis in Diploid and Tetraploid *Primula sinensis*," in *Journal of Genetics*, **24** (1931), 64–96, esp. 89.
6. C. D. Darlington, *Recent Advances in Cytology* (London, 1932), viii (referred to hereafter as *Recent Advances*).
7. *Ibid.*, 488.
8. C. D. Darlington, "The Old Terminology and the New Analysis of Chromosome Behaviour," in *Annals of Botany*, **49** (1935), 579–586, see 579.
9. *Ibid.*, 584–585.
10. C. D. Darlington, Royal Society personal record, 21 November 1941 (copy in Darlington, *Papers*, A4).
11. These three papers were published in the *Journal of Genetics*, **21** (1929): "Meiosis in Polyploids I. Triploid and Pentaploid Tulips," 1–16, written with W. C. F. Newton; "Meiosis in Polyploids. II. Aneuploid Hyacinths," 17–56; and "Chromosome Behaviour and Structural Hybridity in the Tradescantiae," 207–286.
12. C. D. Darlington, "Ring-Formation in *Oenothera* and Other Genera," in *Journal of Genetics*, **20** (1929), 345–363.
13. John Belling and A. F. Blakeslee, "On the Attachment of Non-homologous Chromosomes at the Reduction Division in Certain 25-Chromosome *Daturas*," in *Proceedings of the National Academy of Sciences*, **12** (1926), 7–11.
14. Edmund B. Wilson, *The Cell in Development and Heredity*, 3rd ed. (New York, 1925), 565; Ralph E. Cleland, Oenothera: *Cytogenetics and Evolution* (London and New York, 1972), 54.
15. Wilson, *The Cell*, 566 ff.
16. C. D. Darlington, "Meiosis," in *Biological Reviews*, **6** (1931), 221–264.
17. *Recent Advances*, 448–449.
18. *Ibid.*, 452.
19. *Ibid.*, 478.
20. C. D. Darlington, Royal Society personal record.
21. A. H. Sturtevant, "Chromosome Mechanics," in *Nature*, **131** (1933), 5–6.
22. Hampton L. Carson, "Cytogenetics and the Neo-Darwinian Synthesis," in Ernst Mayr and William B. Provine, eds., *The Evolutionary Synthesis: Perspectives on the Unification of Biology* (Cambridge, Mass., 1980), 91.
23. Theodosius Dobzhansky, *Genetics and the Origin of Species* (New York, 1937); Michael J. D. White, *Animal Cytology and Evolution* (Cambridge, 1945); H. L. K. Whitehouse, *Towards an Understanding of the Mechanism of Heredity* (London, 1965).
24. R. E. Cleland, Oenothera.
25. C. D. Darlington, "My Approaches to Genetics and Evolution," in *Papers*, A2, 10.
26. C. D. Darlington, "J. B. S. Haldane, R. A. Fisher, and William Bateson," in Ernst Mayr and William B. Provine, eds., *The Evolutionary Synthesis*, 430.
27. C. D. Darlington, "In the Evolutionary Soup," *Times Literary Supplement* (4 February 1977), 126.

BIBLIOGRAPHY

I. Original Works. A comprehensive bibliography of Darlington's publications is in D. Lewis' bibliographical memoirs (see below). Darlington assembled and published his own guide to his published works in *C. D. Darlington and Collaborators: Bibliography 1926–1971* (Oxford, 1971). Of his thirteen books the most important is *Recent Advances in Cytology* (London and Philadelphia, 1932; 3rd ed., 1965). The most interesting of his pamphlets is his Conway Memorial Lecture of 20 April 1948, *The Conflict of Science and Society* (London, 1948). The best textbook is the one he wrote with Kenneth Mather, *The Elements of Genetics* (London and New York, 1949; repr. 1952). The most successful of his popular texts is the historical study *The Facts of Life* (London, 1953; repr. 1956); the 2nd edition is titled *Genetics and Man* (London, 1964; repr. 1967). Both editions have been translated into numerous languages. His *Evolution of Man and Society* (New York, 1969; London, 1970) has been translated into seven languages.

Darlington's personal account of the events leading to the synthesis of Mendelism and Darwinian evolution is in "The Evolution of Genetic Systems: Contributions of Cytology to Evolutionary Theory" and "J. B. S. Haldane, R. A. Fisher, and William Bateson," in Ernst Mayr and William B. Provine, eds., *The Evolutionary Synthesis: Perspectives on the Unification of Biology* (Cambridge, Mass., 1980), 70–80, 430–432.

II. Secondary Literature. For a fine overview of Darlington's many contributions to science and to the profession of genetics, see D. Lewis, "Cyril Dean Darlington," in *Biographical Memoirs of Fellows of the Royal Society*, **29** (1983), 113–157. Also invaluable are the introduction and notes to the catalog of his papers compiled by Jeannine Alton and Peter Harper, *Catalogue of Papers and Correspondence of Cyril Dean Darlington (1903–1981), Deposited in the Bodleian Library, Oxford* (Oxford, 1985). Darlington's study of the chemical aspects of chromosome behavior and his participation in multidisciplinary discussions of the nature of the gene are described in Robert C. Olby, *The Path to the Double Helix* (London and Seattle, 1974), chap. 7.

Robert C. Olby

DAVENPORT, HAROLD (*b.* Huncoat, Lancashire, England, 30 October 1907; *d.* Cambridge, England, 9 June 1969), *mathematics*.

Davenport was the only son of Percy Davenport, first office clerk and later company secretary of Perseverance Mill, and of Nancy Barnes, the mill owner's daughter. From Accrington Grammar School he won a scholarship in 1924 to Manchester University, from which he graduated with first-class honors in 1927, at the age of nineteen. Davenport proceeded to Cambridge University with a scholarship to Trinity College, where he graduated with a B.A. in mathematics in 1929, again with the highest honors in each part of the final examination. He wrote his Ph.D. dissertation under J. E. Littlewood. In 1931 he was Rayleigh prizeman, and in 1932 he became a research fellow at Trinity College.

DAVENPORT

In 1944, while he was professor of mathematics at the University College of North Wales, Bangor, Davenport married a colleague in modern languages, Anne Lofthouse; they had two sons. In 1938 Davenport received the Cambridge Sc.D., in 1940 he was elected a fellow of the Royal Society, and in 1941 he won the Adams Prize of Cambridge University. He won the senior Berwick Prize of the London Mathematical Society in 1954, and from 1957 to 1959 he was president of that society. He was elected an ordinary member of the Royal Society of Sciences in Uppsala in 1964, and in 1967 he was Sylvester medalist of the Royal Society. In 1968 he received an honorary degree from the University of Nottingham.

When Davenport embarked on research in mathematics under the supervision of Littlewood, he was not yet committed to any one branch of the subject. Among the problems from analysis and number theory that Littlewood proposed to him, however, was a question about the distribution of quadratic residues that attracted him; virtually all his subsequent work was devoted to the theory of numbers. The topic of Davenport's first piece of research led him directly to the study of character sums and exponential sums, which were then (and still are) central to many of the most profound inquiries in higher arithmetic, and influenced many of his own later researches.

This work brought Davenport to the attention of Louis J. Mordell and of Helmut Hasse; the latter invited him to Marburg for a long visit during 1931. Davenport learned fluent German during this time, and he and Hasse wrote an important paper that still is influential; also, it might be said that Hasse was led by their association to his proof of the Riemann hypothesis for elliptic curves. The natural culmination of this strand of ideas—Weil's proof in 1948 of the Riemann hypothesis for algebraic curves in general by deep methods from algebraic geometry—was at this time far in the future, and Davenport was probably ready for a change of direction.

During his travels in Germany, Davenport met H. Heilbronn, Edmund Landau's last assistant in Göttingen; the two soon became friends and embarked on several new lines of research, including one in the general area of Waring's problem via the celebrated circle method of Hardy, Littlewood, and Ramanujan. Their association deepened when Heilbronn joined Davenport at Cambridge in 1933; they wrote many joint papers over the years, the last appearing in 1971.

Working with Heilbronn or alone, Davenport made several novel adaptations of the circle method and also developed some important technical refinements that led to improved results for Waring's problem itself. While much of this work retains interest, its main importance derives from the fact that it prepared Davenport for the greatest mathematical achievements of his life: the adaptation, starting in 1956, of the circle method (which was invented to deal with additive problems) to nonadditive problems concerning values taken by quadratic and cubic forms in many variables. A critical feature of this adaptation was the use of ideas and results from the geometry of numbers, a branch of number theory that had originated with Hermann Minkowski.

In 1937, at the termination of his Cambridge fellowship, Davenport was appointed by Mordell to an assistant lectureship at the University of Manchester and, under his influence, took up the study of the geometry of numbers. This subject dominated his mathematical activities until 1956, by which time he had become a dominant figure in number theory, not only in the United Kingdom but also in most of the world. In 1941 Davenport, by then a fellow of the Royal Society, moved to a full professorship at the University College of North Wales at Bangor. In 1945 he was appointed Astor Professor of Mathematics at University College, London. In 1950 he became head of the department.

It was during the period 1945 to 1958 that Davenport achieved his full stature. His mathematical prose had been distinguished from the beginning by unusual grace and lucidity, and he now brought these gifts to the classroom and to the supervision of graduate students. Davenport was a superb teacher and an inspiring director of research, and his number theory seminar at University College became a mecca for aspiring number theoreticians from all over the world. It was here that Freeman Dyson conceived his remarkable proof of Minkowski's conjecture for the product of four nonhomogeneous linear forms, that C. A. Rogers developed his deep researches in the theory of packing space, that K. F. Roth was led to his theorem on rational approximations to algebraic numbers, and that D. A. Burgess reported on his dramatic improvement of Ivan M. Vinogradov's estimate of the least quadratic nonresidue. Many other fine pieces of mathematics, not least many of Davenport's own results, first saw the light of day here.

Davenport had an unusually well-organized mind, and was ever ready to make available to his students and associates the wisdom he had gleaned from his mathematical experiences. Like Littlewood he was punctilious about having research problems available for his students, though he was also quick to encourage promising initiatives of their own. Davenport was always eager to discuss mathematics and more than willing to help fellow mathematicians with difficulties. No query ever went unanswered, and he conducted a voluminous correspondence.

Davenport's research activity never flagged, and therefore his expository writing was not on the scale that his literary powers warranted, although he did produce *The Higher Arithmetic* (1954), a small, elegant book that has gone through several editions.

In 1958 Davenport returned to Cambridge as Rouse Ball Professor. He was at the height of his powers, a worthy successor of Hardy, Littlewood, and Mordell, and the unquestioned leader of the British school of number theory. His research went from strength to strength. The original work on forms in many variables had overlapped in a number of ways with that of D. J. Lewis and B. J. Birch, and he embarked on a vigorous collaboration with both these mathematicians. His association with Lewis was especially close and endured for the rest of his life. He was a frequent visitor to Lewis at Ann Arbor; and his two other books, *Analytic Methods for Diophantine Equations and Diophantine Inequalities* (1962) and *Multiplicative Number Theory* (1967), grew out of graduate courses he presented there.

Davenport's activities as research director continued unabated. Among his students were A. Baker, J. H. Conway, P. D. T. A. Elliott, M. N. Huxley, and H. L. Montgomery. The talented young Enrico Bombieri came to his notice, and Davenport brought him to Cambridge for the first of several fruitful visits. Bombieri reawakened Davenport's interest in prime number theory, and they wrote several important joint papers. Baker and Bombieri both won Fields medals in later years. In this last period Davenport also collaborated with A. Schinzel of the Polish Academy of Sciences on properties of polynomials and with W. Schmidt of the University of Colorado on Diophantine approximation. Many achievements and honors seemed still ahead of him when lung cancer set in with awful suddenness and brought his life to an untimely end.

Davenport was shy and reserved, and his outlook on life was at all times conservative. Despite this his organizational gifts, and his willingness to help all who came to him for advice and to render service to the institutions and learned societies with which he was associated, made him a natural academic leader and one of the most influential mathematicians of his time.

BIBLIOGRAPHY

I. ORIGINAL WORKS. A bibliography of Davenport's works is in *Acta arithmetica*, **18** (1971), 19–28. His writings were brought together as *Collected Works of Harold Davenport*, B. J. Birch, H. Halberstam, and C. A. Rogers, eds., 4 vols. (New York, 1977). His writings include "Die Nullstellen der Kongruenzzetafunktionen in gewissen zyklischen Fällen," in *Jahrbuch für Mathematik*, **172** (1934), 171ff., written with Helmut Hasse; *The Higher Arithmetic* (London, 1954; 5th ed., Cambridge and New York, 1982); *Analytic Methods for Diophantine Equations and Diophantine Inequalities* (Ann Arbor, 1962); and *Multiplicative Number Theory* (Chicago, 1967), rev. ed., Hugh L. Montgomery, ed. (New York, 1980).

II. SECONDARY LITERATURE. Articles on Davenport and his work are L. J. Mordell, "Harold Davenport" and "Some Aspects of Davenport's Work," in *Acta arithmetica*, **18** (1971), 1–4 and 5–11; C. A. Rogers, "A Brief Survey of the Work of Harold Davenport," *ibid.*, 13–17; and C. A. Rogers, B. J. Birch, H. Halberstam, and D. A. Burgess, "Harold Davenport," in *Biographical Memoirs of Fellows of the Royal Society*, **17** (1971), 159–192.

H. HALBERSTAM

DAVIS, D. (DELBERT) DWIGHT (*b.* Rockford, Illinois, 30 December 1908; *d.* Chicago, Illinois, 6 February 1965), *evolutionary morphology*.

Davis was the oldest of five children of James Walter Davis, a Methodist minister, and of Ada Ione Fager, who died when Davis was sixteen. His boyhood interest in natural history was encouraged by his relatives. In 1926 he entered North Central College at Naperville, Illinois, but left before graduating (January 1930) to accept a position as assistant in the Division of Osteology at the Field Museum of Natural History, Chicago. (He eventually graduated from North Central College in 1942.) In early 1931 he married Charlotte M. Davis. In 1941 he became curator of the Field Museum's Division of Anatomy, and made this department a center of activity in comparative and functional anatomy.

Davis was no mere anatomist; he also studied the life histories and systematics of reptiles and amphibians and coauthored a study on periodical cicadas. In 1941 he was coauthor (with Karl P. Schmidt) of a widely used field book of snakes in the United States and Canada. After 1937 he worked almost exclusively on mammals—at first on their anatomy, but as he became increasingly aware of the need for an evolutionary interpretation of his findings, he became greatly interested in the living mammal. During this phase of his research, Davis took part in a number of expeditions, notably one in 1950 to North Borneo, and published an excellent taxonomic analysis of the mammals that were collected.

While at the Field Museum, Davis took graduate courses at the University of Chicago Medical School. In 1950 he was appointed lecturer at the University of Chicago, and he supervised the graduate training of a number of students at the museum. In 1954 he held a visiting professorship at the California Institute of Technology, and during the fall and winter of the academic year 1962–1963, he served as acting chairman of the department of zoology of the University of Malaya. In 1963, North Central College awarded him an honorary doctorate.

Davis was a member of the American Society of Mammalogists; the American Society of Ichthyologists and Herpetologists; the Society for the Study of Evolution, which he served from 1961 as managing editor of *Evolution*; and the American Society of Zoologists, which he served as chairman of the Division of Vertebrate Morphology (1961–1962). The latter society honored his memory by establishing the D. Dwight Davis Prize, to be awarded for outstanding morphological studies.

In all, Davis published some fifty papers, about two-thirds dealing with mammals. Among the latter is his magisterial monograph *The Giant Panda* (1964), on which he had worked off and on for more than a dozen years. Stephen Jay Gould has referred to this monograph as "our century's greatest work of comparative anatomy." He also did some popular writing, including regular columns for the *Naperville* (Illinois) *Sun* under the title "This Is Life." He was an outstanding nature photographer and won many awards for his photography.

In Davis's scientific development, the decisive influence was apparently his colleague Karl P. Schmidt, curator of herpetology at the Field Museum. For many years Schmidt sponsored a daily luncheon meeting that other colleagues, such as Rainer Zangerl, Robert Inger, and Bryan Patterson, regularly attended. All were naturalists who were somewhat dissatisfied with the extreme reductionism and one-sidedness of contemporary evolutionary genetics. They searched through the evolutionary and taxonomic literature, particularly that of the Continent, for unorthodox alternatives to the current American orthodoxy. In the process they translated several works from the German, with Schmidt and Zangerl doing the translating and Davis the typing.

Davis was highly critical of the intellectual contributions made in the field of comparative anatomy from Carl Gegenbaur to A. S. Romer. He called for a new evolutionary morphology that would study the hows and whys of structural changes. He was greatly impressed by the essential constancy of the limited number of morphological types of organisms. The most rewarding approach to the analysis of evolutionary change, he suggested, was to study an aberrant type like the giant panda, which is nothing but a modified bear, and correlate its structural peculiarities with those of habit and behavior. In other words, while Gegenbaur tried to reconstruct phylogeny back to the common ancestor, Davis started with the ancestral type and asked how—and owing to what forces of selection—it was changed in its descendants.

Davis rejected Sewall Wright's and George G. Simpson's inadaptive phase in evolution, but also all saltationism. Morphological changes in evolution are ultimately due to changes in genes controlling the rate of growth. Only some of the seemingly adaptive changes are the result of selection-controlled epigenetic mechanisms; all others reflect pleiotropic correlations with the directly selected features.

By these maxims Davis gave a new direction to evolutionary morphology and thus started a school represented by Walter Bock, Carl Gans, David B. Wake, and other morphologists. He himself had been greatly influenced by the writings of Hans Böker (minus his Lamarckism), the physiological genetics of Richard B. Goldschmidt, the adaptationist program of Richard Hesse and Franz Doflein (in *Tiergeographie auf oekologischet Grundlage*, which he translated in part with Schmidt as *Ecological Animal Geography*, New York, 1937), and such new functional morphologists as Walter Moller, Tage Lakjer, and W. L. Engels.

Although he enjoyed excellent health throughout most of his life, Davis was a smoker and died of lung cancer at the age of fifty-six. He was survived by his wife.

BIBLIOGRAPHY

I. ORIGINAL WORKS. *Field Book of Snakes of the United States and Canada* (New York, 1941; repr. 1962), written with Karl P. Schmidt; "Comparative Anatomy and the Evolution of Vertebrates," in Glenn L. Jepsen, Ernst Mayr, and George G. Simpson, eds., *Genetics, Paleontology, and Evolution* (Princeton, 1949), 64–89; *The Baculum of the Gorilla* (Chicago, 1951); and *The Giant Panda* (Chicago, 1964).

II. SECONDARY LITERATURE. Rainer Zangerl, "D. Dwight Davis," in *Bulletin of the Chicago Natural History Museum*, **36**, no. 5 (1965), 6–7.

ERNST MAYR

DE BEER, GAVIN RYLANDS (*b*. Malden, Surrey, England, 1 November 1899; *d*. Alfriston, Sussex, England, 21 June 1972), *zoology, embryology, evolution*.

De Beer was the son of Herbert Chaplin de Beer and Mabel Chaplin Rylands de Beer, whose mothers were first cousins. Soon after Gavin's birth his peripatetic father moved the family to Paris, where he was correspondent of the Exchange Telegraph Company. De Beer thus grew up and went to school in Paris and Versailles. He owed to this experience his exceptional linguistic ability, for he was bilingual in French and English and fluent in Swiss German and Italian. After his schooling in France, he enrolled at Harrow, where he was attracted to zoology. After graduating from Harrow, he entered Magdalen College, Oxford, in 1917, but remained for only a term. De Beer served in the British Army during World War I, and in 1919 he returned to Magdalen to read zoology. After graduation in 1922, he remained at Oxford to teach and do research in zoology; in 1923 he was elected fellow of Merton College, where he stayed until 1938. On 2 March 1925 he married Cicely Glyn Medlycott.

De Beer's earliest research followed in the footsteps of Julian Huxley, then the most influential teacher of zoology at Oxford. Like Huxley, De Beer would dabble on the edges of experimental embryology without becoming deeply involved. He performed simple experiments; he visited Hans Spemann at Freiburg to learn the techniques of operating on embryos; and he published a series of synoptic embryological and zoological texts (*Growth*, 1924; *An Introduction to Experimental Embryology*, 1926; *Vertebrate Zoology*, 1928). His two works of most enduring importance appeared in the early 1930's. The earlier is his best-known work, *Embryology and Evolution* (1930; titled *Embryos and Ancestors* in later editions). It is a short, controversial, and influential assault on the theory of recapitulation. That theory, according to which the successive developmental stages of an individual represent the successive adult ancestral stages in the phylogeny of the species, was then no longer widely accepted by biologists. But its influence persisted, both among a reactionary minority of biologists (such as the anti-Mendelian Ernest William MacBride) and especially among those unfamiliar with the latest trends of biological thought. De Beer wrote *Embryology and Evolution* to make naive recapitulationism inexcusable, and in this he was highly successful.

In the theory of recapitulation, there is only one kind of relation between embryology and evolution. All evolutionary change takes place by adding new evolutionary stages to the end of the animal's ontogeny, through modification of the adult stage. De Beer, following Walter Garstang and others, maintained that the real relation of development and evolution is much broader. Any stage of the life cycle can be modified; evolution can take place by altering the relative rates of development of different organs, at any stage in development. De Beer could improve on Garstang by giving a genetic account of the process. In 1927 Huxley and Edmund B. Ford discovered the Mendelian rate-genes affecting eye color in the amphipod *Gammarus*: different alleles caused melanin pigment to be laid down at different rates, according to the ambient temperature. This was not the first discovery of rate-genes—Richard B. Goldschmidt had priority—but it was the *Gammarus* work that influenced the Oxford zoologists. There it was widely realized—by Huxley, Ford, and J. B. S. Haldane, as well as by De Beer—that changes in rate-genes could cause changes in the time at which different organs developed. For instance, if a rate-gene speeded up the timing of sexual maturity, organs characteristic of larval stages would become adult organs and would stand the process of recapitulation on its head. In De Beer's account, evolutionary changes could take place through the speeding up or the slowing down of the development of organs at any stage. Comparisons among species seemed to demand this kind of general relationship rather than the particular one in the theory of recapitulation. De Beer classified the various possible relations of embryonic and evolutionary changes.

The Elements of Experimental Embryology (1934), De Beer's other well-known book, was concerned not with the phylogeny of species but with the causes of development. Whereas *Embryology and Evolution* was a short polemic, *Elements* was a long synthesis. Written with Huxley, it interprets development in terms of the gradient theory of positional information of Charles M. Child, of which Huxley had always been an enthusiast. It is no longer thought that development can be explained only in those terms, but Huxley and De Beer's remains the classic attempt.

Elements represents the end of one phase in De Beer's career; *Embryology and Evolution*, the beginning of another. In the 1920's he had been interested mainly in causal and descriptive embryology; now he would become increasingly interested in comparative and phylogenetic embryology. He embarked on a major project: describing the development of the skull and head of the main vertebrate groups. It resulted in a large work, *The Development of the Vertebrate Skull* (1937), which he intended to serve as a solid descriptive basis for subsequent analytical work. As with most attempts to provide solid descriptive bases, not much use has been made of it.

In 1938 De Beer moved to London, as reader in embryology at University College, London. His work was interrupted by World War II, during which he again served in the British Army; after the war he became increasingly involved with administration. He was elected fellow of the Royal Society in 1940 and was knighted in 1954. He was professor of embryology at University College, London, from 1946 to 1950, and then became director of the British Museum (Natural History) from 1950 to 1960. In 1958 he organized the celebrations of the Darwin-Wallace centenary. His most important zoological contribution during the 1950's was his descriptive morphological monograph *Archaeopteryx lithographica* (1954).

De Beer retired from full-time zoological work in 1960. He then became interested in Darwinian scholarship, editing and publishing the transmutation notebooks. He also wrote a biography, *Charles Darwin* (1963). Another work was his splendidly illustrated *Atlas of Evolution* (1964), a pioneer in the medium of large-format, color-illustrated works. He continued writing after moving to Switzerland in 1965; in 1972 he returned to England, where he died of a heart attack.

De Beer was a widely read man. He wrote books on a remarkable range of subjects—on travelers in the Alps, the route of Hannibal's passage, and the lives of Sir Hans Sloane, Gibbon, and Rousseau—as well as works on evolution, all in a pure, clear prose. Many of his books continued to be reprinted long after his death.

BIBLIOGRAPHY

I. Original Works. De Beer's most important biological works are *The Comparative Anatomy, Histology and Development of the Pituitary Body* (London, 1926); *Embryology and Evolution* (Oxford, 1930); *The Elements of Experimental Embryology* (Cambridge, 1934; repr. 1963), with Julian Huxley; *The Development of the Vertebrate Skull* (Oxford, 1937); *Archaeopteryx lithographica* (London, 1954); *Charles Darwin* (London and Garden City, N.Y., 1963); and *Atlas of Evolution* (London, 1964).

II. Secondary Literature. E. J. W. Barrington, "Gavin Rylands de Beer," in *Biographical Memoirs of Fellows of the Royal Society of London*, **19** (1973), 65–93, with an extensive bibliography; Stephen Jay Gould, *Ontogeny and Phylogeny* (Cambridge, Mass., 1977); Mark Ridley, "Embryology and Classical Zoology in Great Britain," in T. J. Horder, J. H. Witkowski, and C. C. Wylie, eds., *A History of Embryology* (Cambridge, 1986), 35–67.

Mark Ridley

DE GROOT, JOHANNES (*b*. Garrelsweer, Netherlands, 7 May 1914; *d*. Rotterdam, Netherlands, 11 September 1972), *mathematics*.

De Groot was the son of Maria Margaretha (née Kuylman) and Johannes de Groot, a minister who later became professor of Near Eastern languages at the University of Groningen and subsequently professor of theology at the University of Utrecht. De Groot attended Christelijk Gymnasium in The Hague for two years, and Willem Lodewijk Gymnasium in Groningen for two years. In 1933 he enrolled at the University of Groningen to study mathematics, with minors in physics and philosophy. He was awarded the doctorate in 1942 with a dissertation titled *Topologische Studiën*. His supervisor was G. Schaake.

For the next several years de Groot taught mathematics at secondary schools in Coevorden and The Hague. In 1946 he became a researcher at the Mathematical Center in Amsterdam. In 1947 and 1948 he was lecturer of mathematics at the University of Amsterdam, from 1948 to 1952 professor of mathematics at the Delft Technical University, and from 1952 until his death he was professor of mathematics at the University of Amsterdam. From 1960 to 1964 he was head of the Pure Mathematics Department of the Mathematical Center.

De Groot held guest positions at several universities in the United States: in 1959 and 1960 at Purdue University, in 1963 and 1964 at Washington University in St. Louis, and in 1966 and 1967 at the University of Florida at Gainesville. From 1967 he combined his professorship at Amsterdam with a position of graduate research professor at the University of Florida at Gainesville. In 1969 de Groot became a member of the Royal Dutch Academy of Sciences.

De Groot's early research dealt with general topology and with problems in algebra, mainly in group theory. His early topological research was strongly influenced by Hans Freudenthal. This is noticeable in his dissertation, in which he obtained a number of results in compactification theory and introduced the interesting cardinal invariant's compactness deficiency and compactness degree. In group theory de Groot obtained results on equivalence of Abelian groups and effective construction of indecomposable abelian groups and rigid groups. Rigid structures, that is, structures with only trivial automorphisms, interested him; he later constructed rigid graphs and rigid spaces.

Subsequently, de Groot concentrated his attention on topology, although around 1960 he made good use of his group-theoretical expertise in several papers on groups of homeomorphisms of topological spaces. In the same period he introduced some new topological cardinal invariants, height and spread, which have turned out to be quite useful in set-theoretic topology.

In the last ten years of his life, de Groot was involved with new general approaches, mainly in set-theoretic topology. His early work in compactness and its generalizations, combined with new interest in Baire spaces, led him to introduce concepts like subcompactness, cocompactness, and cotopological properties in general. He elaborated a new compactification method through the use of so-called superextensions. He introduced and studied topological operators that change a given topology into another one by assigning as a closed subbase for the new topology all sets in the first topology with a specific property (for instance, all compact sets, or all connected closed sets). In his last few years, de Groot also began to do research in infinite-dimensional topology and in the topology of manifolds.

De Groot had many students and coworkers. Twelve dissertations were prepared under his supervision, and many of his papers were written with others. A number of his ideas and several of his conjectures were later taken up by others, while he himself, after launching them, went on to new fields. De Groot's influence on his coworkers and students, and through them on the development of general topology, was considerable.

BIBLIOGRAPHY

I. Original Works. De Groot published about ninety papers, including *Topologische Studiën* (Assen, Netherlands, 1942), his dissertation; "Decompositions of a Sphere," in *Fundamenta mathematicae*, **43** (1956), 185–194, written with T. J. Dekker; "Rigid Continua and Topological Group Pictures," in *Archiv der Mathematik*, **9** (1958), 441–446, written with R. J. Wille; "Groups Represented by Homeomorphism Groups I," in *Mathematische Annalen*, **138** (1959), 80–102; "Discrete Subspaces of Hausdorff Spaces," in *Bulletin de l' Académie polonaise des sciences*, ser. Sciences Mathématiques, astronomiques et physiques, **13** (1965), 537–544; and "Inductive Compactness as a Generalization of Semicompactness," in *Fundamenta mathematicae*, **58** (1966), 201–218, written with T. Nishiura.

II. Secondary Literature. A complete list of de Groot's papers and a survey of his scientific work is in P. C. Baayen and M. A. Maurice, "Johannes de Groot, 1914–1972," in *General Topology and Its Applications*, **3** (1973), 3–32. De Groot's contributions to algebra are more fully discussed in L. C. A. Leeuwen, "Some Problems and Results in Abelian Group Theory," in *Niuw archief voor wiskunde*, **22**, no. 2 (1974), 143–155.

P. C. Baayen

DELSARTE, JEAN FRÉDÉRIC AUGUSTE (*b.* Fourmies, France, 19 October 1903; *d.* Nancy, France, 28 November 1968), *mathematics*.

Delsarte was the oldest of three children of the head of a textile factory. After the German invasion in 1914, his father remained in Fourmies to preserve what he could of the factory, while the rest of the family fled to unoccupied France.

Delsarte entered the École Normale Supérieure in 1922. After graduating in 1925 and completing his military service, he was granted a research fellowship, and in a little more than one year he had written his dissertation. Although university teaching jobs were very scarce at that time, Delsarte obtained one at the University of Nancy immediately after receiving the doctorate in 1928; he remained there for the rest of his life, and was dean of the Faculty of Sciences from 1945 to 1949. In 1929 he married Thérèse Sutter; they had two daughters.

In mathematics Delsarte had a predilection for what has been called "hard" analysis, in the tradition of Leonhard Euler, Carl Jacobi, and G. H. Hardy. Like them, he was a superb calculator with a remarkable talent for seeing his way through a maze of computations. His ideas were strikingly original, and his work was completely uninfluenced by the mathematical fashions of his time. Since Delsarte had very few students, he did not receive the recognition he deserved; several of his pioneering ideas were rediscovered much later without his being given credit. Such was the case with his extension of the Möbius function to abelian groups and its use in enumeration problems, and with the generalization to Fuchsian groups of the formula giving the value of the number of lattice points within a ball.

Delsarte's main results in mathematical analysis stem from a common theme: the expansion of a function f by a series

$$S(f) \sim \sum_i L_i(f)\phi_i \qquad (1)$$

where the ϕ_i are functions independent of f and the L_i linear functionals. The Taylor series with $\phi_k(x) = x^k/k!$ and $L_k(f) = D^k f(O)$ is the "classical" example, but many other types were known and were studied in George Watson and Edmund Whittaker's *A Course of Modern Analysis* and Watson's *Treatise on the Theory of Bessel Functions*, which were Delsarte's most cherished books. These works had convinced him that a good understanding of the formal properties of such expansions was necessary to a fruitful study of their domain of definition and their mode of convergence. This was the course he followed with remarkable success, opening up new fields of research that are still far from having been thoroughly explored.

The starting point is a vector space E of complex valued functions, which for simplicity one may assume to be defined in a neighborhood I of 0 in **R**, and to be C^∞.

(A) There is given an endomorphism D of E that has a continuous spectrum. This means that for every $\lambda \in \mathbf{C}$, there is a function $j_\lambda \in E$ for which

$$Dj_\lambda = \lambda j_\lambda. \qquad (2)$$

(B) Each j_λ has a formal expansion

$$j_\lambda \sim \sum_{n=0}^{\infty} \lambda^n \phi_n, \qquad (3)$$

where the ϕ_n are polynomials belong to E and $\phi_0 = 1$, so that

$$D\phi_n = \phi_{n-1} \quad \text{for} \quad n \geqslant 1. \qquad (4)$$

(C) Delsarte introduces the formal series depending on a parameter $y \in I$

$$T^y f \sim \phi_0(y)f + \phi_1(y)Df + \cdots + \phi_n(y)D^n f + \cdots; \qquad (5)$$

these operators satisfy the relation

$$T^y T^z = T^z T^y \qquad \text{for} \quad y, z \in I, \qquad (6)$$

and if $\Phi_f(x, y) = (T^y f)(x)$ for $x, y \in I$,

$$D_x \Phi_f(\cdot, y) = D_y \Phi_f(x, \cdot). \qquad (7)$$

This is in general a partial differential equation, and its integration with the initial condition $\Phi_f(x, 0) = f(x)$ yields $T^y f$. With the initial condition $\Psi(x, 0) = 0$, the equation

$$D_y \Psi(x, \cdot) - D_x \Psi(\cdot, y) = D_x^{n+1} f(x) \phi_n(y) \qquad (8)$$

similarly yields the "remainder," the difference between $T^y f$ and the first n terms in (5). In the classical case, $\Phi_f(x, y) = f(x + y)$. A more interesting example (among many others studied by Delsarte) corresponds to

$$D = \frac{d^2}{dx^2} + \frac{1}{x}\frac{d}{dx}, j_\lambda(x) = J_0(ix\sqrt{\lambda}), \phi_n(x) = \frac{1}{(n!)^2}\left(\frac{x}{2}\right)^n, \qquad (9)$$

for which Delsarte shows that

$$\Phi_f(x, y) = \frac{1}{\pi}\int_0^\pi f(\sqrt{x^2 + y^2 - 2xy\cos\phi})\,d\phi. \qquad (10)$$

(D) Delsarte's main interest in these expansions was in the study of linear endomorphisms U of E that commute with all the T^y for $y \in I$. In the classical case with $E = D(\mathbf{R})$, this condition and minimal continuity properties characterize the convolution operators $f \mapsto f * \mu$, where μ is a measure or distribution with compact support. In general,

$$(U \cdot f)(x) = \langle T^x f, \alpha \rangle \qquad (11)$$

where α is a linear form on E; this implies

$$U \cdot j_\lambda = X(\lambda) j_\lambda \text{ with } X(\lambda) = \langle j_\lambda, \alpha \rangle. \qquad (12)$$

(E) The introduction of these generalized convolution operators U leads to the subclasses J_U of functions $f \in E$ such that $U \cdot f = 0$.

In the classical case, in which μ is the difference of two Dirac measures $\mu = \delta_a - \delta_0$, J_U consists of the functions of period a. For an arbitrary μ, Delsarte called J_U a space of mean-periodic functions; in the general case, he spoke of "J-mean periodic" functions. In many cases, $X(\lambda)$ is an entire function of λ with a sequence of zeroes $\lambda_1, \lambda_2, \cdots, \lambda_n, \cdots$; with each J-mean periodic function is associated a formal expansion

$$f(x) \sim \sum_n c_n j_{\lambda_n}(x), \qquad (13)$$

for which Delsarte gave a general method of computing the c_n; it is similar to the Fourier series, to which it reduces for $U \cdot f = f * (\delta_a - \delta_0)$. If $X(0) \neq 0$, the Taylor series $j_\lambda(x)/X(\lambda)$ introduces polynomials that generalize the Bernoulli polynomials. Delsarte also gave a generalized Euler-Maclaurin formula with explicit remainder using these polynomials.

Earlier, Delsarte had made a thorough study of the case $U \cdot f = f * \mu$; then X is the Fourier transform of μ, and therefore an entire function, and the j_{λ_n} have the form $P_n(x) \exp(2\pi i \lambda_n x)$, with P_n a polynomial. Delsarte investigated the convergence of such expansions, and his results were extended in the later work of Laurent Schwartz and Jean-Pierre Kahane. For the operator D defined in (9), he also showed how, for different choices of U, the expansions (13) became various known expansions in Bessel functions, where usually the coefficients c_n were computed by ad hoc devices but Delsarte's general process applied in every case.

After 1950 Delsarte greatly enlarged the scope of these ideas. The classical translation operators generalize naturally to the operators S^y and T^y defined by $S^y f: x \mapsto f(yx)$ and $T^y f: x \mapsto f(xy)$ in any Lie group. Delsarte considered the more general situation of a Lie group G with a compact group A of automorphisms of G. The operators S^y and T^y are defined by

$$S^y f: x \mapsto \int_A f(y^\sigma x) d\sigma, T^y f: x \mapsto \int_A f(xy^\sigma) d\sigma \quad (14)$$

and reduce to translations when A reduces to the identity element 1_G. In general those operators still satisfy $S^y T^z = T^z S^y$ and depend only on the orbit $\bar{y}$ of y under A, so that they operate on functions on the space of orbits G/A. When G is the additive group $\mathbf{R}^2$ and A the group of rotations, G/A can be identified with the positive reals, and the function on $(G/A) \times (G/A)$,

$$\Phi_f(\bar{x}, \bar{y}) = (T^y f)(x) = (S^x f)(y), \quad (15)$$

is given by (10). When $A = \{1_G\}$, the relation (7) for Φ_f on $G \times G$ still holds when D is an invariant vector field (element of the Lie algebra of G). Delsarte showed that such equations still exist in general, but for invariant differential operators, which in general will be of order > 1, as (9) shows. He considered this result as the beginning of an analog of Lie theory for G/A. This point of view was developed by B. Lewitan, and probably more remains to be done.

The fact that the operators (5) may be obtained by integrating the partial differential equation (7) led Delsarte to consider more generally, for two different differential operators D and D', the equation

$$D_x F(x, y) = D'_y F(x, y), \quad (16)$$

which yields what he called "transmutation" operators, changing D into D'. This was developed by Delsarte in collaboration with J.-L. Lions, and later was used by Lewitan and others for the Sturm-Liouville problem.

The same ideas enabled Delsarte to prove one of his most elegant and unexpected results. It has been known since Gauss that if a C^∞ function f in $\mathbf{R}^n$ has at each point x a value equal to its mean value on every sphere of center x, it is a harmonic function. Delsarte showed that the conclusion still holds if $f(x)$ is equal to its mean value on only two spheres of center x and radii $a > b > 0$, provided the ratio a/b is not a number in a finite set depending only on n.

From 1950 on, Delsarte often lectured at institutes and universities in India and North and South America. After 1960 his eyesight, which had always been poor, began to deteriorate. At the end of his life he could read and write only with great difficulty. From 1962 to 1965 he was director of the Franco-Japanese Institute in Tokyo.

BIBLIOGRAPHY

Delsarte's works are collected in *Oeuvres de Jean Delsarte*, 2 vols. (Paris, 1971), which includes brief essays on his life and work. His writings also include *Lectures on Topics in Mean Periodic Functions and the Two-Radius Theorem* (Bombay, 1961).

JEAN DIEUDONNÉ

DEMEREC, MILISLAV (*b.* Kostajnica, Austria-Hungary [now Yugoslavia], 11 January 1895; *d.* Cold Spring Harbor, New York, 12 April 1966), *genetics*.

The son of Ljudevit Demerec, a schoolteacher and school inspector, and Ljubica Dumbovic Demerec, Demerec graduated from the College of Agriculture at Krizevci in 1916, worked as an adjunct at the Krizevci Experiment Station until 1919, and then came to the United States to commence graduate study with Rollins A. Emerson at Cornell University. After receiving his Ph.D. in 1923, Demerec became a resident investigator in the Department of Genetics of the Carnegie Institution of Washington at Cold Spring Harbor, New York. In 1921 he married Mary Alexander Ziegler; they had two daughters, Rada, who became an anthropologist, and Zlata, who became a microbial geneticist. Demerec was assistant director of the Department of Genetics between 1936 and 1941, acting director between 1941 and 1943, and director between 1943 and 1960. In 1941 he also became director of the adjoining Biological Laboratory of the Long Island Biological Association. Under his administration the two laboratories collaborated closely and became in effect the Cold Spring Harbor Biological Laboratory. Demerec became an American citizen in 1931.

The Biological Laboratory had ample facilities for summer investigators, and it organized the Cold Spring Harbor Symposia in Quantitative Biology.

Under Demerec the Biological Laboratory built a permanent staff, and the symposia, previously physiological, shifted largely to genetics and cytogenetics, population genetics and evolution, biochemistry and molecular biology, and the rapidly developing fields of virus and bacterial genetics. Summer courses were established that gained Cold Spring Harbor a worldwide reputation as the place for scientists to learn how to "do" bacteriophage genetics (from 1945) or bacterial genetics (from 1955). At Cold Spring Harbor one could meet future Nobel Prize winners and geneticists of international stature. Somehow Demerec found sufficient funds from donors to keep the Cold Spring Harbor laboratory growing in staff and reputation, and constantly breaking new ground.

Demerec provided the stimulus for much of the advance in genetics by means of publications to aid investigators. With his colleague Berwind P. Kaufmann he wrote the *Drosophila Guide*, which went through eight editions between 1940 and 1969 and was an indispensable aid to work with the fruit fly in high school, college, and university courses. Demerec edited the compendium, *The Biology of Drosophila* (1950). He started the *Drosophila Information Service* in 1934, which he edited with Calvin B. Bridges, and which was a prototype of the scientific newsletters that enable investigators using a particular organism to exchange unpublished information, list available genetic stocks, and describe new technical aids. Demerec continued editing the newsletter until 1939. He started the first *Drosophila* stock center at Cold Spring Harbor, maintaining and distributing desired experimental stocks throughout the world. He founded and edited through its first nine volumes (1947–1958) *Advances in Genetics*, a series of technical reviews. He supported Bridges in the herculean task of making, at Cold Spring Harbor, his accurate maps of the giant banded salivary chromosomes of *Drosophila melanogaster*.

Demerec also was deeply involved in the international aspects of organized science. He played a significant role in the organization, program arrangements, and financing of the international congresses of genetics, from the sixth (1932) to the tenth (1958). He was vice president of the Seventh International Genetics Congress (1939) and a member of the Permanent International Committee of the International Genetics Congress from 1939 to 1953. He was a member of both the organizing and the program committees for the Tenth Congress.

In spite of these heavy administrative burdens, Demerec was highly productive in scientific research. His first important studies at Cold Spring Harbor continued his interest in the variegated, or mosaic, genetic characters he had worked on as a graduate student studying Indian corn, or maize. In the 1920's Demerec discovered frequently mutating genes and mosaic characters in *Drosophila virilis*, and later in delphiniums. His studies of the effects upon mutation of environmental agents such as temperature and X rays, and of internal conditions such as sex, were classic in their field.

In the 1930's Demerec shifted his interest to problems of radiation-induced mutation. Using *Drosophila melanogaster*, he obtained an approximate answer to the question of whether all X-ray-induced lethals are small chromosome deficiencies, and determined the frequency of cell-lethal mutations, that is, mutations lethal to a single cell in a surrounding of normal tissue. He analyzed the differences in spontaneous mutability of genes in different stocks of *D. melanogaster* and established the existence of mutator genes. His interest in unstable genes led him to explore the "position effects" of genes shifted from euchromatic to heterochromatic portions of the chromosomes. In many of these studies he collaborated with a succession of research assistants, colleagues, and visiting investigators at Cold Spring Harbor.

During World War II Demerec shifted his research to bacteria, first to *Escherichia coli* and subsequently to *Staphylococcus aureus* and *Salmonella typhimurium*. Work also was done on the mold *Neurospora crassa*. Most important were Demerec's studies of induced mutations in bacteria that confer specific kinds of resistance to penicillin, Aureomycin, and streptomycin. Not reported until after the war was the success in producing a mutant strain of *Penicillium* that would grow abundantly when submerged in a vat of nutrient fluid, rather than only when floating on the surface. This achievement multiplied enormously the production of penicillin from a given amount of nutrient. In the 1950's the mutagenic effects on bacteria of many salts, organic chemicals, and especially carcinogens were detected. The fine structure of the gene and the linkage relationships of the genes of *S. typhimurium* were the subjects of another classic study, carried forward largely with the collaboration of Demerec's daughter Zlata and his son-in-law Philip Hartman.

In 1960 Demerec retired as director of the joint Cold Spring Harbor laboratories. Although at first he intended to remain there to do research, unpleasant relations with his successor led Demerec to transfer his experimental work in 1961 to the Brookhaven National Laboratory, where as a senior staff member he continued to investigate problems of mutation, transduction by phage, and linkage in *Salmonella*. Especially significant was his discovery of the clustering of functionally related genes in the bacterial chromosome, and of the genetic homologies between *Salmonella* and *E. coli*. Demerec retired from Brookhaven in 1965 and then assumed a post as research professor of biology at C. W. Post College of Long Island University. He died of a heart attack while organizing his new laboratory.

Among Demerec's many honors were the presidencies of the Genetics Society of America (1939) and the American Society of Naturalists (1954), as well as membership in the National Academy of Sciences, the American Academy of Arts and Sciences, and the American Philosophical Society. He received the Order of St. Sava from Yugoslavia in 1935, and the Kimber Genetics Gold Medal from the National Academy of Sciences in 1962. He held many visiting appointments and lectureships, and served on the genetics panel of the National Academy of Sciences' Committee on the Biological Effects of Atomic Radiation, which issued important reports in 1956 and 1960. His influence as organizer, administrator, and investigator was probably unsurpassed in genetics in the period from 1930 to 1960.

BIBLIOGRAPHY

For a fuller treatment of the life and work of Demerec, and for a complete bibliography of his publications, see Bentley Glass, "Milislav Demerec," in *Biographical Memoirs. National Academy of Sciences*, **42** (1971), 1–27, and *A Guide to the Genetics Collections of the American Philosophical Society* (1988), 25–30. See also Joseph Fruton, *Bio-bibliography of the History of the Biochemical Sciences Since 1800* (Philadelphia, 1982), 159, which lists other biographical references. Primary source material, papers, and 56 volumes of research notes (about 9,500 items in all) are in the Library of the American Philosophical Society, Philadelphia.

BENTLEY GLASS

DENJOY, ARNAUD (*b*. Auch, Gers, France, 5 January 1884; *d*. Paris, France, 21 January 1974), *mathematical analysis*.

Denjoy was the son of Jean Denjoy, a wine merchant in Perpignan, and of a woman surnamed Jayez, who was from Catalonia. After secondary education at Auch and Montpellier, in 1902 he entered the École Normale Supérieure in Paris, where he studied under Émile Borel, Paul Painlevé, and Charles Picard, and graduated first in his class. In 1905 he received a Fondation Thiers fellowship for a three-year period. Denjoy completed his dissertation ("Sur les produits canoniques d'ordre infini") in 1909 and was named *maître de conférences* at Montpellier University, where he taught until 1914. Poor eyesight kept him from military service during World War I. In 1917 he received a professorship at Utrecht, and in 1922 he accepted a position at the University of Paris, where he remained until his retirement in 1955.

On 15 June 1923 Denjoy married Thérèse-Marie Chevresson; they had three sons.

Denjoy led a quiet life, working most of the day at home. During the summers he enjoyed cycling on forest trails or walking along rivers and lakes. His death resulted from a fall in his home.

Elected to the Académie des Sciences in Paris on 15 June 1942, Denjoy was also a member of several other academies, including the Academy of Sciences of Amsterdam, the Société des Sciences et Lettres of Warsaw, and the Société Royale des Sciences of Liège. He was vice president of the International Mathematical Union in 1954. Although he did not write joint papers, he maintained contact with most of the great mathematical analysts, especially those from Russia, Poland, the Netherlands, and Germany.

Denjoy's weak voice and bad eyesight did not make him a notable lecturer, but in private talks he was very entertaining. His written work displayed his gift for brilliant metaphors to convey mathematical discoveries. He coined many illuminating mathematical terms, such as *clairsemé*, *gerbe*, *résiduel*, *plénitude*, and *épaisseur*.

Denjoy was an atheist, but tolerant of others' religious views; he was very interested in philosophical, psychological, and social issues. Throughout his life he wrote about them in his (unpublished) diary. In 1964 he published some of his thoughts in *Hommes, formes et le nombre*, which deals mainly with mathematical discoveries and concepts and with men of science. He liked neither the Bourbaki approach to mathematics nor its style, and at the time of his death he was planning a sharp account of his criticisms.

His activity in the Radical-Socialist Party, at the time headed by Édouard Herriot, led to Denjoy's election to the Montpellier town council in 1912 and to the Gers county council in 1920, and he served on the latter until 1940. From 1949 to 1950, just before François Duvalier became dictator, Denjoy held a cultural diplomatic position in Haiti. In 1960 he joined the Comité d'Honneur de l'Union Rationaliste, which aims to spread the spirit of science and the experimental method, and to fight dogmatism and fanaticism.

Denjoy was the youngest of the prestigious quartet of French mathematicians—the others were Émile Borel, René Baire, and Henri Lebesgue—that devised the theory of functions of real variables at a time when analytic functions $f(z)$ were more commonly studied.

By the time Denjoy completed his dissertation in 1909, Borel had introduced a theory of countably additive measure, divergent series, scales of growth and zero measure, and monogenic functions. Baire had initiated a new approach to real functions by introducing semicontinuity sets of the first category and the transfinite scale of (Baire) functions. Lebesgue had firmly established the notion of measure and his theory of integration. With those ingredients Denjoy realized a synthesis combining topological and metric tools: topology to reduce problems to basics, and metric notions for the final blow.

Denjoy had a strong classical background in complex function theory, differential equations, and continued fractions that permeated all his work. His dissertation, for instance, presented results concerning the series $\Sigma A_n/(z - a_n)$, canonical Weierstrass products for integral functions, asymptotic values of integral functions of finite order, and boundary behavior of conformal representation. Although some of these results are now considered among his best contributions, he wrote in 1934 that he considered his achievements to be (1) the integration of derivatives, (2) the computation of the coefficients of any converging trigonometric series for which the sum is given, (3) his theorem on quasi-analytic functions, and (4) differential equations on a torus.

Posterity has not always confirmed this hierarchy. For instance, the fourth achievement, which completely clarifies the "last Poincaré theorem," has grown into a vast field involving dynamical systems. His theorem concerning quasi-analytical functions (related to Borel monogeneity) and the Denjoy conjecture were the source of many subsequent studies (for example, Benoit Mandelbrot's). The first two are sometimes considered more feats of intellectual strength than sources of practical applications. It remains true, however, that they both had profound implications and that they were needed. It is likely that nobody but Denjoy could have achieved the computation of the coefficients, the results of which were first published (as were most of Denjoy's theorems) as notes in the *Comptes rendus* of the Academy of Sciences (1921). When confronted with growing skepticism, he published the complete proofs in five volumes (1941–1949). They contain much more than was required for the proofs and are an explosion of beautiful theorems and examples. Denjoy's *Leçons sur le calcul des coefficients d'une série trigonométrique* was for a long time recommended reading for research students at Moscow University. It contains, in addition to the Denjoy integral, a wealth of tangential properties of continuous functions that have inspired considerable research by Andrew Bruckner. Some of the properties were generalized by Frédéric Roger and Gustave Choquet.

BIBLIOGRAPHY

The best guide to Denjoy's works is "Arnaud Denjoy, évocation de l'homme et de l'oeuvre," in *Astérisque*, nos. 28–29. See also his *Leçons sur le calcul des coefficients d'une série trigonométrique*, 5 vols. (Paris, 1941–1949); *L'énumération transfinie*, 5 vols. (Paris, 1946–1954); *Mémoire sur la dérivation et son calcul inverse* (Paris, 1954); *Articles et mémoires*, 2 vols. (Paris, 1955); *Un demi-siècle de notes*, 2 vols. (Paris, 1957); and *Hommes, formes et le nombre* (Paris, 1964).

GUSTAVE CHOQUET

DENNISON, DAVID MATHIAS (*b.* Oberlin, Ohio, 26 April 1900; *d.* Ann Arbor, Michigan, 3 April 1976), *physics*.

Dennison was the son of Walter Dennison, a professor of classics at Oberlin College, and of the former Anna L. Green. From 1902 to 1910 the Dennisons lived in Ann Arbor, Michigan, where David's father taught Latin at the University of Michigan. The family then moved to Swarthmore, Pennsylvania, where in 1921 David received his A.B. from Swarthmore College. Dennison was interested in science from an early age, but his first course in physics did not arouse his interest and he majored in mathematics. It was a casual conversation with H. C. Hayes, a Swarthmore physicist, and a summer job in 1920 at General Electric, working with Irving Langmuir, that drew Dennison to physics. He enrolled at the University of Michigan, where he was the first student to submit a theoretical dissertation to the department of physics. He received his Ph.D. in 1924 and, on 18 August of that year, married Helen Lenette Johnson. They had two sons.

Dennison's doctoral research, on the structure of the methane molecule, was directed by Oskar Klein, who had earlier been assistant to Niels Bohr. When Bohr came to the United States in the fall of 1923, he visited Klein at Ann Arbor and met Dennison.

Bohr's impression was favorable, and he invited Dennison to Copenhagen. Dennison was awarded a two-year fellowship by the National Education Board and with it went to Bohr's Institute for Theoretical Physics in Copenhagen.

Dennison was in Europe at a most opportune time. He arrived in Copenhagen in late summer 1924, during the waning months of the old quantum theory. During his three years in Europe, Dennison witnessed the creation of quantum mechanics, starting with Werner Heisenberg's version (June 1925), through Erwin Schrödinger's and P. A. M. Dirac's versions, and ending in the spring of 1927, when he attended Heisenberg's first seminar on the uncertainty principle. "To all of us who heard him," Dennison recalled, "there was never any question but that this was a really wonderful development."

Dennison's first work at Copenhagen continued the analysis of molecular spectra along the lines he had developed in his doctoral dissertation. Although some of his interpretations—made prior to the creation of quantum mechanics—were wrong, the methods Dennison established set the pattern of the subsequent research on molecular structure. After Heisenberg created quantum mechanics, Dennison applied the new mechanics to a symmetric-top molecule, using matrix methods to calculate the rotational energy states, selection rules, and intensities of a symmetric-top rotator. This work, published in 1926, was the first application of matrix mechanics to appear in *Physical Review*.

A fellowship from the University of Michigan allowed Dennison to stay in Europe a third year; he went to Zurich, where Schrödinger was the attraction. There, Dennison worked on homopolar diatomic molecules; later, at Cambridge, during the spring of 1927, this line of work led to Dennison's most noteworthy contribution. Ralph Fowler had invited Dennison to give three lectures to his graduate class, and in preparing the third of these lectures, Dennison resolved the disparity between calculated and measured values for the specific heat of the hydrogen molecule. In this work he was influenced by Heisenberg's 1926 quantum mechanical treatment of the helium atom. Heisenberg explained the puzzling spectrum of helium by recognizing that the wave function for helium can be either symmetric or antisymmetric with respect to the exchange of electrons. In a similar fashion Dennison recognized that the rotational states can be either symmetric or antisymmetric with respect to an exchange of nuclei. Transitions between these symmetric and antisymmetric states are forbidden unless the proton has a spin.

Dennison assumed that the proton, like the electron, had a spin; however, he recognized that the magnetic moment of the proton was so small that transitions between the symmetric and antisymmetric states were so slow that hydrogen gas at low temperatures could be regarded as a mixture of a gas with symmetric rotational states and a gas with antisymmetric ones. When hydrogen was considered as a mixture of two gases, each with its own specific heat, theory agreed with experiment. Dennison's work on the specific heat of hydrogen, published in *Proceedings of the Royal Society of London*, provided the first quantitative evidence for the spin of the proton.

In 1927 Dennison returned to Ann Arbor, where he had received an instructorship in physics at the University of Michigan. He rose rapidly through the ranks and became a full professor in 1935. During this period the University of Michigan became a center for theoretical physics in the United States. In addition to Dennison, S. A. Goudsmit, G. E. Uhlenbeck, and Otto Laporte were full-time members of the faculty. The influence of the University of Michigan physics department was further enhanced by the Summer Symposia in Theoretical Physics, which began in 1928 and continued until 1940. These seminars attracted theoretical physicists from both the United States and leading centers of physics in Europe.

Throughout his career Dennison's primary interest was the application of quantum mechanics to the structure of molecules and the interpretation of molecular spectra. In large part owing to Dennison, the University of Michigan became a world center for both theoretical and experimental molecular studies. An early example of the fruitful interaction between Dennison and the experimentalists occurred in 1932 when Dennison, with George Uhlenbeck, solved the quantum mechanical two-minimum problem that involves the quantum mechanical effect of tunneling. This theory, when applied to the pyramid-shaped ammonia molecule (NH_3), predicted that the nitrogen atom, at the peak of the pyramid, could "tunnel" through the plane of the three hydrogen atoms, thereby inverting the pyramid. This nonclassical effect had spectroscopic consequences, and Dennison persuaded a colleague to perform the experiment. The result agreed with Dennison's predictions; the experiment was the first in what became microwave spectroscopy. Throughout the 1930's Dennison studied the vibrational and rotational behavior of molecular systems and brought his theoretical studies to bear on infrared spectroscopic data.

During World War II, Dennison worked on proximity fuses and was cited for this work by the U.S. Navy. After the war an electron accelerator was built at the University of Michigan. Dennison's colleague H. R. Crane designed the accelerator to have both curved and straight portions. Dennison, with Theodore Berlin, established the general conditions for the stability of such orbits; their paper became a basic reference for accelerator builders. But the accelerator, the glamour tool of postwar physics, could not keep Dennison from his first love: molecules.

Dennison's fascination with molecules was shared by a number of first-rate American physicists and chemists: E. U. Condon, Robert Mulliken, J. Robert Oppenheimer, Linus Pauling, John Slater, J. H. Van Vleck, and H. C. Urey. These scientists did not participate in the creation of quantum mechanics, but with the new theory in place, they immediately started to apply the new theoretical formalism to atoms, molecules, and solids. Their work played a significant part in the emergence of the United States as a leading center for physics during the 1930's.

The field of molecular physics, due in large part to Dennison's contributions, became a highly refined area of research after the war. Until he retired as Harrison M. Randall professor of physics in 1971, Dennison worked on such problems as vibrational-rotational interactions, centrifugal distortion effects, and hindered rotations in molecules. Among his honors were election to the National Academy of Sciences in 1953 and fellowship in the American Physical Society.

BIBLIOGRAPHY

I. Original Works. A complete list of Dennison's publications follows the article by H. Richard Crane in *Biographical Memoirs. National Academy of Sciences*, **52** (1980), 139–159. A personal account of his early years in physics is given in "Recollections of Physics and Physicists During the 1920s," in *American Journal of Physics*, **42** (1974), 1051–1056.

Dennison's papers are deposited in the Bentley Library, University of Michigan.

II. Secondary Literature. Dennison's physical research is described in many treatises on molecular spectroscopy. See, for example, Walter Gordy and Robert L. Cook, *Microwave Molecular Spectra* (New York, 1970); and Gerhard Herzberg, *Molecular Spectra and Molecular Structure*, J. W. T. Spinks, trans., 2 vols., 2nd ed. (Princeton, 1950).

John S. Rigden

DIETRICH, WILHELM OTTO (*b.* Senden, near Ulm, Germany, 30 July 1881; *d.* Berlin, German Democratic Republic, 26 March 1964), *mammalian paleontology*.

Dietrich was the third of the six sons and one daughter of Otto Dietrich, the director of a mill, and of Maria Kramer. He graduated from the gymnasium in Ulm in 1899 and then entered the Technical University in Stuttgart; in 1901 he transferred to the University of Tübingen, where he studied geology and paleontology. In December 1903 Dietrich received the doctorate under E. Koken, with the dissertation "Ältester Donauschotter auf der Strecke Immendingen-Ulm" (1904). He subsequently studied at Freiburg im Breisgau in order to enhance his knowledge of petrography.

By the autumn of 1904, Ernst A. Wülfing had appointed Dietrich an assistant at the Mineralogical-Geological Institute of the newly founded Danzig Technical University, the development of which demanded all of his time and energy. Before this activity could bear scientific fruit, however, otosclerosis forced Dietrich to relocate in the spring of 1907. A stay in the Jura Mountains of Switzerland failed to bring about the desired improvement in his health, and Dietrich suffered a progressive hearing loss, in the last decades of his life becoming completely deaf.

Through Koken's efforts, in 1908 Dietrich secured an assistantship with Eberhard Fraas in the geology department of the Royal Natural History Collection in Stuttgart, where he spent three years. Besides his museum duties he was entrusted with paleontological fieldwork in Steinheim an der Murr, where complete skeletons of the great mammals of the Pleistocene were discovered. In 1909 and 1910 Dietrich reported on the remains of giant deer and other Cervidae. Under his direction the skeleton of a mammoth (*Elephas primigenius*) was salvaged and installed in the museum in 1910. He meticulously described this Swabian mammoth in a thoroughly researched work characterized by careful argumentation and clear presentation—the hallmarks of all of Dietrich's subsequent projects. By the time the work was published (1912), its author was no longer in Stuttgart. Despite Fraas's recommendation and Dietrich's evident promise as a scientist, the Royal Württemberg Ministry of Religion and Education could not bring itself to assure him a secure post.

At Fraas's suggestion W. von Branca (formerly Branco) hired Dietrich in May 1911 as an assistant at the Royal Geological-Paleontological Institute and

Museum of the Friedrich Wilhelm University in Berlin. He served there until March 1945 as assistant and chief assistant and then as curator under Branca, J. F. Pompeckj, and H. Stille. He retired in 1959.

In 1921 Dietrich married Lotte Trendelenburg, daughter of a privy councillor. Their only son was killed on the eastern front in March 1943. This was not the only sacrifice the war demanded of them; in the same year their house was destroyed by bombing. They found modest quarters in two basement rooms at the institute. For Dietrich this close proximity of residence and work place was ideal; even after his official retirement he continued to live in the basement of the institute. In the fall of 1963 he contracted pneumonia, as a consequence of which he died, after months of pain, on 26 March 1964.

Dietrich's only ambition at the Berlin institute was to be a good assistant to the professor and a good manager of the collection. He tended and brought order to the great majority of the rich scientific holdings of the institute. At the same time, his post inspired him to undertake many scientific works and prevented one-sidedness. Besides mammals, he concerned himself with foraminifers, corals, mussels, and snails. He had a masterful command of paleontology, which he always linked with stratigraphy.

Dietrich's specialty was the study of mammals of the early Tertiary and the Quaternary, as well as their biostratigraphic evaluation. He was not interested in European finds; rather, he prized those of Africa. He worked with some of the material yielded by the Tendaguru expedition, as well as the finds of the expeditions of Reck and Kohl. His favorite forms were elephants, hoofed animals, and predators, but primates and rodents also commanded his attention. Predominantly analytically inclined, Dietrich did not pursue theories or grand syntheses. For decades he discussed new works on fossil mammals in the *Neues Jahrbuch für Mineralogie, Geologie und Paläontologie*. His reviews were impartial, reliable, clear, and constructively critical.

The Paleontological Society of German-Speaking Regions named Dietrich an honorary member in 1942, "in acknowledgment of his exemplary researches." He was also an honorary member of the National Association for Natural Science in Württemberg (1956) and the Geological Association of the Upper Rhine (1959). In 1957 the German Geological Society recognized him as an outstanding paleontologist by bestowing on him its highest award, the Hans Stille Medal.

BIBLIOGRAPHY

I. Original Works. Dietrich's publications are listed by Daber (see below). His writings include Ältester Donauschotter auf der Strecke Immendingen-Ulm," in *Neues Jahrbuch für Mineralogie und Geologie*, supp. **19** (1904), 1–39, his dissertation; "Neue Riesenhirschreste aus dem schwäbischen Diluvium," in *Jahreshefte des Vereins für vaterländische Naturkunde. Württemberg*, **65** (1909), 132–161; "*Elephas primigenius Fraasi*, eine schwäbische Mammutrasse," *ibid.*, **68** (1912), 42–106; "Die Gastropoden der Tendaguruschichten der Aptstufe und der Oberkreide im südlichen Deutsch-Ostafrika (wissenschaftliche Ergebnisse der Tendaguru-Expedition 1909–1912)," in *Archiv für Biontologie*, **3,** no. 4 (1914), 101–152; "*Elephas antiquus Recki* n. f. aus dem Diluvium Deutsch-Ostafrikas (wissenschaftliche Ergebnisse der Oldoway-Expedition 1913)," *ibid.*, **4** (1915), 1–80; "Gastropoda mesozoica; Familie Nerineidae," in *Fossilium catalogus*, I (Berlin, 1925), 31; "Beitrag zur Kenntnis der Bohnerzformation in Schwaben. 2. Über die Nager aus den Spaltenablagerungen der Umgebung Ulms," in *Neues Jahrbuch für Mineralogie, Geologie und Paläontologie*, supp. **62,** Abt. B (1929), 121–150; "Beitrag zur Kenntnis der Bohnerzformation in Schwaben. 3. Raubtiere aus den Bohnerzablagerungen der Ulmer und der Eichstätter Alb," *ibid.*, supp. **63,** Abt. B (1930), 451–474; "Die Huftiere aus dem Obereozän von Mähringen auf der Ulmer Alb," in *Palaeontographica*, **A83** (1936), 163–209; "Ältestquartäre Säugetiere aus der südlichen Serengeti, Deutsch-Ostafrika," *ibid.*, **A94** (1942), 77–133; "Stetigkeit und Unstetigkeit in der Pferdegeschichte," in *Neues Jahrbuch für Mineralogie, Geologie und Paläontologie*, Abt. B, Abhandlungen, **91** (1949), 121–148; "Fossile Antilopen und Rinder Äquatorialafrikas (Material der Kohl-Larsschen Expedition)," *Palaeontographica*, **A99** (1950), 1–62; and "Geschichte der Sammlungen des Geologisch-Paläontologischen Institutes und Museums der Humboldt-Universität zu Berlin," in *Berichte der Geologischen Gesellschaft der DDR*, **6** (1962), 247–289.

II. Secondary Literature. R. Daber, "Wilhelm Otto Dietrich," in *Berichte der Geologischen Gesellschaft der DDR*, **10,** no. 1 (1965), 99–106, with portrait; W. Gross, "Wilhelm Otto Dietrich (1881–1964)," in *Neues Jahrbuch für Geologie und Paläontologie*, Monatshefte, 1964, no. 7, 385–387, with portrait; E. Hennig, "Wilhelm Otto Dietrich. Paläontologischer Forscher," in *Jahreshefte des Vereins für vaterländische Naturkunde. Württemberg*, **120** (1965), 55–58, with portrait; K. Staesche, "Wilhelm Otto Dietrich, 1881–1964," in *Jahrbuch und Mitteilungen des Oberrheinischen geologischen Vereins*, n.s. **46** (1964); and H. Wehrli, "Wilhelm Otto Dietrich's Leben und Wirken," in *Geologie*, **5,** no. 4/5 (1956), 261–265, with bibliography 266–270 and portrait.

Longer memorial works are W. D. Heinrich, "Wilhelm Otto Dietrich, 1881–1964," in *Zeitschrift für geologische Wissenschaft*, **10,** no. 7 (1982), 883–1051, and, as editor,

Wirbeltier-Evolution und Faunengeschichte im Kaenozoikum (Berlin, 1983).

EMIL KUHN-SCHNYDER

DIRAC, PAUL ADRIEN MAURICE (*b.* Bristol, England, 8 August 1902; *d.* Miami, Florida, 20 October 1984), *quantum mechanics, relativity, cosmology.*

Dirac was one of the greatest theoretical physicists in the twentieth century. He is best known for his important and elegant contributions to the formulation of quantum mechanics; for his quantum theory of the emission and absorption of radiation, which inaugurated quantum electrodynamics; for his relativistic equation of the electron; for his "prediction" of the positron and of antimatter; and for his "large number hypothesis" in cosmology. Present expositions of quantum mechanics largely rely on his masterpiece *The Principles of Quantum Mechanics* (1930), and a great part of the basic theoretical framework of modern particle physics originated in his early attempts at combining quanta and relativity. Not only his results but also his methods influenced the way much of theoretical physics is done today, extending or improving the mathematical formalism before looking for its systematic interpretation.

Dirac spent most of his academic career at Cambridge and received all the honors to which a British physicist may reasonably aspire. He became a fellow of St. John's College at the age of twenty-five, a fellow of the Royal Society in 1930, Lucasian professor of mathematics in 1932, a Nobel laureate in 1933 for his "discovery of new fertile forms of the theory of atoms and for its applications," a Royal Medalist in 1939, and a Copley Medalist in 1952. He was frequently invited to lecture or to do research abroad. For instance, he traveled around the world in 1929, visited the Soviet Union several times in the 1930's, and was a fellow at the Institute for Advanced Studies, Princeton, in the years 1947–1948 and 1958–1959. In 1973 he was made a member of the Order of Merit. Dirac retired in 1969 but resumed his scientific career in 1971 at Florida State University. In January 1937 Dirac married Margit Wigner, the sister of Eugene Wigner; they had two daughters.

Dirac made his mark through his scientific writings. He had few students: the fundamental problems that he tackled were not for beginners. Unlike many of his colleagues, he was little involved in war projects.

Bristol. Dirac's mother, Florence Hannah Holten, was British; his father, Charles Adrien Ladislas Dirac, was an émigré from French Switzerland. His father did not receive friends at home and forced Paul to silence by imposing French as the language spoken at the dinner table. From childhood Dirac was a loner, enjoying the contemplation of nature, long walks, or gardening more than social life. He was not much inclined to collaboration and did his best thinking by himself. At the Merchant Venturer's Technical College, where his father taught French, he excelled in science and mathematics, and neglected literary and artistic subjects.

From 1918 to 1921 Dirac trained to be an electrical engineer at Bristol University. This background, he explained later, strongly influenced his way of doing physics: he learned how to tolerate approximations when trying to describe the physical world and how to solve problems step by step. He also developed a nonrigorous constructive conception of mathematics, beautifully articulating symbols before precisely defining them, very much as the British physicist Oliver Heaviside did in his calculus.

In 1921 the postwar economic depression prevented Dirac from finding a job, so he accepted two years of free tuition from the mathematics department at Bristol. During this period he was influenced by an outstanding professor of mathematics, Peter Fraser, who convinced him that rigor was sometimes useful and imparted to him his love for projective geometry, with its derivations of complicated theorems by means of simple one-to-one correspondences.

At Bristol, Dirac also attended Charlie Dunbar Broad's philosophy course for students of science, in which Broad criticized the fundamental concepts of science on the basis of Alfred North Whitehead's principle of extensive abstraction and argued that the ideal objects of mathematics must be constructed from the mutual relations—not the inner structure—of the roughly perceived objects of nature. This genesis was supposed to explain the relevance of geometrical concepts when they were applied to the physical world, particularly the success of Einstein's theory of relativity. For Broad, theorists were best when they were their own philosophers. Dirac also read John Stuart Mill's *System of Logic* (1843), but derived the opposite conclusion: that philosophy was "just a way to think about discoveries already made."

Broad's lectures included a serious account of the theory of relativity, which immediately fascinated Dirac. Arthur S. Eddington's *Space, Time and Gravitation* (1920), written in the euphoric period after the British eclipse expedition confirming Albert Einstein's theory in 1919, made a further impression on Dirac. Evidence of epistemological comments by "the fountainhead of relativity in England" can be found in several places in Dirac's work.

Cambridge. In the fall of 1923, Dirac entered St. John's College, Cambridge, as a research student, thanks to an 1851 Exhibition studentship and a grant from the department of scientific and industrial research for work in advanced mathematics. He hoped to study relativity with Ebenezer Cunningham, but was assigned Ralph Fowler as his adviser. Fowler was not only a preeminent specialist in statistical mechanics but also the enthusiastic leader of quantum theoretical research at Cambridge. As a correspondent of Niels Bohr, he regularly got information about the latest advances or failures in atomic theory. As the son-in-law of Ernest Rutherford, he took a strong interest in the experimental work at the Cavendish Laboratory (at Cambridge, theoretical physics was part of the Faculty of Mathematics).

Because of his retiring personality and the relative isolation of the various colleges, Dirac did not have any regular scientific interlocutor but Fowler. To compensate, he joined two physicists' clubs, the $\nabla^2 V$ Club and the more casual Kapitza Club, where theorists and experimenters discussed recent problems and welcomed foreign visitors. He also attended the colloquia at the Cavendish and, to keep up with developments in fundamental mathematics, took part in the tea parties of the distinguished Cambridge mathematician Henry Frederick Baker, who was concerned primarily with projective geometry.

Before arriving in Cambridge, Dirac did not know about the Bohr atom. This gap in his knowledge was quickly and excellently filled by Fowler's detailed lectures. Dirac also read Arnold Sommerfeld's textbook *Atomic Structure and Spectral Lines* (English ed., 1923), Bohr's *On the Application of the Quantum Theory to Atomic Structure* (1923), and Max Born's *Vorlesungen über Atommechanik* (1925). These three fundamental texts involved advanced techniques of Hamiltonian dynamics (to derive the most general expression of the rules of quantization), which Dirac learned from Edmund T. Whittaker's standard text, *A Treatise on the Analytical Dynamics of Particles and Rigid Bodies* (1904). Perhaps more than anything in quantum theory he enjoyed reading Eddington's *Mathematical Theory of Relativity* (1923), which developed the tensor apparatus of Einstein's and Hermann Weyl's theories of gravitation. They became his models of beauty in mathematical physics.

Fowler was quick to detect the qualities of his new student and began to encourage his originality. Only six months after arriving in Cambridge, Dirac started to publish substantial research papers. Whenever his subject had not been imposed by Fowler, he tried to clarify and to generalize in a relativistic way points that he had found obscure in his readings—for instance, the definition of a particle's speed according to Eddington, or the covariance of Bohr's frequency condition, or the expression of the collision probability in the then-fashionable "detailed balancing" calculations. The main characteristics of Dirac's style showed through in this early work: directness, economy in mathematical notation, and little reference to past work.

At the end of 1924, following suggestions by Fowler and Darwin, Dirac focused on the more fundamental problem of generalizing the application of Paul Ehrenfest's adiabatic principle in quantum theory. According to this principle, the quantum conditions for a complicated system could be obtained by infinitely slow ("adiabatic") deformation of a simpler system for which one knew to which variables q the Bohr-Sommerfeld rule $\int p dq = nh$ applied.

Another method, introduced by Karl Schwartzschild and systematized by Johannes Burgers, applied to the so-called multiperiodic systems, the configuration of which can be expressed in terms of s periodic functions with s incommensurable frequencies $\omega_1, \omega_2, \ldots \omega_\alpha, \ldots \omega_s$. One had only to introduce the "angle" variables $w_\alpha = \omega_\omega t$ and the corresponding Hamiltonian conjugates, the "action" variables J_α. In the nondegenerate case for which s is also the number of degrees of freedom, the quantum conditions can simply be written $J_\alpha = n_\alpha \hbar$, where $2\pi\hbar$ is Planck's constant. Burgers showed that this procedure was equivalent to the adiabatic principle because the J's are adiabatic invariants. Dirac increased both the rigor of the demonstration and its scope, including magnetic fields and degeneracy. He also tried to remove the restriction of multiperiodicity and to calculate the energy levels of the helium atom, but he failed. Presumably he believed that a good part of the difficulties of quantum theory could be solved by extension of the adiabatic principle without facing the basic paradoxes emphasized by Bohr and the Göttingen school. Bohr's correspondence principle did not trigger Dirac's interest as a hint toward a fundamentally new quantum mechanics. His only consideration of it was purely operational, as a set of rules to derive intensities of emitted radiation in the action-angle formalism.

Commutators and Poisson Brackets. In 1925 Bohr and Werner Heisenberg both brought their revolutionary spirit to Cambridge. Bohr lectured in May after being distressed by the results of Walther Bothe and Hans Geiger's experiment confirming the light-quantum explanation of the Compton effect and making the paradoxical features of light more obvious than ever. According to Bohr, Pauli, and Born, the crisis in quantum theory had reached its climax. The world needed a new mechanics that would preserve the quantum postulates and agree asymptotically with classical mechanics. Heisenberg came to Cambridge in July 1925 with what soon proved to meet this expectation. He lectured at the Kapitza Club, on "term zoology and Zeeman botanics"—that is, on his latest theory of spectral multiplets and anomalous Zeeman effects. It is not known how much of this talk dealt with more recent ideas, nor if Dirac in fact attended it. Fowler certainly heard of Heisenberg's brand-new "quantum kinematics" in private conversations, and asked to be kept informed.

In late August or early September, Fowler gave Dirac the proof sheets of Heisenberg's fundamental paper, "A Quantum-Theoretical Reinterpretation (*Umdeutung*) of Kinematics and Mechanical Relations." Heisenberg had replaced the position x of an electron by an array $x_{nm}e^{i(E_n - E_m)t/\hbar}$ representing the amplitudes of virtual oscillators directly giving the observable properties of scattered or emitted radiation corresponding to the energy levels E_m and E_n. To keep the new kinematics as analogous as possible to the classical one, he guessed the multiplication law of two arrays x_{nm} and y_{nm} from the corresponding rule for the Fourier coefficients of x and y, and obtained $(xy)_{nl} = \sum_m x_{nm}y_{ml}$. In the same way he guessed the quantum version of the quantization rule $\int pdq = nh$ as $\sum_m |q_{nm}|^2(E_m - E_n) = \hbar^2/2\mu$ (μ being the electron mass). The dynamics—the equation of evolution for x—was taken over from classical dynamics. At that point the most advanced quantum problem that Heisenberg could solve was the weakly anharmonic oscillator. The "essential difficulty," he noticed, was the fact that, according to the new multiplication rule, $xy \neq yx$.

Since there was no familiar Hamitonian formalism in Heisenberg's paper, it was about ten days before Dirac realized that the new multiplication law might solve the difficulties of quantum theory. He first looked for a relativistic generalization of Heisenberg's scheme, but this proved premature. More successfully, he tried to connect it to a Hamiltonian formalism. The difference $xy - yx$, once evaluated for high quantum numbers and in terms of action-angle variables J and w, gave $i\hbar \sum_\alpha \frac{\partial x}{\partial w_\alpha}\frac{\partial y}{\partial J_\alpha} - \frac{\partial y}{\partial w_\alpha}\frac{\partial x}{\partial J_\alpha}$, that is, the classical Poisson bracket $\{x, y\}$ times $i\hbar$. In other words, Heisenberg's strange noncommutativity had a classical counterpart in the Poisson-bracket algebra of Hamiltonian mechanics. Dirac then assumed that the relation $xy - yx = i\hbar\{x, y\}$ held in general (far from the classical limit and for nonmultiperiodic systems) and provided the proper quantum conditions. For canonically conjugate variables p and q it reduced to $qp - pq = i\hbar$, containing Heisenberg's quantization rule.

Dirac was very pleased with this close analogy between classical and quantum mechanics because it allowed him to retain the "beauty" of classical mechanics and to transfer Hamiltonian techniques to quantum mechanics. Hence he could develop very quickly a version of quantum mechanics more elegant than that developed at Göttingen.

q-Numbers. The identity between commutator and Poisson brackets led to the fundamental equations $i\hbar g = gH - Hg$ (for any dynamical variable g evolving with the Hamiltonian H) and $qp - pq = i\hbar$ (for any canonical couple), determining the formalism of quantum mechanics. Dirac thought that Heisenberg's interpretation of the quantum variables in terms of matrices giving the observable properties of radiation was provisional and too restrictive; he preferred a symbolic approach, developing the algebra of abstract undefined "q-numbers" and looking only later for those numbers' representation in terms of observable (ordinary) "c-numbers." The domain of q-numbers had to be extensible, adapting to the further progress of the theory. Some of the axiomatic properties that Dirac imposed on them—for instance, the unicity of the square root and no divisor of zero—had to be dropped later because they cannot be realized in an algebra of operators. Dirac's idea of q-numbers and his axioms for them most probably originated at Baker's tea parties. In Baker's *Principles of Geometry* there is an abstract noncommutative algebra of coefficients for linear combinations of points, which permitted elegant and condensed proofs of theorems in projective geometry (where noncommutativity means dropping Pappus's theorem).

In the case of multiperiodic systems, Dirac could show that his fundamental equations were satisfied by an algebra of matrices with rows and columns corresponding to integral values (times h) of the action variables J. In this representation the energy matrix is diagonal, which suggests that the diagonal elements represent the spectrum of the system. Through a correspondence argument Dirac identified the matrix element $x_{J'J''}$ of the electric polarization with the amplitude of the corresponding transition $J' \rightarrow J''$, in accordance with Heisenberg's original definition of the position matrix. In this representation Dirac could solve the hydrogen atom in early 1926 (a little later than Wolfgang Pauli, but independently). Within a few months he also found the basic commutation and composition rules for angular momentum in multielectron atoms, and he made the first relativistic quantum-mechanical calculation giving the characteristics of Compton scattering. Physicists in Copenhagen were impressed by this achievement, the more so because Dirac treated the field classically, without light quanta.

Dirac assembled all these bright results in his doctoral dissertation, completed in June 1926. At that time he had solved by himself about as many quantum problems as the entire Göttingen group together. In principle his q-numbers were more general and more flexible than the Göttingen matrices, which were rigidly connected to a priori observable quantities. But Dirac had been able to solve the quantum equations only insofar as action-angle variables could be introduced into the corresponding classical problem. To proceed further, he needed a new method of finding representations of q-numbers. That is exactly what Erwin Schrödinger made available in a series of papers submitted for publication between January and June 1926.

The Impact of Schrödinger's Equation. Dirac's first reaction to Schrödinger's equation was negative: Why a second quantum mechanics, since there already was one? Why propose that matter waves were analogous to light waves, since the properties of light waves were already so paradoxical? In a letter written on 26 May 1926, Heisenberg convinced him that Schrödinger's equation $H\left(q, -i\hbar\frac{\partial}{\partial q}\right)\psi_n = E_n\psi_n$ (for one degree of freedom) provided a simple and general method to calculate the matrix elements of a general function F of p and q just by forming the integrals $F_{mn} = \int\psi_m^*(q)F\left(q, -i\hbar\frac{\partial}{\partial q}\right)\psi_n(q)\ dq$. Then, in an astonishingly short time, Dirac accumulated new essential results. The time dependence of the matrix elements could be supplied by the equation $H\psi = i\hbar\partial\psi/\partial t$, suggested by the relativistic substitution $p_\mu \rightarrow i\hbar\partial/\partial x_\mu$. A set of identical particles, following Heisenberg's idea of eliminating unobservable differences from the formalism, had to be represented by either symmetric or antisymmetric wave functions in configuration space, the first corresponding to the Bose-Einstein statistics and the second to Pauli's exclusion principle. Finally, Dirac developed the time-dependent perturbation theory to calculate Einstein's B coefficients of absorption and stimulated emission. He also improved his calculation of the Compton effect. To reach these physical results he did not subscribe to Schrödinger's picture of $|\psi|^2$ as a density of electricity; instead he relied on Heisenberg's interpretation of the polarization matrix or on Born's statistical interpretation of the ψ function.

Interpretation of Quantum Dynamics. Dirac was not satisfied by the provisional and parochial assumptions made to interpret q-numbers and the quantum formalism: according to Heisenberg, the diagonal elements of H and the elements of the polarization matrix had an immediate meaning; according to a paper by Born (June 1926), the coefficients c_n in the development $\psi = \sum_n c_n\psi_n$ over the set of eigenfunctions ψ_n gave the probability $|c_n|^2$ for the system to be in the state n; and, according to Schrödinger's fourth memoir (June 1926), $|\psi|^2$ was "a sort of weight function in configuration space." In Dirac's view, a general interpretation should be based on a transformation theory, as in the theory of relativity (and as emphasized by Eddington).

To arrive at the interpretation, Dirac first worked out the transformations connecting the various matrix representations of his fundamental equations $qp - pq = i\hbar$ and $i\hbar\dot{g} = gH - Hg$. He called ξ and α two maximal sets of commuting q-numbers; ξ' and α', corresponding eigenvalues; and (ξ'/α'), the transformation from the representation where ξ is diagonal to the one where α is diagonal, acting on the representation $g_{\xi'\xi''}$ of g according to $g_{\alpha'\alpha''} = \int(\alpha'/\xi')g_{\xi'\xi''}(\xi''/\alpha'')\ d\xi'\ d\xi''$. In this framework the solutions of the (time-independent) Schrödinger equation were nothing but a particular transformation for which α contains H and ξ contains the position. The notations, introduced for the sake of economy and in obvious analogy to tensor notation, proved to be extremely convenient and spread widely, especially after their later improvement (1939) into the "bra-ket" (or "bra" and "ket") notation. In fact, the symbolic rules were better defined than the mathematical substratum, which was made clear only much later by mathematicians. For instance, the treatment of continuous spectra on the same footing as the discrete ones necessitated singular "δ-functions" (as in $(x'/x'') = \delta(x' - x'')$, perceived by Dirac as limits of sharply peaked functions but raised today to the rank of Schwartz distributions.

To interpret his transformations, Dirac needed only a minimal assumption suggested by the correspondence principle: that for an arbitrary physical quantity g expressed in terms of ξ and the canonical conjugate η, $g_{\xi'\xi'}$ signifies the average of the corresponding classical g for $\xi = \xi'$ and η uniformly distributed. From $\delta(g - g')|_{\xi'\xi'} = |(\xi'/g')|^2$ it follows that $|(\xi'/g')|^2\ dg'$ is proportional to the probability that g is equal to g' within dg' when $\xi = \xi'$. Dirac finished this transformation theory in November 1926 at Copenhagen.

In Göttingen, Pascual Jordan obtained roughly the same results at the same time, though from a different point of view. He defined axiomatically a concept of canonical conjugation at the quantum level and looked for the transformations $(\xi, \eta) \rightarrow (\alpha, \beta)$ from one canonical couple to another. In this more general framework the quantum variables did not necessarily have a classical counterpart, and conjugation did not necessarily correspond to Poisson-bracket conjugation. In other words, Dirac's transformation theory was more constraining than Jordan's, and gave more precise directions for the future extensions of quantum mechanics.

Dirac was also original in his conception of the role of probability in quantum mechanics. He thought that probabilities entered into the description of quantum phenomena only in the determination of the initial state (still described in terms of p's and q's), and not necessarily in the behavior of an isolated system. But, as Bohr had said at the Solvay Conference in 1927, isolated systems were unobservable. Dirac then assumed that the state of the world was represented by its wave function ψ and that it changed abruptly during a measurement, whereupon "nature made a choice."

Dirac retained his basic machinery of transformations in his subsequent lectures on quantum mechanics, but he introduced a substantial change in his fundamental textbook, *The Principles of Quantum Mechanics* (1930). In the original exposition of transformation theory, he had carefully avoided the concept of quantum state, presumably to depart from Schrödinger's idea of ψ as a state. In his *Principles,* however, he presented the principle of superposition and the related concept of space of states as capturing the most essential feature of quantum theory: the interference of probabilities. It seems plausible that this move was inspired by Bohr's insistence on the superposition principle and by John von Neumann's and Hermann Weyl's formulations of quantum mechanics, in which Hilbert spaces played a central role. From this perspective transformations were just a change of base in the space of states. The correspondence with Hamiltonian formalism appeared only in a later chapter of the book.

A New Radiation Theory. Dirac liked his transformation theory because it was the outcome of a planned line of research and not a fortuitous discovery. He forced his future investigations to fit it. The first results of this strategy were almost miraculous. First came his new radiation theory, in February 1927, which quantized for the first time James Clerk Maxwell's radiation in interaction with atoms. Previous quantum-mechanical studies of radiation problems, except for Jordan's unpopular attempt, retained purely classical fields. In late 1925 Jordan had applied Heisenberg's rules of quantization to continuous free fields and obtained a light-quantum structure with the expected statistics (Bose-Einstein) and dual fluctuation properties. Dirac further demonstrated that spontaneous emission and its characteristics—previously taken into account only by special postulates—followed from the interaction between atoms and the quantum field. Essential to this success was the fact that Dirac's transformation theory eliminated from the interpretation of the quantum formalism every reference to classical emitted radiation, contrary to Heisenberg's original point of view and also to Schrödinger's concept of ψ as a classical source of field.

This work was done during Dirac's visit to Copenhagen in the winter of 1927. Presumably to please Bohr, who insisted on wave-particle duality and equality, Dirac opposed the "corpuscular point of view" to the quantized electromagnetic "wave point of view." He started with a set of massless Bose particles described by symmetric ψ waves in configuration space. As he discovered by "playing with the equations," this description was equivalent to a quantized Schrödinger equation in the space of one particle; this "second quantization" was already known to Jordan, who during 1927 extended it into the basic modern quantum field representation of matter. Dirac limited his use of second quantization electromagnetic to radiation: to establish that the corpuscular point of view, once brought into this form, was equivalent to the wave point of view.

The Dirac Equation. An even more astonishing fruit of Dirac's transformation theory was his relativistic equation of the electron. He and many other theorists had already made use of the most obvious candidate for such an equation—$(\hbar^2\partial_\mu\partial^\mu + m^2)\psi = 0$ (Klein-Gordon)—but it did not include the spin effects necessary to explain atomic spectra. More crucially for Dirac, it could not fit into the transformation theory because it could not be rewritten under the form $i\hbar\partial\psi/\partial t = H\psi$. To be both explicitly relativistic and linear in $\partial/\partial t$, the new equation had to take the form $(i\hbar\gamma^\mu\partial_\mu - m)\psi = 0$ or, more explicitly, $i\hbar\partial\psi/\partial t = \vec{\alpha} \cdot \vec{p} + \beta m$. For the spectrum to be limited to values satisfying Einstein's relation $E^2 = p^2 + m^2$, the coefficients $\vec{\alpha}$ and β had to be such that $(\vec{\alpha} \cdot \vec{p} + \beta m)^2 = p^2 + m^2$—that is, $\beta^2 = 1$, $\alpha_i\alpha_j + \alpha_j\alpha_i = 2\ \delta_{ij}$, and $\alpha_i\beta + \beta\alpha_i = 0$.

The simplest entities satisfying these relations are 4×4 matrices, as Dirac noted with the help of Pauli's $\vec{\sigma}$ matrices (such that $[\vec{\sigma} \cdot \vec{p}]^2 = p^2$). Surprisingly, the new equation included spin effects, the value 2 of the gyromagnetic factor, and the correct fine structure formula (Sommerfeld's), as worked out approximately by Dirac and exactly by Darwin and Walter Gordon. Other theorists (Pauli, Darwin, Jordan, Hendrik, Kramers) had been searching for a wave equation integrating spin and relativistic effects, but they all started by assuming the existence of spin, either as an intrinsic particle rotation or as a wave polarization. In contrast, the key to Dirac's success was his persistent adherence to the simplest classical model, the point-electron, as a basis for quantization. Spin effects, as might have been expected from their involving h, were a consequence of relativistic quantization.

Antimatter, Monopoles. The Dirac equation played an essential role not only in atomic physics but also in high-energy physics, through the Klein-Nishina and Møller formulas describing the absorption of relativistic particles in matter. Nevertheless, it presented several strange features that enhanced the "magic" of Dirac's work: a new type of relativistic covariance involving the spinor representations of the Lorentz group, soon elucidated by Göttingen mathematicians; the trembling of the electron imagined by Schrödinger to harmonize the observed electron speed and the expectation value c of the speed operator from Dirac's equation; and, above all, the negative-energy difficulty.

The equation $E^2 = p^2 + m^2$, applying to the spectrum of free Dirac electrons, has two roots: $E = \pm(p^2 + m^2)^{1/2}$; therefore a Dirac electron with an initially positive energy should fall indefinitely by spontaneous emission toward states of lower and lower energy. To avoid this, Dirac imagined in late 1929 that the states of negative energy were normally filled up according to the exclusion principle and that holes in this "sea" would represent protons. If this were true, Dirac had in hand a grandiose unification of the particle physics of his time. But he still had to explain the ratio m_p/m_e between the proton mass and the electron mass. He thought that the disparity in mass might originate in the mutual interaction between the "sea" electrons. The precise numerical value of the ratio would perhaps appear at the same time as the other dimensionless constant, $e^2/4\pi\hbar c$, as suggested by Eddington in a 1928 paper containing a mysterious derivation of this remarkable number.

Eddington believed that electromagnetic interactions could be reduced to the "exchange" interactions, the change of sign of a wave function owing to the permutation of two fermions or to a full rotation being of the same nature as the change of phase following an electromagnetic gauge transformation. At the end of his speculation, he got $e^2/4\pi\hbar c = 1/136$. In his search for a theoretical derivation of m_p/m_e and $e^2/4\pi\hbar c$, Dirac also concentrated on the phase of the wave function.

It is usually assumed that the phase of a wave function is unambiguously defined in space (for a given gauge). But a multivalued phase is also admissible, Dirac noted, as long as the variation of phase around a closed loop is the same for any wave function (to preserve the regular statistical interpretation of ψ based on quantities such as $|\int\psi_1^*(\vec{r})\psi_2(\vec{r})\ d^3r|$). To ensure the continuity of ψ, the variation of phase around an infinitesimal closed loop can only be a multiple of 2π. This determines lines of singularities starting from (gauge) invariant singular points. Now, following the relation between electromagnetic potential and phase implied by gauge invariance, the singular points must be identified with magnetic monopoles carrying the charge $g = n\hbar c/2e$. If there is only one monopole g in nature, every electric charge must be a multiple of $\hbar c/2g$. Dirac always considered this explanation of the quantization of charge in nature as the strongest argument in favor of monopoles.

Unfortunately, no other restriction on e followed from this line of reasoning and Dirac missed his targets, the derivation of $e^2/4\pi\hbar c$ and a subsequent determination of m_p/m_e. But the latter was no longer needed: in 1931 he learned from Weyl that, due to charge conjugation symmetry, the holes in his "sea" theory necessarily carried the charge $-e$. In the same year and in a single paper, he proclaimed the necessity of antielectrons (and also antiprotons) and the possibility of monopoles, and pondered the most efficient method of advance in theoretical physics. As in his quantum-theoretical work, he had first to work out the formalism in terms of abstract symbols denoting states and observables, and next to investigate the symbols' interpretation. This was, Dirac said, "like Eddington's principle of identification," according to which the interpretation of the fundamental tensors of general relativity came after their mathematical justification.

Dirac gave a full quantum-mechanical treatment of his monopoles in 1948 with the help of "nonphysical strings" allowing a Hamiltonian formulation. More recently monopoles have been shown to be necessary in any non-Abelian gauge theory, including electromagnetic interactions. But no experimental evidence has yet been found. On the other hand, the antielectron (or positron) was discovered by Carl Anderson and Patrick Blackett in the years 1932–1933, much earlier than foreseen by Dirac, although its concept faced the general prejudice against a charge-symmetric nature.

The Multitime Theory. After the discovery of the positron, most theorists agreed that the negative-energy difficulty was solved by the "sea" concept. Another fundamental difficulty, also rooted in one of Dirac's early works, his radiation theory, lasted much longer. In 1929, when working out their version of quantum electrodynamics, Heisenberg and Pauli discovered that the second order of approximation involved infinite terms, even when it was related to physical phenomena such as level shifts in atoms. The difficulty looked so serious that in the first edition of his *Principles*, Dirac omitted the quantization of the electromagnetic field and presented only the light-quantum configuration-space approach.

In 1932 Dirac tried to start a new revolution by giving up (for electrodynamics) the most basic requirement of his quantum-mechanical work: the Hamiltonian structure of dynamical equations. Imitating Heisenberg's revolutionary breakthrough, he declared that the new theory should eliminate unobservable things like the electromagnetic fields during the interaction process, and focus on their asymptotic values before and after the interaction. The electromagnetic field, he said, was nothing but a means of observation, and therefore should not be submitted to Hamiltonian treatment. On these lines he derived a set of equations that apparently were quite new; in fact, as Leon Rosenfeld soon pointed out, it differed from the theory of Heisenberg and Pauli only in the use of the interaction representation (for which the quantum fields evolve as free fields) and of a multitime configuration space for electrons (instead of Jordan's quantized waves). Nonetheless, once it had been improved with the help of Vladïmir A. Fock and Boris Podolsky, Dirac's formulation had the great advantage of being explicitly covariant, a feature particularly attractive to the Japanese quantum-field theorists Hideki Yukawa and Sin-Itiro Tomonaga.

The Large-Number Hypothesis. The infinities were still there. The discovery of the positron in 1932 gave some hope that the deformations of Dirac's "sea," the "vacuum polarization," would cure them. But such was not the case (although Wendell Furry and Victor Weisskopf made the infinities "smaller"), and Dirac himself judged the "sea" theory ugly. In 1936, depressed by this state of affairs, he hastily concluded from some experimental results of Robert S. Shankland that the energy principle should be given up in relativistic quantum theory. Needing some diversion, he turned to cosmological speculation following Eddington, who believed in a grand unification of atomic physics and cosmology. Dirac also knew Edward A. Milne, the other famous Cambridge cosmologist, who had been his supervisor for a term in 1925, and he had made friends with the American astronomer Howard P. Robertson, who believed in the expansion of the universe, during a short stay at Göttingen in 1927.

Like Eddington, Dirac focused on dimensionless numbers built from the fundamental constants of both atomic and cosmic phenomena; and he observed that there was a cluster of these numbers around 10^{39}, including the age of the universe in atomic time units and the ratio of electric forces to gravitational ones inside atoms. In 1937 he proposed the "large-number hypothesis," according to which numbers in the same cluster should be simply related. Consequently, the gravitation constant had to vary in time, as in Milne's cosmology and contrary to general relativity.

Milne believed in an "extended principle of relativity," which stipulated that the universe should look the same from wherever it is observed, and completed it by the stricture that the cosmological theory should not include any constant having dimensions. To elaborate his own cosmology further, Dirac provisionally adopted Milne's first principle (he rejected it later, in 1939) but replaced the second hypothesis—which conflicted with Eddington's idea that atomic constants should play a role in cosmology—with his large-number hypothesis. As a result the spiral nebulas (then the furthest objects known, from whose behavior Edwin Hubble had deduced his recession law) had to recede in time according to $t^{1/3}$, and the curvature of the three-dimensional space had to be zero. Relativity did not enter these reasonings; Dirac expected it to play only a subsidiary role in cosmology, since Hubble's law provided a natural speed at any point of space—and therefore a natural time axis. To reconcile this position with his admiration for Einstein's theory of gravitation, Dirac introduced two different metrics for atomic and cosmic phenomena. Only the second one was ruled by Einstein's theory; the first one varied in time according to the large-number hypothesis.

Cosmology was not just a hobby for Dirac. Rather, as he explained in 1939, it embodied his notion of progress in physics—an ever increasing mathematization of the world. In the old mechanistic conception, the equations of motion were mathematical but the initial conditions were given by observation. In the new cosmology the state preceding the initial explosion (posited by Georges Lemaître) was so simple that any complexity in nature pertained to the mathematical evolution. In this context Dirac even expressed the hope that the history of the universe would be only a history of the properties of numbers from 1 to 10^{39}. From the 1970's to the end of his life he often came back to his cosmological ideas. His large-number hypothesis has been seriously considered by several astrophysicists in spite of its speculative character.

Classical Point Electron, Indefinite Metrics. The rest of Dirac's work, from the 1930's on, centered on quantum electrodynamics. Dirac remained true to the research method that he had developed in his early work. He never reached his ultimate aim, a mathematically clean theory, but left interesting by-products of his quest. All the creators of quantum mechanics attempted to deal with the disease of infinite self-energy. One possibility they discussed was a revision of the correspondence basis, the classical theory of electrodynamics, which already involved either ambiguities (dependence on the structure of a finite electron) or infinite self-energy (for point electrons). In 1938 Dirac created a finite theory of point electrons by a convenient "reinterpretation" of the Maxwell-Lorentz equations that canceled the infinite self-mass. In spite of its formal beauty, this theory involved unphysical "runaway" solutions (spontaneously accelerating electrons) that could be eliminated (at the classical level) only at the price of making supraluminal signals possible. Not fully conscious of the latter difficulty, Dirac brought his equations to the Hamiltonian form and quantized them. Unfortunately, only half the divergent integrals of quantum electrodynamics were cured by this procedure. To take care of the other half, Dirac imagined in 1942 a nonpositive (he called it "indefinite") metric in Hilbert space that allowed a new natural representation of the field commutation rules but implied negative probabilities difficult to interpret physically. Pauli admired the new formalism but criticized Dirac's artificial interpretation of it, which involved a "hypothetical world" initially (before collisions occur in the real world) empty of photons and filled up with positrons (to dry out the sea).

In 1946 Dirac realized that his new equations allowed a finite nonperturbative solution; in addition, they could be connected with the regular formalism (with only positive probabilities) by a change of representation, that is, a unitary transformation in Hilbert space. Although not able to explicate this transformation (which presumably would reintroduce infinities), Dirac concluded in 1946 that the difficulties of quantum electrodynamics were purely mathematical. During the next year other theorists realized that the difficulties were connected instead with a proper definition of physical parameters like charge and mass. Nonetheless, the indefinite metric proved to be indispensable in quantum field theory for another reason: a covariant quantization of Maxwell's field requires the introduction of (unobservable) states of negative probability. In the 1960's several theorists, including Heisenberg, also developed Dirac's idea of a finite quantum electrodynamics with indefinite metrics.

Relativistic Ether, Strings. Developed by other physicists in 1947, renormalization, a way to absorb infinities in a proper redefinition of mass and charge, allowed very successful calculations of higher-order corrections to atomic and electrodynamical processes. From this resulted the best numerical agreement ever encountered between a fundamental theory and experiment. Always more concerned with internal beauty than with experimental verdict, Dirac called it a "fluke" and kept searching for a closed quantum electrodynamics purged of infinities at every stage of calculation. His point of view quickly became heterodox as more and more theorists thought that quantum electrodynamics did not have to exist by itself, but only as a part of a more general theory encompassing other types of interactions. As if to stress his originality, Dirac did not show any interest in the growing but messy field of nuclear and particle physics.

Some of Dirac's late attempts at a new quantum electrodynamics brought fundamentally new ideas. For instance, in 1951 he resurrected ether, arguing that quantum theory allowed a Lorentz invariant notion of ether for which all drift speeds at a given point of space-time are equiprobable, in analogy with the *S* states of the hydrogen atom, which are invariant by rotation although the underlying classical model is not. The idea had come to him after the proposal of a new electrodynamics for which the potential is restricted by $A_\mu A^\mu = k^2$, which suggests a natural ether speed $v_\mu = k^{-1}A_\mu$, even in the absence of matter.

In 1955 Dirac proposed strings as the basic representation of quantum electrodynamics, a photon corresponding to a closed string and an electron corresponding to the extremity of an open string. Originally suggested by a manifestly gauge-invariant formulation of quantum electrodynamics in which the electron is explicitly dragging an electromagnetic field with it, this picture "made inconceivable the things we do not want to have," for instance, a physically meaningless "bare" electron.

The Lagrangian in Quantum Mechanics. None of the above-mentioned attempts questioned the basic frame of quantum mechanics that Dirac had established in his younger years. But all through his scientific career he looked for alternative or more general formulations of quantum mechanics that might be more suitable for relativistic applications. Some of the products of this kind of exploration proved to be of essential importance. For instance, in 1933, exploiting a relation discovered by Jordan between quantum canonical transformations and the corresponding classical generating functions, he found that the transformation (q_{t+T}/q_t) from q taken at time t to q taken at time $t + T$ "corresponded" to $\exp i\int_t^{t+T} L(q, \dot{q})\, dt$, where L denotes the Lagrangian and $q(t)$, the classical motion between q_t and q_{t+T}. In the same paper he introduced the "generalized transformation functions," substituting the covariant motion of timelike surface of measurement for the usual hyperplanes "t = constant" in four-dimensional space. The remark about the Lagrangian, generalized by Dirac himself in 1945 to provide the amplitude of probability of a trajectory, inspired Richard Feynman in his discovery of the "Feynman-integrals," now the most efficient method of quantization. The "general transformation function" was adopted by the Japanese school to suggest, in combination with Dirac's multitime theory, Tomonaga's manifestly covariant formulation of quantum electrodynamics (1943).

The Role of Mathematics. Dirac believed in a "mathematical quality of nature." In the ideal physical theory, the whole of the description of the universe would have its mathematical counterpart. Conversely, he claimed around 1924, at one of Baker's tea parties, that any really interesting mathematical theory should find an application in the physical world. After sufficient progress, the field of mathematics would be purified and reduced to applied mathematics, that is, theoretical physics. The foundation of this belief in an asymptotic convergence of mathematics and physics is not easy to trace in Dirac's writings, since he generally avoided philosophical discussion. What could be said mathematically was clear enough to him, and he did not require, as most philosopher-physicists would, a recourse to common language to improve understanding. When circumstances compelled him to epistemological statements—for instance, in the foreword to his *Principles*—he simply borrowed them from physicist-philosophers who were "right by definition": Bohr and Eddington. From both these masters he took the rejection of mental pictures in space-time of the old physics. From Eddington he had the idea of a "nonpicturable substratum" and the recognition, through the development and justification of transformation theories, of "the part played by the observer in himself introducing the regularity that appears in his observations, and the lack of arbitrariness in the ways of nature."

It is doubtful that Dirac regarded these statements as really meaningful. When expressing his personal feelings on the role of mathematics, his leitmotiv was the idea of "mathematical beauty." For him the main reason for the successful appearance of groups of transformations in modern theories was their mathematical beauty, something no more subject to definition than beauty in art, but obvious to the connoisseur. In this perspective the mathematical quality of nature could be just the expression of its beauty. More significantly, Dirac's requirement of beauty materialized into a methodology: one had first to select the most beautiful mathematics and then, following Eddington's "principle of identification," try to connect it to the physical world.

To implement the first stage of this methodology, a more definite notion of beauty is needed. Dirac constantly refers to the museum of his early beautiful mathematical experiences. First comes the magic of projective geometry, exemplifying the power to find surprising relations between picturable mathematical objects through simple, invisible manipulations. Then follows general relativity with the appearance of symmetry transformations, and tensor calculus perceived as a symphony of symbols. At the moment of its introduction, beauty excludes rigor. Exact mathematical meaning comes after a heuristic symbolic stage, as in the introduction of the δ-function or of the q-numbers. It is less difficult, according to Dirac, to find beautiful mathematics than to interpret it in physical terms. Here is perhaps the most creative part of his work, invoking subtle analogies and correspondence with older bits of theories.

On the whole, Dirac's method sounds highly a priori, but he occasionally insisted on the necessity of a proper balance between inductive and deductive methods. A more detailed analysis would also show that where he was the most successful, he always remained securely tied to the empirically solid parts of existing theories.

BIBLIOGRAPHY

I. ORIGINAL WORKS. A list of Dirac's publications is in the biography by Dalitz and Peierls (see below). His main works are "The Fundamental Equations of Quantum Mechanics," in *Proceedings of the Royal Society of London*, **A109** (1925), 642–653; "On the Theory of Quantum Mechanics," *ibid.*, **A112** (1926), 661–677; "The Physical Interpretation of the Quantum Dynamics," *ibid.*, **A113** (1927), 621–641; "The Quantum Theory of Emission and Absorption of Radiation," *ibid.*, **A114** (1927), 243–265; "The Quantum Theory of the Electron, I," *ibid.*, **A117** (1928), 610–624; *The Principles of Quantum Mechanics* (Oxford, 1930); "A Theory of Electrons and Protons," in *Proceedings of the Royal Society of London*, **A126** (1930), 360–365; "Quantized Singularities in the Electromagnetic Field," *ibid.*, **A133** (1931), 60–72; and "The Cosmological Constants," in *Nature*, **139** (1937), 323. Nontechnical writings include "The Relation Between Mathematics and Physics," in *Royal Society of Edinburgh, Proceedings*, **59** (1939), 122–129; "The Evolution of the Physicist's Picture of Nature," in *Scientific American*, **208**, no. 5 (1963), 45–53; *The Development of Quantum Theory* (New York, 1971); and "Recollections of an Exciting Era," in Charles Weiner, ed., *History of Twentieth-Century Physics* (New York, 1977), 109–146.

Some of Dirac's papers have been deposited at the Churchill College Archive, Cambridge. Photocopies of manuscripts and letters, and an interview by Thomas S. Kuhn, are available in the Archive for the History of Quantum Physics (Berkeley, Copenhagen, London, New York, Rome).

II. SECONDARY LITERATURE. Joan Bromberg, "The Concept of Particle Creation Before and After Quantum Mechanics," in *Historical Studies in the Physical Sciences*, 7 (1976), 161–183, and "Dirac's Quantum Electrodynamics and the Wave-Particle Equivalence," in Charles Weiner, ed., *History of Twentieth-Century Physics* (New York, 1977), 147–157; Hendrik Casimir, "Paul Dirac, 1902–1984," in *Naturwissenschaftliche Rundschau*, **38** (1985), 219–223; R. H. Dalitz and Sir Rudolf Peierls, "Paul Adrien Maurice Dirac," in *Biographical Memoirs of Fellows of the Royal Society*, **32** (1986), 137–185; Olivier Darrigol, "La genèse du concept de champ quantique," in *Annales de physique*, **9** (1984), 433–501, and "The Origins of Quantized Matter Waves," in *Historical Studies in the Physical and Biological Sciences*, **16**, no. 2 (1986), 197–253; Michelangelo de Maria and Francesco La Teana, "Schrödinger's and Dirac's Unorthodoxy in Quantum Mechanics," in *Fundamenta scientiae*, **3** (1982), 129–148; Norwood R. Hanson, *The Concept of the Positron* (Cambridge, 1963); Max Jammer, *The Conceptual Development of Quantum Mechanics* (New York, 1966); Helge Kragh, "The Genesis of Dirac's Relativistic Theory of Electrons," in *Archive for the History of Exact Sciences*, **24** (1981), 31–67, "The Concept of the Monopole," in *Studies in History and Philosophy of Science*, **12** (1981), 141–172, "Cosmo-physics in the Thirties: Towards a History of Dirac's Cosmology," in *Historical Studies in the Physical Sciences*, **13**, no. 1 (1982), 69–108, and, as editor, *Methodology and Philosophy of Science in Paul Dirac's Physics*, University of Roskilde text no. 27 (Roskilde, 1979); B. N. Kursunoglu and E. P. Wigner, eds., *Reminiscences About a Great Physicist* (Cambridge, 1987); Jagdish Mehra and Helmut Rechenberg, *The Historical Development of Quantum Theory*, IV, *The Fundamental Equations of Quantum Mechanics* (New York, 1982); Donald F. Moyer, "Origins of Dirac's Electron, 1925–1928," in *American Journal of Physics*, **49** (1981), 944–949, "Evaluation of Dirac's Electron," *ibid.*, 1055–1062, and "Vindication of Dirac's Electron," *ibid.*, 1120–1135; Abdus Salam and Eugene P. Wigner, eds., *Aspects of Quantum Theory* (Cambridge, 1972); and J. G. Taylor, ed., *Tribute to Paul Dirac* (Bristol, 1987).

OLIVIER DARRIGOL

DOBZHANSKY, THEODOSIUS (*b*. Nemirov, Ukraine, Russia, 25 January 1900; *d*. Davis, California, 18 December 1975), *genetics, evolution*.

Dobzhansky was the only child of Sophia Voinarsky and of Grigory Dobrzhansky (the precise transliteration of the Russian family name), a teacher of high school mathematics. In 1910 the family moved to the outskirts of Kiev. During his early gymnasium years, Dobzhansky became an avid butterfly collector. In the winter of 1915–1916, he met Victor Luchnik, a twenty-five-year-old college dropout who was a dedicated entomologist specializing in Coccinellidae beetles. Luchnik convinced Dobzhansky that butterfly collecting would not lead anywhere, that he should become a specialist. Dobzhansky chose to work with ladybugs, which were the subject of his first scientific publication (1918).

Before Dobzhansky graduated in biology from the University of Kiev in 1921, he was hired as an instructor in zoology at the Polytechnic Institute in Kiev. He taught there until 1924, when he became an assistant to Yuri Filipchenko, head of the new department of genetics at the University of Leningrad. Filipchenko had started research with *Drosophila* fruitflies, and Dobzhansky was encouraged to investigate the pleiotropic effects (that is, affecting different features of an organism) of genes.

In 1927, Dobzhansky obtained a fellowship from the International Education Board of the Rockefeller Foundation and arrived in New York on December 27 to work with Thomas Hunt Morgan at Columbia University. In 1928 he followed Morgan to the California Institute of Technology, where he was appointed assistant professor of genetics in 1929 and professor of genetics in 1936. Dobzhansky returned to New York in 1940 as professor of zoology at Columbia University, where he remained until 1962, when he became professor at the Rockefeller Institute (now Rockefeller University), also in New York City. In 1970 Dobzhansky became emeritus at Rockefeller University; in September 1971, he moved to the department of genetics at the University of California, Davis, where he was adjunct professor until his death.

On 8 August 1924 Dobzhansky married Natalia (Natasha) Petrovna Sivertzev, a geneticist in her own right, who at the time was working with the biologist Ivan Schmalhausen in Kiev; they had one daughter. Natasha died in 1969.

During a routine medical checkup in 1968, it was discovered that Dobzhansky suffered from chronic lymphatic leukemia, one of the least malignant forms of leukemia. He was given a prognosis of "a few months to a few years" to live. Over the following seven years, the progress of the leukemia was unexpectedly slow and—even more surprising to his physicians—it had little, if any, noticeable effect on his energy and work habits. However, the disease took a conspicuous turn for the worse in the summer of 1975. In mid-November, Dobzhansky started to receive chemotherapy, but continued living at home and working at the laboratory. He was convinced that the end of his life was near and feared that he might become unable to work and to care for himself. This never came to pass. He died of heart failure on 18 December 1975, as he was being rushed to the hospital. The previous day he had, as usual, worked in the laboratory.

Dobzhansky was a religious man, although he apparently rejected fundamental beliefs of traditional religion, such as the existence of a personal God and of life beyond physical death. His religiosity was grounded on the conviction that there is meaning in the universe. He saw that meaning in the fact that evolution has produced the stupendous diversity of the living world and has progressed from primitive forms of life to mankind. Dobzhansky held that, in mankind, biological evolution has transcended itself into the realm of self-awareness and culture. He believed that somehow mankind would eventually evolve to higher levels of harmony and creativity.

Dobzhansky was one of the most influential biologists of the twentieth century; he also was one of the most prolific. The complete list of his publications has nearly six hundred titles, including a dozen books. The gamut of subject matter is enormous: results of experimental research in various biological disciplines, works of synthesis and theory, essays on humanism and philosophy, and others. These diversified works are nevertheless unified—biological evolution is the theme that threads them together. The place of biological evolution in human thought was best expressed, according to Dobzhansky, in a sentence that he sometimes quoted from the Jesuit paleontologist Pierre Teilhard de Chardin: "Evolution is a light which illuminates all facts, a trajectory which all lines of thought must follow—this is what evolution is."

Dobzhansky's prodigious scientific productivity was made possible by incredible energy and very disciplined work habits. His success as the creator of new ideas and as a synthesizer was, at least in part, based on his broad knowledge, his excellent memory, and an incisive mind able to see the relevance that a new discovery or a new theory might have for other theories or problems. His success as an experimentalist depended on a wise blending of field and laboratory research; whenever possible, he combined both in the study of a problem, using laboratory studies to ascertain or to confirm the causal processes involved in the phenomena discovered in nature. He obtained the collaboration of mathematicians to design theoretical models for experimental testing and to analyze his empirical observations statistically. He was no inventor or gadgeteer, but he had an uncanny ability to exploit the possibilities of any suitable experimental apparatus or method.

Dobzhansky was a world traveler and an accomplished linguist, fluent in six languages and able to read several more. He was a good naturalist and never lacked time for a hike, whether in the California Sierras, the New England forests, or the Amazonian jungles. He loved horseback riding but engaged in no other sports. His interests covered the plastic arts, music, history, Russian literature, cultural anthropology, philosophy, religion, and, of course, science.

Dobzhansky recognized and generously praised the achievements of other scientists; he admired the intellect of his colleagues, even when admiration was alloyed with disagreement. He made many lasting friendships, usually started by professional interactions. Many of Dobzhansky's friends were scientists younger than himself who either had worked in his laboratory as students, postdoctoral fellows, or visitors, or had met him on one of his trips. He was affectionate and loyal toward his friends; he expected affection and loyalty in return. Dobzhansky's exuberant personality was manifest not only in his friendships but also in his antipathies, which he was neither able nor, often, willing to hide.

Dobzhansky was an excellent classroom teacher. More than thirty graduate students obtained the Ph.D. under him, and he had an even greater number of postdoctoral and visiting associates, many of them from foreign countries. Dobzhansky spent long periods of time in foreign academic institutions, and was largely responsible for the establishment or development of genetics and evolutionary biology in Brazil, Chile, and Egypt.

Dobzhansky gave generously of his time to other scientists, particularly to young ones, and to students. On the other hand, he resented time spent on committee activities, which he shunned as much as he reasonably could. Throughout his academic career, Dobzhansky avoided administrative posts; he alleged, perhaps correctly, that he had neither the temperament nor the ability for management. Most certainly, he preferred to dedicate his working time to research and writing rather than to administration.

Dobzhansky received numerous honors and awards. He was elected to the U.S. National Academy of Sciences, to the American Academy of Arts and Sciences, the American Philosophical Society, and many foreign academies, including the Royal Society of London, the Royal Swedish Academy of Sciences, the Royal Danish Academy of Sciences, the Brazilian Academy of Sciences, the Academia Leopoldina, and the Accademia Nazionale dei Lincei. He was president of the Genetics Society of America (1941), the American Society of Naturalists (1950), the Society for the Study of Evolution (1951), the American Society of Zoologists (1963), the American Teilhard de Chardin Association (1969), and the Behavior Genetics Association (1973).

Dobzhansky received the Daniel Giraud Elliot Medal (1946) and the Kimber Genetics Award (1958) from the National Academy of Sciences, the Darwin Medal from the Academia Leopoldina (1959), the Anisfield-Wolf Award (1963), the Pierre Lecomte du Nouy Award (1963), the Addison Emery Verrill Medal from Yale University (1966), the Gold Medal Award for Distinguished Achievement in Science from the American Museum of Natural History (1969), and the Benjamin Franklin Medal from the Franklin Institute (1973). In 1964 he received the National Medal of Science.

The Modern Synthesis of Evolutionary Theory. Dobzhansky's most significant contribution to science was his role in formulating the modern synthesis of evolutionary theory. His *Genetics and the Origin of Species* (1937) is considered by some to be the most important book on evolutionary theory in the twentieth century. The title of the book suggests its theme: the role of genetics in explaining the origin of species, a synthesis of genetic knowledge and Darwin's theory of evolution by natural selection. Considerably revised editions of this book were published in 1941 and 1951. *Genetics of the Evolutionary Process*, published in 1970, was considered by Dobzhansky as the fourth edition of the earlier book, except that the content had changed too much to appear under the same title.

By the early 1930's the work of R. A. Fisher and J. B. S. Haldane in Great Britain, and of Sewall Wright in the United States, had provided a theoretical framework for explaining the process of evolution, particularly natural selection, in genetic terms. This work had a limited impact on the biology of the time because it was formulated for the most part in mathematical language, and it was almost exclusively theoretical, with little empirical support. In *Genetics and the Origin of Species*, Dobzhansky completed the integration of Darwinism and Mendelism initiated by the mathematicians in two ways. First, he gathered the empirical evidence that corroborated the mathematico-theoretical framework. Second, he extended the integration of genetics with Darwinism far beyond the range of issues treated by the mathematicians, and into critical evolutionary issues—such as the process of speciation—not easily subjected to mathematical treatment. Moreover, Dobzhansky's book was written in prose understandable to biologists.

The line of thought of *Genetics and the Origin of Species* is surprisingly modern—in part, no doubt, because the book established the pattern that successive evolutionary treatises would largely follow. The book starts with a consideration of organic diversity and discontinuity. Then, successively, it deals with mutation as the origin of hereditary variation, the role of chromosomal rearrangements, variation in natural populations, natural selection, the origin of species by polyploidy, the origin of species through gradual development of reproductive isolation, physiological and genetic differences between species, and the concept of species as natural units.

Genetics and the Origin of Species was received

with great excitement by the biological community, and it inspired other biologists to bring into the modern synthesis of evolutionary theory the contributions of such fields as systematics (Ernst Mayr, 1942), zoology (Julian Huxley, 1942), paleontology (George G. Simpson, 1944), and botany (G. Ledyard Stebbins, 1950). *Genetics and the Origin of Species* also provided a conceptual framework that stimulated experimental research for many years.

Experimental Population Genetics. Dobzhansky was not only a theorist of evolution but also an experimentalist. During half a century of intensive research and publication, he made fundamental empirical contributions to virtually every major area of population and evolutionary genetics.

Dobzhansky's first contribution to population genetics appeared in 1924—an investigation of local and geographic variation in the color and spot pattern of two Coccinellidae genera, *Harmonia* and *Adalia*. These ladybugs exhibit local polymorphisms, which in some species vary from one locality to another. Dobzhansky explained the genetic variation within and between populations as results of the same fundamental evolutionary processes. Some cardinal themes of his evolutionary theory are already present in this work: the pervasiveness of genetic variation, geographic variation as an extension of local polymorphism and as the first but reversible step toward species differentiation. He continued the study of natural populations of ladybugs until he left Russia in 1927, and on occasion returned to them (for instance, a paper in 1933 and a monograph in 1941).

The beginning of Dobzhansky's studies on the population genetics of *Drosophila* can be traced to 1933, when he published a paper on the sterility of hybrids between *D. pseudoobscura* and *D. persimilis* (then known as *D. Pseudoobscura* races A and B). In a series of papers, he investigated the physiological, developmental, and genetic causes of hybrid sterility. This work developed from the convergence of two independent lines of investigation: the genetics of chromosomal translocations and the study of sex determination. It led in 1935 to a formulation of the concept of (sexually reproducing) species still accepted today: "that stage of the evolutionary process at which the once actually or potentially interbreeding array of forms becomes segregated in two or more separate arrays that are physiologically incapable of interbreeding."

This notion establishes that reproductive isolation is what sets species apart. It is also an evolutionary definition that sees speciation as a dynamic process of gradual change. Dobzhansky introduced in 1935, and formally proposed in 1937, the term "isolating mechanisms" to designate the phenomena that impede gene exchange between species. Throughout his life he identified, classified, and investigated the various kinds of isolating mechanisms.

The experimental contributions of Dobzhansky to population genetics are so numerous and so diversified as to defy the possibility of a brief summary. Following is a discussion of a few major areas of research, with the years when he published some of the major papers in each subject.

Dobzhansky's classical studies on the geographical and temporal variation of chromosomal arrangements in *Drosophila pseudoobscura* and its relatives started with a publication in 1936; in 1938 he published a paper on altitudinal variation; in 1943, a paper on seasonal variation, followed in 1946 by a laboratory study (in collaboration with Sewall Wright) showing adaptive differences (with respect to temperature) between chromosomal arrangements. Numerous other publications on this subject appeared through the 1930's and 1940's, and continued throughout Dobzhansky's life. Starting in the 1950's, the study of geographical variation in chromosomal arrangements was extended to the *D. willistoni* group of tropical species, which exhibit even greater degrees of local polymorphism and geographical variation than *D. pseudoobscura*.

While working with Alfred H. Sturtevant, Dobzhansky realized that the evolutionary phylogeny of chromosomal arrangements can be reconstructed by deciphering the patterns of overlapping chromosomal inversions found in natural populations of *Drosophila;* the first phylogeny was published in 1936. This technique became a major tool in the reconstruction of evolutionary history and was applied to many species by Dobzhansky and by others. A notable example of the success of this method is the reconstruction of the phylogeny of Hawaiian species by Hampton L. Carson and his colleagues.

Originally, Dobzhansky thought that the various chromosomal arrangements of *D. pseudoobscura* were adaptively equivalent (see the 1941 edition of *Genetics and the Origin of Species*), and hence that their geographical and temporal variation was the result of genetic drift. Eventually he became convinced that the chromosomal polymorphisms are adaptive, but remained interested in the roles that migration, mutation, and drift play in the maintenance of variation in natural populations.

Estimates of rates of mutation and of accumulation of lethal genes were first published in 1941 (again in collaboration with Sewall Wright); estimates of the critical parameter *Nm* (the product of effective population size times migration rate) in natural populations appeared in 1942, 1952, and 1954. Dobzhansky developed techniques for the experimental study of migration in nature and published pioneering works in the 1940's; he returned later to this research and spent most of the last few summers of his life at the cabin of the Carnegie Institution research station in Mather, near Yosemite in the Sierra Nevada, measuring the rates of dispersion in *Drosophila*.

Dobzhansky early realized the need to investigate the ecological basis of natural variation. He investigated the nutritional preferences first of *D. pseudoobscura* and later of other species (papers in 1951, 1955, and 1956). Several papers (for instance, 1957, 1959) were devoted to ascertaining—particularly in *D. willistoni*—the relationships between the ecological diversity of the environment and the degree of genetic polymorphism. He also investigated the physiological basis of adaptation, starting with studies of fecundity and rates of oxygen consumption published in 1935.

Genetic variation is a necessary condition for evolution. Dobzhansky probably dedicated more research effort to the study of genetic variation in natural populations than to any other single problem. He studied morphological variations but saw that physiological variation—variation affecting fitness—would be most important in evolution. Taking advantage of genetic methods to produce flies homozygous for full chromosomes, he first investigated the frequency of lethal mutations in nature. In 1942, Dobzhansky published a classical paper showing that variation in fitness is a pervasive phenomenon: Virtually every chromosome found in nature carries genes that are deleterious in the homozygous condition; most individuals in nature are well adapted because they are heterozygous for the deleterious variants—"It is the adaptive level of individuals heterozygous for various chromosomes which is most important" (*Genetics*, **27** [1942], 487).

Dobzhansky pursued the study of this "concealed variation" affecting fitness for two and a half decades. When the techniques of gel electrophoresis were first applied to population genetics in the mid-1960's, he became quite enthusiastic. He appreciated that these studies made it possible to obtain quantitative measures of genetic variation. He also saw that there is a trade-off between electrophoretic studies and the earlier methods of studying concealed variation: the adaptive role of electrophoretic variation is not immediately apparent.

In the 1940's Dobzhansky started work with the *D. willistoni* group of species that resulted in contributions to evolutionary genetics comparable in significance with those derived from the study of *D. pseudoobscura* and its relatives that he had started in the 1930's. The most distinctive results with this group concern the process of speciation and concomitant development of reproductive isolation. The *willistoni* group contains several sibling species. One of these, *D. paulistorum*, is a cluster of semispecies, or species *in statu nascendi*, where varying degrees of hybrid sterility, and particularly sexual isolation, can be observed. He discovered and took advantage of this favorable state of affairs for the experimental study of a fundamental evolutionary problem, speciation. He also used *D. paulistorum* as the organism for laboratory study of sexual isolation by selection. This work brought unsought publicity in such periodicals as the *New York Times* and *Time* magazine.

From around 1960 until his death, Dobzhansky worked on the geotactic and phototactic behavior of *Drosophila*. His interest in this field was only in a small part ascertaining the genetic basis of some simple behavioral traits. His main purpose was, rather, to model the interaction of selection, gene flow, and population size for a behavioral trait with low heritability. There were some unexpected but instructive results, such as the observation of what prima facie appeared to be a case of negative heritability.

Contributions to General Genetics and Other Experimental Work. Dobzhansky made significant contributions to other fields of population biology besides population genetics, particularly to ecology and systematics. Much of his population genetics research had an ecological component: geographical and temporal variation in population characteristics, food resource preferences of *Drosophila* species, rates of dispersion, ecological diversity of environments, and so on. Among his other ecological investigations, two at least deserve mention. One is the study of species' diversity in tropical forests, which led him to a hypothesis to account for the high level of species diversification in the tropics (1950). Then, in the early 1960's, he published several papers on the estimation of the innate capacity for increase in numbers in diverse *Drosophila* populations.

Dobzhansky made significant contributions to "classical" genetics, particularly during the 1920's and 1930's. I shall mention but a few. Using translocations between the second and third chromosomes of *Drosophila melanogaster*, he demonstrated that the linear arrangement of genes based on linkage relationships corresponds to a linear arrangement of genes in chromosomes (1929). This linear correspondence had been postulated before but proof was first provided by Dobzhansky (and independently by Hermann Muller and Theophilus Painter the same year). Also in 1929, Dobzhansky advanced the first sophisticated cytological map of a chromosome—chromosome III of *D. melanogaster*. He showed that the relative distances between genes are different in the linkage and in the cytological map; genes clustered around the center of the linkage map are spread throughout a larger portion of the cytological map. He correctly inferred that the frequency of crossing over is not evenly distributed throughout the chromosome.

Later, Dobzhansky produced cytological maps of the chromosomes II (1930) and X (1932) of *D. melanogaster*, and propounded that the centromere (the "spindle fiber attachment," in the terminology of the time) is a permanent feature of chromosomes. He demonstrated that translocations decrease the frequency of crossing over and advanced a hypothesis to account for this reduction (1931).

Dobzhansky demonstrated that the determination of femaleness by the X chromosome is not due to a single or a few genes, but to multiple factors distributed throughout the chromosome (1931). His publications on the genetic and environmental factors affecting sex determination started in 1928 and extended for more than a decade. These studies included work on bobbed mutants in the Y chromosome and their role in male sterility (1933), as well as numerous publications on gynandromorphs and "superfemales." His publications on developmental genetics began in 1930 and continued for many years.

Working with *D. melanogaster* in Filipchenko's laboratory at the University of Leningrad, Dobzhansky made the first systematic investigation of the pleiotropic, or manifold, effects of genes (1927), a phenomenon that held his interest into the 1940's. His contributions to the study of position effects started in 1932 and continued for several years (a review in 1936).

One distinctive characteristic of Dobzhansky's experimental success is that he selected organisms that provided the best materials to investigate the problems that interested him: The biological particularities of *D. pseudoobscura* and its relatives, and of the *D. willistoni* group, made possible many of his discoveries. Moreover, he always worked at the highest level of genetic resolution possible at any given time: He took advantage of the early methods of genetic analysis, then of various cytological tools, later of the giant polytene chromosomes, and of the techniques to produce chromosomal homozygotes. When gel electrophoresis was developed, Dobzhansky immediately recognized its enormous potential as a tool to study population genetics problems. He felt that it was too late in his life for him to learn the technique, but encouraged his students and collaborators to use it and collaborated in several projects using it.

Human Evolution, Human Individuality, and the Concept of Race. Dobzhansky extended the synthesis of Mendelism and Darwinism to the understanding of human nature in *Mankind Evolving* (1962), a book that some consider to be as important as *Genetics and the Origin of Species*.

Mankind Evolving remains an unsurpassed synthesis of genetics, evolutionary theory, anthropology, and sociology. Dobzhansky stated that human nature has two dimensions: the biological, which mankind shares with the rest of life, and the cultural, which is exclusive to man. These two dimensions result from two interconnected processes, biological evolution and cultural evolution:

> The thesis to be set forth in the present book is that man has both a nature and a "history." Human evolution has two components, the biological or organic, and the cultural or superorganic. These components are neither mutually exclusive nor independent, but interrelated and interdependent. Human evolution cannot be understood as a purely biological process, nor can it be adequately described as a history of culture. It is the interaction of biology and culture. There exists a feedback between biological and cultural processes (*Mankind Evolving*, p. 18).

Two principal topics of *Mankind Evolving* are the interrelated concepts of human diversity and race. Dobzhansky's first major publication on these topics was *Heredity, Race, and Society* (1946), a book written with Leslie C. Dunn that was translated into many languages and sold more than one million copies. The two topics are the main subject of *Genetic Diversity and Human Equality* (1973), the last of Dobzhansky's books published before his death. (Dobzhansky left the completed manuscript for another book, *Evolution*, written with Francisco J. Ayala, G. Ledyard Stebbins, and James W. Valentine, which appeared in 1977.)

Dobzhansky set forth that the individual is not the embodiment of some ideal type or norm but, rather, a unique and unrepeatable realization in the field of quasi-infinite possible genetic combinations. The pervasiveness of genetic variation provides the biological foundation of human individuality and leads to demystification of the much-abused concept of race. Dobzhansky emphasized that populations or groups of populations differ from each other in the frequencies of some genes. These differences may be recognized by distinguishing populations of a given species as races. The number of races and the boundaries between them are largely arbitrary because rarely, if ever, are populations of the same species separated by sharp discontinuities in their genetic makeup. Most important is the fact that races are polymorphic for the same genetic variants that may be used to distinguish one race from another. There is more genetic variation within any human race than there are genetic differences between races. It follows, as Dobzhansky saw it, that individuals should be evaluated by what they are, not by the race to which they belong.

Dobzhansky considered human diversity a fact belonging to the realm of observable natural phenomena: "People are innately, genetically, and therefore irremediably diverse and unlike" (*Genetic Diversity and Human Equality,* p. 4). Biological distinctiveness is not, however, a basis for inequality. Equality—as in equality before the law and equality of opportunity—"pertains to the rights and the sacredness of life of every human being" (*ibid.*). Dobzhansky pointed out that equality of opportunity is the best strategy to maximize the benefits of human biological diversity. "Denial of equality of opportunity stultifies the genetic diversity with which mankind became equipped in the course of its evolutionary development. Inequality conceals and stifles some people's abilities and dissembles the lack of abilities in others. Conversely equality permits an optimal utilization of the wealth of the gene pool of the human species" (*Mankind Evolving,* p. 285). Dobzhansky had little patience with racial prejudice or social injustice, and castigated those who pretended to base them on what he called the "bogus 'science' of race prejudice."

Dobzhansky's lasting interest in the relevance of biology, and particularly evolutionary theory, to human affairs is evident in scores of articles that he wrote on the subject and in the titles of some of his books: *Heredity, Race, and Society* (1946); *Evolution, Genetics, and Man* (1955); *The Biological Basis of Human Freedom* (1956); *Radiation, Genes, and Man* (1959, with B. Wallace); *Mankind Evolving* (1962); *Heredity and the Nature of Man* (1964); *The Biology of Ultimate Concern* (1967); and *Genetic Diversity and Human Equality* (1973).

Humanism. Dobzhansky's interest in the interface between biology and human problems was expressed in numerous publications that, beginning in the mid-1940's, flowed in a continuous stream. His concern was probably kindled by several convergent influences. One factor was the racial bigotry that helped to trigger World War II in Europe; another, Lysenko's suppression of genetics and geneticists in the Soviet Union; a third, his association with Leslie C. Dunn, a colleague at Columbia University and intimate friend, whose compassion for the human predicament was much revered by Dobzhansky, and who became greatly involved in providing shelter in the United States for scientists fleeing from Nazi persecution.

As mentioned above, Dobzhansky published *Heredity, Race, and Society* with Dunn in 1946, and continued publishing on race questions from a biological perspective until the end of his life. Publications criticizing eugenic movements appeared in 1952 and 1964; the subject of eugenics was treated in other papers and several books. In 1946 he translated Lysenko's *Heredity and Its Variability* into English as a way to expose Lysenko's quackery. Dobzhansky criticized Lysenko's "science," and particularly his eradication of genetics and geneticists, in several articles published between 1946 and 1958.

Dobzhansky was concerned with the role of religion in human life and explored the evolutionary basis of religion in several articles published in the 1960's and 1970's, and in his *The Biology of Ultimate Concern* (1967). Yet he did not hesitate to criticize (1953) the antievolutionist stand of Pope Pius XII in the encyclical *Humani generis,* or that of fundamentalist Protestants (1973).

Dobzhansky often expressed his frustration at the limited influence of biology on the thinking of philosophers. He saw that evolutionary biology raises new philosophical problems and throws light on old ones. He wrote several essays on philosophical questions, such as the concepts of determinism and chance (1963, 1966, 1974), transcendent phenomena (1965, 1967, 1977), organismic or compositionist approaches in the philosophy of biology (1967, 1968), and the "creative" character of biological evolution (1954, 1967, 1974).

BIBLIOGRAPHY

I. Original Works. A bibliography of Dobzhansky's publications follows Francisco J. Ayala's memoir in *Biographical Memoirs. National Academy of Sciences,* **55** (1985), 163–213. A bibliography covering Dobzhansky's publications from 1918 to 1969 is in Howard Levene, Lee Ehrman, and Rollin Richmond, "Theodosius Dobzhansky up to Now," in Max K. Hecht and William C. Steere, eds., *Essays in Evolution and Genetics in Honor of Theodosius Dobzhansky* (New York, 1970), 1–41, accompanied by a very brief description of Dobzhansky's scientific work; the bibliography is fairly complete, although there are a few omissions and errors. The same bibliography, with a career summary and supplemented up to 1976, is in *Evolutionary Biology*, **9** (1976), 409–448. Papers published between 1938 and 1976 by Dobzhansky and collaborators under the general title "Genetics of Natural Populations" have been reprinted in R. C. Lewontin, John A. Moore, William B. Provine, and Bruce Wallace, eds., *Dobzhansky's Genetics of Natural Populations I–XLIII* (New York, 1981), which contains introductions about Dobzhansky's scientific work by Provine and Lewontin.

Dobzhansky's books include *Genetics and the Origin of Species* (New York, 1937; 2nd ed., 1941; 3rd ed., rev., 1951); *Heredity, Race, and Society* (New York, 1946; 2nd ed., 1952; 3rd ed., rev and enl., 1956), written with Leslie C. Dunn; *Principles of Genetics* (4th ed., New York, 1949; 5th ed., 1958) written with Edmund W. Sinnott and Leslie C. Dunn (Dobzhansky was not a coauthor of the first three editions of this book); *Evolution, Genetics, and Man* (New York, 1955); *The Biological Basis of Human Freedom* (New York, 1956); *Radiation, Genes, and Man* (New York, 1959), written with B. Wallace; *Mankind Evolving: The Evolution of the Human Species* (New York, 1962); *Heredity and the Nature of Man* (New York, 1964); *The Biology of Ultimate Concern* (New York, 1967); *Genetics of the Evolutionary Process* (New York, 1970); *Genetic Diversity and Human Equality* (New York, 1973); *Evolution* (San Francisco, 1977), written with Francisco J. Ayala, G. Ledyard Stebbins, and James W. Valentine; and *The Roving Naturalist. Travel Letters of Theodosius Dobzhansky*, Bentley Glass, ed. (Philadelphia, 1980).

Dobzhansky's articles include "Description of a New Species of the Genus *Coccinella* from the Neighbourhood of Kiev," in *Materialy dlia fauny iugozapadnoi Rossii*, **2** (1918), 46–47 (in Russian); "Die geographische und individuelle Variabilität von *Harmonia axyridis* Pallas in ihren Wechselbeziehungen," in *Biologisches Zentralblatt*, **44** (1924), 401–421; "Die weiblichen Generationsorgane der Coccinelliden als Artmerkmal betrachtet (*Coleoptera*)," in *Entomologische Mitteilungen*, **13** (1924), 18–27; "Beitrag zur Kenntnis des weiblichen Geschlechtsapparates der Coccinelliden," in *Zeitschrift für wissenschaftliche Insektenbiologie*, **19** (1924), 98–100; "Studies on the Manifold Effect of Certain Genes in *Drosophila melanogaster*," in *Zeitschrift für induktive Abstammungs- und Vererbungslehre*, **43** (1927), 330–388; "The Effect of Temperature on the Viability of Superfemales in *Drosophila melanogaster*," in *Proceedings of the National Academy of Sciences*, **14** (1928), 671–675; "The Reproductive System of Triploid Intersexes in *Drosophila melanogaster*," in *American Naturalist*, **62** (1928), 425–434, written with C. B. Bridges; "Genetical and Cytological Proof of Translocations Involving the Third and the Fourth Chromosomes of *Drosophila melanogaster*," in *Biologische Zentralblatt*, **49** (1929), 408–419; "A Homozygous Translocation in *Drosophila melanogaster*," in *Proceedings of the National Academy of Sciences*, **15** (1929), 633–638.

"Genetical and Environmental Factors Influencing the Type of Intersexes in *Drosophila melanogaster*," in *American Naturalist*, **64** (1930), 261–271; "Cytological Map of the Second Chromosome of *Drosophila melanogaster*," in *Biologische Zentralblatt*, **50** (1930), 671–685; "Interaction Between Female and Male Parts in Gynandromorphs of *Drosophila simulans*," in *Zeitschrift für Wissenschaftliche Biologie, Abt. D, Wilhelm Roux' Archiv für Entwicklungsmechanik der Organismen*, **123** (1931), 719–746; "The Decrease of Crossing-over Observed in Translocations, and Its Probable Explanation," in *American Naturalist*, **65** (1931), 214–232; "Evidence for Multiple Sex Factors in the X-Chromosome of *Drosophila melanogaster*," in *Proceedings of the National Academy of Sciences*, **17** (1931), 513–518, written with J. Schulte.

Cytological Map of the X-Chromosome of *Drosophila melanogaster*," in *Biologische Zentralblatt*, **52** (1932), 493–509; "Geographical Variation in Lady-Beetles," in *American Naturalist*, **67** (1933), 97–126; "On the Sterility of the Interracial Hybrids in *Drosophila pseudoobscura*," in *Proceedings of the National Academy of Sciences*, **19** (1933), 397–403; "Role of the Autosomes in the *Drosophila pseudoobscura* Hybrids," *ibid.*, 950–953; "Deficiency and Duplication for the Gene "Bobbed" in *Drosophila melanogaster*," in *Genetics*, **18** (1933), 173–192, written with Natalia P. Sivertzev-Dobzhansky; "A Critique of the Species Concept in Biology," in *Philosophy of Science*, **2** (1935), 344–355; "Oxygen Consumption of *Drosophila* Pupae. II. *Drosophila pseudoobscura*," in *Zeitschrift für Vergleichende Physiologie*," **22** (1935), 473–478, written with D. F. Poulson; "Position Effects of Genes," in *Biological Reviews*, **11** (1936), 364–384; "Inversions in the Third Chromosome of Wild Races of *Drosophila pseudoobscura*, and Their Use in the Study of the History of the Species," in *Proceedings of the National Academy of Sciences*, **22** (1948), 448–450, written with Alfred H. Sturtevant; "Genetic Nature of Species Differences," in *American Naturalist*, **71** (1937) 404–420; "Genetics of Natural Populations. I. Chromosome Variation in Populations of *Drosophila pseudoobscura* Inhabiting Isolated Mountain Ranges," in *Genetics*, **23** (1938), 239–251, written with M. L. Queal. "Genetics of Natural Populations. V. Relations Between Mutation Rate and Accumulation of Lethals in Populations of *Drosophila pseudoobscura*," in *Genetics*, **26** (1941), 23–51, written with Sewall Wright; "Beetles of the Genus *Hyperaspis* Inhabiting the United States," in *Smithsonian Institution Publication* no. 3642 (1941), 1–94. "Genetics of Natural Populations. VII. The Allelism of Lethals in the Third Chromosome of *Drosophila pseudoobscura*," in *Genetics*, **27** (1942), 363–394, written with Sewall Wright and W. Hovanitz; "Genetics of Natural Populations. VIII. Concealed Variability in the Second and Fourth Chromosomes of *Drosophila pseudoobscura* and Its Bearing on the Problem of Heterosis," *ibid.*, **27** (1942), 464–490, written with A. M. Holz and Boris Spassky; "Genetics of Natural Populations. IX. Temporal Changes in the Composition of Populations of *Drosophila pseudoobscura*," *ibid.*, **28** (1943), 162–186; "Genetics of

Natural Populations. XII. Experimental Reproduction of Some of the Changes Caused by Natural Selection in Certain Populations of *Drosophila pseudoobscura*," *ibid.*, **31** (1946), 125–156, written with Sewall Wright. "Lysenko's 'Genetics,'" in *Journal of Heredity*, **37** (1946), 5–9; "The New Genetics in the Soviet Union," in *American Naturalist*, **80** (1946), 649–651; "The Suppression of a Science," in *Bulletin of the Atomic Scientists*, **5** (1949), 144–146.

"Evolution in the Tropics," in *American Scientist*, **38** (1950), 209–221; "A Comparative Study of Chromosomal Polymorphism in Sibling Species of the *willistoni* Group of *Drosophila*," in *American Naturalist*, **84** (1950), 229–246, written with H. Burla and Antonio Brito da Cunha; "Comparative Genetics of *Drosophila willistoni*," in *Heredity*, **4** (1950), 201–215, written with Boris Spassky; "Some Attempts to Estimate Species Diversity and Population Density of Trees in Amazonian Forests," in *Botanical Gazette*, **111** (1950), 413–524, written with G. Black and C. Pavan; Adaptive Chromosomal Polymorphism in *Drosophila willistoni*," in *Evolution*, **4** (1950), 212–235, written with Antonio Brito da Cunha and H. Burla. "On Food Preferences of Sympatric Species of *Drosophila*," *ibid.*, **5** (1951), 97–101, written with Antonio Brito da Cunha and A. Sokoloff; "Genetics of Natural Populations. XX. Changes Induced by Drought in *Drosophila pseudoobscura* and *Drosophila persimilis*," *ibid.*, **6** (1952), 234–243; "Lysenko's 'Michurinist' Genetics," in *Bulletin of the Atomic Scientists*, **8** (1952), 40–44; "A Comparative Study of Mutation Rates in Two Ecologically Diverse Species of *Drosophila*," in *Genetics*, **37** (1952), 650–664, written with Boris Spassky and N. Spassky; "Two Recent Versions of Eugenics," in *American Naturalist*, **86** (1952), 61–62; "Russian Genetics," Ruth C. Christman, ed., *Soviet Science* (Washington, D.C., 1952), 1–7; "A Comment on the Discussion of Genetics by His Holiness, Pius XII," in *Science*, **118** (1953), 561–563; "Evolution as a Creative Process," in *Atti del IX Congresso internazionale di genetica*, Pt 1, which is supp. to *Caryologia*, **6** (1954), 435–449; "Rates of Spontaneous Mutation in the Second Chromosomes of the Sibling Species, *Drosophila pseudoobscura* and *Drosophila persimilis*," in *Genetics*, **39** (1954), 899–907, written with Boris Spassky and N. Spassky; "Differentiation of Nutritional Preferences in Brazilian Species of *Drosophila*," in *Ecology*, **36** (1955), 34–39, written with Antonio Brito da Cunha; "Studies on the Ecology of *Drosophila* in the Yosemite Region of California. IV. Differential Attraction of Species of *Drosophila* to Different Species of Yeasts," *ibid.*, **37** (1956), 544–550, written with D. M. Cooper, H. J. Phaff, E. P. Knapp, and H. L. Carson; "Genetics of Natural Populations. XXVI. Chromosomal Variability in Island and Continental Populations of *Drosophila willistoni* from Central America and the West Indies," in *Evolution*, **11** (1957), 280–293; "Lysenko at Bay," in *Journal of Heredity*, **49** (1958), 15–17; "Genetics of Natural Populations. XXVIII. Supplementary Data on the Chromosomal Polymorphism in *Drosophila willistoni* in Its Relation to the Environment," *ibid.*, **13** (1959), 389–404, written with Antonio Brito da Cunha, Olga Pavlovsky, and Boris Spassky; "*Drosophila paulistorum*, a Cluster of Species *in Statu Nascendi*," in *Proceedings of the National Academy of Sciences*, **45** (1959), 419–428, written with Boris Spassky.

"Selection for Geotaxis in Monomorphic and Polymorphic Populations of *Drosophila pseudoobscura*," *ibid.*, **48** (1962), 1704–1712, written with Boris Spassky; "Scientific Explanation: Chance and Antichance in Organic Evolution," in Bernard Baumrin, ed., *Philosophy of Science*, I (New York, 1963), 209–222; "Relative Fitness of Geographic Races of *Drosophila serrata*," in *Evolution*, **17** (1963), 72–83, written with L. C. Birch, P. O. Elliot, and R. C. Lewontin; "Human Genetics—an Outsider's View," in *Cold Spring Harbor Symposia in Quantitative Biology*, **29** (1964), 1–7; "The Superspecies *Drosophila paulistorum*," in *Proceedings of the National Academy of Sciences*, **51** (1964), 3–9, written with Lee Ehrman, Olga Pavlovsky, and Boris Spassky; "The Capacity for Increase in Chromosomally Polymorphic and Monomorphic Populations of *Drosophila pseudoobscura*," in *Heredity*, **19** (1964), 597–614, written with R. C. Lewontin and Olga Pavlovsky; "Evolution and Transcendence," in *Main Currents in Modern Thought*, **22** (1965), 3–9; "Determinism and Indeterminism in Biological Evolution," In Vincent E. Smith, ed., *Philosophical Problems in Biology* (New York, 1966), 55–66; "Spontaneous Origin of an Incipient Species in the *Drosophila paulistorum* complex," in *Proceedings of the National Academy of Sciences*, **55** (1966), 727–733, written with Olga Pavlovsky; "Creative Evolution," in *Diogenes*, **58** (1967), 62–74; "Effects of Selection and Migration on Geotactic and Phototactic Behaviour of *Drosophila*. I," in *Proceedings of the Royal Society*, **B168** (1967), 27–47, written with Boris Spassky; "On Some Fundamental Concepts of Darwinian Biology," in *Evolutionary Biology*, **2** (1968), 1–34.

"Polymorphisms in Continental and Island Populations of *Drosophila willistoni*," in *Proceedings of the National Academy of Sciences*, **68** (1971), 2480–2483; written with Francisco J. Ayala and Jeffrey R. Powell; "Genetics and the Diversity of Behavior," in *American Psychologist*, **27** (1972), 523–530; "Effects of Selection and Migration on Geotactic and Phototactic Behaviour of *Drosophila*. III," in *Proceedings of the Royal Society*, **B180** (1972), 21–41, written with Howard Levene and Boris Spassky; "Temporal Frequency Changes of Enzyme and Chromosomal Polymorphisms in Natural Populations of *Drosophila*," in *Proceedings of the National Academy of Sciences*, **70** (1973), 680–683, written with Francisco J. Ayala; "Nothing in Biology Makes Sense Except in the Light of Evolution," in *American Biology Teacher*, **35** (1973), 125–129; "Chance and Creativity in Evolution," in F. J. Ayala and Theodosius Dobzhansky, eds., *Studies in the Philosophy of Biology* (London, 1974), 307–338; and *Humankind—A Product of Evolutionary Transcedence* (Johannesburg, 1977), written with Francisco J. Ayala.

II. SECONDARY LITERATURE. Dobzhansky's work is

reflected in Hampton L. Carson, D. Elmo Hardy, Herman T. Spieth and Wilson S. Stone, "The Evolutionary Biology of the Hawaiian Drosophilidae," in Max K. Hecht and William C. Steere, eds., *Essays in Evolution and Genetics in Honor of Theodosius Dobzhansky* (New York, 1970), 437–543; Julian S. Huxley, *Evolution: The Modern Synthesis* (New York, 1942; repr. 1964); Ernst Mayr, *Systematics and the Origin of Species* (New York, 1942); George G. Simpson, *Tempo and Mode in Evolution* (New York, 1944); G. Ledyard Stebbins, *Variation and Evolution in Plants* (New York, 1950).

FRANCISCO J. AYALA

DORGELO, HENDRIK BEREND (*b.* Dedemsvaart, Netherlands, 9 February 1894; *d.* Eindhoven, Netherlands, 6 March 1961), *physics*.

Dorgelo contributed greatly to the scientific and technical development of the Netherlands, partly as a research physicist and partly as a teacher and organizer of technical physics. His primary field of research was experimental spectroscopy; his name is associated mainly with the spectroscopical investigations of multiplet spectra. These investigations proved important in the transition from the old quantum theory to the new quantum mechanics.

Having completed secondary school, Dorgelo worked in the academic year 1912–1913 as a primary school teacher in his hometown. During World War I he served as a reserve army officer. After the demobilization in 1918, Dorgelo went to Utrecht to study physics at the university. He was supervised by Leonard S. Ornstein, director of the Physical Institute, which Dorgelo joined in 1922 as an assistant. Under Ornstein's guidance he specialized in the study of the multiplet lines recently discovered by the Spanish physicist Miguel A. Catalán. His dissertation, submitted in 1924, dealt with the intensities of the components of multiple spectral lines.

In order to measure the intensity of the multiplet lines with sufficient accuracy, in 1923 Dorgelo developed a new photographic technique that made possible precise determination of the intensity ratios of neighboring spectral lines as well as of lines widely separated in wavelength. He used the improved technique on his own in collaboration with Ornstein and Herman C. Burger, then chief assistant in physics at the University of Utrecht. In 1924 Dorgelo found that the intensity ratios of doublets and triplets can be represented by the ratios of small integers (2:1 for the alkali doublet, 5:3:1 for the triplet of alkaline earths). He also was able to confirm experimentally a suggestion of Arnold Sommerfeld that the scheme might also hold for higher multiplets.

The intensity rules discovered by the Utrecht spectroscopists were relevant to the atomic models of the old quantum theory, especially the vector core model of Sommerfeld and Alfred Landé, and to the correspondence principle of Niels Bohr. Inspired by Sommerfeld, Dorgelo and Burger succeeded, in 1924, in relating the observed intensity ratios to the inner quantum number (j), introduced formally by Sommerfeld in 1920 and used by Landé (as $J = j + \frac{1}{2}$) in his atomic model. They also showed that the intensity ratios of multiplet components arising from atoms in the same initial state do not depend on how the atoms are excited. Dorgelo and Burger arrived at several "sum rules," one of which states that the sum of the intensities of the components of a multiplet associated with the same initial state is proportional to the value of the inner quantum number (J) of the state.

An experimentalist at heart, Dorgelo did not try to justify the sum rules theoretically; Werner Heisenberg did so in 1924. This was a triumph for the old quantum theory and for the correspondence principle. However, it forced Heisenberg to sharpen the arguments based on the correspondence principle and thus contributed to the process that led him to quantum mechanics less than a year later.

After having obtained his doctorate, Dorgelo spent three years at the Philips Incandescent Lamp Company at Eindhoven. He thus had to stop his work on multiplet intensities at a time when it looked most promising. At Eindhoven, Dorgelo investigated the metastable states of inert gases, carrying further the work of Friedrich Paschen and Karl Meissner. He was the first to determine the lifetime of metastable neon, a result obtained in 1925.

In 1927 Dorgelo was appointed professor of physics at the Technical University of Delft, a position he held for twenty-nine years. During this period he worked on spectroscopy, gas discharges, electron optics, and X-ray analysis. Much of the work was done in collaboration with Dutch industrial firms such as Philips and Shell. Though technically valuable, Dorgelo's research at Delft did not reach the level of scientific quality of that done at Utrecht. Much of his time was spent on administration and teaching. He was a popular teacher and wrote several textbooks for students of technical physics. An efficient organizer, Dorgelo was instrumental in the construction of the electron microscope at Delft in the late 1930's, one of the first effective instruments of its kind.

Dorgelo ended his career as rector of the Technical University at Eindhoven, a post he held from 1956 until his death.

BIBLIOGRAPHY

I. ORIGINAL WORKS. "Die Intensität der Mehrfachlinien," in *Zeitschrift für Physik*, **13** (1923), 206–210; "Die Intensität Mehrfacher Spektrallinien," *ibid.*, **22** (1924), 170–177; "Beziehungen zwischen inneren Quantenzahlen und Intensitäten von Mehrfachlinien," *ibid.*, **23** (1924), 258–266, written with Herman C. Burger; "Die photographische Spektralphotometrie," in *Physikalische Zeitschrift*, **26** (1925), 756–794; "Die Lebensdauer der metastabilen s_3- und s_5 Zustände des Neons," in *Zeitschrift für Physik*, **34** (1925), 766–774.

II. SECONDARY WORKS. M. J. Druyvesteyn, "In memoriam Prof. Dr. H. B. Dorgelo," in *Nederlands tijdschrift voor natuurkunde*, **27** (1961), 181–184; A. C. S. van Heel, "H. B. Dorgelo vijfentwintig jaar hoogleraar," *ibid.*, **18** (1952), 26–30; and Jagdish Mehra and Helmut Rechenberg, *The Historical Development of Quantum Theory*, I, pt. 2 (New York, 1982).

HELGE KRAGH

DRYDEN, HUGH LATIMER (*b.* Pocomoke City, Maryland, 2 July 1898; *d.* Washington, D.C., 2 December 1965), *physics, aerodynamics*.

Dryden's parents, Samuel Isaac Dryden and Zenovia (Nova) Hill Culver Dryden, moved their family from Maryland's Eastern Shore to Baltimore in 1907, where his father found work as a streetcar conductor. Dryden was educated in Baltimore public schools and at Johns Hopkins University, completing his B.A. in 1916, his M.A. in 1918, and his Ph.D. in physics in 1919. At the time he was the youngest person to receive a doctorate from Johns Hopkins. On 29 January 1920 he married Mary Libbie Travers; they had a son and two daughters. Throughout his career as a scientist and a public administrator, Dryden was a key figure at Calvary Methodist Church, Washington, D.C., where he was a licensed local preacher.

Dryden took a summer job as an inspector of munitions gauges with the National Bureau of Standards (NBS) in 1918; shortly thereafter he joined the staff of the bureau's new wind tunnel section. At night he completed his doctoral studies under Joseph S. Ames of Johns Hopkins, who taught advanced courses at NBS. Research for his dissertation, "Air Forces on Circular Cylinders," was conducted after hours at the NBS wind tunnel. This work described a series of experiments he had conducted on the drag and distribution of air flowing around cylinders perpendicular to the wind. Dryden was appointed chief of the Aerodynamics Section of NBS in 1920. His research on the problems of wind tunnel turbulence and boundary-layer flow caught the attention of the international engineering and scientific community. The work of Dryden and his colleagues demonstrated the effect of turbulence during the transition period from laminar to turbulent flow in the boundary layer near a solid surface, emphasizing the practicality of maintaining a laminar boundary layer over a large fraction of the surface of an aircraft to reduce drag.

With Lyman J. Briggs, Dryden took some of the earliest high-speed airfoil measurements during the mid 1920's. His work at NBS led him to other engineering projects, such as calculating skyscraper wind loads and ensuring the structural integrity of propeller blades. He also conducted investigations designed to measure the acceleration of gravity.

In 1934 Dryden was named chief of the bureau's Mechanics and Sound Division, which during World War II supported the development of guided glide bombs for the Office of Scientific Research and Development. In cooperation with the U.S. Navy, Dryden's section developed the BAT radar homing missile, which was used in combat. This was the first large-scale research and development project Dryden had directed.

Dryden became associate director of NBS in 1946. The following year he joined the National Advisory Committee for Aeronautics (NACA), of which he had been a member since 1931, as director of research. He was named NACA's director in 1949. During his tenure with the organization, NACA became the leading authority on supersonic flight. High-speed wind tunnel research, flight testing of the X series aircraft, and study of the critical reentry heating problems experienced by high-and-fast-flying missiles and manned vehicles were conducted under Dryden's direction. Dryden had become an administrator of scientists and engineers, and increasingly he devoted his energies to formulating broad research policy rather than pursuing his own research interests.

Concerning the problems of managing research, Dryden said, "Conventional management procedures are well adapted to operations in which the product consists of a series of nearly identical items." But, he wrote, "A research laboratory produces ideas and new knowledge verified by experiment. Its reports are varied in nature and cannot be considered as nearly identical in scope, difficulty, or effort required." Dryden feared that actions taken at high administrative levels, while entirely appropriate for the general operations of government, may "produce unexpected and often harmful results on the efficiency of research activities" ("Science and Public Administration: Viewpoint of a Scientist-Turned-Administrator," 21 March 1958). Fully realizing the difficulties, Dryden was prepared to play a key role in the U.S. response to the orbiting of the first artificial satellite, Sputnik I, by the Soviet Union.

When President Eisenhower established the National Aeronautics and Space Administration (NASA) in 1958, Dryden—along with NACA's other 8,000 employees—was transferred to this new civilian agency. As deputy administrator he participated in the planning of the successful U.S. manned space program and emphasized the importance of international cooperation to space research. Dryden served as NASA's technical negotiator from 1962 to 1965 in an attempt to forge a formal agreement with the Soviet Union.

In 1965, Dryden lost a four-year battle with cancer, bringing to an end his forty-five years of government service. He had been fond of saying that he had grown up with the airplane, but he did much more than passively "grow up" with aeronautics. His work at the National Bureau of Standards, NACA, and NASA had a great impact on the speed, safety, and efficiency of manned flight.

Among the many awards presented to Dryden were the Sylvanus Albert Reed Award (1940), the Presidential Certificate of Merit (1948), the Daniel Guggenheim Medal (1950), the Wright Brothers Memorial Trophy (1955), and the National Medal of Science (1965, awarded posthumously).

BIBLIOGRAPHY

I. Original Works. Dryden's basic work on air flow is summarized in "Turbulence and the Boundary Layer," in *Journal of Aeronautical Sciences*, **6** (1939), 85–105, the second Wright Brothers Lecture to the Institute of Aeronautical Sciences, Columbia University, 12 December 1938. His important technical writings have been collected at the Milton S. Eisenhower Library, Johns Hopkins University. See Richard K. Smith, ed., *The Hugh L. Dryden Papers, 1898–1965; A Preliminary Catalogue of the Basic Collection* (Baltimore, 1974). Other collections of his papers are at the NASA History Office Archives and the National Air and Space Museum Library, both in Washington, D.C.

II. Secondary Literature. An obituary with portrait and bibliography is Jerome C. Hunsaker and Robert C. Seamans, Jr., "Hugh Latimer Dryden," in *Biographical Memoirs. National Academy of Sciences*, **40** (1969), 35–68.

Linda Neuman Ezell

DUBOIS, RAPHAEL-HORACE (*b*. Mans, France, 20 June 1849; *d*. Tamaris-sur-Mer, France, 21 January 1929), *physiology*.

Dubois began his medical education at the medical school of Tours, where he was also demonstrator in botany and chemistry in 1868 and 1869. After serving in the army in the Franco-Prussian War, he returned to study in Paris. He submitted his M.D. thesis in 1876. In 1882 Dubois became *préparateur* for the course in physiology taught at the Sorbonne by Paul Bert and by Albert Dastre. From 1883 to 1886 he was deputy director of the laboratory of physiological optics at the Sorbonne. Dubois took his doctorate in science in 1886, and in February 1887 was elected professor of general and comparative physiology in the Faculty of Sciences at Lyons. He was also director of the marine biology station of Lyons University at Tamaris-sur-Mer.

As a physiologist Dubois's interests were very wide, and he studied a variety of animal types, both vertebrate and invertebrate. He is best remembered for his pioneering studies on hibernation and, especially, for his work on bioluminescence. Henri Regnault and Jules Reiset had previously worked on hibernation, but Dubois made exhaustive studies on the hibernating marmot, measuring rates of metabolism, respiratory changes, temperature, and partial pressure of blood gases. His work led him to believe that the onset of hibernation was due to the accumulation of a fixed amount of CO_2 in the blood.

In 1884 Dubois began work on the bioluminescence of the brilliant tropical firefly *Pyrophorus*. Before his investigations the study of bioluminescence had had a somewhat checkered history. Robert Boyle had demonstrated in 1667 that glowworms cease to emit light when placed in a vacuum and that the glow will return in air. Rudolf A. van Koelliker and Franz von Leydig had studied the fine structure of the luminous organs of the glowworm in 1857, and Louis Pasteur (1864) and E. Ray Lankester (1870) were interested enough in bioluminescence to measure the spectra of the emitted light. It was Dubois, however, who placed the chemical investigation of luminescence on a firm foundation. He disproved the early view that phosphorus played a part in luminescence, and replaced it with evidence that the process was connected with enzyme oxidation of a specific biochemical compound, which he named luciferin. He established the luciferin-luciferase system in the elaterid beetle *Pyrophorus* and in the bioluminescent mollusk *Pholas dactylus*. His detailed work was summarized in two works published in 1886 and (on the mollusk) in 1892. A more popular account of his discoveries, *La vie et la lumière*, was published in 1914, the year of his retirement. After World War I and in his later years, Dubois, like many of his generation, became concerned with the nature of aggression and war. He published several books on these subjects and advocated pacifism.

BIBLIOGRAPHY

I. ORIGINAL WORKS. Among Dubois's scientific works are "Note sur la physiologie des pyrophores," in *Comptes rendus de la Société biologique*, 8th ser., **1** (1884), 661–664, and **2** (1885), 559–562; "Les elaterides lumineaux," in *Bulletin de la Société zoologique de France*, **11** (1886), 1–275; "Sur la production de la lumière chez le *Pholas dactylus*," in *Comptes rendus de la Société biologique*, **40** (1889), 451–453, and **41** (1890), 611–614; "Anatomie et physiologie comparées de la *Pholade dactyle*," in *Annales de l'Université de Lyon*, fasc. 2, pt. (1892); "Étude sur le mécanisme de la thermogenèse et du sommeil chez les mammifères," *ibid.*, fasc. 25, (1896); "Contribution à l'étude des perles fines de la nacre et des animaux qui les produisent," *ibid.*, fasc. 29 (1899); "Production de la lumière et des radiations chimique par les êtres vivants," in *Leçons de physiologie générale et comparée*, II (Paris, 1898), 301–528; and *La vie et la lumière* (Paris, 1914). For his views on war see his *Les origines naturelles de la guerre: Influences cosmiques et théorie anticinétique* (Lyons, 1916).

II. SECONDARY LITERATURE. There is a short note on Dubois by S. Le Tourneur in *Dictionnaire de biographie française*, XI, 970–971. See also H. Cardot, "Aperçu sur l'évolution de la physiologie et sur l'oeuvre des physiologistes lyonnais," in *Revue scientifique*, **66** (1928), 1–9, esp. 6–7.

NEIL DAVIES MORGAN

DUMOND, JESSE WILLIAM MONROE (*b.* Paris, France, 11 July 1892; *d.* Pasadena, California, 4 December 1976), *experimental physics*.

DuMond's long and productive scientific career embraced studies in atomic physics by the use of X rays, nuclear spectroscopy, and measurements of the fundamental physical constants.

He was the only child of Fredrick Melville DuMond, an artist and teacher, and of Louise Adèle Kerr DuMond, also an artist, who came from a well-to-do Philadelphia family living in Paris. After his mother died in 1894, he was brought up by his maternal grandmother and grandaunt in Paris, and then by his father's parents in Rochester, New York, and later in Monrovia, California. DuMond's later passion for designing and building intricate, precision scientific apparatus was sparked by his grandfather, Alonzo Monroe DuMond, a sailor turned sheet-metal artisan. Under his grandfather's guidance, DuMond learned how to design and build mechanical objects using cast-off and homemade materials, from toys to phonographs to telephones.

After graduation from Monrovia High School, DuMond worked at a number of jobs for a year and then, in 1912, entered Throop College (now the California Institute of Technology). He received a B.S. in electrical engineering in 1916. For his thesis DuMond designed and constructed a harmonic analyzer—a mechanical calculator that, he felt, better met the needs of the average laboratory than A. A. Michelson and S. W. Stratton's patented device.

DuMond spent the years 1916 and 1917 as an electrical engineer at the General Electric Company in Schenectady, New York. While there, he took graduate courses at Union College under an arrangement between the company and the school, earning a master's degree (1919) in electrical engineering. The tedious numerical calculations in an advanced course on alternating-current problems inspired his master's thesis, a complex quantity slide rule, a device he once characterized as a substitute for routine work, not intelligence. In 1918, disenchanted with industrial work, DuMond joined the U.S. Army to fight with the American Expeditionary Force in France. He served at the front with a sound-ranging company until the armistice. Upon discharge, DuMond spent one year (1919–1920) as a design draftsman at the French Thomson-Houston Company in Paris and another (1920–1921) at the National Bureau of Standards in Washington, D.C.

DuMond married Irene Gaebel in 1920; they had a son and two daughters. He and Irene were divorced in 1942, and that same year he married Louise Marie Baillet.

In 1921, he returned to Caltech as a graduate student, and he obtained a Ph.D. in physics in 1929. Under physicist Robert A. Millikan's leadership, the school had just embarked on an ambitious research program in physics. In DuMond, Millikan found a relentless, frugal researcher; a producer of ingenious equipment; and an idealist, a man fiercely loyal to the institute. Except for a brief period as visiting associate professor at Stanford University, in 1931, DuMond remained on the Caltech faculty for thirty-four years. He climbed the academic ranks slowly, partly because he voluntarily resigned a graduate teaching fellowship in 1924 to devote more time to research, partly because of the Depression. After spending nine years (1929–1938) as an unpaid research fellow, DuMond became an associate professor. In 1946 he was promoted to full professor, and in 1963 he became professor emeritus.

DuMond's doctoral dissertation dealt with the

Compton scattering of X rays by atoms, and the breadth and structure of the shifted line. In his derivation of the shift in 1922, Arthur Compton had assumed an elastic collision between a photon of light and an electron at rest. But if the atomic electrons had an initial momentum, DuMond reasoned, that momentum could explain the observed broadening of the shifted lines. Using this hypothesis, in 1925 he undertook a detailed spectral study of the Compton line. Now considered one of the classical experiments of atomic physics, DuMond's measurements provided the first direct experimental evidence of the momentum distribution of electrons in atoms, as had been predicted by wave mechanics.

DuMond sought greater intensity and contrast in an X-ray spectroscopic instrument. Therefore, with Harry A. Kirkpatrick (his first graduate student) he conceived, designed, and built a multicrystal spectrometer in 1929. Consisting of fifty small, individual calcite crystals, it was the first of several ingenious spectrometers DuMond developed to study Compton line broadening. (In 1964 DuMond and Kirkpatrick donated this spectrometer to the Smithsonian Institution, Washington, D.C.) Another unique piece of equipment, the curved crystal focusing spectrometer, followed in 1934. It was particularly useful in the high-energy region, where high resolution is hard to secure. Eager to extend the technique developed for high-energy X-ray studies to nuclear physics, DuMond in 1937 designed a large focusing bent-crystal gamma-ray spectrometer. An immediate success when tested after World War II, this new instrument for measuring gamma-ray frequencies became an important research tool in the hands of nuclear spectroscopists at Caltech and elsewhere. DuMond collaborated on a wide range of gamma-ray studies in Pasadena, from analyzing the nuclear energy levels in rare earth nuclei to measuring with extraordinary precision the wavelengths of gamma rays emitted by more than a score of nuclides.

DuMond's work on the fundamental physical constants began in the early 1930's, in response to several independent reports that Millikan's 1917 oil-drop value of the electron charge e was about 0.6 percent too low. Convinced that the discrepancy in measurements was "one of those small errors in our bookkeeping of nature's accounts, behind which there may hide something of great interest and importance" (as he wrote in a letter to L. L. Watters, 11 January 1935), DuMond and his students studied topics ranging from the validity of X-ray methods of determining the charge on the electron to determining the ratio of h/e (where h is Planck's constant) from the short wavelength limit of the continuous X-ray spectrum (1936–1942). His numerous postwar critical studies of the numerical values of the atomic constants, done in collaboration with E. Richard Cohen, remained the definitive work on the subject until a more precise measurement of h/e, using the *ac* Josephson effect in superconductors (1969), led to a new set of values for the fundamental constants.

BIBLIOGRAPHY

I. ORIGINAL WORKS. The starting point for the study of DuMond's life and work is his unpublished "Autobiography of a Physicist," 2 vols. (1972), copies of which are at the Center for History of Physics at the American Institute of Physics, and in the archives of the California Institute of Technology. Volume I includes a complete chronological list of more than 175 scientific papers DuMond wrote between 1915 and 1970. With E. Richard Cohen and K. M. Crowe, DuMond published *Fundamental Constants of Physics* (New York, 1957). On DuMond's role as guardian of the constants, see his "Pilgrims' Progress in Search of the Fundamental Constants," in *Physics Today*, **18** (October 1965), 26–43.

Manuscript sources include extensive correspondence between DuMond and Millikan from 1934 to 1946, in the papers of Robert A. Millikan; between DuMond and DuBridge from 1946 to 1966, in the papers of Lee A. DuBridge; and nine boxes of correspondence and reprints, all in the archives of the California Institute of Technology. Another 157 letters are in the papers of Raymond T. Birge in the Bancroft Library at Berkeley.

II. SECONDARY LITERATURE. W. K. H. Panofsky, DuMond's son-in-law, reviewed his career, treating also DuMond's World War II contributions and inventions, in *Biographical Memoirs. National Academy of Sciences*, **52** (1980), 160–201, with portrait and bibliography. Short obituary notices include E. Richard Cohen, "Jesse W. M. DuMond," in *Physics Today*, **30** (March 1977), 74–75; F. Boehm, "Jesse W. M. DuMond, 1892–1976," in *Engineering and Science*, **40** (March–April 1977), 35–36. On the X-ray determination of the electron charge and Millikan's value of e, see R. H. Kargon, *The Rise of Robert Millikan: Portrait of a Life in American Science* (Ithaca, N.Y., 1982), chap. 6. On the *ac* Josephson effect and the physical constants, the classic paper is B. N. Taylor, W. H. Parker, and D. N. Langenberg, "Determination of e/h, Using Macroscopic Quantum Phase Coherence in Superconductors: Implications for Quantum Electrodynamics and the Fundamental Physical Constants," in *Review of Modern Physics*, **41** (1969), 375–496.

JUDITH R. GOODSTEIN

DUNBAR, CARL OWEN (*b.* Hallowell, Kansas, 1 January 1891; *d.* Dunedin, Florida, 7 April 1979), *geology, paleontology.*

Dunbar's parents, David and Emma Thomas Dunbar, were wheat farmers. He graduated from the University of Kansas in 1913 and received the Ph.D. from Yale University in 1917. On 18 September 1914 he married Lora Beamer; they had a son and a daughter.

Dunbar was closely associated with Charles Schuchert at Yale and did his doctoral work under him, studying the Devonian rocks and fossils of western Tennessee. In 1917 and 1918 he held a Dana resident fellowship that enabled him to prepare his dissertation for publication. He spent the summer of 1918 with Schuchert in Newfoundland. Their geological investigations and determinations of fossils, made during a number of field seasons, were published as the first memoir of the Geological Society of America (1934).

After teaching for two years (1918–1920) at the University of Minnesota, Dunbar returned to New Haven and remained at Yale for the rest of his career, succeeding Schuchert as professor of paleontology and stratigraphy in 1930. Because he did not believe in sabbaticals, Dunbar taught for thirty-nine consecutive years until his retirement in 1959. He trained many of the leading twentieth-century American stratigraphers and paleontologists. Probably his greatest single contribution lay in his teaching abilities. Through his students and his textbooks, Dunbar's influence spread far beyond Yale. He published more than a hundred scientific articles.

Dunbar contributed to the third revision of Louis Pirsson and Schuchert's *Textbook of Geology*. Dunbar and Schuchert's *Historical Geology* was the main text used in the second course in geology throughout most of the United States, going through numerous printings and a major revision in 1941. It was rewritten by Dunbar in 1949 and in 1960. In 1969 he and Karl Waage modified it further. In 1966 Dunbar wrote a popular work on geology, *The Earth*.

For many years Dunbar maintained a series of paleogeographic maps that he had inherited from Schuchert and kept revising as new data appeared. In 1955, twenty-three years after Schuchert's death, the latest revisions were published under his name with an introduction by Dunbar. As chairman of the Committee on Stratigraphy of the National Research Council (1934–1953), Dunbar was charged with preparing a series of charts correlating the formations of the continent deposited during each geologic period. He was also chairman of the subcommittee that prepared the Permian chart and was an active member of three other subcommittees. The charts, published by the Geological Society of America in its *Bulletin* (1941–1960), remain fundamental references.

In addition to being a member of the geology faculty, Dunbar was appointed assistant curator of the Peabody Museum at Yale in 1920. His first task was to pack up the collections prior to demolition of the old building. Dunbar reinstalled the collections in the new building in 1925 and the following year was promoted to curator. In 1942, in addition to his teaching, he became director of the Peabody Museum, a post he held for seventeen years. He initiated a major revision of the exhibits, rapidly bringing the Peabody to the fore as one of the principal public natural history museums in the United States. Under him the museum and its facilities were used as a teaching adjunct to the university.

Dunbar is also well known for his work in invertebrate paleontology. The monograph by Dunbar and G. E. Condra, state geologist of Nebraska, on the Pennsylvanian brachiopods of Nebraska (1932) was written almost entirely by Dunbar. He later wrote a monograph on Permian brachiopods of Greenland (1955). His first major work with Condra was a study of Pennsylvanian fusulines of Nebraska (1927). The fusulines are an extinct group of relatively large microfossils—fusiform and commonly one to two centimeters long; they are extremely abundant locally in Pennsylvanian and Permian strata. They are a group that evolved rapidly and, accordingly, are extremely useful for detailed correlation. Dunbar wrote the early major papers on these fossils in the United States and continued to study them for nearly half a century.

Dunbar was one of the twenty civilian scientists selected to observe the atomic bomb test at Bikini atoll during Operation Crossroads in 1946. Dunbar was treasurer of the Paleontological Society from 1923 to 1937 and an associate editor of the *Journal of Paleontology* from 1931 to 1938. In 1940 he was elected president of the Paleontological Society; he also served another stint as treasurer from 1945 to 1946.

Dunbar received numerous honors, including membership in the National Academy of Sciences (1944), the American Academy of Arts and Sciences (1950), and the American Philosophical Society (1942). In 1959 he was awarded the Hayden Medal of the Academy of Natural Sciences of Philadelphia, and in 1967 he was the fourth recipient of the Paleontological Society Medal. In 1978 he was honored with the William H. Twenhofel Award of the Society of Economic Paleontologists and Mineralogists, named for his mentor at the University of Kansas.

Dunbar was the preeminent student of the historical geology of North American sedimentary rocks. He was probably the last American geologist to have a complete grasp of what were the positions of land and sea, and environments of deposition of the rock strata, in the country over the past 500 million years. Like Schuchert before him, and nearly all paleontologists and stratigraphers of his time, Dunbar accumulated and organized his sense of geology both as a derivation from and a support for a geology of permanent continents and ocean basins. The work of others influenced more strongly by structural contributions, such as Hans Stille and Marshall Kay, or those like Emile Argand who accepted the idea of continental drift, was readily modified to the new concept of mobilist geology that came to the fore during Dunbar's last years. Of the theoretical framework of vanished borderlands and intercontinental bridges that was so carefully constructed by Schuchert and Dunbar and nourished mainstream geology in North America for the first two-thirds of the century, little or nothing remains.

BIBLIOGRAPHY

I. ORIGINAL WORKS. Among Dunbar's writings are "The Fusilinidae of the Pennsylvanian System in Nebraska," in *Nebraska Geological Survey Bulletin*, **2** (1927), written with G. E. Condra; "Brachiopoda of the Pennsylvanian System in Nebraska," *ibid.*, **5** (1932), written with G. E. Condra; "Correlation Charts Prepared by the Committee on Stratigraphy of the National Research Council," in *Bulletin of the Geological Society of America*, **53** (1942), 429–434, with others; *Historical Geology* (New York, 1949; 2nd ed., 1960; 3rd ed., 1969), 3rd ed. written with Karl Waage; "Permian Brachiopod Faunas of Central East Greenland," in *Meddelelser om Grønland*, **110**, no. 3 (1955), 1–169; *Principles of Stratigraphy* (New York, 1957), written with John Rodgers; "Correlation of the Permian Formations of North America," in *Bulletin of the Geological Society of America*, **71** (1960), 1763–1800, with others; and *The Earth* (New York, 1966).

II. SECONDARY LITERATURE. Obituaries include Preston Cloud, "Carl Owen Dunbar (1891–1979)," in *American Philosophical Society. Year Book 1980* (1981), 561–567; J. Thomas Dutro, Jr., "Carl Owen Dunbar," in *Journal of Paleontology*, **55** (1981), 695–697; John Rodgers, "Carl Owen Dunbar," in *Biographical Memoirs. National Academy of Sciences*, **55** (1985), 215–245, with portrait and bibliography; and Karl M. Waage, "Memorial to Carl Owen Dunbar, 1891–1979," in *Geological Society of America. Memorials*, **11** (1981).

ELLIS L. YOCHELSON

DUNN, LESLIE CLARENCE (*b.* Buffalo, New York, 2 November 1893; *d.* New York City, 19 March 1974), *genetics.*

Dunn's parents, Clarence Leslie Dunn and Mary Eliza Booth Dunn, both from farming families, were high school graduates. Leslie went to Dartmouth College on a scholarship, and there his interest in zoology was awakened by John H. Gerould, an inspiring teacher with a strong interest in the new science of genetics. Dunn graduated in 1915 (Phi Beta Kappa) and then went to Harvard University, where he worked under William Ernest Castle. For his D.Sc. dissertation (1920) Dunn made linkage studies on certain genes in mice and rats.

Dunn served as an officer in the U.S. Army in France during World War I. He returned to Harvard in 1919. In May 1918 he had married Louise Porter. They had two sons, the younger of whom, Stephen, in spite of a spastic disability, became a collaborator with his father in anthropological and genetic studies of the Jewish community in Rome in the 1950's.

Dunn's first appointment, in 1920, was to the post of poultry geneticist at the Agricultural Experiment Station at Storrs, Connecticut. In the ensuing eight years he did pioneering work on the genetics of chickens. In scientific insight, versatility, and care in experimental design and execution, that work has probably never been surpassed. Dunn determined the relation between egg weight and hatchability, and the susceptibility of egg size to natural as well as artificial selection. He also explored the loss of vigor resulting from inbreeding and the restoration of vigor by outcrosses between different breeds. Much of this work was done in collaboration with a close friend and colleague, Walter Landauer.

While at Storrs, Dunn entered into collaboration with the plant geneticist Edmund Ware Sinnott to write an introductory textbook, *Principles of Genetics* (1925). Through two subsequent editions (1932, 1939) and two further revisions (1950, 1958), of which Theodosius Dobzhansky became a leading coauthor, this text remained one of the foremost in its field. Scarcely any American geneticist active in the five decades after its first publication did not owe it a great debt, either as student or as teacher. Masterly exposition and an extensive use of carefully designed problems for students to solve characterized all revisions.

In 1928 Dunn was appointed professor of zoology at Columbia University, to succeed Thomas Hunt Morgan. The facilities at Columbia were not suitable for work with poultry, so Dunn decided to continue some work with *Drosophila*, in the Columbia tradition, and to resume his studies of the genetics of mice, which he had never abandoned entirely. He was fortunate in recruiting excellent graduate students to work with him, and these were joined by visiting investigators and postdoctoral associates, especially Salome Gluecksohn-Schoenheimer (later Waelsch), who was trained under Hans Spemann in embryology and who from 1938 to 1954 collaborated with Dunn in many studies of abnormal mouse development caused by mutant genes or specific genotypes. Her place was later filled by Dorothea Bennett, who continued with Dunn to analyze the extraordinary genetic variability at the T = t locus (short tail or tailless phenotypes) arising by repeated mutations and widespread in mouse populations. The evolutionary forces that could lead to the spread of such lethal or near-lethal genes in populations fascinated Dunn. This prolonged investigation was probably his greatest original work in genetics.

The *T* mouse carries a dominant gene that produces a short tail. A recessive *t* gene, when homozygous (*tt*), produces taillessness. The compound, surprisingly, is quite viable and manifests only the short-tail characteristic. When *Tt* mice are bred together, neither homozygous *TT* nor *tt* offspring are born; only *Tt* mice are produced. As the *Tt* cross breeds true, *T* and *t* appear to be alleles, located in the same chromosome pair at the same locus.

Dunn suspected that the *Tt* mice bred true because both homozygous types (*TT* and *tt*) are lethal and do not recombine by crossing over. Hence he looked for deaths during embryonic life in the litters produced from *TT* by *tt* matings, and he did find two classes of lethal embryos, so that half of all the embryos were dying. Such a genetic situation had previously been found only in the fruitfly *Drosophila*, by H. J. Muller. But why, if *T* and *t* are alleles, do they show no interaction in effect?

Next a totally unexpected result turned up. In crosses of *Tt* or *tt* males by normal females, there was a great excess of offspring receiving the *t* gene from its male parent, although when females of those types were crossed to normal males, equal transmission of *T* and *t* occurred. Dunn recognized that he had found a situation in which a deleterious mutant gene could maintain itself in a population, by means of its superior ability to be transmitted through the spermatozoa. He therefore began a long search of many laboratory and wild populations of mice to find whether he could detect other cases of *t* alleles with this startling evolutionary propensity to counteract natural selection. Many were found. Thus, some detrimental genes can survive in populations and contribute materially to the "genetic load" burdening the population.

Dunn was later able to show, by means of crosses to new "marker" genes closely linked to *T* (about 6 units), that *t* mutant genes completely suppress recombination between *T* and the markers. In actuality, then, the *Tt* system is not a single locus, but a chromosome segment carrying genes of related function. This was the first discovery of such clusters of genes with similar or associated functions in a mammal. It was a situation of both evolutionary and teratological importance, for such regions, or segments, of related function are now known to be widespread among species, including the human one. They demonstrate, on the one hand, how genes of related function may, by a process of tandem duplication, become multiplied and thereafter diverge somewhat in function. On the other hand, they reveal the similar, if not quite identical, nature of closely linked genes that cause similar developmental abnormalities. In both these respects, Dunn's studies of the *Tt* genes of mice illuminate current studies of the human genome.

From the early 1920's Dunn actively investigated the genetics of human races and populations. Race mixture in Hawaii, the genetic isolate of the Jews in Rome, and blood group research were interests that led him to establish the Institute for the Study of Human Variation (1952–1958). Dunn's contributions to human genetics were recognized by his election in 1961 to the presidency of the American Society of Human Genetics.

During the 1930's the rise of Nazi race policies and anti-Semitism in Germany greatly distressed Dunn. He labored intensively to aid scientist refugees arriving in America. During World War II his social sympathies led him to participate in the American-Soviet Friendship Council, which he helped to found, and the American-Soviet Science Society. Lysenkoism in Russia altered his feelings toward the Soviet Union, but his associations brought him some political persecution in the 1950's. In the scientific community Dunn always stood strongly for academic freedom and scientific integrity, and was a leader upon whom many young scientists looked with reverence.

His friendship with Milislav Demerec led Dunn to make many influential suggestions in the organization of the great Cold Spring Harbor symposia on genetics and evolution in the 1940's and 1950's. He was renowned as an editor; he was the first editor of *Genetics* (1935–1940), and the editor of *American Naturalist* (1951–1960), of *Genetics in the Twentieth Century* (1951), and of *Race and Biology*, prepared for UNESCO (1951; 3rd ed., 1970). His popular books *Heredity and Variation* (1932), *Heredity, Race and Society* (written with Theodosius Dobzhansky, 1946; 4th ed., 1972), *Biology and Race* (1951), and *Heredity and Evolution in Human Pop-*

ulations (1959), together with many articles in encyclopedias and other compendia, exerted a wide influence. Dunn possessed a strong historical sense, which bore fruit in a final book, *A Short History of Genetics 1864–1939* (1965), as well as in the impetus he gave to the formation of a rich archive of geneticists' documentary materials at the American Philosophical Society.

Dunn received a great many honors, including an honorary Sc.D. degree from Dartmouth College, and election to the National Academy of Sciences, the American Philosophical Society, the Accademia Pataviana, the Norwegian Academy of Sciences, and the presidency of the Genetics Society of America (1932) and of the American Society of Naturalists (1960).

BIBLIOGRAPHY

I. Original Works. A full bibliography is appended to the memoir of Dunn by Theodosius Dobzhansky, in *Biographical Memoirs. National Academy of Sciences*, **49** (1978), 79–104. The L. C. Dunn papers, fully indexed, are preserved in the library of the American Philosophical Society.

II. Secondary Literature. See Joseph S. Fruton, *A Bio-bibliography for the History of the Biochemical Sciences Since 1800* (Philadelphia, 1982) for biographical listings. See also Beutley Glass, *A Guide to the Genetics Collections of the American Philosophical Society* (Philadelphia, 1988). Dorothea Bennett, "L. C. Dunn and His Contribution to T-Locus Genetics," in *Annual Review of Genetics*, **11** (1977), 1–12, is an exceptionally rich and full account of that major scientific work.

Bentley Glass

ECKART, CARL HENRY (*b*. St. Louis, Missouri, 4 May 1902; *d*. La Jolla, California, 23 October 1973), *physics, geophysics, acoustics*.

Eckart was the son of William Eckart and Lilly Hellwig Eckart. After receiving the B.S. in engineering from Washington University (St. Louis) in 1922, he held a one-year fellowship for graduate work there, studying under George E. M. Jauncey and receiving a master's degree in 1923. For the next two years he studied under Karl T. Compton at Princeton, supported by an Edison Lamp Works fellowship, and received the Ph.D. in 1925. In order to work with Paul S. Epstein, Eckart moved to the California Institute of Technology, where he held a National Research fellowship. Eckart's first marriage (1926) ended in divorce after eighteen years. He was married again in 1958 to Klara von Neumann, the widow of a friend of his, the mathematician John von Neumann; she drowned in 1963. Both marriages were childless.

A Guggenheim fellowship enabled Eckart to work with Arnold Sommerfeld at Munich in the academic year 1927–1928. In the latter year he returned to the United States and became assistant professor of physics at the University of Chicago. From 1931 to 1946 Eckart was associate professor at Chicago, and from 1942 to 1946 he was assistant director (and then director) of the University of California's division of war research at San Diego. In 1946 he became professor of geophysics at the University of California, Los Angeles. He served as director of the Marine Physical Laboratory, San Diego, from 1946 to 1952; from 1948 (when the Marine Physical Laboratory became an integral part of the Scripps Institution of Oceanography at the University of California, La Jolla) he was also at this latter institution, serving as director from 1948 to 1950; Eckart was vice chancellor for academic affairs at the University of California, San Diego, from 1965 to 1969. He retired in 1971.

The greatest influence on Eckart's scientific career was Frank W. Bubb, professor of applied mathematics at Washington University. Working with Bubb, Eckart became acquainted with descriptive geometry, vector analysis, and the elements of theoretical physics. Ironically, although he planned to do graduate work in mathematics, only his application for a teaching fellowship in the physics department, where Bubb had prepared the ground, was successful.

When Eckart started graduate work with George E. M. Jauncey at Washington University, he was immediately involved in the ongoing discussions between Jauncey and Arthur H. Compton on the scattering of X rays and gamma rays. Although these discussions were very important for Compton's development of the quantum theoretical explanation of the greater coefficient of absorption of gamma rays, the X-ray experiments, which eventually led to the establishment of the Compton effect, were entirely the work of Compton alone. Jauncey and Eckart only performed experiments to confirm the assumption that the modified radiation is incoherent and thus does not produce any interference effects.

Clearly under the influence of Louis de Broglie's wave theory of the electron, which stressed the similarity of a trajectory of a particle and the normal to a wave front, Eckart tried to give a different explanation of the Compton effect by deriving the change in frequency from the calculation of a curved path of an electromagnetic wave in the nonstatic gravitational field of the recoil electron. In 1925, when Max Born lectured at Pasadena, Eckart's interest in the new quantum mechanics was stimulated. Another influence came from Cornel Lanczos, who in 1925 had shown that the Heisenberg-Born-Jordan theory can be expressed not only in terms of matrices but also in terms of integral equations.

Eckart therefore spent the spring of 1926 using the operator calculus to connect the already published Heisenberg-Born-Jordan matrix mechanics and Erwin Schrödinger's wave equation, which has easily intelligible quantum conditions. Deriving the matrices by integrations of the orthonormal eigenfunctions (of the "characteristic" solutions) of the wave equation, he solved the problem in the case of the harmonic oscillator. His result was partly worked out and presented in a preliminary paper submitted to *Proceedings of the National Academy of Sciences* on 31 March 1926. However, Schröedinger, in a paper received by the editor of *Annalen der Physik* on 18 March 1926, had already clarified the relation between the quantum mechanics of Heisenberg, Born, and Jordan and that of Schrödinger. Due to delay in the mails, Eckart's preliminary paper was published before Schrödinger's paper reached him at Caltech. Thus, in his completed paper in *Physical Review*, Eckart had to acknowledge Schrödinger's priority.

Although theoretical physics in the United States matured around 1925, postdoctoral work in Germany was still very attractive. Eckart and William Houston worked with Arnold Sommerfield at Munich, arriving just in time to be involved in the electron gas theory of metals, which was reformulated by Sommerfeld and his school with the help of Fermi statistics; Eckart dealt with special problems of contact phenomena and with thermionics. Later, in Berlin, he had a disappointing encounter with John von Neumann's completely abstract approach to quantum theory and with Schrödinger's search for a representation in easily visualizable three-dimensional space. When Eckart returned to the United States in 1928 and joined the staff of the physics department at the University of Chicago, his work remained in the realm of quantum theory. With Helmut Hönl he wrote an overview article for *Physikalische Zeitschrift* on the principles and results of wave mechanics, and with Frank C. Hoyt he prepared the English edition of Werner Heisenberg's *Physical Principles of Quantum Theory*, lectures that had been delivered in 1929.

Eckart stressed the importance of the statistical interpretation and the linguistic character of the uncertainty principle. His (and Eugen P. Wigner's) work on the application of group theory in quantum theory led to the Wigner-Eckart theorem, which is a means to calculate the resulting angular momenta of complex systems in quantum theory.

Beginning in 1930, Eckart did research in diffraction by gratings, the quantum theory of polyatomic molecules, the theories of the atomic nucleus, and cosmic ray measurements. In the 1940's he focused on the thermodynamics of fluids, on geophysical hydrodynamics, and on wave generation and propagation in fluids. The latter topics were closely related to the development of sonar systems at the University of California division of war research, so his report, "The Principles of Underwater Sound," was only declassified in 1954.

In underwater acoustics Eckart concentrated on the theory of reverberation and scattering of sound from the sea surface. With better approximations he tried to make the available data for short wave sound an additional source of information for the geometry and the kinematics of the sea surface. Also influential in the sonar field was his research on optimal rectifier systems for detecting steady signals in noise. He continued applying the thermodynamics of irreversible processes to the physics of fluids, to the behavior of elastic and unelastic media, and to oceanic phenomena, especially to surface waves. Starting from a wave equation analogous to Schrödinger's he contributed to the theory of internal oscillations of the sea.

Although Eckart essentially brought into play the mathematical tools developed earlier in hydrodynamics, in the solution of wave equations, in statistical thermodynamics, and in the "new" quantum mechanics, he often carried his approximations to such a degree that his analysis was valuable in studying actual phenomena. So he encouraged experimental work on the unexplained high absorption shown by seawater as opposed to fresh water, and on the propagation of sound penetrating significantly into the crust of the earth below the sea.

Eckart was a member of the National Academy of Sciences and of the American Academy of Arts and Sciences. He received the Agassiz Medal of the National Academy of Sciences for his contribution to the hydrodynamics of the ocean, and the American Geophysical Union awarded him the Bowie Medal. On 31 October 1973 the Acoustical Society of America honored him posthumously with the Pioneers of Underwater Acoustics Medal for his contributions to the understanding of underwater sound and acoustic signal processing.

BIBLIOGRAPHY

I. ORIGINAL WORKS. Eckart's writings include "The Wave Theory of the Compton Effect," in *Physical Review*, 2nd ser., **24** (1924), 591–596; "Post-Arc Conductivity and Metastable Helium," *ibid.*, **26** (1925), 454–464, his dissertation; "Operator Calculus and the Solution of the Equations of Quantum Dynamics," *ibid.*, **28** (1926), 711–726; "Über die Elektronentheorie der Metalle auf Grund der Fermischen Statistik, insbesondere über den Volta-Effekt," in *Zeitschrift für Physik*, **47** (1928), 38–42; "The Application of Group Theory to the Quantum Dynamics of Monoatomic Systems," in *Review of Modern Physics*, **2** (1930), 305–380; "Grundzüge und Ergebnisse der Wellenmechanik," in *Physikalische Zeitschrift*, **31** (1930), 89–119, 145–160, written with Helmut Hönl; *The Physical Principles of the Quantum Theory*, by Werner Heisenberg, translated by Eckart and Frank C. Hoyt (Chicago, 1930); and *Hydrodynamics of Oceans and Atmospheres* (New York, 1960).

"Seminar on Wave Theory," notes of lectures delivered at the Marine Physical Laboratory, San Diego, in 1946 and 1947 are at the library of the University of California, Berkeley. "Surface Waves on Water of Variable Depth," notes of lectures delivered at the Scripps Institution of Oceanography in 1950 and 1951 were issued as *Wave Report* no. 100 (1951), held at the library of the U.S. Department of the Interior.

Letters to or from Eckart are at Mudd Library, Princeton University; Nathan Marsh Pusey Library, Harvard University; Fondren Library, Rice University; the Deutsches Museum, Munich; Bancroft Library, University of California, Berkeley; and Archive for History of Quantum Physics, Office for History of Science and Technology, University of California, Berkeley. The last also has the transcript of an interview with Eckart, conducted by John L. Heilbron on 31 May 1962. The collected unpublished works of Carl Eckart are at the library of the Scripps Institution of Oceanography, University of California, San Diego.

II. SECONDARY LITERATURE. A biographical article is Walter H. Munk and Rudolph W. Preisendorfer, "Carl Henry Eckart," in *Biographical Memoirs. National Academy of Sciences*, **48** (1976), 195–219, with a bibliography of his works. On the Wigner-Eckart theorem, see Eugen Wigner, "Einige Folgerungen aus der Schrödingerschen Theorie für die Termstrukturen," in *Zeitschrift für Physik*, **43** (1927), 624–652.

WALTER KAISER

EDELMAN, CORNELIS HENDRIK (*b.* Rotterdam, Netherlands, 29 January 1903; *d.* Wageningen, Netherlands, 15 May 1964), *soil science*.

The son of a schoolteacher, Edelman began studying geology and mining at the Technological University of Delft in 1919 and graduated in 1924. When one of his professors, Hendrik A. Brouwer, was appointed to the chair of geology at the University of Amsterdam in 1929, Edelman followed him there. Edelman was married twice: to Johanna Van Werkhoven, with whom he had six children, and to Alida W. Veam, a historical geographer.

Edelman was a member of the Royal Netherlands Geological and Mining Society, the Royal Netherlands Geographical Society, and the International Soil Science Society. He was rector magnificus of the State Agricultural University at Wageningen in 1946 and 1947. In 1958 he was elected to the Royal Netherlands Academy of Sciences; he was also a corresponding member of the French Academy of Agronomy. He received an honorary doctorate in science from the University of Ghent and was made a knight (Lion) of the Order of the Netherlands.

While at Amsterdam, Edelman established and operated the first consulting laboratory for sedimentary petrology, in which he was assisted by D. J. Doeglas. This was the beginning of extensive sedimentological investigations conducted for Royal Dutch Shell, to aid in oil exploration, a function later taken over by the corporation's research department.

In the following years the laboratory carried out many investigations of this kind on the sedimentary petrology of the Netherlands' Delta area for the Ministry of Transport and Water Management, as a preparation for the construction of dikes and polders. During this period Edelman completed his doctoral dissertation at the University of Amsterdam ("Petrological Provinces of the Netherlands Quaternary," 1933). His improved method for the analysis of heavy minerals in sediments and soils is still generally used, with only minor modifications.

In 1933 Edelman was appointed to the chair of geology, mineralogy, petrography, and agrogeology (later renamed regional soil science) at the State Agricultural University of Wageningen, a post he held until his death. His investigations on sedimentary petrology continued, and he began research on the clay minerals in soils. His structural model of montmorillonite was considered the best approximation until it was superseded by a more exact method after 1945, and his work on halloysite was of fundamental interest. After 1942 he improved the nineteenth-century discipline of agrogeology by applying it to modern soil survey, with strong emphasis on a physiographic approach. This resulted in the establishment, in July 1945, of the Stichting voor Bodemkartering (Soil Survey Institute), the official national institution for soil and landscape surveys and land evaluation, of which he was director until 1955. Throughout the same period, beginning in 1934, Edelman was involved in soil science research in the Netherlands East Indies (now Indonesia).

Immediately after World War II Edelman established many international contacts, in particular with the U.S. Soil Survey and its head, Charles E. Kellogg. This cooperation led to the use of international standards of soil classification in the Netherlands soil survey. On his first mission for the Food and Agriculture Organization (FAO) of the United Nations (1947), Edelman was a member of an international commission to study problems of food and agriculture in Poland. He was frequently a consultant to that organization. These and other contacts led to the reactivation of the International Soil Science Society and to its fourth congress at Amsterdam in 1950.

Within the framework of FAO, as well as bilateral programs between the Netherlands and other countries, Edelman made great contributions to cooperation on development projects with Third World nations. He helped to establish three institutes designed for this purpose: the International Training Center for Aerial Survey (ITC) at Delft (now the International Institute for Aerospace Surveys and Earth Sciences, Enschede), as well as the International Agriculture Center and the International Institute for Land Reclamation and Improvement, both at Wageningen.

Edelman's scientific activities in the postwar period were especially concerned with (1) physiographic soil surveys with land evaluation and land use planning, (2) late Pleistocene periglacial phenomena and processes, (3) fluvial, estuarine, and marine sedimentation and marine transgressions, and (4) the influence of man on the genesis and development of soils, including cooperation with archaeologists and historical geographers. Often with students and former students he made major contributions to these fields that opened up vistas for further scientific developments.

Edelman was an inspiring and stimulating teacher who supervised many master's and doctoral students in the fields of soil survey and land evaluation. In their careers most of them used and developed Edelman's methods and theories in soil surveys and other research in the Netherlands and abroad, particularly in developing countries. His influence is still very strong in soil survey and land evaluation.

BIBLIOGRAPHY

I. Original Works. Edelman's writings include "Petrologische provincies in het Nederlandsche Kwartair" (Ph.D. diss., Amsterdam, 1933); "On the Crystal Structure of Montmorillonite and Halloysite," in *Zeitschrift für Kristallographie*, **102** (1940), 417–431, written with J. C. L. Favejee; *Studiën over de bodemkunde van Nederlandsch Indië* (Wageningen, 1941; 2nd ed., 1947); *Soils of the Netherlands*, J. Verwey, trans. (Amsterdam, 1950); "Soils," in *Soil Science*, **74** (1952), 15–20; "De subatlantische transgressie langs de Nederlandse kust," in *Geologie en mijnbouw*, **15** (1953), 351–364; "Sedimentology of the Rhine and Meuse Delta as an Example of the Sedimentology of the Carboniferous," in *Gedenkboek H. A. Brouwer* (Amsterdam, 1956), 1–12; "Pleistozän-geologische Ergebnisse der Bodenkartierung in den Niederlanden," in *Geologisches Jahrbuch* (Hannover, 1958), 639–684, written with G. C. Maarleveld; and *Applications of Soil Survey in Land Development in Europe*, publication 12 of the International Institute for Land Reclamation and Improvement (Wageningen, 1963).

II. Secondary Literature. An obituary of Edelman written by A. P. A. Vink can be found in Geologische Rundschau, **54**, no. 2 (1964), 1320–1322.

A. P. A. Vink

EDINGER, JOHANNA GABRIELLE OTTILIE (TILLY) (*b*. Frankfurt am Main, Germany, 13 November 1897; *d*. Cambridge, Massachusetts, 27 May 1967), *vertebrate paleontology*.

Edinger was the youngest of four children of Ludwig and Anna Goldschmidt Edinger. Her father, one of the leading neuroanatomists of his time, created the neurological institute that bears his name at the University of Frankfurt. Her mother, who came from a family of bankers, was highly esteemed for her work in social welfare.

From 1916 to 1918 Edinger studied psychology, zoology, and geology in Heidelberg and then in Munich. After returning to Frankfurt, she devoted herself to zoology, as well as geology and paleontology, under the guidance of Friedrich Drevermann, who decisively influenced her interest in the biological interpretation of fossils. In the fall of 1921 Edinger received her doctorate with a work on *Nothosaurus*. In the same year she described the endocast of the cranial cavity of a nothosaur—a highly consequential research that determined her future scientific pursuits. These were a natural consequence of the scientific talent she inherited from her father but were not the result of his influence, for he died young and was opposed to university education for women. Serving as part-time assistant to Drevermann and enjoying financial independence, Edinger was able to devote herself to the study of fossil brains. In 1927 she became unpaid curator of the natural history museum of Senckenberg.

Edinger undertook the methodical collection of literature on fossil brains. In 1929 her "Die fossilen Gehirne" appeared, a piece that laid the foundations of paleoneurology. It was characterized by George G. Simpson as an invaluable review that served as a basis for continuing and systematizing research on brain casts and as an indication of gaps in current knowledge. Edinger's most meaningful subsequent work while in Frankfurt involved the brains of Sirenia (1933), the first evidence of an admittedly incomplete series of brains deriving from a long time span.

At the same time this work appeared, Hitler was coming to power. Edinger, of Jewish extraction, nonetheless refused to emigrate, and Rudolf Richter, then director of the Senckenberg Museum in Frankfurt, protected her. By 1938, however, her situation had become dire, and it was a matter of having to emigrate. Alfred S. Romer, director of the Museum of Comparative Zoology at Harvard, offered her the possibility of working for him. In May 1939 she fled to London, where she found a position as a translator. She reached Cambridge in 1940, supported initially by the Emergency Committee in Aid of Displaced Foreign Scholars. Subsequently she received fellowships from the Guggenheim Foundation (1943–1944) and the American Association of University Women (1950–1951). In the years 1944 and 1945 she taught at Wellesley College, until her hearing loss made it impossible. She became a U.S. citizen in 1945.

The enormous amount of material in the museums of the United States was available for Edinger's paleoneurological research. In 1948 she published the classic study *Evolution of the Horse Brain*. In it she demonstrated that the progression in brain structure, as in other organs, does not proceed at a constant rate within a given family; instead, the rate varies over time. Edinger also showed that the evolution of the brain can be researched only with the aid of fossils. This monograph was responsible for the extraordinarily rapid development of paleoneurology in subsequent decades.

Edinger's interests extended far beyond fossil brains. With three other authors she published *Bibliography of Fossil Vertebrates Exclusive of North America (1509–1927)* in 1961. She hoped to complete her life's work with a new edition of "Die fossilen Gehirne." In the meantime, however, paleoneurology had grown so much that such an edition would have taken years. She therefore concentrated on completing and publishing a bibliography in that field; she did not, however, live to see it issued. On 26 May 1967 she was involved in a traffic accident and died the following day. Fortunately, her bibliography was later put in order and published.

After World War II, Edinger's scientific accomplishments brought her honorary doctorates from Wellesley College (1950) and the University of Giessen (1957) and an honorary medical doctorate from the University of Frankfurt am Main (1964). In this period Edinger renewed old ties, above all with Frankfurt. She made a vital contribution to bringing German paleontologists out of their postwar isolation. Edinger's congenital hearing loss, which began early in life and grew worse with age, resulted in her being substantially isolated from her environment. To many she therefore appeared to symbolize the solitary woman shut off from the world. But in essence she was precisely the opposite. Temperamental, often stubborn, she was always warmhearted, lovable, and nimble of mind.

In her memory, friends established the Tilly Edinger Fund at the Museum of Comparative Zoology, Harvard University. The fund supports the writing of books in vertebrate paleontology.

BIBLIOGRAPHY

I. ORIGINAL WORKS. A bibliography of Edinger's works is included with the article by Hofer (see below). Among her writings are "Ueber Nothosaurus" (1921), her dissertation; "Ueber Nothosaurus. I. Eine Steinkern der Schädelhöhle," in *Senckenbergiana*, **3** (1921), 121–129; "Ueber Nothosaurus. II. Zur Gaumenfrage," *ibid.*, 193–205; "Ueber Nothosaurus. III. Ein Schädelfund im Keuper," *ibid.*, **4** (1922), 37–42; "Die fossilen Gehirne," in *Ergebnisse der Anatomie*, **28** (1929), 1–249; "Ueber Gehirne tertiärer Sirena Aegyptens und Mitteleuropas sowie der rezenten Seekühe. Ergebnisse der Forschungsreise Prof. E. Stromer in den Wüsten Aegyptens. V. Wirbeltiere," in *Abhandlungen der Bayerischen Akademie der Wissenschaften*, mathematisch-naturwissenschaftliche Abteilung, n.s. **20** (1933), 1–36; *Evolution of the Horse Brain, Memoirs of the Geological Society of America*, no. 25 (1948); *Bibliography of Fossil Vertebrates Exclusive of North America 1509–1927*, 2 vols. (1962), with A. S. Romer, N. E. Wright, and R. von Frank; and *Paleoneurology 1804–1966. An Annotated Bibliography* (New York, 1975), with foreword by Bryan Patterson.

II. SECONDARY LITERATURE. Stephen Jay Gould, "Edinger, Tilly," in *Notable American Women* (1980); H. Hofer, "In Memoriam Tilly Edinger," in *Morphologisches Jahrbuch*, **113** (1969), 303–317, with bibliography and portrait; Bryan Patterson, in *Ergebnisse der Anatomie*, **49** (1975), 7–11; A. S. Romer, "Tilly Edinger, 1897–1967," in *News Bulletin of the Society of Vertebrate Paleontologists*, **81** (1967), 51–53, with portrait; and Heinz

Tobien, "Tilly Edinger†, 13.11.1897–27.5.1967," in *Paläontologische Zeitschrift*, **42** (1968), 1–2.

EMIL KUHN-SCHNYDER

EDWARDS, AUSTIN BURTON (*b*. Melbourne, Australia, 15 August 1909; *d*. Rome, Italy, 8 October 1960), *geology*.

Edwards, fifth and youngest child of William Burton Edwards and Mabel Edwards, was born into a comfortable upper-middle-class home, his father having been a senior public servant. Educated at the University of Melbourne (B.Sc. with first-class honors, 1930), Edwards won a scholarship to the Imperial College of Science, London, where in 1934 he received the doctorate with a dissertation on the petrology of the volcanic and dyke rocks of Victoria.

Tall and well built, Edwards maintained his enthusiasm for sport throughout his life, gaining university sporting honors in Australian football at Melbourne and in athletics at Imperial College. In later years he coached football at Melbourne. In 1935, he married Eileen McDonnell; they had three daughters and a son.

Edwards returned to Melbourne and in 1935 joined Frank Leslie Stillwell in the mineragraphic section of the Council for Scientific and Industrial Research (CSIR), which in 1929 had been established within and closely identified with the geology department of the University of Melbourne. Edwards and Stillwell worked together on the mineralogy and petrology of Australian ore bodies until Stillwell's retirement in 1953, when Edwards took charge of the mineragraphic section of the newly constituted Commonwealth Scientific and Industrial Research Organization (CSIRO).

Supported by funding from the mining industry, Stillwell and Edwards published twelve papers (mainly in the *Proceedings of the Australasian Institute of Mining and Metallurgy*) on Australian ore bodies and their mineralogy. These covered the whole range of size and composition, from the lead-zinc deposits of Broken Hill (New South Wales) and Mount Isa (Queensland) to the uranium deposits of Mount Painter (South Australia), and occurrences of submicroscopic gold. Some of these publications—and papers written with, among others, George Baker and Arthur Gaskin—considered the properties of smelter mattes and mill products, as well as those of the mineral deposits.

Edwards's first major work after joining CSIR was a study of the mineragraphy of the iron ores of the Middleback Ranges of South Australia; it was followed by examination of the iron deposits of Yampi Sound, Western Australia. At the time, the latter were believed to be the major occurrences of very limited Australian iron ore resources. Edwards next studied the mineral composition of the Mount Lyell copper ores of Tasmania.

Edwards (and Stillwell before him) had used Hans Schneiderhöhn and Paul Ramdohr's *Erzmikroskopische Bestimmungstafeln* (1931), Rudolf Van der Veen's *Mineragraphy and Ore-Deposition* (1925), and Fairbanks's *Laboratory Investigation of Ores* (1928) in the original studies of the opaque minerals common in Australian ore bodies, but they became aware that the textures and mutual associations of the ore minerals had received inadequate attention, and that these factors were essential to understanding the genesis of the ore bodies and of the behavior of the ores during milling and smelting.

By 1947 Edwards's grasp of this aspect of mineragraphy enabled him to publish his widely acclaimed *Textures of the Ore Minerals and Their Significance*, which made use of the rapidly advancing knowledge of solid solution phenomena, thanks to the many studies of the atomic structure of minerals that had been undertaken during the 1940's.

Despite his concentration on ore body mineragraphy, Edwards's geological interests were catholic, and his intellectual curiosity and capacity led him to investigate many other geological problems. Related to his mineragraphic work were geochemical studies of ore deposits and examination of a number of meteorites (mostly in collaboration with George Baker). Further removed, but of economic importance, was his work on coal, considering both geological and petrographic aspects, carried out at the behest of the Victoria State Electricity Commission.

Next in volume to his mineragraphic publications was Edwards's petrological work. A few of these writings deal with sedimentary petrology, but the bulk are concerned with igneous and metamorphic petrology. These range from his earliest publications (1932) on the dacitic rocks northeast of Melbourne, through papers on tholeiitic basalt, to those on scapolitization and amphibolites. Despite Edwards's reputation as an outstanding ore microscopist, some colleagues believe his work in igneous petrology is his best. Another of Edwards's geological loves was the landscape itself, inspired by field excursions in his student days under Ernest Wellington Skeats. Ten papers, all written alone, cover aspects of shoreline development and structural control of landforms. Edwards was able to pass on his expertise and enthusiasm in a wide range of geological topics when he served as part-time lecturer in mining geology at the University of Melbourne between 1941 and 1953. He also gave some postgraduate courses—one of which was the basis for his *Textures of the Ore Minerals*. His wide knowledge was put to good use when he edited *Geology of Australian Ore Deposits* for the Fifth Empire Mining and Metallurgical Congress in 1953. This book, containing many contributions from Edwards, has become a classic of Australian geological literature.

Edwards's approach to geological problems was to find the genesis of the particular matter under consideration: What causes the particular textures observed in an ore body? What does the character of a granodiorite reveal about its origin? How does coal form? All these fundamental theoretical problems received attention even if the project being undertaken was ostensibly practical. Edwards was awarded a D.Sc. by the University of Melbourne in 1942, and became a corresponding fellow of the Edinburgh Geological Society and a foreign member of the Mineralogical Society of India. In 1958 he was named observer for the Commission on Geochemistry of the International Union of Pure and Applied Chemistry and served for three years.

In 1960, at the height of his career, Edwards set out to visit European mining areas and mineragraphic centers, and to attend several conferences. After traveling in Greece and Italy, he collapsed and died in Rome. He is buried in the English (Protestant) cemetery there, close to the grave of the poet John Keats.

BIBLIOGRAPHY

I. ORIGINAL WORKS. A bibliography of Edwards's almost 130 works is in the second article by Stillwell listed below. Some of the most significant of his works are "Three Olivine Basalt-Trachyte Associations and Some Theories of Petrogenesis," in *Proceedings of The Royal Society of Victoria*, **48** (1935), 13–26; "The Tertiary Volcanic Rocks of Central Victoria," in *Quarterly Journal of the Geological Society of London*, **94** (1938), 243–320; "The Formation of Iddingsite," in *American Mineralogist*, **23** (1938), 277–281; "Storm Wave Platforms," in *Journal of Geomorphology*, **4** (1941), 223–236; "Differentiation of the Dolerites of Tasmania," in *Journal of Geology*, **50** (1942), 451–480, 579–610; "The Mineragraphic Investigation of Mill Products of Lead-Zinc Ores," in *Journal of the Council for Scientific and Industrial Research, Australia*, **15** (1942), 161–174, written with Frank L. Stillwell; *Textures of the Ore Minerals and Their Significance* (Melbourne, 1947; new and enl. ed., 1954; repr., 1960); "Some Occurrences of Supergene Iron Sulphides in Relation to Their Environments of Deposition," in *Journal of Sedimentary Petrology*, **21** (1951), 34–36, written with George Baker; ed., *Geology of the Australian Ore Deposits*, I (Melbourne, 1953); "The Nature of Brown Coal," in Paul L. Henderson, ed., *Brown Coal, Its Mining and Utilisation* (Melbourne, 1953), 19–61; and "The Present State of Knowledge and Theories of Ore Genesis," in *Proceedings of the Australasian Institute of Mining and Metallurgy*, no. 177 (1956), 69–116.

Internal reports by Edwards are in the archives of the Commonwealth Scientific and Industrial Research Organization, Canberra, Australia. Family papers held in Melbourne contain a limited amount of scientific material, including an unpublished manuscript on the nature and origin of ore deposits.

II. SECONDARY LITERATURE. Obituaries include "Austin Burton Edwards," in *Proceedings of the Australasian Institute of Mining and Metallurgy*, no. 196 (1960), 5–8, with portrait; and Frank L. Stillwell, "Austin B. Edwards," in *Australian Journal of Science*, **23**, no. 8 (21 Feb. 1961), 260, and "Memorial of Austin Burton Edwards," in *American Mineralogist*, **46** (1961), 488–496, with bibliography and portrait.

DAVID BRANAGAN

EHRENHAFT, FELIX (*b.* Vienna, Austria, 24 April 1879; *d.* Vienna, 4 March 1952), *physics*.

Ehrenhaft was raised in comfortable, cultured surroundings; his father, Leopold Ehrenhaft, was a physician; his mother, Louise Egger, the daughter of an industrialist in Hungary. Ehrenhaft studied at the Technische Hochschule in Vienna and, after earning his doctorate in 1903 at the University of Vienna, became an assistant of Viktor von Lang and, in 1905, *Privatdozent*. Ehrenhaft was first married, to Olga Steindler, around 1912. They had two children, Johann and Anna Marie. She died in 1933. He was married again in 1935 to Bettina Stein; she died late in 1939. There were no children from this marriage. His early published papers (1902, 1903) dealt with the preparation of metallic colloids and their optical properties, including the anomalous direction of polarization maxima for reflected light. His lifework was launched in these early studies.

The colloidal state, in which particles of dispersed matter are on the order of 10^{-4} to 10^{-7} cm, was then an exciting research frontier for theory as well as for application, not least because of its bearing on the burning question of the "reality" of atoms. With the invention (1902–1903) of the slit ultramicroscope by Henry Siedentopf and Richard Zsigmondy, this "world of neglected dimensions" (so called by Wolfgang Ostwald), long thought to be forever beyond sense perception, became accessible to direct observation. Soon after Albert Einstein's

EHRENHAFT

(1905) and Marian Smoluchowski's (1906) papers on the kinetic molecular theory interpretation of Brownian movement, Ehrenhaft published (1907) the results of a simple but imaginative experiment for which he was awarded the Lieben Prize (1910) of the Vienna Academy of Sciences. It was a semi-quantitative measurement, on ultramicroscopic particles of silver and other substances, of the Brownian motion in air. As Ehrenhaft stressed, this test on gases provided a much more plausible support for atomism itself than had the earlier results on Brownian motion for liquids, including those of Franz Exner (1900) at Vienna, because the mean free path is larger in gases.

It was a natural step for Ehrenhaft, using essentially the same apparatus—soon after a similar attempt by Maurice de Broglie (1908) and at about the same time as Robert A. Millikan—to measure the electric charges such individual particles can carry, and hence to calculate the charge of the electron, *e* (published in March 1909). An accurate value of *e*, of highest importance for all branches of physics, was widely desired. For Ehrenhaft, joining this search was a decisive point in his scientific and personal life. He had entered into the labyrinth of enormously complex ultramicroscopic phenomena, which fascinated him to the end of his career, and he had also stumbled into a scientific controversy with Millikan and others that was to last for decades.

In his work of 1909, Ehrenhaft followed the motion of individual charged ultramicroscopic metal particles in a horizontal electric field. He obtained the value $e = 4.6 \times 10^{-10}$ ESU—far closer to Rutherford's 4.65×10^{-10} (from radioactivity work) and Planck's 4.65×10^{-10} (from blackbody radiation), than to the value (mean, 4.03×10^{-10}) reported in Millikan's first attempt (1908). On the occasion of the British Association for the Advancement of Science meeting of August 1909, Rutherford referred prominently to Ehrenhaft's experiments and noted that his value of *e* was one of the recent measurements, "which are far more reliable than the older estimates." But in his next paper (1910), Millikan, using balanced drops of water and alcohol, dismissed Ehrenhaft's results for *e* because they had been obtained by a method Millikan regarded as inferior, although Ehrenhaft's results were numerically very close to his own. Ehrenhaft responded to the challenge with ferocious energy.

Ehrenhaft subjected Millikan's 1910 treatment of data to a scathing critique and began a series of ever more intricate experimental attacks on the notion that the electric charge on bodies is invariably a multiple of a definite charge *e*. He claimed the discovery of the "subelectron," a concept completely at variance with all current theories of electronic and atomic phenomena, and indeed a repudiation of his own earlier atomism.

Opponents would argue in vain that in his data reduction Ehrenhaft refused to use Stokes's law in the modified form necessary for small particles; that he falsely assumed the density of the small, sponge-like metal fragments, obtained in an electric arc, to be the same as that of the mother material in the electrode; or that using small, jagged particles rather than round drops would cause leakage effects. Ehrenhaft, however, charged that his adversaries were building the existence of their sought-for indivisible electrons into the theory they were using to calculate the charge from their data; that they were using relatively large droplets on which "subelectrons" tended to be clumped together; and that only he proceeded "from the direct facts," avoiding hypotheses and relying as much as possible on the direct observation of natural phenomena.

In fact, experiments of the Millikan and Ehrenhaft type are difficult. Although in time virtually all researchers outside Ehrenhaft's circle came to corroborate Millikan's view of the atomic nature of the electric charge, there usually had to be some hypothesis-guided selection among the raw data. If Ehrenhaft had had access to Millikan's laboratory notebooks, he would have had no difficulty "proving" his case by concentrating on the runs that Millikan had omitted from his final calculations as flawed. Conversely, Ehrenhaft's greatest methodological flaw may have been that he accepted all observations, whether good or bad, having come to embrace a sensationist view of science that owed some allegiance to Ernst Mach's philosophy.

But at the time, the "quarrel about the electron" was engaging large and often distinguished audiences. This was furthered in Europe by Ehrenhaft's energetic defense, in papers and at scientific meetings, coupled with ever-new experiments issuing from his institute. As a consequence of Ehrenhaft's work, a cloud hung over Millikan's claim for years. In the 1916 edition of *The Theory of the Electron*, Hendrik A. Lorentz concluded that "the question cannot be said to be wholly elucidated," and material in the Nobel Prize archives in Sweden show that as late as 1920 Svante Arrhenius noted that while most physicists agreed with Millikan, the dispute with Ehrenhaft was not regarded as resolved, and that Millikan therefore should not be recommended for a Nobel Prize. On 19 November 1940, Albert Einstein wrote to William F. G. Swann: "Concerning his [Ehrenhaft's] results about the elementary charge

I do not believe in his numerical results, but I believe that nobody has a clear idea about the causes producing the apparent sub-electronic charges he found in careful investigations." Like most such controversies, it never came to a definite falsification of Ehrenhaft's point of view by some crucial experiment; rather, the debate faded into obscurity, although Ehrenhaft continued to publish on "subelectrons" into the 1940's.

During the first decade of his publications, Ehrenhaft's credibility as an experimenter was furthered in 1918 by his demonstration of an effect he called photophoresis (the effect of light on the motion of aerosol particles that both absorb and scatter light). Later he continued work on the interaction of ultramicroscopic particles and light, describing what he called transverse photophoresis, magnetophotophoresis, and the rotation of particles in low-pressure gases in light beams. Some of these effects have since been explained in terms of known effects (such as radiometric forces), while others are still not fully understood. Beginning in the mid 1930's, Ehrenhaft claimed to find experimental evidence for the existence of magnetic monopoles, magnetic currents, and the decomposition of liquids by permanent magnets (magnetolysis). Like most of his proposals after about 1910, they combined the observations of surprising behavior of small particles near the limits of perception, his belief in the validity of direct observation in a regime where many effects interact, and his willingness to go far beyond known theories to explain his observations. While the existence of the raw phenomena was rarely challenged successfully, his interpretations of them became progressively more estranged from the main body of scientific understanding.

In 1912 Ehrenhaft had become associate professor, and in 1920 he was named professor of experimental physics and director of the Third Physical Institute (established for him at the University of Vienna). Those who knew him well regarded Ehrenhaft as deeply devoted to his scientific work, an effective lecturer to large classes, and a stubborn and often difficult fighter for his interpretations. Albert Einstein was one of the many scientists who liked to stay with Ehrenhaft when visiting Vienna. After Ehrenhaft's forced emigration, first to England and then to the United States (1940), following the takeover of Austria by Austrian and German Nazis in 1938, he and Einstein kept in correspondence until the interchange was terminated in unresolvable disagreement over Ehrenhaft's more and more extreme scientific theories.

While in the United States, Ehrenhaft found it very difficult to obtain research support. Thus, in 1946, as a U.S. citizen, he returned to the University of Vienna as U.S. Guest Professor and director of the combined First and Third Physical Institutes, holding these offices until his death.

BIBLIOGRAPHY

I. Original Works. A full bibliography of Ehrenhaft's publications is in Poggendorff, V, VI, and VIIa. Another, somewhat less complete one, was published with a brief, apparently "authorized" biography by Lotte Bittner in her dissertation for the Philosophical Faculty of the University of Vienna, entitled "Geschichte des Studienfaches Physik an der Wiener Universität in den letzten hundert Jahren" (1949). Most of Ehrenhaft's articles on the charge of electrons and subelectrons are listed in Holton (see below, 302–323).

The Center for the History of Physics of the American Institute of Physics in New York City has in its archives approximately three feet of original materials, mostly deposited by Ehrenhaft's son, John L. Ehrenhaft (who still holds some of the more personal letters); these include handwritten autobiographical notes; the MS of an unpublished book, "Magnetismus und Licht" (*ca.* 1947); reprints of works by Ehrenhaft and his pupils; newspaper clippings; personal letters; lecture notes; and the typescript of ten unpublished lectures on his experiments (1946).

The Smithsonian Institution's Dibner Library of the History of Science and Technology has correspondence between Ehrenhaft and other scientists, chiefly Einstein (1917–1941), and related papers. The AIP Center has four reels of microfilm containing most of the correspondence. An exchange of letters is in the R. A. Millikan archive at the California Institute of Technology. An unpublished typescript of ten lectures Ehrenhaft delivered in Vienna in 1947 was produced and distributed in 1967 by his student Paul Feyerabend (with J. Ferber) as "Single Magnetic Northpoles and Southpoles."

II. Secondary Literature. See Paul A. M. Dirac, "Ehrenhaft, the Subelectron and the Quark," in Charles Weiner, ed., *History of Twentieth Century Physics* (New York, 1977), 290–293; and Gerald Holton, *The Scientific Imagination: Case Studies* (New York, 1978), 25–83, 302–323. Evaluations of the dispute over the charge of the electron include R. Bär, "Der Streit um das Elektron," in *Die Naturwissenschaften*, **10** (1922), 322–327, 340–350; J. Mattauch, "Zur Frage nach der Existenz von Subelektron," in *Zeitschrift für Physik*, **37** (1926), 803–815, and "Antwort auf die Bemerkungen Herrn Ehrenhafts zu meiner Arbeit: 'Zur Frage nach der Existenz von Subelektron,'" *ibid.*, **40** (1926), 551–556; and E. Wasser, "Über Ladungmessungen an Selenteilchen bei hohen Gasdrucken," *ibid.*, **78** (1932), 492–509. For a summary of the physics involved, see Milton Kerker, "Movement of Small Particles by Light," in *American Scientist*, **62** (1974), 92–98.

Gerald Holton

EHRESMANN, CHARLES (*b*. Strasbourg, France, 19 April 1905; *d*. Amiens, France, 22 September 1979), *mathematics*.

Ehresmann's father was a gardener employed by a convent in Strasbourg. Ehresmann's parents spoke only the Alsatian dialect, and until 1918 his schooling was in German. He was educated in the Lycée Kléber at Strasbourg and entered the École Normale Supérieure in 1924. After graduation in 1927 and military service, he taught from 1928 to 1929 at the French Lycée in Rabat, Morocco. In the years 1930 and 1931 he did research at Göttingen, and between 1932 and 1934 at Princeton, earning a doctorate in mathematics at the University of Paris in 1934. From 1934 to 1939 he conducted research at the Centre Nationale de la Recherche Scientifique.

Ehresmann became a lecturer at the University of Strasbourg in 1939, and after the German invasion in 1940, he followed that university when it relocated to Clermont-Ferrand. After 1952 he traveled extensively to many countries, where he often was invited to give courses. He became a professor at the University of Paris in 1955, where a chair of topology was created for him. After his retirement in 1975, he taught at the University of Amiens (where his wife was a professor of mathematics) in a semi-official position. He was married twice and had one son by his first wife. He died of kidney failure.

Ehresmann was one of the creators of differential topology, which explores the topological properties (in homotopy and homology) of a differential manifold, in relation to its differential structure. In his dissertation and subsequent papers between 1935 and 1939, he explicitly described the homology of classical types of homogeneous manifolds, such as Grassmannians, flag manifolds, Stiefel manifolds, and classical groups. His methods were based on decomposition of these manifolds into cells, even before the general definition of CW complexes had been given. His results later became a useful tool in the theory of characteristic classes.

Between 1939 and 1956 Ehresmann participated in the creation and development of fundamental notions in differential topology: fiber spaces, connections, almost complex structures, jets, and foliations. Fiber spaces, first considered in special cases by Seifert in 1933 and Hassler Whitney in 1935, became a focus of topological research around 1940, when their importance was realized. Ehresmann approached the theory of fiber spaces from an original angle. He had become familiar with the theory of what Élie Cartan called connections (generalizing the Levi-Cività parallelism in Riemannian manifolds) and with the "generalized spaces" on which these connections are defined; very few mathematicians understood Cartan's ideas at that time. Ehresmann realized that beneath Cartan's formulas and constructions were two fundamental fiber spaces whose basis was a differential manifold: the tangent bundle and the space of frames, the mutual relations of which were the key to Cartan's theory. This gave Ehresmann a view of the general theory of fiber spaces somewhat different from that of other mathematicians in that field.

Ehresmann's theory emphasized the importance of a group of automorphisms of a fiber and led him to the general concept of principal fiber space, where the fibers themselves are topological groups isomorphic to a fixed group. This notion has acquired a fundamental importance in differential and algebraic geometry. Ehresmann could then precisely describe what may be called a two-way correspondence between general fiber spaces and principal ones over a fixed base: with any fiber space there is associated a well-determined (up to isomorphism) principal fiber space; conversely, with any principal fiber space with fibers isomorphic to a group G, and with any action of G on a space F, there is associated a well-determined fiber space with fibers isomorphic to F. Thus, the "space of frames" with group $GL(n\mathbf{R})$ is associated to the tangent bundle of a differential manifold of dimension n as principal fiber space.

With the help of these concepts, Ehresmann could, for any fiber bundle E over a differential manifold M, give a definition (generalizing Cartan's) of a connection on E. Geometrically it amounts to defining, for each $x \in M$ and any point u_x in the fiber E_x, a vector subspace H_{u_x} of the tangent space to E at the point u_x, which is supplementary to E_x in that space and therefore projects isomorphically onto the tangent space to M at x.

When a fiber space E is associated to a principal fiber space P with group G, and G is a subgroup of a group H, it is always possible to consider E as associated to a principal fiber space with group H (extension of G to H). But when K is a subgroup of G, it is not always possible to consider E as associated with a principal fiber space with group K (restriction of G to K). Ehresmann showed that a topological condition must be satisfied: the existence of a section over M of the fiber space associated with P and with the natural action of G on the homogeneous space G/K. This explains why there are always Riemannian structures on an arbitrary manifold M. E is then the tangent bundle, $G = GL(n, \mathbf{R})$ and K the orthogonal group $O(n, \mathbf{R})$; the quotient G/K is then diffeomorphic to an $\mathbf{R}^N$; and for fiber spaces with such fibers, sections

EHRESMANN

over the base always exist. But even for pseudo-Riemannian structures, topological conditions on M are necessary.

Ehresmann studied in detail the case in which M has even dimension $2m$, E is the tangent bundle so that $G = GL(2m, \mathbf{R})$, and $K = GL(m, \mathbf{C})$. When the restriction of G to K is possible, he said, the structure it defines on M is an almost complex structure. The latter term comes from the fact that when M is a complex analytic manifold of complex dimension m, the tangent bundle has fibers that are vector spaces over $\mathbf{C}$; an almost complex structure, however, does not always derive from a complex structure, and additional conditions have to be imposed on the differential structure of M. Independently, Heinz Hopf studied almost complex structures, and many other cases of restrictions to classical subgroups of $GL(n, \mathbf{R})$ were considered later.

In 1944, Ehresmann inaugurated the global theory of completely integrable systems of partial differential equations. In his local study of partial differential equations, Cartan had emphasized the advantages that derive from a geometrical conception of such systems, in contrast with their expression in nonintrinsic terms using local coordinates. For ordinary differential equations, this geometrical conception goes back to Henri Poincaré and substitutes for such an equation (with no singularities) on a manifold M a field of tangent lines on M—or, in modern terms, a line subbundle of the tangent bundle $T(M)$. The natural generalization is therefore a vector subbundle L of rank $p > 1$ of $T(M)$, and the generalization of the integral curves of a differential equation are the injective immersions $f: N \rightarrow M$ into M of a manifold N of dimension $q \leqslant p$, such that for $y \in N$, the image by the tangent map $T_y(f)$ of the tangent space $T_y(N)$ is contained in the fiber $L_f(y)$. Completely integrable systems are those for which there are such maps f for manifolds N of maximal dimension p, whose images $f(N)$ may contain arbitrary points of M.

The characterization of these systems by local properties was presented in the work of Rudolf Clebsch and Georg Frobenius. Ehresmann initiated the study of their global solutions, in the spirit of the "qualitative" investigations started by Poincaré for $p = 1$, which have become known as the theory of dynamical systems. There are always maximal connected solutions $f(N)$; they are called the leaves of the system, forming a partition of M, called a foliation. Ehresmann published only a few papers on that topic; the bulk of the basic notions in the theory was developed, under his guidance, in the dissertation of his pupil Georges Reeb. It was Reeb who obtained the first significant results, in particular the remarkable "Reeb foliation" of the sphere $\mathbf{S}_3$, with a single compact leaf, which later played an important part in the general theory. Until 1960 these papers of Ehresmann and Reeb did not attract much attention, but since then the theory has enjoyed a vigorous and sustained growth that has made it a main branch of differential geometry and differential topology, with recently discovered and surprising links with the theory of C^*-algebras.

The next theory pioneered by Ehresmann was what he called the theory of jets. Two C^∞ maps f,g of a manifold M into a manifold N have a contact of order k at a point $x \in M$ where $f(x) = g(x)$ if, in local coordinates around x and $f(x)$, their Taylor expansions coincide up to order k. This is independent of the choice of local coordinates and defines an equivalence relation in the set $E(M, N)$ of C^∞ maps of M into N. Ehresmann called the equivalence class of such a map f the kth jet of f at the point x. He developed the main properties of that notion in a series of notes. It has since been recognized that this notion provides the best frame for an intrinsic conception of general systems of partial differential equations and for Lie pseudogroups (formerly called infinite Lie groups), free from cumbersome computations in local coordinates.

After 1957 Ehresmann became one of the leaders in the new theory of categories, to which he attracted many younger mathematicians and in which his fertile imagination introduced a large number of concepts and problems. Over the next twenty years, he published his papers in that field and those of his school in a periodical of which he was both editor in chief and publisher, *Cahiers de topologie et de géométrie différentielle*.

Ehresmann's personality was distinguished by forthrightness, simplicity, and total absence of conceit or careerism. As a teacher he was outstanding, not so much for the brilliance of his lectures as for the inspiration and tireless guidance he generously gave to his research students, including Reeb and Jacques Feldbau; throughout his career he supervised a large number of doctoral dissertations.

BIBLIOGRAPHY

Charles Ehresmann, *Oeuvres complètes et commentées*, Andrée Charles Ehresmann, ed., 3 vols. (Amiens, 1982–1984).

JEAN DIEUDONNÉ

EIMER, THEODOR GUSTAV HEINRICH (*b*. Stäfa, near Zurich, Switzerland, 22 February 1843; *d*. Tübingen, Germany, 29 May 1898), *zoology*.

Eimer's father, Heinrich, was a political refugee from the aborted coup that in 1833 had attempted to dissolve the Diet of the German Confederation in Frankfurt. He escaped to Switzerland, where he practiced medicine in the town of Stäfa. There he met and married Albertine Pfenniger, who came from a prominent family of the area. In 1845 the family moved to Lahr, on the western edge of the Black Forest, which Theodor always considered his native town. Subsequently they lived in the Baden towns of Donaueschingen, Langenbrücken, and Freiburg, where Heinrich served as a regional physician.

Until the age of twelve, Eimer received private tutoring; thereafter he attended gymnasiums in Bruchsal and in Freiburg. Following in his father's footsteps, he studied medicine. In 1862 he matriculated at Tübingen, where he was particularly influenced by the histologist Franz von Leydig. He then spent the year 1863–1864 at Freiburg and the year 1864–1865 at Heidelberg. After taking examinations in the natural sciences, Eimer returned to Tübingen for the winter semester of 1865–1866. Between 1866 and 1868 he worked in Rudolf Virchow's laboratory in Berlin, receiving his medical degree in 1867. It was Virchow who turned Eimer from his original interest in anthropology to zoology. In early 1868 Eimer went to Karlsruhe to take his state medical examinations, then spent the next twelve months studying zoology at Freiburg under August Weismann. A three-month winter interlude in Paris rounded out his studies.

In 1869 Eimer was hired as prosector of zootomy (comparative anatomy) by Albert von Kölliker at Würzburg, where he received his doctorate for histological and experimental work on fat absorption in the intestine. On 18 July 1870 he married Anna Lutteroth, the daughter of a Hamburg banker, and the following day was habilitated in zoology and comparative anatomy. Immediately thereafter Eimer volunteered for military service as a field surgeon. He saw action at the siege of Strasbourg and was joined by his wife, who served as an army nurse. After being decorated for service, he was forced to retire from active duty because of illness. Early in 1871 the Eimers journeyed to Capri, where Theodor became familiar with marine organisms. He returned to the island in 1872, 1876, 1877, and 1879. The coelenterates of the Bay of Naples and the lizards of Capri formed the subjects of Eimer's first book-length monographs in zoology.

In 1874 Eimer became the inspector of the grand-duke's collections in Darmstadt and associate professor of zoology at the Technische Hochschule. In 1875 he succeeded Leydig as a full professor of zoology and comparative anatomy at the University of Tübingen. Over the next twenty-three years Eimer developed and taught an array of courses in zoology with only a one-semester interruption. He was known for clear and lively lectures, and he attracted a coterie of loyal, even adulatory, advanced students. An invitation in 1888 to become the director of the Natural History Museum in Hamburg gave Eimer the leverage to convince the Württemberg minister to erect a new building for the zoological institute at Tübingen. He also promoted the study of veterinary medicine at Tübingen and persuaded the university to offer a doctorate in this subject.

From the beginning of his residence in Tübingen, Eimer participated vigorously in the affairs of the Württemburg Society for Natural History and of its Black Forest branch. In 1879 he became a member of the prestigious Leopoldina. Highly patriotic, Eimer was a leader of the Württemberg branch of the National Liberal party (Deutsch-Nationale Partei[1]) until the mid 1880's and continued to practice in a general clinic in order to be prepared to serve again as a military surgeon in the event of national mobilization.

The Eimers were fond of traveling during the summer. They often visited Italy, where Eimer worked at the zoological stations in Naples and Rovigno; they visited the Balkans, Constantinople, and the North and Baltic seas; and during the winter of 1878–1879 they traveled up the Nile as far as Nubia. In 1897 Eimer purchased a small estate near Lindau on Lake Constance, where he intended to spend vacations and to pursue zoological studies on the lake. He fell seriously ill during the fall semester but continued to perform his offical duties at the university until the following May, when he underwent an operation for a severe intestinal disorder. Eight days later, on 29 May 1898, he died at the age of forty-five. He was survived by his wife, two sons, and two daughters.

The first phase of Eimer's scientific career was histophysiological in nature. Under the tutelage of Virchow he studied fat absorption in the small and large intestines. From 1872 to 1875 he studied the nucleus of the cell, but his microscopic technique was considerably inferior to that of Friedrich Anton Schneider, Eduard Strasburger, and Walther Flemming, who were in the process of transforming our knowledge of the nucleus. Eimer also studied the nature of the reptilian egg and made the minor discovery that sponges possess nematocytes and produce spermatozoa. His most important discovery during this period involved a description of the life cycle of a coccidian parasite (a spore-producing protozoan) in the mouse. This became a classic study that was often reproduced in textbooks; the genus of this particular coccidian was later named *Eimeria* by Schneider.

While on Capri during his convalescence of 1871 and again in 1872, Eimer began two series of investigations that were to reorient his biological interests. The first consisted of regeneration experiments with marine organisms. Beginning with the ctenophore *Beroë* and extending the investigation to the true jellyfish, Eimer amputated parts of the organism in order to determine the minimal center of activity of the primitive nervous system. With the medusae of the scyphozoans, he discovered that a single marginal sense organ, known as a lithocyst, was sufficient to initiate a rhythmic pulsation throughout the entire umbrella. He further discovered that a medusa with all the marginal lithocysts excised could gradually reorganize itself and develop new centers of stimulation.

Many of Eimer's experiments paralleled and independently confirmed the contemporaneous physiological studies on the medusa's nervous system done by George John Romanes. The philosophical implications of Eimer's experiments, however, became clear only in 1883, when he delivered the address "Über den Begriff des tierischen Individuums" to the Versammlung der Deutschen Naturforscher und Ärzte in Freiburg. At that time he drew upon a range of nineteenth-century biological assumptions, including the biogenetic law, the principle of the division of labor, a recognition of the alternation of generations, and a belief in the inheritance of acquired characters, to question whether nature consisted of isolated organic individuals. With specific reference to the early-nineteenth-century *Naturphilosoph* Lorenz Oken, Eimer insisted that the individual merged into the species and that both implied "the totality of the animal kingdom."[2]

The other series of investigations begun on Capri pursued the converse of this generalization. If the organic world consisted of a unity, why do we find separate species and individuals? The answer was found on the picturesque Faraglioni cliffs, which lie as isolated promontories jutting into the sea at the southeast end of Capri. There Eimer discovered a race of the common wall lizard (*Lacerta muralis coerulea*) that was markedly darker and bluer than the species on the rest of Capri and elsewhere. After comparing the unique race with members of the species throughout its European range, Eimer concluded that the environment, including selective pressures, interacted with chemically restricted internal growth patterns. "The inherited characters," he explained, "produce the projected line [of development], the direction of which may be altered prior to or after birth only by the influence of the environment. For this resulting alteration can be nothing other than the necessary crystallization product of the changed makeup of the organism."[3] The isolation of the Faraglioni cliffs allowed the consequent changes to accumulate to form a separate race of wall lizard.

Eimer later extended his observations to the markings of birds, mammals, and especially butterflies. He consolidated his findings in *Die Entstehung der Arten* (1888–1901). In a highly speculative way, he argued that the formation of the organism was determined by the operation of four laws of growth: (1) that a directed evolutionary process is preceded by changes in the ontogeny of the individual; (2) that new characters first appear in the mature males and may be transmitted to the rest of the species through heredity—this is what Eimer called the law of male preponderance; (3) that these new characters usually appear at the posterior end or on the inferior side of the male and work their way forward and superiorly as the individual grows older (the law of wavelike development); and (4) that varieties are simply sequential stages in the development of the species.

It was the interplay of these growth patterns that explained for Eimer the division of nature's unity into a multiplicity of forms. The local environment and other external factors, he argued, interacted with the constitution of an organism to render the peculiar characteristics of local varieties. These varieties advanced along particular lines of development while their neighbors remained static in an "Entwicklungsstillstand" or genepistasis. Speciation eventually followed. Eimer also explained atavism, degeneration, and saltations in phylogeny in terms of an environmental impact on local constitutions and the consequent growth patterns. He contrasted his mature evolutionary ideas with those Carl Naegeli presented in his *Mechanisch-physiologische Theorie der Abstammungslehre* (1884), for unlike Naegeli, Eimer did not invoke an internal perfecting principle to explain orthogenetic lines of evolution. He also distinguished his theory from the functionally oriented neo-Lamarckian movement, although he shared its commitment to the inheritance of acquired characters. Like the neo-Lamarckians, after 1883 Eimer found himself increasingly in direct conflict with the results of nuclear cytology and the neo-Darwinian movement.

In his later researches Eimer concentrated on demonstrating the laws of development in butterflies. The results appeared in *Die Artbildung und Verwandtschaft bei den Schmetterlingen* (1888–1895), illustrated by his wife, which presented a detailed taxonomy of the worldwide genus *Papilio* (the swallowtails). Eimer used the longitudinal stripes on the wings to show that the changes from one species to another and from one subgenus to another consisted solely in minor alterations in the development of pigment in accordance with his developmental laws. His argument was based on his belief that such differences must reflect predetermined growth patterns rather than the useful traits required for the operation of natural selection. The wing patterns, which neo-Darwinians explained in terms of mimicry or protective coloration, were understood by Eimer in terms of similar growth patterns directed by slightly different external conditions.

As Eimer became more convinced of the correctness of his ideas, he became increasingly acerbic in his comments on the ideas of others. Weismann, the preeminent neo-Darwinian of the age, with whom Eimer had studied for a year in the late 1860's and to whom he had dedicated a book of his experiments on medusae in 1878, became by the mid 1880's the object of severe criticism in Eimer's published works.

The thrust of Eimer's lifework was to provide an explanation for evolution that did not rely on the utilitarian assumptions of natural selection, on functional adaptation, or on any vitalistic assumptions. Borrowing a term coined by Wilhelm Haacke in 1893, Eimer referred at the end of his life to his evolutionary explanation as "orthogenesis" or directional evolution. Orthogenesis was the subject of his major address delivered to the Congress of Zoologists at Leiden in 1895 and of his last book, *Orthogenesis der Schmetterlinge*, published a year before his death. Eimer remained adamant to the end that his evolution theory utilized only the physicochemical processes associated with his laws of growth.

NOTES

1. This name for Eimer's political affiliation is given in Klunzinger's obituary, but the author was unable to find a record of such a regional party. Klunzinger probably meant the Deutsche Partei, which was the Württemberg branch of the Nationalliberale Partei.
2. "On the Idea of the Individual in the Animal Kingdom," 433.
3. *Zoologische Studien auf Capri. II*, Lacerta muralis coerulea (Leipzig, 1874), 42.

BIBLIOGRAPHY

I. Original Works. A complete bibliography of Eimer's scientific writings is in Klunzinger's obituary (see below). An expansion of his dissertation and his earliest papers on fat-absorbing goblet cells appeared in *Virchow's Archiv* between 1867 and 1869; his microscopical studies on nuclear structure, on reptilian eggs, and on spermatozoa appeared in *Archiv für mikroskopische Anatomie* between 1871 and 1877. The first accounts of Eimer's studies on ctenophores and observations on the Capri wall lizard were published as *Zoologische Studien auf Capri. I. Über* Beroë ovatus. *Ein Beitrag zur Anatomie der Rippenquallen* (Leipzig, 1873) and *Zoologische Studien auf Capri. II.* Lacerta muralis coerulea. *Ein Beitrag zur Darwin'schen Lehre* (Leipzig, 1874). His regeneration experiments on *Beroë* are described in "Versuche über künstliche Teilbarkeit von *Beroë ovatus* (angestellt zum Zweck der Kontrolle seiner morphologischen Befunde über das Nervensystem dieses Tiers)," in *Archiv für mikroskopische Anatomie*, **17** (1879), 213–240. Eimer's address "Über den Begriff des tierischen Individuums" is in the *Amtlicher Bericht der Versammlung der deutschen Naturforscher und Ärzte* for 1883 and in *Humboldt. Monatschrift für die gesammten Naturwissenschaften*, **2** (1883), 437–440. An English translation, "On the Idea of the Individual in the Animal Kingdom," appeared as the appendix to *Organic Evolution*, 413–435 (see below).

The most elaborate discussion of Eimer's laws of growth and their application to evolution is *Die Entstehung der Arten auf Grund von Vererben erworbener Eigenschaften nach den Gesetzen organischen Wachsens. Ein Beitrag zur einheitlichen Auffassung der Lebenwelt*, 3 vols. (Jena and Leipzig, 1888–1901). The first volume was translated by Joseph T. Cunningham as *Organic Evolution as the Result of the Inheritance of Acquired Characters According to the Laws of Organic Growth* (London and New York, 1890). The second volume (1897), written with the assistance of C. Fickert, bore the secondary title *Orthogenesis der Schmetterlinge. Ein Beweis bestimmt gerichteter Entwickelung und Ohnmacht der natürlichen Zuchtwahl bei der Artbildung. Zugleich eine Erwiderung an August Weismann*. The third volume (1901), published posthumously by Fickert and Countess Maria von Linden, bore the secondary title *Vergleichend-anatomisch-physiologische Untersuchungen über das skelett der Wirbeltiere*. Whereas the first volume was published by Gustav Fischer of Jena, the second and third volumes were published by Wilhelm Engelmann of Leipzig.

Eimer presented further explication of the laws of growth with respect to swallowtails in *Die Artbildung und Verwandtschaft bei den Schmetterlingen*, 2 vols. and 2 atlases of 4 colored plates each (Jena, 1889–1895). The first volume bears the subtitle *Eine systematische Darstellung der Abänderungen, Abarten und Arten der Segelfalter-ähnlichen Formen der Gattung* Papilio, the second volume the subtitle *Eine systematische Darstellung der Abänderungen, Abarten und Arten der Schwalbenschwan-*

zählichen Formen der Gattung Papilio. Fickert again assisted Eimer with the second volume.

Eimer's Leiden address on orthogenesis, *Über bestimmt gerichtete Entwicklung, Orthogenesis, und über Ohnmacht der Darwin'schen Zuchtwahl bei der Artbildung* (Leiden, 1896), was reprinted as the first chapter of his *Orthogenesis der Schmetterlinge*, 1–49. An English translation appeared in book form as *On Orthogenesis and the Impotence of Natural Selection in Species-Formation*, Thomas J. McCormack, trans. (Chicago, 1898).

II. SECONDARY LITERATURE. There is no detailed study of Eimer's life. The most important obituary is C. B. Klunzinger, "Theodor Eimer. Ein Lebensabriss mit Darstellung der Eimer'schen Lehren nach ihrer Entwickelung," in *Jahreshefte des Vereins für vaterländische Naturkunde in Württemberg*, **55** (1899), 1–22, which provides a portrait, the most complete bibliography of Eimer's publications, and a brief account of Eimer's most important texts. Countess Maria von Linden, "Professor Dr. Theodor Eimer," in *Biologisches Zentralblatt*, **18** (1898), 721–725, presents a sketch of her former mentor's personality and impact as a teacher. R. von Hanstein, in *Leopoldina*, **34** (1894), 107–108, is brief but contains useful information. Historian of biology Georg Uschmann wrote the standard account in *Neue deutsche Biographie*, IV (Berlin, 1959), 393–394.

Vernon L. Kellogg, *Darwinism To-Day* (New York, 1907), 281–285, presents a capsule summary of Eimer's evolutionary ideas, contrasting them with other orthogenetic theories. A critical summary of Eimer's evolution theory also appears in Yves Delage and Marie Goldsmith, *The Theories of Evolution*, André Tridon, trans. (New York, 1912, 1913), 298–302. While evaluating many contemporary evolutionary theories, Ludwig Plate describes elements of Eimer's ideas throughout his *Selektionsprinzip und Probleme der Artbildung, ein Handbuch des Darwinismus*, 4th ed. (Leipzig and Berlin, 1913). Peter J. Bowler places Eimer's biological accomplishments in the context of Weismann's ideas and the reaction to them by some English and American biologists in "Theodor Eimer and Orthogenesis: Evolution by 'Definitely Directed Variation,'" in *Journal of the History of Medicine*, **34** (1979), 40–73. See also Bowler's *The Eclipse of Darwinism. Anti-Darwinian Evolution Theories in the Decades Around 1900* (Baltimore and London, 1983), esp. 148–160. In both cases the analysis concentrates on Eimer's few translated texts.

FREDERICK B. CHURCHILL

EPSTEIN, PAUL SOPHUS (*b*. Warsaw, Poland [then Russia], 20 March 1883; *d*. Pasadena, California, 8 February 1966), *theoretical physics*.

Epstein's research career spanned the development of quantum theory based on classical mechanics, to which he made fundamental contributions, into quantum mechanics. He was born into a patrician Polish-Russian family of successful businessmen, rabbinical scholars, and philanthropists. His father, Siegmund Simon Epstein, was a road-building contractor and insurance broker. His maternal grandfather, Chaim Lur'ia, was a prominent civic leader in the Jewish community in the city of Minsk, Russia. Epstein's mother, Sarah Sof'ia Lur'ia, who was considerably more ambitious and intellectual than her husband, studied at the medical school for women at St. Petersburg University, corresponded with many literary figures, including Fedor Dostoeevskii, and nurtured her son's scientific aspirations. Epstein's parents divorced when he was a child and he grew up in Minsk, graduating in 1901 from the local high school with a gold medal in mathematics. He spoke Russian and German fluently as a boy and by the age of ten had decided to study physics.

In 1901 Epstein entered the School of Physics and Mathematics at the Imperial University of Moscow. He intended to study mathematical physics, but since there were no professors in this field at Moscow, he worked under Petr N. Lebedev, an experimental physicist best known for measuring the pressure of light. The lectures were available in printed form, so Epstein rarely attended them, aside from the basic two-year lecture demonstration course in physics. The problem courses interested him intensely, however, and the experience of solving problems at the blackboard before the whole class, he later said, gave him a feeling for the subject. An excellent physics library, coupled with the research colloquium that undergraduates could attend but not participate in, directed him to books and journal articles. He earned a bachelor's degree in science in 1906 (the revolution of 1905 having delayed his senior examinations by one year) and then enrolled as a graduate student.

Epstein was a laboratory instructor in physics at the Moscow Institute of Agriculture (1906–1907), and then at the Imperial University of Moscow (1907–1909), while conducting experimental research on the dielectric constant of gases. In 1909 he received a master's degree in physics and became the equivalent of an assistant professor. That December, at a scientific congress in Moscow, he met the theoretical physicist Paul Ehrenfest, who directed his attention to the West.

Having decided that he was not cut out to be an experimentalist ("My hands were not clever enough"), Epstein left Moscow in early 1910 for Munich. There he attended Arnold Sommerfeld's lectures on relativity, then studied the theory of electromagnetic waves, particularly the theory of diffraction, and finally spent, by his own account, four years at Sommerfeld's Institute for Theoretical Physics.

Epstein's first contact with psychoanalysis dates from this period. Plagued by depression and stomach troubles, he spent several months in 1911, and again in 1912, in a sanitarium in Switzerland, undergoing analysis. An enthusiastic and ardent student of Freud's ideas from then on, Epstein went on to cofound the Los Angeles Psychoanalytic Study Group in the late 1920's, the precursor of the Los Angeles Psychoanalytic Society and Institute.

In 1914 Epstein received the Ph.D. in physics, with minors in mathematics and crystallography, from the University of Munich. During World War I, Epstein was classified as an enemy alien by the Germans, and he was interned briefly. After his release, Epstein continued his research, but was not allowed to leave Germany until the war's end. His interest in problems of quantum theory coincided with the publication of Sommerfeld's paper on the fine structure of atomic hydrogen (1916).

Epstein wrote a series of important papers on quantum theory and its applications. In his classic paper on the theory of the Stark effect, the splitting of the spectral lines in a hydrogen atom by a strong electric field (1916), he worked out the quantization rules in an invariant form and then used them to calculate the splitting of the hydrogen lines. The splitting effect, first observed by Johannes Stark in 1913, could not be explained along classical lines. Showing that Niels Bohr's quantum description of the hydrogen atom could solve the problem made Epstein's reputation as a theoretical physicist. The match between his theoretical predictions and Stark's data furnished striking support for the Rutherford-Bohr atomic theory.

Working independently, the German astronomer Karl Schwarzschild publicly announced the solution to the same problem one day after Epstein's paper appeared. (As Epstein relates it, Schwarzchild initially had the wrong formula. He corrected it after seeing an announcement of Epstein's result.) Epstein's other contributions to the development of the quantum theory of atomic structure between 1916 and 1921 ranged from his extension of Bohr's theory to nonperiodic motions, including beta decay and the photoelectric effect, to the application of Bohr's correspondence principle to the interference of spectral lines.

After two years as *Privatdozent* at Zurich (1919–1921) and a short time as assistant to Hendrik A. Lorentz at Leiden, Epstein joined the faculty of the new California Institute of Technology in 1921 as professor of theoretical physics. Except for two years that he spent as an exchange professor (1927, 1929) at the Aachen Institute of Technology in Germany, he served for thirty-two years on the Caltech faculty, becoming professor emeritus in 1953.

Epstein played a significant role in the foundation of Caltech's physics division, introducing and teaching virtually all the theoretical physics courses in the early years. Out of this experience came *Textbook of Thermodynamics* (1937) and a deep interest in Willard Gibbs's statistical mechanics. With Robert A. Millikan he organized and ran the weekly physics research seminar, which became a Thursday afternoon tradition; he also built up the physics library. In 1930 he was elected to the National Academy of Sciences.

At Pasadena, Epstein worked on perturbation theory and the application of Schrödinger's wave mechanics to the Stark effect. In later years his research work mainly fell at the border of physics, acoustics, and hydrodynamics. During World War II he was a consultant for the U.S. Navy Sound Laboratory in San Diego and for the Army Air Corps's Meteorology Project at the Institute.

Outside of physics Epstein's interests ranged over the fields of philosophy, art, history, and psychoanalysis. He married twice. His first wife, Mina, was a concert pianist. They were married in 1909, separated in 1911, and divorced in 1919. Epstein called the marriage "the greatest stupidity that I ever committed in my life" and kept Mina's surname a secret. In 1930 he married Alice Emelie Ryckman; they had one daughter.

BIBLIOGRAPHY

I. Original Works. Epstein's papers referred to in the text are "Zur Theorie des Starkeffektes," in *Annalen der Physik*, 4th ser., **50** (1916), 489–520; "Versuch einer Anwendung der Quantenlehre auf die Theorie des lichtelektrischen Effekts und der β-Strahlung radioaktiver Substanzen," *ibid.*, 815–840; "Über die Interferenzfähigkeit von Spektrallinien vom Standpunkt der Quantentheorie," in *Sitzungsberichte der Bayerischen Akademie der Wissenschaften zu München*, Math.-phys. Kl. (11 January 1919), 73–90; "Problems of Quantum Theory in the Light of the Theory of Perturbations," in *Physical Review*, **19** (1922), 578–608; "The Stark Effect from the Point of View of Schrödinger's Quantum Theory," *ibid.*, **28** (1926), 695–710; "Application of Gibbs' Methods to Modern Problems of Thermodynamics," in *Commentary on the Scientific Writings of J. Willard Gibbs*, **I** (New Haven, 1936), 59–112; and "Critical Appreciation of Gibbs' Statistical Mechanics," *ibid.*, 521–584. One scientific article omitted from the bibliography (see below) is "Ferrite Post in a Rectangular Wave Guide," in *Journal of Applied Physics*, **27** (1956), 1328–1335, written with A. D. Berk.

Epstein's notebooks, autobiographical notes, oral history, correspondence, and manuscripts are deposited in the archives of the California Institute of Technology. For an appreciation of his intellectual development, see "Paul S. Epstein," transcript of an oral interview conducted by John L. Heilbron, Archive for the History of Quantum Physics, Office for History of Science and Technology, University of California, Berkeley, 25 and 26 May and 2 June 1962. Additional source material is listed in Thomas S. Kuhn *et al.*, *Sources for History of Quantum Physics* (Philadelphia, 1967).

II. SECONDARY LITERATURE. See Jesse W. M. DuMond, "Paul Sophus Epstein," in *Biographical Memoirs. National Academy of Sciences*, **45** (1974), 127–152, with portrait and bibliography; and the obituary notice in *Naturwissenschaftliche Rundschau*, **19** (1966), 170. Technical details of Epstein's work on the Stark effect are covered in G. Birtwistle, *The Quantum Theory of the Atom* (Cambridge, 1926), 97–111.

JUDITH R. GOODSTEIN

ERDÉLYI, ARTHUR (*b*. Budapest, Hungary, 2 October 1908; *d*. Edinburgh, Scotland, 12 December 1977), *mathematics*.

Arthur Erdélyi was the first child of Ignác and Frieda (Roth) Diamant. After his father's death, he was adopted by his mother's second husband, Paul Erdélyi. He attended elementary school in Budapest from 1914 to 1926. After studying at the Deutsche Technische Hochschule in Brno, Czechoslovakia, he matriculated at the German University of Prague and was awarded the degree of doctor rerum naturalium in 1938. Forced by the Nazis to flee Czechoslovakia, Erdélyi managed to obtain a research grant from Edinburgh University, where in 1940 he was awarded the degree of doctor of science. In 1942 he married Eva Neuburg, and in 1949 he left Edinburgh to become a professor at the California Institute of Technology. He returned to Edinburgh in 1964 as professor of mathematics, remaining there until his death. In 1975 Erdélyi was elected a fellow of the Royal Society of London, and in 1977 he was awarded the Gunning Victoria Jubilee Prize of the Royal Society of Edinburgh.

Erdélyi began his mathematical career with a study of the confluent hypergeometric function and before arriving in Edinburgh in 1939 had already established himself as a leading expert in the area of special functions. In Edinburgh he continued to pursue his investigations, broadening his interests into generalized hypergeometric functions, classical orthogonal polynomials, and, in particular, Lamé functions, on which he published a series of fundamental papers. His career at Edinburgh University was interrupted, however, by the death at Caltech of Harry Bateman, who left behind voluminous notes on special functions that demanded editing and publication. After due consultation with leading experts, Erdélyi was appointed by Caltech in 1947 to supervise the editing and publication of the Bateman manuscripts. With him came F. G. Tricomi from the University of Turin, W. Magnus from the University of Göttingen, and F. Oberhettinger from the University of Mainz. Together they produced the three-volume *Higher Transcendental Functions* (1953–1955) and the two-volume *Tables of Integral Transforms* (1954). These books became basic reference sources for generations of applied mathematicians and physicists throughout the world, and the most important part of this work, *Higher Transcendental Functions*, remains the most scholarly and comprehensive treatment of the special functions of mathematical physics that is available.

The Bateman Manuscript Project marked a turning point in Erdélyi's development as a mathematician. As the project neared completion, he turned from an investigation of special functions for their own sake to the study of asymptotic expansions of integrals and solutions of differential equations. Erdélyi's most important contribution to this area was in the asymptotic evaluation of integrals. Fundamental to many of his investigations was the idea of an asymptotic scale and generalized asymptotic expansion, an idea that dates back at least to H. Schmidt but that Erdélyi was the first to exploit on a systematic basis. The application of these ideas yielded new theorems on the asymptotic expansion of Laplace integrals involving logarithms and exponential functions, as well as an elegant and unified treatment of Watson's lemma, Darboux's method, and the asymptotic behavior of functions in transition regions.

Erdélyi demonstrated that the Poincaré-type definition of an asymptotic expansion is much too narrow for a satisfactory discussion of the asymptotic behavior of functions depending on more than one parameter. These investigations of asymptotic analysis were influenced by the work then being undertaken in the Guggenheim Aeronautical Laboratory at Caltech on the development of an improved boundary-layer theory for viscous fluid-flow past obstacles, and Erdélyi's lifelong interest in singular perturbation theory can be traced back to this time. His book *Asymptotic Expansions* appeared in 1956 and is now regarded as one of the classic monographs on the subject of asymptotic analysis.

A third major area of Erdélyi's scientific work was in fractional integration and singular partial differential equations. His first major contribution to this area was in 1940, when together with H. Kober he introduced certain modifications of the Riemann-Liouville and Weyl fractional integrals and discussed their connection with the Hankel transform. These generalized fractional integration operators are now called Erdélyi-Kober operators. These results lay dormant for over twenty years until Erdélyi's interest was revived by the publications of Alexander Weinstein on the generalized axially symmetric potential equation. Erdélyi's first paper on this equation appeared in 1956, giving criteria for the location of singularities of solutions, and it laid the foundation for numerous later developments in the analytic theory of partial differential equations. This paper was soon followed by many others on the axially symmetric potential equation and the Euler-Poisson-Darboux equation, as well as further applications of fractional integration to dual integral equations and the theory of generalized functions. He was actively involved with this work at the time of his death.

Arthur Erdélyi was an excellent expositor, and with his broad interests he had something to say in many areas of mathematics. His reputation was based on much more than his published papers, although this alone would have sufficed to make him one of the leading analysts of his time. His combination of mathematical scholarship, an interest and enthusiasm for mathematics, a concern for younger workers, and a willingness to devote his time in aid of the mathematical community won Erdélyi the admiration and respect of an entire generation of mathematicians.

BIBLIOGRAPHY

I. Original Works. *Higher Transcendental Functions*, 3 vols. (New York, 1953–1955), written with W. Magnus, F. Oberhettinger, and F. G. Tricomi; *Tables of Integral Transforms*, 2 vols. (New York, 1954), written with W. Magnus, F. Oberhettinger, and F. G. Tricomi; *Asymptotic Expansions* (New York, 1956); and *Operational Calculus and Generalized Functions* (New York, 1962).

II. Secondary Literature. Obituaries, with bibliographies, include D. Colton, in *Bulletin of the London Mathematical Society*, **11** (1979), 191–207; and D. S. Jones, in *Biographical Memoirs of Fellows of the Royal Society*, **25** (1979), 267–286.

David Colton

ERTEL, HANS (*b*. Berlin, Germany, 24 March 1904; *d*. Berlin, German Democratic Republic, 2 July 1971), *meteorology, geophysics, physics, hydrography, oceanography*.

Ertel studied mathematics, physics, and meteorology at the University of Berlin, where he received the doctorate in 1932. In 1935 he was appointed observer, and in 1938 lecturer, in theoretical meteorology at Berlin. He was professor at the Central Institute of Meteorology and Geodynamics at Vienna in 1942, and from 1943 to 1945 he was professor of geophysics at the University of Innsbruck. Ertel was professor of geophysics at the University of Berlin (German Democratic Republic) and director of its Institute of Meteorology and Geophysics from 1946 to 1969, and was director of the Institute of Physical Hydrography of the German Academy of Sciences, also in Berlin, from 1949 to 1969.

Ertel is a seminal figure in the development of modern meteorology and geophysics. His research, culminating in his general vorticity theorem (Ertel's potential vorticity theorem), was a key element in the transition from classical to modern meteorology.

Through Ertel's work Humboldt University and the Academy of Sciences at Berlin, with its Institute of Meteorology and Geophysics and Institute of Physical Hydrography, became international centers of meteorology, geophysics, and fluid dynamics. Ertel's special field was theoretical meteorology and the application of fluid dynamics to meteorological and geophysical problems. In his earlier years he published on Guilbert's rule (1930) and the curvature of the surface of discontinuity in the atmosphere and ocean (1931). After 1931 he worked on geostrophy, barotropy, fronts and cyclones, and the role of midlatitude disturbances. In 1936 Ertel developed his advective-dynamic theory of air pressure and its periodicities (1936). Later he investigated the instability of the wind field at the troposphere (1936). He analyzed energy content, friction, and turbulence in the atmosphere and obtained a thermodynamic basis for the turbulence criteria (1939).

In 1939 Ertel reduced the equations of state to Hamilton's general dynamic principle. In 1942 he found the most important result, establishing the general vorticity theorem (see Table I) that contained the Bjerknes circulation theorem as a particular case. In 1955 he published a generalization that included the Helmholtz vorticity theorem.

From 1949 to 1971, Ertel derived from the vorticity equations a symmetrical deformation tensor that made possible linear transformation of the vorticity components into their total time derivatives (1962). His other work (some with Carl-Gustav Rossby at Cambridge and Hilding Koehler at Uppsala) included differential equations of fluid dynamics and description of steady convection (1949). Ertel's further studies included his hydrodynamic commutation

TABLE I. Selected results of Ertel's hydrodynamic research with meteorological or geophysical applications.

Year	Topic
1939	Variation principle of hydrodynamics
1939	General variation principle of atmospheric dynamics
1942	Ertel's general vortex theorem
1943	Discovery of planetary waves independently
1949	Ertel-Rossby convection theorem
1950	Ertel's theorem of circulation motions
1952	Ertel's potential theorem (1954)
1955	Ertel's general hydrodynamic theorem
1960	General representation of the equation of continuity (also 1965)
1962	Hydrodynamic equations of motion and vortex equations in corresponding forms
1969	Differential equations of fluid dynamics for an autobarotropic vertical flow
1970	Transformation of the differential form of the Weber hydrodynamic equations in relation to the earth's rotation

formula (1964, 1965). In 1971 he proved that the Jacobian functional determinant of the Eulerian components' velocity with respect to the Langrangian initial coordinates must remain invariant following the motion of a fluid element in inertial currents.

Ertel was a member of several scientific societies and academies, including the academies of sciences at Berlin, Halle, Uppsala, and Vienna. He was editor or coeditor of many scientific journals, including Gerland's *Beitrage zur Geophysik, Acta hydrophysica, Zeitschrift für Meteorologie, Forschungen und Fortschritte, Idöjárás, and Geofisica pura e applicada.*

BIBLIOGRAPHY

I. ORIGINAL WORKS. Ertel's papers are listed in *Idöjárás* 75 (1971), 263–270; Poggendorff, VIIa; and *Zeitschrift für Meteorologie*, **22** (1971), 319–328. His chief books are *Methoden und Probleme der dynamischen Meteorologie*, (Berlin, 1938; repr. 1972); and *Elemente der Operatorenrechnung mit geophysikalischen Anwendungen*, (Berlin, 1940).

Among his articles are "Theoretische Begründung einiger Guilbertscher Regeln," in *Berichte über die Tätigkeit der Preussisches Meteorologisches Institut Berlin* (1930), 114–118; "Die Krümmung der Diskontinuitätsflächen in der Atmosphäre und im Ozean," *ibid.* (1931), 147–152; "Advektiv-dynamische Theorie der Luftdruckschwankungen und ihrer Periodizitäten," in *Veröffentlichungen der Meteorologisches Institut Berlin*, **1**, no. 1 (1936); "Die Arten der Unstetigkeiten des Windfeldes an der Tropopause," in *Meteorologische Zeitschrift*, **53** (1936), 450–455; "Ein allgemeines Variationsprinzip der atmosphärischen Dynamik," *ibid.*, 169–171; "Thermodynamische Begründung des Richardsonschen Turbulenzkriteriums," *ibid.*, **56** (1939), 109–111; "Über ein allgemeines Variationsprinzip der Hydrodynamik," in *Abhandlungen der Preussische Akademie der Wissenschaften*, Physikalisch-mathematische Klasse (1939), no. 7; "Ein neuer hydrodynamischer Wirbelsatz," in *Meteorologische Zeitschrift*, **59** (1942), 277–281; "Über das Verhältnis des neuen hydrodynamischen Wirbelsatzes zum Zirkulationssatz von V. Bjerknes," *ibid.*, 385–387; "Über hydrodynamische Wirbelsätze," in *Physikalische Zeitschrift*, **43** (1942), 526–529; "Über stationäre oszillatorische Luftströmungen auf der rotierenden Erde," in *Meteorologische Zeitschrift*, **60** (1943), 332–334.

"Ein neuer Erhaltungs-Satz der Hydrodynamik," in *Sitzungsberichte der Deutsche Akademie der Wissenschaften zu Berlin*, Mathematisch-naturwissenschaftliche Klasse (1949), no. 1, with Carl Gustaf Rossby; "Ein Theorem über die stationäre Wirbelbewegung kompressibler Flüssigkeiten," *Zeitschrift für angewandte Mathematik und Mechanik*, **29** (1949), 109–113, with Hilding Köehler; "A New Conservation-Theorem of Hydrodynamics," in *Geofisica pura e applicada*, 14, fasc. 3–4 (1949), with Carl-Gustaf Rossby; "Ein Theorem über asynchron-periodische Wirbelbewegungen kompressibler Flüssigkeiten," in *Miscellanea Academica Berolinensia* (1950), 62–68; "Ein Theorem über die Feldstärke in Potentialfeldern," in *Sitzungsberichte der Deutsche Akademie der Wissenschaften zu Berlin*, Klasse fur Mathematik und allgemeine Naturwissenschaften (1954); no. 2; "Kanonischer Algorithmus hydrodynamischer Wirbelgleichungen," *ibid.* (1955), no. 4; "Ein neues Wirbel-Theorem der Hydrodynamik," *ibid.*, no. 5; "Teorema sobre invariantes sustanciales de la hidrodinámica," in *Gerlands Beiträge zur Geophysik*, **69** (1960), 290–293; "Relación entre la derivada individual y una cierta divergencia espacial en hidrodinámica," *ibid.*, 357–361; "Ein System von Identitäten und seine Anwendung zur Transformation von Wirbelgleichungen der Hydrodynamik," in *Monatsberichte der Deutschen Akademie der Wissenschaften zu Berlin*, **4** (1962), 292–296.

"Vertauschungs-Relationen der Hydrodynamik," in *Monatsberichte der Akademie der Wissenschaften zu Berlin*, **6** (1964), 838–841; "Hydrodynamische Vertauschungs-Relationen," in *Acta hydrophysica*, **9** (1965), 115–123; "Theorie der Strömung um Seebuhnen," in *Gerlands Beiträge zur Geophysik*, **77** (1968), 251–256; "Hydrodynamische Theorie der litoralen Sandriffe," *ibid.*, **78** (1969), 245–250; "Eine Differenzengleichung der Hydrodynamik autobarotroper Wirbelströmungen," *ibid.*, 414–418; "Ein Satz zur Kinematik nichtstationärer Stromfelder," *ibid.*, **79** (1970), 147–151; "Transformation der Differentialform der Weberschen hydrodynamischen Gleichungen unter Berücksichtigung der Erdrotation," *ibid.*, 421–424; "Eine Relation zwischen kinematischen Parametern horizontaler Strömungsfelder in der Atmosphäre," in *Idöjárás*, **74** (1970), 98–102; "Analytische Approximation der bodennahen Advektion," *ibid.*, spec. iss., 497–499; "Eine Betrachtung zur geomorphologisch wirksamen. Arbeit der Brandungswellen an Flachküsten," in *Acta hydrophysica*, **16** (1971), 5–10; "Eine Differentialinvariante der Trägheitsbewegungen in der Atmosphäre und im Ozean," in *Zeitschrift für Meteorologie*, **22** (1971), 339–341; and "Quellen und Senken des universellen Schwerefeldes," in *Annalen der Physik*, **26** (1971), 23–28, with Hans-Jürgen Treder.

II. SECONDARY LITERATURE. Obituary notices include F. Dési, in *Idöjárás*, **75** (1971), 261; P. Mauersberger, in *Zeitschrift für Meteorologie*, **22** (1971), 315–317; Wilfried Schröder, in *Wetter und Leben*, **23** (1971), 244; and Heinz Stiller, in *Gerlands Beiträge zur Geophysik*, **81** (1972), 161–163. See also P. Kahlig, in *Archives of Geophysics*, Bioklimatologie, **A19** (1970), 125; H. Pichler, in *Pure and Applied Geophysics*, **65** (1966), 180; and C. Truesdell, "Recent Advances in Rational Mechanics," in *Science*, **127** (1958), 729–739.

WILFRIED SCHRÖDER

ESCHER, BEREND GEORGE (*b.* Gorinchem, Netherlands, 4 April 1885; *d.* Arnhem, Netherlands, 11 October 1967), *geology, minerology, crystallography, volcanology*.

The son of George Arnold Escher, chief engineer and director of the State Public Works Department, and of Charlotte Marie de Hartitzsch, Escher was educated in Switzerland, where a college teacher, F. Mühlberg, inspired him to choose a career in geology. He completed his studies at the Federal Institute of Technology in Zurich under Albert Heim, with a dissertation on the pre-Triassic folding in the western Alps (1911). The thorough schooling in Swiss field geology and the influence of Heim's drawings marked Escher's work. He married Emma Brosy. They had three children.

Escher began his career as an assistant to Eugene Dubois, the discoverer of *Pithecanthropus erectus*, at the University of Amsterdam, then became curator of the geological collection of the Technological University of Delft. He broadened his experience by joining the Bataafse Petroleum Company as a geologist in Batavia (now Djakarta), Java. Here he became acquainted not only with petroleum geology and the petroleum industry but also with recent volcanic activity (dealt with in several of his publications). At Escher's urging the first governmental volcanological survey (Vulkaan Bewakingsdienst) was established in 1919 in the Netherlands East Indies.

In 1922 Escher was appointed professor of geology and director of the State Museum of Geology and Mineralogy in Leiden. With L. M. R. Rutten at the University of Utrecht, he was one of the first to introduce geology as an academic subject in the Netherlands. On his own he produced a curriculum covering the earth sciences, and before 1916 he had written a popular book on physical geology. He adapted the latter work and completely revised it three times, the last edition dating from 1951. Besides geology Escher taught mineralogy and crystallography, and wrote textbooks on these subjects.

At Leiden, Escher again took up the subject of his dissertation, concentrating on the area around Lugano in southern Switzerland, characterized by Permo-Carboniferous sedimentation and volcanism. He undertook the mapping of the area with his students, expanding it eastward to cover the Bergamo Alps of northern Italy. In 1932 he damaged his knee in a fall and was forced to delegate all field activities to his former pupil L. U. de Sitter, who completed the survey.

Escher's inaugural lecture in 1922 dealt with the causes of and the relation between internal geological forces. Of importance to this interest in geophysics is his later work in which he tried to use the sparse geophysical data then available to find a solution to the many conflicting hypotheses regarding the causes of orogenesis. The discovery of the zone of negative gravity anomaly in the East Indies led to discussions regarding its interpretation with Felix Vening Meinesz, P. H. Kuenen, and J. H. F. Umbgrove. Escher's contribution, concerning the place of volcanism in the developing model, introduced Arthur A. Holmes's ideas of underlying convection currents. Escher speculated on the possibility that the moon was derived from the earth and on its possible composition. After studying the moon's morphology, he concluded that its craters were likely of volcanic origin.

Escher was the first geologist in the Netherlands to use laboratory experiments for the study of geological phenomena. His experiments in connection with salt domes were carried out with Kuenen (1929), who later used experimental methods in his sedimentological research. Experimental facilities designed by Escher served engineering purposes in the reclamation of the Zuider Zee.

Escher's outspoken integrity caused his confinement as a hostage by the Germans in 1942. Afterward he went into hiding until the end of hostilities. In 1945 he became the first postwar rector of the University of Leiden, in which capacity he stimulated the development of a modernized student society.

Escher retired in 1955. His deteriorating sight prevented his continuing scientific work.

BIBLIOGRAPHY

I. ORIGINAL WORKS. Escher published some 110 books and papers. A chronological list is appended to his obituary in *Geologie en Mijnbouw*, **46** (see below). The titles of his textbooks are *Grondslagen der Algemene Geologie* (from 7th ed. [1948] onward; Amsterdam, 1916–1951), and *Algemene Mineralogie en Kristallografie* (The Hague, 1935; rev. ed., Gorinchem, 1950). The results of geological work in the southern Alps carried out under his guidance from 1926 to 1939 are compiled by L. U. de Sitter and C. M. de Sitter-Koomans, "The Geology of the Bergamasc Alps, Lombardia, Italy," in *Leidse Geologisch Medelingen*, **14B** (1949), 9–257.

Several of Escher's publications, mostly in Dutch, deal with different East Indian volcanoes; for example, see "Krakatoa," in *Handelingen Eerste Ned. Ind. Natuurwet. Congr. Weltevreden*, (1919), 28–35 and 198–219; the article also contains the excursion guide. This led him to study the caldera problem in "On the Formation of Calderas," in *Leidse Geologisch Medelingen*, **3** (1929), 183–219; also in *Proc. Fourth Pac. Sci. Congr. Java*, (1929), 571–587. Perhaps his most significant contribution is "On the Relation Between the Volcanic Activity in the Netherlands East Indies and the Belt of Negative Gravity Anomalies Discovered by Vening Meinesz," in *Proc. Kon. Ned. Akad. Wet.*, **36** (1933), 677–685.

For his interest in the moon, see "Moon and Earth," in *Proc. Kon. Ned. Akad. Wet.*, **42** (1939), 127–138; and "Origin of the Asymmetrical Shape of the Earth's Surface and Its Consequences upon Volcanism on Earth and Moon," in *Bulletin of the Geological Society of America*, **60** (1949), 353–362. For the experiments on salt domes with Kuenen, see *Leidse Geologisch Medelingen*, **3** (1929), 151–181. A summary of most of Escher's geological experiments is "Eine Übersicht der im geologischen Institut in Leiden von 1920 bis Ende 1937 ausgeführten geologischen Experimente," in *C. R. Congr. Int. de Géogr. Amsterdam*, **2** (1938), 273–278.

II. SECONDARY LITERATURE. An obituary of Escher in Dutch, with a list of most of his publications, is in *Geologie en Mijnbouw*, **46**, 417–422. For further secondary literature, see "Escher" in *Biografisch woordenboek van Nederland*, I (1979). Also of value is M. Neumann van Padang, *History of Volcanology in the East Indies*, Scripta Geologica 71 (1983), which mentions Escher as the initiator of the Vulkaan Bewakingsdienst.

J. J. DOZY

EVANS, GRIFFITH CONRAD (*b.* Boston, Massachusetts, 11 May 1887; *d.* Berkeley, California, 8 December 1973), *mathematics*.

Evans was the son of George William Evans, author of such textbooks as *Algebra for Schools* and teacher of mathematics at English High School in Boston, and of Mary Taylor Evans. After graduating from English High School in 1903, he entered Harvard. He earned the A.B. degree there in 1907, the A.M. in 1908, and the Ph.D. in mathematics in 1910. Among his professors at Harvard were William Fogg Osgood, Julian Coolidge, and Maxime Bôcher. Evans was an instructor in mathematics in the academic years 1906–1907 and 1909–1910. His dissertation, "Volterra's Integral Equation of the Second Kind with Discontinuous Kernel," appeared in two parts in *Transactions of the American Mathematical Society* (1910–1911).

After receiving the doctorate, Evans went to Europe on a Sheldon Traveling Fellowship from Harvard. There he studied at the University of Rome with Vito Volterra, who exerted a lasting influence on his work; in line with his interest in the applications of mathematics, Evans also spent a summer in Berlin, working with the physicist Max Planck.

Upon his return from Europe in 1912, Evans had job offers from M.I.T., the University of California at Berkeley, and the newly founded Rice Institute in Houston. Because of its opportunities he chose Rice and, as assistant professor, began teaching there in 1912. In 1916 he was promoted to full professor of mathematics. On 20 June 1917 Evans married Isabel Mary John; they had three sons.

During World War I, Evans was commissioned a captain in the U.S. Army Signal Corps. His scientific assignments concerning bomb trajectories and sights, and antiaircraft defenses took him to England, France, and Italy. With the help of Volterra, Evans facilitated the enrollment of U.S. military personnel in special wartime courses at Italian universities.

Evans was elected vice president of the American Mathematical Society for the years 1924–1926 and held the same office in the Mathematical Association of America; he was twice elected vice president of the American Association for the Advancement of Science (for mathematics, 1931–1932, and for economics, 1936–1937). He served as an editor of the *American Journal of Mathematics* from 1927 to 1936. During his years at Rice, Evans traveled widely in the United States and Europe. He spent half a year in Belgium, France, and Italy in the period 1929–1930, and the summers of 1921 and 1928 at the University of Minnesota and at Berkeley. While at Minnesota in 1921, discussions with the British statistician R. A. Fisher encouraged Evans to promote the study of mathematical statistics in the United States. He also brought such mathematicians as

Szolem Mandelbrojt, Karl Menger, and Tibor Rado as visiting professors to enrich the program at Rice.

Other institutions bid for Evans's services during these years. Harvard made several offers, and the University of California renewed its efforts to lure him to Berkeley in 1927. He declined these offers, but by 1933 he had changed his mind. The Berkeley administration had long planned a reorganization of its mathematics department upon the retirement of the department chairman, and approved the search for a new chairman and other faculty members in 1933 despite the financial difficulties confronting the university during the Great Depression. Evans was viewed as an exceptionally strong leader, and Berkeley negotiated long and hard to secure his services. Having been elected to the National Academy of Sciences in that year, he was offered a generous salary (subject to cuts applicable to all faculty salaries) and assured that he would enjoy considerable latitude in making new appointments. He extricated himself from his obligations at Rice and moved to Berkeley in the summer of 1934.

Evans's fifteen years as chairman at Berkeley marked a period of change and growth for the Mathematics Department. He had high expectations for Berkeley. As R. G. D. Richardson put it, Evans hoped "to build up [there] a great center in our subject comparable to Princeton, Harvard, and Chicago" (Richardson to E. R. Murrow, 3 May 1935, Emergency Committee Papers, box 109, New York Public Library). In building the Berkeley program, Evans at first argued against favoring displaced foreigners over unemployed Americans; as the economic picture improved, he changed his mind and added Hans Lewy, Jerzy Neyman, and Alfred Tarski to the department. Altogether Evans brought fifteen new faculty members to Berkeley between 1933 and 1949, and engineered such innovations as courses and seminars in mathematical economics (one of his own research interests), a statistical laboratory (headed by Neyman), and greater attention to the applications of mathematics. Evans later described his first few years at Berkeley as "an opportune time" for expanding and strengthening the department there (Evans to Henry Helson, 12 February 1966, Evans Papers, The Bancroft Library).

Evans assumed the presidency of the American Mathematical Society in 1938. In this capacity he encouraged the formation of the American Mathematical Society-Mathematical Association of America War Preparedness Committee (later the War Policy Committee) to guide research on problems of importance to national defense and to design mathematical training programs for the military. As president of the society until 1940, Evans served on the National Research Council committee charged with compiling a national scientific roster; after Pearl Harbor he addressed wartime issues as a member of the Mathematics Committee of the National Academy of Sciences and of the Applied Mathematics Panel of the National Defense Research Council. Between 1943 and 1947 he served as consultant for the Office of the Chief of Ordnance on gun design.

Evans continued as department chairman at Berkeley until 1949. After his retirement at the end of the 1954–1955 academic year, he continued to write and lecture for many years. In 1971 the new mathematics building on the Berkeley campus was named Evans Hall, in recognition of his contributions to mathematics and his dedication to building a world-class center of mathematical sciences at Berkeley.

Much of Evans's work built upon mathematical innovations introduced by Henri Lebesque, Vito Volterra, Maurice Frechet, and Henri Poincaré during his student years. His first paper, published in 1909, while he was still in graduate school, dealt with functional analysis, a field to which he would contribute much in the next decade. Evans's principal results concerned integrodifferential equations and integral equations with singular kernels. The American Mathematical Society invited him to deliver the Colloquium Lectures on this subject in 1916; they were published in 1918 as *Functionals and Their Applications*. These lectures illustrated the utility of integral expressions: for example, replacing second-order partial differential expressions by integral expressions for variable domains with first-order terms permitted derivation of a theorem for integral expressions analogous to Green's theorem.

Another aspect of Evans's work concerned surfaces of minimum capacity. Solution of the problem of minimal surfaces, the so-called plateau problem, depends on local properties. Evans was able to prove, however, in a series of papers beginning in 1920 that, among the surfaces with a given boundary s, there exists a surface of minimum (electric) capacity. In this work he used comparisons of energy integrals. The generalization of such problems led Evans to extensive research, especially in his later years, on multiple-valued harmonic functions.

Beginning with his paper on Kirckhoff's law, written while he was a graduate student and published in 1910, Evans concerned himself with the applications of mathematics. His work during both world wars evinces this interest, as does his innovative work in applying mathematics to economic theory.

Evans formulated a model of the economy as a whole and posed the problem of defining an aggregate variable in terms of microeconomic components. His 1924 paper on the dynamics of monopoly, which introduced time derivatives in demand relations, was recognized as the beginning of dynamic theories of economics. Evans's textbook on mathematical economics was published in 1930 and formed the basis of his pioneering courses at Rice and Berkeley.

BIBLIOGRAPHY

I. ORIGINAL WORKS. Evans's writings include "The Integral Equation of the Second Kind of Volterra, with Singular Kernel," in *Bulletin of the American Mathematical Society*, 2nd ser., **16** (1909), 130–136; "Note on Kirchoff's Law," in *Proceedings of the American Academy of Arts and Sciences*, **46** (1910), 97–106; "Volterra's Integral Equation of the Second Kind with Discontinuous Kernel," in *Transactions of the American Mathematical Society*, **11** (1910), 393–413, and **12** (1911), 429–472; *Functionals and Their Applications* (Providence, R.I., 1918; repr., New York, 1964); "Fundamental Points of Potential Theory," in *The Rice Institute Pamphlet*, **7** (1920), 252–329; "Problems of Potential Theory," in *Proceedings of the National Academy of Sciences*, **7** (1921), 89–98; "The Dynamics of Monopoly," in *American Mathematical Monographs*, **31** (1924), 77–83; *The Logarithmic Potential. Discontinuous Dirichlet and Neumann Problems, American Mathematical Society Colloquium Publications*, 6 (Providence, R.I., 1927); and *Mathematical Introduction to Economics* (New York, 1930); "Potentials of General Masses in Single and Double Layers. The Relative Boundary Value Problems," in *American Journal of Mathematics*, **53** (1931), 493–516, written with E. R. C. Miles; "Complements of Potential Theory," *ibid.*, **54** (1932), 213–234; "Correction and Addition to 'Complements of Potential Theory,'" *ibid.*, **57** (1935), 623–626; "On Potentials of Positive Mass., Parts I and II," in *Transactions of the American Mathematical Society*, **37** (1935), 226–253, and **38** (1935), 201–236; "Potentials and Positively Infinite Singularities of Harmonic Functions," in *Monatshefte für Mathematik und Physik*, **43** (1936), 419–424; "Modern Methods of Analysis in Potential Theory," in *Bulletin of the American Mathematical Society*, **43** (1937), 481–502; "Surfaces of Minimal Capacity" and "Surfaces of Minimum Capacity," in *Proceedings of the National Academy of Sciences*, **26** (1940), 489–491, 662–667; and "Continua of Minimum Capacity," in *Bulletin of the American Mathematical Society*, **47** (1941), 717–733.

Evans's papers at Bancroft Library, University of California, Berkeley, include twenty cartons of notebooks, journals, correspondence, course notes, and drafts and reprints of writings, as well as some papers of his father and of his son George William. Noteworthy among the unpublished materials is correspondence with American and European mathematicians and a typescript of a calculus textbook by Evans and H. E. Bray. The Bancroft collection also documents Evans's wartime activities and participation in professional organizations and university committees. The University Archives at Berkeley contain additional correspondence concerning Evans's appointment and the growth of the mathematics department under his leadership.

II. SECONDARY LITERATURE. The principal biographical notice on Evans is by a colleague at Berkeley, Charles B. Morrey, "Griffith Conrad Evans," in *Biographical Memoirs. National Academy of Sciences*, **54** (1983), 127–155. It contains a summary of Evans's major contributions to mathematics and a ninety-five-item bibliography of his publications. It is supplemented by Charles B. Morrey, Hans Lewy, R. W. Shephard, and R. L. Vaught, "Griffith Conrad Evans," in *University of California, In Memoriam* (1977), 102–103. Evans's success in building the mathematics program at Berkeley is explored in Robin E. Rider, "An Opportune Time: Griffith C. Evans and Mathematics at Berkeley," in Peter Duren, ed., *A Century of Mathematics in America*, II (Providence, R.I., 1989).

ROBIN E. RIDER

EWALD, PAUL PETER (*b.* Berlin, Germany, 23 January 1888; *d.* Ithaca, New York, 22 August 1985), *physics, crystallography.*

Ewald was the son of Paul Ewald, a *Privatdozent* in history at the University of Berlin, and of Clara Philippson Ewald, an internationally known portrait painter. His father died of appendicitis shortly before Paul was born. His mother raised Ewald, and as a result of their travels he learned to speak English and French at a very early age. He was educated at the Königliches Wilhelmsgymnasium in Berlin and the Königliches Victoriagymnasium in Potsdam. Ewald graduated from the latter in 1905 and then began to study chemistry at Gonville and Caius College, Cambridge. In 1906, after one semester, he entered the University of Göttingen to continue studying chemistry. He was, however, disappointed by the lack of consistent theoretical connection of the various facts of chemistry and soon changed to mathematics, which he studied for three semesters (1906–1907). Although he found his first mathematics courses at Göttingen not very helpful, he was compensated during the following semesters, when he worked with David Hilbert and Ernst Hellinger on differential and integral calculus.

Ewald transferred in 1907 to the University of Munich, where his two semesters of mathematical studies included Alfred Pringsheim's lectures on functional analysis. He also attended Arnold Sommerfeld's lectures on hydrodynamics. Fascinated by the interplay of theory and experiment in physics, he became Sommerfeld's student.

In 1910 Ewald chose as the subject of his doctoral dissertation the problem of how to find the optical properties of an anistropic arrangement of isotropic resonators, an area in which Sommerfeld could offer little help. Ewald's approach was quite original. Instead of investigating the reaction of the dipoles on the incident light, he focused on the electromagnetic wave field that is "dynamically possible" in the interior of a lattice arrangement of oscillators. With the help of boundary considerations and especially by explicit calculations, in the reprint of his dissertation (1916) he could explain the "compensation [extinction] of the incident wave," the "dynamically closed" state of the refracted waves (the oscillation modes in the crystal), and the existence of reflected waves outside the crystal. Among the difficulties Ewald had to overcome in working out his thesis was the calculation of the electromagnetic field exciting the dipole oscillations of any one atom. The problem was how to subtract from the entire field the one that originated from the atom itself. This amounted to subtracting infinity from infinity.

Fruitful in another way was Ewald's attempt to discuss some details of his dissertation with Max von Laue in February 1912. His dissertation, with its underlying assumption of a regular spatial arrangement of particles in a crystal, stimulated Laue to think of the phenomena produced by light of very short wavelength (comparable with atomic distances) in the space lattice of a crystal. The personal recollections of Laue and Ewald stress the unique situation in Munich, where Leonhard Sohncke and Paul von Groth allegedly had kept alive the space lattice theory of crystals and Sommerfeld advocated the wave theory of X rays.

At Laue's initiative, Sommerfeld's assistant Walter Friedrich and Röntgen's doctoral student Paul Knipping in April 1912 succeeded in taking photographs of what was soon recognized as X-ray diffraction in a copper sulfate crystal. After receiving his doctorate in that year, Ewald applied its "dynamical theory" to the diffraction of X rays. In this context he invented the "sphere of reflection" (1913). The possible intersections of the "sphere of reflection" with certain points in the reciprocal lattice furnish the directions of constructive interference.

Ewald did postdoctoral work at the University of Göttingen, where he was assistant to David Hilbert. In 1913, soon after his marriage to Elisa Berta (Ella) Philippson, he returned to Munich, at first sharing the post of assistant to Sommerfeld with Wilhelm Lenz. He had learned to operate X-ray equipment for medical purposes, so during World War I he was a field X-ray technician on the northern Russian front. By the autumn of 1915 fighting had practically ceased there, so he had time to continue his work on the diffraction of X rays. In 1916 Ewald published a series of papers on the foundations of crystal optics (one of them was an abbreviation of his dissertation), and in 1917 his *Habilitationsschrift*, on the crystal optics of X rays, which contains the elaborate dynamical theory of X-ray diffraction in perfect crystals.

Ewald's theory predicted a deviation from the simple Bragg equation, which explained the X-ray diffraction geometrically by reflections of the incident rays on the internal atomic net-planes and their subsequent interference. Indications of this deviation could be seen in experimental investigations performed in Manne Siegbahn's institute at Lund. A confirmation of Ewald's theory was, however, possible only with the perfect crystals produced by the semiconductor industry many decades later. Today the perfection of crystals can be tested with the help of Ewald's theory. His dynamical theory stimulated Hans Bethe (who later became his assistant and his son-in-law), who in 1928, in his doctoral dissertation (under Sommerfeld), dealt with electron diffraction by crystals. Another application was in the field of neutron diffraction.

Ewald's lifelong research in the field of X-ray diffraction caused him to develop the ideas of his dissertation and of his *Habilitationsschrift*. The dissertation furnished exact solutions only for two beams. Ewald hoped eventually to solve the *n*-beam problem. According to Bethe's recollections, Ewald wanted thus to finish his dissertation.

In 1918 Ewald became *Privatdozent* at the University of Munich. He was named extraordinary professor of theoretical physics at the Technische Hochschule (now the University) of Stuttgart in 1921, and was appointed professor in 1922. While there, following the work of Richard Glocker, who as early as 1919 had done intensive experimental work on X-ray analysis of metal structures, Ewald helped to create a center of X-ray research and solid state physics.

Ewald was appointed rector at Stuttgart in 1932. In his inaugural address, he urged his audience to strive for social and political harmony, and he ended pathetically with the famous verse of the national anthem "Deutschland über alles." The following year, however, the National Socialists' "law for the restoration of the civil service" caused him to resign the rectorship because his wife was Jewish and he was part Jewish. However, his service at the front in World War I and the Nuremberg redefinition of "part Jewish," and a succession of nominal National Socialists as rectors, allowed him to continue his work as professor. In 1936, when a young Nazi teaching corps leader read a government paper denying the value of an "objective" science, Ewald walked out of the assembly. The new rector, an ardent National Socialist, urged Ewald to resign and had him pensioned off three weeks later.

In 1937 Ewald left Germany, a step he had been considering since 1933. With the help of William Lawrence Bragg, he was able to continue research at Cambridge, supported by a grant. In 1939 he was appointed lecturer, and later professor, of mathematical physics at Queen's University, Belfast. The financial circumstances of the family—there were four children—improved when Ewald accepted the latter post and, from 1949, when he was professor of physics and head of the department at the Polytechnic Institute of Brooklyn. He retired in 1959.

Of outstanding importance was Ewald's contribution to the formation and growth of crystallography. From 1924 to 1940 he was coeditor of *Zeitschrift für Kristallographie*, founded by Paul von Groth. With his pupil Carl Hermann he published the first volume of *Strukturbericht. 1913–1928* as a supplement to *Zeitschrift für Kristallographie* (1931). Six volumes covering the period to 1939 were to follow. After the war Ewald was instrumental in continuing this review work as *Structure Reports*. Through his initiative and with Bragg's help, the International Union of Crystallography was founded in 1947 with Bragg as its president and Ewald as its vice president. From 1948 to 1958 Ewald served as one of the editors of its journal, *Acta crystallographica*.

Ewald was a member or fellow of the American Academy of Arts and Sciences, the Royal Society, and the Deutsche Akademie der Naturforscher (Leopoldina). He received honorary doctorates from the University of Stuttgart in 1954, from the University of Paris in 1958, and from the University of Munich in 1968. In 1978 the Deutsche Physikalische Gesellschaft awarded him the Max Planck Medal, and the following year he received the first Gregori Aminoff Medal of the Royal Swedish Academy.

BIBLIOGRAPHY

I. ORIGINAL WORKS. Ewald's writings include "Dispersion und Doppelbrechung von Elektronengitten (Kristallen)" (Ph.D. dissertation, University of Munich, 1912; published at Göttingen, 1912); "Zur Theorie der Interferenzen der Röntgenstrahlen in Kristallen," in *Physikalische Zeitschrift*, **14** (1913), 465–472; "Zur Begründung der Kristalloptik, Teil I: Theorie der Dispersion," in *Annalen der Physik*, **49** (1916), 1–38, an abbreviated version of his dissertation; "Zur Begründung der Kristalloptik, Teil II: Theorie der Reflexion und Brechung," *ibid.*, 117–143, translated as *On the Foundation of Crystal Optics*, Air Force Cambridge Research Laboratories, Translations no. 84 (Bedford, Mass., 1970); "Zur Begründung der Kristalloptik, Teil III: Die Kristalloptik der Röntgenstrahlen," in *Annalen der Physik*, **54** (1917), 519–597, his *Habilitationsschrift*; *Kristalle und Röntgenstrahlen* (Berlin, 1923); *The Physics of Solids and Fluids, With Recent Developments*, written with T. Pöschl and L. Prandtl, J. Dougall and W. M. Deans, trans. (London and Glasgow, 1930; 2nd enl. ed., 1936); "Die Erforschung des Aufbaues der Materie mit Röntgenstrahlen," in Hans Geiger and Karl Scheel, eds., *Handbuch der Physik*, 2nd ed., XXIII, pt. 2 (Berlin, 1938); "Some Personal Experiences in the International Coordination of Crystal Diffractometry," in *Physics Today*, **6**, no. 12 (1953), 12–17; as editor, *Fifty Years of X-Ray Diffraction* (Utrecht, 1962); "William Henry Bragg and the New Crystallography," in *Nature*, **195** (28 July 1962), 320–325; "The Myth of Myths; Comments on P. Forman's paper on 'The Discovery of the Diffraction of X-Rays in Crystals,'" in *Archive for History of Exact Sciences*, **6** (1969/1970), 72–81; "Physicists I Have Known," in *Physics Today*, **27** (September 1974), 42–47; and "Remembering Peter Debye in Munich," *ibid.*, **38** (January 1985), 9, 122.

Letters from or to Ewald are in the following archives: American Institute of Physics; Olin Library, Cornell University; Landesbibliothek, Stuttgart; Bayerische Staatsbibliothek, Munich; Deutsches Museum, Munich; Perkins Library, Duke University; Bodleian Library, Oxford University; Library of Congress, Washington, D.C.; Bibliothek und Archiv, Max Planck Gesellschaft, Berlin; Staatsbibliothek Preussischer Kulturbesitz, Berlin; and Archive for History of Quantum Physics, Office for History of Science and Technology, University of California, Berkeley. The last also has interviews with Ewald, 29 March 1962, conducted by G. Uhlenbeck with T. S. Kuhn and Mrs. Ewald, and 8 May 1962, conducted by T. S. Kuhn. Personnel records are at the Rektoramt of the University of Stuttgart.

II. SECONDARY LITERATURE. On Ewald or his work, see Michael Eckert, Willibald Pricha, Helmut Schubert, and Gisela Torkar, *Geheimrat Sommerfeld—theoretischer Physiker* (Munich, 1984); Paul Forman, "The Discovery of the Diffraction of X-Rays by Crystals; a Critique of the Myths," in *Archive for the History of Exact Sciences*, **6** (1969/1970), 38–71; G. Hildebrandt, "Zum Tode von Paul Peter Ewald," in *Physikalische Blätter*, **41** (1985), 412–413; and David Phillips, "William Lawrence Bragg," in *Biographical Memoirs of Fellows of the Royal Society*, **25** (1979), 75–143, which on pp. 115–116 provides details on the founding of the International Union of Crystallography.

WALTER KAISER

EWING, WILLIAM MAURICE (*b*. Lockney, Texas, 12 May 1906; *d*. Galveston, Texas, 4 May 1974), *geophysics, oceanography, seismology*.

Ewing hardly ever used the name William, preferring to go by the name Maurice. His parents were Floyd Ford Ewing, a farmer who was also a dealer in hardware and farm implements, and Hope Hamilton Ewing. The fourth of ten children, Ewing grew up as the eldest of seven because his three older siblings died at very early ages. Hope Ewing wanted all her children to be educated, and all but one obtained a university education. John, the youngest, became a well-known geophysicist in his own right. He worked with Maurice at Lamont Geological Observatory and later moved to Woods Hole Oceanographic Institution, where he chaired the department of geology and geophysics.

Ewing was one of the most important exploratory geophysicists who took on the task of finding out about the ocean floor. He was extremely ambitious and had the energy and intelligence to match. He began taking seismic refraction measurements in the ocean in 1935. During World War II, while at Woods Hole Oceanographic Institution, he did pioneering work on the transmission of sound in seawater. In 1949 Ewing was appointed director of Lamont Geological Observatory, which became Lamont-Doherty Geological Observatory in 1969. He remained the director until 1972, when he returned to Texas to become the founding chief of the earth and planetary sciences division of the Marine Biomedical Institute of the University of Texas at Galveston.

Ewing turned Lamont into one of the leading, if not the leading, institutes of marine geology and geophysics. He and the institution were extremely productive in gathering all sorts of data about the seafloor. Ewing was a master at getting the most out of his co-workers and students, as well as his ships and equipment. His work on interpreting surface waves, done with Frank Press, led to the first well-founded estimate of the depth of the Mohorovičić discontinuity under the ocean floor; and the data bank he established at Lamont-Doherty played a crucial role in the confirmation of seafloor spreading and plate tectonics in the late 1960's and early 1970's.

Ewing became a member of the National Academy of Sciences in 1948 and of the American Philosophical Society in 1959, and a foreign member of the Royal Society of London in 1972. He was also an honorary or foreign member of the Geological Society of London (1964), the Royal Astronomical Society (1964), the American Association of Petroleum Geologists (1968), and the Royal Society of New Zealand (1970). Among the numerous scientific prizes Ewing received were the Arthur L. Day Medal of the Geological Society of America (1949), the Agassiz Medal of the National Academy of Sciences (1955), the William Bowie Medal of the American Geophysical Union (1957), the Vetlesen Prize (1960), the Cullum Geographical Medal of the American Geographical Society (1961), the gold medal of the Royal Astronomical Society (1964), the Wollaston Medal of the Geological Society of London (1969), and the Walter H. Bucher Medal of the American Geophysical Union (1974).

Ewing was vice president of the Geological Society of America from 1953 until 1956, president of the American Geophysical Union from 1956 to 1959, and vice president (1952–1955) and president (1955–1957) of the Seismological Society of America.

Following his mother's wishes, Ewing received a sound public school education in Lockney and in 1922 was awarded the Hohenthal Scholarship to Rice Institute in Houston, Texas. Originally he was denied the scholarship; but his mathematics teacher at Lockney High School wrote to Rice, explaining that Ewing was the best mathematics student he had yet encountered.

At Rice, Ewing had to work in order to support himself. At first he worked in an all-night drugstore, then assisted in classes and worked in the library. He also found time to play the trombone in the marching band. While performing one day, he was noticed by a coed named Avarilla Hildenbrand; married 31 October 1928, they had one son and were divorced in 1941. Despite his heavy work load outside of class, he did well at Rice.

Ewing first majored in electrical engineering but switched to physics and mathematics, for he found both the subjects and instructors more exciting than the engineering courses or faculty. He was greatly influenced by H. A. Wilson, a rather unorthodox physicist who had been at the Cavendish Laboratory. Wilson was able to draw prestigious scientists to Rice for his weekly colloquia series, and thus Ewing had the opportunity to meet a number of important physicists. He took no undergraduate courses in geology; indeed, he never took a geology course. He obtained field experience during his summer vacations as part of a crew prospecting for oil in the shallow lakes of Louisiana. This was his first exposure to underwater exploratory geophysics.

In 1926, while still an undergraduate, Ewing published his first scientific paper, "Dewbows by Moonlight." He received the B.A. in 1926, and that fall he became a graduate student in the physics department at Rice. He obtained the M.A. in 1927 and the Ph.D. in 1931. Part of his dissertation, "Calculation of Ray Paths from Seismic Travel-Time Curves," was published in two papers written with Lewis D. Leet, then director of the seismological station at Harvard.

In 1929 Ewing was hired as an instructor in physics at the University of Pittsburgh and a year later left Pittsburgh for a similar position at Lehigh. During his first several years at Lehigh, he had to teach a number of elementary physics courses but still managed to do research in geophysics and present papers at meetings of the American Geophysical Union (1931, 1934). Although this work did not deal with important topics, he developed a fuller understanding of the techniques of explosive seismology. In November 1934 Ewing's first real opportunity to engage in potentially significant research came when Richard Field and William Bowie visited Lehigh. Bowie was chief of the division of geodesy of the Coast and Geodetic Survey; and Field, professor of geology at Princeton, was the major force behind the founding of the American Geophysical Union's Committee on the Geophysical Study of the Ocean Basins.

Most likely knowing of Ewing's work through hearing him speak at meetings of the American Geophysical Union, Bowie and Field suggested that Ewing use explosive seismology to investigate the continental shelf. He agreed, and they decided that he should investigate the region between Cape Henry, Virginia, and the edge of the continental shelf. Although the first attempt did not yield any geological results, it convinced Ewing that he had the technical knowledge to use explosive seismology at sea. Through the influence of Field, he got the use of the R.V. *Atlantis*, which belonged to the Woods Hole Oceanographic Institution. Ewing undertook the project in October 1935, and his study indicated that the seafloor from just off Virginia to the edge of the ocean depths was covered by sediments 3,800 meters thick.

Ewing continued seismic shooting in shallow water throughout the remainder of the 1930's and by 1939 extended his seismic refraction surveys to the shallow waters off Bermuda. In all, Woods Hole provided him with forty-five days of shared time aboard the *Atlantis* during this period. It was on one of these expeditions that Ewing met Edward Bullard. He also, probably through Field's influence, worked with Harry Hess aboard the submarine U.S.S. *Barracuda*. In 1936 they traced the negative gravity anomaly associated with the Puerto Rico Trench around the island arc of the Lesser Antilles. The anomaly had been discovered in 1926 by the Dutch geophysicist Felix Vening Meinesz, who invented a pendulum apparatus for taking gravity measurements at sea from aboard submarines. Obtaining gravity measurements of the seafloor became one of Ewing's major interests, and he continued to work in the field for most of his life.

During World War II, Ewing, on a leave of absence from Lehigh, went to Woods Hole Oceanographic Institution, where he was a research associate from 1940 to 1944. He took Allyn Vine and J. Lamar (Joe) Worzel, two of his former graduate students, with him; and they and Columbus Iselin, director of Woods Hole, wrote a manual for the U.S. Navy entitled *Sound Transmission in Sea Water*. Iselin was impressed with Ewing and his students. Not only were they intelligent but they also worked night and day, seven days a week.

During this time Ewing and his co-workers discovered a low-velocity sound channel in the ocean. It is at a depth of 700–1,300 meters and is called the SOFAR (Sound Fixing and Ranging) channel. The SOFAR channel traps sound waves, and as a result sounds can be transmitted over tremendous distances within this low-velocity tunnel. Ewing found that he could record the sound from the explosion of a small charge dropped off the west coast of Africa as far away as the Bahamas. Needless to say, the U.S. Navy found this discovery important. Ewing met his second wife, Margaret Kidder, at Woods Hole. They were married 19 February 1944 and had two sons and two daughters.

In 1944 Ewing was offered an assistant professorship in the geology department at Columbia University. He accepted and moved to New York City in June 1946, and became a full professor in 1947. The following year the widow of Thomas Lamont, a well-known New York financier, gave Columbia their 155-acre estate, located north of New York City on the west bank of the Hudson River in Palisades, New York. Columbia offered Ewing the use of the estate along with the $250,000 included in the original gift. Around the same time the Massachusetts Institute of Technology asked Ewing to join its staff and offered to provide him with a similar estate and financial support. Ewing and some of his group went to MIT, listened to the offer, and unanimously voted to stay at Columbia.

Even though Ewing had his own oceanographic

institute and a corps of excellent researchers, he still needed his own ship. Although he had some difficulty convincing Columbia University to purchase a ship, in 1953 it bought *Vema* for $150,000. Having his own ship made it much easier for Ewing to design instruments and secure outside funding for oceanic exploration.

Early in 1949, after he settled into Lamont but before the purchase of *Vema*, Woods Hole gave Ewing the simultaneous use of two ships. This was a long-standing desire of Ewing's, since it would allow him to record seismic refraction shots of requisite power and separation to determine the depth of the Mohorovičić discontinuity (Moho) between the crust and the mantle under the Atlantic Ocean. At the time, it was generally felt that the depth of the discontinuity under the oceans was less than under the continents, but the issue was unresolved.

Ewing and his co-workers found the Moho to be five kilometers beneath the ocean floor. There were, however, two problems with the study. One was practical: the results were, strictly speaking, indicative of the depth of the Moho in one small area of the Atlantic Basin. It would take years to get enough shots to determine if the five-kilometer depth was typical of the whole Atlantic and other ocean basins. The other was that the five-kilometer depth was considerably less than previous estimates that had been drawn primarily from time-travel studies of the dispersion of earthquake surface waves over oceanic paths. These studies indicated values more in concert with Beno Gutenberg's estimates.

Ewing and Frank Press, one of his students, found a single solution to both problems. They realized that they had to begin their own analysis of earthquake surface waves. Perhaps they could resolve the conflict and, if successful, use analyses of surface waves as a quick way of finding out the depth of the Moho under any ocean without having to go to sea. As a first step they constructed and installed seismographs at Lamont that were capable of getting reliable travel times for surface waves. They also examined previous analyses and realized that they failed to take into account the effect of water and seafloor sediments on surface waves. Correcting the mistake, they found that their analysis of surface wave distribution yielded the five-kilometer value for the depth of the Moho. He and Press submitted their first article on the subject on 1 August 1949.

This discovery of a fundamental difference between oceanic and continental crust was Ewing's most important specific piece of work. It was unexpected, and its consequences for geological thought were enormous. It also eliminated the need to take thousands of ship-to-ship seismic shootings in order to determine, point by point, the thickness of the oceanic crust—a task that would have taken years to complete.

Ewing remained director of Lamont for twenty-three years, resigning in 1972. His greatest professional achievement was molding Lamont into a superb oceanographic data-gathering institution. By deciding to keep his laboratory, Lamont's ships, at sea for almost the entire year, he built up an enormous data base. Rather than require that the data from one voyage be analyzed before beginning another, Ewing decided to stockpile data concerning all sorts of parameters. He also insisted that any data collected by Lamont ships be available to anyone at Lamont. The policy at most other institutions was to let the person directly responsible for collecting the data or the chief scientist of the cruise view the data as his private property. Ewing created a much more efficient and useful communal data base. Moreover, Ewing encouraged and demanded the most from his students and co-workers, and he taught by example. This provided Lamont with the largest oceanographic data base in the world while he was director and enabled its personnel to confirm, in 1965 and 1966, the key hypothesis that led to plate tectonics and the revolution in the earth sciences during the late 1960's and early 1970's.

It is somewhat ironic, but to Ewing's credit, that the data gathered by Lamont played such a crucial role in the confirmation of seafloor spreading and plate tectonics. Ewing was strongly opposed to the concepts of continental drift and seafloor spreading, although he reluctantly accepted them at the very end of 1966.

The variety of innovative data-gathering techniques developed or improved at Lamont, and the resulting discoveries, were astounding. Ewing, interested in discovering the nature of seafloor sediments, developed seismic reflection shooting at sea and began using it as early as 1949. Seismic reflection allows the determination of small discontinuities in seafloor sediment. Ewing and others at Lamont developed powerful echo sounders that were capable of differentiating layers of sediment. At first a single hydrophone was lowered over the side of the ship to record the shots. Later, hydrophones were towed behind the ship and 0.2-kg. charges were thrown overboard. Around 1962 Ewing began using "sparkers" and "air guns" to produce the sounds. The sparker produces sound by an underwater spark, and the air gun is a container filled with air, kept at a pressure of 150 atmospheres, that is suddenly released through a valve. Ewing and his brother

John, who was responsible for much of this work, mapped the various sediment layers, especially in the northern Atlantic, throughout the 1960's.

During the early 1950's Ewing and two of his co-workers, Bruce Heezen and David Ericson, confirmed the existence of turbidity currents, which had been hypothesized by American geologist Reginald A. Daly in 1933. Daly proposed that underwater currents—he called them "density currents"—caused the formation of submarine canyons, which were known to cut through the continental slope at the mouths of major rivers. In 1938 the Dutch sedimentologist Philip H. Kuenen developed a laboratory model of density currents (he called them "turbidity currents") to show that they could cut through channels and deposit sediment evenly over the bottom of his laboratory tank.

In 1947 Ewing and his co-workers had discovered the existence of a great abyssal plain between Bermuda and the Mid-Atlantic Ridge. They were astonished at the flatness of the plain, and the few cores they took indicated that the plain was covered by relatively recent sediment, nothing older than the Eocene. Two years later they traced the Hudson Canyon 200 miles beyond the continental slope and 15,900 feet beneath the surface. This convinced them of the existence of turbidity currents, which in turn explained the extent of the canyon, the flatness of the abyssal plain, and the recency of the deposits. In 1952 Ewing and Heezen cited turbidity currents in explaining why the 1929 earthquake on the Grand Banks had sequentially broken more than a dozen submarine cables.

Lamont was also responsible for the discovery of the rift valleys that run along the axes of most of the midoceanic ridges. In 1952 Marie Tharp, a cartographer who worked primarily with Heezen, discovered a depression in three profiles of the Mid-Atlantic Ridge and suggested to Heezen the existence of a valley in the center of the ridge. By 1953 Heezen was convinced of the correctness of Tharp's analysis. He had a contract with a cable company to determine why submarine cables often failed and had become interested in the location of earthquakes on the seafloor. He noticed that many of them occurred in the place where Tharp had hypothesized the existence of the rift valley. Ewing also was convinced, and in 1956 he and Heezen presented the discovery at a meeting of the American Geophysical Union. The discovery of the central rift valley and associated earthquakes was extremely important. Not only did it require that any theory about the origin of oceanic ridges explain the presence of the rift valley and shallow earthquakes, it also allowed oceanographers to trace the location of oceanic ridges throughout the world by examining seismological maps of oceanic ridges.

Ewing also thought it would be worthwhile to gather information about submarine magnetic fields, and in 1952 he decided to have his ships tow an airborne magnetometer that had been designed by Victor Vacquier for the detection of submarines. His work inspired other oceanographic institutions to undertake extensive geomagnetic surveys. Given Ewing's plan to collect as much data as possible, however, the Lamont group continued to make magnetic profiles; and its profiles of the Pacific-Antarctic Ridge and the Reykjanes Ridge, analyzed by such Lamont personnel as Walter Pitman III and James Heirtzler, played a central role in the confirmation of seafloor spreading and the development of plate tectonics.

Ewing also stressed the collection of core samples of sediment on the ocean floor. He and his co-workers collected more cores than any other oceanic institution, and work on them led not only to the development of a theory about the cause of ice ages by Ewing and W. L. Donn, one of his students (1956) but also allowed Neil Opdyke, a paleomagnetist hired by Ewing, to trace the time scale of polarity reversal in seafloor sediments ten years later. This latter accomplishment was extremely important in confirming the existence of seafloor spreading and plate tectonics.

Ewing's commitment to his work and the development of Lamont put a tremendous strain upon his private life. Although he cared deeply about his family, his second marriage ended in divorce in 1965. Later that year he married Harriett Green Bassett, who had been his secretary at Lamont. She continued to work at Lamont, and also worked for him after he went to the University of Texas at Galveston in 1972. Ewing died of a massive cerebral hemorrhage.

BIBLIOGRAPHY

I. ORIGINAL WORKS. Ewing was author or coauthor of more than 350 publications. The most comprehensive list is in the memoir by Bullard (see below). Among writings relevant to the account above are "Dewbows by Moonlight," in *Science*, **63** (1926), 257–258; "Geophysical Investigations in the Emerged and Submerged Atlantic Coastal Plain. Part IV. Cape May, New Jersey, Section," in *Bulletin of the Geological Society of America*, **51** (1940), 1821–1840, written with George P. Woollard and A. C. Vine; "Crustal Structure and Surface-Wave Dispersion," in *Bulletin of the Seismological Society of*

America, **40** (1950), 271–280, written with Frank Press; "Turbidity Currents and Sediments in the North Atlantic," in *Bulletin of the Association of Petroleum Geologists*, **36** (1952), 489–511, written with D. B. Ericson and Bruce Heezen; "Turbidity Currents and Submarine Slumps, and the 1929 Grand Banks Earthquake," in *American Journal of Science*, **250** (1952), 849–873, written with Bruce Heezen; "A Theory of Ice Ages," in *Science*, **123** (1956), 1061–1066, written with W. L. Donn; "Mid-Atlantic Ridge Seismic Belt," in *Transactions of the American Geophysical Union*, **37** (1956), 343 (abstract), written with Bruce Heezen; *The Floors of the Oceans: I. The North Atlantic*, Geological Society of America, Special Paper no. 65 (New York, 1959), written with Marie Tharp and Bruce Heezen; "Seismic-Refraction Measurements in the Atlantic Ocean Basins, in the Mediterranean Sea, on the Mid-Atlantic Ridge, and in the Norwegian Sea," in *Bulletin of the Geological Society of America*, **70** (1959), 291–318, written with John Ewing; and "Sediment Distribution on the Mid-ocean Ridges with Respect to Spreading of the Sea Floor," in *Science*, **156** (1967), 1590–1592, written with John Ewing, which contains Ewing's first endorsement, albeit a mild one, of seafloor spreading. The paper that best illustrates Ewing's earlier views on continental drift, one not included in Bullard's bibliography, is "The Atlantic Ocean Basin," part of "The Problem of Land Connections Across the South Atlantic, with Special Reference to the Mesozoic," which is article 3 in *Bulletin of the American Museum of Natural History*, **99** (1952), 87–91.

Harriet Ewing collected many of her husband's private papers and gave them to the University of Texas. They are housed at the Harry Ransom Humanities Research Center at the University of Texas at Austin.

II. SECONDARY LITERATURE. The best commemorative articles are Sir Edward Bullard, "William Maurice Ewing," in *Biographical Memoirs of Fellows of the Royal Society of London*, **21** (1975), 269–311, expanded in *Biographical Memoirs. National Academy of Sciences*, **51** (1980), 119–193; and William L. Donn, "Memories of (William) Maurice Ewing: The Little Boy in the Candy Shop," in *EOS*, **66** (1985), 129–130. Bullard's is more comprehensive, but Donn's offers an intimate and lively account of Ewing, of his and Donn's development of their hypothesis about the cause of ice ages, and his reluctance to accept seafloor spreading and continental drift. William Wertenbaker, *The Floor of the Sea* (Boston, 1974), has captured Ewing's own account of much of his life and gives the reader a feel for Ewing and his work; however, I believe that he sometimes fails to give Ewing's co-workers proper credit for their contributions.

See also Henry Frankel, "The Development, Reception, and Acceptance of the Vine-Matthews-Morley Hypothesis," in *Historical Studies in the Physical Sciences*, **13** (1982), 1–39; William Glen, *The Road to Jaramillo* (Stanford, Calif., 1982); Xavier Le Pichoon, "The Birth of Plate Tectonics," in *Lamont-Doherty Geological Yearbook for 1985/1986* (New York, 1985), 53–61; and Neil Opdyke, "Reversal of the Earth's Magnetic Field and the Acceptance of Crustal Mobility in North America: A View from the Trenches," in *EOS*, **66** (1985), 1177.

HENRY FRANKEL

EYRING, HENRY (*b.* Colonia Juárez, Mexico, 20 February 1901; *d.* Salt Lake City, Utah, 26 December 1981), *physical chemistry*.

Eyring's father, Edward Christian Eyring, was a cattle rancher whose father had emigrated in 1853 from Germany to the United States. His mother, Caroline Cottam Romney, was a descendant of the talented English families of Romney and Cottam, members of which had emigrated from England to Utah in the 1830's and 1840's. Eyring was a Mexican citizen until 1935, when he took out U.S. citizenship.

Following his ancestors on both sides, Eyring was an active Mormon throughout his life. He attended school first in Colonia Juárez and later in Pima, Arizona, where the family moved in 1914. He attended high school in Thatcher, Arizona. During his school years Eyring did much work on the family farm, yet managed to excel academically.

In 1919 Eyring entered the University of Arizona to study mining engineering; his courses were mainly in engineering and mathematics, and he studied little chemistry. Having obtained his bachelor's degree in mining engineering, he transferred to metallurgy, obtaining his master's degree in 1924. Then, after working as a metallurgist in a copper smelter, he decided to study chemistry and in 1925 went on a fellowship to the chemistry department of the University of California at Berkeley, where he came under the influence of such men as G. N. Lewis and Wendell Latimer. His research director was George E. Gibson, and his work was on the radiation chemistry of hydrogen. He received his Ph.D. degree in 1927.

Eyring's first appointment was as an instructor at the University of Wisconsin, where, under the influence of Farrington Daniels, he developed his lifelong interest in chemical kinetics. He carried out some experimental work with Daniels, but because of some friction with the chairman of the department he left in 1929 to work with Michael Polanyi in Berlin on a National Research Foundation fellowship. In 1930 he took a one-year lectureship in chemistry at the University of California at Berkeley. On the recommendation of Hugh S. Taylor, he was then appointed an assistant professor at Princeton University, where he remained until 1946, having become a full professor in 1938. In 1946 he went to the University of Utah in Salt Lake City as the first dean of its graduate school, with the responsibility of establishing a program of graduate study and research. He remained at Utah to the end of his life, continuing an active research program until shortly before he died.

In 1928 Eyring married Mildred Bennion, and they had three sons, Edward Marcus, Henry Bennion, and Harden Romney, all of whom had highly successful careers. Edward followed in his father's footsteps and had a distinguished career at the University of Utah as a physical chemist, with a special interest in chemical kinetics. Mildred Eyring died in 1969, and two years later Eyring married Winifred Brennan Clark, who had been born in Scotland and was a convert to Mormonism.

After his experimental work with Farrington Daniels on the decomposition of nitrogen pentoxide, Eyring's work was largely theoretical, although he always had a deep interest in experimental matters and designed his theoretical work so as to interpret experimental results. There is no doubt that his association with Michael Polanyi in Berlin strongly influenced the direction of his research. He and Polanyi developed for the first time a potential-energy surface for a chemical reaction; they chose the simple process $H + H_2 \rightarrow H_2 + H$, and for the energy of a triatomic species they made use of a semiempirical and very approximate formula that had been given by Fritz London in 1928. This surface contained two valleys meeting at a col or saddle point. In the final part of their published paper Eyring and Polanyi considered the course of the reaction in terms of the motion of a representative point over the surface. This pioneering and highly influential paper was undoubtedly one of the most important contributions made by Eyring and by Polanyi.

At Princeton, Eyring continued to work on a variety of problems, mainly in kinetics. One early investigation that attracted much attention, and that won him an award from the American Association for the Advancement of Science in 1932, was on reactions involving conjugate double bonds, to which he applied quantum-mechanical methods. He also collaborated with H. S. Taylor on some research, including the investigation of reactions brought about by high-energy radiations. This work was important for being the first to demonstrate the mechanistic similarity between radiation-chemical reactions and photochemical reactions. In the 1930's Eyring also became particularly interested in the separation of isotopes, and worked on the theory of the electrolytic separation of heavy water. In 1937 he published a significant paper, with E. U. Condon and W. Altar, on optical activity, in terms of a one-electron model. This work was extended three years later in a paper with W. J. Kauzmann and J. E. Walter. Also during the 1930's Eyring and his students investigated the theory of liquids, and introduced the useful idea of treating a liquid in terms of "holes"; just as a gas is assumed to consist of molecules moving about in empty space, so a liquid may be regarded as made up of "holes" moving around in matter. On this basis Eyring was able to explain in a simple way the near constancy of the sum of the densities of coexisting vapors and liquids. This treatment of liquids became the subject of some controversy, although it was generally conceded to have some usefulness as a simple approach to the problem.

Probably the most important work done by Eyring and his students during the 1930's was on the fundamental theory of the rates of chemical reactions. Following his work with Polanyi on the construction of a potential-energy surface for the H_3 system, Eyring and his associates constructed potential-energy surfaces for a number of other systems by the application of quantum-mechanical and empirical procedures. At that time, in the absence of computers, such calculations entailed considerable labor, even for the H_3 system, and were much more difficult for systems involving more electrons unless considerable empiricism was involved. In spite of this, J. O. Hirschfelder, Eyring, and N. Rosen carried out in 1936 a purely variational calculation for the H_3 system. To avoid an impossible amount of labor, however, they had to use a rather simple variational eigenfunction, and the calculated activation energy was far from the experimental value. It was not until over forty years later that it became possible to treat this problem in a manner that provided a reliable value for the activation energy of the $H + H_2$ reaction.

At the same time that this work was being done on the construction of potential-energy surfaces, Eyring and his students were beginning to make dynamical calculations of the movement of a representative point on the surfaces. The construction of such trajectories had to be done by solving Newton's equations numerically by use of mechanical calculators. The trajectory was produced point by point, and the work was extremely tedious. In spite of the fact that a wide range of conditions could not be employed, the results illustrated important principles as to how chemical processes proceed.

During the early 1930's Eyring gave considerable thought to the formulation of a general treatment of reaction rates, one that would avoid the tedious calculation of trajectories. An important paper in 1932 with H. Pelzer and E. Wigner included a treatment of the rate with which systems pass through the col or saddle point of a potential-energy surface. This work helped Eyring to realize the usefulness of focusing attention on the saddle point, and he came to the conclusion that systems at the saddle point, which he called "activated complexes," are in a state of quasi-equilibrium with the reactants. Also, he concluded, once systems have reached this region of the potential-energy surface, they have reached a point of no return and must continue on their way to become products. This led him to formulate what he called the theory of absolute reaction rates but which is now called transition-state theory. In the first version of this theory the concentration $X\ddagger$ of activated complexes is calculated on the basis of statistical mechanics, and the rate of reaction is this concentration multiplied by the frequency with which the activated complexes pass over the barrier. The resulting expression for the rate is

$$v = \frac{kT}{h}(X\ddagger)$$

where T is the temperature, k the Boltzmann constant, and h the Planck constant. The calculation of $X\ddagger$ requires some knowledge of the structure of the activated complexes, and if this is known $X\ddagger$, and hence the rate, can be calculated satisfactorily.

In November 1934 Eyring submitted his important paper dealing with this theory to the *Journal of Chemical Physics*; the editor, Harold C. Urey, at first rejected it outright on the basis of a referee's report. However, H. S. Taylor and E. Wigner intervened, and as a result the paper, one of the most important ever written in chemical kinetics, was published in the February 1935 issue of the journal. A paper published later that year by Eyring and W. F. K. Wynne-Jones presented an equivalent treatment, but in the language of thermodynamics rather than statistical mechanics. Another 1935 paper, by M. G. Evans and M. Polanyi, presented a very similar treatment, arrived at quite independently.

The original version of transition-state theory, now often referred to as "conventional transition-state theory," involved four basic assumptions, all of which had been made in previous treatments, but which were successfully brought together in the 1935 formulation. The assumptions are: (1) Systems that have reached the activated state are bound to form product molecules; (2) the various types of motion in the activated state can be regarded as separate motions; (3) the motion through the activated state can be treated as classical; and (4) the activated complexes can be treated as being in quasi-equilibrium with the reactant molecules. At first the last of these assumptions was the most controversial, but later work has established that in many cases it is close to the truth. In the original version of the theory, assumption (3) was dealt with in terms of a "transmission coefficient" for reaction, but later treatments have dealt with some success with quantum effects, particularly with quantum-mechanical "tunneling" through the energy barrier. Assumption (2) rarely leads to much error, while inaccuracies arising from assumption (1) have been minimized by more recent treatments, such as variational transition-state theory.

Transition-state theory had considerable impact and is still widely used by those who work on the rates of chemical, physical, or biological processes. Its value is not so much in providing a way of making exact calculations of rates, for there the conventional theory has severe limitations. Its value is rather in providing a conceptual framework with the aid of which one can gain insight into how processes occur. It provides both a statistical-mechanical and a thermodynamic insight, and it leads to useful qualitative predictions, with no need for calculations, of such matters as solvent effects, kinetic-isotope ratios, and pressure influences. Later refinements of the theory, some made by Eyring himself, have allowed it to calculate rates more quantitatively.

Subsequent to the formulation in 1935 of transition-state theory, much of Eyring's effort went into the application of the theory to a variety of problems, both chemical and physical. In an important paper published in 1935, H. Gershinowitz and Eyring applied the theory to trimolecular reactions and showed that it was successful in interpreting them. This was a significant result; no previous theory had dealt satisfactorily with collisions between three molecules, but transition-state theory, by focusing attention on the equilibrium between reactants and activated complexes, was able to treat the problem in a very simple manner. A little later Eyring dealt successfully with physical processes such as viscosity, diffusion, and plasticity in terms of transition-state theory. The arrival in Princeton in 1939 of the electrochemist Samuel Glasstone aroused in Eyring a further interest in electrochemical problems, and in that year, with Glasstone and Laidler, he formulated a treatment of overvoltage. At about the same time he and his co-workers for the first time applied transition-state theory to reactions on surfaces, and were able to obtain reasonable agreement with experiment.

In about 1942 Eyring developed a particular interest in biological systems, largely as a result of a problem in bioluminescence to which he had been introduced by F. H. Johnston of Princeton's department of biology, with whom he was to collaborate for the next three decades. Bioluminescent materials, such as luminescent bacteria, exhibit some curious temperature and pressure effects, and Eyring was able to interpret these on the basis of his transition-state theory. Shortly afterward he worked on the theory of action of such drugs as sulfanilamide and formulated a treatment of the synergism and antagonism effects.

In 1944 Eyring, while retaining his position at Princeton, was appointed acting director of the research program of the Textile Research Foundation, which in that year had established laboratories at Princeton. Eyring had little previous knowledge of textiles, but his work on physical processes and his insight into reaction rates allowed him in a very short time to develop an active research program. With a small group of associates he studied the physical properties of fibers and fabrics, and the effects brought about by such processes as spinning, stretching, and weaving. On the basis of transition-state theory, mathematical models were formulated in order to explain the observed effects. In 1945 and 1946 a series of eleven papers appeared, with George Halsey as coauthor of most of them, under the general title "Mechanical Properties of Textiles."

Eyring's decision to leave Princeton in 1946 for the University of Utah was based on his and his wife's belief that the family's religious and social needs would be better served if they lived in a Mormon community. He was dean of the graduate school at Utah from 1946 until 1966, after which he was appointed distinguished professor of chemistry and metallurgy. During the entire period that he was in Salt Lake City, aside from his own research with his students and colleagues, he exerted a profound influence on the university, greatly strengthening its teaching and research functions.

Eyring's research in Salt Lake City to some extent continued his work at Princeton, but it opened up exciting new avenues and continued to be diverse and original. It included work on processes occurring in the mass spectrometer, on the theory of liquids, on optical rotation, on the mechanical properties of fibers, and on deformation kinetics. In addition, a good deal of his effort went into continuing his research on biological systems.

In the 1950's Eyring and his students made an important contribution to the field of mass spectrometry by developing the theory of the interaction of electrons with atoms and molecules. They constructed potential-energy surfaces, applied transition-state theory, and predicted the fragmentation patterns that occurred with electron beams of various velocities. Little had previously been known of the processes that occur in the mass spectrometer, and a 1950 paper by H. M. Rosenstock, M. B. Wallenstein, A. L. Wahrhaftig, and Eyring greatly clarified the problem and suggested further approaches to it.

In 1957 Eyring resumed his work on the theory of liquids, and his most important work on that problem was done during the next few years. This later work was based on three general ideas: (1) that liquids contain some solidlike regions; (2) that liquids contain holes of molecular size into which molecules can move; and (3) that molecules in these holes show some of the characteristics of gas molecules. On the basis of this model and certain experimental data, Eyring and his students constructed partition functions for a number of substances, including argon, xenon, krypton, hydrogen, chlorine, water, some organic substances, metals, and fused salts. These partition functions were able to interpret the behavior of the solid, liquid, and gaseous phases and to lead to calculated properties that were in good agreement with experimental ones. The properties considered included melting points, critical points, heat capacities, and viscosities. One of the most important accomplishments of this approach was the satisfactory treatment of liquid water, into which two solidlike structures, resembling ice-I and ice-III, were incorporated.

In the 1960's Eyring and his associates worked on the theory of optical rotation and circular dichroism, developing what has been called the octant theory of optical activity. They explored many practical aspects of the problem, applying the theory to determine the conformations of many compounds, some of them important in cancer chemotherapy.

Eyring's work at the University of Utah on biological systems included further work on luminescence and studies of nerve conduction and of diffusion through biological membranes. He worked with T. F. Dougherty in the 1950's on the mechanism of stress and inflammation. They proposed a theory in which stress sets off a chain reaction among body cells, with histamine acting as the destructive agent leading to inflammation. During the next few years these physiological studies were extended to such matters as sodium transport, the functioning of the heart, and the nature of nerve action.

In about 1969 Eyring began to work, particularly with Betsy Jones Stover, on other biological problems, including the nature of mutations. They developed a theory of the probability of survival of an individual in a homogeneous population. The basic equation they obtained was $ds/dt = -ks(1 - s)$, where s is the probability of survival, t is the time, and k is the appropriate rate constant. This equation integrates to $s = 1 + \exp\{-k(t_{\frac{1}{2}} - t)\}$, where $t_{\frac{1}{2}}$ is the half-life of the process. This treatment was applied to mutation rates and death rates, and the constant k was interpreted by the use of transition-state theory. This work with Stover was described in a series of papers that appeared in 1970 with the general title "The Dynamics of Life." A little later Eyring and his students worked on the molecular and kinetic basis of anesthesia.

At the University of Utah, Eyring continued theoretical work started at Princeton on the mechanical properties of fibers and other materials. He considered natural fibers such as cotton, wool, keratin, and collagen, and synthetic fibers like nylon and saran. This work, besides being of great theoretical interest, proved to be of considerable practical value, for example, in the wool industry. These studies were later extended to the physiological problem of the stretching of tendons.

One matter to which Eyring and his associates gave considerable attention was deformation kinetics. When a solid is placed under stress, plastic flow occurs, and this is associated with the making and breaking of intermolecular bonds. Transition-state theory provides a powerful tool for investigating such processes. The matter is complicated, since plastic flow has to be treated in terms of parallel and sequential processes, but Eyring was successful in arriving at useful treatments. After 1970, with Alexander S. Krausz, he gave special attention to the theory of deformation processes, which are important in metallurgical engineering. This collaboration led to a book on deformation kinetics that summarizes the scientific principles relating to the constitutive laws of plastic deformation. An extension of these concepts proved important to the understanding of time-dependent fractures.

Eyring's research style was unusual. He was always brimming with ideas, which he enjoyed sharing with anyone willing to listen. An important function of his students and colleagues was to sift the good ideas from the bad, of which there were many. His work was always stimulating and often controversial. His transition-state theory had its severe critics in its earlier years, and for a period was not taken seriously. Since about 1960, however, it has been realized that in spite of some weaknesses on the quantitative side it does provide, at the cost of very little labor, important insights. Eyring's treatments of the liquid state and of optical rotation have also had their critics. However, even the most controversial aspects of his work have been valuable in stimulating the research of others.

Eyring's devotion to science was matched by his deep devotion to his Mormon faith and to his family. He played an active role in his church, for many years as a member of its General Sunday School Board. He published a number of articles on his faith and on educational topics. The Mormon church is somewhat fundamentalist in its beliefs, and during the 1950's Eyring had a mild clash with church authorities because of his opinions, expressed in a scientific paper, on the origin of life in the universe. The problem was resolved, however, in a dignified manner.

Eyring received many distinctions and awards, including fifteen honorary doctorates and membership in the National Academy of Sciences. He served as president of the American Chemical Society and received its Peter Debye Award, Irving Langmuir Award, and Joseph Priestley Medal. He also received the National Medal of Science and the Berzelius Gold Medal of the Swedish Academy of Sciences.

BIBLIOGRAPHY

I. Original Works. A complete list of Eyring's publications is in S. H. Heath, "Henry Eyring, Mormon Scientist" (M.A. thesis, University of Utah, 1980), and in *Journal of Physical Chemistry*, **87** (1983), 2642–2656. Listed are ten books and over six hundred research articles. Heath also lists a number of articles on religion, education, and other topics.

Eyring's first book, with S. Glasstone and K. J. Laidler, was *The Theory of Rate Processes* (New York, 1941); this was the first detailed presentation of transition-state theory. This work was later brought up to date in Eyring, S. H. Lin, and S. M. Lin, *Basic Chemical Kinetics* (New York, 1980). Other influential books are Eyring, G. E. Kimball, and J. Walter, *Quantum Chemistry* (New York, 1944); F. H. Johnston, Eyring, and B. J. Stover, *The Theory of Rate Processes in Biology and Medicine* (New York, 1974); and A. S. Krausz and Eyring, *Deformation Kinetics* (New York, 1975).

An early paper of great importance was Eyring and M. Polanyi, "Über einfache Gasreaktionen," in *Zeitschrift für physikalische Chemie*, Abt. B, **12** (1931), 279–311, which presented the first construction of a potential-energy surface. Eyring's most important publication was probably "The Activated Complex in Chemical Reactions," in *Journal of Chemical Physics*, **3** (1935), 107–115; this was the first paper on what is now known as conventional transition-state theory.

The mass spectrometry work is presented in H. M. Rosenstock, M. B. Wallenstein, A. L. Wahrhaftig, and Eyring, "Absolute Rate Theory for Isolated Systems and the Mass Spectra of Polyatomic Molecules," in *Pro-*

ceedings of the National Academy of Sciences, **38** (1952), 667–678. Optical activity is covered in L. L. Jones and Eyring, "A Model for Optical Rotation," in *Journal of Chemical Education*, **38** (1961), 601–606, and in more detail in D. Caldwell and Eyring, *The Theory of Optical Activity* (New York, 1971). The work on liquid structure is reviewed in Eyring and M. S. Jhon, *Significant Liquid Structures* (New York, 1969).

II. SECONDARY LITERATURE. A detailed account of Eyring's life and work is to be found in Heath's thesis, cited above. Other valuable sources are S. H. Heath, "The Making of a Physical Chemist: The Education and Early Researches of Henry Eyring," in *Journal of Chemical Education*, **62** (1985), 93–98; J. O. Hirschfelder, "Henry Eyring, 1901–1981," in *American Philosophical Society Year Book 1982* (Philadelphia, 1983), 482–489; and D. W. Urry, "Henry Eyring (1901–1981): A 20th-Century Architect of Cathedrals of Science," in *Proceedings of the International Symposium on Quantum Biology and Quantum Pharmacology*, 9th (1982), 1–3, and "Henry Eyring (1901–1981): A 20th-Century Physical Chemist and His Models," in *Mathematical Modelling*, **3** (1982), 503–522. A volume in honor of Eyring's seventieth birthday, J. O. Hirschfelder and D. Henderson, eds., *Chemical Dynamics* (New York, 1971), contains appreciations of Eyring's contributions and many articles by his former students and other associates.

Interesting reminiscences on Eyring's early work on the calculation of potential-energy surfaces are included in two articles by J. O. Hirschfelder, "A Forecast for Theoretical Chemistry," in *Journal of Chemical Education*, **43** (1966), 457–463, and "My Fifty Years of Theoretical Chemistry: I, Chemical Kinetics," in *Berichte der Bunsen-Gesellschaft für physikalische Chemie*, **86** (1982), 349–355; the first of these articles is reprinted in *Chemical Dynamics*, cited above. See also D. Henderson, "My Friend, Henry Eyring," in *Journal of Physical Chemistry*, **87** (1983), 2638–2640.

An account of some of the work that led Eyring to formulate transition-state theory, with some discussion of the reception of the theory, is in K. J. Laidler and M. Christine King, "The Development of Transition-State Theory," *ibid.*, 2657–2664. On recent extensions of the theory, see D. G. Truhlar, W. L. Hase, and J. T. Hynes, "Current Status of Transition-State Theory," *ibid.*, 2664–2682.

KEITH J. LAIDLER

FAJANS, KASIMIR (*b.* Warsaw, Russian Poland, 27 May 1887; *d.* Ann Arbor, Michigan, 18 May 1975), *chemistry*.

The son of Herman Fajans, a merchant, and Wanda Wolberg, Kasimir Fajans graduated from high school in Warsaw (1904), and then left Russian Poland to study at the universities of Leipzig (bachelor's degree, 1907) and Heidelberg (doctorate, 1909). Then came a year of research at Zurich, followed by a second research year (1910–1911) in Ernest Rutherford's laboratory at the University of Manchester. In 1910 Fajans married Salomea Kaplan, a physician. They had two sons, Edgar and Stefan.

Rutherford and his laboratory attracted some of the best researchers in radioactivity. Since the subject straddled the line between physics and chemistry, generally at least one radiochemist was to be found there. Fajans' interest in radioactivity had been ignited in 1909 by Philipp Lenard, who suggested that he report on the subject for his physics colloquium. This was far afield from the physical-organic topic in stereochemical catalysis that Fajans pursued under Georg Bredig for his doctorate. During the year with Richard Willstätter in Zurich, Fajans became interested in the binding forces in carbon compounds. But he also became convinced that he was more interested in physics, particularly radioactivity, than in organic chemistry. The lectures by Albert Einstein which he attended may also have influenced him. Rutherford's positive response to his request to come to Manchester led to a major redirection in his career.

While in England, Fajans found that the radium decay series branched at radium C, a concept that was not universally accepted at the time. He also collaborated with Henry Moseley in determining the very short half-lives of thorium A (0.14 sec.) and actinium A (0.002 sec.). Most important, however, was the exciting environment created by such colleagues as James Chadwick, Hans Geiger, C. G. Darwin, and others, and the opportunity to learn the status of research into radioactivity's unsolved problems. One of the reasons Fajans left organic chemistry was its empirical approach, which could not satisfy his strong theoretical leanings. In radioactivity he similarly found the disorder of some thirty radioelements, which were meant to fit into only a dozen boxes of the periodic table of elements, too inharmonious to persist.

Fajans continued his study of radioactivity at the Technische Hochschule in Karlsruhe, where he became an assistant in 1911 and a *Privatdozent* for physical chemistry in 1913. The key to rationalization of the three known radioactive decay series was the identification of a number of products with short half-lives. Using chemical and especially electrochemical data, Fajans gained some crucial insights which enabled him to announce the group displacement laws in February 1913: alpha-particle emission moves the daughter product two boxes to the left in the periodic table, while beta-particle emission signifies a jump one box to the right.

Fajans was not alone in working on this theory, which, along with the explanation of the phenomenon of radioactivity by Rutherford and Frederick Soddy a decade earlier, became a cornerstone of the science. Other gifted radiochemists, such as Soddy, György Hevesy, and Alexander Russell, also were approaching a solution. Soddy, in fact, published the same theory, admittedly after seeing Fajans' paper in print, and without having the chemical proof needed to draw the correct conclusions. Despite this apparent plagiarism, and to Fajans' distress, Soddy received the lion's share of credit in the English-speaking world, and his term "isotopes," for the bodies with different radioactive properties but identical chemical characteristics that fit into a single box of the periodic table, easily won out over Fajans' preferred "pleiades" (named after the star cluster).

Rutherford, in an effort to soothe Fajans' frustration, assured him that it is not the idea but its proof that is most important. With his student Oswald Göhring, Fajans thereupon searched for and soon found the first isotope of element 91, which the theory predicted. (This was uranium X_2, initially called brevium, but changed to protactinium in 1918 when Otto Hahn and Lise Meitner, and, independently, Soddy and John Cranston, found the longest-lived isotope.) Another student, Max Lembert, was sent to Harvard to work with the world's leading expert in atomic-weight determinations, Theodore W. Richards. These two found the weight of inactive lead at the end of the uranium decay series to differ from that of natural lead by far more than experimental error would allow. Richards had built his career on the concept of fixity of elements; the paper he wrote expressed amazement that species of the same element could have different weights. By giving an explanation of the way radioelements fit into "normal" chemistry, Fajans provided the key that solved virtually all radiochemical problems. The consequence was that the science of radiochemistry ceased to exist. Applications such as tracer techniques were later investigated, but few questions of basic science. It took the discovery of artificial radioactivity to resurrect the field in the 1930's, when it came to be called nuclear chemistry.

Willstätter had moved to Munich and in 1917 invited Fajans there as an associate professor, to inaugurate teaching and research in physical chemistry. Fajans became full professor in 1925, and director of his own new Institute for Physical Chemistry in 1932, built with funds from the Rockefeller Foundation. Investigations begun in Karlsruhe on the identity and quantity of nonweighable amounts of radioelements led to the Fajans-Paneth-Hahn coprecipitation and adsorption rules. These gave quantitative analysis strong new tools, among which Fajans' adsorption indicators were a vital component. This work in turn drew Fajans' attention once again to the question of chemical binding, for electrical charges proved to be a dominant factor in the phenomena.

Hitler's rise to power forced Fajans from Munich in 1935 and, after some months in Cambridge, he accepted a professorship at the University of Michigan (1936–1957). In Ann Arbor some of his students in nuclear chemistry worked closely with James Cork's cyclotron group, and Fajans himself participated in the discovery of a few new radioisotopes, but his primary interest remained in binding. Although never widely accepted, his quanticule theory, developed from the 1940's onward to replace the classical concept of valence bonds between neutral atoms, proposed that the electrons of a molecule or crystal are subdivided into groups of definite quantization (quanticules) and that all interactions result from the electric forces acting between nuclei and quanticules.

Fajans was coeditor of the *Zeitschrift für Kristallographie* (1924–1939) and associate editor of the *Journal of Physical and Colloid Chemistry* (1948–1949). He was elected to numerous societies and academies of science, among them the academies in Cracow, Leningrad, and Munich, and the Royal Institution of Great Britain. He also received several awards, including the medal of the University of Liège (1948).

BIBLIOGRAPHY

I. ORIGINAL WORKS. Fajans' correspondence and miscellaneous papers are preserved in the Michigan Historical Collection, University of Michigan, Ann Arbor. Among his most significant published papers are "Über die komplexe Natur von Radium *C*," in *Physikalische Zeitschrift*, **12** (1911), 369–378; "Radio-active Products of Short Life," in *Philosophical Magazine*, **22** (1911), 629–638, written with H. G. J. Moseley; "Über eine Beziehung zwischen der Art einer radioaktiven Umwandlung und dem elektrochemischen Verhalten der betreffenden Radioelemente," in *Physikalische Zeitschrift*, **14** (1913), 131–136; "Die Stellung der Radioelemente im periodischen System," *ibid.*, 136–142; "Über das Uran X_2—das neue Element der Uranreihe," *ibid.*, 877–884, written with O. Göhring; "Das Verhalten der Radio-Elemente bei Fällungsreaktionen," in *Berichte der Deutschen chemischen Gesellschaft*, **46** (1913), 3486–3497, written with P. Beer; "Adsorptionsindikatoren für Fällungstitrationen," in *Neuere massanalytische Methoden* (Stuttgart, 1956), 313–369; "Quantikel-Theorie der chemischen Bindung," in *Chimia*, **13** (1959), 349–366.

Fajans authored a popular text entitled *Radioaktivität und die neueste Entwicklung der Lehre von den chemischen Elementen* (Braunschweig, 1919). The fourth edition was translated into English as *Radioactivity and the Latest Developments in the Study of the Chemical Elements*, T. S. Wheeler and W. G. King, trans. (New York, 1923). Other books are *Radioelements and Isotopes: Chemical Forces and Optical Properties of Substances* (New York, 1931); and *Newer Methods of Volumetric Chemical Analysis*, W. C. Böttger, ed., Ralph E. Oesper, trans. (New York, 1938), with E. Brennecke, N. H. Furman, H. Stamm, and R. Lang.

Personal reminiscences by Fajans appeared on the occasion of his Pioneer Lecture on Otto Hahn in *Journal of Nuclear Medicine*, **7** (1966), 402–404; and, in Polish, on the occasion of the centennial of Marie Curie's birth, in *Problemy*, **24** (1968), 392–403.

II. SECONDARY LITERATURE. Józef Hurwic is currently working on a biography of Fajans. Shorter works are by E. Lange, on the occasion of Fajans' seventieth birthday, in *Zeitschrift für Elektrochemie*, **61** (1957), 773–774, and, on his eightieth birthday, in *Jahrbuch der Bayerischen Akademie der Wissenschaften* (1967), 171–173; I. M. Frank in *Uspekhi fizicheskikh nauk*, **99** (1969), 337. Obituary notices are by Thomas M. Dunn in *Nature*, **259** (1976), 611; and by J. Hurwic in *L'actualité chimique*, no. 1 (1976), 28–32. Works dealing specifically with the group displacement laws and atomic weight of lead isotopes are O. U. Anders, "The Place of Isotopes in the Periodic Table: The 50th Anniversary of the Fajans-Soddy Displacement Laws," in *Journal of Chemical Education*, **41** (1964), 522–525; L. Badash, "The Suicidal Success of Radiochemistry," in *British Journal for the History of Science*, **12** (1979), 245–256; J. B. Conant, "Theodore William Richards and the Periodic Table," in *Science*, **168** (1970), 425–428; and Alfred Romer, ed., *Radiochemistry and the Discovery of Isotopes* (New York, 1970).

LAWRENCE BADASH

FEDOROV, EVGENII KONSTANTINOVICH (*b.* Bendery, Bessarabia [now Moldavian S.S.R.], Russia, 10 April 1910; *d.* Moscow, U.S.S.R., 30 December 1981), *geophysics, hydrometeorology, polar exploration.*

Fedorov's father, Konstantin Nikolaevich, was an officer in the Russian Army and, after 1917, in the Red Army; his mother, Sabina Akimovna, was a seamstress. Fedorov completed secondary school at Gorkii and entered Leningrad University in 1928. After graduation in 1932 from the geophysics department of the Faculty of Physics, he joined the Hydrometeorological Service. During the winter of 1932–1933 he worked at a polar station in Tikhaia Harbor. At the same time he conducted magnetic and astronomical observations in Franz Josef Land.

On 23 November 1933 Fedorov married Anna Viktorovna Gnedich, one of the first Soviet women polar explorers and geophysicists; they had two sons and a daughter. During the winter of 1934–1935 Fedorov worked at a polar station on Cape Cheliuskin. Besides routine hydrometeorological observations, he made a number of magnetic determinations, traveling more than 1,600 kilometers (1,000 miles) while surveying an area of almost 15,000 square kilometers (about 5,790 square miles). In the years 1937 and 1938 Fedorov was one of the four workers at the first Severnyi Polius (North Pole) drifting station. The nine-month undertaking, headed by the well-known polar explorer Ivan Dmitrievich Papanin, was exceptional in conception, execution, and scientific results. It paved the way for extensive exploration of the vast central polar basin. All the participants were awarded the title Hero of the Soviet Union and received doctorates in geographical sciences.

In 1939 Fedorov became head of the Hydrometeorological Service. As its director until 1947, and again from 1962 to 1974, he helped set up a well-equipped establishment that served the needs of the national economy, and it was prominent among the national services of the World Meteorological Organization.

In 1939 Fedorov was elected corresponding member of the Soviet Academy of Sciences. Between 1947 and 1955 he worked in the academy, first as head of the laboratory, and from 1949 as assistant director, of the Geophysical Institute. Later he held a number of prominent posts within the presidium of the academy (deputy chief scientific secretary, 1959–1960; chief scientific secretary, 1960–1962). In 1960 he was elected full member. At his initiative the Institute of Applied Geophysics was set up in 1956 within the Hydrometeorological Service; he headed the institute until 1968, and again from 1974 until his death. Under his leadership it made a considerable contribution to environmental studies.

Fedorov concentrated on a broad spectrum of problems related to hydrometeorology. He carried out geophysical investigations in the Arctic, studying magnetic anomalies and variations of the magnetic field. These investigations gave insight into the geophysical processes of this virtually unstudied area and contributed to the development of polar aviation and safe ship passage in the northern seas, thus stimulating exploration of the Arctic.

Fedorov's observations made in the years 1937

and 1938 at Severy Polius number 1 disproved the then current belief that the central Arctic Ocean was under the constant influence of a vast anticyclone with its maximum at the pole. The data Fedorov collected were highly significant for long-range forecasting of ice conditions and stimulated new theories on ice movement in the central Arctic region.

Fedorov's interests included weather and climate, water resources, the seas and oceans, and the ionosphere. He studied the earth's magnetic and radiation fields, and also conducted theoretical and experimental research on the control of hydrometeorological processes, pollution of the environment, and the interaction between society and the environment.

In the 1950's Fedorov embarked on studies of the control of meteorological phenomena that proved to be of great practical value. He directed studies on methods of controlling thunderstorm clouds aimed at preventing hail, believing that geophysical processes must be influenced in such a way that they serve the needs of humanity. He argued that it was possible to use their instability for such purposes when changes in the processes would require relatively small amounts of energy. Traditional description, continuous logging, and forecasting should be combined with the analysis of large-scale experiments.

Fedorov's research stimulated, and contributed to, broad experimentation in geophysics. He particularly studied the physics of clouds, aerosols, and precipitation; radiation geophysics; and the physics of the upper atmosphere. He initiated research in the upper atmospheric layers with the help of missiles and satellites, which soon became the new field of satellite meteorology. This research led to the establishment of a hydrometeorological service that was to provide immediate information on the state of the magnetosphere, ionosphere, and atmosphere as a whole, thereby ensuring reliable weather forecasts.

Because Fedorov's activities required information on the state of the atmosphere and the hydrosphere on a continental, or even global, scale, he became increasingly aware of the close interaction between natural geophysical phenomena and human activities. He concluded that at that time, the progress of geophysical processes on earth was conditioned not only by natural laws but also by human activities that influenced the ways in which the workings of those laws were manifested.

Fedorov considered it essential to study the impurities in the atmosphere and in water, initiating such methods as the gamma-ray survey of snow and aerophotometric techniques for the assessment of pasture vegetation. Later in his career Fedorov devoted much of his efforts to protection of the environment, particularly against radioactive pollution of the biosphere. He introduced methods of determining chemical and radioactive pollution, setting up a system for the continuous monitoring of radioactivity and the extent of pollution of the environment. Fedorov stressed the importance of calculating the existing resources for the transformation of nature, which he called "the earth's capacity," focusing on the expansion of this "capacity."

Fedorov's contributions to hydrometeorology and the other earth sciences, starting with his first expeditions, received much recognition. In 1946 and in 1969, he was awarded the State Prize. In the years 1957 and 1958 Fedorov took part in the Geneva conference of technical experts who agreed that nuclear testing could be monitored by the individual nations. This led to the agreement banning nuclear tests in the atmosphere, in space, and under the seas. From 1963 to 1971 Fedorov was vice president of the World Meteorological Organization, and in 1977 he was awarded its gold medal.

Fedorov was a prominent political and public figure; three times he was elected deputy of the Supreme Soviet of the U.S.S.R.; he was also a member of its Presidium. An enthusiastic peace activist, he chaired the Soviet Peace Committee beginning in 1965 and was vice president (1973–1978) and then president (1978–1981) of the World Peace Council.

Fedorov died at the age of seventy-one after a brief illness and was buried in Moscow.

BIBLIOGRAPHY

Chasovye pogody: Sovetskaia gidrometeorologicheskaia sluzhba ("Sentinels of the Weather: Soviet Hydrometeorological Service"; Leningrad, 1970); *Ekologicheskii krizis i sotsialnyi progress* (Leningrad, 1977), translated into English as *Man and Nature: The Ecological Crisis and Social Progress* (New York, 1981); "Emkost zemli" ("Capacity of the Earth"), in *Nauka narodnomu khoziaistvu* (Moscow, 1979), 56–91; "From the Description of Nature to Its Planning," in *Soviet Geography Today*, I, *Aspects of Theory* (Moscow, 1981), 132–156; *Poliarnye dnevniki* ("Polar Diaries"; Leningrad, 1979; 2nd ed., Moscow, 1983).

V. V. TIKHOMIROV

FELLER, WILLIAM (*b*. Zagreb, Yugoslavia, 7 July 1906; *d*. New York, New York, 14 January 1970), *mathematics*.

Feller was the son of Eugene V. Feller, a wealthy

owner of a chemical factory, and Ida Perc Feller. William was the tenth of twelve children, the youngest of six boys. He was educated by private tutors until he entered the University of Zagreb in 1923, from which he received the equivalent of an M.S. degree in 1925. He received his Ph.D. degree in 1926 from the University of Göttingen, where he remained until 1928. In 1928 he moved to the University of Kiel, where he headed the applied mathematics laboratory until he moved to Copenhagen in 1933, after Hitler came to power. After a year in Copenhagen, he moved to the University of Stockholm to be research associate in the probability group headed by Cramér.

During his Stockholm stay he married Clara Mary Nielsen (27 July 1938), a student of his at Kiel, who as a Danish schoolgirl had bicycled with her friends across the German border carrying anti-Nazi pamphlets. There were no children of this marriage. In 1939 the Fellers immigrated to the United States, where William became a professor at Brown University and the first executive editor of *Mathematical Reviews*. This international review, founded in 1939 because the German review had come under Nazi control, has been an invaluable mathematical tool, and much of its success is due to the policies set by Feller. In 1945 he accepted a professorship at Cornell University; he remained there until 1950, when he moved to Princeton University as Eugene Higgins professor of mathematics.

Although Feller's research was almost entirely in pure mathematics, he had more than an amateur's interest in and knowledge of several scientific fields, including statistics and genetics. He took an excited delight in applications of pure theory, and nothing pleased him more than finding new applications. He wrote several papers applying probability theory to genetics and spent the academic years 1965–1966 and 1967–1968 at Rockefeller University, where he held an appointment as permanent visiting professor and enjoyed close contacts with geneticists.

Before Kolmogorov's measure theoretic formulation (1933) of the basic concepts of probability theory, this theory was a barely respectable part of mathematics, with little interaction with other parts. Probability results were solutions to isolated mathematical problems suggested by a certain nonmathematical context. After 1933 these results took their places in an overall mathematical framework, and probability theory began a rapid development. A host of researchers, with a few great leaders such as Kolmogorov (Soviet Union), Lévy (France), and Feller, transformed mathematical probability into one of the liveliest branches of contemporary mathematics, contributing as much to the newest aspects of other branches as it drew from them.

Feller's first probability paper appeared in 1935. In this and many later papers, he discussed the properties of successive sums $S_1, S_2, \ldots$ of a sequence of independent random variables. Under what conditions does suitably normalized S_n have a nearly Gaussian distribution when n is large? What are asymptotic bounds for S_n when n is large? One of Feller's first results was a set of necessary and sufficient conditions answering the first question. One of his deepest papers answers the second, under appropriate conditions.

In 1931 Kolmogorov gave the first systematic presentation of the intimate relations between parabolic partial differential equations and the probabilistic processes now called Markov processes. Feller completely transformed this subject. First he refined and extended Kolmogorov's work. For example, he proved that the equations in question (in a more general framework than Kolmogorov's) have probabilistically meaningful solutions. Later, he put the analysis in a functional analysis framework, applying semigroup theory to the semigroups generated by Markov process transition probabilities of very general types. He linked the boundary conditions for the differential equations with the domains of the semigroup infinitesimal generators and with the conduct of the process sample paths at the boundaries of the process state spaces, incidentally defining new abstract boundaries when necessary. In particular, Feller found a beautiful form for the infinitesimal generator of the most general one-dimensional diffusion. In much of his work, Feller was a pioneer, yet he frequently obtained definitive results.

One of Feller's greatest legacies, containing research at every level, is his two-volume work *An Introduction to Probability Theory and Its Applications* (1950–1966). He never tired of revising the material in these volumes, in finding new approaches, new examples, and new applications. No other book on the subject even remotely resembles these volumes, with their combination of purest abstract mathematics and interesting applications, employing a dazzling virtuosity of analytical techniques and written in a style betraying the bubbling enthusiasm of the author. The style has made the book popular even among nonspecialists, just as its elegance and breadth have made it an inspiration for specialists.

Mathematical statistics is based on probability theory. Feller made a caustic critique of extrasensory perception experiments, and kept in touch with such statistical controversies as the effect of cigarette smoking on health. It is typical of him that what roused his ire more than the issues in such controversies was the attempt by some statisticians to strengthen weak statistics with irrelevant emotional appeals.

Those who knew him personally remember Feller best for his gusto, the pleasure with which he met life, and the excitement with which he drew on his endless fund of anecdotes about life and its absurdities, particularly the absurdities involving mathematics and mathematicians. To listen to him lecture was a unique experience, for no one else could lecture with such intense excitement.

Feller was a president of the Institute of Mathematical Statistics. He was a member of the National Academy of Sciences and of the American Academy of Arts and Sciences, a foreign associate of the Royal Danish Academy and of the Yugoslav Academy of Sciences, and a fellow of the (British) Royal Statistical Society. He was named to receive the 1969 National Medal of Science shortly before his death but died before the awards ceremony; his widow accepted the medal on his behalf.

BIBLIOGRAPHY

I. Original Works. Some of Feller's early research was published under the name Willy instead of William. His writings included: "Zur Theorie der stochastischen Prozesse (Existenz- und Eindeutigkeitssätze)," in *Mathematische Annalen*, **113** (1937), 113–160; "The General Form of the So-Called Law of the Iterated Logarithm," in *Transactions of the American Mathematical Society*, **54** (1943), 373–402; "The Fundamental Limit Theorems in Probability," in *Bulletin of the American Mathematical Society*, **51** (1945), 800–832; "Diffusion Processes in One Dimension," in *Transactions of the American Mathematical Society*, **77** (1954), 1–31; "On Boundaries and Lateral Conditions for the Kolmogorov Differential Equations," in *Annals of Mathematics*, 2nd ser., **65** (1957), 527–570; and "On the Influence of Natural Selection on Population Size," in *Proceedings of the National Academy of Sciences*, **55** (1966), 733–738.

II. Secondary Literature. See *Annals of Mathematical Statistics*, **41**, no. 6 (1970), iv–xiii, for an obituary, photograph, and complete bibliography. See *Proceedings of the Sixth Berkeley Symposium on Mathematical Statistics and Probability*, II (Berkeley, 1972), xv–xxiii, for obituaries and a very youthful picture; and see *Revue de l'institut international de statistique*, **38** (1970), 435–436, for another obituary.

J. L. Doob

FENN, WALLACE OSGOOD (*b*. Lanesboro, Massachusetts, 27 August 1893; *d*. Rochester, New York, 20 September 1971), *physiology*.

The second of five children of William Wallace Fenn, a Unitarian minister who after 1900 was Bussey professor of theology at Harvard, and of Faith Huntington Fisher, Fenn expected to follow his father's career until he was diverted to physiology as an undergraduate at Harvard. There he received the A.B. (1914), the A.M. (1916), and, after a year (1917–1918) in the army as a camp nutrition officer, the Ph.D. (1919), with a dissertation in plant physiology on colloid chemistry and the problems of salt antagonism. In September 1919 he married Clara Bryce Comstock; they had two sons and two daughters. From 1919 until 1922 he was instructor in applied physiology at Harvard, where his studies of phagocytosis showed that leukocytes ingest carbon particles more readily than quartz particles, a finding that clarified the pathogenesis of pulmonary stilicosis. As a traveling fellow of the Rockefeller Institute for Medical Research between 1922 and 1924, Fenn studied in England, principally in the laboratory of Archibald Vivian Hill. In 1924 he joined the newly created University of Rochester School of Medicine and Dentistry as professor and chairman of the department of physiology; he held both posts until 1959, continued as professor until 1961, and from then until his death was distinguished university professor of physiology. Fenn was elected to the National Academy of Sciences in 1943 and to the American Academy of Arts and Sciences in 1948.

When Fenn arrived in Hill's laboratory in 1922, he began working on muscle physiology, extending the research program for which Hill received the Nobel Prize in that year. Using measurements of heat produced by a contracting frog muscle, Fenn found in 1923 that when a muscle shortens, it liberates more energy than during an isometric contraction. This extra energy production, moreover, is proportional to the work done in shortening. These observations suggested that muscle shortening is an active process, not a passive one like the shortening of a prestretched spring, and what Hill termed the Fenn effect brought the young physiologist to prominence.

After moving to Rochester, Fenn continued the work he had started under Hill's guidance. Early in the 1930's he took measurements from high-speed motion pictures of sprint runners to assess the work performed. He also used measurements made on isolated muscles lifting different loads in the laboratory to establish the relationship between muscle force and the speed of shortening, and the nonlinear

FENN

force-velocity curve for muscle he described in 1935 supported his earlier contention that muscle cannot be regarded simply as an elastic body. However, from the time of his return to the United States in 1924, Fenn's most vigorous research program focused on muscle metabolism. In particular he investigated oxygen consumption in muscle and nerve during stimulation and recovery and in 1926 demonstrated that the conduction of a nerve impulse is accompanied by an increase in oxygen uptake, which he quantified in 1927. Similar studies on contracting muscle drew Fenn's attention to the relationship between oxygen consumption and electrolyte levels and to the role of electrolytes in muscle and nerve functioning.

The investigations of electrolyte metabolism that ensued between 1933 and 1941 marked perhaps the most productive phase of Fenn's research career. It was generally recognized that muscle is rich in potassium, which Fenn showed can pass through the cell membrane. By 1935 his experiments demonstrated that muscle, when it is stimulated, loses intracellular potassium and gains sodium, changes reversed in recovery. To explain this electrolyte exchange, in 1936 Fenn suggested that the loss of potassium from muscle during activity represents an increase in the permeability of the muscle membrane that permits sodium, but not chloride, to enter. This sodium, he further proposed, displaces an equivalent amount of potassium. Fenn went on to suggest in 1939 that the lost potassium tends to follow the carbohydrate cycle from muscle to liver and back again, and elucidated the mechanism by which potassium is reconcentrated in muscle during recovery. After radioactive potassium became available, Fenn used it in 1941 to show further the permeability of cells to potassium. His observations on potassium-sodium exchange during muscle contraction established the foundation for the later understanding of the initiation and propagation of nerve and muscle impulses.

World War II abruptly curtailed Fenn's research program, and although he returned briefly to electrolyte physiology in the late 1950's, he did little further research on muscle physiology. Instead, starting in 1941, Fenn directed his efforts to problems of respiratory physiology of interest to the U.S. Air Force. He took up the problem of pressure breathing, a method for increasing altitude tolerance in non-pressurized airplanes by pressurizing an aviator's lungs to raise the partial pressure of oxygen. To simulate high-altitude conditions, he built a low-pressure chamber. The concepts Fenn developed for dealing with the pressure breathing problem and for displaying graphically the relationships involved were important for respiratory physiology in general, particularly his 1946 descriptions of the pressure-volume diagram of the thorax and lungs and the oxygen-carbon dioxide diagram of the composition of alveolar air. Fenn's group at Rochester continued to work on respiratory physiology under government contract after the war, and Fenn's study of airway resistance helped him describe the mechanics and work of breathing in 1951. He continued to investigate respiratory physiology for the rest of his life.

From the mid 1950's through the 1960's Fenn became increasingly concerned with the standing and integrity of physiology as a discipline. He celebrated the growing calls that government made on physiologists during and after the war as evidence that physiology had "found recognition for itself in the marketplace," but lamented its low prestige compared with the physical sciences. Fenn had been an active leader in the American Physiological Society since 1933, and in the International Union of Physiological Sciences since 1956. In the 1960's he underscored the significance of such organizations to the discipline by writing on their history. His role as a spokesperson for American physiology, sustained by the many awards and honors he received, allowed less time for research. Nevertheless, by the late 1950's, work on aviation physiology had led Fenn to important research on the biological effects of extreme pressure conditions, especially on the ocean floor and in space. From 1962 to 1966 Fenn was founding director of the University of Rochester's Space Science Center, and until the end of his life he vigorously proselytized the promise of the space program for biological research.

BIBLIOGRAPHY

I. Original Works. Lists of some 267 published works authored or coauthored by Fenn are in Rahn's biographical memoir and Dustan's *In Memoriam* (see below). Fenn's books include *History of the American Physiological Society . . . 1937–1962* (Washington, D.C., 1963). Among the most significant of his scientific papers are "A Quantitative Comparison Between the Energy Liberated and the Work Performed by the Isolated Sartorius Muscle of the Frog," in *Journal of Physiology*, **58** (1923), 175–203; "Muscular Force at Different Speeds of Shortening," *ibid.*, **85** (1935), 277–297, with B. S. Marsh; "Electrolyte Changes in Muscle During Activity," in *American Journal of Physiology*, **115** (1936), 345–356, with Doris M. Cobb; and "A Theoretical Study of the Composition of the Alveolar Air at Altitude," *ibid.*, **146** (1946), 637–653, with Hermann Rahn and Arthur B. Otis.

A major collection of Fenn's papers is deposited in the

History of Medicine Collection, Edward G. Miner Library, University of Rochester School of Medicine and Dentistry. The collection (about 30 linear feet) includes Fenn's correspondence, materials relating to his work for scientific societies, and his laboratory notebooks.

II. SECONDARY LITERATURE. Hermann Rahn provides the fullest sketches of Fenn's life and work: "Wallace O. Fenn," in *Respiration Physiology*, **5** (1968), vii–xii, written with P[ierre] D[ejours]; "Wallace O. Fenn, President of the American Physiological Society, 1946–1948," in *The Physiologist*, **19** (1976), 1–10; and his memoir in *Biographical Memoirs. National Academy of Sciences*, **50** (1979), 141–173. These and twenty-three other biographical notices on Fenn, principally from professional journals, are reprinted in Augusta Dustan, ed., *In Memoriam. Wallace O. Fenn, 1893–1971* (n.p., [1979]). The American Physiological Society, Bethesda, Md., has a volume of nearly 100 typescript letters, solicited by Clara B. C. Fenn, in which colleagues and students present their recollections of him, *Wallace O. Fenn, 1893–1971. Memories and Facts from Friends Here and Abroad* (Rochester, N.Y., 1976); and a printed program, *A Memorial Service for Wallace Osgood Fenn, 1893–1971* (Rochester, 1971), with attached remarks delivered at the service and a copy of Fenn's curriculum vitae.

JOHN HARLEY WARNER

FIESER, LOUIS FREDERICK (*b*. Columbus, Ohio, 7 April 1899; *d*. Cambridge, Massachusetts, 25 July 1977), *organic chemistry*.

Fieser, son of Louis Frederick Fieser and Martha Victoria Kershaw, attended Columbus schools and then enrolled at Williams College, where he received a B.A. in 1920. He won letters in varsity football, basketball, and track. Harvard University awarded him a doctorate in 1924. After a year of postdoctoral research in Germany and England, he taught at Bryn Mawr from 1925 to 1930. Among his students was Mary Peters, a premedical student who became a chemist under his influence; they married on 21 June 1932. In 1930 Fieser accepted a Harvard appointment as assistant professor, becoming full professor in 1937 and Sheldon Emery Professor of Organic Chemistry in 1939. He retired in 1968.

Famous as a preparative chemist, Fieser served on the editorial boards of *Organic Syntheses* and *Organic Reactions* and was one of the most prolific contributors to the former. Among his many honors was election to the National Academy of Sciences in 1940. Because of his work with chemical carcinogens, the surgeon general in 1963 appointed Fieser to his committee to investigate the relation between smoking and health. Its 1964 report concluded that smoking is a major cause of lung cancer, chronic bronchitis, pulmonary emphysema, and coronary artery disease. Ironically, Fieser was a heavy smoker and in 1965 underwent surgery for removal of a lung tumor; he also suffered from emphysema and bronchitis. Following his surgery he became a campaigner against smoking.

At Harvard, Fieser studied under James B. Conant, several of whose students became leaders in physical organic chemistry. Fieser, however, loved the active life of the experimenter and became primarily a chemist of natural products and their synthesis. His work with Conant resulted in six joint publications (1922–1925). His dissertation, on the reduction-oxidation potentials of quinones, set new standards of accurate experimentation and is a classic in its field. It had an addendum, the result of independent work in Gregory Baxter's gas analysis course. Fieser used his knowledge of quinones to invent "Fieser's solution," a quinone-containing reagent used to absorb oxygen from other gases and the subject of his first solo publication (1924).

Under Conant's direction Fieser prepared many known and unknown quinones; measured the reduction potentials as a means to uncover relationships between structures and energy of reactions; and determined the effect of ortho-, meta-, and para-directing substituents on the potentials. This work quickly gave him a reputation as an expert in quinone chemistry.

While at Bryn Mawr, Fieser extended his studies, becoming increasingly interested in problems of synthesis, as he prepared more and more quinones. During his five years there, he wrote outstanding papers that placed him in the top rank of American organic chemists. He studied quinones in relation to aromatic systems, the nature of aromaticity, tautomeric forms, and reduction potentials as a measure of the stability of quinonoid structures. In 1927 he achieved the first synthesis of the yellow pigment lapachol, a naphthoquinone with an isoprenoid side chain that was significant for his later work on vitamin K and antimalarials.

The lapachol work stemmed from Fieser's association with Samuel Hooker, an English-born chemist employed at a sugar refinery in Philadelphia. Between 1889 and 1896 Hooker had isolated and elaborated the structures of many natural quinones, including lapachol, and possessed many unique samples. Fieser needed types of quinones that only Hooker had, and he collaborated with Hooker from 1926 until the latter's death in 1935. Hooker bequeathed his samples to Fieser, who wrote or experimentally completed eleven of Hooker's posthumous papers and used the samples in his own researches.

Mary Fieser was a member of her husband's research group at Harvard. She began graduate study there, but the negative attitude of some members of the chemistry department toward women chemists dissuaded her from continuing. Her position in her husband's research group was secure, however, and she engaged in much of the quinone and steroid research, including the development of a new synthetic route to the anthraquinones (1935). Her name appears as coauthor of thirty-six papers.

The best-known synthesis associated with Fieser was achieved in 1939. In that year Henrik Dam, P. Karrer, Edward Doisy, and others had isolated vitamin K_1 as a water-insoluble yellow oil. Fieser thought that it might be a naphthoquinone related to lapachol. Using absorption spectra, composition, and degree of unsaturation, he inferred a structure of 2-methyl-3-phytyl-1,4-naphthoquinone, the phytyl being a large hydrocarbon side chain of four five-carbon isoprenoid units and identical to the hydrocarbon part of the alcohol phytol (a constituent of alfalfa and green leaf pigment). Fieser then developed a simple synthesis by directly combining the methylhydroquinone with phytol and oxidizing the product to the quinone. Although the reaction produced a considerable amount of by-product, he easily isolated the vitamin, showing that it corresponded to natural vitamin K_1 in every way, including biological activity. Fieser protected the synthesis with a patent assigned to the Research Corporation, which in turn licensed Merck to manufacture the vitamin.

His studies of phenanthrene quinones in the 1920's led Fieser to polycyclic systems and the steroids. He developed a new route to polynuclear aromatic types containing the phenanthrene ring system. Soon after moving to Harvard in 1930, he learned that the Royal Cancer Institute in London had found that one of the compounds synthesized by him in 1929, 1,2,5,6-dibenzanthracene, induced cancerous growth in mice; the compound thus became the first known pure carcinogen. Fieser moved quickly into research on polycyclic hydrocarbons and was soon a leader in this area, making several substances available for medical study. His syntheses involved many steps and set such high standards of experimentation that his work laid the foundation for Harvard's preeminence in organic synthesis.

The group at the Royal Cancer Institute also tested methylcholanthrene, first isolated in Germany in 1933 as a bile acid derivative. It proved to be the most potent carcinogen yet known. In 1935 Fieser synthesized it, using the then little-used Elbs reaction, a pyrolytic cyclodehydration of diaryl ketones. He had first used this reaction in the 1920's and thereafter used it to prepare cholanthrenes and various carcinogenic hydrocarbons, finding it to be the best means to obtain important polycyclic hydrocarbons. The awareness that polynuclear hydrocarbons can evoke tumors in animals opened a promising new field of investigation in cancer research. Fieser synthesized carcinogens in the benzpyrene, phenanthrene, and benzanthracene series, and in 1941 received the Katherine Berkan Judd Prize for cancer research.

His researches on condensed ring systems made Fieser the most qualified person to compile a major reference work for the American Chemical Society's Scientific and Technologic Monographs, *The Chemistry of Natural Products Related to Phenanthrene* (1936), of which the confused and undeveloped field of steroid chemistry constituted a large part. He elucidated and critically reviewed the field, along with sections on alkaloids and resin acids. The second and third editions followed in 1937 and 1949. By the 1950's, research in steroid chemistry had become so extensive—with many important studies on hormones, vitamins, and animal and plant sterols—that the field completely took over the fourth edition, titled *Steroids* (1959). Like the third edition it was written with Mary Fieser and became an indispensable reference work.

The Fiesers had an intimate knowledge of steroids and contributed to steroid chemistry in terms of its stereochemistry, its reaction mechanisms, the partial synthesis of steroid derivatives, the isolation of sterols associated with cholesterol, the synthesis of the first *i*-steroids in the ergosterol series, and the nomenclature used to designate the conformation of substituents in the steroid nucleus.

During World War II, Fieser suspended his teaching and research program to engage in the war effort. Working under contract with the National Defense Research Committee, he led a large group in research on explosives, incendiaries, and antimalarials. The Harvard group was intact for four years and produced a number of key weapons.

Fieser first synthesized a number of nitro compounds and evaluated them as explosives. He then turned his attention to divinylacetylene, a Du Pont substance used as a drying oil that often exploded during its manufacture. Divinylacetylene changed from a liquid to a gel on standing and retained its sticky consistency on burning. Fieser thought it might be useful as an explosive or as a spontaneously flammable substance. He soon found, however, that rubber dissolved in hydrocarbons made a superior incendiary. He prepared rubber-gasoline gels and tested them behind Harvard Stadium.

When Japan gained control of the Pacific sources of rubber, America had to find an alternative source. The need was for a gel stable over a wide temperature range and tough enough to withstand the blast of an explosive that would scatter burning globs of gel over an area. An empirical search led Fieser to naphthenic acid, a commercial mixture of cycloparaffin acids from petroleum refining. The aluminum soap made a good gel with gasoline. Napalm, a jellied gasoline made from aluminum *na*phthenate and aluminum *palm*itate, quickly became a component of incendiary bombs and was used in flamethrowers during World War II and in Vietnam. In 1967, in the face of student criticism, Fieser declared that he felt no guilt for having developed the chemical and would do it again if called upon.

The Fieser group developed napalm for a variety of applications, including use in incendiary bombs employed in both the European and Pacific war theaters by the millions, in flamethrowers, in gel-filled grenades, and as small incendiaries carried by saboteurs to light fires and ignite oil slicks.

Fieser's war research was the subject of *The Scientific Method* (1964), his only book written from a personal viewpoint. It consists of twenty-three chapters on napalm, weapons, relations with military officers, trips to manufacturing and testing facilities, and his medical-related projects.

Fieser also headed a large group engaged in a program to develop an antimalarial drug. Since the cinchona tree plantations were in Japanese hands and troops were fighting in malarial regions, the need for alternatives to quinine seemed likely. The Office of Scientific Research and Development's Committee on Medical Research funded about forty programs at universities and drug firms. Fieser worked on sulfa drugs, which the Rockefeller Foundation had shown combated parasites in some cases of malaria.

Various university groups prepared different types of substances, including all possible isomers, derivatives, and analogues of both known and potential drugs, and sent them to Abbott Laboratories for screening. The results changed the course of Fieser's research. Three derivatives of lapachol selected from his Hooker samples had definite antimalarial activity. Fieser switched the work of his group from sulfas to quinones and prepared naphthoquinones with various straight, branched, and cyclic side chains that were numbered and tested.

In 1943 the Rockefeller Foundation reported that the naphthoquinones both suppressed and cured malaria in chickens, and a review panel concluded that they were the most promising of all antimalarials. With top priority the Harvard group worked day and night in a crash program. Some of the substances tested proved to be more potent than quinine, and Fieser felt certain that a superior drug would be developed. At war's end, however, his samples still had not undergone the lengthy and difficult testing needed to provide information on their biological activity in the body.

The best of Fieser's antimalarials was lapinone—potent, effective, and resistant to metabolic deactivation. Its practical value remained unrealized, and Fieser's new drugs never received full evaluation. The research, however, was not wasted, for twenty-two papers appeared on the drugs, beginning in 1948, with details of their preparation, chemistry, and metabolic activity. The government resumed interest in them at the time of the Vietnam war.

Another wartime project with medical aspects was Fieser's work on cortisone, first isolated by Edward Kendall in 1936. The therapeutic value of the hormone was unknown because of inadequate supply. The Committee on Medical Research sought a synthesis in order to explore its medical applications. A major difficulty lay in the fact that cortisone has an oxygen atom at the 11 position in the third ring. No natural steroid served as a starting material. A practical synthesis did not emerge during the war, only complex ones with many steps and small yields, including a method developed at Harvard.

The first batch of synthetic cortisone became available for clinical testing in 1948, and Robert Woodward at Harvard achieved its total synthesis in 1951. That synthesis depended on Fieser's use of compounds prepared by Adolf Windaus in Germany with double bonds that might be used to add oxygen to the 11 position. Fieser synthesized an 11-keto steroid in a four-step sequence. With those steps established, Woodward synthesized the hormone, starting with a common coal tar product.

Fieser's greatest impact lay less in his research than in his writing of important scientific textbooks and reference works. He regarded his success as a teacher to be his greatest achievement. His introductory organic chemistry course, tough and demanding, was taught to large numbers of premedical students. They regarded Fieser as a dynamic, inspirational teacher and were captivated by his enthusiasm for the subject.

Fieser's concern for effective teaching manifested itself in several innovative projects. The earliest was a laboratory manual, *Experiments in Organic*

Chemistry (1935). It and its successor edition, retitled *Organic Experiments* (1964), were possibly the best available manuals for three decades. He continually revised them, always looking for new methods, equipment, and reagents to test and modify for student use. In its successive editions the manual incorporated new methods of separation and isolation, such as chromatography, and new syntheses, such as Fieser's own of vitamin K_1. It differed from other manuals in its diversity of reactions and preparations involving natural products and in being written by a man versed in research who used methods and reactions taken from actual laboratory work, often that of his own group.

A second project aimed at more effective teaching was the film *Techniques of Organic Chemistry*, produced in the 1950's as an aid to laboratory work. Fieser often used novel setups in class demonstrations. His large classes led him to seek a more effective method, and he thought of making a film. He developed the script, obtained a grant from the Ford Foundation, and had the film made by a professional company in his laboratory with himself demonstrating the techniques. He also created and found a manufacturer for a set of inexpensive plastic molecular models of the Dreiding design that students could use to study the spatial relationships in organic chemistry. The American Chemical Society recognized his colorful and innovative teaching with several awards, the last being its Award in Chemical Education (1967), for his contributions as a teacher and writer.

The first of the Fieser's textbooks, *Organic Chemistry* (1944), was so widely adopted that it and its succeeding editions and translations into fifteen languages dominated the market for over a decade. The project began with the time Fieser spent on his travels from coast to coast on his war-related projects: many hours on trains, in hotels, and at airports while delayed by weather. He hated to be idle. Mary Fieser thought writing a book might help to keep her husband's mind on his eventual return to teaching and offered him her notes on her reading, hoping he would write a text on organic chemistry. He wrote while on his journeys, keeping a record of where and when he wrote each page.

Mary Fieser collected material faster than her husband could process it, and at his suggestion she began to write some chapters. Thereafter she wrote part of every textbook and reference work he published: *Textbook of Organic Chemistry* (1950); *Introduction to Organic Chemistry* (1957); *Basic Organic Chemistry* (1959); *Advanced Organic Chemistry* (1961), an expansion of the textbook for both beginning and advanced students, with historical and biographical material; and *Topics in Organic Chemistry* (1963) and *Current Topics in Organic Chemistry* (1964), both of which include material omitted from the 1961 book in order to keep the latter to a reasonable size.

The dominance of *Organic Chemistry* is owing to its excellence. Well written, sound, and authoritative, it uses information and examples of organic reactions that are experimentally real and incorporates historical and biographical material that provide a personal dimension for the reader. The succeeding editions included new developments in antibiotics, vitamins, enzymes, chemotherapy, reaction mechanisms, orbital theory, and aspects of organic chemistry related to technology. These were departures from current texts, and teachers welcomed such innovations.

After *Steroids* (1959) the Fiesers published *Style Guide for Chemists* (1960), devoted to effective communication. Originally based on notes prepared by Mary Fieser on grammar, rhetoric, and style, and incorporated into a pamphlet for contributors to *Organic Reactions*, the *Guide* expanded the notes into a small book; it was still in print and selling well in the 1980's.

Fieser was able to continue his writing following his surgery in 1965. He was by then on his final major project, the *Reagents for Organic Synthesis* series. Six large volumes appeared between 1967 and 1977, as well as several more after his death (under the direction of Mary Fieser). Essential to the practicing chemist, the reagents series provided a source of information on new reagents, which was especially vital as synthesis became increasingly sophisticated and new reagents appeared in the literature at a rapid rate.

By 1967 the Fiesers' textbooks no longer dominated the market; newer ones, reflecting advances in theoretical and physical organic chemistry, were in demand as teachers stressed the intricate details of reactions and structures in terms of reaction intermediates, and mechanisms at the expense of more descriptive synthetic and natural products chemistry. The Fiesers transferred their zeal for preparative chemistry into a new channel: an exhaustive account of substances used in organic synthesis. Each succeeding volume covered the literature up to the date of publication. Listing reagents alphabetically, each volume gave the structure, composition, physical constants, method of preparation, purification, uses, flow sheets, reaction conditions, yields, problems, and references for each, thereby freeing chemists from making lengthy literature searches. The volumes were reliable and thorough, representing an extraordinary effort to search the world's chemical literature, no matter how obscure the journal. The Fiesers were at their best here, utilizing their devotion and skill in finding, collecting, and conveying essential material for the practicing chemist. Thereby, they may be said to have influenced the course of future research through their legacy of indispensable reference works.

BIBLIOGRAPHY

I. ORIGINAL WORKS. A list of Fieser's books and papers is in Poggendorff, VIIb (1968), 1389–1395. Among the more significant papers are the following: "Reduction Potentials of Quinones, I, The Effect of the Solvent on the Potentials of Certain Benzoquinones," in *Journal of the American Chemical Society*, **45** (1923), 2194–2218, written with James B. Conant; "Reduction Potentials of Quinones, II, The Potentials of Certain Derivatives of Benzoquinone, Naphthoquinone and Anthraquinone," *ibid.*, **46** (1924), 1858–1881, written with James B. Conant; "The Alkylation of Hydroxynaphthoquinone, III, A Synthesis of Lapachol," *ibid.*, **49** (1927), 857–864; "The Tautomerism of Hydroxy Quinones," *ibid.*, **50** (1928), 439–465; "The Synthesis of Methylcholanthrene," *ibid.*, **57** (1935), 228–229, 942–946, written with Arnold M. Seligman; "A New Diene Synthesis of Anthraquinones," *ibid.*, 1679–1681, written with Mary Fieser; "Syntheses in the 1,2-Benzanthracene and Chrysene Series," *ibid.*, **61** (1939), 1647–1654, written with William S. Johnson; "Synthesis of Vitamin K_1," *ibid.*, 3467–3475; "Napalm," in *Industrial and Engineering Chemistry*, **38** (1946), 768–773; and "A New Route to 11-Keto Steroids by Fission of a $\Delta^{9(11)}$-Ethylene Oxide," in *Journal of the American Chemical Society*, **73** (1951), 5252–5265, written with Hans Heymann.

Fieser's account of his work on napalm, vitamin K, cortisone, antimalarials, and the writing of his textbook is in *The Scientific Method: A Personal Account of Unusual Projects in War and Peace* (New York, 1964). His reflections on the surgeon general's report are in "More on Smoking and Health," in *Chemistry*, **37** (March 1964), 18–19. He provided an autobiographical sketch for *McGraw-Hill Modern Scientists and Engineers*, I (New York, 1980), 372–374.

II. SECONDARY LITERATURE. On Fieser's life and work, see C. J. W. Brooks, in *Nature*, **270** (1977), 768–769; Hans Heymann, in *Journal of Organic Chemistry*, **30** (1965), insert before 1693; and William S. Johnson, in *Organic Syntheses*, **58** (1978), xiii–xvi. An article based on interviews with Mary Fieser is Stacey Pramer, "Mary Fieser: A Transitional Figure in the History of Women," in *Journal of Chemical Education*, **62** (1985), 186–191.

ALBERT B. COSTA

FILDES, PAUL GORDON (*b.* London, England, 10 February 1882; *d.* London, 5 February 1971), *microbiology*.

The son of Sir Luke Fildes, an illustrator and portrait painter, and of Fanny Woods, Fildes attended Margate Preparatory School and Winchester College before entering Trinity College, Cambridge, in 1900. Upon graduation in 1904, Fildes entered London Hospital Medical College, where he worked with William Bulloch in the Department of Bacteriology and received his medical degree in 1909. Subsequently he received honorary degrees from the University of Cambridge (1948) and the University of Reading (1959). For his contributions to the war effort he was awarded the military O.B.E. in 1919 and was knighted in 1946. His scientific awards included election as a fellow of the Royal Society in 1934, its Royal Medal in 1953, and its Copley Medal in 1963. In 1950 Fildes gave the first Leeuwenhoek Lecture of the Royal Society. With James McIntosh, J. A. Murray, and W. E. Gye, in 1920 he founded the *British Journal of Experimental Pathology*; he served as general editor for many years.

After receiving his degree, Fildes continued to serve as assistant to Bulloch at London Hospital. Between 1909 and 1914, together with James McIntosh, he undertook an investigation of syphilis, about which knowledge was rapidly expanding. They began with a carefully controlled study that demonstrated the effectiveness of Salvarsan (606) in the treatment of syphilis. Salvarsan (606), or arsphenamine, an organic arsenic compound, was developed by Paul Ehrlich and introduced for clinical trials in 1910. Almost immediately controversy arose, in part because of high demand and in part owing to questions about the effectiveness and safety of the drug. In response to questions about the effectiveness of Salvarsan (606), Fildes traced part of the problem to a lack of standardization in the procedures for the Wasserman test and himself developed a standardized procedure. Salvarsan (606) was also alleged to produce a toxic reaction, which Fildes and McIntosh traced not to the drug but to the saline solution used to reconstitute it. This discovery led to the generalized use of "pyrogen-free" distilled water for reconstituting drugs and vaccines.

During World War I, Fildes' attention was directed to problems resulting from battle wounds. Fildes developed a spray of malachite green and mercuric chloride that was very successful in treating deep infected wounds. Also during the war Fildes and McIntosh investigated gangrene and traced it to an anaerobic bacterium. It was then very difficult to study anaerobes, so they designed a special oxygen-free jar with which they investigated several anaerobes isolated from war wounds.

Fildes is best known for his work on bacterial nutrition, research that began between 1918 and 1924 with a study of influenza undertaken with McIntosh. In the course of this work, Fildes showed that two factors from blood, which he labeled X and V, were necessary for the growth of *Haemophilus influenzae*. He proceeded to classify hemoglobulinophilic bacteria by whether they required just one of these factors or both. The X factor was identified as hematin, and in 1936 the V factor was identified by Andre Lwoff as Otto Warburg's coenzyme I, nicotinamide adenine dinucleotide. In subsequent work on nutritional requirements of bacteria, Fildes established that several bacteria required tryptophan as a nutrient, and that other bacteria, which grew without tryptophan, synthesized it. He interpreted this as showing that tryptophan is an essential component of bacterial protoplasm (he later coined the term *essential metabolite* for such substances) and argued that when the bacterium could not synthesize it, it had to be supplied as a nutrient.

Through this work Fildes was led to develop connections between microbiology and chemistry, and sought the cooperation of the biochemist H. D. Kay. Another project during the late 1920's led Fildes to recognize the potential for employing techniques developed in physical chemistry, especially William Mansfield Clark's use of indicator dyes for determining the oxidation-reduction potential of various media, in studying metabolic phenomena. Fildes recognized the need for using these techniques in the course of studying why *Clostridium tetani* did not grow in healthy tissues. After developing evidence that refuted the claim that the tetanus bacillus failed to grow because it was destroyed by phagocytes, Fildes proposed that it succumbed when overoxygenated. Fildes pursued this idea using Clark's redox dyes and established a threshold potential required for germination of tetanus spores. The use of Clark's dye technique proved too laborious for generalized use, and to continue this work Clark began to collaborate with the physical chemist Bert Knight, who developed a glass electrode technique for measuring and regulating oxidation-reduction potentials.

Having come to appreciate the potential for cooperative research involving bacteriology and chemistry, Fildes proposed the establishment of an interdisciplinary research laboratory staffed by bacteriologists and chemists. With funding from the Medical Research Council, the Leverhulme Trustees, and the Halley Stuart Trust Fund, the Medical Research Council Unit in Bacterial Chemistry was created in 1934 at the Bland Sutton Institute of Pathology of Middlesex Hospital. This group continued Fildes' early investigations on the nutritional requirements of bacteria, establishing the amino acid and B vitamin requirements for several (for instance, nicotinamide for *Proteus vulgaris* and glutamine for *Streptococcus hemolyticus*).

While Fildes' work illuminated bacterial nutrition requirements and metabolic processes, his interest remained that of a bacteriologist. He studied how to control the growth and the production of toxins by bacteria. He was responsible for an important advance in this area when he made a connection with work on antibacterial drugs. Sir Lionel Whitby's laboratory at the Bland Sutton Institute was investigating the antibacterial action of sulfonamide. It was proposed that such an antibacterial substance might function by competing with essential metabolites. Fildes' suggestion that sulfonamide might compete with glutamine was not supported; but subsequently Daniel D. Woods, working in Fildes' laboratory, showed that the naturally occurring compound *p*-aminobenzoic acid could compete with and reverse the effects of sulfonamide. Woods and Fildes surmised that *p*-aminobenzoic acid was an essential metabolite and soon confirmed this suggestion. This provided the basis for the Woods-Fildes hypothesis that chemotherapeutic agents can operate by competing with essential metabolites.

The work of the research institute was interrupted by World War II, during which time Fildes moved to the Chemical Warfare Experimental Station at Porton. There he played an instrumental role in developing Britain's germ warfare effort and in getting the United States to develop its own effort. After the war Fildes reconstituted the Medical Research Council Unit in Bacterial Chemistry at the Lister Institute. His own work at the new location focused on the tryptophan requirements of bacteria and on possible tryptophan analogues that might serve as chemotherapeutic agents. After his retirement from the Medical Research Council unit in 1949, Fildes moved to a laboratory in the Sir William Dunn School of Pathology at Oxford, where, for thirteen years, he investigated the interaction of viruses with bacterial hosts by carefully controlling environmental factors. In one of these investigations he examined the effect of tryptophan on the adsorption of a virus to bacterial cells.

Fildes pioneered in the integration of biochemical studies with microbiology. Working from the biochemical side, Marjorie Stephenson was another contributor to the development of a systematic study of bacterial enzymes and metabolism, but among bacteriologists Fildes was unique during his time in recognizing the importance of biochemical studies in developing an understanding of the physiology and function of bacteria.

BIBLIOGRAPHY

I. ORIGINAL WORKS. Among Fildes' most important writings are *Syphilis from the Modern Standpoint* (London, 1911), written with James McIntosh; "A New Antiseptic Mixture for the Treatment of Gunshot Wounds and General Surgical Application," in *Lancet* (1915), **2**, 165–170, written with L. W. Rajchman and G. L. Cheatle; "The Aetiology of Influenza," in *British Journal of Experimental Pathology*, **1** (1920), 119–126, 159–174, written with James McIntosh; "The Growth Requirements of Haemolytic Influenza Bacilli and the Bearing of These upon the Classification of Related Organisms," *ibid.*, **5** (1924), 69–74; "Tetanus VI. The Conditions Under Which Tetanus Spores Germinate *in Vivo*," *ibid.*, **8** (1927), 387–393; "Tetanus VIII. The Positive Limit of Oxidation-Reduction Potential Required for the Germination of Spores of *B. tetani in vitro*," *ibid.*, **10** (1929), 197–204; "Oxidation-Reduction Studies in Relation to Bacterial Growth. III. The Positive Limit of Oxidation-Reduction Potential Required for the Germination of *B. tetani* Spores *in Vitro*," in *Biochemical Journal*, **24** (1930), 1496–1502, written with B. C. J. G. Knight; "The Nitrogen and Vitamin Requirements of *B. typhosus*," in *British Journal of Experimental Pathology*, **14** (1933), 189–196, written with G. P. Gladstone and B. C. J. G. Knight; "Tryptophan and the Growth of Bacteria," *ibid.*, 343–349, written with B. C. J. G. Knight; "Some Medical and Other Aspects of Bacterial Chemistry. President's Address," in *Proceedings of the Royal Society of Medicine*, **28** (1934), 79–90; "Inhibition of Bacterial Growth by Indoleacrylic Acid and Its Relation to Tryptophan: An Illustration of the Inhibitory Action of Substances Chemically Related to an Essential Metabolite," in *British Journal of Experimental Pathology*, **22** (1941), 293–298; "The Evolution of Microbiology. Leeuwenhoek Lecture," in *Proceedings of the Royal Society of London*, **B138** (1951), 65–74; and "Tryptophan as a Bacteriophage Adsorption Factor," in *British Journal of Experimental Pathology*, **38** (1957), 563–572, written with D. Kay.

II. SECONDARY LITERATURE. On Fildes' life and career, see G. P. Gladstone, B. C. J. G. Knight, and Sir Graham Wilson, "Paul Gordon Fildes," in *Biographical Memoirs of Fellows of the Royal Society*, **19** (1973), 317–347; and Robert Kohler, "Bacteriological Physiology: The Medical Context," in *Bulletin of the History of Medicine*, **59** (1985), 54–74.

WILLIAM BECHTEL

FILIPCHENKO [PHILIPTSCHENKO], IURII ALEKSANDROVICH (*b.* Zlyn', Bolkhovskii district, Orlovskii Province, Russia, 13 February 1882; *d.* Leningrad, U.S.S.R., 19/20 May 1930), *genetics, eugenics, zoology*.

Filipchenko was of Moldavian, Ukrainian, Swedish, and Belorussian extraction. His unique family name (which he pronounced "fee-LEEP-chen-ko") probably dates from an eighteenth-century conflation of Filipov and Pilipenko. His grandfather, Efim Ivanovich (1805–1861), was a physician and naturalist. His father, Aleksandr Efimovich (1842–1900), was a landowner and agronomist. His younger brother, Aleksandr Aleksandrovich (1884–1940), became a parasitologist and physician.

Filipchenko's active interest in zoology began at the age of eight, when he developed a fascination with natural history books, started an insect collection, and began keeping annual summer diaries of his entomological observations. He received his secondary education at the Second St. Petersburg Classical Gymnasium. In 1897 he read Darwin's *Origin of Species* and *Sexual Selection*, and two years later he worked his way through Carl Naegeli's *Mechanisch-physiologische Theorie der Abstammungslehre* (1884). These works had an important formative influence on his scientific thinking and on his choice of zoology as a career.

In 1900 Filipchenko graduated from the gymnasium with a silver medal. That same year his father died, and financial difficulties led Filipchenko to enter the Military Medical Academy, but the next year he transferred into the natural science division of the physicomathematical faculty of St. Petersburg University. As a student he spent his summers doing fieldwork, first on a scientific expedition to Kuban (1903), then at the Borodin biological station (1904–1906), and later at the Murmansk station (1908).

During his student years Filipchenko participated in various scientific, popular, and political organizations. In the summer of 1905 Filipchenko's brother was arrested "for political crimes" and sentenced to six years of hard labor; he escaped in 1908 and fled to Italy, returning only in 1917. In early December 1905 Filipchenko himself was arrested while attending a meeting of the Soviet of Workers' Deputies, but was released in a few days. He helped to organize workers in the Aleksandr Nevskii region of the city, and at the end of December was again arrested and spent four months in prison, where he studied philosophy and prepared for his government examinations. Although he subsequently served as a member of the Schlisselburg Committee for aiding political prisoners and gave occasional help to the Social Revolutionary Party, he largely gave up political activity after 1906 and devoted himself to science.

Immediately after Filipchenko's release from prison in the spring of 1906, he passed his government examinations and was accepted for specialized study in V. T. Sheviakov's laboratory of invertebrate zoology at St. Petersburg University. Concurrently he assisted M. N. Rimskii-Korsakov in entomology (1907–1909) at the Stebut Agricultural Courses and taught at the Shaffe, Tagantsev, and Obolenskii women's gymnasiums and at a commercial high school. In May 1907 he was elected a full member of the Russian Entomological Society.

In 1910 Filipchenko was accepted into the master's degree program in zoology and comparative anatomy at St. Petersburg University and was sent abroad to Munich for a year (1911–1912) to work with Richard Hertwig, who was interested in the problem of sex determination. There Filipchenko met Richard Goldschmidt (1878–1958), then a privatdocent in Hertwig's department. Goldschmidt had just recently taken up genetics, switching from studies of the development of *Ascaris* to genetic research on moths and the problem of intersexuality. Because of their closeness in age, their common expertise in invertebrate development, and their mutual interest in insects, the two developed a close relationship. Principally under Goldschmidt's influence, Filipchenko became converted to experimental zoology and especially to genetics.

In the spring of 1912 Filipchenko worked at the Naples Zoological Station, collecting material on crustacean embryology. Later that year he returned to Petersburg and defended his master's thesis on the development of *Apterygota* and the genealogical relation between insects and the millipedes and centipedes. His discussion emphasized the nonhomologous particularities of embryonic layers in different groups. This research convinced Filipchenko that higher systematic taxa differ in their embryological development in ways that are qualitatively different from the kinds of variation occurring within species. The work was published as *Razvitie izotomy* (The development of isotomes) in 1912. That year he was awarded a master's degree in zoology and comparative anatomy from St. Petersburg University, became preparator of its zootomical cabinet, and was appointed to its faculty as a privatdocent.

Genetics and Evolution. Upon his return to Russia, Filipchenko gave up his work on invertebrate embryology and devoted himself to experimental zoology and genetics. On 18 September 1913 he opened a course at St. Petersburg University entitled "The Study of Evolution and Heredity." In 1914 he published anthologies of the latest Western literature on hybridization and sex determination that included his own translations of works by de Vries, Plate, Lotsy, Hertwig, Correns, and others, supplemented with his own essay reviews. During World War I he published a number of popular articles on the new biology in leading contemporary journals and two books based on his lectures, *Izmenchivost' i evoliutsii* (Variation and evolution; 1915) and *Nasledstvennost'* (Heredity; 1917).

Filipchenko's views on evolution were influenced by those of von Baer, Naegeli, and Hertwig. Although he regarded himself as a Darwinian in the sense that he believed in evolution, Filipchenko denied that natural selection could play the evolutionary role that Darwin had assigned it. In the thesis for his master's dissertation, written in late 1912, he emphasized that "the process of organic evolution will be explained neither by the so-called factors of Lamarck, nor by selection, but is one of the fundamental properties of living matter" (Aleksandrov, p. 5). For Filipchenko, as for Herbert Spencer and Naegeli, the evolution of animals and plants was a developmental process analogous to the development of a chick embryo or the solar system.

In his 1915 book Filipchenko criticized the evolutionary mechanisms proposed by de Vries, Lotsy, the English biometricians, and the Lamarckians as inadequate, incomplete, improbable, or unsubstantiated. Citing as a "general law" that "each whole develops primarily under the influence of its own internal causes and impulses that may be affected only secondarily by external ones" (a view referred to as "autogenesis" in Russian scientific literature), Filipchenko concluded with the assertion that "in general organic evolution originates primarily under the influence of causes lying within organisms, which the action of their surrounding environment can affect only in a purely secondary way" (pp. 78–80).

Filipchenko continued to hold this view throughout his career, although it gained sophistication in later presentations. In 1923 he published a 288-page history entitled *Evoliutsionnaia ideia v biologii* (The evolutionary idea in biology) that critically evaluated the theories of Lamarck, Darwin, Wallace, Huxley, Cope, Naegeli, Eimer, Weismann, Korzhinskii, de Vries, Severtsov, Berg, and others. The same year, in his textbook on biometrics and variability, he extended his notion of "group variability" *(gruppovaia izmenchivost')*, asserting that the traits varying within a species are different in kind from those characterizing genera and higher taxa, which exhibit "less variation" and "appear significantly earlier during individual development." Granting that traits characterizing species are carried by genes localized in the chromosomes of the sex cells, he considered it likely that traits of a generic character depend on "entirely special carriers located not in the nucleus, but in the plasm of the sex cells" (p. 213).

Because Filipchenko regarded intraspecific "individual variation" as qualitatively different from interspecific "group variation," he held that evolution comprised two fundamentally different processes. In his 1927 German monograph *Variabilität und Variation*, he distinguished between *microevolution* (the evolution of biotypes, Jordanons, and Linnaean species), which could be elucidated by genetics, and *macroevolution* (the evolution of higher systematic taxa), which lay outside the scope of genetics. First formulated before World War I, Filipchenko's interrelated concepts of autogenesis and group variability dominated his thinking for the rest of his career and underlay each successive stage of his evolving research program.

Genetics, Hybridization, and Craniometry. Filipchenko took up the study of heredity in mammals in 1913, when he was appointed assistant to I. I. Ivanov at the physiological division of the Veterinary Laboratory of the Ministry of Internal Affairs, where he worked through 1916. A student of Pavlov and a pioneering researcher in artificial insemination, Ivanov had been appointed to head the division when it was founded on 19 May 1908. In July 1910 it organized a zootechnical station at Askaniia-Nova, the large estate of F. E. Fal'ts-Fein on the southern Russian steppes that had been donated in 1904 as a wildlife park. There Ivanov conducted important hybridization work on various domesticated and wild varieties of bison, cattle, and other ungulates. At that time, Ivanov was interested in Richard Hertwig's theories of sex determination and probably chose Filipchenko as his assistant for that reason.

Together with Ivanov, Filipchenko taught a course at the Veterinary Laboratory beginning in 1913, lecturing on Mendelian genetics, biometrics, the mutation theory, cytogenetics, and sex determination. Under Ivanov's direction he investigated the effects of mammalian sperm on the sex determination of offspring and studied hybrids of rats and mice produced by artificial insemination. He also mastered the various measures and indices for studying skull characteristics and sought to apply them to the study of cattle.

In 1916 Filipchenko published a popular volume on the origin of domesticated animals. His doctoral dissertation of 1917, "Izmenchivost' i nasledstvennost' cherepa u mlekopitaiushchikh" (Skull inheritance and variation in mammals), led to joint publications with Ivanov in German and Russian and was awarded the Von Baer Prize of the Russian Academy of Sciences for 1919. Filipchenko's expertise in the new biology and his study of mammalian crania led to contacts with V. M. Berkhterev's Psychoneurological Institute, where Filipchenko taught a course on vertebrate zoology (1914–1917). He was elected professor of vertebrate zoology at the institute on 11 November 1915 and served as its academic secretary through 1920.

Because of the effects of world war, revolution, and civil war, the period 1917–1922 was a time of both hardship and opportunity for young Russian scientists. Filipchenko was awarded a doctoral degree in zoology and comparative anatomy from Petrograd University in 1917. On 18 December he was appointed a salaried docent in zoology. In 1918 he became a professor of zoology at the city's women's college, the Advanced Courses for Women; following its merger with the university in 1919, he became a full professor at Petrograd University. In 1918 he organized and directed the university's Laboratory of Genetics and Experimental Zoology; in 1919 the laboratory became a university department and Filipchenko became its chairman.

At this time most academics had to supplement their incomes with extra jobs and popular writing. Filipchenko worked as senior zoologist at the Zoological Laboratory of the Russian Academy of Sciences (1918–1921) and as professor at the Chemical Pharmaceutical Institute (1919–1922), where he lectured on zoology. During this period he also wrote drafts of many of his books. By 1923 he had published a technical book on biometrics and a historical book on evolutionary theory, and had reworked his lectures on general biology for the women's courses into *Obshchedostupnaia biologiia* (Biology for the general reader; 1923), which went through fifteen editions in his lifetime. During the famine, Filipchenko also worked closely with Maxim Gorky on the Commission to Improve the Living Conditions of Scientists (KUBU).

Eugenics and Genetics. Filipchenko's readings on genetics, craniometry, the inheritance of quantitative characters, and neurology brought him into contact with the eugenics work being developed in the United States and Europe. He began giving popular lectures on eugenics in 1917 and published his first popular article on the subject in 1918. In 1919 he became aware of the chromosomal theory of heredity developed by the Morgan school and published an influential popular article that was the first generally available summary of the theory in Russian.

Filipchenko helped to organize the Russian Eugenics Society, founded in Moscow on 19 November 1920. Earlier that month he had decided to develop an independent eugenics research institution in Petrograd, and his Bureau of Eugenics was established on 14 February 1921 under the auspices of the Commission on the Study of Natural Productive Forces of Russia (KEPS) of the Russian Academy of Sciences. He appointed to the bureau's staff Dmitrii M. D'iakonov (1893–1923), Denis Karl Lepin (1895–1964), and Jan Arnold (Janovich) Lusis (1897–1969), three of his senior students at the time. The group undertook vast genealogical studies of Petrograd students, artists, writers, musicians, scientists, and academicians, based not only on archival research but also on widely distributed questionnaires. The studies were published in the bureau's research volumes, edited by Filipchenko, and were the basis of a number of his articles, notably "Intelligentsiia i talanty" (Talent and the intelligentsia; 1925).

In addition to his more technical articles on eugenics, Filipchenko published several pamphlets and books in Russian for the general reader, including *Chto takoe evgenika?* (What is eugenics?; 1921), *Kak nasleduetsia razlichnye osobennosti cheloveka* (How various human traits are inherited; 1921), *Puti uluchsheniia chelovecheskogo roda* (Ways of improving the human race; 1924), and *Gal'ton i Mendel'* (Galton and Mendel; 1925). Although Filipchenko felt that human heredity was a vital area for further research, he believed that humans should never be used as experimental subjects and opposed sterilization and all other immediate social applications, arguing that eugenic progress should be brought about by education rather than legislation.

Beginning in 1924, Marxist ideological discussions began to focus on inheritance, evolution, and eugenics. At the time many Marxists favored Lamarckian inheritance because it appeared more consistent with a materialist philosophy that emphasized the social transformation of nature, society, and man. Genetics, on the other hand, seemed to embody an idealist concept of heredity based on genes impervious to human control and incompatible with rapid human betterment. As a leading geneticist and eugenicist, Filipchenko was drawn into these discussions. In 1925 he forcefully argued against the plausibility of Lamarckism in a booklet entitled *Nasledstvenny li priobretennye priznaki?* (Are acquired characteristics inherited?). Its most controversial passage argued that Lamarckism was actually incompatible with socialism: if acquired characteristics were inherited, the oppression of the lower classes over thousands of years would have made them hereditarily inferior to the privileged classes.

For various reasons, including the hostile reaction of Marxists to his booklet, Filipchenko lost interest in eugenics in the mid 1920's, although he continued to contribute occasional popular articles and book reviews on the subject. Beginning in 1925, he instead focused his considerable energies on experimental work with animals and plants and on the cultivation of genetics as a Soviet discipline. In late 1925 his Bureau of Eugenics was renamed the Bureau of Genetics and Eugenics; in 1927 it became simply the Bureau of Genetics; in April 1930, shortly before his death, it became the Laboratory of Genetics of the U.S.S.R. Academy of Sciences.

Genetics and Agriculture. In the spring of 1920, together with six colleagues from the Advanced Courses for Women, Filipchenko founded the Peterhof Natural Science Institute in Old Peterhof (now Petrodvorets) at "Sergievka," the former estate of the duke of Leichtenberg. There he created the institute's Laboratory of Genetics and Experimental Zoology and became its director, and the institute's scientific secretary, for the next decade. Working principally at Peterhof, Filipchenko conducted research on the genetics of agriculturally important plants and animals.

Filipchenko's chief assistant, D. M. D'iakonov, died of complications from tuberculosis on 30 September 1923. To replace him Filipchenko recruited, from Kiev, Theodosius G. Dobzhansky, who had come to his attention through a piece of research on the genetics of *Drosophila,* the principal research object of the Morgan school. Dobzhansky accepted and became a docent in Filipchenko's department at the university, arriving in Petrograd in January 1924, within hours of the announcement of Lenin's death; the city was renamed Leningrad the next day. Eight months later Filipchenko's assistant at the Peterhof laboratory, V. M. Isaev (an outspoken Communist), was killed by a White guerrilla band in September 1924 while mountain climbing in the northwest Caucasus. In 1925, then, Dobzhansky became assistant director of both the Peterhof laboratory and the KEPS bureau.

Over the next three years Filipchenko and Dobzhansky became very close. As head of the KEPS bureau, Filipchenko organized a series of summer expeditions to survey local breeds of hoofed animals in remote regions: the Semirechie (1926) and Semipalatinsk (1927) regions of Kazakhstan, Kirghizia (1928), and Turkmenistan (1929, 1930). As Filipchenko's chief assistant, Dobzhansky led the first two such expeditions, which also included Lusis and two new students in the department, Nikolai Nikolaevich Medvedev and Iulii Iakovlevich Kerkis.

In 1926 Filipchenko nominated Dobzhansky for an International Education Board fellowship to work in T. H. Morgan's laboratory; he was accepted and left for New York in the fall of 1927.

In the mid 1920's Filipchenko conducted a systematic survey of the genetics literature that was published as a two-volume text dealing with plants (1927) and animals (1928). For practical and theoretical reasons he became especially interested in the inheritance of quantitative characters in soft wheats, focusing on the size of the ears and seed kernels in grains and grasses. Working with Lepin on experimental plots at Peterhof, he conducted regular studies of seeds and samples collected by Vavilov's worldwide expeditions. Although these investigations on soft wheats did not entirely confirm his expectations on group variability, they proved useful to other researchers and were posthumously published as a book in 1934.

By 1929 Filipchenko had become convinced that genetics had to be reconstituted as a discipline. He believed that its traditional, strictly morphological character had to be replaced by a physiological approach that emphasized the development of traits. Drawing on the recent work of Spemann and other embryologists to support his idea of group variability, Filipchenko called for the integration of developmental mechanics and genetics. In 1929 he began teaching a new course embodying these ideas and systematically set forth its subject matter in a textbook (published posthumously in 1932 as *Eksperimental'naia zoologiia* (Experimental zoology). In May 1930 he laid out the argument and agenda for physiological genetics in his paper "Morfologiia i fiziologiia nasledstvennosti" (The morphology and the physiology of heredity), presented to the VIth All-Union Congress of Zoologists in Kiev.

During the period of the first Five-Year Plan, ideological disputes and institutional reorganizations disrupted academic and scientific life. Beginning in 1929, Filipchenko was publicly castigated for his autogenetic view of evolution and his earlier work in eugenics. He was relieved of all university teaching duties in early 1930, but he was never arrested. Forced out of the university, Filipchenko devoted himself to the organization of applied research in animal genetics.

Since 1928 Filipchenko had been involved in plant and animal breeding work at the State Institute of Applied Agronomy (GIOA). In 1929, at the invitation of Nikolai Vavilov, he agreed to organize the Genetics Division of its Institute of Animal Breeding, which was slated to become the core of the new Institute of Animal Breeding of the Lenin All-Union Academy of Agricultural Sciences. In the spring of 1930, now free from university duties, Filipchenko devoted himself to planning the division and started work on the genetics of swine breeding. On 4 April 1930 Filipchenko's Bureau of Genetics became the Laboratory of Genetics of the U.S.S.R. Academy of Sciences, and he looked forward to expanding his academy research base.

Filipchenko's death was natural but sudden. A few days after returning from the Kiev meeting, while sowing experimental wheat at Peterhof, he developed a severe headache and returned to Leningrad to be attended by his physician brother. After a three-day illness he died of streptococcal meningitis during the night of 19/20 May 1930, around midnight. His head was donated to Bekhterev's Brain Institute for research and his remains were buried in Smolensk Cemetery in Leningrad following a large public funeral procession. His wife, Nadezhda Pavlovna, and his only son, Gleb, a physicist, died during World War II in the blockade of Leningrad.

Impact. Filipchenko was a founder of Soviet genetics and experimental biology. Fluent in German, French, English, and Latin, and a member of the American Genetic Association, the Société de Morphologie (Paris), the Deutsche Gesellschaft für Vererbungswissenschaft, and the Deutsche Gesellschaft für Zuchtungskunde, he played a central role in introducing the latest Western biological developments into Russia. His energetic personality, precise mind, and broad encyclopedic knowledge in a wide range of biological fields exercised a profound influence on Theodosius Dobzhansky and other members of the Leningrad school. His review articles, textbooks, and popular tracts educated a generation of Soviet biologists.

Despite his early death Filipchenko had a lasting impact on the institutionalization of Soviet genetics. Founded in 1919, Filipchenko's department at Petrograd University was the first university genetics department in Russia and one of the first in Europe; despite subsequent organizational, ideological, and political disruptions, it remains one of the leading Soviet centers of genetics. The Laboratory of Genetics and Experimental Zoology of the Peterhof Natural Science Institute, which he founded in 1920, was one of Russia's earliest research centers for the new experimental biology; renamed the Peterhof Biological Institute in 1930, the complex remains the university's principal base for biological research and field work. The KEPS bureau that Filipchenko founded in 1921 became the central genetics institution of the U.S.S.R. Academy of Sciences under the leadership of N. I. Vavilov, who took over the

Laboratory of Genetics upon Filipchenko's death. In 1933 it became the academy's Institute of Genetics (IGEN). Although IGEN moved to Moscow in 1934, Filipchenko's students continued to form the core of its research staff and that of its mutation laboratory, directed by H. J. Muller (1933–1937), until Lysenko forced them out in the early 1940's.

In the Soviet Union the recognition of Filipchenko's historical importance was eclipsed by the rise of Lysenkoism. Ironically, his principal detractor was the Leningrader Isaak Izrailovich Prezent. In 1932 Prezent edited Filipchenko's *Eksperimental'naia zoologiia* for posthumous publication, and his extensive annotations indicated only minor quibbles with Filipchenko's views. Five years later Prezent had become Lysenko's chief ideologist and was leading the attack on Filipchenko for his "reactionary" support of evolutionary autogenesis, his "idealist" concept of the gene, and his "bourgeois" eugenic views. Filipchenko's importance began to be recognized in the Soviet Union only after the repudiation of Lysenkoism and the rebirth of Soviet genetics (1964–1967). In the succeeding years two of Filipchenko's articles were reprinted (1968), his book on the history of evolutionary theory was reissued (1977), and a brief biography by his student N. N. Medvedev was published (1978).

In the West, however, Filipchenko's distinction between microevolution and macroevolution had an impact on the development of evolutionary theory in the 1930's and 1940's. The distinction was employed by Filipchenko's protégé Theodosius Dobzhansky in *Genetics and the Origin of Species* (1937), a key work in the development of the synthetic theory of evolution. The same distinction became the organizing principle of the most outspoken attack on that theory, *The Material Basis of Evolution* (1940), by Filipchenko's friend Richard Goldschmidt. By the 1940's the relationship between microevolution and macroevolution had become a central issue in evolutionary theory. In challenging the relevance of population genetics to macroevolution and highlighting the importance of morphological, embryological, physiological, and developmental factors, Filipchenko helped to shape the problematics and discourse of modern evolutionary biology.

BIBLIOGRAPHY

I. Original Works. Filipchenko's archives are Fund 813 of the Manuscript Division of the Saltykov-Shchedrin Public Library in Leningrad. He published more than 100 works in Russian, several of which include summaries in Western languages. Spelling his name "J. A. Philiptschenko," he also published approximately 20 works in German and 4 in French.

Filipchenko was the author of twenty pamphlets, books, and texts: *Razvitie izotomy* (The development of isotomes; St. Petersburg, 1912); *Izmenchivost' i evoliutsiia* (Variation and evolution; Petrograd and Moscow, 1915; 2nd ed., Petersburg, 1921); *Proiskhozhdenie domashnykh zhivotnykh* (The origin of domesticated animals; Petrograd, 1916; 2nd ed., Leningrad, 1924); *Nasledstvennost'* (Heredity; Moscow, 1917; 2nd ed., 1924; 3rd ed., 1926); *Chto takoe evgenika?* (What is eugenics?; Petrograd, 1921); *Kak nasleduetsia razlichnye osobennosti cheloveka* (How various human traits are inherited; Petrograd, 1921); *Izmenchivost' i metody ee izucheniia* (Variation and methods for its study; Petrograd, 1923; 2nd ed., Leningrad, 1926; 3rd ed., 1927; 4th ed., Moscow and Leningrad, 1929); *Obshchedostupnaia biologiia* (Biology for the general reader; Petrograd, 1923; 15th ed., 1930); *Evoliutsionnaia ideia v biologii* (The evolutionary idea in biology; Moscow, 1923; 2nd ed., 1926; 3rd ed., 1977); *Puti uluchsheniia chelovecheskogo roda (evgenika)* (Ways of improving the human race [eugenics]; Leningrad, 1924); *Frensis Gal'ton i Gregor Mendel'* (Francis Galton and Gregor Mendel; Moscow, 1925); *Besedy o zhivykh sushchestvakh* (Conversations about living substances; Leningrad, 1925); *Nasledstvenny li priobretennye priznaki?* (Are acquired characteristics inherited?; Leningrad, 1925), translation of T. H. Morgan's essay of that title, translated by Filipchenko, plus an essay of his own; *Variabilität und Variation* (Berlin, 1927); *Chastnaia genetika* (Systematic genetics), I, *Rasteniia* (Plants; Leningrad, 1927), II, *Zhivotnye* (Animals; Leningrad, 1928); *Genetika* (Genetics; Leningrad, 1929); *Genetika i ee znachenie dlia zhivotnovodstva* (Genetics and its significance for animal breeding; Moscow and Leningrad, 1931); *Eksperimental'naia zoologiia* (Experimental zoology; Leningrad and Moscow, 1932); and *Genetika miagkikh pshenits* (The genetics of soft wheats; Moscow and Leningrad, 1934), with T. K. Lepin.

Filipchenko also published numerous articles on invertebrate embryology, vertebrate hybridization, genetics, and eugenics, notably: "Anatomische Studien über *Collembola*," in *Zeitschrift für wissenschaftliche Zoologie*, **85** (1906), 270–304; "Beiträge zur Kenntnis der *Apterygoten*," 3 parts, *ibid.*, **88** (1907), 99–116, **91** (1908), 93–111, and **103** (1912), 519–660; "O vidovykh gibridakh" (On species hybrids), in *Novye idei v biologii*, no. 6 (1913), 122–149; "Opisanie gibridov mezhdu bizonom, zubrom i rogatym skotom v zooparke 'Askaniia Nova' F. E. Fal'ts-Feina" (Description of hybrids between bison, aurochs and percora in the "Askaniia Nova" Zoological Park), in *Arkhiv veterinarnykh nauk*, no. 2 (1915), 1–33, written with I. I. Ivanov, published the next year in German as "Beschreibung von Hybriden zwischen Bison, Wisent und Hausrind," in *Zeitschrift für induktiv Abstammungs- und Vererbungslehre*, **16** (1916), 1–48; "Izmenchivost' i nasledstvennost' cherepa u mlekopitaiushchikh" (Variation and heredity of mammalian skulls), 2 parts, in *Russkii arkhiv anatomii, gistologii i embriologii*,

1, no. 2 (1916), 311–404, and no. 3 (1917), 747–818; "Evgenika" (Eugenics), in *Russkaia mysl'*, 1918, no. 3–4, 69–95; "Khromozomy i nasledstvennost'" (Chromosomes and heredity), in *Priroda*, 1919, no. 7–9, 327–350; "Zakon Mendelia i zakon Morgana" (Mendel's law and Morgan's law), *ibid.*, 1922, no. 10–12, 51–66; "Nashi vydaiushchiesia uchenye" (Our leading scientists), in *Izvestiia Biuro po evgenike Akademii nauk SSSR*, no. 1 (1922), 21–38; "Rezul'taty obsledovaniia leningradskikh predstavitelei iskusstva" (Results of a survey of Leningrad artists), *ibid.*, no. 2 (1924), 5–28; "Deistvitel'nye chleny b. imperatorskoi, nyne Rossiiskoi, akademii za poslednie 80 let" (Members of the former imperial, now Russian, Academy for the past eighty years), *ibid.*, no. 3 (1925), 3–82, with T. K. Lepin and Ia. Ia. Lus; "Intelligentsiia i talanty" (Talent and the intelligentsia), *ibid.*, 83–101; "Izmenchivost' kolichestvennykh priznakov u miagkikh pshenits" (Variation of quantitative characters in soft wheats), *Izvestiia Biuro po genetike i evgenike*, no. 4 (1926), 5–58; "O parallelizme v zhivoi prirode" (On parallelism in living nature), in *Uspekhi eksperimental'noi biologii*, **3** (1925), no. 3–4, 242–258; "Uspekhi genetiki za poslednie 10 let (1918–1927) v SSSR" (Achievements of genetics in the U.S.S.R. over the past ten years, 1918–1927), in *Trudy Leningradskogo obshchestva estestvoispytatelei*, **57**, no. 1 (1927), 3–11; and "Morfologiia i fiziologiia nasledstvennosti" (The morphology and physiology of heredity), in *Trudy VI Vsesoiuznogo s"ezda zoologov* (Proceedings of the Sixth All-Union Congress of Zoologists; Kiev, 1930), 15–30.

II. SECONDARY LITERATURE. In Russian, see the obituaries by T. K. Lepin, in *Priroda*, 1930, no. 7–8, 683–698; by A. Zavarzin, in *Trudy Leningradskogo obshchestva estestvoispytatelei*, **60**, no. 2 (1930), 3–16; and by his brother, A. A. Filipchenko, in *Trudy Laboratorii genetiki*, no. 9 (1932), 1–11. See also the works by his student N. N. Medvedev, *Iurii Aleksandrovich Filipchenko 1882–1930* (Moscow, 1978), and "Iurii Aleksandrovich Filipchenko 1882–1930," in *Vydaiushchiesia sovetskie genetiki* (Outstanding Soviet geneticists; Moscow, 1980), 88–100. On his role in the development of Soviet genetics, see A. E. Gaisinovich, *Zarozhdenie i razvitie genetiki* (The birth and development of genetics; Moscow, 1988).

On Filipchenko's evolutionary views, see Mark B. Adams, "La génétique des populations était-elle une génétique évolutive?," in *Histoire de la génétique*, Jean-Louis Fischer, ed. (Paris, 1989); and D. A. Aleksandrov, "Iurii Aleksandrovich Filipchenko kak genetik-evoliutsionist: Formirovanie nauchnykh interesov i vzgliadov" (Iurii Aleksandrovich Filipchenko as a geneticist and evolutionist: The formation of his scientific interests and views), in *Evoliutsionnaia genetika (k 100-letiiu so dnia rozhdeniia Iu. A. Filipchenko)* (Evolutionary genetics [in honor of the 100th anniversary of Iu. A. Filipchenko's birthday]; Leningrad, 1982), 3–21.

In English, see Mark B. Adams, "Eugenics in Russia 1900–1940," in *The Wellborn Science*, Mark B. Adams, ed. (New York, 1988), 153–216. See also the following essays in *The Evolutionary Synthesis*, Ernst Mayr and William B. Provine, eds. (Cambridge, Mass., 1980): Mark B. Adams, "Severtsov and Schmalhausen: Russian Morphology and the Evolutionary Synthesis," 193–225, and "Sergei Chetverikov, the Kol'tsov Institute, and the Evolutionary Synthesis," 242–278; and Theodosius Dobzhansky, "The Birth of the Genetic Theory of Evolution in the Soviet Union in the 1920s," 229–242. See also A. E. Gaissinovitch, "The Origins of Soviet Genetics and the Struggle with Lamarckism, 1922–1929," Mark B. Adams, trans., in *Journal of the History of Biology*, **13**, no. 1 (1980), 1–51.

MARK B. ADAMS

FOK, VLADIMIR ALEKSANDROVICH (*b*. St. Petersburg, Russia, 22 December 1898; *d*. Leningrad, U.S.S.R, 27 December 1974), *physics*.

Fok (often spelled Fock) was the son of Aleksandr Aleksandrovich Fok, a forestry specialist whose works were well known at the beginning of the twentieth century, and of Nadezhda Alexeevna Fok. He graduated from the Practical School of the Reformation Churches in Petrograd with a gold medal in 1916 and then enrolled in the Faculty of Physics and Mathematics of Petrograd University. The following year he was called to military service and, after graduating from an accelerated course at the Artillery School, was sent to the front. He was back at the university in 1919, and that same year began work as a calculator at the State Optical Institute, which was directed by Dmitri Rozhdestvenskii. In 1922 Fok graduated from the university. In 1934 he received a doctor of science degree. The beginning of his scientific activity coincided with the dawn of Soviet physics and took place in poor conditions because there were almost no physics scholars in the country. Fok worked simultaneously in various scientific and educational institutions. In addition to his research in the department of theoretical physics of the State Optical Institute (1919–1923, 1928–1941), Fok was a researcher at the Leningrad Institute of Physics and Technology (1924–1936). He also taught at Leningrad University (1924–1974), becoming a professor there in 1932, and during the same period he conducted a seminar in experimental mathematics at the Faculty of Physics and Mechanics, Leningrad Polytechnical Institute. At this seminar students tried to solve actual problems of applied physics. Fok did research at the Lebedev Physics Institute (1944–1953) and at the Institute for Physical Problems (1954–1964), both part of the Soviet Academy of Sciences in Moscow.

Fok's first paper, written while he was at the

State Optical Institute (1924), concerned the theory of luminosity of surfaces of arbitrary shape. Fok was also involved in investigating the mathematical aspects of the process of optical-glass preparation. At the Institute of Physics and Technology, Fok worked out a rigorous theory of thermal breakdown of dielectrics and calculated the thermal resistance of a multicore cable. While serving as a consultant for various institutions, Fok solved the problem set by Joseph Lagrange (which proved to be too hard for G. F. Bernhard Riemann): the problem of gas pressure within an artillery barrel before the ejection of the shell. Fok was also involved in theoretical aspects of geological surveys and electrical methods of prospecting for minerals, and he carried out investigations in the theory of elasticity and other problems of mathematical physics. During World War II Fok performed ballistics calculations.

A number of Fok's papers are purely mathematical (concerning integral equations, Bessel functions, and the theory of analytic functions and Airy functions, the latter tabulated by Fok in 1945). These papers are complemented by a series of investigations on the theory of diffraction and approximate solutions to some particular problems of radio-wave diffractions. He worked out how to determine the field in the region of half-shadow, or the so-called principle of locality. In his rigorous theory of diffraction of radio waves around the earth's surface (1946), Fok took into account the nonhomogeneity of the atmosphere and the earth's surface. The mathematical techniques developed by the Italian physicist Tullio Regge for the description of the scattering of elementary particles (the method of complex momenta) resemble those constructed by Fok for the solution of similar problems (spherical functions with a complex index used for the solution of diffraction problems).

Fok's reputation rests mainly on his work on quantum mechanics, begun in the late 1920's. In 1926 he generalized the Schrödinger wave equation to the relativistic case by replacing the momentum in the relativistic equation linking energy, momentum, and mass by the gradients. (The problem was solved independently by the Swedish physicist Oskar Klein, and the relativistic scalar equation describing particles with spin in an electromagnetic field was named the Klein-Fok equation.) Fok was also involved in quantum mechanics of multiparticle systems. Developing Douglas Hartree's approach (1927) and using Wolfgang Pauli's principle, Fok worked out the technique called the Hartree-Fok method (1930). The wave function of electrons in the atom is represented as a determinant of one-electron functions, the latter being determined by a standard variational procedure. The resulting self-consistent solution automatically takes into account correlations of the orbital electrons associated with their mutual exchange. Fok's system of equations of a self-consistent field with exchange is successfully used not only for calculations of multielectron atoms but also for all multiparticle problems of quantum mechanics, including the theory of superconductivity (for the latter, Fok's equations have been generalized by Nikolai Bogolubov).

In 1931 Fok published his book *Fundamentals of Quantum Mechanics*, which played an important role in familiarizing Soviet physicists with a domain of physics already being explored in this country.

Between 1928 and 1934, Fok obtained important results in the quantum field theory. He completed the mathematical method of secondary quantization proposed by Paul A. M. Dirac and developed by Pascual Jordan and Eugene Wigner. It has been shown that this method does not go beyond the traditional framework of quantum mechanics (as Jordan argued) and that the two ways of description are completely equivalent. As early as 1934 Fok suggested describing a system with a variable number of particles (bosons) in the representation of the secondary quantization with the aid of a generating functional ("Fok's functional"). In his papers Fok introduced a number of new notions, such as "the Fok space" (a Hilbert space in the representation of secondary quantization). In papers written with Dirac and Boris Podolsky (1932), Dirac's results on interaction of charged particles by the exchange of virtual photons were extended. The one-dimensional problem solved by Dirac was generalized to the realistic three-dimensional case; the multitime formalism of Dirac, Podolsky, and Fok was introduced; and quantum electrodynamics was formulated in its modern form.

Fok also studied problems of the general theory of relativity. In 1939 he solved the problem of motion of a many-body system within the framework of this theory. Fok demonstrated that for the finite masses (not "point masses") the general equations of the theory of gravitation (specifically, their compatibility conditions) make it possible to obtain the law of universal gravitation and Newton's equations of motion without any additional considerations. Fok took into account the corrections to Newton's equations of gravitation in the second approximation of Einstein's theory. He summarized the results of his investigations in a special monograph (1955).

Fok paid a great deal of attention to the philosophical aspects of quantum mechanics (especially from the late 1940's on) and of the theory of relativity. He was among those Soviet physicists (including A. Ioffe, Ya. Frenkel, L. Landau, I. Tamm, and S. Vavilov) who struggled against the vulgarization of both these great theories by some Soviet philosophers. In the postwar period Fok developed his own ideas on the interpretation of quantum mechanics and discussed them with Niels Bohr.

Fok loved literature and poetry, and wrote many facetious poems that became popular among physicists, including some dedicated to Bohr and Born that were written in Russian and in German. Fok wrote other lyrical poetry as well.

In 1932 Fok was elected as corresponding member, and in 1939 as full member, to the Soviet Academy of Sciences. In 1936 he received the Mendeleev Prize; in 1946, the State Prize; and in 1960, the Lenin Prize. In 1968 Fok was awarded the title Hero of Socialist Labor. He was a foreign member of the Norwegian and Danish Academies of Sciences and the Deutsche Akademie der Wissenschaften in Berlin, and held honorary degrees from the universities of New Delhi, Leipzig, and Michigan.

BIBLIOGRAPHY

I. ORIGINAL WORKS. "Zur Berechnung der Beleuchtigungsstärke," in *Zeitschrift für Physik*, **28**, no. 2 (1924), 102–113; "Über die invariante Form der Wellen- und die Bewegungsgleichungen für einen geladenen Massenpunkt," *ibid.*, **39**, nos. 2–3 (1926), 226–232; "Zur Wärmetheorie des elektrischen Durchschlages," in *Archiv für Elektrotechnik*, **19**, no. 1 (1927), 71–81; "Näherungsmethode zur Lösung des quantenmechanischen Mehrkörperproblems," in *Zeitschrift für Physik*, **61**, nos. 1–2 (1930), 126–148; "On Quantum Elektrodynamics," in *Physikalische Zeitschrift der Sowjetunion*, **2**, no. 6 (1932), 468–479, written with P. A. M. Dirac and Boris Podolsky; and "Zur Quantenelektrodynamik," *ibid.*, **6**, no. 5 (1934), 425–469.

Nachala kvantovoi mekhaniki (Moscow, 1931; 2nd ed., Moscow, 1976), trans. and rev. by Eugene Yankovsky as *Fundamentals of Quantum Mechanics* (Moscow, 1978); *Teoriya katorazha* ("Theory of [Well] Logging"; Moscow and Leningrad, 1933); "Sur le mouvement des masses finies d'après la théorie de gravitation Einsteinienne," in *Journal of Physics* (Moscow), **1**, no. 2 (1939), 81–116; "Diffraction of Radio Waves Around the Earth's Surface," in *Journal of Physics* (Moscow), **9**, no. 4 (1945), 255–266; "New Methods in Diffraction Theory," in *Philosophical Magazine*, 7th ser., **39**, no. 289 (1948), 149–155; *Teoriya prostranstva, vremeni i tyagoteniya* (Moscow, 1955), trans. by N. Kemmer as *The Theory of Space, Time and Gravitation* (London and New York, 1959; 2nd rev. ed., Oxford and New York, 1964); and "Quantum Physics and Philosophical Problems," in *Foundations of Physics*, **1**, no. 4 (1971), 293–306.

II. SECONDARY LITERATURE. "Sbornik statei posvyashchennikh 80-letiyu so dnya rozhdeniya V. A. Foka" ("Collection of Essays Dedicated to V. A. Fok on the Occasion of the 80th Anniversary of His Birth"), in *Trudy GOI*, **43**, no. 177 (1978), includes a bibliography of Fok's work; M. G. Veselov, "Vladimir Aleksandrovich Fok," in *Uspekhi fizicheskikh nauk*, **66**, no. 4 (1958), 695–699; M. G. Veselov, G. F. Drukarev, and Yu. V. Novozhilov, "Vladimir Aleksandrovich Fok," in *Uspekhi fizicheskikh nauk*, **96**, no. 4 (1968), 741–743, trans. by E. Bergman in *Soviet Physics Uspekhi*, **11**, no. 6 (1969), 921–923; and M. G. Veselov, P. L. Kapitsa, and M. A. Leontovich, "Pamyati Vladimira Aleksandrovicha Foka," in *Uspekhi fizicheskikh nauk*, **117**, no. 2 (1975), 375–376, trans. as "In Memory of Vladimir Aleksandrovich Fok" by R. W. Bowers in *Soviet Physics Uspekhi*, **18**, no. 10 (1975), 840–841.

V. J. FRENKEL

FOSHAG, WILLIAM FREDERICK (*b.* Sag Harbor, New York, 17 March 1894; *d.* Westmoreland Hills, Maryland, 21 May 1956), *geology, mineralogy.*

Foshag was the son of William Frederick Foshag and Joanna Eva Riegler. The family moved to California in his early youth, and he entered the University of California with chemistry as his major. He received the A.B. in chemistry in 1919 and the Ph.D. in 1923. While at the university he came under the influence of the distinguished mineralogist Arthur S. Eakle, who aroused in him an intense and abiding interest in minerals. In 1917 and 1918, while still an undergraduate, Foshag worked in the laboratory of the Riverside Portland Cement Company at Crestmore, California, whose limestone quarry was an outstanding mineral locality and provided the material for several of his early publications on mineralogy.

In 1919 Foshag was appointed assistant curator in the Division of Mineralogy and Petrology at the Smithsonian Institution, where he remained for the rest of his life. On 5 September 1923 he married Merle Crisler; they had one son.

In 1929 he was promoted to curator, and in 1948 to head curator, of the Department of Geology, a position he held until his death.

Foshag was known as a mineralogist, but he was a geologist, chemist, volcanologist, gemologist, and student of meteorites as well. During his many years of service to the Smithsonian Institution the mineral collection was enormously enriched, both by his work in the field and by some notable bequests that raised it from comparatively undistinguished stature to one of the world's great collections. Most notable of these bequests was the collection (16,000 carefully selected specimens) of Colonel Washington A. Roebling (1926); this bequest was undoubtedly the fruit of the friendly relationship Foshag had with Roebling. Roebling's son, John A. Roebling, established an endowment fund of $150,000, the income of which has been used for the continued growth of the Roebling collection.

Foshag's own collecting and research activities were largely devoted to the study of mineral and mining localities in the western United States and in Mexico. His knowledge of Mexico was utilized by the U.S. Geological Survey during World War II, when he headed a cooperative project with the Mexican authorities for the discovery and development of strategic mineral deposits in that country. In 1943 a new volcano, Paricutín, erupted in central Mexico, and Foshag followed the course of the activity from its inception until Paricutín became extinct in 1952, a notable first in volcanology.

In 1946 Foshag and his colleague Edward P. Henderson spent more than four months in Japan supervising the grading, classifying, and evaluation for the U.S. government of some $25 million worth of diamonds that the Japanese people had given to their government to aid in the war effort.

Foshag had a deep interest in the archaeology of Latin America and made a collection of artifacts and related materials that was later acquired by the National Gallery of Art. He acted as a consultant to Robert Woods Bliss in matters relating to jade and other archaeological materials, and he prepared the foreword for the catalog of the Bliss collection. He also had a keen interest in Oriental porcelains, especially those from China and Japan.

Foshag was a fellow of the Geological Society of America, a charter fellow of the Mineralogical Society of America, and president of the latter in 1940. He was an honorary member of the Geological Society of Mexico. The mineral foshagite, $Ca_4Si_3O_9(OH)_2$, from Crestmore, California, was named in his honor by A. S. Eakle.

BIBLIOGRAPHY

I. ORIGINAL WORKS. Foshag's bibliography comprises more than 100 publications, including descriptions of thirteen new minerals. The following is a small selection: "The Origin of the Colemanite Deposits of California," in *Economic Geology*, **16** (1921), 199–214; "Saline Lakes of the Mojave Desert Region," *ibid.*, **21** (1926), 56–64; "Gems and Gem Minerals," in *Minerals from the Earth and Sky*, pt. 2, Smithsonian Scientific Series, no. 3 (1929), 169–332, written with George P. Merrill; "The Ore Deposits of Los Lamentos, Chihuahua, Mexico," in *Economic Geology*, **29** (1934), 330–345; "Problems in the Study of Meteorites," in *American Mineralogist*, **26** (1941), 137–144; "Tin Deposits of the Republic of Mexico," in *U.S. Geological Survey Bulletin* 935-C (1942), 99–176; "Exploring the World of Gems," in *National Geographic*, **98** (1950), 770–810; "Mineralogical Studies on Guatemalan Jade," in *Antropología e historia de Guatemala*, **6**, no. 1 (1954); and "Birth and Development of Paricutín Volcano, Mexico," in *U.S. Geological Survey Bulletin* 965-D (1956), 355–489.

II. SECONDARY LITERATURE. On foshagite, see Arthur S. Eakle, in *American Mineralogist*, **10** (1925), 97–99. For an assessment of Foshag's work, see E. H. Kraus, "Presentation of the Roebling Medal of the Mineralogical Society of America to William Frederick Foshag," in *American Mineralogist*, **39** (1954), 293–295. For an obituary and bibliography, see W. T. Schaller, "Memorial of William Frederick Foshag," in *American Mineralogist*, **42** (1957), 249–255.

BRIAN MASON

FOURMARIER, PAUL (*b.* La Hulpe, Belgium, 25 December 1877; *d.* Liège, Belgium, 20 January 1970), *geology*.

In 1895 Fourmarier enrolled in the mining engineering course at Brussels University and then at Liège University, where he studied with Max Lohest. He received a degree in mining engineering in 1899 and in geological engineering in 1901, both from Liège.

Soon after graduation Fourmarier joined the Bureau of Mines as a mine inspector, thereby gaining firsthand knowledge of the tectonic structures of coalfields. At the same time he was appointed a teaching assistant at Liège University, where he showed a talent for teaching, and was invited to join the Geological Commission, where he worked on the revision of André Dumont's geological map. Fourmarier was responsible for revising the Hamoir-Ferrières, Seraing, and Chenée map sheets of the Ardennes.

In 1901 Fourmarier published a revised geological map of the Theux region, where a Devonian-Carboniferous basin, surrounded on three sides by a curved fault, was supposedly sunk into Lower Devonian rocks. However, a new excavation for a second railroad track between Pépinster and Theux exposed the superposition of Lower Devonian upon Carboniferous beds. This Theux fault was obviously a thrust, but it dipped to the north. Having carefully described the new exposure, Fourmarier was puzzled by this thrust apparently opposite to the usual direction. If the displacement was northward, as elsewhere, could this be an underthrust? But it would have to be an odd kind of underthrust, engulfing the Carboniferous on three sides, under the Lower Devonian.

This idea illustrates the uncertainty that still existed, even among the best workers in tectonics. Henri de Dorlodot concurred, after having conceived and rejected the interpretation that was later shown to be correct: that the Theux fault forms the frontal northern limit of a *tectonic window* (the concept had hardly been defined, and no term had yet been proposed for it). Dorlodot used the term "eyelet" (*oeillet* in French) to describe this still hypothetical structure. But, relying on field evidence, he rejected his hypothesis because of the very meager evidence regarding the Marteau fault, which should have closed the window to the southeast. This is indeed a weak point. The southeastern closure of the Theux circular fault was to remain a problem until 1950, when Fernand Geukens traced it further south, well into the monotonous pre-Devonian basement.

That the medieval town of Theux lies in the middle of a tectonic window was established by Fourmarier's research in 1905. The demonstrative criterion emerged from a detailed study of the facies variations in the Devonian limestones within the window and along the Ourthe Valley, in the nappe. As confirmed by a revision by Marie Coen-Aubert in 1974, this tracing of isofacies lines demonstrates a 12-kilometer (7.5-mile) horizontal displacement of the nappe northward with regard to the window.

If this hypothesis was correct, the Carboniferous should underlie the Lower Devonian, and a new coalfield might be discovered. Accordingly, two test boreholes were drilled near Pépinster, north of the window. They reached the coal measures—unfortunately barren—underneath the Lower Devonian of the nappe. This added the final point to the demonstration.

Fourmarier's detailed mapping also had disclosed two smaller windows within the larger window at Oneux. More complex than might have been thought at first, the Theux window, one of the first and best-documented examples of a tectonic window, thus acquired wide renown.

On a regional scale, the tectonic structure of the Ardenne is somewhat symmetrical. In the east it must be related to the Eifel overthrust, which extends east toward Aachen. In the west the La Tombe klippe relates to the Midi overthrust, which has been traced westward, south of the concealed coalfields of northern France, to the Boulonnais inlier on the Channel coast. Between these two thrust segments, the intermediate "Condroz" segment was interpreted by Dorlodot as a major overfold. Fourmarier advocated that this structure be replaced by a *charriage du Condroz* connecting the western and eastern thrusts. This major overthrust, the frontal structure of the Hercynian chain in Western Europe, is well documented by seismic prospection and confirmed by several deep boreholes.

This synthesis brings to the fore the two kinds of tectonics proposed by Pierre Termier. Albrecht Heim had drawn tectonic nappes in the Alps in the form of recumbent anticlines with stretched inverted limbs, an image that Marcel Bertrand had employed in Provence and that, after acceptance of Maurice Lugeon's Alpine synthesis, was taken by Termier as the general model of nappe structure. This model has since gained a wide, if rather undue, acceptance. It had inspired Dorlodot to put forth his recumbent Condroz anticline; nevertheless, he and (now) Fourmarier offered as evidence against this "stretching" style the clear-cut *shearing* across the limbs of the folds, as seen both in outcrop and underground in the mine profiles. With Fourmarier's tectonic synthesis the Ardennes thus became the type locality of Termier's tectonic "second genre." The 1922 meeting of the International Geological Congress in Belgium provided the opportunity for demonstrating these structures in the field.

In fact, the Assynt slices (northwestern Scotland), the Appalachian thrust sheets, those of the northern Rockies, the Katanga (now Shaba) thrust belt, and many others all exhibit this shearing style. Even the Helvetic Alps, where the Morcles overfold remains an exception, belong to this "second genre," which thus represents the prime type of overthrusting. Fourmarier's analyses and syntheses of 1907, 1934, and 1954 exemplify the tangential style of tectonics, except perhaps for his overemphasis of the idea that thrusting develops only after folding instead of the contrary, as is generally accepted today.

In 1919, he was elected to the Royal Academy of Belgium, over whose committees for geology and geography he later presided. In 1920, Fourmarier was appointed professor of mining geology, hydrogeology, and mining and industrial geography, and in 1927 professor of physical geology, at Liège University. As was usual in Belgium, he linked geomorphology to geology. An enthusiastic teacher and superb speaker, for almost three decades he taught many young geologists, mining engineers, and geographers. He wrote textbooks on physical geology and hydrogeology, and also edited the *Prodrome*, on the regional geology of Belgium.

In 1913, on a mission to central Africa, Fourmarier

traveled, mostly on foot, through bush and forest to the shores of Lake Tanganyika. There he mapped the Lukuga coalfield (now in Zaïre) and explored the Malagarassi Valley (now in Tanzania).

Having become acquainted with the intricate correlation of the mostly nonfossiliferous formations of central Africa, Fourmarier proposed a synthesis of the sometimes divergent proposals by drawing, in 1924, the first geological map of the Congo Basin, on the scale of 1:4,000,000; it was revised and amplified in 1930 on the scale of 1:2,000,000. The success of this work brought him the further task of organizing, in 1929, the Geological Commission (within the Colonial Office), which he headed for thirty years. He was coauthor of the International Geological Map of Africa, and honorary president of the Association of African Geological Surveys. He was also one of the founding members, in 1929, of the Royal Colonial Institute, which later became the Royal Overseas Academy of Sciences.

In the realm of theory, Fourmarier considered the megatectonic pattern of crustal architecture on a planetary scale. Ideas about the symmetry and permanence of continents and oceans, and the parallelism of folded belts tightly wrapped around the pre-Cambrian shields, underlie the three fundamental rules that he proposed to explain the evolution of the earth's crust. This was an advance on the ideas of Eduard Suess, Émile Haug, and Hans Stille, but Fourmarier was nevertheless strongly opposed, as were most geologists of the period, to the then misunderstood intuitions of Alfred Wegener.

These speculative endeavors did not prevent Fourmarier from returning frequently to the field. Short of build and seemingly frail, he was nevertheless an energetic and tireless walker. Throughout his teaching career, and long after his retirement in 1948, he continued to revise attentively the structures of the Paleozoics south and east of Liège, mapping in detail the fault traces, appraising their importance, and evaluating the possible role of cross faults.

His wide travels through many continents had brought him in touch with a variety of geological settings where his keen eye promptly recognized the structural pattern. This explains why among his many interests he took the trouble to bring together field observations concerning minor structural types such as slaty cleavage from far and wide. He pursued this endeavor with mixed success—he was not a microscopist—but with unabated enthusiasm until the very last years of his life.

Faithful to his motto "He who devotes heart and soul to scientific pursuits will keep working up to the threshold of the grave," he published six papers in 1969 and, despite his failing eyesight, left on his desk an unfinished manuscript when he died in January 1970 at the age of 92.

He was an inspiration to many younger geologists. Colleagues from all over the world heaped fame and honors upon him. An associate of the Institut de France, of the Spanish Academy at Madrid, and of many other academies and institutions, he received the Wollaston, Penrose, Gaudry, and van Watershoot van der Gracht medals and was awarded honorary doctorates by the universities of Paris, Lille, Grenoble, Caen, and Geneva. However, it is likely that he took greater pride in the British War Medal and the Médaille de la Résistance, awarded for his actions during World War I and World War II, respectively.

BIBLIOGRAPHY

I. Original Works. Only a few works by Fourmarier can be mentioned here; most were published in *Annales de la Société géologique de Belgique (ASGB)*, *Bulletin* or *Mémoires de la classe des sciences de l'Académie royale de Belgique* (*BARB* or *MARB*), or *Proceedings* of several International Geological Congresses.

On the Theux window, see "Le bassin dévonien et carboniférien de Theux," in *ASGB*, **28** (1901), 27–53; "La structure du massif de Theux," *ibid.*, **33** (1906), M109–139; and the map sheet Louveigné-Spa (with "Notice explicative") of *Carte géologique de la Belgique à l'échelle du 25.000-ème* (Brussels, 1958).

The Ardennes tectonic synthesis was first proposed in "La tectonique de l'Ardenne," in *ASGB*, **34** (1907), M15–123, and restated in "Vue d'ensemble sur la géologie de la Belgique," *ibid.*, in *Mémoires*, **8** (1933–1934), and in the chapter on tectonics in the *Prodrome* (see below). See also the *13th IGC Guidebooks* A1, B2, and C2 (1922); and *planche* 10—*tectonique* (with "Commentaire"), in *Atlas national de Belgique* (Brussels, 1953).

His *Principes de géologie* went through three editions (Paris, 1933; 2nd ed., 1944; 3rd ed., 2 vols., 1949–1950). His textbook *Hydrogéologie: Introduction à l'étude des eaux destinées à l'alimentation humaine et à l'industrie* went through two editions (Paris, 1939; 2nd ed., 1958). *Prodrome d'une description géologique de la Belgique* (Liège, 1954) remains the best source on the geology of Belgium.

A bibliography of his works on Africa is given in full in Denaeyer's obituary notice (see below). His geological map of the Belgian Congo was published in *Revue universelle des mines*, 7th ser., **4**, no. 4 (15 November 1924).

On the three fundamental rules of crustal evolution, see "Les idées actuelles sur les déformations de l'écorce terrestre et la dérive des continents," in *Revue universelle des mines*, 7th ser., **18** (1928), 39; "Essai sur la probabilité de l'existence d'une règle de symétrie dans l'architecture de l'écorce terrestre," in *MARB*, 2nd ser., **11**, fasc. 2 (1930), 3–46; "Recherches sur l'existence d'une règle de symétrie . . . ," in *World Engineering Congress, Tokyo* (1929); *Trois règles fondamentales de l'architecture de l'écorce terrestre* (Paris, 1932); "Recherches complémentaires sur l'existence d'une règle de symétrie . . . ," in *Proceedings of the Sixteenth International Geological Congress* (1935), 925–936; "La dérive des continents et la règle de symétrie" in *BARB*, 5th ser., **22** (1938), 1391–1414; "La règle de symétrie appliquée à la géomorphologie," in *Bulletin de la Société belge d'études géographiques*, **8** (1938), 20–40; "Efforts tangentiels et efforts verticaux dans la tectonique," in *ASGB*, **49** (1946), B88–182; "Le problème de l'origine des continents," in *BARB*, 5th ser., **48** (1961), 1368–1426; "L'arrangement systématique des continents et des océans," in *Mélanges de géographie physique, humaine, économique, appliquées offerts à M. Omer Tulippe* (Gembloux, 1967) 5–25; "Le problème de la dérive des continents," in *MARB*, 2nd ser., **17**, no. 2 (1969).

Papers relating to slaty cleavage include "Le clivage schisteux dans les terrains paléozoïques de la Belgique," in *Proceedings of the Thirteenth International Geological Congress* (1923), 517–530; "De l'importance de la charge dans le développement du clivage schisteux," in *BARB*, 5th ser., **9** (1923), 454–458; "Observations sur le développement de la schistosité dans les séries plissées," in *BARB*, 5th ser., **18** (1932), 1048–1053; "Schistosité, foliation et microplissement," in *Archives des sciences* (Geneva), **4** (1951), 5–23; "Schistosité et grande tectonique," in *ASGB*, **76** (1953), B275–301; "Le granite et les déformations mineures des roches," in *MARB*, **31** (1959), 3–101; "Les déformations mineures des roches . . . ," in *Memoire FALLOT*, Société géologique de France (1961), 57–81; "L'intérêt de l'étude des déformations mineures des roches . . . ," in *Revue des questions scientifiques*, **26** (1965), 438–517; and "La montée systématique des fronts de schistosité en rapport avec la granitisation," in *MARB*, **39** (1969).

II. SECONDARY LITERATURE. An obituary notice appeared in *Annuaire de l'Académie royale de Belgique* (Brussels) by Paul Michot, who has published a short obituary notice in *ASGB*, **93** (1970), 425–429. Notices also are in *Bulletin de la Société géologique de France*, 7th ser., **13** (1971) 205–213, with an extensive bibliography, by Léon Calembert, and in *Bulletin des séances de l'Académie des sciences d'Outremer à Bruxelles* (1971) 71–85, with a complete bibliography of works relating to Africa, by Marcel E. Denaeyer.

PIERRE DE BÉTHUNE

FRÉCHET, RENÉ MAURICE (*b*. Maligny, France, 10 September 1878; *d*. Paris, France, 4 June 1973), *mathematics*.

Maurice Fréchet (he never used the name René) was the fourth of six children born to Protestant parents of modest means. His father, Jacques Fréchet, was director of a Protestant orphanage. The family moved to Paris while Maurice was still a boy. There his father was a schoolteacher and his mother, Zoé, ran a boardinghouse for foreigners. At the Lycée Buffon in Paris the young Fréchet was taught mathematics by Jacques Hadamard, then in his twenties, who perceived his pupil's precocity and gave him special attention. This tutelage extended to encouragement well beyond the ordinary student-teacher relationship and continued (by correspondence) even after Hadamard's appointment as professor at Bordeaux in 1894. Fréchet entered the École Normale Supérieure in 1900, graduating in 1903. In the winter of 1903–1904 he wrote up the lectures of Émile Borel that were turned into a book, *Leçons sur les fonctions de variables réelles et les développements en séries de polynômes* (1905). This work was part of a long and close relationship between Borel and Fréchet that continued as long as Borel lived.

Fréchet wrote his doctoral thesis (completed in 1906) as a student of Hadamard, by that time a professor in Paris. Under the influence of Hadamard and as a result of reading the work of Vito Volterra, Fréchet had become interested in what was then called the "functional calculus," the study of numerically valued functions (Hadamard called them functionals) defined on a class of ordinary functions (or, perhaps, curves). Fréchet made the bold jump to numerical functions defined on an abstract class. In order to be able to speak about limits and continuity, he introduced several ways of developing a theory of point set topology in an abstract space. The most important of these involved what came to be called a metric space, a concept invented by Fréchet.

In his thesis Fréchet was the first to demonstrate conclusively the feasibility and fruitfulness of developing in abstract spaces an effective generalization of the point set topology that had been developed for Euclidean spaces. Following Georg Cantor, Fréchet took as fundamental the concept of a derived set of a given set. Fréchet introduced the concepts of compactness, separability, and completeness. He established the connection between compactness and what later came to be known as total boundedness. He also established, under certain conditions, the linkage in metric spaces between two properties of a set: that of being closed and compact, and that of having the property concerning open coverings that is expressed in the Borel-Lebesgue (also called Heine-Borel) theorem for closed and bounded sets in Euclidean space.

Hadamard, reporting to the Académie des Sciences in 1934 on the work of Fréchet (in connection with the latter's candidacy for election to the Section de Géométrie of the academy), said that the boldness of Fréchet's effort of abstraction was without precedent since the work of Évariste Galois. (The report by Hadamard, as well as one by Borel, can be found in the archives of the academy in Paris.) Jean Dieudonné, in his *History of Functional Analysis* (1981), wrote that the blend of algebra and topology that has come to be known as functional analysis emerged largely because of a sudden crystallization of ideas that was essentially due to the publication of four fundamental papers, of which one was Fréchet's thesis (1906), the others being works by Ivar Fredholm (1900), Henri Lebesgue (1902), and David Hilbert (1906).

From 1907 to 1910 Fréchet held teaching positions at lycées in Besançon and Nantes and at the University of Rennes. He married Suzanne Carrive in 1908, and they had four children. He held a professorship at Poitiers from 1910 to 1918, but was on leave in military service throughout World War I, mainly as an interpreter with the British army. From 1919 to 1928 he was head of the Institute of Mathematics at the University of Strasbourg.

The most important further accomplishments of Fréchet in his pre-Paris years were: (1) the representation theorem (1907) for continuous linear functionals on the vector space of functions now known as L^2 (a theorem discovered independently at almost exactly the same time by F. Riesz); (2) the formulation of the concept of the differential of a function (1911 and 1925), generalizing the concept from ordinary calculus, in a way suitable for use in the general analysis of functions in abstract normed vector spaces; (3) the formulation in 1915 of an imporant generalization of the work of J. Radon, showing how to extend the work of Lebesgue and Radon to the integration of real functions on an abstract set without a topology, using merely a generalized measurelike set function; (4) a theory of an abstract topological space, now known as a T_1-space, which Fréchet called an *H*-class, or, alternatively, an accessible space, which he first defined in 1918 using axioms about derived sets, and in 1921 using axioms about neighborhoods. Such a topological space is more general than the type of topological space introduced in 1914 in a book by Felix Hausdorff.

Fréchet moved to the University of Paris in 1928, where he taught until his retirement in 1949 (he was made honorary professor in 1951). His book *Les espaces abstraits* was published before his move. It is devoted almost exclusively to his work on general topology. Fréchet's early influence as the pioneer of an effective theory of topology in abstract spaces was substantial, but in time his influence was superseded by that of Hausdorff, whose book became an important resource for students and scholars, and in the 1920's by the brilliant work of two young Russians, Paul Aleksandrov and Paul Urysohn, who were notably influenced by both Fréchet and Hausdorff.

Fréchet's career in Paris was mainly devoted to probability theory and its applications. This interest had developed while he was at Strasbourg. Following Borel's wishes, Fréchet wrote expositions of probability theory in two books (1937–1938). Although he published prolifically on probability and statistics, bringing functional analysis to bear, his contributions in these fields did not match in originality and importance his early work on topology and general analysis.

Apart from his standing with Borel and Hadamard and certain other good friends among French mathematicians, Fréchet did not secure as much approbation in France as he did abroad, notably in America and Poland. He lost out repeatedly to more classical analysts in running for election to the Académie des Sciences. He was finally elected in 1956 to the vacancy created by the death of Borel. He was a chevalier of the Légion d'Honneur, was elected to the Polish Academy of Sciences in 1929, and was an honorary member of the Royal Society of Edinburgh and a member of the International Institute of Statistics.

BIBLIOGRAPHY

I. ORIGINAL WORKS. Fréchet's thesis was published as "Sur quelques points du calcul fonctionnel," in *Rendiconti del Circolo matematico di Palermo*, **22** (1906), 1–74. Two works dealing with the representation theorem for functionals on L^2 are "Sur les ensembles de fonctions et les opérations linéaires," in *Comptes rendus de l'Académie des sciences, Paris*, **144** (1907), 1414–1416; and "Sur les opérations linéaires," in *Transactions of the American Mathematical Society*, **8** (1907), 433–446. Other works include "Sur la notion de différentielle," in *Comptes rendus de l'Académie des sciences, Paris*, **152** (1911), 845–847; "Sur l'intégrale d'une fonctionelle étendue à un ensemble abstrait," in *Bulletin société mathématique de France*, **43** (1915), 248–264; "Sur la notion de voisinage dans les ensembles abstraits," in *Bulletin des sciences mathématiques*, **42** (1918), 138–156.

"Sur les ensembles abstraits," in *Annales scientifiques de l'École normal supérieure*, 3rd ser., **38** (1921), 341–388 (includes a characterization of *H*-classes, which are also known as T_1-spaces); "La notion de différentielle dans l'analyse générale," *ibid.*, **42** (1925), 293–323; *Les espaces abstraits et leur théorie considérée comme introduction à l'analyse générale* (Paris, 1928); *Notice sur les travaux scientifiques de Maurice Fréchet* (Paris, 1933); *Recherches théoriques modernes sur le calcul des probabilités*, 2nd ed., I, *Généralités sur les probabilités: Éléments aléatoires*, II, *Méthode des fonctions arbitraires: Theorie des événements en chaîne dans le cas d'un nombre fini d'états possibles* (Paris, 1950–1952).

II. SECONDARY LITERATURE. Obituaries are by Daniel Dugué in *International Statistical Review*, **42** (1974), 113–114; by Szolem Mandelbrojt in *Comptes rendus de l'Académie des sciences, Paris, Vie académique*, **277** (19 November 1973), 73–76; and by F. Smithies in *Year Book of the Royal Society of Edinburgh, 1975* (1975), 31–33. See also L. C. Arboleda, "Les débuts de l'école topologique soviétique: Notes sur les lettres de Paul S. Alexandroff et Paul S. Unysohn à Maurice Fréchet," in *Archive for History of Exact Sciences*, **20** (1979), 73–89; Jean Dieudonné, *History of Functional Analysis* (Amsterdam, 1981); Angus E. Taylor, "A Study of Maurice Fréchet, I, His Early Work on Point Set Theory and the Theory of Functionals," "II, Mainly About His Work on General Topology, 1909–1928," and "III, Fréchet as Analyst, 1909–1930," in *Archive for History of Exact Sciences*, **27** (1982), 233–295, **34** (1985), 279–380, and **37** (1987), 25–76.

ANGUS E. TAYLOR

FREUDENBERG, KARL JOHANN (*b.* Weinheim, Germany, 29 January 1886; *d.* Weinheim, 3 April 1983), *organic chemistry*.

The third of ten children of Hermann Ernst Freudenberg, co-owner of one of the biggest German tanneries and leather plants, and Helene Siegert, daughter of a professor at the Düsseldorf Academy of Arts, Freudenberg was educated in Weinheim and Frankfurt am Main. In 1904 he enrolled at Bonn to study science and botany. He soon concentrated on chemistry, which he pursued at the University of Berlin beginning in 1907. In 1910, Freudenberg completed his doctoral dissertation on Chinese gallotannin under Emil Fischer, then the unquestioned leader of Germany's organic chemists. In July of the same year he married Doris Nieden; they had three daughters and two sons.

Freudenberg's early work on tannins and the stereochemistry of organic acids followed very much in the footsteps of Fischer, whose assistant he was from 1910 to 1914. Eventually he felt the need to escape from his teacher's overwhelming influence and moved to the University of Kiel in April 1914, becoming *Privatdozent* there in July. The war soon interrupted his career; and for some time Freudenberg, like many other chemists, served as an adviser on gas warfare. After returning to Kiel, he received the title professor in 1919. In the autumn of 1920 he went to Munich to work with Richard Willstätter at the State Chemical Laboratory. In October 1921 he was appointed extraordinary professor of organic chemistry at the University of Freiburg im Breisgau, and one year later became full professor and director of the Chemical Institute at the Technische Hochschule in Karlsruhe. In April 1926, Freudenberg succeeded Theodor Curtius in the chemistry chair at the University of Heidelberg. He retained this position until his retirement in 1956, except for a sabbatical spent at the University of Wisconsin, Madison, and the Johns Hopkins University, in 1931. Until 1969 he continued working in Heidelberg as head of the Research Institute for the Chemistry of Wood and Polysaccharides, which had been created for him in 1938.

Most of Freudenberg's early work sprang from the foundation laid by Emil Fischer, but he soon extended the classical approach to include the more complicated and less easily obtainable carbohydrates. The combination of chemical and spectroscopic methods enabled him to assign absolute configurations to carbohydrates, terpenes, and steroids. Much effort was spent on the preparation of acetone sugars and their use in syntheses. Freudenberg's studies on tannins and their relations to catechins, dating back to his dissertation and taken up again after World War II, were first summarized in *Die Chemie der natürlichen Gerbstoffe*, which he wrote while a *Privatdozent* at Kiel. The comprehensive handbook *Stereochemie*, which he contributed to and edited in 1933, was a lasting success.

In 1916 the wartime need for a substitute for vegetable tannins made Freudenberg familiar with the main components of wood tissue—cellulose and lignin—on which he later did the work that brought him fame as a chemist. As early as 1921 he concluded from quantitative acetolysis that cellulose is a linear polymer composed of glucose residues linked throughout by covalent glycosidic bonds. In 1928 he published the first correct formula. In the same year the controversy between Hermann Staudinger and Mark and Meyer about the macromolecular concept emerged. Freudenberg, however, bred in the preparative tradition and not overly concerned with priority claims, was little inclined to enter this dispute. Instead he proceeded to extend the idea of a highly ordered cellulose structure to other natural polymers, and in 1930 he was able to confirm that starch is a chain-type macromolecule. For the next

β H α	β H α	β H α
C	C	C
HCOH	HCOH	HCOH
HOCH	HOCH	HOCH
HCO—	HCO—	HCO—
HCO—	HCO—	HCO—
CH_2OH	CH_2OH	CH_2OH

Source: K. Freudenberg and E. Braun, "Methycellulose," *Annalen der Chemie*, **460** (1928), 296.

ten years Freudenberg studied Schardinger dextrines produced from starch, and in 1939 he inferred that they are cyclic saccharides with a helical structure, which explained the starch-iodine test result as an inclusion compound. In the 1930's this research was complemented by extensive studies on the spectroscopic behavior of carbohydrates that were carried out in close cooperation with the physical chemist Werner Kuhn.

Freudenberg's method of pursuing different analytical and synthetical pathways simultaneously revealed its value when he brought together this accumulated expertise in order to determine the structure of lignin, an extremely complicated natural polymer responsible for the stability of vascular plants and therefore one of the most abundant natural products. However, its separation and chemical treatment presented major problems. New analytical methods, degradation procedures, and model syntheses had to be developed. They supported Freudenberg's assumption that coniferyl alcohol was a major precursor of lignin, an assumption he was able to confirm by simulating the enzymatic lignification in the laboratory, interrupting the polymerization, and identifying the oligomeric intermediates. For almost three decades, however, progress on the lignin problem was disappointingly slow. In 1952 Freudenberg achieved the first real breakthrough when he succeeded in isolating the oligomers, determining their structure, and explaining the mechanism of their formation through intermediate radicals. Lignin then became the main field of his research, which from 1962 to 1965 culminated in the design of a formula scheme for the constitution of spruce lignin, involving eighteen C_6—C_3 units.

Over a period of sixty years Freudenberg published more than 450 papers as well as several books and received international recognition, including foreign membership in the Royal Society in 1963. His talent also was applied in academic and civic affairs. He served as rector of Heidelberg University in 1949 and 1950 and as a member of the City Council from 1951 to 1956, primarily engaging in city planning and adult education. He also devoted a considerable amount of time and energy to local and family history, as well as to the history of chemistry.

BIBLIOGRAPHY

I. ORIGINAL WORKS. The most comprehensive bibliography (on microfiche) is in the memoir by Stevens (see below). Freudenberg's books include *Die Chemie der natürlichen Gerbstoffe* (Berlin, 1920), revised as *Tannin, Cellulose, Lignin* (Berlin, 1933); *Stereochemie, eine Zusammenfassung der Ergebnisse, Grundlagen und Probleme* (Leipzig and Vienna, 1933); and *Constitution and Biosynthesis of Lignin* (New York, 1968), written with Arthur C. Neish, which contains Freudenberg's most detailed account of his research. For an earlier presentation, see Freudenberg's "Lignin im Rahmen der polymeren Naturstoffe," in *Angewandte Chemie*, **68** (1956), 84–92.

Freudenberg's personal papers, including an unpublished autobiography, are held by his family.

II. SECONDARY LITERATURE. The most comprehensive biography is T. S. Stevens, "Karl Johann Freudenberg," in *Biographical Memoirs of Fellows of the Royal Society*, **30** (1984), 167–189, with bibliography on microfiche. *Angewandte Chemie*, **68** (1956), 81–120, is dedicated to Freudenberg and contains articles by B. Helferich, W. Kuhn, O. T. Schmidt, and F. Cramer, on various aspects of his work. Another account is A. Wacek, "Karl Freudenberg zum 70. Geburtstag," in *Österreichische Chemiker-Zeitung*, **57** (1956), 33–38. His life and personality are portrayed in Friedrich Cramer, "Leben und Werk von Karl Freudenberg," in *Heidelberger Jahrbücher*, **28** (1984), 57–72.

CHRISTOPH MEINEL

FRISCH, KARL RITTER VON (*b*. Vienna, Austria, 20 November 1886; *d*. Vienna, 12 June 1982), *experimental zoology, comparative physiology, ethology*.

Karl von Frisch is remembered for his pathfinding studies of the sensory capabilities and behavior of animals, particularly fish and honeybees. Through ingenious experiments of classical simplicity, informed by an evolutionary biologist's sense of function and a physiologist's concern with biological mechanisms, he greatly enlarged the scientific understanding of animal behavior. His remarkable discovery of the dance "language" of the honeybee is generally regarded by ethologists as the single most important contribution to the study of animal behavior of the twentieth century.

Frisch was the youngest of four sons of Anton Ritter von Frisch and Marie Exner von Frisch. Three generations of physicians preceded him on his father's side of the family. Careers in academia predominated on his mother's side of the family. His father was a surgeon and urologist who conducted significant research in anatomy and bacteriology, and became professor of surgery at the Vienna General Polyclinic. His paternal grandfather, also named Anton, was a surgeon in the Austrian Imperial Army who received the Order of the Iron Cross, and with it a hereditary knighthood, for his work in reorganizing the Army Medical Corps.

Karl's mother, a woman of keen intellect, was the daughter of Franz Exner, professor of philosophy at the University of Prague. All four of her brothers became professors. The most important of these for Karl's development as a scientist was Sigmund Exner, professor of physiology and director of the Physiological Institute at the University of Vienna. Another of Karl's uncles, Franz Serafin Exner, was professor of physics and director of the Physics Institute, also at the University of Vienna. This scholarly tradition continued among Karl's brothers: Hans von Frisch became a professor of constitutional law (initially at Basel, eventually at Vienna); Otto von Frisch became assistant professor of surgery and director of the Rudolfinerhaus clinic in Vienna; and Ernst von Frisch became director of the Salzburg Studienbibliothek.

Frisch attended the Benedictine Schottengymnasium in Vienna and then enrolled in 1905 as a medical student at the University of Vienna, following his father's wish that he become a physician. Zoology was his primary interest, however, and in 1908 he transferred to the Zoological Institute at Munich to study experimental zoology under Richard Hertwig. In 1909 he returned to Vienna to write his doctoral dissertation under Hans Przibram at the Biological Experimentation Institute. He received his doctorate from the University of Vienna in 1910. Frisch went back to Munich in the fall of 1910 as an assistant to Hertwig and became a *Privatdozent* at Munich in 1912. During World War I he worked at the Rudolfinerhaus, the hospital directed by his brother Otto. He was appointed assistant professor at the University of Munich in 1919, professor of zoology and director of the Zoological Institute at Rostock University in 1921, and professor of zoology at the University of Breslau in 1923. Upon Hertwig's retirement in 1925, Frisch succeeded him as professor of zoology and director of the Zoological Institute at the University of Munich.

In the early 1930's Frisch secured funds from the International Education Board of the Rockefeller Foundation and from the state of Bavaria to replace the old Zoological Institute at Munich with a new one. The new institute, completed in 1932, was the most up-to-date zoological institute in Europe. After the coming to power of the Nazis, however, Frisch was unable to choose his own staff freely, and in 1941 he learned that the Ministry of Education was planning to remove him from his professorship because his maternal grandmother, Charlotte Dusensy Exner, had been of Jewish ancestry. The intended dismissal did not come to pass, however, because the economic and political influence of powerful friends, as reported by Frisch in his autobiography, was brought to bear at the Ministry of Food and at Nazi party headquarters. Frisch's expertise with bees stood him in good stead at the Ministry of Food, which valued him because an epidemic was devastating the bee populations of Europe and causing serious damage to German agriculture. The ministry gave Frisch an official assignment to study what could be done to combat the problem.

In July 1944 Frisch's house in Munich was destroyed and the Zoological Institute was severely damaged by Allied bombing. Frisch had already moved much of his equipment to his summer home in Brunnwinkl, Austria, and he continued his research there. In 1946 he accepted the professorship of zoology and directorship of the Zoological Institute at the University of Graz (Austria). In 1950, however, he returned to Munich, where the Zoological Institute had been largely rebuilt. He remained at Munich as professor of zoology and director of the Zoological Institute until his retirement in 1958.

Frisch married Margarethe Mohr on 20 July 1917. They had three daughters and one son. In the course of his long and distinguished career, Frisch received many honors, including the Nobel Prize for Medicine or Physiology, which he shared with Konrad Lorenz and Nikolaas Tinbergen in 1973. He was elected a foreign member of the Royal Society of London and belonged to numerous other scientific academies in Europe and the United States. He received honorary degrees from six universities. His other prizes included the Order of Merit for Sciences and Arts (1952), the Kalinga Prize for the popularization of science (1959), the Austrian Medal of Honor for Science and Art (1960), the Balzan Prize for Biology (1963), and the Distinguished Service Cross with Star and Ribbon of the Order of Merit of the German Federal Republic (1974). He was president of the German Zoological Society from 1928 to 1929.

Early Scientific Work. Like most of the main

contributors to the biological study of animal behavior in the first half of the twentieth century, Frisch was exceptionally interested in animals as a child. His mother, who possessed a strong love of nature, permitted him to maintain a sizable menagerie, which in his secondary school days included, by his later count, nine different species of mammals, sixteen species of birds, twenty-six species of cold-blooded terrestrial vertebrates, twenty-seven species of fish, and forty-five species of invertebrates. He also took great pleasure in collecting animals at his family's summer home at Brunnwinkl, on Lake Wolfgang in the Salzkammergut region of Austria.

In Frisch's three years of medical studies at Vienna, the teacher who influenced him most was his uncle Sigmund Exner. Exner, a pioneer in the study of comparative physiology, impressed upon his nephew the importance not only of designing careful experiments but also of being cautious in drawing conclusions. Under his uncle's direction, Frisch investigated how the position of the pigment cells in the compound eyes of butterflies, beetles, and shrimp shifted in response to light and darkness. These investigations were the subject of his first major scientific paper, which appeared in *Biologisches Zentralblatt* in 1908.

Experimental zoology was the forte of Richard Hertwig, under whom Frisch studied at Munich. Hertwig stressed that zoology had to go beyond the mere description of forms to an understanding of causal relationships. Such an understanding was to be achieved primarily through experimentation. Frisch also benefited greatly from studying with Richard Goldschmidt, who was Hertwig's chief assistant, and Franz Doflein, who, as assistant professor of the systematics and biology of animals, led Frisch and his fellow students on numerous excursions into the field.

When Frisch returned to Vienna to write his doctoral dissertation under Hans Przibram at the Biological Experimentation Institute, Przibram selected a morphological topic for him. Frisch chose instead to study nerve pathways and color changes in fish, a topic in which he had become interested after observing the ability of minnows to adapt their coloration to light or dark backgrounds. His *Habilitationsschrift* at Munich was also on color adaptation in fish. Published in 1912, it included the results of research he had carried out in the spring of 1911 at the Naples Zoological Station.

Crucial for the direction of Frisch's researches as a young scientist was the stimulus he received from the work of Carl von Hess, director of the Munich Eye Clinic. A prolific researcher, Hess studied phototactic reactions in a wide range of animals. He found that fish and invertebrates appeared to perceive the brightness values of the colors of the spectrum in the same way as a totally color-blind human did: colors in the gold-green to green region appeared brightest, and the spectrum was shortened at the red end. On the basis of this correspondence, Hess concluded that fish and invertebrates were totally color-blind.

Having studied the ability of fish to adapt themselves to the color of their surroundings, Frisch could not believe that they were unable to perceive colors. In June 1911, at the annual meeting of the German Zoological Society, held in Basel, he delivered a lecture entitled "On the Color Sense of Fish," in which he challenged Hess's thesis. The fact that many fish possess an ability to change color and many of them assume splendid colors at spawning time made it difficult for Frisch to believe that fish are color-blind, but he wished to judge the issue primarily on experimental grounds. His experiments showed that fish can adapt not only their lightness or darkness but also their color to their surroundings. He was able to show further that fish did not respond simply to the brightness of colors but also to the colors themselves. The gold and red pigment cells expanded in response to gold or red backgrounds but not to a gray background of equal brightness. That this response was mediated by the fish's eyes was shown by the fact that fish that had been blinded did not display color changes. For Frisch these results strongly confirmed the idea that fish possess a color sense.

Hess responded by calling Frisch's results "completely mistaken," and an acrimonious dispute ensued between the two that continued for more than a decade. Though the dispute began over the color sense of fish, it soon opened up on a second front: the question of the color sense of bees.

Frisch's dispute with Hess over the color sense of bees was not without precedent. As early as 1793, in a book entitled *Das entdeckte Geheimniss der Natur im Bau und in der Befruchtung der Blumen*, Christian Konrad Sprengel proposed that the bright colors and special "honey guides" of flowers were contrivances designed to bring about the fertilization of flowers by insects. Charles Darwin supported Sprengel's idea that the colors of flowers served to attract insect pollinators, but unlike Sprengel he explained this in terms of natural selection rather than design on the part of a wise Creator. Darwin was more interested in the special mechanisms that assured cross-fertilization in plants than he was in how insects are attracted to flowers in the first place, and he did not test whether bees are capable of distinguishing colors.

Darwin's contemporary Sir John Lubbock did precisely that. He succeeded in training bees to associate food with certain colors. His experiments led him to conclude not only that bees can distinguish colors but also that they have color preferences. Lubbock's work was challenged by the French biologist Félix Plateau, who maintained, on the basis of his own experiments, that insects are guided to flowers not by colors but by the smell of nectar or pollen. Between 1900 and 1910 the issue was taken up by scientists in Germany (Albrecht Bethe and Hugo von Buttel-Reepen), Switzerland (Auguste-Henri Forel), and the United States (Charles Henry Turner).

Frisch's debate with Hess over the color sense of bees led him to the first of his classical experiments on honeybee behavior. Beginning in the summer of 1912, he arranged on a feeding table a series of gray squares representing a gradation in brightness from white to black. Among the gray squares he placed a blue square of equal size. He then placed a watch glass on each square and filled the watch glass on the blue square with sugar water. The bees were allowed to feed on the sugar water. New, clean squares and watch glasses were then substituted for the old ones, the relative positions of the squares were changed, and the process was repeated. Frisch found that after a training period of one to two days, the bees would congregate on a blue square when it, like the gray squares among which it was placed, had only a clean and empty watch glass on it. This showed, he believed, that bees are able to distinguish a blue square from a gray square of a comparable brightness, and that bees thus have a color sense. That the bees were not attracted to a special smell associated with the blue square was demonstrated by the fact that the bees were still attracted to the blue square when it and the gray squares were all covered with a large glass plate.

In 1914 Frisch published a 188-page monograph, *Der Farbensinn und Formensinn der Biene*. He reported that while bees could distinguish the color blue from all shades of gray, they confused red with black and blue-green with gray. Similarly, they mistook red-orange for yellow or green, and blue for violet or purple. He concluded that bees have a color sense, but that it is like that of a red-green color-blind man rather than a man with normal vision. He noted, further, that the colors of flowers and the color sense of bees are closely correlated: colors that are seldom seen in flowers are colors that bees cannot distinguish. This supported the idea that the colors of flowers have developed as adaptations to the color sense of their pollinators. In the following decade, investigators using more refined apparatus than Frisch's soon revised and extended his findings (identifying, among other things, that bees perceive blue-green and ultraviolet as colors), and in the 1950's and 1960's Frisch's student Karl Daumer further extended the study of the bee's color sense and the biological significance of floral colors.

Frisch also reported on his efforts to train bees to respond to different shapes. He was unsuccessful in his attempts to get them to distinguish between different geometric shapes, and suggested that they could not distinguish between forms that were foreign to them. This interpretation was modified in the 1920's and 1930's by the researches of Mathilde Hertz, who found that what seemed to matter most for the form sense in bees was the degree of "brokenness" of the patterns.

Hess's response to Frisch's experiments on the color sense of fish and bees was to maintain, without looking at the experiments, that Frisch's work was flawed or incorrectly reported. Hertwig supported Frisch by attending the experiments and attesting to the accuracy of the reported results. The debate had the positive effect of bringing attention to Frisch's work, and it stimulated additional studies on color vision in animals by others. In a paper of 1923 reviewing the debate on the color sense in animals over the previous decade and a half, Frisch was able to list not only eight papers of his own and twenty-four of Hess's but an additional thirty-three publications produced by twenty-eight other authors.

The sensory physiology of bees offered Frisch further topics for study. During a brief leave from his wartime hospital duties, he began using food-training techniques to test the honeybee's sense of smell. In 1919 he published a 238-page monograph on the bee's sense of smell and its significance for the biology of flowers. Through a wealth of experiments he found not only that bees could be trained to particular odors but also that their olfactory reactions appeared to be very similar to those of humans, in that substances that smelled similar to humans evidently smelled similar to bees, and substances that had no smell for humans had no smell for bees. He conducted experiments that established that the bee's sense of smell is located on its antennae. He also undertook comparative experiments that showed that although the color of flowers can attract bees from a greater distance, it is the scent that bees use to identify flowers at close range.

First Studies of the "Language" of the Bees. In

the course of his studies on the senses of color and smell in bees, Frisch observed that a food source set out in a new location often remained undiscovered by scout bees for hours or even days, but that once one bee discovered the food, there would soon be dozens of bees feeding upon it. He also observed that after a troop of bees that had been trained to feed at a particular spot exhausted the food there, the number of bees at the spot quickly dropped off to zero; but as soon as the food was replenished, and a scout bee discovered the renewed supply and returned to the hive laden with nectar, the whole troop would soon be back at the feeding place. It thus appeared to Frisch that bees somehow communicated to their hive mates the existence of worthwhile food sources.

In the spring of 1919 Frisch set out to discover how this communication took place. Using a queen-breeding cage with a single honeycomb that could be observed through glass on both sides, he was able to witness what he later described in his autobiography as "the most far-reaching observation" of his life. He saw a scout bee return from a freshly replenished feeding station and perform on the honeycomb a lively *Werbetanz* (courting dance), which excited the bees around her and caused them to fly out to the feeding station. He spent the next three years studying this communication system. He set up an observational hive in which the honeycombs were arranged side by side so they could all be observed through a glass window. He also devised a system of numbering bees by painting them with spots of different colors so they could be individually identified.

Frisch first reported on these researches in a lecture to the Society for Morphology and Physiology at Munich in January 1920. He followed this with two more reports in 1921 and 1922, and in 1923 he published a 186-page monograph entitled "Über die 'Sprache' der Bienen: Eine tierpsychologische Untersuchung." It turned out that bee "dances" had been described by observers in the eighteenth and nineteenth centuries. Frisch's special contribution was his analysis of what the dances signified.

In speaking of a "language" of the bees, Frisch was responding to Buttel-Reepen's suggestion of 1915 that bees make perceptibly different sounds—a flight tone, a stinging tone, a hunger tone, and so forth—that constitute a kind of language for them. Frisch found no evidence for what Buttel-Reepen called the *Lautsprache* (audible language) of the bees. He concluded instead that the "language" bees employed to communicate the existence of food sources involved two special dances. The first of these was a "round dance" in which the forager tripped around in a tiny circle on a cell of the honeycomb, alternately to the left and to the right, continuing this for up to half a minute or longer. The second was a *Schwänzeltanz* (waggle or tail-wagging dance), in which the returning forager performed half-circle turns, alternately to the left and the right, between which she did a short, straight-line run while wagging her abdomen rapidly from side to side.

In each case, neighboring bees trooped excitedly along behind the dancer. Frisch interpreted the round dance as a dance performed by nectar collectors and the tail-wagging dance as a dance performed by pollen collectors. Each served, he believed, to recruit additional bees to go out and collect food of the same sort as that brought back to the hive by the dancer. Early in his researches Frisch thought there was a third dance as well, but his later observations failed to confirm this.

It appeared to Frisch that the means by which new bees were recruited to seek a food supply and the means by which they proceeded to find the supply were two different things. His observations showed that the new recruits did not simply accompany the dancer back to the source. Instead, the recruits flew out from the hive in all directions, searching for the particular flower scent that had clung to the dancer. If the source happened to be a flower that was odorless, the recruits were aided in their quest by the scent of their predecessors, for bees in the presence of an excellent food source will discharge their scent organs, thereby impregnating the air with a special aroma that serves to attract their fellow workers.

Frisch acknowledged at a meeting of the German Society for Natural Science and Medicine at Innsbruck in 1924 that at the beginning of his work he had actually contemplated the possibility that the dancer might use a "secret sign" to indicate the direction and distance of the food. This was not the conclusion to which his initial studies on the "language" of the bees had led him. Nonetheless, what he had found was quite remarkable. He had found that honeybees, through their symbolic dances and the comparatively simple means of touch and smell, were able to communicate to bees back in the hive not only the existence of profitable food sources but also the species of flowers with which the food was associated. In this way they could effectively exploit the food sources in their surroundings as different species of plants came into flower. Not until two decades later did he discover that bees can indeed communicate to their hive mates the direction and the distance of their feeding places. Work done primarily in the 1950's established further that the dances can serve to communicate information not only about food but also about water, resinous substances, and potential nest sites.

Other Experimental Studies. Frisch's researches on bees were necessarily restricted to the warm months of the year, when bees are active. For his laboratory work in the colder months, fish were his favorite, though not exclusive, experimental subjects. He also conducted a study on color changes in salamanders, confirming Paul Kammerer's report that raising salamanders on backgrounds of different colors affected the animals' coloration. In 1923 Frisch took up the long-controverted question of whether fish, which lack the basic auditory structures of the higher vertebrates, are capable of hearing. Using food-training techniques, he was able to teach a catfish to respond to a whistle.

Frisch proceeded to explore in detail the sensory physiology of fish. Combining histological examinations of the retina of fish with color-training experiments on fish in bright and dim light, he was able in 1925 to provide strong support for the duplicity theory of vertebrate vision. He demonstrated that the rods take over for the cones in the fish retina at just the time that the fish in semidarkness becomes unable to distinguish colors. Later, performing surgical operations on the inner ears of fish, he was able to show that one part of the labyrinth serves for hearing while another part serves the sense of balance. He established that minnows cannot localize sound. Along different lines, he discovered that when a minnow's skin is punctured, a chemical substance is released that serves as a warning to other minnows. This was a fundamental discovery for subsequent work by others on the role of alarm substances in animal social behavior.

These various studies of Frisch's, like his work on bees, illustrate the kind of comparative physiological investigations at which he excelled. Most zoologists in his day, as he saw it, were preoccupied with the study and description of form. The pioneers of comparative physiology tended to be physiologists working in medical faculties. Perceiving the need for a journal of comparative physiology, Frisch and Alfred Kühn persuaded the publisher Dr. Ferdinand Springer to start the *Zeitschrift für vergleichende Physiologie*. Frisch served as an editor of this journal, now called *Journal of Comparative Physiology*, from its inception in 1924 until his death in 1982.

The "Language" and Orientation of the Bees Revisited. Frisch continued his honeybee studies in the 1920's and 1930's, devoting his attention in particular to a long course of investigations on the bee's sense of taste that laid the foundations for work on the subject thereafter. In the 1940's two factors led him to take up the question of the bee's "language" again. One was a doctoral dissertation written at Bonn in 1938 by Christoph Henkel, who maintained that Frisch had been mistaken in supposing that the bee's round dance signified a nectar source while the waggle dance signified a pollen source. Henkel believed that round dances simply indicated an unnatural abundance of food. The second was the reported successes of Russian scientists in increasing clover production by training bees to clover-flavored sugar water. Frisch's initial testing of Henkel's view seemed only to confirm the original distinction between the dances of nectar gatherers and pollen gatherers, but in the course of trying to train bees to visit distant feeding places, he made a discovery that opened up a new set of researches on the dance language and orientation of bees.

The discovery was made in August 1944 at Brunnwinkl, where Frisch found it still possible to work despite the dislocations caused by the war. For the first time in the course of his researches, he set up a feeding station at a distance of more than 100 meters from the hive. Previously he had always set the feeding station close enough to the hive so that both could be watched at the same time. He was surprised to find newly recruited bees not searching for food in ever-widening circles from the hive but instead flying to the immediate proximity of the feeding place. Doubting that the direction of a food source could be communicated by a scout to her hive mates, in September 1944 he tested the more likely explanation that the scout bees were emptying their scent organs in flight, thus providing a trail that the other bees followed. This, however, proved not to be the case, for shellacking the bees' scent organs shut did not keep new recruits from arriving at the feeding station.

Frisch then constructed an observation hive to see whether bees returning to the hive from different distances behaved differently from one another. He set up feeding stations at 12 and 280 meters from the hive, and marked the bees that fed at the two stations with blue and red spots, respectively. He found, to his amazement, that all the bees that had been feeding at the nearer source performed "round" dances when they returned to the hive, while all those that had been feeding at the more distant source performed "waggle" dances. He made these observations on 6 October 1944. In the short time that remained before cold weather called a halt to his researches for the rest of the season, he was able to establish that the round dance merged into the waggle dance at a distance of 50 to 100 meters from the hive. He had not resolved the question, however, of whether bees had a means of communicating direction as well as distance.

Eagerly resuming these studies in June 1945, Frisch observed that the direction of the bees' waggle runs on the honeycomb in the hive changed during the course of the day, though the location of the feeding place remained fixed. He discovered that the dances were oriented with respect to the position of the sun. Thus, for example, a waggle run directly upward on the honeycomb in the dark hive indicated that the food source was in the direction of the sun, while a run 60 degrees to the left of vertical indicated a food source 60 degrees to the left of the sun. When bees danced outside the hive on a horizontal surface—as opposed to inside the dark hive on a vertical honeycomb—they pointed directly toward the food source. In other words, the sun appeared at the same angle and on the same side to them as it had on their outward flight to the food.

Continuing his analysis of how bees communicate distance, Frisch increased the distance of the feeding places from the hive to see if the form of the dance would change again. Perceiving a difference in the rhythm of the dances of bees returning from 100 meters away compared with those of bees returning from unknown distances, he used a stopwatch to record the frequency of the dance turns of bees returning from feeding stations at 100-meter intervals from 100 to 1,500 meters away from the hive. He found that the number of round dance movements per unit time decreased with distance, while the number of waggle movements in the waggle runs and the duration of the runs increased.

Having established that nectar collectors performed waggle dances as well as round dances, Frisch tested to see if pollen collectors also performed both dances. This proved to be the case. Contrary to his earlier conclusion, then, it was not the nature of the food source but the distance of the source from the hive that determined the form of the dance, with round dances performed at distances of less than 100 meters or so from the hive, and waggle dances at greater distances.

Frisch reported his major new discoveries in two papers published in 1946. He also discussed them in his book of 1947 entitled *Duftgelenkte Bienen im Dienst der Landwirtshaft und Imkerei*, in which he described the results of his own experiments on how training bees to particular scents, such as that of red clover, could increase crop production.

Frisch's work, which had attracted considerable attention before the war, became the subject of intense interest. The English ethologist W. H. Thorpe visited Frisch at Brunnwinkl in September 1948 to witness and repeat for himself Frisch's key experiments. In an article in *Nature* in 1949, Thorpe confirmed the accuracy of Frisch's remarkable findings. In the United States, Donald R. Griffin of Cornell University repeated Frisch's experiments and invited Frisch to lecture at Cornell, an invitation that developed into a three-month lecture tour of the United States in the spring of 1949.

Frisch's new discoveries of 1945 and 1946 on the dance language of the bees opened up a host of new research topics for him. Having learned that in long-range flights bees orient themselves with respect to the position of the sun, he asked himself how bees orient themselves when the sun is not visible. A series of experiments led him to conclude—mistakenly, it now seems—that the bees' perception of ultraviolet rays allows them to see the sun through a cloud cover that obscures the sun to the human eye. Of more lasting and fundamental importance, he made the remarkable discovery that bees perceive polarized light and are able to orient their dances according to the plane of polarization. Thus, with the sun obscured but a patch of blue sky visible to them, they are able to retain their sense of direction.

In the 1920's Frisch's student Ingeborg Beling had conducted pioneering studies on the temporal sense of bees. Frisch in the 1950's proceeded to establish that thanks to their "internal clock," bees are able to take into account the changes in the sun's position during the course of the day and are thus able to use the sun (as well as conspicuous objects in the landscape) for orientation on long flights. This was another performance by bees that Frisch found almost too extraordinary to be true, but carefully designed "displacement" experiments showed that bees, by adjusting for the time of day and the sun's diurnal course, can indeed use the sun's position to find a given compass direction.

With his student Martin Lindauer, Frisch undertook a comparative study of the dance language of different varieties and species of bees. This revealed different "dialects" in the dance language of bees, and provided a basis for evaluating how the complex dance language exhibited by the black Austrian honeybee (*Apis mellifera carnica*), the original subject of Frisch's researches, could have evolved from a more primitive form of dance, such as that discovered by Lindauer in the Indian dwarf honeybee *Apis florea*.

In his autobiography Frisch acknowledged that throughout his career he was plagued by periods when he doubted that he would have any more new ideas for research. At the beginning of his career he also worried about whether he would be able to think of interesting dissertation topics for his students. The vast majority of the topics he investigated, however, proved immensely fruitful for additional studies of his own and for work conducted by his students and other researchers. His early studies on pigmentation and color adaptation led to his studies of color vision. His experiments on the color sense of bees led to his later studies of bees' senses of smell and taste, their "language," and their orientation.

In his 1965 book *Tanzsprache und Orientierung der Bienen*, which describes in detail the results of the work of Frisch and others on bees for the twenty years preceding the book's publication, he was able to list forty-four doctoral students who had written dissertations on bees under his direction. Among them was Martin Lindauer, whose own students by then represented a third generation working in the tradition Frisch had established. Frisch's skill in devising simple experiments under natural conditions was taken as a model by other scientists—for example, the Dutch ethologist Nikolaas Tinbergen.

Frisch not only was exceptionally skilled as an experimenter but also was unusually gifted in his ability to write lucidly about biology. In addition to an introduction to bacteriology that he wrote for nurses during World War I, he produced a number of books of an introductory or semipopular nature, several of which have gone through many editions. Among these (with titles of the English translations and dates of the original edition and translations) were *Aus dem Leben der Bienen* (*The Dancing Bees*; 1927, 1954); *Du und das Leben* (*Man and the Living World*; 1936, 1962); *Bees: Their Vision, Chemical Senses, and Language* (1950, rev. ed. 1971); *Biologie* (*Biology*; 2 vols., 1952, 1964); *Erinnerungen eines Biologen* (*A Biologist Remembers*; 1957, 1967); and, written with his son Otto, *Tiere als Baumeister* (*Animal Architecture*; 1974, 1983).

Frisch was always careful to point out, in speaking of a "language" of the bees, that the language metaphor had its limits, and that the dances of bees, though variable and adapted to different conditions, were mainly instinctive. The 1970's witnessed considerable discussion of the issue of whether it is appropriate to attribute a "language" to bees, and whether the dances of bees actually serve to communicate information regarding the location of food sources to new recruits. Ethologists in general, however, regard the symbolic, communicative function of the bee dances as well established and, indeed, regard Frisch's discovery of the dance language of the bees as the single most impressive discovery of classical ethology.

BIBLIOGRAPHY

I. Original Works. A nearly complete bibliography of Frisch's scientific writings, with relatively few flaws in the citation of page and volume numbers, is in *Journal of Comparative Physiology*, **147** (1982), 420–422. For Frisch's authoritative summing up of his own and other researches on bees, see his *Tanzsprache und Orientierung der Bienen* (Berlin, 1965), translated as *The Dance Language and Orientation of Bees* (Cambridge, Mass., 1967).

In addition to Frisch's magnum opus of 1965, the following are among his most important scientific writings: "Studien über die Pigmentverschiebung im Facettenauge," in *Biologisches Zentralblatt*, **28** (1908), 662–671, 698–704; "Über die Beziehungen der Pigmentzellen in der Fischhaut zum sympathischen Nervensystem," in *Festschrift zum sechzigsten Geburstag Richard Hertwigs*, III (Jena, 1910), 17–28; "Beiträge zur Physiologie der Pigmentzellen in der Fischhaut," in *Pflügers Archiv für die gesamte Physiologie des Menschen und der Tiere*, **138** (1911), 319–387; "Über den Farbensinn der Fische," in *Verhandlungen der Deutschen zoologischen Gesellschaft*, **21** (1911), 220–225; "Über farbige Anpassung bei Fischen," in *Zoologische Jahrbücher. Abteilung für allgemeine Zoologie und Physiologie*, **32** (1912), 171–230; "Sind die Fische farbenblind?" *ibid.*, **33** (1912), 107–126; "Der Farbensinn und Formensinn der Biene," *ibid.*, **35** (1914), 1–182, also separately reprinted (Jena, 1914); "Zur Streitfrage nach dem Farbensinn der Bienen," in *Biologisches Zentralblatt*, **39** (1919), 122–139; "Über den Geruchsinn der Biene und seine blütenbiologische Bedeutung," in *Zoologische Jahrbücher. Abteilung für allgemeine Zoologie und Physiologie*, **37** (1919), 1–238; "Über den Einfluss der Bodenfarbe auf die Fleckenzeichnung des Feuersalamanders," in *Biologisches Zentralblatt*, **40** (1920), 390–414; "Über die 'Sprache' der Bienen, I–III," in *Münchener medizinische Wochenschrift*, **67** (1920), 566–569, **68** (1921), 509–511, **69** (1922), 781–782; "Über die 'Sprache' der Bienen. Eine tierpsychologische Untersuchung," in *Zoologische Jahrbücher. Abteilung für allgemeine Zoologie und Physiologie*, **40** (1923), 1–186; and "Das Problem des tierischen Farbensinnes," in *Naturwissenschaften*, **11** (1923), 470–476.

These writings were followed by "Ein Zwergels, der kommt, wenn man ihm pfeift," in *Biologisches Zentralblatt*, **43** (1923), 439–446; "Sinnesphysiologie und 'Sprache' der Bienen," in *Naturwissenschaften*, **12** (1924), 981–987; "Farbensinn der Fische und Duplizitätstheorie," in *Zeitschrift für vergleichende Physiologie*, **2** (1925), 393–452; "Neue Versuche über die Bedeutung von Duftorgan und Pollenduft für die Verständigung im Bienenvolk," *ibid.*, **4** (1926), 1–21, with G. A. Rösch; "Versuche über den Geschmacksinn der Bienen, I–III," in *Naturwissenschaften*, **15** (1927), 321–327, **16** (1928), 307–315, **18** (1930),

169–174; "Über die Labyrinth-Funktionen bei Fischen," in *Verhandlungen der Deutschen Zoologischen Gesellschaft*, **33** (1929), 104–112; "Untersuchungen über den Sitz des Gehörsinnes bei der Elritze," in *Zeitschrift für vergleichende Physiologie*, **17** (1932), 686–801, with H. Stetter; "Über den Geschmacksinn der Biene," *ibid.*, **21** (1934), 1–156; "Über den Gehörsinn der Fische," in *Biological Reviews*, **11** (1936), 210–246; "Psychologie der Bienen," in *Zeitschrift für Tierpsychologie*, **1** (1937), 9–21; and "Zur Psychologie des Fisch-Schwarmes," in *Naturwissenschaften*, **26** (1938), 601–606.

Frisch's publications of the 1940's include "Die Tänze und das Zeitgedächtnis der Bienen im Widerspruch," in *Naturwissenschaften*, **28** (1940), 65–69; "Die Bedeutung des Geruchsinnes im Leben der Fische," *ibid.*, **29** (1941), 321–333; "Über einen Schreckstoff der Fischhaut und seine biologische Bedeutung," in *Zeitschrift für vergleichende Physiologie*, **29** (1941), 46–145; "Die Werbetänze der Bienen und ihre Auslösung," in *Naturwissenschaften*, **30** (1942), 269–277; "Versuche über die Lenkung des Bienenfluges durch Duftstoffe," *ibid.*, **31** (1943), 445–460; "Die 'Sprache' der Bienen und ihre Nutzanwendung in der Landwirtschaft," in *Experientia*, **2** (1946), 397–404; "Die Tänze der Bienen," in *Österreichische zoologische Zeitschrift*, **1** (1946), 1–48, translated as "The Dances of the Honey Bee," in *Bulletin of Animal Behaviour*, spec. no. (December 1947), 5–32; *Duftgelenkte Bienen im Dienst der Landwirtschaft und Imkerei* (Vienna, 1947); "Gelöste und ungelöste Rätsel der Bienensprache," in *Naturwissenschaften*, **35** (1948), 12–23, 38–43; and "Die Polarisation des Himmelslichtes als orientierender Faktor bei den Tänzen der Bienen," in *Experientia*, **5** (1949), 142–148.

Among Frisch's later publications were "Die Sonne als Kompass im Leben der Bienen," in *Experientia*, **6** (1950), 210–221; "Orientierungsvermögen und Sprache der Bienen," in *Naturwissenschaften*, **38** (1951), 105–112; "Gibt es in der 'Sprache' der Bienen eine Weisung nach oben oder unten?" in *Zeitschrift für vergleichende Physiologie*, **35** (1953), 219–245, with H. Heran and M. Lindauer; "Himmel und Erde in Konkurrenz bei der Orientierung der Bienen," in *Naturwissenschaften*, **41** (1954), 245–253, with M. Lindauer; "Über die Fluggeschwindigkeit der Bienen und über ihre Richtungsweisung bei Seitenwind," *ibid.*, **42** (1955), 377–385, with M. Lindauer; "The 'Language' and Orientation of the Honey Bee," in *Annual Review of Entomology*, **1** (1956), 45–58, with M. Lindauer; "Lernvermögen und erbgebundene Tradition im Leben der Bienen," in Mario Autuori *et al.*, *L'instinct dans le comportement des animaux et de l'homme* (Paris, 1956), 345–386; "Über die 'Missweisung' bei den richtungsweisenden Tänzen der Bienen," in *Naturwissenschaften*, **48** (1961), 585–594, with M. Lindauer; "Dialects in the Language of the Bees," in *Scientific American*, **207** (1962), 78–87; "Honeybees: Do They Use Direction and Distance Information Provided by Their Dancers?" in *Science*, **158** (1967), 1072–1076; and "Decoding the Language of the Bee," *ibid.*, **185** (1974), 663–668.

II. SECONDARY LITERATURE. On Frisch's life and career the best source is his autobiography, *Erinnerungen eines Biologen* (Berlin, 1957), translated by Lisbeth Gombrich as *A Biologist Remembers* (Oxford, 1967). W. H. Thorpe, "Karl von Frisch," in *Biographical Memoirs of Fellows of the Royal Society*, **29** (1983), 197–200, is a brief notice that borrows heavily from the autobiography. For a sampling of current work directly or indirectly in the tradition of research established by Frisch, and for other comments on Frisch as a scientist, see Bert Holldobler and Martin Lindauer, eds., *Experimental Behavioral Ecology and Sociobiology: In Memoriam Karl von Frisch 1886–1982* (New York, 1985).

The author gratefully acknowledges the helpful comments of Donald R. Griffin and Martin Lindauer on an earlier draft of this paper.

RICHARD W. BURKHARDT, JR.

FRISCH, OTTO ROBERT (*b.* Vienna, Austria, 1 October 1904; *d.* Cambridge, England, 22 September 1979), *physics*.

Frisch grew up in Vienna. His grandfather, Moriz Frisch, a Polish Jew from Galicia, had settled there and started a printing business in 1877. His father, Justinian Frisch, was a doctor of law and held senior positions in a number of firms, mainly in printing and publishing. His mother, Auguste Meitner, was the daughter of a lawyer and at one time a concert pianist. The small family (there were no other children) seems to have been a very happy one, and Frisch always retained a strong sense of family and of community. Much of his scientific work was collaborative.

With his father's help Frisch developed a strong interest in mathematics but also enjoyed making things, and at Vienna University he elected to read physics, with mathematics as a second subject. After receiving a doctorate (1926) under the supervision of Karl Przibram, he worked for a year in the laboratory of a small instrument firm, then for three years, in Berlin, at the Physikalisch-Technische Reichsanstalt. This brought him into contact with the strong school of physics at the University of Berlin, and in his last year there he was able to work with Peter Pringsheim. In 1930 Frisch moved to the University of Hamburg, as assistant to the great experimentalist Otto Stern; and over the next three years he worked both with Stern, on diffraction experiments with molecular beams, and with the young Emilio Segrè.

In 1933 the racial laws of the National Socialist government compelled Stern, Frisch, and a host of other Jewish physicists to leave the country. Frisch then worked for a short time with Patrick Blackett at Birkbeck College, London, before being invited to join Niels Bohr's Institute for Theoretical Physics in Copenhagen in 1934. In Copenhagen, Frisch, working largely with Hans von Halban, began to concentrate on the field of experimental nuclear physics recently opened up by Enrico Fermi. Despite his strong experimental bias, however, it was for the joint authorship of two theoretical papers that Frisch was to become famous.

The first of these contributions was written over the Christmas holidays in late 1938. Earlier that year Frisch's aunt, Lise Meitner, had fled from Germany and, after spending time in the Netherlands and Denmark, had joined the Nobel Institute in Stockholm. Meitner was a physicist of distinction who had collaborated for about thirty years with Otto Hahn. She had been very kind to Frisch while he was in Berlin, and they had since spent Christmas together regularly. During his visit she received a letter from Otto Hahn that reported the apparent splitting of the uranium nucleus under neutron bombardment. Frisch and Meitner then worked out a simple theoretical interpretation of this phenomenon, which they christened "fission." Back in Copenhagen, Frisch confirmed the fission interpretation by repeating the experiment and detecting the fission fragments.

Meanwhile, as war threatened to engulf much of Europe, even Denmark was becoming uncomfortably close to the German state for a Jew. In the summer of 1939, Mark Oliphant responded to Frisch's clearly expressed desire to move to England by inviting him to Birmingham University for a summer vacation, with a view to a more permanent appointment thereafter. Before the vacation was over, however, war had broken out; Frisch was promptly appointed to a temporary teaching assistantship at Birmingham. While there, he completed with Rudolf Peierls two theoretical papers relating to the possibility of a nuclear chain reaction. In the first, a general paper on subatomic phenomena written for the annual report of the Chemical Society, they concluded, as Bohr had earlier, that there was no possibility of an explosive chain reaction in natural uranium. Before writing this, Frisch had tried to verify experimentally the theoretical claim underlying Bohr's argument: that the observed fissions were due entirely to the isotope uranium 235, which is less than 1 percent of natural uranium. He was unable to separate out the isotope for experiment, but the attempt got him thinking about what would happen if the isotope could be isolated. Also with Peierls, who had been investigating the mathematics of the chain reaction, Frisch worked out an order-of-magnitude estimate for the critical size of a sphere of pure uranium 235 (about one pound) needed to sustain a chain reaction. Another calculation gave an estimate for the proportion of fission energy that would be released by such a sphere in an explosion before the sphere split too much to sustain the reaction. The results suggested that a pure uranium 235 bomb might be feasible with only a few pounds of that isotope.

These important investigations were written up in April 1940 in the famous Frisch-Peierls memorandum; and though they did not at first dispel skepticism about the possibilities of a bomb (the separation task seemed impracticably enormous), they led directly to the British wartime atomic energy project and indirectly provided a strong stimulus to the American project. Despite the problems caused by his German background, Frisch made valuable contributions to both projects, first at Liverpool, where he worked from August 1940 to late 1943, and then at Los Alamos, New Mexico, where the combination of his practical resourcefulness and his theoretical command made him a useful troubleshooter. Early in 1946, the war over, he returned to England to become head of the nuclear physics division of the new Atomic Energy Research Establishment at Harwell. The following year he accepted the Jacksonian Professorship of Natural Philosophy at Cambridge.

Frisch was not a natural administrator or a great leader. He delegated the administration of Harwell and lived a relatively quiet life at Cambridge, teaching, supervising research (and undertaking a little of his own), and writing a large number of popular articles and books. His great gift, whether in experimental or theoretical physics, had always been to see things clearly, usually through some simple analogy. His experiments were elegant and straightforward, and his theoretical contributions had the same character. Popular exposition thus came naturally to him.

In 1951 Frisch married Ursula Blau, an artist; they had a son and a daughter. As in his own childhood, art (in which his father had been passionately interested) and music (inherited from his mother's side) dominated family life. He received the O.B.E. in 1946 and was elected a fellow of the Royal Society in 1948. At Cambridge he was a fellow of Trinity College. Frisch died after an accident, just before his seventy-fifth birthday.

BIBLIOGRAPHY

I. Original Works. A full bibliography of Frisch's writings is in the obituary by Peierls (see below). The two most significant papers are "Disintegration of Uranium by Neutrons: A New Type of Nuclear Reaction," in *Nature*, **143** (1939), 239–240, written with Lise Meitner; and "On the Construction of a 'Super-Bomb'; Based on a Nuclear Chain Reaction in Uranium," written with Rudolf Peierls, part I in Margaret M. Gowing, *Britain and Atomic Energy, 1939–1945* (London, 1964), 389–393, and part II in Ronald W. Clark, *Tizard* (Cambridge, Mass., 1965), 215–217. On the discovery of fission, see "The Discovery of Fission," in *Physics Today*, **20** (November 1967), 43–52, written with J. A. Wheeler.

Frisch's papers, cataloged by the Contemporary Scientific Archives Centre in Oxford (catalog no. 87), are in Trinity College Library, Oxford, and comprise an extensive collection of research notebooks, personal diaries and notebooks, correspondence, and memorabilia, including articles written by and about him.

II. SECONDARY LITERATURE. The two principal accounts of Frisch's life are his autobiography, *What Little I Remember* (Cambridge, 1979); and Sir Rudolf Peierls, "Otto Robert Frisch," in *Biographical Memoirs of Fellows of the Royal Society*, **27** (1981), 282–306. His wartime work is covered in Margaret M. Gowing, *Britain and Atomic Energy, 1939–1945* (London, 1964). On the discovery of fission, see Hans G. Graetzer and David L. Anderson, eds., *The Discovery of Nuclear Fission: A Documentary History* (New York, 1971).

JOHN HENDRY

FRUMKIN, ALEKSANDR NAUMOVICH (*b.* Kishinev, Russia, 24 October 1895; *d.* Tula, U.S.S.R. [buried in Moscow], 27 May 1976), *physical chemistry*.

Frumkin was the son of Naum Yefimovich Frumkin, an official in the insurance business, and of Margarita L'vovna. After graduating in 1912 from the Practical School in Odessa, he continued his studies for two years in Germany and Switzerland. In Bern he studied inorganic chemistry under F. Kaltenschütter. On his return to Odessa, Frumkin graduated as a nonresident student from the Faculty of Physics and Mathematics of Novorossiiskii (Odessa) University (1915). After a year as a chemist in the analytical laboratory of a metallurgical plant, he taught at Novorossiiskii University from 1917 to 1922, in 1920 becoming a professor. In 1919 Frumkin defended his master's thesis, and in 1934 he received his doctorate. Frumkin married Amalia Davidovna Bogdanovskaia in June 1926. After her death he married Emilia Georgievna Perevalova. There were no children from either marriage.

In Frumkin's master's thesis he anticipated the program of his subsequent scientific research associated with the theory of surface phenomena and the theory of electrochemical processes. Frumkin is justly considered the founder of this area of research. He proved the applicability of Willard Gibbs's thermodynamic equation to the phenomena of adsorption and derived the state equation of an adsorbed layer (Frumkin's isotherm, which takes into account the interaction between molecules absorbed at adjacent lattice centers). He worked out the quantitative theory of the effects of an electric field on the adsorption of molecules and (through measurement of potential jumps along the gas-solution boundary) obtained information about polarization of organic substances.

In 1932 and in the following years Frumkin developed a theory of the quantitative relationship between the rate of an electrochemical reaction and the structure of the electric double layer on the metal-electrolyte interface. Having studied the electrical characteristics of this layer, he investigated the dependence of the shape of the electrocapillarity curves on the presence of various molecules within the boundary layer; he also worked out methods of studying adsorption phenomena, for example, the rise of electrode potentials and the capacity of an electrical double layer. These works led Frumkin to elaborate the theory of an electromotive force intimately connected with chemical current sources. This theory led to the conclusion that the value of the electromotive force of an electrochemical element is the sum of the difference of potential jumps in the ionic double layer on both electrodes and of the contact potentials of the electrode metals.

In 1927 Frumkin introduced the idea of zero-charge potential (the value of a metal electrode's potential at which there are no free electrical charges on the metal's surface) as the fundamental characteristic of metallic electrodes. Another field of his research laid the foundations of modern polarography as one of the means of monitoring—by noninterfering instantaneous investigation—the kinetics of chemical reactions, including some that are very difficult to study (such as simultaneous polymerization of several monomers of various types of polarity). This method was based on investigations of concentrational polarization. Concentrational polarization is the difference ΔE between the values of electrode potentials at equilibrium and at the flow through the electrode of an external electrical current i. ΔE is due to the deviation, by electrode concentrations, of the substances in reaction from their values in the electrolyte solution volume (as the result of hindered diffusion of those substances).

$$\Delta E = \frac{RT}{nF} ln\left(1 - \frac{i}{i_d}\right),$$

where F is the Faraday number, n is the number of electrons involved in the process, and i_d is the limiting diffusion current.

Frumkin clarified the mechanism of many electrode reactions, for instance, those of regeneration of oxygen and a number of anions. Many of his research papers concern the corrosion of metals and various methods of protecting against it. He studied wetting of metals by electrolytes. These investigations resulted in an important contribution to the mining industry—the development of the theory of flotation processes, which makes it possible to separate the metals from minerals.

Frumkin's theoretical ideas have found applications in investigations of chemical sources of current, commercial electrolysis, polarography, heterogeneous catalysis, colloid chemistry, and bioelectrochemistry. In his last works, published between 1965 and 1975, he revised some of the fundamental ideas of electrochemistry concerning the charge of electrodes and elaborated a thermodynamic theory of the metal-electrolyte interface of catalytically active electrodes.

Frumkin devoted much effort to the organization of Soviet science. In 1924 he became head of the department of surface phenomena at the L. Ia. Karpov Institute of Physical Chemistry in Moscow, which he had joined in 1922 at the invitation of A. N. Bach; he remained there until 1946. From 1939 to 1949 he was director of the Institute of Physical Chemistry of the U.S.S.R. Academy of Sciences, and from 1958 to 1976 he was head of the Institute of Electrochemistry of the Academy of Sciences.

From 1930 to 1976 Frumkin was head of the department of electrochemistry at Moscow University. In the year 1928–1929 he lectured on colloidal chemistry at the University of Wisconsin at Madison. His students and followers at Novorossiiskii University and in a number of laboratories of the Academy of Sciences' institutes formed a school that gave rise to a number of satellites, the best-known of which are the school of physicochemical mechanics founded by P. A. Rebinder and the school of two-dimensional liquids led by B. V. Deryagin. With N. N. Semenov, Frumkin organized international meetings on physical chemistry held in the Soviet Union in the 1920's and mid 1930's. These meetings played an important role in the formation of physical chemistry as a separate field of research in the Soviet Union.

Frumkin's scientific achievements received wide recognition. In 1932 he was elected full member of the U.S.S.R. Academy of Sciences. In 1931 he was awarded the Lenin Prize. He was an honorary member of the American Academy of Arts and Sciences, the Hungarian Academy of Sciences, and the French and Belgian chemical societies; and a foreign (or full) member of the Bulgarian, Dutch, and Polish academies, the German Academy of Sciences (Leopoldina), and the U.S. National Academy of Sciences.

An admirer of poetry and art, Frumkin amassed a very valuable collection of paintings and a sizable library of Russian poetry.

BIBLIOGRAPHY

By gracious permission of Prof. V. E. Kazarinov, this essay draws upon his article on Frumkin in the third edition of the *Great Soviet Encyclopedia*.

I. ORIGINAL WORKS. *Elektrokapilliarnie iavleniia i elektrodnia potentsiali* (Electrocapillary phenomena and electrode potentials; Odessa, 1919); "Wasserstoffüberspannung und Struktur der Doppelsicht," in *Zeitschrift für physikalische Chemie*, ser. A, **164**, nos. 1–2 (1933), 121–133; *Kinetika elektrodnykh protsessov* (Kinetics of electrode processes; Moscow, 1952), written with V. S. Bagotskii, Z. A. Jofa, and B. N. Kabanov; "K voprosu ob energii aktivatsii razriada iona vodoroda" (Activation energy of the discharge of hydrogen ions), in *Zhurnal fizicheskoi khimii*, **29**, no. 8 (1955), 1513–1526, written with M. I. Temkin; "O kolcevom diskovom elektrode" (On the ring disk electrode), in *Doklady Akademii nauk SSSR*, **126**, no. 1 (1959), 115–118, written with L. N. Nekrasov; "Ob opredelenii zariada reagiruiushchei chastitsi iz zavisimosti kinetiki elektrovosstanovleniia ot potentsiala i kontsentratsii fona" (On the determination of the influence of a charge of a reactive particle from the dependence of the Kinetics of electroreduction on the potential and environmental concentration), *ibid.*, **147**, no. 2 (1961), 418–421, written with O. A. Petrii.

"Hydrogen Overvoltage and Adsorption Phenomena," in *Advances in Electrochemistry and Electrochemical Engineering*, **1** (1961), 65–121, and **3** (1963), 287–391; "Vliianie adsorbtsii neutralnikh molekul i organicheskikh kationov na kinetiku elektrodnikh protsessov" (The influence of the adsorption of neutral molecules and organic cations on the kinetics of the electrode processes), in *Trudy 14-go Soveshchaniia mezhdunarodnogo komiteta po elektrokhimicheskoi termodinamike i kinetike* (Moscow, 1965), 302ff.; "Hydrogen Evolution from Alkaline Solutions on Metals of High Overvoltage," in *Electrochimica Acta*, **15**, no. 2 (1970), 289–301; *Potentsiali nulevogo zariada* (The zero-charge potentials; 2nd ed., Moscow, 1982); *Izbrannie trudi. Elektrodnii protsessi* (Selected works. The electrode processes), B. P. Nikol'skii, ed. (Moscow, 1987).

II. SECONDARY LITERATURE. N. A. Bakh, "Raboty A. N. Frumkina" (A. N. Frumkin's works), in *Uspekhi khimii*, **6**, no. 11 (1937), 1572–1582; and B. N. Kabanov and O. V. Isakova, "Aleksandr Naumovich Frumkin," in *Materiali K biobibliografii uchenikh SSSR*, ser. Khimicheskikh nauk, fasc. 44 (Moscow, 1955; 2nd ed., 1970). In the latter, in both editions Kabanov wrote the text; the bibliography in 1955 was by Isakova and in 1970 by R. I. Goriachova and I. A. Makhrova.

V. J. FRENKEL

GABOR, DENNIS (*b*. Budapest, Hungary, 5 June 1900; *d*. London, England, 9 February 1979), *physics, electrical engineering*.

Dennis Gabor was awarded the Nobel Prize for physics in 1971, for the invention and development of the method of holography, to date the most influential product of his restlessly inventive intellect. By his own admission, Gabor lived on and for the compulsion to invent. Never the mechanic or, strictly speaking, an experimentalist, he saw his role more as that of conceiving future possibilities, both technical and social.

Gábor Dénes, to give the Hungarian form of his name, was the eldest of three boys. His childhood was a period of intense and liberal intellectual stimulation within a culturally rich Jewish middle-class society. His father, Bertalan, was director of Hungary's largest industrial firm, a coal mining company. His mother, Adrienne Kálmán, was a former actress. The family adopted the Lutheran faith in 1918, and although Gabor nominally remained true to it, religion appears to have had little influence in his life. He later acknowledged the role played by an antireligious humanist education in the development of his ideas and stated his position as being that of a "benevolent agnostic." By the age of fifteen, he was performing home experiments with his younger brother, George, and was set on a scientific career.

School in Budapest was followed by military training toward the end of World War I and then a course in mechanical engineering at the city's technical university, where he enrolled in 1918. Gabor left during the third year because of his distaste for registering for military service under a reactionary government. His formal education was completed at the Technische Hochschule in Berlin, where he received the diploma in electrical engineering in 1924 and the doctorate in 1927. His technical ingenuity found employment in industry in Germany, briefly in Hungary, and, from 1934, in England.

On 8 August 1936 Gabor married Marjorie Louise Butler, a fellow employee of the British Thomson-Houston Company; they had no children. From 1949 until his retirement in 1967, he worked at Imperial College, London. By the latter year, his interests were dominated by his concern for the future of industrial societies, which found expression through his involvement in the Club of Rome, of which he was a founding and very active member. Elected a fellow of the Royal Society in 1956, he was also the first émigré to be elected an honorary member of the Hungarian Academy of Sciences in 1964.

Short and sturdy, Gabor enjoyed good health for most of his adult life, apart from a serious attack of thrombosis-phlebitis in the legs in 1961. In 1974, however, he suffered a severe cerebral hemorrhage, following which he could neither read nor write.

Besides his work on holography, Gabor's scientific and technical contributions include developments in the theory of communications, plasma theory, and technical ideas embodied in more than a hundred commercial patents. Like anglers, however, inventors are haunted by the ones that get away. Much of Gabor's scientific work was influenced by the fact that he had not invented the electron microscope.

In 1931 Ernst Ruska and Max Knoll developed the first two-stage electron microscope. Within a few years Ruska had improved "this wonderful instrument," as Gabor more than once described it, to an extent that its resolving power improved on optical instruments. Gabor could never forgive himself for not having done it himself and spent years trying to make up for what he considered his missed opportunity. The cause of this deep regret—indeed, envy—lay in the possibilities presented by his own doctoral research at the Technische Hochschule in Berlin, carried out between 1924 and 1926, on the measurement of fast surges in high-voltage power lines. These surges, caused by lightning, often resulted in considerable damage.

In order to record these surges, Gabor developed a fast-response cathode-ray oscilloscope of advanced design. Particularly significant was the replacement of the conventional long focusing solenoid by a short coil, encased in iron, intended to confine the field within the coil and prevent the intrusion of stray magnetic fields. Without realizing it, Gabor had constructed in crude form the first iron-shrouded magnetic electron lens, the forerunner of the modern high-resolution magnetic lens of an electron microscope.

The full significance and correct explanation of the working of the magnetic electron lens were soon grasped by Hans Busch (1926). After this, the opportunity was clear. As Gabor himself later commented: "How could anybody capable of putting two and two together not think of an electron microscope?" He recalled suggesting in 1928 to his compatriot and friend Leo Szilard that he had the expertise within his grasp to build such a microscope. His cathode-ray oscilloscope provided, in a crude embryonic form, the basic technology. To his lasting regret, he made no move. The completion of his doctorate had made him "temporarily allergic" to electrons, he later explained. In addition, the usefulness of an electron microscope was not immediately clear.

After completing his doctoral dissertation, Gabor found work in the physics laboratory of Siemens and Halske at Siemensstadt, where he studied lamp technology. His interest in the improvement of mercury vapor lamps had developed from his private research aimed at detecting the "mitogenic rays" claimed to be emitted from growing onion roots. The lamps were used in these experiments to attempt to induce mitosis. He remained at Siemens until 1933. Within weeks of Hitler's rise to power, Gabor's contract with the company was terminated. Returning briefly to Hungary, he used the patents of a new lamp design to negotiate an inventor's agreement to go to England and work on its improvement. Though never a success, his plasma lamp gave him a foothold in the British Thomson-Houston Company at Rugby in Warwickshire. With the end of the agreement in 1937, he was appointed to the permanent staff, where he remained until 1948. Gabor was the only refugee to be employed by the B.T.H. Research Laboratory. His vivid personality enriched the establishment, and his scientific output remained considerable. Nevertheless, Gabor found this period in his life, during which he made acclaimed contributions to the theory of communications, and at the end of which he established the principles of holography, sterile and at times depressing.

His time in Berlin had been one of the great periods of Gabor's life. He had gone to the university as often as possible to witness at first hand the work of Albert Einstein, Max Planck, Walter Nernst, and Max von Laue. He moved in a circle of expatriate Hungarians that included John von Neumann, Eugene Wigner, Leo Szilard, and Michael Polanyi, as well as his close friend Peter Goldmark. In contrast, from 1939 until 1945 Gabor was isolated from the mainly classified work at Rugby, accommodated in a hut built for him on the fringe of the restricted area that could be reached only by a specified route. He was virtually cut off from the scientific literature, *Nature* being the only journal he could obtain regularly. In 1942 his father, who had been a great influence on his life, died in Hungary. About 1943 Gabor began to give more serious thought to the electron microscope.

It is clear that the electron microscope had never been far from Gabor's mind from the moment he realized what he had missed in its initial development. Now he had an opportunity to make what he thought of as a comeback in the field. The real prize, as he saw it, would be gained for producing an instrument that could "see" individual atoms. Yet despite gradual improvements in the resolving power of electron microscopes, a theoretical barrier set by a compromise between diffraction effects at the aperture edge and spherical aberration placed crucial practical limits short of the resolution needed to focus atomic lattices. After Otto Scherzer first pointed out the limitation posed by spherical aberration in 1936, several authors attempted to calculate the limit to resolving power set by various imperfections, and in 1943 the exact rule was elaborated by Walter Glaser.

For Gabor this provided just the kind of challenge his inventive mind needed. The barrier to progress was of a technical nature, yet formidable. At the time, his main work, on developing his ideas on infrared detection, was going poorly. His initial efforts at thinking of ways around the theoretical limits placed on the electron microscope brought no ready solutions. His attempts to interest others in the problem were unsuccessful. In 1946 the theoretical limit was virtually reached in practice by James Hillier and Edward Ramberg. By then Gabor was seriously considering offers from the United States, frustrated by what he felt to be a lack of support for his ideas at B.T.H., in particular for his latest project, to develop a 3-D projection system. A number of factors conspired to keep Gabor in England. In 1947 he became a British citizen. In the same year he made "my luckiest find yet," the one he hoped would enable him to get his own back on the electron microscope, and so reveal the atomic lattice.

Gabor had been searching for the "trick" to get around the barrier of the theoretical limit to resolution. During the Easter holiday in 1947, he was sitting on a bench at the local tennis club when an idea suddenly came to him: Why not take an electron picture distorted by lens imperfections and correct it by optical means? A few calculations convinced him he was right.

Gabor was proposing a two-stage process. In the first stage an interference pattern produced by the interaction of electrons diffracted by the object and a separate but coherent reference beam of electrons would be photographically recorded on transparent film. Gabor argued that this interference pattern, or "hologram," as he later called it, would carry the complete information needed to reconstruct an image of the object, using an optical system free from the limitations of electron optics. In the second stage, the hologram would be scaled up by a factor in the ratio of the wavelength of the light used in the reconstruction to the wavelength of the electron beam. The new hologram would then be illuminated with a light wave of the same aberration as the electron wave to, in theory, reveal an exact replica of the original object, magnified by the scaling factor.

In July 1947 Gabor, assisted by Ivor Williams, began experiments at B.T.H. to establish the principle by using a purely optical model—that is, using visible light instead of electrons—with a mercury vapor lamp as a source of coherent light, to produce the interference photographs of simple two-dimensional images. Five months later he was able to show his close confidant Lawrence Bragg his first successful wavefront reconstructions: hazy images of simple printed words used as objects. Even when Bragg fully understood the theory, he still stated that it was a miracle it should work. The first public indication of Gabor's success came with a preliminary note to *Nature*, published on 15 May 1948. The following year he wrote a more complete theoretical treatment, for the *Proceedings of the Royal Society*, in which he introduced the word "hologram" and indicated possible applications in light optics. Among these was the ability, using the same method, to record the data associated with 3-D objects in one interference photograph.

Gabor's private correspondence was full of enthusiasm. To Max Born, he declared that his holographic reconstructions had made him happier than anything he had done in the last twenty years. And he told Arthur Koestler: "I missed inventing the electron microscope when I was 27, and I have every intention to make a comeback, with a fresh start, at 48." For the moment he was very happy. Within a few years he again felt the inventor's frustration as everything seemed to turn sour.

In his Nobel Lecture, Gabor acknowledged the influence of both Lawrence Bragg and Frits Zernike on his ideas in wavefront reconstruction, as he called his new principle. The idea of a two-step imaging process came directly from Bragg's X-ray microscope, described in 1942, in which holes were drilled in a brass plate that corresponded to the photographically recorded image of an X-ray diffraction pattern produced by a crystal lattice. When the plate was illuminated with monochromatic light, an image of the crystal structure could be viewed through the microscope. The double diffraction process explicit in the X-ray microscope is crucial to holography.

At the time, neither Gabor nor Bragg was aware that the method had been suggested by the work of Mieczislav Wolfke in 1920. Bragg's method was limited, however, to cases where both amplitude and phase of the wave were preserved. In 1950 Martin Buerger extended the principle to crystals producing known phase changes by using "phase shifters." Gabor's crucial addition to the double-diffraction concept was the preservation of phase information through the introduction of the coherent reference wave. The whole information (hence "hologram," from the Greek *holos*, meaning "whole") would thus be preserved in the recorded interference pattern, allowing a complete reconstruction.

A coherent background wave had been used with great success by Zernicke for his investigation of lens aberration by phase contrast. To this Gabor added the concept of reconstruction. The electron shadow microscope of Hans Boersch (1939) was similar to the first stage of Gabor's process, except in the use of coherent illumination. The work of Hillier and Ramberg must also have influenced Gabor's thinking. In early 1947 they published work showing that electrons, penetrating through membranes, could interfere with the illuminating wave passing the edge. Gabor had advance notice of these results and had suggested adapting the concept of Zernicke's phase-contrast microscope for use with electrons. He was a strong believer in the rational cumulative development of knowledge and was generous to those he recognized as having influenced his work. He had stood on the shoulders of Bragg and Zernicke, he claimed in his Nobel Lecture.

Early in 1949 Gabor moved to an academic appointment at Imperial College of Science and Technology, London, as reader in electronics, although at least until 1952 he was still debating whether to accept offers of a permanent move to the United States. Final acceptance that his marriage would be childless probably clinched the issue. At the same time, his new post relieved him of some of the frustrations he had felt at B.T.H., as a succession of postgraduate students allowed him to put many of his ideas into practice. Together they built a Wilson cloud chamber, an analog computer, a flat television tube, and many other devices.

Meanwhile, attempts were under way to build a working electron microscope, based on Gabor's holographic principle, in the new research laboratories of Associated Electrical Industries, at Aldermaston. A.E.I. was the parent company of both

B.T.H. and Metropolitan-Vickers, the latter a pioneer in the development and manufacture of electron microscopes. In 1950, with government funding, this program of holographic electron microscopy got under way with the participation of Michael Haine, James Dyson, and Tom Mulvey. Gabor remained closely involved as a consultant. By 1953 they had demonstrated that reconstituted images were possible with use of the technique, although overall results emphasized the limitations. They were unable to reduce the resolution to the point that there was any real advantage over more traditional methods. The work was held in abeyance from 1953 until a decision was made to close it down in 1955.

Gabor took little consolation in the modest improvements made and let his disappointment be known. With feelings running high, he declared to Haine: ". . . it was a very ill wind which I let out now almost eight years ago which blew nobody any good, least of all to myself." The holographic principle, on which Gabor had set such high hopes, appeared to be receding to the status of a scientific curiosity.

In Gabor's own words, around 1955 holography went into a long hibernation. Among other early attempts to apply the technique, Gabor, with Walter P. Goss, constructed a holographic interference microscope that failed to arouse any interest from the optical industry. In California, Hussein El-Sum, Paul Kirkpatrick, and Alberto Baez attempted to produce X-ray holograms, and Gordon Rogers in Britain worked with radio waves. In general, however, interest in the subject seemed to have petered out. Far from Gabor's intended use of holography, only the recognition that it could be successfully applied to radar stimulated the research that kept an interest alive.

Doing classified radar research at the Willow Run Laboratory of the University of Michigan, Emmett N. Leith first became aware of Gabor's work in late 1956, although not until 1960 did he and his collaborator, Juris Upatnieks, initiate a research program. In that year they duplicated Gabor's optical experiments. Using simple means, they succeeded in overcoming one of the major limitations experienced by Gabor and others, the elimination of a spoiling "twin" image produced in the reconstruction. In 1962 they produced the first laser holograms.

The availability of the powerful coherence provided by lasers is often seen as the key factor in the revival of holography. This is only partly true. The invention of the laser coincided with a clear increase of interest in holography, and many of the achievements of this period could have been, and in some cases were, produced without the use of lasers. It is true, however, that the dramatic three-dimensional holographic images could be produced only by diffuse reflection using lasers.

In 1963 and later, Gabor claimed he had the complete idea for a laser back in 1950, but could not find a student willing to take up the project. Optical holography, meanwhile, was enough to make Gabor a celebrity and truly give him his "comeback," even if it was not exactly in the way he intended. He experienced the inventor's dream of seeing an idea grow into concrete applications. These came in abundance, and many further applications of the general principles of wavefront reconstruction are still possible in the future.

Concern for the future came to dominate Gabor's later years. In the same year that Leith and Upatnieks revealed their laser holograms, Gabor published the book he had written in "a year of Saturdays," *Inventing the Future*, in which he eloquently expounded his concern for the future of industrial civilization, faced with the triple threat of overpopulation, nuclear weapons, and the "leisure society." It expanded on themes first developed in his inaugural address following election to a personal chair of applied electron optics at Imperial College in 1958 (published in an abridged form in *Encounter* in 1960). His major concern was his perception of a mismatch between technology and social institutions, and the necessity for inventive people to turn to "social inventions" as a first priority—indeed, this became his own priority. He promoted the mobilization of "a force of thinkers" to provide visions thirty or forty years ahead and found opportunity for such a project when, in 1968, he became a founding member and active participant in the Club of Rome, an elite drawn from diverse backgrounds who were interested in the study of global problems. His views are reflected in *Inventing the Future* and its sequel, *The Mature Society*. In the former he confesses, "Any book on the future will tell more of its author than about things to come." Gabor's beliefs were essentially those of a conservative and benevolent humanist. Although his books have an optimistic outlook, in private he was a pessimist and held a sneaking longing for the past. An avid reader of science fiction, he also, by contrast, admired the more conservative writings of Evelyn Waugh. One of the greatest influences on him, however, came from Aldous Huxley. Haunted by the Malthusian threat of overpopulation, Gabor shared many of the views of the British eugenicists. Writings on the subject by Julian Huxley and Charles Galton Darwin also influenced him. Politics and religion had little part in Gabor's view of things. His pessimism came more from what he saw as the "irrationality" of human behavior. He opposed the Vietnam war and saw the space program as "the last collective folly of mankind." At the same time, and in contrast, he viewed the student unrest of the late 1960's as a symptom of social malaise and the decline of morality.

In his later years, Gabor still found time for scientific interests. With the revival of holography, he was much in demand as a speaker on the subject. He also made further important contributions of his own, including introduction of the application of holography to computer data processing. After retiring he remained a research fellow and professor emeritus of Imperial College. Much of his time, however, was divided between the CBS Laboratories in the United States, where he did much of his later work as a part-time consultant, and a summer home in Italy.

In December 1971 came the Nobel Prize. Gabor had made his comeback. Yet by the time of his Nobel Lecture, he had virtually given up any idea of the realization of his original aim, holographic electron microscopy. One year later, Lawrence Bartell, at the University of Michigan, realized how it would be possible to devise a holographic electron microscope capable of forming images of electron clouds in gas-phase atoms. Several undergraduate students developed the method to a point where atomic and molecular photographic images of a quality allowing bond lengths to be measured with a ruler were produced. In April 1974, Bartell informed Gabor of these developments, which immediately set him to designing his own holographic electron microscope.

During the summer of that year, Gabor suffered a severe stroke, after which he could neither read nor write, and later almost totally lost the power of speech, although his intellect remained unimpaired. Four years later, after a summer spent at his Italian home, he was confined to his bed, and, during the following winter, he died peacefully in a London nursing home. In 1977 he had visited the newly created Museum of Holography in New York City.

Gabor's inventions may yet hold greater future significance. The basic optical hologram, however, will remain one of those things that stir the imagination. As Emmett Leith remembered about duplicating Gabor's original optical experiments: "The results were only what we had expected, yet the physical realization was rather awesome."

BIBLIOGRAPHY

I. Original Works. A complete list of Gabor's publications and references to many of his patents are in Allibone (see below). His works include *Inventing the Future* (London, 1963); "Holography 1948–71," in *Le Prix Nobel en 1971* (Stockholm, 1972), 169–201; *The Mature Society* (London, 1972); and "The Principle of Wavefront Construction, 1948," in Ezio Camatini, ed., *Optical and Acoustical Holography* (New York, 1972), 9–14.

Gabor's private papers and correspondence are in the archives of Imperial College of Science and Technology, South Kensington, London. Autobiographical notes are at the Royal Society of London.

II. Secondary Literature. T. E. Allibone, "Dennis Gabor," in *Biographical Memoirs of Fellows of the Royal Society*, **26** (1980), 107–147, the most comprehensive outline of Gabor's life and work; Michael Edward Haine and Vernon Ellis Cosslett, *The Electron Microscope* (London, 1961; repr. 1962); Emmett N. Leith, "Dennis Gabor, Holography and the Nobel Prize," in *Proceedings of the IEEE*, **60** (1972), 653–654, and "The Legacy of Dennis Gabor," in *Optical Engineering*, **19** (1980), 633–635; and Tom Mulvey, "Fifth Years of High Resolution Electron Microscopy," in *Physics Bulletin*, **34** (1983), 274–278.

S. T. Keith

GAEDE, WOLFGANG (*b.* Lehe, near Bremerhaven, Germany, 25 May 1878; *d.* Munich, Germany, 24 June 1945), *physics*.

Gaede, the founder of high-vacuum technology, through his invention of the mercury rotary pump (1905), the molecular pump (1912), and the diffusion pump (1915), made possible numerous new applications in technology and physics—semiconductor technology and nuclear physics, to mention only two.

The son of Karl Wilhelm Gaede, a Prussian army captain, and of Amalie Ruef, Gaede spent his youth in Freiburg im Breisgau, his mother's native city. There he attended the university. From childhood he had shown a remarkable ability to visualize and analyze complicated moving systems, a skill obviously of invaluable help in his later inventions. He gave his technicians little more than a rough sketch, since he told them every detail of the pump under construction from memory.

At the University of Freiburg, Gaede first studied medicine but soon changed to physics. He received the Ph.D. in 1901 with a dissertation titled "Über die Änderung der spezifischen Wärme der Metalle mit der Temperatur." He remained at Freiburg, as assistant to Franz Himstedt, until 1908. Continuing the work of his dissertation with experiments on

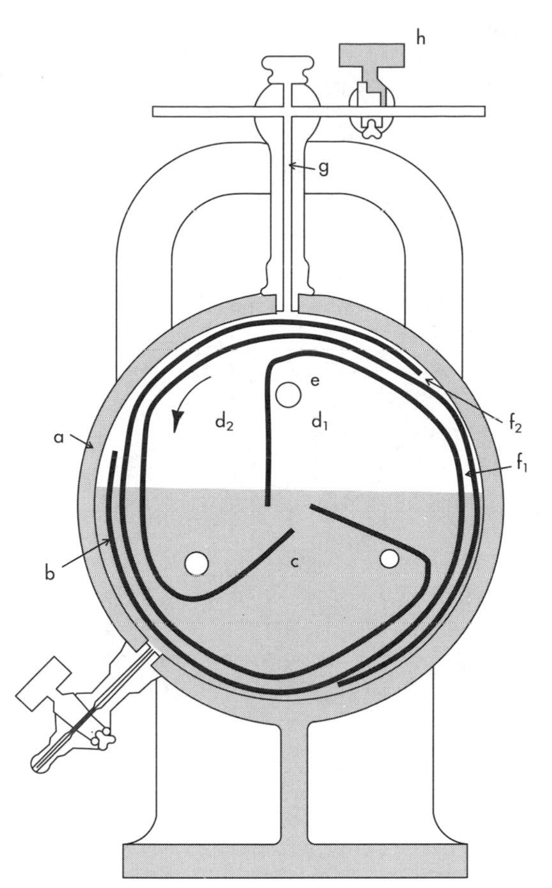

FIGURE 1. Schematic Diagram of a Rotary Mercury Pump.
a housing
b porcelain drum
c mercury pool
d_1, d_2 chambers
e intake port
f_1, f_2 exhaust channels
g connection to vacuum vessel
h stopcock

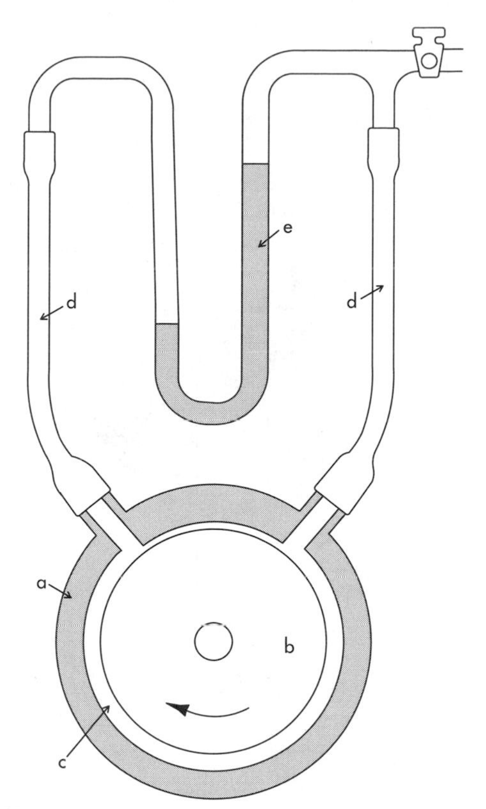

FIGURE 2. Schematic Diagram of a Molecular Pump.
a housing
b rotating cylinder
c slot
d hose
e manometer

the Volta effect in a vacuum, Gaede was led to his first invention, the mercury rotary pump. It was the first continuously working air pump and had a much higher pumping speed than previous pumps—about 0.2 cubic decimeters per second.

Gaede presented the pump to the scientific world in 1905, at the Naturforscherversammlung in Meran. The efficient pump immediately was in great demand, and Gaede commissioned the small firm E. Leybolds Nachfolger to produce it. Gaede maintained a close connection with this establishment, which became a leader in vacuum technology.

Between 1905 and 1909 Gaede spent much of his time investigating the external friction of gases and its possible usefulness for a new method of creating a high vacuum. He qualified to teach at the University of Freiburg with the *Habilitationsschrift* "Die äussere Reibung der Gase" (1911).

One consequence of Gaede's work on external friction of gases was the construction of the molecular pump in 1912. It was the first pump to be based on a method different in principle from that used by Otto von Guericke. The gas molecules adhere to a fast-rotating surface, thus gaining additional momentum away from the vessel to be exhausted. The pumping speed was about 2 cubic deciliters per second, ten times greater than the mercury rotary pump. The ultimate vacuum pressure was less than 10^{-6} millimeters of mercury. In 1913 this invention brought Gaede an associate professorship at Freiburg and the Elliot Cresson gold medal of the Franklin Institute.

Another new pumping principle discussed in Gaede's *Habilitationsschrift* was that of the diffusion pump. In it the gas molecules diffuse into a rapidly moving vapor jet and thus move away from the evacuated space and toward the backing pump. With this apparatus Gaede obtained a new record low pressure. The prototype had a pumping speed of 0.2 cubic deciliters per second. In 1923 the efficiency

GAEDE

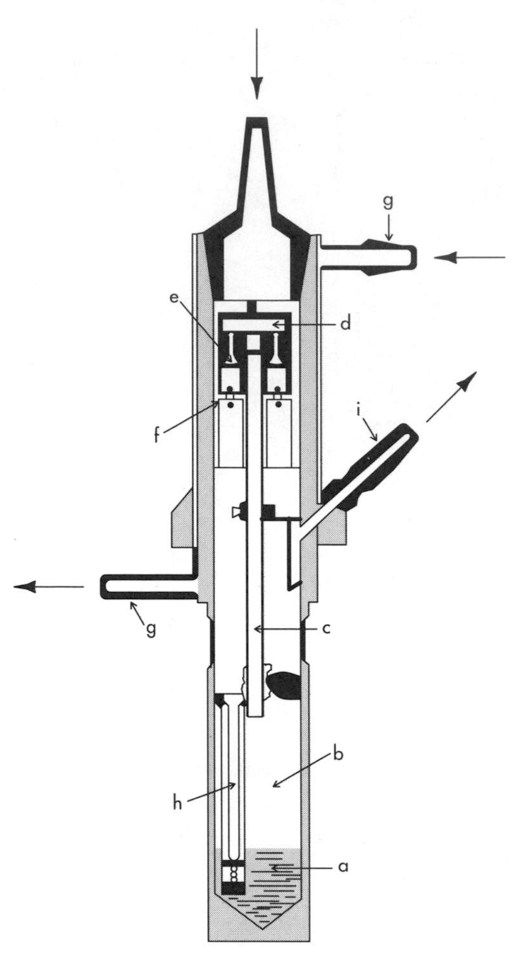

FIGURE 3. Schematic Diagram of a Diffusion Pump.
a mercury pool
b vapor compartment
c conduit for mercury vapor
d distributor
e nozzles
f slit
g cooling water intake and outlet, respectively
h return pipe for liquid mercury
i vacuum pump connection

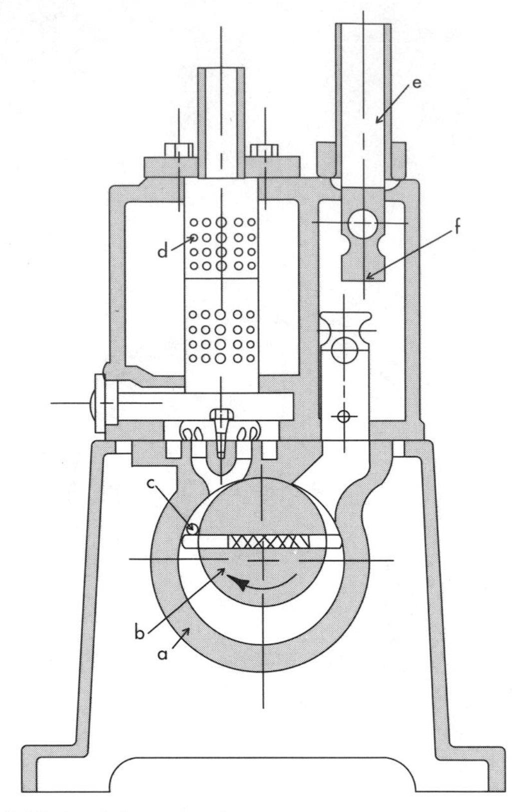

FIGURE 4. Schematic Diagram of a Gas Ballast Pump.
a housing
b rotor with dampers
c gas ballast valve
d oil trap
e intake port
f dirt trap

was improved to 150 cubic deciliters per second and in 1944 to 1,500 cubic deciliters per second.

In 1917 Gaede was offered a professorship at the Technical University in Berlin-Charlottenburg. He declined because of the political instability in that city. In 1919 he decided to go to the Technical University of Karlsruhe, where he accepted the professorship once held by Heinrich Hertz. The Physical Society of London awarded Gaede the Dudell medal in 1933; in the same year he was denounced by some of his assistants as being "politically unreliable," and the medal made him an object of persecution by the Nazi regime. Although he was not Jewish, Gaede was dismissed from his professorship at the end of 1933. However, the Siemens-Ring Foundation voted to give him the Siemens-Ring, the highest award a German inventor could receive. He transferred his work to a private laboratory in Karlsruhe that was equipped and sponsored by Leybold.

In 1935 Gaede invented the gas ballast device that he applied to an oil-filled rotary vane pump (gas ballast pump). This device, which prevents pumped vapors from being condensed in the pump during the compression phase, injects a permanent gas, usually air, into the compression chamber of the pump so that the compression ratio of the vapor is reduced when the exhaust valve of the pump opens, and thus no condensation occurs.

In 1939 Gaede went to Munich, where he established a private laboratory. In his contact with Arnold Sommerfeld and his co-workers he found an acknowledgment he had not earlier enjoyed. The laboratory was destroyed during an air raid in 1944. A year later Gaede contracted a fatal case of diphtheria.

Throughout his life Gaede was an outsider. He could afford to be, since he had no financial problems after the invention of the mercury rotary pump. He never married, and his sister Hannah always took care of him. Contacts with his colleagues were rare, and his students testified that he was a bad teacher. He seems to have been an amiable, eccentric man, as ingenious inventors are sometimes said to be.

BIBLIOGRAPHY

I. ORIGINAL WORKS. "Über die Änderung der spezifischen Wärme der Metalle mit der Temperatur" (Ph.D. dissertation, Freiburg im Breisgau, 1902); "Demonstration einer rotierenden Quecksilberluftpumpe," in *Physikalische Zeitschrift*, **6** (1905), 758–760; "Demonstration einer neuen Verbesserung an der rotierenden Quecksilberluftpumpe," *ibid.*, **8** (1907), 852–853; "Die äussere Reibung der Gase," in *Berichte der Naturforschenden Gesellschaft zu Freiburg im Breisgau*, **18**, no. 2 (1911), 133–197, his *Habilitationsschrift*; "Die äussere Reibung der Gase und ein neues Prinzip für Luftpumpen: Die Molekularluftpumpe," in *Physikalische Zeitschrift*, **13** (1912), 864–870, and in *Annalen der Physik*, **41** (1913), 289–336; "Die Molekularluftpumpe," in *Annalen der Physik*, **41** (1913), 337–380; "Die Diffusion der Gase durch Quecksilberdampf bei niederen Drucken und die Diffusionsluftpumpe," *ibid.*, **46** (1915), 357–392; "Die Entwicklung der Diffusionspumpe," in *Zeitschrift für technische Physik*, **4** (1923), 337–369; "Die Öldiffusionspumpe," *ibid.*, **13** (1932), 210–212; and "Gasballastpumpen," in *Zeitschrift für Naturforschung*, **2A** (1947), 233–238.

Gaede wrote articles on pumps and vacuum technology for Felix Auerbach and W. Hort, eds., *Handbuch der physikalischen und technischen Mechanik*, VI (Leipzig, 1928), 90–121; and Wilhelm Wien and F. Harms, eds., *Handbuch der Experimentalphysik*, IV, pt. 3 (Leipzig, 1930), 413–461.

II. SECONDARY LITERATURE. Tributes are E. Justi, in *Elektrotechnische Zeitschrift*, **64** (1943), 285–287; K. May, in *Zeitschrift für technische Physik*, **24** (1943), 65; and W. Molthan, *ibid.*, **19** (1938), 153–154.

Obituaries include Arnold Sommerfeld, "In Memoriam. Wolfgang Gaede," in *Zeitschrift für Naturforschung*, **2A** (1947), 240; and Franz Wolf, in *Physikalische Blätter*, **3** (1947), 384–386.

Biographical material is in Manfred Dunkel, *Geschichte der Firma E. Leybolds Nachfolger: 1850–1966* (Cologne, 1963); Hannah Gaede, *Wolfgang Gaede, der Schöpfer des Hochvakuums* (Karlsruhe, 1954); Poggendorff, V (1926), 407, and VIIa, pt. 2 (1958), 154; and Franz Wolf, in *Karlsruher akademische Reden*, n.s. **3** (1947), and *Neue deutsche Biographie*, VI (1964), 16–17.

A. BRACHNER

GARBASSO, ANTONIO GIORGIO (*b.* Vercelli, Italy, 16 April 1871; *d.* Arcetri, near Florence, Italy, 14 March 1933), *physics*.

Nothing seems to be known of Garbasso's family background. In 1900 he married Bianca Ventura, who was born in 1879; they had one son, Giorgio, born in 1918.

Garbasso attended *liceo* in Turin, then took his doctorate in physics at the university there in 1892. In the years 1893 and 1894, on a government traveling fellowship, he went to Bonn to work with Heinrich Hertz and, after Hertz's death in 1894, to the University of Berlin Physics Institute. He qualified for the *libera docenza* (equivalent to the *Privatdozentur*) in experimental physics in 1894 and in mathematical physics in 1895, unusual for an Italian physicist. Garbasso taught mathematical physics at the University of Pisa from 1895 to 1897 and at Turin from 1897 to 1903. In 1902, in national competitions (*concorsi*) for university chairs, he became the first physicist in two decades to be able to choose between experimental and mathematical physics; he opted for experimental physics at Genoa in order to have a laboratory, which he believed even mathematical physics required. Tenured in 1906, Garbasso was promoted to full professor in 1909. In 1913 he transferred to the Istituto di Studi Superiori in Florence (after 1924 the University of Florence), where he oversaw the building of the new physics institute, opened in 1920. On Italy's entry into World War I in 1915, he enlisted in the army. He rose to the rank of major in the Corps of Engineers and was creator and director of the artillery's sound-ranging service.

Garbasso's research interests were shaped by his unusual university experience. At Turin his teachers had been the experimental physicist Andrea Naccari and the mathematical physicist Giuseppe Basso. The latter was a physicist rather than the usual mathematician; also atypically, he coupled theoretical and experimental approaches in his research. Basso was perhaps the first in Italy to teach and use Maxwell's electromagnetic theory of light, in the 1880's. All Garbasso's work showed the influence of Basso in its integration of theory and experiment.

Much of Garbasso's research derived from the work of Hertz and Augusto Righi on the optics of electromagnetic waves. At Turin, in the years 1892 and 1893, Garbasso devised an apparatus composed of insulating screens covered with a uniform network of metal strips. The screens behaved like a series of resonators, and electromagnetic waves aimed at them exhibited selective absorption and reflection as if the screens were colored bodies. Hertz praised the experiment as extremely beautiful and stated that he wished he had conceived and performed it himself. In Berlin, Garbasso collaborated with Emil

Aschkinass on adapting his apparatus to refract Hertzian radiation as a prism refracts light. Heinrich Rubens and Ernest Fox Nichols at the Physikalisch-Technische Reichanstalt, also in Berlin, soon used the same arrangement to demonstrate the reflection, refraction, and dispersion of infrared radiation. Thus, Garbasso's work forms a link in the chain of experimental and theoretical studies that led to Planck's radiation law in 1900.

Garbasso's later research built upon his interest in mechanical models of electrodynamic phenomena: he devised models for the emission of light and the explanation of spectra, for electron behavior, and for atomic structure. His theoretical and experimental researches on the mirage and variable refraction in general were highly praised.

Although he never had a chair to match his inclinations, Garbasso could be considered Italy's first German-style theoretical physicist; indeed, he referred to himself as a professor of theoretical physics in the information he submitted in 1902 for Poggendorff IV. He has also been called the precursor of atomic physics in Italy. Garbasso began propagandizing for Maxwell's electrodynamics in Italy upon receiving his doctorate, publishing two series of lectures in 1893 and 1897, and he published a course of lectures on theoretical spectroscopy in German (1906). In 1913 he welcomed the publication of Bohr's atomic theory. Garbasso's research career was terminated by World War I and by the efforts to build the new physics institute; but he became, after Orso Mario Corbino, the second key patron of theoretical physics in Italy in the 1920's and early 1930's and supported efforts to establish the first chairs in the field in the mid 1920's.

Through Garbasso's Florence institute passed many young physicists interested in modern topics, including Enrico Fermi and Franco Rasetti before they moved to Rome, Enrico Persico, Bruno Rossi, Gilberto Bernardini, Giuseppe Occhialini, and Giulio Racah. Like Corbino, Garbasso saw the need for closer collaboration between theorists and experimentalists in science, as well as for new institutions designed to exploit scientific results for industrial and public purposes. Italy's lack of precision optical instruments during World War I led him to promote a national institute for optics and precision mechanics and then to house in his institute at Arcetri the Istituto Nazionale di Ottica. But Garbasso's proposed solutions to Italy's scientific and technical backwardness were more piecemeal than Corbino's, and they explicitly rejected the channeling of "pure" research or the organization of "pure" scientists.

Also in contrast with Corbino, Garbasso was early attracted to the Fascist message of order and renewal. He was elected mayor (1920–1927) and then appointed first Fascist *podestà* (1927–1928) of Florence, and became a member of the Fascist party in 1923. For his success in keeping Florentine coalition partners of the Fascists loyal to the government during the crisis following the murder of the Socialist deputy Giacomo Matteotti in 1924, he received several honors from Mussolini.

Early in his career, Garbasso held a view on models similar to those of Hertz and Henri Poincaré: that the many possible mechanical models of a phenomenon do not tell us about the underlying reality. He esteemed science as the privileged way to knowledge in a manner that at times appeared to border on scientism, yet he was also a professing Roman Catholic. These two factors, coupled with his antipathy toward both positivism and idealism, led him to collaborate with the Roman Catholic modernist reformers of the journal *Il Rinnovamento* (1907–1909), for which he prepared a series of articles on contemporary issues in the physical sciences. Here he modified his earlier position to incorporate the idea that the advance of science narrows the choice among models, which come increasingly to resemble reality. Garbasso did not contribute to *Il Rinnovamento* after the blanket excommunication of its collaborators in November 1907, which followed the papal condemnation of modernism earlier that year. Within a few years he proclaimed the experimental sciences to be "realistic," consistent with the moderate realism of the neo-scholastics dominant in the church until the Second Vatican Council.

In his later years Garbasso held the secular neo-idealist philosophy represented by Benedetto Croce, responsible for the cultural undervaluation of science and the weak support for it in Italy. Against Croce, he argued for the aesthetic and moral value of science, and as director of the Museum of Antique Instruments and promoter of the National Museum of the History of Science, he spoke often about Italy's past scientific greatness.

Among Garbasso's many honors were the gold medal of the Madrid Academy of Sciences (1905), the Matteucci Medal of the Società Italiana delle Scienze (1908), the Royal Prize for physics for 1918 of the Accademia dei Lincei (awarded in 1920), life appointment as senator of the kingdom (1924), and the presidencies of the National Social Insurance Fund and the Committee for Astronomy, Applied Mathematics, and Physics of the Italian National Research Council.

BIBLIOGRAPHY

I. ORIGINAL WORKS. Extensive lists of Garbasso's scientific and popular scientific writings are found in Poggendorff and in the obituary by Brunetti noted below. Early lectures on Maxwell's electromagnetic theory of light given at the University of Turin in 1892 and 1893 were published as "La teoria di Maxwell dell'elettricità e della luce," in *Rivista di matematica* **3** (1893), 149–169. The book *Quindici lezioni sperimentali su la luce considerata come fenomeno elettromagnetico* (Milan, 1897; repr. Milan, 1898) was especially important in the diffusion of the knowledge of Maxwell in Italy. *Vorlesungen über theoretische Spektroskopie* was written and published in German (Leipzig, 1906) because no Italian publisher was interested in such a work. The book *Fisica d'oggi, filosofia di domani* (Milan, 1910) was composed of articles on developments and problems in contemporary physics originally planned for, and in part published in, the short-lived (1907–1909) Roman Catholic modernist journal *Il Rinnovamento*.

After Garbasso's death twenty-two of his nontechnical biographical, historical, and philosophical lectures and essays were collected in *Scienza e poesia*, edited by Jolanda de Blasi, with a preface by Benito Mussolini (Florence, 1934).

In his long essay "Perchè, quando, e come nacque l'Istituto Nazionale di Ottica di Arcetri," in *Atti della Fondazione Giorgio Ronchi*, **31**, no. 2 (1976) through **32**, no. 2 (1977), and also available as a book by the same title (Florence, 1977), Vasco Ronchi reprinted two reports Garbasso coauthored with Luigi Pasqualini on the need for and the work of a national laboratory for optics and precision mechanics, "Per un laboratorio e una rivista di 'Ottica e Meccanica di Precisione': Promemoria [1917]," **31** (1976), 180–184; and "Ai fondatori del Laboratorio di Ottica e Meccanica di Precisione [c. 1925]," 363–367; and two related speeches of Garbasso's, "Per l'inaugurazione del Laboratorio di Ottica e Meccanica di Precisione [1918]," 200–213, and "Sulle condizioni attuali dell'insegnamento dell'ottica pratica in Italia [1928]," 673–677.

Letters written by Garbasso are included in the Vito Volterra Papers and the Tullio Levi-Civita Papers, both at the Accademia Nazionale dei Lincei, Rome, and the Augusto Righi Papers, at the Accademia Nazionale dei Quaranta, Rome. A 1914 exchange with Niels Bohr on the Stark–Lo Surdo effect is located in the Archive for History of Quantum Physics.

II. SECONDARY LITERATURE. The major obituary is by Garbasso's longtime assistant at Florence, Rita Brunetti, "Antonio Garbasso: La vita, il pensiero e l'opera scientifica," in *Nuovo cimento*, new ser., **10** (1933), 129–152. A second important obituary, not listed in Poggendorff VIIb, is Luigi Puccianti, "Commemorazione del Presidente dell'Accademia Antonio Garbasso," in *Rendiconti dell'Accademia dei Lincei classe di scienze fisiche*, 6th ser., **17** (1933), 988–995. Another not listed in Poggendorff is Gilberto Bernardini, "Antonio Garbasso," in *Ricerca scientifica*, 4th ser., **1** (January–June 1933), 441–446, which reprints the sections relating to Garbasso of the 1920 report of the Commission for the Royal Prize in Physics of the Accademia dei Lincei. An acerbic treatment of Garbasso's philosophical and historical approaches is Sebastiano Timpanaro, "La scienza di Garbasso," in *Scritti di storia e critica della scienza* (Florence, 1952), 260–269. Useful for contemporary notice of Garbasso's public and political activities is Alessandro Martelli, "Antonio Garbasso," in *Illustrazione toscana e dell'etruria*, 2nd ser., **11** (May 1933), 4–8; see also Marco Palla, *Firenze nel regime fascista (1929–1934)* (Florence, 1978), 93–96. For Garbasso's views on the organization of Italian physics and their relation to Corbino's, see Carlo Tarsitani, "La fisica italiana tra vecchio e nuovo: Orso Mario Corbino e la nascita del gruppo Fermi," in Giovanni Battimelli, Michelangelo De Maria, and Arcangelo Rossi, eds., *La ristrutturazione delle scienze tra le due guerre mondiali*, vol. 1, *L'Europa* (Rome, 1984), 323–346.

BARBARA J. REEVES

GARROD, ARCHIBALD EDWARD (*b*. London, England, 25 November 1857; *d*. Cambridge, England, 28 March 1936), *medicine, medical chemistry, genetics*.

Garrod was the fourth and youngest son of Sir Alfred Baring Garrod, physician to King's College Hospital, and his wife, Elisabeth Ann Colchester. In 1848 his father, deeply interested in chemistry, discovered that in patients suffering from gout, the blood contains an elevated concentration of uric acid, an observation that enabled him to distinguish gout from rheumatoid arthritis.

Garrod was educated at Marlborough, whence he went to Christ Church, Oxford. In 1880 he was graduated B.A., with first-class honors in natural science, and began to study medicine at St. Bartholomew's Hospital, London. Four years later he received the M.R.C.S. diploma of the Royal College of Surgeons and was graduated M.B. at Oxford. In 1886 he proceeded to the Oxford M.D. degree, and the same year married Laura Elisabeth Smith, daughter of Sir Thomas Smith, baronet, surgeon to St. Bartholomew's Hospital.

After receiving his medical qualification in 1884, Garrod remained at St. Bartholomew's Hospital, first as a house physician (1884–1887) and then as casualty physician, assistant demonstrator in practical medicine, medical registrar, demonstrator of morbid anatomy, and lecturer on chemical pathology. While engaged in the daily work of a medical school teacher and the care of patients in the hospital, Garrod also pursued research. In 1886 he published his first book, a manual on the use of the laryngoscope, and in 1888 became a member of the Royal College of Physicians of London. In 1890 Garrod published a work on rheumatism and rheumatoid arthritis, diseases on which he had already written several papers pointing out the close connection among rheumatic fever, chorea, and endocarditis. Garrod argued that rheumatoid arthritis was essentially a disease of the joints and should be distinguished clearly from rheumatism, which he viewed as a single, specific disease that might manifest itself as rheumatic fever, rheumatic heart disease, or chorea. In 1891 he was elected a fellow of the Royal College of Physicians.

In his studies of rheumatic diseases, Garrod began in the 1890's to investigate both normal and abnormal pigments in the urine for the purpose of detecting changes in metabolism induced by disease. In 1892 he found an unusually colored urine in a patient suffering from chorea and showed that it was caused by the pigment urohematoporphyrin.[1] He worked out methods for the separation of the pigment and showed that it was present in small amounts in normal urine.[2] In 1896 Garrod began, in collaboration with the biochemist Frederick Gowland Hopkins, to study the urinary pigment urobilin, describing its spectroscopic properties.[3]

In addition to his posts at St. Bartholomew's Hospital, Garrod served on the visiting staffs of the West London Hospital and of the Hospital for Sick Children in Great Ormond Street. In 1897, at the Hospital for Sick Children, Garrod examined a three-month-old boy, Thomas P., brought to the outpatient department because his urine was staining his diapers a deep reddish-brown. Garrod recognized the phenomenon as alkaptonuria, a condition described in 1822 by Alexandre Marcet and named in 1858 by the pharmaceutical chemist Carl Heinrich Detlev Bödeker.[4] In 1891 Michajl Volkov and Eugen Baumann found that the pigment in alkaptonuric urine was a derivative of the amino acid tyrosine, and they named it homogentisic acid.[5] From his study of Thomas P., Garrod developed an improved method for the extraction of homogentisic acid from the urine.[6] He went on to determine the levels of homogentisic acid in four other patients with alkaptonuria. The second patient, a boy of fourteen, had a brother with alkaptonuria.

In 1899 Garrod drew together data on thirty-nine cases of alkaptonuria, either recorded in the literature or observed by himself or other London physicians.[7] He noted the tendency of the anomaly to occur among siblings, but at that time he knew of no instance of the transmission of alkaptonuria from one generation to another.

In March 1901 a fifth child born to the parents of Garrod's former patient Thomas P. proved to have alkaptonuria. Three of the five children in the family were alkaptonuric, and several months later Garrod thought to ask the mother whether she and her husband were related. He learned that they were first cousins; their mothers were sisters. When Garrod inquired about other families possessing one or more children with alkaptonuria, he found that in each instance the parents were first cousins.

In accordance with the germ theory of disease, then strongly in the ascendant, Garrod had first thought that alkaptonuria must result from an intestinal infection by some organism that converted tyrosine to homogentisic acid. But one afternoon, as he was walking home from the hospital, thinking about the question, it occurred to him that alkaptonuria might result from some error in the metabolism more or less comparable with a structural malformation. It should not be considered a disease, but simply an alternative course of metabolism, inherited and harmless.[8]

Garrod saw, too, that Mendel's laws of inheritance, presented to English readers by the biologist William Bateson in his book *Mendel's Principles of Heredity* (1902), provided an explanation of the puzzling phenomenon of alkaptonuria. The mating of first cousins created the conditions that enabled a rare, recessive character to appear. Garrod found that some twelve alkaptonuric children, described by himself and others, all resulting from the marriages of first cousins, had a total of thirty-six normal siblings—a ratio of three normal children to one alkaptonuric in families containing alkaptonurics. Such a perfect 3:1 Mendelian ratio showed that alkaptonuria was inherited as a simple Mendelian recessive factor.

To reinforce his view that alkaptonuria was simply an alternative mode of metabolism, Garrod studied such similar familial anomalies as albinism and cystinuria. Like alkaptonurics, albinos tended to be the offspring of marriages of first cousins. Garrod suggested that in addition to the known anomalies he had studied, others might exist still unrecognized. Just as no two individuals of a species were exactly identical in bodily structure, neither might their internal chemistry be exactly the same.

In June 1908 Garrod delivered the Croonian Lectures before the Royal College of Physicians of London on the various metabolic anomalies he had studied and published them as a book entitled *Inborn Errors of Metabolism*. By 1908 he was able to suggest that the anomaly in alkaptonuria derived from the lack of a specific enzyme that catalyzed a specific chemical change within living tissue. The lack of such an enzyme would result in the accumulation of an intermediate substance that would then be excreted. The metabolic error in alkaptonuria was an inability to break down the aromatic portion of the molecules of phenylalanine and tyrosine. Since the inborn error was inherited and caused by a specific Mendelian factor (in modern parlance, a gene), Garrod was introducing a link between a specific enzyme and a specific gene. He thus foresaw the one gene–one enzyme hypothesis that in the 1940's would be developed so effectively by George Wells Beadle and Edward L. Tatum in their study of the fungus *Neurospora*, and that has been central to the development of theories of gene action.

From a historical standpoint, Garrod's work represents a convergence of clinical medicine and medical chemistry with the new science of genetics that developed so rapidly after 1900. In England during the 1890's the presuppositions of both medical chemistry and clinical medicine were undergoing subtle but profound change under the influence of Charles Darwin's theory of natural selection, an influence especially great upon Garrod and Hopkins. In Garrod's work, clinical medicine and medical chemistry joined with genetics to produce the concept of inborn errors of metabolism—a concept containing the views that metabolism consisted of a series of chemical reactions and intermediate substances and that the production of the enzyme catalyzing each metabolic reaction was governed by a specific inherited factor.

In recognition of his contributions to science, Garrod was elected a fellow of the Royal Society in 1910. In 1912 he was appointed a full physician to St. Bartholomew's Hospital. Following the outbreak of World War I in 1914, Garrod joined the staff of the First London General Hospital at Camberwell. In 1915 he was promoted to temporary colonel in the Army Medical Service and went to Malta, where he remained until 1919 as consulting physician to the Mediterranean forces. In 1918 he was knighted. Two of Garrod's three sons were killed in action during the war, and one died of influenza after the armistice. In 1919 Garrod returned to St. Bartholomew's Hospital as director of a new medical unit that included a clinical laboratory close to hospital wards.

In 1920 Garrod succeeded Sir William Osler as Regius professor of medicine at Oxford University. At Oxford he worked to develop clinical investigation. He emphasized that in disease the clinician saw unique phenomena, experiments of nature, that he alone could study. For that reason the clinician needed at hand his own laboratories for the scientific study of the problems that arose in the care of patients.

A tall, handsome man of distinguished appearance and kindly manner, Garrod taught medical students to think biochemically. He was also a supremely effective bedside clinician who could elicit the signs and symptoms of disease with great skill and simplicity. In 1923 he brought out a second edition of *Inborn Errors of Metabolism*, in which he added discussions of two more inherited anomalies, steatorrhea and hematoporphyria. In 1924 Garrod delivered the Harveian Oration to the Royal College of Physicians, "The Debt of Science to Medicine," in which he pointed out the many contributions of clinical observation to the growth of medical science.

Garrod remained at Oxford until his retirement in 1927, at the age of seventy. After living for a short time at their country house at Woodbridge in Suffolk, Sir Archibald and Lady Garrod moved to Cambridge, where their daughter, the archaeologist Dorothy A. E. Garrod, was a fellow of Newnham College. Garrod died at his house in Cambridge.

NOTES

1. A. E. Garrod, "On the Presence of Urohaematoporphyrin in the Urine of Chorea and Articular Rheumatism," in *Lancet* (1892), **1**, 793.
2. A. Garrod, "Haematoporphyrin in Normal Urine," in *Journal of Physiology*, **17** (1894), 349–352.
3. A. Garrod and F. Gowland Hopkins, "On Urobilin," *ibid.*, **20** (1896), 112–144, and **22** (1897–1898), 451–464.
4. [Carl Heinrich Detlev] Boedeker, "Über das Alcapton; ein neuer Beitrag zur Frage: Welche Stoffe des Harns können Kupferreduction bewirken?" in *Zeitschrift für rationale Medizin*, **7** (1859), 130–145.
5. M. Volkov [Wolkow] and E. Baumann, "Über das Wesen der Alkaptonuria," in *Zeitschrift für physiologische Chemie*, **15** (1891), 228–285.
6. A. Garrod, "Alkaptonuria: A Simple Method for the Extraction of Homogentisic Acid from the Urine," in *Journal of Physiology*, **23** (1898–1899), 512–514.
7. Archibald E. Garrod, "A Contribution to the Study of Alkaptonuria," in *Medico-Chirurgical Transactions published by the Royal Medical and Chirurgical Society of London*, **82** (1899), 367–391.
8. Archibald E. Garrod, "About Alkaptonuria," *ibid.*, **85** (1902), 69–77; and "The Incidence of Alkaptonuria: A Study in Chemical Individuality," in *Lancet* (1902), **2**, 1616–1620.

BIBLIOGRAPHY

I. Original Works. In addition to about 100 papers published in various medical and scientific journals, Garrod published the following books: *An Introduction to the Use of the Laryngoscope* (London, 1886); *A Treatise on Rheumatism and Rheumatoid Arthritis* (London, 1890); *A Handbook of Medical Pathology for the Use of Students in the Museum of St. Bartholomew's Hospital* (London, 1894); *Inborn Errors of Metabolism* (London, 1909; 2nd ed., 1923); *The Debt of Science to Medicine* . . . (Oxford, 1924); and *The Inborn Factors in Disease* . . . (Oxford, 1931).

Among his many articles in journals the more significant include: "On the Relation of Chorea to Rheumatism, with Observations of Eighty Cases of Chorea," in *Medico-Chirurgical Transactions*, **72** (1889), 145–163; "On the Occurrence and Detection of Haematoporphyrin in the Urine," in *Journal of Physiology*, **13** (1892), 598–620; "Concerning Cystinuria," in *Journal of Physiology*, **34** (1906), 217–223, with W. H. Hurtley; "The Croonian Lectures on the Inborn Errors of Metabolism," in *Lancet* (1908), **2**, 1–7, 73–79, 142–148, 214–220; "Congenital Family Steatorrhea," in *Quarterly Journal of Medicine*, **6** (1912–1913), 242–258; "Medicine from the Chemical Standpoint," in *British Medical Journal* (1914), **2**, 228–235; "On Congenital Porphyrinuria, Associated with Hydro Aestivale and Pink Teeth," in *Quarterly Journal of Medicine*, **15** (1921–1922), 319–330, with L. Mackey; and "The Lessons of Rare Maladies," in *Lancet* (1928), **1**, 1055–1059.

II. SECONDARY LITERATURE. There is no biography of Garrod. Obituaries include "Sir Archibald Garrod," in *British Medical Journal* (1936), **1**, 731–736; and "Sir Archibald Edward Garrod," in *Lancet* (1936), **1**, 807–809. See also F. G. Hopkins, "Archibald Edward Garrod, 1857–1936," in *Obituary Notices of Fellows of the Royal Society of London*, **2** (1936–1939), 225–228.

LEONARD G. WILSON

GARSTANG, WALTER (*b*. Blackburn, England, 9 February 1868; *d*. Oxford, England, 23 February 1949), *zoology*.

Garstang, the son of Walter Garstang, a doctor, intended to read medicine when he entered Oxford in 1884. Influenced by Henry N. Moseley, professor of zoology, he soon changed his mind and took his degree in zoology. In 1888 he was appointed a founder member to the staff of the Marine Biological Laboratory at Plymouth, where he worked off and on until 1901. In 1891 he spent a year at Owens College, Manchester, with Arthur Milnes Marshall, an embryologist and recapitulationist of the Balfour school; and in 1893 he was elected fellow of Lincoln College, Oxford, where he taught zoology until 1897 while continuing to carry out his research at Plymouth during vacations. In 1895 he married Lucy Ackroyd, who shared his biological interests; they had a son and five daughters.

Garstang's research in this early phase concerned marine biology. His first paper, for instance, was on the warning coloration of nudibranch mollusks; he also wrote descriptive papers on the natural history and morphology of various marine organisms, particularly the invertebrate group called tunicates, which was a lifelong interest. By this time he had probably formulated his original views on the relation of embryology and evolution; they are implied, for instance, in a paper of 1894 on the ancestry of the vertebrates; but their main publication would come later.

From 1897 to 1907 Garstang worked full time on the fisheries of the British Isles. He was at Plymouth until 1901, when the International Council for the Exploration of Sea was established; he then founded its Fisheries Laboratory at Lowestoft. He sought the kind of biological knowledge that was necessary for the scientific exploitation of the sea. He thus considered the relation of physical conditions in the sea and plankton numbers, and at Plymouth he investigated variation in size among the different stocks of mackerel. This work was a forerunner of a larger study of plaice at Lowestoft. By mark and recapture studies, Garstang worked out the growth rates and migrations of this species. Having noticed that young plaice grow faster on the Dogger Bank (in the North Sea) than off the Dutch coast, he tried catching, marking, and transplanting fry from the latter site to the former. The transplanted fry duly grew faster. Garstang argued that mass transplantations could be commercially possible.

By 1907 the British government had designs upon the laboratory at Lowestoft. Garstang was not the kind of man to take direction from civil servants, and with the douceur of the new professorial chair in zoology at Leeds University, his resignation from Lowestoft was secured. At Leeds he had less time for research: he was setting up a new school, establishing two laboratories (one at Leeds and a marine station on the coast at Robin Hood Bay), and—as was then customary among professors—taking an active interest in his undergraduate students. He helped to found the Leeds Boat Club, assisted with the O.T.C., wrote and produced student entertainments, and kept open house. Teaching the subject, however, enabled him to return to his interest in fundamental zoology. He turned back to the phylogenetic speculations that had been so important a part of zoology in the late nineteenth century and, in the 1920's, published his most important work.

In the 1880's the most influential idea in phylogenetic reconstruction was the theory of recapitulation—Haeckel's "biogenetic law"—according to which "ontogeny recapitulates phylogeny": the phylogenetic relations of a species can be read off by observing, with suitable interpretation, the successive developmental stages of an individual. In the decade before World War I, the vanguard of embryology had moved away from phylogenetic reconstruction to investigate, experimentally, the causes of development itself; and in the 1920's experimental embryology continued to dominate.

Experimental embryology and genetics implicitly cut away both the factual support and the theoretical coherence of the biogenetic law, but they produced no influential critique of it. A critique was what Garstang (who had little interest in modern experimental biology) provided, together with a more general theory of the relation of evolution and development to replace it. In his "critical re-statement" (1922) of the biogenetic law, he pulled no punches. The law still appealed to some, such as the powerful E. W. MacBride, and Garstang's work was controversial. The fact that, despite his obvious deserts, he was not elected a fellow of the Royal Society, has been attributed to the opposition of the minority who were interested in phylogenetic questions and the indifference of the majority who felt that kind of work was no longer to be counted real science. Nevertheless, there are two important general papers—the critique of recapitulation (1922) and a study of larval forms (1928)—and two particular applications—to the ancestry of vertebrates (1928) and of siphonophores (1946).

According to the theory of recapitulation, evolutionary changes are confined to the adult stage, new forms being added to the end of the organism's ontogeny. Garstang maintained that changes take place at all stages, and he urged that the larval forms of the marine groups he knew well are adaptive dispersal stages, not recapitulated ancestral forms. Some important phylogenetic transitions, he argued, had taken place through the modification of the ancestral larval form into a descendant adult form, with the loss of the ancestral adult stage. He called this process paedomorphosis. Although the idea was not original with him, he was highly influential in giving it currency.

Garstang applied his theory of paedomorphosis in both of his major phylogenetic papers. The ancestry of the vertebrates was the standard problem in late-nineteenth-century phylogenetic speculation; almost no invertebrate group had been left uninvestigated during that imaginative age in the search for the ancestor. Garstang pointed to the similarity of certain echinoderm larvae and the larvae of tunicates, which in turn are similar to the main chordate groups. The vertebrates, in this theory, originated by paedomorphosis. The theory is still as widely accepted as any other account of vertebrate origins. Garstang likewise suggested that the siphonophores—a group of free-floating colonial marine organisms—originated by paedomorphosis, in this case from a larval form, such as the actinula of the common hydroid *Tubularia*.

But Garstang's main influence then and later was due not so much to his scientific papers as to his delightful verse. He found time to work out few of his broad-ranging phylogenetic speculations in detail, and he instead wrote down the main ideas amusingly and memorably, in light verse. The verse was written for friends, and little of it appeared in his lifetime; but it no doubt encouraged his graver colleagues not to take him seriously. Garstang's ideas, however, have aged well, and his verse is still read with pleasure by generations that have forgotten the productions of his more fashion-conscious contemporaries.

BIBLIOGRAPHY

I. Original Works. "The Theory of Recapitulation: A Critical Re-statement of the Biogenetic Law," in *Journal of the Linnean Society, Zoology*, **35** (1922), 81–101; "The Origin and Evolution of Larval Forms," *British Association*, Glasgow, 23 pages, (1928); "The Morphology of the Tunicata, and Its Bearing on the Phylogeny of the Chordata," in *Quarterly Journal of Microscopical Science*, **72** (1928), 51–187; and "The Morphology and Relations of the Siphonophora," *ibid.*, **87** (1946), 103–193.

II. Secondary Literature. Obituaries of Garstang by his son-in-law Sir Alister C. Hardy appeared in *Journal of the Marine Biological Association of the United Kingdom*, **29** (1951), 561–566, and in *Proceedings of the Linnean Society of London*, **162** (1950), 99–105. Hardy gives an account of Garstang's ideas in his introduction to Garstang's verse anthology, *Larval Forms* (Oxford, 1951), which also contains a list of Garstang's main publications.

More generally, see Stephen Jay Gould, *Ontogeny and Phylogeny* (Cambridge, Mass., 1977); and Mark Ridley, "Embryology and Classical Zoology in Great Britain," in T. Horder, J. W. Witkowski, and C. C. Wylie, eds., *A History of Embryology* (Cambridge, 1986), 35–67.

Mark Ridley

GIAUQUE, WILLIAM FRANCIS (*b.* Niagara Falls, Ontario, Canada, 12 May 1895; *d.* Berkeley, California, 28 March 1982), *physical chemistry, chemical physics.*

Giauque received the 1949 Nobel Prize in chemistry for his research on chemical thermodynamics, particularly his pioneering and exhaustive investigations on entropy and low-temperature chemistry. He contributed greatly to establishing the third law of thermodynamics as a fundamental scientific law, invented the adiabatic demagnetization cooling process, and demonstrated the natural occurrence of oxygen's O^{17} and O^{18} isotopes and molecular hydrogen's ortho and para forms. Giauque's experimental researches were meticulous, most of them definitive, with improvements on his results coming only from refinements in technique. In his long and productive career, all of which he spent at the University of California (Berkeley), Giauque published 183 papers and trained 51 graduate students.

Giauque was the eldest of two sons and a daughter. His parents, William Tecumseh Sherman Giauque and Isabella Jane Duncan, were American citizens, which, according to the citizenship laws in effect, automatically made their children American citizens. Giauque received his elementary school education mainly in Michigan, where his father was a weightmaster and station agent for the Michigan Central Railroad. Upon his father's death in 1908, the family returned to Niagara Falls, and despite their opposition, Giauque enrolled in a two-year commercial course at the Niagara Falls Collegiate and Vocational Institute, intending to acquire the training he needed to help support them. By this time, Giauque's mother had become a part-time seamstress and tailor for J. W. Beckman, a chemist with American Cyanamid Company. Her employment with Beckman proved a very fortunate development because convincing Giauque to switch to the five-year general (college preparatory) course the next year took the combined efforts of his mother and the Beckman family. Giauque selected electrical engineering, but lacking both finances and engineering experience, he planned to work for a short time in one of the power generating plants at Niagara Falls. Unable to find any engineering openings, he accepted a position with the Hooker Electro-Chemical Company across the river in Niagara Falls, New York. Hooker's well-organized laboratory impressed Giauque greatly, and the two years he spent there convinced him to study chemical engineering.

By 1916 Beckman had been transferred to Berkeley; hearing of Giauque's new interest in chemical engineering, he suggested that Giauque attend the University of California at Berkeley. Gilbert Newton Lewis had arrived there in 1912 to serve as the chemistry department's chairman and dean of the College of Chemistry, which included chemical engineering. He also attracted a first-rate faculty, among them Joel Hildebrand, George E. Gibson, William Bray, and Gerald E. K. Branch. Beckman spoke highly of Lewis's research on the electron valence theory, thermodynamics, and free energy, and praised the research program Lewis had established. His recommendation of Berkeley's program, combined with its ten-dollar total semester fee, easily persuaded Giauque to enroll in the College of Chemistry rather than attend the more expensive Massachusetts Institute of Technology or Rensselaer Polytechnic Institute.

Giauque graduated with highest honors in 1920, receiving a B.S. degree in chemistry for a program of study that contained 25 percent engineering courses. Hildebrand, who taught Giauque, described him as outstanding. Two years later Giauque earned the Ph.D. in chemistry with a minor in physics. Gibson directed his dissertation although Giauque worked closely with the physicist Raymond Birge during his graduate studies. Because of Giauque's obvious promise, Lewis immediately offered him a faculty position. Giauque still had hopes of an engineering career, but the excellent research environment that Lewis had created finally led him, after several months of ambivalence, to pursue a career in chemistry. Giauque remained at Berkeley for the rest of his life, moving from assistant (1922–1927) to professor (1934) in twelve years. On 19 July 1932, Giauque married Muriel Frances Ashley. They had two sons.

Giauque's earliest investigations were on low-temperature entropy and the third law of thermodynamics. The German chemist Walther Nernst first stated the third law, then called the Nernst heat theorem, in 1906. According to Nernst, in any reaction involving only solids and liquids (including solutions), the change in entropy approached zero as the temperature approached absolute zero. Five years later Max Planck, in the third edition of his *Thermodynamik*, argued that Nernst's theorem did not hold for solutions and required modification. He suggested assigning zero entropy to each element at absolute zero and interpreted the third law to mean that all pure solids and liquids had zero entropy at absolute zero. But Lewis and Gibson pointed out that entropy measured randomness of a macroscopic state, and even in a pure solid or liquid some randomness existed in its structure. Only a perfect crystal of a pure solid or liquid lost its entropy at absolute zero. In 1920 Lewis and Gibson gave the definition of the third law accepted today: The entropy of a perfect crystal is zero at absolute zero.

Giauque demonstrated the correctness of Lewis and Gibson's third-law interpretation in his doctoral dissertation and in his first publication (1923). He showed experimentally, from heat-capacity and heat-of-fusion measurements, that glycerol glass (supercooled glycerol) at 70 K had considerably more entropy than crystalline glycerol (about 5.6 cal mol^{-1} K^{-1}) and concluded that this difference remained even at absolute zero. His third-law demonstrations continued in the 1920's and early 1930's with a series of investigations on diatomic gases in which he calculated their entropies theoretically from spectroscopic data and compared them with experimental entropies determined calorimetrically.

Giauque and several graduate students, among them R. Wiebe, H. L. Johnston, and J. O. Clayton measured low-temperature heat capacities and changes of state to obtain the calorimetic entropies for molecules such as hydrogen chloride, hydrogen bromide, hydrogen iodide, oxygen, nitrogen, nitric oxide, and carbon monoxide. For the spectroscopic entropies they used quantum-statistical equations developed within the last twenty years from band spectra studies on gaseous molecules. These included Otto Sackur's and Hans Tetrode's equations for gaseous volumes and translational energies (1911–1913), equations that Richard Tolman and Harold Urey had independently derived for rotational energies (1923), and Hervey Hicks and Allan Mitchell's equations for vibrational energies (1926).

All of these experiments, which Giauque began publishing in 1928, showed the thoroughness and high degree of accuracy that characterized his research. Not only did the close agreement of his spectroscopic and calorimetric entropy values clearly support the third law of thermodynamics, but Giauque provided further confirmation when he showed that his experimental entropies and those calculated from equations for the entropy of formation were in excellent agreement. At the same time his experiments verified the use of quantum statistics and the partition function in calculating entropy.

As a result of calorimetric studies in 1852 by Julius Thomsen at Copenhagen and later by Marcelin Berthelot at Paris, most chemists at that time believed the quantity of heat evolved in an exothermic reaction measured the reactants' affinity for one another. While this correlation often seemed to hold, some reactions that were clearly endothermic occurred spontaneously. Confusion also surrounded the meaning of entropy after introduction of the concept in 1854 by Rudolf Clausius at Bonn, and persisted until J. Willard Gibbs at Yale University, in a brilliant series of publications (1875–1878), showed that both the heat (enthalpy) and the entropy changes were necessary to establish a reaction's spontaneity at constant temperature and pressure. To measure spontaneity, Gibbs in 1875 defined a new function called the "potential" and expressed its dependence on enthalpy, entropy, and temperature with the equation $d\zeta = d\chi - t\ d\eta$, where ζ is the potential, χ the enthalpy, η the entropy, and t the temperature. In 1882 Hermann von Helmholtz at Berlin derived a similar relation in which he introduced the term "free energy" instead of Gibbs's "potential." For this reason chemists and physicists today speak of the Gibbs-Helmholtz equation, writing it as

$$\Delta G = \Delta H - T\Delta S,$$

where ΔG is Gibbs's free-energy change; ΔH is the enthalpy change; and $T\Delta S$ is the product of the entropy change and absolute temperature.

Gibbs (1876), Helmholtz (1882), and Jacobus van't Hoff (1886) showed that equilibrium constant (K) and electromotive force (EMF) measurements provided additional ways of calculating free energy. At Berkeley, Lewis, Gibson, and Merle Randall, carried out numerous free-energy determinations from chemical equilibrium studies and EMF measurements, while Giauque and his students calculated them from enthalpies and entropies measured thermodynamically, and from spectroscopic data.

Giauque's research on the entropy of gases from 1928 to 1932 showed that the hydrogen molecule, H_2, and molecules with similar ends, such as CO, NO, and N_2O, crystallized with a definite amount of residual entropy as the temperature approached absolute zero. The entropy of solid hydrogen results from a disordered nuclear spin alignment of the molecule's two protons. This alignment is parallel or in the same direction in the ortho form, and antiparallel or in opposite directions in the para form. Disordered crystalline arrangements cause the entropy found in the other molecules. Werner Heisenberg, at Niels Bohr's Institute for Theoretical Physics in Copenhagen, first suggested in 1927 that hydrogen and other elementary diatomic molecules existed in symmetrical (para) and antisymmetrical (ortho) forms. That same year Friedrich Hund, also in Copenhagen, pointed out that nuclear spin accounted for an element's hyperfine spectrum and that spectral analysis therefore provided information on the two spin states.

Nuclear spins, like electron spins, are difficult to reverse; and when a molecule interacts with electromagnetic radiation, the resulting electronic band spectrum contains two different sets of lines, one for each spin alignment. Molecular hydrogen's spectrum showed that the more intense set belonged to ortho-hydrogen and corresponded to odd rotational levels (odd rotational quantum numbers), while the fainter set represented para-hydrogen and even rotational levels. The two forms produced a regular pattern of lines that alternated in intensity, and from the intensities it appeared that molecular hydrogen contained a three:one ortho:para mixture at ordinary temperatures. Edward Condon, at that time a National Research fellow in Germany, had carried out theoretical calculations on the hydrogen molecule in 1927 showing that the ortho-para equilibrium was temperature dependent though established only slowly. In a letter to the Berkeley laboratory he suggested that evidence of the two forms might result from keeping hydrogen at liquid air temperatures for two to three months. Condon expected to see a marked difference in hydrogen's heat capacity if a transition occurred between the two forms.

Giauque began the suggested experiments late in 1927. He and H. L. Johnston obtained twenty grams of pure hydrogen by electrolyzing water, and after keeping it in a steel container at the temperature of liquid air (85 K) for 197 days (19 October 1927–3 May 1928), they observed a decrease of 0.04 cm (0.4 Torr.) in its vapor pressure at the triple point. The change occurred because as the temperature decreased, fewer and fewer molecules had sufficient rotational energy to remain in the higher ortho (odd rotational) energy levels. More and more molecules reversed their nuclear spins to assume the lower-energy para form and occupied the lower (even) rotational levels. Giauque and Johnston's experiment indicated an entropy difference of 4.39 cal mol^{-1} K^{-1} for the two forms, which clearly supported the third law interpretation that Lewis and Gibson gave in their 1920 paper.

Karl Bonhoeffer and Paul Harteck, at the Kaiser Wilhelm Gesellschaft in Berlin, first separated ortho- and para-hydrogen in 1929, using ten grams of charcoal to adsorb a small amount of hydrogen gas. The charcoal acted as a catalyst in establishing the equilibrium between ortho- and para-hydrogen. They succeeded in keeping the charcoal at the temperature of liquid hydrogen (20 K) for about twenty minutes, and after pumping off the gas they showed, from its higher thermal conductivity, that it consisted of 99.7 percent para-hydrogen. That same year Arnold Eucken and Kurt Hiller, at the Technische Hochschule in Breslau, measured a change in the heat capacity of hydrogen they had cooled to 90 K for periods of four to fourteen days. By the 1930's low-temperature research established beyond doubt that molecular hydrogen existed in two forms and that a catalytic separation produced almost pure parahydrogen.

To account for the entropy values of CO, NO, and N_2O, Giauque believed that in the crystal state some of the molecules had heat-to-head arrangements (CO · · CO, NO · · NO, N_2O · · N_2O) and others had head-to-tail arrangements (CO · · OC, NO · · ON, N_2O · · O_2N). Every molecule's orientation in the crystal was not the same as required for zero entropy. Such disorder could account for the small entropy at 0 K. For CO the entropy was 1.1 cal mol^{-1} K^{-1}. In a completely random crystalline arrangement of CO, NO, or N_2O, the entropy increased to a maximum of $S = R\ ln\ 2$, or 1.38 cal mol^{-1} K^{-1}. While the behavior of these compounds seemed to fall outside the third law, Giauque's results proved the law's validity only for perfect crystalline order in the lowest energy state.

Giauque's third-law investigations resulted in his most significant accomplishment, the invention in late 1924 of cooling by adiabatic demagnetization. His new cooling method enabled scientists to understand better the principles and mechanisms of electrical and thermal conductivity, to determine heat capacities, and to investigate the behavior of superconductors at extremely low temperatures. Michael Faraday in London had conducted the first systematic low-temperature research beginning in 1823, when he used compression and cooling with ice-salt mixtures to liquefy such gases as chlorine, sulfur dioxide, ammonia, and carbon dioxide. In 1877 Carl von Linde at Munich developed a commercially practical refrigeration process based on the expansion of ammonia gas, and in 1877 and 1878 Louis Cailletet at Châtillon-sur-Seine reached temperatures lower than 80 K and liquefied the "permanent gases," oxygen, nitrogen, nitrogen dioxide, carbon monoxide, and acetylene. Linde and Cailletet used the Joule-Thomson effect (1852), in which a compressed and cooled gas, after expansion through a small opening, cools further because the expanding gas expends some of its kinetic energy in overcoming intermolecular attractions.

Almost simultaneously, in 1877 Raoul Pictet at Geneva developed a cascade process in which each gas in a group of several gases with decreasing critical temperatures and triple points, such as sulfur dioxide, carbon dioxide, and oxygen, liquefied the group's next member. The process liquefied a gas by compression at the critical temperature, the highest temperature at which it existed as a liquid, and cooled it to its triple point, the lowest temperature at which it existed as a liquid, by boiling under reduced pressure. Because no liquids have critical temperatures and triple points between nitrogen's boiling point (77 K), hydrogen's critical point (33.3 K) and triple point (14 K), and helium's critical point (5.2 K), the cascade process failed to liquefy hydrogen and helium. A solution to the problem of reaching these temperatures finally appeared near the turn of the century. In 1895 Linde considerably improved Joule-Thomson cooling with the invention of his regenerator or heat-interchanger cyclic cooling. In 1898 James Dewar at the Royal Institution in London combined the Joule-Thomson, Linde, and cascade processes to liquefy hydrogen (20.4 K), and in 1908 Heike Kamerlingh Onnes at the Cryogenic Laboratory in Leiden used the combined process to liquefy helium (4.2 K). The low-temperature study of matter (cryogenics) made available laboratory temperatures of 5.0–0.8 K and led in 1911 to Kamerlingh Onnes' discovery of superconductivity in metals such as mercury, tin, and lead.

Kamerlingh Onnes also studied the magnetic susceptibility of the paramagnetic compound gadolinium sulfate octahydrate, $Gd_2(SO_4)_3 \cdot 8H_2O$, at liquid helium temperatures. These measurements became Giauque's starting point in 1924 when he calculated the effect of a magnetic field on the octahydrate's entropy and showed theoretically that application and subsequent adiabatic removal of the field at liquid helium temperatures produced additional cooling. Adiabatic demagnetization suggested a new method of reaching temperatures near absolute zero. Two years later, in a theoretical paper published at Zurich on 11 December 1926, Peter Debye used the same gadolinium compound to describe in detail the principle of adiabatic demagnetization cooling. Debye's paper appeared eight months after Wendell Latimer at Berkeley publicly discussed Giauque's work for the first time, at the California Section meeting of the American Chemical Society (9 April 1926).

In 1924 it was well known from Nernst's heat theorem and the Gibbs-Helmholtz equation that the heat capacities of substances become very small and approach zero at temperatures below 10–15 K. At these temperatures a substance loses practically all its thermal entropy, and magnetization/demagnetization should produce no further significant cooling. But paramagnetic compounds, such as gadolinium, cerium, and dysprosium salts, have thermal and magnetic entropy. In the absence of a magnetic field at low temperatures, they no longer have appreciable thermal entropy but still possess magnetic entropy because their atomic magnets have an irregular arrangement. Application of a powerful magnetic field forces the atomic magnets to line up with the field, reducing the magnetic entropy. A cooling bath removes the heat generated by the entropy decrease. Giauque recognized that if he insulated the compound thermally and removed the field under adiabatic conditions, the total entropy must remain constant. By removing the field, the atomic magnets return to their random arrangement and increase the magnetic entropy. Temperature measures thermal motion; therefore, the accompanying decrease in thermal entropy that corresponds to a decrease in motion results in the temperature's lowering. Because magnetic entropy is a factor in cooling only at low temperatures, Giauque pointed out that cooling by adiabatic demagnetization is most effective at temperatures produced by the evaporation of liquid helium (1 K). He compared the sequence of steps in magnetic cooling with the three steps in the refrigeration process, using an idealized expansion engine.

When Giauque began calculating the low temperatures achievable with magnetic cooling, he had neither the expensive large-scale equipment to conduct experiments nor the thermometer to record the readings. He planned not merely to measure low temperatures but also to use magnetic cooling in his low-temperature thermodynamics research. Paramagnetic salts were ideal for use because of their high heat capacities at low temperatures, though Giauque later experienced difficulty in making good thermal contact with the cooled salt. The equipment he required included a magnet with a strong homogeneous field (8,000–20,000 gauss); a hydrogen and a helium compressor; a purification system for removing oil, air, and other gases from the helium; and vacuum pumps for hydrogen recovery and reduction of liquid helium's temperature. Because this apparatus was not immediately available, Giauque and his graduate student D. P. MacDougall succeeded in carrying out the first adiabatic demagnetization cooling that produced temperatures below 1 K (0.53 K) only on 19 March 1933. Nine years had passed since Giauque first conceived of cooling by adiabatic demagnetization. The Leiden group had known all along of the research, but to Giauque's astonishment they never attempted the cooling experiments before he did.

In the experiments, Giauque and MacDougall placed a sixty-one-gram sample of paramagnetic gadolinium sulfate octahydrate in a copper calorimeter tube. A vacuum jacket filled with helium gas to conduct heat from the compound surrounded the tube. The tube and jacket rested inside a copper-lead Dewar flask to which Giauque and MacDougall added liquid helium through a vacuum-jacketed transfer tube to a height of one meter and then placed the flask within the copper coils of a solenoid magnet. Low-viscosity cooling oil (kerosene) pumped rapidly over bare copper conductors removed heat and promoted efficient heat transfer, which, Giauque found, was the principal problem in designing solenoid magnets. An inductance bridge measured the gadolinium sulfate's magnetic susceptibility.

The cooling progressed in three stages: (1) the

electric current through the copper coils caused the atomic magnets to line up, releasing heat and decreasing the entropy of the paramagnetic compound; (2) when the cooling stopped, the compound was insulated against heat flow by evacuating helium gas from the surrounding jacket; (3) the electric current was turned off, quickly demagnetizing the compound, which did magnetic work by inducing an electric current in the copper coils. Because no heat entered, the atomic magnets absorbed energy and cooled the compound.

Measuring such low temperatures presented a problem. The commonly used constant-volume gas thermometer, even one containing helium, deviated from ideal behavior at these temperatures and was in error. An alternative was to measure the salt sample's magnetic susceptibility, which, according to Curie's Law (1905), varied inversely with the absolute temperature. Giauque made the first susceptibility measurements with a coil of several thousand turns of fine copper wire around the insulating vacuum jacket. As the temperature decreased, the alternating current flowing through the coil also decreased while the salt's magnetic susceptibility increased. From the relation between current and susceptibility, Giauque obtained magnetic susceptibility values and then calculated the absolute temperature from Curie's law. For gadolinium compounds the Curie constant C is 7.880, giving the Curie-law equation

$$T = 7.880/\text{magnetic susceptibility}.$$

The Curie equation provided good low-temperature values. But upon approaching absolute zero (1 K) it failed because it allowed entropy values to decrease asymptotically during magnetization and conflicted with the accepted view that entropy was finite. Giauque obtained true thermodynamic temperatures by plotting the change in enthalpy between zero field and some constant field (H) against entropy, or $d\mathrm{H}_0^{\mathrm{H}}\ dS = T$. The graph showed the variation of T with magnetic field strength and Giauque's calculation of the absolute temperature from the equation $T = dH/dS$ at constant magnetic field.

By 1938 Giauque had improved temperature measurement by inventing an extremely sensitive amorphous carbon (lampblack) resistance thermometer for work below 1 K. It consisted of a single layer of glass-lens paper (which he had chosen for its very loose open structure) applied to the sides of a twelve-inch-long glass sample tube. Lampblack mixed with a large amount of ethyl alcohol was painted on the paper and then coated with a collodion-ethyl ether-alcohol solution. Two platinum wires connected the carbon layer to tungsten terminals sealed into the glass wall. The thermometer measured temperatures accurately and precisely and was suitable for low temperatures because its resistance had a high temperature coefficient and changed little with magnetic field strength. By this time Giauque had decreased temperatures from 1 K, obtained by evaporating liquid helium, to 0.004 K with adiabatic demagnetization.

The absorption band spectra from which Giauque had calculated the entropies of diatomic gases led unexpectedly to the discovery of oxygen's two isotopes. A band spectrum contains many strong and weak lines. By applying quantum statistics to their distribution pattern, Tolman at the California Institute of Technology in 1923, Birge at Berkeley in 1926, and others determined the molecular energy levels of these gases. Giauque had used the same quantum statistical distribution patterns of the molecular energy levels to calculate entropies and had found excellent agreement with values obtained from the third law and from entropies of formation. In 1928, while examining atmospheric oxygen band spectrum photographs provided by Harold D. Babcock of Mount Wilson Observatory, Giauque and Johnston noticed some unaccounted-for weak lines in the spectrum. In 1928 Robert Mulliken at New York University had shown that the spectrum's strong doublets belonged to molecular oxygen (O^{16}-O^{16}). Giauque also knew that many of the weak lines that Babcock had discovered and measured belonged to the molecule's higher energy states. Yet he and Johnston could not account for all the weak lines. Giauque never left anything of significance unexplained and considered his entropy calculations unsatisfactory because they failed to account for the additional weak lines. The origin of these lines remained unknown until early 1929, when Giauque recalled awaking one morning and suddenly realizing that the lines came from oxygen isotopes.

Francis W. Aston, J. J. Thomson's assistant at Cambridge and the world's authority on mass spectroscopy, had established the existence of neon, sulfur, chlorine, and silicon isotopes by 1929, but had not found any isotopes of oxygen. Detailed frequency calculations by Giauque and Johnston in January 1929 proved that the O^{16}-O^{18} molecule generated one set of weak lines. By May their calculations identified another set of very weak lines that Babcock had reported but had not associated with the oxygen spectrum. They belonged to the O^{16}-O^{17} molecule. Their presence confirmed the existence of the O^{17} isotope that P. M. S. Blackett and others in 1925 had reported to result from collisions between alpha particles and nitrogen nuclei. By assigning masses of 18 and 17 to the two isotopes, Giauque and Johnston succeeded in calculating accurately all of the oxygen spectrum's weak lines from the positions of the strong lines. Articles announcing the discovery of O^{18} and O^{17} appeared in the 2 March and 1 June 1929 issues of *Nature*.

Giauque's unexpected discovery that normal atmospheric oxygen contained small amounts of the isotopes O^{17} (0.037 percent) and O^{18} (0.204 percent), in addition to the abundant O^{16} (99.759 percent), caused a problem with the chemists' and physicists' atomic mass values. Chemists arbitrarily had assigned an exact relative atomic mass of 16 to oxygen's unsuspected three-isotope mixture as their standard for atomic mass determinations and continued to use this value even after 1929. Physicists now based their atomic mass standard on the lightest and most abundant oxygen isotope, O^{16}. The mean atomic mass of atmospheric oxygen was then 16.0044 on the mass spectrometric scale. To convert from the physicists' to the chemists' scale required dividing the physicists' atomic mass by 1.00027 (16.0044/16). The two scales remained in use until 1961, when the International Union of Pure and Applied Chemistry and the International Commission of Atomic Weights abandoned O^{16} and adopted the carbon^{12} isotope with C = 12.0000 as the new standard. On this scale the chemists' atomic mass of atmospheric oxygen decreased to 15.9994 from 16.0000. The change in 1961 was the first since the 1860's, when the precise atomic mass determinations made by Jean-Servais Stas at Brussels had made 0 = 16.0000 the accepted atomic mass standard. Prior to that time chemists had assigned various masses to oxygen: Thomas Thomson's 0 = 1, William H. Wollaston's 0 = 10, and Jöns Jacob Berzelius' 0 = 100.

Of the two oxygen isotopes Giauque discovered in 1929, O^{18} provided scientists with an isotopic tracer that enabled them to study the photosynthesis and respiration mechanisms and led Harold Urey, George Murphy, and Ferdinand Breckwedde in 1931 to the spectroscopic identification of hydrogen's isotopes. Giauque's isotopic research also demonstrated Heisenberg's earlier prediction that a molecule retains a half-quantum unit of vibrational energy (zero point or residual energy) even in its lowest energy quantum state. Thus vibration motion within an atom did not stop at 0 K.

In his nearly sixty-year career at Berkeley, Giauque interrupted his research on low-temperature entropy, adiabatic demagnetization, and oxygen isotopes only once. This occurred between 1939 and 1944, when he directed a classified engineering program that designed and built a mobile liquid-oxygen generating plant. The government required liquid oxygen for medical and survival purposes and for use in rocket fuel. The heat exchangers designed in the program were prototypes of the large units constructed later for the liquefaction of natural gas.

When G. N. Lewis became chairman of Berkeley's chemistry department in 1912, his policy required all faculty members to teach introductory chemistry. From the time of his appointment as instructor in 1922, and continuing every semester for thirty-four consecutive years, Giauque taught a discussion-laboratory section. Until retirement in 1962, his teaching duties also included advanced physical chemistry and chemical thermodynamics. For fifteen years (1945–1960) Giauque served as adviser for Letters and Science students majoring in chemistry. Despite a no-nonsense, strictly business image that students sometimes found forbidding, Giauque was at heart a storyteller and a humorous personality who often developed close relations with his graduate students.

In addition to winning the 1949 Nobel Prize in chemistry, Giauque earned many other honors in his lifetime. They included two honorary degrees—an Sc.D. from Columbia University (1936) and an LL.D. from the University of California, Berkeley (1963)—the American Chemical Society's J. Willard Gibbs Medal (1951) and its California Section's G. N. Lewis Award (1955), the Charles Frederick Chandler Foundation Medal (1936), and the Franklin Institute's Elliott Cresson Medal (1937). Giauque was elected to the National Academy of Sciences in 1936, and he held membership in the American Philosophical Society from 1940. He died of complications from a fall.

BIBLIOGRAPHY

I. ORIGINAL WORKS. Giauque's papers are in the Bancroft Library's Archives, University of California, Berkeley. There is a sixty-one-page oral history (1974) on Giauque in the archives, but because Giauque never signed a release, it is not available for examination. The most comprehensive summary of Giauque's work is his Nobel Lecture, "Some Consequences of Low Temperature Research in Chemical Thermodynamics," in *Nobel Lectures: Chemistry 1942–1962* (New York, 1964), 227–250. There is a collection of his articles, *The Scientific Papers of William F. Giauque: Low Temperature Chemical and Magneto Thermodynamics*, I, *1923–1949* (New York, 1969).

A bibliography of Giauque's articles and reviews (1923–1978) is included in his papers. Many of them prior to 1962 appeared in *Journal of the American Chemical So-*

ciety. After 1962 Giauque published increasingly in *Journal of Physical Chemistry* and *Journal of Chemical Physics*. Some of his important articles are "The Third Law of Thermodynamics. Evidence from the Specific Heats of Glycerol That the Entropy of a Glass Exceeds That of a Crystal at the Absolute Zero," in *Journal of the American Chemical Society*, **45** (1923), 93–104, with G. E. Gibson; "Paramagnetism and the Third Law of Thermodynamics. Interpretation of the Low-Temperature Magnetic Susceptibility of Gadolinium Sulfate," *ibid.*, **49** (1927), 1870–1877; "The Entropy of Hydrogen Chloride. Heat Capacity from 16° K. to Boiling Point. Heat of Vaporization. Vapor Pressures of Solid and Liquid," *ibid.*, **50** (1928), 101–122; "Symmetrical and Antisymmetrical Hydrogen and the Third Law of Thermodynamics. Thermal Equilibrium and the Triple Point Pressure," *ibid.*, 3221–3228, with H. L. Johnston; "An Isotope of Oxygen, Mass 18. Interpretation of the Atmospheric Absorption Bands," *ibid.*, **51** (1929), 1436–1441, with H. L. Johnston; "An Isotope of Oxygen, Mass 17, in the Earth's Atmosphere," *ibid.*, 3528–3534, with H. L. Johnston; "The Entropy of Hydrogen and the Third Law of Thermodynamics. The Free Energy and Dissociation of Hydrogen," *ibid.*, **52** (1930), 4816–4831; "The Calculation of Free Energy from Spectroscopic Data," *ibid.*, 4808–4815; "The Conditions for Producing Temperatures Below 1° Absolute by Demagnetization of $Gd_2(SO_4)_3 \cdot 8H_2O$). Temperature-Magnetic Field Isentropics," *ibid.*, **54** (1932), 3135–3142, with C. W. Clark; "Experiments Establishing the Thermodynamic Temperature Scale Below 1° K. The Magnetic and Thermodynamic Properties of Gadolinium Phosphomolybdate as a Function of Field and Temperature," *ibid.*, **60** (1938), 376–388; with D. P. MacDougall; and "Amorphous Carbon Resistance Thermometer-Heaters for Magnetic and Calorimetric Investigations at Temperatures Below 1° K," *ibid.*, 1053–1060, with C. W. Clark.

II. SECONDARY LITERATURE. In addition to the oral history, other accounts of Giauque are D. N. Lyon and K. S. Pitzer, "William Francis Giauque," in University of California, *In Memoriam* (Berkeley, 1985), pp. 152–156; William Jolly, *From Retorts to Lasers* (Berkeley, 1987), which has a chapter on Giauque and also gives a valuable historical account of the Berkeley chemistry department; "Giauque Awarded Nobel Prize for Low Temperature Research," in *Chemical and Engineering News*, **27** (28 November 1949), 3571; and "William Francis Giauque," in *McGraw-Hill Modern Scientists and Engineers* (New York, 1980), 191–193. Giauque's papers also contain four short unpublished biographical memoirs, written in 1948 and 1949, totaling ten pages (no authors given).

ANTHONY N. STRANGES

GIGNOUX, MAURICE (*b*. Lyons, France, 19 October 1881; *d*. Grenoble, France, 20 October 1955), *geology*.

Gignoux was born into a family of Swiss origin. A brilliant student, he was accepted by both the École Polytechnique and the École Normale Supérieure. He chose the latter, which offered courses closer to his interests in research and teaching, and passed the *agrégation* in natural sciences with honors. Gignoux married Marie Garel in 1909; they had five sons and a daughter. In Lyons, under the supervision of Charles Depéret, he prepared a dissertation on the Mediterranean Pliocene; presented in 1913, it remains the basis of more recent studies on the subject.

Gignoux's career was interrupted by World War I, during which, because of his delicate health, he served in the meteorological research department of the army. After the war he was appointed professor of geology at the University of Strasbourg, now under French jurisdiction, where he taught stratigraphy, placing sediments in their structural frame for the first time. In 1925 he published *Géologie stratigraphique*, a highly acclaimed work that ran to five editions and was translated into English, Russian, and Polish.

While at Strasbourg, Gignoux collaborated with the École Supérieure du Pétrole and thus was able to pursue his interest in the tectonics of the evaporite-bearing formations often associated with petroleum ores. He extended his tectonic findings to the genesis of some Alpine structures, notably to the emplacement of the large *nappes de charriage*, and demonstrated, conversely, that the assumed remains of a gigantic nappe (the Suzette-Gigondas massif in the southern Rhône Valley) were only rooted diapirs.

During his summer holidays Gignoux spent as much time as possible in the Alps, both for reasons of health and for the pleasure of being close to the mountains. In 1926 he accepted the chair of geology at the University of Grenoble in the Alps, succeeding Wilfrid Kilian, whose temporary assistant he had been in 1909. This was the beginning of the most productive period of his life, devoted to the innermost and most complex Alpine zones, particularly the Briançon region. This tremendous undertaking lasted until World War II and produced the fundamental study *Description géologique du bassin supérieur de la Durance*, published in 1938 in collaboration with Léon Moret. This study set the precedent for geological research in the Durance area for the next forty years.

The strength of Gignoux and Moret was that they outlined the existence of great structural zones, homogeneous in geological evolution and style of deformation, and applied to these zones a special terminology (Briançonnais zone, sub-Briançonnais zone, Piedmont zone, etc.) that clarified the synthesis and became a model easily applied to other great mountain chains.

Among the series that Gignoux explored, one was an important turning point in his research: the Embrunais flysch. Under the term *flysch* geologists class thick accumulations of sediments arranged in very regular beds, rhythmically repeated, that provide regularly stratified mountains. Furthermore, these flysch series are folded in a manner that suggested to Gignoux deformations arising from the sliding or "flowing" of the plastic mass of flysch down a slope, propelled by its own weight. This gravity tectonics theory had been put forward by the Swiss geologist Hans Schardt in 1893 but did not receive much attention until Gignoux proved its validity.

Although later studies did not fully consider Gignoux's theory, it was recognized between 1948 and 1955 as far away as the United States. In 1955 Gignoux was awarded the Penrose Medal of the Geological Society of America, rarely awarded outside the United States.

Like most true scientists, Gignoux never separated pure research from its applications. He was interested in the geological problems encountered in hydroelectrical engineering, upon which he was frequently consulted. This work led to the publication of *Géologie des barrages*, written with Reynold Barbier, in 1955. This highly acclaimed work was not republished due to Gignoux's premature death at Grenoble in 1955.

Throughout his life, Gignoux's work was outstanding in its simplicity and directness. Above all else, he retained the ability to arrive at an accurate interpretation of the observed facts. This persistent interest is well expressed by his choice of Pirandello's words in the preface to his *Géologie stratigraphique*: "Un fatto è come un sacco, che vuoto non si regge. Perchè si regga, bisogna prima farci entrar dentro la ragione" ("A fact is like a bag that will stand up only if filled with common sense").

Throughout his life Gignoux worked ceaselessly, despite his chronic asthma, without regard for personal gain or honor. He received many prestigious appointments and honors; he was dean of the Faculty of Sciences at Grenoble, an officer of the Legion of Honor, a member of the French Academy of Sciences (1932), and a recipient of the Prix Gaudry of the Geological Society of France.

Gignoux was not only a famous geologist but also a fascinating man of extreme personal charm who was devoted to his family, friends, and students. He was also a knowledgeable musician. Through his professional and literary talents, his personal integrity, and his intellect, he made a great impression on those who knew him. Gignoux is remembered not only as the founder of the Alpine geological school that is still active, but also as a leader of European geology from 1938 to 1955.

BIBLIOGRAPHY

I. Original Works. Among Gignoux's writings are *Géologie stratigraphique* (Paris, 1925; 5th ed., 1960), translated by Gwendolyn G. Woodford as *Stratigraphic Geology* (San Francisco, 1955); *Description géologique du bassin supérieur de la Durance* (Grenoble, 1938), written with Léon Moret; *Géologie dauphinoise: Initiation à la géologie par l'étude des environs de Grenoble* (Paris, 1944; 2nd ed., 1952), written with Léon Moret; *Méditations sur la tectonique d'écoulement par gravité* (Grenoble, 1948); "La notion de temps en géologie et la tectonique d'écoulement par gravité," in *International Geological Congress, Report of the 18th Session, Great Britain*, pt. XIII (London, 1952); and *Géologie des barrages et des aménagements hydrauliques* (Paris, 1955), written with Reynold Barbier.

II. Secondary Literature. A biography of Gignoux is Léon Moret, "Maurice Gignoux (1881–1955)," in *Bulletin de la Société Géologique de France*, 6th ser., **6** (1956), 289–317.

JACQUES DEBELMAS

GILLULY, JAMES (*b*. Seattle, Washington, 24 June 1896; *d*. Denver, Colorado, 29 December 1980), *geology*.

Gilluly's father, Charles Elijah Gilluly, was descended from a disciple of Robert Emmett who fled from Ireland to the United States in 1793 after Emmett was arrested and hanged. Louisa Elizabeth Briegel Gilluly, his mother, was a member of a Württemburg family that found refuge in the United States in 1830, after participating in an abortive attempt to convert that German state to an autonomous republic. Thus the habit of bold and independent thought was a part of his background and his upbringing.

Gilluly was a scholar and a leader; in high school he was elected captain of the football team as well as class valedictorian. He entered the University of Washington in 1915, but straitened financial circumstances compelled him to work both before entering college and between his college years. Gilluly served in the U.S. Navy during 1917 and 1918 and then returned to the university. His courses followed an erratic path from civil engineering to economics and finally to geology. He received the B.S. degree in 1920.

After graduation Gilluly worked for a short time as a junior geologist for the National Refining Company in Montana, then moved to the U.S. Geological Survey, where he remained, with brief interruptions, until his retirement in 1966. The first interruption allowed him to take graduate courses part-time at Johns Hopkins until he was accepted in 1922 as a doctoral candidate at Yale. He received the Ph.D. in 1925 with a dissertation on the sedimentary rocks of the San Rafael Swell in Utah. Gilluly married Enid Frazier on 30 June 1925. They had two daughters. After his return to the U.S. Geological Survey, he conducted field studies in Alaska, Utah, Oregon, Arizona, New York, and the Panama Canal Zone. In 1932 Gilluly went to Europe to work at Innsbruck, Austria, with Bruno Sander on the newly developed field of petrofabrics. Before returning to the United States he traveled extensively in Eastern Europe to familiarize himself with global geological problems.

In 1938 Gilluly joined the faculty of the University of California at Los Angeles, where he became known as an inspiring but demanding professor. The entry of the United States into World War II interrupted his university career, and he transferred to the Geological Survey's Military Geology Unit. He was assigned to the Southwest Pacific Command, where he studied the geology of possible invasion sites. He made a landing with the U.S. Marines on the island of Leyte, at a location he had recommended.

At the end of the war Gilluly returned to UCLA to resume his teaching career. During the McCarthy era, however, the university was forced to require loyalty oaths from the faculty; Gilluly, whose service and loyalty to the United States were unquestioned, resigned rather than accept this demeaning condition of employment.

In 1950 he returned to full-time work with the Geological Survey for what may have been his most productive years. In a 1949 paper he expressed his conviction that orogenies were not periodic events of short duration, interspersed with periods of quiescence, but a continuing process that could be observed through much of geologic time. This conclusion, bold for its time, had a significant impact on geological thought. In the words of a later biographer:

> He questioned among other items: (1) then-accepted ideas on systematic connections among plutonism, volcanism, and tectonism; (2) the supposed universal association of orogeny and plutonism; and (3) the idea that radiometric ages of plutons also date the deformation of their wall rocks. He proposed that subcrustal flow or a transfer of subcrustal and basal crustal material is needed to meet the requirements of isostasy in changes like the uplift of the Colorado Plateaus province, and that other subcrustal flow was required to drag sialic material beneath the western continental margin of the United States to be available for remobilization into the overlying plutonic belt. (Smith, "Memorial to James Gilluly," p. 3)

Gilluly wrote a memoir, *The Origin of Granite* (1962), published by the Geological Society of America, and discussed the origins of plutonic granite rocks in the William Smith Lecture that he presented to the Geological Society of London. The origin of granite was important in the dispute between the followers of James Hutton and Abraham G. Werner over volcanism and neptunism at the end of the eighteenth century.

With A. O. Woodford and Aaron C. Waters, Gilluly wrote a textbook, *Principles of Geology*, that ran through several editions and was used in many American colleges. Although he was a field geologist, Gilluly was a dedicated student of the processes of the earth and the principles that guided them. He was firm in his belief, however, that the processes and their inducted principles must be studied and tested in the field.

Gilluly received many honors, and many invitations to lecture at both American and European universities. He served successively as vice president and president of the Geological Society of America in 1947 and 1948 and received its Penrose Medal in 1958. The following year he was awarded the Distinguished Service Medal, the highest award given by the U.S. Department of the Interior. He was a member of the National Academy of Sciences and the American Academy of Arts and Sciences, and an honorary member of the Geological Society of London.

BIBLIOGRAPHY

I. ORIGINAL WORKS. Gilluly's writings include *A Reconnaissance of the Point Barrow Region, Alaska, U.S. Geological Survey Bulletin* 772 (1925), written with Sidney Page and W. T. Foran; "Analcite Diabase and Related Alkaline Syenite from Utah," in *American Journal of Science*, 5th ser., **14** (1927), 199–211; "Sedimentary Rocks of the San Rafael Swell and Some Adjacent Areas in Eastern Utah," in *U.S. Geological Survey Professional Paper* 150 (1928), 61–110, written with John B. Reeside; *Geology and Oil and Gas Prospects of Part of the San Rafael Swell, Utah, U.S. Geological Survey Bulletin* 806-C (1929); *Copper Deposits near Keating, Oregon*, *ibid.*, 830-A (1931); *Geology and Ore Deposits of the*

Stockton and Fairfield Quadrangles, Utah, U.S. Geological Survey Professional Paper 173 (1932); *Replacement Origin of the Albite Granite Near Sparta, Oregon, ibid.*, 65–81; "Mineral Orientation in Some Rocks of the Shuswap Terrane as a Clue to Their Metamorphism," in *American Journal of Science*, **228** (1934), 182–201; "Keratophyres of Eastern Oregon and the Spilite Problem," *ibid.*, **229** (1935), 225–252, 336–352; *Geology and Mineral Sources of the Baker Quadrangle, Oregon, U.S. Geological Survey Bulletin* 879 (1937); "Emplacement of the Uncle Sam Porphyry, Tombstone, Arizona," in *American Journal of Science*, **243** (1945), 643–666; "*The Ajo Mining District, Arizona, U.S. Geological Survey Professional Paper* 209 (1946); "Distribution of Mountain Building in Geologic Time," in *Bulletin of the Geological Society of America*, **60** (1949), 561–590; and "Subsidence in the Long Beach Harbor Area, California," *ibid.*, 461–529, written with U.S. Grant IV.

Later works are *Principles of Geology* (San Francisco, 1951, 4th ed., 1975), written with A. O. Woodford and Aaron C. Waters; "Geologic Contrasts Between Continents and Ocean Basins," in Arie Poldervaart, ed., *Crust of the Earth (A Symposium), Geological Society of America Special Paper* 62 (1955), 7–18; "The Tectonic Evolution of the Western United States," in *Quarterly Journal of the Geological Society of London*, **119** (1963), 133–174, the William Smith Lecture; "The Scientific Philosophy of G. K. Gilbert," in C. C. Albritton, ed., *The Fabric of Geology* (Reading, Mass., 1963), 218–224; "Atlantic Sediments, Erosion Rates, and the Evolution of the Continental Shelf: Some Speculations," in *Bulletin of the Geological Society of America*, **75** (1964), 483–492; "Orogeny and Geochronology," in *American Journal of Science*, **264** (1966), 97–111; "The Role of Geological Concepts in Man's Intellectual Development," in *Texas Quarterly*, **11**, no. 2 (1968), 11–23; "Plate Tectonics and Magmatic Evolution," in *Bulletin of the Geological Society of America*, **82** (1971), 2383–2396; and "American Geology Since 1910: A Personal Appraisal," in *Annual Review of Earth and Planetary Science*, **5** (1977), 1–12.

II. Secondary Literature. On Gilluly's life and work, see Thomas Nolan, "James Gilluly," in *Biographical Memoirs, National Academy of Sciences*, **56** (1987), 119–132; J. Fred Smith, Jr., "Memorial to James Gilluly," in *Memorials of the Geological Society of America*, **12** (1982); and Aaron C. Waters, "Portrait of a Scientist: James Gilluly, Pioneer of Modern Geological Ideas," in *Earth Science Reviews*, **5** (1969), A19–A27.

J. J. Lloyd

GLEY, MARCEL-EUGÈNE-ÉMILE (*b.* Épinal, France, 16 January 1857; *d.* Paris, France, 25 October 1930, *physiology*.

After obtaining his Bachelor of Arts at the University of Nancy in 1878, Gley studied medicine in Montpellier and then returned to Nancy, where he became a medical doctor in 1881. His flair for scientific research was evident early in his career: even as a student he was fascinated by physiological experimentation. After moving to Paris in 1880, Gley worked until 1883 in Jules Marey's laboratory at the Collège de France. From 1883 to 1889 he was assistant in physiology at the Faculty of Medicine and from 1886 to 1893 was director of the clinical laboratory of the Hôtel-Dieu. He became assistant professor of physiology at the Faculty of Medicine in 1889, assistant to the professor of physiology at the Museum of Natural History in 1893, and deputy to Charles Richet in 1895. During this period Gley achieved a felicitous combination of university instruction and pure research; his inventive genius had free rein. He also participated enthusiastically in various projects of the French scientific community.

Editor in chief of *Journal de physiologie et de pathologie générales*, secretary (1899) and vice president (1897) of the Société de Biologie, and member of the Academy of Medicine from 1903 (president for the year 1927), Gley was considered one of the major continuators of the work of Claude Bernard. He received official recognition in 1908, when he was named to the newly created chair of general biology of the Collège de France. From then on, he devoted more time to his professorial activity than to experimental work, his boldness of experimental imagination supplanted by a broader consideration of the great problems of physiology.

Gley's initial discoveries had to do with blood clotting, in particular with the mechanisms of anticlotting factors. He collaborated with Charles Richet in the latter's neurophysiological experiments, notably on the role of the central nervous system in regulation of temperature, discovering, in 1884, that stimulation of the anterior lobe of the cerebral cortex in rabbits provoked a rise in temperature. In 1890 he was awarded the Prix Montyon for his study of the nervous system. Like Richet, Gley studied immunization against the hemolytic action of serums. In April 1891—independent of similar experiments conducted in Italy in 1890 by Gustavo Pisenti and Giulio Vassale—he showed that dogs and rabbits whose thyroids had been surgically removed were able to survive for some time if aqueous extracts from this gland were administered. This experimental demonstration was followed almost immediately by its practical application to humans. In October 1891 George B. Murray obtained the first success in the treatment of myxedema through thyroid medication.

Gley noted that the ablation of the thyroid alone

was less serious than the simultaneous extirpation of the small attached glands. After his first publication in 1891, he described not only the substitutive effects of the total removal of the thyroid in thyroidectomized animals but also the existence of parathyroids (unaware that a description of them had already been published in Sweden by Ivar Sandström), and he was the first to discuss their physiological role (1891–1893). In 1891 he induced glycosuria in dogs by tying the efferent veins of the pancreas.

Gley was one of the founders of organotherapy and modern endocrinology, both in his experimental research and in his theoretical study of "functional and interrelational humoral correlations." At the center of his interest was the notion of internal secretion, of which he produced an excellent historical analysis, highlighting the role of Claude Bernard and of Charles-Édouard Brown-Séquard. His rules of epistemological criticism in experiments with extracts of organs were strict, and he was highly cautious in the therapeutic administration of these extracts. In fact, the severity of these rules caused him to abandon his efforts to treat diabetes, and he published nothing about those experiments. In December 1922, however, after learning of the discovery of insulin, Gley asked that a sealed folder that had been deposited at the Société de Biologie in 1905 be opened. Its contents showed that in 1894 and 1900 he had obtained encouraging results in treating diabetic dogs with extracts from a pancreas whose ducts previously had been obstructed.

Gley's other research concerned the presence of iodine in the thyroid and in blood, the relations between suprarenal secretion and the activity of the autonomic system, acquired immunity after the injection of toxic doses of an organ extract (in particular the phenomenon he called tachyphylaxis), the physiology of the liver (most especially its antitoxic role), the physiology of the cardiac muscle, the sense of taste, and the relation between physiological phenomena and mental activity. Gley had a keen appreciation of historical precedents and strove for elegance in literary expression. His publications on philosophical questions and on the history of science stood in the positivist tradition and were outstanding for their clarity and good sense.

BIBLIOGRAPHY

I. Original Works. Gley's principal discoveries were published in "Sur les fonctions due corps thyroïde," in *Comptes rendus de la Société de biologie*, **43** (1891), 841–847; "Sur les troubles consécutifs de la destruction du pancréas," in *Comptes rendus de l'Académie des sciences*, **112** (1891), 752–755; and "Action des extraits de pancréas sclérosé sur des chiens diabétiques," in *Comptes rendus de la Société de biologie*, **87** (1922), 1322–1325. Among his articles on the history of science are "The Theory of Internal Secretion: Its History and Development," in *The Practitioner*, **94** (1915), 2–15; and his articles on the work of F. X. Bichat, Claude Bernard, A. Vulpian, and M. Schiff. Other publications include *Traité élémentaire de physiologie* (Paris, 1910; 8th ed., 1934), with Mathias Duval; a manual of endocrinology entitled *Les sécrétions internes* (Paris, 1914; 3rd ed., 1925), trans. and ed. by Maurice Fishberg as *The Internal Secretions* (New York, 1917); and the following collections of essays and lectures: *Essais de philosophie et d'histoire de la biologie* (Paris, 1900); *Études de psychologie physiologique et pathologique* (Paris, 1903); *Quatre leçons sur les sécrétions internes* (Paris, 1920); and *Les grands problèmes de l'endocrinologie* (Paris, 1926).

II. Secondary Literature. There is no satisfactory monograph on Gley's life and work. A subjective but very well documented overview of his early research is in *Notice sur les titres et travaux d'Eugène Gley* (Paris, 1902; reiss. with an appendix, 1907). There are obituary notices by an anonymous author in *Archives internationales de pharmacodynamie*, **38** (1930), vii–xxx; by A. d'Arsonval in *Revue scientifique*, **68** (1930), 577–580; by J. Jolly in *Presse médicale*, **38** (1930), 1565–1566; by P. Portier in *Bulletin de l'Académie de médecine*, 3rd ser., **104** (1931), 392–398; and by C. Richet in *Journal de physiologie et pathologie générales*, **29** (1931), 1–6. A more recent assessment of his work in endocrinology is M. Pestel, "Le cinquantenaire de la découverte de l'insuline: E. Gley, précurseur de F. J. Banting et C. H. Best," in *Nouvelle presse médicale*, **1** (1972), 1527–1528.

M. D. Grmek

GÖDEL, KURT FRIEDRICH (*b.* Brünn, Moravia [now Brno, Czechoslovakia], 28 April 1906; *d.* Princeton, New Jersey, 14 January 1978), *mathematical logic, set theory, general relativity.*

The character and achievement of Gödel, the most important logician since Aristotle, bear comparison with those of the most eminent mathematicians. His aversion to controversy is reminiscent of Newton's, while his relatively small number of publications—each quite precise and almost all making a major contribution—echo Gauss's motto of "few but ripe." Like Newton, he revolutionized a branch of mathematics—in this case, mathematical logic—giving it a structure and a Kuhnian paradigm for research. His notebooks, like those of Gauss, show him to be well in advance of his contemporaries. While Newton made substantial efforts in theology as well as in mathematical physics, Gödel contributed to philosophy as well as to mathematical logic. But during the formative period of his career, unlike either Gauss or Newton, he was torn between two cultures, the Austrian and the American, and he spent most of his life on foreign soil.

Life. Gödel's family belonged to the German-speaking minority in Brünn, a textile-producing city in the Austro-Hungarian province of Moravia, where they had lived for several generations. His grandfather, Josef Gödel, after marrying a Viennese wife, Aloisia Keimel, moved from Brünn to Vienna, where he worked in the leather industry. He allegedly committed suicide in the 1890's. Subsequently his wife left their son, Rudolf, with a sister-in-law, "Aunt Anna," who raised him in Brünn and later lived with Kurt's family until her death. Rudolf was first sent to a gymnasium, where the classical education bored him, and then to a weavers' school, from which he graduated with honors. An energetic and practical man intrigued by weaving machinery, he worked until his death for the Friedrich Redlich textile factory in Brünn, eventually becoming manager and part owner of it—a man of property who belonged to the upper middle class. He married Marianne Handschuh, the daughter of a weaver originally from the Rhineland. She had been educated at a French institute in Brünn and had broad cultural interests. They had two children; the elder, Rudolf, became a radiologist in Vienna; the younger was Kurt Gödel.

According to his brother, Kurt had a happy childhood, though he was shy and easily upset. His family dubbed him Herr Warum (Mr. Why) because of his continual questions. When two weeks old, he was baptized a Lutheran, his mother's religion, with Friedrich Redlich as his godfather and as the source of his middle name, Friedrich; his father remained nominally Old Catholic. At the age of five, according to his brother, he experienced a mild anxiety neurosis, from which he recovered completely. Three or four years later he contracted rheumatic fever, which, despite his recovery, left him convinced that his heart had been permanently damaged. World War I did not directly affect him, since Brünn was far from the fighting, but the establishment of Czechoslovakia as a nation in 1918 tended to isolate the German-speaking minority there. In 1929 he was to renounce his Czechoslovakian citizenship and officially become an Austrian.

Gödel's education began in September 1912, when he enrolled in the Evangelische Privat-Volks- und Bürgerschule, a Lutheran school in Brünn, from which he graduated in 1916. That fall he entered the Staats-Realgymnasium, a German-language high school in Brünn, where he excelled in mathematics, languages, and religion. (He took religious questions more seriously than the rest of his family, later describing his belief as "theistic.") As an adolescent he became interested first in foreign languages, then in history, and finally, around 1920, in mathematics; about a year later he was attracted to philosophy as well, and in 1922 he read Kant. According to his brother, he had mastered a good deal of university mathematics before graduating from the gymnasium in 1924.

That year Gödel matriculated at the University of Vienna, intending to study mathematics, physics, and philosophy, and to take a degree in physics. He had become interested in science by reading Goethe's theory of colors. At the university he attended lectures on theoretical physics by Hans Thirring, who had just published a book on the theory of relativity, and his interests focused on the foundations of physics. But about 1926, influenced by the number theorist Philipp Furtwängler, he changed to mathematics. During 1927 he attended Karl Menger's course in dimension theory. About this time his philosophical interests were greatly stimulated through a course in the history of philosophy given by Heinrich Gomperz.

Soon Hans Hahn, an analyst intrigued by general topology and logic, became his principal teacher and introduced him to the group of philosophers around Moritz Schlick, the logical positivists later known abroad as the Vienna Circle. He wrote in the curriculum vitae submitted with his *Habilitationsschrift*: "At that time [1926–1928], stimulated by Prof. Schlick, whose philosophical circle I frequently attended, I was also occupied with modern works in epistemology." In 1928 he grew less involved with the Vienna Circle and attended its meetings sporadically until 1933, when he stopped going altogether. This occurred, despite their common interest in logic, because Gödel's Platonism conflicted with their positivism. "I only agreed with some of their tenets," he later noted; "I never believed that mathematics is syntax of language." All the same, he remained in contact with one member of the Vienna Circle, Rudolf Carnap, whose 1928 lectures on mathematical logic and the philosophical foundations of arithmetic strongly influenced the future direction of his research.

In February 1929, Gödel's father died from an abscess of the prostate, a medical condition that Kurt later experienced. Soon his mother moved with her two sons to Vienna. The three of them often went to the theater together, since Kurt was greatly interested in it at that time; his taste in music ran to light opera and Viennese operettas. In 1937 his mother returned to the family villa in Brünn; she died three decades later, in Vienna.

After Gödel completed his dissertation in the summer of 1929, it was approved by his supervisor Hahn and by Furtwängler. He received a doctorate in mathematics from the University of Vienna in February 1930. During the academic year 1931–1932 he was assistant in Hahn's seminar in mathematical logic, selecting much of the material discussed.

By invitation, in October 1929 Gödel began attending Menger's mathematics colloquium, which was modeled on the Vienna Circle. There in May 1930 he presented his dissertation results, which he had discussed with Alfred Tarski three months earlier, during the latter's visit to Vienna. From 1932 to 1936 he published numerous short articles in the proceedings of that colloquium (including his only collaborative work) and was coeditor of seven of its volumes. Gödel attended the colloquium quite regularly and participated actively in many discussions, confining his comments to brief remarks that were always stated with the greatest precision.

At a conference in September 1930, Gödel announced his startling first incompleteness theorem: there are formally undecidable propositions in number theory. He sent a paper on his incompleteness results to *Monatshefte für Mathematik und Physik* in November 1930; it was published two months later. The theorem constituted his *Habilitationsschrift*, which he submitted to the University of Vienna in June 1932 and for which Hahn served as referee. In December 1932 Hahn wrote an evaluation of it for the university, praising Gödel's submission as

> a scientific achievement of the first rank which . . . will find its place in the history of mathematics. . . . The work submitted by Dr. Gödel surpasses by far the standard usually required for *Habilitation*. Today Dr. Gödel is already the principal authority in the field of symbolic logic and research on the foundations of mathematics.

The following March, Gödel was made *Privatdozent*, and during the summer he gave his first course, on the foundations of arithmetic. Extremely shy as a teacher, he always lectured to the blackboard; he had few students. During the next seven years he gave only two more courses at the University of Vienna.

The most obvious reason for this situation was Gödel's increasing association with the Institute for Advanced Study at Princeton, of which he was a member for three years during the 1930's; a less obvious reason was the state of his mental health. When the institute first began operation in the fall of 1933, he was a visiting member for the academic year, thanks to the efforts of Oswald Veblen, and from February to May 1934, he lectured on the incompleteness results. There he made Einstein's acquaintance, but came to know him well only a decade later. Lonely and depressed while at Princeton, he had a nervous breakdown after returning to Europe in June 1934, and was treated for this condition by the eminent psychiatrist Julius von Wagner-Jauregg. That fall he again stayed briefly in a sanatorium (the first stay had been in 1931, for suicidal depression), postponing an invitation to return to the Institute for Advanced Study for the spring term of 1935 and informing Veblen that the delay was due to an inflammation of the jawbone. At Vienna, during the summer semester of 1935, he gave a course on topics in mathematical logic, then traveled to Princeton in September. Suffering from depression and overwork, he resigned suddenly from the institute in mid-November, returned to Europe in early December, and spent the winter and spring of 1936 in a sanatorium. Veblen who had seen him to the boat, wrote to Paul Heegaard (who was on the organizing committee for the 1936 International Congress of Mathematicians), urging that Gödel "be invited to give one of the principal addresses. There is no doubt that his work on the foundations of mathematics is the most important which has been done in this field in our time."

In 1935 Gödel had made the first breakthrough in his new area of research: set theory. During May and June 1937 he lectured at Vienna on his striking result that the axiom of choice is relatively consistent. That summer he obtained the much stronger result that the generalized continuum hypothesis is relatively consistent; and in September 1937 John von Neumann, an editor of the Princeton journal *Annals of Mathematics*, urged him to publish his new discoveries there. Yet Gödel did not announce them until November 1938, and then not in the *Annals* but in a brief summary communicated to the *Proceedings of the National Academy of Sciences*.

Two weeks after marrying Adele Porkert Nimbursky, a nightclub dancer, on 20 September 1938, Gödel left Austria to work for a term at the Institute for Advanced Study. At Menger's invitation he spent the first half of 1939 as visiting professor at Notre Dame. Both at the institute and at Notre Dame, he gave a lecture course on his relative consistency results in set theory; he also presented them in December 1938 at the annual meeting of the American Mathematical Society. In February 1939, Gödel wrote to Veblen, asking him to submit the manuscript containing the proof to the *Proceedings of the National Academy of Sciences*, and promising a lengthy article with a detailed proof for the *Annals of Mathematics*. No such article ever appeared in the *Annals*, and the standard source for his proof remained notes of his lectures taken by George W. Brown in the fall/winter of 1938–1939 and published in 1940.

When he returned to Vienna in June 1939, both Gödel's personal situation and the global one were darkening. After the *Anschluss* of March 1938, when Nazi Germany forcibly annexed Austria, the position of *Privatdozent* had been abolished and replaced by that of *Dozent neuer Ordnung*. Most lecturers at the University of Vienna were quickly transferred to the new position, but Gödel was not. He formally applied for it in September 1939. Within a week he received a letter responding to his application and noting that he moved in Jewish-liberal circles, although he was not known to have spoken for or against the Nazis. Finally, in June 1940, he was granted the position of *Dozent neuer Ordnung*, when it was no longer of any use to him. Ironically, from 1941 to 1945 he was listed at the University of Vienna as *Dozent für Grundlagen der Mathematik und Logik*.

During the summer of 1939 Austria, as a part of Germany, was preparing for war. Gödel, ordered by the military authorities to report for a physical examination, was declared "fit for garrison duty" in September and feared that he would soon be called into service. In late November he wrote to Veblen, urgently seeking an American nonquota immigrant visa. During the summer, in Vienna, ultrarightist students had physically assaulted him. While attempting to obtain German exit permits for himself and his wife, he lectured at Göttingen in December on the continuum problem. By early January 1940, thanks to the vigorous efforts of Veblen and of Abraham Flexner, those permits had arrived, as had the U.S. visas and the Soviet transit visas. Since the war made it dangerous to cross the Atlantic, he took the Trans-Siberian Railway, then a ship from Yokohama to San Francisco, finally reaching Princeton, where he was appointed an ordinary member of the Institute. He never returned to Europe.

In Princeton, Gödel had a quiet social life, his closest friends being Albert Einstein and, later, Oskar Morgenstern, who was also from Vienna. Both of them served as witnesses when he became an American citizen in 1948. The gregarious Einstein and the reclusive Gödel often walked home together from the institute. During 1942 Gödel had begun to be close to Einstein, and their conversations generally concerned philosophy, physics, or politics. Einstein regularly informed him of advances in unified field theory, but Gödel did not collaborate with Einstein because he remained skeptical of this theory.

Gödel had no formal duties at the Institute for Advanced Studies, and thus was free to pursue his research. Several of his working notebooks from that time (in the now archaic Gabelsberger shorthand) record the development of his ideas. In particular, they chronicle his attempts around 1942 to prove the independence of the continuum hypothesis and of the axiom of choice. Slightly earlier, he found some essential errors in Jacques Herbrand's proof that there is a quantifier-free interpretation of first-order logic; these errors were independently rediscovered two decades later by Peter Andrews and others.

In 1943, however, Gödel began to turn his research from mathematics to philosophy—at first, philosophy of mathematics, and later, philosophy in general. He expressed his Platonist views publicly in two well-known articles, one on Russell's mathematical logic in 1944 and the other on Cantor's continuum problem in 1947. Later he wrote that "the greatest philosophical influence on me came from Leibniz, whom I studied about 1943–1946," adding that Kant influenced him to some degree. According to Gödel, his Platonism had earlier led him to give a philosophical analysis that culminated in the discovery of both his completeness theorem and his incompleteness results.

Moreover, it was Kant who stimulated Gödel, during the period 1947–1951, to consider new cosmological models, in some of which "time travel" into the past was theoretically possible. These models, so far removed from mathematical logic, echoed his work on differential geometry in Menger's colloquium in the early 1930's.

Late in his career, honors were showered upon Gödel in abundance. In 1946 he was finally made a permanent member of the Institute for Advanced Study. In 1950 he gave an invited address on relativity theory to the International Congress of Mathematicians, and the following year he was one of two recipients of the first Einstein Award. He was also asked to deliver the annual Gibbs Lecture to the American Mathematical Society, and did so in December 1951, arguing against mechanism in the philosophy of mind. That same year Yale University granted him an honorary D.Litt., and Harvard followed with an honorary Sc.D. in 1952. (Later there were two more honorary doctorates, one from Amherst College in 1967 and the other from Rockefeller University in 1972.) In 1955 Gödel was elected a member of the National Academy of Sciences. In

1957 he was elected a fellow of the American Academy of Arts and Sciences and, in 1967, an honorary member of the London Mathematical Society. He was made a foreign member of the Royal Society a year later, and a corresponding member of the Institut de France in 1972. The United States honored him in 1975 with a National Medal of Science.

Gödel repeatedly refused honors from Austrian academic institutions (for instance, in 1966, honorary membership in the Vienna Academy of Sciences), apparently because he was still perturbed by his treatment after the *Anschluss*. He could not, however, refuse the honorary doctorate that the University of Vienna awarded him posthumously.

After his promotion to professor of mathematics in 1953, Gödel spent a great deal of his time on institute business, particularly to ensure that aspiring young logicians were made visiting members. His promotion had been considerably delayed for fear of burdening him with administrative duties (and for fear that he would be overzealous in carrying them out).

Gödel's personality was idiosyncratic. Shy and solemn, short and slight of build, he was very courteous but lacked warmth and sensitivity. Since he was very much an introvert, he and his wife had guests infrequently; he found the excitement tiring. A hypochondriac whose health actually was relatively poor, he often complained of stomach trouble but remained distrustful of doctors, believing throughout his life that his medical judgment was better than theirs. (In the 1940's he delayed treatment of a bleeding duodenal ulcer for so long that blood transfusions were required to save his life.) Keenly sensitive to the cold, he was often seen in Princeton wearing a large overcoat, even on warm summer days.

In 1976 Gödel retired from the Institute for Advanced Study as professor emeritus. Soon illness and death visited those near to him. In 1977 his wife underwent major surgery, and Oskar Morgenstern, to whom he had become especially close since Einstein's death, died. His own health, uncertain throughout his life, had turned worse near the end of the 1960's. At that time he experienced a severe prostate condition, but he refused to have surgery performed, despite the urgings of his doctor and friends. During the last year of his life, he suffered from depression and paranoia. Late in December 1977, at his wife's insistence, he was hospitalized; two weeks later he died. According to the death certificate, this was due to "malnutrition and inanition" caused by a "personality disturbance." Fearing that his food would be poisoned, he refused to eat, and thus starved himself to death. His wife died three years later. They had no children.

Work. When, in 1928, Gödel began to do research in mathematical logic, it was not a well-defined field. It was still a part of the "foundations of mathematics"—a subject that belonged more to philosophy than to mathematics. As Ernst Zermelo described the situation in logic about that time, apropos of his axiomatization of set theory two decades earlier, "A generally accepted 'mathematical logic' . . . did not exist then, any more than it does today, when every foundational researcher has his own logistical system." The state of foundations was epitomized by the 1930 conference at Königsberg at which Gödel announced his incompleteness result. The conference was primarily devoted to a discussion of the three competing foundational schools—formalism, intuitionism, and logicism—that had dominated the subject for more than a decade.

Formalism, developed by David Hilbert, treated mathematics as a purely formal and syntactical subject in which meaning was introduced only at the metamathematical level. "Hilbert's program," as it was called, was concerned primarily with proving the consistency of classical mathematics and the existence of a decision procedure for it. Intuitionism, developed by L. E. J. Brouwer, stressed the role of intuition as opposed to the formal aspect, and rejected much of both classical logic and classical mathematics. Logicism, developed by Bertrand Russell and embodied in his *Principia Mathematica*, asserted that mathematics was a part of logic and could be developed as a logical system without any further recourse to intuition. All three schools were to influence Gödel, who followed none of them but established a new paradigm: mathematical logic as a part of classical mathematics, answering questions of genuine mathematical interest and only indirectly of philosophical import.

Gödel's dissertation, completed in 1929 and published the following year, grew out of a problem that Hilbert and Wilhelm Ackermann had posed in their book *Grundzüge der theoretische Logik*. (He became interested in this problem in 1928, a year before he first read *Principia Mathematica*.) Hilbert and Ackermann remarked that it was not known whether every valid formula of first-order logic is provable and, moreover, whether each axiom of first-order logic is independent. In his dissertation Gödel solved both problems, considering only a countable set of symbols, since uncountable languages had not yet been introduced. His solution to the first problem is now known as the completeness theorem for first-order logic. In addition to showing that every valid first-order formula is provable, and then extending this result to countably infinite sets of first-order formulas, he gave a different form of the theorem: A first-order formula (or a countable set of such formulas) is consistent if and only if it has a model. In this form the theorem made rigorous a long-standing fundamental belief of Hilbert's.

In 1930 a revised version of Gödel's dissertation was published, now supplemented by a theorem that became very important two decades later, the compactness theorem for first-order logic: A countably infinite set A of first-order formulas has a model if and only every finite subset of A has a model. The published version omitted the philosophical remarks found at the beginning of his dissertation, where he cast doubt on Hilbert's program (even hinting at possible incompleteness) at the same time that he furthered the program by his proof of the completeness theorem. Apparently his doubts about Hilbert's program had been nurtured by Brouwer's lectures, given at Vienna in March 1928, on the inadequacy of consistency as a criterion for mathematical existence.

Hilbert and Ackermann had made the theory of types their framework for logic, considering first-order logic and second-order logic as subsystems. Hence it is not surprising that Gödel, after publishing his completeness results, soon turned to considering higher-order logic. He began by an attack on Hilbert's problem of finding a finitist consistency proof for analysis (that is, second-order number theory). His attack first divided the problem into two parts: (1) to establish the consistency of number theory by means of finitist number theory, and (2) to show the consistency of analysis by means of (the truth of) number theory. But he found that truth in number theory cannot be defined within number theory itself, and so his plan of attack failed. Thereby he was led to his first incompleteness theorem, which he announced privately to Carnap on 26 August 1930.

On 7 September, Gödel made his first public announcement at the Königsberg conference. By 23 October, when he submitted an abstract of the result to the Vienna Academy of Sciences, he had obtained the second incompleteness theorem: The consistency of a formal system S cannot be proved in S if S contains elementary number theory, unless S is inconsistent. (Von Neumann, who had heard him discuss the first incompleteness theorem at Königsberg and was keenly interested in it, wrote to him on 20 November with an independent discovery of the second theorem.) Although Gödel proved his incompleteness results for higher-order logic—in particular, for the theory of types—he pointed out that they applied equally to axiomatic set theory. Adding finitely many new axioms would not change the situation, nor would adding infinitely many new axioms, as long as the resulting system was omega-consistent.

Although Gödel's incompleteness theorems were eventually recognized as the most important theorems of mathematical logic, at first they had a mixed reception. At Princeton, where von Neumann soon lectured on them, Stephen Kleene was enthusiastic about them but Alonzo Church believed, mistakenly, that his new formal system (which included the lambda calculus) could escape incompleteness. In Europe, Paul Bernays corresponded with Gödel about the results and, after some initial reservations, came to accept them. By contrast, Zermelo was skeptical both in correspondence and in person. As late as 1934, Hilbert denied in print that the incompleteness results refuted his program. During the correspondence with Bernays and Zermelo, Gödel showed that for a language truth cannot be defined in the same language, a theorem that was found independently by Tarski in 1933.

In 1932 Gödel published his formulation of the incompleteness results from the standpoint of first-order logic. If number theory is regarded as a formal system in first-order logic, then the above results about incompleteness and unprovability of consistency apply to S. If, however, S is extended by variables for sets of numbers, for sets of sets of numbers, and so on (together with the corresponding comprehension axioms), then we obtain a sequence of systems S_n; the consistency of each system is provable in all subsequent systems. But in each subsequent system there are undecidable propositions. Going up in type in this way, he noted, corresponds in a type-free system of set theory to adding axioms that postulate the existence of larger and larger infinite cardinalities. This was the beginning of Gödel's interest in large cardinal axioms, an interest that he elaborated in 1947 in regard to the continuum problem.

One consequence of Gödel's research was that in 1931 the editors of *Zentralblatt für Mathematik* invited him and Arend Heyting to prepare a joint report on the foundations of mathematics. Although Gödel labored for some time at this report, he eventually withdrew from the venture, and Heyting published his version alone in 1934. Gödel's fragmentary draft survives in his *Nachlass*.

From 1932 to 1936 Gödel published a variety of brief but substantial papers on logic as well as on differential and projective geometry, primarily for

Menger's colloquium. The articles on geometry, all published in 1933, dealt with curvature in convex metric spaces, projective mappings, and coordinate-free differential geometry.

Articles on logic were more numerous and diverse, treating certain cases of Hilbert's decision problem (*Entscheidungsproblem*), the propositional calculus, intuitionistic logic and arithmetic, and speed-up theorems. In 1933 Gödel's results on the decision problem extended work by Ackermann and by Thoralf Skolem to obtain a sharp boundary between decidable and undecidable first-order formulas. In the propositional calculus he solved a problem of Hahn's by showing that some independence results in this calculus require infinite truth tables, and he answered a question of Menger's by formulating the propositional calculus so as to have uncountably many symbols. He did not, however, introduce a first-order language with uncountably many symbols, as Anatolii Ivanovich Maltsev was to do in 1936.

As for intuitionism, Gödel established that if only finitely many truth-values are permitted, then there is no completeness theorem for the intuitionistic propositional calculus; moreover, there are infinitely many systems of logic between the intuitionistic and the classical propositional calculi, each stronger than the previous system. In another paper he proved the philosophically important result that if intuitionistic number theory is consistent, then so is classical number theory; this was done by interpreting the latter theory within the former. Then Gödel reversed direction, showing that the intuitionistic propositional calculus can be given a "provability" interpretation within the classical propositional calculus—an interpretation that, by his second incompleteness theorem, cannot represent provability in a formal system.

What turned out to be an important contribution occurred in Gödel's 1934 lectures on incompleteness. There, refining a suggestion made by Herbrand three years earlier, he introduced the notion of a general recursive function. But he did not believe at the time that this notion captured the general informal concept of computability, and was convinced that it did so ("Church's thesis") only by Alan Turing's work in 1936.

In 1936 Gödel published the first example of a speed-up theorem—pointing out that if one goes from a logic S_n of order n to a logic S_{n+1} of next higher order, there are infinitely many theorems of S_n, each of whose shortest proof in S_{n+1} is vastly shorter than its shortest proof in S_n. Such speed-up theorems later were studied extensively in computer science.

Gödel's next major accomplishment occurred in set theory. Around 1930 he began to think about the continuum hypothesis and learned of Hilbert's attempt, during the period 1925–1928, to prove it. In contrast with Hilbert, he felt that one should not build up the sets involved in a strictly constructive way. Then he reconsidered the question from the standpoint of relative consistency and of models of set theory, in which his discoveries in time became just as famous as his incompleteness theorems. The first breakthrough came in 1935. In October, while at the Institute for Advanced Study, he informed von Neumann of his new result, obtained by means of his "constructible" sets, that the axiom of choice is consistent relative to the other axioms of set theory. When he gave a course on axiomatic set theory at Vienna in 1937, he used what was later called Bernays-Gödel set theory; this is a slight variant of Bernays' system (about which Bernays had informed him in a 1931 letter) used to identify a set with the corresponding class. One of those attending the course was Andrzej Mostowski, who later recalled that Gödel "constructed a model in which the axiom of choice was valid; at that time, I am sure that he did not have the consistency proof for the continuum hypothesis."

During this period Gödel was trying to prove that the generalized continuum hypothesis is true in the model of constructible sets. On 14 June 1937 he found the crucial step in establishing this fact. By September he had communicated his new result to von Neumann, but refrained from publishing it for over a year. When it appeared, late in 1938, it was clear that he had taken Zermelo's cumulative type hierarchy in set theory and had treated it from the standpoint of Russell's ramified theory of types, extending the latter theory to transfinite orders. In particular, Gödel stated that the generalized continuum hypothesis can be proved consistent relative to von Neumann's system of set theory, Zermelo-Fraenkel set theory, or the system of *Principia Mathematica*. At that time he did not mention first-order logic as the basis of his notion of "constructible set" but merely regarded it as excluding impredicative definitions.

When he communicated his second brief paper on the subject, in February 1939, Gödel spelled out in detail how to prove that the generalized continuum hypothesis holds in the model. The hierarchy of constructible sets was introduced by transfinite recursion in such a way that its definition differed from Zermelo's cumulative hierarchy only at successor ordinals, where $M_{\alpha+1}$, the next level after M_α, was defined as the set of all subsets of M_α that are first-order definable from parameters in M_α. The critical step, later called the condensation lemma, used a form of the Löwenheim-Skolem theorem to show that each constructible subset of $M\omega_\alpha$ is $\aleph_\alpha$. Then the generalized continuum hypothesis holds in the model, since the cardinality of $M\omega_\alpha$ is $\aleph_\alpha$. Finally, Gödel gave two set models, $M\omega_\omega$ and M_Ω, where Ω was the first inaccessible cardinal. The first of these was a model of Zermelo set theory, while the second was a model of Zermelo-Fraenkel set theory.

In the 1938 paper Gödel introduced the axiom of constructibility, which stated that every set is a constructible set. He asserted that this proposition, "added as a new axiom, seems to give a natural completion of the axioms of set theory, in so far as it determines the vague notion of an arbitrary infinite set in a definite way." The axiom of constructibility, as he showed in the 1939 paper and in the 1940 monograph, implies both the axiom of choice and the generalized continuum hypothesis. The last step was to establish that the axiom of constructibility holds in the model. To do so, he introduced and developed the critical notion of being "absolute." That is, a formula is absolute for a model if it holds in the model precisely when it is true.

Gödel's 1938 paper also asserted that the axiom of constructibility implies that there is a nonmeasurable set of real numbers (as well as an uncountable set of real numbers having no perfect subset) that occurs low in the projective hierarchy. These results followed from an observation of Stanislaw Ulam, who noticed that in the model there is a projective well-ordering of the real numbers.

Gödel's results on constructible sets became known largely through his 1940 monograph, based on lectures delivered at the Institute for Advanced Study during October–December 1938. There he did not use the hierarchy M_α but presented the constructible sets as built up by transfinite recursion from eight fundamental operations on sets, formulated within Bernays-Gödel set theory. Later set theorists tended to find this second approach less intuitive than the first.

It is uncertain why Gödel refrained from submitting for publication his results on the relative consistency of the axiom of choice and the generalized continuum hypothesis, known by the summer of 1937, until November 1938. But a clue can be found in a letter he wrote to Menger in December 1937:

> I have continued my work on the continuum problem last summer and I finally succeeded in proving the consistency of the continuum hypothesis (even the generalized form) with respect to general set theory. But for the time being please do not tell anyone of this. So far, I have communicated this, besides to yourself, only to von Neumann. . . . Right now I am trying to prove also the independence of the continuum hypothesis, but do not yet know whether I will succeed with it. (Wang, *Reflections*, p. 99)

Gödel persisted in his attempts to prove that the continuum hypothesis is independent. During the summer of 1942, while on vacation in Maine, he obtained a proof of a related result, the independence of the axiom of choice relative to the theory of types as well as the independence of the axiom of constructibility. But he did not succeed in showing the independence of the continuum hypothesis.

Gödel never published his result on the independence of the axiom of choice. According to comments he made later to John Addison and others, he feared that such independence results would lead research in set theory "in the wrong direction." What he considered to be the right direction became apparent in his expository paper of 1947 on the continuum problem.

It was very likely, he observed in that paper, that the continuum hypothesis would eventually be proved independent of the axioms of set theory; to seek such a proof was the best way to attack the continuum problem. But, he insisted, even such a proof of independence would not solve the continuum problem because, he argued Platonistically, the cumulative hierarchy of sets forms a well-determined reality, and thus it should be possible to establish the truth or falsity of the continuum hypothesis. The way to proceed was to search for new true axioms (especially large cardinal axioms) that would settle its truth or falsity. It was very likely, he believed, that the continuum hypothesis was false.

In 1946 Gödel had delivered a paper on set theory at the Princeton Bicentennial Conference on Problems of Mathematics. There he introduced the notion of ordinal-definable set (which was related to, but distinct from, his notion of constructible set), and conjectured that the ordinal-definable sets would provide a model of set theory in which the axiom of choice held, thereby providing a new proof for its relative consistency. Moreover, Gödel asserted, it would be impossible to prove that the continuum hypothesis held in that model. Because this paper circulated only in manuscript at the time, the ordinal-definable sets were rediscovered independently in 1962 by John Myhill and Dana Scott, who used them in the way proposed by Gödel.

During the 1940's Gödel turned increasingly to

philosophy. The first fruits were found in his 1944 article on Russell's mathematical logic. Like almost all of his publications from this date on, it was requested by an editor—in this case by Paul Schilpp for the volume *The Philosophy of Bertrand Russell*. The article, solicited in November 1942, was sent to Schilpp six months later. In September 1943 Gödel wrote to Russell, urging him to reply in detail to the criticisms contained in the article. But Russell, who had not worked in logic for three decades, declined to answer the criticisms and merely granted that they had some merit.

This article was Gödel's first public defense of his Platonism, which was unfashionable in philosophy of mathematics at the time. He put forward suggestions for further research in foundations, especially in regard to the theory of types, and considered (though he was somewhat dubious) the possibility of infinitely long logical formulas. Such infinitary logics were fully developed only a decade later by Alfred Tarski and others.

In July 1946, Schilpp asked Gödel to write an article (completed three years later) for the volume *Albert Einstein: Philosopher-Scientist*. This request prompted Gödel to return to his early interest in the foundations of physics, and thereby led him to publish two technical papers on the general theory of relativity (1949). Highly original and eventually quite influential, these papers gave the first rotating solutions to Einstein's cosmological equations. The original solution put forward by Gödel permitted an observer, in principle, to travel into the past, while his later solutions (for an expanding universe) did not allow this. The article for Schilpp's volume dealt with the relationship between relativity theory and Kantian philosophy. Gödel elaborated on this subject in a still unpublished paper, in which he argued that relativity justified certain aspects of Kant's view of time. "Einstein told me," Oskar Morgenstern wrote in May 1972, "that Gödel's papers were the most important ones on relativity theory since his own [Einstein's] original paper appeared. On the other hand, other cosmologists such as Robertson did not like Gödel's work at all."

In December 1951, Gödel delivered to the American Mathematical Society the Gibbs Lecture, "Some Basic Theorems on the Foundations of Mathematics and Their Implications," which remains unpublished. In it he discussed the implications of his incompleteness theorems for mathematics and philosophy. His chief result was that "either mathematics is incompletable in the sense that its evident axioms can never be comprised in a finite rule, i.e. the human mind (even within the realm of pure mathematics) infinitely surpasses the powers of any finite machine, or else there exist absolutely unsolvable Diophantine problems." From this time on Gödel made it a policy to refuse any invitation to lecture, turning down many important ones.

In May 1953 Schilpp requested a third article from Gödel, this time for the volume *The Philosophy of Rudolf Carnap*. Gödel worked for an extended period on this article, "Is Mathematics Syntax of Language?" which was devoted to showing why the answer was no. He was, however, not satisfied with any of his six versions of the article, and it remains unpublished. Indeed, because he published so little after the 1940's, Gödel appeared rather unproductive to many of the mathematicians interested in his work.

In 1958 there appeared the last of Gödel's published papers, solicited two years earlier for a volume honoring Bernays' seventieth birthday. Known as the "Dialectica Interpretation" (after the journal *Dialectica*, in which it was published), the paper supplied a new quantifier-free interpretation for intuitionistic logic by using primitive recursive functionals of any finite type. The paper extended Hilbert's program by not confining the notion of "finitary," as Hilbert had, to concrete objects, but permitting abstract objects as well. This was an instance, however, of a result that was first published decades after Gödel found it. For he discovered it in 1941 and lectured on it that year at both Princeton and Yale. Late in the 1960's he wrote an expanded version of this paper, but never published it—despite Bernays' repeated pleas.

In 1963, when Paul Cohen discovered the method of forcing and used it to prove the independence of the axiom of choice as well as that of the continuum hypothesis, he went to Princeton to seek Gödel's *imprimatur*. At Cohen's request, Gödel submitted Cohen's article containing these results to the *Proceedings of the National Academy of Sciences*, in which Gödel's relative consistency results had appeared a quarter-century earlier. Their correspondence makes it clear that Gödel made many revisions in Cohen's paper.

That correspondence also reveals Gödel's concern with showing the existence of a "scale" of length $\aleph_1$ majorizing the real numbers (treated as the set of all sequences of natural numbers). He regarded this question as, "once the continuum hypothesis is dropped, the key problem concerning the structure of the continuum." In 1970 Gödel drafted a paper, intended for the *Proceedings*, on this problem and gave in it some axioms on scales (the square axioms) that, he claimed, imply that the continuum hypothesis is false and that the power of the real numbers is $\aleph_2$. Gödel submitted this paper to Tarski for his judgment and, when D. A. Martin found an error in it, withdrew it. Thus his final contribution to Cantor's continuum problem ended inconclusively.

During the 1970's, and perhaps earlier, in his philosophical research Gödel pursued the ideal of establishing metaphysics as an exact axiomatic theory, but he never achieved what he regarded as a satisfactory treatment. His work in this area was stimulated by Leibniz's *Monadology*. Leibniz also influenced his version, which began to circulate about 1970, of the ontological argument for the existence of God.

Since much on Gödel's philosophical work remains unpublished, his philosophical influence will likely increase as more of his work becomes available. By contrast, his mathematical work (essentially complete three decades before his death) was, and is, in the words of von Neumann, "a landmark which will remain visible far in space and time."

BIBLIOGRAPHY

I. ORIGINAL WORKS. Gödel's *Collected Works* are being prepared by Solomon Feferman (editor in chief), John W. Dawson, Jr., Stephen C. Kleene, Gregory H. Moore, Robert M. Solovay, and Jean van Heijenoort. Vol. I, containing publications from 1929 to 1936, was published in 1986; the second volume, containing his publications after 1937, appeared in 1988. A third volume, consisting of correspondence and previously unpublished manuscripts, is in preparation.

Gödel's scientific *Nachlass* is in the Firestone Library at Princeton University. Part of his personal *Nachlass*, including about 1,000 letters (mainly to his mother), is in Vienna at the Nueu Stadtbibliothek.

II. SECONDARY LITERATURE. An excellent account of Gödel's life and work is in the biography by Solomon Feferman in vol. I of Godel's *Collected Works* (1986), 1–36. For more personal accounts, see G. Kreisel, "Kurt Gödel," in *Biographical Memoirs of Fellows of the Royal Society*, **26** (1980), 149–224 (corrections *ibid.*, **27**, p. 697, and **28**, p. 718); and Hao Wang, *Reflections on Kurt Gödel* (Cambridge, Mass., 1987). On Gödel's life, see also Curt Christian, "Leben und Wirken Kurt Gödels," in *Monatshefte für Mathematik*, **89** (1980), 261–273; John W. Dawson, Jr., "Kurt Gödel in Sharper Focus," in *Mathematical Intelligencer*, **6**, no. 4 (1984), 9–17; Stephen C. Kleene, "Kurt Gödel," in *Biographical Memoirs. National Academy of Sciences*, **56** (1987), 135–178; Willard V. Quine, "Kurt Gödel," in *Year Book of the American Philosophical Society 1978* (1979), 81–84; and Hao Wang, "Kurt Gödel's Intellectual Development," in *Mathematical Intelligencer*, **1** 1978), 182–184, and "Some Facts About Kurt Gödel," in *Journal of Symbolic Logic*, **46** (1981), 653–659. Studies of his work are in the "Introductory Notes" in the *Collected Works*, as well as in Martin Davis, "Why Gödel Didn't Have Church's Thesis," in *Information and Control*, **54** (1982), 3–24; John W. Dawson, Jr., in *PSA 1984: Proceedings of the 1984 Biennial Meeting of the Philosophy of Science Association*, **2** (1985), 253–271; Stephen C. Kleene, "The Work of Kurt Gödel," in *Journal of Symbolic Logic*, **41** (1976), 761–778 (addendum, *ibid.*, **43**, 613) and in articles by Stephen C. Kleene, G. Kreisel, and O. Taussky-Todd in *Gödel Remembered* (1987), edited by P. Weingartner and L. Schmetterer. On Gödel's contributions to philosophy, see Hao Wang, *From Mathematics to Philosophy* (1974). Finally, there is a biographical video, 'Dr. Kurt Gödel: Ein mathematischer Mythos," by P. Weibel and W. Schimanovich.

GREGORY H. MOORE

GOLDRING, WINIFRED (*b.* Kenwood, near Albany, New York, 1 February 1888; *d.* Albany, 30 January 1971), *paleontology, geology*.

Goldring's father, Frederick, trained as an orchid grower at Kew Gardens before emigrating from England in 1879 to assume charge of the orchid collections at the Erastus Corning estate near Albany. Her mother was Mary Grey, a teacher and daughter of Corning's head gardener. Subsequently her father opened his own floral business in Slingerlands, an Albany suburb.

Winifred showed unusual talent, and in 1905 graduated as class valedictorian from the high school of the New York State Normal School. After taking her A.B. at Wellesley College in 1909, Goldring was assistant to Elizabeth F. Fisher, professor of geology. She studied as well with the Harvard geographer William Morris Davis, who supervised the thesis that earned her an A.M. from Wellesley in 1912. From 1912 to 1914 she remained at Wellesley as instructor in petrology and geology, and assistant in geography and field geology. In the summer of 1913 Goldring studied at Columbia University with systematic paleontologist Amadeus W. Grabau, and in 1921 with paleobotanist Edward W. Berry at Johns Hopkins University.

In 1914 Goldring accepted a temporary appointment as "scientific expert" to develop exhibits on fossils at the New York State Museum in Albany, which had recently opened a hall of invertebrate paleontology. The museum provided her with lifetime employment as she moved from assistant paleontologist through the ranks to become, in 1939, state paleontologist. All of Goldring's major work was done in the New York area, a result of her appointment and of the fact that the strata between

Lake Erie and the Hudson River, exposed by deep-cut rivers and the Finger Lakes but otherwise tectonically undisturbed, had become a classic reference for Devonian paleontology and stratigraphy. Under James Hall the New York State Geological Survey had produced the monumental thirteen-volume *Palaeontology of New York* (1847–1894) and had amassed an outstanding museum collection (despite the fact that Hall had been allowed to sell a substantial number of specimens in lieu of being paid a sufficient salary as state paleontologist).

Charles Abiathar White, working under Hall at midcentury, had found an extensive "colony of crinoids" in Hamilton shales that were described individually in subsequent years in New York State publications by Hall and his successor, John M. Clarke. The "sea lilies" were an important, primitive class of echinoderms. Goldring took up the study of crinoids after 1915; the resulting volume, *Devonian Crinoids of the State of New York*, revised earlier descriptions and detailed important characteristics of specimens held at the New York State Museum and the United States National Museum. The oversized volume had sixty plates produced by illustrator George S. Barkentin. Goldring's work helped to clarify the controversial boundaries between the Silurian and Devonian periods (and within the Devonian as well) by further refining the evolution of the organisms. Goldring continued her investigation of crinoids and fossil plants while she met the other demands of her position.

Her initial museum assignment had been to "fill the cases," and Goldring used the opportunity to explore technical and educational alternatives. The resulting displays "What Is a Fossil?" and "What Is a Geological Formation?" were considered models and were copied elsewhere. Most famous was the representation of a large Gilboa (Middle Devonian) fossil forest of seed ferns, designed to represent the oldest known "petrified forest," investigated by the museum staff in the early 1920's. Assigned to exhibit the exciting new material, Goldring juxtaposed an impressionistic background painting (by Henri Marchand and sons) depicting the ancient forest and Schoharie Creek with life-size restorations of the fern trees and fossil stumps. The display, opened in 1925, also revealed three series of forests that had successively flourished, and subsequently been submerged, destroyed, and fossilized, at the Gilboa site.

In the late 1920's state funds were more restricted, and the museum staff concentrated on educational programs, mapping, and applied geology. Goldring spent the summers between 1928 and 1937 largely doing fieldwork on the classic and complicated Helderberg Mountain region, which was essential to her geological maps and discussion of the Berne (1935) and Coxsackie (1943) quadrangles. Her responsibility to maintain public programs was met through several ambitious handbooks and exhibitions. Part 1 of her *Handbook of Paleontology for Beginners and Amateurs*, on fossils (1929), was simple enough for general readers but also was used as an introductory college textbook; it is still in print. Part 2, on formations (1931), updated the nomenclature and provided new correlation charts of the stratigraphy of the New York Devonian. Her field abilities, thoroughness, and incisive mind made even her *Guide to the Geology of John Boyd Thacher Park* (1933) and its setting a "case study" used by college faculty.

Goldring never pursued an academic appointment, but she worked actively with colleagues and their students (such as Charles Schuchert's student G. Arthur Cooper) who studied New York fossils and sedimentary formations, even arranging for their summer support. She did not much like public lecturing, but she regularly attended professional meetings and was an active member of the major geological, paleontological, and museum associations. Although her research was done exclusively in New York and adjacent states and provinces, Goldring took extended trips throughout the United States, Canada, and Alaska, and lived for one summer in Cuba. When the International Geological Congress met at Washington, D.C., in 1933, for example, she was principal guide for the ten-day tour through New York State and joined a thirty-day transcontinental trip conducted mainly for foreign geologists.

Goldring's appointment as the first woman state paleontologist received considerable public attention. A quiet yet forceful woman who never married, she could be very direct in her opinions regarding opportunities for women in paleontology (not very good) and evaluating the quality of work produced by her peers. The latter recognized her merit by electing her president of the Paleontological Society in 1949 (another female first) and vice president of the Geological Society of America in 1950. Goldring retired in 1954 and spent the next sixteen years at her family home in Slingerlands, reading and walking in the Helderbergs.

BIBLIOGRAPHY

I. Original Works. Goldring's writings include *The Devonian Crinoids of the State of New York*, New York

State Museum Memoir 16 (1923); *Handbook of Paleontology for Beginners*, I, *The Fossils* (Albany, N.Y., 1929; 2nd ed., 1950); "The Oldest Known Petrified Forest [Gilboa, New York]," in Smithsonian Institution, *Annual Report . . . for 1928* (1929), 315–341; *Handbook of Paleontology for Beginners*, II, *The Formations* (Albany, N.Y., 1931); "Some Upper Devonian Crinoids from New York," in *Annals of the Carnegie Museum*, **24** (1934–1935), 337–348; "New and Previously Known Middle Devonian Crinoids of New York," *ibid*., 349–368; *Geology of the Berne Quadrangle*, *New York State Museum Bulletin*, 303 (1935), with a chapter on glacial geology by John H. Cook; and *Geology of the Coxsackie Quadrangle*, *New York State Museum Bulletin*, 332 (1943), with a chapter on glacial geology by John H. Cook.

Official correspondence is in the New York State Archives in Albany, in the Winifred Goldring and John M. Clarke MSS. A brief summary of Goldring's personal activity is recorded in intermittent alumnae reports to Wellesley College. There are a significant number of letters in the Charles Schuchert MSS in the Yale University Archives and in the Division of Marine Invertebrate MSS, J. Brookes Knight MSS, and Ray S. Bassler MSS at the Smithsonian Institution Archives.

II. Secondary Literature. The only biographical article of any length is Donald W. Fisher, "Memorial to Winifred Goldring, 1888–1971," in *Memorials of the Geological Society of America*, **3** (1974), 96–102, with a list of her publications. See also *American Men and Women of Science*, I (1955), 705; and *Notable American Women: The Modern Period*, IV (1980), 282–283.

Sally Gregory Kohlsted

GOODSPEED, THOMAS HARPER (*b*. Springfield, Massachusetts, 17 May 1887; *d*. Calistoga, California, 17 May 1966), *plant genetics, botany*.

Goodspeed is remembered by botanists for two achievements. First, he helped to establish and develop the botanical garden of the University of California, which under his direction became one of the leading university botanical gardens in the United States and the source from which numerous ornamental plants native to South America were introduced to the horticulturists and amateur gardeners of California. Second, he was the author of a systematic monograph on the tobacco genus *Nicotiana* that combined traditional with modern methods.

Goodspeed was the son of George S. Goodspeed, professor of comparative religion and ancient history at the University of Chicago, and of Florence Duffy Mills Goodspeed. After schooling in Chicago, followed by a year at Gaillard College in Lausanne, Switzerland, he entered Brown University, where he became interested in plant science and graduated with the A.B. degree in 1909.

In the same year, Goodspeed was appointed assistant in the department of botany at the University of California at Berkeley, where he remained until his retirement in 1957, rising to the rank of professor in 1928.

In 1911 Goodspeed married Florence Beman; they had a son and a daughter.

During World War I, Goodspeed took part in a survey of plants that might provide an emergency source of rubber. He also cooperated with the chairman of the department, the distinguished algologist William A. Setchell, in developing the university's botanical garden. He was largely instrumental in selecting the present site in Strawberry Canyon, about a mile above the present campus, and establishing a new garden there. Under Goodspeed's direction, the garden became a valuable source of plant material for teaching and research, a location for testing and propagating previously ignored or unknown species that enriched the gardens of Northern California, and a quiet spot of beauty that attracted many visitors.

When Goodspeed arrived in Berkeley, Setchell had already established a collection of related species, including cultivated tobacco (*Nicotiana tabacum*) and was doing research on them. Goodspeed joined him in this effort, along with Roy E. Clausen, a graduate student who later took his Ph.D. under Setchell. Until 1928 they collaborated in cytogenetic investigations of *Nicotiana*, becoming the first botanists to demonstrate the origin of a new plant species (*N. glutinosa-tabacum*) via the combination of interspecific hybridization and doubling of the chromosome number. They also contributed valuable data that led to experimental demonstration of a similar origin for cultivated tobacco itself.

Goodspeed and Clausen then continued separate lines of research. After a brief period of investigations of X-ray-induced mutations and of plants having an extra chromosome (trisomics) in *N. sylvestris*, Goodspeed continued his main line of research: the cytogenetics, systematics, and distribution of the approximately sixty species belonging to the genus *Nicotiana*. With collaborators he wrote a monograph on the genus that was a synthesis of taxonomic, morphological, cytogenetic, and phytogeographical evidence that his and other laboratories had obtained. His account of evolution and the origin of species reflected his deep knowledge and wide experience with *Nicotiana*. Published in 1954, when he was sixty-seven, it was his final important research production.

His desire to see and collect as many species of *Nicotiana* as possible in their native environment led Goodspeed to organize and lead two botanical collecting and exploring expeditions to the Andes, where the genus is most diverse, the first in 1935–1936 and the second in 1938–1939. With Mrs. Goodspeed and eleven associates who either accompanied him or went to areas that he did not visit, he gathered for the university's botanical garden not only a nearly complete collection of living plants and herbarium specimens of *Nicotiana* species, including five that are described for the first time, but also numerous other ornamentals native to temperate South America that were tested in Berkeley and have enriched the gardens of California. His account of these expeditions, *Plant Hunters in the Andes*, was of interest both botanically and in its account of climate and vegetation.

Goodspeed was secretary of the Save-the-Redwoods League (1917–1918), a director of the Golden Gate International Exposition (1939–1940), and president of the American-Scandinavian Foundation (1944–1945). He was honored by the Seventh and Eighth International Botanical Congresses with an honorary vice presidency and presidency of the Section on Experimental Taxonomy (Seventh, Stockholm, 1950), and an honorary vice presidency (Eighth, Paris, 1954). He was also president of a section of the International Scientific Tobacco Congress (Paris, 1955), at which he gave the opening address. He received honorary degrees from Brown University (1940), the University of La Plata (Argentina, 1943), and the University of Cuzco (Peru, 1957). Among his other distinctions were election as a foreign member of the Swedish Royal Academy of Sciences and receipt of the Chilean government's highest decoration, commander of the Order of Merit Bernardo O'Higgins (1953).

Goodspeed was a distinguished botanist and an able botanical garden director, and he did much to promote good relationships between scientists of North and South America.

BIBLIOGRAPHY

I. Original Works. Goodspeed's writings include "Parthenocarpy and Parthenogenesis in *Nicotiana*," in *Proceedings of the National Academy of Sciences*, **1** (1915), 341–346; "Interspecific Hybridization in *Nicotiana*, II, A Tetraploid *glutinosa-tabacum* Hybrid: An Experimental Verification of Winge's Hypothesis," in *Genetics*, **10** (1925), 278–284, written with Roy E. Clausen; "Interspecific Hybridization in *Nicotiana*, VIII, The *sylvestris-tomentosa-tabacum* Triangle and Its Bearing on the Origin of *tabacum*," in *University of California Publications in Botany*, **11** (1928), 245–256, written with Roy E. Clausen; "Nature and Significance of Structural Chromosomal Alterations Induced by X-Rays and Radium," in *Cytologia*, **1** (1930), 308–327, written with P. Avery; "Induced Chromosomal Alterations," in *Biological Effects of Radiation*, **2** (1936), 1281–1295; "Trisomic and Other Types in *Nicotiana sylvestris*," in *Journal of Genetics*, **38** (1939), 381–458, written with P. Avery; *Plant Hunters in the Andes* (New York, 1941; 2nd ed., rev. and enl., Berkeley, 1961); "Cytotaxonomy of *Nicotiana*," in *Botanical Review*, **11** (1945), 533–592; and "The Genus *Nicotiana*: Origins, Relationships and Evolution of Its Species in the Light of Their Distribution, Morphology and Cytogenetics," in *Chronica Botanica*, **16** (1954), written with H.-M. Wheeler and Paul C. Hutchison.

II. Secondary Literature. See *In Memoriam* (University of California), June 1967.

G. Ledyard Stebbins, Jr.

GORTANI, MICHELE (*b*. Lugo, Spain, 16 January 1883; *d*. Tolmezzo, Italy, 24 January 1966), *geology, paleontology*.

Gortani was the son of Luigi Gortani, an engineer, and of Angela Grassi. His parents were from Friuli in northeast Italy; his father was working in Spain when he was born. Gortani attended the secondary school in Udine, the main town of Friuli, and then entered Bologna University, from which he graduated in natural sciences in 1904. His first publication appeared in 1902, while he was still a student.

Gortani deeply loved Friuli, particularly Carnia, an area of mountains and valleys near the Austrian border. He did work in geology and paleontology, as well as entomology and botany (in the latter fields he published a guide to the Coleoptera of Friuli in 1905 and to the Friulian flora, the latter with his father, in 1905–1906). The Carnian Alps are formed mainly of Paleozoic rocks; paleontology and stratigraphy of that geologic era were the main field of Gortani's research.

Gortani was assistant in geology at Perugia University (1905), at Bologna (1906–1910), and at Turin (1911–1913); he was assistant and professor in charge at Pisa University (1913–1922). In addition he was a deputy in the Italian Parliament from 1913 to 1919. About 1910 he married Maria Gentile, who was from Friuli; they had no children.

During World War I, Gortani was a volunteer in the Alpine troops of the Italian army; he helped refugees from the territories, including Friuli, occupied by the Austro-Hungarian army in 1917 and was involved in the reconstruction after the war.

Gortani became full professor of geology at Cagliari University, in Sardinia, in 1922. His stay in Sardinia, though short, influenced his further geological research, the island being the other Italian region besides Carnia with extensive Paleozoic outcrops. He moved to Pavia later in 1922. In 1924 Gortani succeeded Giovanni Capellini at Bologna. He completely reorganized the department of geology there and was editor of the magazine *Giornale di geologia* from 1926 to 1953.

Gortani conducted numerous field investigations and laboratory examinations of the Paleozoic fossil flora and fauna of the Carnian Alps, achieving brilliant results in geologic age determinations and a major advance in Paleozoic stratigraphy. His most important stratigraphic synthesis of the region was published in 1921; a review on the Paleozoic of Sardinia followed in 1922. He continued to study Paleozoic fauna, especially graptolites, as well as Paleozoic fossils from Karakorum (1921).

In 1936 and 1938 Gortani led scientific missions to Ethiopia and Eritrea, then governed by Italy; his petrographer was Angelo Bianchi, from the University of Padua. In later years Gortani studied and described the fossils and the stratigraphic sequence of this region, not yet scientifically explored.

In structural geology Gortani pointed out the importance of gravitational tectonics and the imbricate structure in the Carnian Alps (1957).

He did work in hydrogeology and geomorphology, as well as petroleum geology. He was president of the International Symposium on the Gas Fields of Western Europe (Milan, 1957).

Gortani's textbook of geology, *Compendio di geologia* (1946–1948), has a lengthy section devoted to recovery of groundwater for human needs. Gortani solved practical problems of coal mining (at Valdarno, Tuscany, in 1943), of geopedology, and of soil conservation. He was president of several commissions for applied geology questions, such as earthquakes, landslides, subsidence, and thermal springs. His studies in geomorphology emphasized the glacial phenomena in the central Apennines (1930–1931) and in Friuli (1959), and the Quaternary terraces throughout Italy (1929–1952). He founded the Italian Institute of Speleology in 1929.

Among Gortani's many publications a note on Italian pioneers in geology and mineralogy is worthy of mention (1962). In 1960 he published the geological bibliography of Friuli.

Gortani was a member of the Constituent Assembly of Italy after World War II and then a member of the Italian Senate, as a Christian Democrat (1948–1953). He was responsible for several laws connected with mountain populations and with geological problems, as well as for the geological map of Italy on the scale 1:100,000 (completed in 1970).

Gortani's love for his native region was expressed in the Museum of Carnian Traditions, which he established in Tolmezzo with the help of his wife. The Museo delle Arti e delle Tradizioni Popolari Carniche collects typical furniture and crafts.

Gortani was honorary member of the geological societies of England, France, Austria, and Germany; national member of the Accademia Nazionale dei Lincei, of the Accademia Benedettina di Bologna, and of several other regional academies; and president of the Italian Geological Society, of the Friulian Alpine Society, and of the Carnian Community.

An Alpine shelter in Friuli and the Naturalist Society of Portogruaro (Venezia) have been named for Gortani; the Acts of the Friulian Museum of Natural History are titled *Gortania*.

BIBLIOGRAPHY

I. Original Works. Gortani's more than 300 publications appeared from 1902 to 1965. Among them are "Nuovi fossili raibliani della Carnia," in *Rivista italiana di paleontologia*, **8** (1902), 76–94; "Coleotteri del Friuli," in *Rivista coleotterologica italiana*, **3** (1905); *Flora friulana, con speziale riguardo alla Carnia*, 3 vols. (Udine, 1905–1906), with Luigi Gortani; "La serie paleozoica delle Alpi carniche," in *Rendiconti della R. Accademia dei Lincei*, 5th ser., **30** (1921), 100–103; "Ordoviciano nella catena del Caracorum orientale," *ibid.*, 183–188; "Osservazioni sul Paleozoico della Sardegna," in *Bollettino della Società geologica italiana*, **41** (1922), 356–369; "Relazione sui terrazzi fluviali e marini d'Italia," in *Bollettino della Società geografica italiana*, 6th ser., **6** (1929), 1–18; "Sulla glaziazione quaternaria nell'Appennino abruzzese," in *Rendiconti della R. Accademia delle scienze dell'Istituto di Bologna*, **35** (1931), 34–39; "Sull'origine della lignite del Valdarno," in *Memorie della R. Accademia delle scienze di Bologna*, 9th ser., **9** (1943), 175–201; *Compendio di geologia*, 2 vols. (Udine, 1946–1948); "Graptoliti di Rigolato (Carnia)," in *Memorie dell'Istituto geologico e mineralogico della Università di Padova*, **16** (1950), 3–27; "Carta della glaciazione würmiana in Friuli," in *Atti della R. Accademia delle scienze di Bologna*, 11th ser., **6** (1959); *Bibliografia geologica d'Italia*, VI, *Friuli* (Naples, 1960); "I pioneri italiana della geologia e della mineralogia," in *Giornale di geologia*, 2nd ser., **29** (1962), 1–17; and "Le doline alluvionali," in *Natura e montagna*, **5** (1965), 120–128.

II. Secondary Literature. Obituaries include Ardito Desio, in *Atti dell'Accademia di scienze, lettere ed arti di Udine*, **9** (1968), 1–23; Raimondo Selli, "Michele Gortani. Discorso commemorativo . . . ," in *Accademia nazionale dei Lincei, Celebrazioni Lincee*, **9** (1968), 1–23, with bibliography on pp. 11–23; and Luigi Usoni, in *Atti del Symposium internazionale sui giacimenti minerari delle Alpi* (Trent, 1968), 5–12.

Giuliano Piccoli

GOUDSMIT, SAMUEL ABRAHAM (*b*. The Hague, Netherlands, 11 July 1902; *d*. Reno, Nevada, 4 December 1978), *physics*.

Goudsmit was born into comfortable circumstances. His father, Isaac, was a well-to-do wholesaler of bathroom fixtures. His mother, Marianne Gompers, ran a fashionable millinery shop. When Goudsmit was ten, his mother began consulting him on the design of new hats. She and her designers often followed his advice. Goudsmit later said that he found the task of guessing six months in advance what kinds of hats would be in demand exciting and challenging. If not for the fact that poor health forced his mother to abandon her business in 1918, during Goudsmit's last year in high school, he might well have followed her into the business.

Goudsmit's interest in science first manifested itself when, at age eleven, he read an account of spectrographic phenomena in the elementary physics textbook of his older sister Rosemarie. While he found it fascinating, his decision to study physics when he entered the University of Leiden was based almost solely on the fact that in high school his best grades had been in science and mathematics, and on the further fact that he had no interest in following his father into the business world.

At the University of Leiden, Goudsmit came under the influence of Paul Ehrenfest, who quickly recognized that Goudsmit had a keen intuition for making sense out of empirical data. Under Ehrenfest's tutelage, Goudsmit's mild interest in physics developed into a deep and committed passion. The problems to which Ehrenfest directed Goudsmit were related to analyzing and finding order in the "fine structure" of atomic spectra. Two years after beginning his physics studies, Goudsmit published his first paper (1921). He was convinced that he had found a relativistic formula for doublets in alkali metals that was similar to Arnold Sommerfeld's formula for X-ray doublets. Though Ehrenfest was skeptical, he encouraged Goudsmit to proceed with publication. Goudsmit's results did not gain much attention even though the problem of understanding doublets, multiplets, and the anomalous Zeeman effect was the focus of attention for those attempting to develop a comprehensive theory of the atom. One of the reasons Goudsmit's work may not have gotten more attention was the fact that his relativistic treatment was eclipsed by the popularity of the so-called rump theory first introduced by Werner Heisenberg in early 1922.

Between 1922 and 1925 Goudsmit continued to work on complex spectra and the Zeeman effect. In 1924, Ehrenfest arranged for him to spend half of each week working as an experimental spectroscopist in Pieter Zeeman's laboratory in Amsterdam. With Ehrenfest's group in Leiden for the rest of the week, Goudsmit concentrated on manipulating formulas for accounting for the location and intensity of X-ray lines. There were no general rules. Getting the right formula was a matter of intuitive manipulation of quantum-number rules. It was the kind of detective work in which Goudsmit reveled. During this period he was author or coauthor of several papers on complex spectra and the intensity of Zeeman components.

Early in 1925 physicists struggling to develop a comprehensive quantum theory to account for the data of atomic spectra were faced with several riddles and conundrums. The doublet riddle had emerged in early 1924. The rump model accounted for optical multiplets by ascribing them to the magnetic interaction between a single-valence electron and the rest of the atom, or rump. X-ray multiplets were explained by using relativistic precession of electron orbits. The riddle arose from the fact that the two approaches are, at base, incompatible with each other, and yet each was able to partially explain phenomena thought to lie within the range of the other. A second riddle was associated with the assumptions of the Bohr *Aufbauprinzip* (buildup principle). Starting with hydrogen, Bohr proposed a one-by-one adiabatic addition of electrons. Each atom was treated as the core of the next to be built, it being assumed that the quantum numbers of electrons in the core would be invariant. But Alfred Landé had shown, early in 1923, that the quantum number was not an invariant when an electron was added to the atom, and that in fact the so-called core must have half-integer angular momentum. These were only two of a wealth of difficulties. For example, no model or theory was able to account for the anomalous Zeeman effect, the splitting of each line of a multiplet in the presence of a weak magnetic field.

Near the end 1924 Wolfgang Pauli had completed a paper in which he proposed replacing the various models that had thus far been put forward with an abstract conceptual system in which each electron in the atom was assigned a total of four (instead of the usual three) quantum numbers. He was able to do this by asserting the exclusion principle—no two electrons in an atom could have the same four quantum numbers. The paper was published in early 1925, and almost immediately it caught Goudsmit's attention. In May 1925 he published a paper suggesting a simplification of the exclusion principle in which the fourth quantum number was always

$\pm \frac{1}{2}$. But in the spirit of the work that Goudsmit had been doing for the previous four years, his suggestion was purely formal and did not carry with it any physical interpretation.

In June 1925 George Uhlenbeck, another of Ehrenfest's students, returned from Rome, where he had served for three years as personal tutor to the son of the Dutch ambassador. Ehrenfest suggested that Uhlenbeck spend the summer with Goudsmit, learning about recent developments in the theory of atomic spectra. They were an excellent match. Goudsmit supplied the intuitions necessary to recognize and summarize regularities not immediately obvious in the data provided by spectral analysis. Uhlenbeck was more analytically oriented, more readily able to make connections between formal systems and traditional physical concepts. When Goudsmit explained Pauli's proposals, Uhlenbeck immediately jumped to the premise that each of the quantum numbers is associated with a degree of freedom for each of the electrons, and that therefore the fourth quantum number must correspond to the spin of the electron. Between them they quickly worked out the result that if the angular momentum of the electron was $\hbar/2$, then the interpretation of the alkali doublets was the two orientations of the electron spin relative to its orbital motion. Assuming that the magnetic moment was one Bohr magneton, all of the Zeeman effects could be accounted for. Goudsmit and Uhlenbeck immediately wrote up the result in a short note they gave to Ehrenfest. Although he was skeptical, Ehrenfest submitted it for publication. Later, when they had second thoughts and asked Ehrenfest to withdraw the paper, he refused, on the grounds that they were "both young enough to be able to afford a stupidity." The paper was published in mid November.

Serious objections were raised to their electron spin hypothesis from two quarters. H. A. Lorentz quickly pointed out that, using the standard radius for the electron, $r_0 = e^2/mc^2$, if the electron possessed an angular momentum of $\hbar/2$, then the surface velocity would be about ten times the speed of light. On the day the paper was published, Goudsmit received a letter from Heisenberg in which Heisenberg noted that the doublet separation predicted by their formulation was too large by a factor of two. In February 1926, L. H. Thomas showed that the factor of two could be explained if one took into account relativistic precession of the electron orbit. By that time the issue had already all but been decided. In December 1925, in a series of informal discussions at Leiden with Ehrenfest, Goudsmit, and Uhlenbeck, both Niels Bohr and Albert Einstein became convinced of the essential soundness of the concept of spin. Bohr in turn convinced Heisenberg. Pauli held out until he was satisfied of the correctness of Thomas' explanation of the factor of two. Still, the notion of spin was not completely assimilated until 1928, when Paul Dirac developed the complete relativistic wave equation of the electron.

Goudsmit and Uhlenbeck had not been the only people stimulated by Pauli's exclusion principle paper to consider introducing the concept of electron spin. The idea was definitely in the air. On a visit to Tübingen in January 1925, the American Ralph Krönig, who earlier had collaborated with Goudsmit in Zeeman's laboratory, used the concept of spin to work out the relativistic doublet formulas. Because of the skepticism and even derision he encountered from (among others) Heisenberg, Pauli, and Ernst Jordan, he suppressed the idea. Earlier, in 1921, Arthur Compton had suggested that a spinning electron should be the ultimate magnetic particle, but he never worked out the consequences.

The discovery of electron spin and its acceptance propelled Goudsmit into the inner circle of the world physics community. He was invited to Copenhagen to work with the group at Bohr's institute, and in 1926 he was awarded a Rockefeller fellowship that enabled him to travel to Tübingen, where he collaborated with Ernst Back in measuring the hyperfine splittings in the spectrum of bismuth and in showing that one could account for this structure by assuming that the bismuth nucleus had a spin and a magnetic moment.

Some, such as I. I. Rabi, were later bemused that Goudsmit and Uhlenbeck were never awarded the Nobel Prize for the discovery of electron spin. In 1953 the discovery did bring them the Research Corporation Award, and in 1965 they shared the German Physical Society's Max Planck Medal. But the most immediate significant effect of the discovery for both Goudsmit and Uhlenbeck was the fact that it led to career opportunities that might not otherwise have materialized.

During the mid 1920's, under the direction of Harrison M. Randall, the University of Michigan physics department sought to strengthen its theoretical group in order to provide a balance to its experimental program, which was especially strong in infrared spectroscopy. In 1926 the head of Michigan's theoretical group, Walter Colby, offered Goudsmit and Uhlenbeck positions in the department. Goudsmit was hesitant. He hated the thought of leaving Ehrenfest and was reluctant to give up his only recently acquired niche in the inner circle of European physics. Ehrenfest, who had recommended Goudsmit and Uhlenbeck to Colby, urged that Goudsmit take the offer, on the grounds that American physics was becoming more and more important, and the chances for rapid career advancement in America were far greater than in Europe. The superior financial prospects for a university physicist in America were of no small consideration, since Goudsmit had only recently become engaged. In the summer of 1927, after he had received his Ph.D., he married Jeanne (Jaantje) Logher, who had been a designer in his mother's shop; they had a daughter, Esther. With his new wife, he moved to Ann Arbor, Michigan, where he was to be an instructor in the Michigan physics department.

Goudsmit's fears that by moving to the University of Michigan he would be removing himself from the inner circle of the international physics community were unfounded. Beginning in the summer of 1928, what had been a summer school for experimental physics primarily for the students and faculty at Michigan was transformed into an annual summer symposium on theoretical physics. Leaders in the field from all over the world were invited to participate, and their lectures attracted physicists from far and wide. As a result Goudsmit remained at the center of both the formal and the informal network of the world's leading physicists. He had an excellent reputation as a teacher, serving as professor from 1932 to 1946. In 1929, with his first graduate student, Robert Bacher, he published two articles on hyperfine structure. In 1932 they collaborated on *Atomic Energy States, as Derived from the Analyses of Optical Spectra*. In 1930 Goudsmit had collaborated with Linus Pauling on *The Structure of Line Spectra*. Both works were important source books for many years. In the mid 1930's Goudsmit's research interests shifted somewhat from work on atomic spectra to studies in neutron physics and problems in electron scattering.

The degree to which Goudsmit became assimilated to his American surroundings can be judged from the fact that in 1938, while on a trip to the Netherlands, he was offered a professorship at Amsterdam as successor to Zeeman. Even though the appointment had earlier seemed to Goudsmit to be an unattainable dream, he decided against the move. Had World War II not come along, Goudsmit most likely would have lived out his life learning and teaching at the University of Michigan. It was not to be.

In 1941 Goudsmit joined the Radiation Laboratory at the Massachusetts Institute of Technology, where initially he worked on signal-to-noise ratio problems. In the summer of 1943 he was sent to England to investigate the source of dissatisfaction among American air crews with radar sets Royal Air Force crews found perfectly satisfactory. After several months of study, Goudsmit concluded, in part on the basis of many hours of interviews with flight crews, that the source of the difficulty lay in the fact that the radars had been designed with low-flying tactical airplanes in mind and were ill suited for the high-flying strategic bombers of the American Eighth Air Force. It was another example of Goudsmit's talent as a detective and a sleuth.

Goudsmit's interest in puzzle-solving and mysteries, which had led him to take a course in forensics and to study Egyptology (to the point of being proficient in deciphering hieroglyphics) further manifested itself in a memo he wrote in June 1943 to Lee A. Dubridge, head of the Radiation Laboratory, in which he recommended himself as a person who would be well suited for getting information on the work of scientists in Europe: "I have very close personal contacts with most of the physicists in Italy, France, Belgium, Holland, and even Germany. I think there are even some German physicists who still believe I am their friend." Three months later, in September 1943, Goudsmit was offered and accepted the position of scientific head of Project ALSOS. The name ALSOS, derived for the Greek word for "grove," was chosen by the army because the project was conceived of by General Leslie R. Groves, the head of the American atom bomb program. The purpose of Project ALSOS was to follow directly behind advanced elements of Allied forces in Europe in order to uncover firsthand information on war-related German scientific research, especially research in nuclear physics and atom bomb development.

Between April 1944 and May 1945 the ALSOS unit hopscotched through France, Belgium, the Netherlands, and Germany, investigating significant physics laboratories, seizing and assessing documents and apparatus, capturing and interrogating scientists—especially physicists and chemists thought likely to be working on problems of nuclear physics and nuclear engineering. When Goudsmit arrived at the University of Strasbourg in November 1944, he found what he considered incontrovertible evidence in the files of the physicist Carl Friedrich von Weizsäcker that German work on an atomic bomb was not very far advanced. He also discovered that his old colleague and friend Heisenberg and his laboratory had been evacuated from Berlin to the small Bavarian town of Hechingen. Goudsmit was of the opinion that Heisenberg was in overall charge of German atomic bomb work. In the spring of 1945, elements of the ALSOS team captured the

Hechingen laboratory and uncovered the German uranium heavy-water reactor that had been under construction in a camouflaged cave in the nearby village of Haigerloch.

Shortly after the end of hostilities in Europe, while Goudsmit was still attached to ALSOS, he took a trip to Holland to visit his parents' house in the Hague. He had last heard from his parents, who were Jewish, in March 1943, in a letter sent from a Nazi concentration camp. The house was in shambles, but the framework still stood. Subsequently, from records kept by the Germans, Goudsmit confirmed that his parents had been put to death in the gas chamber of a concentration camp.

By all accounts these experiences had a deep and lasting effect on Goudsmit's psyche. He was stunned when he realized that his colleagues in what has been termed "scientific intelligence" did not seem to have been repelled, as he had been, by evidence he uncovered of the atrocities committed by German scientists and physicians in the name of scientific research. He returned to the United States determined not to bow to the expediency of looking the other way in the interests of exploiting the talents of German scientists who had so recently been part of the Nazi war effort. He was aided in this effort by the fact that in 1948 his erstwhile colleague at Michigan, Walter Colby, who had worked with him in Project ALSOS, became first director of the Office of Intelligence for the Atomic Energy Commission.

Goudsmit and Colby were very close. On the one hand, Colby lent Goudsmit the money to buy a house. On the other hand, Colby, who had extremely serious bouts of self-doubt, relied heavily and sometimes totally on Goudsmit's advice with regard to whom to hire, how to approach different government intelligence agencies, how to organize the office, how to address individual personnel problems, how and when to organize official trips to various European establishments, whom to see on such trips, what to say, and what kinds of information would be important to obtain. Colby kept Goudsmit closely informed on the latest intelligence information regarding particular individual German and Eastern European scientists. It was an extraordinary arrangement.

Goudsmit's personal vigilance on the question of the treatment and reception of German scientists has to be seen against a background in which his own career took a very sharp turn. At the end of the war, Goudsmit realized that he did not want to return to what he now perceived to be the relatively cloistered environment of Michigan. Immediately after the war he moved from Michigan to Northwestern but stayed for only a year. The same restlessness that prevented him from returning to Michigan drove him to seek a position that, in his judgment, was more in the center of the postwar action in physics. At the end of 1947 he accepted an offer to join the staff of the Brookhaven National Laboratories (BNL) as senior physicist. BNL director Philip Morse expected that Goudsmit would want to pursue his own research, but he expressed the hope that Goudsmit could play the role of adviser and guide to younger members of the staff.

Goudsmit took up residence at Brookhaven in early 1948 and remained there until his retirement in 1970. In fact, his only research work was in the development of a time-of-flight mass spectrometer, a device in which ions of an atomic species are given a known momentum and then measurement of the time of flight of the ions over a known distance is used to give accurate information on the mass of the species. But there can be no question that he was a force and a presence at BNL. In 1952 he became chairman of the physics department.

By the time Goudsmit moved to Brookhaven in 1948, he was well known not just to the physics community but also to the general public for his outspoken attacks against the easy assimilation of German scientists back into the world community of physicists. With regard to individual German scientists, Goudsmit concentrated much of his energy speaking out against the opinions of Werner Heisenberg. He suggested that Heisenberg and those working with him in nuclear research had not sorted out the most basic principles necessary to develop an effective nuclear engineering program. He wrote a scathing public denouncement of Heisenberg's claim that during the war he had actively dragged his heels in developing the atomic bomb. He engaged in direct exchanges with Heisenberg and openly battled with newspaper people who showed any inclination to treat the Heisenberg claims seriously.

Whether individuals agreed with his judgments or not, Goudsmit's outspoken, even brutal, attacks on German scientists were widely interpreted as an indication of the fact that his wartime experiences had left him embittered and depressed, and without his prewar zeal for research. There is no question that the war had deeply affected Goudsmit. But his outspoken opposition to the postwar accounts of Heisenberg and others with regard to their motives and accomplishments during the war was a reflection of a more general concern about the effects of the war on the world's physics communities, the quality of life one could expect as a practicing physicist, the reestablishment of prewar international camaraderie, and the freedom to pursue interesting questions for no other reason than their intrinsic appeal. It was these concerns that underpinned his postwar attitudes toward German science and individual German scientists.

At first, immediately after the war, Goudsmit was convinced that the only form of government under which it was possible to have a thriving, robust research community was a democracy. His convictions on these matters had been reinforced by the public struggle over the question of military versus civilian control of atomic energy and the exact character that would be prescribed by law for the Atomic Energy Commission. In Goudsmit's view, the United States had come perilously close to making "the Nazi mistake." It had been saved from that fate by the open nature of its governmental structures, which had provided organizations such as the American Federation of Scientists direct, meaningful access to the deliberative process. With the passing of time, Goudsmit worried more and more about the prospects for American science. He feared that government regulation and the influence of the concerns of the military would have disastrous consequences for American science. After the announcement of the detonation of the first Soviet atomic bomb in September 1949, Goudsmit shifted his analysis from consideration of the political forms under which science was being performed to the social, cultural, and material circumstances in which the activities of scientists were embedded.

By the end of his life, Goudsmit had a very simple, very clear picture of how physics had been affected by the war. Prior to the war, according to his analysis, there had been a relatively small number of physicists, no more than a hundred, at the forefront of research. They were in close communication with each other and were protected from the day-to-day events of the world by "the thick walls of academia." The war changed all that. Physicists now found themselves in the midst of the turmoil of world problems. The destruction of their isolation had meant the end of freedom of action for physicists and for scientists in general, and might even signal the onset of an era in which physicists were no longer free to pursue the questions they deemed important not just in Nazi Germany, or the Soviet Union, but even in the United States.

In 1952 Goudsmit opened a new phase in his career by becoming editor of *The Physical Review*. At the time *The Physical Review* was a single journal, appearing twice a month and publishing about 5,000 pages per year. During the next twenty-three years, in his tenure first as editor of *The Physical Review* and then, after 1961, as managing editor and editor in chief, Goudsmit oversaw the growth of the journal as it split first into two parts and then into five, publishing over 25,000 pages per year at the time that Goudsmit retired in 1974. Of course the discipline of physics was going through a period of intense growth, and the encouragement of such growth would not have been in the interests of creating the kind of ambience in physics for which Goudsmit yearned. Goudsmit's ability to manage the growth of *The Physical Review* reflects the degree to which, even as he pined for what he remembered as a simpler, kinder time, Goudsmit recognized the realities of the present.

It was in the interest of seeking ways to return the physics community to something of its prewar character that, in 1951, Goudsmit proposed the creation of a "letters journal." This was just at the time he was expressing deep distress over the loss of the spirit of what he characterized as the prewar, worldwide physics community. Goudsmit's proposal did not find favor with the Council of the American Physical Society (APS). Whether or not anyone else shared Goudsmit's philosophic concerns about the changing character of the world community of physicists, there were other, more practical considerations that eventually allowed Goudsmit's proposal to prevail. As the size of the physics community expanded, the time between submission, acceptance, and publication of research papers grew to what practitioners considered unacceptably long. As a result, individual researchers would distribute preprints of their publications, using mailing lists that were rather narrowly and exclusively defined. There was much concern about the fact that physicists in small universities and colleges didn't get information critical to their research, and hence to their own advancement.

With his usual persistence, Goudsmit kept the issue before the Council of the APS meeting after meeting, year after year. In April 1958 the Council finally approved going forward with the journal on a trial basis. The first issue of *Physical Review Letters* was published on 1 July 1958. The journal was an instant success. At the time of Goudsmit's death in 1978, it was the mostly widely read and emulated journal in physics. Typically, once this journal began, Goudsmit did not simply sit by and let nature take its course. For the sixteen years he served as editor, he published regular editorials, exhorting his readers and authors to do better.

After his retirement in 1974 as editor in chief of the APS, Goudsmit accepted a position as distinguished visiting professor at the University of Nevada, Reno, where he lectured on what he called "physics appreciation" to a large undergraduate audience. On 4 December 1978 he was found dead of a heart attack in his car in the university parking lot. Later, when officials entered his office, they found a safe owned by the Department of Energy. Inside the safe Goudsmit had kept, among other things, much of the intelligence material he had collected during his tenure as scientific head of Project ALSOS. Thirty-five years after World War II, detective to the end, he had still been keeping track of "his Germans."

Goudsmit was a recipient of the O.B.E. (1948), of the Medal of Freedom, of the Karl Taylor Compton Medal of the American Institute of Physics (1974), and of the U.S. National Medal of Science (1976). He was a fellow of the American Physical Society, the National Academy of Science (elected 1947), and the Netherlands Physical Society. He was a member of the American Philosophical Society and the American Academy of Arts and Sciences.

BIBLIOGRAPHY

I. Original Works. Goudsmit's most significant contributions to physics are "Relativistische Auffassung des Dubletts," in *Naturwissenschaften*, **9** (1921), 995; "Les doublets dans les spectres visibles," in *Archives Neerlondaisas dos Sciences Exactos et Naturalles*, **6** (1922), 116–126; "The Intensities of the Zeeman-Components," in *Koninklijke Akademie van Wetenschappen Amsterdam, Afdeeling voor de Wis- en Natuurkundige Wetenschappen*, **28** (1925), 418–422; "Über die Komplexstruktur der Spektren," in *Zeitschrift für Physik*, **32** (1925), 794–798; "Ersetzung der Hypothese vom unmechanischen Zwang durch eine Forderung bezüglich des innern Verhaltens jedes einzelnen Elektrons," in *Naturwissenschaften*, **13** (1925), 953–954, with G. E. Uhlenbeck; "Die Kopplungsmöglichkeiten der Quantenvektoren im Atom," in *Zeitschrift für Physik*, **35** (1926), 618–625, with G. E. Uhlenbeck; "Spinning Electrons and the Structure of Spectra," in *Nature*, **117** (1926), 264–265, with G. E. Uhlenbeck; "Over het roteerende electron en de structuur der spectra," in *Physica*, **6** (1926), 273–290, with G. E. Uhlenbeck; "Die Koppelung der Quantenvektoren bei Neon, Argon, und einigen Spektren der Kohlenstoffgruppe," in *Zeitschrift für Physik*, **40** (1926), 530–538, with E. Back; *Atoommodel en structuur der spectra* (Amsterdam, 1927); "Kernmoment und Zeemaneffekt von Wismut," in *Zeitschrift für Physik*, **47** (1928), 174–183, with E. Back; "Separations in Hyperfine Structure," in *Physical Review*, 2nd ser., **34** (1929), 1501–1506, with R. F. Bacher; "Zur Hyperfeinstruktur des Wismuts," in *Zeitschrift für Physik*, **66** (1930), 1–12, with P. Zeeman and E. Back; *The Structure of Line Spectra* (New York, 1930), with L. Pauling; "Theory of Hyperfine Structure Separations," in *Physical Review*, 2nd ser., **37** (1931), 663–681; *Atomic Energy States, as Derived from the Analyses of Optical Spectra* (New York, 1932), with R. F. Bacher; "Nuclear Magnetic Moments," in *Physical Review*, 2nd ser., **43** (1933), 636–639; "Diffusion of Slow Neutrons," *ibid.*, **50** (1936), 461–463, with D. S. Bayley, B. R. Curtis, and E. R. Gaerttner; "Introduction to the Problem of the Isochronous Hairspring," in *Journal of Applied Physics*, **11** (1940), 806–815, with Ming-chen Wang; "Multiple Scattering of Electrons," in *Physical Review*, 2nd ser., **57** (1940), 24–29, and **58** (1940), 36–42, with J. L. Saunderson; and "Mass Measurements with a Magnetic Time-of-Flight Mass Spectrometer," *ibid.*, **84** (1951), 824–829, with E. E. Hays and P. I. Richards. Goudsmit published several informal reminiscences: "The Discovery of Electron Spin," typescript of a lecture given on Goudsmit's acceptance of the Max Planck Medal (1965), held by the Niels Bohr Library at the American Institute of Physics, a German translation of which was published in *Physikalische Blätter*, **21** (1965), pp. 445–453; "It Might as Well Be Spin," in *Physics Today*, **29** (June 1956), 40–43, followed by George Uhlenbeck, "Personal Reminiscences," 43–48; and "The Michigan Symposium in Theoretical Physics," in *Michigan Alumnus Quarterly Review* (20 May 1961), 178–182. The following material, written by Goudsmit, concerns his wartime experiences in Project ALSOS or his views on German science during or after World War II: "How Germany Lost the Race," in *Bulletin of the Atomic Scientists*, **1** (1946), 4–5; "Secrecy or Science?" in *Science Illustrated*, **1** (1946), 97–99; "War Physics in Germany," in *Review of Scientific Instruments*, **17** (January 1946), 49–52; *ALSOS* (New York, 1947; 2nd enl. ed., 1983); and "Our Task in Germany," in *Bulletin of the Atomic Scientists*, **4** (1948), 106.

The bulk of Goudsmit's papers are deposited at the archives of the American Institute of Physics. A small collection of material is now being processed by Military Reference at the National Archives.

II. Secondary Literature. The best sources for investigating the history of the introduction of the concept of electron spin are P. Forman, "The Doublet Riddle and Atomic Physics *circa* 1924," in *Isis*, **59** (1968), 156–174; Daniel Sewer, "*Unmechanische Zwang:* Pauli, Heisenberg, and the Rejection of the Mechanical Atom, 1923–1925," in *Historical Studies in Physical Science*, **8** (1977), 189–256; David C. Cassidy, "Heisenberg's First Core Model of the Atom: The Formation of a Professional Style," *ibid.*, **10** (1979), 187–224; Abraham Pais, *Inward Bound: On Matter and Force in the Physical World* (Oxford and New York, 1986), chaps. 10 and 13; Angelo Barraca, "Early Proposals of an Intrinsic Magnetic Moment of the Electron in Chemistry and Magnetism (1915–1921) Before the Papers of Goudsmit and Uhlenbeck," in *Rivista di storia della scienza*, **3** (1986), 353–374. No documentary history of Project ALSOS exists. The following account of the project is by its military commander: Boris Pash, *ALSOS Mission* (New York, 1969). Goudsmit's activities with regard to German scientists after the war has been analyzed in Mark Walker, "Uranium Machines, Nuclear Weapons and National Socialism: The German Quest for Nuclear Power, 1939–1949" (Ph.D. diss., Princeton, 1987); S. Goldberg, "Between Old and New: Goudsmit at Brookhaven," in *The Restructuring of Physical Sciences in Europe and the United States, 1945–1960* (in press). The following are biographical sketches of Goudsmit: Daniel Lang, "A Farewell to String and Sealing Wax, I," in *The New Yorker* (7 November 1953), 47–72, and "A Farewell to String and Sealing Wax, II," *ibid.* (14 November 1953), 46–67; "Goudsmit, Samuel Abraham," in *Current Biography, 1954,* 304–306; and "Goudsmit, Samuel Abraham," in *McGraw-Hill Modern Scientists and Engineers,* I (New York, 1980), 452–453.

STANLEY GOLDBERG

GREGORY, WILLIAM KING (*b.* New York City, 19 May 1876; *d.* Woodstock, New York, 29 December 1970), *vertebrate paleontology, comparative anatomy.*

Gregory was the son of George Gregory and Jane King Gregory. His father was a printer who lived and worked in lower Manhattan, and there young William spent his boyhood, living with his parents in a small house, the front of which was occupied by his father's shop. He attended St. Luke's Primary School, for a short time a public school, and then Trinity School, where from 1894 to 1895 he took the science course to prepare for Columbia University. At Columbia he first studied in the School of Mines, but he soon shifted to Columbia College, where he obtained his bachelor's degree in 1900. Gregory then earned his master's degree there in 1905 and his Ph.D. in 1910. The five-year intervals between his degrees were due in part to the fact that while still an undergraduate he became scientific assistant to Henry Fairfield Osborn, the famous paleontologist who was first dean of the graduate faculty at Columbia and then for many years president of the American Museum of Natural History. During all of his adult life, Gregory was closely associated with those two institutions. He retired from the American Museum of Natural History in 1944 and from Columbia University in 1945.

Gregory married Laura Grace Foote on 4 December 1899; she died in 1937. In 1938 he married Angela DuBois. There were no children.

Gregory was not a robust person, yet he enjoyed good health to such a degree that he lived well into his nineties. He was always active in his own way—that is, attending to his multitudinous affairs—but he did not participate in any sports; his was a contemplative life. He lived for many years across the street from the American Museum of Natural History in Manhattan, where he spent so many of his waking hours.

Although much of his work was in paleontology, a profession that commonly demands rigorous field trips to collect fossils, Gregory was not a field man. He made a few short visits to paleontological field parties in the western states, but his most serious field adventures were of a zoological nature. In 1921 and 1922 he went with Henry C. Raven, of the Museum of Natural History, to Australia to collect marsupials; in 1925 he participated in the *Arcturus* expedition, led by William Beebe, to collect and study marine life in the Sargasso Sea; and in 1929 and 1930 he went to Africa with Raven, James H. McGregor, and Earle T. Engel to study gorillas in their habitat and to collect specimens for study. On the whole, his adventures were in the laboratory and the library; he was a superb comparative anatomist and an outstanding student of vertebrate osteology and evolution.

During the first two decades of his scientific career, Gregory was research assistant to and collaborator with Osborn, who in 1910 turned his professorial duties over to Gregory. From then until his retirement, Gregory trained and was mentor to a host of graduate students, many of whom became famous zoologists and paleontologists. About 1920 Gregory had become so outstanding and involved in his chosen fields that he could no longer devote time to Osborn's projects; that work was taken over by others. Nevertheless, Gregory did work closely with zoologists and paleontologists, notably W. D. Matthew, Walter Granger, George Gaylord Simpson, Henry Raven, and Milo Hellman.

Perhaps no other scientist of his time had such extensive knowledge of the vertebrates as did Gregory. He was a preeminent authority on fishes, both fossil and recent. He was noted for his work on lower tetrapods, notably fossil reptiles and especially mammal-like reptiles.

Although his studies of birds were limited, Gregory accomplished a formidable amount of research on both fossil and recent mammals. Indeed, his Ph.D. dissertation, "The Orders of Mammals," has remained a standard reference on the relationships of all the mammals. He was a leading authority on the evolution of the marsupials.

Gregory's interests developed along definite lines. He did important studies of origins—of the development of the tetrapod limb from the fin of crossopterygian fishes, for example—and he devoted years of work to evolutionary sequences. His studies on the series from fish to man were justly famous.

Even more noted, perhaps, was his work on the evolution of primates, particularly his studies of the origins of human lines of evolution. In this connection Gregory was a leader in the study of mammalian dentition; one of his important books was on *The Origin and Evolution of the Human Dentition* (1922).

Gregory developed some important evolutionary concepts, such as that of "heritage and habitus"—the principle that all animals possess a combination of heritage characters, derived from their ancestors, and habitus characters, developed as evolutionary responses to the world in which they live. Another concept was that of polyiseromerism and anisomerism, the contrast of numerous primitive duplicate or similar elements in an organism with more advanced differentiation and reduction of such elements. Gregory also developed his "palimpsest" concept of evolutionary development—the idea that organisms may show a faint record of primitive features beneath their dominant adaptations.

All of these ideas, and many others, were brought together in Gregory's final great work, *Evolution Emerging*, a two-volume treatise on vertebrates published in 1951, the first volume devoted to text and the second to hundreds of illustrations, many of them arranged to show evolutionary comparisons and developments.

As might be expected of so eminent a scientist, Gregory was a member of many scientific societies, including the National Academy of Sciences. At the time of his death he was one of its oldest members.

BIBLIOGRAPHY

I. Original Works. Gregory's writings include "The Orders of Mammals," in *Bulletin of the American Museum of Natural History*, **27** (1910), 1–524; "On the Structure and Relations of *Notharctus*, an American Eocene Primate," in *Memoirs of the American Museum of Natural History*, **3** (1920), 49–243; *The Origin and Evolution of the Human Dentition* (Baltimore, 1922); *Our Face from Fish to Man* (New York, 1929); "Fish Skulls: A Study of the Evolution of Natural Mechanisms," in *Transactions of the American Philosophical Society*, **23** (1933), i–vii, 75–481; "A Half Century of Trituberculy. The Cope-Osborn Theory of Dental Evolution, with a Revised Summary of Molar Evolution from Fish to Man," in *Proceedings of the American Philosophical Society*, **73** (1934), 169–317; and *Evolution Emerging: A Survey of Changing Patterns from Primeval Life to Man*, 2 vols. (New York, 1951).

II. Secondary Literature. A memoir of Gregory that includes a complete bibliography is by Edwin H. Colbert, in *Biographical Memoirs: National Academy of Sciences*, **46** (1975), 90–133.

Edwin H. Colbert

GROSS, WALTER ROBERT (*b.* Katlakaln, near Riga, Latvia, 20 August 1903; *d.* Tübingen, Germany, 9 June 1974), *vertebrate paleontology*.

Both Gross and his work were profoundly influenced by the Baltic region, where he was born. The son of a country pastor, Erwin Gross, and his wife, Maria, he continued the Baltic-German tradition in natural sciences of Karl Ernst von Baer and especially the paleontological work of Christian Heinrich Pander. His interest in natural sciences was awakened by observing the surrounding plants and animals, specifically butterflies. His detailed observations went unpublished except in a posthumous work about his place of birth. While still in secondary school in Riga, Gross studied fossil fishes that he collected nearby. These studies set the course of his later scientific work. While at the University of Marburg, he decided to give up zoology and focus on vertebrate paleontology. Studies on antiarchs, a specialized group of armored fishes from the Devonian period, which Gross began during secondary school, became the subject of the dissertation for which he received the Ph.D. in 1931 from the University of Berlin.

During the next years Gross was supported by the Notgemeinschaft der Deutschen Wissenschaft and produced many publications. His publications encompassed all agnathans and fishes from the Silurian and the Devonian found in deposits in the Baltic region and the Rhineland. He amazed his colleagues with an almost unbelievable ability to identify and correctly associate small remnants of these fishes. Following the advice of a colleague who had studied complete specimens of one genus, Gross added the latter's corrections to a paper in press (1936); today we know that his original determinations were correct and the corrections in error. All Gross's publications demonstrate a considerable ability for detailed and correct description of a wide variety of agnathans and fishes. Perhaps his main contribution to paleontology was the descriptions of various representatives of the placoderms and their classification. Nevertheless, he preferred to take a broader approach and, when possible, to encompass the entire fish fauna where he worked.

Gross married Ursula Wolff, a scientific illustrator, in 1935; they had three children, Roland in 1937, Sabine in 1940, and Harro in 1943. He spent most of his career at the University of Berlin (after 1945 called Humboldt University) until 1961, but he also served as assistant and associate professor (1934–1937) at the University of Frankfurt am Main. His work on the histology of bones of fossil amphibians and reptiles (1934) formed the basis for the work of de Riqles more than thirty years later.

Gross considered paleontology to be a biological science with additional dimensions in time and space (1960). To him paleontology represented the actual sequence of fossils—their true history—while biology allowed only the deduction of a hypothetical phylogeny or genealogy from recent forms. Biology always had priority for the explanation of processes, so Gross opposed special explanations for the evolutionary process derived from paleontology—concepts such as the macroevolution and typostrophism of Otto Heinrich Schindewolf (1943, 1956). He agreed that the rate of evolution varied over time and from one group to another, but he saw this as only an acceleration in time of processes acting in recent forms as well. Gross published on accelerated rates of evolution in what became known as his "rocket scheme" (1964). All supposedly large jumps in evolution (Schindewolf's typogenesis) are artifacts of the theoretical framework of the supposer—he specifically attacked the "type" concept—while evolution progresses continuously by small steps.

Gross served in the German army during World War II and then was interned (1943–1949). During this time he compiled a long paper on the origin of vertebrates, which opposed the generally accepted idea of their freshwater origin; he argued for a marine origin (1951, as in 1933) based on the fossil record. His views are generally accepted today. He returned to Humboldt University in 1949 and became director of the section of paleontology in 1950.

Gross seldom wrote general overviews or dealt extensively with subjects outside his main interest. An exception was his work on conodonts, initiated when he felt that prevailing opinion had strayed far from observable facts. He wrote three articles on the histology of conodonts (1954, 1957, 1960), microscopic forms first described by Pander. Conodonts were considered gill rakers of fishes at that time. Gross demonstrated through histology the impossibility of that interpretation and identified the then unknown conodont-bearing animal as a chordate of unknown systematic position. These papers became classics in the histology of conodonts. The building of the Berlin Wall in 1961 stranded Gross and his family in West Germany; they were returning from a visit to a fossil excavation in southern Switzerland. He was left without his personal and scientific material and without property owned by his family. Nevertheless, Gross continued his productive scientific work as full professor and temporary director of the department of paleontology and geology at the University of Tübingen until his retirement in 1969.

After 1961 Gross concentrated more on the histology of scales and teeth and became a leader in the field of paleohistology. He described in detail the histology of teeth, dermal bones, and especially scales from different groups of fish and agnathans. His approach might be called histomorphology because he took into consideration the histological differences between hard tissues, but his interest focused on the arrangement and pattern of canals, growth lines, and cell spaces in these hard tissues. He demonstrated that these structures were valuable features that could be used for systematic and taxonomic determination of fossil fishes and agnathans. One of his most comprehensive and instructive papers on scales and their histology in agnathans and fishes was published in 1966.

Gross was a corresponding member of the Senckenbergische Naturforschende Gesellschaft, Frankfurt am Main. He received an honorary doctorate from the University of Munich and became an honorary member of the Paläontologische Gesellschaft in 1972. Throughout his life he worked alone, but his influence was far-reaching, particularly in the Soviet Union, where studies similar to his continue today.

BIBLIOGRAPHY

I. ORIGINAL WORKS. A complete list of Gross's papers is in *Paläontologische Zeitschrift*, **48** (1974), 145–148. They include "Die phylogenetische Bedeutung der altpaläozoischen Agnathen und Fische," in *Paläontologische Zeitschrift*, **15** (1933), 102–137; "Die Typen des mikroskopischen Knochenbaues bei fossilen Stegocephalen und Reptilien," in *Zeitschrift für Anatomie und Entwicklungsgeschichte*, **103** (1934), 731–764; "Neue Crossopterygier aus dem baltischen Oberdevon," in *Zentralblatt für Mineralogie, Geologie, und Paläontologie*, Abt. B (1936), 69–78; "Paläontologische Hypothesen zur Faktorenfrage der Deszendenzlehre: Über die Typen- und Phasenlehre von Schindewolf und Beurlen," in *Die Naturwissenschaften*, **31** (1943), 237–245; "Die paläontologische und stratigraphische Bedeutung der Wirbeltierfaunen des Old Reds und der marinen altpaläozoischen Schichten," in *Abhandlungen der Deutschen Akademie der Wissenschaften zu Berlin*, math.-naturwiss. Kl., Jahrgang 1949 (1951), 1–130; "Zur Conodonten-Frage," in *Senckenbergiana lethaea*, **35** (1954), 73–85; "Über die 'Watsonsche Regel,' " in *Paläontologische Zeitschrift*, **30** (1956), 30–40; "Über die Basis der Conodonten," *ibid.*, **31** (1957), 78–91; "Über die Basis bei den Gattungen Palmatolepis und Polygnathus (*Conodontida*)," *ibid.*, **34** (1960), 40–58; "Über die Bedeutung der Paläontologie im Rahmen der Biologie," in *Forschen und Wirken. Festschrift zur 150-Jahr-Feier der Humboldt-Universität zu Berlin*, II (Berlin, 1960), 297–308; "Polyphyletische Stämme im System der Wirbeltiere?" in *Zoologischer Anzeiger*, **173** (1964), 1–22; "Kleine Schuppenkunde," in *Neues Jahrbuch für Geologie und Paläontologie, Abhandlungen*, **125** (1966), 29–48; "Christian Heinrich Pander 1794–1865 und seine Bedeutung für die Paläontologie," in *Münstersche Forschungen in Geologie und Paläontologie*, no. 19 (1971), 101–183, with P. Siegfried; and *Kirchspiel und Pastorat Roop in Südlivland 1907–1917* (1974), i–iii, 1–58 (unpublished manuscript, 150 copies circulated).

II. SECONDARY LITERATURE. Obituaries appeared in *Attempto. Nachrichten für die Freunde der Universität Tübingen*, nos. 51/52 (1974), 110–111; in *Paläontologische Zeitschrift*, **48** (1974), 143–148; and in *News Bulletin of the Society of Vertebrate Paleontology*, no. 102 (1974), 47. His evolutionary viewpoint is defended in Wolf-Ernst Reif, "The Search for a Macroevolutionary Theory in German Paleontology," in *Journal of the History of Biology*, **19** (1986), see 119–120.

HANS-PETER SCHULTZE

GULICK, JOHN THOMAS (*b.* Waimea, Kauai, Sandwich Islands [now Hawaii], 13 March 1832; *d.* Honolulu, Oahu, Hawaii, 14 April 1923), *zoology, evolution*.

A noted student of geographical distribution, a systematist, and a participant in late-nineteenth-century debates on evolution theory, Gulick was the third of eight children born to Peter Johnson Gulick, a Presbyterian missionary, and Fanny Hinckley Thomas Gulick. Born in the small native village where Captain Cook had landed fifty-four years earlier, Gulick grew up in modest circumstances. In early childhood he suffered an eye inflammation that left him with deficient sight for the rest of his life. He attended a small mission school year Honolulu, where as of 1846 his studies included Greek, surveying, and natural philosophy as well as reading and spelling. He traveled to Oregon in 1848 to improve his health, joined the gold rush to California, and returned to Hawaii in 1850. Gulick began a journey to the United States in 1853 to continue his education; he graduated from Williams College in 1859 and attended Union Theological Seminary, in New York City, in 1860 and 1861.

Barred from Civil War service because of his poor eyesight, Gulick sailed for Japan in 1862 with the first U.S. minister to that nation, Robert H. Pruyn. He became one of the first American missionaries, and one of the first photographers, in that country. Discouraged by official resistance to Christianity, he left Japan for China in 1863. On 3 September 1864 he married Emily De La Cour, an Englishwoman who taught school in Hong Kong. He was ordained a Congregationalist minister on 22 August 1864, and after less than a year of Chinese language study in Peking (now Beijing) moved with his wife to Chang-Kia K'ou (Kalgan), a city of about 75,000 located 140 miles northwest of Peking. The rigorous climate and living conditions, his own poor health, and the death of his wife forced Gulick to withdraw from this mission in 1875.

In 1876 Gulick was posted to Japan, where he remained as a missionary until his retirement in 1899. Following a second marriage on 31 May 1880 to Frances Amelia Stevens, an American teacher in a missionary school, the couple settled in Osaka; they had a son and a daughter. After his retirement Gulick lived for five years in Oberlin, Ohio, where his children attended college, then moved to Honolulu, where he spent his last years. Although his health was poor all his life, he lived to the age of ninety-one.

Gulick's scientific avocation was closely associated with his religious vocation and the life circumstances it entailed. His earliest memories were of marine and forest animals near his Hawaiian home, and his intense curiosity about natural objects and appreciation of their beauty found expression in snail collecting and in frequent references to natural history in his boyhood journals. These enthusiasms provoked the first, and to all appearances the only, religious crisis of his long life. The austere and single-minded Calvinism of his upbringing led Gulick to perceive his preoccupation with nature as a diversion from sacred things, and thus a breach of Christian piety. By 1852, however, he was beginning to find release from this tension in the view that study of nature is a means to study God's character through his works. This attitude, derived from natural theology, stayed with him throughout his long career and was unshaken by his adoption and vigorous advocacy of evolutionary views.

In 1853 Gulick's youthful interests began to turn to science proper. Having returned from a missionary trip to the Caroline Islands, he suddenly perceived the relative richness in number and variety of the Hawaiian land snail populations. Reading Darwin's account of the *Beagle* voyage led him to reflect on the phenomena of island life, and Hugh Miller's *Footprints of the Creator* (1847) exposed him to pre-Darwinian evolutionary views. Gulick began close study of the Hawaiian snails, especially the genus *Achatinellae*, carefully recording all data on their geographical distribution. He emphasized the local centers of production of species and their adaptation to physical environment and to one another.

During his stay in the United States, Gulick participated for the first time in the scientific community, broadened his knowledge of literature (including that of science), and extended his systematic work, using collections he had brought from Hawaii. He met James Dwight Dana at Yale, became a leader of the Williams College Lyceum of Natural History, and presented several papers on the systematics of *Achatinellae* to the New York Lyceum of Natural History, which elected him to membership. His reading included Tayler Lewis' *The Bible and Science*, which argued for a saltationist version of evolution, and Darwin's *Origin of Species*.

From the time that Gulick began his missionary activity in 1862 until his retirement in 1899, his religious responsibilities had first claim on his time and energies, and he lived for long periods in relative isolation from the scientific community. He never ceased work on his collections or reflection on the bearing of their interpretation on the problems of evolution, however, and on two occasions—during the years 1871–1873 and 1888–1889—furloughs made possible valuable contacts with the scientific world.

Part of the first of these furloughs was spent in England. Gulick took his collections to the British Museum, where he made progress in sorting, classifying, and labeling. During a visit to Down (Kent), Darwin encouraged him to write up his interpretations of the data. He presented his results in a talk to the British Association meeting at Brighton, a paper published by the Linnean Society, and a letter to *Nature*. These works reveal both his commitment to evolution in an essentially Darwinian form and his questioning of the sufficiency of natural selection as defined by Darwin to explain all cases of specific divergence.

Gulick pointed out that the Hawaiian Islands possessed a very high number of species of land snails within a small geographical area. These species were almost all found only in the Hawaiian Islands, some in only a part of one island. Most species were confined to a very narrow range, and those found in any given local area were connected by varieties displaying minute gradations of form and color. Given the evidently uniform climate, soil, and other circumstances in which the different species were found, differential fitness as defined by natural selection did not seem a likely explanation for the differences. Gulick concluded that the evolution of the different forms of Hawaiian land snails required two conditions: the division of a given population into two or more noninterbreeding parts (for example, by migration of one part) and an inherent tendency to variation in each part strong enough to lead to cumulative divergence over time.

Gulick's relatively settled living circumstances after his move to Japan in 1876 permitted sustained development of his ideas. In 1887 and 1889 he presented to the Linnean Society of London two long papers, "Divergent Evolution Through Cumulative Segregation" and "Intensive Segregation, or Divergence Through Independent Transformation," which elaborated and generalized the conclusions of his 1872 papers. The central problem of evolution, he now emphasized, was the explanation of specific divergence. Natural selection, even comprising Darwin's principle of "advantage of divergence of character," was not sufficient to split one species into two or more, Gulick argued. Segregation or isolation, the separation of a parent stock into mutually exclusive breeding populations, was also required. Once effected, Gulick maintained, such separation would eventually lead to divergence even in the absence of natural selection. Much of the bulk of the papers was taken up with specification of the various forms of isolation and their effects with or apart from selection.

Gulick's papers gave new life to a debate already under way among the British Darwinists that centered on George John Romanes's theory of "physiological selection." In a paper published in 1886, Romanes had argued for the insufficiency of natural selection and had proposed a supplementary principle that amounted to the spontaneous appearance of mutual infertility between two or more portions of a species population. In Gulick's scheme Romanes's physiological selection figured as one possible mode of varietal isolation. Romanes quickly embraced Gulick's more general formulation, and until his death in 1894 argued for recognition of isolation as an essential component of the evolutionary process. The debates occasioned by the work of Romanes and Gulick were largely responsible for the wide recognition of the problems of specific divergence and of the necessary role played by isolation that was evident among English and American naturalists by the first decade of the twentieth century.

After his retirement Gulick wrote out a larger statement of his evolutionary ideas, *Evolution, Racial and Habitudinal* (1905). This work incorporated his efforts to apply the mathematical theory of probability to his evolutionary problems and introduced principles for understanding the development of human civilization analogous to, but distinct from, those applicable to biological evolution.

Born in the year of Cuvier's death, Gulick lived to see—and applaud—the introduction of Mendelism in biology. His career spanned and exemplified first the age of the naturalist explorers, then the period of ambitious and often contentious evolutionary theorizing that marked the decades around 1900. Viewed as a revisionist by some Darwinians, Gulick nevertheless remained loyal to Darwinian natural selection in the face of challenges from de Vries's mutation theory and other quarters after 1900. His emphasis on isolation as a necessary complement to selection helped open the way for studies that formed part of the synthetic theory of evolution.

BIBLIOGRAPHY

I. Original Works. A listing of Gulick's publications is in Addison Gulick, *Evolutionist and Missionary: John Thomas Gulick* (Chicago, 1932), 511–514. This work also prints much of Gulick's correspondence as well as extracts from his journals. His descriptive and systematic efforts are represented by "Descriptions of New Species of *Achatinellae*," in *Proceedings of the Zoological Society of London* (1873), 73–89, written with Edgar A. Smith; and "On the Classification of the *Achatinellae*," *ibid.*, 89–91. Early statements on evolution include "On the Variation of Species as Related to Their Geographical Distribution, Illustrated by the *Achatinellae*," in *Nature*, **6** (1872), 222–224; and "On Diversity of Evolution Under One Set of External Conditions," in *Journal of the Linnean Society, Zoology*, **11** (1873), 496–505. The most important of Gulick's later articles are "Divergent Evolution Through Cumulative Segregation," *ibid.*, **20** (1888), 189–274; and "Intensive Segregation [or Divergence Through Independent Transformation]," *ibid.*, **23** (1891), 312–380. His final and most comprehensive statement is *Evolution, Racial and Habitudinal* (Washington, D.C., 1905).

II. Secondary Literature. The principal biographical source is Addison Gulick, *Evolutionist and Missionary: John Thomas Gulick* (see above). See also Addison Gulick, "John T. Gulick, a Contributor to Evolutionary Thought," in *Scientific Monthly*, **18** (1924), 83–91; and David Starr Jordan, "John Thomas Gulick, Missionary and Darwinian," in *Science*, **58** (1923), 509. Gulick's connection with George John Romanes and the debates occasioned by both men's work are discussed in John E. Lesch, "The Role of Isolation in Evolution: George J. Romanes and John T. Gulick," in *Isis*, **66** (1975), 483–503.

John E. Lesch

HADORN, ERNST (*b.* Forst, Switzerland, 31 May 1902; *d.* Wohlen, Switzerland, 4 April 1976), *developmental biology*.

Hadorn was the son of Christian Hadorn, a farmer, and of Elisabeth Lehner Hadorn. His ancestors had been farmers in the same place for centuries. He studied at the teachers' college at Muristalden from 1918 to 1922, then taught in the primary school of a village close to his home. When he had saved some money, Hadorn entered the University of Bern (1925), where he studied zoology under the embryologist Fritz Baltzer, himself a student of Theodor Boveri. On 21 October 1930 he married Marie Daepp; they had a son and two daughters. He received the doctorate the following year and returned to teaching, this time at the secondary school level. At the same time, in the basement of his home he continued experimental work on the topic he had treated in his dissertation. With this work he qualified as *Privatdozent* in 1935 with a *Habilitationsschrift* on nucleocytoplasmic interactions.

In 1936 Hadorn was awarded a Rockefeller fellowship, which enabled him to do research at the University of Rochester. His original plan was to work with the distinguished embryologist Benjamin Willier on the chicken, but he soon switched to *Drosophila*, working first with Curt Stern, then by himself. After returning to Switzerland in 1937, he again taught school, this time at the gymnasium in Biel. In 1939 Hadorn was appointed professor at the University of Zurich, and in 1942 he was named head of the Institute of Zoology in Zurich. He held this position until his retirement in 1972, building the institute from a small, provincial institution to one of the leading centers of developmental genetics. He served from 1962 to 1964 as rector of the university of Zurich and was a member of the Swiss Council of Science. After his retirement from the university he lived in Wohlen, not far from the village of his birth.

Hadorn's first scientific experiments were extensions of Baltzer's work, using hybrids between two species of newts. "Hybrid merogonic" embryos were produced, that is, embryos containing the haploid nucleus of one species in the cytoplasm of the other. Such embryos undergo normal cleavage and gastrulation but die in early embryonic stages, in part because they are haploid. Diploid hybrids between these two species die also, but much later in development. Haploid embryos from the same species die later than those containing nucleus and cytoplasm from different species. By an ingenious use of transplantation and explantation experiments,

Hadorn could show that the primary damage occurs in the head mesoderm. The difference in the damage between homogenic and hybrid merogones must be due to a disharmony between the nucleus (that is, the genes) and the cytoplasm of the two species in the latter.

At this point Hadorn saw that continuing this work with newts would be futile because they are not suitable for the genetic analysis of development. Hence he gave up his work with amphibians and, at Rochester, intended to work on chickens, an organism on which both embryological and genetic experiments had been carried out. Not that he gave up amphibians altogether: throughout his life he published papers on various aspects of amphibian development, notably a paper dealing with the influence of the notochord on the breakdown of yolk (1951). He also published a popular book on the experimental embryology of amphibians that went through two German editions.

At Rochester, Hadorn started to work with Willier, whose reputation was based on his work on the development of the chicken feather. The chicken feather seemed a suitable object of research because many genes affecting it were known. Hadorn was coauthor with Willier and Mary Rawles of a paper on transplantation experiments with the wings of chicken embryos that demonstrated the autonomy of feather pigmentation. But at this time (1936) the classical work of George Beadle and B. Ephrussi appeared that developed the technique of transplantation in *Drosophila*. Hadorn saw that this technique was exactly what he was looking for and, with Curt Stern, who had just finished his fundamental work on somatic crossing over, started to apply the transplantation method to developmental problems. They demonstrated that the sterility of XO males is an autonomous character, and that the pigmentation of the male genital apparatus is also autonomous at the cellular level, though individual pigmented cells can migrate from the testis sheath to the sperm ducts.

From this time on, transplantation in *Drosophila* remained Hadorn's primary experimental technique, and he applied it to a large number of different topics. The first concerned lethal mutations that kill the animal in the late larva or early pupa. Around this time a great many developmental geneticists were interested in the action of lethal genes—for instance, the work of D. F. Poulsen on embryonic lethals in *Drosophila*, of H. Grüneberg and S. Gluecksohn-Waelsch in the mouse, of W. Landauer in the chicken. But this work was mostly morphological, while Hadorn used transplantation experiments to analyze late lethals.

Hadorn's first lethal mutation was "lethal giant larvae" (*lgl*), a stock obtained from Beadle. Larvae homozygous for this gene develop normally but are unable to undergo pupation. In this transplantation work on *lgl*, Hadorn asked James V. Neel, a graduate student of Stern's, to collaborate with him, saying that two people work not twice, but four times, as efficiently as a single person. This was his method of work ever after: His transplantation experiments were carried out with one or two assistants working at the same table, one partner preparing the tissues while the other injected the grafts into the host. This concern with the efficiency of experimental techniques contributed greatly to the large amount of experimental work Hadorn accomplished.

It was known from earlier experiments by Gottfried Fraenkel (1935) that a center for the induction of pupation in flies must be located in the head region. Hadorn accordingly transplanted several organs from the heads of normal larvae into *lgl* larvae and found that implantation of one organ of unknown function called Weismann's ring (described by August Weismann in 1864) was able to induce puparium formation in *lgl* larvae. With Berta Scharrer, Hadorn established the glandular nature of this structure, now called the ring gland, and showed that in *lgl* larvae these glands are reduced in size. But this reduction of the ring gland was not the only effect of the mutant gene: In the induced puparia, development did not reach completion because the imaginal disks and the testes degenerated, and even the ovaries, which can develop further after transplantation into wild-type hosts, were unable to produce normal structures. Hadorn concluded that reduction of the ring glands was only one of several effects of the *lgl* gene.

These experiments immediately placed Hadorn in the first rank of developmental geneticists. The induction of insect metamorphosis by hormones was a very popular topic at the time. Since 1935, experiments with several insects had suggested such hormonal effects, and in some insects glandular structures had been implicated as sources of these hormones. Hadorn was the first, however, to prove that a puparium-inducing hormone is secreted by the ring gland in dipterans, and that genes are involved in the growth and functioning of this gland. This experiment formed the basis for his great scientific reputation.

Hadorn continued for some time to study a great number of late lethal mutations in *Drosophila*. He attracted to Zurich a large number of graduate stu-

HADORN

dents who participated in this research. He summarized his research on lethal mutations in a review article in 1951 and in a book in 1955. He observed in all cases that each mutation produces a specific pattern of damage while other organs remain unaffected, and he accounted for this finding by assuming a pleiotropic pattern of gene manifestation. Genes are not active in all cells, but are called into action gradually during development, and only in some tissues. Hadorn was able, in some cases, to demonstrate that some or all of the effects in a pattern of damage go back to a unitary primary damage: All lethal effects of the gene "lethal-meander" are consequences of starvation due to inability of the larva to digest or absorb proteins. But, probably because of his experience with *lgl* gonads, Hadorn did not exclude the possibility of "primary pleiotropy," different activities of the same gene in different tissues. Consequently, he did not try to establish "pedigrees of causes" going back to one primary cause, as Grüneberg had done for lethals in the mouse.

In 1950 Hadorn was awarded another Rockefeller fellowship, which enabled him to work at the California Institute of Technology with Herschel K. Mitchell. In these experiments they devised a remarkably simple and sensitive method to study pleiotropic gene action at the biochemical level: paper chromatography of isolated organs of *Drosophila* and the study of the chromatogram in ultraviolet light. In this way a number of fluorescent spots could be identified in different organs at different stages and under the influence of specific mutant genes. After his return to Zurich, Hadorn continued this work in close collaboration with Alfred Kühn and Albrecht Egelhaaf of the Max Planck Institute for Biology at Tübingen. They devised a method to determine the amounts of fluorescent substances in the spots by fluorometry and applied it to eye color mutations in *Drosophila* and in the moth *Ephestia*.

At the same time, Hadorn collaborated with the chemist M. Viscontini in identifying the fluorescent substances chemically. It turned out that most of them, including red eye pigments, are pteridines, and the aberrations induced by mutant genes were used to elucidate pteridine metabolism. Of great importance is the demonstration, with Ilse Schwinck (1956), that the mutant gene "rosy" (*ry*) blocks the synthesis of isoxanthopterine and that this character is nonautonomous. This work forms the basis of extensive studies of the *ry* locus, the gene determining the structure of the enzyme xanthine dehydrogenase. In particular the work of A. Chovnick and his collaborators has made the rosy locus one of the best-investigated genes of *Drosophila*.

At the same time as he was conducting these studies in biochemical pleiotropism, Hadorn initiated another important series of investigations, the developmental study of imaginal disks in *Drosophila*. The imaginal disks are clusters of undifferentiated cells present in the larva that after pupation develop into adult structures. Some of the main conclusions about this system emerged from the basic work of Hadorn, Bertani, and Gallera (1949) on the male imaginal disk, a group of cells that gives rise to all internal and external sexual structures except the testes. It could be shown by transplantation of different parts of this disk that it represents a mosaic of different domains, each of which is determined to develop into specific structures; in this way a map indicating domains of specific developmental fates could be established for the mature imaginal disk. While each of these domains gives rise, after transplantation, to specific adult structures, parts of these domains have the ability to regulate into a complete structure; they possess the properties of a "morphogenetic field." The occurrence of regulation depends in turn on a certain amount of growth of the piece of imaginal disk implanted: Pieces transplanted to adult larvae shortly before pupation cannot regulate, since the pupation hormone stops the tissue's ability to grow.

The study of the behavior of imaginal disks and their parts after transplantation was the main tool for the analysis of development by Hadorn and his collaborators and students in the later part of his life. One aspect of this work was serial transplantation of pieces of imaginal disks into larvae. The pieces were first transplanted from one larva into another, and this process was repeated several times, some pieces being permitted to undergo metamorphosis by remaining in the host. It appeared that the pieces that were serially transplanted continued to produce the structures for which they had been determined; on further serial transplantation, however, there appeared a tendency to produce only part of the structures for which the disk had been determined, possibly because of the dilution of determined cells induced by continued growth.

These experiments on serial transplantation in larvae led to transplantation of parts of disks into adults. It was found that in adult hosts the grafts grew and proliferated, but did not differentiate. They behaved, thus, as cell cultures. If pieces of these cultures were transplanted into larvae and permitted to undergo metamorphosis, they developed according to their original determination. But if disks cultured in adults were further subcultured in adults, then tested for their state of determination at intervals, sooner or later the subcultures gave rise to foreign structures: For instance, in cultures that originally had been male genital disks, legs and antennae appeared. In still further subcultures, wing structures, and still later, thoracic structures, could appear.

This finding, called "transdetermination," constitutes an important discovery. In classical embryology it had been assumed that once a group of cells has been determined, it can differentiate only into the structures to which it has been determined, and that this state is irreversible. The state of determination is transmitted by cellular heredity. The serial transplantations in larvae supported this prediction, and only prolonged subculture in adults could show that determination is not a final state but one that can change and, in the long run, be lost. After this discovery, Hadorn thoroughly investigated the rules of transdetermination—which imaginal disks can be transformed into which other structures and in which order they can be transformed. The discovery of transdetermination must be regarded as Hadorn's most important contribution.

Hadorn introduced a large number of other techniques into the study of insect development. He showed that one could dissociate imaginal disks by means of the enzyme trypsin. The cells could then be mixed with similarly dissociated cells from another genetically marked strain, so that they could form mosaic structures. He also showed that embryonic cells could be grown in adult hosts. All these discoveries and techniques opened up a very fruitful field of investigation that Hadorn himself was not able to explore completely. The developmental physiology of imaginal disks continues to be a very active field, and it is due mainly to Hadorn that *Drosophila* has become one of the most widely investigated subjects of embryological studies.

Hadorn was primarily a brilliant experimenter. He constantly devised new techniques and adapted techniques from other branches of science to his problems. In this way he left a large and varied amount of experimental work. The theoretical formulation of conclusions followed the experiments and led to further experiments, always carefully designed to answer specific questions.

Hadorn was a man of great physical strength and working capacity. This enabled him to carry out his transplantation work while being involved in many other activities. He enjoyed teaching and lectured regularly. He carried a considerable administrative load as head of his institute and, later, as rector of the University of Zurich. He provided much initiative in international scientific undertakings. Foremost among these was the journal *Developmental Biology*, which he founded with Paul Weiss and Jean Brachet. The three men constituted the editoral board in the journal's first decade (1959–1969) and gave it its specific scientific character.

Hadorn was informal and jovial in his contacts with friends and students. He proved helpful and generous to his numerous students and colleagues and was extremely popular with them. He practiced severe self-discipline; his life was tightly regulated, and the schedule of his activities strictly determined. He went to church regularly and had no difficulty in combining his faith with his scientific way of thinking.

In the history of biology Hadorn represents a bridge between classical embryology and modern developmental biology. Classical experimental embryology is based mainly on experiments in amphibians. It lost favor among the younger generation of biologists in the late 1930's and 1940's, partly because it appeared that most of the important problems had been solved. But more important was the disregard of many investigators of the genetic basis of development and the rejection of the reductionist approach taken by geneticists. Hadorn, in his experiments with larval lethals, introduced genetic thinking into developmental studies, and he defended the reductionist approach throughout his life, the last time in a speech given a few weeks before his death to a meeting of chemists in Zurich. His original work on imaginal disks in *Drosophila* led further toward the modern approach to development in object, techniques, and concepts. His discovery of transdetermination was a major starting point for a new line of developmental research that has not yet been exhausted.

BIBLIOGRAPHY

I. ORIGINAL WORKS. Hadorn's writings include "Über die Entwicklungsleistungen bastardmerogonischer Gewebe von *Triton palmatus* (♀) × *Triton cristatus* (♂) im Ganzkeim and als Explantat in vitro," in *Wilhelm Roux Archiv für Entwickl.-Mech. d. Organismen*, **131** (1934), 238–284; "An Accelerating Effect of Normal 'Ring-Glands' on Puparium-Formation in Lethal Larvae of *Drosophila melanogaster*," in *Proceedings of the National Academy of Sciences*, **23** (1937), 478–484; "The Structure of the Ring-Gland (*corpus allatum*) in Normal and Lethal Larvae of *Drosophila melanogaster*," *ibid.*, **24** (1938), 236–242, written with Berta Scharrer; "The Relation Between the Color of Testes and Vasa Efferentia in *Drosophila*," in *Genetics*, **24** (1939), 162–179, written with Curt Stern; "Die Auswirkung eines Letalfaktors (*lgl*) bei *Drosophila melanogaster* auf Wachstum und Differenzierung der Gonaden," in *Revue suisse de zoologie*, **49** (1942), 228–236, written with Hans Gloor; "Zur Pleiotropie der Genwirkung," in *Archiv Julius Klaus-Stiftung für Vererbforschung*, **20**, spec. iss. (1945), 82–95; "Regulationsfähigkeit und Feldorganisation der männlichen Genital-Imaginalscheibe von *Drosophila melanogaster*," in *Wilhelm Roux Archiv für Entwickl-Mech. d. Organismen*, **144** (1949), 31–70, written with Giuseppe Bertani and J. Gallera; "Developmental Actions of Lethal Factors in *Drosophila*," in *Advances in Genetics*, **4** (1951), 53–85; "Properties of Mutants of *Drosophila melanogaster* and Changes During Development as Revealed by Paper Chromatography," in *Proceedings of the National Academy of Sciences*, **37** (1951), 650–665, written with Herschel K. Mitchell.

See also "Chromatographische und fluorometrische Untersuchungen zur biochemischen Polyphänie von Augenfarb-Genen bei *Ephestia kühniella*," in *Zeit. Naturforsch.*, **8B** (1953), 582–589, written with Alfred Kühn; *Letalfaktoren in ihrer Bedeutung für Erbpathologie und Genphysiologie der Entwicklung* (Stuttgart, 1955), trans. by Ursula Mittwoch as *Developmental Genetics and Lethal Factors* (New York, 1961); "Patterns of Biochemical and Developmental Pleiotropy," in *Cold Spring Harbor Symposia on Quantitative Biology*, **21** (1956), 363–373; "Fehlen von Isoxanthropterin und Nicht-Autonomie in der Bildung der roten Augenpigmente bei einer Mutante (rosy^2) von *Drosophila melanogaster*," in *Zeitschr. ind. Abstamm-u. Vererb.-Lehre*, **87** (1956), 528–553, written with Ilse Schwinck; *Experimentelle Entwicklungsforschung, im besonderen an Amphibien* (Berlin, 1961, 2nd enl. ed., 1970), trans. by David Turner as *Experimental Studies of Amphibian Development* (New York, 1974); "Weitere Untersuchungen über Musterbildung in Kombination aus teilweise dissoziierten Flügel-Imaginalscheiben von *Drosophila melanogaster*," in *Developmental Biology*, **4** (1962), 40–66, written with Heinrich Ursprung; "Differenzierungsleistungen wiederholt fragmentierter Teilstücke männlicher Genitalscheiben von *Drosophila melanogaster* nach Kultur *in vivo*," *ibid.*, **7** (1963), 617–629; "Genetics on Its Way," presidential address, in S. J. Geerts, ed., *Proceedings of the XI International Congress of Genetics: Genetics Today*, II (Oxford and New York, 1965), lxiii–lxxii; and "Konstanz, Wechsel und Typus der Determination und Differenzierung in Zellen aus männlichen Genitalanlagen von *Drosophila melanogaster* nach Dauerkultur *in vivo*," in *Developmental Biology*, **13** (1966), 424–509.

II. SECONDARY LITERATURE. Hadorn's life and work are discussed in Dietrich Bodenstein, "For the 70th Birthday of Ernst Hadorn," in Heinrich Ursprung and Rolf Nöthiger, eds., *The Biology of Imaginal Discs* (New York, 1976), vii–ix; Walter J. Gehring, "In Memoriam Ernst Hadorn," in *Developmental Biology*, **53**, no. 1 (1976), iv–vi; and Rolf Nöthiger, "Nachruf. Ernst Hadorn (31. Mai 1902–4. April 1976)," in *Wilhelm Roux's Archives of Developmental Biology*, **180**, no. 2 (1976), i–vi, and "Ernst Hadorn (1902–1976)," in *Genetics*, **86** (1977), 1–4.

ERNST CASPARI

HALLIBURTON, WILLIAM DOBINSON (*b.* London, England, 21 June 1860; *d.* Exeter, England, 21 May 1931), *biochemistry*.

William Halliburton was the only son of Thomas Halliburton and his wife, Mary Homan. An attack of polio left his right arm paralyzed when he was very young, and this doubtless played its part in the course of his future career, making it impossible for him to do clinical medicine. After being educated privately and at University College School in London, Halliburton enrolled as a medical student at University College in 1878. He qualified for a B.Sc. in physiology in 1879 and obtained membership in the Royal College of Physicians in 1883. He became an M.D. (with gold medal) in 1884. He succeeded Gerald Yeo as professor of physiology at King's College, London, in 1889, and was elected a fellow of the Royal Society two years later. In 1896 he married Annie Dawes; they had no children. Halliburton was elected a fellow of the Royal College of Physicians in 1892, and he delivered both the Goulstonian Lectures (1893) and the Croonian Lecture (1901). Halliburton was elected to the secretaryship of the Royal Society, but this decision was later altered. The reason for this change of mind remains obscure; it might have been because the Royal Society wanted a secretary with a broader biological background. He received honorary LL.D. degrees from the universities of Aberdeen and Toronto. In 1923, he resigned his chair because of ill health.

It was early perceived that Halliburton was a founding influence on the discipline of biochemistry in Great Britain. Frederick Gowland Hopkins wrote in Halliburton's obituary notice: "He was the first in this country by his works and his writing to secure for biochemistry general recognition and respect." On his retirement Halliburton was elected the first honorary member of the British Biochemical Society, an honor that was not extended again until fifteen years later, to Arthur Harden.

Halliburton's education at University College and its medical school coincided almost precisely with a period of resurgence of British physiology in London and in Cambridge. Its leading spirits were Michael Foster, John Burdon-Sanderson, and Halliburton's mentor, Edward Albert Schäfer (later Sharpey-Schäfer). During the decade 1880 to 1890 the preclinical faculty at University College was one of the most scientifically active in Britain. Halliburton became part of its gifted young elite, and he formed friendships with Sidney Martin, Victor Horsley, and Frederick Mott. All were to become leaders in their chosen spheres of scientific medicine.

Under Schäfer's guidance, Halliburton began his physiological career by familiarizing himself with heat fractionation, a new technique for protein separation. Such developments brought "old" physiological issues within the ambit of "newer" chemical investigations. In the hands of Ernst Schulze and Heinrich Ritthausen, separation techniques were a fresh impetus to investigation of protein composition and heterogeneity.

In 1887 Halliburton published a major paper on muscle proteins that secured his reputation not only in Britain but on the Continent as well, adding to Wilhelm Kühne's works on muscle contraction of 1859 and 1864. Kühne had obtained "myosin" from muscle plasma, the fluid that can be pressed from fresh muscle. Halliburton, using heat fractionation, claimed to show that the myosin could be further separated into two proteins that he called paramyosinogen and myosin. He also showed that muscle plasma coagulates more rapidly if an extract of muscle prepared in the same way as fibrin ferment from blood is added. Finck has drawn attention to Halliburton's having, in these experiments, actually prepared the second myofibrillar protein, actin, and with it having precipitated a much less soluble complex, actomyosin.

In a series of follow-up papers Halliburton attempted to apply these techniques to differentiated animal tissues. By studying kidney, liver, and other tissues, he tried to assess the relationship between the functions of different organs and the various proteins found in the cells of each tissue. With Frederick Mott, Halliburton turned to research in the chemical pathology of the nervous system and of mental illness, showing that choline is produced in true nerve degeneration but not in functional diseases such as neurasthenia, depression, and hysteria.

Throughout his career Halliburton attempted to consolidate the status of chemical physiology in England by encouraging the research of others. In 1898 King's College spent £20,000 to improve its anatomy and physiology departments, including greatly expanded facilities for chemical physiology. As a result more people became involved in the field, although the department maintained its conservative continuity with medical practice and problems.

Among Halliburton's early research students was Christine Tebb, one of the first female biochemists. In 1907, while working under Halliburton, she showed that protagon, a substance that was claimed to represent a basic component of brain cells, was in fact an unresolved mixture of other biochemicals. She later married Otto Rosenheim, another scientist who was strongly influenced by being in Halliburton's physiology department but who, as a chemist, brought different attitudes to his work.

After the turn of the century, as Halliburton's own research career waned, he continued to influence younger workers to move toward biochemistry, an increasing number of whom were trained in chemistry. For example, Jack Drummond, professor of biochemistry at University College, London, from 1922 to 1945, came to Halliburton's department as chemical assistant in 1913. R. T. Hewlett, later professor of bacteriology at King's College, London, and Charles J. Martin, later director of the Lister Institute for Preventive Medicine, learned from Halliburton to apply chemistry to bacteriological problems.

Within his own subject, Halliburton moved cautiously from physiology into the newer, unexplored area of biochemistry but returned to physiology for his overall perspective on the new science. Nevertheless, his emphasis on chemical separation and resolution marks him as an important and transitional figure in the development of British biochemistry.

BIBLIOGRAPHY

I. Original Works. "On Muscle-Plasma," in *Journal of Physiology*, **8** (1887), 133–202; *A Textbook of Chemical Physiology and Pathology* (London, 1891); *Essentials of Chemical Physiology* (London, 1893; 13th ed., 1936); "On the Chemical Physiology of the Animal Cell," in *British Medical Journal* (1893), 501–506, 512–517, 627–632, the Goulstonian lectures; *The Chemical Side of Nervous Activity* (London, 1901), the Croonian lecture; and *Biochemistry of Muscle and Nerve* (London, 1904), with a selected bibliography of Halliburton's most important scientific papers.

Halliburton took over William Senhouse Kirkes's *Handbook of Physiology* (1st ed., London, 1848), beginning in 1896 and continuing until his death. The last edition to bear his name was *The Handbook of Physiology and Biochemistry* (37th ed., London, 1942), written with R. J. S. McDowall, who subsequently took over the project.

The Wellcome Institute for the History of Medicine, London, has forty-eight notebooks of medical lecture notes and early laboratory notes. The Wellcome Contemporary Archive Collection has correspondence on Halliburton's troubled election as secretary of the Royal Society in 1903.

II. SECONDARY LITERATURE. J. A. H., obituary in *Biochemical Journal*, **26**, pt. 1 (1932), 269–271; Frederick G. Hopkins, in *British Medical Journal*, **1** (1931), 1006, and *Lancet* **1** (1931), 1263; R. J. S. McDowall, obituary in *British Medical Journal*, **1** (1931), 957–958; and Neil D. Morgan, "William Dobinson Halliburton FRS (1860–1931), Pioneer of British Biochemistry," in *Notes and Records of the Royal Society of London*, **38** (1983), 129–145.

NEIL MORGAN

HÄNTZSCHEL, WALTER HELMUT (*b.* Dresden, Germany, 16 November 1904; *d.* Hamburg, Federal Republic of Germany, 10 May 1972), *geology, paleontology*.

Häntzschel was the son of Theodor Johannes Häntzschel, headmaster at an elementary school, and of Minna Müller. After graduating from the Dresden Realgymnasium in 1924, Häntzschel enrolled at the Technical University of Dresden, where he studied mineralogy, geology, chemistry, botany, zoology, and geography. Upon completing his courses he received the state certificate for high school teachers. From 1930 to 1934 he taught at a local high school, and he received the doctorate with high honors in 1932. Two years later he became chief of the Forschungsanstalt für Meeresgeologie und Meerespaläontologie Senckenberg at Wilhelmshaven.

In 1938 Häntzschel was appointed curator of paleontology in the department of geology and paleontology of the State Museum for Mineralogy and Geology at Dresden. He was drafted in 1942 and served in an engineering unit. The Russians captured him in 1945, and he was a prisoner of war until 1948. After four months as an assistant at the Institute of Geology and Paleontology of the University of Halle, in 1949 Häntzschel became curator at the Geologisches Staatsinstitut in Hamburg. He remained there until he retired in 1969.

Häntzschel married Marianne Krausse on 23 May 1936; they had two daughters. He was a member of most German scientific societies. From 1963 until his death he edited *Paläontologische Zeitschrift*, and from 1950 until 1969 *Mitteilungen aus dem Geologischen Staatsinstitut in Hamburg*. In recognition of his scientific achievements, Häntzschel was awarded the title of honorary professor in 1964 by the University of Hamburg. In 1958 he was invited to attend the First International Conference on Salt Marshes at Sapelo Island, Georgia.

While still a student, and to a greater extent while a teacher, Häntzschel wrote a number of papers on paleontological and sedimentological topics. Most important was his doctoral dissertation, "Das Cenoman und die Plenus-zone der sudetischen Kreide" (The Cenomanian and the plenus-zone in the Cretaceous system of the Sudeten Mountains), which caused Rudolf Richter, director of the Institute of Geology at the University of Frankfurt, to offer him a position there. Shortly afterward, Häntzschel took over the Senckenberg am Meer research station at Wilhelmshaven, which was devoted to investigating marine geology and paleontology in the tidal flats of the North Sea.

The foundation of this station as an offshoot of the University of Frankfurt can be credited to the perspicacity of Richter, who realized the necessity to study geology and paleontology as and where it is in process. In 1928 he coined the terms "actuogeology" and "actuopaleontology." Häntzschel spent four years at the station, investigating what interested him most: sedimentary structures and trace fossils (*Lebensspuren*) in the making, and their relationship.

A less happy chapter in Häntzschel's career started in 1938, with his appointment as curator in his home town of Dresden. Here he applied his actualistic understanding to the fossil outcrops. The war soon interrupted this activity, however, and after six years he came back to find Dresden and the museum destroyed.

The Hamburg State Geological Institute, at which Häntzschel became a curator in 1949, also had been completely destroyed, so his and the entire staff's energy went into the rebuilding of collections and library, teaching and administration. Nevertheless, Häntzschel managed to continue in his chosen field. His main area of concern was the same as it had been at Wilhelmshaven: how marine life came to be transformed into recognizable sedimentary structures (trace fossils). What are now known as trace fossils formerly were mostly considered to be enigmatic "problematica" and described under numerous names, such as the overcrowded genus *Fucoides*, which was considered a kind of seaweed before it became evident that most of the fossils placed in that genus arise from sedimentary processes or by locomotion of various benthic organisms.

Häntzschel realized that these structures were not body fossils in the usual sense, although they were caused by organisms. Adolf Seilacher showed that they usually cannot be assigned to specific producers but to certain functions like "grazing" or "burrowing," irrespective of the taxonomy of the individual producer. Häntzschel soon saw the necessity to review the existing knowledge of trace fossil morphology as well as of the various names previously given to them. He collected the widely scattered pertinent data from the literature, an immense task that, when published in 1962, made trace fossils accessible to further research and started a worldwide boom in trace fossils. It takes some direct experience with trace-fossil literature fully to acknowledge the accomplishment hidden in this book (and even more so in its second edition, which appeared after Häntzschel's death). Another important book on trace fossils is *Vestigia invertebratorum et problematica* (1965).

BIBLIOGRAPHY

I. Original Works. Among Häntzschel's works are "Das Cenoman und die Plenus-zone der sudetischen Kreide," in *Abhandlungen der Preussischen Geologischen Landesanstalt*, n.s. **150** (1933), his dissertation; "Trace Fossils and Problematica," in Raymond C. Moore, ed., *Treatise on Invertebrate Paleontology*, pt. W (New York and Lawrence, Kan., 1962), W177–W245, 2nd ed. (1975), W1–W269; and *Vestigia invertebratorum et problematica*, in F. Westphal, ed., *Fossilium catalogus*, I, *Animalia*, pt. 108 (The Hague, 1965).

II. Secondary Literature. On Häntzschel's life and work, see Günther Hertweck, "Walter Häntzschel," in *Paläontologische Zeitschrift*, **46**, no. 3/4 (1972), 105–112, with complete bibliography; U. Lehmann, "Walter Häntzschel: Ein Nachruf," in *Mitteilungen aus dem Geologisch-paläontologisches Institut Universität Hamburg*, **41** (1972), 6–14, with a complete bibliography; and Adolf Seilacher, "Walter Häntzschel (1904–1972) and the Foundation of Modern Invertebrate Ichnology," in Robert W. Frey, ed., *The Study of Trace Fossils* (New York, 1975), v–vii.

Ulrich Lehmann

HARINGTON, CHARLES ROBERT (*b.* Llanerfyl, North Wales, 1 August 1897; *d.* London, 4 February 1972), *biochemistry, medical research administration.*

Harington was the elder son of Rev. Charles Harington and Audrey Emma Bayly. His grandfather was Sir Richard Harington, eleventh baronet of Ridlington; Harington was thus a member of one of the ancient English families, tracing his ancestry back to the twelfth century, and grew up in a long tradition of service in the law, church, and army. At preparatory school in Malvern Wells, he contracted tuberculosis of the hip, which left him with a severe limp and kept him out of school for six and a half years. Nevertheless, he enrolled at Malvern College when he was seventeen. From there Harington won a mathematics exhibition to Magdalene College, Cambridge, in 1916. Family sentiment, to which was added medical advice, dissuaded him from an initial intention to follow a career in engineering. He switched to reading natural sciences and took a first in Part I of the tripos examinations before transferring to Edinburgh in 1920. There he did research with the pharmacologist George Barger and took a Ph.D. in 1922.

Harington was then appointed research assistant in the department of therapeutics at the Royal Infirmary but in 1932 was lecturer in pathological chemistry at University College Hospital Medical School, London. Here he was subsequently reader (1928) and professor (1931). The first year of his appointment was spent in New York, where he worked with D. D. van Slyke at the Rockefeller Institute; and for shorter periods he worked with H. D. Dakin and Otto Folin. It was in London, though, that he did the very important work on the biochemistry of the thyroid gland that secured him election to the Royal Society in 1931. In 1933 he published *The Thyroid Gland: Its Chemistry and Physiology,* which remains a classic of endocrinology.

A member of the Biochemical Society from 1921, Harington was its secretary from 1930 until 1942. From 1938 he served on the Medical Research Council, and from 1941 to 1945 on the Agricultural Research Council as well. Harington married Jessie McCririe, a physician, in 1923; they had a son and two daughters. The second part of Harington's career is intimately associated with his directorship of the National Institute for Medical Research (NIMR), where he succeeded Sir Henry Dale in 1942. In 1944 he was awarded the Royal Medal of the Royal Society. In 1962 he retired from the NIMR. He was awarded the Gold Medal in Therapeutics from the Society of Apothecaries in the same year. Harington held honorary doctor of science degrees from Paris (1945), London (1962), and Cambridge (1949). He was knighted in 1948 and made K.B.E. in 1962. The first nonmedically qualified director of the NIMR, he received another token of respect for the way in which he fused medical and chemical research when he was elected an honorary fellow of the Royal College of Physicians in 1963.

The symptoms of thyroid disorder were first recognized in the latter part of the nineteenth century. In 1871 C. H. Fagge associated sporadic cretinism (impaired neurological development) in infants with thyroid atrophy. Shortly afterward W. M. Ord reported atrophy of the thyroid in cases of the then newly described Gull's disease, which he renamed myxedema. Goitrous patients whose thyroid was removed improved at first, but they later developed symptoms of myxedema. At the turn of the century confirmation that the thyroid contained iodine supported early use of iodine, and explained the fact that burned sponge had for centuries been a recognized remedy for goiter. E. C. Kendall first purified such a substance from the thyroid in December 1914 and named it thyroxine. It was Harington who determined the structure of thyroxine and suggested how it was synthesized, in a series of papers in the late 1920's. Harington suggested that thyroxine was formed by the oxidative coupling of two di-iodotyrosine molecules with the loss of one alanine side chain:

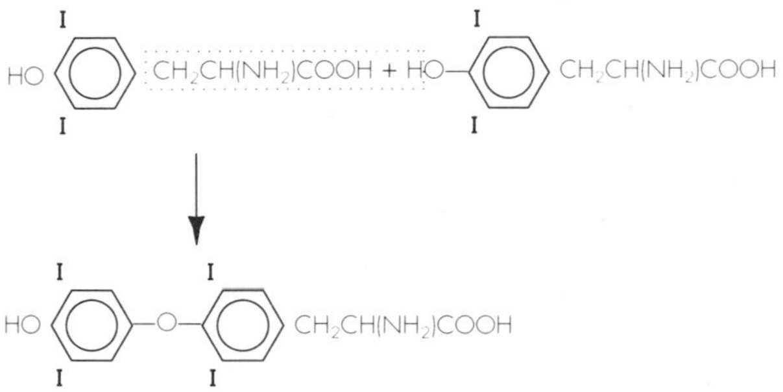

HO–C₆H₂I₂–$CH_2CH(NH_2)COOH$ + HO–C₆H₂I₂–$CH_2CH(NH_2)COOH$

↓

HO–C₆H₂I₂–O–C₆H₂I₂–$CH_2CH(NH_2)COOH$

FIGURE 1. From Harold Himsworth and Rosalind Pitt-Rivers, "Charles Robert Harington," p. 279.

This process is associated with transformations within the protein molecule thyroglobulin. Harington synthesized thyroxine (T_4) and thyronine, and reported that the synthetic thyroxine had been used clinically in the cure of hypothyroidism.

In addition to his research on the thyroid gland, Harington worked on the immunology of the endocrine system. In 1937 he produced antisera to thyrotrophin by prolonged injection of anterior pituitary extracts into rabbits. In 1946 he elucidated the hypersensitivity mechanism of contact dermatitis in munitions workers handling tetryl via its reaction, and antigenicity, in coupling with tissue protein.

Harington's perspective on his own work was always broad, as demonstrated by the synthetic treatment of physiology, pathology, and biochemistry in his 1933 book on the thyroid gland. He brought this ability to see the scope and potential of very different aspects of medical research to his role as director of the NIMR. Initially head of the division of biochemistry, he gave full attention to administration from 1955 on, and prided himself on an active role in leading and coordinating the institute's various programs of research. These principles were expressly portrayed in a number of lectures on research strategy, such as the Linacre Lecture, "The Place of the Research Institute in the Advance of Medicine" (1958), and the Shattuck Lecture (1951).

BIBLIOGRAPHY

I. ORIGINAL WORKS. A full list of Harington's scientific papers is in Himsworth and Pitt-Rivers (see below). His works include *The Thyroid Gland: Its Chemistry and Physiology* (London, 1933); "Thyroxine: Its Biosynthesis and Its Immunochemistry," in *Proceedings of the Royal Society of London*, **B132** (1944), 223–230, the Croonian Lecture; "Twenty-five Years of Research on the Biochemistry of the Thyroid Gland," in *Endocrinology*, **49** (1951), 401–416; "The Role of the Basic Sciences in Medical Research," in *New England Journal of Medicine*, **244** (1951), 777–785; *Leadership in Scientific Research: Sir David Russell Memorial Lecture* (London, 1958); and "The Place of the Research Institute in the Advance of Medicine," in *Lancet* (1958), **2,** 1345–1351.

II. SECONDARY LITERATURE. Harold Himsworth and Rosalind Pitt-Rivers, "Charles Robert Harington," in *Biographical Memoirs of Fellows of the Royal Society*, **18** (1972), 267–308.

NEIL MORGAN

HARISH-CHANDRA (*b.* Kanpur, India, 11 October 1923; *d.* Princeton, New Jersey, 16 October 1983), *mathematics.*

Harish-Chandra was one of the most profound mathematicians of his time. Although he started his scientific career as a theoretical physicist, he was always preoccupied with the purely mathematical aspects of physical theories, and eventually worked entirely in mathematics. His work, in algebra and analysis, is concerned almost exclusively with the theory of representations of groups and their harmonic analysis. This area of mathematics is at the interface of such varied disciplines as physics, number theory, and geometry; and some of the greatest mathematicians—such as Georg Frobenius, Élie Cartan, Hermann Weyl, I. M. Gel'fand, Atle Selberg, and Robert Langlands—have been attracted to it. Harish-Chandra's work is widely regarded as a brilliant and enduring part of it. That representation theory and harmonic analysis are among the most central and active areas of interest in present-day mathematics is partly due to his lifelong efforts dating from the 1940's. He had his early scientific training in India but settled in the United States, where almost all of his work was done. No mathematician of comparable stature has arisen from India in the second half of the twentieth century.

Harish-Chandra was born into an upper-middle-class family. His father, Chandra Kishore, was a civil engineer in the government of Uttar Pradesh (then known as the United Provinces); his mother, Satyagati Seth Chandrarani, was the daughter of a lawyer. Harish-Chandra was a brilliant student, and graduated with an M.Sc. from the university of Allahabad in 1943. Although the educational system under the British colonial administration was mostly sterile, there were some exceptions here and there, mainly due to the efforts of a few remarkable but isolated individuals, such as C. V. Raman in Bangalore and P. C. Mahalanobis in Calcutta. Another was K. S. Krishnan, a physicist at the University of Allahabad. Harish-Chandra studied under Krishnan, whose influence stayed with him throughout his life.

From Allahabad, Harish-Chandra went to Bangalore in southern India as a research student with H. J. Bhabha, a leading theoretical physicist who later created the Tata Institute of Fundamental Research and developed it into a world-class research institution in mathematics and physics. In 1945 Harish-Chandra went to England to work with P. A. M. Dirac at Cambridge. His thesis, on the classification of irreducible representations of the Lorentz group, gave clear indications that he was already becoming fascinated with the purely mathematical. In the year 1947–1948 he was at the Institute for Advanced Study at Princeton as an assistant to Dirac, who was a visiting professor there. In 1950 he went to Columbia University, remaining there until 1963. He then returned to the Institute for Advanced Study as a permanent member, a position he held until his death. In 1968 Harish-Chandra was named the I.B.M.–von Neumann professor of mathematics at the institute. He was elected a fellow of the Royal Society in 1973 and a member of the U.S. National Academy of Sciences in 1981. He was awarded honorary doctorates by Delhi University in 1973 and Yale University in 1981.

In 1952 Harish-Chandra married Lalitha Kale; they had two daughters.

Harish-Chandra's health was always fragile, and his health problems were compounded by the relentlessly ascetic and overwhelmingly intense nature of his work patterns. He suffered a heart attack in 1969, but continued to work on the central questions, as he saw them, of representation theory. But the damage to his heart was irreversible and his problems grew worse, resulting in a fatal heart seizure as he was walking in the woods near the institute at Princeton.

The origins of the theory of group representations go back at least to the work of Georg Frobenius in the nineteenth century; its main problem is to represent abstractly given groups by concrete matrices in such a way that the group operation corresponds to multiplication of matrices. Originally only finite groups and matrices of finite size were considered. However, with the rise of quantum mechanics it became increasingly clear that infinite groups and matrices have to be included. Indeed, one of the fundamental techniques for understanding very complex interacting systems of highly energetic elementary particles is to study the groups of symmetries that leave the system unchanged, and to use this study to limit the possibilities.

In order to apply this technique, it is indispensable to have a complete knowledge of the representations of the symmetry groups. The groups that are important in these problems, such as the rotation and Lorentz groups and their variants, are examples of what mathematicians call simple groups. The simple groups were classified by Cartan and Wilhelm Killing in the nineteenth century, and by 1925 all their finite dimensional representations were classified by Cartan and Weyl. What Harish-Chandra did was to construct the most fundamental infinite dimensional representations of all the simple groups and use them to carry out the harmonic analysis of functions and generalized functions in these groups. The technical problems that he had to overcome in completing this project involved entirely novel aspects of the theory of differential equations on group manifolds.

High-energy physics is not the only area where representation theory of simple groups is a crucial ingredient. Modern number theory is another, and current formulations of the problem of understanding the arithmetic of Galois extensions of algebraic number fields make essential use of the language and results of the representation theory. The classical number theoretic investigations, culminating in the work of David Hilbert, Teiji Takagi, Emil Artin, Helmut Hasse, and others, had obtained a marvelous description of the Abelian extensions of number fields. One of the great achievements of modern number theory is the discovery that similar descriptions of *all* extensions (Abelian or not) of number fields ultimately depend on establishing certain natural correspondences between representations of the Galois groups and appropriate representations of certain simple groups associated to the number fields.

Although there had been many attempts to study infinite-dimensional representations of groups before Harish-Chandra, it was he who began the systematic theory of representations of all simple groups. Unlike his predecessors, who worked with special groups and ad hoc methods, Harish-Chandra realized immediately that it was essential to develop an approach that integrated the algebraic, geometric, and analytic aspects of simple groups. For instance, his papers were the first to consider representations of nontrivial infinite-dimensional associative algebras. Quite early in his work he made the major discovery that one can associate to any irreducible representation of a simple group a character that is the trace of the matrices that represent the group elements. Characters are not new, of course; both Frobenius and Weyl had worked with them and obtained beautiful formulas for them, but they worked with matrices of finite size and there was no conceptual difficulty in defining the characters and developing their theory. But the Harish-Chandra character is more subtle and singular, being the trace of an infinite matrix; indeed, it is a nontrivial matter even to define it, and the only tools available for its study are the differential equations satisfied by it.

The cornerstone of Harish-Chandra's work is a profound and astonishingly complete study of the solutions of these differential equations that led him to a complete description of all the fundamental characters. Complicating the development was the fact that the characters are a priori functions only in a very generalized sense, and Harish-Chandra had to prove ex post facto that they are functions in the classical sense. This result, known as the Harish-Chandra regularity theorem, is one of his greatest achievements and its influence pervades the entire subject. In spite of repeated efforts by many people, it has not been possible to deduce this theorem as a consequence of general principles from the theory of differential equations.

In order to carry out the harmonic analysis of functions on the simple groups in terms of the characters, it is necessary to understand how the elementary functions that arise from the characters behave at infinitely distant parts of the group manifold. The second major achievement of Harish-Chandra was his complete solution to this problem and the explicit harmonic analysis that flowed from this asymptotic theory. Roughly speaking, these elementary functions can be expanded in an infinite series at infinity, and the leading terms of these expansions determine in a very simple and explicit manner the weights with which the various characters enter the harmonic analysis of the delta function on the group.

BIBLIOGRAPHY

I. ORIGINAL WORKS. Harish-Chandra's works were brought together in his *Collected Papers*, V. S. Varadarajan, ed., 4 vols. (New York and Berlin, 1984).

II. SECONDARY LITERATURE. Robert P. Langlands, in *Biographical Memoirs of Fellows of the Royal Society*, **31** (1985), 199–225; and V. S. Varadarajan, "Harish-Chandra (1923–1983)," *The Mathematical Intelligencer*, **6**, no. 3 (1984), 9–19.

V. S. VARADARAJAN

HARVEY, EDMUND NEWTON (*b*. Germantown, Pennsylvania, 25 November 1887; *d*. Woods Hole, Massachusetts, 21 July 1959), *physiology*.

The son of William Harvey and Althea Ann Newton, Harvey was raised by his mother and three older sisters after his father's early death. In 1909 he received the B.S. in general science from the University of Pennsylvania and in 1911 the Ph.D. from Columbia University with a dissertation on cell permeability. In 1911 Harvey was appointed instructor in Edwin Grant Conklin's department of biology at Princeton, where he became assistant professor in 1915, full professor in 1919, and Henry Fairfield Osborn research professor in 1933. Harvey occupied this post until he retired in 1956. On 12 March 1916 he married Ethel Nicholson Browne, who had received her Ph.D. in zoology from Columbia in 1913; they had two sons. The couple shared laboratories during much of their lives.

Harvey never lost the interest in natural history fieldwork developed in his youth, but as an undergraduate he was drawn to cell physiology and the experimental laboratory by the teachings of Ralph Lillie. Both he and Lillie, Harvey later asserted, were influenced by Jacques Loeb and his program for the physicochemical analysis of life, and early in his career Harvey looked to Loeb's work on artificial parthenogenesis as a model of what biological experimentation should be. Alfred Goldsborough Mayor invited Harvey to be a collector at the Dry Tortugas Marine Biological Laboratory of the Carnegie Institution in the summer of 1909 and, later that season, at the Marine Biological Laboratory at Woods Hole, where Harvey investigated membrane formation and artificial parthenogenesis.

Harvey's visit to Woods Hole initiated a lifelong association with marine biological stations and an emphasis on marine organisms in much of his research. From 1909 until the end of his life he spent part of every summer doing research at Woods Hole. It was in 1913, while on an expedition with Mayor to Australia's Great Barrier Reef, that Harvey's interest was directed to bioluminescence, the phenomenon that would occupy the greatest part of his research career. In that year he began publishing a series of studies on light emission in luminous bacteria, but his first important paper on the mechanism of light production in animals did not appear until 1916. In it Harvey first drew attention to the work of Raphael Dubois, who in the 1880's had shown that the luminous organs of *Pyrophorus moctilucus* contained two separable substances, *luciférine* and *luciférase*, that, when mixed, produced light. Harvey proceeded to demonstrate the existence of a luciferin-luciferase system in three other organisms. With bold optimism he asserted that "the problem of bioluminescence has been solved at least in its broad aspects" and predicted that the exact chemical nature of luciferin soon would be determined.

While visiting Japan in 1916 to study a luminous squid, Harvey was introduced to the organism most central to his research from then on, the ostracod crustacean *Cypridina hilgendorfii*. What made this tiny luminescent organism especially intriguing is that when it is dried after collecting, its luminescence system not only remains stable but can be easily activated, even after years of storage, by aqueous solutions containing molecular oxygen. Between 1916 and 1919, working with dried material shipped to his Princeton laboratory, Harvey found that *Cypridina* also possesses a two-component light-emitting system and developed techniques for separating and purifying its substrate, luciferin, and enzyme, luciferase. Luciferin emits light when oxidized, he showed, and the oxidation product can be reduced to luciferin again. *Cypridina* proved exceedingly useful in kinetic studies of its enzyme-substrate system as well, for bioluminescence provided a visible indicator of its own reaction velocity. Harvey continued to pursue chemical analyses of animal light throughout his career, after the late 1920's focusing particularly on how bioluminescent activity is affected by alterations in the chemical and physical environment.

By the late 1920's Harvey's elucidation of bioluminescence had established him as doyen of the field, yet the expectations so confidently proclaimed in 1916 remained discouragingly unfulfilled. He systematically sought the *Cypridina*-type luciferin-luciferase system in every kind of luminescent organism he could examine but found the reaction in only a few, which frustrated his hopes of disclosing a general pattern. This, along with the seeming intractability of the precise biochemistry of luminescence, helped rechannel his energies toward other physiological interests. Harvey had never set aside his research on cell permeability, and between 1930 and 1934 he actively investigated cell surface tension. He was also managing editor, from 1932 to 1939, of *Journal of Cellular and Comparative Physiology*.

Harvey was encouraged to diversify his research interests about 1927 by his friendship with the physicist Alfred Lee Loomis, who had just opened a private laboratory at Tuxedo Park, New York, within driving distance of Princeton. At the Loomis Laboratory they conducted experiments on the biological effects of high-frequency sound waves and about 1930 constructed the centrifuge microscope, which Harvey used to study the effects of centrifugal force on cells. In 1935 they entered the new field of electroencephalography, and for the remainder of the 1930's studied the electrical potentials of the human brain during sleep.

The outbreak of World War II redirected Harvey's research efforts. At the Loomis Laboratory radio communications research supplanted the biophysics work of interest to Harvey, while at Princeton he was among those scientists who diverted their energies to military research. In 1942 he started working for the Committee on Medical Research of the Office of Scientific Research and Development. Of the two major programs he directed, the more important investigated bubble formation in animal blood and tissues and the pathogenesis of decompression sickness; the other, which involved firing high-velocity missiles at anesthetized animals, studied the mechanisms of wounding. For his work on decompression sickness Harvey was awarded the Armed Forces Certificate of Merit in 1948.

After the war Harvey published many reports on his military research, but his laboratory activity slackened. He was vice president of the corporation of the Marine Biological Laboratory at Woods Hole from 1942 to 1952, and helped shape its reorganization and growth after the war. Harvey continued his research on bioluminescence, for which he received the Rumford Medal of the American Academy of Arts and Sciences in 1947, and published the book *Bioluminescence* in 1952. Perhaps in part out of frustration at never finding the simple physicochemical solutions to the problems of bioluminescence he had once anticipated, however, he increasingly shifted his attention to broader questions of the evolution of luminescent organisms, and he wrote on the history of luminescence.

BIBLIOGRAPHY

I. ORIGINAL WORKS. Harvey's published works are listed in Poggendorff, VI, 1038–1040, and VIIb, 1875–1878. His books include *The Nature of Animal Light* (Philadelphia and London, 1920); *Living Light* (Princeton, 1940; repr. New York, 1965); *Bioluminescence* (New York, 1952); and *A History of Luminescence from the Earliest Times to 1900* (Philadelphia, 1957). Among more than 250 scientific papers Harvey authored or coauthored, particularly significant are "The Mechanism of Light Production in Animals," in *Science*, **44** (1916), 208–209; "Potential Rhythms of the Cerebral Cortex During Sleep," *ibid.*, **81** (1935), 597–598, with Alfred L. Loomis and Garret Hobart; and "Decompression Sickness and Bubble Formation in Blood and Tissues," in *Harvey Lectures*, **40** (1944–1945), 41–76.

The American Philosophical Society Library, Philadelphia, has a collection of Harvey's papers (1923–1959, about 5,600 items plus 19 volumes), which is described in Margaret Miller, "The Papers of Edmund Newton Harvey (1887–1959)," in *Survey of Sources for the History of Biochemistry and Molecular Biology*, no. 6 (November 1977), 8–9.

II. SECONDARY LITERATURE. The fullest biographical sketch of Harvey is Frank H. Johnson, "Edmund Newton Harvey," in *Biographical Memoirs. National Academy of Sciences*, **39** (1967), 193–266, which contains extracts from an unpublished autobiographical memoir. Other obituary notices that supplement this are by Elmer Grimshaw Butler, in American Philosophical Society, *Year Book 1959* (1960), 127–130; and by Aurin M. Chase, in *Biological Bulletin*, **119** (1960), 9–10.

JOHN HARLEY WARNER

HASSE, HELMUT (*b.* Kassel, Germany, 25 August 1898; *d.* Ahrensburg, near Hamburg, Federal Republic of Germany, 26 December 1979), *number theory*.

One of the most important mathematicians of the twentieth century, Helmut Hasse was a man whose accomplishments spanned research, mathematical exposition, teaching, and editorial work. In research his contributions permeate modern number theory; particularly noteworthy are his "local-global principle," which established Kurt Hensel's p-adic numbers as indispensable tools of number theory, and his proof of the "Riemann hypothesis" for elliptic curves. In exposition his report on class field theory in the 1920's made the work of Teiji Takagi, Philipp Furtwängler, Emil Artin, Hasse, and others available to a wide audience (and, like any good exposition, contained a great deal of Hasse's own reworking of the material). Later books and monographs confirmed Hasse's reputation as a writer who could be counted on to present the most difficult subjects with great clarity. In teaching, the long list of his students and their descriptions of his inspiring lectures give ample testimony to his excellence. In his editorial work the continuation of "Crelle's Journal"—*Journal für die reine und angewandte Mathematik*—as one of the world's foremost mathematical periodicals during the fifty years of his editorship was largely a result of his painstaking efforts, high standards, and editorial ability.

Hasse's parents were Paul Reinhard Hasse, a judge, and Margaretha Quentin (born in Milwaukee, Wisconsin, but raised from the age of five by an aunt in Kassel). His secondary education was in gymnasiums in the Kassel area until the family moved to Berlin, where his father had received a high judicial appointment in 1913. After two years in the Fichte-Gymnasium there, he took the exit examination early (a *Notabitur*) in order to volunteer for the navy. Hasse had evidently decided on a career in mathematics while still at gymnasium, because while he was stationed in the Baltic, he studied, on the advice of his gymnasium teacher, the Dirichlet-Dedekind lectures on number theory. During the last year of his naval service he was stationed in Kiel, where he attended classes in mathematics under Otto Toeplitz. Upon leaving the navy in December 1918, Hasse went to Göttingen to begin his mathematical studies in earnest. The teacher at Göttingen who made the greatest impression on him was Erich Hecke, who left Göttingen to go to Hamburg in the spring of 1919. Hasse left the following year. Hasse did not, however, follow Hecke. Greatly impressed by Kurt Hensel's book *Zahlentheorie* (1913), which he had found while browsing in a Göttingen bookstore, he decided to go to Marburg to study with Hensel. What made Hensel's book special was his introduction of p-adic numbers, and it was in order to study this new tool of number theory that Hasse went to Marburg.

In October 1920 Hasse discovered his "local-global principle," which transformed the p-adic numbers from a curiosity that the Göttingen establishment regarded as a fruitless sidetrack into a natural tool of number theory. In 1975, in a *Geleitwort* to the first volume of his mathematical papers, Hasse recounted his discovery of the local-global principle, emphasizing Hensel's role in it. At Hensel's suggestion, Hasse had investigated the necessary and sufficient conditions for a rational number to be representable by a rational quadratic form. He found the key to the answer in a reduction procedure of Lagrange for ternary quadratic forms that he had learned from his reading of Dirichlet-Dedekind. However, to his disappointment, his solution did not appear to call for the use of p-adic numbers. He communicated this to Hensel, whose reply—a postcard that Hasse kept all his life—opened Hasse's eyes, as he later said, to the true significance of what he had proved: a ternary quadratic form that represents 0 nontrivially over the p-adic numbers for all p (including $p = \infty$, which corresponds to the real numbers) represents 0 nontrivially over the rationals. That is, if there is a solution "locally" for all p, then there is a "global" (rational) solution. Hasse soon expanded this principle to a wide range of problems dealing with the equivalence of quadratic forms, work that allowed him to complete his doctoral dissertation in 1921 and his *Habilitationsschrift* in 1922.

Hasse left Marburg in 1922 to accept a paid teaching appointment as *Privatdozent* in Kiel. Hasse married Clara Ohle on 11 April 1923. They had a daughter, Jutta, and a son, Rüdiger.

In 1925 he was appointed professor at Halle, and in 1930 he returned to Marburg to assume the chair made vacant by Hensel's retirement. This succession, which was the realization of Hensel's fondest wish, was of short duration. The strong center of German mathematics, Göttingen, was demolished by the firings and resignations that followed the coming to power of the Nazis in 1933. Hasse, who appeared to be politically acceptable to the Nazis and yet was a mathematician of the highest caliber, was a natural choice as a successor to the deposed director of the Mathematics Institute at Göttingen, Richard Courant. Even Courant, in the interest of the continuation of mathematics in Göttingen, favored Hasse's appointment. (The directorship of the institute was not linked to any one chair. Courant was still "on leave" in 1934, and his chair was not vacant. Hasse was appointed to the more prestigious chair left vacant by Hermann Weyl's resignation, the chair that had formerly been David Hilbert's.) Hasse became the director at Göttingen in 1934 and remained in that post formally—although he was on leave and engaged in naval research in Berlin from 1939 to 1945—until he was dismissed by the British occupation authorities in September 1945.

During the years in Kiel, Hasse explored the subject of norm residue symbols and explicit reciprocity laws by means of p-adic techniques. The proximity of these interests to those of Emil Artin, and Artin's physical proximity in Hamburg, led to friendly collaboration and frequent meetings to exchange ideas. (A joint paper, written in 1925, says the work was drafted by "the younger" of the two authors. Few readers could have known that Hasse was younger—by less than half a year.) During this period Hasse undertook to prepare for the Deutsche Mathematiker Vereinigung a report on recent developments in class field theory, particularly the advances of the Japanese mathematician Teiji Takagi. The report was delivered at Danzig in the summer of 1925, but the published *Klassenkörperbericht*, as it was called, took longer to produce. It was published in two parts, the second part not appearing until 1930; by this time Artin had succeeded in proving his general law of reciprocity (1927), and Hasse included a full account of it in the second part of the *Bericht*.

In the early 1930's Hasse completed a thoroughgoing revision of class field theory—including reciprocity laws and norm residue symbols—in terms of the theory of noncommutative algebras applied to p-adic number fields, an approach that has been called "the high point of the local-global principle."

Soon after this success, Hasse began work in another area that was to become one of the central subjects of modern number theory. Challenged by his English colleague Harold Davenport to bring his algebraic methods to bear on a problem Davenport and Louis Joel Mordell had been studying concerning the number of solutions of a congruence of the form $y^2 = f(x) \bmod p$ for large primes p when $f(x)$ is a fixed cubic polynomial, Hasse succeeded magnificently. He first perceived that the problem Davenport and Mordell had been considering was in fact equivalent to a problem that Artin had formulated in an altogether different form as an analogue of the Riemann hypothesis, and he then proved this analogue in the case Davenport and Mordell considered, which was the "Riemann hypothesis" in the case of a function field of an elliptic curve over a finite field. To establish connections between such apparently disparate subjects in mathematics and, more than that, to solve a problem using techniques that were developed for altogether different purposes is an achievement of the highest sort in pure mathematics. Great progress has since been made in generalizing Hasse's solution. It is now known that the "Riemann hypothesis" is true in a vast range of cases, provided that algebraic techniques apply; the actual Riemann hypothesis, which is transcendental rather than algebraic, is still an unsolved problem.

Hasse's political views and his relations with the Nazi government are not easily categorized. On the one hand, his relations with his teacher Hensel, who was unambiguously Jewish by Nazi standards, were extremely close, right up to Hensel's death in 1941, and his relations with the Hensel family remained close and warm throughout his life. One of his most important papers was a collaboration with Emmy Noether and Richard Brauer, both Jewish, published in 1932 in honor of Hensel's seventieth birthday (which occurred at the end of 1931). Also in 1932 he dedicated an extremely important paper to Emmy Noether on the occasion of her fiftieth birthday. Hasse did not compromise his mathematics for political reasons. In his years as director of the Mathematics Institute in Göttingen, he brought to Göttingen as difficult a figure as Carl Ludwig Siegel, and he struggled against Nazi functionaries who tried (sometimes successfully) to subvert mathematics to political doctrine. Hasse never published in the journal *Deutsche Mathematik*. On the other hand, he made no secret of his strongly nationalistic views and of his approval of many of Hitler's policies. As an academic administrator and, during the war, as a military officer, he was of course a participant in the regime. He did apply, in 1937, for membership in the National Socialist Party, but the application was refused because he had a remote Jewish ancestor. (However, Siegel reported in a letter to Courant in March 1939 that he had seen Hasse wearing Nazi insignia.) After the war, many emigrant mathematicians condemned Hasse's activities during the Nazi period; invitations to the United States were few, considering his scientific eminence, and what invitations there were aroused controversy.

The occupation authorities revoked Hasse's right to teach in September 1945. He refused to remain in Göttingen in a purely research capacity, on the grounds that his inability to teach would be detrimental to his research. However, after a difficult year, he finally decided to accept an appointment as a research professor at the Berlin Academy, and moved to Berlin in September 1946. During the next two years he gave private lectures to a small group of students, many of whom later had successful careers. In 1948, Hasse's rating in the "denazification" program was improved to the point where he was allowed to give public lectures at Humboldt University in Berlin, and in 1949 he was named to a professorship there. One year later he accepted an appointment as professor at the University of Hamburg. He retired in 1966, but continued to make his home in Ahrensburg, outside Hamburg, for the rest of his life. In the postwar period he remained active as a teacher, as a writer of both research and expository works, and as an editor, although in his last years his work was curtailed by ill health.

Among the many honors Hasse received were the German National Prize, First Class, for Science and Technology (1953), an honorary doctorate from the University of Kiel (1968), and the Cothenius Medal of the Academia Leopoldina in Halle (1968). He was a member of the academies of science of Berlin, Halle, Göttingen, Helsinki, Mainz, and Madrid.

BIBLIOGRAPHY

I. Original Works. *Mathematische Abhandlungen*, Heinrich Wolfgang Leopoldt and Peter Roquette, eds., 3 vols. (Berlin and New York, 1975). A complete list of Hasse's scientific publications is at the end of vol. III.

II. Secondary Literature. G. Frei, "Helmut Hasse (1898–1979)," in *Expositiones Mathematicae*, **3** (1985), 55–69; Heinrich Wolfgang Leopoldt, "Zum wissenschaftlichen Werk von Helmut Hasse," in *Journal für die reine und angewandte Mathematik*, **262/263** (1973), 1–17, and "Helmut Hasse," in *Journal of Number Theory*, **14** (1982), 118–120; and S. L. Segal, "Helmut Hasse in 1934," in *Historia Mathematica*, **7** (1980), 46–56.

Harold M. Edwards

HEEZEN, BRUCE C. (*b*. Vinton, Iowa, 11 April 1924; *d*. at sea, south of Iceland, 21 June 1977), *oceanography*.

Heezen was the son of Charles Christian Heezen and Esther Shirding. When he was still young, the family moved to Muscatine, Iowa. As the only child of affluent parents, he was able to pursue his many interests in the outer world about him and the inner world of the laboratory and the library. Heezen was also called upon, in his younger years, to help his father and, in his father's absence, to manage the family turkey farm.

Heezen enrolled at the University of Iowa in 1942, and as a geology major he worked closely with his professor, A. K. Miller, and Professor Arthur Trowbridge. After he spent a summer in Nevada as an assistant to Walter Youngquist, who was collecting cephalopods for his Ph.D. dissertation, Heezen's future career as a paleontologist seemed assured. However, in April 1947, after hearing a Sigma Xi lecture by Maurice Ewing of the Woods Hole Oceanographic Institute on the perils and adventures of deep-sea exploration, Heezen went up to meet him. Ewing said to him, "Young man, would you like to go on an expedition to the Mid-Atlantic Ridge? There are some mountains out there, and we don't know which way they run." Heezen went, and spent the next thirty years mostly at Lamont Geological Observatory, under the direction of Ewing, working in the new, wide open, and rapidly expanding field of oceanography.

Heezen had a long, productive collaboration with Ewing. The team of Ewing, Heezen, and David Ericson was particularly effective in finding both ships and time to collect data to solve both specific and general problems in the analysis of deep-sea cores and the distribution patterns of sediments in the abyssal plains. Heezen's work as expedition leader was a great challenge and joy to him, particularly in the early days, when every cruise was a voyage of discovery and there was freedom for the chief scientist to explore features of interest. His ability as a teacher developed through work with his graduate students, many of whom later occupied outstanding positions in academe, governmental service, and industry.

Heezen's contributions to knowledge of the ocean floor included seafloor processes, structure and trends of the Mid-Oceanic Ridge, and the visual observation of the seafloor from both near and far. His first major work, undertaken with Ewing and Ericson soon after he arrived at Lamont, was concerned with turbidity currents. The Grand Banks earthquake of 1929 and the resulting series of cable breaks documented on a grand scale the existence of turbidity currents, a dense mixture of sand and seawater, and the speed with which they can race downslope and erode, transport, and deposit sands far out to sea. These coarse-grained sands formed abyssal plains on the ocean basin floor between the continental margins and the Mid-Oceanic Ridge. Turbidity currents are transient, and their deposits represent only a small percentage of deep-sea sediments; Heezen was ever concerned that there should be other processes involved in forming the seafloor.

An unexpected opportunity to study this problem arose in 1965 when a disabled winch and lost coring rig on an early cruise of the R. V. *Eastward* forced Heezen to revise his program completely. He did so, using a bottom camera with a compass and a short corer. The results of this study revealed the role that deep geostrophic contour currents, controlled by the Coriolis effect, play in the erosion, transportation, and deposition of sediments along the continental margins. These great wedges of sediment, smoothed out by deep currents flowing parallel to the contours, are now known as drifts.

Students and colleagues of Heezen—Charles Hollister, Leonard Johnson, and Anthony Laughton—further identified and surveyed many drifts in the North Atlantic and Pacific oceans, showing that these deposits are a major feature of the seafloor. Throughout his life Heezen continued to be interested in sedimentation as it related to topography and to the underlying geological structure.

A study of the topography of the North Atlantic using data obtained from the three Mid-Atlantic Ridge expeditions of 1947–1948 was begun by Heezen and Marie Tharp in 1952; they continued to work together until Heezen's death in 1977. A simultaneous study of earthquakes as a cause of cable failures resulted in the discovery by Marie Tharp of the seismically active rift valley as a median feature in the 40,000-mile, world-encircling Mid-Oceanic Ridge. This discovery of the rift valley was announced by Ewing and Heezen at the 1956 meeting of the American Geophysical Union in Toronto. The rift valley, which is tensional, has associated shear features or fracture zones. Although fracture zones were discovered in the Pacific by M. F. Maury in 1939, they were not known to exist in the Atlantic until 1961, when Heezen and Tharp discovered the equatorial fracture zones and recognized that they were not mere fracture zones but a major offset between two ridge segments.

Meanwhile, Heezen, knowing that the rift valley was tensional in nature, proposed that the earth expanded as new crust arose from within the rift valley. He also suggested that this expansion of the earth provided a mechanism for continental drift. Heezen first presented his idea in 1958 in a paper, "Géologie sous-marine et déplacements des continents," delivered at Nice, France. He continued to propose his expanding-earth hypothesis throughout the early 1960's. Although Heezen's idea turned out to be incorrect, his idea of the creation of new seafloor material at the ridge axis was correct and a fundamental aspect of Harry Hess's idea of seafloor spreading.

Heezen's concern with the visual aspects of the seafloor covered a broad range of scales: from the distant view of the physiographic diagrams, 1:10 million (1957–1971), the panoramas of the several oceans (1967–1970), and the final "World Ocean Floor" panorama, 1:23 million (1977), to the bottom photographs taken from a surface ship (1971), and finally to actually observing the seafloor from manned submersibles.

In order to better portray the features of the seafloor, Heezen and Tharp adapted the sketching technique developed by A. K. Lobeck of Columbia University to portray land topography. The first of their physiographic diagrams of the North Atlantic was accompanied by a text. A succession of physiographic diagrams of the South Atlantic, Indian Ocean, and west central Pacific followed. These diagrams, plus others covering the world's oceans, served as a basis for the panoramas prepared by the artist Heinrich Berann, who worked closely with Heezen and Tharp. The panoramas were published in *National Geographic* and brought a vivid and realistic picture of the seafloor to the general public. The "World Ocean Floor," a panorama that was sponsored by the Office of Naval Research, was completed just before Heezen left on his last cruise.

Heezen's interest in the near visual aspects of the seafloor had originated in 1947, during his first field season at sea. He had collected many bottom pictures and cores, which he published with John Northrop in 1951. His lifelong interest in the visual aspects of the floor of the ocean resulted in a book, *The Face of the Deep* (1971), written with Charles Hollister, in which the description of submarine topographic provinces and seafloor processes was accompanied by nearly 600 photographs of the deep-sea floor. The publication of this book led directly to Heezen's involvement as a principal participant in the U.S. Navy program investigating the deep-sea floor with submersibles. This work helped assuage Heezen's unquenchable curiosity about the actual appearance of the seafloor.

Heezen, who never married, died of a heart attack while he was aboard the U.S. Navy nuclear submarine *NR-1* to observe the crest of the Reykjanes Ridge. His death cut short further publication of his visual observations on tidal currents as an additional process in the initial formation of submarine canyons, descriptive bench and talus geology of the continental slope and canyon gorges, submarine karst topography in the Bahamas, volcanism at Puna Ridge off Hawaii, and subduction in the Middle America Trench.

Heezen received his Ph.D. from Columbia University in 1957. He was research associate (1956–1958) and senior research scientist (1958–1960) at the university's Lamont-Doherty Geological Observatory and then assistant professor (1960–1964) and associate professor (from 1964) in the department of geology. He served as a consultant to the U.S. Navy, submarine cable companies, and the oil industry, and was an adviser on the Law of the Sea. Among the many national and international organizations in which he took an active part were the Commission for Marine Geology of the International Union of Geological Sciences (president) and the Commission on Marine Geophysics of the International Association of the Physical Sciences of the Ocean (president). He was coordinator of the IUGGI Atlantic, Indian and Pacific Oceans of the Geological World Atlas.

Heezen was the recipient of the Henry Bryant Bigelow Medal (1964), the Cullum Geographical Medal of the American Geographical Society (1973), the Francis P. Shepard Medal of the Society of Economic Paleontologists and Mineralogists (1975), the Walter H. Bucher Medal of the American Geophysical Union (1977), and the Gardiner-Greene Hubbard Medal of the National Geographic Society, awarded posthumously to Heezen and to Marie Tharp (1978).

BIBLIOGRAPHY

I. Original Works. "An Outcrop of Eocene Sediment on the Continental Slope," in *Journal of Geology*, **59** (1951), 396–399, written with John Northrop; "Turbidity Currents and Submarine Slumps, and the 1929 Grand Banks Earthquake," in *American Journal of Science*, **250** (1952), 849–873, written with Maurice Ewing; "Mid-Atlantic Ridge Seismic Belt," in *Transactions of the American Geophysical Union*, **37** (1956), 343, an abstract written with Maurice Ewing; *Physiographic Diagrams of the North Atlantic; the South Atlantic; the Indian Ocean; and the Western Pacific* (New York, 1957–1971), with Marie Tharp; "Oceanographic Information for Engineering Submarine Cable Systems," in *Bell System Technical Journal*, **36** (1957), 1047–1093, written with C. H. Elmendorf; *The Floors of the Oceans*, I, *The North Atlantic*, Geological Society of America Special Paper no. 65 (New York, 1959), written with Marie Tharp and Maurice Ewing; "Géologie sous-marine et déplacements des continents," in *La topographie et la géologie des profondeurs océaniques* (1959), 295–304; "The Rift in the Ocean Floor," in *Scientific American*, **203**, no. 4 (1960), 98–110; "Equatorial Atlantic Fracture Zones," in *Special Papers of the Geological Society of America*, **68** (1962), 195–196, written with Marie Tharp and Robert D. Gerard; "Chain and Romanche Fracture Zones," in *Deep-Sea Research*, **11** (1964), 11–33, written with E. T. Bunce, J. B. Hersey, and M. Tharp; "Shaping of the Continental Rise by Deep Geostrophic Contour Currents," in *Science*, **152** (1966), 502–508, written with Charles D. Hollister and William F. Ruddiman.

Additional works include *Quaternary History of the Ocean Basins* (New York, 1967); *Panoramas of the Indian Ocean, the Atlantic Ocean, the Pacific Ocean;* and *The Arctic Ocean*, inserts in *National Geographic* (1967–1970), with Marie Tharp; *The Face of the Deep* (New York, 1971), written with Charles D. Hollister; *World Ocean Floor* (Washington, D.C., 1977), with Marie Tharp; as editor, *Influence of Abyssal Circulation on Sedimentary Accumulations in Space and Time* (Amsterdam, 1977); and "Visual Evidence for Subduction in the Western Puerto Rico Trench," in *Géodynamique des Caraïbes* (Paris, 1985), 287–304, written with Wladimir D. Nesteroff, Michael Rawson, and R. P. Freeman-Lynde.

Heezen's professional papers and underwater films are at the Smithsonian Institution Archives.

II. SECONDARY LITERATURE. A commemorative volume is R. A. Scrutton and M. Talwani, eds., *The Ocean Floor: Bruce Heezen Commemorative Volume* (New York, 1982). See also Marie Tharp and Henry Frankel, "Mappers of the Deep," in *Natural History*, **95** (October 1986), 48–62.

MARIE THARP

HEILBRON, IAN MORRIS (*b.* Glasgow, Scotland, 6 November 1886; *d.* London, England, 14 September 1959), *organic chemistry*.

The younger son of David Heilbron, a wine merchant, and Fanny Jessel was named Isidor but altered the name to Ian during his youth. During his studies at Glasgow High School he developed a fascination for chemistry that met resistance from his father, who expected his son to follow him into the family business. Heilbron persisted and was permitted to enroll in the Royal Technical College of Glasgow (now Strathclyde University), where the influence of George Gerald Henderson fortified his resolve to become a chemist. With the aid of a Carnegie fellowship he did research with Arthur Rudolf Hantzsch at Leipzig, taking his Ph.D. in 1909. Heilbron was married in 1924 to Elda Marguerite Davis of Liverpool; they had two sons. His wife's death in 1954 was a serious blow from which he never fully recovered.

After completing his formal education at Leipzig, Heilbron returned to Glasgow as a lecturer at the Royal Technical College, where, after a wartime interruption, he was made professor of organic chemistry. His reputation in research brought successive professorships at Liverpool (1920), Manchester (1933), and Imperial College of Science and Technology of London University (1938).

During World War I, Heilbron served in Salonica, Greece, as assistant director of supplies, holding the ranks of lieutenant and lieutenant colonel. He received several awards for distinguished service. When the war ended, he spent a short period with the British Dyestuffs Corporation, but his interest in the academic life soon took him back to Glasgow. However, he continued to serve the company, which became a part of Imperial Chemical Industries, Ltd., as a consultant until 1949. Heilbron's chemical interests always showed a strong bent for industrial application. When he retired from Imperial College in 1949, he became the first director of the Brewing Industry Research Foundation. He was not a figurehead officer but actively developed the foundation as a sound agency for promotion of better scientific understanding of fermentation. He retired in 1958.

Heilbron's organizational capacity was utilized in World War II in his capacity as a scientific adviser in the Ministry of Supply (1939–1942) and in a similar role in the Ministry of Production (1942–1945). He strongly supported the new insecticide DDT, and was a leader in developing production on a large scale for wartime use in North Africa, the Mediterranean basin, and the Far East. After the war he vigorously promoted its routine use in agriculture.

Heilbron's most lasting contributions, however, were in organic chemical research, where he quickly became one of the leaders. He combined a keen mind, driving energy, and meticulous attention to detail with a rare capacity for organization. The result was an unusual success in exploring structural problems, particularly in the realm of natural products.

Heilbron was a leader in the application of physical methods in the investigation of structural problems and was effective in turning organic chemists away from the tedious structural approaches of the nineteenth century. He was among the first to use spectroscopic techniques effectively in a full range of wavelengths; he pioneered in the revival of chromatography as a separatory and diagnostic tool; and he was one of the pioneers in use of the molecular still. He quickly adopted microanalysis in his laboratory.

Heilbron pursued a broad range of organic structural problems, from early work with Henderson on terpenes and with students on semicarbazones, to studies on dyes, on photocatalysis with Edward C. C. Baly, on reactions of polyunsaturated ketones, and ultimately to studies on a variety of compounds found in nature and having biochemical significance.

When he moved to Liverpool, Heilbron worked for a time with Baly, studying the hypothesis that in photosynthesis carbon dioxide is degraded to formaldehyde in the presence of colored materials such as chlorophyll, the formaldehyde becoming a building block in the synthesis of sugars. In extending the studies to reactions involving the formation of nitrogenous compounds in plants, it was learned that traces of simple organic compounds, including formaldehyde, frequently arose from rubber tubing. Although Baly was reluctant to abandon the formaldehyde hypothesis, Heilbron was not inclined to pursue the matter.

About the time that Thomas Percy Hilditch left the oil-and-soap industry to become the first James Campbell Brown professor of industrial chemistry at Liverpool (1925), where he pursued a quarter of a century of research on the composition of fatty oils, Heilbron and Hilditch undertook a study of the unsaponifiable matter in the oils of elasmobranch fish. Heilbron immediately became interested in the structure of squalene (from shark-liver oil). He was not only able to establish the formula as $C_{30}H_{56}$ and to propose the probable structure as an isoprene polymer, but also to demonstrate that it showed a structural relation to the sterols; he suggested a biosynthetic pathway from squalene to the steroid nucleus. His speculation was later substantiated by studies of Konrad Bloch and David Rittenberg at Columbia, who used isotopic tracers.

Heilbron's subsequent research dealt almost entirely with problems related to natural materials of biological significance and compounds produced in their degradation. While always focusing on the main problem, he and his research students and associates uncovered a body of new chemistry related to the degradation products.

Heilbron's studies on squalene led quite naturally to studies of other compounds in unsaponifiable matter. An interest in the sterols converted to vitamin D by irradiation resulted in extensive work on cholesterol and to the characterization of ergosterol. The squalene work continued into work on substituted naphthalenes and was a factor in working toward the structure of vitamin A. A final proof of structure by Heilbron and his associates was anticipated by Paul Karrer and his collaborators.

The work in Heilbron's laboratory on both sterols and carotenoids led to characterization of a number of naturally occurring compounds in seaweeds. The degradation of such molecules led to studies on biphenyls and pyridylquinolenes, and the development of substances with antispasmodic properties.

During World War II, Heilbron continued a vigorous research program in London, concurrent with his governmental commitments. He began to explore the chemistry of penicillin at a time when concentration of the material was laborious and difficult. The structural studies in his laboratory were important in establishing the synthesis of penicillin F and G, and in elucidation of the chemistry of heterocyclic five-membered rings containing nitrogen and sulfur.

BIBLIOGRAPHY

I. ORIGINAL WORKS. A bibliography of Heilbron's nearly 300 scientific papers is in *Biographical Memoirs of Fellows of the Royal Society*, **6** (1960), 75–85. They include "The Chemical Society: A Mid-century Review," in *Journal of the Chemical Society* (January–June 1950), 1641–1653; "Carotenoids and Vitamin A: The End of a Chapter," in *Endeavour*, **10** (1958), 175–182, with A. H. Cook; "Expanding Horizons in Organic Chemistry," in *Journal of the Royal Society of the Arts*, **106** (1958), 861–871; and "Reflections on Science in Relation to Brewing," in *Journal of the Institute of Brewing*, **65** (1959), 144–154.

II. SECONDARY LITERATURE. A. H. Cook, "Ian Morris Heilbron (1886–1959)," in *Biographical Memoirs of Fellows of the Royal Society*, **6** (1960), 65–85, deals primarily with his chemical contributions; Cook's sketch in *Dictionary of National Biography, 1951–1960*, 469–470, summarizes his career and lists his many honors. Also see an obituary by R. Robinson in *Nature*, **184** (1959), 767–768.

AARON J. IHDE

HEILBRONN, HANS ARNOLD (*b*. Berlin, Germany, 8 October 1908; *d*. Toronto, Canada, 28 April 1975), *mathematics*.

Heilbronn was born the son of Alfred and Gertrud Heilbronn, middle-class Jews who were cousins. After graduating from the Realgymnasium at Berlin-Schmargenhof in 1926 and studying at the universities of Berlin, Freiburg, and Göttingen, in 1930 he became assistant to Edmund Landau at Göttingen. Heilbronn's work on the distribution of primes rapidly brought him renown. He received the doctorate in 1933.

The advance of Hitler to power forced Heilbronn to emigrate to England in 1933, and until 1935 he was supported at the University of Bristol by refugee organizations. In this period he rose to prominence with his proof of Gauss's conjecture that the class number $h(d)$ of the imaginary quadratic field of discriminant $d < 0$ tends to infinity as d tends to minus infinity. Erich Hecke had already shown in 1918 that this would follow from a generalized Riemann hypothesis. Heilbronn showed that Gauss's conjecture also follows from the falsity of the same hypothesis. By the "law of the excluded middle" Gauss's conjecture holds, but this argument is ineffective in the logical sense: for given h_0 it does not provide a d_0 such that $h(d) > h_0$ for all $d < d_0$. Heilbronn's theorem was strengthened and generalized by C. L. Siegel (1935), Richard Brauer (1950), and others, but the first effective version was obtained only in 1983 by Gross and Zagier, using a conditional result of Goldfeld. In 1934 Heilbronn showed (with E. H. Linfoot) that there are at most ten imaginary quadratic fields with class number $h(d) = 1$. Again the proof was ineffective, and although nine fields with this property were known, the proof gave no means for deciding whether there actually was a tenth. Its existence was claimed to be disproved in 1952 by K. Heegner in a controversial paper (later seen to be basically correct) and by H. M. Stark and A. Baker in 1966–1967.

From 1935 to 1940 Heilbronn was at Cambridge as a fellow of Trinity College. This period saw the beginning of his close collaboration with Harold Davenport, which lasted until the latter's death in 1971. In 1936 Heilbronn published a paper simplifying, strengthening, and making generally accessible the methods by which I. M. Vinogradov had greatly improved the estimates of Godfrey Hardy and John Littlewood for such additive problems as Waring's problem; and in a series of papers (1936–1937) with Davenport he considered specific problems of this type. In another area, Heilbronn showed that there are only finitely many real quadratic fields in which the euclidean algorithm holds. His method was not suited to determining them all, which was done only in the 1950's by Davenport and others, using the geometry of numbers. At that time Heilbronn returned to the problem; he showed that only finitely many cyclic cubic fields have a euclidean algorithm and obtained similar results in other cases to which the methods of the geometry of numbers do not apply.

In 1940, at the outbreak of war, Heilbronn was interned for a period as an "enemy alien" but then served with the British services. He was demobilized in 1945 and, after a brief period at University College, London, returned in 1946 to Bristol as reader and became a British citizen; in 1949 he became professor and head of department. He rapidly built up a strong department, particularly a school of number theory. One of his pupils was A. Fröhlich.

At the beginning of the Bristol period, Heilbronn considered the approximations to an arbitrary real number θ by rational fractions whose denominator is a perfect square. He showed that for every $\eta > 0$ there is a $C(\eta) > 0$ with the property that for every real θ and every integer $N \geq 1$ there are integers n,g such that

$$1 \leq n \leq N : |n^2\theta - g| \leq C(\eta)N^{-1/2+\eta}.$$

A weaker form with $2/5$ instead of $1/2$ had been given by Vinogradov (1927). Heilbronn asked whether $1/2$ is the best possible constant, but no further progress has been made; there have, however, been a large number of generalizations. The work on euclidean algorithms in cyclic fields was also done at Bristol, but increasingly Heilbronn's preoccupation with policy problems and administrative irritations left him little time for mathematics.

In 1963, feeling that the university was not supporting his department as it should, Heilbronn resigned his chair at Bristol and in 1964 accepted one in Toronto. He moved to Canada with his wife, Dorothy Greaves, whom he had married that same year. In 1970 he became a Canadian citizen. In Toronto he built up a thriving research school and took an active part in Canadian mathematics. In what was to be his last published paper, he showed that the question of whether there is a real zero s_0 in $1/2 < s_0 < 1$ of the zeta-function $\zeta_L(s)$ of an algebraic number field L reduces to the case when L is quadratic. This was subsequently used by Stark to show that the Siegel-Brauer theorem about the class numbers of algebraic number fields (which generalizes Heilbronn's) can be made effective in many cases.

In 1973 Heilbronn had a heart attack and did not recover completely. He died during an operation to implant a pacemaker. During his life Heilbronn was elected to the Royal Society (1951) and to the Royal Society of Canada (1967). He was president of the London Mathematical Society from 1959 to 1961.

BIBLIOGRAPHY

I. Original Works. Heilbronn's papers are brought together in *The Collected Papers of Hans Arnold Heilbronn*, Ernst J. Kani and Robert A. Smith, eds. (New York, 1988). A complete bibliography of his works is in Cassels and Fröhlich (see below).

II. Secondary Literature. A full account of Heilbronn's life and work is J. W. S. Cassels and A. Fröhlich, "Hans Arnold Heilbronn," in *Biographical Memoirs of Fellows of the Royal Society*, **22** (1976), 119–135, reprinted in *Bulletin of the London Mathematical Society*, **9** (1977), 219–232. See also *Proceedings of the Royal Society of Canada*, 4th ser., **13** (1975), 53–56.

For the Gross-Zagier-Goldfeld effective estimate, see D. Zagier, "L-Series of Elliptic Curves, the Birch-Swinnerton-Dyer Conjecture, and the Class Number Problem of Gauss," in *Notices of the American Mathematical Society*, **31** (1984), 739–743. For recent work on the existence of euclidean algorithms in number fields, see H. W. Lenstra, Jr., "Euclidean Number Fields of Large Degree," in *Inventiones Mathematicae*, **38** (1977), 237–254.

J. W. S. Cassels

HEINROTH, OSKAR AUGUST (*b.* Kastel-Kostheim, near Mainz, Germany, 1 March 1871; *d.* Berlin, Germany, 31 May 1945), *ornithology, ethology*.

Oskar Heinroth came from a family of musicians and academics on his father's side and of day laborers on his mother's. His father, August Heinroth, earned a Ph.D. from the University of Jena and then for many years was tutor to the sons of a rich lumber merchant in St. Petersburg, who rewarded him so handsomely at the end of his service that when he returned to Germany, he was independently wealthy. August and his wife, Katharina Bodenmüller, had two children, the elder of whom was Oskar.

Heinroth graduated from the Holy Cross Gymnasium in Dresden in 1890 and went on to study medicine and zoology at Leipzig, Halle, and Kiel. He was licensed as a physician in 1895. The same year he received the doctorate from the Physiological Institute at Kiel. From 1896 to 1899 Heinroth studied zoology at Friedrich Wilhelm University in Berlin. He worked as a volunteer at the Berlin Zoo until 1904, when he was named assistant to the director. In 1904 he married Magdalene Wiebe, who became his invaluable co-worker. In 1910 Heinroth was put in charge of establishing an aquarium at the zoo, and when it was completed in 1913, he became its first director. He served in this capacity for the rest of his career. His wife died in 1932, and Heinroth married Dr. Katharina Berger Rösch the following year; both marriages were childless. She collaborated with him in his ornithological studies, and after his death she wrote the major biographical study of his life and career.

Heinroth was twice the president of the German Ornithological Society and once of the German Society for the Study of Mammals. He was a corresponding or honorary member of numerous other societies, and his researches were honored with the Leibniz and Goethe medals.

From his earliest years, Heinroth was fascinated by the behavior of animals, especially birds. He is said to have learned to walk in the family hen house, where he loved to watch the hens and to imitate the noises they made. Though his family moved to Dresden so that he could attend the highly regarded Holy Cross Gymnasium, he took no interest in the classical education offered there, much preferring to spend his time at the Dresden Zoological Garden.

Heinroth earned his doctorate at Kiel with a study on urine formation in fish. When he went to Berlin in 1896, he divided his time between his university and museum studies, and working at the zoo. At the Zoological Museum he studied avian systematics under Anton Reichenow. In 1898 he published an important analysis of the course of molting in birds. In 1900 and 1901 Heinroth participated as zoologist and physician on a scientific expedition to the South Seas. Though the high hopes for this expedition were destroyed when the party was attacked by natives in New Guinea and the expedition's leader, Bruno Mencke, was killed, Heinroth managed to return to Berlin with a valuable collection of living animals and animal skins. He wrote up the ornithological results of the expedition and continued his observations on animal behavior at the zoo.

Heinroth was especially concerned with what he called the "finer details" of bird behavior. His approach to behavior studies was elaborated in a major paper he delivered at the Fifth International Ornithological Congress, held at Berlin in 1910. In this paper, "Beiträge zur Biologie: Nämentlich Ethologie und Psychologie der Anatiden," he argued that behavior patterns are highly instructive for reconstructing the phylogenies of ducks and geese. He claimed that *arteigene Triebhandlungen* (species-specific instinctive actions) could be used like morphological features to determine the genetic affinities of species, genera, and subfamilies. He focused in particular on behavioral displays through which individuals of a species signal to other members of the same species. He also described, among other things, aspects of the phenomenon that his disciple, Konrad Lorenz, later named *imprinting* (*Prägung*). Heinroth found that greylag goslings do not instinctively recognize adult greylag geese as conspecifics. To the contrary, goslings hatched in an incubator and then exposed to a human being, even for a short period of time, will follow the human as if he or she were the birds' parent. However, it was not until the 1930's, when Lorenz insisted upon the theoretical significance of this phenomenon as a process distinct from normal association learning, that imprinting became an important focus of research for ethologists and other students of animal behavior.

With his wife, Heinroth undertook the systematic study of the instinctive behavior patterns of different bird species. The methodology they developed involved rearing baby birds at home, in isolation from other birds. The Heinroths saw these isolation or deprivation experiments as a means of distinguishing between those species-specific actions which were innate or genetically determined and those which were acquired through imitating or learning from conspecifics. Essentially, they defined as innate all those species-specific behavior patterns which emerged in a developing bird independently of any contact with other individuals of its own species. Emphatically empirical in their approach, they proceeded on the assumption that what was innate and what was learned in the different bird species had to be determined through experiments conducted on a species-by-species basis. Magdalena Heinroth gave an early report on this work at the International Ornithological Congress of 1910. Over the course of more than two decades, the Heinroths proceeded to rear nearly all the bird species of Central Europe in isolation. The results of this work were detailed in their four-volume classic, *Die Vögel Mitteleuropas in allen Lebens- und Entwicklungsstufen photographisch aufgenommen und in ihrem Seelenleben bei der Aufzucht vom Ei ab beobachtet* (1924–1933).

Heinroth's studies, in the opinion of the distinguished German ornithologist Erwin Stresemann, constituted the first great step toward closing the long-standing gap between avian systematics and avian biology, the two major branches of ornithology. At first Heinroth's work attracted few disciples or emulators. At the beginning of the 1930's, however, Heinroth was sought out by the young Austrian naturalist Konrad Lorenz, and over the next decade they developed a strong friendship, and exchanged observations and ideas. This interaction was of great importance for the subsequent history of ethology. Heinroth's love of animals and concern with establishing the study of animal behavior on a firm inductive foundation were shared by Lorenz, though the younger man proved significantly less reluctant to generalize about animal behavior than Heinroth had been.

It was primarily Lorenz, first with Heinroth's encouragement and then with the very able support and collaboration of the young Dutch naturalist Nikolaas Tinbergen and others, who between 1935 and 1960 played the leading role in establishing ethology, the comparative study of behavior, as a modern scientific discipline.

BIBLIOGRAPHY

I. Original Works. For a complete bibliography of Heinroth's works, encompassing 484 items, see Katharina Heinroth, *Oskar Heinroth, Vater der Verhaltensforschung* (Stuttgart, 1971), 204–227.

II. Secondary Literature. The major biographical study of Heinroth is the book by Katharina Heinroth cited above. Other discussions of Heinroth's life and career include Konrad Herter, "Zur Erinnerung an Oskar Heinroth (1871 bis 1945)," in *Sitzungsberichte der Gesellschaft Naturforschender Freunde zu Berlin*, **3** (1963), 117–122; Otto Koehler, "Oskar Heinroth zum 70. Geburtstag," in *Die Naturwissenschaften*, **29** (1941), 169–171; Konrad Lorenz, "Nachruf auf Oskar Heinroth, 1871–1945," in *Der zoologische Garten*, **24** (1958), 264–274; Kurt Priemel, "Zum 70. Geburtstag von Oskar Heinroth," *ibid.*, **13** (1941), 133–140; Karl Max Schneider, "Dr. Oskar Heinroth, 1871–1945," in *Zeitschrift für Säugetierkunde*, **19** (1951[1954]), 57–65; and Erwin Stresemann, *Ornithology from Aristotle to the Present*, Hans J. Epstein and Cathleen Epstein, trans., G. William Cottrell, ed. (Cambridge, Mass., 1975), 272–273, 345–347, 359, 363, 364.

Richard W. Burckhardt, Jr.

HEISENBERG, WERNER KARL (*b.* Würzburg, Germany, 5 December 1901; *d.* Munich, Federal Republic of Germany, 1 February 1976), *quantum theory, nuclear physics.*

Heisenberg was the younger son of August and Anna Wecklein Heisenberg. His father taught ancient languages at the Altes Gymnasium in Würzburg and Greek philology at the university. In 1910 he moved his family to Munich, where he had been appointed professor of Greek philology at the university. Heisenberg attended primary school in Würzburg and Munich. He began piano lessons early, and by the age of thirteen he was playing master compositions. He remained an excellent and avid player throughout his life. On 29 April 1937 Heisenberg married Elisabeth Schumacher, daughter of the noted Berlin professor of economics Hermann Schumacher. They had seven children.

In 1911, Heisenberg entered the Maximilians-Gymnasium, of which his maternal grandfather was rector. There he displayed an outstanding talent for mathematics. By the time of his final examinations (*Abitur*), he had taught himself calculus, had explored the properties of elliptic functions, and had attempted to publish a paper on number theory. Heisenberg entered the University of Munich in 1920, intending to study pure mathematics. Following the refusal of mathematics professor Ferdinand von Lindemann to admit Heisenberg to his seminar for advanced students, Heisenberg's father arranged an interview with the professor of theoretical physics, Arnold Sommerfeld, who tentatively accepted the ambitious student into his advanced seminar. Heisenberg received his doctorate under Sommerfeld in 1923—over the objections of Wilhelm Wien, the professor of experimental physics, who found the candidate deficient in his field. Heisenberg's dissertation involved an approximate solution of the complicated equations governing the onset of hydrodynamic turbulence.

When Heisenberg entered the University of Munich, theoretical physics, although it had attained recognition through the efforts of the older generation, was overshadowed by the work of experimentalists like Wien. Sommerfeld's institute was one of the few, mainly in Germany, where theoretical atomic physics was pursued, and the only German institute concerned with the entire quantum theory of atomic spectroscopy.

By 1920, atomic structure, the properties of light, and spectroscopy had become focuses of research on quantum atomic theory. This theory, formulated by Niels Bohr in 1913, regarded atomic motions as governed by integral quanta of energy and momenta. Transitions between quantum states involved the emission or absorption of monochromatic radiation of frequency proportional to the change of energy. During the early 1920's the mechanical properties of quantized models of atoms and molecules consisting of more than two particles disagreed with observed properties, and atomic spectra and their behavior in applied electric and magnetic fields displayed numerous inexplicable regularities and anomalies. The discovery of the Compton effect at the end of 1922 lent support to the light-quantum hypothesis, contradicting the well-established wave theory of light and raising the wave-particle dualism to a fundamental problem.

During his studies and research at Munich, and subsequently at Göttingen with Max Born and at Copenhagen with Niels Bohr, Heisenberg became familiar with each of the above difficulties, as well as with the limitations of quantum theory and of the methods employed by each of his mentors. He also made the acquaintance of such young and brilliant theoretical physicists as Wolfgang Pauli, Enrico Fermi, and Paul Dirac, who would dominate atomic physics for at least a decade. Heisenberg, a leading representative of this group, is best known scientifically for his contributions to the creation and development of quantum mechanics.

As early as his first semester at Munich, Heisenberg displayed the audacity, optimism, and independence of thought and action that characterized his physics as well as his personal life during and immediately following World War I. While his father, a reserve infantry officer, was away from home for nearly the entire war, his sons were left increasingly on their own. A severe shortage of fuel and food forced the occasional closing of school and encouraged Heisenberg to educate himself. Weak from lack of food, he helped to bring in the harvest with schoolmates on a Bavarian farm in the summer of 1918.

The loss of the war and the abdication of the monarchy caused revolutionary unrest throughout Germany. In Bavaria a socialist republic came to power in 1918, then was replaced in 1919 by a Bolshevik-oriented republic that was suppressed by troops dispatched from Berlin. During the sometimes heavy street fighting and the subsequent restoration of democratic socialist rule, Heisenberg supported the invading forces as part of a unit composed of boys from his gymnasium. Soon after the restoration of democratic socialism, Heisenberg was elected leader of a small group of younger boys from the earlier unit who became associated with the Bund Deutscher Neupfadfinder (New Boy Scouts). The New Boy Scouts strove for a renewal of supposedly decadent German personal and social life through the direct experience of nature and the uplifting beauties of Romantic poetry, music, and thought. Heisenberg's comrades from these years remained among his closest friends throughout his life, and outdoor activities, together with music, remained among his favorite pastimes.

During his studies with Sommerfeld, Heisenberg became acquainted not only with Wolfgang Pauli, a fellow student who thereafter was his closest collaborator and severest critic, but also with the intricacies of the anomalous Zeeman effect, the inexplicable splitting of atomic spectral lines into more than the expected three components in a magnetic field. In his first semester, Heisenberg offered a classification of the anomalous lines using thoroughly unorthodox half-integral quantum numbers. In 1922 he publicly displayed his audacity—and his intuition—in his first published paper, which offered a model for the Zeeman effect that described all of the known data in terms of the couplings between valence electrons and the remaining atomic "core" electrons. The model, however, violated many of the basic principles of quantum theory and classical mechanics. It thus served both as the basis for most of the subsequent work on the Zeeman effect until the advent of electron spin and as the first indication of the radical changes required for solving the quantum riddle.

The core model brought its author to the attention of established theoreticians. Sommerfeld had already written to his colleagues about Heisenberg's work when, in June 1922, he brought his student to Göttingen for a series of lectures on quantum atomic physics presented by Bohr. Heisenberg's audacious criticism of one of Bohr's assertions and a subsequent confrontation between the two over the core model resulted in a mutual admiration and the beginning of a lifelong collaboration that was as important for Heisenberg as his collaboration with Pauli.

During the Göttingen meeting Sommerfeld arranged for his students to continue their studies at Göttingen with Max Born while Sommerfeld traveled. A lecture on the core model, delivered soon after he arrived in Göttingen for the winter semester of

1922–1923, brought Heisenberg a private assistantship with Born.

Except for semester-long visits to Munich and Copenhagen, Heisenberg remained in Göttingen until May 1926. The period was one of his most productive scientifically. With his colleagues there, he developed the matrix form of quantum mechanics, progressed toward an interpretation of the new formalism, and applied the quantum theory, along with electron spin, to the Zeeman effect, the helium atom, and other old problems. In July 1924 Heisenberg qualified to teach on the university level by presenting to the Göttingen faculty a modification of the quantum rules for the Zeeman effect. The modification foreshadowed the notions of what Born was now calling a future "quantum mechanics."

After Pauli had indicated the inadequacy of quantum theory for the hydrogen molecule ion in 1922, attention turned to an exact calculation of the orbits of helium. While Pauli grew increasingly skeptical of any approach that assumed the existence of electron orbits in atoms, Born and Heisenberg managed to support Bohr's building up of the periodic table, which explicitly assumed the existence of orbits. But the celestial mechanics of atoms clearly failed to reproduce the properties of helium, and in 1923 Born and Heisenberg officially pronounced the premises of quantum atomic theory inadequate.

During a seven-month visit to Bohr's institute at Copenhagen beginning in the fall of 1924, Heisenberg turned to the nature of light. Earlier in 1924 Bohr, Hendrik A. Kramers, and John C. Slater had attempted to resolve the wave-particle duality by assuming the statistical conservation of energy and momentum in the absorption and emission of radiation. The proposal was already in doubt and was soon refuted, but it indicated again the belief that radical change was necessary. With Bohr and Kramers, Heisenberg attempted to account for optical fluorescence and dispersion by a "sharpened" version of Bohr's correspondence principle between classical and quantum physics. In the Kramers-Heisenberg scheme, which has proved fruitful ever since, emission and absorption were treated as if they occurred via quantized harmonic oscillators whose frequencies and amplitudes were related to the Fourier components of the supposed electron motions in atoms. From his visit to Copenhagen, Heisenberg also gained an appreciation of what Pauli called Bohr's "philosophical thinking," a concern for the physical and conceptual nature of the difficulties in quantum theory, in addition to their mathematical expression. All of these elements contributed to his discoveries soon after his return to Göttingen in 1925.

In May 1925, Heisenberg took on a new and difficult problem, the calculation of the line intensities in the hydrogen spectrum. Just as he had done with Kramers and Bohr, Heisenberg began with a Fourier analysis of the electron orbits. When the hydrogen orbits proved too difficult, he turned to the anharmonic oscillator. With a new multiplication rule relating the amplitudes and frequencies of the Fourier components to observed quantities, Heisenberg succeeded in quantizing the equations of motion for this system in close analogy with the classical equations of motion.

A severe attack of hay fever in early June forced Heisenberg's retreat to the island of Helgoland. There he completed the calculation of the anharmonic oscillator, determined the constants of motion, and obtained from his multiplication rule the Thomas-Kuhn summation rule for spectral lines. After nearly two weeks on Helgoland, Heisenberg returned to Göttingen, where he drafted his fundamental paper "Über die quantentheoretische Umdeutung kinematischer und mechanischer Beziehungen," which he completed in July. In this paper Heisenberg proclaimed that the quantum mechanics of atoms should contain only relations between experimentally observable quantities. The resulting formalism served as the starting point for the new quantum mechanics, based, as Heisenberg's multiplication rule implied, on the manipulation of ordered sets of data forming a mathematical matrix.

Born and his assistant, Pascual Jordan, quickly developed the mathematical content of Heisenberg's work into a consistent theory with the help of abstract matrix algebra. Their work, in collaboration with Heisenberg, culminated in their "three-man paper" that served as the foundation of matrix mechanics. Confident of the correctness of the new theory, Heisenberg, Pauli, Born, Dirac, and others began applying the difficult mathematical formalism to the solution of lingering problems. But most physicists soon welcomed a rival theory propounded by Erwin Schrödinger in 1926.

Schrödinger offered a quantum "wave mechanics" that purported to replace electron orbits, banish quantum jumps, and require only the familiar methods of partial differential equations. His recognition of the mathematical equivalence of the rival theories and his claim that his theory superseded matrix mechanics caused a flurry of activity among the matrix mechanicians.

In May 1926 Heisenberg succeeded H. A. Kramers as Bohr's assistant in Copenhagen. The acceptance of electron spin, Born's interpretation of the square of Schrödinger's wave function as a probability density, and the formulation of the Dirac-Jordan transformation theory led to intensive studies in Copenhagen of the "perceptual content" of the rival quantum formalisms. In extensive correspondence with Pauli, Heisenberg analyzed the measurement of individual quantum events in the context of transformation theory. This led him in early 1927 to the formulation of his principle of uncertainty, or indeterminacy, the contribution for which Heisenberg is perhaps best known.

According to Heisenberg, the precise, simultaneous measurement of canonically conjugate variables, such as the position (q) and the momentum (p), or the energy (E) and time (t), of a particle, are excluded in principle. Instead, reciprocal relationships exist between the indeterminacies in the measurements of position (Δq) and momentum (Δp), or energy (ΔE) and time (Δt). These can be represented by Heisenberg's famous uncertainty relations:

$$\Delta p \cdot \Delta q \approx h$$
$$\Delta E \cdot \Delta t \approx h,$$

where h is Planck's constant. The implications were enormous. Heisenberg declared in particular that traditional notions of causality are weakened by the indeterminacy formula, since all initial conditions of a particle's motion cannot be known with enough precision to predict its subsequent motion with certainty. The difficulty was one of principle, not of technique. Strict determinism, he insisted, will never be reestablished.

Heisenberg formulated his principle in Copenhagen, where he was teaching, after long conversations with Bohr. Bohr was not satisfied with Heisenberg's formulation, among other versions, because it did not adequately treat the simultaneous use of the physical pictures of particles and waves. Bohr's views, expressed in the principle called "complementarity," provided a deeper understanding of the uncertainty principle in terms of the actual measuring processes in any experiment. These two principles combined with Born's probabilistic interpretation of the wave function to form the basis of the "Copenhagen interpretation" of quantum mechanics. Although this interpretation, despite its revolutionary implications for measurement, causality, and objectivity, was soon accepted by the majority of physicists, a vocal minority that included Einstein, Schrödinger, and Planck was never fully reconciled to it. Physics that renounces knowledge of objective processes in principle cannot, they argued, be a complete account of physical reality.

In 1927, as Heisenberg, Bohr, and Born presented the Copenhagen interpretation, Heisenberg accepted a call to Leipzig as professor of theoretical physics, at the age of twenty-five Germany's youngest full professor.

Heisenberg and his Leipzig colleagues transformed the Physics Institute into a leading center for research on atomic and quantum physics. Among Heisenberg's early students and collaborators were Felix Bloch, Rudolf Peierls, Edward Teller, Peter Debye, Frederick Hurd, and Carl Friedrich von Weizsäcker. Heisenberg traveled widely during his early Leipzig years, carrying the message of quantum mechanics to the United States, Japan, India, and Italy.

Heisenberg's research in Leipzig concentrated upon applications and extensions of quantum mechanics. In 1928 he showed that a quantum-mechanical exchange integral that had played a crucial role in his earlier solution of the helium problem could account for the strong molecular magnetic field in the interior of ferromagnetic materials. Bloch, then Heisenberg's assistant, supplemented the theory with his notion of spin waves, and Peierls offered a theory of the anomalous Hall effect.

Heisenberg also explored the philosophical implications of quantum theory. With Weizsäcker and Grete Hermann, a student of Kantian philosophy, he considered the problems of language in quantum theory and the applicability of neo-Kantian notions of causality. With Bohr, Heisenberg pushed the Copenhagen view into chemistry, biology, and even into social and ethical phenomena. To many lay persons, however, the new ideas seemed as strange and disturbing as the revisions in world view brought about earlier by the relativity theory. Consequently, Heisenberg addressed himself increasingly to popular audiences. As theoretical physics and physicists fell into increasing disfavor with the rise of Hitler, the task of popularization became more pressing.

Heisenberg's foremost scientific concern after 1927 involved the search for a consistent extension of the quantum formalism that would yield a satisfactory unification of quantum mechanics and relativity theory. This required the formulation of a covariant theory of interacting particles and fields that accounted for elementary processes at high energies and small distances. In 1929, drawing upon the work of Dirac, Jordan, Oskar Klein, and others, Heisenberg and Pauli succeeded in formulating a general gauge-invariant relativistic quantum field theory by treating particles and fields as separate entities interacting through the intermediaries of field quanta.

The formalism led to the creation of a relativistic quantum electrodynamics, equivalent to that developed by Dirac, which, despite its puzzling negative energy states, seemed satisfactory at low energies and small orders of interaction. But at high energies, where particles approach closer than their radii, the interaction energy diverged to infinity. Even at rest, a lone electron interacting with its own field seemed to possess an infinite self-energy, much as it did in classical electrodynamics. Attention was directed to the resolution of such difficulties for more than two decades.

Soon after the discovery of the neutron in 1932, Heisenberg developed a neutron-proton model of the nucleus by introducing the concept of the nuclear exchange force and the formalism of isotopic spin. Nonrelativistic quantum mechanics could be applied to the nucleus, Heisenberg showed, as long as one did not consider the structure of nucleons. Heisenberg's work served as the basis for contemporary nuclear physics, much of which was pursued by his assistants, especially Weizsäcker. Heisenberg preferred to continue the search for a consistent quantum physics of fields. In 1935 Heisenberg and his assistant Hans Euler discovered that nonlinear interactions in positron theory, which yielded photon-photon scattering, could be represented by treating the electron as possessing a minimum size, below which the interferences predominated.

These studies were slowed by the deterioration of German politics. The takeover of German student bodies, such as Leipzig's, by Nazi students preceded the Nazi control of German society. Heisenberg was thirty-one when Hitler came to power in 1933. At the end of the year Heisenberg was awarded the Nobel Prize for physics (for 1932) for his contributions to quantum mechanics. His scientific renown and his ties to leading members of the older generation of German quantum physicists—Max Planck, Max von Laue, Arnold Sommerfeld—rendered the young man a leading spokesman for German physics.

It needed defense. The attack upon academic professions, including physics, had three phases before the outbreak of war: the dismissal of Jewish and leftist teachers and professors, the dismissals of those protected during the first round of dismissals, and an attempt to force the remaining Germans and German institutions into acquiescence with, if not overt support of, the dictatorship. Heisenberg's response was perhaps typical of many educated Germans. There was little chance that he would emigrate voluntarily, despite numerous opportunities and invitations to do so. The reasons included his attachment to Germany, his agreement with its resurgence as a nation, and his perceived duty to his profession, made more acute by the unsavory politicians in control. As for many Germans, nation and politics were separable for Heisenberg, and, like many, he believed the Nazis would not be in power long.

Planck, Heisenberg's main political mentor, counseled optimism for the future and quiet diplomacy for the present. Public protest or direct confrontation was not compatible with their temperaments or their strategy of administrative diplomacy. At no time, however, was Heisenberg an overt supporter of Nazi ideals; nor do Nazi party records show that he was ever a member. He was a man dedicated to his country and to the preservation of the best of its culture, even at personal risk. His were the dilemma and the tragedy of many German scientists under Hitler.

Heisenberg found support and protection in his efforts to counter the effects of the regime through circles of numerous friends and colleagues of like mind, extending from his family acquaintances to the university faculty, to the Saxon Academy of Sciences in Leipzig, to the entire non-Nazi academic community of Germany. Optimism, perseverance and attempts to utilize bureaucratic channels characterized Heisenberg's early efforts on behalf of colleagues dismissed in 1933. The dismissal of four Jewish veterans of World War I from the Leipzig science faculty two years later occasioned perhaps the strongest verbal protests from the Leipzig physicists and an implied threat to resign. A strong reprimand of the Leipzig physicists resulted, but the four victims were never reinstated.

Beginning in 1936, antiscientific political attention focused upon theoretical physics and upon Heisenberg in particular. Sommerfeld, who had by then reached retirement age, wanted Heisenberg as his successor in Munich. That touched off a campaign by Nazi physicists against the theoretical physics establishment. Early in 1936 the Nobel Prize–winning physicist Johannes Stark and his followers unleashed a newspaper assault against "Jewish [that is, theoretical] physics" and contrasted it with a supposedly experimentally oriented "German physics." Heisenberg responded in the same party newspaper with a defense of teaching and research in contemporary theoretical physics. With the experimentalists Max Wien and Hans Geiger, he authored and circulated a petition among German physicists protesting the attack and the effects of Nazi policy on teaching and research. It was submitted to the Reich Education Ministry with seventy-five signatures. Although the petition did have some immediate ef-

HEISENBERG

fect, it exerted little influence upon the overall political course of the regime or its policies.

While engaged in this political fight, Heisenberg vigorously pursued his search for a consistent quantum field theory. His tenacious adherence to what he believed to be the beginning of a new quantum revolution is in part attributable to his concern for the vitality of German research. In 1935 Heisenberg's research began to focus on high-energy collisions of elementary particles in cosmic rays, the highest-energy phenomena then known. Examining the Fermi (weak) interaction in early 1936, Heisenberg discovered a mathematical minimum length, about the size of elementary particles, that appeared to trigger the onset of "explosion showers" of cosmic rays. The minimum length, a notion that he had earlier considered in the context of quantum electrodynamics, marked, he believed, the boundary of quantum mechanics and the frontier of a wholly new and revolutionary physics.

Heisenberg's revolutionary notions were challenged soon afterward by the alternative quantum electrodynamics of "cascade showers," generated by *Bremsstrahlung* and pair production. A controversy ensued, mainly between Heisenberg and several American physicists, over the existence of explosion showers and over allegiances to the two types of theories and their implications for the future course of physics. Fermi's weak-field theory soon proved inapplicable to the problem, but in 1939 Heisenberg extended his notions to Yukawa's (strong) meson theory of nuclear forces, revitalizing the controversy into the war years. A universal minimum length remained a permanent feature of Heisenberg's physics. Although explosion showers, later called "multiple processes," were discovered after the war in cosmic-ray events, the invention of renormalization techniques and the experimental confirmation of quantum electrodynamics to the highest energies left Heisenberg's physics with only minority support.

In 1937 a new attack in the Nazi press interrupted Heisenberg's work on quantum field theory. When Heisenberg accepted a renewed call to succeed Sommerfeld in Munich, *Das Schwarze Korps*, the journal of the S.S., published an article signed by Stark declaring that Heisenberg and other theoreticians were "white Jews." It accused Heisenberg in particular of actions amounting to overt opposition to the totalitarian regime. Not only was Heisenberg's appointment canceled, but his personal safety and future ability to work and teach in Germany seemed endangered. With the support of his colleagues, he decided to write directly to the head of the S.S., Heinrich Himmler, requesting an official disavowal of the charges and a termination of the campaign against him and his physics. A year of investigations and interrogations, some in the Berlin office of the S.S., brought partial relief. Himmler disavowed the charges and placed the proponents of "German physics" on the defensive, but Heisenberg never did succeed his former teacher.

The outbreak of world war in September 1939 profoundly affected Heisenberg and his career. Still of military age, he was ordered to report to the Army Weapons Bureau (Heereswaffenamt) in Berlin. There the authorities asked him and other leading German nuclear physicists to investigate whether nuclear fission, discovered in Berlin a year earlier, could be used for large-scale energy production. Within two months Heisenberg completed a comprehensive report on the theory of chain reactions and their uses, including their use in an atomic bomb. The report made Heisenberg the leading specialist on nuclear energy in Germany.

In order to continue the promising research, the Army Weapons Bureau designated the Kaiser Wilhelm Institute for Physics in Berlin the center of German fission research. After the departure of the institute's Dutch director, Peter Debye, who chose emigration over German citizenship, Heisenberg was named adviser, and later acting director, of the institute and its nuclear research. At the same time, Heisenberg supervised preliminary reactor experiments in Leipzig. He also continued with high-energy interactions. In papers written between 1942 and 1944, Heisenberg developed a theory of particle collisions based, as in 1925, only upon the observable properties of the colliding particles. The resulting "S-matrix" theory of particle scattering, especially in its later analytic forms, enjoyed considerable attention after the war, then again during the 1960's, but renormalized field theories eventually found more followers.

Heisenberg consciously pursued his assumed role of protector of German physics after 1938. The chance that he might emigrate accordingly declined, despite the various offers and pleas from his American colleagues during his tour of the United States a year later. Heisenberg was later much criticized for his decisions to remain in Germany and to work on nuclear energy under Hitler. He always responded that Germany needed him.

The aims of German nuclear research have also been debated. The theoretical possibility of an atomic bomb was known among nuclear physicists before the outbreak of the war. In 1942 a decision was made at a meeting between German government officials and atomic scientists, calling for the construction of a reactor, the first step toward atomic energy and, if so desired, an atomic bomb. During a private meeting a year earlier between Bohr and Heisenberg in German-occupied Copenhagen, Bohr received the distinct impression that the Germans were indeed aiming for a bomb. Heisenberg later insisted that a misunderstanding had occurred; however that may be, a very disturbed Bohr conveyed his impressions to Allied scientists when he fled Denmark in 1943.

The German reactor effort never succeeded. Technical and scientific errors, lack of financial and material support, and the effects of Allied bombing prevented the Germans from achieving a chain reaction. By war's end the German project and its scientists had moved to small towns in southern Germany in order to escape the bombing of the larger cities. The ALSOS Mission, a secret American intelligence unit, captured Heisenberg and nine other atomic scientists there during the last chaotic days of the war. Heisenberg and his colleagues were eventually turned over to the British, who held them for six months at a country estate near Cambridge.

The British returned the German scientists to Göttingen in early 1946. In addition to developing themes of superconductivity, hydrodynamic turbulence, and particle physics, Heisenberg set out to revitalize German science. While reestablishing in Göttingen the Kaiser Wilhelm Institute for Physics, renamed the Max Planck Institute and placed under his direction, Heisenberg devoted enormous energy to realizing his conception of government science policy. He sought a direct role for the new West German federal government in forming a national science policy and a direct role for science advisers to the chancellor. Such conceptions found expression in 1949 in the German Research Council (Deutscher Forschungsrat), composed of fifteen leading scientists, including Heisenberg as president. With the support of the scientific establishment and of Chancellor Konrad Adenauer, the German Research Council represented German science in international affairs and directly in the chancellor's office. Among the council's successes were the acquisition of Marshall Plan money for German research, the admission of the Federal Republic into the International Union of Scientific Councils, and the statement of federal responsibility for science in the West German constitution.

As head of both the research council and the leading West German physics institute, Heisenberg became a leading spokesman for German science in the international arena. Seeking to reestablish international relations, he headed (beginning in 1953) the Alexander von Humboldt Foundation for the support of foreign scholars in the Federal Republic, and he traveled and lectured widely abroad. In 1954 Heisenberg served as the West German delegate to the conference Atoms for Peace in Geneva. After the founding of the European Council for Nuclear Research in 1952, Heisenberg, as head of the German delegation, participated in the decision to locate the European research center for high-energy physics (CERN) in Geneva and later served as chairman of its scientific policy committee.

But at home Heisenberg's research council went against the German tradition that support for science fell to the cultural minister of each state. The research council thus came into increasing conflict with the Emergency Association of German Science (Notgemeinschaft der Deutschen Wissenschaft), revived in 1949 by the cultural ministers and the university rectors. In order to avoid further friction, Heisenberg reluctantly allowed the amalgamation of the research council with the emergency association in 1951 to form the German Research Association (Deutsche Forschungsgemeinschaft). Representation of German science was assumed by the senate of the new body, composed mainly of the old research council. Heisenberg became a member of the presidential committee of the research association and chairman of its committee for atomic physics, which coordinated nuclear research.

Heisenberg's influence and that of his colleagues is evidenced by their twofold impact on the important field of West German nuclear policy: support of nuclear energy and opposition to nuclear weapons. In 1955 the Western allies granted the Federal Republic full sovereignty and full membership in the NATO alliance, removing all restrictions upon West German research. Heisenberg and his colleagues immediately launched a public campaign for a crash program in nuclear energy development. Under Heisenberg's direction, Germany's first nuclear reactor, a research model, was set up at Garching (near Munich) in 1957. At the same time, a major nuclear research section was established at Heisenberg's Max Planck Institute under the direction of Karl Wirtz; it eventually relocated in Karlsruhe.

While Heisenberg energetically argued for nuclear energy production, he equally energetically opposed Adenauer's plans to equip the West German army with tactical nuclear weapons. With strong support from an aroused public, Heisenberg and other leading German scientists launched a broadly based political campaign against nuclear weapons. It culminated in 1957 in a public declaration, formulated mainly by Weizsäcker and Heisenberg, and signed by seventeen prominent nuclear scientists opposed to research on or possession of nuclear weapons by West Germany. This time the effort met with success; the West German army has remained nonnuclear.

Despite the enormous demands of his political involvement, Heisenberg continued to pursue his search for a consistent quantum field theory. After an unsuccessful attempt at a nonlocal theory, he turned in 1952 to the investigation of nonlinear field equations in which the mathematical state space was extended beyond that used in quantum mechanics. Heisenberg and his collaborators discovered that with these equations, results could be obtained without the introduction of supplementary subtraction or renormalization techniques. After demonstrating the consistency of the theory in 1957, Heisenberg entered into an extremely intense collaboration with Pauli that a year later yielded a nonlinear spinor equation designed to describe the properties and the behavior of all known elementary particles. The resulting equation, dubbed the "world formula" by journalists, caused immense excitement among physicists, but Pauli, ever the critic, withdrew his support later that year.

Saddened by Pauli's renunciation and by his death soon afterward, Heisenberg nevertheless continued research on the nonlinear spinor theory after moving his institute to Munich from Göttingen in 1958. A year later he and his colleagues published a long paper enunciating the new theory and obtaining its resonance states, one of which, a new particle (the eta-meson), was discovered in 1960. Although the theory did not enjoy wide popularity, Heisenberg retained his belief in it to the end of his life. In his view the prevailing method of inventing ever more elementary material objects, such as quarks, missed the main objective of a fundamental theory: to explain the very existence and behavior of matter itself by fields. In Heisenberg's opinion, his theory agreed with Plato's representation of the structure of matter by simple geometrical forms. The symmetries of Heisenberg's field equation seemed to him to be analogous to Plato's forms. Matter and its properties would follow from the symmetries of the nonlinear field equation and the conditions imposed upon it. While attempting to realize this plan in the complicated mathematics of nonlinear equations, Heisenberg extended his Platonic notions to the rest of his world view, merging it in his later years with his views on the philosophy of physics, on religion, on society, and on his role in the history of quantum physics.

Heisenberg continued to work tirelessly into the 1970's on his nonlinear spinor theory, on philosophy, on international relations, and on the direction of his Max Planck Institute. In the middle of 1973 he fell seriously ill. He slowly improved and appeared to have fully recovered, but in July 1975 he suffered a severe relapse. He died six months later.

BIBLIOGRAPHY

I. ORIGINAL WORKS. Heisenberg published over 500 independent works, of which some 100 may be considered original scientific contributions. The others concern philosophical, cultural, political, and popular subjects. All of Heisenberg's published writings have been reissued in facsimile in *Werner Heisenberg: Gesammelte Werke/Collected Works*, 9 vols., Walter Blum, Hans-Peter Dürr, and Helmut Rechenberg, eds. (Berlin, Munich, and New York, 1984–). The volumes are divided into three series issued by two publishers. Series A (Berlin and New York, Springer-Verlag) contains the original scientific writings, which are arranged by topic with an introductory survey by a distinguished figure in the field. Series B (Munich: Piper-Verlag) contains review articles, lectures, and books. Series C (Munich: Piper-Verlag) contains the philosophical, political, and popular writings. The *Collected Works* also contains a number of significant, previously unpublished items, among them a long manuscript titled (by the editors) "Ordnung der Wirklichkeit" (1942) and all available wartime nuclear research reports authored or coauthored by Heisenberg. A complete bibliography, including references to subsequent reprintings, translations, and excerpts, has been compiled by David Cassidy and M. Baker, *Werner Heisenberg: A Bibliography of His Writings* (Berkeley, 1984). References to translations of Heisenberg works cited below are in the *Bibliography*.

Heisenberg summarized his own physics in various periods in series of lectures that were subsequently published as textbooks. The most important of these include *Die physikalischen Prinzipien der Quantentheorie* (Leipzig, 1930, repr. Mannheim, 1958), trans. by Carl Eckart and Frank C. Hoyt as *Physical Principles of the Quantum Theory* (New York, 1930); *Kosmische Strahlung*, which Heisenberg edited (Berlin, 1943; 2nd ed., 1953), translated by T. H. Johnson as *Cosmic Radiation* (New York, 1946); *Die Physik der Atomkerne* (Brunswick, 1943; 2nd ed., 1949), also in English, *Nuclear Physics* (New York, 1953); and *Introduction to the Unified Field Theory of Elementary Particles* (New York, 1966).

Nearly all of Heisenberg's nontechnical writings originated as lectures. Some of the more widely distributed collections, published before the *Collected Works*, include *Wandlungen in den Grundlagen der Naturwissenschaft* (Leipzig, 1935; 11th ed., Stuttgart, 1980); *Das Naturbild der heutigen Physik* (Hamburg, 1955; 2nd ed., 1961); *Physics and Philosophy: The Revolution in Modern Science* (New York, 1958; repr. 1962); *Schritte über Grenzen: Gesammelte Reden und Aufsätze* (Munich, 1971; 2nd ed.,

1973); and *Tradition in der Wissenschaft: Reden und Aufsätze* (Munich, 1977).

Extensive unpublished interviews of Heisenberg, focusing mainly upon his contributions to quantum physics, were conducted in the early 1960's by the Sources for History of Quantum Physics (SHQP). Transcriptions are available in the repositories of the project. Heisenberg's memoirs, written largely as a series of dialogues, appeared as *Der Teil und das Ganze: Gespräche im Umkreis der Atomphysik* (Munich, 1969), trans. by Arnold J. Pomerans as *Physics and Beyond: Encounters and Conversations* (New York, 1971).

Portions of Heisenberg's extensive correspondence have been published in *Wolfgang Pauli: Wissenschaftlicher Briefwechsel mit Bohr, Einstein, Heisenberg u.a.*, I, Armin Hermann *et al.*, eds. (New York, 1979), II, Karl von Meÿenn, ed. (New York, 1984); in *Niels Bohr: Collected Works*, 10 vols., Erik Rüdinger, general ed. (Amsterdam, 1972–); and in many other places, for which see Bruce R. Wheaton and John L. Heilbron, *An Inventory of Published Letters to and from Physicists 1900–1950* (Berkeley, 1982).

Heisenberg's unpublished papers and correspondence are considerable, although many early items were not preserved or were lost during World War II. The main body of material, currently in Munich, at the Werner Heisenberg Archives in the Werner Heisenberg Institute of the Max Planck Institute for Physics and Astrophysics, will be deposited in the library and archives of the Max Planck-Gesellschaft, Berlin. The papers include administrative documents, lecture course notes, and extensive correspondence. A survey of all of the available extant archival papers through 1950 is provided by the Inventory of Sources for History of Twentieth-Century Physics, Office for History of Science and Technology, University of California, Berkeley. Some of the scientific correspondence and manuscripts are available on microfilm in the SHQP repositories. Copies of German wartime documents, interviews, and other materials pertaining to the fission project, many of which refer to Heisenberg, are preserved in David Irving's collection *Records and Documents Relating to the Third Reich*, group 2, microfilms DJ-29 to DJ-31 (East Ardsley, Wakefield, England, 1966).

II. Secondary Literature. There is, as yet, no comprehensive account of Heisenberg's life and work in print. Surveys are provided by Armin Hermann, *Werner Heisenberg in Selbstzeugnissen und Bilddokumenten* (Reinbek bei Hamburg, 1976) and *Die Jahrhundertwissenschaft: Werner Heisenberg und die Physik seiner Zeit* (Stuttgart, 1977). Appreciations of Heisenberg's life and work by his friends and colleagues are in the *Collected Works*, ser. A, and in Fritz Bopp, ed., *Werner Heisenberg und die Physik unserer Zeit* (Brunswick, 1961); Hans-Peter Dürr, ed., *Quanten und Felder: Physikalische und philosophische Betrachtungen zum 70. Geburtstag von Werner Heisenberg* (Brunswick, 1971); Heinrich Pfeiffer, ed., *Denken und Umdenken: Zu Werk und Wirkung von Werner Heisenberg* (Munich, 1977); and Carl-Friedrich von Weizsäcker and Bartel L. van der Waerden, *Werner Heisenberg* (Munich, 1977).

Heisenberg's political life is treated in Elisabeth Heisenberg, *Das politische Leben eines Unpolitischen: Erinnerungen an Werner Heisenberg* (Munich, 1980), translated by S. Cappellari and C. Morris as *Inner Exile: Recollections of a Life with Werner Heisenberg* (Boston, 1984); Sir Nevill Mott and Sir Rudolf Peierls, "Werner Heisenberg, 5 December 1901–1 February 1976," in *Biographical Memoirs of Fellows of the Royal Society*, **23** (1977), 213–251; and Samuel A. Goudsmit, "Werner Heisenberg (1901–1976)," in *Year Book of the American Philosophical Society for 1976* (1977), 74–80, which revises Goudsmit's evaluation in *ALSOS* (New York, 1947; 2nd ed., 1983). A history of the "Heisenberg affair" and the broader problems of academic physicists in the Third Reich is offered by Alan D. Beyerchen, *Scientists Under Hitler: Politics and the Physics Community in the Third Reich* (New Haven, 1977).

Portrayals of Heisenberg's participation in German wartime nuclear research include Jost Herbig, *Kettenreaktion: Das Drama der Atomphysiker* (Munich, 1976); Robert Jungk, *Heller als tausend Sonnen: Das Schicksal der Atomforscher* (Bern, 1956), translated by James Cleugh as *Brighter Than a Thousand Suns: A Personal History of the Atomic Scientists* (New York, 1958); David Irving, *The German Atomic Bomb: The History of Nuclear Research in Nazi Germany* (New York, 1967, repr. 1983); and Mark Walker, "The German Quest for Nuclear Power, 1939–1949" (Ph.D. dissertation, Princeton University, 1987). Studies of postwar science policy, especially nuclear policy, with reference to Heisenberg include Armin Hermann *et al.*, *History of CERN*, I, *Launching the European Organization for Nuclear Research* (New York, 1987); Joachim Radkau, *Aufstieg und Krise der deutschen Atomwirtschaft 1945–1975* (Hamburg, 1983); Hans Karl Rupp, *Ausserparlamentarische Opposition in der Ära Adenauer: Der Kampf gegen die Atombewaffnung in den fünfziger Jahren* (Cologne, 1970); Thomas Stamm, *Zwischen Staat und Selbstverwaltung: Die deutsche Forschung im Wiederaufbau 1945–1965* (Cologne, 1981); and Kurt Zierold, *Forschungsförderung in drei Epochen: Deutsche Forschungsgemeinschaft. Geschichte, Arbeitsweise, Kommentar* (Wiesbaden, 1968).

Historical and philosophical studies of Heisenberg's physics differ widely in interpretation, methodology, and use of sources. Some of the better-known works include Mara Beller, "The Genesis of Interpretations of Quantum Physics, 1925–1927" (Ph.D. dissertation, University of Maryland, 1983), and "Matrix Theory Before Schrödinger: Philosophy, Problems, Consequences," in *Isis*, **74** (1983), 469–491; Joan Bromberg, "The Impact of the Neutron: Bohr and Heisenberg," in *Historical Studies in the Physical Sciences*, **3** (1971), 307–341; David C. Cassidy, "Cosmic Ray Showers, High Energy Physics, and Quantum Field Theories: Programmatic Interactions in the 1930s," *ibid.*, **12** (1981), 1–39; Olivier Darrigol, "Les débuts de la théorie quantique des champs (1925–1948)" (doctoral dissertation,

University of Paris I [Panthéon-Sorbonne], 1982); Peter Galison, "The Discovery of the Muon and the Failed Revolution Against Quantum Electrodynamics," in *Centaurus*, **26** (1983), 262–316; Patrick A. Heelan, *Quantum Mechanics and Objectivity: A Study of the Physical Philosophy of Werner Heisenberg* (The Hague, 1965); John Hendry, *The Creation of Quantum Mechanics and the Bohr-Pauli Dialogue* (Hingham, Mass., 1984); Herbert Hörz, *Werner Heisenberg und die Philosophie* (Berlin, 1968); Max Jammer, *The Philosophy of Quantum Mechanics: The Interpretations of Quantum Mechanics in Historical Perspective* (New York, 1974); Edward MacKinnon, "Heisenberg, Models, and the Rise of Matrix Mechanics," in *Historical Studies in the Physical Sciences*, **8** (1977), 137–188, and *Scientific Explanation and Atomic Physics* (Chicago, 1982); Jagdish Mehra and Helmut Rechenberg, *The Historical Development of Quantum Theory*, II–IV (New York, 1982–1987); and Daniel Serwer, "*Unmechanischer Zwang*: Pauli, Heisenberg, and the Reception of the Mechanical Atom, 1923–1925," in *Historical Studies in the Physical Sciences*, **8** (1977), 189–256. Further references are in John L. Heilbron and Bruce R. Wheaton, *Literature on the History of Physics in the 20th Century* (Berkeley, 1981).

DAVID C. CASSIDY

HEISKANEN, WEIKKO ALEKSANTERI (*b.* Kangaslampi, Finland, *ca.* 23 July 1895; *d.* Helsinki, Finland, 23 October 1971), *geodesy.*

The ninth and youngest child of Heikki Heiskanen and Riikka Jurvanen, Heiskanen grew up on a small farm in eastern Finland. He was exceptionally energetic, generous in his opinions of others, and of a deep religious faith. In 1922 he married Kaarina Levanto; they had one daughter. From 1933 to 1936 Heiskanen was a member of the Finnish Diet, where he worked to improve the legal status of the Finnish language (Swedish was still the dominant language in some quarters). He also translated and wrote popular works on astronomy with the intent of widening the cultural sphere of his Finnish-speaking countrymen. His work in geodesy was recognized by memberships in seven different academies of science, including the American Academy of Arts and Sciences, and he was an honorary member of the Council of the International Association of Geodesy. The University of Bonn, the Helsinki University of Technology, the University of Uppsala, and Ohio State University awarded him honorary doctorates.

After three years of study, Heiskanen graduated with an M.S. degree from the University of Helsinki in 1919, with top honors in physics, mathematics, astronomy, political economy, and theoretical philosophy. With opportunities more promising in geodesy than in astronomy, his first love, he joined the Finnish Geodetic Institute in 1921 and produced a doctoral dissertation under Ilmari Bonsdorff, "Untersuchungen über Schwerkraft und Isostasie," in 1924. Bonsdorff's own studies on the isostatic equilibrium of the earth's crust may well have provided the initial inspiration that led Heiskanen to become an expert in isostasy and a major figure in physical geodesy.

Heiskanen's computational gravity calculations contributed to the development of methods for the isostatic reduction of gravity measurements, which were then used to compute the undulations of the geoid and eventually a worldwide geodetic system, the Columbus geoid. Outside of geodesy his work was important to the successful inertial guidance of the first United States satellites and missiles, which depended upon detailed knowledge of the gravity field, and important to geophysical studies of the structure of the earth's crust. Heiskanen's scientific career may be divided into three phases: his computations in the 1920's in connection with the international gravity formula; his development and application of the Airy-Heiskanen hypothesis in isostasy in the 1930's and 1940's, most notably associated with the Isostatic Institute in Helsinki; and his program for the creation of a worldwide geodetic system, carried out after 1950 at Ohio State University.

The calculation of the geoid—the equipotential surface of the gravity field—was the central problem of classical geodesy. The reference surface was provided by the ellipsoid approved by the International Association of Geodesy in 1924, and the comparison of observed gravity values with it was based on the international gravity formula approved at the Stockholm congress of the association in 1930. Heiskanen first achieved international recognition when his value for gravity at the equator, $\gamma = 978.049\,(1 + 0.0052884 \sin^2 \phi - 0.0000059 \sin^2 2\phi)$ cm./sec.^{2}, was adopted by the congress as the first term, or idealized constant, of the formula. He had arrived at this figure on the basis of several thousand isostatically reduced gravity stations in all parts of the world.

From the beginning Heiskanen advocated the importance of the isostatic correction of gravity stations. His development of G. B. Airy's less-favored hypothesis, according to which land masses float on a fluid base, like icebergs in the ocean, was to prove immensely fruitful in his own work; and present seismological findings would suggest that the Airy-Heiskanen hypothesis prevails.

In 1928 Heiskanen began a twenty-one-year association with the Helsinki University of Technology. A research group of gifted students quickly gathered around him, and in 1936 that group was officially recognized by the International Association of Geodesy as the Isostatic Institute. Under Heiskanen's guidance a simple cartographic method was developed for computing the topographic-isostatic effect, tables and maps for isostatic reduction were prepared, actual reductions of thousands of stations were carried out, the undulations of the geoid were calculated over wide areas, and the problems of isostatic equilibrium and structure of the earth's crust were studied.

After World War II several forces converged to bring Heiskanen to the United States. It was difficult to get the maps necessary for isostatic research in postwar Finland; and Heiskanen, appreciating the potential of improved gravimeters for extended gravity measurements, dreamed of a truly universal geodetic system that could replace the disparate regional and national systems of the time. Also at that time the United States government needed accurate gravity measurements to provide support to its rocket and satellite programs. In 1950 Heiskanen accepted a post at Ohio State University and for the next fifteen years spent most of the year in America, holding the post of director of the Finnish Geodetic Institute (1949–1962) mostly in absentia. For the first time courses in advanced geodesy were offered at an American university; they were heavily attended by U.S. Air Force officers. The Institute of Geodesy, Photogrammetry and Cartography, founded at Ohio State in 1951, was headed by Heiskanen from 1953 until 1965.

The program for the work at Ohio is laid out in Heiskanen's *On the World Geodetic System* (1951). Gravity measurements from different systems, isostatically reduced, were to be used with George Stokes's formula for computing the undulations of the geoid, and with Felix Vening Meinesz's formulas for computing the absolute deflection of vertical components ξ and η. With this "gravimetric" method the absolute direction of the plumb line could be calculated for every point on earth; and thus every geodetic measurement, even measurements on different continents, could be incorporated into a uniform system. The result of the program, "The Columbus Geoid" (1957), was based on measurements collected from thirty-five countries, reduced for the first time by computer. Ironically, perhaps, this work appeared in the same year as the first satellite observations, which were to pave the way for the era of satellite geodesy. Heiskanen's abilities as a teacher and administrator were such that Ohio State University quickly became the leading center for geodetic education in the United States, and perhaps half of the most prominent geodesists today were once his students.

BIBLIOGRAPHY

I. ORIGINAL WORKS. Heiskanen's "On the World Geodetic System" is in *Suomen geodeettisen laitoksen julkaisuja* ("Publications of the Finnish Geodetic Institute"), no. 39 (Helsinki, 1951), repr. in *Publications of the Ohio State University, Institute of Geodesy, Photogrammetry and Cartography*, no. 1 (Columbus, 1951). His other works include *Gravity Survey of the State of Ohio* (Columbus, 1956), written with U. A. Uotila; "The Columbus Geoid," *Transactions of the American Geophysical Union*, **38**, no. 6 (1957), 841–847; *Size and Shape of the Earth: Symposium Held at the Ohio State University . . . November 13–15, 1956* (Columbus, 1957), as editor; *The Earth and Its Gravity Field* (New York, 1958), written with Felix Vening Meinesz; *Assembly of Gravity Data* (Columbus, 1959); *Symposium on Geodesy in the Space Age . . . February 6–8, 1961* (Columbus, 1961), edited with Simo H. Laurila; and *Physical Geodesy* (San Francisco, 1967), written with Helmut Moritz.

II. SECONDARY LITERATURE. A complete bibliography of Heiskanen's writings is in "The Finnish Geodetic Institute 1918–1968," in *Suomen geodeettisen laitoksen julkaisuja*, no. 65 (1969), 129–138. An account of his life and work is R. A. Hirvonen, in *Proceedings of the Finnish Academy of Sciences and Letters* for 1972 (1974), 75–83.

KATHLEEN AHONEN

HELLAND-HANSEN, BJØRN (*b.* Christiania [now Oslo], Norway, 16 October 1877; *d.* Bergen, Norway, 7 September 1957), *physical oceanography.*

Helland-Hansen's parents were Kristofer Hansen, a stenographer to the Norwegian parliament, and Nikoline Mathilde Helland. He attended the Aars og Vos High School in Christiania, where one of his teachers was Kristian Birkeland, who was also a professor at the Royal Frederik University (later called the University of Christiania).

Helland-Hansen entered the Royal Frederik University with the intention of studying law but soon changed to medicine. He assisted Birkeland at lectures and in early 1898 accompanied him on the first aurora borealis expedition to northernmost Norway. During that trip his hands were frozen and his fingers had to be totally or partially amputated. This loss required him to give up a medical career, so he turned to oceanography.

Helland-Hansen entered the young field of ocean-

HELLAND-HANSEN

ography at a significant time. An interest in the erratic movements of food fishes had led to research support by the governments of the Scandinavian nations. The programs, which involved biologists, meteorologists, and physicists, led to considerable international scientific exchange. In 1899, with the encouragement of Johan Hjort, director of the Norwegian Board of Sea Fisheries, Helland-Hansen studied with the physicist Martin H. C. Knudsen in Copenhagen.

Fridtjof Nansen, at the Royal Frederik University, was invited by Hjort to carry out hydrographic studies on the first cruise of the *Michael Sars* in the Norwegian Sea (1900). Nansen took Helland-Hansen as his assistant, thereby beginning a lifelong association. From 1900 to 1906 Helland-Hansen was an employee of the Norwegian Board of Sea Fisheries and participated in several expeditions of the *Michael Sars* with Nansen. He had obtained a sample of standard water (of known salinity) from Knudsen for purposes of comparison, and the scientists made remarkably precise measurements of temperature and salinity from which they could plot the positions of specific water masses. In 1909 Nansen and Helland-Hansen published a monograph on the Norwegian Sea, a classic in physical oceanography.

The publication of Vilhelm Bjerknes on the circulation of fluids of different densities (1898) was followed by that of Nansen's assistant Vagn Walfrid Ekman on wind-driven ocean currents (1902). From these Helland-Hansen and Johan Sandström began working out equations for calculating ocean currents that were based on both the varying density of seawater and the Coriolis force; these became standards in the field.

Helland-Hansen never received a college degree. In 1902 he married Anna-Marie Krag, the daughter of a Lutheran church official in Denmark. They had six children.

From 1906 to 1921 Helland-Hansen directed the biological station at the Bergen Museum, and from 1914 to 1946 he was also professor of oceanography at that museum (which became the University of Bergen). For the biological station he designed and had built the seventy-six-foot vessel *Armauer Hansen*, launched in 1913, which was in service for more than forty years. Helland-Hansen participated in several expeditions on that ship with Ekman, using the latter's current meter, in the Norwegian Sea in 1923 and 1924, and in the area from the Canary Islands to Portugal and Gibraltar in 1930.

In later years Helland-Hansen considered that his major contribution to oceanography had been "the demonstration that work at sea can be undertaken from a small vessel which can be operated inexpensively." However, his scientific accomplishments included significant papers on adiabatic temperature changes in the sea, on variations of the Gulf Stream and the effect of the Gulf Stream on the Norwegian Sea, and on temperature-salinity relationships in the upper layers of the ocean.

At Helland-Hansen's urging the Bergen Museum established the Geophysical Institute, which he directed from 1917 until his retirement in 1947. It included a chair in hydrography, one in dynamical meteorology (held by Bjerknes), and one in geomagnetism and cosmic physics. Helland-Hansen also was instrumental in the founding of the Norwegian Geophysical Commission and the Norwegian Geophysical Association, both in 1917. He was active in the International Council for the Exploration of the Sea and especially in the International Union for Geodesy and Geophysics, of which he was president from 1945 to 1948. The citizens of Bergen subscribed a new building for the Geophysical Institute in 1928.

Helland-Hansen became closely involved with shipping magnate and statesman Christian Michelsen in the planning of an institute devoted to science and intellectual freedom. When the Christian Michelsen Institute was established in 1930, Helland-Hansen became chairman of the board and director, serving until 1955. The broad scope of its interests occupied most of his later years.

A man of great personal charm, Helland-Hansen was often honored for his scientific achievements and appreciated for his social and intellectual interests.

BIBLIOGRAPHY

I. Original Works. Helland-Hansen's writings include *Current Measurements in Norwegian Fiords, the Norwegian Sea, and the North Sea in 1906* (Bergen, 1908); *Croisière océanographique accomplie à bord de la Belgica dans la mer du Grönland, 1905* (Brussels, 1909); *The Norwegian Sea* (Christiania, 1909), written with Fridtjof Nansen; "Physical Oceanography," in Sir John Murray and Johan Hjort, eds., *Depths of the Ocean* (London, 1912), 210–306; *The Ocean Waters: An Introduction to Physical Oceanography* (Leipzig, 1912), written with Adolph H. Schröder *et al.*; *Temperature Variations in the North Atlantic Ocean and in the Atmosphere* (Washington, D.C., 1920), written with Nansen; *The Eastern North Atlantic* (Oslo, 1926), written with Nansen; and *Bericht über die ozeanographischen Untersuchungen im zentralen und östlichen Teil des Nordatlantischen Ozeans im Frühsommer 1938* (Berlin, 1939), written with Albert Defant.

II. SECONDARY LITERATURE. *Festskrift til professor Bjørn Helland-Hansen* (Bergen, 1956) includes a biography by Olaf Devik and a bibliography. Other biographical accounts, by Hakon Mosby, are in *Journal du Conseil*, **33**, no. 1 (1957), 321–323, and in *Norwegian Academy of Sciences Yearbook* for 1958, 37–43.

ELIZABETH NOBLE SHOR

HELLY, EDUARD (*b*. Vienna, Austria, 1 June 1884; *d*. Chicago, Illinois, 28 November 1943), *mathematics*.

Helly was the only son of Sigmund Helly, a civil servant, and of Sara Necker Helly. He and his sister, Anna, grew up in a sheltered, middle-class home. Helly married Elise Bloch, a mathematician, on 4 July 1921; their only child, Walter Sigmund, became professor of operations research at the Polytechnic Institute of New York in Brooklyn.

Helly entered the Maximilians-Gymnasium in Vienna in 1894, passing his final school examination in 1902; he then took up the study of mathematics, physics, and philosophy at the University of Vienna. Five years later he presented his (handwritten) dissertation, "Beiträge zur Theori der Fredholm'schen Integralgleichung," to the Department of Philosophy, receiving the doctorate on 15 March 1907 (the referees were Wilhelm Wirtinger and Franz Mertens). With the help of a fellowship, Helly then spent two semesters (winter of 1907 to 1908 and summer of 1908) at the University of Göttingen, studying primarily under David Hilbert, Felix Klein, Hermann Minkowski, and Carl Runge.

Upon his return to Vienna, Helly was confronted with the problem of earning a living. He began by giving private lessons in mathematics, and from 1910 on, he also taught at a gymnasium. In 1908 he became a member of the Viennese Mathematical Association (VMA), to which he delivered a total of seventeen lectures at its sessions.

Helly's first paper, "Über lineare Funktionaloperationen," appeared in 1912. His work basically consists of five items: his first work and "Über Systeme linearer Gleichungen . . ." rank as landmarks in the history of functional analysis; "Über Mengen konvexer Körper . . ." and "Über Systeme von abgeschlossenen Mengen . . ." are concerned with his intersection theorem of convex analysis; and his paper "Über Reihenentwicklungen . . ." deals with several convergence criteria for a general class of orthogonal expansions.

Inspired by work of Friedrich Riesz, "Über lineare Funktionaloperationen" is a contribution to the moment problem that played a fundamental role in the development of functional analysis. Using the gliding hump method, Helly gave a first functional analytic proof of a particular case (linear functionals on the space of functions, continuous on a compact interval) of the uniform boundedness principle. Concerning the Hahn-Banach extension theorem, it is his proof, still used in today's courses, by which the matter is extended to one further dimension. This method is to be seen in connection with the particular case $n = 1$ of Helly's intersection theorem: A family of compact, convex sets of the n-dimensional euclidean space possesses a nonempty intersection provided any $n + 1$ of the sets have a common element. Among the many important concepts and assertions developed in Helly's first work is his selection theorem, attributed to the foundations of real analysis and probability theory (see Wintner). Given a set of functions that are bounded and of bounded variation, both uniformly on a compact interval, one may select a subsequence that converges pointwise to a limit function of bounded variation.

In 1913 Helly presented his intersection theorem in a VMA lecture. Further projects he had announced were put aside with the outbreak of World War I. Helly volunteered for the army in 1914, was called up in 1915, and was shot in the chest while serving on the Austrian-Russian front in September 1915. Subsequently taken prisoner, he was deported to eastern Siberia and not released until 1920.

Helly immediately resumed his studies, his goal being the *Habilitation*. Titles of two lectures he delivered at VMA meetings in 1914 ("Über unendliche Gleichungssysteme und lineare Funktionaloperationen"; "Einiges Geometrische über den Raum von unendlich vielen Dimensionen") indicate that he had a concrete conception of his *Habilitationsschrift* even before the war. In fact, he presented his thesis, "Über Systeme linearer Gleichungen," early in 1921 (the referee was Hans Hahn). Emphasizing connections with Minkowski's work, Helly studied general sequence spaces and included an axiomatic introduction of normed linear spaces that parallels the treatments given by Stefan Banach and Norbert Wiener (see Bernkopf, p. 67).

Helly was appointed *Privatdozent* at the University of Vienna in August 1921, remaining in this rank throughout his time in Vienna. Because the position was unpaid, he had to earn a living outside the university. From 1921 to 1929 he was employed in a bank; when it failed in 1929, he became an actuary at the Viennese life insurance company Phönix from 1930 to 1938. This may be the reason why Helly wrote only two further papers: his proof in 1923 of his intersection theorem and a long paper in 1930 in which he showed that the intersection theorem is a particular case of a general, purely topological theorem. (Recent concepts include Helly number, Helly space, and Helly hypergraph [graph theory].) In fact, entry 52A35 in *Subject Classification Scheme 1979* of the American Mathematical Society is devoted to "Helly type theorems."

After the occupation of Austria in 1938, Helly could not teach the course he had announced for the summer semester; he also was dismissed by the insurance company because he was a Jew. In September 1938 the Hellys emigrated to the United States. The first years were difficult ones. Helly was a lecturer at several small colleges in New Jersey, even though he was recommended to more prominent schools by Oswald Veblen, Hermann Weyl, and Albert Einstein. But he was relatively old, had not held a "regular" university position in Europe, and was one of many highly qualified immigrants seeking a position in the United States. His situation seemed to improve in September 1943, when he was appointed a visiting lecturer at the Illinois Institute of Technology in Chicago. He did not long enjoy this first more substantial university position; he died of heart failure two months later.

BIBLIOGRAPHY

I. ORIGINAL WORKS. "Über lineare Funktionaloperationen," in *Österreichische Akademie der Wissenschaften Mathematische-naturwissenschaftliche Klasse*, IIa, **121** (1912), 265–297; "Über Reihenentwicklungen nach Funktionen eines Orthogonalsystems," *ibid.*, 1539–1549; "Über Systeme linearer Gleichungen mit unendlich vielen Unbekannten," in *Monatshefte für Mathematik und Physik*, **31** (1921), 60–91; "Über Mengen konvexer Körper mit gemeinschaftlichen Punkten," in *Jahresbericht der Deutschen Mathematiker-Vereinigung*, **32** (1923), 175–176; and "Über Systeme von abgeschlossenen Mengen mit gemeinschaftlichen Punkten," in *Monatshefte für Mathematik und Physik*, **37** (1930), 281–302.

II. SECONDARY LITERATURE. Michael Bernkopf, "The Development of Function Spaces with Particular Reference to Their Origin in Integral Equation Theory," in *Archive for History of Exact Sciences*, **3** (1966/1967), 1–96; P. L. Butzer, S. Gieseler, F. Kaufmann, R. J. Nessel, and E. L. Stark, "Eduard Helly (1884–1943): Eine nachträgliche Würdigung," in *Jahresbericht der Deutschen Mathematiker-Vereinigung*, **82** (1980), 128–151; P. L. Butzer, R. J. Nessel, and E. L. Stark, "Eduard Helly (1884–1943): In Memoriam," in *Resultate der Mathematik*, **7** (1984), 145–153; Ludwig Danzer, Branko Grünbaum, and Victor Klee, "Helly's Theorem and Its Relatives," in Victor Klee, ed., *Convexity* (Providence, R.I., 1963), 101–180; Jean Dieudonné, *History of Functional Analysis* (Amsterdam and New York, 1981); Eugene Lukacs, *Characteristic Functions*, 2nd ed., rev. and enl. (London, 1970); A. F. Monna, *Functional Analysis in Historical Perspective* (New York, 1973); H. M. Mulder and A. Schrijver, "Median Graphs and Helly Hypergraphs," in *Discrete Mathematics*, **25** (1979), 41–50; I. Netuka and J. Veselý, "Eduard Helly: Konvexita a funkcionální analýza," in *Pokroky mathematiky, fyziky astronomie*, **29** (1984), 301–312; Jan van Tiel, *Convex Analysis* (Chichester, West Sussex, and New York, 1984); and Aurel Wintner, *Spektraltheorie der unendlichen Matrizen* (Leipzig, 1929).

PAUL L. BUTZER
ROLF J. NESSEL
EBERHARD L. STARK

HENNIG, EMIL HANS WILLI (*b*. Dürrhennersdorf, near Zittau, Germany, 20 April 1913; *d*. Ludwigsburg, near Stuttgart, Federal Republic of Germany, 5 November 1976), *entomology, theory of taxonomy*.

The eldest of three sons of Emil Hennig, a railway official, and of Emma Gross Hennig, Willi Hennig attended the elementary school of his native village and learned English, French, and Latin from a local doctor. His secondary education was obtained at the Höhere Schule in Dresden. In 1932 he entered the University of Leipzig to study natural history (zoology, botany, geology); in 1936 he submitted a doctoral dissertation devoted to the genitalia of Diptera. After some months of personal research at the Staatliches Museum für Tierkunde in Dresden, Hennig joined the staff of the Deutsches Entomologisches Institut (DEI) in Berlin in January 1937. He married Irma Wehnert in 1939; they had three sons. Except during the war, he worked at the DEI until August 1961 as scholar, assistant director, and vice-director, publishing a great many of his dipterological writings in its periodicals (*Arbeiten über morphologische und taxonomische Entomologie aus Berlin-Dahlem, Arbeiten über physiologische und angewandte Entomologie aus Berlin-Dahlem, und Beiträge zur Entomologie*).

During the first half of World War II, Hennig was an infantryman in Poland, France, Denmark, and Russia; he was wounded in Russia in 1942. In the following years he served in mosquito control groups in Germany, in Greece, and in Italy, where he was taken prisoner of war. After the war he began the redaction of his *Grundzüge einer Theorie der phylogenetischen Systematik* (published in 1950), the first draft of which was written while he was a prisoner of war.

In August 1961, although working at the DEI in East Berlin, Hennig resided in West Berlin. The erection of the Berlin Wall forced him to resign from the DEI at the very time he was to succeed the retiring director. He remained for two years in West Berlin as a professor at the Technische Universität. During this difficult period he completed the German manuscript of what became *Phylogenetic Systematics* (1966).

In spite of employment offers from foreign institutions, Hennig did not wish to leave Germany. In April 1963 he accepted appointment as director of the Abteilung für Phylogenetische Forschung at the Ludwigsburg branch of the Staatliches Museum für Naturkunde in Stuttgart. He spent the rest of his life there, publishing numerous dipterological contributions in *Stuttgarter Beiträge zur Naturkunde*. Although he found in West Germany all the opportunities for the continuation of his work and for international research travels to Canada, the United States, and Australia, Hennig never forgot Berlin and the Middle European way of science of his youth.

Hennig received an honorary doctorate from the Free University of Berlin. He was honorary professor at the University of Tübingen, member of the Deutsche Akademie Leopoldina der Naturforscher in Halle, foreign member of the Royal Swedish Academy of Sciences, and honorary member of the Society of Systematic Zoology. He received the Fabricius Medal of the Deutsche Entomologische Gesellschaft and the Gold Medal of the American Museum of Natural History (New York) and of the Linnean Society of London.

Although Hennig's first scientific publication (1932) was devoted to the taxonomy of the snake genus *Dendrophis* and was followed by other herpetological contributions, he was nevertheless an entomologist; his first publication on insects, which dealt with the copulatory organs of *Diptera Tylidae*, appeared in 1934. His doctoral dissertation on the copulatory apparatus of the higher Diptera was published in 1936 in *Zeitschrift für Morphologie und Ökologie der Tiere*. The majority of his 150 later publications concern the genitalia, wings, larvae, taxonomy, biogeography, paleontology, and phylogeny of Diptera. Besides original descriptive papers and revisions published in the periodicals of the DEI and of the Stuttgart Museum, this corpus includes monographs on major topics *(Larvenformen der Dipteren)*, regions (New Zealand), or families (in Lindner's *Die Fliegen* . . . he treated fourteen families, including the difficult and enormous Muscidae and Anthomyiidae). In 1973 he revised the Diptera section in Kükenthal's *Handbuch der Zoologie*.

It is impossible to analyze this immense body of achievements here. There are, however, two important points: (1) the profusion of species in a large and homogeneous order of insects with complete metamorphosis, such as the Diptera, was an ideal background for the birth and the testing of an extensional methodology of taxonomy; (2) the discussion among dipterists of Hennig's first ideas (1936) on the taxonomic congruence gave impetus to his research on larvae and on a consequent methodology of classification.

In spite of the impressive volume of his dipterological works, Hennig deserves more attention than would usually be paid to a classical taxonomist. This attention is required by five major books and some preliminary sketches, all devoted to the theory of method in phylogenetic taxonomy.

It is not strictly correct to say that the Hennigian revolution in taxonomy is equivalent to the Darwinian one in biology. Darwin concentrates on the processes of evolution, whereas Hennig, discarding the various models of processes grounded on utility, success, and adaptation, focuses on a taxonomy of results, that is, objects and attributes. Considering without any restriction that living taxa are genealogically linked, he restored the taxonomic prevalence of lineages over characters. This constitutes his improvement on Darwin. Indeed, Hennig refuted the false assumption of an equivalence between the relationships of descent and those of similarity, and refused the phenetic compromise that Darwin accepted as inescapable.

Hennig's requisite for genealogical objectivity of taxa implied an extensional judgement (that is, among numerous taxa) of the phylogenetic recency of the characters; such an appreciation is the inverse of an intensional measure (that is, involving numerous characters) of the phenetic distance between the taxa. This is hardly compatible with the pheneticism of "numerical" taxonomy. However, the conflict is not, as often stated, between qualitative and quantitative appraisals; it is between extensional and intensional ways of thinking.

An extensional appraisal of the relative recency of characters must decide whether they represent apomorphies—that is, evolutive states—proper to some smaller recent taxa or plesiomorphies—that is, evolutive states—shared by many members of large older taxa. This decision makes it possible to construct monophyletic taxa (taxa that contain a given ancestor plus all its codescendants). Taxa that share the same apomorphy are considered *Schwestergruppe* (sister groups). Starting from the smallest of them, the monophyletic arrangement is constructed by continuous chaining of synapomorphies of rising "rank," that is, by nesting of sister taxa of growing magnitude. This method breaks with the extensional method of Ernst Haeckel, which, starting from symplesiomorphies, cannot guarantee the completeness of taxa.

The key terms of Hennig's operational logic number only four: apomorphy, plesiomorphy, sister group, and monophyly. This logic has been called "cladism" by its opponents, a designation rejected by Hennig but easily adopted by his followers.

The reversals in taxonomy from an intensional logic to an extensional one, and from symplesiomorphic constructs to synapomorphic ones were not immediately perceived as a true revolution. Initially, Hennig's methodology was almost totally overlooked; from 1966 to 1975 it stirred considerable opposition from both the process taxonomists (G. G. Simpson, Ernst Mayr, A. J. Cain), who claimed they were the "evolutionary" taxonomists par excellence, and the intensional-pattern taxonomists (P. H. A. Sneath, Robert R. Sokal), who called themselves the true "numerical" taxonomists.

Although he received much recognition from his peers, Hennig was disappointed at the slow reception of his *Grundzüge* of 1950; he regretted the poor English version of his *Phylogenetic Systematics* of 1966, which was printed without his consent; and he suffered from the attacks of his former compatriot Ernst Mayr, to which he replied in 1974. All these disappointments, controversies, and dissatisfactions have been studied in their historical and epistemological details by Dupuis (1979, 1984).

Since 1975, Hennig's methodology has received increasing attention in many countries and fields of biology (zoology, paleontology, botany, molecular biology). This is well illustrated by the fact that the on-line citational bank SCISEARCH, although very incomplete, indicates a growing number of citations for Hennig's major books (1950, 1966): for the years 1974–1976, 79 references; for 1977–1979, 155; for 1980–1982, 158; for 1983–1985, 212. The Willi Hennig Society, established in 1981, publishes the journal *Cladistics*.

The price of the growing success of Hennig's ideas is not small. Saether (1986) denounces "post-Hennigian deviations" that prove to be neither extensional (the various "cladistifications" of intensional procedures) nor even genealogical (the "transformed cladism," born in part from the vicariance biogeography). In such conditions, the edition by Wolfgang Hennig of some of his father's works—particularly the 1961 text of *Phylogenetische Systematik* (1982)—is a well-timed service for biologists.

BIBLIOGRAPHY

I. Original Works. A reasonably complete bibliography of Hennig's publications is in *Beiträge zur Entomologie*, **28** (1978), 169–177. His books or monographs include *Die Larvenformen der Dipteren*, 3 vols. (Berlin, 1948–1952, repr. 1968); *Grundzüge einer Theorie der phylogenetischen Systematik* (Berlin, 1950; repr. Königstein, 1981); *Muscidae*, 2 vols. (Stuttgart, 1955–1964), pt. 63b of Erwin Lindner's *Die Fliegen des paläarktischen Region; The Diptera Fauna of New Zealand*, Petr Wygodsinsky, trans. (Honolulu, 1966); *Phylogenetic Systematics*, D. Dwight Davis and Rainer Zangerl, trans. (Urbana, Ill., 1966; repr. 1979), a revision of *Grundzüge* . . . and a rather poor translation of a German manuscript of 1961, also available in an excellent Spanish translation of the manuscript, *Elementos de una sistemática filogenética* (Buenos Aires, 1968), and in German, edited by Hennig's son Wolfgang, *Phylogenetische Systematik* (Berlin, 1982); *Anthomyiidae* (Stuttgart, 1966–1976), pt. 63a of Erwin Lindners' *Die Fliegen der paläarktischen Region; Die Stammesgeschichte der Insekten* (Frankfurt, 1969), translated by Adrian C. Pont as *Insect Phylogeny* (Chichester, England, and New York, 1981); "Diptera (Zweiflügler)," in Kükenthal's *Handbuch der Zoologie*, IV.2, pt. 2, no. 31 (Lieferung 20) (1973); and *Aufgaben und Probleme stammesgeschichtlicher Forschung* (Berlin and Hamburg, 1984).

Hennig's papers include "Die Schlangengattung *Dendrophis*," in *Zoologischer Anzeiger*, **99** (1932), 273–297, with W. Meise; "Zur Kenntnis der Kopulationsorgane der Tyliden (Micropeziden, Dipt. Acalypt.)," *ibid.*, **107** (1934), 67–76; "Beiträge zur Kenntnis der Kopulationsapparates der cyclorrhaphen Dipteren," in *Zeitschrift für Morphologie und Oekologie der Tiere*, **31** (1936), 328–370; "Beziehungen zwischen geographischer Verbreitung and systematischer Gliederung bei einiger Dipterenfamilien . . . ," in *Zoologischer Anzeiger*, **116** (1936), 161–175; "Ein Beitrag zum Problem der 'Beziehungen zwischen Larven- und Imaginalsystematik,'" in *Arbeiten über morphologische und taxonomische Entomologie aus Berlin-Dahlem*, **10** (1943), 138–144; "Probleme der biologischen Systematik," in *Forschung und Fortschritte*, **21–23** (1947), 276–279; "Zur Klärung einiger Begriffe der phylogenetischen Systematik," *ibid.*, **25** (1949), 136–138; "Kritische Bemerkungen zum phylogenetischen System der Insekten," in *Beiträge zur Entomologie*, **3** (spec. iss., Festschrift Sachtleben) (1953), 1–85; "Phylogenetic Systematics," in *Annual Review of Entomology*, **10** (1965), 97–116; "Kritische Bemerkungen zur Frage 'Cladistic Analysis or Cladistic Classification,'" in *Zeitschrift für zoologische Systematik und Evolutionsforschung*, **12** (1974), 279–294, translated by C. D. Griffiths and edited by G. Nelson as "Cladistic Analysis or Cladistic Classification? A Reply to Ernst Mayr," in *Systematic Zoology*, **24** (1975), 244–256—Mayr's article is "Cladistic Analysis or Cladistic Classification?" in *Zeitschrift für zoologische Systematik und Evolutionsforschung*, **12** (1974), 94–128.

II. SECONDARY LITERATURE. Peter Ax, "Willi Hennig, 20.4. 1913 bis 5.11. 1976," in *Verhandlungen der Deutschen zoologischen Gesellschaft*, **70** (1977), 346–347, and *Das phylogenetische System* (Berlin, 1984), translated by R. P. S. Jefferies as *The Phylogenetic System* (Chichester, England, and New York, 1987); George Byers, "In Memoriam. Willi Hennig (1913–1976)," in *Journal of the Kansas Entomological Society*, **50** (1977), 272–274; Claude Dupuis, "The Hennigo-Cladism: A Taxonomic Method Born of Entomology," in *Abstracts of the Sixteenth International Congress of Entomology* (Kyoto, 1980), 15, "Permanence et actualité de la systématique: La systématique . . . de . . . Hennig," in *Cahiers des naturalistes*, **34** (1979), 1–69 (about 400 references), "La volonté d'être entomologiste . . . ," in *Bulletin de la Société entomologique de France*, **88** (1983), 18–38, "Haeckel or Hennig? The Gordian Knot of Characters, Development, and Procedures in Phylogeny," in *Human Development*, **27** (1984), 262–267, "Willi Hennig's Impact on Taxonomic Thought," in *Annual Review of Ecology and Systematics*, **15** (1984), 1–24, and "Darwin et les taxinomies d'aujourd'hui," in *L'ordre et la diversité du vivant* (Paris, 1986); S. Kiriakoff, "Nécrologie: Willi Hennig (1913–1976)," in *Bulletin et annales de la Société royale belge d'entomologie*, **113** (1977), 240–243; Ole Saether, "The Myth of Objectivity—Post-Hennigian Deviations," in *Cladistics*, **2** (1986), 1–13; D. Schlee, "In Memoriam. Willi Hennig 1903–1976," in *Entomologica germanica*, **4** (1978), 377–391; R. T. Schuh and P. Wygodzinsky, "Willi Hennig," in *Systematic Zoology*, **26** (1977), 104–105; and E. O. Wiley, *Phylogenetics* (New York, 1981).

CLAUDE DUPUIS

HENRI, VICTOR (*b.* Marseilles, France, 6 June 1872; *d.* La Rochelle, France, 21 June 1940), *physics*.

Henri's birth record contains the statement "parents unknown." According to French custom, the child was given an arbitrary surname. In fact, his parents were Russian aristocrats, forbidden to marry by orthodox church law because they were cousins. His uncle was Admiral Alexis Krylov, a famous naval engineer. Krylov's mother was related to the Lyapunovs, the family of the great mathematician.

Henri studied at the German Gymnasium in St. Petersburg and received the *baccalauréat* in Paris in October 1889. He studied mathematics, physics, and chemistry in the Classes Préparatoires aux Grandes Écoles, then at the Sorbonne, where he was enrolled from 1893 to 1894. From 1892 to 1894 he worked in the laboratory of physiological psychology at the Sorbonne under Alfred Binet and attended the psychology course of Théodule Ribot at the Collège de France. He contributed chapters on sensation, memory, and psychometry to Binet's *Introduction à la psychologie expérimentale* (1894). From 1895 to 1904, he was an editor of the journal *Année psychologique*.

From October 1894 to March 1896, Henri was a student in the philosophy department of the University of Leipzig, where he worked under Wilhelm Wundt, and from May 1896 to May 1897 at Göttingen, where he worked under Georg Elias Müller. On 5 June 1897, he presented a dissertation at Göttingen that concerned space perception in the sense of touch, a question little studied by psychophysicists. It was published under the title *Über die Raumwahrnehmungen des Tastsinnes* (1898). At the end of the work, Henri sketched a biological interpretation of the extensional aspects of touch, which in his view should depend on cutaneous physiological structures.

Henri's physiological orientation to experimental psychology is revealed in his monograph *La fatigue intellectuelle* (1898), written with Alfred Binet. The work was undertaken to study the mental strain experienced by students. Numerous physiological parameters—cardiac and respiratory rhythm, blood pressure, body temperature, cutaneous sensitivity—were studied during intellectual work. Significant variations were observed, compared with the resting state.

Pursuing his physiological interests, Henri entered the experimental physiology laboratory of the Sorbonne, headed by Albert Dastre, who had been a pupil of Claude Bernard. Henri became technical assistant to Dastre, who encouraged him to study the physicochemical basis of physiology. Henri began to study quantitative aspects of enzymatic action and went to Wilhelm Ostwald's laboratory of physical chemistry in Leipzig to study catalysis. His science dissertation was presented at the Paris faculty of sciences on 20 February 1903, and published under the title *Lois générales de l'action des diastases*. This work laid the theoretical foundations of enzyme kinetics. Relying on Emil Fischer's concept of a complex between the enzyme and its substrate, Henri treated the formation and dissociation of this complex by the mass action law, thereby deriving fundamental equations. Thus the activity of biological agents was reduced to the basic laws of physical chemistry. Henri opposed vitalist conceptions and the search for laws that would be peculiar to biology. Ten years later, in their classical work on enzyme kinetics, Michaelis and Menten acknowledged Henri's work.

Henri's treatment of enzymatic catalysis is based on the concept that the catalyst acting on an equilibrium reaction does not change the equilibrium value. Thus, if a is the total amount of substrate, x the substrate (or product) combined with the enzyme, $(a - x)$ the free substrate, c the total amount of the enzyme, m the amount of enzyme combined with the substrate, $(c - m)$ the free enzyme, then the mass action law allows the following equation: $(a - x)(c - m) = Km$, where K is the equilibrium constant for the formation of the complex between the enzyme and its substrate. Henri then assumes that the rate of formation of the product x is proportional to the quantity of combined catalyst m, whose value is given by:

$$m = \frac{c(a - x)}{K + a - x}.$$

Thus, the rate of formation of the product $x, \frac{dx}{dt}$, is given by the following equation: $\frac{dx}{dt} = K_1 \frac{c(a - x)}{K + a + x}$, where K_1 is a proportionality constant. The value of x given by integrating this equation combines a logarithmic and a linear factor. In their 1913 paper, Michaelis and Menten used the same basic assumptions but introduced some simplifications and gave a simpler expression for the rate equation.

In 1907 Henri was appointed *maître de conférences* at the Sorbonne, in charge of a course on physical chemistry, and did research in colloid chemistry. He started using absorption spectra of various organic molecules in the infrared, ultraviolet, and X-ray spectra, and he oriented the activity of his research group toward photochemistry, ranging from the basic properties of rays to their chemical and biological actions and their possible technical and industrial applications. The following year Henri went to Jena in order to study new techniques of microscopy at the Carl Zeiss Foundation. For three years he gave a course on photochemistry at the Sorbonne, and worked on a treatise on photochemistry with René Wurmser. During this period he was appointed adjoint director of the physiological laboratory at the École Pratique des Hautes Études (1913–1914). In 1910, Henri married his second wife, Pauline Cernovodeanu, a Romanian bacteriologist at the Pasteur Institute who was part of his research group. They were later divorced.

Many results in the photochemistry and ultraviolet spectra of molecules of organic compounds such as hemoglobin were obtained, and work on photosynthesis was envisaged when World War I broke out. In 1915, Henri collaborated in a research program on gas warfare for national defense. In September of that year, the French government sent him to Russia to inform the Russians of the French work and help them organize their chemical industry for military purposes. After the October Revolution, he was given a chair of physiology at the Choniawski University in Moscow (1917–1918) and a laboratory at the Moscow Scientific Institute; he also became the scientific secretary of the Moscow section of the Academy of Sciences Committee for the Study of Natural Resources. Henri resumed his studies on photochemistry, completing calculations, discovering general laws, and formulating hypotheses on the structure of molecules and mechanisms of light absorption. The work, completed in 1918, was published under the title *Études de photochimie* (1919). While in Moscow, he met a cousin, Vera de Lyapunov, the daughter of Princess Elizabeth Khoviansky and of Vasily Lyapunov. Married in 1923, they had four children.

Henri left Moscow, hoping to resume work at the Sorbonne laboratory. But Dastre had died in 1917 and had been replaced by Louis Lapicque. The laboratory was empty, and the position of adjoint director at Hautes Études was not a steady one. In 1920 Henri left Paris for Zurich, where he held a chair of physical chemistry at the university. Among his colleagues were Hermann Weyl, Erwin Schrödinger, and Peter Debye. While there he made his most important discovery, the predissociation of molecules in an activated state, a significant contribution to molecular physics.

Looking carefully at the absorption spectra of gaseous molecules, Henri discovered narrow and wide bands, depending on the frequency of the light used, for the same molecule. Wide bands were obtained with a higher frequency. Henri was deeply puzzled by this difference in the width of the bands, a phenomenon neglected by other spectroscopists, who were more concerned with the position of the bands. Henri was able to give a theoretical interpretation in terms of the stability of molecular structure. Wide bands are the sign of a "predissociated" state of the molecule, a short-lived state of enhanced chemical reactivity. They were attributed to rotatory states of the whole molecule that were not quantified—as they are in the normal state of the molecule. Electronic activated states and atomic vibratory states also were considered in the explanation of molecular spectra. A basic concept of photochemical action was thus established, and was presented in the monograph *Structure des molécules* (1925). Henri continued to speculate on the relationships between predissociation and photochemistry for the rest of his life.

In *Structure des molécules*, Henri proposed more controversial ideas on structural chemistry. From the study of absorption spectra and other experimental evidence, and for reasons of symmetry, he deduced unexpected structural models instead of classical ones for molecules such as benzene and its derivatives. In his view, benzene was a distorted octahedron. This was before the Heitler-London concept of the homopolar chemical bond.

In 1927 Henri and René Wurmser published another important contribution to photochemistry, in which they claimed that Einstein's concept of photochemical equivalence does not hold true for photochemical reactions, since for most of them the quantum yield is not equal to unity. Henri and Wurmser found that Einstein's law is valid only for elementary photochemical processes, activations, predissociations, and dissociations or ionizations of atoms and molecules. But because other energy exchanges take place within the reaction, the kinetics does not follow Einstein's law.

In 1930, Henri left the University of Zurich to head a research laboratory to be established by an oil company near Marseilles. He planned to work on the basic mechanisms of oil cracking. Unfortunately, the project never came to fruition, and Henri accepted a chair of physical chemistry at the University of Liège in December 1931.

At Liège, Henri continued his work on spectroscopy, turning to polyatomic molecules, and began to work on thermal activation, a natural complement to photochemical activation. In 1934, Peter Debye was invited by the Francqui Foundation to give a series of lectures at the University of Liège. During this year, Debye and Henri organized regular meetings, which were a great source of excitement for those who attended. One of Henri's pupils, the neuropharmacologist Zénon Bacq, in his book *Les transmissions chimiques de l'influx nerveux* (1974), showed that during this period Henri's interest in biophysics was as strong as ever. Henri helped Bacq, who was searching for the sympathetic neurotransmitter, to use ultraviolet spectroscopy in order to characterize chemically the molecules being investigated. Henri studied the spectra of hormone and vitamin molecules in order to establish structural features as well as to devise measurement and dosing methods. His Liège lectures were published under the title *Physique moléculaire: Matière et énergie* (1933). He also published numerical tables of molecular spectra and, with William A. Noyes and Fritz London, edited the section on general chemistry of the International Meeting on Physics, Chemistry, and Biology that was held at Paris in 1938.

When World War II broke out, Henri went to Paris to do military research at the French National Center of Scientific Research (CNRS). He left Paris in June 1940, suffering from a lung infection, and died at La Rochelle on the twenty-first of that month.

BIBLIOGRAPHY

I. Original Works. *La fatigue intellectuelle* (Paris, 1898), written with Alfred Binet; *Über die Raumwahrnehmungen des Tastsinnes. Ein Beitrag zur experimentellen Psychologie* (Berlin, 1898); *Lois générales de l'action des diastases* (Paris, 1903); *Cours de chimie physique* (Paris, 1907); "Absorption des rayons ultra-violets et action photochimique," in *Journal de physique théorique et appliquée*, 5th ser., **3** (1913), 305–323, written with René Wurmser; *Études de photochimie* (Paris, 1919); *Structures des molécules*, Publications de la Société de Chimie Physique 12 (Paris, 1925), rev. ed. by H. Brasseur and C. Corin (Liège, 1934); "Le mécanisme élémentaire des actions photochimiques," in *Journal de physique et le radium*, 6th ser., **8** (1927), 289–310, written with René Wurmser; "Structure des molécules et spectres de bandes," in *L'activation et la structure des molécules* (Paris, 1928), 96–105; "Experimentelle Grundlagen der Prädissoziation der Moleküle," in Peter Debye, ed., *Leipziger Vorträge 1931. Molekülstruktur* (Leipzig, 1933), 131–154; *Physique moléculaire. Matière et énergie* (Paris, 1933); "Étude de l'adsorption de CO_2 par différents cokes et de leur pouvoir réducteur," in *Chimie et industrie*, **34** (1935), 1485, written with G. Perlmutter and E. Gevers-Orban; "Étude du spectre d'absorption ultraviolet de la vapeur de pyridine: Relation avec le spectre de Raman," in *Journal de chimie physique*, **33** (1936), 641–665, written with A. Angenot; *Spectres moléculaires. Structure des molecules*, nos. 11 and 12 of *Tables annuelles de constantes et données numériques* (Paris, 1937); *Chimie générale* (Paris, 1938), edited with William A. Noyes and Fritz London; and "The Ultraviolet Absorption Spectra of 1-3 Cyclohexadiene," in *Journal of Chemical Physics*, **7** (1939), 439–440, written with Lucy W. Pickett.

II. Secondary Literature. Léon Brillouin, "Hommage à Victor Henri," in *Volume commémoratif Victor Henri. Contribution à l'étude de la structure moléculaire* (Liège, 1947), vii–ix; Jules Duchesne, "Avant propos," *ibid.*, xi–xiii, "La structure moléculaire," in *Journal de chimie physique et de physicochimie biologique*, **50** (1953), 608–610, "Victor Henri," in Robert Demoulin, ed., *L'Université de Liège de 1936 à 1966. Liber memorialis*, II, *Notices biographiques* (Liège, 1967), 471–477, and "Victor Henri," in *Biographie nationale publiée par l'Académie royale des sciences, des lettres et des beaux-arts de Belgique*, XLII (Brussels, 1981), 346–354; M. Letort, "La cinétique chimique dans l'oeuvre scientifique de Victor Henri," in *Journal de chimie physique et de physicochimie biologique*, **50** (1953), 604–607; B. Rosen, "La prédissociation," *ibid.*, 601–603; and René Wurmser, "La théorie des enzymes," *ibid.*, 611–612.

CLAUDE DEBRU

HERBST, CURT ALFRED (*b.* Meuselwitz, near Altenburg, Thuringia, Germany, 29 May 1866; *d.* Heidelberg, Germany, 9 May 1946), *embryology*.

Herbst was the son of Heinrich Herbst, a manufacturer, and of Henriette Martin. He decided when young to become a biologist. In 1886 and 1887 he studied at Geneva with Carl Vogt, then went to Jena, where Ernst Haeckel was among his teachers. He was awarded the Ph.D. at Jena in 1889 under the sponsorship of Arnold Lang. He passed his examination magna cum laude; his dissertation was a morphological study of a myriapod, *Scutigera coleoptrata*.

Herbst became acquainted with Hans Driesch in 1887 while they were students at Jena. They were very close friends; Richard Goldschmidt describes Herbst as Driesch's "alter ego." Both Herbst and Driesch had independent means, and, slow to settle into academic life, they traveled widely together for a number of years. In 1889 they went to the Mediterranean for the Easter holidays, and from November 1889 through April 1890 they traveled to Ceylon, Java, and India. Herbst spent the summer of 1890 as an assistant to Lang in Jena, then briefly attended the Polytechnical Institute in Zurich to increase his knowledge of chemistry. Then his travels with Driesch continued—throughout Europe, Algeria and Tunis, Palestine, Syria and Greece, Scandinavia, Egypt (twice), and India (twice)—until Driesch married in 1899, and even shortly after. After attending the International Zoological Congress held at Berlin in August 1901, the Driesches and Herbst took the long way around to Naples, traveling through Russia and Turkestan for two months.

The voyagers did not abandon biological investigation for the pleasures of cosmopolitan travel. The latter were interrupted regularly by the performance of experimental investigations by both men on the development of marine eggs, occasionally at Trieste or Rovigno, more often at the zoological station in Naples. Herbst qualified as *Privatdozent* in zoology at Heidelberg in 1901, under the tutelage of Otto Bütschli. After several years as *Privatdozent* there, he became an assistant professor in 1906, and he succeeded Bütschli in the professorial chair of zoology in 1919. The Driesches settled in Heidelberg in 1900 and spent nineteen years there. In 1935 Herbst became emeritus professor, and he remained in Heidelberg for the rest of his life. He was strongly and openly anti-Nazi, and his later years were very difficult.

Herbst was elected to membership in the Heidelberg Academy of Sciences in 1919. From 1914 to 1919 he was a corresponding member of the Kaiser Wilhelm Institute for Biology in Berlin-Dahlem. Beginning around 1914 there was discussion of Herbst's becoming an assistant to Hans Spemann there, but the plans were not carried through.

Herbst was held in great respect by his contemporaries for his inventive studies in experimental embryology, and his work was of great influence in providing impetus for the development of this new form of investigation. In 1891, at Trieste, he began a series of investigations of the development of sea urchin eggs in seawater of altered composition. In 1889 G. Pouchet and Laurent Chabry, in France, had reported that sea urchin larvae developed abnormally in seawater from which calcium had been precipitated out by the addition of potassium or sodium oxalate. They observed that as a result of calcium deficiency in the medium, the skeletons of the larvae were defective or absent, and that when the calcareous skeletal rods were not present, the arms of the pluteus larvae in which they are normally located failed to form.

Pouchet and Chabry postulated a mechanical explanation, and thought that the arms failed to form because they were not pushed out by skeletal rods—as if they were normally pushed out by the skeleton in the way that a portion of glove might be pushed out by the protrusion of a finger. Herbst repeated the experiments in 1891 and confirmed their results, but he interpreted the effect differently, in terms of morphogenetic stimuli. He thought that the arms were absent because of the lack of the stimulus to growth that the rods would have exerted upon the cells that would have formed the arms.

Herbst published these experimental results and his interpretation of them in 1892. He continued to perform experiments that were related to them, directly or indirectly, until 1943. Herbst had diminished the concentration of calcium ions in the seawater in which the embryos were maintained in the experiments described in 1892. He later (1900) extended this investigation to demonstrate that when cleaving sea urchin eggs or larvae were raised in calcium-free seawater, the cells completely separated from one another. At a time when important experiments involved blastomeres' being separated from each other mechanically by shaking, or by other methods that could introduce serious experimental errors, this was a technical advance of great significance; the method remained useful for many decades.

Herbst not only removed ions from seawater, he also added them. He added calcium chloride, as well as other salts, to the seawater in which he maintained sea urchin embryos. He increased the ion concentration of sodium, potassium, or magnesium, which he knew to be normally present in seawater. He also added salts that increased the concentration of ions of lithium, cesium, or rubidium, all elements found in the same column of the periodic table as sodium and potassium.

The most interesting and surprising effects were those of increased lithium ion concentration. Some larvae maintained in seawater to which lithium salts had been added lacked skeletal rods and arms in the pluteus stage. In some of the gastrulas raised in this medium, the layer of the blastula destined to invaginate within the cap that was to become ectoderm, to form the endoderm and the digestive system, instead turned outward to form an empty endodermal bag. These gastrulas, instead of gastrulating, "exogastrulated." Furthermore, in a number of the exogastrulas, the amount of the endoderm was considerably increased at the expense of the ectoderm.

This was a great blow to the doctrine of germ layer specificity, which was then so rigid that it threatened to stifle progress in embryology. Furthermore, since the endoderm is situated at the pole of the egg opposite to that where the ectoderm is located, the use of the lithium ion provided a method of altering the polar gradients of the egg. For decades it was applied, by followers of Herbst, to the most sophisticated morphological, physiological, and biochemical studies of polarity and gradients that embryologists have yet accomplished. Herbst's studies on the effects of altered ionic constitution on development may be considered the first important steps in the origins of modern chemical embryology.

Later in his life Herbst became interested in the effect of such simple chemical substances as carbonic acid and dilute nitric acid on sex determination in *Bonellia*, a marine worm. In one of Hans Spemann's Silliman Lectures on embryonic induction, begun at about the time he was awarded the Nobel Prize (1935), Spemann referred to the work of Herbst on *Bonellia* as supporting the concept of a chemical inductive stimulus. At the same time, Spemann drew attention to the possibility that "the first inductions . . . were discovered, or even suspected, by . . . Herbst . . . in his theoretical discussions on the role of formative stimuli in animal development."

As an outcome of his new interpretation of the results of the Pouchet-Chabry experiments, Herbst published (1894, 1895) two long articles on the significance of stimulus physiology for the causal interpretation of processes of animal development, and in 1901 his monograph on formative stimuli in animal development appeared. In these he extended the tropism theories of such plant physiologists as Julius von Sachs and Wilhelm Pfeffer to causal explanations of the growth and development of plants and animals. He first considered the effects of external stimuli on plant growth, then those of stimuli from one part of an organism upon the growth or development of another.

Herbst acknowledged that it was the pathologist Rudolf Virchow who had introduced the concept of formative stimuli. He adduced much evidence demonstrating the existence and actions of such stimuli, however, and because of his high standing among his contemporaries, his elaboration of the theory was widely known. Also, in his *Analytical Theory of Organic Development* (1894), Driesch had independently devoted considerable attention to contact and chemical induction and to releasing effects in development. Of particular importance was Herbst's double interest in induction as a releasing effect, and in its evocation of the development of something qualitatively new and different from what would have been formed in the absence of the evoking influence. His emphasis on the actions and significance of morphogenetic stimuli provided strong support for the concepts of embryonic induction that have dominated embryological theory ever since.

Except in his association with Driesch, Herbst led a withdrawn life. He was socially aloof and, although amiable, he seemed unapproachable. He had few friends other than Driesch. He never married. In part because of his Saxon accent, he seemed a rather comic figure to his students; nevertheless, they greatly admired him for his knowledge and wisdom. One of Herbst's seminars inspired an interest in experimental embryology on the part of Viktor Hamburger, later a codiscoverer of the nerve growth factor and a figure of enormous influence in the establishment of developmental neurobiology as a separate discipline during the latter half of the twentieth century. Walter Landauer, who under Herbst's guidance wrote his doctoral dissertation on the effect of lithium ions on echinoderm hybridization, came to America in 1924 and became an influential innovator in the application of genetics and biochemistry to the study of vertebrate teratology.

Thus the work of Herbst, in terms of both technical

advance and theoretical insight, greatly influenced the development of embryology and deserves far more attention than it has received.

BIBLIOGRAPHY

I. ORIGINAL WORKS. Herbst's monograph is *Formative Reize in der tierischen Ontogenese* (Leipzig, 1901). Articles related to work discussed in this article include "Experimentelle Untersuchungen über den Einfluss der veränderten chemischen Zusammensetzung des umgebenden Mediums auf die Entwicklung der Tiere, I. Teil, Versuche an Seeigeleiern," in *Zeitschrift für wissenschaftliche Zoologie*, **55** (1892), 446–518; ". . . II, Weiteres über die morphologische Wirkung der Lithiumsalze und ihre theoretische Bedeutung," in *Mitteilungen der zoologischen Station zu Neapel*, **11** (1893), 136–220; "Über die Bedeutung der Reizphysiologie für die kausale Auffassung von Vorgängen in den tierischen Ontogenese, I," in *Biologisches Centralblatt*, **14** (1894), 657–666, 689–697, 727–744, 753–771, 800–810; "Ueber die Bedeutung der Reizphysiologie für die kausale Auffassung von Vorgängen in der tierischen Ontogenese, II," *ibid.*, **15** (1895), 721–745, 753–772, 792–805, 817–831, 849–855; "Über das Auseinandergehen von Furchungs- und Gewebezellen in kalkfreiem Medium," in *Wilhelm Roux Archiv für Entwicklungsmechanik der Organismen*, **9** (1900), 424–463; "Untersuchungen zur Bestimmung des Geschlechtes, I und II," in *Sitzungsberichte der Heidelberger Akademie der Wissenschaften*, math.-wiss. Kl. (1928), 1–19, and (1929), 1–43; ". . . III," in *Naturwissenschaften*, **20** (1932), 375–379; ". . . IV," in *Wilhelm Roux Archiv für Entwicklungsmechanik der Organismen*, **132** (1935), 337–383; ". . . V," *ibid.*, **134** (1936), 313–330; ". . . VI," *ibid.*, **135** (1936), 178–201; ". . . VII," *ibid.*, **136** (1937), 147–168; ". . . VIII," *ibid.*, **138** (1938), 451–464; ". . . IX," *ibid.*, **139** (1939), 282–302; ". . . X," *ibid.*, **140** (1940), 252–284; and "Die Bedeutung der Salzversuche für die Frage nach der Wirkungsart der Gene," *ibid.*, **142** (1943), 319–378.

II. SECONDARY LITERATURE. Because of the state of disorder in Germany at the time of Herbst's death, no obituaries relating to him were ever published. Thirteen years later a brief, unsigned note about him appeared in *Sitzungsberichte der Heidelberger Akademie der Wissenschaften, Jahreshefte 1943/55* (1959), 41–42. A short article about him by Hans Querner is in *Neue deutsche Biographie*, VIII (Berlin, 1969), 593. Hans Driesch, *Lebenserinnerungen* (Basel, 1951), describes Driesch and Herbst's travels together; a few details about Herbst's life are in Richard B. Goldschmidt, *Portraits from Memory* (Seattle, 1956); and in Georg Uschmann, *Geschichte der Zoologie und der zoologischen Anstalten in Jena 1779–1919* (Jena, 1959). Hans Spemann, *Embryonic Development and Induction* (New Haven, 1938), discusses Herbst's work on 222–224. See also Frederick B. Churchill, "From Machine-Theory to Entelechy: Two Studies in Developmental Teleology," in *Journal of the History of Biology*, **2** (1969), 165–185; and Jane M. Oppenheimer, "Some Diverse Backgrounds for Curt Herbst's Ideas about Embryonic Induction," in *Bulletin of the History of Medicine*, **44** (1970), 241–250.

Valuable information about Herbst's character and personality was provided by Hans Querner and Viktor Hamburger. I learned of the possibility that there was discussion of Herbst's joining Spemann through the kindness of Klaus Sander of Freiburg, who had received photocopies of relevant documents held by the Max-Planck Gesellschaft zur Förderung der Wissenschaften E. V., Bibliothek und Archiv zur Geschichte.

JANE OPPENHEIMER

HESS, HARRY HAMMOND (*b.* New York City, 24 May 1906; *d.* Woods Hole, Massachusetts, 25 August 1969), *geology, geological geophysics, mineralogy, oceanography*.

Hess was the son of Elizabeth Engel Hess and of Julian S. Hess, a member of the New York Stock Exchange. He attended Asbury Park High School in New Jersey, where he failed to distinguish himself academically. Nevertheless, he entered Yale University in 1923, planning to become an electrical engineer. He became bored with electrical engineering, however, and in 1925 switched to geology. He took courses from Alan Bateman, Adolph Knopf, Chester Longwell, and Carl Dunbar. After receiving his B.S. degree in 1927, Hess spent two years in Rhodesia as an exploration geologist for Loangwa Concessions, Ltd., an experience, he later reported, that gave him a profound respect for fieldwork. In 1929 he returned to the United States to attend graduate school.

Hess eventually decided to study at Princeton. His major professors there were A. F. Buddington (petrology), A. H. Phillips (mineralogy), R. M. Field (oceanic structure), and Edward Sampson (mineral deposits). He was quite close to Buddington and, except for Sampson, he eventually coauthored articles with all of them. While still a graduate student he worked with the renowned Dutch geophysicist Felix Vening Meinesz, helping him obtain gravity measurements in the West Indies and the Bahamas. Vening Meinesz taught him the rudiments of geophysics, and they became lifelong friends. Hess obtained the Ph.D. from Princeton in 1932 with a dissertation on serpentinization of a large peridotitic intrusive located in Schuyler, Virginia.

Hess taught at Rutgers during the 1932–1933 academic year and spent several months in 1933 and 1934 at the Geophysical Laboratory of the Carnegie Institution of Washington, then returned to Princeton in 1934 to teach in the geology department. On 15 August 1934 Hess married Annette Burns, daughter of George Plumer Burns, a professor of botany at the University of Vermont; they had two sons.

A reserve officer in the navy at the time of the attack on Pearl Harbor, 7 December 1941, Hess initially was stationed in New York City, where he headed an operation charged with estimating the daily positions of German submarines in the North Atlantic. He volunteered for active sea duty and eventually took over the command of the attack transport U.S.S. *Cape Johnson*. He took part in four major combat landings, and at the close of the war he returned to Princeton with the rank of commander. Hess remained active in the navy reserve and was on call for advice during the Cuban missile crisis, the loss of the submarine *Thresher*, and the *Pueblo* affair. At his death he held the rank of rear admiral.

Except for visiting professorships at Capetown University in South Africa (1949–1950) and Cambridge University (1965), Hess remained at Princeton, chairing the geology department from 1950 to 1966. In 1964 he was appointed to the Blair professorship of geology. Hess received numerous scientific honors, and he devoted a considerable amount of time to scientific organizations. In 1952 he was elected to the National Academy of Sciences, and in 1960 to the American Philosophical Society. He was an honorary foreign member of the Geological Society of London, the Geological Society of South Africa, and the Sociedad Venezolana de Geólogos. In 1966 he received the Penrose Medal of the Geological Society of America and the Feltrinelli Prize of the Accademia Nazionale dei Lincei. He was president of two sections of the American Geophysical Union, Geodesy (1951–1953) and Tectonophysics (1956–1958); of the Mineralogical Society of America (1955); and of the Geological Society of America (1963). In 1969 he was awarded (posthumously) the Distinguished Public Service Award of the National Aeronautics and Space Administration.

Hess was one of the major figures in the American Miscellaneous Society, an informal group of scientists from various fields formed to consider new ideas that might be worth considering. He played a crucial role in bringing to life one of these ideas, the drilling beneath the ocean into the mantle. Labeled "Project Mohole," this project was originally suggested to Hess by Walter Munk in 1957. Hess pushed for the project, and the National Science Foundation supported it from 1958 to 1966. He chaired the panel charged with determining where to drill, and the first core sample was obtained in 1958.

From 1962 until his death, Hess chaired the Space Science Board of the National Academy of Sciences, the function of which was to advise the National Aeronautics and Space Administration on its scientific program. He was chairing a Space Science Board conference at Woods Hole, Massachusetts, organized by him to reformulate the scientific objectives of lunar exploration, when he consulted a doctor about chest pains he had experienced and died in the doctor's office.

Hess was a devoted family man. Although somewhat reserved, he possessed a forceful personality and was known for his courage—as naval officer, department chair, or defender of scientists or ideas. Throughout his career he was not afraid to hypothesize solutions to major problems and was usually the first to find fault with his earlier solutions.

The range of Hess's research accomplishments was extraordinary. He wrote detailed mineralogic studies about pyroxenes, devoted much of his life to the origin and significance of peridotite, and combined his work on peridotite with his treatment of large-scale problems about the origin of island arcs and oceanic trenches, ridges, and crust, in which he continually showed his ability to utilize data from exploratory geophysics to develop hypotheses about their origin. These concerns eventually led to his hypothesis of seafloor spreading—the most important conceptual innovation leading to the plate tectonics revolution in the earth sciences during the late 1960's and early 1970's.

From 1932 through 1938 Hess published a number of papers in which he offered solutions to several diverse problems: the presence and formation of magnetic serpentine belts in island arcs and Alpine-type mountain ranges, the presence of gravity anomalies near trenches in island arc regions, and the formation of island arcs and mountain ranges. By 1937 he had developed a unifying solution to all of these problems. He (along with Vening Meinesz and others) supposed downbuckling of the earth's crust, which resulted in the formation of negative gravity anomolies around trench regions, accompanying island arcs containing serpentinized peridotite intrusions of magma, and eventual alteration of sediments squeezed upward through continued action of the downbuckle. Hess supported this downbuckling hypothesis throughout his early career, and he extended its problem-solving effectiveness in 1940 when he offered an explanation for the generation of the Hawaiian Islands and accompanying swell that was derivative from the downbuckling hypothesis. In 1950 he explicitly coupled the downbuckling hypothesis with convection currents in the earth's mantle in order to explain how the crust could remain downbuckled for an extended period of time, and he utilized the conjunction to explain the rather complicated pattern of deep-foci earthquakes typically located on the continental side of island arcs.

Hess found a way to combine defense of his country with his ongoing research. While commanding the U.S.S. *Cape Johnson*, he managed to take numerous soundings of the Pacific seafloor. Through these soundings he discovered a number of submerged, reefless, flat-topped seamounts, which he named guyots in honor of the Swiss geologist Arnold Guyot, the first professor of geology at Princeton. In 1946 Hess presented an ingenious hypothesis for their development and formation. The unique and puzzling characteristic of guyots is their absence of reefs; consequently he had to construct a solution for their origin and development that would prohibit reef development. Hess argued that guyots originally were Precambrian islands that, by the time lime-secreting organisms had emerged, were too far below sea level for those organisms to survive. Central to his analysis was the claim that sea level has risen with respect to oceanic structures because of the continual deposition of continental sediment upon the seafloor. The sediments have raised the seafloor; and therefore, assuming a relative constancy of oceanic water since the Precambrian era, sea level has risen.

In 1953 and 1955 Hess wrote speculative papers in which he expanded his downbuckling theory; suggested a new solution to the origin of guyots; and devoted much of his attention to the nature, formation, and development of oceanic ridges. All of these pursuits were undertaken in light of his new analysis of the oceanic crust and its layer of covering sediments, which was based upon new studies of the ocean floor by Maurice Ewing and his co-workers at the Lamont Geological Observatory that had begun after the end of World War II. Hess's new model of the oceanic crust, like many others at that time, was as follows: The oceanic crust was taken to be only 5 kilometers thick with an average of 0.7 kilometer of unconsolidated sediments above it, and the mantle was considered to be made of peridotite. Major differences between this model and previous ones, including Hess's former model, were elimination of any granitic material from the oceanic crust and a drastic thinning of the basalt layer along with the consequent raising of the Moho discontinuity. Moreover, the 0.7 kilometer of unconsolidated sediments reduced by a factor of three to five times the size of former estimates, which had been based upon extrapolation of present rates back to the Precambrian.

With this model of the oceanic crust, Hess turned to the problem of the origin of oceanic ridges. The relevant, new ridge data he had to explain were that almost all ridges were associated with basalt volcanism, had no sediments older than Cretaceous atop them, appeared to be ephemeral, and exhibited little folding in their formation. In 1953 Hess, beginning with a thin basaltic crust and peridotite upper mantle, envisioned the formation of a less dense layer of crust through magmatic intrusion of basalt mixed with peridotite. Since the basaltic intrusion would be less dense than the surrounding peridotite, the column would rise as a result of isostatic adjustment. All this could occur without crustal folding; and the resulting surface materials, basalt mixed with peridotite, matched most of the samples that had been collected from the Mid-Atlantic Ridge. The subsequent cooling of the rising magma and cessation of the convection currents provided Hess with a solution to the problem of how ridges decrease in height after initial formation.

In 1955 Hess rejected his 1953 hypothesis for the origin of oceanic ridges. The genesis of this new model is found in his continued concern with serpentinite. Central to his 1955 hypothesis was the reaction

olivine + water = serpentinite + heat.

Both olivine rock and serpentinite are forms of the rock type called peridotite. At temperatures above approximately 500°C the reaction proceeds to the left, while below 500°C it moves to the right. Because the reactive equivalents of serpentinite are less dense than those of olivine, when the reaction proceeds to the right, the resulting serpentinite has a greater volume than the reactive olivine rock. This reaction, consequently, offered a mechanism for the formation and subsequent disappearance of oceanic ridges.

Aware of new data indicating high heat flows in the oceanic crust, Hess applied this reaction to the formation of oceanic ridges. He placed the crucial 500°C isotherm at a depth well below the Moho. Assuming the upward movement of water from the earth's interior, he proposed transformation of olivine peridotite to serpentine peridotite at the 500°C isotherm. With the addition of more water, continued serpentinization would occur; at the same time the whole mass of serpentinized peridotite would rise, if the 500°C isotherm migrated upward.

Hess suggested two possible causes for the isotherm rise: either convective overturn in the mantle or intrusion of basalt. Once the rising isotherm reached the crust, where there was no more olivine, the deserpentinization below, brought about by the rising isotherm, would lead to a net loss of serpentinized material. Rising of the oceanic surface would occur whenever the serpentinization was greater than the deserpentinization, while subsequent lowering of the surface would result when deserpentinization exceeded serpentinization.

By 1953 Hess realized that his former solution to the development of guyots was untenable. The difficulty with his earlier solution was that it utilized a mechanism for submergence of guyots that was twenty-five times too slow. Because of the discovery of Upper Cretaceous shallow-water fossils atop guyots, Hess realized that guyots could not have begun sinking until the Late Cretaceous. This was five times more recent than he had formerly supposed. Moreover, he decided that his former estimate of the rate of sediment deposition upon the seafloor was five times too rapid in light of the new estimates of 0.7 kilometer of unconsolidated sediment upon the seafloor. Rather than revise his former solution, Hess applied the serpentinization reaction to the problem. He proposed that guyots formed above a well of serpentinized peridotite. Once the mass of serpentinite began to deserpentinize, the resulting decrease in volume of the mass would cause the required downward movement of the guyot.

In 1959 Hess further developed his 1955 hypothesis for the origin and development of midocean ridges. He presented a new model of the oceanic crust and continued to recognize the importance of the fact that seismic profiles of accumulated oceanic sediment consistently yielded lower values than those established through extrapolation of present sedimentary rates. When Hess reproposed his solution to the problem of midocean ridge development, he opted for convection currents rather than rising basalts as the driving force behind the serpentinization-deserpentinization transform because of two new discoveries: higher-than-normal heat-flow data over several Pacific ridges and association of the central rift valley along the Mid-Atlantic Ridge with shallow earthquakes.

Hess presented his new model of the oceanic crust in December 1959. It differed in two major respects from his 1953/1955 model. He replaced the five-kilometer basalt layer with a five-kilometer layer of serpentinized peridotite. The overall reason for this switch was that seismic velocity data indicated either basalt or serpentinized peridotite. But the newly discovered fact that this layer of oceanic crust (layer 3), the bottom layer, has uniform thickness indicated to Hess that it had to have been formed on location—it was just too even to have been formed, say, in the mantle and then transported piecemeal to its present location. Hess then suggested that layer 3 had been formed by the serpentinization of peridotite because that would give the uniform thickness, and he now believed that the mantle was partly composed of peridotite. He supposed that ascending water serpentinized the peridotite down to a level of five kilometers, the depth at which he placed the critical 500°C isotherm, in order to account for the five-kilometer thickness of layer 3.

The other major difference between this model and Hess's former one was his more detailed account of the upper two layers. He argued that layer 1, the top layer, was unconsolidated sediment, while layer 2, the middle layer, was consolidated sediment or volcanic rocks or both. He remained puzzled over the lack of seafloor sediment and suggested three solutions without opting for any one of them.

> The most obvious alternatives are: (1) The oceans are relatively young. . . . (2) The pre-Cretaceous sediments have in some manner been removed; for example, by incorporation into the continents by continental drift. (3) Nondeposition of any sediment over much of the ocean floor was a common attribute of the past. ("The AMSOC Hole . . . ," p. 343)

All that he was sure of was that the sedimentary-rock sections of layers 1 and 2 were very incomplete, although he was "rooting against" such a prediction. In the year that followed, Hess chose an option that was not extremely obvious to him in 1959: that the ocean floors but not the oceans are very young, since they are continually being created and destroyed through seafloor spreading. However, in 1959 Hess had not developed his seafloor-spreading hypothesis.

In December 1960 Hess, in a preprint, proposed his seafloor-spreading hypothesis—the name "seafloor spreading" was given to Hess's hypothesis by Robert Dietz, an American earth scientist who, with Hess's preprint in hand, published the first article on seafloor spreading in 1961, one year before Hess's version was published. With this hypothesis Hess became a proponent of continental drift, for he realized that it offered a solution to the number-one problem faced by continental drift: how to move the continents through the seafloor without having them break up.

Hess proposed that the continents do not plow their way through the seafloor, as formerly suggested by Alfred Wegener, the German earth scientist who presented the first detailed account of continental drift during the second decade of the twentieth century, but are carried passively atop the spreading seafloor. (Arthur Holmes, one of the leading British earth scientists of the twentieth century, proposed a hypothesis of ocean basin formation that was a forerunner of Hess's seafloor spreading in the 1930's.)

The central aspect of Hess's hypothesis was its new solution to the problem of the origin and development of midocean ridges. He realized that his 1955/1959 solution was inadequate and that if he were to propose a slightly different solution incorporating many of the same elements, he could avoid difficulties with his former view and even solve the problem of how layer 3 of oceanic crust forms. He also saw that his new hypothesis solved the problem of guyot formation and explained why no sediments on the ocean floor are older than Cretaceous. Hess claimed that young midocean ridges are located on upward-moving convection currents and are the sites for generation of new seafloor. That is, they are where layer 3 of the oceanic crust, composed of serpentinized peridotite, is created—the place where the peridotite is serpentinized.

Hess explained the five-kilometer thickness of layer 3 by positioning the critical 500°C isotherm five kilometers below the surface, arguing that the rising convection currents would elevate the isotherm to such a level. Once the serpentinized peridotite is created, it is forced outward from ridge axes by the movement of parting convection currents. Eventually the convection cell subsides, and the ephemeral ridge disappears. Meanwhile, the outward-moving new seafloor ultimately sinks into the mantle on the backs of descending convection currents. This creates ocean trenches. Continents, pushed along by the convection currents, cease moving when convection stops; and their leading edges become deformed as they impinge upon the downward-moving limbs of convecting material. The continents do not sink into the mantle because of their relatively low density, and sometimes seafloor sediments are metamorphosed and added onto continents at ocean trenches. In addition, since the spreading seafloor moves away from both sides of a ridge axis at equal rates, ridges have a median position between drifting continents. Because of the continual destruction of its material, the seafloor is always young, and sediments are never older than Cretaceous. Guyots form on ridges, where they are truncated by wave erosion. Once guyots move off ridges, they drown themselves well below sea level.

Hess correctly categorized his 1960 preprint as an "essay in geopoetry." By 1966, however, it began to gain acceptance with the confirmation of the Vine-Matthews-Morley hypothesis. This hypothesis, independently proposed in 1963 by Fred J. Vine and Drummond H. Matthews at Cambridge University and by Lawrence W. Morley, a geophysicist working for the Geological Survey of Canada, was a direct corollary of Hess's idea of seafloor spreading and the notion that the geomagnetic field undergoes reversals in polarity. They reasoned that if seafloor spreading occurs and the geomagnetic field reverses its polarity, the seafloor should be made up of strips of alternately normal and reversely magnetized material running parallel to midocean ridges.

Hess learned of the Vine and Matthews version of the Vine-Matthews-Morley hypothesis in late 1963 or early 1964—the article was published in *Nature* in September 1963, and Vine wrote to Hess about the idea in 1964. Hess was quite excited about it. He eventually worked with Vine and Matthews at Cambridge in 1965, and he arranged for Vine to come to Princeton during the 1965–1966 academic year.

Although many aspects of Hess's idea of seafloor spreading have become somewhat problematic, his basic idea that the seafloor is created at ridges and sinks into trenches has become accepted background knowledge in the earth sciences.

BIBLIOGRAPHY

I. ORIGINAL WORKS. Hess published more than 110 monographs, articles, and discussions. The most complete listing of his publications is in Harold L. James, "Harry Hammond Hess," in *Biographical Memoirs. National Academy of Sciences*, **43** (1973), 109–128. His two most important works in mineralogy are "Pyroxenes of Common Mafic Magmas," in *American Mineralogist*, **26** (1941), 515–535 and 573–594, and *Stillwater Igneous Complex, Montana: A Quantitative Mineralogical Study*, Geological Society of America Memoir no. 80 (New York, 1960).

Some key articles, particularly relevant to his work on serpentinization, peridotites, the structure of the seafloor, and his idea of seafloor spreading, are "Interpretation of Gravity-Anomalies and Sounding-Profiles Obtained in the West Indies by the International Expedition to the West Indies in 1932," in *Transactions of the American Geophysical Union, 13th Annual Meeting* (1932), 26–33; "Island Arcs, Gravity Anomalies, and Serpentine Intrusions: A Contribution to the Ophiolite Problem," in *17th International Geological Congress, Report*, II (Moscow, 1937), 263–283; "Geological Interpretation of Data Collected on Cruise of USS *Barracuda* in the West Indies—Preliminary Report," in *Transactions of the American Geophysical Union, 18th Annual Meeting* (1937), 69–77; "A Primary Peridotite Magma," in *American Journal of Science*, **235** (1938), 321–344; and "Gravity Anomalies and Island Arc Structure with Particular Reference to the West Indies," in *Proceedings of the American Philosophical Society*, **79** (1938), 71–96.

Later works are "Drowned Ancient Islands of the Pacific Basin," in *American Journal of Science*, **244** (1946), 772–791; "Comment on Mountain Building," in "Colloquium on Plastic Flow and Deformation within the Earth," *Transactions of the American Geophysical Union*, **32** (1951), 528–531; "Geological Hypotheses and the Earth's Crust under the Oceans," in *Royal Society of London Proceedings*, **A222** (1954), 341–348; "Serpentines, Orogeny, and Epeirogeny," in A. W. Poldervaart, ed. *Crust of the Earth*, Geological Society of America, Special Paper 62 (New York, 1955), 391–407; "The Oceanic Crust," in *Journal of Marine Research*, **14** (1955), 423–439; "Nature of the Great Oceanic Ridges," in *Preprints of the First International Ocean Congress* (Washington, D.C., 1959), 33–34; "The AMSOC Hole to the Earth's Mantle," in *Transactions of the American Geophysical Union*, **40** (1959), 340–345; "History of Ocean Basins," in A. E. J. Engel, Harold L. James, and B. F. Leonard, eds., *Petrologic Studies: A Volume in Honor of A. F. Buddington* (New York, 1962), 599–620; and, with Fred Vine, "Seafloor Spreading," in *The Sea*, vol. 4, part 2, Arthur E. Maxwell, A. C. Bullard, E. Goldberg, and J. L. Worzel, eds. (New York, 1970).

Annette Hess collected many of her husband's private papers and gave them to Princeton. They are presently in the Firestone Library at Princeton University.

II. SECONDARY LITERATURE. Besides the memoir by James, see A. F. Buddington, *Geological Society of America Memorials*, **1** (1973), 18–26; and William W. Rubey, "Harry Hammond Hess (1906–1969)," in *Year Book of the American Philosophical Society* (1970), 126–129. Two articles on the development and presentation of Hess's idea of seafloor spreading and its eventual confirmation are Henry Frankel, "Hess's Development of His Seafloor Spreading Hypothesis," in Thomas Nickles, ed., *Scientific Discovery: Case Studies* (Boston, 1980), 345–366; and "The Development, Reception and Acceptance of the Vine-Matthews-Morley Hypothesis," in *Historical Studies in the Physical Sciences*, **13** (1982), 1–39.

HENRY FRANKEL

HEYTING, AREND (*b.* Amsterdam, Netherlands, 9 May 1898; *d.* Lugano, Switzerland, 9 July 1980), *logic, mathematics.*

Heyting was the eldest child of Johannes Heyting and Clarissa Kok. Both parents were schoolteachers; his father, a man of considerable intellectual gifts, later became principal of a secondary school.

Originally Heyting was to become an engineer, but later it was decided that he should go to the university. In 1916 he enrolled as a student of mathematics at the University of Amsterdam. The funds to pay for his studies were earned by Heyting and his father by supervising the homework of high school students. Two of his teachers at the university, L. E. J. Brouwer and, to a lesser extent, Gerrit Mannoury, shaped and determined his future scientific interests: the greater part of Heyting's work is devoted to intuitionism, Brouwer's philosophy of mathematics, although at certain points his views are closer to the ideas of Mannoury.

After receiving the equivalent of the M.Sc. in 1922, Heyting became a teacher at two secondary schools in Enschede, an industrial town in the eastern Netherlands, far removed from any of the Dutch universities. In his leisure hours he worked on his dissertation, which dealt with the axiomatics of intuitionistic projective geometry. In 1925 he received his doctorate under Brouwer, cum laude.

Heyting's reputation grew rapidly, and in 1937 he was appointed lector at the University of Amsterdam, having been admitted the year before as *Privatdocent*. In 1948 he became a full professor, and in 1968 professor emeritus. In 1942 he was elected a member of the Royal Dutch Academy of Sciences.

Heyting was retiring and modest, lacking all ostentation. His interests were very wide-ranging and varied: music, literature, linguistics, philosophy, astronomy, and botany; he also was fond of walking, cycling, and gardening. As a teacher and lecturer he impressed his students and his international audiences at congresses with his exceptionally clear presentations. In 1929 Heyting married Johanne Friederieke Nijenhuis; they had eleven children. The couple were divorced in 1960, and in 1961 he married Joséphine Frédérique van Anrooy.

In 1927 the Dutch Mathematical Association published a prize question that asked for a formalization of Brouwer's intuitionistic theories. Heyting's answer was awarded the prize early in 1928; a revised and expanded version of his essay was published in 1930. This work made Heyting's name well known among logicians and philosophers of mathematics. It also marked the beginning of a lifelong friendship with Heinrich Scholz at Münster, not far from Enschede, who put his extensive library at Heyting's disposal. Scholz held the only chair of mathematical logic in Germany.

In Brouwer's intuitionism, mathematics consists in the mental construction of mathematical systems, an activity that is supposed to be carried out in the mind of an idealized mathematician, in principle without the use of language; language enters only in attempts to suggest similar constructions to other persons. Something is true in intuitionistic mathematics only if it can be shown to hold by means of a construction.

Brouwer's presentation of his views was deliberately antiformal, in a highly personal style, and often difficult to understand. Heyting's formalization, partially anticipated by V. Glivenko and Andrei N. Kolmogorov, made comparison with formalized traditional mathematics possible. Though Heyting's work has led some into the mistake of identifying intuitionism with his formalization, in the long run the study of intuitionistic formalisms has greatly helped the understanding of the basic intuitionistic concepts.

Around 1930 Heyting also formulated his explanation of the meaning of the intuitionistic logical operations, based on constructive proof or construction as a primitive notion. Though the germs of such an explanation can already be found in Brouwer's writings, this was an important step forward. (Kolmogorov independently gave, in 1932, a closely related interpretation of intuitionistic logic as a calculus of problems.)

Heyting also continued his work, begun with his dissertation, on Brouwer's program of the reconstruction of actual pieces of mathematics along intuitionist lines; in 1941 he published a pioneering paper on intuitionistic algebra, and in the 1950's he investigated the intuitionistic theory of Hilbert space. Some of his students who wrote dissertations under his direction also contributed to the program: J. G. Dijkman (theory of convergence, 1952), B. van Rootselaar (measure theory, 1954), D. van Dalen (affine geometry, 1963), Ashvinikumar (Hilbert space, 1966), A. S. Troelstra (general topology, 1966), and C. G. Gibson (Radon integral, 1967). In the period of his professorship Heyting also published textbooks in projective geometry, one of which was *Axiomatic Projective Geometry* (1963).

Heyting always saw the creation of a better understanding and appreciation of Brouwer's ideas as one of his principal tasks, and thus many of his talks at international meetings and his published writings are devoted to expositions and defense of the basic ideas of intuitionism, in a style that was never dogmatic or polemical.

There are differences in outlook between Brouwer and Heyting, however; in particular, Heyting frankly recognized the formal-theoretical element introduced into intuitionistic mathematics by the (in practice) inescapable use of language, an aspect suppressed in most of Brouwer's writings although Brouwer was aware of it. He also did not share Brouwer's pessimistic views on language as a means of communication, and accordingly he valued positively the use of formalization and axiomatization for intuitionism.

In 1934 Heyting wrote a short monograph titled *Intuitionism and Proof Theory*, a concise and well-written survey in which the viewpoints of intuitionism and formalism are clearly described and contrasted. In 1956 Heyting published his very successful *Intuitionism: An Introduction*, from which many logicians and mathematicians learned about intuitionism. It is certainly in large measure due to Heyting that intuitionism is still very much alive today; without his efforts the "intuitionistic revolution" might well have dwindled away and Brouwer's ideas would have become part of the past.

BIBLIOGRAPHY

I. Original Works. Heyting's writings include "Die formalen Regeln der intuitionistischen Logik," in *Sitzungsberichte der Preussischen Akademie der Wissenschaften*, Phys.-math. Kl. (1930), 42–56; "Die formalen Regeln der intuitionistischen Mathematik II, III," *ibid.*, 57–71, 158–169; *Mathematische Grundlagenforschung: Intuitionismus, Beweistheorie* (Berlin, 1934; repr. 1974), enl. French translation, *Les fondements des mathématiques: Intuitionisme, théorie de la démonstration*, P. Fevrier, trans. (Paris, 1955); "Untersuchungen über intuitionistische Algebra," in *Verhandelingen der Nederlandsche akademie van wetenschappen*, Afd. Natuurkunde, sec. 1, **18**, no. 2 (1941); *Intuitionism: An Introduction* (Amsterdam, 1956; 2nd, rev. ed., 1966; 3rd, rev. ed., 1971); *Axiomatic Projective Geometry* (New York, 1963; 2nd ed., 1980); "Intuitionistic Views on the Nature of Mathematics," in *Synthèse*, **27** (1974), 79–91.

II. Secondary Literature. Information on Heyting's life and work is in A. S. Troelstra, "Arend Heyting and His Contribution to Intuitionism," in *Nieuw archief voor wiskunde*, 3rd ser., **29** (1981), 1–23, and "Logic in the Writings of Brouwer and Heyting," in V. N. Abrusci, E. Casari, and M. Mugnai, eds., *Atti del Convengo internazionale di storia della logica, San Gimignano 4–8 dicembre 1982* (Bologna, 1983), 193–210; and J. Niekus, H. van Riemsdijk, and A. S. Troelstra, "Bibliography of A. Heyting," in *Nieuw archief voor wiskunde*, 3rd ser., **29** (1981), 24–35, with errata *ibid.*, 139.

A. S. Troelstra

HILL, MAURICE NEVILLE (*b.* Cambridge, England, 29 May 1919; *d.* Cambridge, 11 January 1966), *marine geophysics.*

Hill was the younger son of Archibald Vivian Hill, an eminent physiologist and Nobel laureate, and Margaret Keynes, sister of John Maynard Keynes, the Cambridge economist. When he was four, the family moved to Highgate, in North London. Much of Hill's formal education was obtained as a day student at Highgate Junior and Senior Schools. At the age of eleven, however, after an undistinguished performance in the Junior School, he was sent to Avondale, a boarding school at Clifton, near Bristol, for two years. In the Senior School he did well at mathematics and physics, and displayed an aptitude for leadership.

Summer holidays during his boyhood were often spent at Ivybridge, near Plymouth, where Hill's father worked at the Marine Biological Association's (MBA) laboratory. From here Hill and a cousin, Richard Keynes, went to sea in the MBA's ship *Salpa*; it was on these occasions that he began to acquire the skills of seamanship and a love of the sea. In the summer of 1938, on leaving school, Hill sailed to Bermuda on the Royal Society's research ship *Culver* (a ninety-foot ketch) as second officer, an experience that probably turned his thoughts to oceanography.

In October 1938 Hill entered King's College, Cambridge, to read natural sciences: specifically, mathematics, physics, geology, and physiology. Following the outbreak of war, however, he did not return to Cambridge in the fall of 1939, applying instead for a technical post in the Royal Navy. Initially he was posted to H.M.S. *Osprey*, a shore establishment at Portland Bill devoted to antisubmarine warfare. After two years Hill was transferred to the sweeping division of the mine design department at Edinburgh, where he ultimately became a group leader and specialized in countermeasures to the German acoustic homing torpedo. While at Portland he regularly visited family friends who lived nearby and, through them, met his future wife, Philippa Pass. Married in 1944 at Edinburgh, they had three daughters and two sons.

In 1945 Hill returned to Cambridge and did a further year in physics. These two years as an undergraduate enabled him to qualify for a B.A. degree under wartime regulations. In the autumn of 1946 he started as a research student in the department of geodesy and geophysics at Cambridge. His Ph.D. project was to develop a method of making deep-sea seismic observations in order to determine the thickness of the sediments and the nature of the rocks beneath them. Before the war a two-ship method for use in shallow water had been developed, initially by Maurice Ewing in the United States. Hill's great achievement was to devise a single-ship method for use in deep (and shallow) water in which both shots and receivers were placed at comparatively shallow depths in the water column. In Hill's method neutrally buoyant hydrophones were suspended beneath unmoored buoys and the signals transmitted by radio to the ship, where they were recorded. The new technique was first tested from a weather ship about three hundred miles (five hundred kilometers) west of Ireland and was a great success, seismic layers 1 (the sediments), 2, and 3 being recognized for the first time. A disadvantage of the method was that the radio range was limited to about nineteen miles (thirty kilometers) or so, and hence the Moho (or base of the crust) could not be detected. This deficiency was remedied in 1952 when, during work in the northeastern Atlantic and on the shelf around southwestern Britain, a second ship was used to fire shots and lines up to 62.5 miles (100 kilometers) and depths to the Moho were obtained.

In the late 1940's Hill employed the method in the English Channel south of Plymouth, using an MBA vessel. This work provided some of the first evidence of the great thicknesses of Mesozoic and Tertiary sediments that characterize the shelf areas around the British Isles (in contrast with the land area) and provide such important hydrocarbon prospects. This work, together with the initial experiment west of Ireland, constituted the dissertation for Hill's Ph.D., which was awarded early in 1951. In 1949 he had been elected a fellow of King's College and appointed a research assistant in the department of geodesy and geophysics.

From 1950 to 1953 H.M.S. *Challenger* was on a round-the-world scientific cruise during which 48 Hill-type seismic lines were shot, Hill supervising the work personally in the northeastern Atlantic. These lines confirmed the existence of layer 2, which had been somewhat controversial, and revealed that in general layer 1, the sedimentary layer, is very thin, usually one hundred yards to about six-tenths of a mile (one hundred meters to one kilometer).

During the 1950's Hill developed and became adept at deploying a number of other techniques of the new and rapidly expanding science of marine geology and geophysics: notably coring (of sediments), dredging (for hard rocks), and measurements of the geomagnetic field, using the newly developed proton-precession magnetometer. The importance of the latter two techniques in paving the way for the formulation of the Vine-Matthews hypothesis in 1963 cannot be overestimated. The hypothesis maintained that the anomalies in the geomagnetic field observed around midocean ridge crests could be explained in terms of seafloor spreading and reversals of the geomagnetic field. By 1967 this idea had led to confirmation of the theories of seafloor spreading and continental drift, and the formulation of the concept of plate tectonics. By the early 1960's, and as a result of Hill's inspiration and leadership, a large and probably unique collection of rocks had been dredged by the Cambridge group and a considerable amount of magnetic data acquired from the ridges in the northern Atlantic and northwestern Indian oceans.

Thus two researchers in Hill's group, D. H. Matthews and F. J. Vine, had access to new and unpublished information, in part magnetic data from the early phase of the International Indian Ocean Expedition (IIOE), in which the Cambridge group was enthusiastically participating, as well as the crucial knowledge that the measured remanent magnetization of the dredged basalts was very much greater than that of basalts extruded on land. However, as an experimentalist rather than as a theoretician, Hill did not have much time for grand theories such as those of continental drift and seafloor spreading, and there is very little speculation in his papers.

The very considerable contributions of the Cambridge group to the IIOE were presented at a Royal Society discussion meeting organized by Hill and held in November 1964. In the meantime, work in the northeastern Atlantic had not been neglected, and included a magnetic survey of the English Channel and Western Approaches that revealed a Hercynian structural grain trending at right angles to the shelf edge; the measurement of geomagnetic variations from moored buoys on the shelf that documented an unforeseen ocean-edge effect; studies of the median valley of the mid-Atlantic ridge to explore its continuity; and the integrated study of a number of seamounts.

Hill was one of the very small number of research group leaders who in the 1950's developed the techniques of marine geology and geophysics that were to produce such startling results and profound implications for the whole of the earth sciences in the 1960's. He was clearly a dedicated scientist and a born leader. He had a friendly nature and great charm, and was very popular with his research students and associates. He was greatly attached to King's College, to which he contributed as a director of studies in natural sciences, a tutor, and a member of various committees. In 1962 he was elected fellow of the Royal Society; in 1963 he was awarded the Chree Medal of the Physical Society; and in 1965 he was made reader in marine geophysics at Cambridge. Despite all this, in the mid 1960's he began to experience depression and to feel that in some way he was inadequate. In 1966, after a short spell in the hospital, he took his own life.

BIBLIOGRAPHY

A detailed account of the life and work of M. N. Hill is given by E. C. Bullard in *Biographical Memoirs of Fellows of the Royal Society*, **13** (1967), 193–203. This includes a complete bibliography of Hill's writings.

The first three volumes of *The Sea* (New York, 1963), for which he was general editor, were subtitled *Ideas and Observations on Progress in the Study of the Seas*. Volume IV, *New Concepts of Sea Floor Evolution* (New York, 1970), produced under the general editorship of Arthur E. Maxwell, was dedicated to Hill's memory.

F. J. VINE

HÖBER, RUDOLF OTTO ANSELM *(b*, Stettin, Prussia [now Poland], 27 December 1873; *d.* Philadelphia, Pennsylvania, 5 September 1953), *physiology*.

Höber was born into a family of shopkeepers and merchants. His mother, Elise Köhlau, was descended from one of the oldest merchant families in the city. After graduating from the Königlich Mariensstiftsgymnasium in Stettin, Höber enrolled in 1892 at the University of Freiburg im Breisgau, where he attended August Weismann's lectures on zoology and studied chemistry with Baumann. His father's financial difficulties soon forced Höber to leave Freiburg and continue his medical education at the University of Erlangen, where his uncle, the noted physiologist Julius Rosenthal, offered him financial assistance and a room in his home. Höber received his medical degree in 1897. In the following year he accepted a position as assistant at the Physiological Institute in Zurich.

In Zurich, Höber met Josephine Marx, Rosenthal's niece. After their marriage in 1901, she studied medicine and became an accomplished physiologist. In the years prior to her death in 1941, she wrote several papers with her husband. Höber remained in Zurich until 1909, when he moved to the University of Kiel as a *Privatdozent*. In 1915 he was promoted to professor of physiology and director of the Institute for Physiology at Kiel. In addition to his teaching and extensive research and publication activities, Höber was coeditor with Albrecht Bethe of *Pflügers Archiv für die gesammte Physiologie des Menschen und der Tiere* from 1918 to 1933. In the latter years, as a result of the anti-Jewish laws, Höber was forced to surrender his position at Kiel and emigrate from Germany. He settled at the University of Pennsylvania, where he was visiting professor in the department of physiology from 1936 until his death.

Rosenthal's suggestion that he make a detailed study of Walther Nernst's textbook of physical chemistry led Höber to his major scientific achievements. Before completing his studies in Erlangen, he had decided to investigate general physiological processes from the standpoint of physical chemistry. By 1902 the first fruits of his program were recorded in his textbook *Physikalische Chemie der Zelle und Gewebe*, a work that went through six German editions and long remained the standard work on cellular physiology. An English edition of the book was published in 1945.

Höber's scientific research was devoted almost exclusively to investigating physiological transport mechanisms, particularly through elucidating the properties, organization, and functional characteristics of cellular membranes, especially their permeability. The concepts of osmotic pressures, concentration gradients, and ionic and potential gradients derived from physical chemistry provided the basis for this enterprise, but Höber also coupled them with work on colloidal chemistry, a rapidly developing research area at the turn of the twentieth century. Work on the lipoid solubility of various staining agents, as well as microscopic investigation of their differential concentration in tissues and cell components, were part of his investigation of the structure and permeability of cell membranes.

These research tools enabled Höber to propose a modification of Charles Overton's lipoid model for cellular membranes. In order to explain how dissolved substances in the extracellular medium enter the protoplasm, Overton had proposed in 1895 that lipoids are the chief architectural components of cellular membranes and that the rate of penetration of organic compounds is determined by their solubility in the lipoids present in the membranes. Confirmation of this model was provided by experiments on the solubility of certain dyestuffs in a variety of solventlike membranes constructed from different organic fluids in the laboratory, such as solutions of olive oil, oleic acid, and diamylamine, or of cholesterol or lecithin with benzene or oil of turpentine.

Although the solubilities of the dyestuffs tested were comparable with their ability to penetrate and stain organic cells intravitally, Höber showed that the process was more complicated through experiments that revealed a variety of lipoid-insoluble substances that, at variance with the Overton solvent-membrane concept, were nonetheless capable of entering cells at remarkable speeds. He demonstrated that these lipoid-insoluble substances have a low molecular volume and interpreted this as evidence in support of a porous, sievelike architecture of the cell membrane. Substances of high lipoid solubility that easily entered the cell turned out to have high molecular volume. From such experiments Höber proposed a mosaic as the general architecture of cell membranes consisting of at least two kinds of structural elements existing side by side at the surface of the cells: lipoid areas, which can vary in their chemical and physicochemical properties, and porous areas, which can vary in size and shape of the pores. Each occupies a different percentage of the total surface area.

Höber successfully utilized the "lipoid-sieve" model of the membrane in a wide variety of experimental researches elucidating the physiology of glandular secretion. His work on kidney function established the activity of tubular secretion and reabsorption in the lumen and tubular epithelium of the nephron, thereby adding support to the ultrafiltration theory of the formation of the glomerular filtrate first proposed by Carl Ludwig. The lipoid-sieve concept also proved fruitful in Höber's many investigations on the secretory activity of the liver.

Perhaps the most important area in which Höber applied his generalized concept of the membrane was investigations on the generation of bioelectric potentials. Early in his career he contributed significantly to elaborating models of muscle and nerve cell membranes essential to establishing Julius Bernstein's theory of the action potential. Bernstein conceived the action potential wave as a self-propagating depolarization resulting from the local breakdown of selectively permeable membranes surrounding nerve and muscle cells. The mechanisms controlling the transitory local alterations of the cell membranes were assumed to be chemical in nature.

In order to elaborate these mechanisms, Höber studied the effect of local application of neutral inorganic salts on the resting potential of the muscle. This work was later supplemented with experiments on synthetically produced colloid sieve membranes with varying pore sizes. Interfacial changes of ion concentrations in relation to pore size were studied by applying different electrolytes on each side of these artificial membranes. Comparing the results of studies on model membranes with experiments on physiological membranes, Höber hypothesized that as a result of polar changes in the concentration and composition of hydrogen, potassium, calcium ions, the hydroxyl ion, and aromatic sulfonates, the membranes undergo alterations of their colloidal structure that produce condensing or swelling effects combined with changes of resistance due to increased or decreased ion permeability. Following this chemical wave with concomitant loosening of the membrane, potassium ions concentrated in the cell interior would be permitted to diffuse across the membrane, leading to a depolarization of the nerve or muscle.

Höber's willingness to adapt his concepts to new findings is illustrated by doubts he began to harbor toward the end of his career concerning his generalized model of cellular membranes. His lipoid sieve model employed passive transport mechanisms derived from physical chemistry. In his very earliest experimental researches on absorption in the intestines (1898, 1899), however, Höber pointed out that some substances moved against their concentration gradients and that some active transport system might be playing a role, but he did not explore this observation further at the time. He began to consider active transport mechanisms more seriously after experiments done in 1937 established that the absorption of amino acids through the walls of the intestine is in no way proportional to their concentration. Finally, in his publications after the mid 1940's, Höber began to feel that the passive electrostatic mechanisms he had proposed for producing changes in membrane permeability were inadequate; and a decade before Hodgkin and Keynes's revolutionary work on the sodium pump in 1955, he began to stress the need for exploring active, energy-consuming mechanisms for transporting ions in and out of cells that, he speculated, might be fueled by a cycle of phosphorylation and dephosphorylation.

BIBLIOGRAPHY

I. ORIGINAL WORKS. Höber's *Habilitationsschrift* is *Über Konzentrationsänderungen bei der Diffusion zweier gelösten Stoffen gegeneinander* (Zurich, 1899); his inaugural lecture at Zurich is "Über die Bedeutung der Theorie der Lösungen für Physiologie und Medizin," in *Biologisches Zentralblatt*, **19** (1899), 271–285. His books include *Physikalische Chemie der Gewebe* (Leipzig, 1902; 6th ed., 1926), also in English as *Physical Chemistry of Cells and Tissues* (Philadelphia, 1945), with the collaboration of David I. Hitchcock, J. B. Bateman, David R. Goddard, and Wallace O. Finn; and a general textbook, *Lehrbuch der Physiologie des Menschen* (Berlin, 1919; 8th ed., 1939).

In addition to his many research articles, the following offer a general overview of Höber's developing views on the architecture of cellular membranes: "Membranen als Modelle physiologischer Objekte," in *Die Naturwissenschaften*, **24** (1936), 196–202; "Membrane Permeability to Solutes in Its Relation to Cellular Physiology," in *Physiological Review*, **16** (1936), 52–102; "The Influence of Organic Electrolytes and Non-electrolytes upon the Membrane Potentials of Muscle and Nerve," in *Journal of Cellular and Comparative Physiology*, **13** (1939), 195–218, written with Marie Andersh, Josephine Höber, and Bernard Nebel; and "The Membrane Theory," in *Annals of the New York Academy of Sciences*, **47** (1946), 381–392.

Among Höber's works on secretory functions of the kidney and the liver are "Studies Concerning the Nature of the Secretory Activity of the Isolated Ringer-Perfused Frog Liver, 1, Differential Secretion of Pairs of Dyestuffs," in *Journal of General Physiology*, **23** (1939), 185–190; "Studies Concerning the Nature of the Secretory Activity of the Isolated Ringer-Perfused Frog Liver, 2, Inhibitory and the Promoting Influence of Organic Electrolytes and Non-electrolytes upon the Secretion of Dyestuffs," *ibid.*, 191–202, written with Elinor Moore; and "Conditions Determining the Selective Secretion of Dyestuffs by the Isolated Frog Kidney," in *Journal of Cellular and Comparative Physiology*, **15** (1940), 35–46, written with Priscilla M. Briscoe-Woolley.

II. SECONDARY LITERATURE. Biographical material on Höber is in Albrecht Bethe, "Rudolph Höber," in *Pflügers Archiv*, **259** (1954), 1–3. An article on his work is H. Netter, "Rudolph Höbers wissenschaftliches Werk," *ibid.*, 4–13. See also J. C. Poggendorff, *Biographisch-literarisches Handwörterbuch der exacten Naturwissenschaften*, VIIa, pt. 2 (Berlin, 1958), 505.

TIMOTHY LENOIR

HODGE, WILLIAM VALLANCE DOUGLAS (*b.* Edinburgh, Scotland, 17 June 1903; *d.* Cambridge, England, 7 July 1975), *mathematics*.

Hodge came from a solid middle-class background. His father, Archibald James Hodge, was a searcher of records (an office concerned with land titles), and his mother, Janet Vallance, the daughter of a prosperous proprietor of a confectionery business. He was educated at George Watson's Boys College, Edinburgh University (1920–1923, M.A.), and St. John's College, Cambridge (1923–1926, B.A.). On 27 July 1929 he married Kathleen Anne Cameron; they had one son and one daughter. A fellow of the Royal Society since 1938, Hodge was physical secretary between 1957 and 1965 and was a foreign associate of many academies, including the U.S. National Academy of Sciences. He was a vice president of the International Mathematical Union (1954–1958) and master of Pembroke College, Cambridge (1958–1970). He also was responsible for organizing the International Congress of Mathematicians at Edinburgh in 1958. A cheerful and energetic person, he was both popular and effective in his numerous administrative roles.

On leaving Cambridge, where he had already

started to specialize in algebraic geometry, Hodge took up his first teaching appointment at Bristol in 1926, and, influenced by a senior colleague, Peter Fraser, he made strenuous efforts to master the work of the Italian algebraic geometers. In the course of his reading, he soon came across a problem that determined the course of his work for years to come.

In one of his papers, Francesco Severi mentioned the importance of knowing whether a nonzero double integral of the first kind could have all its periods zero. By a stroke of luck, Hodge saw a paper by Solomon Lefschetz in the 1929 *Annals of Mathematics* in which purely topological methods were used to obtain the period relations and inequalities for integrals on a curve. It was clear to Hodge that Lefschetz's methods could be extended to surfaces to solve Severi's problem, and he found it incredible that this should have escaped Lefschetz's notice. It did not take Hodge long to work out the details and write "On Multiple Integrals Attached to an Algebraic Variety." This was the turning point in his career, and therefore it is perhaps appropriate to describe the essence of his proof and its relation to Lefschetz's paper.

If $\omega_1, \cdots, \omega_g$ are a basis of the holomorphic differentials on a curve X of genus g, we have $\omega_i \wedge \omega_j = 0$ for simple reasons of complex dimension. If $[\omega_i]$ denotes the 1-dimensional cohomology class defined by ω_i (that is, given by its periods), it follows that the cup product $[\omega_i] \cup [\omega_j] = 0$. If we spell this out in terms of the periods, we obtain the Riemann bilinear relations for the ω_i. This is the modern approach and the essential content of Lefschetz's 1929 paper, though it must be borne in mind that cohomology had not yet appeared and that the argument had to be expressed in terms of cycles and intersection theory. It was Lefschetz himself who led the developments in algebraic topology that enable us to express his original proof so succinctly.

Similarly, the Riemann inequalities arise from the fact that, for any holomorphic differential $\omega \neq 0$, $i\omega \wedge \overline{\omega}$ is a positive volume and thus $i \int_x \omega \wedge \overline{\omega} > 0$. In cohomological terms this implies in particular that $[\omega] \cup [\omega] \neq 0$, and hence that $[\omega] \neq 0$; in other words ω cannot have all its periods zero. If we now replace X by an algebraic surface and ω by a nonzero holomorphic 2-form (a "double integral," in the classical terminology), exactly the same argument applies and disposes of Severi's question.

In addition to the primitive state of topology at this time, complex manifolds (other than Riemann surfaces) were not conceived of in the modern sense. The simplicity of the proof indicated above owes much to Hodge's work in later years, which made complex manifolds familiar to the present generation of geometers. All this must have had something to do with Lefschetz's surprising failure to see what Hodge saw. In fact, however, this was no simple omission on Lefschetz's part, and he took a great deal of convincing on this point. At first he insisted publicly that Hodge was wrong, and he wrote to him demanding that the paper be withdrawn. In May 1931 Lefschetz and Hodge had a meeting in Max Newman's rooms at Cambridge. There was a lengthy discussion leading to a state of armed neutrality and an invitation to Hodge to spend the next academic year at Princeton. After Hodge had been at Princeton for a month, Lefschetz conceded defeat and, with typical generosity, publicly retracted his criticisms of Hodge's paper. Thereafter Lefschetz became one of Hodge's strongest supporters and fully made up for his initial skepticism. His support proved crucial when in 1936 Hodge was elected to the Lowndean chair of astronomy and geometry at Cambridge.

The publication of "On Multiple Integrals . . ." opened many doors for Hodge. In November 1930 he was elected to a research fellowship at St. John's College, Cambridge, and shortly afterward was awarded an 1851 Exhibition Scholarship. He was thus in a position to take up Lefschetz's invitation to spend a year in Princeton.

In 1931 Princeton was a relatively small university with a very distinguished academic staff. In mathematics, pioneering work was being done in the new field of topology by Oswald Veblen, James W. Alexander, and Lefschetz. Although Hodge never became a real expert in topology, he regularly attended the Princeton seminars and picked up enough general background for his subsequent work.

Lefschetz was, without doubt, the mathematician who exerted the strongest influence on Hodge's work. In Bristol, after his early encounter with Lefschetz's *Annals* paper, Hodge had proceeded to read his Borel tract, *L'analyse situs et la géométrie algébrique*, and was completely won over to the use of topological methods in the study of algebraic integrals. At Princeton, Lefschetz propelled Hodge further along his chosen path. They had frequent mathematical discussions in which Lefschetz's fertile imagination would produce innumerable ideas, most of which would turn out to be false but the rest of which would be invaluable.

Hodge in due course became Lefschetz's successor in algebraic geometry. Whereas the Princeton school inherited and developed Lefschetz's contributions in topology, his earlier fundamental work on the homology of algebraic varieties was somewhat neglected, probably because it was ahead of its time. Hodge's work was complementary or dual to that of Lefschetz, providing an algebraic description of the homology instead of a geometric one. The fact that Lefschetz's theory has now been restored to a central place in modern algebraic geometry is entirely due to the interest aroused by its interactions with Hodge's theory.

Recognizing that he was no longer an expert on algebraic geometry, Lefschetz persuaded Hodge to spend a few months at Johns Hopkins, where Oscar Zariski was the leading light. The visit had a great impact on Hodge's future. In the first place he and his wife formed a close lifelong friendship with the Zariskis. He also was impressed with the new algebraic techniques that Zariski was developing, and in later years he devoted much time and effort to mastering them. Although their technical involvement with algebraic geometry was different, Zariski and Hodge felt a common love for the subject and had serious mathematical discussions whenever they met. They also maintained an intermittent correspondence for more than forty years.

Princeton. By the time Hodge came to Princeton, his mathematical ideas, arising from "On Multiple Integrals . . . ," had already progressed very significantly. In studying integrals in higher dimensions, he soon put his finger on the crucial point. Whereas for the Riemann surface of a curve the number of holomorphic 1-forms is half the first Betti number, there is no corresponding relation in higher dimensions for holomorphic p-forms with $p > 1$. Hodge discussed this point on a number of occasions with Peter Fraser until one day Fraser pointed out Georges de Rham's dissertation, which had just arrived in the Bristol University library. In later years Hodge described this as a stroke of good fortune; although it did not solve his problem, it helped him to see what was involved. In de Rham's theory, valid on any real differentiable manifold, the main result is that there always exists a closed p-form ω with prescribed periods, and that ω is unique modulo derived forms. On a Riemann surface there are natural choices given by the real and imaginary parts of the holomorphic differentials, and Hodge was looking for an appropriate generalization to higher-dimensional algebraic varieties. He saw that the real and imaginary parts of a holomorphic 1-form on a Riemann surface are in some sense duals of one another, and he had a hunch that there should be an analogous duality in general. More precisely, for each p-form ω there should be an $(n\text{-}p)$-form $^*\omega$ (n being the dimension of the manifold); the preferred forms, later called harmonic, would be those satisfying $d\omega = 0$ and $d(^*\omega) = 0$. The main theorem to be proved would be the existence of a unique harmonic form with prescribed periods.

Once established at Princeton, Hodge tried to clarify and develop these vague ideas. He soon realized that the relationship of ω to $^*\omega$ was a kind of orthogonality and was able to make this precise in euclidean space and, more generally, on a conformally flat manifold. He then attempted to prove the existence theorem by generalizing the classical Dirichlet methods. The next stage was to try to remove the restriction of conformal flatness, but for this Hodge needed to become familiar with classical Riemannian geometry. This was his major preoccupation and achievement during his stay in Baltimore. At the same time he also came across a paper by the Dutch mathematician Gerrit Mannoury in which an explicit and convenient metric was introduced on complex projective space, and hence on any projective algebraic manifold. This metric proved of fundamental importance for all of Hodge's subsequent work.

Cambridge, 1932–1939. On returning to Cambridge, Hodge continued his efforts to prove the existence theorem for harmonic forms on a general Riemannian manifold. His first version was, in his words, "crude in the extreme"; Hermann Weyl found it hard to judge whether the proof was complete or, rather, how much effort would be needed to make it complete. Nevertheless, Hodge was convinced that he was on the right track, and his next step was to apply his theory in detail to algebraic surfaces. Using the Mannoury metric, he proceeded to study the harmonic forms and found the calculations much simpler than he had expected. Finally, to his great surprise, he discovered that his results gave a purely topological interpretation of the geometric genus (the number of independent holomorphic 2-forms).

This was a totally unexpected result; and when it was published ("The Geometric Genus of a Surface as a Topological Invariant," 1933), it created quite a stir among algebraic geometers. In particular it convinced even the most skeptical of the importance of Hodge's theory and became justly famous as "Hodge's signature theorem." Twenty years later it played a key role in Friedrich Hirzebruch's work on the Riemann-Roch theorem, and it remains one of the highlights of the theory of harmonic forms. It is intimately involved in the spectacular results (1983–1988) of Simon Donaldson on the structure of four-dimensional manifolds. Essentially Donaldson's work rests on a nonlinear generalization of the Hodge theory.

After his success with the signature theorem, Hodge worked steadily, polishing his theory and developing its applications to algebraic geometry. He also began to organize a connected account of all his work as an essay for the Adams Prize. He was awarded that prize in 1937, but the magnum opus took another three years to complete and finally appeared in book form (*The Theory and Applications of Harmonic Integrals*) in 1941. In the meantime he had published another approach to the existence theorem that had been suggested to him by Hans Kneser. It involved the use of the parametrix method of F. W. Levi and David Hilbert and was, as Hodge said, superior in all respects to his first attempt. Unfortunately this version, reproduced in his book, contained a serious error that was pointed out by Bohnenblust. The necessary modifications to provide a correct proof were made by Hermann Weyl at Princeton and independently by Kunihiko Kodaira in Japan.

Hodge freely admitted that he did not have the technical analytical background necessary to deal adequately with his existence theorem. He was only too pleased when others, better-qualified analysts than himself, completed the task. This left him free to devote himself to the applications in algebraic geometry, which was what really interested him.

In retrospect it is clear that the technical difficulties in the existence theorem required not significant new ideas but a careful extension of classical methods. The real novelty, which was Hodge's major contribution, was in the conception of harmonic integrals and their relevance to algebraic geometry. This triumph of concept over technique is reminiscent of a similar episode in the work of Hodge's great predecessor Bernhard Riemann.

Wartime Cambridge. By the spring of 1940, Hodge had completed the manuscript of his book on harmonic integrals and felt he had exhausted his ideas in that area for the time being. He was therefore looking for a new field. On the other hand, his increasing administrative commitment to Pembroke College and Cambridge University left him less time and energy to devote to mathematics. These two factors help to explain the shift in his interests over the next decade. For some time he had been aware of the powerful algebraic techniques that had been introduced into algebraic geometry by Bartel L. van der Waerden and Zariski. These ideas had had little impact on British geometers, so Hodge felt a duty to interpret and explain the new material to his colleagues. In this he was motivated by a desire to make amends to the Baker school of geometry at Cambridge for the sharp change of direction that his work on harmonic integrals had produced. He thus conceived the idea of writing a book that would replace Henry Baker's *Principles of Geometry*. Although not as demanding as original research, this task soon proved too much for his unaided effort. He therefore enlisted the assistance of Daniel Pedoe, thus beginning a collaborative enterprise that lasted for ten years and led to their three-volume *Methods of Algebraic Geometry*.

Although this book discharged Hodge's obligations to classical algebraic geometry and contained much useful material, it did not achieve its main objective of converting British geometers to modern methods, principally because it was overtaken by events. By the time it appeared, algebraic geometry was exploding with new ideas, and entirely new foundations were being laid. In addition, Hodge did not have the elegance and fluency of style that make algebra palatable. He recognized his limitations as an algebraist, however, and despite his admiration for, and interest in, Zariski's work, he eventually returned to his "first love," the transcendental theory.

Postwar Cambridge. Stimulated by a visit to Harvard in 1950, Hodge directed his research interests back to harmonic integrals. He wrote a number of papers that, though not sensational, were a steady development of his original ideas. In particular his paper "A Special Type of Kähler Manifold" led, a few years later, to Kodaira's final characterization of projective algebraic manifolds. The manifolds singled out by Hodge were, for a few years, known as Hodge manifolds; ironic ally, Kodaira's proof that Hodge manifolds are algebraic led to their disappearance.

Hodge also took an interest in the theory of characteristic classes and wrote a paper ("The Characteristic Classes on Algebraic Varieties") to bridge the gap between the algebraic-geometric classes of John A. Todd and the topological classes introduced by Shiing Shen Chern. He clearly saw the significance of this work at an early stage, and subsequent developments have fully justified him.

The early 1950's saw a remarkable influx of new topological ideas into algebraic geometry. In the hands of Henri Cartan, Jean-Pierre Serre, Kodaira, Donald Spencer, and Hirzebruch, these led in a few years to spectacular successes and the solution of many classical problems, such as the Riemann-Roch theorem in higher dimensions. The great revival of transcendental methods provided by sheaf theory and its intimate connection with harmonic forms naturally aroused Hodge's interest. He made strenuous efforts to understand the new methods and eventually saw that they could be applied to the study of integrals of the second kind. At this time (early 1954) he was busy preparing a talk to be delivered at Princeton in honor of Lefschetz's seventieth birthday, so he suggested that the present author, one of his research students, might try to develop the ideas further and see if they led to a complete treatment of integrals of the second kind. It did not take long to see that one obtained a very elegant and satisfactory theory in this way, and Hodge was given a complete manuscript a few days before his departure for Princeton. He was thus able to describe the results at the Princeton conference ("Integrals of the Second Kind on an Algebraic Variety").

Mathematical Assessment. Hodge's mathematical work centered so much on the one basic topic of harmonic integrals that it is easy to assess the importance of his contributions and to measure their impact. The theory of harmonic integrals can be roughly divided into two parts, the first dealing with real Riemannian manifolds and the second dealing with complex, and particularly algebraic, manifolds. These will be considered separately.

For a compact Riemannian manifold (without boundary), Hodge defined a harmonic form as one satisfying the two equations $d\phi = 0$ and $d^*\phi = 0$, where d is the exterior derivative and d^* its adjoint with respect to the Riemannian metric. An equivalent definition, proposed later by André Weil, is $\Delta\phi = 0$, where $\Delta = dd^* + d^*d$. Hodge's basic theorem asserts that the space $\mathcal{H}^q$ of harmonic q-forms is naturally isomorphic to the q-dimensional cohomology of X (or dual to the q-dimensional homology). The beauty and simplicity of this theorem made a deep impression. As mentioned earlier, Hermann Weyl, the foremost mathematician of the time, was so impressed that he assisted Hodge with the technicalities of the proof. At the International Congress of Mathematicians in 1954, Weyl said that in his opinion, Hodge's *Harmonic Integrals* was "one of the great landmarks in the history of science in the present century."

As an analytical result in differential geometry, one might have expected Hodge's theorem to have been discovered by an analyst, a differential geometer, or even a mathematical physicist (since in Minkowski space the equations $d\omega = d^*\omega = 0$ are simply Maxwell's equations). In fact Hodge knew little of the relevant analysis, no Riemannian geometry, and only a modicum of physics. His insight came entirely from algebraic geometry, where many other factors enter to complicate the picture.

The long-term impact of Hodge's theory on differential geometry and analysis was substantial. In both cases it helped to shift the focus from purely local problems to global problems of geometry and analysis "in the large." Together with Marston Morse's work on the calculus of variations, it set the stage for the new and more ambitious global approach that has dominated much of mathematics ever since.

One of the most attractive applications of Hodge's theory is to compact Lie groups, in which, as Hodge showed in his book, the harmonic forms can be identified with the bi-invariant forms. Another significant application, due to Salomon Bochner, showed that suitable curvature hypotheses implied the vanishing of appropriate homology groups, the point being that the corresponding Hodge-Laplacian Δ was positive definite.

If we turn now to a complex manifold X with a Hermitian metric, we can decompose any differential r-form ϕ in the form

$$\phi = \sum_{p+q=r} \phi^{p,q},$$

where $\phi^{p,q}$ involves, in local coordinates $(z_1, \cdots, z_n)$, p of the differentials dz_i and q of the conjugate differentials $d\bar{z}_i$ (and is said to be of type (p, q)). In general, if $\Delta\phi = 0$, so that ϕ is harmonic, the components $\phi^{p,q}$ need not be harmonic. It was one of Hodge's remarkable discoveries that for the Mannoury metric on a projective algebraic manifold (induced by the standard metric on projective space), the $\phi^{p,q}$ are in fact harmonic. This property of the Mannoury metric is a consequence of what is now known as the Kähler condition: that the 2-form ω (of type (1, 1)) associated to the metric ds^2 by the formula

$$ds^2 = \sum g_{ij} dz_i d\bar{z}_j, \quad \omega = \frac{i}{2\pi} \sum g_{ij} dz_i \wedge d\bar{z}_j$$

is closed (that is, $d\omega = 0$).

As a consequence we obtain a direct sum decomposition

$$\mathcal{H}^r = \sum_{p+q=r} \mathcal{H}^{p,q}$$

of the space of harmonic forms and consequently, by Hodge's main theorem, a corresponding decomposition of the cohomology groups. In particular the Betti numbers $h^r = \dim \mathcal{H}^r(X, C)$ are given by

$$h^r = \sum_{p+q=r} h^{p,q},$$

where $h^{p,q} = \dim \mathcal{H}^{p,q}$. Moreover, as Hodge showed, the numbers $h^{p,q}$ depend only on the complex structure of X and not on the particular projective embedding (which defines the metric).

In this way Hodge obtained new numerical invariants of algebraic manifolds. As the $h^{p,q}$ satisfy certain symmetries, namely

$$h^{p,q} = h^{q,p} \quad \text{and} \quad h^{p,q} = h^{n-p,n-q},$$

various simple consequences are immediately deduced for the Betti numbers. Thus h^{2k+1} is always even, generalizing the well-known property of a Riemann surface but also showing that many even-dimensional real manifolds cannot carry a complex algebraic structure.

The fact that one obtains in this way the new intrinsic invariants $h^{p,q}$ is a first vindication of the Hodge theory as applied to algebraic manifolds. It shows that the apparently strange idea of introducing an auxiliary metric into algebraic geometry does in fact produce significant new information. For Riemann surfaces the complex structure defines a conformal structure, and hence the Riemannian metric is not far away; but in higher dimensions this relation with conformal structures breaks down and makes Hodge's success all the more surprising. Only in the 1950's, with the introduction of sheaf theory, was an alternative and more intrinsic definition given for the Hodge numbers, namely

$$h^{p,q} = \dim H^q(X, \Omega^p),$$

where Ω^p is the sheaf of holomorphic p-forms.

A further refinement of Hodge's theory involved the use of the basic 2-form ω, which is itself harmonic. Hodge showed that every harmonic r-form ϕ has a further decomposition of the form

$$\phi = \sum_s \phi_s \omega^s,$$

where ϕ_s is an "effective" harmonic $(r - 2s)$ form. This decomposition is the analogue of the results of Lefschetz that relate the homology of X to the homology of its hyperplane sections, ω playing the role of a hyperplane.

The theory of harmonic forms thus provides a remarkably rich and detailed structure for the cohomology of algebraic manifolds. This "Hodge structure" has been at the basis of a vast amount of work since the mid-1930's, and it has become abundantly clear that it will in particular play a key role in future work on the theory of moduli.

One problem that Hodge recognized as of fundamental importance is the characterization of the homology classes carried by algebraic subvarieties of X. For divisors (varieties of dimension $n - 1$) this problem had been settled by Émile Picard and Lefschetz, and Hodge saw what the appropriate generalization should be. For many years he attempted to establish his conjecture, but it eluded all his efforts. The conjecture is that a rational cohomology class in $H^{2q}(X, Q)$ is represented by an algebraic subvariety if and only if its harmonic form is of type (q, q). The necessity of this condition is easy; the difficulty lies in the converse, which asserts the existence of a suitable algebraic subvariety. This "Hodge conjecture" has achieved a considerable status, almost on a par with the Riemann hypothesis or the Poincaré conjecture. Its central importance is fully recognized, but no solution is in sight.

BIBLIOGRAPHY

I. Original Works. Hodge's writings include "On Multiple Integrals Attached to an Algebraic Variety," in *Journal of the London Mathematical Society*, **5** (1930), 283–290; "The Geometric Genus of a Surface as a Topological Invariant," *ibid.*, **8** (1933), 312–319; "A Special Type of Kähler Manifold," in *Proceedings of the London Mathematical Society*, 3rd ser., **1** (1951), 104–117; "The Characteristic Classes on Algebraic Varieties," *ibid.*, 138–151; *The Theory and Applications of Harmonic Integrals* (Cambridge, 1941; 2nd ed., 1952; repr. 1959); *Methods of Algebraic Geometry*, 3 vols. (Cambridge, 1947–1954; vol. I repr. 1959; vol. II repr. 1968), written with Daniel Pedoe; and "Integrals of the Second Kind on an Algebraic Variety," in *Annals of Mathematics*, **62** (1955), 56–91, written with Michael F. Atiyah.

II. Secondary Literature. Michael F. Atiyah, "William Vallance Douglas Hodge," in *Biographical Memoirs of Fellows of the Royal Society*, **22** (1976), 169–192, includes an extensive bibliography. Hodge's work is treated in Phillip Griffiths and Joseph Harris, *Principles of Algebraic Geometry* (New York, 1978).

Michael Atiyah

HOFMEISTER, FRANZ (*b.* Prague, Austria-Hungary [now Czechoslovakia], 30 August 1850; *d.* Würzburg, Germany, 26 July 1922), *physiological chemistry, pharmacology*.

Hofmeister was named after his father, a well-respected, prosperous physician in Prague; his mother's maiden name was Anna Hess. During his medical studies at the German university in Prague, Hofmeister showed aptitude in chemistry and was advised by the physiologist Ewald Hering to work with Hugo Huppert, the newly appointed professor of applied medical chemistry. Huppert had been a student of Carl Gotthelf Lehmann, and specialized in clinical chemistry. Upon completing work for the medical degree in 1872, Hofmeister became an assistant in Huppert's laboratory, and in 1879 received his *Habilitation* in physiological chemistry for work done there on peptones. The Prague medical faculty then decided to make Hofmeister head of a new institute for experimental pharmacology, so he had to obtain (in 1881) a *Habilitation* in that subject. In 1883, after six months in Strassburg, where Oswald Schmiedeberg headed the leading German institute of pharmacology, Hofmeister was appointed *ausserordentlicher Professor* of pharmacology at Prague; two years later he was made *ordentlicher Professor*, and in 1887 pharmacognosy was added to his professorial title. In 1896 Hofmeister succeeded Felix Hoppe-Seyler as professor of physiological chemistry at Strassburg; he remained there until 1919, after the reversion of the city to French rule. His final years were spent in Würzburg, where he was appointed honorary professor of physiological chemistry. In 1891 he married Johanna Gröger.

In Huppert's laboratory, Hofmeister showed that the sugar excreted in the urine during pregnancy is not glucose, as had been thought, but lactose. He also examined various analytical methods for the identification of amino acids and for the detection of proteins in biological fluids. He then conducted a series of studies on peptones, the products of the gastric digestion of food proteins. During the latter half of the nineteenth century, many physiologists believed that peptones were the immediate metabolic precursors of blood proteins. In 1882 Hofmeister showed that peptones largely disappear in the intestinal mucosa, but his evidence was not sufficient to disprove the so-called peptone theory; this came in 1901 with the discovery by Otto Cohnheim of the enzyme he named erepsin.

The experimental work for which Hofmeister is most remembered dealt with the precipitation of proteins by inorganic salts. Beginning in 1887, he and his students (Siegmund Lewith, Rudolf von Limbeck, and Egmont Münzer) published a series of papers in which they reported that different salts could be placed in a regular order with respect to their salting-out effect on proteins, the order remaining essentially the same for different proteins. This relationship later came to be known as the "Hofmeister series" or "lyotropic series." Hofmeister noted that the effect of the acidic and basic components of different salts is essentially additive, and he recognized that the differences among them are a general function of their hydration. He did not use the terms "anions" and "cations," however; Svante Arrhenius' theory of electrolytic dissociation was not widely accepted until the 1890's. An important consequence of this systematic study of the salting-out phenomenon was Hofmeister's use of ammonium sulfate (introduced into protein chemistry by Camille Méhu in 1878) to effect the crystallization of egg albumin; this approach to the purification of proteins was actively pursued in his Strassburg laboratory: in the crystallization of human serum albumin (Hans Theodor Krieger, 1899) and of the Bence-Jones protein (Adolf Magnus-Levy, 1900), in the isolation of thyroglobulin (Adolf Oswald, 1899), and in the demonstration by Otto Porges and Karl Spiro (1903) that at least three types of globulins are present in human serum. Further work on the physical chemistry of proteins was largely conducted by Spiro; among his important publications in this field was one in 1914 (with Max Koppel) on buffers. Initially an organic chemist, Spiro became a pharmacologist, and then was associated with Hofmeister until they both left Strassburg in 1919.

In 1902, at the Karlsbad meeting of the Gesellschaft der Deutscher Naturforscher und Ärzte, Hofmeister delivered a plenary lecture on the structure of proteins. This was also the subject of his lengthy review article that appeared shortly afterward in *Ergebnisse der Physiologie*. After considering various earlier proposals, Hofmeister presented several arguments in favor of the view that, in proteins, the amino acid units are joined largely by amide bonds. He attached special significance to the biuret reaction, a color test given by proteins, peptones, and glycine derivatives that had been synthesized during the 1880's by Theodor Curtius. He also cited, in support of his theory, physiological studies on the enzymatic cleavage of proteins and of hippuric acid. At the Karlsbad meeting the renowned organic chemist Emil Fischer, who had recently entered the protein field, espoused the same theory of protein structure, and gave the name "peptide" to the compounds formed by the linkage of amino acids through amide bonds. In subsequent accounts of the history of protein chemistry, these two lectures mark the appearance of the so-called Fischer-Hofmeister peptide theory of protein structure. This theory received strong support from Fischer's extensive work between 1903 and 1909 on the chemical synthesis of peptides, but his success fell short of his hope to synthesize a natural protein.

Although Hofmeister is best known for his personal research and writings on proteins, he also concerned himself with other major biochemical problems. Among them was the question of the role in intracellular metabolism of enzymes, the term given by Wilhelm Friedrich (Willy) Kühne in 1876 to so-called unorganized or soluble ferments such as pepsin. Hofmeister worked in Prague on the newly discovered process of biological methylation and on the chemical pathway of biological urea formation; in 1901 he published a book on the chemical organization of the cell, and discussed the integration of enzyme action in metabolic processes. By that time, many intracellular enzymes had been identified, although some prominent physiologists (for example, Max Verworn and William Dobinson Halliburton) still adhered to the opinion that cellular respiration and biosynthesis are the expression of the activity of a protoplasm not susceptible to chemical dissection. Hofmeister's 1901 book may therefore be considered to reflect the emergence of a modern biochemistry based on the enzyme theory of life. It should be noted, however, that in 1914 Hofmeister wrote another article in which he modified his views in line with the current ideas about the colloidal nature of living matter.

After 1900 Hofmeister made no personal experimental contributions to the subject of his 1901 essay, but during the succeeding decade his junior associates at the Strassburg institute did much important work in this field. From their subsequent accounts it is evident that Hofmeister played a large role in suggesting problems and in guiding the research. Perhaps the greatest single success was the discovery (1904) by Franz Knoop of the β-oxidation pathway for the metabolic breakdown of fatty acids. Shortly afterward (1908) Ernst Friedmann extended Knoop's work to the study of the metabolic breakdown of the carbon chains of amino acids and of the formation of ketone bodies in this process. Other notable achievements included perfusion experiments on carbohydrate metabolism (1904) performed by Gustav Embden and the discovery of aldehyde mutase (1910) by Jacob Parnas; both men later became leaders in the elucidation of the so-called Embden-Meyerhof-Parnas pathway of carbohydrate breakdown in muscle. Among the more important enzyme studies were also those of Otto von Fürth on tyrosinase (1902) and of Julius Schütz on the kinetics of pepsin action (1900). In addition to these contributions, significant work was done on other biochemical problems; especially noteworthy are Fürth's study of adrenaline (1903), the work of Friedrich Bauer on the structure of inosinic acid (1907), and the discovery by Wilhelm Stepp of what later came to be called fat-soluble vitamin A (1909). From 1914 to 1918, Hofmeister contributed to the German war effort through such studies as an investigation of the nutritional value of army bread. He also began personal research on vitamins, and his final publications dealt with the purification of the anti-beriberi factor.

The productivity of Hofmeister's Strassburg institute, especially during the years 1900 to 1910, attests to his role in the development of modern biochemistry and his influence on many young people who were to achieve distinction in the medical sciences. The Germans or Austrians included not only biochemists (Franz Knoop, Gustav Embden, Ernst Friedmann, Otto von Fürth, Wilhelm Stepp), but also pharmacologists (Otto Loewi, Ernst Peter Pick, Alexander Ellinger, Walter Siegfried Loewe), the physiologist Albrecht Bethe, and several clinical investigators (Friedrich Kraus, Gustav von Bergmann, Adalbert Czerny, Adolf Magnus-Levy, Hans Eppinger, and Ernst Freudenberg). There were also many foreign guests, most of them from Japan, Italy, and Russia; in addition to Jacob Parnas from Poland, they included Lawrence J. Henderson from the United States and Henry Stanley Raper from Great Britain. Moreover, in the face of the tension between the German rulers and the French population of Alsace, Hofmeister won the affection of his Alsatian students and made special efforts to promote their interests. Among them were Léon Blum and Gustave Schickelé, who after 1918 became professors of medicine and of obstetrics and gynecology, respectively, at Strassburg.

In 1901 Hofmeister founded the journal *Beiträge zur chemischen Physiologie und Pathologie: Zeitschrift für gesamte Biochemie*. It ceased publication in 1908 when he decided to merge it with *Biochemische Zeitschrift*, established two years before by Carl Neuberg. Hofmeister's journal included many important papers from his own and other biochemical laboratories.

From the foregoing it is evident that Hofmeister's role as the leader of his research group was that of a senior counselor in the independent work of his junior associates on a great variety of scientific problems rather than that of the director of closely supervised teamwork along one, or only a few, lines of research identified with his name. This style of leadership was appropriate in the emergent field of biochemistry, whose many problems came largely from mammalian physiology, and where all available methods, including those of organic chemistry, physical chemistry, and biology, had to be applied to solve these problems. In this respect, Hofmeister imitated his predecessor Hoppe-Seyler, whose laboratory provided the seedbed for the subsequent development of physiological chemistry in many countries.

Hofmeister's personal ambitions appear to have been modest. He had a happy family life and, until shortly before his death, enjoyed good health; there were leisure hours spent in doing watercolors or listening to music. A gracious Austrian, brilliant in conversation, Hofmeister also appears to have been at root a solitary person who regarded the world around him with skepticism, and often with scorn. He seems to have avoided scientific meetings, except when he was invited to present a lecture. He received few public honors, and sought none. Hofmeister's greatness lay not only in his personal research achievements but even more in the fact that, given the opportunity to head an important center of biochemical research, he chose to further his discipline through the education of the next generation of its leaders.

BIBLIOGRAPHY

I. ORIGINAL WORKS. A list of publications from Hofmeister's laboratories is appended to the biographical memoir by J. Pohl and K. Spiro, "Franz Hofmeister: Sein Leben und Wirken," in *Ergebnisse der Physiologie*, **22** (1923), 1–50. Among Hofmeister's scientific articles, of special interest are "Zur Lehre von der Wirkung der Salze, II, Über Regelmässigkeiten in der eiweissfällenden Wirkung der Salze und ihre Beziehung zum physiologischen Verhalten derselben," in *Archiv für experimentelle Pathologie und Pharmakologie*, **24** (1888), 247–260; "Über die Darstellung von kristallisiertem Eieralbumin und die Kristallisierbarkeit kolloider Stoffe," in *Zeitschrift für physiologische Chemie*, **14** (1890), 165–172; *Die chemische Organisation der Zelle* (Brunswick, 1901); "Über Bau und Gruppierung der Eiweisskörper," in *Ergebnisse der Physiologie*, **1** (1902), 759–802; and "Über qualitativ unzureichende Ernährung," *ibid.*, **16** (1918), 1–39, 510–589.

II. SECONDARY LITERATURE. The most important available source of biographical information is the article by Pohl and Spiro (above). Other obituary notices include those by Karl Spiro, in *Archiv für experimentelle Pathologie und Pharmakologie*, **95** (1922), i–vii; and by Gustav Embden, in *Klinische Wochenschrift*, **1** (1922), 1974–1975. See also John Leo Abernethy, "Franz Hofmeister: The Impact of His Life and Research on Chemistry," in *Journal of Chemical Education*, **44** (1967), 177–180; and Joseph S. Fruton, "Contrasts in Scientific Style. Emil Fischer and Franz Hofmeister: Their Research Groups and Their Theory of Protein Structure," in *Proceedings of the American Philosophical Society*, **129** (1985), 313–370.

JOSEPH S. FRUTON

HUBBS, CARL LEAVITT (*b.* Williams, Arizona, 18 October 1894; *d.* La Jolla, California, 30 June 1979), *ichthyology*.

Hubbs was considered one of the last general naturalists and was the dean of American ichthyology for half a century. His father, Charles Leavitt Hubbs, was variously a prospector, assayer, and land developer. Hubbs's early years were spent in San Diego, California, then a city of fewer than twenty thousand people and considerable open country. He became acquainted with the local wildlife and began collecting seashells. After the divorce of his parents in 1907, he and his brother, Leonard, lived with their mother, Elizabeth Goss Johnson Hubbs, who from 1908 ran a private school in Redondo Beach, California. There he devoted his spare time to roaming the beaches and barren hills. When his mother married Frank Newton, the family moved to the San Joaquin Valley near Turlock, California, where he attended high school and, in his words, "plunged into nature study with a vengeance," with a special interest in birds. An indifferent student in his early years, he became a keen one in high school and considered becoming a chemist.

Hubbs's final year of high school was in Los Angeles, and he then attended junior college, where George Bliss Culver interested him in the local freshwater fishes. Hubbs entered Stanford University, then the center of American ichthyology under the leadership of David Starr Jordan and Charles Henry Gilbert; the latter was his mentor. He received his A.B. in 1916 and his M.A. in 1917, by which time he had become assistant curator of fishes, amphibians, and reptiles at the Field Museum in Chicago. In 1918 he married Laura Clark, a fellow student at Stanford who had majored in mathematics. They had one daughter and two sons.

In 1920 Hubbs became curator of the fish division of the museum of zoology at the University of Michigan and an instructor on the faculty. He was awarded a Ph.D. by that university in 1927 and in 1940 was promoted to full professor. In 1944 Hubbs became professor of biology at the Scripps Institution of Oceanography, a graduate school of the University of California in La Jolla. He remained there for the rest of his life, as professor emeritus from 1969.

Among the various honors that Hubbs received were election to the National Academy of Sciences (1952) and the Joseph Leidy Award and Medal of the Academy of Natural Sciences of Philadelphia (1964).

The primary work of this indefatigable scientist was in ichthyology. When he entered it, the study of fishes in the United States was advancing from the descriptive stage to broader aspects of ecology and behavior. Noted from his student days to the end of his life as a keen and meticulous worker, Hubbs was an outstanding taxonomist of fishes. He was also an intense collector who enlarged the holdings at the University of Michigan and the Scripps Institution enormously. While at Michigan, Hubbs and his wife conducted studies on hybridization in various fishes in nature and in the laboratory. In an investigation of geographic variation, Hubbs concluded that differences in temperature during growth could lead to variation in numbers of vertebrae in fishes (1922).

Hubbs wrote many papers on the systematics, distribution, and habits of cyprinodont fishes. Other groups to which he devoted considerable effort were lampreys, catastomid fishes, sculpins, and hagfishes. He also published regional studies, most notably on the fishes of the Great Lakes region (1941) and of California (1979). The latter, a publication that had been in preparation for thirty-five years, was a shortened version of his ideal goal: to record and define taxonomically all fishes in the fresh and marine waters of that state. Hubbs was also involved in fisheries management, with the Michigan Institute for Fisheries Research in the 1930's and with the California Cooperative Oceanic Fisheries Investigation in the 1950's.

One of Hubbs's major interests was the relict fishes, survivors from an earlier epoch of extensive river systems in the western Great Basin. He first observed these in 1915 with John Otterbein Snyder of Stanford while conducting a survey of the fishes of the Bonneville Basin in Utah. He returned to the western basins many times to trace the distribution of surviving fishes and its relationship to the hydrographic history of the region (1948, 1974). He proposed the common name "pupfish" for the genus *Cyprinodon* and pushed successfully for the inclusion of the freshwater spring at Devil's Hole, Nevada, in Death Valley National Monument, in order to save a species of pupfish.

Studies begun on fishes often led Hubbs down other scientific paths. When he moved to California and began collections of local fishes, he soon became involved in directing a program during the 1950's on the total ecology of the kelp beds. His earlier interest in the effects of temperature variations on fishes was continued in California. It led him into a consideration of past climatic changes, derived from measurements of oxygen isotopes in mollusk shells of known ages. In the 1960's he conducted an extensive study of aboriginal middens in Southern California and Baja California, Mexico.

Hubbs's first trip to the islands off the coast of western Mexico in 1946 was to collect fishes in a little-studied area and to determine island endemism. He returned to Guadalupe Island many times, chiefly to record and tally the endangered Guadalupe fur seal and the northern elephant seal. For a decade Hubbs conducted an annual count of gray whales on their migration and at the calving lagoons in Baja California. He became a valued spokesman for the conservation of marine mammals.

Hubbs, an intense worker and prodigious writer, kept many projects going simultaneously. The participation of his wife was an important factor in his accomplishments; she worked with him always, for many years as a volunteer. He devoted considerable time to each of his many graduate students, who were awed by his energy and attention to detail. Hubbs's goals were to carry out scientific research as steadily as he could, and to encourage students and colleagues to do likewise.

BIBLIOGRAPHY

I. Original Works. Hubbs's bibliography runs to more than seven hundred entries, of which half are on fishes. The others are on marine mammals, birds, archaeology, and climatology, plus book reviews and obituaries. His early temperature studies were summarized in "Variations in the Number of Vertebrae and Other Meristic Characters of Fishes Correlated with the Temperature of Water During Development," in *American Naturalist*, **56** (1922), 360–372. His Ph.D. dissertation was "The Structural Consequences of Modifications of the Developmental Rate in Fishes, Considered in Reference to Certain Problems of Evolution," originally published in *American Naturalist*, **60** (1926), 57–81. The first edition of the work on Great Lakes fishes was "Guide to the Fishes of the Great Lakes and Tributary Waters," in *Bulletin of the Cranbrook Institute of Science*, no. 18 (1941), written with Karl F. Lagler; the last edition was *Fishes of the Great Lakes Region* (Ann Arbor, Mich., 1964). The long-pursued California project was "List of the Fishes of California," in *Papers of the California Academy of Sciences*, **133** (1979), 1–51, written with W. I. Follett and Lillian J. Dempster; Hubbs saw it in published form just three weeks before he died.

Work on relict fishes was first summarized by Hubbs and his son-in-law, Robert Rush Miller, in "The Zoological Evidence: Correlation Between Fish Distribution and Hydrographic History in the Desert Basins of Western United States," in *The Great Basin, with Emphasis on Glacial and Postglacial Times*, which is *Bulletin of the University of Utah*, **38** (1948), 17–166. A later summary, which Hubbs called his "opus," was "Hydrographic History and Relict Fishes of the North-Central Great Basin," in *Memoirs of the California Academy of Sciences*, **7** (1974), 1–259, written with Robert Rush Miller and Laura C. Hubbs.

With George S. Bien and Hans H. Suess, Hubbs wrote "La Jolla Natural Radiocarbon Measurements I–V," in *American Journal of Science Radiocarbon Supplement*,

2 (1960), 197–223, **4** (1962), 204–238, **5** (1963), 254–272, **7** (1965), 66–117, and **9** (1967), 261–294 (with Bien only).

Hubbs gave his large personal library and his extensive correspondence and scientific files to Scripps Institution of Oceanography; his files include many biographical and autobiographical items.

II. SECONDARY LITERATURE. An account of Hubbs's life is Elizabeth N. Shor, Richard H. Rosenblatt, and John D. Isaacs, "Carl Leavitt Hubbs," in *Biographical Memoirs, National Academy of Sciences*, **56** (1987), 214–249, which includes a partial bibliography. A very warm account is Kenneth S. Norris, "To Carl Leavitt Hubbs, a Modern Pioneer Naturalist, on the Occasion of His Eightieth Year," in *Copeia* (1974), 581–610, which includes a selected bibliography by Elizabeth N. Shor and a list of doctoral students by Laura C. Hubbs. A complete bibliography is in Frances Hubbs Miller, *The Scientific Publications of Carl Leavitt Hubbs: Bibliography and Index, 1915–1981*, Special Publication no. 1, Hubbs–Sea World Research Institute (La Jolla, 1981).

ELIZABETH NOBLE SHOR

HUENE, FRIEDRICH, FREIHERR VON (*b*. Tübingen, Germany, 22 March 1875; *d*. Tübingen, 4 April 1969), *vertebrate paleontology*.

Over a period of seven decades, Friedrich von Huene decisively deepened and advanced knowledge of fossil reptiles. He was the son of Alexandra, Baroness Stackelberg of Estonia, and Baron Johannes Hoyningen of Livonia. Two younger brothers died at an early age. In 1876 Huene's father became a professor at the Evangelical Divinity School in Basel, Switzerland. Huene attended gymnasium in Basel, passing the final examination in 1895. In his first semester at the University of Lausanne, he studied theology and geology. Under the influence of Eugène Renevier, he began to concentrate on geology. He subsequently spent three semesters at the University of Basel. Attracted by the work of E. Koken, he then studied at the University of Tübingen. After conducting research on Silurian Craniacea, published in 1899, he received the doctorate at Tübingen in 1898. He then completed a geological description, begun in 1896, of the Liestal region in the Jura Mountains of Switzerland.

On 1 April 1899 Huene became an assistant to Koken, who encouraged him to study Saurischia of the Upper Triassic. Huene set about this new assignment in the broadest possible terms, visiting many European museums. In 1902 he submitted his *Habilitation* thesis, a comprehensive overview of Triassic fossil reptiles. At this time he decided to dedicate himself exclusively to research and to treat lecturing as a secondary activity.

On 17 March 1904 Huene married Theodora Lawson; they had four daughters. In 1907–1908 his monograph on the dinosaurs of the European Triassic, considered in light of non-European events, appeared. In it he expounded the position of Harry Govier Seeley (1887) that dinosaurs comprised two independent orders, Saurischia and Ornithischia. In 1908 Huene was promoted to assistant professor, and in 1910 he became a German citizen.

In 1914 Huene began excavations in the Stuben sandstone of Trossingen (southern Germany) that yielded several *Plateosaurus* skeletons. While he was in the midst of this work, World War I broke out. In August 1914 Huene enlisted as a volunteer and became an officer. After seeing service at the front, he assumed the post of military geologist in late 1916. After the war he worked on thecodonts and Sauropterygia. American support enabled him to continue his excavations in Trossingen during the summers of 1921 to 1923. Besides unearthing the *Plateosauri,* he discovered ichthyosaurs of the Lias in Holzmaden.

Huene traveled to Argentina at the conclusion of the summer semester in 1923, in order to study sauropod *(Titanosaurus)* remains. Even more important to him than South America, however, was South Africa, to which he went in the spring of 1924. He collected at many sites in the Karroo region and returned to Tübingen with a great amount of material. Vertebrate finds from Santa Maria in southern Brazil led him to organize a collecting expedition there (1928–1929). These journeys, as well as the scientific treatment of the Karroo and Gondwanaland saurians of Africa and South America, demanded strenuous efforts that were crowned in August 1939 with the opening of a new hall in the Tübingen museum. Huene also unearthed many other paleontological finds before World War II. In 1936 he discovered a specialized placodont, *Henodus,* from the Upper Triassic in the vicinity of Tübingen. After that he undertook to shed light on the osteology and systematics of the *Mesosaurus* (1940). Huene's long and intensive preoccupation with fossil reptiles led him early on to investigate phylogenetic relationships. He brought together his fifty-five years of experiences in *Paläontologie und Phylogenie der niederen Tetrapoden* (1956). It is interesting to compare his classification with contemporary classifications, as A. S. Romer has done.

As the diagram shows, Huene divided lower tetrapods into the small group of Urodelidia and the large group of Eutetrapoda. This division relied on the views of E. Jarvik, which had then been rejected by most paleontologists. Huene divided Eutetrapoda

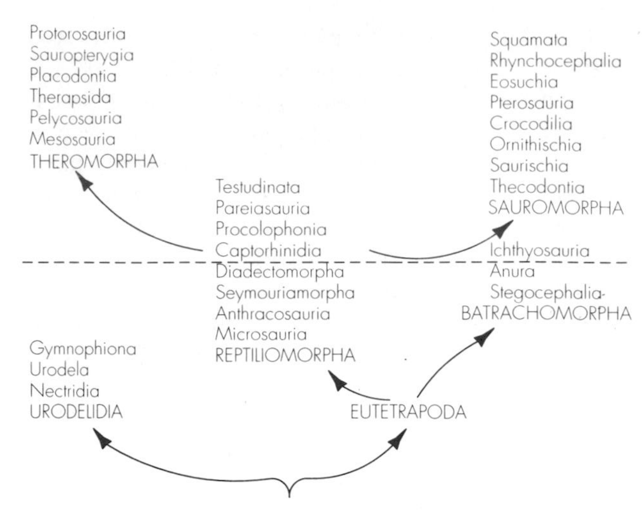

FIGURE 1. Huene's 1956 classification represented in diagrammatic form. All forms above the horizontal broken line are definitely reptilian. A. S. Romer (1967, p. 830).

into Reptiliomorpha and Batrachomorpha. He also assigned Ichthyosauria, which were undoubtedly reptiles, to the category of Batrachomorpha. Huene counted certain forms among the primitive Reptiliomorpha that today are placed among the amphibians. He derived both the line of Sauromorpha and that of Theromorpha from the Captorhinidia. Among the Theromorpha are both Pelycosauria and Therapsida, which lead to the mammals. Mesosauria, Placodontia, Sauropterygia, and Protorosauria bear no relation to either Pelycosauria or Therapsida.

By the time it appeared, Huene's classification was already obsolete. Rigidly fixated on the morphology of fossil reptiles, he paid little or no attention to the totality of vertebrates, to their comparative anatomy and embryology. The value of his publications lies in the profusion of individual bits of knowledge about fossil lower tetrapods.

Huene did not see the question of God and the world or of religion and nature as a great problem. In the origins of vertebrates he perceived a single, systematic course of life. Generally speaking, he reckoned on the "guidance" and "planning" of the Creator, as he said when speaking or writing to those in religious circles.

Huene worked indefatigably and briskly, and thus was able to master the huge amount of material that he collected and that was sent to him from many sources. As a result, much that he published was hastily conceived and often ill considered. Ironically, in his last publication, "Ein sehr junger und ungewöhnlicher Ichthyosaurier" (1966), he fell victim to the hoax of an artificial fossil.

Huene lived an ascetic life. He was modest and helpful, and never acted against his religious precepts. He formed friendships with numerous vertebrate paleontologists, many of whom visited him. He was a member of the German Academy of Natural Scientists, Leopoldina (1944), and an honorary member of the Paleontological Society (1956). In 1956 he received the Buch Plaque of the German Geological Society.

BIBLIOGRAPHY

I. Original Works. A complete list of Huene's publications is in Reif and Lux (see below), 118–140. Huene's works include "Die silurischen Craniaden der Ostseeländer, mit Ausschluss Gotlands," in *Verhandlungen der Kaiserlichen russischen mineralogischen Gesellschaft zu St. Petersburg*, 2nd ser., **36** (1899), 181–359, also published separately (St. Petersburg, 1899); "Geologische Beschreibung der Gegend von Liestal im Schweizer Tafeljura an Hand von Blatt 30 des Siegfried-Atlas," in *Verhandlungen der Naturforschenden Gesellschaft in Basel*, **12** (1900), 295–373; "Uebersicht über die Reptilien der Trias," in *Geologische und paläontologische Abhandlungen*, n.s. **6**, no. 1 (1902), 1–84; "Die Dinosaurier der europäischen Triasformation mit Berücksichtigung der aussereuropäischen Vorkommnisse," *ibid.*, supp. 1 (1907–1908), 1–149; "Die Cotylosaurier der Trias," in *Palaeontographica*, **59** (1912), 69–102; *Die Ichthyosaurier des Lias und ihre Zusammenhänge* (Berlin, 1922); "Vollständige Osteologie eines Plateosauriden aus dem schwäbischen Keuper," in *Geologische und Paläontologische Abhandlungen*, n.s. **15**, no. 2 (1926), 1–45; *Die fossilen Reptilien des südamerikanischen Gondwanalandes*, 3 pts. (Munich, 1935–1942); "*Henodus chelyops*, ein neuer Placodontier," in *Palaeontographica*, **A84** (1936), 97–148; "Osteologie und systematische Stellung von *Mesosaurus*," *ibid.*, **A92** (1940), 45–58; "Die Anomodontier des Ruhu-Gebietes in der Tübinger Sammlung," *ibid.*, **A94** (1943), 154–188; "Arbeitserinnerungen von Dr. Friedrich Frhr. von Huene," in *Selbstbiographie von Naturforschern*, no. 2 (Halle an der Saale, 1944); "Die Theriodontier des Ruhu-Gebietes in der Tübinger Sammlung," in *Neues Jahrbuch für Mineralogie, Geologie und Paläontologie*, Abhandlungen, **92** (1950), 47–136; *Paläontologie und Phylogenie der niederen Tetrapoden* (Jena, 1956; supp., 1959); and "Ein sehr junger und ungewöhnlicher Ichthyosaurier aus dem oberen Lias von Holzmaden," in *Neues Jahrbuch für Geologie und Paläontologie*, Abhandlungen, **124** (1966), 53–54.

II. Secondary Literature. Edwin H. Colbert, *Men and Dinosaurs* (New York, 1968); Walter Gross, "Friedrich Freiherr von Huene, 22.3.1875–4.4.1969," in *Paläontologische Zeitschrift*, **43** (1969), 111–112; H. Hölder, "Geschichte der Geologie und Paläontologie der Universität Tübingen," in *Mineralogie, Geologie und Paläontologie an der Universität Tübingen von den Anfängen bis zur Gegenwart* (Tübingen, 1977), 87–284; Wolf-Ernst Reif, "Evolutionary Theory in German Paleontology," in Marjorie Grene, ed., *Dimensions of Darwinism* (Cambridge and New York, 1983), 173–203; "Paleoecology and Evolution in the Work of Friedrich von Huene," in W.-E. Reif and F. Westphal, eds., *Third Symposium on Mesozoic Terrestrial Ecosystems, Short Papers* (Tübingen, 1984), 193–197; "The Search for Macroevolutionary Theory in German Paleontology," in *Journal of the History of Biology*, **19** (1986), 79–130; and "Evolutionstheorie und religiöses Konzept im Werk des Wirbeltierpaläontologen Friedrich Freiherr von Huene (1875–1969)," in *Werkschriften des Universitätsarchivs Tübingen*, 1st ser., no. 12 (Tübingen, 1987), 91, 140, with W. Lux; A. S. Romer, "Early Reptilian Evolution Reviewed," in *Evolution*, **21** (1976), 821–835; A. Seilacher, "Friedrich Freiherr von Huene," in *News Bulletin of the Society of Vertebrate Paleontologists*, **86** (1969), 41; and A. Seilacher and F. Westphal, "Friedrich Freiherr von Huene, 22 März 1875–4 April 1969," in *Jber. u. Mitt. Oberrhein. geol. Ver.*, n.s. **51** (1969), 25–30.

EMIL KUHN-SCHNYDER

HUNT, FREDERICK VINTON (*b.* Barnesville, Ohio, 15 February 1905; *d.* Buffalo, New York, 20 April 1972), *acoustics*.

Hunt was the youngest of the three children of Fred Hunt and Ella Shipley. The family lived in an agricultural town that was on a major railroad line. His father worked for the First National Bank of Barnesville and eventually became its president. His mother came from a substantial farming family. As a child Hunt received considerable encouragement in his studies from his family and teachers. His brother, Spencer, ten years older, who became an electrical engineer, introduced him to electricity and radio. He received bachelor's degrees in physics (1924) and electrical engineering (1925) from Ohio State University.

Hunt then entered Harvard, where he remained throughout his career, working in acoustics, especially electroacoustical transduction. Harvard already had a tradition in acoustics, dating at least from Wallace C. Sabine's studies of reverberation in 1895, that was being carried on by Frederick A. Saunders and George W. Pierce when Hunt arrived as a graduate student. He wrote doctoral dissertations in both physics and engineering. Hunt received the doctoral degree, only one being allowed, in physics (1934) and served on the faculty until he retired in 1971. He held the Gordon McKay professorship of applied physics from 1946 and the Rumford professorship of physics from 1953. He directed the Harvard Underwater Sound Laboratory (1941–1946) and then the Acoustics Research Laboratory.

Hunt's doctoral dissertation in physics, written under Pierce, was in acoustical reverberation: he developed a technique for determining the decay curves for sound in a room and, by using a frequency-modulated sound source, obtained improved accuracy. In the late 1930's Hunt and his first doctoral students investigated topics in the field of room acoustics. This work formed the basis for extensive studies of the acoustics of concert halls that his students later pursued.

Hunt's other doctoral dissertation also dealt with a fundamental topic in acoustics: the measurement of audio frequencies. He developed a "direct-reading" frequency meter in which a current is proportional to the frequency of the source. This meter extended the limits within which frequency indication was not distorted by amplitude and wave form, and it was capable of responding to rapid variations in the frequency of the signal. This made it suitable for studying the variation of frequency in speech, for example. Used in conjunction with heterodyne methods, the meter also had the potential for measuring and controlling radio frequencies, and the General Radio Company manufactured it in great numbers.

Beginning in 1936, Hunt searched for ways to reduce distortion in reproducing sound from phonograph records. He considered primarily the reproducer (stylus and transducer). Pickup weight, the force of the stylus on the record, in the 1930's was more than one hundred grams, because the design at the time required that the stylus should be ground while tracking in order to fit the groove. Hunt and J. A. Pierce developed a reproducer with a pickup weight of only five grams that did not need grinding. At about the same time (1938) they also gave a theoretical analysis of the distortion due to the fact that the reproducing stylus, having a different shape, does not follow the trajectory of the cutting stylus exactly. They suggested ways in which this "tracing" distortion could be reduced, one of which involved the use of light forces and a stylus with a much smaller tip (which would, as they noted, permit long-playing records). This analysis led to further theoretical studies by Hunt's students and others. Over the years Hunt continued to encourage the development of lighter pickups, and in the 1960's he was attempting to make one that weighed only 0.1 gram.

Hunt was particularly influential in the field of underwater acoustics. Extensive development of this field was due to the U.S. Navy's projects to develop sonar. (The acronym, originally derived from "sonic azimuth and range," is attributed to Hunt.) During World War I, at the U.S. Navy's laboratory in New

London, Connecticut, G. W. Pierce had made an important beginning in developing sonar devices. Hunt was named director of the Harvard Underwater Sound Laboratory (HUSL), established during World War II for research related to undersea warfare. At its peak the HUSL included 450 people, and Hunt was actively involved in all its projects. Central developments at the HUSL were the prototype of scanning sonar and the acoustically guided torpedo. The HUSL closed after the war, but Hunt remained in touch with the U.S. Navy.

In 1950 Hunt proposed that entire oceans be scanned with low-frequency sonar and encouraged the organization of the Navy's Artemis underwater surveillance project. His final project, which he called *sic transit sonitus*, was to develop a technique for processing the sound signals received by a monitor to allow an "after-the-fact" detection of a moving source (a propeller-driven ship, for example). With this technique, which Hunt first described in 1965, a sequence of averaged intensity readings is analyzed statistically to discriminate between a moving source and a random fluctuation. By 1972 Hunt was able to detect, with a probability of 90 percent, a source which had a signal-to-noise-level difference no higher than −10.6 dB, with an extremely low false alarm rate.

Hunt's writing style was conversational and lucid, and he produced effective review articles. He had planned to write a textbook on physical acoustics but instead published the monograph *Electroacoustics* (1954). It contains a long historical chapter extensively based on Patent Office publications. He also left a manuscript, *Origins in Acoustics*, part of which, covering the period from antiquity through the seventeenth century, was published in 1978. Hunt wrote some four dozen research papers and technical reports and held about a dozen patents. Thirty-eight students received their doctorates under his direction.

In 1932 Hunt married Katharine Buckingham, a graduate of M.I.T. in architecture; they had one son, Thomas Kintzing, born in 1937. In 1925 Hunt had made a 16,000-mile tour of the West in a Model T Ford, and he continued to pursue automobile touring with his family. He received many honors, including an honorary Sc.D. from Harvard in 1945, the Presidential Medal of Merit in 1947, the U.S. Navy's Distinguished Public Service Medal in 1970, and medals and awards of the Acoustical Society of America and of the Audio Engineering Society. An oceanographic research ship was named for him in 1965. He died of a heart attack while atttending a convention of the Acoustical Society of America.

BIBLIOGRAPHY

I. Original Works. An extensive list of Hunt's publications, technical reports, talks, and patents is in Harry A. Schenck, *Journal of the Acoustical Society of America*, **57** (1975), 1255–1257. The majority of his papers appeared in *Journal of the Acoustical Society of America*, *Journal of the Audio Engineering Society*, and *Review of Scientific Instruments*. A number of his papers are reprinted in H. E. Roys, ed., *Disc Recording and Reproduction* (Stroudsburg, Pa., 1978). There is a collection of Hunt's papers and technical reports in the Hunt Reading Room of the McKay Library at Harvard University. An interview of Hunt by L. L. Beranek is in the oral history collection of the Niels Bohr Library at the Center for the History of Physics, American Institute of Physics.

Hunt's publications include "On Frequency Modulated Signals in Reverberation Measurements," in *Journal of the Acoustical Society of America*, **5** (1933), 127–138; "A Direct-Reading Frequency Meter Suitable for High Speed Recording," in *Review of Scientific Instruments*, **6** (1935), 43–46; "HP6A: A Radical Departure in Phonograph Pick-up Design," in *Electronics*, **11** (March 1938), 9–12, with J. A. Pierce; "On Distortion in Sound Reproduction from Phonograph Records," in *Journal of the Acoustical Society of America*, **10** (1938), 14–28, with J. A. Pierce; "Analysis of Sound Decay in Rectangular Rooms," *ibid.*, **11** (1939), 80–94, with L. L. Beranek and D. Y. Maa; *Electroacoustics: The Analysis of Transduction, and Its Historical Background* (Cambridge, Mass., 1954); "Propagation of Sound in Fluids," in *American Institute of Physics Handbook* (New York, 1957, 1963, 1972), sec. 3c; "Electroacoustics and Transducers," in *Journal of the Acoustical Society of America*, **30** (1958), 375–377; "The Rational Design of Phonograph Pickups," in *Journal of the Audio Engineering Society*, **10** (1962), 274–289; "The Past Twenty Years in Underwater Acoustics: Introductory Retrospection," in *Journal of the Acoustical Society of America*, **51** (1972), 992–993; "Signal Rate Processing for Transit Detection," *ibid.*, 1164–1185; and *Origins in Acoustics*, Robert E. Apfel, ed. (New Haven, 1978).

II. Secondary Literature. There are accounts of Hunt by John V. Bouyoucos, "Pioneers of Underwater Acoustics Medal, 1965: Frederick V. Hunt," in *Journal of the Acoustical Society of America*, **39** (1966), 765–769; and by Laurence Batchelder, "Frederick V. Hunt," *ibid.*, **52** (1972), 52–54. The Hunt memorial issue, *ibid.*, **57**, no. 6, pt. I (1975), 1239–1401, includes articles about Hunt and his work by colleagues and students. The collection *Epistolae pro Frederick V. Hunt* (n.p., n.d.), written in connection with the award of the Underwater Acoustics Medal of the Acoustical Society, describes Hunt's role at the HUSL. The work done at the HUSL is described in the *Summary Technical Reports* of Division 6, National Defense Research Committee (declassified 1960).

Sigalia Dostrovsky

HUNT, REID (*b.* Martinsville, Ohio, 20 April 1870; *d.* Belmont, Massachusetts, 10 March 1948), *pharmacology.*

Hunt was the younger son of Milton L. Hunt, a banker, and Sarah E. Wright. Both of his parents, who were Quakers, had been schoolteachers and encouraged him in his educational goals. Hunt's interest in science was acquired at an early age when he studied chemistry under the direction of the local pharmacist. After graduating from Martinsville High School at the age of sixteen, he spent a year at Wilmington College and a year at Ohio University. In 1888 Hunt transferred to the Johns Hopkins University, where he obtained his B.A. in 1891. The interest in physiology that he had developed during his undergraduate years led him to undertake graduate studies in that field at Hopkins under H. Newell Martin. Early in 1892 he went to Germany for medical studies at the University of Bonn. During this period his interest in pharmacology was aroused by the lectures of Carl Binz, professor of pharmacology. Hunt returned to Hopkins in the fall of 1892 to resume his graduate work in physiology under Martin and later under William H. Howell, receiving a Ph.D. in 1896. In the same year he obtained an M.D. from the College of Physicians and Surgeons in Baltimore.

Hunt spent the next two years as a tutor in physiology at the College of Physicians and Surgeons of Columbia University. During the summers of 1898 and 1899, he accompanied expeditions of Columbia zoologists to Egypt and the Sudan in an attempt to obtain specimens of the African lungfish and its developmental stages. One of the party died of fever, and Hunt and another colleague were stricken. Neither expedition succeeded in obtaining a lungfish specimen.

In the fall of 1898, Hunt returned to Hopkins to become an associate in pharmacology under John J. Abel, the principal founder of modern American pharmacology. He was given the rank of associate professor of pharmacology in 1901. Although still officially connected with Johns Hopkins, Hunt spent much of his time during 1902 and 1903 in Germany, working in the Frankfurt laboratory of Paul Ehrlich, who exercised a strong influence on the direction of Hunt's scientific work.

Hunt left Hopkins in 1904 to organize and head the newly created division of pharmacology of the Hygienic Laboratory, United States Public Health Service. In addition to his research Hunt was responsible for the testing and analysis of vaccines and antitoxins required under the 1902 Biologics Control Act. In 1913 he became professor and head of the pharmacology department at Harvard Medical School, a position he held until his retirement in 1936.

Though Hunt was tall and gave an impression of physical strength, he was basically a shy, modest individual who never sought the limelight. Research rather than teaching was his forte, and it was evidently with mixed feelings that he left government to return to a university post in 1913. Hunt had an impressive command of the medical literature that was widely recognized by his colleagues. He appears to have had few interests outside his work and home life, except for a love of travel. He married Mary Lillie Taylor on 12 December 1908; they had no children.

Hunt was one of the pioneers in the establishment of pharmacology as an independent discipline in the United States, a movement in which his mentor Abel played the pivotal role. He was the first secretary of the American Society for Pharmacology and Experimental Therapeutics, founded in 1908, and one of the drafters of its constitution and bylaws. He later (1916–1919) served as president of the society. Hunt also applied his pharmacological knowledge in a number of other professional commitments. He served as a member of the American Medical Association's Council on Pharmacy and Chemistry for thirty years (president 1927–1936); as president of the Committee of Revision for the United States Pharmacopeia (1920–1930); and as chairman of the Hunt Committee, an ad hoc board of scientists created by the United States secretary of agriculture in 1927 to determine tolerable levels of lead and arsenic on fruits sprayed with lead arsenate pesticide. Hunt was elected to the National Academy of Sciences in 1919.

One of Hunt's most important research contributions was his study of the biological activity of choline compounds and its relationship to chemical structure. In Abel's laboratory in 1899, Hunt had noted that suprarenal extracts free of epinephrine caused a lowering of blood pressure. He identified choline as one of the substances responsible, but recognized that it did not account for the full effect. Further work suggested that suprarenal extract might contain derivatives of choline more potent than choline itself.

Apparently influenced by Ehrlich's structure-activity studies while in Frankfurt, Hunt later began to study the physiological action of a series of choline derivatives and analogues. His first publication on the subject (1906), with R. deM. Taveau, reported the extraordinary activity of one of these substances, acetylcholine. He was profoundly impressed by the fact that acetylcholine was 100,000 times more active than choline in lowering blood pressure. The biological significance of this discovery became clear only in 1914, when Henry Dale and Otto Loewi identified the role of acetylcholine in chemical transmission of nerve impulses, work for which they received the Nobel Prize in 1936. Choline and analogous compounds continued to interest Hunt throughout his career, and a long series of papers on the subject issued from his laboratory, contributing to knowledge of the pharmacology of the autonomic nervous system and of structure-activity relationships in onium compounds.

Another important area of research that occupied a significant amount of Hunt's time was the physiology and pharmacology of the thyroid gland. His interest in the subject stemmed from his work with Ehrlich on the toxicity of the nitriles. Based on the assumption that the toxicity of acetonitrile was due to a conversion to hydrocyanic acid by oxidation, Hunt reasoned that administration of thyroid gland should enhance the toxicity of acetonitrile by increasing basal metabolism. He found that this was indeed the case for rats and guinea pigs, but that mice reacted in the opposite way and became more resistant to acetonitrile. For two decades following the publication of his first paper on the subject (1905), Hunt used the "acetonitrile reaction" to study thyroid physiology and pharmacology. He was able to show that thyroid gland preparations have a physiological activity parallel to their iodine content, thus making it possible to standardize thyroid preparations by measuring the iodine content. He also demonstrated the influence of different diets on thyroid function.

BIBLIOGRAPHY

I. ORIGINAL WORKS. A good bibliography of Hunt's publications appears on pp. 38–44 of the obituary by E. K. Marshall, Jr. (see below). A key paper reporting the biological activity of acetylcholine is "On the Physiological Action of Certain Cholin Derivatives and New Methods for Detecting Cholin," in *British Medical Journal* (1906), **3**, 1788–1791, with R. deM. Taveau. See also "The Relation of Iodin to the Thyroid Gland," in *Journal of the American Medical Association*, **49** (1907), 1823–1829. The Manuscripts Division of the Countway Library, Boston, holds one box of Hunt manuscript materials, including his lectures for a pharmacology course and some correspondence on professional matters.

II. SECONDARY LITERATURE. The best biographical sketch is E. K. Marshall, Jr., "Reid Hunt, 1870–1948," in *Biographical Memoirs, National Academy of Sciences*, **26** (1949), 25–44. Also useful is Otto Krayer, in *Dictionary of American Biography*, supp. IV, 410–412. Several obituaries are listed in the bibliography to the brief biographical article on Hunt in John Parascandola and Elizabeth Keeney, *Sources in the History of American Pharmacology* (Madison, Wis., 1983), 38–40. Hunt's participation in the Hunt Committee is described in James Whorton, *Before Silent Spring: Pesticides and Public Health in Pre-DDT America* (Princeton, 1974), 154–160. On his role in the American Society for Pharmacology and Experimental Therapeutics, see K. K. Chen, ed., *The American Society for Pharmacology and Experimental Therapeutics, Incorporated: The First Sixty Years, 1908–1969* (Bethesda, Md., 1969).

JOHN PARASCANDOLA

HUREWICZ, WITOLD (*b*. Lodz, Russian Poland, 29 June 1904; *d*. Uxmal, Mexico, 6 September 1956), *topology*.

Although well acquainted with the topology of the Polish school, Hurewicz, the son of an industrialist, began the study of topology under Hans Hahn and Karl Menger in Vienna, where he received the Ph.D. in 1926. After being a Rockefeller fellow at Amsterdam in 1927–1928, he was *Privatdozent* and assistant to L. E. J. Brouwer at the University of Amsterdam from 1928 to 1936. In the latter year Hurewicz took a year's leave of absence to visit the Institute for Advanced Study at Princeton. He decided to stay in the United States, first at the University of North Carolina and then, from 1945, at the Massachusetts Institute of Technology. He died after the International Symposium on Algebraic Topology at the National University of Mexico. While on an excursion he fell from a pyramid he had climbed.

Hurewicz was a marvelously clear thinker, a quality reflected by his style of oral and written communication. This clarity characterizes his early work in set-theory topology. By grasping the essentials and putting them into a larger context, he simplified approaches and generalized theorems and theories. For instance, the switch in dimension theory from subsets of Cartesian to general separable metric spaces is due to Hurewicz. The so-called Sperner proof of the invariance of dimension was independently and simultaneously found and published by Hurewicz.[1] A remarkable result of this first period is his topological embedding of separable metric spaces into compact spaces of the same (finite) dimension.[2]

The next period of Hurewicz's scientific life started with the recognition of spaces of mappings (of one space into another) as a powerful means of topological research; the extensive use of the principle that in complete metric spaces the intersection of overall dense open subsets is itself overall dense[3] is quite characteristic of Hurewicz's thought. In this period he also developed beautiful theorems on dimension-raising mappings,[4] as well as theorems and lucid proofs on embedding finite-dimensional into Cartesian spaces.[5]

For a long time combinatorial methods were belittled as a useful but ugly tool in topology—a necessary evil, as it were. In the early 1930's the desire to change the homological into a homotopical approach was reinforced by Heinz Hopf's homological classification of the mappings of *n*-dimensional polyhedra into the *n*-dimensional sphere.[6] Hurewicz was particularly impressed by Karol Borsuk's homotopic characterization of closed sets dividing *n*-space by essential mappings into the (*n*-1) sphere.[7] In this respect the last paragraph of one of his papers deserves to be quoted (in translation):

> . . . the part played by mappings on the *n*-sphere in the topological research of the last few years (in particular, in investigations by Hopf and Borsuk). One may expect that a closer study of these mappings (especially by group theory means) will lead to clarifying the relation between homology and homotopy, which would create the possibility to apply set theory methods in those domains which are at present exclusively dominated by combinatorial methods. Among others one should consider that an essential mapping of an *n*-dimensional closed set of a space *R* on the *n*-sphere is in a sense the set theory analogue of the combinatorial concept of *n*-cycle, whereby mappings that can be continued to the whole *R* correspond to "bordering" cycles.[8]

As valuable as it may be, Hurewicz's work, as reported so far, is entirely overshadowed by the discoveries made during a short period in the year 1934–1935, which in due course assured him of a place among the greatest topologists: the discovery of the higher homotopy groups and their foremost properties. In hindsight, it all looks so simple: replacing, in Henri Poincaré's definition of the fundamental group, the circles by spheres of any dimension. In fact, the idea was not new, but until Hurewicz nobody had pursued it as it should have been. Investigators did not expect much new information from groups, which were obviously commutative, and in this respect no better than the commutative homology groups. The paragraph quoted may explain why this did not bother Hurewicz, yet the experience acquired in dealing with spaces of mappings gave him a head start.

The wealth of results displayed in the four papers on the topology of deformations[9] is overwhelming. There is little need to go into detail, since most of them are now among the rudiments of homotopy theory, the creation of Hurewicz. Even homological algebra is rooted in this work: In "aspherical" spaces the homology groups are uniquely determined by the fundamental group. Other theorems include the following: If the first *n*-1 homotopy groups are trivial, the *n*th homology and homotopy groups are isomorphic. A polyhedron with only trivial homotopy groups is in itself contractible to a point.

Hurewicz's second great discovery (1941) is exact sequences, an almost imperceptible abstract[10] that generated an enormous literature. His work, with others, on fibre spaces has been of lasting importance.[11]

Surprisingly, Hurewicz's bibliography shows a relatively small number of items. His personal influence, however, cannot be overestimated. His knowledge of mathematics went far beyond topology; he lectured in a rich variety of fields. Hurewicz, who never married, was a highly cultured and charming man, and a paragon of absentmindedness, a failing that probably led to his death.

NOTES

1. "Über ein topologisches Theorem," in *Mathematische Annalen*, **101** (1929), 210–218.
2. "Theorie der analytischen Mengen," in *Fundamenta mathematicae*, **15** (1930), 4–17.
3. "Dimensionstheorie und Cartesische Räume," in *Koninklijke Akademie van wetenschappen te Amsterdam, Proceedings*, **34** (1931), 399–400.
4. "Über dimensionerhöhende stetige Abbildungen," in *Journal für die reine und angewandte Mathematik*, **169** (1933), 71–78.
5. "Über Abbildungen von endlich dimensionalen Räumen auf teilmengen Cartesischer Räume," in *Sitzungsberichte der Preussischen Akademie der Wissenschaften* (1933), 754–768.
6. Heinz Hopf, "Die Klassen der Abbildungen der *n*-dimensionalen Polyeder auf die *n*-dimensionale Sphäre," in *Commentarii mathematici helvetici*, **5** (1933), 39–54.
7. Karol Borsuk, "Über Schnitte der *n*-dimensionalen Euklidischen Räume," in *Mathematische Annalen*, **106** (1932), 239–248.
8. "Über Abbildungen topologischer Räume auf die *n*-dimensionale Sphäre," in *Fundamenta mathematicae*, **24** (1935), 144–150.
9. "Höher-dimensionale Homotopiegruppen," in *Koninklijke Akademie van wetenschappen, te Amsterdam, Proceedings*, **38** (1935), 112–119; "Homotopie und Homologiegruppen," *ibid.*, 521–528; "Klassen und Homologietypen von Abbildungen," *ibid.*, **39** (1936), 117–126; and "Asphärische Räume," *ibid.*, 215–224.
10. "On Duality Theorems," *Bulletin of the American Mathematical Society*, abstract 47-7-329.
11. "Homotopy Relations in Fibre Spaces," in *Proceedings of the National Academy of Sciences*, **27** (1941), 60–64, with N. E. Steenrod; "On the Concept of Fiber Space," *ibid.*, **41** (1955), 956–961; and "On the Spectral Sequence of a Fiber Space," *ibid.*, 961–964, with E. Fadell.

BIBLIOGRAPHY

Solomon Lefschetz, "Witold Hurewicz: In Memoriam," in *Bulletin of the American Mathematical Society*, **63** (1957), 77–82, includes a complete bibliography of Hurewicz's works.

HANS FREUDENTHAL

HYMAN, LIBBIE HENRIETTA (*b*. Des Moines, Iowa, 6 December 1888; *d*. New York City, 3 August 1969), *invertebrate zoology*.

Hyman was the third of four children born to Joseph Hyman and Sabina Neumann Hyman, who were recent Jewish immigrants. As the only daughter, Libbie was expected to do much of the housework and felt her early years were "devoid of affection and consideration." Although she received no encouragement from her family, she found pleasure in schoolwork and was always at the top of her class. Her fascination with the natural world began with a love for wildflowers, which she hunted in the woods and then meticulously categorized according to the scientific classifications in a botany text.

After graduating from high school in 1905, Hyman worked in a factory, gluing labels on cereal boxes. A chance meeting with her high school German teacher (shocked to find her prize student so occupied) led to a scholarship to the University of Chicago. In 1906 she began a course of study to major in botany, but the anti-Semitism of a laboratory assistant caused her to switch to zoology. Hyman received her B.S. degree in 1910, then continued at Chicago under the direction of Charles Manning Child, receiving her Ph.D. in 1915. She worked as Child's assistant from 1915 until 1931. Her research centered on the physiology of planarian flatworms and other lower invertebrates. Although Hyman published more than forty articles during her years with Child, she never considered this work to be outstanding and did not classify herself as a "research type."

As a graduate student Hyman supported herself by assisting in the introductory zoology courses, and she soon realized the need for good laboratory manuals. She published *A Laboratory Manual for Elementary Zoology* in 1919 and *A Laboratory Manual for Comparative Vertebrate Anatomy* in 1922. Both books were very successful and were immediately adopted for use by many colleges. Indeed, from the time of her resignation as Child's research assistant in 1931 to her death, Hyman never held a paid position, living entirely on the royalties from her books.

With the success of these laboratory manuals, Hyman began to contemplate writing a similar book on invertebrates to stimulate the teaching of her preferred subject. She was advised by colleagues that a more advanced textbook was needed, and planned a monograph covering the invertebrate phyla. In 1931 she left Chicago and traveled through Europe, where she visited the marine station at Naples. She returned in 1933, settling in New York City to begin writing her invertebrate treatise. She lived near the American Museum of Natural History in order to use its library, and was made an honorary research associate of the museum in 1937. She maintained this affiliation for the rest of her life.

Volume 1 of *The Invertebrates, Protozoa Through Ctenophora*, was published in 1940. In the preface Hyman states that her intent is to "furnish a reasonably complete and modern account of the morphology, physiology, embryology, and biology of the invertebrates." She goes on to say, "It is obviously impossible for any one person to have a comprehensive first-hand knowledge of the entire range of invertebrates, and consequently a work of this kind is essentially a compilation from the literature." Yet Hyman did much more than survey the world literature; she drew almost all of the figures herself from both live and preserved specimens and was an excellent histologist, preparing tissue sections herself. She traveled extensively to collect animals and spent many summers at the Marine Biological Laboratory at Woods Hole. She never had any secretarial help, typing all of the volumes on an old Underwood typewriter.

Hyman spent the rest of her life attempting to complete the McGraw-Hill series on the invertebrates. Volumes 2 (*Platyhelminthes and Rhynchocoela*) and 3 (*Acanthocephala, Aschelminthes, and Entoprocta*) were published in 1951; volume 4 (*Echinodermata*) in 1955; volume 5 (*Smaller Coelomate Groups*) in 1959; and volume 6 (*Mollusca I*) in 1967.

Hyman maintained an active research career and was considered the foremost authority on the taxonomy of North American turbellarian flatworms. She was outspoken in her belief that the identification of new species based on a single (often sexually immature) specimen was inadequate. Although she began her career by emphasizing small differences and creating new species based on them, as she continued with her work, she was "disposed to think that at least many planarian species tend to show extensive geographical variation and to exist as numerous races," and she became an ardent taxonomic "lumper."

Hyman never married. In 1941 she bought a house in the country in order to enjoy a large garden and daily commuted to the Museum of Natural History. She moved back to New York City in 1952 when she felt the need to spend more time on her writing. She became progressively debilitated by Parkinson's disease and was confined to a wheelchair while working on volume 6.

Hyman belonged to many scientific societies, held office in the American Society of Zoologists and the Society of Systematic Zoology, and was elected to the National Academy of Sciences. She edited *Systematic Zoology* from 1959 to 1963. Hyman was awarded the Elliot Medal of the National Academy of Sciences in 1951, the Gold Medal of the Linnean Society of London in 1960 (only the third American to receive this award), and the American Museum of Natural History's Gold Medal for Distinguished Achievement in Science in 1969, just months before she died.

BIBLIOGRAPHY

I. Original Works. A complete bibliography of Hyman's works, compiled by William K. Emerson, is in Nathan W. Riser and M. Patricia Morse, eds., *Biology of the Turbellaria* (New York, 1974), vol. VII of the Invertebrate Zoology series.

II. Secondary Literature. A more complete biography of Hyman, written by Horace W. Stunkard, is included as a memorial preface to the above volume.

Rachel D. Fink

ILLING, VINCENT CHARLES (*b*. Jalandar, Punjab, India [now Pakistan], 24 September 1890; *d*. London, England, 16 May 1969), *geology*.

Illing was the younger son of Thomas Illing and Annie Payton. His father was serving with the British Army, and Illing's early life was spent in the foothills of the Himalayas of India and on the island of Malta.

When his father retired and the family returned to England, Vincent was thirteen. He was enrolled in the King Edward VI Grammar School at Nuneaton, where his education was shown to be somewhat deficient by British standards. He soon made up this deficiency and went on to win scholarships that enabled him to enter Sidney Sussex College, Cambridge, in 1909. Inspired by W. G. Fearnsides, a fellow of the college, Vincent made geology his major subject. He completed parts 1 and 2 of the general science tripos in three years, winning first-class honors in each. He was awarded a Harkness Scholarship, which allowed him while still an undergraduate to map the Cambrian inlier near his home in Hartshill. He discovered a rich Cambrian fauna, near Birmingham in Warwickshire, that proved to be the link between the fauna of Newfoundland on the west and of Scandinavia on the east, thus laying the foundation for the framework of geological history in these areas. This work was recognized by the Geological Society of London when he received its Lyell Fund in 1918.

In 1913 Professor W. W. Watts at Imperial College of Science and Technology in London invited Illing to start a course on petroleum geology at the Royal School of Mines, which was part of Imperial College. This was the beginning of a long association with Imperial College; in 1914 he was demonstrator in petroleum and lecturer by 1915. Illing was appointed assistant professor in 1921 and professor in 1935; he retired in 1955. He built the course in petroleum geology mainly on the literature and emphasized the achievements of the American oil industry. The core of his courses related to the geological occurrence of oil fields and their relationship to stratigraphy. His first publication on this topic was in 1919, when he wrote "The Search for Subterranean 'Oil-pools' in the British Isles," a pessimistic view of the economic prospects that was provoked by exaggerated ideas held by some politicians. Illing thought that Britain's Carboniferous strata were too strongly folded, faulted, and eroded to have retained economic oil or gas fields. He also criticized the use of percussion drills, which yielded poor rock samples, and recommended methods of coring that would provide more important data. In this opinion he was ahead of his time.

Originally Illing's philosophy about oil geology was that an oil geologist could concentrate on exploration or move into oil development, and he modified his course to accommodate this belief. Later he recognized no essential difference between these fields.

In 1915 Illing's teaching was interrupted while he studied the oil geology of Trinidad. At this time he decided to combine academic and consulting work in order to assist in the building of a viable oil technology department at the Royal School of Mines. He held a lifelong conviction that knowledge and scientific discoveries should be applied to the welfare of mankind.

After returning to London, Illing met Frances Jean Leslie, a teacher. They were married on 20 December 1919, and had a son and four daughters. Illing then took a two-year absence from Imperial College to investigate the geology of the Naparima area of Trinidad for the Naparima Oil Company. During this study he initiated the use of heavy-mineral suites as a correlation aid in the highly disturbed strata of Trinidad. He is credited by later writers with laying the foundation for this correlation technique.

Upon returning to teaching and research at the Royal School of Mines, Illing stressed the need for good field geologists with a sound geological training and a good knowledge of chemistry and physics. During World War I, Britain had established an aeronautical inspection department at the Royal School of Mines; after working on matters concerning the physics and chemistry of petroleum and its products, Illing acknowledged the importance of geophysical work in the search for and development of oil reservoirs.

After the war Illing's geological studies during summer vacations took him to the oil fields of Poland and Romania, and the oil mine at Pechelbronn, France; these investigations stimulated an interest in how to obtain maximum recovery from oil deposits. In the course of this work he set up a research project to study the entire process of oil, gas, and water movement in porous media; this led him to speculate on the origin, accumulation, and preservation of hydrocarbons—the field in which he was an early writer and in which he made his greatest contributions to petroleum geology.

His paper "The Migration of Oil and Natural Gas" was a pioneer work in which Illing clearly enumerated many ideas and principles that were new and today are as valid as when they first appeared. Illing's experiments demonstrated that the main cause of oil movement out of the source rocks is compaction, with hydraulics and gravity important in secondary migration. Primary migration is determined by rock texture; oil enters coarse rocks preferentially and stops at the boundary with fine rocks until the pressure difference overcomes the resistance. Illing followed this paper with seven articles in Volume I of *Science of Petroleum* (1938), including one on the origin, one on the migration, and an introduction to the principles of accumulation, of petroleum. Some of his ideas on these subjects were revised and restated in "Role of Stratigraphy in Oil Discovery" (1945), which was followed by the five-part article "Geology Applied to Petroleum" (1946). He rounded out his major published contributions as editor of and contributor to Volume VI of *Science of Petroleum* (1955).

Starting in 1921 and for part of every summer from 1928 to 1939, Illing did fieldwork in Venezuela. These studies led him to recommend the acquisition and development of properties on which the Mercedes and associated fields in the state of Guárico are located. During World War II he did further work in Trinidad and Venezuela related to Britain's expanding oil needs. This led to an expansion of Trinidad's oil supplies and, after the war, to the full development of the Mercedes area of Venezuela.

As a result of Illing's enthusiasm, several of his banking friends were persuaded to set up an organization, later known as Caracas Petroleum Corporation, to further explore Venezuela. Illing advised the corporation on the selection of concessions and directed gravity surveys, refraction and reflection surveys, and surface geological mapping. In the late 1930's Caracas Petroleum Corporation was merged with Ultramar Company; Illing continued to contribute to Ultramar for many years.

In 1947 the British government asked Illing to arbitrate with the Mexican government over the payment due to the Mexican Eagle (Shell) Oil Company as compensation for the properties nationalized before World War II. Since the Mexican government insisted that the negotiations be strictly confidential, Illing took on the task alone. As a result of his efforts, a settlement was arranged and Illing gained the confidence and respect of the Mexican authorities. So well did he succeed in this endeavor that he later became adviser to Pemex, the government-owned oil company of Mexico.

These and other activities firmly established Illing's reputation as an oil authority throughout the world, and his services were much sought after. In 1950 he established the firm of V. C. Illing and Partners, with offices close to Imperial College. Illing and his partners were consultants to the Gas Council of Great Britain on geological matters and to the government of Nigeria in connection with the development of the oil and gas fields of that country.

Illing joined the Geological Society of London in 1913 and served on its council from 1927 to 1928. In addition to the Lyell Fund (1918), he received many honors, including the Murchison Medal of the Geological Society of London in 1944 and election to the Royal Society in 1945. He took a leading part in the organization and functioning of the petroleum geology sessions of the XVIIIth International Geological Congress at London in 1948. Illing joined the American Association of Petroleum Geologists in 1934 and became the first geologist from England to be granted honorary membership (1961). He was elected an honorary associate of the Royal School of Mines in 1951 and a fellow of Imperial College in 1958. Upon his retirement in 1955, the senate of London University conferred on him the title of professor emeritus of geology (oil technology) at the University of London.

Illing's ideas and principles of petroleum genesis and migration are accepted and quoted in numerous books and articles. He is reported to have had seemingly inexhaustible energy in the field, often walking twenty or thirty miles a day. His former students at Imperial College remember him as an inspiring lecturer and teacher of geology. Many of them went on to occupy positions of prominence in teaching and in the petroleum industry.

BIBLIOGRAPHY

I. Original Works. Illing's writings include "Paradoxidian Fauna of a Part of the Stockingford Shales," in *Quarterly Journal of the Geological Society*, **71** (1915), 386–450; "The Search for Subterranean 'Oil-Pools' in the British Isles," in *Geological Magazine*, **6** (1919), 290–301; "The Migration of Oil and Natural Gas," in *Journal of the Institute of Petroleum Technology*, **19** (1933), 229–274; seven articles in Albert E. Dunstan *et al.*, eds., *The Science of Petroleum*, **I** (Oxford, 1938): "The Origin of Petroleum," 32–38, "Eastern Venezuela and Trinidad," 106–110, written with Hans Kugler, "The Migration of Oil," 209–215, "An Introduction to the Principles of the Accumulation of Petroleum," 218–219, "The Origin of Pressure in Oilfields," 224–229, "The Role of Faulting in the Accumulation of Oil and Gas," 252–254, and "The Significance of Surface Indications of Oil," 294–296; "The Role of Stratigraphy in Oil Discovery," in *Bulletin of the Association of Petroleum Geologists*, **29** (1945), 872–884; and "Geology Applied to Petroleum," in *Oil Weekly*, **122** (1946).

II. Secondary Literature. Obituaries of Illing are G. D. H., "Vincent Charles Illing, F.R.S.," in *Journal of the Institute of Petroleum*, **55** (1969), 422–424; N. L. Falcon, "Vincent Charles Illing," in *Biographical Memoirs of Fellows of the Royal Society*, **16** (1970), 365–384; and Hans G. Kugler, "Vincent Charles Illing," in *Bulletin of the American Association of Petroleum Geologists*, **54** (1970), 542–544, and *Proceedings of the Geological Society of London*, no. 1664B (1971), 357–360.

GERALD M. FRIEDMAN

JEPSEN, GLENN LOWELL (*b*. Lead, South Dakota, 4 March 1903; *d*. Princeton, New Jersey, 15 October 1974), *paleontology*.

Jepsen was the second son of Victor Theodore Jepsen and Kittie Gallup Jepsen. He received his schooling in Rapid City, South Dakota, where his family moved when he was four years old. After a year's study at the University of Michigan, he returned to Rapid City and taught English at the South Dakota School of Mines (1923–1925) while continuing his studies there. An early interest in fossils was nourished by visits to the American Museum of Natural History's excavations at Agate, Nebraska, and by collecting in the Badlands. On one of these trips Jepsen met Prof. William J. Sinclair, who persuaded him to enroll at Princeton University, from which he graduated with highest honors in geology in 1927 and received his Ph.D. in 1930. He was appointed instructor in geology at Princeton in 1930; successive promotions led to the Sinclair professorship of vertebrate paleontology in 1946. Jepsen was named curator of vertebrate paleontology in 1935 and director of the Princeton Museum of Natural History in 1940; he became professor emeritus in 1971.

Jepsen married Janet E. Mayo on 14 June 1934; they had one daughter and were divorced in 1953. He held memberships and offices in numerous scientific societies, including presidency of the Society of Vertebrate Paleontology in 1944 and 1945. In 1962 he received the Addison Verrill Medal of the Peabody Museum of Natural History at Yale.

While still an undergraduate at Princeton, Jepsen published on fossil collecting in the Badlands of South Dakota and on the oldest known cat; other papers on Oligocene saber-toothed cats followed in 1927 and 1933; in 1934 he described *Sinclairella*, a specialized insectivore from the White River beds, and separated the apatemyid insectivores from plesiadapid primates. From 1936 to 1941 Jepsen edited, with William B. Scott, a series of monographs on the mammalian fauna of the White River Oligocene, to which he contributed the section on Insectivora and Carnivora. These early studies of Oligocene mammals influenced his attitudes toward the interpretation of phylogenetic series of fossils, the recognition of species in paleontology, and the significance of fossil vertebrates for stratigraphic correlation.

In 1927, under Sinclair, Jepsen began intensive studies of the geology and vertebrate faunas of Tertiary deposits of the Bighorn Basin in Wyoming, which he continued for more than forty years. From the Polecat Bench he described four successive mammalian assemblages below the well-known Lower Eocene fauna, the most complete sequence of fossiliferous Paleocene continental deposits known at that time in North America. The lowest mammals came from sandstones that had been mapped as Cretaceous but contained no remains of dinosaurs. In later years he commented on the inadequacy of the "absence of dinosaurs" as a criterion for the Cretaceous-Tertiary boundary.

Continued fieldwork produced additional specimens and previously unknown forms. Jepsen's 1940 study of the multituberculates from this area proposed new methods to classify these mostly mouse-sized mammals. Other treasures of his Paleocene collections remained undescribed in his closely guarded cabinets at Princeton, while each year the patient search in the sparsely fossiliferous beds continued during the field seasons. Jepsen was a field paleontologist par excellence who recognized how our limited knowledge of the history of life could be improved only by a better knowledge of the fossil record. His desire for the most complete possible representation of the fauna of this important epoch in the early history of mammals kept him and his students at the difficult and tedious search for the tiny fossils.

Jepsen also was alert to exploit unusual opportunities, such as the discovery of a rich bed of Triassic fishes during excavation for the Firestone Library on the Princeton campus in 1946 and the Horner paleo-Indian buffalo-hunt site near Cody, Wyoming (1948–1962). Many accounts of the trials and pleasures of collecting fossils flowed from his pen. His most prized find was the complete articulated skeleton of an early Eocene bat (1966), to which he devoted thousands of hours of painstaking preparation under the binocular microscope.

Concern with problems of evolution and the fossil record is discernible in Jepsen's systematic papers of the 1930's. His 1944 essay "Phylogenetic Trees" foreshadows the synthetic theory of evolution that emerged at the 1947 Princeton Bicentennial Conference on Genetics, Paleontology, and Evolution, which he organized with Ernst Mayr and George G. Simpson. "Selection, 'Orthogenesis,' and the Fossil Record" (1949) traces the history of conflicting uses of the term "orthogenesis" and its relationship to concepts of irreversibility and evolutionary momentum, and concludes that fossil vertebrates provide no valid examples of the supposed phenomenon.

Jepsen was widely known and admired in the profession as a friendly, urbane, and witty gentleman, a meticulous scholar, and an inspiring undergraduate teacher. His elegant lectures attracted many excellent students to the field of vertebrate paleontology. He placed great stress on laboratory work and fieldwork, and demanded high standards of performance. He allowed his graduate students little freedom and was secretive about his specimens, which he guarded in locked cabinets. At times he could be ferocious toward his assistants and co-workers. For many years he stood in adversarial relationship to his curator.

In later years he published very little about his main collection from the Paleocene of Wyoming and often appeared despondent and withdrawn. At times he emerged from this mood to deliver animated and provocative lectures on diverse aspects of vertebrate history. His emphasis was on the incompleteness of our knowledge of past life, problems of extinction ("Dinosaurs never intended to become extinct"), and the fallacious explanations too often offered to smooth over difficulties or ignorance.

BIBLIOGRAPHY

I. Original Works. Jepsen's publications on fossil vertebrates are listed in Charles L. Camp *et al.*, comps., "Bibliographies of Fossil Vertebrates," in *Special Papers of the Geological Society of America*, **27** (1940) and **42** (1942); and *Memoirs of the Geological Society of America*, **37** (1949), **57** (1953), **84** (1961), **92** (1964), **117** (1968), **134** (1972), and **141** (1973). Other articles are in Jepsen's "Princeton University Museum of Natural History 1964" (Princeton, 1964), appendix C (mimeographed). A selected list of titles is in Erling Dorf, "Memorial to Glenn Lowell Jepsen," in *Geological Society of America, Memorials*, **6** (1977).

On evolution, see "Phylogenetic Trees," in *Transactions of the New York Academy of Sciences*, 2nd ser., **6** (1944), 81–92; "Selection, 'Orthogenesis,' and the Fossil Record," in *Proceedings of the American Philosophical Society*, **93** (1949), 479–500; and, edited with Ernst Mayr and George G. Simpson, *Genetics, Paleontology and Evolution* (Princeton, 1949). See also *Futures in Retrospect* and "Riddles of the Terrible Lizards" (below).

His major descriptive works on fossil vertebrates are "New Vertebrate Fossils from the Lower Eocene of the Bighorn Basin, Wyoming," in *Proceedings of the American Philosophical Society*, **69** (1930), 117–131; "Stratigraphy and Paleontology of the Paleocene of Northeastern Park County, Wyoming," *ibid.*, 463–528; "A Revision of the American Apatemyidae and the Description of a New Genus, *Sinclairella*, from the White River Oligocene of South Dakota," *ibid.*, **74** (1934), 287–305; "The Mammalian Fauna of the White River Oligocene," in *Transactions of the American Philosophical Society*, **28** (1936–1941), edited with William B. Scott; and "Paleocene Faunas of the Polecat Bench Formation, Park County, Wyoming. Part I," in *Proceedings of the American Philosophical Society*, **83** (1940), 217–340.

Jepsen's philosophy of paleontological research is presented in *Futures in Retrospect* (New Haven, 1962) and "Riddles of the Terrible Lizards," in *American Scientist*, **54** (1964), 227–246.

See also "Fossil Collecting in the Badlands," in *Black Hills Engineer*, **14** (1926), 77–86; "A Natural Library," in *Bulletin of the New Jersey State Museum*, no. 3 (1949); "Ancient Buffalo Hunters," in *Wyoming Archaeologist*, **5**, no. 3 (1962), 2–7; "Early Eocene Bat from Wyoming," in *Science*, **154** (1966), 1333–1338; and "Bat Origins and Evolution," in William A. Wimsatt, ed., *Biology of Bats*, I (New York, 1970), 1–64.

II. SECONDARY LITERATURE. Besides the Dorf memorial cited above, see Farish A. Jenkins, Jr., "Obituary, Glenn Lowell Jepsen 1903–1974," in *News Bulletin of the Society of Vertebrate Paleontology*, no. 103 (1975), 89–92; and Sheldon Judson, "Biographical Memoir: Glenn Lowell Jepsen (1903–1974)," in *American Philosophical Society Yearbook 1976* (1977), 85–99. The collection of fishes unearthed during construction of the Firestone Library is described in Bobb Schaeffer, "The Triassic Coelacanth Fish *Diplurus*, with Observations on the Evolution of the Coelacanthini," in *Bulletin of the American Museum of Natural History*, **99** (1952), 25–78.

JOSEPH T. GREGORY

JONES, WALTER JENNINGS (*b.* Baltimore, Maryland, 28 April 1865; *d.* Baltimore, 28 February 1935), *biochemistry*.

The thirteenth child of Levin Jones, a successful ship's chandler, and of Zeanette Jane Bohnen, Walter grew up in a devout and wealthy Methodist family. Educated at Baltimore City College and at Johns Hopkins University (A.B., 1888; Ph.D., 1891), he married Grace Crary Clarke on 1 September 1891; they had one daughter, Marion. After teaching at Wittenberg College in Springfield, Ohio (1891–1892) and at Purdue University (1892–1895), he began his lifelong professional association with Johns Hopkins. Ill health caused him to retire in 1927.

The major figure who determined Jones's choice of research topics was Albrecht Kossel, an authority on nucleoproteins, under whom he worked during a seven-month period of study at Marburg in 1899. There, too, he came to know the nucleic acid chemist P. A. Levene, who was to become his rival and the acknowledged leader in his field in the United States.

In the 1890's the relation of the nucleins (nucleoproteins) to other organophosphorus compounds and to the proteins was still unclear. Kossel hoped that the study of the breakdown products of the nucleins would furnish clarification. Among these products he had recognized the amino-purine bases adenine and guanine, and the corresponding oxypurine bases hypoxanthine and xanthine. He traced their source to the nucleic acid component of nuclein, as he did the pyrimidine bases thymine and cytosine. His conception of nucleic acid structure, however, was vague, and the precise identity of thymine uncertain. By the preparation of a bromine derivative of thymine, Jones, working in Marburg, was able to confirm Kossel's opinion that it was distinct from the known 4-methyl uracil, with which it shared the same empirical formula.

After returning to Baltimore, Jones taught physiological chemistry and toxicology under the pharmacologist John Abel. From the establishment of the Johns Hopkins medical department in 1893 on, the importance of chemistry in the curriculum had been recognized. When physiological chemistry was established as an independent department in 1908, Jones became full professor. The confusion in the subject of nucleic acid chemistry and physiology attracted and held him in this field. His mission was to bring clarification, and this he did, though his claim in 1914 that "the nucleic acids constitute what is possibly the best understood field of Physiological Chemistry" was surely an overstatement. Levene at the Rockefeller Institute certainly considered that Jones had also added some confusion.

Jones's chief contributions to physiological chemistry concerned the enzymatic and hydrolytic breakdown of the nucleic acids. It was by such studies that the nucleic acids were shown to be built from four different nitrogenous bases—two purines and two pyrimidines—each base being attached to a sugar to constitute a "nucleoside," and each nucleoside being phosphorylated to constitute a "nucleotide"; all four were linked to produce a "tetranucleotide." Although subsequent research showed that the nucleic acids were much larger than a tetranucleotide, this structure did embody the correct constituents in the right arrangement. Before the tetranucleotide was established, controversy continued as to whether the nucleic acids were mixtures of different mononucleotides or "polymers" of all of them (what we would call oligonucleotides). As a result of such controversy, much of the literature seems very confused to the modern reader.

Jones's first contribution to clarification concerned the purine products of nucleic acid decomposition. He showed that the tissues contain deaminating enzymes that convert the guanine and adenine bases in nucleic acids into their corresponding oxypurines, xanthine and hypoxanthine. These bases were therefore secondary products of the degradation and not present as such in the nucleic acids. Jones's next contribution concerned the products of the partial breakdown of yeast nucleic acid. He demonstrated the production of dinucleotides and of guanylic and adenylic acids. These acids were therefore constituent nucleotides of yeast nucleic acid. Guanylic acid was not an independent nucleic acid. These results supported Leven and Jacob's opinion that the constituents of nucleic acid were nucleotides, the structure of each of which corresponded to that of inosinic acid. They showed that in the latter the phosphoric acid and the purine base hypoxanthine were linked through the pentose sugar ribose.

The manner of linkage of the nucleotides to each other in the nucleic acid molecule of yeast remained a subject of dispute until the late 1930's. Levene, Robert Feulgen, and Jones supported different structures. The linkages involved were either (1) ribose-ribose, (2) phosphoric acid-phosphoric acid, or (3) phosphoric acid-ribose. In 1919 Leven concluded that all the links were between phosphoric acid and ribose, as in the structure accepted today. Jones, however, presented evidence first in favor of only ribose-ribose links, and later (in 1923) in favor of both ribose-ribose and ribose-phosphoric acid links (a structure that Levene had suggested in 1912 but later rejected). Testily, Levene wrote:

> During the period when Jones believed in the existence of the cytosine-uracil dinucleotide, this formulation might have been warranted, but after Levene had shown in 1917 that the dinucleotide of Jones and Read consisted of a mixture of two mononucleotides . . . , there was much less justification for referring to a structural scheme that had been abandoned earlier. (*Nucleic Acids*, pp. 274–275)

Perhaps the most intriguing of Jones's enzymatic studies was that indicating the presence of an enzyme in an extract of pig pancreas that cleaved internucleotide bonds. Although Levene failed to confirm this observation, René Dubos and R. H. S. Thompson succeeded when they described an RNA depolymerase (RNase) in 1938.

The two most popular sources of nucleic acid were the thymus gland and yeast, hence the names for the two fundamental types: thymonucleic acid (DNA) and yeast nucleic acid (RNA). Jones, in common with other authorities, believed the former to be confined to animal tissues and the latter to plants. His study of the β nucleoprotein of the pancreas, however, caused him to change his opinion and conclude: "It thus seems more than probable that the distinction between animal and plant nucleic acids will in the future not be so definitely drawn."

Jones accepted the identification of the sugar in thymonucleic acid as a hexose and in yeast nucleic acid as ribose. At one time he doubted that guanylic acid was present in thymonucleic acid, and he retained a skeptical view on the uracil found in yeast nucleic acid, for he suspected it was the product of the deamination of cytosine. But he accepted Levene's assertion that both nucleic acids were constructed on the same general plan.

The state of analytical chemistry and the knowledge of extractive procedures in Jones's day made the study of nucleic acid chemistry a difficult one. Jones was at times hasty and overconfident. Levene's frequent criticisms of his work were to be expected. Always forthright, a gifted conversationalist, and an enthusiastic controversialist, Jones won wide respect. He was a member of the National Academy of Sciences and twice president of the American Society of Biological Chemists.

BIBLIOGRAPHY

I. Original Works. Jones and his co-workers published sixty-eight articles, the majority of them in the *Journal of Biological Chemistry*. Among them is "The Occurrence of Plant Nucleotides in Animal Tissue," in *Journal of Biological Chemistry*, **62** (1924–1925), 291–300, with M. E. Perkins. As well as contributing to the literature of toxicology in his early days, Jones wrote *Nucleic Acids: Their Chemical Properties and Physiological Conduct* (London, 1914; 2nd ed., 1920).

II. Secondary Literature. The fullest account of Jones, with portrait and bibliography, is William Mansfield Clark, "Walter (Jennings) Jones, 1865–1935," in *Biographical Memoirs, National Academy of Sciences*, **20** (1939), 79–139. Clark also wrote an obituary of Jones in *Science*, n.s. **81** (1935), 307–308. His work is criticized in P. A. Levene, *Nucleic Acids* (New York, 1931).

Robert Olby

JORDAN, ERNST PASCUAL (*b.* Hannover, Germany, 18 October 1902; *d.* Hamburg, Federal Republic of Germany, 31 July 1980), *theoretical physics*.

Jordan's father, Ernst Pascual Jordan, the descendant of a Spaniard, was a painter of portraits, landscapes, and buildings. His interest in perspective drawings and his acquaintance with the elements of projective geometry seem to have influenced the inclinations of his son. Jordan's mother, Eva (née Fischer), was responsible for his permanent concern with biological problems and his early dedication to numerical computations. Through readings of the popular science literature of his time, such as the influential works of Wilhelm Bölsche, Ernst Haeckel, Ernst Mach, and Friedrich Lange, Jordan became interested in the discussion of the problem of neovitalism. As a result he was later an ardent defender of positivism.[1]

At the age of sixteen, when he attended the so-called reformed gymnasium at Hannover, Jordan became acquainted with higher mathematics through Walther Nernst and Arthur Schönflies' *Einführung in die mathematische Behandlung der Naturwissenschaften*. Against the wishes of his father, who wanted him to study architecture, Jordan decided to study physics and mathematics. In the spring of 1921 he matriculated at the Technische Hochschule of his hometown, Hannover, where he attended the mathematical lectures of Heinrich Müller and Georg Prange. The only lectures on physical topics he found of interest were courses on electrical engineering by Friedrich Kohlrausch and on physical chemistry by Max Bodenstein. The low level of the other lectures on physics led him to study by himself. He learned atomic physics through Arnold Sommerfeld's *Atombau und Spektrallinien* and relativity through Moritz Schlick's *Raum und Zeit in der gegenwärtigen Physik*.

After two semesters Jordan moved to the University of Göttingen, one of the most prominent centers of theoretical physics in postwar Germany. Two weeks after his arrival, he attended Niels Bohr's celebrated Wolfskehl lectures on atomic physics, an event of great importance for the future development of the field in German-speaking countries.

In Göttingen, Jordan soon was introduced to Richard Courant and Max Born. Having recently taken the chair for theoretical physics vacated by Peter Debye, Born was initiating his research program for a more rational foundation of quantum theory. So he welcomed Jordan, whose exceptional talents—despite a speech defect that plagued him all his life—soon became apparent.

During the course of his lectures and seminars Born applied the perturbation methods developed in celestial mechanics to problems in atomic physics. Born had worked out these procedures with the help of Jordan's predecessors, Wolfgang Pauli and Werner Heisenberg. Jordan was quickly introduced into the most pressing problems of quantum physics. He increasingly participated in the research as Born's closest collaborator. In Göttingen, Jordan also attended Alfred Kühn's lectures on biology, a subject that continued to hold his interest all his life.

During his student years Jordan assisted his teachers in the writing of an article on lattice dynamics for the famous *Encyklopädie der mathematischen Wissenchaften*. He helped Richard Courant in the editing of the first volume of Courant and Hilbert's *Methods of Mathematical Physics* (1924), a volume written to a great extent as a response to the mathematical demands of the rapidly developing quantum mechanics.

In October 1924 Jordan was officially employed as Born's assistant. Among his other duties was assisting James Franck in the preparation of his handbook article *Anregungen von Quantensprüngen durch Stösse* (1926); it was eventually coauthored by Jordan and Franck.

After finishing his regular courses in physics, mathematics, and zoology, the twenty-two-year-old Jordan presented a highly original doctoral thesis on the light-quantum problem,[2] which aroused considerable interest in view of the discovery of the Compton effect. But Jordan's suggestion that the momentum distribution of scattered light quanta may also be continuous—in opposition to Einstein's original idea of needlelike radiation—was immediately disproved by the latter.[3] The attention that Einstein had paid to his first publication, however, made a strong impression on the young man.

Jordan now began to participate in Born's research program on a new formulation of quantum theory based on correspondence arguments and the use of observable magnitudes only. These and similar procedures developed independently by John van Vleck showed clearly that Einstein's earlier probabilistic treatment of the radiation phenomena could be brought into line with Bohr's work. In a joint publication Born and Jordan[4] emphasized the importance of the concept of "transition amplitudes," which later proved to be decisive for the emergence of matrix mechanics.

After pursuing extensions of Pauli's investigations on the Zeeman effect and studies of the thermal equilibrium between atoms and radiation,[5] Jordan became involved in the formal development of matrix mechanics advanced by Heisenberg at the end of July 1925. Jordan's familiarity with the methods of matrix calculus from his collaboration with Courant made him an invaluable collaborator for Born, and together they developed the general foundations of the new theory. Only two months after the submission of Heisenberg's fundamental paper the two authors submitted their results for publication to the *Zeitschrift für Physik*.[6] Whereas Heisenberg had supplied the basic idea of how the new formalism could be obtained from classical theory by means of correspondencelike arguments, Born and Jordan worked out in detail a proof of this procedure and its mathematical framework. They found that quantum physics could be formulated adequately in matrix language and discovered that the fundamental basis of the new quantum mechanics could be expressed by the famous commutation relation $pq - qp = (h/2\pi i)\,1$. Because of Born's ill health, Jordan carried out most of the detailed elaborations and calculations, as well as the writing of this important paper.[7]

A thorough study of quantum mechanics, including the extension of the formalism to systems with many degrees of freedom and the inclusion of angular momentum, was the subject of Jordan's next paper, the famous "three-man paper," elaborated in collaboration with Max Born and Werner Heisenberg.[8] An improvement in the definition of matrix differentiation in accordance with the product rule allowed an especially simple formulation of the equations of motion of quantum mechanics. The authors also generalized the concept of canonical transformation in order to preserve the commutation relation in such an operation. One of the most important insights of their joint work was the recognition that the solution of a quantum mechanical problem was equivalent to a canonical transformation of a matrix into its diagonal form. After treating the consequences of the spinning-electron hypothesis in matrix theory in a paper together with Werner Heisenberg,[9] Jordan turned to the more fundamental questions of the new theory.

By the middle of 1926 four different formulations of quantum mechanics existed: Heisenberg's matrix mechanics, Dirac's q-number formalism, Schrödinger's wave mechanics, and the operator formalism of Born and Norbert Wiener. The necessity to clear up the relations between them came to the fore. The all-embracing formulation finally was supplied in December 1926 independently by Dirac and Jordan with the so-called statistical transformation theory, which also paved the way for comprehending the physical content of the new formalism.[10] In general there are infinitely many different possibilities for representing the quantum mechanical magnitudes by operators connected by canonical transformations, Heisenberg's and Schrödinger's representations being only special cases. Whereas Dirac started incorporating Schrödinger's theory into his q-number formalism, Jordan was guided by Pauli's suggestion that according to the statistical interpretation of quantum theory, interference between probabilities must also occur. In spite of their different methods of attack, both authors solved the fundamental problem of determining the probability amplitude of two arbitrary mechanical magnitudes, concluding that Schrödinger's eigenvalue functions constitute just those elements of the canonical transformation matrix that render the Hamiltonian diagonal.

As Born's collaborator, Jordan soon became one of the strongest adherents of the new indeterministic world picture based on Born's statistical interpretation of Schrödinger's wave function. Rejecting Schrödinger's attempts to return to a classical description of atomic processes in terms of continuous changes, Jordan participated in the debates concerning the fundamental question of the meaning of the new quantum magnitudes, relying on his transformation theory.[11] In his *Habilitationsvortrag*,[12] delivered in February 1927, Jordan could state simultaneously with Heisenberg the following implicit formulation of the principle of indeterminacy: "If certain coordinates of a quantum mechanical system are empirically observable magnitudes . . . then the corresponding momenta to these coordinates are in principle nonobservable magnitudes." A visit to Copenhagen made him especially apt to accept Bohr's complementarity principle.

The application of the new formalism to the radiation field contained in the last section of the Born-Heisenberg-Jordan paper was considered by Jordan one of his most important contributions to physics because it served as the beginning of quantum field theory. It was especially satisfying to Jordan that the quantization of the vibrations of an elastic continuum according to the quantum-rules supplied from first principles the energy fluctuations derived thirty years earlier by Einstein using thermodynamical arguments from Planck's radiation law. In Einstein's second fluctuation law the mean square fluctuation of the cavity radiation energy E in the frequency interval ν, $\nu + d\nu$ is given by $(E - \overline{E})^2 = h \nu E + (c^3/8\pi\nu^2\Delta\nu) E^2/V$, V being the volume of the cavity. This expression was the most elegant early formulation of the wave-particle dualism of light quanta. The striking fact that all the particle aspects of light could in 1925 be obtained directly by quantization of the electromagnetic field made Jordan consider the possibility of applying the same procedure to matter waves in order to obtain, by "second quantization," the material particles in a natural way.[13]

Convinced that the many-body problem in quantum mechanics can be stated correctly only in the context of quantized matter waves ("repeated" or "second" quantization, as it was called later by Léon Rosenfeld), Jordan started working out his ideas together with Wolfgang Pauli, Oskar Klein, and Eugene Wigner. During his stay in Copenhagen in the summer 1927 Jordan established, together with Klein, the first nonrelativistic formalism of second quantization for a system of interacting Bose particles.[14] The more complicated case of particles obeying the Pauli exclusion principle could be solved only in collaboration with his colleague Eugene Wigner when Jordan returned in October 1927 to Göttingen. Their work led to the commutation rules for Fermi-Dirac particles.[15] Because in Dirac's theory (second) quantization was applied only to the components of the electromagnetic field and not to the matter waves associated with the material particles, the Jordan-Wigner theory accounts only for the creation and destruction of photons, but not for the material particles.

To incorporate transformations between matter and radiation, Jordan devised a theory with waves and corpuscles treated in a more symmetrical fashion. As a first indication of the appropriateness of his view, Jordan used this method to offer a derivation of Einstein's first fluctuation theorem, which states that the probability W to encounter all the radiation energy E of frequency ν in a subvolume V' of V is equal to $W = (V'/V)^{E/h\nu}$, where $E/h\nu = n$ is interpreted as the number of light quanta in the radiation field. Jordan claimed that, although he had proposed in the winter of 1925–1926 a full program to develop these ideas, it was only after Dirac's success with the radiation problem early in 1927 that his ideas gained the acceptance of the scientific community.[16]

According to Jordan's conception, interactions between light and matter should be described by interacting three-dimensional quantized wave fields, the occurrence of discrete electrified particles and of the light quanta merely being manifestations of the quantum laws. As a counterpart of Heisenberg's and Dirac's treatment of the many-body problem, Jordan had first formulated the same problem in the context of wave quantization, before going on to solve the general case. Since, as in Dirac's radiation theory, space and time coordinates are not given in their covariant form, the relativistic invariance of Jordan's theory was not obvious. On the other hand, the method of second quantization revealed its practical usefulness, particularly in the relativistic domain, where creation and annihilation of particles take place. At that time no method was known to handle the change of particle number in the configuration-space approach. Further, it was expected that the feared retardation problem, present when moving particles interact, would automatically vanish as soon as the theory was expressed in a relativistically covariant formulation. Such a relativistically invariant formalism for the charge-free radiation field was developed first by Jordan and Pauli with the introduction of relativistically invariant commutation relations as well for the field variables at different space-time points.[17]

The more ambitious program of a general relativistic field theory of interacting spinorial and electromagnetic fields was accomplished finally in 1929 and 1930 by Heisenberg and Pauli. This difficult task had been made possible only by utilizing the more sophisticated mathematical tools of functional analysis, required for the treatment of the nonlinear field equations. But the basic difficulties of quantum field theory, such as the infinite self-energy of elementary particles and transitions to the negative energy states first noted in Dirac's theory of the electron, also were present in the new formalism. So the final goal of Jordan's program never could be achieved.[18]

In the meantime Jordan had become *Privatdozent* in Göttingen. In 1928 he succeeded Pauli as Wilhelm Lenz's assistant in Hamburg, before attaining in 1929 a more permanent position as extraordinary professor at the University of Rostock. Soon after his appointment he married Hertha Stahn in 1930; they had two sons. In 1935 he was promoted to ordinary professor, a position he held until 1944, when he was called to take the directorship of the Institute of Theoretical Physics at the University of Berlin as Max von Laue's successor.

During those years, Jordan continued to do research in fundamental problems. But his different attempts to change the foundations of quantum physics in order to get a more consistent relativistic theory[19] never obtained general acceptance. In this context he developed new nonassociative algebraic forms[20] that instead founded a new branch of mathematical investigation, the so-called Jordan algebras.[21]

When, at the beginning of the thirties, progress in physics became slower, many physicists tried to extend the applicability of the quantum theory beyond their own disciplines. After Niels Bohr's famous lecture "Light and Life" in the summer of 1932, Jordan was one of the first to search for further manifestations of the complementarity principle in biology, color vision, and psychology.[22] In a series of controversial articles, disapproved by Bohr, Jordan put forward the idea that the background of biological phenomena are individual quantum processes (quantum jumps), adequately amplified by the biological organism to produce indeterministic effects at the macroscopic level. He also paid much attention to genetics and especially to the problem of radiation-induced gene-mutation (topics then of great interest in Germany) as studied by the Russian-born scientist Nikolai Timoféeff-Ressovsky, K. G. Zimmer, and Max Delbrück. For many years in the late 1930's quantum biology became Jordan's main field of research.[23] But his intention after the war to found a large institute devoted to pure research in quantum biology was never realized.[24]

In spite of his sympathies for the National Socialist movement, Jordan never broke with the tenets of modern theoretical physics, which were then under attack by a group of physicists sympathetic to the

Nazi leaders.[25] During the war he served on the meteorological staff of the Luftwaffe in Hamburg-Fulsbüttel.

Jordan had an unfortunate disposition to put science at the service of political rulers. In June 1936, attending the Copenhagen conference on theoretical physics and philosophy, he informed the Nazi authorities in a secret report about the activities of various participants. In spite of the scientific character of the meeting, he claimed, there was often a definitely materialistic and political worldview involved in many of the reports presented to the philosophical section.[26]

In his popular writings Jordan liked to use strange political analogies, comparing a cell to the state and the nucleus to the Führer, thereby offending many of his foreign colleagues. The enthusiastic disquisitions on military power and armaments he frequently gave in his scientific and philosophical explanations were often almost comic. This behavior, even if interpreted as a sign of opportunism—and in spite of his outstanding scientific contributions—prevented his being reinstalled in a full academic professorship right after the war. But after some inquiries by foreign authorities, Jordan was reinstated as a visiting professor in 1947 and as a full professor in 1953 in Hamburg—a position he held until his retirement in 1971. It would be incorrect, Pauli commented at the time, "if West Germany chooses to ignore a person like P. Jordan."[27]

During the Adenauer era Jordan again became involved in politics. He was a supporter of the notion that only Western atomic armaments could guarantee a peaceful world order.[28] From 1957 to 1961 Jordan was a member of the German Bundestag, contributing to the elaboration of the laws regulating the peaceful use of atomic power. In this last period, Jordan's scientific activities centered on general relativity, astrophysics, cosmology, and pure mathematics.

Motivated by Stanley Eddington's number speculations about the connection of fundamental physical constants and Dirac's conjectures on a slowly decreasing gravitational constant on a cosmological scale, Jordan and his collaborators attempted after 1944 to incorporate these ideas into the framework of Einstein's general relativity theory using the five-dimensional formalism as developed earlier by Theodor Kaluza, Oskar Klein, and Oswald Veblen. As one of the practical conclusions of this generalized relativity theory, Jordan suggested explaining Wegener's continental drift phenomenon as a result of an expansion of the earth.[29] Even more unconventional were Jordan's cosmological speculations, including a theory of the formation and evolution of stars.[30]

Jordan's books on scientific and philosophical subjects addressed to lay audiences[31] as well as his textbooks on physics[32] found wide readerships. Many of them were translated into several foreign languages and appeared in many editions. In 1942 Jordan was awarded the Max Planck Medal and in 1955 the Gauss Medal.

NOTES

1. P. Jordan, "Über den positivistischen Begriff der Wirklichkeit," in *Naturwissenschaften*, **22** (1934), 485–490, and "Positivismus in der Naturwissenschaft," in *Glaube und Forschung*, II (Gütersloh, 1950), 93–112.
2. P. Jordan, "Zur Theorie der Quantenstrahlung," in *Zeitschrift für Physik*, **30** (1924), 297–319.
3. A. Einstein, "Bemerkungen zu P. Jordans Abhandlung 'Zur Theorie der Quantenstrahlung,'" *ibid.*, **31** (1925), 784–785.
4. M. Born and P. Jordan, "Zur Quantentheorie aperiodischer Vorgänge," *ibid.*, **33** (1925), 479–505.
5. P. Jordan, "Über das thermische Gleichgewicht zwischen Quantenatomen und Hohlraumstrahlung," *ibid.*, 649–655.
6. M. Born and P. Jordan, "Zur Quantenmechanik, I," *ibid.*, **34** (1925), 858–888.
7. In response to an inquiry from B. L. van der Waerden, Jordan recalled in October 1964, "During Born's stay at Silvaplana I was in Hannover, in my parents' house, thinking about a part of the material, which was then explained in the paper by Born and myself. I was in correspondence with Born, to whom I naturally reported my progress. I remember that he suggested after some time to stop our exchange of letters, because the double demands of the exhausting treatment in the sanatorium and our conversations by letter about this exciting theme had a bad effect on him. So it could in fact have been as you suppose, that I had already written most of the work in a first draft when we met again in Göttingen."
8. M. Born, W. Heisenberg, and P. Jordan, "Zur Quantenmechanik, II," in *Zeitschrift für Physik*, **35** (1925), 557–615.
9. W. Heisenberg and P. Jordan, "Anwendung der Quantenmechanik auf das Problem der anomalen Zeemaneffekte," *ibid.*, **37** (1926), 263–277.
10. P. Jordan, "Über Kanonische Transformationen in der Quantenmechanik, I, II," *ibid.*, **37** (1926), 383–386, and **38** (1926), 513–517, and "Über eine neue Begründung der Quantenmechanik," *ibid.*, **40** (1927), 809–838.
11. P. Jordan, "Anmerkung zur statistischen Deutung der Quantenmechanik," *ibid.*, **41** (1927), 797–800, "Philosophical Foundations of Quantum Theory," in *Nature*, **119** (1927), 566–569, and "Reply to N. C. Campbell," *ibid.*, 779. See also M. Beller, "Pascual Jordan's Influence on the Discovery of Heisenberg's Indeterminacy Principle," in *Archive for History of Exact Science*, **33** (1985), 337–349.
12. P. Jordan, "Kausalität und Statistik in der modernen Physik," in *Naturwissenschaften*, **15** (1927), 105–110.
13. P. Jordan, "Zur Quantenmechanik der Gasentartung," in *Zeitschrift für Physik*, **44** (1927), 473–480, and "Über Wellen und Korpuskeln in der Quantenmechanik," *ibid.*, **45** (1927), 766–775. See also J. Bromberg, "The Concept of Particle Creation Before and After Quantum Mechanics," in *Historical Studies in the Physical Sciences*, **7** (1977), 161–191.
14. P. Jordan and O. Klein, "Zum Mehrkörperproblem der Quantentheorie," in *Zeitschrift für Physik*, **45** (1927), 751–765.
15. P. Jordan and E. Wigner, "Über das Paulische Äquivalenzverbot," *ibid.*, **47** (1928), 631–651.

16. P. Jordan, "Die Lichtquantenhypothese. Entwicklung und gegenwärtiger Stand," in *Ergebnisse der exakten Naturwissenschaften*, **7** (1928), 158–208.
17. P. Jordan and W. Pauli, "Zur Quantenelektrodynamik ladungsfreier Felder," in *Zietschrift für Physik*, **47** (1928), 151–173.
18. The equivalence of both methods later was cleared up by different authors, particularly by V. Fock, "Konfigurationsraum und zweite Quantelung," *ibid.*, **75** (1932), 622–647. See also P. Jordan, "Zur Methode der zweiten Quantelung," *ibid.*, 648–653.
19. P. Jordan, "Über die Multiplikation quantenmerchanischer Grössen, I, II,"*ibid.*, **80** (1933), 285–291, and **87** (1934), 505–512.
20. P. Jordan, "Eine Klasse nichtassoziativer hyperkomplexer Algebren," in *Nachrichten aus der Gesellschaft der Wissenschaften zu Göttingen*, **33** (1932), 569–575; and P. Jordan, J. von Neumann, and E. Wigner, "On an Algebraic Generalization of the Quantum Mechanical Formalism," in *Annals of Mathematics*, **35** (1934), 29–64.
21. H. Braun and M. Koecher, *Jordan-Algebren* (Berlin, 1966); and N. Jacobson, *Structure and Representations of Jordan Algebras* (Providence, 1968).
22. P. Jordan, "Die Quantenmechanik und die Grundprobleme der Biologie und Psychologie," in *Naturwissenschaften*, **20** (1932), 815–821, "Quantenphysikalische Bemerkungen zur Biologie und Psychologie," in *Erkenntnis*, **4** (1934), 215–252, "Positivistische Bemerkungen über die parapsychologischen Erscheinungen," in *Zentralblatt für Psychotherapie*, **9** (1936), 3–17, "Quantenphysik und Biologie," in *Naturwissenschaften*, **32** (1944), 309–316, "Theorie des Farbensehens," in *Physikalische Zeitschrift*, **45** (1944), 327, and "Zur Biophysik des Farbensehens," in *Optik*, **2** (1947), 169–189.
23. An exposition of quantum biology is also in his book *Die Physik und das Geheimnis des organischen Lebens* (Braunschweig, 1941).
24. P. Jordan, "Zukunftsaufgaben quantenbiologischer Forschung," in P. Jordan, A. Meyer-Abich, and H. Petersen, eds., *Physis* (Stuttgart, 1942).
25. P. Jordan, "Naturwissenschaft im Umbruch," in *Deutschlands Erneuerung*, **25** (1941), 452–458. Compare also S. Balke, "Laudatio auf Prof. Dr. Pascual Jordan," unreferenced printed booklet (after 1969) in the Jordan-Nachlass at the Staatsbibliothek Preussischer Kulturbesitz in Berlin.
26. D. Hoffman, "Zur Teilnahme deutscher Physiker an den Kopenhagener Physiker Konferenzen nach 1933," in *Schriftenreihe für Geschichte der Naturwissenschaften, Technik, und Medizin*, **25** (1988), 49–55.
27. In a letter of 8 May 1952 to the dean of the Faculty of Mathematics and Science of the University of Hamburg. Later, in 1979, Jordan was also proposed by Wigner for the Nobel Prize.
28. P. Jordan, *Der gescheiterte Aufstand. Betrachtungen zur Gegenwart* (Frankfurt am Main, 1956). See also the critical remarks by W. Kliefoth, "Forschung in veränderer Umwelt," in *Physikalische Blätter*, **13** (1957), 23–32.
29. P. Jordan, "Zum Problem der Erdexpansion," in *Naturwissenschaften*, **48** (1961), 417–425, "Geophysical Consequences of Dirac's Hypothesis," in *Reviews of Modern Physics*, **34** (1962), 596–600, and *The Expanding Earth*, Arthur Beer, trans. and ed. (Oxford, 1971).
30. P. Jordan, "Entstehung der Sterne, I, II," in *Physikalische Zeitschrift*, **45** (1944), 183–190, 233–244, "Zur Theorie der Sternentstehung," in *Physikalische Blätter*, **3** (1947), 97–106, and *Die Herkunft der Sterne* (Stuttgart, 1947). See also J. Singh, *Great Ideas and Theories of Modern Cosmology* (New York, 1961).
31. P. Jordan, *Physics of the 20th Century*, Eleanor Oshry, trans. (New York, 1944), *Die Physik und das Geheimnis des organischen Lebens* (Braunschweig, 1941), *Der Naturwissenschaftler vor der religiösen Frage* (Oldenburg, 1963), *Albert Einstein* (Frauenfeld und Stuttgart, 1969), *Begegnungen* (Oldenburg and Hamburg, 1971), and *Erkenntnis und Besinnung* (Oldenburg and Hamburg, 1972).
32. J. Franck and P. Jordan, *Anregung von Quantensprüngen durch Stösse* (Berlin, 1926); M. Born and P. Jordan, *Elementare Quantenmechanik* (Berlin, 1930); and P. Jordan, *Statistische Mechanik auf quantentheoretischer Grundlage* (Braunschweig, 1933), *Anschauliche Quantentheorie* (Berlin, 1936), and *Schwerkraft und Weltall* (Braunschweig, 1952).

BIBLIOGRAPHY

I. Original Works. There is no complete bibliography of Jordan's writings, but his most important scientific publications can be found in the corresponding volumes of Poggendorff. Most of his early scientific work is published in the *Zeitschrift für Physik* and in the *Nachrichten aus der Gesellschaft der Wissenschaften zu Göttingen*. Jordan's review articles and a great number of book reviews are contained in *Naturwissenschaften*. Beginning in the 1930's he published also in journals of a more general nature, such as *Erkenntnis, Forschungen und Fortschritte, Radiologica*, and, after 1945, *Universitas, Optik, Zeitschrift für Naturforschung*, and *Physikalische Blätter*. As vice president (1950–1963) and president (1963–1967) of the Akademie der Wissenschaften und der Literatur in Mainz, he contributed more than twenty-five papers on mathematics and on general relativity to the *Abhandlungen* of the academy.

Jordan's unpublished papers and literary remains, which include twenty-one boxes and twenty-two files, are deposited at the Staatsibliothek Preussischer Kulturbesitz in Berlin. More than eighty letters from his correspondence with physicists during the 1920's are cited by T. S. Kuhn, *et al., Sources for History of Quantum Physics: An Inventory and Report* (Philadelphia, 1967). Two hundred twenty-four letters from the correspondence with his main publisher are preserved in the archives of the Vieweg-Verlag in Wiesbaden. Thirty-three letters from Jordan's correspondence with Wolfgang Pauli are kept in the Pauli letter collection at the Centre Européen Pour la Recherche Nucléaire in Geneva and will be published in the forthcoming edition of Wolfgang Pauli's *Wissenschaftlicher Briefwechsel mit Bohr, Einstein, Heisenberg, u.a.*, edited by Karl von Meyenn. A more complete list containing also Jordan's later scientific correspondence is provided by the "Inventory of Sources for History of Twentieth-Century Physics" (ISHTCP), available to researchers at the Office for History of Science and Technology of the University of California, Berkeley. The transcripts (101 pages) of four interviews conducted by T. S. Kuhn on 17–20 June 1963 in Hamburg are available at the repositories of the material listed in the above-cited *Sources for History of Quantum Physics*.

II. Secondary Literature. There are no major biographical studies of Jordan. E. Brüche provided a short notice on the occasion of Jordan's sixtieth birthday in *Physikalische Blätter*, **18** (1962), 513; and J. Ehlers, Jordan's collaborator during the 1960's, wrote an appreciation on his seventieth birthday, *ibid.*, **28** (1972), 468–469.

Jordan's philosophical convictions are discussed by H. Laitko, "Zur philosophischen Konzeption des Physikers Pascual Jordan. Versuch einer kritischen Analyse" (Ph.D. diss., Berlin, 1964). Jordan's neopositivistic views in the 1930's aroused much opposition from members of the Vienna circle, such as O. Neurath, "Jordan, Quantentheorie und Willensfreiheit," in *Erkenntnis*, **5** (1935), 179–181; H. Reichenbach, "Metaphysik bei Jordan?" *ibid.*, 178–179; M. Schlick, "Ergänzende Bemerkungen über P. Jordans Versuch einer quantentheoretischen Deutung der Lebenserscheinungen," *ibid.*, 181–183; and E. Zilsel, "P. Jordans Versuch, den Vitalismus quantenmechanisch zu retten," *ibid.*, 56–65.

Since Jordan was an active member of the Bundestag from 1957 until 1961, his political actions were also discussed thoroughly in the press. See S. Nowak, "Der Anti-Göttinger," in *Rheinischer Merkur*, **12**, no. 35 (1957), 6. Concerning Jordan's collaboration with Max Born, see Born, *My Life* (New York, 1978), and *The Born-Einstein Letters*, Irene Born, trans. (New York, 1971). Jordan's contributions to quantum mechanics are described in E. Bagge, "Pascual Jordan und die Quantenphysik," in *Physikalische Blätter*, **34** (1978), 224–228; M. Jammer, "Pascual Jordan und die Entwicklung der Quantenphysik," in *Naturwissenschaftliche Rundschau*, **37** (1984), 1–9; J. Mehra and H. Rechenberg, *The Historical Development of Quantum Theory*, III (New York, Heidelberg, and Berlin, 1982); and in the introduction to B. L. van der Waerden's *Sources of Quantum Mechanics* (Amsterdam, 1967).

More detailed historical studies have been carried out on Jordan's work on quantum field theory by J. Bromberg, "The Concept of Particle Creation Before and After Quantum Mechanics," in *Historical Studies in the Physical Sciences*, **7** (1976), 161–191; and by O. Darrigol, "The Origin of Quantized Matter Waves," *ibid.*, **16** (1986), 197–253.

KARL VON MEYENN

JORDAN, (HEINRICH ERNST) KARL (*b.* Almstedt, near Hildesheim, Germany, 7 December 1861; *d.* Tring, Hertfordshire, England, 12 January 1959), *entomology, evolution.*

From the 1890's to the 1930's Jordan was the world's foremost animal taxonomist. He was the youngest of seven children of a farmer who died when Karl was five. The generosity of an uncle enabled Jordan to attend high school at Hildesheim and the University of Göttingen, where he received a training in the best tradition of German botany and zoology. Among his teachers was Ernst Ehlers, a distinguished comparative anatomist of invertebrates, who brought him into contact with the battles concerning the causes of evolutionary change then raging among August Weismann, Ernst Haeckel, Theodor Eimer, Wilhelm Haacke, and others. In 1886 Jordan received his Ph.D. summa cum laude.

After a year in the army, Jordan acquired a teaching diploma and in 1888 began to teach in the gymnasium at Münden. In 1891 he married Minna Brünig; she died in 1925. At Münden he was in active scientific contact with several local biologists, one being the ornithologist Count Hans Berlepsch. In Berlepsch's house he met the future director of the Rothschild Museum, Ernst Hartert, and a close friendship soon developed between them. It was Hartert who in 1892 persuaded Lord Walter Rothschild to engage Jordan for his private museum at Tring, Hertfordshire, England, which Jordan joined in 1893, beginning a brilliant career extending over more than sixty years. During the next forty years Hartert was the outstanding ornithologist in the world, and Jordan the premier entomologist.

Since his early school days Jordan had been an enthusiastic naturalist and beetle and butterfly collector. Relieved of all teaching duties, he could now devote himself entirely to his taxonomic researches. Although his Ph.D. dissertation dealt with butterflies, his main activity from 1886 to 1895 was devoted to beetles. Among his bibliography of more than 450 papers, more than a third are devoted to the Coleoptera, particularly the family Anthribidae. His first major contribution to the Lepidoptera, part of a paper written with Lord Walter Rothschild, dates from 1895. This was followed by monographs on *Papilio*, *Charaxes* (1898–1903), and the Sphingidae (1903). The latter has been characterized as "possibly the finest example of a taxonomic monograph that has ever been produced in Lepidoptera." In contrast with other contemporary authors, who based their classification primarily on color pattern, Jordan took numerous anatomical and behavioral characteristics into consideration, producing classifications that have stood the test of time remarkably well. Among the more than one hundred publications on Lepidoptera in the ensuing fifty years are many important monographs on genera and subfamilies, but none as lavishly illustrated as those on *Charaxes* and the sphingids.

Beginning in 1906, Jordan published papers (up to 1923, mostly in joint authorship with N. Charles Rothschild) on fleas (Siphonaptera). He was soon the world's outstanding authority on this group. The more than 1,280 figures in 140 publications on fleas were all drawn by Jordan himself. His reclassification of this previously chaotic order of insects is now largely adopted. As in other parts of his taxonomic research, Jordan constantly asked profound biological questions. How are specializations in certain species and genera of fleas connected with the structure of hair or feathers of the hosts? Among the abundance of variable characters, which best indicate close relationship? What contribution do fleas make to zoogeography? Taxonomy for Jordan was always a means to an end.

Jordan's taxonomic publications were in a class by themselves. The accuracy of the descriptions and figures was matched by a careful population analysis of large series of specimens; allopatric "species" were carefully studied for the possibility of conspecificity; synonymies were conscientiously compiled; and, most of all, the totality of characters was carefully evaluated in order to determine relationship.

Jordan's greatest contribution, however, was his development of the principles of population systematics in a series of papers (1895, 1896, 1903, 1905) issued as a deliberate challenge to the typological taxonomy prevalent—indeed, quite universal—at the time (Mayr 1955). What he promoted in 1895, in an introduction to a monograph on Oriental *Papilio* by Rothschild, was the gist of the "new systematics." If systematics has by now regained some of the prestige it had in former centuries, it is largely due to Jordan's pioneering contributions. Eventually most evolutionists agreed with his claim "Sound systematics are the only safe bases upon which can be built up sound theories as to the evolution of the diversified world of live beings" (1910, p. 385).

Jordan was particularly emphatic about the distinction between individual and geographical variation. Being a strict believer in allopatric speciation, he considered only subspecies ("geographical varieties") as incipient species. In a series of papers he attempted to refute the claims of sympatric speciation frequently made by other entomologists. Just as he believed in comprehensive polytypic species, so he also believed in large genera. Like most distinguished taxonomists, Jordan was definitely a "lumper." As a taxonomist, he frequently had to make decisions as to the relationship of a species or genus and the best way to subdivide a higher taxon. He was almost a genius in finding the best solution, inducing Walter Rothschild once to remark, half admiringly, half despairingly: "Oh, he is always right."

Jordan was very modest, never pushing himself forward, but his merits were so fully appreciated that he was eventually made honorary life president of the International Congresses of Entomology and of the International Commission on Zoological Nomenclature, and a fellow of the Royal Society, to mention only a few of his many honors. He was immensely helpful to other workers and played an important role after World War I in reviving good relations among the entomologists of the formerly warring nations. Also notable is the help that he gave Lord Rothschild in building up the world's greatest collections of butterflies and fleas.

Jordan enjoyed remarkably good health into his old age. Well into his eighties he collected butterflies and fleas in the Alps, and at age ninety-three he studied the complex genitalic structures of fleas through the microscope and drew the illustrations for his papers with a steady hand. He died at ninety-seven after a short illness.

BIBLIOGRAPHY

I. Original Works. Many of Jordan's monographs, although published in the Tring Museum journal, *Novitates zoologicae*, are virtually books—for example, the monograph on *Charaxes* (1898–1903) and the revision of the Sphingidae (1903), both written with Walter Rothschild. Other important papers are "On Mechanical Selection and Other Problems," in *Novitates zoologicae*, **3** (1896), 426–525; "Der Gegensatz zwischen geographischer und nicht-geographischer Variation," in *Zeitschrift für wissenschaftliche Zoologie*, **83** (1905), 151–210; "The Systematics of Some Lepidoptera . . . and Their Bearing on General Questions of Evolution," in *Proceedings of the International Entomological Congress*, I (Brussels, 1910), 385–404; and "Notes on Arctiidae," in *Novitates zoologicae*, **23** (1916), 124–150

II. Secondary Literature. There is a festschrift for Jordan in *Transactions of the Royal Entomological Society of London*, **107** (December 6, 1955), 1–402, with a biography by Miriam Rothschild and evaluations of Jordan's contribution to the knowledge of fleas (Robert Traub), anthribid beetles (E. C. Zimmerman), lepidopterans (E. G. Munroe), and systematics and evolution (Ernst Mayr), and a bibliography to 1954. See also N. D. Riley, "Heinrich Ernst Karl Jordan," in *Biographical Memoirs of Fellows of the Royal Society*, **6** (1960), 106–133, with bibliography; and Miriam Rothschild, *Dear Lord Rothschild* (London, 1983), with an account of Jordan's role at Rothschild's museum.

Ernst Mayr

JULIAN, PERCY LAVON (*b.* Montgomery, Alabama, 11 April 1899; *d.* Waukegan, Illinois, 19 April 1975), *organic chemistry*.

Percy Lavon Julian was the oldest of the six children of James Sumner and Elizabeth Lena (Adams) Julian. His mother was a schoolteacher, and his father, originally a teacher, took a job as a railway mail clerk shortly after Percy was born. Although public education for Alabama blacks at the time rarely extended beyond eighth grade, in 1916 Julian graduated from the State Normal School for Negroes in Montgomery, which his parents had attended, and he entered DePauw University in Greencastle, Indiana. When he received his A.B. degree in 1920, Julian was elected to Phi Beta Kappa and named valedictorian of his class. For two years he served as an instructor in chemistry at Fisk University, then began graduate study at Harvard University, where he received his A.M. degree in chemistry in 1923. He remained at Harvard as a research fellow until 1926. He then served as professor of chemistry, for one year at West Virginia State College for Negroes and for two years at Howard University.

In 1929 Julian received a fellowship from the Rockefeller Foundation to continue his graduate work at the University of Vienna. His research there under the direction of Ernst Späth on the alkaloids of *Corydalis cava* earned him the Ph.D. degree in 1931. He returned to Howard University as head of the chemistry department for one year, then was a research fellow and taught organic chemistry at DePauw University from 1932 to 1936. On 24 December 1935 Percy Julian married Anna Johnson, a sociologist. They had two children, Percy Lavon, Jr., and Faith Roselle.

When he was denied a professorship at DePauw because of prejudice against blacks, Julian became director of research of the Soya Products Division of the Glidden Company, Chicago, Illinois. Later he also was named director of research for Glidden's Durkee Famous Foods Division and manager of the Fine Chemicals Division. In 1954 Julian started his own chemical firm, Julian Laboratories, which manufactured intermediates for industrial production of steroids. After selling his company in 1961 he continued as president for several years. In 1964 he founded the Julian Research Institute, where he served as director until his death from cancer at age seventy-six.

Julian's financial success and fame enabled him to be a major force in the civil rights movement. In 1967 he was appointed cochairman of a group of successful black Americans who raised money for the Legal Defense and Education Fund of the National Association for the Advancement of Colored People. This fund enabled lawsuits to be brought to enforce civil rights laws. He was active in civic affairs and many national organizations and was elected a member of the National Academy of Sciences in 1973.

Percy Julian's significant research in the chemistry of natural products began with the total synthesis of physostigmine, an alkaloid isolated from Calabar beans and important in the treatment of glaucoma. This he accomplished in 1935 with the assistance of Josef Pikl, formerly a fellow student at the University of Vienna whom Julian had assisted in coming to the United States to work with him. Their total synthesis confirmed the structural formula assigned to physostigmine. Julian also at this time extracted from Calabar beans stigmasterol, a phytosterol that reportedly could serve as raw material for synthesis of male and female hormones.

Julian's principal work during his eighteen years directing research at the Glidden Company was the development of processes for preparation of substances of commercial value from soybeans. When he arrived in 1936, the firm had purchased a new plant in Germany for the extraction of oil from soybeans for the production of refined oils for paints and other uses; he supervised the assembly of this plant at Glidden. Julian's first laboratory task was to develop processes for further separation and purification of the various components of the extracted, oil-free soybean meal and to determine uses for them. Julian's laboratory prepared refined protein products that could replace milk casein in industrial applications, such as coating and sizing of paper and in manufacture of paints and other Glidden products. Julian designed and supervised the construction of a plant to produce these soya proteins; its eventual daily output of forty tons of commercially important products made the Soya Products Division one of Glidden's most profitable divisions. Soy protein was also developed as the base of an oxygen-impenetrable, fire-fighting foam used by the U.S. armed forces in World War II. The remaining meal, following the separation of these protein products, was shown to be of value as a supplement in livestock and poultry feeds. Julian and his laboratory workers also developed derivatives of the crude oil fraction, including edible oils for preparation of margarine, salad oils, and dressings, and soy lecithin, used widely as an emulsifier in manufacture of foods. As the scope of research expanded, Julian hired additional chemists, some of whom were his former students at DePauw University, and within ten years his research program developed into an important industrial laboratory.

A change in the direction of Julian's research occurred in 1940 when he began recovering sterols from soybean oil. At that time clinicians were discovering many uses for the newly discovered sex hormones. However, only minute quantities of progesterone, testosterone, and the estrogens could be produced from the extraction of hundreds of pounds of spinal cords, testicles, or ovaries. Soybean oil was known to contain a mixture of sterols, and Julian perfected means by which soya sterols could be isolated and converted to these hormones. At one stage in processing the oil a porous, solid, foamlike soap could be produced, from which the soya sterols were extracted with solvents. Glidden manufactured large quantities of progesterone and testosterone, utilizing on an industrial scale the chemical reactions for the modification of sterols devised in Julian's laboratory, which now included over fifty chemists. Julian and his co-workers published many articles on the chemistry of sterols in the *Journal of the American Chemical Society*, and they obtained patents on key processes in preparation of steroids from soybeans.

In 1949 the importance of cortisone for treatment of rheumatoid arthritis produced a rush by many pharmaceutical firms to produce cortisone and its analogs, yet most of the syntheses were dependent upon scarce bile acid starting materials. Julian undertook the synthesis of cortisone from pregnenolone, available in abundance from soya sterol, and devised a multistep process for conversion of pregnenolone to cortexolone (Reichstein's Substance *S*), which differs from cortisone in lacking an oxygen at the *C*-11 position. The industrial production of cortisone was revolutionized in 1952 when the Upjohn Company announced that oxygenation at *C*-11 could be accomplished in high yield by microbiological oxidation. This made progesterone, pregnenolone, and cortexolone valuable intermediates for the production of a variety of corticoid drugs, and Upjohn sought manufacturers capable of supplying the ton quantities needed.

It was at this time that Percy Julian left his position at Glidden to start Julian Laboratories in Franklin Park, Illinois. Julian realized that an even more abundant plant source of hormones was a Mexican *Diosorea*, commonly called wild yam, rich in diosgenin. In a subsidiary in Mexico City, Laboratorios Julián de México, 16-dehydropregnenolone oxide was prepared from diosgenin and shipped to Julian Laboratories in the United States. There Julian and his team of chemists transformed "oxide" into many derivatives of cortexolone through a complex series of reactions they devised. These valuable intermediates were sold to pharmaceutical firms for production via microbiological oxidation of a wide variety of steroids of the cortisone family. In 1961 Julian sold his firm to Smith, Kline, and French Laboratories for 2.3 million dollars. In later years he continued his research on steroid chemistry as director of the Julian Research Institute. Percy Julian's skill in guiding industrial research and his creativity in novel chemical syntheses are well illustrated in his publications and patents dealing with chemical manipulations of the structures of steroids.

BIBLIOGRAPHY

I. Original Works. A list of Julian's scientific publications and patents is in Bernhard Witkop, "Percy Lavon Julian, 1899–1975," in *Biographical Memoirs. National Academy of Sciences*, **52** (1980), 223–266. His most important articles include "Studies in the Indole Series. V. The Complete Synthesis of Physostigmine (Eserine)," in *Journal of the American Chemical Society*, **57** (1935), 755–757, written with Josef Pikl; "Sterols. IX. The Selective Halogenation and Dehalogenation of Certain Steroids (Part I)," *ibid.*, **72** (1950), 362–366, written with William J. Karpel; "Sterols. XIII. Chemistry of the Adrenal Cortex Steroids," in *Recent Progress in Hormone Research*, **6** (1951), 195–214; "Sterols. XV. Cortisone and Analogs. Part I. 16α-Hydroxy and 16α,17α-Epoxy Analogs to Cortisone," in *Journal of the American Chemical Society*, **77** (1955), 4601–4604, written with Wayne Cole, Edwin W. Meyer, and Bernard M. Regan; and "Sterols. XVI. Cortisone and Analogs. Part 2. 17α,21-Dihydroxy-4-pregnene-3,12,20-trione," *ibid.*, **78** (1956), 3153–3158, written with Chappelle C. Cochrane, Arthur Magnini, and William J. Karpel.

Addresses and written works for general audiences include "The Chemist as Scholar and Humanist," in *Chemist* (American Institute of Chemists), **42** (1965), 101–104; "On Being Scientist, Humanist, and Negro," in Stanton L. Wormley and Lewis H. Fenderson, eds., *Many Shades of Black* (New York, 1969), 147–157; and "Science, an Ally of the Humanities," in *American Scientist*, **63** (1975), 13–15.

Some personal papers are in the possession of Dr. Anna J. Julian.

II. Secondary Literature. W. Montague Cobb, "Percy Lavon Julian," in *Journal of the National Medical Association*, **63** (1971), 143–150, and "Onward and Upward—Percy Lavon Julian, 1899–1975," in *Crisis*, **85** (1978), 166–171; Louis Haber, *Black Pioneers of Science and Invention* (New York, 1970), 86–101; and Max Tishler, "Percy L. Julian, the Scientist," in *Chemist* (American Institute of Chemists), **42** (1965), 105–113. The text of the addresses given at a symposium to honor Julian on 12–13 May 1972 at MacMurray College is in the library of that college in Jacksonville, Illinois.

Daniel P. Jones

KALMÁR, LÁSZLÓ (*b*. Edde, Hungary, 27 March 1905; *d*. Mátraháza, Hungary, 2 August 1976), *mathematics, mathematical logic, philosophy of mathematics, computer science.*

László Kalmár was the youngest child of Zsigmond

Kalmár and Róza Krausz Kalmár. His father was an estate bailiff on a manor situated in Transdanubia, about 30 kilometers from Lake Balaton. About the beginning of World War I, Kalmár moved with his widowed mother to Budapest, where he attended secondary school. His outstanding mathematical abilities were already evident; he had read and understood Cesàro's calculus text when he was only thirteen. He studied mathematics and physics at the University of Budapest between 1922 and 1927. Despite unhappy circumstances (his mother had died earlier), he was very successful in his studies and was considered by his fellow students as their master in mathematics. During his university years he studied under such eminent mathematicians as József Kürschák and Lipót Fejér. After obtaining a Ph.D. he accepted a faculty position at the University of Szeged, where he remained until his retirement in 1975. He was an assistant to Alfréd Haar and Frigyes Riesz from 1930 to 1947. He was promoted in 1947 to full professor. He married Erzsébet Árvay in 1933. Three of their four children survived him.

Kalmár was elected a corresponding member of the Hungarian Academy of Sciences in 1949 and as a full member in 1961. He was awarded the highest orders in Hungary for scientific activity: the Kossuth Prize in 1950 and the State Prize in 1975. He was honorary president of the János Bolyai Mathematical Society and the John von Neumann Society for Computer Science. In spite of his age, he continued his research with full energy until the last day of his life.

One of the fields in which his contribution is of greatest importance is mathematical logic. His interest in logic was aroused on a visit to Göttingen in 1929. He gave simplified proofs of several fundamental results: Bernays and Post's theorem on the completeness of the propositional calculus, Gentzen's theorem on the consistency of elementary number theory, Löwenheim's theorem on the satisfiability of any first-order sentence in a countable set, and Post and Markov's theorem on the algorithmic unsolvability of the word problem of associative systems. He analyzed carefully the possibilities for stating generally and proving straightforwardly Gödel's celebrated incompleteness theorem. He studied the interrelations and significance of the incompleteness results of Church and Gödel. Concerning Church's famous thesis in which the heuristic concept of effective calculability is identical to the precise notion of general recursivity, he advocated the view that the limits of effective calculability become ever broader, and that therefore this concept cannot be identified permanently with an unalterably fixed notion. He wrote, partly with János Surányi, a series of articles on the reduction theory of the so-called *Entscheidungsproblem* (the decision problem of mathematical logic).

Kalmár also was extensively involved in theoretical computer science. He concerned himself from the mid 1950's with the mathematics of planning and programming electronic computers. He dealt with adapting the usual mathematical formula language and the programming languages to each other and with questions of mathematical linguistics. In addition, he wrote papers on defining the field of cybernetics, the use of computers, and the applicability of cybernetical ideas in various sciences.

It is common knowledge that mathematics can be applied widely in more practical fields. Kalmár often expressed his conviction that the connection with other domains of science is important to both sides, because the influence of more empirical sciences may be the source of permanent inspiration for the development of mathematics. He wrote a large number of articles popularizing mathematics. Some of his articles, written in Hungarian, are so constructed that the paper begins with a broad survey of a branch of logic before concluding with his own results.

Kalmár's scholarly personality was vivid and well rounded. His work cannot be discussed adequately by considering only his published works. He was enthusiastically inclined toward various sorts of personal contacts in his profession: regular teaching of university students, informal discussions with colleagues, lectures to general audiences. He taught primarily calculus, beginning with integral and then continuing with differential calculus, and foundations of mathematics (set theory and mathematical logic). His ideas on teaching calculus were explained in a posthumous book compiled by his pupils.

In the Department of Mathematics at the University of Szeged, he was the founder and first occupant of the Chair for Foundations of Mathematics and Computer Science. He also founded the Cybernetical Laboratory, which bears his name, and the Research Group for Mathematical Logic and Automata Theory at the university.

He was member of several scientific committees and editorial boards of scientific journals. The journals *Acta cybernetica* and *Alkalmazott matematikai lapok* were founded by him.

The names of Hungarian mathematicians whose scientific activity was essentially promoted by Kalmár would fill a long list. Of these, the following four persons are most noteworthy. Rózsa Péter was Kalmár's contemporary. Her basic contributions to the theory of recursive functions was close to one of the research areas of Kalmár. Among his younger colleagues, he was the teacher of the set theorist Géza Fodor. The algebraists Tibor Szele and Andor Kertész were also extensively encouraged and guided by him.

From the viewpoint of the development of the sciences in Hungary, Kalmár will probably be remembered most for his ceaseless effort in promoting the development of computer science and the use of computers in his country.

BIBLIOGRAPHY

I. ORIGINAL WORKS. "A Hilbert-féle bizonyításelmélet célkitűzései, módszerei és eredményei" (The purposes, methods, and results of the Hilbertian Proof Theory), in *Matematikai és fizikai lapok*, **48** (1941), 65–119; "Quelques formes genérales du théorème de Gödel," in *Comptes rendus de l'Academie des sciences* (Paris), **229** (1949), 1047–1049; "Ein direkter Beweis für die allgemein-rekursive Unlösbarkeit des Entscheidungsproblems des Prädikatenkalküls der ersten Stufe mit Identität," in *Zeitschrift für mathematische Logik und Grundlagen der Mathematik*, **2** (1956), 1–14; "An Argument Against the Plausibility of Church's Thesis," in A. Heyting, ed., *Constructivity in Mathematics* (Amsterdam, 1959), 72–80; "Über einen Rechenautomaten, der eine mathematische Sprache versteht," in *Zeitschrift für angewandte Mathematik und Mechanik*, **40** (1960), T64–T65; "A kvalitatív, információelmélet problémái" (The problems of qualitative information theory), in *Magyar tudományos akadémia Matematikai és fizikai*, **12** (1962), 293–301; "Foundations of Mathematics—Whither Now?" in Imre Lakatos, ed., *Problems in the Philosophy of Mathematics* (Amsterdam, 1967), 187–207; "Meaning, Synonymy, and Translation," in *Computational Linguistics*, **6** (1967), 27–39; *Bevezetes a matematikai analízisbe* (Introduction to mathematical analysis; Budapest, 1982).

II. SECONDARY LITERATURE. For a systematic treatment of the reduction theory of the *Entscheidungsproblem*, including a number of Kalmár's results in this area, see János Surányi, *Reduktionstheorie des Entscheidungsproblems im Prädikatenkalkül der ersten Stufe* (Budapest, 1959). See also R. Péter, "Kalmár László matematikai munkássága" (The mathematical activity of László Kalmár), in *Matematikai lapok*, **6** (1955), 138–150; A. Ádám, "Kalmár László matematikai munkásságáról" (On the mathematical activity of László Kalmár), *ibid.*, **26** (1975), 1–10, which includes a detailed bibliography. Brief obituaries appeared in *Acta scientiarum mathematicarum*, **38** (1976), 221–222, and *Alkalmazott matematikai lapok*, **2** (1976), 151–155 (with bibliography).

ANDRÁS ÁDÁM

KARGIN, VALENTIN ALEKSEEVICH (*b.* Ekaterinoslav [now Dnepropetrovsk], Russia, 23 January 1907; *d.* Moscow, U.S.S.R., 21 October 1969), *chemistry*.

Kargin began his scientific activities unusually early: having finished secondary school in 1922, the fifteen-year-old youth became a laboratory assistant. By 1927 he had written nine articles on problems in analytical chemistry and electrochemistry. From 1925 to 1927 he worked as a practical student in the laboratory of analytical chemistry at the L. Ia. Karpov Institute of Physical Chemistry, as an assistant to the chemist of the Rudmetalltorg Trust, as a chemist for the Group on Radioactive Ores (directed by Aleksandr Evgenievich Fersman), and as senior chemist of the Russian Gems Trust.

From 1925 to 1930, while continuing to work, Kargin studied at the Physical-Mathematical Faculty of Moscow University, from which he graduated in 1930. In 1927 he was hired as a scientific worker at the Karpov Institute of Physical Chemistry. In 1936 he was awarded the doctorate in chemistry without having defended a dissertation.

At the Karpov Institute, Kargin worked from 1927 to 1937 on developing a theory of the coagulation of colloids by electrolytes. His works were devoted to the creation of new types of electrodes for potentiometry, and in the 1920's these works played a major role in spreading high-efficiency potentiometric methods in the Soviet Union. In the 1920's and 1930's Kargin revealed the physicochemical nature of fundamental differences between lyophilic and lyophobic colloids. His work was distinguished by its combination of practical direction with major theoretical achievements. His research promoted the creation of new methods of analysis and purification of substances, as well as the strengthening with liquid glass of bore holes and soils.

In 1937 Kargin headed the Laboratory of Macromolecular Compounds of the Karpov Institute. By the mid 1930's he had become interested in a new area of physical chemistry of macromolecular compounds, the production of artificial fibers. This interest was also stimulated by practical interest—an urgent demand for the development of the scientific basis for new methods of producing artificial fibers and films formed from polymer liquids. The study of the nature and properties of polymer liquids led to the conclusion that they were thermodynamically reversible systems. As early as the 1930's Kargin showed the correctness of applying many concepts proposed for colloid solutions of low-molecular-weight substances to such polymer liquids at the same time that the principal differences were being discovered.

The clarification of the nature of polymer liquids permitted Kargin to begin studying the structure of polymers and the processes of forming artificial fibers and films from their solutions. This work, as well as the thermodynamic investigations of the molecular nature of polymer bodies, permitted the disclosure of regularities of the mechanical and thermomechanical properties of polymers, and the connection between the physicochemical properties of polymer materials and the structure on the molecular and submolecular levels. These works led to the structural-chemical and physical modification of chemical fibers, plastics, and rubbers. This research was closely connected with the development of problems on the deterioration of polymers and the fatigue of elastic polymers (rubbers) that began at the end of the 1940's. Kargin also conducted a study of the structural characteristics of the reaction medium in the formation of macromolecules.

During the war years, 1941–1945, Kargin worked on practical problems related to increasing the defense and economic potential of the Soviet Union. He developed and introduced a new method for processing protective materials (paper) that was awarded the State Prize in 1943. In construction he was responsible for a new method for the artificial strengthening of sandy soil saturated with water. He received the State Prize for this work in 1947.

In 1959 Kargin organized the journal *Vysokomolekuliarnye soedineniia* (Macromolecular compounds) and became its editor in chief. He was also involved in pedagogical activities: in 1954 he was appointed professor at Moscow State University, and in 1955 he organized and chaired the department of macromolecular compounds of the Faculty of Chemistry at the university. His research in the physical chemistry of macromolecular substances brought him the State Prize in 1950 and the Lenin Prize in 1962. In 1953 he was elected full member of the Academy of Sciences of the U.S.S.R.

During the last years of his life, Kargin devoted much of his time to studying the nature of the elasticity of polymers and the submolecular mechanism of deformation. With Viktor Aleksandrovich Kabanov and Nikolai Al'fredovich Plate he elaborated on the synthesis of polymers on freshly formed surfaces of hard particles. He also conducted experiments on submolecular deformation with V. A. Goldanskii. Kargin was also the founder and editor of *Entsiklopediia polymerov* (Encyclopedia of polymers).

BIBLIOGRAPHY

I. Original Works. *Kratkie ocherki po fiziko-khimii polimerov* (Short articles on the physical chemistry of polymers; 2nd ed., Moscow, 1967), with G. L. Slonimskii; *Kolloidnye sistemy i rastvory polimerov. Izbrannye trudy* (Colloid systems and polymer solutions. Collected works; Moscow, 1978); and *Struktura i mekanicheskie svoistva polimerov* (The structure and mechanical properties of polymers; Moscow, 1979).

II. Secondary Literature. Akademiia Nauk SSSR, *Valentin Alekseevich Kargin*, Materialy k biobibliografii uchenykh SSSR, ser. khimicheskikh nauk, no. 29 (Moscow, 1960).

A. N. Shamin

KARPECHENKO, GEORGII DMITRIEVICH (*b.* Vel'sk, Vologda (modern Arkhangel'sk) Province, Russia, 3 May 1899; *d.* in Soviet prison, location unknown, 17 September 1942), *cytogenetics, botany.*

Karpechenko was born into the family of a surveyor. He completed his secondary education at the Vologda gymnasium in 1917 and entered the natural science division of Perm University. In 1918 he transferred to the Petrovskii Agricultural Academy in Moscow (renamed the K. A. Timiriazev Agricultural Academy on 23 December 1923), where he supported himself by working for an agronomist and teaching science in local schools. He received his diploma in 1922 and went into graduate work in the academy's Department of Selection of Agricultural Plants, headed by Sergei I. Zhegalov, where he studied plant cytology with A. G. Nikolaeva.

Karpechenko's earliest work (1922–1925) was carried out at the department's Petrovskoe-Razumovskoe plant breeding station near Moscow, headed by Zhegalov. Following up the hypothesis of the Danish plant cytologist Otto Winge that natural polyploid plant species had originated historically through hybridization, attempts were under way to cross the radish with common cabbage, savoy cabbage, brussels sprouts, and kohlrabi. Karpechenko began his research career by studying the karyology of clovers (*Trifolium*), beans (*Phaseolus*), and cruciferous vegetables (Cruciferae). His cytological study of two crucifers, the radish (*Raphanus sativus* L.) and the common cabbage (*Brassica oleracea* L.), demonstrated that they had the same number of chromosomes (nine pairs, $2n = 18$) and that their chromosome sets appeared morphologically similar (1922). This finding led Karpechenko to hybridization experiments on the two species that brought him to international attention.

In 1922, aided by his teachers Zhegalov, Nikolaeva, and I. N. Sveshnikova, Karpechenko began his study of the hybrids produced from a cross between *Raphanus sativus* and *Brassica oleracea*. Both species are biennial. Typically, during the first vegetative season cytological investigations of the hybrids were conducted, and crosses were made in the second year. In 1923, 202 pollinated flowers produced 123 hybrids that were cultivated and studied. Although these F_1s each had 18 chromosomes, reduction division was abnormal and the plants appeared sterile. After the stumps were stored for the winter and replanted in the spring of 1924, they began flowering. These F_1s were studied cytologically, and the findings were reported in an article published in England (1924).

In 1925 Nikolai Vavilov was made director of the new All-Union Institute of Applied Botony and New Cultures (VIPBiNK [Vsesoiuznyi institut prikladnoi botaniki i novykh kul'tur]) in Leningrad. He invited Karpechenko, then a twenty-six-year-old graduate student at his alma mater, to head the institute's Laboratory of Plant Genetics located outside the city at Detskoe Selo (renamed Pushkin in 1937). Karpechenko accepted and held the post from 1925 until October 1940. His experimental work was transferred from Moscow to Leningrad in the autumn of 1925, and the second and subsequent generation of hybrids were cultivated and studied with the help of laboratory members S. A. Shchavinskaia, A. N. Lutkov (assistant director of the laboratory from 1925 to 1941), O. N. Sorokina, and E. P. Gogeisel. Karpechenko traveled to Europe (1925–1926) and continued his cytological investigation of the hybrids in the laboratories of Otto Winge (Den Kgl. Veterinaer- og Landbohøjskole's Arvelighedslaboratorium, Copenhagen), Erwin Baur (Institut für Vererbungsforschung, Berlin-Dahlem), and William Bateson (John Innes Horticultural Institution, London-Merton).

On the plot at Petrovskoe-Razumovskoe, where the plants were not isolated from radishes, 13 tetraploid hybrids produced 209 seeds, from which 91 F_2 hybrid plants were grown, of which 73 were investigated cytologically: 61 were triploids (27 chromosomes) that proved sterile, and it was concluded that they had resulted from promiscuous crossing of the tetraploid ($n = 18$) and the radish parental form ($n = 9$). By contrast, 6 F_1 tetraploids grown in isolation from radishes on plots at Gribovo yielded 612 seeds, from which 361 F_2 plants were grown, of which 229 were investigated cytologically: 213 were tetraploid ($2n = 36$), and subsequently proved fertile and constant. In all, 302 plants were examined cytologically. Although the other polyploid hybrid forms tended to be infertile or unstable (including triploids, and occasional pentaploids and hexaploids), the tetraploid hybrid, which combined the full diploid complement of both the radish and the cabbage, underwent normal meiosis and proved constant, fertile, morphologically distinct, and reproductively isolated from both parental species. Karpechenko concluded that it was an experimentally produced "species nova" and indeed a new genus, since it combined the genetic material from two existing genera. He named the new form *Raphanobrassica* and reported his findings in 1927 at the Vth International Congress of Genetics (Berlin), and in extensive publications in Russian (1927) and English (1928).

This work brought Karpechenko international recognition. In 1929 he won a Rockefeller Foundation stipend to study in the United States and spent October 1929 through February 1931 based at Berkeley in the laboratory of Ernest Brown Babcock. He also spent time in Pasadena at the laboratory of Thomas Hunt Morgan, where he became close friends with his fellow countryman Theodosius Dobzhansky, who was beginning his cytogenetic comparisons of various *Drosophila* species. He was invited to give a major address on distant hybridization at the plenary session of the VIth International Congress of Genetics (Ithaca, New York) in 1932. In 1934 he was made vice president of the genetics section of the VIth International Botanical Congress in Amsterdam.

Karpechenko's work also brought him prominence in the Soviet Union as one of that country's leading geneticists. In 1929 he was appointed general secretary of the All-Union Congress of Genetics, Selection, Seed Culture, and Animal Husbandry in Leningrad. In 1930 VIPBiNK was transformed into the All-Union Institute of Plant Industry (VIR [Vsesoiuznyi institut rastenievodstva]) and became the central research institution of the newly formed Lenin All-Union Academy of Agricultural Sciences (VASKhNIL [Vsesoiuznaia akademiia sel'sko-khoziaistvennykh nauk imeni V. I. Lenina]). As director of its laboratory of plant genetics (1930–1941) Karpechenko became prominent in Soviet agriculture. In 1932 he was made a member of the presidium of the all-union conference on planning research in genetics and selection for the second five-year plan. On 2 November 1934, on Vavilov's recommendation, the presidium of VASKhNIL awarded Karpechenko a doctorate of biological science without his having to defend a dissertation.

After Iurii A. Filipchenko's death in 1930, his department of experimental zoology and genetics at Leningrad University was reorganized: A. P. Vladimirskii was appointed to head the new department of animal genetics, and in 1932 Karpechenko was asked to organize the department of plant genetics. He recruited its faculty from the VIR, including Grigorii A. Levitskii (cytology), Mariia A. Rozanova (experimental systematics), Leonid I. Govorov (selection, specializing in legumes), and Doncho Kostov (*chastnaia genetika*, specialized genetics of various plant species). From 1932 through 1940, Karpechenko chaired the department, gave an advanced course on topics in plant genetics, directed the work of graduate students, and headed the Laboratory of Plant Genetics at the Peterhof Biological Institute associated with the university.

In 1935, at the age of thirty-six, Karpechenko was widely regarded as one of the world's leading authorities on polyploidy and distant hybridization in plants. Vavilov invited him to write two chapters for the three-volume collection *Theoretical Foundations of Plant Selection* (1935), "The Theory of Distant Hybridization" and "Experimental Haploidy and Polyploidy." The first of these chapters was also published as a separate booklet (1935); drawing on the Soviet and international literature, it constituted his most general theoretical statement on the subject.

Karpechenko divided distant hybridization into two categories. "Congruent" or compatible crosses are those in which "the parent forms, despite great differences in their genes, have 'corresponding' chromosomes that can combine in hybrids without lowering their viability or fertility" (1935, p. 293). Because they contain many different genes as a result of their evolution in different environmental conditions, such crosses allow the breeder to select for desirable combinations of traits (such as disease or insect resistance, growing properties, and adaptation to particular climates) that may have been lost in particular domesticated breeds and can be reintroduced from wild or local varieties. By contrast, "incongruent" or incompatible crosses are those in which the "parent forms have 'non-corresponding' chromosomes or a different number, or differences in the cytoplasm, or both, with the result that the hybrids . . . commonly show disrupted meiosis, partial or total sterility, and not infrequently developmental abnormalities" (*ibid.*). Karpechenko focused on incongruent crosses, analyzed the causes of lowered viability and fertility, set forth ways to overcome the difficulties through polyploidy, and evaluated the potentials of such crosses for the breeder.

In the 1930's Karpechenko extended his researches on *Raphanobrassica*. Although it did not easily cross with its parental forms, he found that it could be hybridized with *Brassica carinata* and *Brassica chinensis*. By crossing a tetraploid *Brassica oleracea* (4n = 36) and *Brassica chinensis* (2n = 20), Karpechenko created a fertile hexaploid (2n = 56) in 1937. In the late 1930's he worked on tetraploid barleys and investigated the artificial induction of polyploidy by use of X rays, ultraviolet light, colchicine, and other chemicals. His work led to that of A. N. Lutkov on polyploid sugar beets, of Mikhail I. Khadzhinov on hybrid corn, of Boris L. Astaurov on artificial polyploidy in the silkworm, and to many other studies on plant polyploidy and hybridization throughout the VIR's system of plant-breeding laboratories and stations.

Beginning in the year 1935, Karpechenko's positions brought him into increasing conflict with Lysenkoism. As one of the Soviet Union's leading experts on hybridization, he had to contend with the cult growing up around the popular elderly breeder Ivan V. Michurin, who had claimed spectacular and hereditarily permanent results from grafting, so-called vegetative hybridization. Following Michurin's death in 1935, Lysenko and his philosophical partner I. I. Prezent claimed that their own techniques and theories extended Michurin's legacy into a distinctly Russian Marxist science that they called "Michurinist biology." When the second edition of Morgan's *Scientific Basis of Evolution* appeared in 1935, Karpechenko and his wife, Galina Sergeevna, made the authorized translation into Russian; it was published after some delay in late 1936. At the December 1936 meeting of VASKhNIL, where Lysenko and his supporters castigated Morgan and attacked his Soviet followers, Karpechenko defended genetics and spoke out against the assertion by Prezent and Lysenko that somatic cells are involved in the formation of germinal cells and that acquired characteristics can thereby be inherited.

These skirmishes intensified over the next four years. In 1935 Karpechenko had been appointed a member of the organizing committee for the VIIth International Congress of Genetics (scheduled for 1937 in Moscow), but Lysenkoists were added, geneticists were removed, and the congress was twice postponed until it was finally held at Edinburgh in 1939 without any Soviets attending. In February 1937 the Leningrad University newspaper criticized Karpechenko as a Morganist and suggested that his views were subversive. By April, Karpechenko was complaining that the moral atmosphere at the university was making it difficult to teach genetics.

 KARPECHENKO

Nonetheless, on 28 December 1938 he was promoted to full professor and confirmed as chairman of the department of plant genetics.

At the VIR, however, the situation began to deteriorate. After Lysenko became president of VASKhNIL in 1938, he appointed his supporters to the institute staff, and they began to harass Vavilov and his protégés. In particular, a man without scientific qualifications, Stepan Shundenko, was appointed deputy director of the VIR for science; as later became clear, he was an officer of the secret police (NKVD). At a meeting organized by the journal *Pod Znamenem Marksizma* in 1939, Karpechenko tried to defend genetics and his own work. That same year, as president of VASKhNIL, Lysenko refused to permit the work of Karpechenko or members of his university department to be displayed at the All-Union Agricultural Exhibit.

In his last letter to Dobzhansky, sent from Paris as his final trip abroad was drawing to a close, Karpechenko had predicted that his return to the Soviet Union would almost certainly mean prison and death. On 6 August 1940 Vavilov was arrested by NKVD agents. In subsequent months, as the case against Vavilov was being assembled, a number of his colleagues were also seized. Karpechenko was arrested in October 1940. According to a recent account, at his very first prison interrogation he "confessed" that his artificial induction of polyploidy in vegetables was anti-Soviet in character, hoping that his judges would see the absurdity of such a claim (Popovsky, p. 148).

Karpechenko was not heard from again. In 1948 he was publicly referred to as one of those who had been eliminated as "enemies of the people." Although he was posthumously rehabilitated in the mid 1950's, only after 1965 were his works republished and the exact date of his death made known. The details of the last twenty-three months of his life and the place, cause, and circumstances of his death have yet to be revealed.

BIBLIOGRAPHY

I. Original Works. Karpechenko published some fifty works. For the papers on *Raphanobrassica* that won him international fame in the 1920's, see "Chislo khromosom i geneticheskie vzaimootnosheniia u kul'turnykh Cruciferae" (The number of chromosomes and genetic interrelationships in cultivated cruciferae), in *Trudy po prikladnoi botanike i selektsii*, **13**, no. 2 (1922), 4–14; "Hybrids of ♀ *Raphanus sativus* L. x ♂ *Brassica oleracea* L.," in *Journal of Genetics*, **14**, no. 2 (1924), 375–394; "The Production of Polyploid Gametes in Hybrids," in *Hereditas*, **9** (1927), 349–368; "Poliploidnye gibridy *Raphanus sativus* L. x *Brassica oleracea* L. (K probleme eksperimental'nogo vidoobrazovaniia)," in *Trudy po prikladnoi botanike, genetike, i selektsii*, **17**, no. 3 (1927), 305–410, published in English as "Polyploid Hybrids of *Raphanus sativus* L. *Brassica oleracea* L. (On the Problem of Experimental Species Formation)," in *Zeitschrift für induktive Abstammungs- und Vererbungslehre*, **48**, no. 1 (1928), 1–85.

In the 1930's Karpechenko published several works of a more general character, notably "Teoriia otdalennoi gibridizatsii" (The theory of distant hybridization), in *Teoreticheskie osnovy selektsii rastenii* (Theoretical foundations of plant selection), I (Moscow and Leningrad, 1935), 293–354 (also published as a separate pamphlet); and "Eksperimental'naia poliploidiia i gaploidiia" (Experimental polyploidy and haploidy), *ibid.*, 398–434. Karpechenko also published the authorized translation of Thomas Hunt Morgan, *The Scientific Basis of Evolution* (2nd ed., New York, 1935), as *Eksperimental'nye osnovy evoliutsii* (The experimental basis of evolution; Moscow and Leningrad, 1936). For his comments at the December 1936 VASKhNIL meeting, see O. M. Targul'ian, ed., *Spornye voprosy genetiki i selektsii* (Issues of genetics and selection; Moscow and Leningrad, 1937), 281–284.

Two of Karpechenko's papers are reprinted in *Klassiki sovetskoi genetiki 1920–1940* (Classics of Soviet genetics 1920–1940; Leningrad, 1968), 461–538; and fifteen in G. D. Karpechenko, *Izbrannye trudy* (Selected works), G. S. Karpechenko, O. N. Sorokina, and V. V. Svetozarova, comps., A. N. Lutkov and D. V. Lebedev, eds. (Moscow, 1971), which also includes a brief biography and a list of his published works.

II. Secondary Literature. See V. N. Lebedev's biography in *Vydaiushchiesia sovetskie genetiki* (Leading Soviet geneticists; Moscow, 1980), 37–48. Of special interest is E. S. Levina, "Iz istorii otechestvennoi genetiki: N. I. Vavilov i G. D. Karpechenko" (From the history of genetics of our country: N. I. Vavilov and G. D. Karpechenko), in *Genetika*, **23**, no. 11 (1987), 2007–2019, which reprints eight unabridged letters between Karpechenko and Vavilov (1925–1938) and provides valuable information on their institutional and personal relationship.

For discussions of Karpechenko's scientific work, see the studies by Z. M. Rubtsova: "Znachenie rabot G. D. Karpechenko dlia razvitiia evoliutsionnoi tsitogenetiki" (The significance of G. D. Karpechenko's work for the development of evolutionary cytogenetics), in *Iz istorii biologii* (From the history of biology), IV (Moscow, 1973), 148–159; *Razvitie evoliutsionnoi tsitogenetiki rastenii v SSSR (1920–1940-e gody)* (The development of evolutionary cytogenetics of plants in the U.S.S.R., 1920–1940's; Leningrad, 1975); and her chapters in S. R. Mikulinskii and Iu. I. Polianskii, eds., *Razvitie evoliutsionnoi teorii v SSSR (1917–1970-e gody)* (The development of evolutionary theory in the U.S.S.R., 1917–1970's; Leningrad, 1983), 92–128.

In English, see the occasional references to Karpechenko

in David Joravsky, *The Lysenko Affair* (Cambridge, Mass., 1970); Zhores A. Medvedev, *The Rise and Fall of T. D. Lysenko*, I. Michael Lerner, trans. (New York, 1969); and Mark Popovsky, *The Vavilov Affair* (Hamden, Conn., 1984).

MARK B. ADAMS

KAY, MARSHALL (*b*. Paisley, Ontario, Canada, 10 November 1904; *d*. Englewood, New Jersey, 3 September 1975), *geology*.

Marshall Kay was the older of two sons born to George Frederick and Bethea Kay, both natives of Ontario, but of Presbyterian Scottish and Yorkshire derivation. The father, who was educated at Toronto and Chicago and became a distinguished geologist, moved his family to the United States in 1904 when he accepted a position on the faculty of the University of Kansas. In 1907 he transferred to the University of Iowa as professor of Pleistocene geology, but soon became department head and state geologist (1911) and then dean of liberal arts (1917–1941). Marshall became a United States citizen at the age of five and grew up in Iowa City, but he retained a lifelong affection for Canada and Scotland. Indeed, he did considerable research in eastern Canada. He was originally christened "George Marshall Kay" and in his early years used "G. Marshall," but he later dropped "George" entirely to minimize professional confusion with his father.

Besides the model of an illustrious geologist father, Marshall Kay was influenced as a student by such other Iowa faculty members as J. J. Runner, A. O. Thomas, A. C. Trowbridge, and C. K. Wentworth. He was also in the company of several exceptionally promising geology classmates, such as R. E. King, P. B. King, H. S. Ladd, A. Pabst, and M. Stainbrook. These nascent young scientists amused themselves by solving mathematical puzzles and practicing memory training with railroad timetables and baseball statistics. Marshall received the B.S. degree cum laude from Iowa in 1924 and the M.S. in 1925. He shared the Lowden Prize with P. B. King and received a special citation for his M.S. thesis. In 1924 he attended a University of Chicago field geology course taught at Ste. Genevieve, Missouri, which complemented his 1923 University of Iowa field course conducted at Baraboo, Wisconsin. Kay always considered himself to be a field geologist, and he continued to do active field research up to his death.

Kay chose Columbia University for further postgraduate study, and after receiving the Ph.D. in 1929 commenced a forty-four-year teaching career at that institution. He began as lecturer and then instructor at Barnard College (1929–1931), soon became instructor at Columbia College (1931–1937), and subsequently advanced through the professorial ranks to his ultimate appointment as Newberry Professor of Geology in 1967. Marshall frequently represented Columbia on the faculty of the University of Wyoming's Summer Science Field Camp, where he met his wife-to-be, Inez Clark, who was a botany student from the University of Michigan. They were married in 1935 and raised four children, three of whom studied earth science.

Geology at Columbia University had its beginning in 1792 with the appointment of Samuel L. Mitchill as professor of chemistry and natural history in Columbia College. Edinburgh-trained Mitchill, who was a well-known contemporary of William Maclure, comte de Volney, and Amos Eaton, was one of the first persons to teach geology in an American university. Stratigraphic geology, which was Kay's specialty, was not identifiable as a distinct study at Columbia until the creation of the department of geology and paleontology and the appointment of John S. Newberry as its first professor (1866–1892). Stratigraphy continued under the guidance of Amadeus W. Grabau (1901–1919) and J. J. Galloway (1919–1931). Kay became the fourth in this distinguished lineage when he took the reins in 1931 from his mentor, Galloway, who had resigned to move to Indiana.

Kay's midwestern roots led to an early emphasis upon Paleozoic stratigraphic paleontology. His M.S. thesis director at Iowa, A. O. Thomas, had himself been a student of paleontologist Stuart Weller at the University of Chicago. Weller's approach emphasized faunal assemblages more than phylogeny, and it stressed the ability to identify fossils in the field. The Weller stamp was clearly transferred by Thomas, for Kay both practiced and taught this approach at Columbia. Kay's thesis research on the stratigraphy and paleontology of the Ordovician Decorah Formation in Iowa began his lifelong love affair with lower Paleozoic strata and their invertebrate fossils; only rarely did he stray very far from the Ordovician.

Stratigraphic paleontology or biostratigraphy began in America a century before Kay's student days with Lardner Vanuxem (1792–1848), who introduced the concept of index fossils and faunal correlation developed first in Europe by William Smith and Georges Cuvier. This field flourished in the latter part of that century in the hands of James Hall (1811–1898), one of Vanuxem's many protégés and younger colleagues on the newly organized New

York Geological Survey. By the 1920's and 1930's, stratigraphic paleontology was a maturing specialty in which heated debates often developed over issues that assumed a degree of refinement of the fossil record not always justified by the available evidence. Some of these exchanges were heated and personal.

Young Kay was himself involved in some of the acrimony, especially with the opinionated E. O. Ulrich, who had proposed in 1911 a whole new geologic system, the Ozarkian, to include uppermost Cambrian and lower Ordovician strata of the conventional time scale. Ulrich discounted Kay's early work on the Ordovician because it did not support his own prejudices. This was not to be the last time that Marshall would be involved in controversy.

Kay's own research interests and teaching mirrored the evolution of stratigraphy in America during the middle twentieth century. Although he never abandoned his early training in biostratigraphy, already during the 1930's he was becoming increasingly interested in physical aspects of the subject, that is, the fossil-containing strata themselves and their complex variations both laterally and vertically. He acknowledged that it was J. J. Galloway who had inspired his own interest as a graduate student by "taking the dry bones of stratigraphy and giving them flesh and life." Kay, likewise, gave life to a subject that, in the hands of less skilled teachers, could be the dullest of all geology courses. He stressed the distribution of strata in time and space as records of tectonic evolution and used the stratigraphic record to illustrate general principles rather than to present it descriptively as an end in itself. The book *Stratigraphy and Life History* (1965), coauthored with E. H. Colbert, provides a hint of the approach. One of his pedagogical techniques was to insist that students extract many of the regional relationships themselves from stratigraphic data provided on countless posterlike diagrams drawn from the literature for teaching aids.

During the 1940's and 1950's the preeminence of the classic biostratigraphic emphasis was being supplemented by a new, rapidly evolving physical approach to stratigraphy, which received great impetus from both the needs of and the subsurface data from the petroleum industry. Together with other Americans such as L. L. Sloss, W. C. Krumbein, E. C. Dapples, H. E. Wheeler, and A. I. Levorsen, Kay emphasized the importance of lithofacies as well as biofacies, stressed the inherent limitations of index fossils, and refined the concepts of chrono- (or time-) stratigraphy. Following World War II, stratigraphy became increasingly concerned with the analysis of regional patterns of thickness, facies, and unconformities, as well as the relationships of these to major tectonic elements. Marshall pursued such analyses primarily in New York, Vermont, Ontario, Pennsylvania, and Virginia (1930's–1940's), Nevada and Utah (1950's), and Newfoundland (1960's–1970's), but he was always very well versed in other regions through the literature. His papers "Analysis of Stratigraphy" and "Isolith, Isopach, and Palinspastic Maps" were harbingers of the new physical stratigraphy.[1] Other workers, including Sloss and Wheeler, broke more completely with the old stratigraphy, however, and developed regional stratigraphic analysis even further. Present-day emphasis on sequence stratigraphy and the importance of global (eustatic) sea-level changes as a major determiner of the sequences is the latest outgrowth of the "new" stratigraphy.

Beginning early in his career, Kay emphasized that one of the ultimate goals of stratigraphy should be the refinement of paleogeographic restorations. His 1945 paper "Paleogeographic and Palinspastic Maps," although received at first with skepticism, soon proved to be a fundamental step forward for the study of complexly deformed orogenic belts.[2] Although Kay was always more stratigrapher than petrologist, he urged attention to such details as conglomerate pebble compositions as an important clue to the histories of orogenic belts. In this he anticipated in some measure an important phase of sedimentologic studies that was to mature in the 1960's and 1970's, namely, detailed attention to sedimentary petrography and paleocurrent measurements in paleogeographic studies. This approach was pioneered in America primarily by sedimentary petrologists P. D. Krynine and F. J. Pettijohn.

After coming to Columbia University, Marshall had become increasingly concerned with tectonics, which led to the contributions for which he was to be most famous. Because of Columbia's location within the classical Appalachian orogenic belt, development of this interest might seem inevitable, but its seeds actually were sown back on the midwestern prairies before he arrived in New York City. In his early Iowa work Kay had encountered an Ordovician altered volcanic ash layer and wondered about its origin. In two papers he argued that this Hounsfield Metabentonite extended from New York State all the way to Iowa and Minnesota.[3] Because the Hounsfield thickened toward the southeast and because more such Ordovician ash layers appeared in the Appalachian region, Kay inferred that a volcanic source lay somewhere in the southeastern United States. On a paleogeographic map published in 1935 he showed such a source superimposed upon the orthodox Appalachia borderland.[4] This was the first step in an assault upon conventional wisdom about the evolution of mountain belts, which already for more than half a century had been regarded as the locus of generation of continental crust, and so was of truly fundamental global importance.

An enormously influential concept of borderlands, that is, long-lived highlands lying along the edges of continents and composed of Precambrian crystalline rocks, had dominated American thought beginning with publications in the 1870's by J. D. Dana of Yale University. Dana's ideas, in turn, had been stimulated by James Hall's celebrated but incomplete theory of mountains (1857–1859). Erosion of the borderlands for hundreds of millions of years was thought by Dana to have supplied great volumes of sediments to adjacent, long, subsiding belts named geosynclines by Dana. These subsiding belts lay between the presumed borderlands and the more stable continental interior. After hundreds of millions of years of subsidence and thick sedimentation, geosynclines apparently were crumpled to form mountain ranges, which henceforth were converted to stable, deeply eroded additions or accretions to continents.

Dana's venerable scheme of Precambrian borderlands was threatened by the 1934 discovery by M. P. Billings in New Hampshire of marine Devonian fossils in metamorphosed sedimentary and volcanic rocks within the very heart of the presumed Precambrian borderland. Kay's next paleogeographic map, drawn in 1935 but published in 1937, now showed a marine trough in New England rather than the old borderland.[5] This was the first major break with orthodoxy. In 1936 Kay saw for himself the important New Hampshire rocks with Billings and in 1937, on an International Geological Congress field trip in the Ural Mountains, he again observed volcanic rocks associated with Paleozoic sedimentary ones.

For his graduate course in stratigraphy Kay compiled the distribution of major rock types on continental-scale maps for different segments of geologic time. From his compilations as well as the critical field observations noted above, he was beginning to see in the late 1930's a consistent pattern of distribution of ancient volcanic rocks along the continental margins.

Kay's illumination was bolstered by an encounter with the important writings of H. Stille and H. H. Hess. Hans Stille contrasted geosynclines with stable continental interiors, which he called "cratons." He also recognized two parallel subdivisions within what had been termed geosynclines since Dana's time. To formalize his distinction Stille coined the term *orthogeosyncline* (straight geosyncline) for the entire linear zone of thick strata first denoted by Hall and Dana. This he then subdivided into a belt of thick strata with associated volcanic and other igneous rocks, which he named *eugeosyncline* (truly or wholly geosyncline), and a complementary nonvolcanic zone, which he named *miogeosyncline* (lesser geosyncline).

Because Stille's distinction fit the pattern that Kay himself had discovered independently, Kay adopted the terminology and subsequently added several new categories. Meanwhile Hess had begun in 1939 to draw the attention of Americans to analogies between modern volcanic island archipelagoes and ancient mountain belts, a comparison that had begun in Europe half a century earlier. In particular Hess had noted that zones of serpentine rocks in ancient mountain belts occupied positions similar to such rocks in modern volcanic arcs. Kay was impressed with this comparison as well as the Alpine geologists' comparison of their ophiolite suite (consisting of serpentine, gabbro, basalt, and chert) with oceanic volcanic rocks. Moreover, he noted that a number of mountain belts located along continental margins pass oceanward directly into volcanic archipelagoes, his favorite example being the Aleutian arc extending into the North Pacific from southern Alaska.

Marshall Kay's most important contribution to the understanding of ancient mountain or orogenic belts developed during the 1940's and culminated in the publication in 1951 of his famous Geological Society of America Memoir 48, *North American Geosynclines*. Kay concluded that the nature of strata of geosynclines depends upon the complex interrelationships of uplift, adjacent subsidence, sedimentation, and presence or absence of contemporary volcanism. Most important, geosynclines, though generally subsiding, have at different times and places contained ancient volcanic arcs or tectonic lands, or both, raised within them. Kay argued that such internal lands were the major sources of clastic geosynclinal sediments rather than hypothetical borderlands external to the geosyncline. His tectonic lands were a special innovation that allowed for cannibalism of much geosynclinal sediment from but-slightly-older rocks of varied types. Thus was completed the overturn of the paleogeography and tectonic history of mountain belts. By the late 1940's A. J. Eardley (who had studied with Hess at Princeton) and others were also comparing ancient volcanic belts with modern arcs.

Kay was always conscious of great complexity and ambiguity in rocks, especially in those of deformed mountain belts. Although a master of synthesis and generalization himself, he regarded the principle of simplicity as a dangerous tool when carelessly applied. "Much of nature is not simple," he frequently observed. It was characteristic for him to draw attention to differences of pattern and exceptions to simplifications. In fact it was to emphasize regional differences of stratigraphic and tectonic patterns that he proposed in the 1940's his well-known and much criticized classification of geosynclines. Given his own extensive early training in stratigraphic paleontology and his admiration for Stille's insight, it seems natural that Kay would propose additional taxonomic categories of depositional basins besides Stille's orthogeosyncline and its two subdivisions. It was also natural that he would extend Stille's binomial approach by coining a Greek prefix appended to the suffix geosyncline for each new category, as in the following list:

GEOSYNCLINES ASSOCIATED WITH TECTONICALLY ACTIVE CONTINENTAL MARGINS

Orthogeosyncline (from Stille)	
Eugeosyncline	Volcanic-bearing zone
Miogeosyncline	Nonvolcanic zone
Epieugeosyncline	Nonvolcanic troughs on former eugeosynclines
Taphrogeosyncline	Fault-bounded troughs

GEOSYNCLINES ASSOCIATED WITH TECTONICALLY PASSIVE CONTINENTAL MARGINS

Paraliageosyncline	Coastal-plain wedges of sediments

DEPOSITIONAL BASINS WITHIN CONTINENTAL CRATONS

Autogeosyncline	Subsidence without complementary highlands
Zeugogeosyncline	Subsidence yoked to complementary highlands
Exogeosyncline	Subsidence with sediment derived from beyond the craton (that is, from an uplifted orthogeosyncline)

One may wonder why, during a period of still great taxonomic emphasis in geology, this classification aroused so much antipathy. Two common criticisms suggest a partial answer. First, it was confusing to have "geosyncline" now applied both to a first-order feature (Stille's orthogeosyncline) and also to subdivisions thereof (eu- and miogeosyncline), and, second, Kay's extension of "geosyncline" to include cratonic basins blurred the whole geosynclinal concept, which until then had been linked with major mountain belts. Attention to the important patterns that Kay wished to emphasize by establishing his categories was obscured by semantics and emotional reaction for more than two decades. Indeed, his classification was a frequent butt of jokes. Ironically it was overlooked that "miogeocline," which was coined by others in the 1960's and quickly received wide currency, was a redundant synonym of Kay's "paraliageosyncline." With the advent of the sea-floor spreading hypothesis in the early 1960's and the revolutionary theory of plate tectonics in the late 1960's, the significance of tectonic distinctions that Kay had emphasized finally came to be appreciated, but his terms remained generally unpopular. While a few workers strained to fit his terminology to the new paradigm, Kay recognized that the older terminology was no longer very appropriate. It had served its purpose by drawing attention to diverse tectonic patterns that could now at last be mostly explained by the new theory. He put the situation in perspective in 1967 when he wrote, "The concern about the terms has resulted in more penetrating analyses of the history in the rocks; the very endeavour to classify has been rewarding."[6]

Marshall Kay was also active in several significant investigations of lesser scope than global tectonics. For example, beginning with early work on Lower Paleozoic rocks along the New York–New England boundary, he developed a long-standing interest in overthrust faulting, especially where different sedimentary facies have been juxtaposed. He carried this interest to Nevada in the 1950's and still later was among the first to suggest large-scale overthrusting in western Newfoundland. He also was early to recognize the significance of exotic carbonate-rock boulders contained in Ordovician shales in Quebec and Newfoundland. And with characteristic insight he recognized in his later years, in light of plate tectonics, that the Ottawa-Bonnechere graben of southeastern Ontario and adjacent Quebec, which he had mapped in the late 1930's, represented an aulacogen.

The year 1967 was a landmark in geology, for it was the year that the theory of plate tectonics was first announced, that Kay was named Newberry professor, and that he convened the International

Gander Conference on Stratigraphy and Structure Bearing on the Origin of the North Atlantic. Even in his sixty-third year the man was still at the forefront of his field with a perfect sense of timing to bring together in Gander, Newfoundland, over 100 leading authorities on Western European and eastern North American geology. Unlike most symposia, which are soon forgotten, this conference had exceptional impact. At a critical time in the history of geology, as nothing before it focused attention upon the compelling evidence for continental separation across the North Atlantic ocean basin. Many fruitful collaborations by workers from both sides of the Atlantic were forged here, and a momentous symposium volume of 1,082 pages appeared two years later, thanks to the careful and efficient attention of its editor, Marshall Kay.

It is poignant that the American Association of Petroleum Geologists, which published the Gander volume, had also sponsored in 1928 a famous symposium in New York City that was very uncomplimentary to continental drift, in keeping with the characteristic American antipathy for that idea. Kay was then a Ph.D. candidate at Columbia University and was thoroughly steeped in the American tradition of permanently fixed continents. Thirty-nine years later, however, his open-mindedness and firsthand knowledge of Lower Paleozoic rocks in both eastern North America and the British Isles allowed him to see clearly the merit of continental separation across the Atlantic in spite of that intellectual legacy.

Marshall Kay's many honors included election to Phi Beta Kappa in 1924, receipt of the Kunz Prize of the New York Academy of Sciences in 1941, and election as honorary foreign member of the Geological Society of London in 1964 and honorary correspondent of the Geological Society of Stockholm in 1968. He received the prestigious Penrose Medal of the Geological Society of America in 1971, the Distinguished Service Award of the University of Iowa in 1971, and an honorary degree from Middlebury College in 1974. He was honored by a special "Conference on Modern and Ancient Geosynclinal Sedimentation" in 1972 and another on "Paleozoic Margins of Paleo-American and Paleo-Eurafrican Plates" in 1975.

Kay was a member of many professional societies, including the Geological Society of America, Paleontological Society, Geological Association of Canada, American Association of Petroleum Geologists, New York Academy of Sciences, Society of Economic Paleontologists and Mineralogists, American Association for the Advancement of Science, American Geophysical Union, Paleontological Society of Japan, Paleontological Association, and the Iowa Academy. He was a member of the board of managers of the New York Botanical Garden, the American Commission of Stratigraphic Nomenclature, and the International Commission on Stratigraphy, and was a delegate to four international congresses. During 1966 and 1967 he was invited to lecture in the U.S.S.R., Sweden, Canada, and Great Britain.

NOTES

1. In *Bulletin of the American Association of Petroleum Geologists*, **31** (1947), 162–168, and **38** (1954), 916–917.
2. *Ibid.*, **29** (1945), 426–450.
3. "Age of the Hounsfield Bentonite," in *Science*, **72** (1930), 365, and "Stratigraphy of the Ordovician Hounsfield Metabentonite," in *Journal of Geology*, **39** (1931), 361–376.
4. In "Distribution of Ordovician Altered Volcanic Materials and Related Clays," in *Bulletin of the Geological Society of America*, **46** (1935), 243.
5. In "Stratigraphy of the Trenton Group," *ibid.*, **48** (1937), pl. 7.
6. "On Geosynclinal Nomenclature," in *Geological Magazine*, **104** (1967), 315.

BIBLIOGRAPHY

I. ORIGINAL WORKS. Kay wrote almost 200 publications, the most important of which are listed in the Geological Society of America's *Memorial Volume*, **7** (1977). A dozen or so more papers on stratigraphy and paleontology published in the 1930's first established his reputation in America. The 1945, 1947, and 1954 papers cited in the article were major contributions to stratigraphic concepts. Papers published in 1930 and 1931 on the Hounsfield metabentonite and papers on Ordovician paleogeography in 1935 and 1937 (also cited in the article) were the precursors of his contributions on geosynclines and mountain belts. "Geosynclines in Continental Development," in *Science*, **99** (1944), 461–462, was his first major paper on this theme. It was followed by "Geosynclinal Nomenclature and the Craton," in *Bulletin of the American Association of Petroleum Geologists*, **31** (1947), 1289–1293; and his single most important publication, *North American Geosynclines*, Geological Society of America Memoir 48 (1951).

In 1955 he was invited to write "The Origin of Continents" for *Scientific American*, **193** (September 1955), 62–66. The Gander conference symposium volume, *North Atlantic: Geology and Continental Drift* (Tulsa, Oklahoma, 1969), was edited by Kay and contains five articles by him. This was his last major contribution on large-scale tectonics, but the following two papers provide important retrospective insights by Kay on his involvement with geosynclines and tectonics: "On Geosynclinal Nomenclature," in *Geological Magazine*, **104** (1967), 311–316; and "Reflections: Geosynclines, Flysch, and Melanges," in *Modern and Ancient Geosynclinal Sedimentation* (Tulsa, Oklahoma, 1974), a symposium volume dedicated to Kay.

While all of these contributions to the literature on tectonics were appearing, Kay continued to publish many papers dealing with stratigraphy, paleontology, and structure of the regions in which he was doing fieldwork. Many of these were important contributions even though of more restricted geographic interest. An example is his last publication, "Dunnage Melange and Subduction of the Protacadic Ocean, Northeast Newfoundland," in Geological Society of America Special Paper 175 (1976). Another major contribution of still a different sort was the textbook *Stratigraphy and Life History* (New York, 1965), written with E. H. Colbert.

II. SECONDARY LITERATURE. Short biographical treatments of Kay have been published by P. A. Chenoweth in *Bulletin of the American Association of Petroleum Geologists*, **60** (1976), 1129–1130; and R. H. Dott, Jr., in *Memorial Volume*, **7** (1977), 1–9, published by the Geological Society of America. The latter contains a list of Kay's most important publications. The introductory paper in the *Modern and Ancient Geosynclinal Sedimentation* volume, "The Geosynclinal Concept," by R. H. Dott, Jr., provides an overview of changing ideas about geosynclines from Hall and Dana to plate tectonics and contains an assessment of Kay's contributions. K. J. Hsü, "The Odyssey of Geosyncline," in Robert N. Ginsburg, ed., *Evolving Concepts in Sedimentology* (Baltimore, 1973), provides a similar overview from a different perspective. Dott presented further historical analysis illustrated by reproductions of several paleogeographic maps (including three of Kay's) in "Tectonics and Sedimentation a Century Later," in *Earth-Science Reviews*, **14** (1978), 1–34.

R. H. DOTT, JR.

KETTLEWELL, HENRY BERNARD DAVIS (*b.* Howden, Yorkshire [now Humberside], England, 24 February 1907; *d.* Steeple Barton, near Oxford, England, 11 May 1979), *evolution, ecological genetics, entomology.*

Kettlewell was the only son of Kate Davis and of Henry Kettlewell, a merchant and member of the Corn Exchange. He was educated at Old College, Windermere, and then at the renowned English public school Charterhouse (1920–1924); studied briefly in Paris; read medicine (with zoology) at Gonville and Caius College, Cambridge, from 1926; and undertook his clinical training at St. Bartholomew's Hospital, London, from 1929. In 1935, after graduating in medicine, he joined a general medical practice in Cranleigh, Surrey. The following year he married Hazel Margaret Wiltshire. They had a daughter who died young and a son.

During World War II Kettlewell worked for the Emergency Medical Service at the Working War Hospital. He was also for a time an anesthetist at a hospital in Guildford. With the inauguration of the National Health Service, Kettlewell left general medical practice, and in 1949 he moved to South Africa to carry out research at the International Locust Control Centre at Cape Town University, making several expeditions to the Kalahari, the Knysna Forest, the Belgian Congo, and Mozambique. In 1951 he was appointed to a senior research fellowship in the laboratory of Edmund Brisco Ford in the zoology department at Oxford, dividing his time between England and South Africa until 1954, when he was appointed senior research officer and finally settled in Oxford.

It was there that he carried out his best-known work, on the evolution of melanism in moths, over a period of two decades, leading his small research team, which sometimes included his wife and son, with great panache. Although he had the true comedian's deep streak of sadness, Bernard Kettlewell was a big man, with a personality larger than life. Kind, charming, and irascible, he had a huge and infectious ebullience and energy, could be the life and soul of any party, and was much loved by his friends. In 1952 he returned to England by driving from Cape Town to Alexandria, surely with the same abandon with which he used to drive the English lanes. He retired in 1974 and died five years later, from an apparently accidental overdose of the medication he was using to relieve the pain of a back injury sustained during fieldwork.

Kettlewell belonged to that English tradition which sees no sharp boundary between amateur naturalist and professional biologist. As perhaps the last great exponent of that tradition, it was appropriate that he saw his work as confirming the work of its greatest exponent of all time, Charles Darwin, and that his first profession, medicine, was also Darwin's. From his youth Kettlewell had shown an extraordinary facility for fieldwork with insects, in his teens finding species which had completely eluded experienced collectors, and later in life establishing in three days the life cycle of a rare species which had defeated a large team of enthusiasts. It was this ability which he exploited so superbly in the work he carried out at Oxford, where, at some financial loss, he converted his lifelong hobby of entomology into a profession. The ingenuity with which he could combine science and natural history was shown by his demonstration that a moth caught in his trap had migrated direct from North Africa—it was carrying a radioactive particle from a French nuclear test explosion.

Because of a long friendship with Edmund Ford, whom he had met in the Black Wood of Rannoch, a famous ecological site, in 1937, Kettlewell had agreed to join the team Ford was building up in Oxford for the "study of evolution by observation and experiment" to investigate the most spectacular and rapid evolutionary change ever witnessed: industrial melanism in moths. It was the ideal experimental situation, in which all variables but one have been manipulated into insignificance. The normal richness and complexity of the coadaptations of organisms with each other and with their physical environment had been almost obliterated by one overwhelming factor: the industrial and domestic soot which blanketed the more populous parts of Britain and Europe.

The extermination of the lichens that otherwise clothed the trunks of trees had left those moths which spent the day exposed there, relying only on stillness and invisibility, deprived of the camouflage they had enjoyed. Forced to rest on the soot-blackened bark, some eighty species of them had darkened during the nineteenth century. This was well enough known, and had been subjected to various neo-Lamarckian, mutationist, and (notably by J. B. S. Haldane) Darwinian interpretations, but it was Kettlewell who used it to provide spectacular evidence of the competence of natural selection to produce evolutionary change.

The two personal qualities that led to this success were Kettlewell's great facility with insects in the field, and his status as an amateur entomologist. Although walking with the kings of population genetics, he had not lost the common touch; his knowledge of natural history and his enthusiasm for the insects themselves were readily appreciated by the amateur lepidopterist, who found the work of the more rigorously trained of Kettlewell's colleagues excessively recondite. Kettlewell put this power to communicate with the amateur world to good effect in organizing a national survey (1952–1970), conducted by 170 entomologists, of the distribution of the black and the lichen-patterned forms of one of the more prominent industrial melanics, the peppered moth (*Biston betularia*). This showed clearly that the highest frequencies of the black forms were found in the heavily polluted industrial heartlands of Britain, and that populations in the western and northern rural areas were still entirely of the lichen-camouflaged form.

By using a massive field experiment, replicated in polluted and unpolluted woodlands in 1953 and 1955, and with the help of Niko Tinbergen in observing and filming the insectivorous birds which attacked the moths during the day, Kettlewell showed that twice as many of the black form as of the pale form survived in polluted woodland, the situation being reversed in the unpolluted wood. This constituted what he described as "Darwin's missing evidence," that evolutionary changes could be produced in natural populations through the force of natural selection acting on inheritable variation. It was a problem that could have been solved only in this way, by direct work in the field by a field naturalist. Kettlewell's demonstration was of such elegance and simplicity as to make it one of the great classics of evolutionary biology.

During his amateur and his professional work, Kettlewell built up a large collection of Lepidoptera, rich in specimens demonstrating the genetics of melanism and other variations, which he combined with the collection of A. E. Cockayne, another medical amateur entomologist with a distinguished reputation in genetics. United with Lord Rothschild's collection, this became the Rothschild-Cockayne-Kettlewell collection of the British Museum, a unique resource for the study of natural variation and its inheritance.

The fame his research brought him led to more popular scientific commissions: an expedition to Brazil in 1958 to commemorate the centennial of Darwin's *Origin of Species*, which Kettlewell converted into a study (with a film) of insect camouflage, and a biography of Darwin (jointly with Julian Huxley), which was a pioneering example of the now standard genre of scientific biographies decorated with contemporary illustrations.

In the work that immediately followed his classic experiments, Kettlewell extended his interest to species which were melanic, and had perhaps been so for a long time, in rural rather than industrial environments: his most notable study was of *Amathes (Paradiarsia) glareosa*, which becomes darkened in the Shetland Isles as a result of flying in subarctic twilight against black, peaty soil. Melanism, according to his book *The Evolution of Melanism*, was therefore a "recurring necessity," arising in many and varied circumstances when moths were predated on a darkened background, industrial pollution providing but the latest of these. This view opened a way of comparing the coadaptation of the whole genome to genes which had been present for a long period with those that were newly arisen. This synthesis attracted relatively little attention, in part because it entailed Ronald A. Fisher's theory of the evolution of dominance, always dear to members of Ford's school of ecological genetics but deeply controversial among geneticists.

Although some of Kettlewell's students extended his work to additional species, research on melanism showed signs of withering within five years of his death. Three of its best practitioners (Philip M. Sheppard, E. Robert Creed, James A. Bishop) had died young; Kettlewell's personal style, on which the research greatly depended, was inimitable; and funding from the scientific establishment was sparse. The Royal Society declined to elect him to a fellowship, and it was significant that Kettlewell's formal honors were all awarded in the Soviet bloc (Darwin Medal, Soviet Union, 1959; Mendel Medal, Czechoslovakia, 1965) and that in Darwin's own country, Kettlewell had found the "missing evidence" not with state funds but with a grant from the privately administered Nuffield Foundation.

BIBLIOGRAPHY

I. ORIGINAL WORKS. Kettlewell's major book is *The Evolution of Melanism: The Study of a Recurring Necessity* (Oxford, 1973). The bibliography contains an extensive, but not exhaustive, compilation of his scientific papers; the preface and acknowledgments contain some autobiographical material. Kettlewell's two more popular works are *Your Book of Butterflies and Moths*, which communicates clearly his enthusiasm for natural history (London, 1963), and *Charles Darwin and His World* (London and New York, 1965), with Julian Huxley. The chief papers describing his experiments on *Biston betularia* were published in *Heredity*: "Selection Experiments on Industrial Melanism in the Lepidoptera," **9** (1955), 323–342; "Further Selection Experiments on Industrial Melanism in the Lepidoptera," **10** (1956), 287–301; and "A Survey of the Frequencies of *Biston betularia* L. (Lep.) and Its Melanic Forms in Britain," **12** (1958), 51–72. For a wider audience he wrote "Darwin's Missing Evidence," in *Scientific American*, **200** (March 1959), 48–53. His two films, later to look amateur beside the work of professional wildlife moviemakers, but pioneering classics in their day, were *Evolution in Progress* (cinematographer Niko Tinbergen, 1956) and *Darwin and the Insects of Brazil* (with the Shell Film Unit, 1958) (both distributed by British Universities Film and Video Council, London). The annotated Rothschild-Cockayne-Kettewell National Collection of British Lepidoptera is in the British Museum (Natural History), London.

II. SECONDARY LITERATURE. Obituaries are Cyril A. Clarke, in *Antenna*, **3**, no. 4 (1979), 125, and *Entomologist's Record and Journal of Variation*, **91**, no. 10 (1979), 253–255; R. F. Demuth, *ibid.*, 255–257; E. B. Ford, in *Nature*, **281** (1979), p. 166; and E. P. W. [Wiltshire], in *Proceedings of the British Entomological and Natural History Society*, **12**, no. 3/4 (1979), 101–103. For a scientific review of Kettlewell's work in context, see Edmund B. Ford, *Ecological Genetics* (London, 1964; 5th ed., 1975), ch. 14. A copy of the Granada TV film about his Shetland work is held by the biology department of the University of York.

JOHN R. G. TURNER

KHOKHLOV, REM VICTOROVICH (*b*. Livny, U.S.S.R., 15 July 1926; *d*. Moscow, U.S.S.R., 8 August 1977), *physics*.

Khokhlov's father, Victor Khristoforovich, was a professor of technical sciences. His mother, Marya Yakovlevna Vassil'eva, was a physicist at Moscow State University.

Khokhlov spent practically all his life in Moscow. He started working during school vacations, initially because he wanted to do manual work and later, in wartime, because he wished to contribute to the struggle against fascism. Working as an automobile locksmith, he graduated from school and in 1943 entered the Moscow Aviation Institute, from which he transferred in 1945 to the department of physics of the University of Moscow.

Khokhlov's subsequent life was closely connected with the university. He graduated from it in 1948, and in 1952 he defended his candidate's thesis. Later he served as an assistant and a reader in the department in the area of the theory of oscillations. From September 1959 to June 1960, he worked at Stanford University. In 1961 he defended his doctoral thesis, and in 1967 he became a professor. Beginning in 1973, Khokhlov served as a rector of the University of Moscow.

Khokhlov began his scientific activity as a pure theoretician (for example, his work on associated Laguerre functions). Very quickly he became a leader of investigations on nonlinear optics at the University of Moscow. In 1965 he organized a unit specializing in this area within the department of physics.

During his first years of work at the university, Khokhlov concentrated his efforts on radiophysics. He belonged to the third generation of the Soviet school of the physics of oscillations, founded by L. I. Mandelshtam and N. D. Papaleski. (Khokhlov entered the university one year after Mandelshtam's death.) Khokhlov's work has greatly influenced the development of many trends in radiophysics, optics, and acoustics. Between 1954 and 1960 he worked out in detail an effective procedure for analyzing oscillations and waves in nonlinear systems—the method of stepwise simplification of the reduced equation (Khokhlov's method). With this method Khokhlov obtained fundamental results in the theory of the interaction of waves in highly dispersive media and in the theory of weak shock waves in dissipative media.

The approaches formulated in this series of works were widely used by Khokhlov and his coauthors and pupils in their theoretical work on nonlinear optics, nonlinear acoustics, and laser physics. After 1960 Khokhlov and his collaborators became more and more involved in experimental investigations. Their most important works completed during the period between 1960 and Khokhlov's death in 1977 may be divided into four groups. First, in nonlinear optics there was the start-up of tuned parametric generators of light (systems in which the parametric excitation of light waves in a nonlinear medium is realized). The theoretical foundations were derived from nonlinear wave optics (1962–1964). Powerful optical frequency multipliers were built and important investigations conducted on self-focusing and self-defocusing of light (1966). Second, in nonlinear acoustics Khokhlov worked out the theoretical procedure of slowly varying the profile of a propagated wave, and formulated the equations of nonlinear acoustics of plane waves (1960–1961). He also formulated and solved the equations of nonlinear acoustics of diffracted beams (the equations of Khokhlov-Zabolotskaya, 1969).

Third, in X-ray optics Khokhlov developed the theory of gamma lasers, provided the analyses of schemes of contraction of the lines of gamma resonance, and built up the theory of dynamic diffraction in amplifying media (1972–1975). Last, in the physics of the interaction of laser radiation and matter, Khokhlov put forward the idea of control of the chemical reactions by a resonance photoaction (1970), worked out the theory of control of the surface phenomena with the aid of laser radiation, and performed significant investigations in the theory of nuclear synthesis of elements.

Apart from his intensive scientific and teaching activity, Khokhlov was active as an organizer of scientific work. He was a member of the Presidium of the U.S.S.R. Academy of Sciences, and during the last years of his life was vice president of the academy. He also served as the head of a scientific council of the academy on the problem of noncoherent and nonlinear optics, and organized the All-Union and international meetings on these problems. He was a member of the Central Revision Commission of the Communist party and a member of the Supreme Soviet of the U.S.S.R. From 1965 Khokhlov headed the wave processes department at Moscow State University, and from 1973 to 1977 he was rector of the university.

Khokhlov's achievements have been widely recognized. In 1966 the U.S.S.R. Academy of Sciences elected him a corresponding member, and in 1974 a full member. In 1970 he was awarded the Lenin Prize, and in 1985 (posthumously) the State Prize of the U.S.S.R. Khokhlov was a full member of the Bulgarian Academy of Sciences, a corresponding member of the Portuguese Academy, and an honorary doctor of a number of universities.

BIBLIOGRAPHY

I. Original Works. "K teorii zakhvativaniia pri maloi amplitude vneshnei sily" (On the theory of locking-in at a small value of the amplitude of an external force), in *Doklady Akademii nauk SSSR*, **97**, no. 3 (1954), 411–414; "K teorii udarnykh radiovoln v nelineinikh liniyakh" (On the theory of shock radio waves in nonlinear lines), in *Radiotekhnika i elektronika*, **6**, no. 6 (1961), 917–925; "O raspostranenii voln v nelineinikh dispergiruyushchikh sredakh" (On the propagation of waves in nonlinear disperse lines), *ibid.*, no. 7, 1116–1127; "Ob odnoi vozmozhnosti usileniia svetovikh voln" (On some possibility of the amplification of light waves) in *Zhurnal eksperimentalnoi i teoreticheskoi fiziki*, **43**, no. 1 (1962), 351–353, written with S. A. Akhmanov; "K teorii prostikh magnitogidrodinamicheskikh voln konechnoi amplitudi v dissitipativnikh sredakh" (On the theory of simple magnetohydrodynamic waves of finite amplitude in a dissipative medium), *ibid.*, **41**, no. 2 (1961), 534–543, written with S. I. Soluian.

"Ob upravlenii khimicheskimi reakciiami putem rezonansnogo fotovozdeistviia na molekuli" (On the control of chemical reactions by means of resonance photo attack influence on molecules), *ibid.*, **58**, no. 6 (1970), 2195–2201, written with V. T. Platonenko and N. D. Artamonova; "Problemi teorii nelineinoi akustiki" (Problems of nonlinear acoustical theory), in *Akusticheskii zhurnal*, **20**, no. 3 (1974), 449–457, written with O. V. Rudenko and S. I. Soluian; "K voprosu o vozhmozhnosti sozdania gamma-lazera na osnove radioaktivnikh kristallov" (On the possibility of the creation of the laser on the basis of radioactive crystals), in *Pis'ma v zhurnal eksperimentalnoi i teoreticheskoi fiziki*, **15**, no. 9 (1972), 580–583; *Problems of Nonlinear Optics* (New York, 1972), written with S. A. Akhmanov and A. P. Sukhorukov; and "Self-focusing, Self-defocusing and Self-modulation of Laser Beams," in *Laser Handbook*, II (Amsterdam, 1972), 1151–1228.

II. Secondary Literature. V. I. Grigoriev, *Rem Viktorovich Khokhlov (R. V. Khokhlov)* (Moscow, 1981), and *Akademik Rem Viktorovich Khokhlov* (Moscow, 1982). For an obituary see S. A. Akhmanov *et al.*, *Uspekhifizicheskikh nauk*, **124**, no. 2 (1978), 355–358.

V. J. Frenkel

KING, HAROLD (*b.* Llanegan, Caernarvonshire, Wales, 24 February 1887; *d.* Wimbourne, England, 20 February 1956), *organic chemistry, chemotherapy, pharmacology*.

King was the first of four children born to Herbert and Ellen Elizabeth Hill King, both originally from Lancashire. From 1891 until their retirement in 1923, King's parents were head teachers of the St. James's Church School in Bangor. King's early education there was followed by five years at the Friars' Grammar School, Bangor. In 1905 he entered University College, Bangor, on two scholarships. The teaching of K. J. P. Orton led King to chemistry, and in 1909 he graduated with first-class honors. With additional university support he remained for two more years at Bangor, training for research in Orton's laboratory.

In 1911 King received an industrial bursary and took an appointment in the analytical laboratory of the Gas Light and Coke Company's Tar and Ammonia Products Works at Beckton. Seeing no prospects for serious research, he moved six months later (1912) to a temporary post in the Wellcome Physiological Research Laboratories at Herne Hill. Here, in the environment created by Henry Hallett Dale and George Barger, and in collaboration with A. J. Ewins, King found an exhilarating research atmosphere in which biology and chemistry were brought together to mutual advantage. Six months later King accepted a position in the experimental department of the Wellcome Chemical Works at Dartford. In Frank Lee Pyman's laboratory he worked on the glycerophosphates and improved his experimental technique.

During World War I, King contributed to Wellcome's urgent efforts to create a British synthetic drug industry. In Pyman's laboratory he carried out research on hyoscine that expanded his knowledge of alkaloidal chemistry, an interest already stimulated by his association with Ewins. In 1919 King accepted a position as chemist on the scientific staff of the Medical Research Council, which was then establishing an institute for laboratory work. He remained with the Medical Research Council, at the National Institute for Medical Research, until his retirement in 1950, concentrating his research in areas where organic chemistry might inform medicine or biology: chemotherapy, alkaloidal chemistry, and the structure of steroids. In 1923 King married Elsie Maud Croft; they had one son.

King was well suited by temperament and intellect to the life of a research institute. Quiet and retiring, he thrived on long hours at the laboratory bench and on freedom from teaching or administrative responsibilities. He found sufficient intellectual stimulus in his own curiosity and in contacts with colleagues. Though personally reserved, he readily shared his knowledge with associates, and his interest in work proceeding in other laboratories of the institute prompted him on several occasions to initiate fruitful collaborative efforts.

An active member of the Chemical Society, King served on both its Council (1928–1931) and its Publication Committee. He was a member of the Alkaloids Committee and the Chemical Revision Committee of the Pharmacopoeia Commission. For the Medical Research Council he prepared *Chemotherapy Abstracts*, served as chemical secretary of the Chemotherapy Committee, and during World War II was secretary of the Committee on the Synthesis of Penicillin. Apart from his work as a chemist, King had a serious amateur interest in entomology, a recreation to which he devoted the six years of his retirement.

King owed his position with the Medical Research Council in part to the council's interest in chemotherapy in the wake of Paul Ehrlich's success with Salvarsan, and chemotherapy remained a central focus of his research throughout his career. In his first publications after joining the council, King reported his isolation of the principal impurity in commercial preparations of Salvarsan, a substance responsible for part of the drug's toxic side effects. Organic arsenicals also figured in King's next project. Through the 1920's and early 1930's he attempted to produce an arsenical drug analogous to suramin (Bayer 205, a compound similar to the dye trypan red). The idea was to combine the antitrypanosomal therapeutic power of the arsenicals with the long-lasting therapeutic effect of suramin. No clinically useful drug came out of this effort, but another set of investigations on the thioarsenites, formed by the reaction between arsenious oxide derivatives and thiol compounds, yielded insights into the mode of action of arsenical drugs on spirochetes and trypanosomes.

A by-product of King's work on arsenicals was his study of the constitution of sulfarsphenamine and neoarsphenamine, published between 1933 and 1935. In the late 1930's and 1940's King investigated the effects of structural modifications of the cinchona alkaloids on their antiplasmodial action in malaria. Identifying the traits of the molecules essential for biological activity, he was able to synthesize several analogues of these alkaloids, one of which had marked antiplasmodial action. Although no clinically useful drug emerged immediately from these studies, King's work found an important place in the U.S. wartime search for synthetic antimalarials. In the 1940's King, working with E. M. Lourie and Warrington Yorke, demonstrated that long aliphatic straight-chain diguanidines, diisothioureas, and diamidines showed antitrypanosomal activity, a result that led to the development by A. J. Ewins of stilbamidine for the treatment of kala-azar. King's finding, with James Walker and C. H. Andrewes, also in the 1940's, that p-sulfonamidobenzamidine had antirickettsial activity was the first indication that rickettsial infections were vulnerable to chemotherapy.

King's early work with A. J. Ewins and then F. L. Pyman on alkaloidal chemistry opened to him a second field of lifelong research that yielded significant results for chemistry, pharmacology, and therapeutics. Of greatest importance was his work on curare, prompted by Dale's work on chemical transmission of nerve impulses and the resulting interest within the institute in substances affecting neuromuscular transmission. In studies that he began to publish in 1935, King succeeded in isolating a specific alkaloid, d-tubocurarine chloride, from the South American arrow poison tube curare, and in showing that it was the active principle. He went on to demonstrate its botanical origin and to determine its chemical constitution. Access to the pure alkaloid was an immediate boon to surgeons, who could safely control its dosage for use as a muscle relaxant. Consideration of the structure of d-tubocurarine chloride prompted King to attempt synthesis of simpler molecules that might reproduce its effects. This work, done in King's laboratory by Eleanor Zaimis, did yield such a compound, decamethonium iodide. Still more important medically was the unexpected finding that other members of the series of methonium compounds acted as powerful sympathetic blocking agents, producing a drop in blood pressure. Follow-up of this derivative result of King's curare studies resulted in the first effective drugs for treatment of hypertension.

More theoretical was King's proposal, with institute colleague Otto Rosenheim, of new structural formulas for cholesterol and the bile acids. Existing formulas in the early 1930's suffered from several problems, including failure to conform to J. D. Bernal's X-ray crystallographic studies. Beginning in 1932, King and Rosenheim published a series of papers arguing that the ring structure of the sterols consisted of four fused six-membered rings. Quickly accepted in slightly modified form—one of the rings was shown to have five members—this proposal was of major significance in that it opened up many opportunities for investigation in the biologically important field of steroids.

The value of King's work was well recognized by the scientific community. Elected fellow of the Royal Society in 1933, he received the Hanbury Medal of the Pharmaceutical Society in 1941, and the Addingham Gold Medal of the William Hoffman Wood Trust (Medical) in 1952. In 1950, the year of his retirement, King was appointed C.B.E. in consideration of his work for the Medical Research Council.

BIBLIOGRAPHY

I. Original Works. King's publications are listed at the end of C. R. Harington, "Harold King, 1887–1956," in *Biographical Memoirs of Fellows of the Royal Society*, **2** (1956), 157–171. Representative of his work in chemotherapy is "Trypanocidal Action and Chemical Constitution," published with various collaborators in *Journal of the Chemical Society*, **125** (1924), 2595–2611; **127** (1925), 2632–2651, 2701–2714; **129** (1926), 817–831, 1355–1370; (1927), 1049–1060, 3068–3097; (1928), 2426–2447; (1930), 669–694; (1931), 3043–3056, 3236–3257; and (1932), 2505–2510, 2866–2872. The main work on curare was published in "Curare Alkaloids," in *Journal of the Chemical Society* (1935), 1381–1389; (1936), 1276–1279; (1937), 1472–1482; (1939), 1157–1165; (1940), 737–746; (1947), 936–937; (1948), 265–266, 1945–1949; and (1949), 955–958, 3263–3271. King's principal joint publication with Otto Rosenheim was "The Ring-System of Sterols and Bile Acids," in *Journal of the Society of Chemical Industry*, **51** (1932), 464–466, 954–957; **52** (1933), 299–301; and **53** (1934), 91–92, 196–200. King's biographical notice on Otto Rosenheim includes remarks on their collaboration: "Sigmund Otto Rosenheim, 1871–1955," in *Biographical Memoirs of Fellows of the Royal Society*, **2** (1956), 257–267.

II. Secondary Literature. The fullest available biographical source is C. R. Harington (see above). See also T. S. Work, "Dr. Harold King, C.B.E., F.R.S.," in *Nature*, **177** (1956), 604–605.

John E. Lesch

KING, HELEN DEAN (*b*. Owego, New York, 27 September 1869; *d*. Philadelphia, Pennsylvania, 7 March 1955), *genetics*.

King was the daughter of William A. and Lenora Dean King. Her father was a prosperous businessman, president of the King Harness Company. She attended Owego Free Academy, and at the age of eighteen, she entered Vassar College, where she received the A.B. degree in 1892. In 1894 King returned to Vassar as a graduate student in biology and assistant demonstrator in the biology laboratory. She began graduate studies at Bryn Mawr College in 1895 and received a Ph.D. in 1899 under Thomas Hunt Morgan. She continued postdoctoral work under Morgan and assisted in the Bryn Mawr biology laboratory from 1899 to 1904. At the same time she taught science at Miss Florence Baldwin's School from 1899 to 1907.

Following two years (1906–1908) as a university fellow for research in zoology at the University of Pennsylvania, King joined the research staff of the Wistar Institute of Anatomy and Biology in Philadelphia. She remained there until her retirement in 1950. King became a full professor of embryology in 1927 and a member of Wistar's advisory board in 1928. She never married.

King had the best education a woman could receive in the 1890's. Vassar led all other women's colleges in its expenditures on scientific apparatus in 1887, the year before King entered. At Bryn Mawr her major subject was morphology, which she studied under Morgan, and she was exposed to Darwinian theory in the paleontology classes of Florence Bascom. Bryn Mawr was an ideal spot for an aspiring woman biologist, for its biology department had been established by Edmund B. Wilson in 1885 and had been taken over by Morgan in 1891. Morgan's brilliant researches in embryology and Wilson's continued ties with Morgan guaranteed that the women were exposed to the latest research in embryology, cytology, and the newly emerging field of genetics.

Morgan suggested the topic of King's dissertation: "The Maturation and Fertilization of the Egg of *Bufo lentiginosus*" (the common toad). After King published her dissertation in 1901, Morgan continued as her mentor, suggesting new problems until 1909, when she changed the subject of her research from amphibians to rats and Morgan abandoned embryology to work exclusively on the genetics of *Drosophila*.

King's interest in sexual fertilization, the maturation of the egg, and the determination of sex reflect the influence of Morgan and the *Entwinklungsmechanik* school. She believed that embryological questions could be solved only through rigorous experimentation and a mechanistic interpretation.

In her work on sex determination, King emphasized the influence of environmental factors on altering the sex ratio. She was inclined to accept the view advanced by Nettie M. Stevens, Wilson, and Clarence E. McClung that there are two kinds of spermatozoa, those carrying the accessory element (the Y chromosome) and those from which it is absent. However, King opposed the prevailing theory advanced by Wilson in 1910 that any egg is capable of fertilization by any spermatozoon that happens to come in contact with it. She persisted in her hypothesis that ova exercise a kind of selection, accepting one kind of sperm and rejecting the other kind, on the basis of different environmental conditions.

In "Studies on Sex Determination in Amphibians" (1909–1912), King worked on toads' eggs and increased the proportion of females by slightly drying the eggs or by withdrawing water from them by placing them in solutions of salts, acids, or sugars. Edwin G. Conklin singled out her work as important evidence for environmental influence on the alteration of the sex ratio. In other studies King found that hybridizing altered the sex ratio, as did the age of the female, and she suggested that the germ plasm undergoes changes due to age.

During her first ten years of inbreeding albino rats (1909–1919), King found that she could select for the tendency to produce an excess of males or an excess of females. She concluded that in the rat the sex ratio is, to a certain extent, amenable to selection and that the female has more influence in determining sex than does the male. King suggested that factors such as heredity, environment, and nutrition act on ova in such a way as to render them more easily fertilized by one kind of sperm than by the other. Her conclusion contained unmistakable feminist overtones. She wrote in 1919: "In the female element, as in the female organism, resides the power to select that which is for the best interests of the species." In order to explain seasonal changes in the sex ratio of the rat, King suggested in 1927 that ova select some sperm and reject others because they have a greater chemical attraction or repulsion for one kind of sperm than for the other, depending upon seasonal changes in body metabolism.

King's major research focused on the effects of inbreeding on albino rats and the effects of captivity on wild gray Norway rats. Before joining the Wistar Institute, she had devoted herself to cytological studies, performing painstaking operations on amphibian eggs. In 1909, however, she abandoned cytology and embarked on a lifelong project of animal breeding when she mated two males and two females from a single litter of albino rats. King would continue this dynasty for more than 130 generations, to develop what would later be called the King colony, a part of the famous Wistar Institute stock of white rats that figured in innumerable research projects throughout the world.

In 1919 King published the results of fifteen generations of mating brother and sister albino rats. Her study soundly refuted Darwin's assertion that inbreeding inevitably leads to degeneration in animals. King concluded that the closest form of inbreeding in mammals is not necessarily inimical to body growth, fertility, or constitutional vigor, provided that one selects the best animals from a large population to breed. (After fifty generations of brother-sister matings, King noted their superiority over stock controls in terms of fertility, growth, size, and longevity.) Her results supported the contention of Edward M. East and H. K. Hayes (1912) that a completely homozygous strain can stand up forever if it has a good gametic constitution and a natural inherent vigor. King found that adverse environmental conditions, such as malnutrition, set back the rats only temporarily and did not alter their genetic constitution.

King suggested that some factors tended to be inherited together: for instance, large body size, early sexual maturity, high fecundity, superior vigor, and longevity. But she did not think these factors were linked, as were many of the genes in *Drosophila*, according to work being done by Morgan and his researchers at Columbia.

King's work followed the inbreeding experiments of William E. Castle, who found no decrease in fertility in *Drosophila* or, in his later work, in rats. Her data supported Castle's conclusion that selection from the productive pairs more than offsets the effects of inbreeding. King's results differed sharply from those of Sewall Wright, however, who noted a deterioration in vigor among guinea pigs inbred for more than twenty generations. Wright had not bred selectively, whereas King had allowed only the most vigorous individuals to reproduce. Much of King's work on inbreeding pointed to the difficulty of establishing complete homozygosity in an animal stock and contributed to the growing recognition among biologists in the 1920's and 1930's that there is a tremendous genetic diversity even within a local population of a single species.

In 1927 King and Leo Loeb of the Washington University Medical School began some of the first tissue transplants to determine the extent of genetic similarity in two populations of inbred animals: King's albino rats and Wright's guinea pigs. Although King's rats had been inbred for thirty-seven to thirty-nine generations, and Wright's guinea pigs for only fifteen to twenty-three generations, the researchers found the host's reaction to donor tissue much more severe between inbred rats than between guinea pigs. A transplant from one inbred rat to another led to just as severe a reaction as a transplant between noninbred rats. In order to explain the unexpected results, Loeb and King suggested that selection of the strongest individuals during many generations of inbreeding had caused heterozygosity to persist. They cited the greater number of spontaneous mutations in the rat than in the guinea pig as another possible explanation. However, when Loeb, King, and Blumenthal (1943) did tissue transplants between individual rats that had been inbred for 102 to 106 generations, they noted that homozygosity in the strain had gradually increased since the earlier tissue transplants. Still, a completely homozygous condition had not been attained even after more than one hundred generations of brother-sister matings.

In another lengthy breeding experiment to determine the effects of captivity on wild animals, King bred twenty-five generations of gray Norway rats originating from six pairs of wild rats trapped in the Philadelphia streets. Reporting on changes from the eleventh to the twenty-fifth generation, King found that the rats grew more rapidly and attained a much larger size than their wild prototypes. This result agreed with earlier studies on the domestication of wild animals. At the same time Calvin Bridges was breeding special stocks of *Drosophila* mutants at Columbia, King was discovering a number of mutations in her wild rats affecting coat color, structure of the fur, and eye color. With controlled mating she determined whether the mutations were autosomal or sex-linked, dominant or recessive, the result of a single gene mutation or of selection that affected modifying genes.

King inbred mutants to distinguish between homozygotes and heterozygotes. The inbreeding of rat mutants produced more mutants: cinnamon rats, waltzing rats, and stub rats (runts with shortened tails). King rejected the hypothesis that the genes for all the mutations were present in the original six pairs of wild rats and remained latent until chance matings resulted in their phenotypic expression. She speculated that mutations appeared after several years of captivity because the very young and the very old females were allowed to reproduce, protected from the competition of natural selection. Thus, variability of the germ plasm due to age led to an increase in mutant individuals, King concluded.

In 1932 King presented a paper before the Sixth International Congress of Genetics summarizing the kinds of mutations that had occurred in twenty-eight generations of captive Norway rats, comprising over forty-five thousand individuals. She suggested that captivity, with its altered conditions of environment and nutrition, tends to produce diversity in a wild race, not to render it more homogeneous.

In the 1930's King collaborated with Castle on four linkage studies of the Norway rat in which they mapped the genes for several newly discovered mutations. King did the matings, while Castle provided the genetic analysis of the experimental data.

In her seventies, King performed a series of ex-

KING

periments with Margaret Reed Lewis, a Bryn Mawr classmate and Wistar colleague, on tumor transplants in rats. They were successful in extracting a fat-soluble, tumor-produced substance from rat sarcomas and inducing tumors by injecting the substance into rats of the same inbred strain. In a later experiment King and Lewis were able to inhibit the growth of tumors and make rats immune to the further growth of grafts of the same tumor by injecting a substance directly into the tumor. The substance was made by grinding tumors, both primary and induced, to a pulp, extracting them in 95 percent alcohol, and distilling the extract.

Like many early geneticists, King favored the use of eugenics to improve the human race. Although she found nothing wrong with consanguineous marriages if the original stock was good, she thought laws forbidding such marriages were necessary because one out of every fourteen persons carries some hidden defect in the germ plasm. King foresaw a time when marriage would be based on the physical fitness of the individuals and on their recorded pedigree for several generations. In the future, people would realize the great value of favorable genetic combinations that produce unusual ability, King predicted in 1935.

In 1932 King shared the Ellen Richards Prize with the astronomer Annie Jump Cannon. The prize had been awarded by the Association to Aid Women in Science since 1901 for the best piece of experimental work submitted by a woman in science.

King served as associate editor of the *Journal of Morphology and Physiology* from 1924 to 1927, and as vice president of the Society of Zoologists in 1937. She was a fellow of the American Association for the Advancement of Science and a member of the Society of Experimental Biology and Medicine, the Eugenics Research Association, the American Association of Anatomists, the American Society of Naturalists, and the American Genetics Association. She also was a member of Phi Beta Kappa and Sigma Xi.

BIBLIOGRAPHY

I. Original Works. Scientific articles published during King's years at Bryn Mawr are in the bound reprint series of *Bryn Mawr College Monographs*, **1**, no. 1 (1901), **1**, no. 3 (1904), and **6** (1906). Other works include "The Maturation and Fertilization of the Egg of *Bufo lentiginosus*," in *Journal of Morphology*, **17**, no. 2 (1901), 293–350; "Notes on Regeneration in *Tubularia crocea*," in *Biological Bulletin*, **4** (1903), 287ff.; "The Effects of Heat on the Development of the Toad's Egg," *ibid.*, **5** (1908), 218ff.; "Studies on Sex Determination in Amphibians, II," *ibid.*, **16** (1909), 27ff.; "The Effects of Various Fixatives on the Brain of the Albino Rat, with an Account of a Method of Preparing This Material for a Study of the Cells of the Cortex," in *Anatomical Record*, **4** (1910), 213–244; "The Effects of Pneumonia and of Postmortem Changes on the Percentage of Water in the Brain of the Albino Rat," in *Journal of Comparative Neurology*, **21**, no. 2 (1911), 147–154.

"The Effects of External Factors, Acting Before or During the Time of Fertilization, on the Sex Ratio of *Bufo lentiginosus*," in *Biological Bulletin*, **20** (1911), 205ff.; "The Sex Ratio in Hybrid Rats," *ibid.* **21** (1911) 104ff.; "The Effects of Some Amido-Acids on the Development of the Eggs of *Arbacia* and of *Chaetopterus*," *ibid.*, **22** (1912), 273ff.; *Dimorphism in the Spermatozoa of* Necturus maculosus (Baltimore, 1912); "Some Anomalies in the Gestation of the Albino Rat (*Mus norvegicus albinus*)," in *Biological Bulletin*, **24** (1913), 377–391; "The Effects of Formaldehyde on the Brain of the Albino Rat," in *Journal of Comparative Neurology*, **23** (1913), 283–314; "On the Weight of the Albino Rat at Birth and the Factors That Influence It," in *Anatomical Record*, **9** (1915), 213–231; "On the Normal Sex Ratio and the Size of the Litter in the Albino Rat," *ibid.*, **9** (1915), 403–420, written with J. M. Stotsenburg; "The Growth and Variability in the Body Weight of the Albino Rat," *ibid.*, **9** (1915), 751–776.

"On the Postnatal Growth of the Body and of the Central Nervous System in Albino Rats That Are Undersized at Birth," *ibid.*, **11** (1916), 41–52; "The Relation of Age to Fertility in the Rat," *ibid.*, **11** (1916), 269–287; "Studies in Inbreeding," in *Journal of Experimental Zoology*, **26**, nos. 1 and 2 (1918), **27**, no. 1 (1918), and **29**, no. 1 (1919); *Studies on Inbreeding* (Philadelphia, 1919); "Linkage Studies of the Rat," pts. 1–4, 9–10, in *Proceedings of the National Academy of Sciences*, **21** (1935), 390–399, **23** (1937), 56–60, **26** (1940), 578–580, written with W. E. Castle; *ibid.*, **27** (1941), 250–254, written with Castle and Amy L. Daniels; *ibid.*, **27** (1941), 394–398, **30** (1944), 79ff., **34** (1948), 135ff., **35** (1949), 545–546, written with Castle; and in *Journal of Heredity*, **38** (1947), 341–344, written with Castle; and *Life Processes in Gray Norway Rats During Fourteen Years in Captivity* (Philadelphia, 1939).

Five volumes of King's scientific papers are in the archives of the Wistar Institute of Anatomy and Biology, Philadelphia. The institute's annual reports from 1924 to 1956 contain correspondence between King and the director regarding her research. A file on King's graduate work under Morgan is in the Bryn Mawr College archives. For a discussion on the deliberations over the Ellen Richards Prize by the Association to Aid Women in Science, see the Mary Thaw Thompson papers at Vassar College, the Florence Sabin papers at the American Philosophical Society, and the Ida Hyde papers at the American Association of University Women headquarters in Washington, D.C.

II. SECONDARY LITERATURE. References to King are in Jane M. Oppenheimer, "Thomas Hunt Morgan as an Embryologist: The View from Bryn Mawr," in *American Zoologist*, **23** (1983), 845–854; and Margaret Rossiter, *American Women in Science* (Baltimore, 1982).

MARY BOGIN

KING, WILLIAM BERNARD ROBINSON (*b*. West Burton, Yorkshire, England, 12 November 1889; *d*. 23 January 1963), *geology (stratigraphy), paleontology, military geology (hydrology)*.

King, the descendant of a long line of Yorkshiremen, was the younger son of William Robinson King, Jr., a solicitor, and Florence Muriel Theed King. He attended a local preparatory school, then Uppingham. In 1908 he entered Jesus College, Cambridge, and read the natural sciences tripos. In 1912 he graduated in the first-class in geology and was awarded the Harkness Scholarship. King then worked for two years as a field surveyor for the Geological Survey of Great Britain, mapping Ordovician and Silurian rocks in the Oswestry area and the Lake District until the outbreak of World War I. In 1914 he was commissioned second lieutenant in the Seventh Battalion of the Royal Welch Fusiliers and was sent to France to provide geological advice on ensuring an adequate water supply for the army in trench warfare. On 7 June 1916, King married Margaret Amy Passingham; they had two daughters, Margaret and Cuchlaine, the latter a noted geographer/oceanographer in her own right.

For his work with the Royal Engineers during the war, King was awarded the O.B.E. In 1920 he accepted the posts of demonstrator and assistant to Professor J. E. Marr at Cambridge. King was elected a fellow of Jesus College, and in 1922 he was elected a fellow of Magdalene College and held the Charles Kingsley lectureship. In 1931 he accepted the Yates-Goldsmid chair of geology at University College, London, but kept his association with Magdalene College. In 1936 Cambridge awarded him the higher doctoral degree, Sc.D., upon submission of published works. While he was a professor in London, World War II started and King, now a major (later promoted to lieutenant colonel), was again called to serve as a geologist in France. He was awarded the Military Cross for bravery, for driving the lead truck of a convoy carrying high explosives through enemy lines during the Dunkirk withdrawal. King was elected to the Woodwardian chair at Cambridge, once held by the illustrious geologist Adam Sedgwick, in 1943.

King received many honors in his lifetime. He was on the Council of the Geological Society of London for seventeen years and served as president from 1953 to 1955. He was president of the Yorkshire Geological Society in 1949 and 1950. He was elected a fellow of the Royal Society in 1949 and was foreign correspondent of the Palaeontological Society of India, the Geological Society of America, and the Geological Society of France. King received the Wollaston Fund Award in 1920 and the Murchison Medal in 1951 from the Geological Society of London; the Gosselet Medal in 1923 from the Société Géologique du Nord; and the Prix Prestwich in 1945 from the Geological Society of France. The French universities of Lille and of Rennes conferred honorary doctorates upon him in 1947 and 1952, respectively. King died of thrombosis after a relatively minor operation.

Military geology as a branch of geology has long been ignored by the scientific community and by historians of earth science. The application of geology to military action, however, has had untold effect on the outcome of political/military disputes, aside from contributing to the development of the science itself. For example, during World War I, the neutral Dutch government was made to stop allowing German transport of materials through Dutch canals into Belgium after King proved that captured concrete aggregate contained Niedermendig basalt fragments from the Rhineland (Shotton, 1963). In addition, King's pioneering work (1921) on hydrogeology, or hydrology, based on his wartime borings in the region of the Somme River, was a significant contribution. He was highly successful in finding water supplies for the troops, and his deductions on the fluctuations of the water table and his knowledge of the geology of the area provided invaluable advice to the army on the effect of tunneling and countertunneling beneath enemy lines to plant explosives.

King's work as a military geologist during World War II was instrumental in the selection of the Normandy coast for the Allied invasion. The original choice for the landing site was the Cotentin peninsula, with Cherbourg to be used as a port to receive supplies. According to F. W. Shotton (personal communication), student and colleague of King, the plan to set up several airfields for fighters to counter enemy strikes was not carried out because King advised that the rugged Lower Paleozoic rocks there would afford no flat ground suitable for the quick construction of airfields. At his suggestion, the invasion area was changed to Normandy, where the flat Middle Jurassic limestone, covered usually by loess, could support the quick construction of small airfields, and its extension below low tide allowed the setting up of blockship and pontoon Mulberry harbors. King had further alerted the Allied Command regarding the possibility of more obstacles in the invasion beaches in Normandy "than were dreamed of by the Intelligence Corps" (Shotton, personal communication). These obstacles were met by Allied troops when they landed on the beachheads. King retired from the army in 1943 upon his election to the Woodwardian chair.

In his stratigraphic work, King did not limit his interest to a narrow period in geologic history, although his main contributions were on the Lower Paleozoic. He also published on the stratigraphy of the Pleistocene, and on the geology of the floor of the English Channel. Simultaneously he made contributions in the field of paleontology, having described Cambrian fauna of the Salt Range of India (1941) and trilobites of Persia (1930), and having named some new genera and species. He also appears to have been the first to recognize the distinctive *Hippopotamus* fauna as being characteristic of the last major interglacial period, although this observation is usually attributed to A. Sutcliffe (Andrew Currant, personal communication).

King was considered a first-rate field geologist in the classical mold. His students appreciated his encouragement and keen interest in their work. His Lower Paleozoic work included publications (1923, 1928) on the Carodocian and the Ashgillian and their relation to each other, and to the unconformable Silurian in the Berwyn Hills and the Meifod district, respectively. He recognized that the black shale known as the Pen-y-garnedd shales, containing a scarce benthic fauna as well as abundant graptolites, had been laid down in anoxic conditions in shallow offshore lagoons. In 1932 and (with W. H. Wilcockson) in 1934, King published numerous observations on these formations in his home county of Yorkshire, including the description of a band of fossiliferous Ashgillian limestone that proved to be a Neptunian dike in the highly inclined Ingletonian rocks, proving the older age (Precambrian) of the latter. The age of the Ingletonian at that time was in dispute, as some thought it was Upper Ordovician. In 1948, King (with Alwyn Williams) published his last paper on the British Lower Paleozoic. The authors reclassified the Ashgillian of King's mentor, J. E. Marr, and subdivided it into three faunal zones.

King's detailed and farsighted observations on the Pleistocene succession were eclipsed by later work by others who based their biostratigraphy on palynological evidence. His presidential address (1955) to the Geological Society of London, however, embodied "a wealth of integrated knowledge that is still of direct value today" (Andrew Currant, personal communication). At present a revival of King's type of approach, involving detailed fieldwork on a regional scale, seems to be taking place in Great Britain, so that a thorough reexamination of his work is indicated.

Long before the explosion of interest in oceanographic geological exploration, King became interested in the application of geophysical techniques to the study of the geology of the English Channel (1949), and showed that the en echelon pattern of folding observed on both the English and the French coasts is continuous across the floor of the Channel. He published (1953, with M. N. Hill; 1954) on the interpretation of seismic data in the English Channel. He also collaborated with Edward Bullard (1954) in a discussion of the floor of the Atlantic Ocean.

To add further variety to his professional life, King contributed geologic evidence to the reports on the Mid-Pleistocene Swanscombe finds of human skull remains and associated animal bones and implements.

King's career, therefore, was highly varied. He made valuable contributions in stratigraphy and paleontology, and his contributions in military geology were outstanding.

BIBLIOGRAPHY

I. ORIGINAL WORKS. King's publications are listed in a complete bibliography by F. W. Shotton, "William Bernard Robinson King, 1889–1963," in *Biographical Memoirs of Fellows of the Royal Society*, **9** (1963), 171–182. King's archival materials covering the period 1946–1955 are in the Cambridge University Archives, file no. 32/40.

II. SECONDARY LITERATURE. Besides the Shotton memorial, see a memorial by T. C. Nicholas in *Proceedings of the Geological Society of London* (1962–1963), nos. 1603–1611, pp. 150–153; and an obituary by F. W. Shotton, "Prof. W. B. R. King, O.B.E., F.R.S.," in *Nature*, **198** (1963), 244. In addition, a lengthy obituary appeared in *The Times* (London), 26 January 1963, 10; and a short announcement (with photo) in *Illustrated London News*, 2 February 1963, 167.

Personal communications with King's daughter, Cuchlaine, with A. M. King, and with F. W. Shotton were of great value in writing this essay. In addition, the following individuals provided insight in the evaluation of King's contributions: Art Boucot, Andrew Currant, Desmond Donovan, Angharad Hills, C. H. Holland, J. D. Lawson, H. B. Whittington, and Alwyn Williams.

ELLEN T. DRAKE

KLEIN, OSKAR BENJAMIN (*b.* Mörby, Sweden, 15 September 1894; *d.* Stockholm, Sweden, 5 February 1977), *theoretical physics.*

Klein was the third and youngest child of Gottlieb Klein, Sweden's first rabbi, who immigrated from Homonna, a small town in the southern Carpathians, via Eisenstadt (near Vienna) and Heidelberg, and of Toni Levy. From an early age his interest in the natural sciences was apparent; he was given to collecting small animals, observing the stars, and reading popular scientific books. Later he performed chemical experiments and studied such works as Darwin's *On the Origin of Species.*

On the occasion of a peace conference held in the summer of 1910 at Stockholm, fifteen-year-old Oskar was introduced by his father to the chemist Svante Arrhenius, who subsequently invited the boy to work in his laboratory at the Nobel Institute on the solubility of salts with radioactive indicators. The results of this investigation were published the following year. Under the guidance of Arrhenius, Klein was introduced to more specialized scientific literature when he began his university studies. Klein considered Hendrik A. Lorentz's *Leerboek der differentiaal- en integraalrekening*, which was also available in German, as having been important to his scientific development.

In the spring of 1914, Klein arranged to stay in Germany and France. However, the outbreak of war in August 1914 forced him to return to Sweden, where he fulfilled his military service. He continued his studies at the Stockholm Högskola and at the same time worked as scientific assistant at the Nobel Institute, where between 1917 and 1919 he published three papers on the dielectric properties of dipolar molecules and on electrolytes. In his doctoral dissertation, submitted in May 1921, he presented an important study of the statistical theory of suspensions and solutions based on Josiah Willard Gibbs's statistical methods.

From 1918 on, Klein frequently visited Copenhagen, and finally remained there from September 1921 until September 1922. In June 1922 he accompanied Niels Bohr when he lectured at Göttingen. From then on, Klein and Hendrik Kramers were considered Bohr's closest collaborators. Meanwhile, his interest had shifted from physical chemistry to quantum theory. He and the Norwegian physicist Svein Rosseland, with whom he shared a grant from the Danish Rask-Ørsted Foundation (created to promote international scientific relations after the war), investigated the collisions of electrons with atoms, which were then of great interest for the theoretical development of quantum theory. In the course of these investigations, they introduced, as the counterparts of normal collisions, "collisions of the second kind," in which the colliding electron gains energy instead of losing it and the interacting atomic system undergoes a transition to a lower stationary state (1921). The successful application of this concept to various areas of atomic, molecular, and celestial physics contributed to Klein's increasing reputation and aided the favorable reception of quantum theory in Sweden. But at the time, academic posts for theoretical physicists in Europe were few, and in 1922 Klein had to accept low-paid teaching positions at the universities of Stockholm and Lund.

In September 1923, Klein joined the spectroscopist Harrison M. Randall to help build up the physics department at the University of Michigan in Ann Arbor. In August 1923, before leaving for the United States, he married Gerda Koch; they had six children.

From 1923 to 1925, relatively isolated from European developments, Klein pursued his own attempts to formulate a unified relativistic theory embracing electromagnetism, gravitation, and quanta. Inspired by the optical-mechanical analogy underlying William Rowan Hamilton's mechanics, he achieved independently of Theodor Kaluza and others a five-dimensional generalization of Einstein's general relativity theory that included electromagnetism.[1] In this work the wave-particle duality of matter was anticipated through the imposition of a quantum condition of periodicity on the wave function associated with the motion of a charged particle in five-dimensional space. Since these considerations were not published until April 1926, after the advent of wave mechanics, they had no influence on the early development of that area of physics.

There was a special interest in molecular spectra at Ann Arbor. Randall had been working for some years with Friedrich Paschen at Tübingen. Participating in these activities, Klein studied molecular interactions in terms of perturbation methods used by Bohr for the Stark effect.[2] The outcome was a paper on the simultaneous action on a hydrogen atom of crossed homogeneous electric and magnetic fields (1924), which revealed a fundamental difficulty of the older quantum theory: that in this situation, transitions became possible from "allowed" to "prohibited" orbits. In his review of the state of quantum theory in 1925 (1926), Wolfgang Pauli concluded that this difficulty could be avoided only by a fundamental revision of the basic principles.

Klein returned to Europe in the summer of 1925 and resumed his post as docent in theoretical physics at the University of Lund. During a visit to Copenhagen in March 1926, he was informed of Erwin

Schrödinger's wave mechanics, and after introducing some changes in his five-dimensional approach, he published his results, which included the relativistic wave equation (discovered independently by various authors) known as the Klein-Gordon equation (1926):

$$\Delta\psi + \frac{4\pi^2}{c^2h^2}\{(h\nu - eV)^2 - m^2c^4\}\psi = 0.$$

Attempts to relate the fifth dimension to known physical quantities remained unsuccessful, but some interest was aroused by Klein's presentation at Leiden in June 1926. Under the influence of P.A.M. Dirac's relativistic theory of the electron in 1928, Klein temporarily gave up his five-dimensional theory of quantum phenomena.

During 1926 Klein became deeply involved in Bohr's work on correspondence and complementarity, which evolved into the Copenhagen interpretation of the quantum theory, offering a unified conception of the particle and wave character of microparticles. He obtained a relativistic extension of Schrödinger's expressions for the electric charge and current density associated with the wave field (1927). The point of view of correspondence allowed him to establish a rule to determine the atomic transition probabilities before Dirac did so in a more satisfactory way by quantization of the electromagnetic field. However, Klein's procedure remained accepted for many years.

Other important contributions to quantum field theory emerged in collaboration with Pascual Jordan, who visited the Bohr Institute in 1927. Klein had succeeded Kramers and Heisenberg as lecturer at the Copenhagen Institute of Theoretical Physics. He and Jordan became interested in the quantum theoretical treatment of the many-particle problem as a result of Dirac's radiation theory. For the case of particles without spin (Bose particles), they found a method of obtaining the number of particles associated with the three-dimensional wave field by the introduction of commutation rules between the corresponding field operators (1927). This procedure, known as second quantization, is specially suited to describe processes involving particle creation and annihilation, and hence was important for the development of particle physics. Later the method was extended to fermions by Jordan and Eugene Wigner.

Dirac's relativistic theory of the electron, published at the beginning of 1928, offered a new possibility for discussing the interaction between radiation and free electrons in a consistent manner. In a joint work with Yoshio Nishina, a Japanese guest at the Bohr Institute, Klein faced the laborious enterprise of calculating the angular intensity distribution of Compton scattering according to the new theory (1929). The Klein-Nishina formula improved the earlier intensity formula obtained by Dirac and Walter Gordon with the method of electrodynamic potentials.

The experimental verification of the Klein-Nishina formula convinced many physicists of the soundness of Dirac's relativistic equation, in spite of fundamental difficulties. One such difficulty is known as Klein's paradox. By direct calculation Klein could show that, in appropriate potential fields, transitions to states with negative kinetic energy became possible. The interpretation of particles that appeared to accelerate in the opposite direction to an applied force was a major problem before the discovery of antiparticles.

In the last year of his stay at Copenhagen, Klein completed a paper containing an elegant method, still used in microwave spectroscopy, of determining rotational energy levels. Later he helped to develop a procedure, of great utility in modern molecular research, for calculating the interaction potentials directly from the energy levels of a diatomic molecule (1932).

In 1927 the chair of mechanics and mathematical physics at the Stockholm Högskola, occupied until then by the mathematician Ivar Fredholm, became vacant. Klein applied for the post and won it, but because of the complicated Swedish nomination procedure, he was not appointed until 1930. Once in Stockholm, in addition to his scientific work and teaching obligations, Klein developed an active cultural and political life. Worthy of mention in the latter line is his help to refugees during World War II, and, in the former, his engagement in historical and philosophical studies, and in the diffusion of the new physical ideas.

In his first scientific paper at Stockholm (1931), Klein gave, in the tradition of Gibbs's statistical mechanics, an explanation of the thermodynamic irreversibility paradoxically generated by mechanically reversible systems. For this purpose he used an inequality (later known as Klein's lemma) for the statistical probability given by the density matrix introduced by Dirac. In 1933 he proposed an approximate method for solving the wave equation by a recursion procedure very similar to the Wentzel-Kramers-Brillouin method. In 1935 Klein used a generalized Dirac equation for an approximate description of a nuclear system. In 1938 he presented some procedures for solving the Schrödinger equation in a periodic force field and also for describing electron interactions in a crystal. Also in 1938, at a conference on new theories in physics held in Warsaw, Klein presented a formulation of nonabelian vector field interactions that anticipated some aspects of the formalism of Chen Ning Yang and Robert Lawrence Mills. This work returned him to the fifth dimension.

The original five-dimensional theory of 1926 was an elegant attempt to unify the then known physical interactions, gravity and electromagnetism. In 1947 Klein tried to incorporate meson forces into this theory. In order to confine the charge of the mesons to integral positive and negative multiples of the elementary charge, the wave function must be a periodic function of the fifth coordinate, and this periodicity introduced a fundamental length scale into the theory. Klein was the first to recognize the similarity of the newly discovered muon decay to nuclear beta decay. This was the first indication of the universal Fermi interaction for weak interactions, and was discovered independently by E. Clementel and G. Puppi.

At about the same time, Klein also made important contributions to the theory of superconductivity. In order to obtain a quantum mechanical microscopic model of this phenomenon, he computed the diamagnetic properties of an electron gas (1944). Then, in 1945, in collaboration with Jens Lindhard, he pointed out that wave functions extending over large regions of space are favorable for large diamagnetism. In 1952, using the Bloch method, he reached a tentative description of superconductivity. The definitive solution was achieved in 1957 by John Bardeen, Leon N. Cooper, and John Robert Schrieffer.

In 1946, with the Swedish physicists Göran Beskov and Lars Treffenberg, Klein published a work on the distribution of chemical elements. They intended to solve the basic problems connected with the so-called hypothesis of frozen equilibrium in accounting for the distribution of the chemical elements. This hypothesis assumes a special kind of chemical equilibrium among atomic species at high temperature and pressure, frozen by a rapid expansion. The simple equilibrium formula resulting from the hypothesis, however, does not agree with observation for large atomic masses. To correct this, Klein and his collaborators suggested that different atoms could be created at different places and at different times, in the interior of the primitive stars. Although ingenious, their solution is not widely accepted. Also in relation to cosmology, Klein pursued the problem of finding exact solutions of Einstein's field equations (1947). In 1954 he presented a solution that provided the relativistic analogue to the polytropic gas spheres.

Cosmology further stimulated Klein's interest in statistical mechanics. In 1944 he derived the laws of thermodynamic equilibrium for a fluid in a static gravitational field. He showed that, in addition to the Tolman condition, one must introduce the condition $\alpha(g_{\mu\nu})^{\frac{1}{2}} = cte$, where α is the chemical potential of the fluid and $g_{\mu\nu}$ is the metric tensor. This relation constitutes the relativistic generalization of Gibbs's equilibrium condition.

Klein's cosmology culminated in the mid 1950's, in a supergalactic system regarded as an ordinary, but very large, superstellar system, condensed from a thin, cold, hydrogen cloud. With some further assumptions, Klein deduced the Eddington relations and established the basis for the cosmological model that he later constructed with Hannes Alfvén. In 1956 and 1957 Klein proposed an extension of the five-dimensional theory of combined gravitational and electromagnetic fields including quantum fields, claiming that the inclusion of gravitational effects might eliminate the problem of the divergences of electron theory.

At the Solvay Conference on the Structure and Evolution of the Universe, held in 1958 at Brussels, Klein presented some speculative conjectures against the cosmological constant introduced by Einstein. In his report, "Some Considerations Regarding the Earlier Development of the System of Galaxies" (1959), he offered an alternative description of the evolution of a hypothetical gas cloud in the framework of general relativity.

As professor emeritus of the University of Stockholm in 1962, Klein pursued his studies of the classification schemes for the growing number of elementary particles. In 1959 he had presented an extension of the isobaric spin scheme to include K mesons and hyperons, suggesting also a scheme for weak interactions. In 1966 he proposed a classification of the baryons in terms of baryon and lepton numbers and strangeness, without using hypercharge or supercharge.

The experimental discovery of the antiproton by Owen Chamberlain and Emilio Segrè in 1955 confirmed the supposed matter-antimatter symmetry. In 1963, Klein and Alfvén presented a cosmological model based on a perfect symmetry between matter and antimatter on a cosmological scale. This model was inspired by Klein's earlier speculations on the metagalactic structure of the universe, originating from a sphere of very dilute plasma containing equal quantities of matter and antimatter. When the density of the gravitationally contracting sphere reached a critical value, annihilation processes became dominant. Then the collapse stopped and reversed to the outward motion connected with the red shift of galaxies. The Klein-Alvén cosmology adopts the concept of a globally nonexpanding universe, considering the observed red shift only as a local phenomenon. Difficulties with this cosmological model were soon recognized.

Also in connection with his cosmological model, Klein scrutinized the foundations of general relativity and concluded that many difficulties could be overcome by keeping close to the original ideas of Einstein. He claimed that the principle of equivalence between gravitation and inertial forces is incompatible with the idea proposed by Mach, and accepted by Einstein, to describe the universe by an analogy in three dimensions to the closed surface of a sphere. Along the same line, he derived an extension of Einstein's principle of equivalence to include the case of particles described through the Dirac equation.

In 1970 Klein presented "A Tentative Program for the Development of Quantum Field Theory as an Extension of the Equivalence Principle of General Relativity Theory," which starts with his five-dimensional formulation of 1926:

> Although the periodicity as an interpretation of the elementary quantum of electricity occurred to me almost in the beginning (in 1925) and later led me to the consideration of the corresponding group, from early times until recently I oscillated between this way and a formalism related to isospin, obtained by cutting down the higher harmonics of the Fourier expansion. This, however, led to two difficulties: the linearity in the momenta in the Lagrangian density could not be maintained unless the group was limited in analogy to the transformation group; and that the equivalence claim could not be satisfied. With the full periodicity these difficulties are absent. (p. 253)

Klein's last papers are devoted to this program.

In 1951 Klein received an honorary doctorate from the University of Oslo, and in 1965 from the University of Copenhagen. In 1954 he became a member of the Nobel Committee on Physics.

NOTES

1. Since there is no documentary evidence of this early work, we must rely on Klein's statements given in his autobiographical account on six interviews conducted in 1962 and 1963. See Kuhn *et al.*, 103f.
2. See also interview 3 with Klein (Kuhn *et al.*) from 25 February 1963, p. 8f. of the transcript.

BIBLIOGRAPHY

I. Original Works. A bibliography of Klein's most important scientific articles is in Poggendorff, VI, pt. 2, 1330, and VIIb, pt. 4, 2492–2494. He also wrote many popular scientific essays and books on atomic and nuclear physics, on relativity, and on cosmology, most of them in Swedish: *Den Bohrsk a atomtheorien* (Stockholm, 1922); "Vad vi veta om ljuset," in *Natur och kultur*, **41** and **42** (1925): *Orsak och verkan i den nya atomteoriens belysning* (Stockholm, 1935); *Entretiens sur les idées fondamentales de la physique moderne*, L. Rosenfeld, trans. (Paris, 1938); and *Les processus nucléaires dans les astres* (Liège, 1953).

Only a small portion of Klein's scientific manuscripts and correspondence deposited at the Niels Bohr Institute, Copenhagen, is listed in Thomas S. Kuhn, John L. Heilbron, Paul Forman, and Lini Allen, *Sources for History of Quantum Physics* (Philadelphia, 1967), 55. Some letters from the correspondence with Bohr and Kramers are in Niels Bohr's *Collected Works*, Klaus Stolzenburger, ed., V (1984), 382–394, and IX, 592–596; and letters from the correspondence with Wolfang Pauli are in Pauli's *Wissenschaftlicher Briefwechsel*, Karl von Meyenn, Armin Hermann, and Victor F. Weisskopf, eds., I, 488–492, 494–495, and II, 3–4, 7, 43–46, 51, 422–431, and 534.

"Über die Löslichkeit von Zinkhydroxyd in Alkalien," in *Zeitschrift für anorganische Chemie*, **74** (1912), 157–169; "Zur statistischen Theorie der Suspensionen und Lösungen," in *Arkiv för matematik, astronomi och fysik*, **16** (1921), no. 5, 1–51; "Über Zusammenstösse zwischen Atomen und freien Elektronen," in *Zeitschrift für Physik*, **4** (1921), 46–51, with Svein Rosseland; "Über die gleichzeitige Wirkung von gekreuzten homogenen elektrischen und magnetischen Feldern auf das Wasserstoffatom. I," *ibid.*, **22** (1924), 109–118; "Quantentheorie und fünfdimensionale Relativitätstheorie," *ibid.*, **37** (1926), 895–906; "Elektrodynamik und Wellenmechanik vom Standpunkt des Korrespondenzprinzips," *ibid.*, **41** (1927), 407–442; "Zur fünfdimensionalen Darstellung der Relativitätstheorie," *ibid.*, **46** (1927), 188–208; "Zum Mehrkörperproblem der Quantentheorie," *ibid.*, **45** (1927), 751–765, with Pascual Jordan; "Über die Streuung von Strahlung durch freie Elektronen nach der neuen relativistischen Quantendynamik von Dirac," *ibid.*, **52** (1929), 853–868, with Yoshio Nishina; "Die Reflexion von Elektronen an einem Potentialsprung nach der relativistischen Dynamik von Dirac," *ibid.*, **53** (1929), 157–165; "Zur Frage der Quantelung des asymmetrischen Kreisels," *ibid.*, **58** (1929), 730–734.

"Zur quantenmechanischen Begründung des zweiten Hauptsatzes der Wärmelehre," in *Zeitschrift für Physik*, **72** (1931), 767–775; "Zur Berechnung von Potentialkurven für zweiatomige Moleküle mit Hilfe von Spektraltermen," *ibid.*, **76** (1932), 226–235; "Zur Frage der quasimechanischen Lösung der quantenmechanischen Wellengleichung," in *Zeitschrift für Physik*, **80** (1933), 792–803; "Quelques remarques sur le traitement approximatif du problème des électrons dans un réseau cristallin par la mécanique quantique," in *Journal de physique et le radium*, 7th ser., **9** (1938), 1–12; "Philosophy and Physics," in *Theoria* (Gotebrog-Lund) (1938), 59–61; "Sur la théorie des champs associés à des particules chargées," in *Les nouvelles théories de la physique* (Paris, 1939), 81–98; "On the Magnetic Behaviour of Electrons in Crystals," in *Arkiv för matematik, astronomi och fysik*, **31A** (1944–1945), no. 12; "Some Remarks on the Quantum Theory of the Superconductive State," in *Reviews of Modern Physics*, **17** (1945), 305–309, with J. Lindhard; "On the Origin of the Abundance Distribution of Chemical Elements," in *Arkiv för matematik, astronomi och fysik*, **33B** (1946–1947), no. 1, with G. Beskov and L. Treffenberg; "Meson Fields and Nuclear Interaction," *ibid.*, **34A** 1947–1948), no. 1; "On a Case of Radiation Equilibrium in General Relativity Theory and Its Bearing on the Early Stage of Stellar Evolution," *ibid.*, no. 19; and "On the Thermodynamical Equilibrium of Fluids in Gravitational Fields," in *Reviews of Modern Physics*, **21** (1949), 531–533.

"Theory of Superconductivity," in *Nature*, **169** (1952), 578–579; "On a Class of Spherically Symmetric Solutions of Einstein's Gravitational Equations," in *Arkiv för fysik*, **7** (1954), 487–496; "Generalizations of Einstein's Theory of Gravitation Considered from the Point of View of Quantum Field Theory," in *Helvetica physica acta*, supp. **4** (1956), 58–71; "Some Remarks on General Relativity and the Divergence Problem of Quantum Field Theory," in *Nuovo Cimento*, supp. **6** (1957), 344–348; "On the Systematics of Elementary Particles," in *Arkiv för fysik*, **16** (1960), 191–196; "Matter-Antimatter Annihilation and Cosmology," *ibid.*, **23** (1963), 187–194, with Hannes Alfvén; "Remark Concerning the Basic SU(3) Triplets," in *Physical Review Letters*, **16** (1966), 63–64; "A Tentative Program for the Development of Quantum Field Theory as an Extension of the Equivalence Principle of General Relativity Theory," in *Nuclear Physics*, **B21** (1970), 253–260; "Arguments Concerning Relativity and Cosmology," in *Science*, **171** (1971), 339–345; "The Equivalence Principle and the Dirac Equation with Gravitation," in *Physica Norvegica*, **5** (1971), 145–147; "Ur mitt liv i fysiken," in *Svensk naturvetenskap* (1973), 159–172; "Generalization of Einstein's Principle of Equivalence so as to Embrace the Field Equations of Gravitation," in *Physica scripta*, **9** (1974), 69–72; "Einstein's Principle of Equivalence Used for an Alternative Relativistic Cosmology Considering the System of Galaxies as Limited and not as the Universe," in *Kongelige danske videnskabernes selskabs matematisk-fysiske meddelelser*, **39** (1974), 3–18; and "Electromagnetic Theory Treated in Analogy to the Theory of Gravitation," in *Nuclear Physics*, **B92** (1975), 541–546.

II. Secondary Literature. Obituaries are by Stanley Deser, in *Physics Today*, **30** (June 1977), 67–68; Inga Fischer-Hjalmars and Bertel Laurent, in *Kosmos* (1978), 19–29; and Christian Møller, in *Fysisk tidsskrift*, **75** (1977), 169–171.

Authoritative descriptions of Klein's contributions to quantum theory are in Wolfgang Pauli, "Quantentheorie," in *Handbuch der Physik*, XXIII (Berlin, 1926), 1–278, and "Die allgemeinen Prinzipien der Wellenmechanik," in *Handbuch der Physik*, XXIV, pt. 1 (Berlin, 1933), 83–272.

Klein's activity during his Copenhagen period is described in Max Dresden, *Hendrik Anthony Kramers* (New York, 1987); and in Peter Robertson, *The Early Years. The Niels Bohr Institute 1921–1930* (Copenhagen, 1979). A good survey of Klein's contribution to the development of the relativistic wave equation is Helge Kragh, "Equation with Many Fathers. The Klein-Gordon Equations in 1926," in *American Journal of Physics*, **52** (1984), 1024–1033. Klein's cosmological speculations are summed up in Hannes Alfvén, "Antimatter and Cosmology," in *Scientific American* (April 1967), 106–113; and by Jagjit Singh, *Great Ideas and Theories of Modern Cosmology* (New York, 1961), 273–279. For the new formulations of the Kaluza-Klein theories, especially in the context of supergravity, see Venzo De Sabbata and Ernst Schmutzer, eds., *Unified Field Theories of More Than Four Dimensions* (Singapore, 1983).

Karl von Meyenn
Mariano Baig

KLENK, ERNST (*b*. Pfalzgrafenweiler, near Freudenstadt, Germany, 14 October 1896; *d*. Cologne, Federal Republic of Germany, 29 December 1971), *physiological chemistry*.

The son of Johannes Klenk, a brewer, and of Katharina Grossman, Klenk attended gymnasium at Tübingen, and after serving in World War I, he entered the University of Tübingen in 1918 to study chemistry. He took his doctorate in 1923 and worked under Hans Thierfelder, and later Franz Knoop, the discoverer of the β-oxidation of fatty acids. With Thierfelder he wrote a book on the chemistry of the cerebrosides (1930). In his early research under Knoop, Klenk worked on the oxidation of fatty acyl benzene derivatives, thereby laying the foundation for his interest in lipid and fatty acid chemistry. He became associate professor at Tübingen in 1931, and in 1936 he accepted the chair of physiological chemistry at Cologne, a position he held until his retirement in 1967. In 1937 Klenk married Margarete Aldinger, a physician; they had three sons.

Klenk made important contributions to lipid chemistry and more specifically to neurochemistry, a field that had been opened up by Johann Ludwig Wilhelm (John Louis William) Thudichum at the turn of the century. Thudichum summarized his lifework on the sphingolipids of the brain in his *Die chemische Konstitution des Gehirns des Menschen und der Tiere* (1901). Klenk did a considerable amount of work on the cerebrosides. He was the first to isolate nervon, cerebron, and other cerebrosides, and to describe in detail the structures of their fatty acid constituents. In 1929 he demonstrated for the first time the correct chain length (18C) of sphingosine; later he contributed to the elucidation of its structure. He also worked on the composition of polyenolic fatty acids, and he isolated a C_{22} fatty acid that was more highly unsaturated than the well-known arachidonic acid.

In 1935 Klenk discovered a new class of sphingolipids, the gangliosides. This result came from the investigation of the biochemistry of various neurolipodystrophies, including Niemann-Pick disease, and especially from the study of Tay-Sachs disease, otherwise known as amanrotic idiocy. The latter is a familiar disease of infancy, characterized by a progressive degeneration of nerve cells, eventuating in weakness of the muscles and paralysis. Klenk suggested the name *ganglioside* for the type of glycolipid isolated in this disease, in direct analogy to the already named cerebrosides, and because this kind of glycolipid accumulated in the ganglia cells of the brain in Tay-Sachs disease.

Whereas cerebrosides are localized in white matter, gangliosides were initially thought to be unique to the gray matter of the brain. Because of their high concentration in the gray matter, they were initially regarded as characteristic components of nerve cells. Although gangliosides may be said generally to possess greater complexity than most other sphingolipids, it is now known that they are not exclusively localized either in neurons or in the brain.

In 1941 Klenk isolated neuraminic acid, in the form of the crystallized methylglycoside, from brain gangliosides. Later, with other groups, he went on to elucidate the structure of several gangliosides. He described the acceptor function of neuraminic acid for myxoviruses of the mumps-influenza group and found that neuraminic acid is the determinant group of the MN blood group system.

In the last years of his life, Klenk's major effort was directed to the further purification and elucidation of ganglioside structures. His life's work acted as a strong stimulus to others interested in the structure of sphingolipids and their chemical and enzymatic synthesis and metabolism.

Klenk was a founding member of the editorial board of the *Journal of Neurochemistry* (1957), and served as president of the Deutsche Gesellschaft für Biologische Chemie. Among his honors were the Norman Medal of the Deutsche Gesellschaft für Fettwissenschaft in 1953, the Oil Chemist Award in 1958, the Heinrich Wieland Prize in 1964, the American Oil Chemists Society Award in 1965, the Stouffer Prize in 1966, and the Otto Warburg Medal of the Deutsche Gesellschaft für Biologische Chemie in 1971.

BIBLIOGRAPHY

I. ORIGINAL WORKS. A bibliography of Klenk's scientific papers is in J. C. Poggendorff, *Biographisch-literarisches Handwörterbuch der exakten Naturwissenschaften*, VI, 1333, and VIIa, pt. 2, 781–872. His writings include *Die Chemie der Cerebroside und Phosphatide* (Berlin, 1930), with Hans Thierfelder; and "On the Discovery and Chemistry of Neuraminic Acid and Gangliosides," in *Journal of the Chemistry and Physics of Lipids*, **5** (1970), 193–197.

II. SECONDARY LITERATURE. Hildegard Debuch, "Ernst Klenk, 1896–1971," in *Journal of Neurochemistry*, **21** (1973), 725–727.

NEIL MORGAN

KNIGHT, SAMUEL HOWELL (*b.* Laramie, Wyoming, 31 July 1892; *d.* Laramie, 1 February 1975), *geology*.

S. H. Knight was the son of Wilbur C. and Emma Howell Knight. His father was the first professor of geology at the University of Wyoming, and his mother was dean of women at the same institution from 1911 to 1921. In his youth, Knight spent considerable time working as an assistant to both his father and Professor William C. Reed, paleontological curator at the university. With his brother, Oliver, he followed his father into a career in geology. He was educated at the University of Wyoming (B.A. 1913) and Columbia University (Ph.D. 1929). In 1916 he married Edwina Hall, whom he met while at Columbia. They had two children, Wilbur and Eleanor, both of whom studied geology.

After several years as a graduate student at Columbia, Knight returned to Wyoming in 1916 as an assistant professor of geology and by 1917 was full professor and head of the one-man department of geology. During the early years (1917–1928) he established the undergraduate curriculum and initiated a graduate program that began awarding the master's degree in geology in 1926. By 1931 the department had grown to three members and after World War II to seven. The graduate program included doctoral students by the early 1950's. During much of this time "Doc" Knight wore two hats: head of the department and Wyoming state geologist (1933–1941).

Knight's emphases as a career academic geologist were the education of undergraduate students, the development of exceptional field geologists (Wyoming Geology Science Camp), and, using his considerable artistic abilities, the popularization of geology for the general public in a geologically blessed state. The latter efforts are best represented by the university geology museum that houses his reconstruction of a Wyoming *Brontosaurus* skeleton and murals depicting the paleogeography of the Rocky Mountain West. His full-scale copper *Tyrannosaurus rex* now stands before the S. H. Knight Geology Building at the University of Wyoming.

Interests in unraveling sedimentary paleoenvironments, the relationships between intermontane basin formation and episodic mountain growth, the evolution of landscapes, and history of geology, published as abstracts and field guides, provide insight into his active and sharing mind. That experience and enthusiasm were brought to students and to lay and professional audiences throughout the West using a "chalk talk" approach in which the geologic history of the area "evolved" through perspective drawings before their very eyes.

Although the bulk of his time was spent for the benefit of students, faculty, and the university, Knight made significant contributions to the field of geology. Two papers, one sedimentological (1929), the other paleontological (1968), illustrate his unique geologic insight and his recognition of the importance of three-dimensional representation of geologic features. His doctoral work on the sedimentology of the Pennsylvanian Fountain and Casper formations in the Laramie Basin characterizes his background and scientific style: The work was field based and tied intimately to the present Laramie River basin as a modern analogue. Quantitative grain-size, mineral, and chemical analyses of representative samples from carefully measured stratigraphic sections of the Fountain and Casper formations, from mechanically weathered Precambrian Sherman granite, and from Laramie River sediments provided a unique application of the principle of uniformitarianism. He concluded that the geologic conditions now are much like those that existed during Fountain time and many primary sedimentary structures of the Fountain Formation are common in the present drainage system.

In addition to his interpretation of the lithic character and distribution of the Fountain Formation as a Pennsylvanian alluvial fan complex, he described and defined unusual festoon cross-lamination, some with penecontemporaneous folds. More fully developed in the laterally equivalent Casper Formation, these primary sedimentary structures were elegantly documented in plan, in cross section, and in three-dimensional sketches. Commonly reproduced in modern sedimentology texts, his diagrams and explanation that festoon cross-lamination forms "as a result of shifting currents in a large shallow body of water" are easily translated to modern features in nearshore tidal channels where strong currents constantly change and rework clastic sediments. In the same work, interestingly, he recognized the possibility that sandstone dikes in the Fountain and Casper sandstones resulted from earthquake activity during the Pennsylvanian—a paleoseismic indicator receiving renewed attention.

Toward the end of his academic career and after his 1963 retirement, Knight completed a detailed study of Precambrian stromatolites, remarkably well-preserved in the Nash Formation in the Medicine Bow Mountains, Wyoming. With his grandson (now a professional geologist) as an assistant, he mapped the distribution of stromatolites, bioherms, and reefs by plane table and alidade (1:1200 scale) and individual heads by net and protractor methods (1:60 to 1:120 scale) in such detail that entire algal communities could be reconstructed. To distinguish the three basic Proterozoic algal forms, he cut oriented samples into four-inch cubes, resectioned the sides into thin, polished translucent slabs, photographed, and reassembled the cubes with an internal light source to examine their full three-dimensional character. His very clever treatment of this Precambrian occurrence is a fully illustrated, descriptive classic.

S. H. Knight became an honorary member of the American Association of Petroleum Geologists in 1959. He was an honorary life member of the Wyoming Geological Association (1952) and a fellow of the Geological Society of America. His published work and the time, thought, and effort he devoted to his students and his university provide a living legacy in those whom he touched and in their students. His philosophy that geology begins in the field and that theoretical and experimental advances must be returned to a field setting for the science to flourish is no less true today than during the fifty years of "Doc" Knight's tenure in and for Wyoming.

BIBLIOGRAPHY

I. Original Works. Knight was the author of thirty-four scientific papers, including abstracts of work presented at professional meetings and field guides to the geology of Wyoming. A complete bibliography is in the memorial by J. D. Love and Jane M. Love in *Geological Society of America, Memorials*, 7 (1977). His most significant works are "The Fountain and the Casper Formations of the Laramie Basin: A Study on Genesis of Sediments," in *University of Wyoming Publications in Science: Ge-*

ology, **1** (1929), 1–82; and "Precambrian Stromatolites, Bioherms, and Reefs in the Lower Half of the Nash Formation, Medicine Bow Mountains, Wyoming," in *Contributions to Geology*, **7** (1968), 73–116.

II. SECONDARY LITERATURE. R. S. Houston and D. W. Boyd, memorial, in *Bulletin of the American Association of Petroleum Geologists*, **60** (1977), 1130–1132, and H. D. Thomas, introduction, in *Contributions to Geology*, **2** (1963), 1–6, an issue dedicated to Knight in honor of his retirement that contains a bibliography complete through 1963. There is also a brief treatment of Knight's impact on the University of Wyoming in *Time* (12 July 1963).

WALLACE A. BOTHNER

KNOPF, ADOLPH (*b.* San Francisco, California, 2 December 1882; *d.* Palo Alto, California, 23 November 1966), *petrology, economic geology*.

Adolph Knopf, son of German immigrants Anna Greisel and George Tobias Knopf, became one of the most influential geologist-educators in the United States. Born and raised in California, he entered the University of California at Berkeley and received a B.S. in 1904, an M.S. in 1905, and a Ph.D. in 1909. His mentor was Andrew Cowper Lawson, a master of field observations and a leading thinker in structural geology, petrology, and the origin of mineral deposits. From Lawson, Knopf learned the value of experience gained through field studies; he followed that lesson throughout his life and passed it along to his own students.

The summer of 1905 found Knopf in Alaska as a temporary employee of the U.S. Geological Survey; in the fall of 1906 he joined that institution as a permanent employee, continuing to work in Alaska until 1911 and submitting one of his studies, on the tin deposits of the Seward Peninsula, for his Ph.D. dissertation. In 1911 Knopf was assigned to study the region between Butte and Marysville in Montana; this brought him into contact with the Boulder Batholith, a major geological feature he continued to study for the rest of his life.

Knopf married Agnes Burchard Dillon in 1908; they had three daughters and one son. The family lived in Washington, D.C., site of the headquarters of the U.S. Geological Survey. The months when fieldwork was possible found Knopf in the West; winter months were spent in Washington writing reports and completing petrographic studies. Through his combined field and laboratory studies, Knopf became distinguished both as an economic geologist and as a petrologist. In 1915 he commenced what was to become probably his most important work, a study of the Mother Lode district of California. The result appeared as Professional Paper 157 of the U.S. Geological Survey (1929) and was immediately recognized as a classic work. It is still one of the most important studies ever made of the geometry and petrology of an intricate system of faults and veins in a metamorphic complex.

Agnes Knopf died of influenza in 1918; in January 1920 Knopf started teaching at Yale and in June of the same year he married Eleanora Frances Bliss, a geologist with whom he had worked at the Geological Survey since 1912. No children were born of the second marriage. Among the faculty members at Yale when Knopf arrived was Bertram Borden Boltwood, who had, in earlier years, collaborated with Ernest Rutherford on studies of radioactivity, and who had demonstrated through chemical analysis that lead is the daughter product derived by radioactive decay of both uranium and thorium. Boltwood's demonstration in 1907 that the proportions between uranium and thorium and the associated lead might be used to determine mineral ages had, by 1920, been strongly supported by a number of influential people, in particular Arthur Holmes in England and Joseph Barrell of the Yale faculty. Knopf recognized the importance of radiometric dating, became a spokesman in seeking support for development of the technique, and published a number of important papers concerning the age of Earth. He did not carry out any of the measurements himself, but through the 1930's and 1940's he was the most influential scientist in the country supporting radiometric dating.

Knopf was a legendary teacher among graduate students. His lecturing style was not good and his successes in teaching undergraduates were few, but his close, interactive teaching with graduate students at Yale left a lasting impact, not only on the students themselves but also on the institutions they later joined, and thereby on the entire geological profession. A combination of careful field and rigorous laboratory observations underlay Knopf's work, and he passed on a belief in this combination to students. Among those who have written of Knopf's influence on them are James Gilluly and William W. Rubey. Knopf continued working after retirement from Yale, first as visiting professor, and then as consulting professor, at Stanford.

Knopf was elected to the National Academy of Sciences in 1931. He was a fellow of the Geological Society of America (president, 1944) and a member of the Society of Economic Geologists. He was awarded the Penrose Medal of the Geological Society of America in 1959. Perhaps the greatest honor paid

Adolph Knopf is the frequency with which his former students mention in their published reminiscences that he played an essential role in their education and scientific development.

BIBLIOGRAPHY

I. ORIGINAL WORKS. "Geology of the Seward Peninsula Tin Deposits, Alaska," *Bulletin of the U.S. Geological Survey*, no. 358 (1908); "The Seward Peninsula Tin Deposits," *ibid.*, no. 345 (1908), 251–267; "Some Features of the Alaskan Tin Deposits," in *Economic Geology*, **4** (1909), 214–223; "The Eagle River Region, Southeastern Alaska," *Bulletin of the U.S. Geological Survey*, no. 502 (1912); "Ore Deposits of the Helena Mining Region, Montana," *ibid.*, no. 527 (1913); "The Tourmalinic Silver-Lead Type of Ore Deposit," in *Economic Geology*, **8** (1913), 105–119; "Mineral Resources of the Inyo and White Mountains, California," *Bulletin of the U.S. Geological Survey*, no. 540 (1914), 81–120; "A Geologic Reconnaissance of the Inyo Range and the Eastern Slope of the Southern Sierra Nevada, California," *Professional Papers of the U.S. Geological Survey*, no. 110 (1918); "Geology and Ore Deposits of the Yerington District, Nevada," *ibid.*, no. 114 (1918).

"Ore Deposits of Cedar Mountain, Mineral County, Nevada," in *Bulletin of the U.S. Geological Survey*, no. 725 (1921), 361–382; "Geology and Ore Deposits of the Rochester District, Nevada," *ibid.*, no. 762 (1924); "The Mother Lode System of California," *Professional Papers of the U.S. Geological Survey*, no. 157 (1929); "Age of the Earth; Summary of Principal Results," in *Bulletin of the National Research Council*, **80** (1931), 3–9; "Age of the Ocean," *ibid.*, 65–72; "Igneous Geology of the Spanish Peaks Region, Colorado," in *Bulletin of the Geological Society of America*, **47** (1936), 1727–1784; "The Geosynclinal Theory," *ibid.*, **59** (1948), 649–669; "Bathyliths in Time," in Arie Poldervaart, ed., *Crust of the Earth*, Geological Society of America Special Paper no. 62 (1955), 685–702; and "The Boulder Bathylith of Montana," in *American Journal of Science*, **255** (1957), 81–103.

II. SECONDARY LITERATURE. Robert G. Coleman, "Memorial of Adolph Knopf," in *American Mineralogist*, **53** (1968), 567–576; and Chester R. Longwell, "Adolph Knopf," in *Biographical Memoirs. National Academy of Sciences*, **41** (1970), 235–249.

BRIAN J. SKINNER

KNORR, LUDWIG (*b.* Munich, Germany, 2 December 1859; *d.* Jena, Germany, 4 June 1921), *chemistry.*

Knorr, one of five sons of Angelo Knorr, a merchant, attended gymnasium in Munich. He began to study chemistry at the university there in 1878, and also attended the universities of Erlangen and Heidelberg; at the latter he spent a term working under Robert Bunsen. In 1879 the organic chemist Emil Fischer accepted a professorship at Munich, and offered Knorr an assistantship in the department of chemistry. In 1882 Knorr moved with Fischer to Erlangen and took his doctorate there with a dissertation on piperyl-hydrazine. He married the daughter of the artist C. Piloty before moving in 1885, again with Fischer, to the University of Würzburg, where he became assistant professor. In 1889 Knorr was elected to the chair of chemistry at the University of Jena.

During the last decades of the nineteenth century many of the "name reactions" of synthetic organic chemistry were developed and extended as German organic chemistry responded to the challenge of producing in the laboratory both natural and novel organic complexes, and contributed to the needs of a growing chemical industry for drugs and dyestuffs. Early in his career Knorr began important work on heterocyclic compounds, especially the pyrazolones (1883–1911). He worked on several syntheses, and by condensing acetoacetic acid and phenylhydrazine, he produced the compound 3-methyl, 1-phenyl pyrazolone, which, when methylated, produced antipyrine, a chemical that was shown to reduce fever. He also synthesized by condensation reactions other complicated heterocyclic compounds, such as quinoline (1886), pyrazole derivatives (1887), and derivatives of pyrrole (1902). (The pyrrole ring is the basic unit of the porphyrin system that occurs, for example, in hemoglobin and chlorophyll.) In the 1890's Knorr also did much work on the complicated structure of the vegetable alkaloid morphine and discovered morpholine.

Besides his addition to synthetic organic chemistry, Knorr contributed significantly to the theoretical problem of tautomerism. In 1863 acetoacetic ester (ethyl β-ketobutyrate) was isolated. It became evident that it reacts as though it possesses two different structural formulas. In fact it exhibits keto-enol tautomerism, in which the structures differ in the point of attachment of a hydrogen atom:

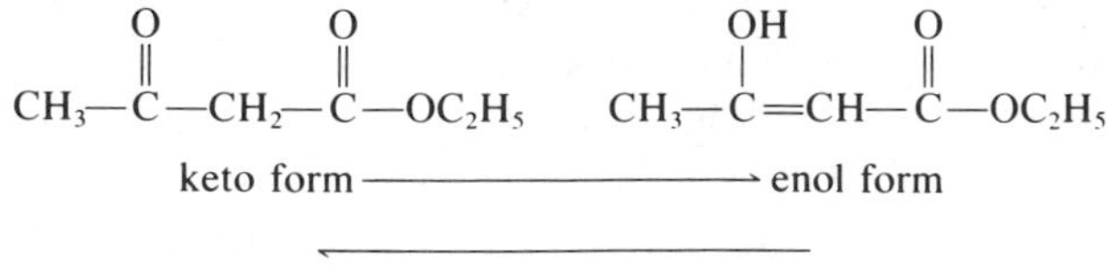

$$CH_3-\overset{\overset{O}{\|}}{C}-CH_2-\overset{\overset{O}{\|}}{C}-OC_2H_5 \qquad CH_3-\overset{\overset{OH}{|}}{C}=CH-\overset{\overset{O}{\|}}{C}-OC_2H_5$$

$$\text{keto form} \; \rightleftharpoons \; \text{enol form}$$

Acetoacetic Acid

Along with other German chemists such as Ludwig Claisen and Johannes Wislicenus, Knorr became interested in the phenomenon, and in the 1890's he worked on the keto-enol allelotropic mixtures of diacetylsuccinic ester, acetylacetone, and acetoacetic ester. In 1911 he prepared the two tautomeric forms of acetoacetic ester separately, each free of contamination by the other. The keto form was obtained by cooling a solution of the ordinary ester in alcohol and ether to −78°C. The enol form was obtained by the action of dry hydrogen chloride gas on a suspension of sodium acetoacetic ester at the same temperature. By measuring the refractive indexes of the enol and keto forms, and also that of the equilibrium mixture, Knorr was able to calculate that the latter contains only 7 percent of the enol form. K. H. Meyer, working independently, succeeded in obtaining pure keto and enol isomers at about the same time.

BIBLIOGRAPHY

I. ORIGINAL WORKS. A list of Knorr's scientific papers is in J. C. Poggendorff, *Biographisch-literarisches Handwörterbuch der exakten Naturwissenschaften*, IV, 768–769, V, 645–646, and VI, 1345.

II. SECONDARY LITERATURE. P. Duden and H. P. Kaufmann, "Ludwig Knorr zum Gedächtnis," in *Berichte der Deutschen Chemischen Gesellschaft*, **60A** (1927), 1–34; H. W. Flemming, ed., *Ludwig Knorr, Dokumente aus Höchster Archiven* no. 31 (Frankfurt, 1967); and R. Scholl, "Ludwig Knorr," in *Sitzungsberichte der Sächsischen Akademie der Wissenschaften zu Leipzig*, **75** (1923), 155–165.

NEIL MORGAN

KNUDSEN, VERN OLIVER (*b.* Provo, Utah, 27 December 1893; *d.* Los Angeles, California, 13 May 1974), *physics*.

Knudsen was the son of Andrew Knudsen and the former Chesty Sward, who had emigrated from Norway and Sweden, respectively, in the middle of the nineteenth century. They were farmers and members of the Church of Jesus Christ of Latter-Day Saints (Mormon). They raised seven children, of whom Vern was the youngest. As a boy he worked on the family farm until it was decided that because of his interest and success in schoolwork, he should be permitted to attend Brigham Young University in Provo, from which he graduated A.B. in 1915. His first plans were for a career in engineering, but in his later undergraduate years he came under the influence of Harvey Fletcher, who was teaching physics at Brigham Young University at the time. Fletcher persuaded Knudsen to go into physics. After graduation Knudsen served for three years as a Mormon missionary in the Chicago area. It was in connection with his missionary work that he met his wife, Florence Telford. Married in December 1919, they had two sons and two daughters.

Toward the end of World War I, Knudsen joined Fletcher in the research division of the American Telephone and Telegraph Company (which later became the Bell Telephone Laboratories) and pursued military research on electronics.

In 1919 Knudsen entered the graduate school of the University of Chicago, where he studied under Albert A. Michelson, Robert A. Millikan, and Henry G. Gale. For his doctoral research Millikan suggested that Knudsen work on the electron theory of the specific heats of metals, but Knudsen had already decided that acoustics was to be his field. His dissertation was based on the study of the sensitivity of the ear to small differences in intensity and frequency. In his dissertation research Knudsen profited greatly from his collaboration with the otologist George E. Shambaugh. The Ph.D. degree was awarded to Knudsen magna cum laude in 1922.

Though offered positions at both the University of Chicago and Bell Laboratories, Knudsen chose to accept an instructorship in physics at what was then called the University of California, Southern Branch (later renamed the University of California at Los Angeles). This institution had only recently been established and provided rather primitive conditions for both teaching and research. However, Knudsen showed great resourcefulness in making do with the available equipment and embarked at once on a research program in architectural acoustics. Here he was much aided by the presence in the Los Angeles area of many auditoriums and assembly rooms, church sanctuaries, and other spaces that badly needed acoustical treatment. He also continued his interest in otology through his association with a distinguished Los Angeles otologist, Dr. Isaac H. Jones. During the period 1922 to 1932 Knudsen published many papers on audition problems in collaboration with Jones. There was also a considerable output of papers on the history of special auditoriums. Much of this was summarized in lectures given in 1930 at the Twelfth International Congress of Architects in Budapest.

Around 1930 Knudsen began to devote attention to the purely physical problems associated with the absorption of sound in the air in rooms, particularly the effect of humidity on this absorption. In his first paper published in the *Journal of the Acoustical Society of America* (July 1931), Knudsen showed conclusively that the absorption of sound in a room does not take place solely in the materials in the walls, the floor, the ceiling, and the clothing of the audience but also in the air, and depends especially on the humidity of the latter. Through collaboration with the German physicist Hans O. Kneser in Stuttgart, Knudsen's experimental results were interpreted on the basis of relaxation processes of the molecules of oxygen and nitrogen among themselves as well as with the molecules of water vapor. A paper by Knudsen summarizing his results was presented at a meeting of the American Association for the Advancement of Science in the early 1930's and was awarded a prize of $1,000 as the outstanding paper at the meeting.

Knudsen's interest in this physical side of room acoustics continued for many years and led to the production of ten papers on the acoustical relaxation process in oxygen, nitrogen, and other gases during the period 1931 to 1955.

In the meantime Knudsen consulted on the acoustical renovation and design of auditoriums in the Los Angeles area. This led to the publication in 1932 of his fundamental text *Architectural Acoustics*, a volume that epitomized both the theoretical side of the subject and the contemporary practical applications, such as radio broadcasting and sound moving pictures. Altogether, some thirty-five professional articles on architectural acoustics were published by Knudsen from 1925 to 1970. He became one of the most sought-after consultants in the field.

Knudsen's second book, written in collaboration with Cyril Harris, was *Acoustical Designing in Architecture* (1950). This incorporated the newer developments in room acoustical theory, such as the concept of diffusivity, the role of normal modes in sound production, reverberation control, sound reinforcement, and newer and more accurate absorption coefficients for acoustical materials. The volume, which emphasizes the proper design of rooms and auditoriums, was reprinted in paperback (1973).

As he grew older, Knudsen became increasingly concerned with the problem of noise in human society and was active in movements to control it.

A prime mover in the organization of the Acoustical Society of America in 1929, Knudsen served as its third president (1933–1935). The society honored him with its honorary fellowship in 1954, with the first presentation of the Wallace Clement Sabine Medal in 1957, and with its gold medal in 1967.

Knudsen was responsible for much of the development of the University of California at Los Angeles, not only in the department of physics, of which he was chairman for several years, but also in its general academic program. He successfully promoted graduate studies, becoming the first dean of its graduate school (1954–1958). He was appointed vice-chancellor of UCLA in 1956, and from 1959 to 1960 he was chancellor.

During World War II, Knudsen was very active in war research. He helped to organize what became the Naval Undersea Research and Development Center in San Diego and served as its first research director. He also monitored NDRC contracts on research in artillery sound ranging in air.

In addition to the honors mentioned above, Knudsen received honorary degrees from Brigham Young University and UCLA. The new physics building erected at UCLA in 1964 was named in his honor.

BIBLIOGRAPHY

I. ORIGINAL WORKS. A bibliography of Knudsen's principal technical papers is included in *Vern O. Knudsen: Collected Papers from the Journal of the Acoustical Society of America*, I. Rudnick and T. Bomba, comps. and eds. (New York, 1975). This volume contains the complete texts of Knudsen's articles in *Journal of the Acoustical Society of America*.

II. SECONDARY LITERATURE. Leo P. Delsasso: "The Gold Medal 1967. Vern Oliver Knudsen," in *Journal of the Acoustical Society of America*, **42** (1967), 535–536; and Isadore Rudnick, "Vern Oliver Knudsen," *ibid.*, **56** (1974), 712–715.

ROBERT LINDSAY

KOEBE, PAUL (*b.* Luckenwalde, Germany, 15 February 1882; *d.* Leipzig, Germany, 6 August 1945), *mathematics*.

The son of Hermann Koebe, a factory owner, and Emma (née Kramer) Koebe, Paul Koebe attended a Realgymnasium in Berlin. He started university studies in 1900 in Kiel, which he continued at Berlin University (1900–1905) and Charlottenburg Technische Hochschule (1904–1905). He was a student of H. A. Schwarz; the other referee of his thesis was F. H. Schottky. His habilitation as a *Privatdozent* at Göttingen University took place in 1907. He was appointed a professor extraordinary at Leipzig University in 1910, a professor ordinary at Jena University in 1914, and again at Leipzig in 1926. His numerous papers are all concerned with one chapter of the theory of complex functions, which is best characterized by the headings "conformal mapping" and "uniformization." In fact, in 1907, simultaneously with and independently of H. Poincaré, he accomplished the long-desired uniformization of Riemann surfaces.

The strange story of uniformization has still to

be written. Its origin is the parametrization of the algebraic curves (z,w) with $w^2 = (z - a_1)(z - a_2)(z - a_3)(z - a_4)$ by means of elliptic functions, achieved by Abel and Jacobi. The attempt to use abelian integrals for general algebraic curves of genus p in the same way as elliptic ones had been used in the case of genus one led Jacobi to reformulate the problem: parametrization of the p-th power of the Riemann surface by means of a p-tuple of functions of p variables—the Jacobi inversion problem, which was solved by B. Riemann. By Poincaré's intervention in 1881 and 1882, history took a quite unexpected turn. When studying differential equations, Poincaré discovered automorphic functions, which F. Klein was investigating at the same time. If in $\mathbf{C}$ the group G is generated by rotations with centers $a_1, a_2, \ldots, a_{n+1}$ and corresponding rotation angles $2\pi/k_i$ (integral k_i, $i = 1, \ldots, n$), an automorphic function F of G is easily constructed by Poincaré series. Such an F maps its domain conformally on a Riemann surface branched at the $F(a_i)$ with degrees k_i. Its inverse achieves the "uniformization" of that Riemann surface. Counting parameters and applying "continuity" arguments, Klein (1882) and Poincaré (1884) stated that the scope of this method included Riemann surfaces of all algebraic functions, although their proofs were unsatisfactory. It was not until 1913 that the "continuity method" of proof was salvaged by L. E. J. Brouwer.

At present, if a Riemann surface is to be uniformized, it is wrapped up with, rather than cut up into, a simply connected surface, which is then conformally mapped upon a standard domain (circular disk, plane, or plane closed at infinity), and in this framework automorphic functions are an a posteriori bonus. The idea of uniformizing the universal wrapping rather than the Riemann surface itself goes back to Schwarz. As early as the 1880's various methods were available to solve the boundary value problem of potential theory for simply connected (even finitely branched) domains, or, equivalently, to map them conformally upon the standard domain, as long as the boundary was supposed smooth, say piecewise analytic; and as early as 1886 Harnack's theorem had made convergence proofs for sequences of harmonic functions easy. Using these tools, uniformization could have been achieved in the 1880's were it not for the blockage of this access by automorphic functions.

In 1907 the lock was opened. Koebe and Poincaré simultaneously noticed that if an arbitrary simply connected domain (the universal wrapping of the Riemann surface) has conformally to be mapped on the standard domain, it suffices to exhaust it by an increasing sequence of smoothly bounded ones. As a matter of fact, because of Harnack's theorem, it was preferable to use Green's functions of the approximating domains, with the $\log \frac{1}{r}$ – singularity at P, which form an increasing sequence u_n. If the $u_n - \log \frac{1}{r}$ are bounded at P, it converges toward Green's function u for the prescribed domain, which together with its conjugate v solves the mapping problem by $\exp\,[-(u + iv)]$. If not (the case of mapping upon the whole plane) the goal is attained by noticing that all *schlicht* images of the unit circle by f with $f(0) = 0$, $f'(0) = 1$ contain a circle with center $f(0)$ and a radius independent of f (this "Koebe constant" has been proved to be 1/4). The latter remark is a weak form of Koebe's distortion theorem, which for *schlicht* mappings f of the unit circle states for $|z| < r$ an inequality of the form

$$Q(r) \leqq |f^1(z_1)/f^1(z_2| \leqq Q(r)^{-1}$$

with $Q(r)$ independent of f.

Koebe's most influential contribution to conformal mapping on the unit circle was his 1912 proof by *Schmiegung*, which has become so common that textbooks are silent about its authorship. It rests on the remark that $z \rightarrow z^2$ for $|z| < 1$ increases the distances from the boundary, and consequently the square root reduces them. To use the square root univalently the given domain D is transformed by linear fractions such as to lie on one sheet of the square root surface, with the branching outside D. Then the square root operation brings the boundary of D nearer to that of the unit circle. This process is repeated with suitable branching points such as to deliver a sequence of mappings converging to that of the unit circle. This square root trick goes back to Koebe's 1907 paper; it seems that Carathéodory suggested it be applied that fundamentally.

Koebe's mathematical style is prolix, pompous, and chaotic. He tended to deal broadly with special cases of a general theory by a variety of methods so that it is difficult to give a representative selective bibliography. Koebe's life-style was the same; Koebe anecdotes were widespread in interbellum Germany. He never married.

BIBLIOGRAPHY

I. Original Works. See Poggendorff. The following papers by Koebe are particularly important: "Über die Uniformisierung beliebiger analytischer Kurven," in *Nachrichten von der Köninglichen Gesellschaft der Wis-*

senschaften zu Göttingen, Math.-Phys. Kl. (1907), 191–210, 633–669; "Über eine neue Methode der konformen Abbildung und Uniformisierung," *ibid.* (1912), 844–848; "Über die Uniformisierung der algebraischen Kurven, I," in *Mathematische Annalen*, **67** (1909), 145–224; "Allgemeine Theorie der Riemannschen Mannigfaltigkeiten," in *Acta Mathematica*, **50** (1927), 27–157.

II. SECONDARY LITERATURE. Ludwig Bieberbach, "Das Werk Paul Koebe," and H. Cremer, "Erinnerungen an Paul Koebe," in *Jahresbericht der Deutschen Mathematiker-Vereinigung*, **70** (1968), 148–158, 158–161; L. E. J. Brouwer, *Collected Works*, H. Freudenthal, ed., II (Amsterdam, 1976), 572–576, 583, 585–586; H. Freudenthal, "Poincaré et les fonctions automorphes," in *Le livre du centenaire de la naissance de Henri Poincaré* (Paris, 1955), 212–219, and "A Bit of Gossip: Koebe," in *Mathematical Intelligencer*, **6**, no. 2 (1984), 77; Reiner Kühnau, "Paul Koebe und die Funktionentheorie," in Herbert Beckert and Horst Schuman, eds., *100. Jahre Mathematisches Seminar der Karl-Marx-Universität* (Leipzig, 1981), 183–194; Otto Volk, "Paul Koebe," in *Neue Deutsche Biographie*, XII (1980).

HANS FREUDENTHAL

KÖHLER, WOLFGANG (*b.* Reval [now Tallinn], Estonia, 21 January 1887; *d.* Enfield, New Hampshire, 11 June 1967), *Gestalt psychology, animal behavior.*

Köhler was the son of Franz Eduard Köhler, director of the German-language gymnasium in Reval, and of Wilhelmine (Minni) Girgensohn Köhler, both of whom were offspring of ministers. The family moved to Wolfenbüttel, Germany, when Wolfgang was six years old. He attended the gymnasium there and then went on to the universities of Tübingen, Bonn, and Berlin. He received his Ph.D. from Berlin in 1909 after studying philosophy and psychology under Carl Stumpf, and physical chemistry and physics under Walther Nernst and Max Planck.

Köhler was appointed assistant at the Psychological Institute of Frankfurt am Main in 1909 and *Privatdozent* in 1911. In 1913 he was named director of the anthropoid research station established by the Prussian Academy of Sciences on Tenerife in the Canary Islands. Isolated on Tenerife by World War I, he did not return to Germany until May 1920. He became acting director of the Psychological Institute at the University of Berlin in August 1920. In 1921 he was named professor at Göttingen. He held this position for only one semester; in 1922 he was appointed professor of philosophy and director of the Psychological Institute at Berlin, the positions previously occupied by Stumpf.

When the Nazis came to power in 1933, Köhler was outspoken in his criticism of the new regime. In 1935 he resigned his positions in Berlin and emigrated to the United States, where he became professor of psychology (1935–1946) and then research professor of philosophy and psychology (1946–1958) at Swarthmore College. Earlier he had been a visiting professor at Clark University (1925–1926) and Chicago (1935), and William James lecturer at Harvard (1934–1935). Upon retiring from Swarthmore he settled in New Hampshire and held the title of research professor at Dartmouth. He became a U.S. citizen in 1946.

Köhler was married twice: in 1912 to Thekla Gelb, with whom he had two sons and two daughters; and in 1927 to Lili Harleman, with whom he had one daughter. Among his many honors were *Ehrenbürger* of the Free University of Berlin, honorary degrees from eight colleges and universities in the United States and Europe, presidency of the American Psychological Association (1959), and membership in the American Academy of Arts and Sciences, the American Philosophical Society, and the National Academy of Sciences.

Köhler's doctoral dissertation under Stumpf at Berlin and his *Habilitationsschrift* at Frankfurt were studies of acoustic perception. His identification of vowel character as an attribute of tone, with the principal vowels separated by octaves, stimulated research among experimental psychologists. In addition, his experimental results concerning the physical and psychological dimensions of the perception of tones led him to challenge the common assumption that perception was a straightforward function of peripheral stimulation. At Frankfurt from 1910 to 1913 Köhler was closely associated with Max Wertheimer and Kurt Koffka. With Wertheimer providing the initial lead, the three men proceeded to lay down the foundations of Gestalt psychology.

In 1913, in his first major theoretical contribution to Gestalt theory, Köhler attacked what he called the "constancy hypothesis," the idea of a one-to-one relation between peripheral stimulation and perceptual sensation. He maintained that modern investigators would have to give up the traditional distinction between (1) peripheral, physiologically produced sensations and (2) central, psychological operations upon these sensations, because organization was a basic, underived feature of perception.

The work that first brought Köhler international fame was that on chimpanzees on Tenerife (1913–1917). In experiments now regarded as classics, he showed that instead of learning simply by "trial and error," which was the most that E. L. Thorndike would attribute to infrahuman animals, chimpanzees were able to solve problems by grasping the relations between means and ends. They displayed what Köhler called "insight"—"the appearance of a complete solution with reference to the whole layout of the field" (*The Mentality of Apes* [1925], 198). Köhler also investigated what kinds of problems chimpanzees were unable to solve. His challenge to associationist psychology, his observation of primate social behavior, his attention to individual differences, and his use of films were all major contributions to the study of animal behavior in this period.

On Tenerife, Köhler also did important experiments demonstrating relational learning in chimps and chickens, and he composed his most fundamental philosophical work: *Die physischen Gestalten in Ruhe und im stationaren Zustand* (1920). In this book he sought to model psychological theory on the field theory of Faraday and Maxwell. Using examples from electrostatics, he argued that in the physical world there were Gestalt phenomena, the qualities of which could not be described as merely sums of parts. He then maintained that the same characteristics could be found in brain processes, and that the whole "somatic field" of the brain could be treated as a single physical system.

In the 1920's Gestalt theory became one of the leading schools of German academic psychology, with Köhler as one of its leading spokesmen. He helped found and was coeditor of *Psychologische Forschung* (1921–1938), he debated such prominent critics of Gestalt theory as Georg Elias Müller and Eugenio Rignano, and he wrote *Gestalt Psychology* (1929), a book designed to introduce Gestalt ideas to American psychologists, whose views were dominated by behaviorism, an approach Köhler found philosophically naive in its attempt to remove consciousness from science and atomistic in its reduction of behavior to reflexes. At the same time he sought to extend the principles derived from studying perception to other areas of psychology. He developed Gestalt explanations of association, memory, and the perception of other people. Köhler also pursued the problem of the relation between psychology and neurophysiology and laid the basis for a theory of organisms as "open systems." His work in the United States on figural aftereffects and cortical currents in the brain grew out of his belief in an isomorphism between brain processes and perceptual structures. His theoretical conclusions from research on cortical currents, though challenged by Karl Lashley and others, remain open to investigation.

Köhler, like his co-workers, did not see Gestalt psychology as simply a response to technical questions in the psychology of perception. They felt that in demonstrating the structured nature of experience they were providing an alternative to the "atomistic" and "mechanistic" world view that dominated philosophic accounts of psychological reality and that was, in their view, threatening modern culture. Köhler, who throughout his career was interested in relating psychological experience both to philosophy and to physics and physiology, was especially keen to urge that one could reject atomism and mechanism without sacrificing one's allegiance to natural science. His philosophical position, presented in the books of 1920 and 1929, was developed further in *The Place of Value in a World of Facts* (1938). He again promoted his field theory of perception in *Dynamics in Psychology* (1940). The posthumously published *The Task of Gestalt Psychology* (1969) reviewed the achievements of Gestalt psychology and identified problems for the future.

BIBLIOGRAPHY

I. ORIGINAL WORKS. The main collection of Köhler's personal papers and manuscript correspondence is in the American Philosophical Society Archives, Philadelphia. A complete listing of Köhler's published scientific writings and reviews, compiled by Edwin B. Newman, is in Mary Henle, ed., *The Selected Papers of Wolfgang Köhler* (New York, 1971), 437–449. Henle's volume provides an excellent selection of Köhler's papers.

II. SECONDARY LITERATURE. Köhler's work and its historical context through 1920 are analyzed in Mitchell G. Ash, "The Emergence of Gestalt Theory: Experimental Psychology in Germany, 1890–1920" (Ph.D. dissertation, Harvard University, 1982). Ash has also treated aspects of Köhler's work after 1920 in "Gestalt Psychology: Origins in Germany and Reception in the United States," in Claude E. Buxton, ed., *Points of View in the Modern History of Psychology* (Orlando, Fla., 1985), 295–344. On Köhler's behavior toward the Nazis see Ash, "Ein Institut und eine Zeitschrift. Zur Geschichte des Berliner psychologischen Instituts und der Zeitschrift 'Psychologische Forschung' vor und nach 1933," in Carl Friedrich Graumann, ed., *Psychologie im Nationalsozialismus* (Berlin, 1985), 113–137; and Mary Henle, "One Man Against the Nazis—Wolfgang Köhler," in *American Psychologist*, **33** (1978), 939–944. Other valuable treatments of Köhler's work and/or career include Solomon E. Asch, "Wolfgang Köhler, 1887–1967," in *American Journal of Psychology*, **81** (1968), 110–119; Rudolf Bergius, "Wolfgang Köhler zum Gedenken," in *Psychologische Forschung*, **31** (1967), i–v; Mary Henle, "Wolfgang Köhler (1887–1967), in *American Philosophical Society Year Book, 1968* (1969), 139–145—also, with minor changes, in Mary Henle, ed., *The Selected Papers of Wolfgang Köhler* (New York, 1971), 3–10; W. C. H. Prentiss, "The Systematic Psychology of Wolfgang Köhler," in Sigmund Koch, ed., *Psychology: A Study of a Science*, I (New York, 1959), 427–455; Hans Lukas Teuber, "Wolfgang Köhler zum Gedenken," in *Psychologische Forschung*, **31** (1967), vi–xiv; and Carl B. Zuckerman and Hans Wallach, "Köhler, Wolfgang," in David Sills, ed., *International Encyclopedia of the Social Sciences*, VIII (New York, 1968), 438–442.

RICHARD W. BURKHARDT, JR.

KOWARSKI, LEW (*b*. St. Petersburg, Russia, 10 February 1907; *d*. Geneva, Switzerland, 27 July 1979), *physics, administration*.

Kowarski was the younger son of a prosperous Jewish businessman, Nicholas Kowarski, and a Christian singer, Olga Vlassenko; he was born out of wedlock, as might be expected for the son of a couple so mismatched by Russian standards of the time. After a few years his parents drifted apart. He was raised as a Christian and a liberal, but retained fewer religious convictions than political ones. Kowarski recalled that all this "contributed to create a fundamental feeling of borderline life, not quite belonging to the core of things, which characterized my whole career later on."

In 1917 the father took his sons to Vilnius (Wilno), another borderline area. Subsequently, like many others, Kowarski left Poland to study in Belgium and France. Taking a degree in chemical engineering at Lyons in 1928, he moved to Paris and found a part-time job with a company that manufactured gas mains.

Determined to strive for greater things, he entered Jean Perrin's laboratory of physical chemistry and prepared a doctor's thesis. Kowarski's great opportunity came when Frédéric Joliot hired him as a part-time assistant at the Radium Institute and later at the Collège de France. This opened the way to a brief period of highly significant work on nuclear chain reactions, followed by a distinguished administrative career, marked less by running established organizations than by setting up new ones. Meanwhile, Kowarski married Dora Heller and had a daughter. The marriage subsequently dissolved, following the pattern of separations that would persist through his life, but he found stability in a second marriage, in 1948, to Kathe (Kate) A. Freundlich. Though from middle age Kowarski was burdened with severe circulatory and digestive problems, physically he was a big man. He impressed all who knew him with his restless, inexhaustible energy and his rare precision of thought and language.

Kowarski's early work on practical questions of gas diffusion and on crystal growth (his thesis was an important early look at the way atoms move across a crystal surface before binding to a permanent site) at first seemed poor preparation for the nuclear physics that became his specialty under Joliot's tutelage. But the physical chemist's approach to reactions en masse served Kowarski well after the discovery of uranium fission was announced. Joliot and Hans von Halban, Jr., immediately took up the subject, and in early 1939 they asked Kowarski to join them. The team quickly determined and published the fact that uranium fission releases neutrons. Now the crucial question was how many neutrons were released per fission; the answer would tell whether nuclear chain reactions were feasible. From measurements of the distribution of neutrons in a tank full of uranium oxide and water, the team concluded that roughly 3.5 neutrons are emitted per fission, more than enough to sustain a chain reaction. (In fact their theory for analyzing the experiments was too crude; the true value is about 2.5, just barely enough.)

In April 1939, after turning down a suggestion from Leo Szilard that fission work should be kept secret, the Paris team published their conclusion (*Nature*, **143** [1939], 680). This publication convinced many physicists that uranium chain reactions were a serious possibility. Joliot, Halban, and Kowarski themselves, joined by Francis Perrin, secured aid from the government and private industry and launched a concerted effort to attain a chain reaction. They worked out that the reaction could be furthered by using a moderator, especially heavy water, inhomogeneously intermixed with the uranium; they devised simple but powerful equations that analyzed the approach to a chain reaction through easily measured quantities; and they demonstrated a limited chain reaction in piles of uranium mixed with paraffin or water. The German invasion of France in May 1940 cut short a program that might otherwise have been the world's first to achieve a self-sustained chain reaction.

Kowarski and Halban fled to England, taking with them most of the world's supply of heavy water. At the Cavendish Laboratory in December they studied a sphere containing a mixture of uranium and heavy water and demonstrated that the mixture could reach the critical condition; the only thing they lacked for a working reactor was a few tons more of each substance. Their result was one of various influences that persuaded British leaders to launch an atomic bomb project.

Meanwhile, irreconcilable differences of programmatic choice and administrative approach developed between Halban and Kowarski. At the end of 1942 the British sent Halban to Montreal to establish an independent laboratory, while Kowarski remained in Cambridge with a few helpers. By 1944 Halban's effort had come to a standstill; he was retired from his post and Kowarski was brought to Canada. Aided by information and materials from the United States, he directed construction of the first reactor to operate outside that country, a zero-power, heavy-water pilot plant. Its descendants were a line of successful Canadian reactors.

In 1946 Kowarski returned to Paris to take a post under Joliot as a leader of the French Commissariat à l'Énergie Atomique. He ran a project to build a small heavy-water plant like his Canadian one, began work on larger reactors, and in other ways helped create and administer the programs that would make France inferior to none but the United States and the Soviet Union as a civilian and military nuclear power. But when Joliot was removed from his post in 1950, Kowarski felt increasingly out of place.

Meanwhile he had been an important early supporter of the plan to build a joint European high-energy physics facility, CERN, and beginning in 1952 his activity was gradually transferred to the new laboratory in Geneva. As director of scientific and technical services, Kowarski set up many elements of the laboratory's infrastructure, such as health physics and information services. With that accomplished his role might have ended, but beginning in 1956 he took an interest in the linked, embryonic technologies of liquid-hydrogen bubble chambers, bubble-chamber film processors, and especially computers; he began yet another career by pushing CERN to the forefront of work with these devices. He also advised the European Nuclear Energy Agency and other organizations that dealt with nuclear policy. He remained active into the 1970's, when he joined the nuclear power debate by arguing publicly for caution in the development of some reactor types, while remaining a champion of the heavy-water technology he had pioneered.

BIBLIOGRAPHY

Kowarski's papers and an extensive oral history interview by Charles Weiner are at the American Institute of Physics, New York City, with a bibliography of his writings. Other papers are in family hands and at the Radium Institute, Paris, and CERN, Geneva. His chain reaction work is described in Spencer R. Weart, *Scientists in Power* (Cambridge, Mass., 1979). Obituaries include Otto R. Frisch in *Nature*, **282** (1979), 541; Bertrand Goldschmidt in *Physics Today*, **32** (December 1979), 68–70; and a CERN publication, *Lew Kowarski, 1907–1979* (Geneva, 1979).

SPENCER R. WEART

KRAVETS, TORICHAN PAVLOVICH (*b*. Volkovo, Tula guberniia [now oblast], Russia, 22 March 1876; *d*. Leningrad, U.S.S.R., 21 May 1955), *physics*.

Kravets' parents, Pavel Naumovich Kravets and Felicitiana Karpovna Shagina, were physicians. In 1894 he graduated with a medal from the Tula gymnasium, where he received an excellent education that included both ancient and Western languages. (The knowledge of Latin was of great benefit for his future work in history of science.) In 1894 Kravets entered the mathematics section of the Faculty of Mathematics and Physics at Moscow University; he graduated in 1898 with a first-class diploma. While Kravets was at the university, the Moscow school of physics headed by P. N. Lebedev was at its height. In 1907 he married Ekaterina Mikhailovna Svechina; they had no children.

Kravets' scientific activity was conducted in Moscow (1898–1914), Kharkov (1914–1923), Irkutsk (1923–1926), and Leningrad (1926–1955). In Lebedev's Moscow laboratory, Kravets began work on the spectroscopy of colored solutions. The results were summed up in his M.Sc. thesis, "Absorbtsiia sveta v rastvorakh okrashennikh veshchestv" (Absorption of light in solutions of colored substances), defended at St. Petersburg in 1913. This work was both of purely scientific importance (the analysis of interaction of electromagnetic waves in a medium within a broad range of wavelengths) and of applied interest, for it made clear the possibility of chemical analysis of substances by means of absorption of electromagnetic waves. The "Kravets absorption integral," introduced in this work (and since then included in modern physics dictionaries), enables one to determine the charge and mass of an equivalent oscillator corresponding to the given line or band of absorption from the analysis of absorption spectra. This work has become especially important with the appearance of dye lasers.

During his three years in Irkutsk, Kravets worked on problems in geophysics, investigating seiches of Lake Baikal. At approximately the same time these problems interested I. V. Kurchatov, who studied seiches in the Azov and Black seas.) The seiches are free oscillations of the water level in closed (natural or artificial) basins that may influence seismographic readings. Kravets was commissioned to set the work of the Irkutsk seismic station in motion.

In the course of doing so, he developed a model of the free oscillations of the basin that agreed to within 1 percent with the actual process. These investigations were continued at Leningrad in the late 1920's, this time for the Baltic.

The main area of Kravets' physical research in Leningrad was the physics of photographic images. To this end Kravets established a special department of scientific photography at the Vavilov State Optical Institute. He was first of all interested in the problem of latent photographic images, which had attracted the attention of D. F. J. Arago. Kravets also studied absorption of light by haloid silver salts. His interest in these problems had arisen while he was working in Lebedev's laboratory. In the course of research on the physics of latent images, a group of Kravets' pupils and colleagues was formed; this group carried out investigations on applied photography and photographic sensitometry (photographic metrology). Based on results of this research, standards for photographic materials were established in the Soviet Union.

Kravets devoted much attention to the history of science. This interest was stimulated by his contacts with Lebedev. Beginning in 1912 and continuing almost until his death, Kravets studied Lebedev's school, which has produced many outstanding physicists (N. N. Andreev, V. K. Arkadiev, S. I. Vavilov, P. P. Lazarev, P. S. Epstein, and others). In 1949 he edited (and commented on) *Dokumenti po istorii izobretenia fotografii* (Documents in the history of the invention of photography). The classics of natural philosophy had always interested Kravets, and many volumes of scientific papers by Newton, Lomonosov, Lenz, H. A. Lorenz, Mendeleev, Monge, and Faraday were published in the Soviet Union with Kravets (and Vavilov) as editor and commentator. Kravets also wrote reminiscences of P. N. Lebedev, S. I. Vavilov, and D. S. Rozhdestvenskii, among others.

Kravets was a brilliant lecturer. His academic career had begun in 1898, in the department of physics of the Moscow Higher Technical School, which was headed by A. A. Eichenwald. Before the revolution he also taught at Moscow University and the Moscow Pedagogical Institute. In 1914 he was named professor at Kharkov University and worked there until 1923, when he was appointed professor at Irkutsk University. From 1930 Kravets taught in Leningrad higher schools, and from 1938 he was the head of the department of general physics at Leningrad University.

Kravets' hobbies were literature, poetry, history. Here he was able to use his talents as lecturer and as storyteller. His work in the history of science was related to this avocational writing.

Kravets was elected a corresponding member of the Academy of Sciences of the U.S.S.R. in 1943. He received the State Prize of the U.S.S.R. in 1946.

BIBLIOGRAPHY

I. Original Works. *Trudi po fizike* (Works on physics; Moscow, 1959) contains papers by Kravets, including the following: "Absorbtsiia sveta v rastvorakh okrashennikh veshchesty" (Absorption of light in solutions of colored substances; 1912), 33–145; "Predvaritelnaia zametka o prilivakh Baikala" (On tides of Lake Baikal; 1926), 267–270; "Prakticheskoe i teoreticheskoe znachenie sensitometrii" (The practical and theoretical sense of sensitometry; 1939), 245–252; and "Nekotoriie noviie danniie o pogloshchenii sveta v rastvorakh i adsorbirovannikh sloiakh" (Some new data on light adsorption in solutions and in adsorbed [surface] layers; 1950), 191–206. Kravets wrote the introduction to and edited *Dokumenti po istorii izobretenia fotografii* (Documents in the history of the invention of photography; Moscow and Leningrad, 1949). See also his *Ot N'iutona do Vavilova* (From Newton to Vavilov; Leningrad, 1967).

II. Secondary Literature. G. P. Fuerman, "Torichan Pavlovich Kravets" (obituary), in *Uspekhi fizicheskikh nauk*, **58**, no. 2 (1956), 183–192; Iu. N. Gorokhovskii *et al.*, "K 75-letiiu so dnia rozhdeniia T. P. Kravetsa" (On the seventy-fifth birthday of T. P. Kravets), *ibid.*, **44**, no. 2 (1951), 301–310; M. Rodovskii, "K semidesiati letiiu T. P. Kravetsa" (T. P. Kravets—on his seventieth birthday), *ibid.*, **29**, no. 1 (1946), 212–213; and "Torichan Pavlovich Kravets (k 75-letiiu so dnia rozhdeniia" (T. P. Kravets [on his seventy-fifth birthday]), in *Zhurnal tekhnicheskoi fiziki*, **21**, no. 4 (1951), 385–388. Papers on Kravets' work are included in his *Trudi*. See also M. V. Savostianova and V. Iu. Roginskii, *Torichan Pavlovich Kravets* (Moscow, 1979).

V. J. Frenkel

KREBS, HANS ADOLF (*b.* Hildesheim, Germany, 25 August 1900; *d.* Oxford, England, 22 November 1981), *biochemistry.*

Hans Adolf Krebs was the elder son, and second of three children, of Georg Krebs, an otolaryngologist with a flourishing private practice, and Alma Davidson Krebs, the daughter of a banker and member of a close-knit family that had been settled in the Hildesheim area for several centuries. As a boy he was deeply impressed by the intact late-medieval and Renaissance architecture of the old city of Hildesheim, at the edge of which his family lived in a comfortable house with a spacious garden.

Through the diverse interests of his widely cultivated father, Krebs came into contact with music, poetry, art, and clever conversation. On the customary family Sunday hikes into the nearby wooded hills, his father instilled in Hans a strong interest in nature in general and in wildflowers in particular. His father remained for him, however, a somewhat aloof figure who was skeptical of his son's intellectual abilities and made him feel that nothing he did was quite good enough. Shy and somewhat solitary, Hans made no close friends as a boy; but he was industrious and well organized, and he read widely, cycled the surrounding countryside, and pursued hobbies such as botanical collecting and bookbinding. For much of his boyhood, the activity that consumed most of his time was practicing the piano. Although his father and mother both came from Jewish families, Hans and his younger brother Wolf were raised outside the formal faith, because Georg Krebs believed that the best solution to the Jewish question in Germany was assimilation.

In *Mittelschule* Krebs ranked first in his class, but during his years in the classical Gymnasium Andreanum his record was not outstanding. He did well in all of his subjects without showing exceptional talent in any of them. His education was concentrated in Latin and Greek, modern languages, history, literature, and mathematics, with relatively little science. His favorite subject was history, in which he read, beyond his assigned textbooks, major works by eminent German historians such as Theodor Mommsen and Leopold von Ranke.

At the age of fifteen, Krebs decided that he wanted to follow his father into medicine. He envisioned that after completing his training he would probably join his father's practice until he could establish himself on his own.

Along with most other Germans, during the years after the outbreak of the world war, the Krebs family experienced increasing austerity owing to shortages of food, fuel, and other commodities. By the late summer of 1918, Hans was old enough to be drafted into the army. Assigned to a signal corps regiment in Hanover, he had completed only a few weeks of basic training when mutinies by sailors at Kiel and elsewhere precipitated the end of the war. Returning home, Krebs was able to obtain an immediate discharge on the grounds that he intended to enroll as a medical student at the nearby University of Göttingen. Allowed to enter as a veteran midway through the school term, he had to work very hard to learn what he had missed, but his disciplined habits and capacity to absorb large amounts of information enabled him quickly to catch up. After completing the summer term, he transferred to the University of Freiburg in order to listen to the lectures of its outstanding faculty, to broaden his cultural experience, and to enjoy hiking in the Black Forest.

Stimulated by accounts of their own scientific discoveries that some of his teachers gave in their lectures, Krebs became interested in trying his hand at research. In the anatomical institute at Freiburg, Wilhelm von Möllendorff, a leader in the field of vital staining, gave him a project to study comparatively the staining effects of different dyes on muscle tissue. Adopting Möllendorff's general "physicalist" approach to staining, Krebs concluded that the intensity and distribution of the staining in muscle tissue are governed not by the respective chemical properties of the dyes but by the degree of their dispersibility and the varying densities of the tissue structures. Krebs wrote up his results in a well-crafted research paper, and Möllendorff arranged for its publication with the young medical student as sole author.

After passing with high marks the examination that completed his preclinical training in March 1921, Krebs stayed on for one more semester in Freiburg in order to take in the lectures of the famed pathologist Ludwig Aschoff. Then he moved on to the University of Munich because of the general renown of its clinical faculty. After remaining for two semesters in Munich, Krebs spent the winter semester in Berlin in order to hear lectures by some of the leaders in that internationally famous center of medicine. He returned to Munich for the summer semester of 1923, and in the fall of that year passed, again with very good marks, the final medical examination known as the *Ärtzliche Prüfung*.

During the clinical years in medical school, Krebs was too busy studying and attending lectures and demonstrations to undertake any further research projects. The conviction nevertheless grew stronger that he wanted eventually to participate in scientific investigation. Warned by his father and others that it was not possible to make his living by science alone, he planned to enter internal medicine, in which he hoped to be able to combine clinical practice with experimental work.

Krebs went back to Berlin in January 1924 to fulfill his required year of hospital service at the Third Medical Clinic of the University of Berlin. There he carried out preliminary examinations in the outpatient clinic. Encouraged to use his spare time in the laboratory, he undertook on his own a study of the gold sol reaction that was currently in use in some clinical laboratories as a sensitive diagnostic test for syphilis. In collaboration with Annalise Wittgenstein, a member of the clinical staff, he began concurrently experiments on dogs on the passage of foreign substances from the blood into the cerebrospinal fluid, a problem also connected with syphilis, because of difficulties encountered in getting therapeutic drugs such as Salvarsan into the latter fluid to attack the spirochetes lodged in the central nervous system. Drawing on his experience in Möllendorff's laboratory, Krebs decided to employ dyes whose presence in the cerebrospinal fluid could easily be detected colorimetrically. Wittgenstein proved to be untrained as an investigator, so it fell entirely to Krebs to design the experimental attack and to write the papers reporting their results.

In these efforts, carried out with little guidance, Krebs proved himself to be a resourceful independent investigator. The experience reinforced, however, a belief that he had already acquired through the lectures in biochemistry and other subjects that he had heard in medical school: that chemistry was becoming ever more important to medical research, and that he did not have enough systematic training in the subject to conduct investigations more fundamental than those he had so far done. He decided, therefore, after completing his hospital year, to enroll in a special chemistry course offered at the nearby Charité Hospital for doctors like himself who needed to strengthen their chemical backgrounds. There he spent most of 1925 learning to carry out basic methods of qualitative and quantitative analysis.

Through a series of contacts made by Bruno Mendel, a close friend he had met at the Third Medical Clinic, Krebs had a rare opportunity to become, at the beginning of 1926, a paid research assistant to the great biochemist Otto Warburg at the Kaiser-Wilhelm Institut für Biologie in Berlin-Dalheim. There he learned the tissue slice and manometric methods that Warburg had devised in order to measure the rates of respiration and glycolysis of cancer cells and to compare these with the rates in normal tissues. In one of his first assigned projects, Krebs extended the measurements that Warburg had made on rat tumor tissue to tissues obtained from human patients. In 1927 Warburg discovered that carbon monoxide inhibits the respiration of yeast cells, but that illuminating the cells diminishes the extent of the inhibition. He asked Krebs to find an oxidative iron-heme catalyst that would have similar characteristics. By early 1928 Krebs had identified a heme-pyridine compound whose responses to carbon monoxide in darkness and light, as well as its responses to HCN, were so similar to the responses of yeast cells that Warburg could regard the compound as a "model" of the respiratory enzyme *(Atmungsferment)* that he had postulated several years earlier, and could utilize Krebs's results to strengthen his argument that the respiratory enzyme itself is an iron-heme compound.

Otto Warburg was a formidable chief: demanding and authoritarian, skeptical of much of what passed for scientific work in other laboratories, easily angered by opposition or criticism. He was also, however, a man of strong integrity and singleness of purpose, devoted wholly to his research and able to apply great experimental skill as well as theoretical acumen to the problems he pursued. In his small laboratory he required everyone present to work on the problems that interested him, and he required everyone to be present, without fail, from 8:00 A.M. to 6:00 P.M., six days per week. Krebs was both inspired and intimidated by Warburg. What he had previously done on his own now appeared to him narrow and dilettante. He was content to become a loyal apprentice, adopting with little question Warburg's opinions and his approach to scientific investigation. From Warburg, Krebs learned that he should seek out the central questions in his field, rely on precise methods, carry out many experiments without hesitating about whether they were worthwhile, and write up his results in clear, concise papers.

After he had become skilled at the manometric-tissue slice techniques that Warburg used, it occurred to Krebs that the method could be applied to great advantage to study intermediary metabolism. From some of his teachers in medical school, and particularly in the lectures of Franz Knoop in Freiburg, he had heard both that very little was known of the series of reactions between the foodstuffs that enter the body and their final decomposition products, and that it should be a central goal of biochemistry to establish the unbroken sequences of chemical equations that must connect them. When he was bold enough to suggest to Warburg that he might like to apply the manometric methods to these problems, Warburg told him that such experiments would be of no interest to him, and that there was not enough space in his laboratory for anyone to work on problems other than his own. Krebs had no choice but to conform; the idea nevertheless appeared so compelling to him that he kept it firmly in mind as a goal he would pursue when he eventually became free to define his own investigative pathway.

In 1929 Warburg told Krebs that he could not remain as a research assistant beyond 31 March 1930. Since Warburg did not help him to find another research position, Krebs thought that Warburg did not consider him capable of independent scientific work. Doubting his own investigative talents, but still keen to continue if he could, Krebs looked for a position in clinical medicine that would also offer some chance to do laboratory work. After a number of fruitless inquiries, he obtained such a post at the municipal hospital of Altona, near Hamburg, in the Department of Medicine, directed by Leo Lichtwitz, an outstanding physician with an interest in metabolic diseases.

At Altona, Krebs had heavy clinical responsibilities. The medical staff was outstanding, however, and he learned to be a caring physician, sensitive to the individual needs of the patients in the wards of which he had charge. In order to obtain Warburg's support for a grant to purchase manometers so that he could do research, he was obliged to undertake an investigation of proteolysis in tumor cells for Warburg. The project did not much interest him. While dutifully carrying it out, he searched for openings that would enable him to begin a research program of his own. His first successful idea came from the startling recent discovery by Einar Lundsgaard that muscles poisoned with iodoacetate could continue to contract for a short time even though they could not produce lactic acid. This finding upset the dominant current view, formulated by Otto Meyerhof, that the formation of lactic acid was directly connected with muscle contraction. Extending Lundsgaard's results to other animal tissues, Krebs was able to show that lactic acid added to slices poisoned with iodoacetate restored their respiration. In March 1931 he published his results in a short paper that, although minor in itself, marked his emergence as an independent investigator and his entry into the field of intermediary carbohydrate metabolism.

Not long after arriving at Altona, Krebs had already made arrangements to move to Freiburg in April 1931, to become an assistant to Siegfried Thannhauser, an expert on metabolic diseases in the Department of Medicine. There too he spent much of his time on clinical duties; but he had more extensive laboratory facilities, increased support from research grants, independence from Warburg's interests, and a university ethos that encouraged research on fundamental problems. During his first three months there, he continued the line of investigation of carbohydrate metabolism that he had begun in Altona, but in July he embarked on a major new problem: to determine how urea is synthesized in the animal organism.

During the previous sixty years, a succession of investigators had established, through feeding experiments on animals and the perfusion of isolated organs, that amino acids and ammonia give rise to urea in the liver. It was generally assumed that the amino acids were deaminated, yielding ammonia that served in turn as the source of the urea nitrogen. The methods employed in 1930, however, did not appear capable of answering such unresolved questions as whether ammonia was an obligatory intermediate or amino nitrogen could be incorporated directly into urea; whether all amino acids gave rise to urea through a common mechanism; and whether other proposed intermediates such as cyanate took part. Krebs began his investigation with similar questions in mind; more broadly, he hoped to test the suitability of the manometric-tissue slice method to study complex synthetic metabolic processes. To begin, he adapted the manometric urease methods of urea analysis of D. D. van Slyke to use in the Warburg manometer, providing him with an exceptionally accurate and rapid means to measure the very small quantities of urea that would be produced by tissue slices. Shortly after he had begun, a medical student, Kurt Henseleit, came to him for an M.D. thesis problem, and Krebs soon turned most of the daily experimental operations for the urea investigation over to Henseleit. Between August and October 1931, Krebs and Henseleit carried out numerous experiments that only confirmed that amino acids and ammonia can produce urea in isolated liver slices and in no other animal tissues.

During November, Krebs and Henseleit tested the effects of several amino acids and some intermediates of carbohydrate metabolism, seeking to identify substances that might influence the rate of formation of urea in liver slices. In one of these experiments they included, almost as an afterthought, the uncommon amino acid ornithine. Unlike anything else they had tried, ornithine, in combination with ammonia, dramatically increased the rate. During the following weeks they carried out experiments designed to test whether compounds chemically related in various ways to ornithine might exert an analogous effect, and whether ornithine acted as a nitrogen donor or only influenced the formation of urea from ammonia. By February 1932 Krebs had ascertained that the effect was specific to ornithine and that ornithine could act in catalytic quantities. He was also convinced that the effect must be related to the reaction arginine → ornithine + urea, well known to occur enzymatically in the liver; but for some time he could not visualize a specific connection between them.

In late March or early April, Krebs perceived that if ornithine gave rise to arginine by a route *different* from the reaction by which arginine yielded

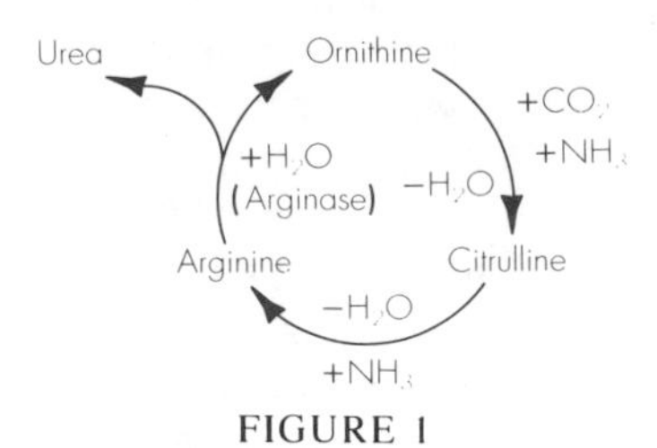

FIGURE 1

ornithine and urea, then the catalytic action of ornithine would be explained, for the ornithine consumed in forming the intermediate would subsequently be regenerated. At first he constructed on paper the simplest possible alternative pathway to give a balanced equation: ornithine + $2NH_3$ + CO_2 → arginine + H_2O. His solution made all of the elements of the problem fit together so coherently that he believed it could not be wrong, and he quickly submitted a preliminary paper to *Klinische Wochenschrift*. He recognized immediately, however, that there must be further intermediates, because the above equation postulated a chemically implausible simultaneous reaction of four molecules. Even before the first paper appeared, he had established that one of the required intermediates is citrulline, and he published a second paper reporting that a further step had been identified in what he called the "*Kreislauf des Ornithins*," a term shortly afterward translated into English as "ornithine cycle." A year later he began to depict these reactions in a form that highlighted their cyclic nature (Figure 1).

The ornithine cycle was rapidly recognized as a major discovery, earning Krebs the praise of leading biochemists in Germany and elsewhere. His achievement led to his appointment in December 1932 as a *Privatdozent* at the University of Freiburg. It caused Warburg to recognize the ability of his former assistant, and it gave Krebs a self-assurance that he never lost. Historically the discovery and the methods used to attain it now appear as the opening of a new era in metabolic biochemistry.

As his reputation spread, Krebs began to attract students to work with him in his well-equipped laboratory, and he appeared to be developing a small research school in intermediary metabolism. His students pursued questions related to his prior work on urea synthesis. Theodor Benzinger in particular applied the same strategies to study the corresponding process in birds, the synthesis of uric acid. Krebs himself attempted to demonstrate conclusively that amino acids are deaminated oxidatively, especially in the kidney. To his surprise, however, it turned out that the "unnatural" optical isomers of the amino acids were deaminated far more rapidly than their "natural" antipodes.

Early in 1933 Krebs took up a broad new research endeavor: to identify intermediates in the metabolic breakdown of foodstuffs, particularly of carbohydrates and fatty acids. He began in part empirically, intending to try out systematically a large number of possible intermediates to see whether they were oxidized in tissue slices and whether they increased the respiration of the tissues. He was, however, strongly influenced by current hypotheses concerning metabolic reactions, especially a widely discussed closed circuit of oxidative reactions known as the Thunberg-Wieland-Knoop scheme that appeared capable, if confirmed, of connecting the metabolism of amino acids, carbohydrates, and fatty acids into a common pathway: 2 acetic acid → succinic acid → fumaric acid → malic acid → oxaloacetic acid → pyruvic acid + CO_2 → acetaldehyde + CO_2 → acetic acid. He also sought to elucidate the final steps of Franz Knoop's long-known β-oxidation theory for fatty acids, and a recently proposed variation that postulated a double-ended oxidation. None of Krebs's early experiments in this area yielded definite conclusions.

Meanwhile, in January 1933 Adolf Hitler became chancellor of Germany, and during the following weeks the Nazis swiftly consolidated their power and subverted democratic protections. On April 12 Krebs was among the numerous Jews who were dismissed from their academic posts in accordance with the newly decreed law for the reform of the civil service. Having learned of the admiration in which the leader of the Cambridge school of biochemistry, Frederick Gowland Hopkins, held his work on urea synthesis, Krebs wrote to him inquiring if he might work in the Cambridge laboratory. Although eager to make a place for him, Hopkins had no financial means available for him. Fortunately the Rockefeller Foundation, which had already supported Krebs's work in Freiburg through a grant to Thannhauser, offered him a one-year fellowship. Krebs left for England in June, and within a month he had resumed his experimental research in Cambridge. Warmly received there, he quickly felt at home in England, even though he had no assurance that he would be permitted to remain there permanently.

In Cambridge, Krebs pursued for six more months the questions concerning the metabolism of carbohydrates, fatty acids, and related substances that he had begun in Freiburg. He included several efforts to connect the respiratory oxidation of citric acid with other metabolic pathways. None of these experiments led him beyond confirming what was already known about substances readily oxidized in animal tissues. Feeling that he was bogged down, he abruptly abandoned this line of investigation in December and turned to other metabolic questions.

In May 1934 Krebs's position in Cambridge was consolidated when he was appointed demonstrator in biochemistry. Of the several dozen German refugee scientists by then in England, he was the first to receive a regular academic post.

After several further starts on problems that did not lead him to significant new findings, Krebs returned in the fall of 1934 to his earlier study of the deamination of amino acids. Pursuing an anomaly that he had first noticed in Freiburg, that the dicarboxylic amino acids glutamic and aspartic acid released much less ammonia than did other amino acids, he found that they absorbed ammonia added to the tissue medium. By the spring of 1935, he had amassed compelling evidence that the ammonia combines with glutamic acid to form glutamine. He thus established a previously unknown metabolic synthesis, a reaction whose significance was at first unclear but that became the starting point for an extensive field of investigation. During the same period he extracted from kidney and liver tissue an enzyme able to catalyze the deamination of the "unnatural" optical isomers of amino acids. Neither he nor others were able to explain why organisms possessed a special enzyme to act upon a class of compounds that they do not normally encounter.

During the summer of 1935, Krebs was invited to apply for a position as lecturer in the Department of Pharmacology at the University of Sheffield, with the understanding that his research and teaching would be in biochemistry. He did so, and took up his post on 1 October. Krebs chose to move from one of the international centers of biochemistry and a milieu that he valued highly, where he benefited from the intellectual stimulation of colleagues working on problems closely related to his own, to a small provincial university in a sooty industrial town, where he would be isolated in his field. He did so mainly because he had the prospect at Sheffield of sufficient laboratory facilities to begin to build a research team. At Cambridge, where research space was crowded, he had already begun to draw more students than he could accommodate. He was attracted also by the beauty of the countryside surrounding Sheffield and by the warmth and generosity of his new chief, Edward Wayne.

In the months before leaving for Sheffield Krebs had undertaken, with two students, a further study of the synthesis of uric acid in birds. In Sheffield he continued to work on this problem, keeping in close touch with the students, who remained in Cambridge. By October they had identified hypoxanthine and xanthine as two intermediates that give rise, by successive oxidations, to uric acid. Krebs hoped that he would then be able to find the precursors that joined to form the molecular skeleton common to these three purines. By the end of the year, however, he had made little progress in this direction, and he turned again to other problems.

During his investigations of deamination, Krebs had several times turned briefly to the long-standing question of the converse process: how are amino acids formed in the organism? Early in 1936 he tested a hypothesis first proposed twenty-five years before by Knoop: that pyruvic acid reacts with ammonia and a second ketonic acid to form an acetylamino acid, which then decomposes to yield the amino acid corresponding to the second ketonic acid. In experiments carried out on rat liver tissue, Krebs initially obtained results that appeared to support this hypothesis; but he soon recognized that the data could equally well fit an oxidoreduction reaction between the two ketonic acids for which the ammonia was unnecessary. This outcome induced him to study the anaerobic reactions of pyruvic and other ketonic acids as potential connections between the recently established Embden-Meyerhof pathway of glycolysis and the further oxidative breakdown of carbohydrates. By the summer of 1936, he was persuaded that he had sufficiently established the existence of a family of such reactions to submit to *Nature* a preliminary paper titled "Intermediate Metabolism of Carbohydrates," in which he claimed to have "found some new chemical reactions in living cells which represent steps in the breakdown of carbohydrates." He presented a sequence of three dismutation reactions that he believed to be the "primary steps of the oxidation of pyruvic acid." Since the latter had long been viewed as one of the key substances linking the various paths of intermediary metabolism, Krebs was putting forth a scheme that could potentially join the fragmentary known reaction chains into a comprehensive network leading from the carbohydrate foodstuffs to their final oxidation products and connecting them as well with fatty acid metabolism.

Meanwhile, in Szeged, Hungary, Albert Szent-Györgyi was investigating respiration in isolated tissue with methods similar to those Krebs used. Instead of tissue slices, however, Szent-Györgyi employed suspensions of tissue coarsely minced so as to leave most of the cells intact. Between 1934 and 1936 Szent-Györgyi and his co-workers showed that fumaric acid added in catalytic quantities to pigeon breast muscle can sustain its respiration at a normal rate. From this and other supporting evidence, he developed a theory that the C_4 dicarboxylic acids long thought to be key intermediates in oxidative metabolism—succinic, fumaric, and oxaloacetic acids—formed instead a catalytic system that transports the hydrogen ions removed from food-stuffs to the respiratory chain of cytochromes and the *Atmungsferment* identified earlier by David Keilin and Otto Warburg, respectively. Szent-Györgyi proposed a cyclic scheme in which succinate was oxidized through fumarate to oxaloacetate, the latter being in turn converted directly to succinate by what he called "overreduction."

Krebs, who knew Szent-Györgyi personally, followed his publications on this subject closely. He was favorably impressed with the experimental work and adopted for some of his own experiments the minced pigeon breast muscle preparation. He was also influenced by Szent-Györgyi's general concept of the catalytic cycle but was dissatisfied with the idea of the overreduction of oxaloacetate. During the summer and fall of 1936, Krebs extended his investigations of coupled oxidation-reduction reactions to include dehydrogenation reactions of fumaric, oxaloacetic, and malic acids. In this work he sought to combine the conception of intermediate reactions of carbohydrate metabolism he had recently presented in his *Nature* article with modifications of the views of Szent-Györgyi.

Citric acid had long been known to be one of the relatively few substances able to accelerate strongly the respiration of isolated tissues, but it had been left out of the various hypotheses that had been proposed to link the steps of oxidative metabolism, probably because it was difficult to envision the chemical steps that might connect it with other intermediates such as the C_4 dicarboxylic acids. From time to time Krebs tested possible reaction schemes experimentally, but he attained no promising results until the fall of 1936. Then he and his first graduate student at Sheffield, William Arthur Johnson, obtained experimental evidence that pyruvic and oxaloacetic acid added together to minced pigeon breast muscle gave rise to significant quantities of citric acid. Believing that he had found the pathway by which citric acid is formed, Krebs was still unable during this time to ascertain how that substance might be further oxidized. In October and November 1936, he discussed in several public lectures the various dismutation reactions he had found that appeared to be involved in the oxidative breakdown of carbohydrates, but he qualified the scheme he presented as provisional and incomplete. He considered that he had shown citric acid to be "an intermediate in the breakdown of pyruvic and malic acid," but emphasized "that it appears to be a side reaction only."

Through the winter and spring of 1937, Krebs continued to gather data on the anaerobic dismutation reactions occurring in animal tissues, as well as in several types of bacteria. He now presented his results cautiously, without claiming that the particular reactions found in one organism or tissue necessarily occurred generally, and without attempting to organize them into sequential reaction schemes. In late April, by which time it probably appeared that this line of investigation was not leading toward broader conclusions, he saw in the latest issue of *Biochemische Zeitschrift* a preliminary paper by Carl Martius and Franz Knoop proposing a new pathway for the physiological decomposition of citric acid:

citric acid $\rightarrow$ cis-aconitic acid $\rightarrow$ isocitric acid $\rightarrow$ oxaloacetic acid $\rightarrow$ α-ketoglutaric acid.

Martius had devised the mechanism from theoretical chemical considerations and verified that liver extract can cause citric acid to give rise enzymatically to α-ketoglutaric acid. It was already well established, as Martius and Knoop pointed out, that α-ketoglutaric acid is decarboxylated to form succinic acid, thus connecting this sequence with the C_4 dicarboxylic acid series.

When Krebs read this paper, he probably realized at once that these reactions might provide the crucial link between the decomposition of citric acid and the main pathways of oxidative metabolism. With the assistance of Johnson, he quickly confirmed that in the presence of suitable blocking agents, citric acid added to pigeon breast muscle gives rise to substantial quantities of α-ketoglutaric and succinic acid. By then he had probably arrived at the general conception that there is a cyclic pathway through which the succinic acid formed from citric acid gives rise to malic and oxaloacetic acid, which regenerate citric acid by means of the dismutation reaction he had found in pigeon muscle tissue during the previous autumn. Soon afterward, however, Johnson's experiments showed that oxaloacetic acid alone produced citric acid anaerobically more rapidly than did oxaloacetic acid + pyruvic acid, so Krebs gave up the idea that a specifically identified dismutation was involved in the synthetic reaction that produced citric acid. He assumed then that oxaloacetic acid reacts with an unknown product of carbohydrate metabolism that he labeled provisionally as "triose."

On 7 June, Johnson performed an experiment

showing that citric acid added in very small quantity to pigeon breast muscle caused the tissue to absorb "extra" O_2 in quantities greater than those necessary to oxidize the citric acid completely. Krebs interpreted this result to mean that the citric acid was acting catalytically, being regenerated in a cyclic process whose net effect was to oxidize the "triose" that entered the cycle. By now—only about six weeks after encountering Martius and Knoop's article—Krebs had amassed what he considered convincing evidence for the existence of a "citric acid cycle" that constituted the "preferential" pathway "through which carbohydrate may be oxidized in animal tissues." By 10 June he had sent to *Nature* the short paper "The Role of Citric Acid in Intermediate Metabolism in Animal Tissues," while he continued to gather supporting data. When the editor of *Nature* informed him that, because of a backlog, the paper would not be published without a substantial delay, Krebs wrote a longer version of the paper and sent it to *Enzymologia*, a recently founded journal that was less prestigious than those in which he normally published, but in which he knew his paper would be accepted immediately and appear rapidly. He was impatient because he knew that he had arrived at a conclusion of fundamental importance in intermediary metabolism. The citric acid cycle that he presented, when coupled with the Embden-Meyerhof glycolytic pathway, enabled for the first time a coherent view of the principal steps in the process that was regarded as the chief source of energy for the vital activities of living organisms.

Late in 1936 Krebs met Margaret Fieldhouse, daughter of a Yorkshire family and a teacher of domestic science at a convent school in Sheffield. Although she was thirteen years younger than he, they quickly established an easy rapport, finding that they shared interests in hiking, botany, and other activities, and laughed at the same things. Shy but blunt and spirited, Margaret adapted to Hans's disciplined style and his relentless absorption in his work, while leavening his life with her spontaneous warmth. They were married in the spring of 1938 and spent their honeymoon in North America, where Margaret had to share his time with his professional colleagues as he attended meetings and engaged in an extensive speaking tour. They then settled in a small but comfortable house in Sheffield, in which they remained for nineteen years and raised three children, Paul, Helen, and John.

In 1938 Krebs was given the title lecturer in biochemistry, and made head of a newly established Department of Biochemistry at Sheffield. The Rockefeller Foundation, which had until then supported his research with annual research grants, began to fund the activity of his department in five-year grants, which were renewed regularly. The laboratory at Sheffield began to attract students and visiting investigators from both England and overseas. Krebs and his associates continued investigating the topics that he had previously pursued. By 1940 he had substantially strengthened the evidence for the citric acid cycle and refuted some criticisms that had been made of it. Early isotopic studies in other laboratories had by 1941 confirmed the main outlines of the theory, but seemed for a time to indicate that citric acid itself lay outside the main pathway. Consequently the cycle was renamed the "tricarboxylic acid cycle," although it was already referred to sometimes as the "Krebs cycle."

During the war Krebs was able to carry on his normal research program. In addition he participated in a wartime project to test the adequacy of special diets intended to stretch food supplies made scarce by the German submarine attacks on British shipping. The experimental subjects were conscientious objectors who volunteered for the project as an alternative to military service. Near the end of the war, the Medical Research Council, planning for a postwar expansion of scientific investigation, offered to establish a research unit under Krebs's leadership at Sheffield. With this support he was able to organize the Unit for Research in Cell Metabolism and assemble a team of workers, some of whom worked directly on projects of interest to him, others of whom developed more independent lines of investigation. In increasing numbers graduate students from Britain and abroad came to Sheffield to earn Ph.D. degrees within the unit.

Krebs had envisioned the citric acid cycle originally as the oxidative phase of carbohydrate metabolism, but it gradually became clear that it forms the final common pathway for the oxidative decomposition of all the major classes of foodstuffs, and that it is also involved in synthetic pathways. As the significance of this particular metabolic cycle expanded, Krebs began to ponder more deeply the general significance of metabolic cycles, of which the ornithine and citric acid cycles offered the paradigm examples. In 1947 he published a meditative paper titled "Cyclic Processes in Living Matter," in which he maintained that "metabolic cycles seem to be a feature peculiar to life, and the question why they have arisen therefore deserves a more detailed enquiry." Although acknowledging that "the specific meaning of cycles is still a puzzle," he suggested that while studying metabolic processes it would be worthwhile to "expect and look for further cycles."

Much of the research of Krebs's laboratory in the postwar period was directed toward further development of investigations based on his brilliant prewar discoveries, especially the question of how widely the citric acid cycle occurs among different tissues and organisms. Until 1953 Krebs and others believed that the cycle did not operate within microorganisms. Later it turned out that the apparent inability of intermediates of the cycle to activate respiration in these organisms resulted from the impermeability of their cell membranes to these substances, and that the cycle is nearly universal. While pursuing that problem, Krebs raised the question of how microorganisms that subsist on acetic acid as the sole organic nutrient can synthesize large molecules; acetic acid enters the citric acid cycle, but the cycle cannot provide a net synthesis of larger molecules from it. He suggested that problem as a research topic to a former student, Hans Kornberg, who discovered in 1957 the glyoxylate cycle, a modification of the citric acid cycle.

In 1953 the paramount importance that the citric acid cycle had attained in biochemistry was recognized in the award of the Nobel Prize jointly to Krebs and to Fritz Lipmann, whose discovery of coenzyme-A had specified the details of the crucial synthetic step in the cycle. Soon afterward Krebs was invited to become Rudolph Peters' successor as professor of biochemistry at Oxford, and to assume responsibility for the large Department of Biochemistry there. After receiving assurances that he could bring his metabolic research unit along with him, he accepted.

By the time Krebs came to Oxford, radioactive isotopes, spectrophotometry, chromatography, and other new methods had greatly facilitated the investigation of the intermediary steps in metabolic pathways. Many laboratories had made contributions, and most of the major pathways had been identified and linked up. Numerous details remained to be elucidated, and Krebs's team continued to play a significant role in that work. He increasingly turned his attention, however, toward the question of how the rates and direction of the reactions are controlled to meet the varying requirements of the organism. In 1957 he sketched out his general approach to the problem. Of the many intermediary steps contained in such processes as anaerobic glycolysis or respiration, only a few serve as "pacemakers," upon which the controlling factors operate. Control points that select between alternative metabolic pathways include the initial reaction at which a particular substrate enters a pathway, and "branching points." Other control points determine the direction in which a particular pathway proceeds. For the regulation of the latter type, Krebs drew attention to the importance of the ratios of reduced and oxidized forms of catalysts, such as the pyridine nucleotides, which are linked to the metabolic pathways as hydrogen acceptors or donors, and the ratio of ATP to ADP and inorganic phosphate. These control systems operated on the principles of feedback mechanisms.

For the next two decades, Krebs directed the work of his group toward more detailed specification of these general ideas. In 1963 he explained how the formation of carbohydrates from noncarbohydrate sources, a process known as gluconeogenesis, is regulated. Two critical steps in the pathway of gluconeogenesis—the conversion of fructose diphosphate to fructose-6-phosphate, and the conversion of pyruvate to phosphopyruvate—were identified as the pacemaker reactions. In order to attain experimental conditions more closely resembling the normal physiological circumstances that affect metabolic rates, Krebs switched from the minced tissue and tissue slice methods upon which he had relied for over twenty years to perfusing organs of small animals, especially the rat liver. Scientific progress itself sometimes displays cyclic patterns. When Krebs began his investigations in 1930, perfused whole organs had been the customary means to study metabolic processes, but they were tedious and did not permit close control of experimental conditions. His introduction of tissue slices quickly replaced whole organs in such research. Thirty years later, more advanced analytical technique made the reintroduction of perfusion methods a progressive move.

During the late 1960's Krebs developed methods to measure the "redox state" of the pyridine nucleotides within cells. He regarded the ratio of the oxidized to the reduced form of these coenzymes as a central factor in controlling the direction of metabolic reactions. The ratio itself was not directly measurable, but he and his colleagues were able to determine it by measuring the concentrations of substrates involved in particular dehydrogenation reactions catalyzed by the pyridine nucleotides, and in this way they could distinguish the redox state in the cytoplasm from that in the mitochondria. During the 1970's he returned to the subject of his earliest major discovery, the synthesis of urea, in order to examine the regulatory mechanisms controlling that process.

In 1967 Krebs reached the mandatory retirement

age at Oxford, and it appeared that he would be forced to end his research career despite the active leadership he still provided for his department. Arrangements were made, however, largely through the support of George Pickering and Paul Beeson, to provide a small laboratory for Krebs and his own unit at the Radcliffe Hospital. Several of the members of his team who had already served with him for a long time—in particular Leonard Eggleston, Reginald Hems, Patricia Lund, and Derek Williamson—came with him. There he continued his work with undiminished energy.

Ever since he formulated the citric acid cycle, Krebs had wondered what was the advantage to organisms of oxidizing their foodstuffs through such a complex, circuitous pathway instead of a simpler, more direct one. Forty-five years later he had still not found a satisfying explanation. Then, in April 1980, Jack Baldwin, an organic chemist at Oxford, suggested to him that acetic acid, the immediate substrate oxidized by the cycle, cannot, on chemical grounds, be directly dehydrogenated. It therefore must become attached to another molecule in order to complete its oxidative decomposition. Stimulated by this idea, Krebs quickly worked out an explanation for the necessity of a cyclic process as the most efficient mechanism for the utilization of the energy available in the foodstuffs. Obviously excited by finding an answer to a question that had for so long puzzled him, he embarked on a general inquiry into the broader question of whether all metabolic pathways can be shown to be the most efficient possible for achieving their particular functions, and whether they can be explained in this way as the expected outcome of evolution. During the following months Krebs participated in celebrations of his eightieth birthday at symposia in Dallas and in Sheffield. At each of these occasions, he invoked his new ideas concerning the evolution of metabolic pathways to illustrate the creative satisfaction he felt in continuing an active scientific life as long as he felt scientifically competent. The suggestion that he could retire and enjoy the things for which he had never had time he repeatedly rejected as "irritating."

Krebs carried on his scientific activity vigorously, until September 1981. Then, after returning from a trip to Germany, he experienced a loss of appetite that lasted for several weeks. Admitted to the hospital for observation, he rapidly weakened, and died on 22 November. His sudden passing was a severe and unexpected loss to his family, his many friends, and his colleagues. There was some consolation in the fact that he had maintained a rich scientific and personal life for over eighty years without finally having to endure that enforced inactivity of retirement that held so little appeal to him.

Hans Krebs combined in an extraordinarily fruitful manner the discipline to carry on a steady investigative pace day after day, year after year, with the imagination to connect the results of his experiments and other work into broadly unifying insights. He was not given to planning very far ahead, and he often shifted from one problem to another when he thought he was getting bogged down in a given investigation. Nevertheless, looking back, one can see that he pursued persistently a coherent set of related problems in intermediary metabolism throughout his long research pathway. His early work was most innovative, but his later investigations and publications added significantly to the breadth and depth of the insights he had acquired. He was more gifted in the design than in the practical execution of experiments, and from the late 1930's onward he relied almost entirely on his collaborators, students, and technicians to carry out the experiments he planned. Krebs was not, for everyone, easy to work for, because he demanded high standards of performance and was sparing in his praise of work well done—he expected good work as a matter of course. But he inspired in many a lifelong loyalty. His integrity was obvious to all who knew him. He did not allow his time to be wasted in idle talk, but he was generous with his help and his concern when it mattered. Accustomed in his later years to the adulation accorded a Nobel laureate, and to the deeper admiration of the many younger biochemists who looked upon him as one of the founders of their field, he was not falsely humble but he retained a genuine modesty. He could mention his "fame" without embarrassment, but he did not dwell on himself or his past achievements. Rather, until the very end he fixed his attention on his science and on what he conceived as his responsibilities toward the world at large.

BIBLIOGRAPHY

I. ORIGINAL WORKS. *Reminiscences and Reflections* (Oxford, 1981), an autobiography written in collaboration with Anne Martin, includes a complete bibliography of Krebs's scientific publications and those of his collaborators. An extensive collection of Krebs's correspondence and papers, cataloged by the Contemporary Scientific Archives Centre, is deposited in the Sheffield University Library. Inquiries concerning this collection can be addressed to The Librarian, Sheffield University Library, Sheffield 510 2TN U.K.

II. SECONDARY LITERATURE. The most comprehensive of the numerous biographical memoirs that appeared after Krebs's death is Sir Hans Kornberg and D. H. Williamson, "Hans Adolf Krebs," in *Biographical Memoirs of Fellows of the Royal Society*, **30** (1984), 351–385, which contains an extensive list of Krebs's publications. Two articles that treat particular aspects of Krebs's scientific work are Steven Benner, "The Tricarboxylic Acid Cycle," in *Yale Scientific Magazine*, **50**, no. 7 (1976), 4–10, 34–36; and Frederic L. Holmes, "Hans Krebs and the Discovery of the Ornithine Cycle," in *Proceedings of the Federation of Biological Sciences*, **39** (1980), 216–225.

FREDERIC L. HOLMES

KRULL, WOLFGANG (*b.* Baden-Baden, Germany, 26 August 1899; *d.* Bonn, Federal Republic of Germany, 12 April 1971), *mathematics*.

Krull was the son of Helmuth Krull, a dentist in Baden-Baden, and Adele Siefert Krull. After graduating from high school in 1919, he studied at the University of Freiburg and the University of Rostock. In 1920 and 1921 he studied at Göttingen, where he became acquainted with Felix Klein, whom he greatly admired. It was Emmy Noether, however, who awakened in Krull an enthusiasm for modern algebra, which at that time was making rapid advances. On his return to Freiburg in 1921, Krull earned his doctor's degree with a dissertation on the theory of elementary divisors.

On 1 October 1922 Krull became an instructor at Freiburg, and in 1926 he was appointed unsalaried associate professor. In 1928 he went to Erlangen as full professor. His early publications were about the theory of rings and the theory of algebraic extensions of fields. In 1925 he had proved a theorem concerning the decomposition of an abelian group of operators as the direct sum of indecomposable groups, the Krull-Schmidt theorem. In a paper published in 1928 Krull applied the fundamental ideas of the Galois theory, at first valid only for finite extensions, to infinite normal separable extensions. In this way the Galois group becomes a topological group with Krull topology (a linear topology), and the classical theorems of Galois theory carry over verbatim to the general case provided we replace the term "subgroup" by "closed subgroup." Later Krull again examined the topological groups he had introduced, especially the compact abelian groups with linear topology and countable bases. The same fundamental thought (introduction of a "natural" topological structure in algebraic systems) is to be found again in a paper published in 1955 with applications to the arithmetic of infinite algebraic number fields and the construction of a generalized multiplicative ideal theory. The theory of groups with subgroup topology, published in 1965, belongs to the same sphere of work.

The years Krull spent as full professor in Erlangen were the high point of his creative life. About thirty-five publications of fundamental importance for the development of commutative algebra and algebraic geometry date from this period. In 1921 Emmy Noether had recognized the importance of rings in which the maximal condition for ideals is satisfied, rings that are now termed Noetherian.

In 1928 Krull introduced the important concept of the Krull dimension of a commutative Noetherian ring. Krull's basic results on dimension (*e.g.*, Krull's principal ideal theorem) mark a turning point in the development of the general theory of Noetherian rings. Previously a Noetherian ring had been a kind of pale shadow of a polynomial ring, but after the publication of Krull's results (1928–1929) the way was open for the introduction of a surprising amount of interesting detail, as shown in the work of D. G. Northcott.

In 1932 Krull introduced the theory of additive valuations. The ideas of this theory are used in the theory of integrally closed rings and in algebraic geometry. Krull also defined rings that today are called Krull rings. The importance of Krull rings lies in the fact that the integral closure of a Noetherian integral domain is not necessarily a Noetherian ring, but it is always a Krull ring. In 1937 Krull proved the main part of the Krull-Akizuki theorem, and in 1938 the Krull-Amazuya lemma, a classical line of reasoning in algebra.

Let p be a proper prime ideal of a Noetherian ring R and put $S = R - p$; then we can form the ring $R_p \subset R$ of quotients of R with respect to S, and R_p has precisely one maximal prime ideal: R_p is a local ring. The name "local ring" has been given to these rings because they are used to study the local properties of algebraic varieties. The German name is *Stellenring*, and the algebraic study of local rings began with Krull's famous investigation of *Stellenringe* (1938). Let R be a Noetherian local ring with maximal ideal m; then $\cap_{n=1}^{\infty}(m)^n = (0)$. This is Krull's intersection theorem (1938), the basis of the Krull topology of the ring R in which R is a metric space. The Cauchy completion $\overline{R}$ of R is a Noetherian local ring and dim $\overline{R}$ = dim R (1938). Krull posed the problem of determining the structure of all complete local rings, which I. S. Cohen solved in 1946.

While at Erlangen, Krull became chairman of the

Faculty of Science. In 1929 he married Gret Meyer; they had two daughters. In 1939 he accepted an appointment to the University of Bonn. During the war he was called up into the naval meteorological service. In 1946 he resumed his work in Bonn. From this time until his death over fifty further publications appeared. These were in part a continuation of his earlier studies, but they also dealt with other fields of mathematics: group theory, calculus of variations, differential equations, Hilbert spaces.

As a mathematician of high international standing he received many invitations and honors. In 1962 the University of Erlangen conferred on him the degree of honorary doctor; he was the only mathematician given this honor. Krull had close professional and human contact with his many students. He directed thirty-five doctoral theses.

Krull described his attitude toward mathematics as that of an aesthete: "For the mathematician it is not merely a matter of finding theorems and proving them. He wants to arrange and group these theorems together in such a way that they appear not only as correct but also as imperative and self-evident. To my mind such an aspiration is an aesthetic one and not one based on theoretical cognition." Krull ascribed great importance to the mathematical imagination of the mathematician and said that it is the possession of this imagination that distinguishes the great scientist from gifted average people.

BIBLIOGRAPHY

I. ORIGINAL WORKS. Important works include "Über verallgemeinerte endliche abelsche Gruppen," in *Mathematische Zeitschrift*, **23** (1925), 161–196; "Galois'sche Theorie der unendlichen algebraischen Erweiterungen," in *Mathematische Annalen*, **100** (1928), 687–698; "Primidealketten in allgemeinen Ringbereichen," in *Sitzungsberichte der Heidelberger Akademie der Wissenschaften, Mathematische-Naturwissenschaftliche Klasse* (1928); "Über einen Hauptsatz der allgemeinen Idealtheorie," *ibid.*, (1929); "Über die ästhetische Betrachtungsweise in der Mathematik," in *Semesterberichte Erlangen*, **61** (1930), 207–220; "Allgemeine Bewertungstheorie," in *Journal für die reine und angewandte Mathematik*, **167** (1932), 160–196; "Galoissche Theorie der ganz abgeschlossenen Stellenringe," in *Semesterberichte Erlangen*, **67/68** (1937), 324–328; "Beiträge zur Arithmetik kommutativer Integritätsbereiche, III, Zum Dimensionsbegriff der Idealtheorie," in *Mathematische Zeitschrift*, **42** (1937), 745–766.

"Dimensionstheorie in Stellenringen," in *Journal für die reine und angewandte Mathematik*, **179** (1938), 204–226; "Allgemeine Modul-, Ring- und Idealtheorie," in *Enzyklopädie der Mathematischen Wissenschaften*, 2nd ed., I (Leipzig, 1939); "Beiträge zur Arithmetik kommutativer Integritätsbereiche, VI, Der allgemeine Diskriminantensatz: Unverzweigte Ringerweiterungen," in *Mathematische Zeitschrift*, **45** (1939), 1–19; "Über separable, insbesondere kompakte separable Gruppen," in *Journal für die reine und angewandte Mathematik*, **184** (1942), 19–48; "Jacobsonsche Ringe, Hilbertscher Nullstellensatz, Dimensionstheorie," in *Mathematische Zeitschrift*, **54** (1951), 354–387; "Jacobsonsche Radikal und Hilbertscher Nullstellensatz," in *Proceedings of the International Congress of Mathematicians, 1950*, II (Cambridge, Mass., 1952); "Charakterentopologie, Isomorphismentopologie, Bewertungstopologie," *Memorias de matematica del Instituto Jorge Juan* (Spain), no. 16 (1955); "Zur Theorie der Gruppen mit Untergruppentopologie," in *Abhandlungen aus dem Mathematischen Seminar Universität Hamburg*, **28** (1965), 50–97; *Idealtheorie* (New York, 1968).

II. SECONDARY LITERATURE. H. J. Nastold, "Wolfgang Krull's Arbeiten zur kommutativen Algebra und ihre Bedeutung für die algebraische Geometrie," in *Jahresberichte der Deutschen Mathematiker-Vereinigung*, **82** (1980), 63–76; H. Schoeneborn, "In Memoriam Wolfgang Krull," *ibid.*, 51–62, and complete bibliography of Krull's works, *ibid.*, 77–80; D. G. Northcott, *Ideal Theory* (Cambridge, England, 1953).

HEINZ SCHOENEBORN

KRUTKOV, IURII ALEKSANDROVICH (*b.* St. Petersburg [now Leningrad], Russia, 29 May 1890; *d.* Leningrad, U.S.S.R., 12 September 1952), *physics*.

Krutkov's father, Aleksandr Fedorovich Krutkov, was a teacher of Russian and ancient languages. Soon after Krutkov's birth the family moved to the Ukraine; there, in the small town of Lubny, his father was the master of the classical gymnasium. In 1906 the family returned to St. Petersburg, where in 1908 Krutkov graduated from the gymnasium with a gold medal and then entered the mathematics section of the Faculty of Mathematics and Physics of the university. The decade 1906–1916 was a remarkable period in the history of this faculty; at that time many brilliant mathematicians (A. S. Besikovich, R. O. Kuz'min, N. I. Muskhelishvili, V. I. Smirnov, Ia. D. Tamarkin, A. A. Friedmann, I. M. Vinogradov) and physicists (L. V. Mysovskii, P. I. Lukirskii, I. V. Obreimov, N. N. Semenov, D. V. Skobel'tsyn, Ia. I. Frenkel, and others) studied there. Krutkov, who belonged to this group, was the first "pure theorist" in Russia—he was never involved in experimental work. He was a pupil of Paul Ehrenfest and an active member of his seminar (1908–1912); in 1913 he continued his studies with Ehrenfest at Leiden.

The first series of Krutkov's scientific publications continued the work of Ehrenfest on adiabatic invariants. Before Krutkov's papers, the adiabatic invariants of a given physical quantity could be found only by inspired guessing; he was the first to work out a general tool. His papers on this subject played a part in preparation of the conceptual construction of quantum mechanics. Also along Ehrenfest's lines, Krutkov confirmed that the hypothesis of the independent quanta $h\nu$ of energy of an oscillator leads to Wien's formula and that only the idea of the association of quanta results in Planck's formula.

Krutkov was the first Soviet physicist to receive a stipend from the Rockefeller Foundation; he spent two years (1922–1923) in Germany and Holland. This time was an important period not only in his life but also in the history of relativistic cosmology: in a number of talks with Einstein in the spring of 1923 Krutkov showed that Einstein's criticism of Friedmann's paper on a nonstationary (expanding) universe was wrong.

The theory of oscillations and statistical mechanics occupied an important place in Krutkov's work. In a series of his papers (1933–1936) related to the theory of Brownian motion, he considered the nonlinear case of rotational Brownian motion of a particle. Using his results on the Brownian motion of an oscillator, Krutkov built up the theory of the rolling motion of a ship on a "random" (nonperiodic) sea. In this way he established the connection between the rolling angle and the sea characteristics, obtaining simple and important expressions for the function of distribution of the rolling angle together with average values of various characteristics of the rolling motion.

Krutkov turned to problems of classical and statistical mechanics, under the influence of A. N. Krylov, a distinguished expert in the theory of ships. This collaboration produced a classical treatise on the theory of gyroscopes presented in both a vector (Krutkov) and a usual (Krylov) form. Some of Krutkov's papers were devoted to the theory of rotation of a solid body. The last (postwar) series of papers concerns the theory of elasticity.

Though Krutkov taught for some time at the Faculty of Physics and Mechanics of the Leningrad Polytechnical Institute, his closest ties were with Leningrad University, where he was a professor (with a short break) from 1921 to 1952. During World War II, though imprisoned, he worked on aviation engineering problems under A. N. Tupolev. In his last years he was the head of the department of mechanics of the Faculty of Mechanics and Mathematics. Among Krutkov's pupils at the university was V. A. Fock, with whom he published a paper on the theory of the Rayleigh pendulum (here one can easily see a connection with Krutkov's interests in the theory of adiabatic invariants).

In 1933 the U.S.S.R. Academy of Sciences elected Krutkov a corresponding member. In 1952 he was awarded the State Prize for a series of papers on mechanics.

Krutkov had married Lidiia Dmitrievna Khudiakova in 1949. They had no children.

BIBLIOGRAPHY

I. Original Works. "Bemerkungen zu Herrn Wolfkes Note: 'Welche Strahlungsformel folgt aus der Annahme der Lichtatome?' " in *Physikalische Zeitschrift*, **15** (1914), 363–364; "O kvantovanii uslovnoperiodicheskikh sistem" (On the quantization of conditionally periodic systems), in *Zhurnal Russkogo fiziko-khimicheskogo obshchestva*, chast' fizicheskaia, **50**, no. 4–6 (1918), 134–142; "Contribution to the Theory of Adiabatic Invariants," in *Proceedings of the Royal Academy of Amsterdam*, **21** (1919), 1112ff.; "On the Determination of Quanta Conditions by Means of Adiabatic Invariants," *ibid.*, **23** (1921), 826ff.; "Über das Rayleigsche Pendel," in *Zeitschrift für Physik*, **13**, no. 3 (1923), with V. A. Fock; "Zur Schwankungstheorie," *ibid.*, 203–205; "Notiz über die mechanischen Grundgleichungen der statistischen Mechanik," *ibid.*, **36** (1926), 623–627.

Obshchaia teoria giroskopov i nekotorikh tekhnicheskikh ikh primenenii (The general theory of gyroscopes and some of its technical applications; Leningrad, 1932), with A. N. Krylov; "Zur Theorie der Brownischen Bewegung," in *Zeitschrift für Physik der Sowjetunion*, **5**, no. **2** (1934), 287–300; "O lineinikh zadachakh teorii brounovskogo dvizheniia" (On linear problems of the theory of Brownian motion), in *Doklady Akademii nauk SSSR*, n.s. **1** (1934), 479–482 (German trans., 482–485), pt. 2, n.s. **3** (1934), 215–217 (German trans., 218–220), pt. 3, n.s. **4** (1934), 120–122 (German trans., 122–124); "Zamechanie o bokovoi kachke korablia na volnenii" (Note on the rolling motion of a ship in heavy seas), *ibid.*, n.s. **2** (1934), 158–159 (German trans., 160–161); "Issledovaniia po teorii brounovskogo dvizheniia" (Studies on the theory of Brownian motion), in B. I. Davydov, ed., *Brounovskoie dvizhenie, A. Einstein, M. Smoluchovski* (Moscow, 1936); and *Tenzor funktsii napriazheniia i obshchii resheniia v statike teorii uprugosti* (The stress function tensor and the general solutions in the static of the elastic theory; Moscow and Leningrad, 1949).

II. Secondary Literature. V. Ia. Frenkel, "Iurii Aleksandrovich Krutkov," in *Uspekhi fizicheskikh nauk*, **102**, no. 4 (1970), 639–654.

V. J. Frenkel

KUENEN, PHILIP HENRY (*b.* Dundee, Scotland, 22 July 1902; *d.* Leiden, Netherlands, 17 December 1976), *sedimentology, marine geology, experimental geology.*

Kuenen grew up in a bilingual household and came from a strong intellectual and scientific background. He was the third of the five children of Johannes Petrus Kuenen, of Dutch heritage, and Dora Wicksteed, who was English. Kuenen's father was professor of physics at Dundee (1896–1907) and then at Leiden until his death (1922); Kuenen's brother Donald Johan became an eminent zoologist at Leiden University.

As a child, Kuenen was a collector: his first acquisition was a human appendix in a jar, and he began collecting rocks at the age of eight. He received virtually all of his schooling in Leiden and was awarded an undergraduate degree in geology in 1922 and a Ph.D. in 1925. As an undergraduate Kuenen had begun research on Tertiary fossils from Timor under the direction of the paleontologist Karl Martin, whom he greatly admired. Taxonomic work did not suit him, however, and he was relieved to turn it over to another student when Martin retired in 1922, and to study under B. G. Escher, a structural geologist. Escher set up an experimental laboratory upon his arrival in 1922, and his approach to solving geological problems had a lasting impact on Kuenen.

Kuenen's graduate research was on the porphyry district of Lugano, in the Alps, but he soon diverged from this area of study. He remained at Leiden from 1926 to 1934 as a geological assistant, and during this time began conducting diverse geological experiments. Escher recommended Kuenen as geologist for the *Snellius* expedition (1929–1930) to the Netherlands Indies (specifically, to the Moluccas) and its associated deep basins and troughs. The emphasis of this expedition was on oceanography, biology, and geology, particularly on collecting data with new instrumentation, such as continuous-recording echo sounding. During this same period and at various other times in the 1920's, the eminent Dutch geophysicist Felix Andries Vening Meinesz was conducting gravity surveys in the Netherlands Indies. These surveys were to have a major impact on tectonic concepts, and Kuenen modeled some of the results in an experiment.

In 1934 Kuenen took a position as undergraduate lecturer and curator of collections at Groningen, where the professorship of geology had been abolished after the retirement of Bonnema in 1932. There were very few students, an arrangement that suited Kuenen, who could therefore devote most of his time to research. He was promoted to reader in 1939 and to professor in 1946 (later backdated to 1943); the latter promotion was held up by the German-influenced administration, ostensibly because of Kuenen's English heritage, though his political views were probably a factor. Kuenen served as an observer in the Dutch air force and was wounded during the German takeover (May 1940); later he was briefly imprisoned by the Germans, and his collection of photographs was confiscated (it was never returned).

When Leiden University was closed early in the war, geology students came to Groningen to study, which at first pleased Kuenen but soon reconfirmed his preference for research rather than teaching and supervision. He tried to discourage undergraduates from pursuing geology as a career, commonly expressing the opinion that there was no future in geology, and he would not take graduate students until they had completed all their graduate coursework (and a first graduate degree) elsewhere. Kuenen did take teaching seriously, however, and considered it his duty to educate. He was a thorough but dry lecturer, and students who came to know him learned to recognize his very subtle humor. He helped to write two textbooks and also wrote nontechnical books and articles. He produced at least one paper on teaching technique, encouraging the use of stereo pairs of photographs; Kuenen sometimes used stereo projections in class, and the students wore special glasses. He was an excellent photographer and used stereo pairs in some publications or made them available to readers.

Kuenen was professor of geology and director of the Geological Institute at the University of Groningen until his retirement in 1972; he briefly held the positions of dean of the faculty (1951–1952) and rector magnificus (1960). Although Kuenen's eminence had resulted in the revival and maintenance of a geological program at Groningen, he did not develop a strong school or a political base in an academic system that was strongly political; the small scale of the program at Groningen was in part a result of Kuenen's reluctance to take on students. Nevertheless, some very prominent young sedimentologists came to do their doctorates at Groningen or to conduct postdoctoral research, and others held state-supported positions in the institute. Around 1970 Kuenen spent considerable time and effort fighting abolition of the geology program at Groningen and elsewhere in the Netherlands; the stress contributed to his second nervous breakdown.

On 2 April 1932 Kuenen married Charlotte Susanne Wilhelmine Pijzel, who became a strong helpmate and counterbalance to Kuenen's rather sober demeanor. They had three daughters and a son. Kuenen's daily life was quite regimented—he bicycled to work, bicycled home for lunch, bicycled back to work. Kuenen did not own a car, nor did he drive. But he traveled extensively to international meetings, usually with his wife, and he typically took advantage of these trips to visit field localities. The death of his wife in 1967 was a great shock to Kuenen, probably causing the first of two nervous breakdowns. This loss, combined with administrative problems, caused a dramatic drop in Kuenen's scientific productivity. He died in 1976 during surgery for an aneurysm.

Kuenen's influence was in three fields: experimental geology, encouraged by his mentor, Escher; marine geology, stimulated by his experience on the *Snellius* expedition; and the combination of these two with field geology (in collaboration with others) to generate a model for the deposition of coarse sediments in the deep sea by the action of turbidity currents. On the latter topic he wrote some fifty papers (about a quarter of his published work), nearly all of which he wrote alone. For his scientific contributions, particularly his work on turbidity currents and turbidites, Kuenen received six major geological medals (Waterschoot van der Gracht, Penrose, Wollaston, Shepard, Dumont, and Steinmann), three honorary doctorates (Dublin, Cracow, and Exeter), and numerous distinguished lectureships, honorary memberships, and other awards.

Over a ten-year period beginning in 1928, Kuenen published a remarkable series of papers in experimental geology (almost all in *Leidsche geologische mededeelingen*), principally relating to structural and tectonic problems. His first major paper (1928) concerned the effects of wind action on pebbles; he reversed the flow of a vacuum cleaner and sandblasted easily erodible artificial pebbles of chalky consistency, in order to speed up the processes. This work confirmed laws of wind faceting formulated by Albert Heim (Escher's former supervisor). He followed this project with an experimental study written with Escher (1929) on the genesis and structure of folds in salt domes; other related work included a paper with L. U. de Sitter on folding (1938) and a paper on ptygmatic folding (1938), a subject to which Kuenen later returned (1968). Spurred by results from the *Snellius* expedition and Vening Meinesz's gravity data, Kuenen also performed experiments on the genesis of volcanic cones (1934) and of negative gravity anomalies by downbuckling of the crust due to horizontal compressive stresses (1936).

As a preface to the latter paper, Kuenen quoted Chester R. Longwell: "The field of tectonics is no place for a prim individual who likes everything orderly and settled and has a horror of loose ends." It was not long after this paper that Kuenen left the field of structural geology and tectonics—partly because he considered it to be "messy" and probably partly because he was being overshadowed by L. U. de Sitter and particularly by J. H. F. Umbgrove.

In a review of experimental geology (1965), Kuenen concluded that structural geology and tectonics ("endogenic geology") were stubborn subjects, and that the most promising area of experimental geology was that of external or surface processes. He stated that transportation by ice, water, and wind could be tested, qualitatively and quantitatively, and that slumping, load casting, mud flows, and turbidity currents offered a wide field for the experimental approach. Kuenen's first experimental work, mentioned above, was on surface processes, and his first experiment on turbidity currents was conducted in 1936 and 1937; he continued experimenting with turbidity currents and related phenomena for the rest of his active career. In addition, in the 1950's he began a protracted series of experiments on abrasion of pebbles and sand by various processes, published in a series of six papers (1955–1964).

Experiments conducted by Kuenen ranged from sophisticated to simple; he had a knack for clever experimental design on a low budget, and he enjoyed tinkering and woodworking—as a hobby he constructed detailed models of sailing ships. Clearly he was led into experimental geology by Escher, but several other factors motivated him. His upbringing and his superior intellect would have inclined him to test critically ideas or hypotheses generated in his or others' studies; he also had strong physical intuition but professed no mathematical ability. (For the time and in his field of sedimentology, Kuenen was considered to be quantitative.) Experimental work suited his life-style as well as his scientific approach; although he traveled extensively, he was not particularly attracted to fieldwork.

Kuenen became a proselytizer for experimental geology, publishing several versions of papers on its value while acknowledging that experimental technique did not seem to suit most geologists, compared with physicists or chemists. He discussed the pitfalls of experiments, such as incorrect scaling or producing something that looked right by processes improbable in nature; but he extolled the productive functions of geological experiments: "to illustrate, clarify, test, or exclude suggested explanations or imagined processes, to provide new concepts, to furnish data, to allow inspection of processes [that is, processes that otherwise could not be observed], [and] to allow study of simplified natural processes" ("The Value of Experiments in Geology," p. 32).

Marine Geology. Kuenen's career as marine geologist started with his participation in the *Snellius* expedition. As had been true of Charles Darwin and James Dwight Dana, this endeavor matured Kuenen, deepened and broadened his experience in a very short time, and provided a fertile ground for the generation of ideas to be tested (and ultimately to be published). Kuenen was a very careful observer, and his original reports include geological mapping and description of islands, discussion of bathymetrical results, discussion of geological significance of gravity data, and geology of coral reefs; Gerda A. Neeb analyzed the sediment samples taken by Kuenen on the expedition.

Oceanic coral reefs had been a focus of geological study at least since Darwin's time, particularly because they were indicators of sea level; yet some were at present dead and submerged, whereas others had survived, apparently by accumulating to great thickness. Theories for the controls of coral-reef life, growth, and death included tectonic subsidence and eustatic (worldwide) sea-level change, particularly during the late Cenozoic glacial ages; temperature changes during glacial ages were also incorporated. Darwin and Dana considered some aspects of this problem in the nineteenth century, and twentieth-century contributors to the discussions included William Morris Davis, Reginald Aldworth Daly, and Umbgrove. Kuenen (1933, 1947) favored a combination of controls on reef growth and pointed out that coral in the Moluccas had survived sea-level changes and ocean-temperature variation in the late Cenozoic.

Other work that emanated from Kuenen's *Snellius* experience included discussions of eustatic changes of sea level, calculations of the total mass of sediments in the deep sea and on Earth, studies of several volcanogenic phenomena, and consideration of submarine currents. Three books were generated during this time, two of them popular works: Kuenen's "diary" of the expedition, *Kruistochten over de Indische diepzeebekkens* (1941), and *De kringloop van het water* (1948; revised and translated into English in 1955), a discussion of the presence and action of water in its various forms on earth. The third was a textbook, *Marine Geology* (1950).

The field of marine geology grew explosively following World War II, and Kuenen's text was influential, though quickly eclipsed by rapid progress in the field. He never revised it. In his book Kuenen tried to summarize the state of the entire field, but the text also presents many of his own studies and contains many fertile ideas. At about the same time (1948), Francis P. Shepard published *Submarine Geology*, which went through two revisions and ultimately supplanted Kuenen's work. Whereas Shepard continued to practice marine geology—to direct scientific cruises—Kuenen never again participated in an oceanographic expedition. However, he kept abreast of developments in marine geology and several times gave papers in which he emphasized the importance to geologists of marine geology and the study of modern processes. He contributed ideas, for example, to the studies of H. Postma and of Kuenen's colleague L. M. J. U. van Straaten concerning sediment transport and deposition on modern Dutch tidal flats (1958). Kuenen firmly believed that experiments and studies of the modern were critical tests of historical geological reconstructions.

Submarine Canyons and Turbidity Currents. The development of continuous echo sounding in the 1930's led to the discovery and detailed mapping of many submarine topographic features, some of the most spectacular of which were submarine canyons. Study of these steep-walled, deep-sea valleys launched the careers of Kuenen and Shepard, but in very different ways. Shepard made a career of the detailed field study of canyons—their topography, the composition of their walls and floors, the currents that flowed within them, and related features. Impressed with their steepness, he originally postulated that they were eroded subaerially when the continents were much more elevated, and by observation he spent much of his career testing and revising this hypothesis.

Kuenen never personally studied a submarine canyon; he entered the field when he read a paper by Daly (1936) postulating that submarine canyons were eroded by the action of subaqueous, sediment-laden density currents (subsequently named "turbidity currents" by Douglas Johnson, 1938). These currents had been observed in lakes and reservoirs, and Daly believed they had been very active at the edge of the continental shelf during glacioeustatic lowstands of sea level, when waves would generate great turbulence there. By his own admission, Kuenen did not take this paper seriously until Vening Meinesz said he thought it presented a viable hypothesis. Kuenen was then spurred to conduct experiments on the generation, nature, and erosive capability of turbidity currents. He concluded that they could erode submarine canyons, siding with

Daly in a debate that has not yet been entirely resolved. At that time, both Daly and Kuenen were considering low-density, mud-laden currents. Kuenen's first two papers on this subject, published in 1937 and 1938, did not generate much interest, acceptance, or rebuttal outside of those directly involved.

During World War II, Kuenen continued to publish papers based on his *Snellius* experience, as well as on a variety of topics that did not require fieldwork. At this time he apparently also was working on his books, as well as a text on earth history with Isaak M. van der Vlerk (the principal author), all of them published in the late 1940's. In 1947 Kuenen published a major summary and discussion of his ideas about coral atolls and submarine canyons; in regard to the latter, he restated his position but put more emphasis on the action of submarine slumps in addition to turbidity currents.

Also at this time Kuenen renewed his experiments, now concentrating on high-density turbidity currents generated in a flume and in a ditch in his yard. He produced a film of these experiments, and presented the results and conclusions at the International Geological Congress at London in 1948. He postulated that submarine slumps and mudflows could transport very coarse sediments out of canyons, and that they could generate or evolve into high-density turbidity currents, transporting sediments coarser than silt (that is, sand) to the deep-sea floor. Although it is clear that previous workers had published some of these ideas, it is widely agreed that Kuenen's paper and film of 1948 started a revolution in the interpretation of graded beds of sand in the geological record.

Attending the meeting was C. I. Migliorini, who in 1944 had suggested that coarse-grained graded beds in Italy were deep-water deposits of submarine landslides and associated muddy currents. In the discussion following Kuenen's presentation, he suggested that Kuenen's high-density turbidity currents were the mechanism that transported coarse sediments and deposited these graded beds. Kuenen had not considered the deposits until this time—he was more concerned with erosion of submarine canyons. Migliorini took Kuenen to the Apennines, and they collaborated on a paper, "Turbidity Currents as a Cause of Graded Bedding" (1950). The paper, combining experimental results and field data, was very well received, helped in particular by a simultaneously published favorable review by Francis J. Pettijohn, one of the most influential sedimentary geologists of the era.

Also present at the meeting was Ian Campbell. He was familiar with the studies of Manley L. Natland (1933), whose paleoecological analysis of Foraminifera in the Cenozoic Ventura Basin in California indicated to him that not only sands but also conglomerates had been deposited in deep water. His ideas had been met with rejection and even ridicule. Upon hearing of Kuenen's paper and as chairman of the Research Committee of the Society of Economic Paleontologists and Mineralogists (SEPM), Natland organized a symposium at the 1950 SEPM annual meeting in Chicago, inviting Kuenen to attend and to visit him in the Ventura Basin; this visit resulted in a collaborative paper presented at the symposium. During his visit, Kuenen also traveled around the United States, lecturing and showing his film. By the time (or shortly after) the symposium proceedings were published (1951), most concerned sedimentary geologists had accepted that graded beds were the products of turbidity currents (Kuenen and Migliorini, 1950), and that these graded beds of coarse sediments (sand and coarser) were deposited in up to thousands of meters of water (Natland and Kuenen, 1951).

The revision in interpretation of coarse-grained, graded beds caused the reinterpretation of numerous sedimentary basins. Kuenen and his students, coworkers, and others went on to refine the interpretation of sedimentary structures generated by turbidity currents and associated processes. The importance of paleocurrent indicators in these sediments was recognized, and new paleogeographic reconstructions of various basins were generated. Kuenen continued to publish numerous papers on aspects of these problems, including results of fieldwork and of newly designed experiments. He was particularly excited to learn more of coarse-grained sediment samples from deep-sea cores, especially as reported by Maurice Ewing, Bruce C. Heezen, and others from Lamont Geological Observatory in New York. He also followed their work and contributed his own to calculation of turbidity-current velocities from submarine cable breaks; it was later shown that some of these breaks were caused by slumps.

Kuenen is considered by many to be the first modern, process-oriented sedimentary geologist, principally for his work on turbidity currents, which integrated marine geology, experiment, and geological fieldwork. Roger G. Walker (1973, p. 3) argued that "the [turbidity current] theory . . . represents the *only* true revolution in thought in this century about clastic rocks." Kuenen's other experimental work, in particular his studies of sedimentary structures and textures, generated significant results; none of Kuenen's students became experimentalists, however, to carry on the tradition. Kuenen was a meticulous observer and a critical thinker who constantly questioned the validity of geological interpretations. A supremely intelligent individual, he could be merciless to those who were not so critical in their thinking; if asked incisive questions, however, he was very helpful. Kuenen could be gracious and witty when the occasion arose, and exhibited a sharp sense of humor. Although at times, especially in later life, Kuenen expressed insecurity about his eminence in sedimentology, it is clear that he was a giant in his field. Many prominent sedimentologists were influenced by him personally, as well as by his papers.

BIBLIOGRAPHY

I. Original Works. No bibliography has been published; Kuenen's curriculum vitae lists 211 publications.

Kuenen's monographs include *Geology of Coral Reefs: The "Snellius" Expedition*, V, pt. 2 (Leiden, 1933); *Geological Interpretation of the Bathymetrical Results, The* Snellius *Expedition*, V, pt. 1 (Leiden, 1935); *Kruistochten over de Indische diepzeebekkens* (The Hague, 1941); *De Kringloop van het water* (The Hague, 1948); *Marine Geology* (New York, 1950); and *Realms of Water* (New York, 1955). In addition he contributed to *Geheimschrift der aarde* (Utrecht, 1948; 6th ed., 1951), Isaak M. van der Vlerk the principal author; and wrote the introductory section to *Bottom Samples: The "Snellius" Expedition*, V, pt. 3 (Leiden, 1943), Gerda Neeb author of the body of the report.

Kuenen's papers include "Experiments on the Formation of Windworn Pebbles," in *Leidsche geologische mededeelingen*, **3** (1928), 17–38; "Experiments in Connection with Salt Domes," *ibid.*, **3** (1929), 151–182, with B. G. Escher; "Experiments on the Formation of Volcanic Cones," *ibid.*, **6** (1934), 99–118; "The Negative Anomalies in the East Indies (with Experiments)," *ibid.*, **8** (1936), 169–214; "Experiments in Connection with Daly's Hypothesis on the Formation of Submarine Canyons," *ibid.*, **8** (1937), 327–351; "Experimental Investigation into the Mechanism of Folding," *ibid.*, **10** (1938), 217–239, with L. U. de Sitter; "Observations and Experiments on Ptygmatic Folding," in *C. R. Soc. géol. Finl.*, **12** (1938), 11–28; "Density Currents in Connection with the Problem of Submarine Canyons," in *Geological Magazine*, **75** (1938), 241–249; "Geochemical Calculations Concerning the Total Mass of Sediments in the Earth," in *American Journal of Science*, **239** (1941), 161–190; "Pitted Pebbles," in *Leidsche geologische mededeelingen*, **13** (1942), 189–201; "Two Problems of Marine Geology: Atolls and Canyons," in *Verhandelingen der Koninklijke Nederlandsche akademie van wetenschappen, afdeeling natuurkunde*, sec. 2, **43**, no. 3 (1947), 1–69; "The Formation of Beach Cusps," in *Journal of Geology*, **56** (1948), 34–40; "Turbidity Currents of High Density," in *Report of the 18th International Geological Congress*, pt. 8 (London, 1950), 44–52; "Turbidity Currents as a Cause of Graded Bedding," in *Journal of Geology*, **58** (1950), 91–127, with C. I. Migliorini.

"Sedimentary History of the Ventura Basin, California, and the Action of Turbidity Currents," in J. L. Hough, ed., *Turbidity Currents and the Transportation of Coarse Sediments to Deep Water*, Society of Economic Paleontologists and Mineralogists Special Publication no. 2 (1951), 76–107, with Manley L. Natland; "Significant Features of Graded Bedding," in *Bulletin of the American Association of Petroleum Geologists*, **37** (1953), 1044–1066; "Origin and Classification of Submarine Canyons," in *Bulletin of the Geological Society of America*, **64** (1953), 1295–1314; "Sea Level and Crustal Warping," in Arie Poldervaart, ed., *Crust of the Earth*, Geological Society of America Special Paper no. 62 (1955), 193–204; "Experimental Abrasion of Pebbles 2. Rolling by Current," in *Journal of Geology*, **64** (1956), 336–368; "Tidal Action as a Cause of Clay Accumulation," in *Journal of Sedimentary Petrology*, **28** (1958), 406–413, with L. M. J. U. van Straaten; "Sand," in *Scientific American*, **202**, no. 4 (1960), 94–110; "Deep-Sea Sands and Ancient Turbidites," in A. H. Bouma and A. Brouwer, eds., *Turbidites* (Amsterdam, 1964), 3–33; "Experimental Abrasion, 6, Surf Action," in *Sedimentology*, **3** (1964), 29–43; "Value of Experiments in Geology," in *Geologie en mijnbouw*, **44** (1965), 22–36; "Experimental Turbidite Lamination in a Circular Flume," in *Journal of Geology*, **74** (1966), 523–545; "Emplacement of Flysch-Type Sand Beds," in *Sedimentology*, **9** (1967), 203–243; "Origin of Ptygmatic Features," in *Tectonophysics*, **6** (1968), 143–158; "Tentative Data on Flow Resistance in Suspension Currents," in *Geologie en mijnbouw*, **50** (1971), 429–442; and "Continental Shelf" and "Continental Slope," in *Encyclopaedia Britannica*, 15th ed. (1974), III, 585.

II. Secondary Literature. Only short memorial notes have been published: A. Brouwer, in *De Leidse geoloog*, **40** (1977), 7–12, a newsletter of Leiden geology students (in Dutch); Francis P. Shepard, in *Geological Society of America. Memorials*, **8** (1978); L. M. J. U. van Straaten, in *Geologie en mijnbouw*, **56** (1977), 1–3 (in Dutch); and E. K. W. (E. K. Walton), in *Geological Society of London Annual Report for 1977*, 35–36.

Kuenen and F. L. Humbert published a 700-entry "Bibliography of Turbidity Currents and Turbidites," in A. H. Bouma and A. Brouwer, eds., *Turbidites* (Amsterdam, 1964), 222–246. Roger G. Walker has written a history of turbidity-current and turbidite concepts: "Mopping Up the Turbidite Mess," in Robert N. Ginsburg, ed., *Evolving Concepts in Sedimentology* (Baltimore, 1973), 1–37.

This biography was prepared from the content of Kuenen's papers, from memorials and medal citations, and from interviews with and questionnaires from Kuenen's students, colleagues, and contemporaries, including Tj. van Andel, G. J. Boekschoten, J. R. Boersma, A. H.

Bouma, A. Brouwer, R. H. Dott, Jr., G. de V. Klein, G. V. Middleton, G. Postma, H. G. Reading, J. E. Sanders, L. M. J. U. van Straaten, J. Terwindt, and R. G. Walker. I was provided with copies of correspondence between some of these people and Kuenen, as well as correspondence between Kuenen and Shepard from the Scripps Institution of Oceanography archives. Kuenen's daughter Madeline Dommering-Kuenen compiled information for me from the Kuenen family.

JOANNE BOURGEOIS

KÜHN, OTHMAR (*b*. Vienna, Austria, 5 November 1892; *d*. Vienna, 26 March 1969), *paleontology*.

Kühn's father was a watchmaker and goldsmith. Because of the long sickness of his mother, Kühn's early years were hard. After finishing secondary school in 1911, he started to learn the brewery business, but in 1914 he matriculated at the University of Vienna, where he studied natural sciences (especially botany), aiming for a career in teaching. His studies were interrupted by military service during World War I. In 1919 he received his Ph.D. in botany and paleontology. From 1921 to 1943 Kühn taught in secondary schools in Vienna. His main interests, however, were his paleontological studies as a voluntary scientific collaborator at the Museum of Natural History in Vienna. Following military service in World War II, Kühn became curator in 1944 and then director in 1951 of the geological-paleontological section of the Museum of Natural History. From 1951 to 1964 he was full professor of paleontology at the University of Vienna. He served as dean of the faculty and as elected rector. Kühn was married and had one daughter.

Kühn's professional and scientific activities resulted in many honors, including membership in the academies of science in Athens, Belgrade, Copenhagen, Ljubljana, Vienna, and Zagreb, and honorary doctorates from the universities of Athens and Bucharest.

Kühn's scientific work covered a wide range of topics. Starting with botanical studies, he soon became interested in systematic paleontology of invertebrate groups, which had not been studied carefully for a long time. For nearly his whole life Kühn was one of the very few leading experts in Mesozoic and tertiary corals and hydrozoans as well as in cretaceous rudist pelecypods. Influenced by his teacher at the university, Carl Diener, Kühn considered systematic descriptions of fossils an absolute necessity for a sound biostratigraphical base of geological studies. Although some results (for example, the biostratigraphical scheme based on rudist evolution) are out of date, Kühn's studies on cretaceous and tertiary fossils are still fundamental works.

A main feature of Kühn's scientific work was the publication of precise and up-to-date reviews of the groups he was studying. He wrote catalogs of fossil hydrozoans and rudists, and he summarized his knowledge of various invertebrate groups in review articles. He founded the *Fossilium catalogus Austriae*, a paleontological treatise that aimed to provide a critical synopsis of all fossils described from Austrian territory. Kühn's work was focused on the integration of paleontology and geology. Although he emphasized the biological aspect, he was able to understand the needs of geoscientists seeking biostratigraphical and paleoecological data. This is shown by his contribution to F. X. Schaffer's *Geologie von Österreich* (Kühn wrote the chapter describing the geology of the southern Alps) and by the volume *Autriche* of the *Lexique stratigraphique international* (he was the editor and also wrote large parts of the volume).

Most of Kühn's more than 160 publications describe fossils from Austria and from southeastern and southern Europe. In this he followed the tradition of paleontological research in Vienna, which was strongly directed to the countries formerly belonging to the Austro-Hungarian monarchy. Kühn's personal relations with colleagues in these countries greatly assisted the reorganization of Viennese paleontology after World War II. Clearly recognizing the future trends in paleontology and paleobiology, Kühn promoted the development of micropaleontology, vertebrate paleontology, paleobotany, and biostratigraphy in the Institute of Paleontology at the University of Vienna by establishing new staff positions and research possibilities. Kühn offered assistance and kindness to many colleagues.

BIBLIOGRAPHY

I. ORIGINAL WORKS. Important papers on the systematic paleontology of rudist pelecypods, published in *Neues Jahrbuch für Mineralogie, Geologie und Paläontologie*, are "Rudistenfauna und Kreidentwicklung in den Westkarpathen," **86B** (1942), with D. Andrusov; "Rudisten aus Griechenland," **89A** (1945); and "Die borealen Rudistenfaunen," **90B** (1949). *Das Danien der äusseren Klippenzone bei Wien* (Jena, 1930) presents the first evidence of lowermost tertiary deposits in Austria. Catalogs include *Hydrozoa* (Berlin, 1928); *Hydrozoa* (Berlin, 1939); and *Rudistae* (Berlin, 1932).

II. SECONDARY LITERATURE. Biographical sketches include F. Bachmayer and H. Zapfe, in *Annalen des Naturhistorischen Museums in Wien*, **74** (1970); F. Steininger, in *Mitteilungen der Geologischen Gesell-*

schaft in Wien, **62** (1969), with complete bibliography; and H. Zapfe in *Almanach der Akademie der Wissenschaften in Wien*, **120** (1970).

ERIK FLÜGEL

KUNITZ, MOSES (*b*. Slonim, Russia, 19 December 1887; *d*. Philadelphia, Pennsylvania, 21 April 1978), *physical biochemistry, enzymology*.

Kunitz was born in Slonim, in Russia. After World War I Slonim was within Poland, some thirty miles west of Baranovichi. Early in World War II Slonim was reincorporated within the USSR, in the state of Belorussia. The town has been described as a center of Talmudic studies, before it was occupied by the German army in 1941. The changes in the fortunes of Slonim and its Jewish community were undoubtedly a matter of concern to Kunitz throughout his life in the United States, and he participated in reunions in New York of emigrants from Slonim. Kunitz emigrated to the United States in 1909, obtained a job in a factory making straw hats, and entered the Cooper Union Evening School of Chemistry in New York in 1910. He then brought his brother and wife-to-be, Sonia Bloom, to America. He married in 1912; a daughter was born in 1914 and a son in 1921.

In 1913 he became the laboratory assistant of Jacques Loeb at the Rockefeller Institute, continuing in this position until 1923. He became a United States citizen in 1915. With Loeb's encouragement Kunitz graduated from Cooper Union in 1916 with a B.S. degree and continued his evening studies for two years at the Electrical Engineering School of Cooper Union and an additional year at the Extension School of Columbia University. Two more years were spent in the Columbia School of Mines, Engineering, and Chemistry, and in 1922 Kunitz became a graduate student in biochemistry at Columbia University. Having worked day and night for almost a decade, he was appointed to the staff of the institute in 1923 with the title of assistant. In that year, Kunitz published two articles with Loeb on the effects of acids and salts on the behavior of proteins. His thesis at Columbia was a physiochemical study of the properties of gelatin in salt solutions, and he received a Ph.D. degree in 1924. He was a competent glassblower, and the first of his independent papers, published in 1924, described an improved microelectrophoretic cell, which he had made. This was followed by a study of the effects of salts on sodium salts of gelatin. Kunitz concluded the latter with an expression of indebtedness to Loeb. Loeb died in 1924 and Kunitz's role and position in the institute then changed significantly.

Loeb's interest in salt effects had grown markedly after his discovery of the artificial parthenogenesis of sea urchin eggs in seawater supplemented with Mg^{++}. The continuing analysis of the different effects of various cations and anions led to the detailed studies that Loeb carried out with the initially meticulous, later skilled and knowledgeable, Kunitz. These studies eventually demonstrated the weakness of theories of colloid chemistry in explaining the behavior of proteins, rejected some commonly held views, and asserted the importance of controlling hydrogen ion concentration (or pH) in determining the binding of ions by proteins. Kunitz's definitive laboratory work in the last years of Loeb's life established the former as a significant figure in studies of the physical chemistry of proteins.

When Loeb's major colleague, John H. Northrop, in the department of general physiology, was appointed to head that division after Loeb's death, he invited Kunitz to stay and to continue these physiochemical studies. Northrop and Kunitz published many papers together on ion binding, swelling, osmotic pressure, and other properties of gelatin, as well as on related topics, from 1924 to 1931. In 1926 Kunitz was promoted to an associateship at the institute at the recommendation of Northrop. He had demonstrated unusual and useful skills in mathematics, as well as in the experimental physical chemistry of proteins.

In 1926 Northrop, who preferred country life, transferred his group to the branch of the institute in Princeton, New Jersey, containing the departments of animal and plant pathology. Kunitz also moved with Northrop to the relatively new laboratories, organized by Theobald Smith in 1916 and 1917. It may be mentioned that some new members of this branch of the institute, including "the physiologists" Northrop and Kunitz, were viewed by Smith as an administrative burden.

As early as 1919 Northrop had begun to work on proteolysis by pepsin, and he now concentrated on proteinases. Following the reported crystallization of urease by J. B. Sumner in 1926, Northrop advanced to the crystallization of pepsin by 1930. Kunitz then began work on the purification of pancreatic enzymes, and in 1931 the collaborators reported the crystallization of trypsin. The skills of Kunitz, honed on the physical chemistry of gelatin, were now brought to bear on enzyme isolation and characterization, and in the next three decades produced numerous landmarks in the development of enzymology. In the transition of the work of the laboratory, gelatin was used to detect the presence of proteolytic gelatinases and to assay such enzymes.

In 1930 Northrop was concerned with the problem of establishing the purity of crystals of pepsin, of determining whether such crystals contained one protein or a mixture of proteins of similar solubility. Application of the theory of the phase rule enabled him to demonstrate that the pepsin crystals contained a single protein. Determination of the solubility of pepsin in concentrated salt solutions at low temperature with saturating and saturated concentrations should, and indeed did, give a single value for the solubility of the protein. In 1930 also, Northrop and Kunitz considered the general theory of solubility curves obtainable from mixtures or solid solutions of proteins. This now became a major approach to the establishment of the purity of the many proteins isolated by these investigators. Kunitz, in collaboration with M. L. Anson and Northrop, also developed methods of establishing the molecular weight, molecular volume, and hydration of various proteins. In 1932 the methodologies were applied to crystalline trypsin, and the range of activity of the enzyme and its molecular properties were described. Kunitz and Northrop studied the irreversible and reversible denaturation and inactivation of the enzyme in 1934. A classical study of the thermodynamics of the reversible denaturation of a trypsin inhibitor was published by Kunitz in 1948.

In 1933 Kunitz and Northrop extracted the proenzyme chymotrypsinogen from pancreas with cold sulfuric acid (0.12 M), crystallized the protein, and demonstrated its conversion by trypsin to the active proteolytic enzyme chymotrypsin, which was also crystallized. This discovery began to explain the long-known fact that fresh pancreas lacked active tissue-digesting proteolytic enzymes. The following year they showed that trypsin is similarly contained in the acid extract as the crystallizable trypsinogen, itself activatable by trypsin. The generality and consequences of the paring of proproteins to functional proteins have become clear only after some thirty years. Many of the peptides removable by proteolysis or other active fragments of the degradation of proteins have more recently been found to possess striking physiological effects in themselves. The presence of leading hydrophobic sequences, subsequently removable, have been shown to be important in the secretion of proteins through membranes. Autodigestible giant proteins have been detected in the multiplication of an RNA virus, such as that of foot-and-mouth disease. The arrest of proteolytic fragmentation is a possible approach to blocking the multiplication of such a virus, a problem of great interest to Theobald Smith.

By 1934 and 1935, as Northrop extended his work in other directions, Kunitz took the leadership in studies of the pancreatic proenzymes and on the activity and inhibition of the enzymes themselves. The startling success of Northrop and Kunitz in the institute in Princeton led their younger colleague, Wendell Stanley, in the nearby department of plant pathology, to exploit similar methods to isolate tobacco mosaic virus. This ambitious effort succeeded in 1935, and a group of young biochemists worked under Stanley's leadership in Princeton until these laboratories were closed at the end of the 1940's. In this period Kunitz was a warm and fatherly figure in assisting the younger people attracted by the laboratories of Northrop and of Stanley.

Prior to American involvement in World War II, Kunitz continued to explore the kinetics of conversion of the zymogens to active enzyme. With Northrop he discovered and crystallized a pancreatic basic protein that combines with and inhibits the appearance of active trypsin. He also turned to the isolation of pancreatic nucleases. In 1939 and 1940, exploiting once again the extraordinary stability of some proteins in acid, he described the isolation and crystallization of a small, heat-stable ribonuclease, whose contamination by proteinase was subsequently eliminated by heat. The availability of this enzyme, as well as of crystalline deoxyribonuclease, whose isolation he reported in 1948, became major tools in the subsequent exploration of the structure and function of the nucleic acids beginning in the 1940's. The specific degradation and inactivation of genetic function by the purified nucleases became crucial elements of proof that pneumococcal DNA was an agent of inheritable transformation in 1944 or that viral RNA or DNA were infectious genetic units in the late 1950's. Almost immediately Kunitz's enzymes, eventually made available commercially, became essential tools in the progress of the biochemistry of the nucleic acids and in molecular biology. Ribonuclease itself became a protein of choice for the analysis of unfolding and inactivation and spontaneous refolding and regeneration of active enzyme. In 1940 Kunitz was promoted to associate membership in the Rockefeller Institute.

Early in the American involvement in World War II, Kunitz was asked by a civilian governmental agency, the Office of Scientific Research and Development, to isolate hexokinase from yeast, as was the laboratory of Carl Cori. Both groups succeeded, Kunitz publishing this work in 1946 with Margaret

McDonald, with whom he had worked on several problems of protein isolation since 1940. Characteristically, in the course of his investigation of hexokinase, he first crystallized three other proteins, only one of which was readily defined as an enzyme. Kunitz and McDonald also crystallized the toxic ricin, which, according to R. Herriott, was recalcitrant to other workers. In later years Kunitz became the crystallizer of last resort for many younger enzyme chemists.

In 1946 Sumner, Northrop, and Stanley were awarded the Nobel Prize for their work in crystallizing proteins, and in 1948 a second edition of the book *Crystalline Enzymes* (first published in 1939), containing the elegant work of Northrop, Kunitz, and Herriott, appeared. The volume is, in fact, a manual summarizing and detailing their most important studies and marks the end of their close departmental association. In 1947 the board of trustees of the Rockefeller Institute decided to close the Princeton branch and to merge it with the larger branch in New York. In the next year a dozen members of the Princeton staff, including Northrop and Stanley, elected to leave the institute. The Nobelists moved to the University of California at Berkeley, and the Princeton laboratories were closed in 1950. Almost a decade later, in 1957, Northrop called upon Kunitz once again, in an effort to define the nature of mutations in mathematical terms and to clarify problems in the development of lysogenic bacteriophage.

With Northrop's move to California and the disruption of a collaboration extending some thirty-five years, Kunitz returned to the institute in New York, with a promotion to full membership in 1949. After the death of his wife, he had married a former Slonimer in 1939. The couple welcomed their return to New York, enjoying the social and cultural opportunities of the city. Kunitz was particularly fond of opera and found amiable colleagues with whom to play chess.

Now in his early sixties, Kunitz turned to several new enzymes and biological materials. In the early 1950's he isolated and characterized an inorganic pyrophosphatase from yeast; in his last paper in 1962 he confirmed the observation that the enzyme also hydrolyzed adenosine triphosphate in the presence of Zn^{++}. In his last working years he had studied two extremely recalcitrant enzymes, the alkaline phosphatase of chicken intestine and an invertase of yeast. Despite his experience and artistry in coaxing protein crystals from solution, both enzymes possess unusual structural features and neither responded felicitously to his touch. It is not clear how Kunitz regarded his lack of success in obtaining pure crystalline products in these challenging final efforts. In 1953 Kunitz had become member emeritus at the institute, and in 1970 he was retired from the Rockefeller University, which awarded him an honorary degree in 1973.

Kunitz did not write reviews that integrated his studies into the developing thought of the time. With few exceptions, his papers do not tell us his thoughts on why enzymes, such as ribonuclease or inorganic pyrophosphatase, warranted his attention. To anyone discussing science with him at the Rockefeller Institute, it was clear that he was not attempting to keep up with the explosive advances of the biochemistry of the time; having decided on a project, he would sift the literature for essential information but would give his time, labor, and thought to the project at hand. He was delighted with the crystals as they appeared, but his deliberate efforts were marked, as Northrop has said, with "imagination, ingenuity, persistence, great technical skill, mathematical facility, and a thorough theoretical knowledge." It is regrettable that in his long career he had so few students, as either postdoctoral fellows or technicians. Nevertheless, the many young fellows abounding at the Rockefeller Institute who were sufficiently courageous to peek into the laboratory of this living legend were given every opportunity to ask questions, to see Kunitz at work, and to learn a method under his friendly surveillance.

Kunitz was extremely modest and self-effacing. In 1957 he was awarded the Carl Neuberg Medal by the American Society of European Chemists and Pharmacists and in 1967 he was belatedly elected to the National Academy of Sciences. He spent many summers working with Loeb and later with Northrop at the Marine Biological Laboratory in Woods Hole, Massachusetts. He enjoyed gardening at his summer home at Falmouth Heights in Massachusetts, now occupied by his son, Jacques Kunitz. Kunitz died in Philadelphia in the care of his daughter, Rosaline Albert.

BIBLIOGRAPHY

I. Original Works. A complete list of Kunitz's publications is obtainable from the archives of the Rockefeller University. See also Poggendorff. Key publications to which references are made in the article include "Valency Rule and Alleged Hofmeister Series in the Colloidal Behavior of Proteins, I, The Action of Acids," and "II, The Influence of Salts," in *Journal of General Physiology*, **5** (1923), 665–691 and 693–707, written with J. Loeb, and "III, The Influence of Salts on Osmotic Pressure, Membrane Potentials, and Swelling of Sodium Gelatinate," *ibid.*, **6** (1924), 547–564; "The Combination of Salts and Proteins, I," *ibid.*, **7** (1924), 25–38, written with J. H. Northrop; "Solubility Curves of Mixtures and Solid Solutions," *ibid.*, **13** (1930), 781–791, written with J. H. Northrop; "Crystalline Trypsin, I. Isolation and Tests of Purity," *ibid.*, **16** (1932), 267–294, written with J. H. Northrop; and "Molecular Weight, Molecular Volume, and Hydration of Proteins in Solution," *ibid.*, **17** (1934), 365–373, written with M. L. Anson and J. H. Northrop.

"Crystalline Chymo-trypsin and Chymo-trypsinogen, I. Isolation, Crystallization, and General Properties of a New Proteolytic Enzyme and Its Precursor," *ibid.*, **18** (1935), 433–458, written with J. H. Northrop; "Isolation from Beef Pancreas of Crystalline Trypsinogen, Trypsin, a Trypsin Inhibitor, and an Inhibitor-trypsin Compound," *ibid.*, **19** (1936), 991–1007, written with J. H. Northrop; "Crystalline Ribonuclease," *ibid.*, **24** (1940), 15–32; "Crystalline Hexokinase (Heterophosphatese [*sic*]): Method of Isolation and Properties," *ibid.*, **29** (1946), 393–412, written with Margaret R. McDonald; *Crystalline Enzymes*, 2nd ed. (New York, 1948), written with J. H. Northrop and R. M. Herriott; "Isolation of Crystalline Desoxyribonuclease [*sic*] from Beef Pancreas," in *Science*, **108** (1948), 19; "The Kinetics and Thermodynamics of Reversible Denaturation of Crystalline Soybean Trypsin Inhibitor," in *Journal of General Physiology*, **32** (1948), 241–269; "Crystalline Inorganic Pyrophosphatase Isolated from Baker's Yeast," *ibid.*, **35** (1952), 423–450; and "Hydrolysis of Adenosine Triphosphate by Crystalline Yeast Pyrophosphatase. Effect of Zinc and Magnesium Ions," *ibid.*, **45** (1962), 31–46.

II. SECONDARY LITERATURE. George W. Corner, *A History of the Rockefeller Institute, 1901–1953: Origins and Growth* (New York, 1965); Joseph S. Fruton, *Molecules and Life* (New York, 1972); and R. Herriott, "A Biographical Sketch of John Howard Northrop," in *Journal of General Physiology*, **45** (1962), 1–16, and "Moses Kunitz," in *Nature*, **275** (1978), 351–352.

SEYMOUR S. COHEN

KUNO, HISASHI (*b.* Tokyo, Japan, 7 January 1910; *d.* Tokyo, 6 August 1969), *petrology*.

The eldest son of Kamenosuke Kuno, a Japanese painter, and Tome, his wife, young Kuno went to Sendai to study at the Second High School, where he was very interested in geology but spent most of the time mountain climbing and skiing. In 1929 he enrolled at the Geological Institute of the University of Tokyo. Under the influence of Professor S. Tsuboi, he decided to concentrate on petrology, and he began his career by studying the petrology of the Izu-Hakone region.

Kuno intensively investigated the crystallization of pyroxenes from magmas and the genesis of basaltic magmas, subjects that were widely discussed in the 1930's. "Petrological Notes on Some Pyroxene Andesites from Hakone Volcano" was outstanding among his earlier papers in that it showed his remarkable ability in field petrology and microscopic observations. Indeed, high-precision optical data of rock-forming minerals were Kuno's hallmark. Recognizing the importance of groundmass minerals, he proceeded to establish the crystallization course of pyroxenes from magmas. The map of Hakone Volcano, appended to Kuno's dissertation (1950a), has remained one of the best geologic maps of volcanoes in Japan.

In 1939 he was appointed associate professor, but he was drafted in 1941 and was in military service in northeastern China for most of World War II. After this difficult time he returned to academic life in 1946 and completed his doctoral dissertation, "Petrology of Hakone Volcano and the Adjacent Areas, Japan," which was published in 1950. Here Kuno elucidated the genesis of two groups of andesites and the role of pyroxenes. With this paper he established his international reputation as a petrologist. During 1951 and 1952 he was invited to study pyroxenes at Princeton University by a grant of the Geological Society of America. At Princeton he collaborated with H. H. Hess, who became his lifelong friend.

For his petrological study of pyroxenes Kuno was awarded the Japan Academy Prize in 1954. His book *Volcanoes and Volcanic Rocks* (1954) was well received as a standard textbook. In 1955 he was promoted to professor of petrology at the University of Tokyo, a position he held until his death. In "Differentiation of Hawaiian Magmas," based on observations in Hawaii, he showed the possibility of generating granitic magma from tholeiitic magma through fractional crystallization, and also that the depth where partial melting of peridotite occurs in the upper mantle determines whether tholeiite or alkali olivine basalt magma is produced.

Extending this view to the Japanese islands, Kuno found a close correlation between the depths of earthquake focuses and the distribution of various basaltic rocks and presented a hypothesis that tholeiite magma is produced at depths of less than 200 kilometers and alkali olivine basalt magma at depths greater than 200 kilometers. With additional petrochemical data he later added high-alumina basalt magma, intermediate between the two magmas in composition and formed at an intermediate depth, as the third primary magma. This model provoked much interest among experimental petrologists, and Yoder and Tilley, Ringwood and Green, Kennedy, and later Kushiro made various experiments at high temperatures and pressures on artificial systems containing olivine, pyroxene, and natural rocks, most of which supported Kuno's model. Thus Kuno made a great contribution to experimental petrology, although he himself approached the problem mainly from the viewpoint of a field petrologist.

In the field of volcanology he wrote many papers on calderas, volcanic eruptions based on the pyroclastic materials, and the origins of andesites and petrographic provinces. His paper "Origin of Cenozoic Petrographic Provinces of Japan and Surrounding Areas" greatly influenced petrogenetic discussions. In his later years Kuno was interested in the petrology of the moon and was registered with NASA as a principal investigator on the returned lunar samples. It is a great pity that he died of cancer shortly before the lunar rocks from Apollo 11 were available for investigation.

Kuno trained many excellent petrologists, among them A. Miyashiro, Y. Seki, S. Aramaki, and I. Kushiro. His activities in academic societies were numerous; among his many honors, he was honorary fellow, Geological Society of America and Mineralogical Society of America; honorary member, Geological Society of London; foreign associate, National Academy of Sciences; president, Volcanological Society of Japan, Geological Society of Japan, and International Association of Volcanology and Chemistry of the Earth's Interior; and vice president, International Union of Geodesy and Geophysics. He was survived by his wife, Kimiko, his son, Takashi, and his daughter, Shizuko.

BIBLIOGRAPHY

I. ORIGINAL WORKS. "On the 'Pyroxene Andesites' from Japan," in *Bulletin of the Volcanological Society of Japan*, **1** (1932), 20–37; "Petrological Notes on Some Pyroxene Andesites from Hakone Volcano, with Special Reference to Some Types with Pigeonite Phenocrysts," in *Japanese Journal of Geology and Geography*, **13** (1936), 107–140; "Fractional Crystallization of Basaltic Magmas," *ibid.*, **14** (1937), 189–208; "Geology of Hakone Volcano and Adjacent Areas, Parts I and II," in *Journal of the Faculty of Science, University of Tokyo*, **7** (1950a), 257–279, 351–402. "Petrology of Hakone Volcano and the Adjacent Areas, Japan," in *Bulletin of the Geological Society of America*, **61** (1950b), 957–1019; "Formation of Calderas and Magmatic Evolution," in *Transactions, American Geophysical Union*, **34** (1953), 267–280; *Iwanami Zensho* ("Volcanoes and Volcanic Rocks," Tokyo, 1954); "Differentiation of Hawaiian Magmas," in *Japanese Journal of Geology and Geography*, **28** (1957), 179–218, written with K. Yamasaki, C. Iida, and K. Nagashima. "Origin of Cenozoic Petrographic Provinces of Japan and Surrounding Areas," in *Bulletin volcanologique*, 2nd ser., **20** (1959), 37–76; "High-alumina Basalt," in *Journal of Petrology*, **1** (1960), 121–145; "Origin of Primary Basalt Magmas and Classification of Basaltic Rocks," *ibid.*, **4** (1963), 75–89; "Lateral Variation of Basalt Magma Type Across Continental Margins and Island Arcs," in *Bulletin volcanologique*, 2nd ser., **29** (1966), 195–222; "Volcanological and Petrological Evidences Regarding the Nature of the Upper Mantle," in T. F. Gaskell, ed., *The Earth's Mantle* (London, 1967), 89–110; "Differentiation of Basalt Magmas," in H. H. Hess and A. Poldervaart, eds., *Basalts: The Poldervaart Treatise on Rocks of Basaltic Composition*, II (New York, 1968), 623–688; "Pigeonite-bearing Andesite and Associated Dacite from Asio, Japan," in *American Journal of Science*, **267A** (1969), 257–268; *Selected Papers by Professor Hisashi Kuno* (Tokyo, 1969).

II. SECONDARY LITERATURE. Memorials were written by H. L. Foster in *Geological Society of America: Memorials*, I (1973), 27–37, and K. Yagi in *American Mineralogist*, **55** (1970), 573–583; both contain bibliographies. See also K. Yagi, "Life and Works of Professor Hisashi Kuno," in *Journal of the Japanese Association of Mineralogists, Petrologists, and Economic Geologists*, **63** (1970), 30–42.

KENZO YAGI

KURATOWSKI, KAZIMIERZ (*b.* Warsaw, Poland, 2 February 1896; *d.* Warsaw, 18 June 1980), *mathematics.*

Kuratowski's father was a well-known Warsaw lawyer. After completing his secondary education in Poland, Kuratowski enrolled as an engineering student at the University of Glasgow in 1913. He was spending his summer vacation of 1914 at home when war broke out, so he had to remain in Poland. In 1915 he began to study mathematics at the University of Warsaw, which had reopened after almost half a century of inactivity. He was a student of Stefan Mazurkiewicz and Zygmunt Janiszewski in the Seminar on Topology (initiated in 1916), and of the philosopher and logician Jan Łukasiewicz. Kuratowski graduated from the university in 1919. He obtained his doctorate in 1921 under the supervision of Wacław Sierpiński, with a dissertation on fundamental questions in set theory that contributed to international recognition of the embryonic Polish school of mathematics.

In 1927 Kuratowski was named to the chair of mathematics at Lwów Technical University. In 1933 he became professor at Warsaw University, where he was in charge of a wide range of academic and administrative functions inside and outside of Poland. He was secretary of the Mathematical Commission of the Council of Exact and Natural Sciences, which formulated the organizational plan of the Polish school of mathematics from 1936 to 1939. The outbreak of the war and the dramatic events of the following years did not put an end to mathematical education, which the scientific elite carried out through the clandestine university network. Kuratowski was active in the restructuring of mathematics in Poland after the liberation in February 1945.

Kuratowski is prominent in the history of mathematics in Poland as a result of his original contributions and his intense activity in mathematical education. A member of the editorial board of *Fundamenta mathematicae* from 1928, Kuratowski replaced Sierpiński as its editor in chief (1952) and held the post until his death. He was one of the founders and editor of the series Monografie Matematyczne (1932), which published the works of well-known Polish mathematicians. Kuratowski was vice president of the Polish Academy of Sciences and founded and directed its Institute of Mathematics. He was also vice president of the International Mathematical Union and a foreign member of the Academy of Sciences of the U.S.S.R., of the Royal Society of Edinburgh, of the Accademia Nazionale dei Lincei, and of the academies of Palermo, Hungary, Austria, the German Democratic Republic, and Argentina.

Kuratowski's mathematical activity focused on the properties and applications of topological spaces. His first contribution to general topology was the axiomatization of the closure operator (1922). He used Boolean algebra to characterize the topology of an abstract space independently of the notion of points. Subsequent research showed that, together with Felix Hausdorff's definition of a topological space in terms of neighborhoods, the closure operator yielded more fertile results than the axiomatic theories based on Maurice Fréchet's convergence (1906) and Frigyes Riesz's points of accumulation (1907).

Another field that interested Kuratowski was compactness, which, along with the metrization of topological spaces, was a pillar of mathematics in the 1920's. Prior to the appearance of the pioneering work of Pavel Aleksandrov and Pavel Urysohn (1923), many workers in this field used the properties of Émile Borel and Henri Lebesgue, of Bernard Bolzano and Karl Weierstrass, and of Georg Cantor, with little discretion. Kuratowski and Sierpiński were the first to publish a comprehensive study of the Borel-Lebesgue property (1921). In this context they made a remarkable presentation of Lindelöf spaces. Kuratowski and his colleagues in the Seminar of Mathematics at Warsaw made important contributions in the field of metric spaces, the spaces of greatest interest in that period. The first volume of his *Topologie* (1933) was the first complete work on metric spaces to appear in several decades.

In the theory of connectedness, the efforts of Bronisław Knaster and Kuratowski to organize the ideas of contemporary mathematicians culminated in an important contribution (1921) that revealed the conditions for connectedness of a subspace and the union of a family of subspaces of a topological space. However, the most remarkable feature of this work was the ingenious construction of a set known as the Knaster-Kuratowski fan, a subspace of the plane, obtained from the Cantor set by the category method, that has a central point that, if removed, causes the fan to become completely disconnected. The appearance of the Knaster-Kuratowski fan and its singular characteristics gave a new impulse to the research in connectedness and dimension.

The topology of the continuum was one of the original areas of activity of the Polish mathematicians. Kuratowski, who made significant contributions in this field, dealt with the problem of the indecomposable continuum and the common frontier. He showed that such a frontier is itself an indecomposable continuum or the union of two indecomposable continua (1924). Knaster proved the second of these propositions (1925). The implicit use of minimal principles in the study of irreducible continua and the need for a theory to deduce a method to eliminate the transfinite ordinals from the topological demonstrations led Kuratowski to formulate the Kuratowski-Zorn lemma, one of the fundamental notions in set theory. Before Max Zorn formulated this lemma (1935), Kuratowski used the axiom of choice to establish a minimal principle in a paper that proposed to generalize certain results of Janiszewski (1910) and L. E. J. Brouwer (1911) with regard to an irreducible continuum containing two given points (1922). The appearance of such a general method represented a historic moment of reaction to the generalized and even artificial use of the well-ordered sets and the transfinites of Cantor. From this time on, it became customary to use the transfinites only when it was absolutely necessary. A detailed account of Kuratowski's impact in this area is in Kuratowski and Andrzej Mostowski's *Set Theory* (1968).

Another concern of the Polish school of mathematics was measure theory. Are there nonmeasurable sets in the theory of a completely additive measure? The study of this question led to the Banach-Tarski paradox (1929). Stefan Banach and

Kuratowski (1929) answered in the affirmative, provided the continuum hypothesis is assumed. In his general formulation of the problem, Banach raised questions about the cardinality of a set on which a measure was desired (1930). Stimulated by Kuratowski, his student Stanisław Ulam provided the solution in the same year. This was a beautiful example of scientific collaboration and understanding, and of the ability to organize and encourage creative activity at its height.

Even though at first the theory of dimension was considered to be a part of point-set topology, the situation changed radically after the publication of the work of Aleksandrov (1926), in which the dimension of a metric space was characterized in terms of the dimension of its polyhedra. The work of Kuratowski in this field reveals his talent and his ability to adapt to new theories. Extending the principal results of Aleksandrov, Kuratowski (1933) devised a method to characterize the dimension of a metric space by means of barycentric mapping into the nerve of an arbitrary finite cover. The generality of this method permitted the extension of various spaces to normal spaces. In addition to the generalization of the theorem of Georg Nöbeling and Lev S. Pontriagin, Kuratowski, working alone or with Karl Menger and Edward Otto, among others, derived many additional results in the theory of dimension.

Mention also should be made of the theory of projective sets and analytic sets, in which Kuratowski extended the results in Euclidean spaces to Polish spaces (complete and separable); the theory of graphs, in which he obtained a rather difficult characterization of planar graphs; and the structure and classification of linear spaces from the point of view of topological range or type of dimension.

BIBLIOGRAPHY

I. Original Works. Most of Kuratowski's papers were written in French and appeared mainly in *Fundamenta mathematicae*. They include "Le théorème de Borel-Lebesgue dans la théorie des ensembles abstraits," in *Fundamenta mathematicae*, **2** (1921), 172–178, with Wacław Sierpiński; "Sur les ensembles connexes," *ibid.*, 206–255, with B. Knaster; "Une méthode d'élimination des nombres transfinis des raisonnements mathématiques," *ibid.*, **3** (1922), 76–108; "Sur l'opération Ā de l'analysis situs," *ibid.*, 182–199; "Sur les coupures irréductibles du plan," *ibid.*, **6** (1924), 130–145; "Sur une généralisation du problème de la mesure," *ibid.*, **14** (1929), 127–131, with Stefan Banach; and "Sur un théorème fondamental concernant le nerf d'un système d'ensembles," *ibid.*, **20** (1933), 191–196. His books include *Topology*, 2 vols., I translated by J. Jaworski and II translated by A. Krikor (New York, 1966–1968), which also appeared in French, Russian, and Polish; and *Set Theory* (Amsterdam and New York, 1968; 2nd, rev. ed., 1976), with Andrzej Mostowski. See also his autobiography, "Zapsiki do autobiografii," in *Kwartalnik historii nauk i techniki*, **24**, no. 2 (1979), 243–289; and *A Half Century of Polish Mathematics: Remembrances and Reflections*, Andrzej Krikor, trans. (Oxford and New York, 1980).

II. Secondary Literature. P. S. Aleksandrov, "In Memory of C. Kuratowski," A. Lofthouse, trans., in *Russian Mathematical Surveys*, **36** (1981), 215–216; J. Krasinkiewicz, "A Note on the Work and Life of Kazimierz Kuratowski," in *Journal of Graph Theory*, **5** (1981), 221–223; and Mary Grace Kusawa, *Modern Mathematics: The Genesis of a School in Poland* (New Haven, 1968).

Luis Carlos Arboleda

LACK, DAVID LAMBERT (*b.* London, England, 16 July 1910; *d.* Oxford, England, 12 March 1973), *ornithology, ecology, ethology.*

David Lack was the eldest of the four children of Harry Lambert Lack, a leading ear, nose, and throat surgeon in London, and of Kathleen Rind, a professional actress before her marriage. He was educated at Gresham's School, Holt, Norfolk, and in 1929 entered Magdalene College, Cambridge, where he completed the natural sciences tripos and graduated in June 1933. He was steered by Julian Huxley to a position as biology teacher at a progressive Devonshire school, Dartington Hall. After a leave of absence spent in the Galapagos (1938–1939), he resigned from the school at the end of the academic year in 1940, having grown disenchanted with its educational philosophy.

From 1940 to 1945 Lack was involved in war-related work as a civilian. From 1945 on, he was director of the Edward Grey Institute of Field Ornithology at Oxford in 1963; he became professorial fellow of Trinity College, Oxford. He was elected to the Royal Society of London in 1951, served as president of the Fourteenth International Ornithological Congress in 1966, and was awarded the Darwin Medal of the Royal Society in 1972.

On 9 July 1949 Lack married Elizabeth Twemlow Silva, who collaborated with him in his research and was still active in ornithology at Oxford in the late 1980's. They had three sons and a daughter. He converted to the Anglican religion in 1948 from an agnostic stance and was confirmed in 1951. In addition to his scientific research, Lack wrote several books on the natural history of birds for a more general audience.

LACK

As an ornithologist and committed Darwinian, Lack was instrumental in forming and promoting a synthesis of evolutionary biology and ecology, and in developing the field of population ecology. From adolescence his dominant enthusiasms were birds and evolution, the latter interest stimulated by W. P. Pycraft's popular books. He also read Huxley's articles on bird courtship while in high school. At Cambridge, Lack was influenced by J. B. S. Haldane and Ronald A. Fisher, though he did not adopt the mathematical methods of these theorists. At Dartington Hall he pursued his research on bird ecology and behavior. As a member of the Scientific Advisory Committee of the British Trust for Ornithology (formed in 1932), he helped to organize a heathland census inquiry in 1933. Through this work he became interested in territorial behavior of the European robin.

During the 1930's Huxley acted as Lack's unofficial supervisor, eventually securing grants from the Royal Society and the Zoological Society of London to finance an expedition by Lack and others to the Galapagos Islands in the fall of 1938. During his leave of absence (1938–1939) Lack investigated speciation in the Galapagos finches, spending five months studying museum collections in the United States. The California Academy of Sciences, owner of the largest collection, published his first monograph on the finches, with a four-year delay after his submission of the manuscript, in 1945. However, it was not this work but his second book on these birds, *Darwin's Finches*, published in 1947, that brought Lack worldwide attention. The book is now regarded as a classic of evolutionary biology.

Darwin's Finches is a lucid study of the importance of ecological factors in promoting divergence among closely related species. It was commonly thought in the 1930's that minor differences between similar species were not adaptive, thereby setting a limit to the role of natural selection in originating species. Using the finches as a paradigm, Lack argued to the contrary: that most of the minor differences between these species were adaptations produced by natural selection, which enabled the species to avoid competing with each other for food. Lack's views in this book were a reversal of what he had written in his first monograph. His change of mind came after reflecting on the implications of a principle discussed by the Russian ecologist G. F. Gause in the 1930's: that two competing species cannot occupy identical ecological niches. Lack's development of what is now known as the "competitive exclusion principle" stimulated a great deal of experimental research on competition and adaptation, and contributed to a greater awareness among biologists in the 1960's of the subtlety of natural selection. Lack's conversion was influenced by conversations with the entomologist George C. Varley on the mechanisms of population regulation. His new interpretations interested Ernst Mayr, his close friend, who invited Lack to present them at a Princeton University conference on evolutionary biology in 1947.

Lack dated his conversion on the Galapagos finches to 1943, when he was touring coastal ship-watching radar stations in connection with his work for the Operational Research Group. His companion, Varley, discovered that birds were detected by radar, and Lack henceforth made effective use of radar in studies of bird migration. Also in 1943 he had his second major scientific idea, which concerned the evolution of reproductive rates in birds. He argued that clutch size was adjusted not to mortality or to the physiological capability of birds, but to the food supply available to the parents to feed their young. He especially objected to the teleological nature of arguments that saw clutch size as compensation for mortality. His theory maintained that natural selection would regulate clutch size so as to produce the greatest number of young surviving to independence. These ideas were developed in the context of debates in the growing field of population ecology over mechanisms of population regulation. The basic choice among causes of regulation was between biotic factors (predation, food supply), which varied in severity in proportion to population density, and abiotic factors (climate), whose severity was largely independent of a population's density. Lack's research, published in 1954 in *The Natural Regulation of Animal Numbers*, supported the primacy of density-dependent regulation. His argument was influenced by the theories of the Australian entomologist A. J. Nicholson. This issue became the focus of intense ecological discussion in the 1960's.

Lack's ideas about natural selection led to a further controversy with V. C. Wynne-Edwards, who in 1962 published *Animal Dispersion in Relation to Social Behaviour*, in which he applied the theory of "group selection," the idea that natural selection could actually override the interests of the individual by working on features that benefited the group as a whole. Lack's response to Wynne-Edwards and his criticisms of group selection appeared in *Population Studies of Birds* (1966).

Lack's conversion to Christianity resulted in his 1957 book *Evolutionary Theory and Christian Belief*; it met with a poor reception among his colleagues. Unlike those who were interested in reconciling science and religion. Lack had no qualms about embracing orthodox religious beliefs that appeared to be in contradiction with his scientific views about evolution. He felt no such contradiction, however.

As director of the Edward Grey Institute, Lack devoted his energies to problems of population regulation, turning his institute into a world center for population ecology. In research his emphasis was always on field observation and comparative surveys, rather than on experiment. He distrusted the artificial conditions imposed by experiments and had a deep aversion to any research that involved killing birds. He often presented an argument by gathering facts from field studies until the sheer weight of the evidence supported whatever position he had adopted. In this respect his style was similar to Darwin's. He also distrusted the often crude mathematical approaches to ecology that had become prominent in the 1960's. His last book, *Island Biology*, published posthumously in 1976, was partly a response to the mathematical theory of biogeography introduced by Robert H. MacArthur and Edward O. Wilson in *The Theory of Island Biogeography* (1967).

Lack was first and foremost a naturalist who sought to combine problems in animal behavior, ecology, and evolutionary biology. He may justly be counted among the small group of naturalists, including Huxley and Mayr, who contributed to the "modern synthesis" in evolutionary biology in the middle decades of the twentieth century.

BIBLIOGRAPHY

I. ORIGINAL WORKS. A nearly complete bibliography is in the memoir by W. H. Thorpe in *Biographical Memoirs of Fellows of the Royal Society of London*, **20** (1974), 271–293. *Darwin's Finches* has been republished with notes and an introduction by Laurene M. Ratcliffe and Peter T. Boag (Cambridge, 1983). Competitive exclusion was the subject of several articles by Lack, the first being "Ecological Aspects of Species-formation in Passerine Birds," in *Ibis*, **86** (1944), 260–286. His first public presentation of his change of view is reported in the symposium on the ecology of closely allied species, in *Journal of Animal Ecology*, **13** (1944), 176–177. See also "The Significance of Ecological Isolation," in G. L. Jepsen *et al.*, eds., *Genetics, Paleontology, and Evolution* (Princeton, 1949), 299–308. These ideas are extended in *Ecological Isolation in Birds* (Oxford and Edinburgh, 1971). Lack's position in the controversies on population regulation and animal dispersion is explained in his appendix to *Population Studies of Birds* (Oxford, 1966). His major original work on reproductive rates is *The Natural Regulation of Animal Numbers* (Oxford, 1954). His posthumous book is *Island Biology, Illustrated by the Land Birds of Jamaica* (Berkeley and Los Angeles, 1976).

Lack's general writings on natural history include *The Life of the Robin* (London, 1943; rev. ed., 1946; paperback ed., 1953; 4th ed., 1965); and *Swifts in a Tower* (London, 1956; repr. 1973).

Lack's memoirs of his life to 1945, "My Life as an Amateur Ornithologist," with some reminiscences from several colleagues, are in *Ibis*, **115** (1973), 421–441. Archival material, including scientific correspondence relating to his publications, is at the Edward Grey Library of Field Ornithology, Oxford University.

II. SECONDARY LITERATURE. The fullest account of his life, based heavily on Lack's own memoir, is by W. H. Thorpe (see above). For contemporary assessments of his work, see Denis Chitty, "Population Studies and Scientific Methodology," in *British Journal for the Philosophy of Science*, **8** (1957), 64–66; and Gordon H. Orians, "Natural Selection and Ecological Theory," in *American Naturalist*, **96** (1962), 257–263. Of further interest are Frank J. Sulloway, "Darwin and His Finches: The Evolution of a Legend," in *Journal of the History of Biology*, **15** (1982), 1–53; and Marjorie Grene, ed., *Dimensions of Darwinism: Themes and Counterthemes in Twentieth-Century Evolutionary Theory* (Cambridge, 1983), essays by William B. Provine, "The Development of Wright's Theory of Evolution: Systematics, Adaptation, and Drift," and by Stephen Jay Gould, "The Hardening of the Modern Synthesis."

SHARON KINGSLAND

LANCEFIELD, REBECCA CRAIGHILL (*b.* New York City, 5 January 1895; *d.* New York City, 3 March 1981), *bacteriology.*

Rebecca Craighill was born at Fort Wadsworth in New York City, where her father, William E. Craighill, was stationed as a colonel in the U.S. Army Engineering Corps. Her mother, Mary Wortly Montague Byram, a native of Mississippi, had met Craighill through her brother, who was his classmate at West Point. After a somewhat peripatetic early life as the daughter of a Regular Army officer, Rebecca attended Wellesley College, graduating in 1916. Her graduate training was an integral part of the development of her scientific career. In 1918, just after receiving a master's degree from Columbia University, she married Donald E. Lancefield, a fellow graduate student in biology who later became a professor of biology at Queens College. They had one daughter, Jane.

Essentially all of Rebecca Lancefield's scientific research, extending over a period of some sixty years, dealt with a single large group of pathogenic bacteria, the streptococci. Her contributions form the basis for the present state of knowledge of this field. She brought order to the classification of this

LANCEFIELD

diverse group of organisms, clarified much of their basic biology, and provided explanations for their disease-producing capacity. This work had its beginning in 1918, shortly after she received her M.A., when she obtained a position as a technician at the Rockefeller Institute for Medical Research. She was assigned to the laboratory of Oswald T. Avery and Alphonse R. Dochez, who were studying a large collection of hemolytic streptococci that had been isolated during serious epidemics at military installations in 1917. It was not known at that time whether a single strain was involved or whether there were many different types of virulent streptococci, as had been shown to occur in the pneumococci. It was Lancefield's task to sort out these strains serologically and determine their degree of diversity. Within a year she identified four distinct serological types that served to classify 70 percent of the strains studied. The paper describing this work marks the beginning of modern streptococcal biology.

Lancefield returned to graduate school in 1919 and later spent a year at the University of Oregon, where both she and Donald taught. She returned to the Rockefeller Institute in 1922, an affiliation that continued until her death. Her new assignment was in the laboratory of Homer F. Swift, which was concerned with streptococcal infections and rheumatic fever. At the outset her work dealt with a group of organisms known as "green" *(viridans)* streptococci; this research provided the dissertation for her Ph.D. from Columbia, which she received in 1925.

Turning her attention again to the hemolytic streptococci, Lancefield made rapid strides in the next few years. Two different antigens were identified and isolated from the organisms in soluble form: the first was type-specific and responsible for the distinctions between different strains that she had first observed in 1919; the other was more generally distributed and was present in all of her strains from cases of human disease. Lancefield determined that the type-specific antigen was a protein, which she called M-protein. Subsequently she demonstrated that it was the principal virulence factor of streptococci and that immunity to streptococcal infection was directed primarily to M-protein (and thus was type-specific). This proved of great importance in the understanding of the epidemiology of streptococcal disease, since there are now known to exist several dozen different types, each with a different M-protein.

Lancefield showed that the other antigen was carbohydrate in nature and that this kind of antigen could be used to divide streptococci into subgroups. Thus, while the strains isolated from strep throat and other human diseases (designated group A) all had the same carbohydrate antigen, a group of strains from bovine mastitis had a quite different carbohydrate, and those from horses with strangles a third. Still other groups were encountered, and Lancefield was able to sort out the many different varieties of hemolytic streptococci that exist in nature by means of serological analysis of the different antigens of this class.

Much of Lancefield's subsequent work dealt with group A streptococci because of their role in human diseases, particularly rheumatic fever and glomerulonephritis. She clarified the nature and biological properties of the antigens found previously, described new protein antigens, and explored the implications of all of them for the production of disease. Her work was not limited to this single group, however, and she defined the properties of group B streptococci, originally thought to be primarily of bovine origin but now of concern because of neonatal infections in humans. As in group A, she found several different serological types, but in this case the type-specific antigens were capsular polysaccharides rather than proteins.

Lancefield's contributions have been widely recognized, and the varieties of streptococci are known throughout the world as the Lancefield groups and types. She received many honors, but the one that best reflects the esteem of her colleagues came when the professional organization concerned with streptococci in this country changed its name to the Lancefield Society in 1977. The international society in this field made the same change in 1981, shortly after her death.

BIBLIOGRAPHY

I. Original Works. Nearly all of Rebecca Lancefield's research was published in *Journal of Experimental Medicine* between 1919 and 1979 (35 papers). Listed here are her initial paper and reviews that will serve as reference to her most important studies: "Studies on the Biology of Streptococcus, I, Antigenic Relationships Between Strains of *Streptococcus hemolyticus*," in *Journal of Experimental Medicine*, **30** (1919), 179–213, with A. R. Dochez and O. T. Avery; "Specific Relationship of Cell Composition to Biological Activity of Hemolytic Streptococci," in *Harvey Lectures*, **36** (1941), 251–290; "Current Knowledge of Type-specific M Antigens of Group A Streptococci," in *Journal of Immunology*, **89** (1962), 307–313; and "Multiple Mouse Protective Antibodies Directed Against Group B Streptococci: Special Reference to Antibodies Effective Against Protein Antigens," in *Journal* *of Experimental Medicine*, **142** (1975), 165–179, with Maclyn McCarty and William N. Everly.

II. SECONDARY LITERATURE. Maclyn McCarty, "Rebecca Craighill Lancefield," in *Biographical Memoirs, National Academy of Sciences*, **57.**

MACLYN MCCARTY

LANE, ALFRED CHURCH (*b.* Boston, Massachusetts, 29 January 1863; *d.* New York City, 15 April 1948), *geology*.

Alfred Church Lane, a great-great-grandson of a Concord minuteman and grandson of an abolitionist, belonged to the eleventh generation of his family in New England. His father, Jonathan Abbot Lane, served some years as president of the Boston Merchants' Association and the Massachusetts State Senate. His mother was Sarah Delia Clarke, a graduate of Mt. Holyoke College. The Lanes were steadfast members of the Congregationalist church and the Republican party. As the only child in this prosperous and prominent Boston family, Alfred developed a wide range of interests, a sense of responsibility toward the local and world communities, and a deep faith in a righteous but loving personal God. At the end of his life he was described as a modern Renaissance man, distinguished as a scientist, humanist, and humanitarian.

Lane attended the Boston Latin School and Harvard College, where he received his bachelor's degree in natural science in 1883. For the next two years he served as an instructor in mathematics at Harvard while working toward a Ph.D. degree in geology. It was common practice at that time for geology students to obtain some of their education at a German university, so Lane spent two years (1885–1887) at Heidelberg studying petrography under Harry Rosenbusch. Returning to Harvard, he received his A.M. and Ph.D. degrees in 1888. Throughout his long career, which spanned sixty active years, Lane emphasized the importance of applying mathematics, chemistry, and physics to geological problems, and his efforts were instrumental in changing geology from a qualitative natural philosophy to a quantitative science. Probably his single greatest contribution to science arose from the catalytic role he played in efforts to develop methods for measuring the age of the earth and of individual geological formations.

Lane's career fell into three separate periods. For the first twenty years (1889–1909) he served as a member of the Geological Survey of Michigan. For the next twenty-seven years he was a professor of geology and mineralogy at Tufts College in Medford, Massachusetts, and for the twelve years following his 1936 resignation he remained actively involved in scientific and public affairs.

In 1889 Lane joined the Michigan Geological Survey as a petrographer. In 1892 he was appointed assistant state geologist, and in 1899 state geologist of Michigan. On 15 April 1896 Lane married Susanne Foster Lauriat, and the couple had two sons and a daughter. During his years in Michigan, Lane initiated studies on essentially every aspect of Michigan geology—the copper deposits, iron ores, water resources, geomorphology, and stratigraphy—and he published papers applying his observations to wider problems, such as the role of eutectics in rock magmas, studies on the grain of igneous intrusives, theories of copper deposition, the nature of geothermal gradients, and the early surroundings of life. In 1908 Lane coined the word "connate," still in use to describe interstitial waters trapped in sediments. His studies of the geology of Isle Royale in Lake Superior are commemorated by Lane Cove, named in his honor.

In 1909 Lane accepted the Pearson professorship of geology and mineralogy at Tufts College. At Tufts he instituted a rigorous type of training that was most uncommon in undergraduate courses then and now. He announced in the college catalog that all students enrolling in his classes needed a working knowledge of elementary chemistry, physics, and mathematics, and the ability to use French or German atlases. His advanced courses required calculus.

During World War I, Lane went to France to do educational work for the Young Men's Christian Association (YMCA), and he stayed on through 1919 as head of the department of mining in the college organized by the American Expeditionary Force at the Université de Beaune. In 1929 he was appointed the first science consultant to the Library of Congress, and in 1931 he was elected president of the Geological Society of America.

In 1936 the Commonwealth of Massachusetts enacted a bill requiring all teachers to take an oath swearing to uphold the constitutions of the United States and of the commonwealth. Lane opposed the oath in legislative hearings and in January 1936 he resigned his professorship rather than sign it. Many teachers publicly opposed the oath but only one other, Earle M. Winslow, chairman of the economics department at Tufts, resigned in protest. Lane was seventy-three years old. In a period when there was no generally agreed-upon retirement age, however, this action by a man of such towering reputation brought widespread acclaim. Regrettably, the same action by Winslow, a much younger man who made a greater sacrifice, passed largely unnoticed.

Lane had a lifelong interest in finding methods to measure geologic time, and he recognized exciting new potentialities when radioactivity was discovered in 1896. In 1924, the National Research Council's Division of Geology and Geography formed the Committee on the Measurement of Geologic Time by Atomic Disintegration, with Lane as chairman. At first the committee focused its attention mainly on radiometric methods, but it soon broadened its scope and shortened its name—deleting "by Atomic Disintegration." To help keep the record straight Lane applied a Dewey decimal system in listing forty-six different methods—isotopic, astronomical, and geological—for measuring some fraction of geologic time. He sought quantitative results, but he never saw science as a quest for certainty.

Lane held the chairmanship for twenty-two years, during which time he oversaw the formal meetings and issued the reports of the committee and acted as a one-man clearinghouse for specimens, data, methods, and ideas. He wrote letters to scientists around the world who were doing research on geologic time and, when feasible, he paid regular visits to their laboratories. He regarded it as especially important for different laboratories to analyze the same samples from well-documented sources, and he personally saw to it that such samples were made available. Lane took a special interest in the work of a young physicist, Alfred O. C. Nier, who joined the laboratory of Kenneth Bainbridge at Harvard in 1936 to build a mass spectrometer capable of measuring isotopic abundances of lead and uranium.

Lane corresponded regularly with Otto Hahn in Berlin, and, in 1938, when Hahn succeeded in splitting uranium and thorium atoms by bombarding them with neutrons, he sent Lane the first report on the process to reach America (*New York Times*, 17 April 1948, p. 15). At that time, however, the event was not seen as a historic breakthrough in the effort to harness atomic energy. The 1939 report of Lane's Committee abstracted Hahn's results, pointing out that the splitting was strictly a laboratory phenomenon, not known to occur in nature, and of no apparent significance in geochemistry.

World War II brought an end to the free exchange of information among atomic scientists and a temporary lull in efforts to measure geologic time. But Lane found other things to do. He was involved with the Boy Scouts of America, and he never ceased actively participating in the Congregational church, where he served as deacon for many years. He was vice president of the American Academy of Arts and Sciences from 1944 to 1946, and he belonged to numerous scientific, social, political, or historical societies, such as the American Civil Liberties Union, the New England Historic Geneological Society, and the Twentieth Century Association.

With the explosion of atomic bombs in 1945, Lane instantly recognized the awesome power and inherent danger. He wrote that we must find ways to put this new power to the service of international law and justice. He said that we can never be sure of peace in the world, but he joined the American Association for the United Nations in the hope that this body would act as a force for peace and worldwide cooperation.

In 1946 Lane resigned his chairmanship of the N.R.C. committee, partly because his eyesight was failing and partly because he wanted free time to pursue his many other interests. That year the committee report cited Arthur Holmes in England as having determined the age of the earth's granitic layer as about 3.35 billion years. Holmes equated that with the age of the earth itself, and many scientists concurred. Ten years later the age of the earth was determined to be about 4.6 billion years. Lane remained keenly interested in dating individual geologic formations, and he anticipated a wealth of data (which he did not live to see) from measurements on newly discovered radioactive isotopes such as potassium 40, rubidium 87, and samarium 147.

Lane published 1,087 papers on science, politics, economics, national and international issues, and religion. His single book was a small volume of poetry, which he had printed anonymously.

Lane appeared to be in good health in April 1948 when he drove with friends to New York to welcome back the Finne Ronne South Polar Expedition, which included Robert L. Nichols, his former student and successor as chairman of the geology department at Tufts. The reunion was held at the home office of the American Institute of Mining and Metallurgical Engineers. There, toward the end of a day full of excitement, storytelling, and singing, Alfred C. Lane died suddenly of a heart attack. A plaque in Lane's honor in the Goddard Chapel at Tufts is inscribed with the following statement, which he had written many years earlier: "Science and Religion aim to know, to share, and to spread the truth freely."

BIBLIOGRAPHY

I. Original Works. No complete list of Lane's 1,087 published titles has been compiled. One hundred and seventeen of the more important publications are listed by Robert L. Nichols in his memorial to Alfred Church

Lane in *Geological Society of America: Proceedings* (1952), 107–118. Lane's own notes on his career may be found in the archives of Harvard University. Most of them are published in the reunion volumes of the Harvard class of 1883. Information on Lane's scientific ideas comes from his articles, particularly the annual *Report of the Committee on the Measurement of Geologic Time*, published by the National Research Council, Division of Geology and Geography, between 1927 and 1946.

II. SECONDARY LITERATURE. No biography of Lane has been written. The chief sources of information are memorials, the most detailed of which is the one by Robert L. Nichols, cited above. Others are by Leonard Carmichael, in *Science*, **108** (1948), 567–568; and by Esper S. Larsen, Jr., in *American Mineralogist*, **34** (1949), 249–252. Articles appeared in the *New York Times* and the *New York Herald Tribune* on 17 April 1948.

URSULA B. MARVIN

LARK-HOROVITZ, KARL (*b.* Vienna, Austria, 20 July 1892; *d.* West Lafayette, Indiana, 14 April 1958), *physics*.

Karl Lark-Horovitz was a son of Moritz Horovitz and Adelle Hofmann. His father was a noted dermatologist and a man of wide-ranging scholarly interests, including botany and classical poetry. Karl Horovitz attended a humanistic high school and entered the University of Vienna in 1911 to study chemistry, physiology, physics, and pre-Socratic philosophy. During World War I he served as an officer in the signal corps of the Austrian Army, acquiring practical experience with the use of crystal detectors. In 1919 he returned to his studies and later that year received his Ph.D. in physics from the University of Vienna, where he stayed to teach and conduct research until 1925. In 1916 he married Betty Friedländer, a printmaker whose pseudonym was Lark, and in 1926 formally changed his family name to Lark-Horovitz.

From the beginning of his scientific career Karl Lark-Horovitz was attracted by newly opened fields of research and, in particular, by interdisciplinary subjects. One of his first publications, in 1914, dealt with the historical development of the principle of relativity, and he always remained interested in the history of science. His other early investigations included the use of radioactive materials in the study of crystals, the physics of image formation by the human eye, and the electrochemistry of glasses.

In 1925 Lark-Horovitz was awarded an International Research Council Fellowship by the Rockefeller Foundation. He first went to the University of Toronto to conduct X-ray diffraction studies at low temperatures and subsequently spent some time at the University of Chicago (1926), Rockefeller Institute (1926–1927), and Stanford University (1927–1928). Invited to Purdue University in the spring of 1928 to deliver a series of lectures, Lark-Horovitz returned to West Lafayette in the fall of 1929 to assume a permanent position at Purdue as the director of the Physical Laboratory. In 1932 he was appointed head of the department, a position he held until his death.

During his tenure at Purdue, Lark-Horovitz was very actively involved in a number of diverse activities that can roughly be divided into four areas. Prior to World War II, he conducted research in a variety of subfields of physics, including X rays, physics of surfaces, and, from 1936 to 1942, nuclear physics (a small cyclotron was constructed in 1938); he created at Purdue a highly regarded graduate program in physics; he actively participated in the creation, development, and fruition of the American Association for the Advancement of Science (AAAS) Cooperative Committee on the Teaching of Science and Mathematics; and he established at Purdue one of the leading centers of solid-state physics in the United States.

It was in this last field of research, in particular in the study of semiconductors, that Lark-Horovitz made his most important contributions to physics. His interest in solid-state physics was prompted by the wartime mobilization of science for national defense. The Radiation Laboratory, established in 1940 at the Massachusetts Institute of Technology to develop the microwave radar for the military, was seeking in early 1942 laboratories that would assume subcontracts to investigate various technical aspects of the new device. The contract Lark-Horovitz brought to Purdue was concerned with the development of crystal rectifiers, and his experience as a chemist led him to the choice of germanium as a material for investigation.

At the time the Purdue group started its wartime project, very little was known about the electrical properties of germanium. Proper methods of purification and controlled doping had first to be developed, and once relatively pure samples of the material became available, Lark-Horovitz and his collaborators pursued research along two lines. First of all, the measurements of the resistivity, Hall effect, and thermoelectric power provided the basis for seeking answers concerning the mechanisms of electric conductivity in germanium. Results of these experiments led Lark-Horovitz to the conclusion that germanium was an intrinsic semiconductor.

The second line of research was aimed directly at the production of operational crystal rectifiers intended for use as microwave radar detectors. Lark-Horovitz and his group developed a high-back-voltage germanium diode, a successful device that found many applications, but went into mass production too late to be used during the war.

A sense of urgency and strong emphasis on application prevented Lark-Horovitz from following up, with appropriate thoroughness, on some of the discoveries made at Purdue while working on the germanium wartime project. Some observed phenomena, such as the spreading resistance anomaly, or negative resistance, as it became clear only a few years later, anticipated the discovery of the transistor effect made by a team of Bell Labs researchers in 1947.

After the end of the war, Lark-Horovitz's interests turned temporarily away from solid-state physics, but after the development of the transistor, he resumed his research in this area with new vigor. His major contribution from this period was his work on the impact of radiation on semiconductors, which threw much light on the mechanism of production and effect of lattice defects in crystals.

Lark-Horovitz was very actively involved in the work of the AAAS. He was general secretary from 1947 to 1949 and a member of the editorial board from 1949 to the time of his death. He was an original member of the AAAS Cooperative Committee on the Teaching of Science and Mathematics, and chairman of the committee from 1945 to 1950.

Karl Lark-Horovitz became a naturalized American citizen in 1936. His children, Caroline Betty and Karl Gordon (who used the family name Lark), were born in 1929 and 1930, respectively. He died of a heart attack on 14 April 1958, while at work in his office at Purdue.

BIBLIOGRAPHY

I. Original Works. A bibliography of Lark-Horovitz' research publications is in Vivian Annabelle Johnson, *Men of Physics: Karl Lark-Horovitz, Pioneer in Solid State Physics* (Oxford and New York, 1969). The list of his scholarly contributions consists of eighty-three entries, and a selection of nineteen of his most important works in physics is reprinted in the book.

II. Secondary Literature. Johnson's book includes a biography of Karl Lark-Horovitz (46 pages long). Brief obituaries appeared in *Science*, **127** (1958), 1487–1488, by Hubert M. James, and in *Proceedings of the Indiana Academy of Sciences*, **68** (1958), 35–37.

Kris Szymborski

LAUB, JAKOB JOHANN (*b.* Rzeszów, Galicia, Austria-Hungary, 7 February 1882; *d.* Fribourg, Switzerland, 22 April 1962), *physics*.

Laub was the son of the manager of an estate near the German border. His turbulent and polyfaceted career is reflected in uncertainties surrounding his origins and identity. At birth he was Jakub; at death, Jacobo-Juan; he referred to his father as Adolf rather than by his probable name, Abraham; his mother's maiden name, Anna Schenborn, may or may not be a corrupt spelling. Beginning with his Argentine period, Laub let it be known that he was born in Jägerndorf, Austria. In 1911 he married Ruth Wendt, daughter of a Hamburg professor. They had one daughter and were divorced sometime after 1928. Wendt later became an organizer of migrant farm workers in California.[1]

Laub attended gymnasium in Rzeszów. In 1902, after studying briefly at the universities of Cracow and Vienna, he entered the University of Göttingen as a student of mathematics and physics. There, taking courses and seminars with David Hilbert and Hermann Minkowski, he became interested in the electron theory. He turned to experiment, and in 1905 he decided to work with Wilhelm Wien at Würzburg. Laub's doctoral dissertation (1907) concerned secondary cathode-ray emission. At his oral defense (1906), he introduced Einstein's special theory of relativity, which Wien had recommended to him in September 1905. For the next several years Laub remained at Würzburg and concentrated on extending Einstein's ideas.

Although by early 1908 Einstein was attracting notice from distinguished physicists, he had not yet received a university appointment. It was an unusual step, then, when in February 1908 Laub wrote to Einstein to ask if he could visit Bern to study relativity with him. Laub became Einstein's first scientific collaborator. Together they published articles criticizing Minkowski's notion of electromagnetic force and suggesting an experiment to decide between Einstein's special relativity and Hendrik Lorentz's electron theory.

In 1909, At Einstein's urging, Laub accepted a post as assistant to Philipp Lenard at Heidelberg. Lenard was jealous of Einstein's revolutionary interpretation of the photoelectric effect. (He had won a Nobel Prize in 1905, in part for his work on that effect, just as Einstein would win one in 1921, in part for *his* work on it.) Lenard set Laub to measure the density of the electromagnetic ether, a peculiar project hatched by Lenard and Vilhelm Bjerknes on which Laub expended little effort. In 1910 Laub published a masterly survey of the experimental evidence for relativity, the appearance of which displeased Lenard. Late in that year Laub began looking for employment outside Germany. He declined an assistant professorship at the University of Illinois to assume a chair of geophysics and theoretical physics at the University of La Plata in Argentina.

Laub's academic career in Argentina was relatively short. He left La Plata for a chair in physics at the Instituto Nacional del Profesorado Secundario, in nearby Buenos Aires, in 1914, after running afoul of the American director of the La Plata observatory, William Joseph Hussey. Laub taught in Buenos Aires for the rest of the decade. In 1920 he joined the Argentine diplomatic corps, a position made possible by his having become an Argentine citizen in 1915.

Except for furloughs in Argentina during the years 1928–1930 and 1939–1947, Laub spent most of the rest of his life in Europe. In the 1920's and 1930's he was stationed at consulates in Munich, Breslau, Hamburg, and Warsaw. In the early 1930's, when the Argentine coup d'état by José F. Uriburu seems to have deprived him of a salary for several years, Laub carried out fundamental research on high-frequency radiotelephone transmission for the Reichspost in Breslau and for the Berlin firm of C. Lorenz.[2] Beginning in 1947, after he retired, Laub was a researcher in the Physics Institute at the University of Fribourg, Switzerland. He returned to an old interest, atmospheric electricity and radiation, which he pursued in collaboration with Friedrich Dessauer.[3] His health declined through the 1950's, parallel with the fall in the value of the Argentine peso. By 1960 he was destitute. To raise cash he sold his unique scientific correspondence.

Jakob Laub, the close collaborator or friend of four Nobel laureates and a host of scientist luminaries in half a dozen countries, died in poverty and obscurity.

NOTES

1. Biographical data from the *Lebenslauf* of Laub's 1907 doctoral dissertation; courtesy of Eduardo L. Ortiz, Imperial College, London; and from correspondence and an interview with the late Ruth Wendt. The Jägerndorf origin appears in Laub's official file at the Ministry of Foreign Affairs, Buenos Aires.
2. F. Budischin, "Entwicklung und Ausbau des hochfrequenten Drahtfunks in Deutschland," in *Fernmeldetechnische Zeitschrift*, **1** (1948), 201–202, where, however, the experiments are inaccurately credited to "Prof. W. Laub."
3. Jakob Laub, "Ueber Schwankungen atmosphärischer Ionen und ihre biologische Wirkung," in *Bulletin der Schweizerischen Akademie der medizinischen Wissenschaften*, **16** (1960), 292–304, where Laub chronicles his forty-year interest in related questions.

BIBLIOGRAPHY

I. ORIGINAL WORKS. Laub's correspondence and interaction with Einstein figure in Carl Seelig's *Albert Einstein: Ein dokumentarische Biographie* (Zurich, 1954); some letters that Laub auctioned about 1960 are extracted in the catalogs of the auctioneer Gerd Rosen of Berlin (available in an annex of the New York Public Library). Most of Laub's German publications from the early years are in vol. V of Poggendorff; there is no systematic record of his Spanish-language publications. In his last years, Laub claimed to have had patent rights.

Most of Laub's extant correspondence with Einstein may be found (in original or in facsimile) in the Einstein papers (Hebrew University, Jerusalem; Boston University). The Smithsonian Institution Archives contain correspondence between Laub and Paul Hertz, as well as a number of letters from the latter part of Laub's career. Laub's activity at Heidelberg is recounted in correspondence between Vilhelm Bjerknes and Philipp Lenard, located at the University of Oslo. Correspondence between Laub and Emil Bose is preserved in the Fundación Walter B. L. Bose, Buenos Aires. Correspondence between Laub and Wilhelm Wien is at the Deutsches Museum, Munich.

II. SECONDARY LITERATURE. Lewis Pyenson, "Laub, Jakob," in *Neue deutsche Biographie*, V., 688–689, *Cultural Imperialism and Exact Sciences: German Expansion Overseas, 1900–1930* (New York and Bern, 1985), 163–170, 199–202, 227–228, *The Young Einstein: The Advent of Relativity* (Bristol and Boston, 1985), esp. 215–246, and "Silver Horizon: A Note on the Later Career of the Physicist-Diplomat Jakob Laub," in *Jahrbuch für Geschichte von Staat, Wirtschaft und Gesellschaft Lateinamerikas*, **25** (1988), 757–766.

LEWIS PYENSON

LAVES, FRITZ H. (*b.* Hannover, Germany, 27 February 1906; *d.* Laigueglia, Italy, 12 August 1978), *chemical crystallography, structural inorganic chemistry, metallurgy, mineralogy.*

Fritz Henning Emil Paul Berndt Laves was the son of Georg Ludwig Eduard Laves, a judge, and Margarethe Hoppe. His father claimed descent from Georg Ludwig Friedrich Laves (1788–1864), court architect of the elector of Hannover (later George I of England). In 1938 Laves married Melitta Druckenmüller, an architect who frequently drew structural diagrams for her husband. The couple had three daughters, Gracia, Charlotte, and Katarina.

Laves was a piano student and built up a large collection of Beethoven and Mozart piano scores. He read literature extensively and particularly admired the works of Hermann Hesse and Thomas Mann. Evidently the young Laves brothers were fascinated by natural history; the two elder brothers collected butterflies and beetles while Fritz dabbled in spiders, rocks, and minerals. Perhaps the experience of discovering an attractive specimen, which Professor Otto Mügge at the University of Göttingen identified as orthoclase feldspar ($KAlSi_3O_8$), was responsible for Laves' interest in this perplexing family of major rock-forming minerals.

Laves' scientific activity can be divided into three professional phases: German (1929–1948), American (1948–1954), and Swiss (1954–1978). In all three phases order-disorder in crystals was his leitmotiv. His German phase focused principally on alloys and intermetallic compounds; the American phase on the order-disorder in, and crystallochemical definition of, feldspars; and the Swiss phase on more elaborate probings into order-disorder in general through infrared absorption, nuclear magnetic resonance, electron spin resonance, and the microscopies. In the last decade of his life, he studied Al, Ga, V, Nb, and Ta alloys with respect to superconductivity.

Laves studied at the universities of Innsbruck and Göttingen. He received his Ph.D. in 1929 after studying with Paul Niggli at Zurich. Niggli, the foremost proponent of theoretical crystallography, introduced the "lattice complex" in 1919, extending this to *Bauverbände* (building units). Intrinsic in each of the 230 crystallographic space groups is a set of equivalent points that are related to each other under the symmetry or equivalence operations of the space group. Connecting these equivalent points together gave, among other topogeometric objects, polyhedra. It was Niggli's objective to arrange the space groups according to each of the derived lattice complexes, and from this to evolve a catalog of crystal structures. Although Laves initially planned to study petrology, Niggli's crystallography pleased him more and he was soon deep into the study of lattice complexes with respect to sphere packings. A strict definition of coordination number was wanting, and Laves pursued *Wirkungsbereiche* (spatial partitionings or domains such as Voronoy polyhedra or Dirichlet domains) and the newly published Wigner-Seitz cells.

Laves remained as a doctoral candidate and assistant at Zurich until 1930, then was assistant and *Privatdozent* at Göttingen until 1944, initially under Victor Moritz Goldschmidt. He remained *Privatdozent* because the Hitler regime suspected Laves, a principled man, of protecting Jews. Out of a sense of justice, in 1933 he tried to rally support from the science faculty to keep Goldschmidt. But Goldschmidt, a Jew, had to leave Germany in 1935, returning to his former position in Oslo, Norway, only to be later imprisoned several times by the Nazis.

Laves was assigned to a group under Reich Marshal Hermann Goering to develop an alloy "stronger than steel and lighter than air," and this was perhaps why much of his research at that time involved alloys of magnesium. His strong background in theoretical crystallography and crystal chemistry, gained from Niggli, led to the study of alloys and intermetallic compounds. In 1934 and 1935 he studied the structure of AB_2 compounds.

The Laves phases play a pivotal role in the chemical crystallography of alloys and intermetallic compounds. In such compounds, of which over 220 discrete types are presently known, the relative sizes of the metal atoms are of key importance. The determination of atomic size usually proceeded from the estimation of interatomic distances from X-ray diffraction experiments. For predominantly ionic structures, such as $Mg^{2+}O^{2-}$, the procedure was relatively straightforward and ionic radius was governed by coordination number and ionic formal charge. From arrays of large numbers of bond distances, effective ionic radii could be established with remarkable replicability for different structures. For alloys the problem was, and still is, more difficult: formal charge as such does not exist and coordination number is frequently confounded by a range of interatomic distance values. Initially, metallic radii were determined from the metals themselves (*e.g.*, Mg, Cu, Au, Pb), and an additivity rule was applied with corrections for coordination number. Laves contributed significantly to retrieving such distances and subsequent metallic radii.

The Laves phases are AB_2 compounds, such as those found in $MgZn_2$, $MgCu_2$, and $MgNi_2$, studied by Witte and Laves. Note here that A has two valence electrons and B belongs to the first transition series of elements. The Laves phases are allied to the σ phases. The coordination number for A is based on $12B + 4A$. Laves recognized their structural basis on dense-packing, the selected omission of nodes that leads to a tetrahedral tridymite-like net (such a net figured heavily in his later work on silica polymorphs). With A corresponding to the central cavity in the tridymite net and B corresponding to the tridymite framework, the desired arrangement was obtained. Although the ideal radius ratio $r_A : r_B = \sqrt{3}/\sqrt{2} = 1.225$ for such arrangements implicit in close-packings, the radius ratios observed in real crystals range from 1.05 to 1.67, demonstrating the wider flexibility of geometrical factors in alloys compared with ionic crystals.

The influence of Niggli can be found throughout

the crystallographic communications of Laves. The topology of crystallography, crystallographic networks, and structure systematology are themes of a publication that appeared during Laves' early period at Göttingen.[1] It appeared after the classic work on silicate structures by W. L. Bragg, who organized them according to increasing condensation of the (SiO_4) tetrahedra through sharing corners with other (SiO_4) tetrahedra. It was already recognized that one oxide anion could receive at most two bonds from tetrahedrally coordinated silicon (4+) atoms, that is, (Si—O—Si). The known structures were arranged according to increasing condensations of the tetrahedra, such as insular units (SiO_4), clusters (Si_2O_7), and rings (Si_3O_9); chains (SiO_3); sheets (Si_2O_5); and frameworks (SiO_2). Laves inquired about the theoretical dimensionality that was possible for each of these stoichiometries. If dimensionality $n = 0$ corresponds to finite units, $n = 1$ to one-dimensional chains, $n = 2$ to two-dimensional sheets, and $n = 3$ to three-dimensional frameworks, he derived for SiO_4, $n = 0$; for SiO_3, $n = 0, 1$; for Si_2O_5, $n = 0, 1, 2, 3$; and for SiO_2, $n = 1, 2, 3$ as topological possibilities. In short, this stimulating work was an alternative, more topological way of comprehending crystal structures both real and hypothetical.

A profound influence on Laves' insights into the chemical crystallography of metals and alloys came from the brilliant studies conducted by Eduard Zintl (1898–1941) during his brief career. At Munich, Zintl perfected methods of potentiometric titration for analysis of elements in the periodic system. This was followed by the study of saltlike compounds of sodium and other metals in liquid ammonia at Freiburg, and finally the evolution of a remarkable chemical model among the elements while at Darmstadt. He concluded that a wall or boundary existed between elements of groups I–III (which formed intermetallic phases and alloys with each other) and elements of groups IV–VII (which formed saltlike compounds with groups I–III). This boundary is known to this day as the Zintl border. (Earlier, Friedrich Adolf Paneth recognized that elements to the right of the border formed liquid or gaseous hydrides, whereas those to the left of the border, with the exception of boron, did not.) Thus the compound Na_4Pb_9 prepared in liquid ammonia was in fact a polyanionic salt $[Na(NH_3)_x]_4^+[Pb_9^{4-}]$ or $[Na(NH_3)_x]_4^+[Pb(Pb_8)]^{4-}$. This "polyplumbide" formed the basis for the enumeration of many other polyanionic clusters comprising groups IV–V, such as polyplumbides, polystannides, polybismuthides, polyantimonides, and polyarsenides. Zintl's discoveries had a profound influence on Laves. Both declared X-ray diffraction the tool of choice for unraveling their crystal chemistries and both sought a focused systematology. Furthermore, Laves drafted a far-reaching appraisal of Zintl's work and the tasks of metallurgists in general.[2] It should be recalled that at that time formal crystal structure analysis on such compounds was hampered by gross differences among atomic numbers of the constituent atoms, and confronted problems that have become structurally solvable at the required level of accuracy only relatively recently, by the use of X-ray techniques.

In 1943 Laves was called to the faculty at Halle, and in 1945 he was made professor at Marburg. During this time he concentrated on problems in one-dimensional disorder. Shortly thereafter, however, began the second phase in his life. In 1948 he arrived at the University of Chicago as a "paper clip specialist" (code for scientists whisked out of Germany at their own consent). There he worked with J. R. Goldsmith on order-disorder in feldspars. Although feldspars are a far cry from alloys, the underlying principles of substructure-superstructure, order-disorder, and twinning in these seemingly very different classes of substances join at the level of the topology and geometry of crystal structure. In addition, interest in feldspars went back to Laves' youth.

Order-disorder is a central theme in feldspar crystal chemistry. A good example is found in anorthite, $CaAl_2Si_2O_8$, and its synthetic gallium (Ga) and germanium (Ge) analogues: $Ca(Al_{1.25}Ga_{.75})Si_2O_8$, $CaAl_2(Si_{1.25}Ge_{.75})O_8$, and $Ca(Al_{1.75}Ga_{.75})(Si_{1.25}Ge_{.75})O_8$. Goldsmith and Laves deliberately studied such synthesized substitutions because the relative intensities of structure factors vary as atomic numbers of substituents over nonequivalent sites in a structure type.[3] Three types of X-ray reflections occur in anorthite: so-called type *a* constitute the reflections of the average structure; type *b* are additional reflections, the ordering reflections over the (Si + Al) cations in the tetrahedral framework; type *c* are yet additional "diffuse" reflections. By comparing single crystal photographs of these various compositions, it was possible to conclude that the type *b* reflections were due to Si-Al ordering, based on the contrast patterns of equivalent reflections for various crystals with Al, Ga, and Ge substitutions. The type *c* reflections hardly varied at all among the different compositions. Since the Ca cation was an invariant in these compositions, it was concluded that the centroid of this relatively large cation was responsible for these diffuse reflections. Order-disorder in such feldspars can lead to structure cells with at least one axial multiple or with a space group of reduced symmetry, usually a subgroup of the parent space group.

Laves and his co-workers probed problems of order-disorder and substructure-superstructure in great detail among many oxides of fundamental importance to mineralogic, petrologic, and ceramic sciences. Included were the fundamental structure types corundum (Al_2O_3) and spinel (Al_2MgO_4), which are based respectively on principles of hexagonal and cubic close-packing, and quartz (SiO_2). Quartz displays many principles in chemical crystallography; over one dozen polymorphs of SiO_2 are known. Geologically important silica polymorphs include tridymite, cristobalite, chalcedony, low quartz, and high quartz. The last two are related by a group-subgroup relationship in their space groups, and low quartz, owing to its relative hardness, is the most frequent constituent of beach sands.

In 1954 M. J. Buerger discussed the stuffed derivatives of the silica structures, thereby opening up an enormous field of applied crystallographic research. Stuffed derivatives of a fundamental structure type admit additional atoms into the crystal structure. For example, nepheline, $KNa_3(Al_4Si_4O_{16})$, is a stuffed derivative of tridymite $\square\square_3(Si_4Si_4O_{16})$, where vacancies (□) in the one are filled by alkalies in the other (also note that Al and Si can be ordered in nepheline). On structural grounds this may lead to an integral multiple of one or more of the crystallographic axes, a subgroup of the parent structure space group, or both. The most frequent cause of this phenomenon is order, or the sequential alternation of two or more ionic species over sites formerly occupied by only one ionic species. Ordering is usually temperature dependent; as atomic thermal vibrations increase with increasing temperature, the ordering or superstructure reflections on X-ray films will become weaker and vanish completely when perfect mixing of the two or more ionic species merges at a crystallographic site. This is one way to use single crystals of minerals as potential geothermometers. Laves and his co-workers assiduously studied stuffed derivatives of high quartz and vigorously pursued the possibility of using feldspars as geothermometers.[4]

The high quartz structure is a three-dimensional framework built of pairwise corner-linkages of (SiO_4) tetrahedral modules. Its symmetry is $P6_222$ or $P6_422$, space groups which are enantiomorphic, that is, right- or left-handed. Thus quartz is piezoelectric. In addition, linking just the Si^{4+} cations together results in the (6 · 3 · 6 · 3) Kagomé 4-connected net. The hexagonal rings in the net are the largest channels in the structure and, like nepheline, can incorporate larger cations. Laves became interested in order-disorder, solid-solution, defect, and ionic conduction problems in the channels. Electrolysis, electrical conductivity, color centers, and the presence of trace elements, particularly Ti^{4+}, in quartz attracted his and his students' attention.[5]

Order-disorder studies continued in the spirit of Laves' research. Interaction with colleagues in physics led him to emphasize infrared spectroscopy, nuclear magnetic resonance, electron spin resonance, and other spectroscopies because these new techniques admitted new and more direct observations of previously inaccessible structural phenomena. Such studies constituted the bulk of his crystallographic research after he accepted the chair of mineralogy at the Eidgenössische Technische Hochschule in Zurich in 1954, succeeding Paul Niggli.

Laves served as chairman of the Deutsche Mineralogische Gesellschaft (DMG) (1956–1958) and was a member of the executive committee of the International Union of Crystallography (1960–1966). Among other honors, he received an honorary doctorate from the University of Bochum, the Abraham Gottlob Werner Medal of the DMG, and the Roebling Medal of the Mineralogical Society of America.

NOTES

1. "Zur Klassifikation der Silikate," in *Zeitschrift für Kristallographie*, **82** (1932), 1–14.
2. "Eduard Zintls Arbeiten über die Chemie und Struktur von Legierungen," in *Die Naturwissenschaften*, **29** (1941), 244–255.
3. "Cation Order in Anorthite ($CaAl_2Si_2O_8$) as Revealed by Gallium and Germanium Substitutions," in *Zeitschrift für Kristallographie*, **106** (1955), 213–235, written with J. R. Goldsmith.
4. "On the Use of Calcic Plagioclases in Geologic Thermometry," in *Journal of Geology*, **62** (1954), 405–408, written with J. R. Goldsmith.
5. "Eigenschaften von Elektrolyse-Farbzentren in Quartzkristallen," in *Die Naturwissenschaften*, **48** (1961), 714, written with P. Schindler and H. E. Weaver.

BIBLIOGRAPHY

I. ORIGINAL WORKS. A list of over 230 professional papers of Laves', and a select few by his former students, is in *Zeitschrift für Kristallographie*, **151** (1980), 9–20.

II. SECONDARY LITERATURE. A short biographical sketch by J. R. Goldsmith is in *American Mineralogist*, **55** (1970), 541–546. For a detailed biographical account of Laves' life, see the memorial by E. Hellner in *Zeitschrift für Kristallographie*, **151** (1980), 1–9.

PAUL BRIAN MOORE